KEY TO EXERCISE SYMBOLS

 Agriculture

 Business/Economics

 Education

 Engineering/Technical

 Environmental Studies

 Medicine

 Political Science

 Psychology

 Sociology

An Introduction to Statistical Methods and Data Analysis

THE DUXBURY SERIES IN STATISTICS AND DECISION SCIENCES

An Introduction to Statistical Methods and Data Analysis

Third Edition

Lyman Ott
MERRELL DOW RESEARCH INSTITUTE

PWS-KENT PUBLISHING COMPANY
BOSTON

PWS–KENT
Publishing Company

20 Park Plaza
Boston, Massachusetts 02116

PWS-KENT Publishing Company is a division of Wadsworth, Inc.

Library of Congress Cataloging-in-Publication Data

Ott, Lyman.
    An introduction to statistical methods and data analysis / Lyman Ott. — 3rd ed.
        p.   cm.
    Bibliography: p.
    Includes index.
    ISBN 0-534-91926-X
    1. Mathematical statistics.   I. Title.
QA276.O77 1988   87–22445
519.5—dc 19   CIP

Printed in the United States of America.

    89 90 91 92 — 10 9 8 7 6 5 4 3

Sponsoring Editor: Michael Payne
Production: Stacey Sawyer, Complete Editorial Production Services
Interior and Cover Design: Ellie Connolly
Cover Photo: Charles Krebs/The Stock Market of NY
Interior Illustration: Folium
Typesetting: Syntax International *and* Typothetae
Cover Printing: New England Book Components
Printing and Binding: R. R. Donnelley & Sons Company

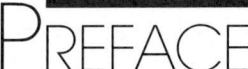PREFACE

An Introduction to Statistical Methods and Data Analysis, Third Edition, was written to give students from many different disciplines a first exposure to statistical methods and data analysis. The availability and widespread use of portable electronic calculators and statistical software systems that are fairly easy to use have enabled us to focus on the methods, the assumptions underlying the methods, and the results from analyses, without requiring students to spend endless hours doing calculations by hand.

The first and second editions of this text have been used very successfully in both one-term and one-year courses. It assumes a background in high-school algebra and no prior knowledge of statistics; the first few chapters provide a good review of introductory probability and statistics.

The focus of the third edition is the same as that of the second edition. Since many computer software systems are available for mainframes and for personal computers and since many of these systems have been made very "user-friendly," a number of us have become more comfortable with analyzing and interpreting data using available software. We're still interested in teaching the statistical methods appropriate for various experimental conditions, but we don't want to spend needless hours with calculations, once we understand the methods. If we can identify an appropriate method for analyzing data from an experiment, we can use available statistical software to do the actual (and sometimes tedious) calculations required by the method. This will enable us to focus more on the interpretation of the results of an analysis rather than on the analysis of the data. In many cases the computer output is provided in the text, so that only a modest amount of calculation is required.

The main new features of this edition are shown here:

- Many additional exercises have been added. Some of these involve straight-forward drills, others are more thought-provoking exercises involving practical experimental situations.

- The regression chapters have been expanded to include more work on residual analysis.
- More computer output is provided to illustrate the results of analyses.
- More attention is paid to the underlying assumptions. Data plots, residual plots, probability plots and other descriptive tools are used to diagnose potential problems with the assumptions.
- Some computer simulations are provided to illustrate the effects of violating certain underlying assumptions.
- A section on nested sampling and split-plot designs has been added to Chapter 16.
- A new chapter, Chapter 17, has been added to introduce the analysis of repeated measures experiments.

In addition to these major changes, numerous other less substantive changes have been made in response to constructive comments from reviewers and users.

ACKNOWLEDGMENTS

I wish to express my appreciation to friends and colleagues who have made many constructive suggestions and criticisms at various stages of development of the original text and during the preparation of the second and third editions. A special thanks to my colleague and fellow author, David Hildebrand, University of Pennsylvania, who provided a long, thoughtful review of the second edition. For this edition, I am deeply appreciative of the constructive comments of Mary Sue Beersman, Northeast Missouri State University; Larry Claypool, Oklahoma State University; Dale O. Everson, University of Idaho; Bruce Johnson, University of Connecticut; Larry J. Stephens, University of Nebraska; Robert F. Strahan, Iowa State University; Peter Westfall, Texas Tech University; and Douglas A. Wolfe, Ohio State University.

Also thanks are due A. Hald, E. S. Pearson, H. O. Hartley, R. E. Kirk, the Biometrika Trustees, the Chemical Rubber Company, Lederle Laboratories, the Editors of the Annals of Mathematical Statistics, D. B. Duncan, R. A. Waller, the Editors of Biometrics, the Editors of JASA, and the American Society of Testing and Materials for permission to reprint tables. A special note of appreciation is extended to my typists for their careful translations of my "drafts" into typed form; to Ruth Campbell, Jeanette Beach, and Ellen Evans for the first edition, to Vicki Mason for the second and to Linda Rabe for her careful attention to detail in typing this edition. I am also indebted to my editor, Michael Payne, and to the many others at PWS-KENT who worked long and hard behind the scenes to transform my manuscript into a finished product. I am also deeply indebted to Pat Hildebrand for generating the simulation studies used in this edition, to Chris Schmid for his help with exercises and preparation of the Solutions Manual, and to Jim Stegeman for proofreading the manuscript and for checking answers and to Victor Prybutok and Steve Bajgier of Drexel University, who provided me with some excellent test questions to be used as part of the exercise sets. Finally, I acknowledge the support and encouragement of my wife Sally and two children Curtis and Kathryn during the long evenings and weekends of work required for this textbook.

CONTENTS

7   INFERENCES ABOUT POPULATION VARIANCES

8   LINEAR REGRESSION AND CORRELATION   300

9   INFERENCES RELATED TO LINEAR REGRESSION AND CORRELATION   338

CONTENTS

1

WHAT IS STATISTICS?

INTRODUCTION

Almost everyone is confronted with statistics in their day-to-day living, but few people have much of a notion about the discipline of statistics. A commonly held opinion is that statistics deals with data tabulations and summarizations in the form of batting averages, divorce rates, incidences of lung cancer for a given sector of the United States, and, in general, the tedious and somewhat boring description of our world by using numbers. While the field of statistics does utilize numbers and statisticians do attempt to describe phenomena in our world, modern-day statistics deals primarily with **statistical inference**.

Suppose that a manufacturer of light bulbs produces roughly a half million bulbs per day. Concerned about customer reaction to its product, the firm wishes to determine the fraction of bulbs produced on a given day that are defective. The firm can solve the problem in two ways. All the half million bulbs could be inserted into sockets and tested, but the cost of this solution would be substantial and could greatly increase the price per bulb. A second method for determining the fraction that are defective is to select 1000 bulbs from the half million produced and to test each one. The fraction of bulbs defective in the 1000 tested could be used to estimate the fraction defective in the entire day's production. We will show in later chapters that the fraction defective in the bulbs tested will probably

1

be quite close to the fraction defective for the entire half million bulbs. Also, we will be able to tell you by how much you might expect this estimate to differ from the fraction of defective bulbs produced on a given day.

A second and similar example of statistics is brought to mind by the frequent use of the Gallup poll, the Harris poll, and other public opinion polls. How can these pollsters presume to know the opinions of more than 100 million Americans? They certainly cannot reach their conclusions by contacting every voter in the United States. Rather, as we have suggested in the light bulb example, they sample the opinions of a small number of voters, perhaps as few as 1000, to estimate the reaction of every voter in the country. The amazing result of this process is that the fraction of those people contacted who hold a particular opinion will match very closely the fraction of voters holding that opinion in the complete population at that point in time. Most students find this assertion difficult to believe; convincing supportive evidence will be supplied in subsequent chapters.

A third example of a statistical problem is taken from the field of medicine. Suppose a research physician wishes to investigate the effect of a new drug on the stimulation of a patient's heart. Note that the physician is really interested in the effect of the drug on all future heart patients who might be treated with the drug. Fifty heart patients are selected and each is treated with the drug. The increase in cardiac index (a measure of the efficiency of the heart) is recorded for each. After observing the effect of the drug on the 50 patients, the physician may infer that the drug will have a similar effect on most such heart patients in the future.

characteristics

measurement

Now let us try to identify from the preceding examples the **characteristics** common to inferential statistical problems. First, each example involved making an observation or **measurement** that cannot be predicted with certainty. Indeed, results of repeated observations are likely to vary in a haphazard or random manner. For example, we cannot say in advance whether a particular light bulb selected from production will work, whether a randomly selected voter will vote for a particular political candidate, or what a new heart patient's cardiac index will be after receiving the drug.

sampling

Second, each example involved **sampling**. A group of light bulbs was selected from the day's production, a group of people was taken from the entire voting population of the United States, and a group of 50 heart patients was obtained from the total of all heart patients.

collection of data

Third, each example involved the **collection of data** or measurements, one measurement corresponding to each element of the sample (group). We realize that observations on the elements of the sample may be quantitative in nature (when we record age, income, cardiac index, and so on) or qualitative in nature (when we record political affiliation, gender, preferences, and so on). But even these qualitative observations can be viewed as measurements if we assign a number to each qualitative category. For example, when a light bulb is tested, we will associate a 1 with a defective bulb and a 0 with a good bulb. Each number in the set will be a 0 or a 1. The total number of defective light bulbs in the sample is the sum of these measurements. Similarly, in a sample of voter intentions, we can assign the measurement 1 to each person who supports a specified candidate and 0 to each person who does not.

common objective

Finally, each example exhibits a **common objective**. That is, the purpose of sampling is to obtain information that can be used to make an inference about a much larger set of measurements called the population. For the manufacturer who wishes to make a statistical inference about (estimate) the fraction of defectives in a day's production, the population of interest is the set of 1s and 0s—roughly a half million in number—corresponding to the defective and nondefective bulbs produced that day. Similarly, the population for the voter problem is the set of 1s and 0s corresponding to the 100 million or more voters in the United States. Each voter is assigned a 1 if the voter intends to vote for the political candidate and a 0 if not. The objective of sampling is to estimate the fraction of eligible voters who favor the candidate—that is, the fraction of 1s in the population.

The population associated with the heart patients is the set of cardiac index values for all heart patients who might be treated by the drug. The sample of 50 cardiac index values for the group of heart patients is presumably representative of this population. The objective of the experiment is to estimate the average cardiac index that might be observed in patients following treatment with the drug. Thus the experimenter wishes to make a statistical inference about a large set of cardiac index values that could be acquired from patients treated by the drug at some time in the future. Note again that populations are collections of measurements and are not collections of people (which is the usual connotation of the term "population"). Note also that populations may exist in fact or may be conceptual in nature. The populations for the first two examples exist even though we do not actually possess the complete collection of 1s and 0s corresponding to defectives and nondefectives, or the entire set of 1s and 0s corresponding to voters favoring or opposing the political candidate. In contrast, the population of cardiac index values measured on all heart patients who could be administered the drug, now or in the future, is referred to as an imaginary or conceptual population.

Putting the four characteristics together—random observations, sampling, numerical **statistics** data, and a common inferential objective—we might define **statistics** to be a theory of information. Information is obtained by experimentation or, equivalently, by sampling; it is employed to make an inference about a larger set of measurements, existing or conceptual, called a population.

Relevant definitions for this section are as follows:

DEFINITION 1.1

A **population** is the set of all measurements of interest to the sample collector.

DEFINITION 1.2

A **sample** is any subset of measurements selected from the population.

DEFINITION 1.3

The **objective of statistics** is to make an inference about a population based on information contained in a sample.

FIGURE 1.1

Population and
Sample

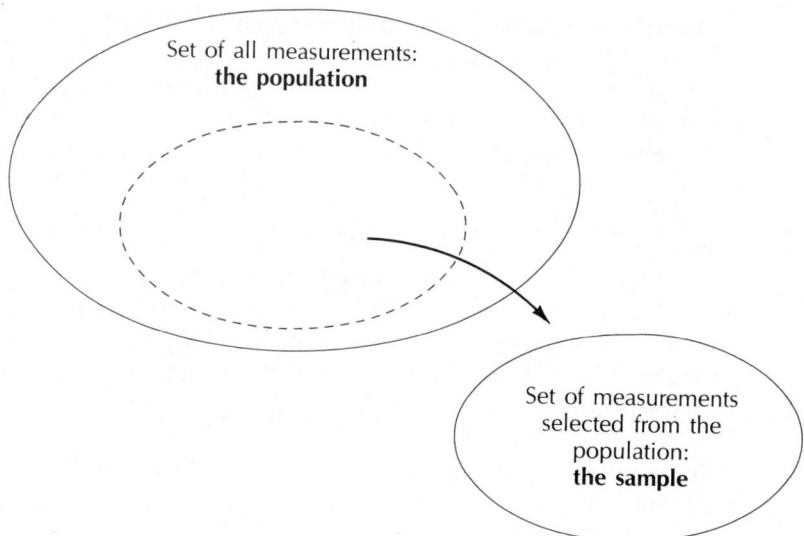

1.2   WHY STUDY STATISTICS?

We can think of two good reasons for taking an introductory course in statistics. First, every person is exposed to manufacturers' claims for products, to the results of sociological, consumer, and political polls, and to the published achievements of scientific research. Many of these results are inferences based on sampling. Some of the inferences are valid; others are invalid. Some are based on samples of adequate size; others are not. Yet all these published results bear the ring of truth. Some people say that statistics can be made to support almost anything (particularly statisticians). Others say it is easy to lie with statistics. Both statements are true. It is easy, purposely or unwittingly, to distort the truth by using statistics when presenting the results of sampling to the uninformed. Thus one reason for studying statistics is that you need to know how to evaluate published numerical facts, when to believe them, when to place tongue in cheek, and when to reject them.

A second reason for studying statistics is that your job or profession may require you to interpret the results of sampling (surveys or experimentation) or to employ statistical methods of analysis to make inferences in your work. For example, practicing physicians receive large amounts of advertising describing the benefits of new drugs. These advertisements frequently display the numerical results of experiments that compare a new drug with an older one. Do such data really imply that the new drug is more effective, or is the observed difference in results due simply to random variation in the experimental measurements?

Recent trends in the conduct of court trials indicate an increasing use of probability and statistical inference in evaluating the quality of evidence. The use of statistics in the social, biological, and physical sciences is essential because all these sciences make use of observations of natural phenomena, through sample surveys or experimentation, to develop and test new theories. Statistical methods are employed in business when sample data are used to forecast sales and profit. In addition, they are used in engineering and manufacturing to monitor product quality. The sampling of accounts is a new and useful tool to assist accountants in conducting audits. Thus statistics plays an important role in almost all areas of science, business, and industry; persons employed in these areas need to know the basic concepts, strengths, and limitations of statistics.

1.3   SOME CURRENT APPLICATIONS OF STATISTICS

ACID RAIN: A THREAT TO OUR ENVIRONMENT

The accepted *causes* of acid rain are sulfuric and nitric acids; the *sources* of these acidic components of rain are hydrocarbon fuels, which spew sulfur and nitric oxide into the atmosphere when burned. The effects of acid rain are many. Some of these effects are listed here:

- Acid rain, which is often formed when snow melts, frequently invades breeding areas for many fish and this prevents successful reproduction. Forms of life that depend on ponds and lakes contaminated by acid begin to disappear.
- In areas surrounded by affected bodies of water, nutrients are leached from the soil.
- Manufactured structures are also affected by acid rain. Experts estimate that acid rain has caused nearly $15 billion damage to buildings and other structures thus far.

Solutions to the problems associated with acid rain will not be easy. The National Academy of Sciences has recommended that we strive for a 50% reduction in sulfur oxide emissions. Perhaps that is easier said than done. High-sulfur coal is a major source of these emissions, but in states dependent on coal for energy, a shift to lower-sulfur coal is not always possible. Rather, better scrubbers must be developed to remove these contaminating oxides from the burning process before they are released into the atmosphere. Fuels for internal combustion engines are also major sources of the nitric and sulfur oxides of acid rain. Clearly, better emission control is needed for automobiles and trucks.

Reducing the oxide emissions from coal-burning furnaces and motor vehicles will require greater use of existing scrubbers and emission control devices as well as the development of new technology to allow us to use available energy sources. Developing alternative, cleaner energy sources is also important if we are to meet the goal of the National Academy of Sciences. Statistics and statisticians will play a key role in monitoring atmosphere conditions, testing the effectiveness of proposed emission control devices, and developing new control technology and alternative energy sources.

DETERMINING THE EFFECTIVENESS OF A NEW DRUG PRODUCT

The development and testing of the Salk vaccine for protection against poliomyelitis (polio) provide an excellent example of how statistics can be used in solving practical problems. Most parents and children growing up in the 1950s and earlier can recall the panic brought on by the outbreak of polio cases during the summer months. Although relatively few children fell victim to the disease each year, the pattern of outbreak of polio was unpredictable and caused great concern because of the possibility of paralysis or death. The fact that very few of today's youth have even heard of polio demonstrates the great success of the vaccine and the testing program that preceded its release on the market.

It is standard practice in establishing the effectiveness of a particular drug product to conduct an experiment (often called a *clinical trial*) with human subjects. For some clinical trials, assignments of subjects are made at random, with half receiving the drug product and the other half receiving an inactive solution or tablet (called a *placebo*) that does not contain the medication. One statistical problem concerns the determination of the total number of subjects to be included in the clinical trial. This problem was particularly important in the testing of the Salk vaccine because data from previous years suggested that the incidence rate might be less than 50 cases for every 100,000 children. Hence a large number of subjects had to be included in the clinical trial in order to detect a difference in the incidence rates for those treated with the vaccine and those receiving the placebo.

With the assistance of statisticians it was decided that a total of 400,000 children should be included in the Salk clinical trial begun in 1954, with half of them randomly assigned the vaccine and the remaining children assigned the placebo. No other clinical trial had ever been attempted on such a large group of subjects. Through a public school inoculation program, the 400,000 subjects were treated and then observed over the summer to determine the number of children contracting polio. Although less than 200 cases of polio were reported for the 400,000 subjects in the clinical trial, more than three times as many cases appeared in the group receiving the placebo. These results together with some statistical calculations were sufficient to indicate the effectiveness of the Salk polio vaccine. However, these conclusions would not have been possible if the statisticians and scientists had not planned for and conducted such a large clinical trial.

The development of the Salk vaccine is not an isolated example of the use of statistics in the testing and developing of drug products. Statistics has played an important role in the development and testing of birth control pills, rubella vaccines, chemotherapeutic agents in the treatment of cancer, and many other preparations. More recently, statistical analyses led to the early conclusion that the drug AZT is effective in the treatment of AIDS, for patients who have previously been afflicted and have survived a bout with a particularly virile strain of pneumonia.

APPLICATIONS OF STATISTICS IN OUR COURTS

Libel suits related to consumer products have touched each one of us; you may have been involved as a plaintiff or defendant in a suit or you may know of someone who was involved in such litigation. Certainly we all help to fund the costs of this litigation indirectly through increased insurance premiums and increased costs of goods. The testimony in libel suits concerning a particular product (automobile, drug product, and so on) frequently leans heavily on the interpretation of data from one or more scientific studies involving the product. This is how and why statistics and statisticians have been pulled into the courtroom.

For example, epidemiologists have used statistical concepts applied to data to determine whether there is a statistical "association" between a specific characteristic, such as the use of a brand-name tampon, and a disease condition, such as toxic shock syndrome. An epidemiologist who finds an association should try to determine whether the observed statistical association from the study is due to random variation or whether it reflects an actual association between the characteristic and the disease. Arguments in courtrooms about the interpretations of these types of associations involve data analyses using statistical concepts as well as a clinical interpretation of the data.

THE ENERGY CRISIS: A SEARCH FOR NEW SOURCES AND A SEARCH FOR OIL

The OPEC oil crisis of 1973–1974 brought to America's attention a problem that will continue to plague us for decades: a shortage of energy. The United States is confronted by staggering annual demands for energy with supplies that may not meet future demands, especially if a major supplier "interrupts" service. Such an interruption by OPEC in 1974 led to an energy rush and subsequent supply problems.

Possible sources of energy needed to supply the present and future requirements of the United States include vast coal and oil shale reserves, nuclear reactors, new oil and natural gas reserves, solar energy, and alternative new fuels. For example, methanol (wood alcohol) and ethanol (grain alcohol) may be major contributors to octane boost as leaded fuels are phased out. These alcohols are also likely candidates to reduce our dependence on foreign crude oil.

Although we can have temporary reprieves from oil and gas shortages, the nation must still address important questions related to energy supplies for future generations. In which of these resources should we, the American public, invest the capital necessary for development? Which source will yield a given amount of energy at minimum cost? What unfavorable impact will each have on the quality of life? Which might yield dangerous side effects? These questions and others must be answered by experimentation. Statisticians will assist in designing experiments and in interpreting experimental data.

OPINION AND PREFERENCE POLLS

Public opinion, consumer preference, and election polls are commonly used to assess the opinions or preferences of a segment of the public for issues, products, or candidates of interest. And we, the American public, are exposed to the results of these polls on a daily basis in newspapers, in magazines, on the radio, and on television. For example, the results of polls related to the following subjects were printed in local newspapers over a two-day period:

- consumer confidence related to future expectations about the economy
- preferences for candidates in upcoming elections and caucuses
- attitudes toward cheating on federal income tax returns
- preference polls related to specific products (for example, foreign vs. U.S. cars, Coke vs. Pepsi, McDonald's vs. Wendy's)
- reactions of North Carolina residents to arguments about the morality of tobacco
- opinions of voters about proposed tax increases and proposed changes in the Defense Department budget

A number of questions can be raised about these polls. How many people were polled? What questions were asked? Was each person asked the same question? How were people chosen or selected for the poll? Can we believe the results of these polls? Do these results "represent" how the general public feels?

Opinion and preference polls are an important, visible application of statistics for the consumer. We will discuss this topic in more detail in Chapter 6. After studying this material, you should have a better understanding of how to interpret the results of these polls.

1.4   WHAT DO STATISTICIANS DO?

What do statisticians do? Statisticians, both in consulting and research, devote their time to two major areas. The first concerns the acquisition of the sample data. Sample surveys and experiments cost money and yield information that is usually in the form of numbers on sheets of paper. By varying the survey or experimental procedure—where you select the data and how many observations you take from each source—you can vary the cost and quantity of information in the experiment. Rather simple modifications in the data selection procedure can reduce the cost of the sample drastically. Thus statisticians study

designing surveys

various methods for **designing sample surveys** and experiments (selecting the sample) and attempt to find the method that will yield a specified amount of information at minimum cost.

selecting the method for analyzing the data; analyze the data

After the data are acquired, the second task facing statisticians is the **selection** of the appropriate method of inference for a given sample survey or experimental design. Some of these methods are good, some are bad, and some seem to be good for most occasions. It is the statistician's job to choose the appropriate method for **analyzing the data**, and then to analyze the data.

evaluating goodness of the inference

The foregoing discussion leads to the most important contribution of statistics to science and business. Anyone can devise a method to make inferences based on the sample data. The major contribution of statistics is in **evaluating the "goodness" of the inference** When predicting, we seek an upper limit to the error of prediction. In reaching a decision concerning a characteristic of the population, we ask the probability of reaching a correct conclusion.

To summarize: first, statisticians design surveys and experiments to minimize the cost of obtaining a specified quantity of information. Second, they seek the best method for analyzing the data and making an inference for a given sampling situation. Third, they summarize and analyze the data. Finally, statisticians provide a measure of the goodness of an inference.

1.5   A NOTE TO THE STUDENT

We think with words and concepts. Thus statistics, a theory of information, requires the memorization of new terms and concepts (as does the study of a foreign language). Commit definitions, theorems, and concepts to memory.

Also, focus on the broader aspects of statistics. What is statistics? How does it work? What are some of the more important applications? Do not let details obscure these general characteristics of the subject. The teaching objective of this text is to identify and amplify these broader concepts of statistics.

EXERCISES

 1.1 Selecting the proper diet for shrimp or other sea animals is an important aspect of sea farming. A researcher wished to estimate the mean weight of shrimp maintained on a specific diet for a period of six months. One hundred shrimp were randomly selected from an artificial pond and each is weighed.
  a. Identify the population of measurements that is of interest to the researcher.
  b. Identify the sample.
  c. What characteristics of the population would be of interest to the researcher?
  d. If the sample measurements are used to make inferences about certain characteristics of the population, why would a measure of the reliability of the inferences be important?

 1.2 Radioactive waste disposal as well as the production of radioactive material in some mining operations was creating a serious pollution problem in some areas of the United States. State health officials decided to investigate the radioactivity levels in one suspect area. Two hundred points were randomly selected in the area and the level of radioactivity was measured at each point. Answer questions (a), (b), (c), and (d) in Exercise 1.1 for this sampling situation.

 1.3 A social researcher in a particular city wished to obtain information on the number of children in households that receive welfare support. A random sample of 400 households was selected from the welfare rolls of the city. A check on welfare recipient data provided the number of children in each household. Answer questions (a), (b), (c), and (d) of Exercise 1.1 for this sample survey.

class exercise **1.4** Search issues of your local newspaper to locate the results of a recent Harris or Gallup survey.
  a. Identify the items that were observed to obtain the sample measurements.
  b. Identify the measurement made on each item.
  c. Clearly identify the population associated with the survey.
  d. What characteristic(s) of the population is (are) of interest to the pollster?
  e. Does the article explain how the sample was selected?
  f. Does the article include the number of measurements in the sample?
  g. What type of inference is made concerning the population characteristics?
  h. Does the article tell you how much faith you can place in the inference about the population characteristic?

2 DATA DESCRIPTION

2.1 INTRODUCTION

The field of statistics can be divided into two major branches: descriptive statistics and inferential statistics. In both branches we will be working with a set of measurements. For situations in which **data description** is our major objective, the set of measurements available to us is frequently the entire population. For example, suppose that we wish to describe the distribution of annual incomes for all families registered in the 1980 census. Since all these data are recorded and are available on computer tapes, we do not need to obtain a random sample from the population; the complete set of measurements is at our disposal. Our major problem is in organizing, summarizing, and describing these data—that is, making sense of the data. Similarly, vast amounts of monthly, quarterly, and yearly data are available on sales for the steel industry broken down by domestic consumption, exports, consumer inventory change, and imports. However, in order to present such data in formats appropriate for audiences of consumers, financial analysts, or managers, it is necessary to organize, summarize, and describe the data. Good descriptive statistics will enable us to make sense of the data by reducing a large set of measurements to a few summary measures that provide a good, rough "picture" of the original measurements.

In situations in which we are concerned with **statistical inference**, a sample is usually the only set of measurements available to us. We use information in the sample to draw conclusions about the population from which the sample was drawn. Of course, in the process of making inferences we need to organize, summarize, and describe the sample data also.

For example, the tragedy surrounding isolated incidents of product tampering has brought about federal legislation requiring tamper-resistant packaging for certain drug products sold over the counter. These same incidents also brought about increased awareness by industry of the rigid standards of product and packaging quality that must be maintained by companies involved with delivering these products to the store shelves. In particular, one company was interested in determining the proportion of packages out of total production that are improperly sealed or have been damaged in transit. Obviously it would be impossible to inspect all packages at all the stores where the product is sold, but a random sample of the production could be obtained and the proportion defective in the sample could be used to *estimate* the actual proportion of improperly sealed or damaged packages.

Similarly, in developing an economic forecast of new housing starts for the next year, it is necessary to use sample data from various economic indicators in order to make such a prediction (inference).

In both of these examples involving an inference, description of the sample data (making sense of the data) is an important step leading toward the inference that we make. Thus no matter what our objective, statistical inference or data description, we must first describe the set of measurements at our disposal.

There are two ways to describe a set of measurements. We can use either graphical techniques or numerical descriptive techniques. Section 2.2 is concerned with graphical methods for describing data on a single variable. In Sections 2.3, 2.4, and 2.6 we discuss numerical techniques for describing data from a single variable. Section 2.5 is optional and deals with coding techniques. The final topics on data description are presented in Section 2.7, where we consider a few techniques for describing (summarizing) data on more than one variable. Section 2.8 of this chapter discusses the role of computers and computer software in doing the business of statistics.

2.2 DESCRIBING DATA ON A SINGLE VARIABLE: GRAPHICAL METHODS

guideline for organizing data

When many measurements are obtained on a variable, the data must first be organized, prior to presentation, by using one of several graphical techniques. As a general rule, the data should be arranged in such a way that **each observation can fall into one and only one category of the variable**. This procedure eliminates any ambiguity that might arise in placing observations into categories and aids in the interpretation of the data. For example, if we are trying to determine how our food dollars are spent, we might use the categories in Table 2.1 to classify each purchased food item.

TABLE 2.1   Categorization of Food Items

Food Item
Dairy products
Cereal and baked goods
Nonalcoholic beverages
Poultry
Seafood
Meat
Fruit and vegetables
Other foods

Assuming these categories are clearly defined, all food items could be placed into one and only one category of the variable.

However, if we asked social researchers to respond to a questionnaire about perceived causes of the Liberty City (Miami) riots, the categories listed in Table 2.1 would not suffice.

Which of the following do you think was the most important cause of the Liberty City disorder?

Criminal elements
Poverty
Lack of jobs
Black nationalism
Poor housing
Police brutality
Don't know
Other

Clearly there are too many overlapping categories. For example, the categories poverty, lack of jobs, and poor housing may be indicators of the same problem. Hence the data from the qualitative variable "major cause of the disorder" could not be organized according to our guideline.

Having organized the data according to the guideline suggested, there are several ways to graphically display the data. The first and simplest graphical procedure for data organized in this manner is the *pie chart*. It is used to display the percentage of the total number of measurements falling into each of the categories of the variable by partitioning a circle (much as one might slice a pie). The data of Table 2.2 represent a summary of a study to determine paths to authority for individuals occupying top positions of responsibility in key public-interest organizations.

Using biographical information, each of 1345 individuals was classified according to how she or he was recruited for the current elite position.

Although you can scan the data in Table 2.2, the results are more easily interpreted by using a pie chart. From Figure 2.1 we could make certain inferences about channels to positions of authority. For example, more people were recruited for elite positions from

TABLE 2.2   Recruitment to Top Public-Interest Positions*

Recruitment from	Number	Percentage
Corporate	501	37.2
Public-interest	683	50.8
Government	94	7.0
Other	67	5.0

Source: Thomas R. Dye and L. Harmon Zeigler, *The Irony of Democracy*, 5th ed. (Monterey, Calif.: Duxbury Press, 1981), p. 130.

* Includes trustees of private colleges and universities, directors of large private foundations, senior partners of top law firms, and directors of certain large cultural and civic organizations.

FIGURE 2.1

Pie Chart for the Data of Table 2.2

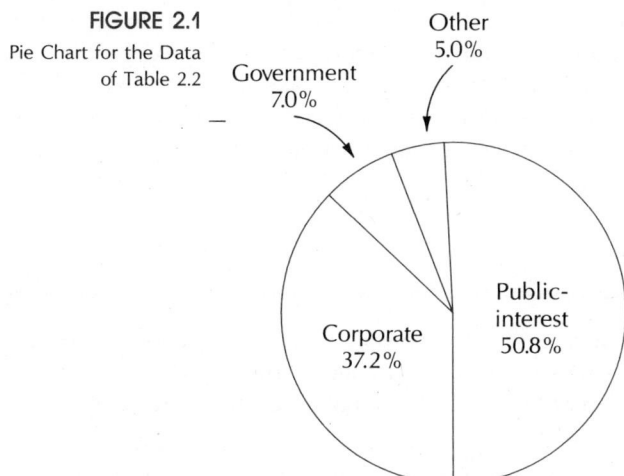

public-interest organizations (approximately 51%) than from elite positions in other organizations.

Over recent years many state legislatures and the federal government have watched with great interest and concern what has been labeled "The Great American Buy-in." Figure 2.2 shows sources of foreign investments in the United States and where the investments were made.

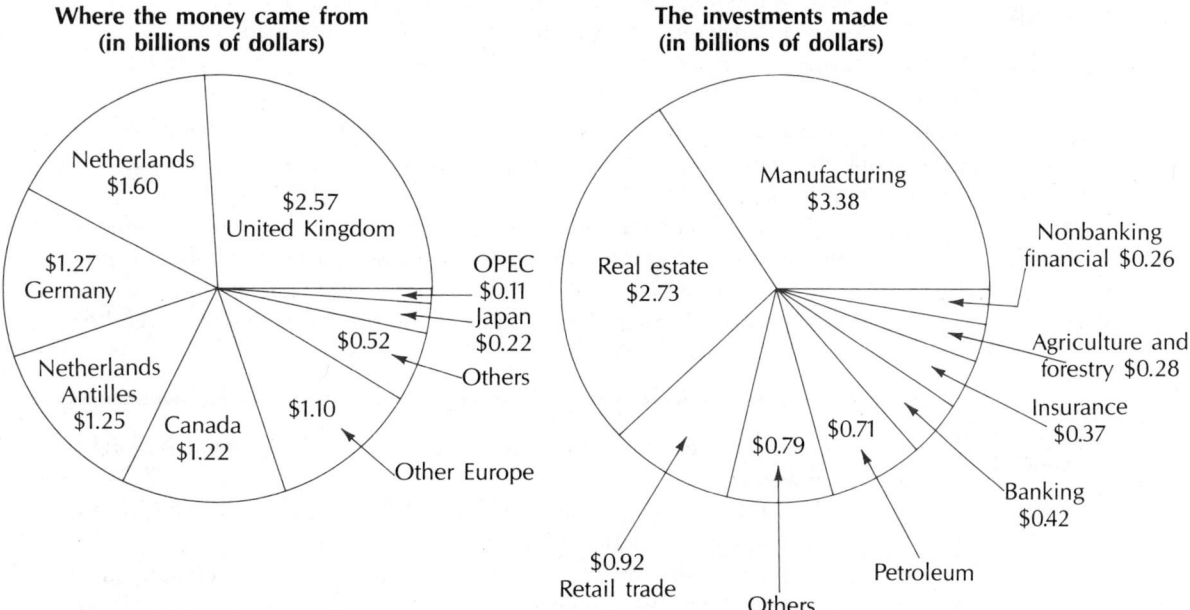

FIGURE 2.2 Foreign Investments in the United States

From Figure 2.2 it is clear that more than half of all foreign investments in the United States came from the United Kingdom, the Netherlands, and Germany. Similarly, over 50% of all the investment money went into manufacturing and real estate.

In summary, the pie chart can be used to display percentages associated with each category of the variable. The following guidelines should help you to obtain clarity of presentation in pie charts.

GUIDELINES FOR
CONSTRUCTING
PIE CHARTS

1. Choose a small number of categories for the variable, preferably about five or six. Too many categories make the pie chart difficult to interpret.
2. Whenever possible, construct the pie chart so that percentages are in either ascending or descending order.

statistical map   A second graphical technique for summarizing data on a single variable is the **statistical map**. Such graphs are frequently used to describe election results where either the Republican or the Democratic candidate wins each state (or district). Figure 2.3 uses four statistical maps to show the results of the 1964, 1972, 1980, and 1984 presidential landslide victories. Note that states won by the Democratic candidate are shaded. The guidelines for constructing good statistical maps are as follows:

GUIDELINES FOR
CONSTRUCTING
A MAP

1. Units on the map must be large enough to be clearly visible.
2. Too many different shadings cannot be used for the map.

In Figure 2.3, the states (units) are visible (except maybe for the smaller states) and there is only one shading to differentiate between states won by the Democrat and those won by the Republican.

bar chart   Another graphical technique for data organized according to the recommended guideline is the **bar chart** or **bar graph**. One such chart, displaying U.S. oil prices from 1971 to 1986, is shown in Figure 2.4.

Bar charts are relatively easy to construct if you use the guidelines given here.

GUIDELINES FOR
CONSTRUCTING
BAR CHARTS

1. Label frequencies along one axis and categories of the variable along the other axis.
2. Construct a rectangle at each category of the variable with a height equal to the frequency (number of observations) in the category.
3. Leave a space between each category to connote distinct, separate categories and to clarify the presentation.

FIGURE 2.3 Recent Presidential Landslides

Statistical Maps
Showing Recent
Presidential
Landslides

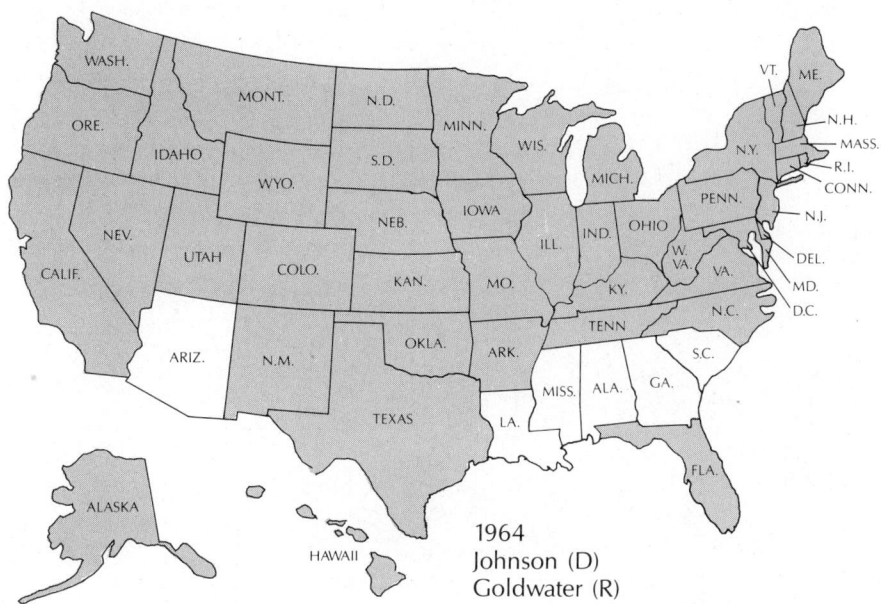

1964
Johnson (D)
Goldwater (R)

1972
Nixon (R)
McGovern (D)

Democrat   Republican

FIGURE 2.3   Recent Presidential Landslides
(continued)

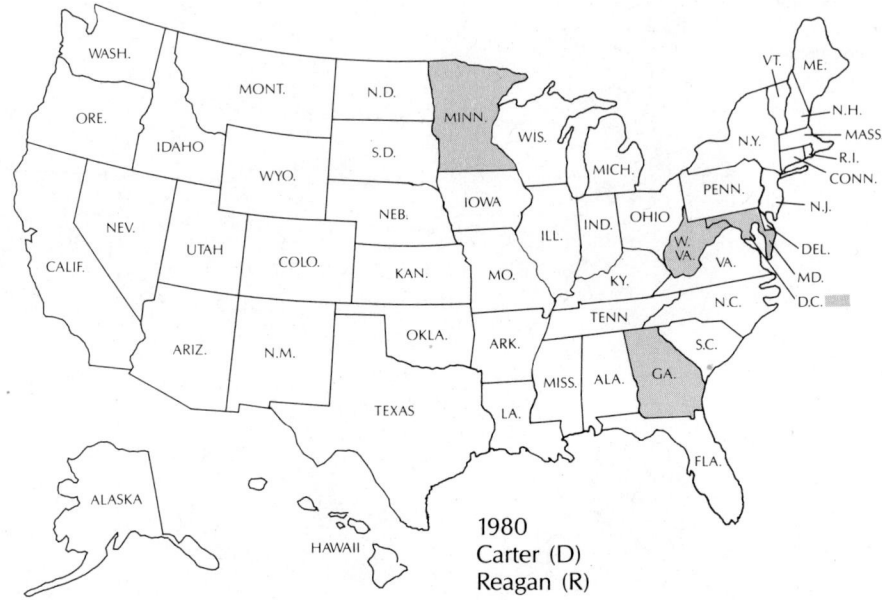

1980
Carter (D)
Reagan (R)

1984
Mondale (D)
Reagan (R)

Democrat   Republican

FIGURE 2.4

U.S. Oil Prices
(Dollars) from 1971
to 1986

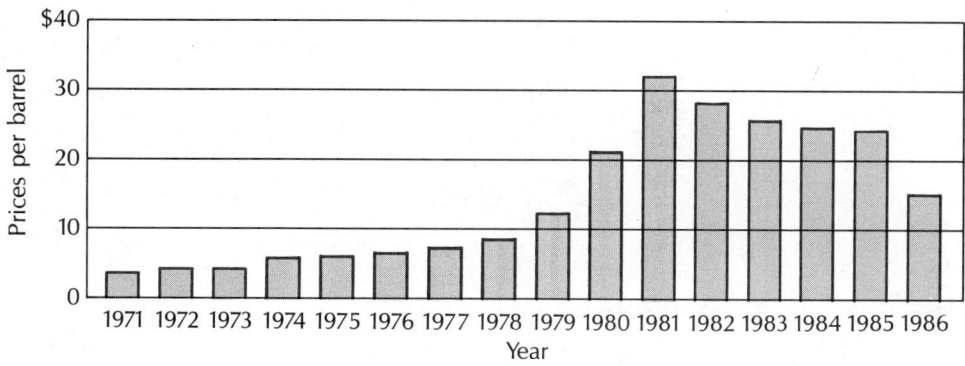

There are many variations of the bar chart, which you may have seen in printed form. Figure 2.5 compares the sources of the world crude oil from 1971 to 1986 while Figure 2.6 shows the number of active drill rigs in the United States during this same period. From Figure 2.5, it is clear that there have been major shifts in the sources of crude oil from 1971 to 1986. Also it appears that the United States is headed toward increased dependence on oil imports. This latter opinion seems to bear additional credence when we see the drop in active drilling rigs over recent years shown in Figure 2.6, the projected increase in U.S. petroleum demand from 1986 to 1990 illustrated in Figure 2.7, and the projected drop in U.S. production shown in Figure 2.8.

frequency histogram, relative frequency histogram
    The next two graphical techniques that we will discuss in this text are the **frequency histogram** and the **relative frequency histogram**. Both of these procedures are applicable only to quantitative (measured) data. As with the previous graphical techniques, we must organize the data before constructing a graph.

    Consider the following kind of measurement: weight gain for each of 100 baby chicks fed on a new diet and observed over an 8-week period. These data are recorded in Table 2.3.

    In trying to describe the set of measurements recorded in Table 2.3, we note that the largest weight gain is 4.9 and the smallest is 3.6. But although we might examine the table very closely, it is difficult to describe how the measurements are situated along the interval from 3.6 to 4.9. Are most of the measurements near 3.6, near 4.9, or are they evenly distributed along the interval? To answer these questions, we summmarize the data

frequency table
in a **frequency table**.

    To construct a frequency table, we begin by dividing the range from 3.6 to 4.9 into an

class intervals
arbitrary number of subintervals called **class intervals**. The number of subintervals chosen depends on the number of measurements in the set, but we generally recommend using from 5 to 20 class intervals. The more data we have, the larger the number of classes we tend to use. The guidelines given here can be used for constructing the appropriate class intervals.

FIGURE 2.5

Sources of World
Crude Oil, 1971 and
1986

Where the World Gets Its Crude Oil
Leading suppliers' proved reserves in billions of barrels

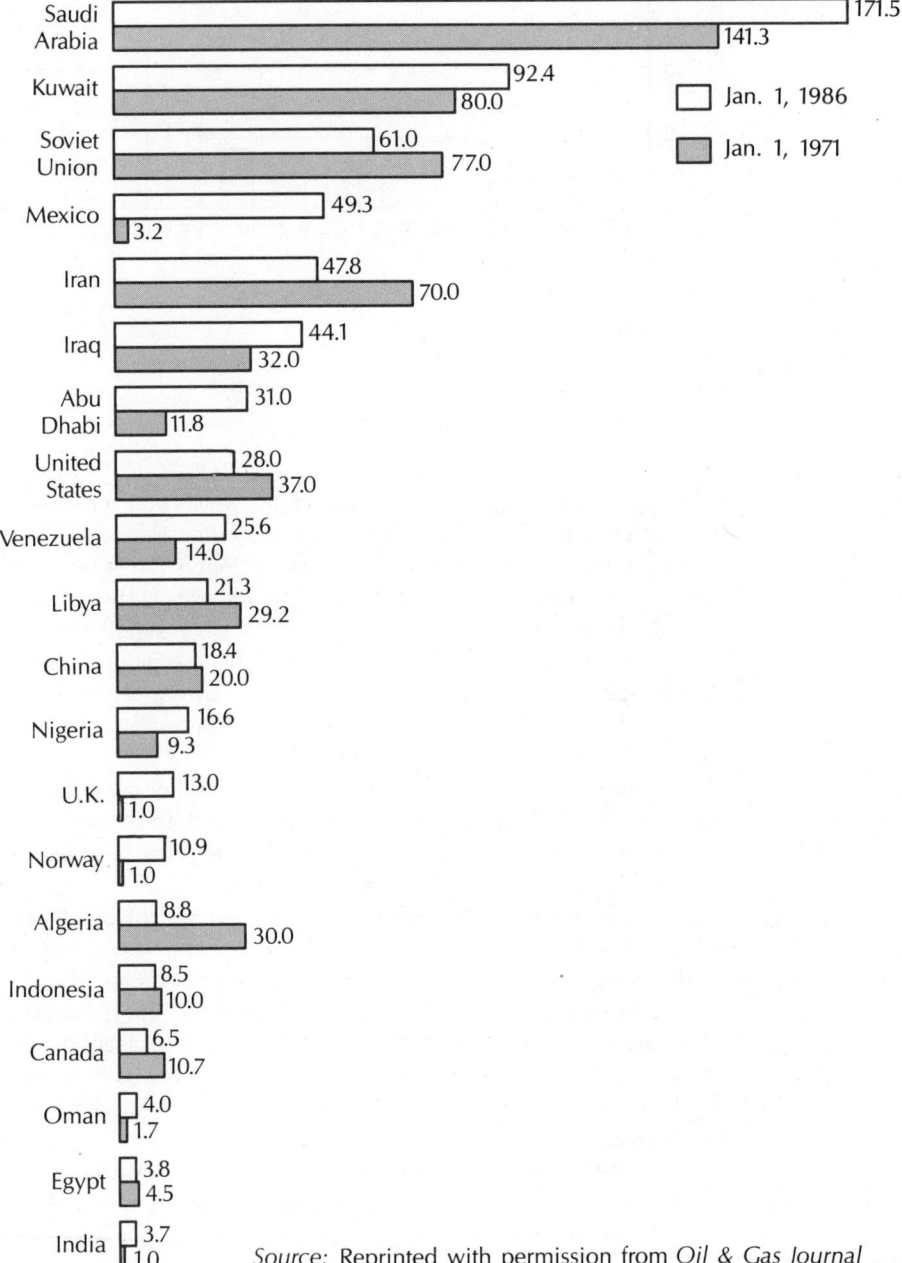

Country	Jan. 1, 1986	Jan. 1, 1971
Saudi Arabia	171.5	141.3
Kuwait	92.4	80.0
Soviet Union	61.0	77.0
Mexico	49.3	3.2
Iran	47.8	70.0
Iraq	44.1	32.0
Abu Dhabi	31.0	11.8
United States	28.0	37.0
Venezuela	25.6	14.0
Libya	21.3	29.2
China	18.4	20.0
Nigeria	16.6	9.3
U.K.	13.0	1.0
Norway	10.9	1.0
Algeria	8.8	30.0
Indonesia	8.5	10.0
Canada	6.5	10.7
Oman	4.0	1.7
Egypt	3.8	4.5
India	3.7	1.0

Source: Reprinted with permission from *Oil & Gas Journal*

FIGURE 2.6

Active Drilling Rigs in the United States, 1971 to 1987

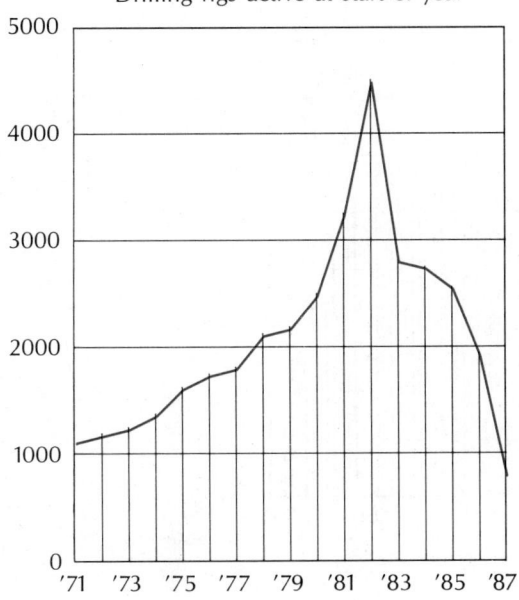

Source: Hughes Tool Co.

FIGURE 2.7

U.S. Petroleum Demand

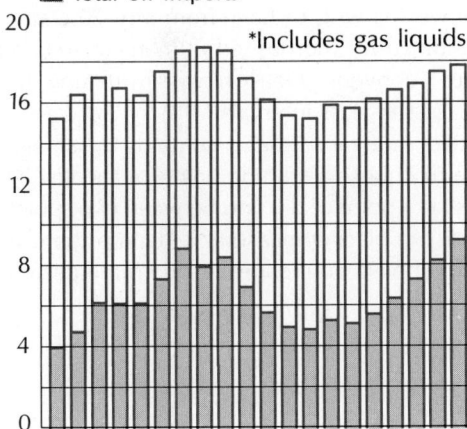

Note: 1986–1990 are projections

Sources: American Petroleum Institute (1971–1985); projections by industry sources

FIGURE 2.8

U.S. Crude Oil
Production

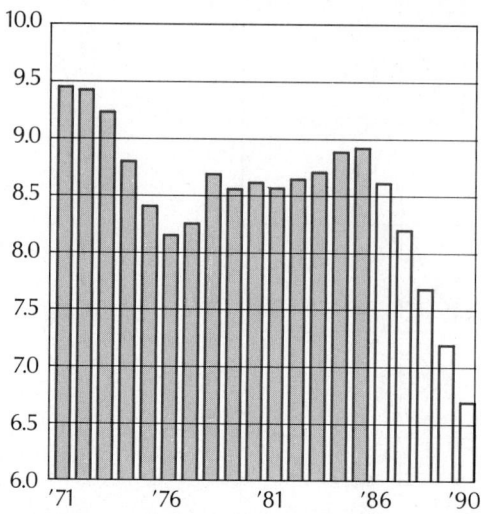

U.S. daily crude output in millions of barrels

Note: 1986–1990 are projections

Sources: Energy Information Administration
(1971–1985); projections by industry sources

GUIDELINES FOR
CONSTRUCTING
CLASS INTERVALS

1. Divide the *range* of the measurements (the difference between the largest and the smallest measurements) by the approximate number of class intervals desired. Generally we will wish to have from 5 to 20 class intervals.
2. After dividing the range by the desired number of subintervals, round the resulting number to a convenient (easy to work with) unit. This unit represents a common width for the class intervals.
3. Choose the first class interval so that it contains the smallest measurement. It is also advisable to choose a starting point for the first interval so that no measurement falls on a point of division between two subintervals. This eliminates any ambiguity in placing measurements into the class intervals. (One way to do this is to choose boundaries to one more decimal place than the data.)

For the data in Table 2.3, the range is

range = 4.9 − 3.6 = 1.3.

Assume that we want to have approximately 10 subintervals. Dividing the range by 10 and rounding to a convenient unit, we have 1.3/10 = .13 ≈ .1. Thus the class interval width is .1.

TABLE 2.3   Weight Gains for Chicks (Grams)

3.7	4.2	4.4	4.4	4.3	4.2	4.4	4.8	4.9	4.4
4.2	3.8	4.2	4.4	4.6	3.9	4.3	4.5	4.8	3.9
4.7	4.2	4.2	4.8	4.5	3.6	4.1	4.3	3.9	4.2
4.0	4.2	4.0	4.5	4.4	4.1	4.0	4.0	3.8	4.6
4.9	3.8	4.3	4.3	3.9	3.8	4.7	3.9	4.0	4.2
4.3	4.7	4.1	4.0	4.6	4.4	4.6	4.4	4.9	4.4
4.0	3.9	4.5	4.3	3.8	4.1	4.3	4.2	4.5	4.4
4.2	4.7	3.8	4.5	4.0	4.2	4.1	4.0	4.7	4.1
4.7	4.1	4.8	4.1	4.3	4.7	4.2	4.1	4.4	4.8
4.1	4.9	4.3	4.4	4.4	4.3	4.6	4.5	4.6	4.0

It is convenient to choose the first interval to be 3.55–3.65, the second to be 3.65–3.75, and so on. Note that the smallest measurement, 3.6, falls in the first interval and that no measurement falls on the endpoint of a class interval (see Table 2.4).

Having determined the class interval, we then construct a frequency table for the data. The first column labels the classes by number and the second column indicates the class intervals. We then examine the 100 measurements of Table 2.3, keeping a tally of the number of measurements falling in each interval. The number of measurements falling in a given class interval is called the **class frequency**. These data are recorded in the third column of the frequency table (see Table 2.4).

class frequency

relative frequency

The **relative frequency** of a class is defined to be the frequency of the class divided by the total number of measurements in the set (total frequency). Thus if we let f_i denote the frequency for class i and n denote the total number of measurements, the relative frequency for class i is f_i/n. The relative frequencies for all the classes are listed in the fourth column of Table 2.4.

TABLE 2.4   Frequency Table for the Chick Data

Class	Class Interval	Frequency f_i	Relative frequency f_i/n
1	3.55–3.65	1	.01
2	3.65–3.75	1	.01
3	3.75–3.85	6	.06
4	3.85–3.95	6	.06
5	3.95–4.05	10	.10
6	4.05–4.15	10	.10
7	4.15–4.25	13	.13
8	4.25–4.35	11	.11
9	4.35–4.45	13	.13
10	4.45–4.55	7	.07
11	4.55–4.65	6	.06
12	4.65–4.75	7	.07
13	4.75–4.85	5	.05
14	4.85–4.95	4	.04
Totals		$n = 100$	1.00

FIGURE 2.9

Frequency Histogram
for the Chick Data of
Table 2.4

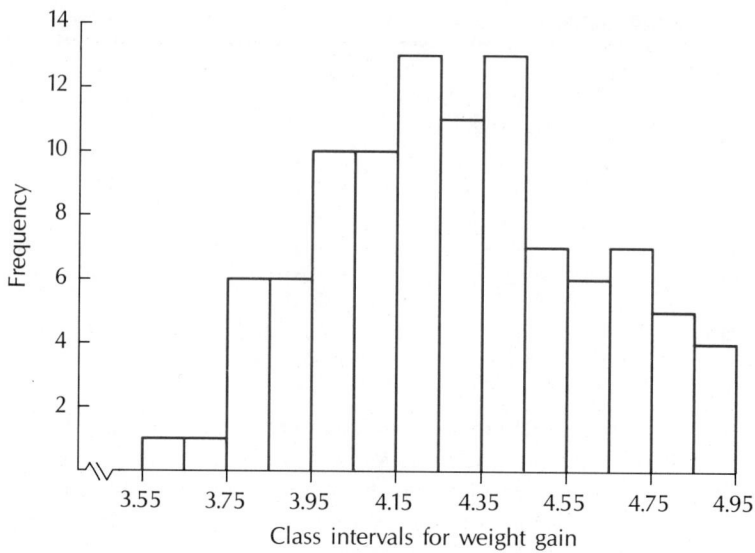

The data of Table 2.3 have been organized into a frequency table, which can now be
histogram   used to construct a *frequency histogram* or a *relative frequency* **histogram**. To construct
a frequency histogram, draw two axes: a horizontal axis labeled with the class intervals
and a vertical axis labeled with the frequencies. Then construct a rectangle over each
class interval with a height equal to the number of measurements falling in a given sub-
interval. The frequency histogram for the data of Table 2.4 is shown in Figure 2.9.

The relative frequency histogram is constructed in much the same way as a frequency
histogram. In the relative frequency histogram, however, the vertical axis is labeled as relative frequency, and a rectangle is constructed over each class interval with a height equal
to the class relative frequency (the fourth column of Table 2.4). The relative frequency
histogram for the data of Table 2.4 is shown in Figure 2.10. Clearly, the two histograms
of Figures 2.9 and 2.10 are of the same shape and would be identical if the vertical
axes were equivalent. We will frequently refer to either one as simply a histogram.

There are several comments that should be made concerning histograms. First, the
distinction between bar charts and histograms is based on the distinction between *qualitative* and *quantitative* variables. Values of qualitative variables vary in kind but not degree,
and hence are not measurements. For example, the variable political party affiliation can be
categorized as Republican, Democrat, or other, and, although we could label the categories
as one, two, or three, these values are only codes and have no quantitative interpretation. In
contrast, quantitative variables have actual units of measure. For example, the variable yield
(in bushels) per acre of corn can assume specific values. *Bar charts are used to display frequency data from qualitative variables; histograms are appropriate for displaying frequency
data for quantitative variables.*

Second, the histogram is the most important graphical technique we will present because of the role it plays in statistical inference, a subject we will discuss in later chapters.
Third, if we had an extremely large set of measurements, and if we constructed a histogram

FIGURE 2.10

Relative Frequency
Histogram for the
Chick Data of Table 2.4

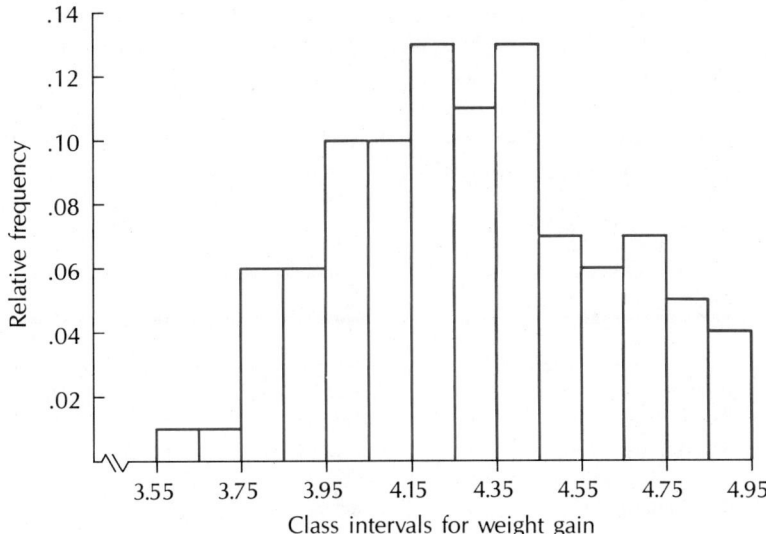

using many class intervals, each with a very narrow width, the histogram for the set of measurements would be, for all practical purposes, a smooth curve. Fourth, the fraction of the total number of measurements in an interval is equal to the fraction of the total area under the histogram over the interval. For example, if we consider the interval 3.75 to 4.35 for the chick data of Table 2.4, we see that there are exactly 56 of the 100 measurements in that interval. Thus .56, the fraction of the total number of sample measurements falling in that interval, is equal to the fraction of the total area under the histogram over that interval, as indicated in Figure 2.11.

FIGURE 2.11

Fraction of
Measurements in the
Interval 3.75 to 4.35 for
the Chick Data

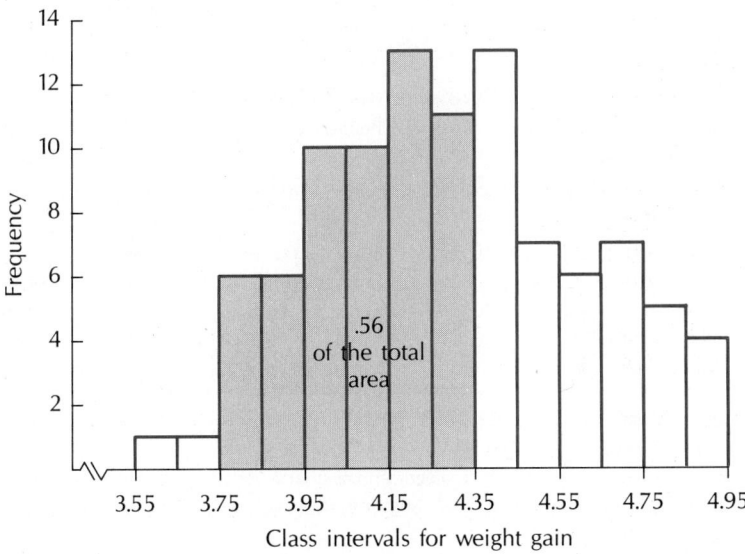

probability   Fifth, if a single measurement is selected at random from the set of sample measurements, the chance, or **probability**, that it lies in a particular interval is equal to the fraction of the total number of sample measurements falling in that interval. This same fraction will be used to estimate the probability that a measurement randomly selected from the population lies in the interval of interest. For example, from the sample data of Table 2.3, the chance or probability of selecting a baby chicken with a weight gain in the interval 3.75

TABLE 2.5   Violent Crime Rates for 90 Standard Metropolitan Statistical Areas Selected from the North, South, and West

South	Rate	North	Rate	West	Rate
Atlanta	843	Albany, N.Y.	270	Bakersfield, Calif.	818
Augusta, Ga.	560	Allentown, Pa.	171	Boise, Idaho	441
Baton Rouge	792	Atlantic City	858	Colorado Springs	451
Beaumont, Tex.	668	Canton, Mass.	336	Denver	640
Birmingham	637	Chicago	585	Eugene, Ore.	351
Charlotte, N.C.	707	Cincinnati	521	Fresno	925
Chattanooga, Tenn.	522	Cleveland	794	Honolulu	322
Columbia, S.C.	953	Detroit	860	Kansas City	869
Corpus Christi, Tex.	592	Evansville, Ind.	444	Las Vegas	1148
Dallas	752	Grand Rapids, Mich.	411	Lawton, Okla.	572
El Paso	621	Johnstown, Pa.	201	Modesto, Calif.	570
Fort Lauderdale	991	Kankakee, Ill.	453	Oklahoma City	618
Greensboro, N.C.	503	Kenosha, Wis.	250	Oxnard, Calif.	480
Houston	719	Lancaster, Pa.	112	Pueblo, Colo.	792
Jackson, Miss.	495	Lansing, Mich.	322	Sacramento	709
Knoxville, Tenn.	404	Lima, Ohio	346	St. Louis	766
Lexington, Ky.	370	Madison, Wis.	212	Salinas, Calif.	574
Lynchburg, Va.	184	Mansfield, Ohio	758	Salt Lake City	390
Macon, Ga.	470	Milwaukee	306	San Diego	629
Mobile, Ala.	832	Newark	1000	San Francisco	1005
Monroe, La.	918	Paterson, N.J.	737	San Jose	514
Nashville, Tenn.	550	Philadelphia	638	Seattle	559
Newport News, Va.	437	Pittsfield, Mass.	305	Sioux City, Iowa	289
Richmond,Va.	499	Racine, Wis.	527	Spokane, Wash.	439
Roanoke, Va.	272	Rockford, Ill.	592	Stockton, Calif.	739
Shreveport, La.	656	South Bend, Ind.	436	Tacoma, Wash.	578
Washington, D.C.	802	Springfield, Ill.	607	Topeka, Kan.	494
West Palm Beach	1068	Syracuse, N.Y.	265	Tucson, Ariz.	616
Wichita Falls, Kan.	830	Vineland, N.J.	482	Vallejo, Calif.	487
Wilmington, Del.	448	Youngstown, Ohio	409	Waco, Tex.	548

Source: Department of Justice, *Uniform Crime Reports for the United States: 1980* (Washington, D.C.: Government Printing Office, 1980), pp. 60–86.

Rates represent the number of violent crimes (that is, murder, forcible rape, robbery, and aggravated assault) per 100,000 inhabitants, rounded to the nearest whole number.

to 4.35 is .56. This value, .56, is an approximation to the probability of selecting a measurement in the interval 3.75 to 4.35 for the population of all weight gains for baby chickens.

Finally, since we use proportions rather than frequencies in a relative frequency histogram, we can compare two different samples (or populations) by examining their relative frequency histograms even if the samples (populations) are of different sizes.

exploratory data analysis

The last graphical technique to be presented in this section is a display technique taken from an area of statistics called **exploratory data analysis (EDA)**. Professor John Tukey (1977) has been the leading proponent of this practical philosophy of data analysis aimed at exploring and understanding data.

stem-and-leaf plot

The **stem-and-leaf plot** is a clever, simple device to construct a histogramlike picture of a frequency distribution. It allows us to use the information contained in a frequency distribution to show the range of scores, where the scores are concentrated, the shape of the distribution, whether there are any specific values or scores not represented, and whether there are any stray or extreme scores. The stem-and-leaf plot does *not* follow the organization principles stated previously for histograms. We will use the data shown in Table 2.5 to illustrate how to construct a stem-and-leaf plot.

The original scores of Table 2.5 are either three- or four-digit numbers. We will use the first or *leading* digit of scores as the stem (see Figure 2.12) and the *trailing* digits as the leaf. For example, the violent crime rate in St. Louis is 766. The leading digit is 7 and the trailing digits are 66. In the case of Las Vegas, the leading digits are 11 and the trailing digits are 48. If our data consisted of 6-digit numbers such as 104,328, we might use the first two digits as stem numbers, the second two digits as leaf numbers, and ignore the last two digits.

For the data on violent crime, the smallest rate is 112, the largest is 1148, and the leading digits are 1, 2, 3, . . . , 11. In the same way that a class interval determines where a measurement is placed in a frequency table, the leading digit (stem of a score) determines the row in which a score is placed in a stem-and-leaf plot. The trailing digits for a score are then written in the appropriate row. In this way each score is recorded in the stem-and-leaf plot. This has been done in Figure 2.12 for the violent crime data.

We can see that each stem defines a class interval and the limits of each interval are the largest and smallest possible scores for the class. The values represented by each leaf must be between the lower and upper limits of the interval.

The plot can be made a bit neater by ordering the data within a row from lowest to highest. The end result is that a stem-and-leaf plot is a graph that looks much like a

FIGURE 2.12
Stem-and-Leaf Plot for Violent Crime Rates of Table 2.5

```
 1 | 84 71 12
 2 | 72 70 01 50 12 65 89
 3 | 70 36 22 46 06 05 51 22 90
 4 | 95 04 70 37 99 48 44 11 53 36 82 09 41 51 80 39 94 87
 5 | 60 22 92 03 50 85 21 27 92 72 70 74 14 59 78 48
 6 | 68 37 21 56 38 07 40 18 29 16
 7 | 92 07 52 19 94 58 37 92 09 66 39
 8 | 43 32 02 30 58 60 18 69
 9 | 53 91 18 25
10 | 68 00 05
11 | 48
```

FIGURE 2.13

Stem-and-Leaf Plot
for the Data of
Table 2.5 and
Figure 2.12

1	12 71 84
2	01 12 50 65 70 72 89
3	05 06 22 22 36 46 51 70 90
4	04 09 11 36 37 39 41 44 48 51 53 70 80 82 87 94 95 99
5	03 14 21 22 27 48 50 59 60 70 72 74 78 85 92 92
6	07 16 18 21 29 37 38 40 56 68
7	07 09 19 37 39 52 58 66 92 92 94
8	02 18 30 32 43 58 60 69
9	18 25 53 91
10	00 05 68
11	48

histogram turned sideways, as in Figure 2.13. The advantage of such a graph over the histogram is that it reflects not only frequencies, concentration(s) of scores, and shape of the distribution, but also the actual scores from which we can determine whether there are any values not represented, and whether there are stray or extreme values.

GUIDELINES FOR
CONSTRUCTING
STEM-AND-LEAF
PLOTS

1. Split each score or value into two sets of digits. The first or leading set of digits is the stem and the second or trailing set of digits is the leaf.
2. List all possible stem digits from lowest to highest.
3. For each score in the mass of data, write down the leaf numbers on the line labeled by the appropriate stem number.
4. If the display looks too cramped and narrow, we can stretch the display by using two lines per stem so that, for example, we place leaf digits 0, 1, 2, 3, and 4 on the first line of the stem and leaf digits 5, 6, 7, 8, and 9 on the second line.
5. If too many trailing digits are present, such as in a 6- or 7-digit score, we drop the right-most trailing digit(s) to maximize the clarity of the display.
6. The rules for developing a stem-and-leaf plot are somewhat different from the rules governing the establishment of class intervals for the traditional frequency distribution and for a variety of other procedures that we will consider in later sections of the text. Class intervals for stem-and-leaf plots are, then, in a sense slightly atypical.

Before leaving graphical methods for describing data there are several general guidelines that can be helpful in developing graphs with an impact. These guidelines pay attention to the design and presentation techniques and should help you make better, more informative graphs.

general
guidelines
for successful
graphics

1. Before constructing a graph, set your priorities. What messages should the viewer get?
2. Choose the type of graph (pie chart, bar graph, histogram, and so on).
3. Pay attention to the title. One of the most important aspects of a graph is its title. The title should immediately inform the viewer of the point of the graph and draw the eye toward the most important elements of the graph.

4. Fight the urge to use many type sizes, styles, and color changes. The indiscriminate and possibly excessive use of different type sizes and styles and of numerous colors will confuse the viewer. Generally, we recommend only two typefaces; color changes and italics should be used in only one or two places.

5. Convey the tone of your graph by using colors and patterns. The more intense, warm colors (yellows, oranges, reds) are more dramatic than the blues and purples and hence help to stimulate enthusiasm by the viewer. On the other hand, pastels (particularly grays) convey a conservative, business-like tone. Similarly, the simple patterns convey a conservative tone, whereas the busier patterns help to stimulate more excitement.

6. Don't underestimate the effectiveness of a simple, straightforward graph.

7. Practice drawing graphs frequently. As with almost anything, practice improves skill.

EXERCISES

2.1 University officials periodically review the distribution of undergraduate majors within the colleges of the university to help determine a fair allocation of resources to departments within the colleges. At one such review the following data were obtained:

College	Number of Majors
Agriculture	1500
Arts and sciences	11,000
Business administration	7000
Education	2000
Engineering	5000

a. Construct a pie chart for these data.
b. Use the same data to construct a bar graph.

2.2 Because of the difficult times many basic industries have endured in recent years, financial analysts have monitored the influx of foreign materials. The data here show steel industry imports (in 1000s of tons) for the years 1979 to 1986.

Year	1979	1980	1981	1982	1983	1984	1985	1986
Import	17,518	15,491	19,898	16,663	17,061	26,171	23,650	19,650

a. Would a pie chart be an appropriate graphical method for describing these data? Explain.
b. Construct a bar graph.

2.3   Graph the data shown here in the allocation of our food dollars to the categories of Table 2.1. Try a pie chart and a bar graph. Which seems better?

Where Our Food Dollars Go	(Percent)
Dairy products	13.4
Cereal and baked goods	12.6
Nonalcoholic beverages	8.9
Poultry and seafood	7.5
Fruit and vegetables	15.6
Meat	24.5
Other foods	17.5

 2.4   The following table shows quarterly gross national product (GNP) and disposable personal income (DPI) for the years 1985 and 1986 (billions of dollars). Choose a form of a single bar graph to display these data on a single graph.

Year/ Quarter	1985				1986			
	I	II	III	IV	I	II	III	IV
GNP	3910	3961	4017	4067	4137	4203	4266	4308
DPI	2505	2532	2503	2533	2536	2555	2579	2589

 2.5   The regulations of the board of health in a particular state specify that the fluoride level must not exceed 1.5 parts per million (ppm). The 25 measurements given below represent the fluoride levels for a sample of 25 days. Although fluoride levels are measured more than once per day, these data represent the early morning readings for the 25 days sampled.

.75	.86	.84	.85	.97
.94	.89	.84	.83	.89
.88	.78	.77	.76	.82
.72	.92	1.05	.94	.83
.81	.85	.97	.93	.79

a. Determine the range of the measurements.
b. Dividing the range by 7, the number of subintervals selected, and rounding, we have a class interval width of .05. Using .705 as the lower limit of the first interval, construct a frequency histogram.
c. Compute relative frequencies for each class interval and construct a relative frequency histogram. Note that the frequency and relative frequency histograms for these data have the same shape.
d. If one of these 25 days was selected at random, what is the chance (probability) that the fluoride reading will be greater than .90 ppm? Guess (predict) what proportion of days in the coming year will have a fluoride reading greater than .90 ppm.

2.6   Construct a bar graph using the unemployment data shown here.

Year	Percentage Unemployed in the Civilian Labor Force
1950	5.3
1955	4.4
1960	5.5
1965	4.5
1970	4.9
1975	8.5
1980	6.6

Source: Bureau of the Census, *Statistical Abstract of the United States: 1980* (Washington, D.C.: Government Printing Office, 1980), p. 396.

2.7   Construct a relative frequency histogram for the data in the accompanying table.

Per Capita Public Welfare Expenses, by State

Dollars	f
50–74	3
75–99	6
100–124	14
125–149	11
150–174	2
175–199	5
200–224	2
225–249	5
250–274	1
275–299	1
Total	50

2.8   Construct a frequency table with suitable class intervals for the data presented here.

32.3	22.8	30.5
31.3	31.3	30.0
29.4	31.1	29.4
31.6	27.6	29.7
30.4	28.8	31.6
31.2	30.7	32.5
29.8	30.3	29.2

2.9   Construct a relative frequency histogram for the data of Exercise 2.8.

2.10   Construct a line plot for the following data.

2.9	3.0	4.4
0.8	2.7	1.6
3.5	3.6	1.2
1.9	3.8	2.2
2.6	3.9	1.5
2.8	4.4	0.9
2.5	4.1	2.3
4.5	3.5	2.5

2.11   Construct a stem-and-leaf diagram for the data in Exercise 2.10. Which plot seems to be more informative for these data?

2.12   Survival times (in months) are shown for patients with severe chronic left-ventricular heart failure. Construct a stem-and-leaf diagram for these data.

4	15	24	10
1	27	31	14
2	16	32	7
13	36	29	6
12	18	14	15
18	6	13	21
20	8	3	24

2.13   Use the data from Exercise 2.12 to construct a frequency histogram. Which plot (the stem-and-leaf diagram or the frequency histogram) describes these data best? Why?

2.14   Below are data from SAT exams for selected years. Plot these data and give some interpretations to the data.

	Year				
Gender, Type	1967	1970	1975	1980	1983
Male, Math	514	509	495	491	493
Female, Math	467	465	449	443	445
Male, Verbal	463	459	437	428	430
Female, Verbal	468	461	431	420	420

Source: College Entrance Examination Board.

2.15   Construct a frequency histogram plot for the telephone data in the accompanying table (telephones per 1000).

State	Telephones	State	Telephones
Alabama	500	Montana	540
Alaska	350	Nebraska	590
Arizona	550	Nevada	720
Arkansas	480	New Hampshire	590
California	610	New Jersey	650
Colorado	570	New Mexico	470
Connecticut	620	New York	530
Delaware	630	N. Carolina	530
Florida	620	N. Dakota	560
Georgia	570	Ohio	550
Hawaii	480	Oklahoma	580
Idaho	550	Oregon	560
Illinois	650	Pennsylvania	610
Indiana	580	Rhode Island	560
Iowa	570	S. Carolina	510
Kansas	600	S. Dakota	540
Kentucky	480	Tennessee	540
Louisiana	520	Texas	570
Maine	540	Utah	560
Maryland	610	Vermont	520
Massachusetts	570	Virginia	530
Michigan	580	Washington	570
Minnesota	560	W. Virginia	450
Mississippi	470	Wisconsin	540
Missouri	570	Wyoming	580

2.16   Construct a stem-and-leaf plot for the data of Exercise 2.15. Interpret the data display.

2.17   Computer output is shown for the data of Exercise 2.15. Compare the stem-and-leaf plot in the output to the one you constructed in Exercise 2.16.

```
      PRINT C1
C1
     500     350     550     480     610     570     620     630     620     570     480
     550     650     580     570     600     480     520     540     610     570     580
     560     470     570     540     590     720     590     650     470     530     530
     560     550     580     560     610     560     510     540     540     570     560
     520     530     570     450     540     580

MTB > STEM-AND-LEAF C1;
SUBC> INCREMENT = 100.

Stem-and-leaf of C1          N  = 50
Leaf Unit = 10

      1     3 5
      7     4 577888
    (33)    5 012233344444555666667777777888899
     10     6 011122355
      1     7 2

MTB > STOP
```

2.18   Construct a frequency histogram for the data of Table 2.5. Then compare the histogram to the stem-and-leaf plot of the data shown in Figure 2.13.

2.3 DESCRIBING DATA ON A SINGLE VARIABLE: MEASURES OF CENTRAL TENDENCY

Numerical descriptive measures are commonly used to convey a mental image of pictures, objects, and other phenomena. There are two main reasons for this: first, graphical descriptive measures are inappropriate for statistical inference, since it is difficult to describe the similarity of a sample frequency histogram and the corresponding population frequency histogram. The second reason for using numerical descriptive measures is one of expediency—we never seem to carry the appropriate graphs or histograms with us, and so must resort to our powers of verbal communication to convey the appropriate picture. We seek several numbers, called *numerical descriptive measures*, that will create a mental picture of the frequency distribution for a set of measurements.

central tendency, variability   The two most common numerical descriptive measures are measures of **central tendency** and measures of **variability**. That is, we seek to describe the center of the distribution of measurements and also how the measurements vary about the center of the distribution. We will draw a distinction between numerical descriptive measures for a

parameter, statistic   population, called **parameters**, and numerical descriptive measures for a sample, called **statistics**. In problems requiring statistical inference, we will not be able to calculate values for various parameters, but we will be able to compute corresponding statistics from the sample and use these quantities to estimate the corresponding population parameters.

In this section we will consider various measures of central tendency, followed in Section 2.4 by a discussion of measures of variability.

mode   The first measure of central tendency we consider is the **mode**.

> **DEFINITION 2.1**
>
> The **mode** of a set of measurements is defined to be the measurement that occurs most often (with the highest frequency).

We illustrate the use and determination of the mode in an example.

EXAMPLE 2.1   Slaughter weights (in pounds) for a sample of 15 Herefords each with a frame size of three (on a one-seven scale) are shown here.

962	1005	1033
980	965	1030
975	989	955
1015	1000	970
1042	1005	995

Determine the modal slaughter weight.

Solution

For these data, the weight 1005 occurs twice and all others once. Hence the mode is 1005.

Identification of the mode for Example 2.1 was quite easy because we were able to count the number of times each measurement occurred. When dealing with grouped data—data presented in the form of a frequency table—we can define the modal interval to be the class interval with the highest frequency. However, since we would not know the actual measurements, but only how many measurements fall into each interval, the mode is taken as the midpoint of the modal interval; it is an approximation to the mode of the actual sample measurements.

The mode is also commonly used as a measure of popularity that reflects central tendency or opinion. For example, we might talk about the most preferred stock, a most preferred model of washing machine, or the most popular candidate. In each case we would be referring to the mode of the distribution.

It should be noted that some distributions have more than one measurement that occurs with the highest frequency. Thus we might encounter bimodal, trimodal, and so on, distributions.

median

The second measure of central tendency we consider is the **median**.

DEFINITION 2.2

The **median** of a set of measurements is defined to be the middle value when the measurements are arranged from lowest to highest.

The median is most often used to measure the midpoint of a large set of measurements. For example, we may read about the median wage increase won by union members, the median age of persons receiving social security benefits, and the median weight of cattle prior to slaughter during a given month. Each of these situations involves a large set of measurements, and the median would reflect the central value of the data.

However, we may use the definition of median for small sets of measurements by using the following convention. The median for an even number of measurements will be the average of the two middle values when the measurements are arranged from lowest to highest. When there is an odd number of measurements, the median is still the middle value. Thus, whether there is an even or odd number of measurements, there is an equal number of measurements above and below the median.

EXAMPLE 2.2

Each of 10 children in the second grade was given a reading aptitude test. The scores were as shown below.

| 95 | 86 | 78 | 90 | 62 | 73 | 89 | 92 | 84 | 76 |

Determine the median test score.

Solution

We must first arrange the scores in order of magnitude.

| 62 | 73 | 76 | 78 | 84 | 86 | 89 | 90 | 92 | 95 |

Since there is an even number of measurements, the median is the average of the two midpoint scores.

$$\text{median} = \frac{84 + 86}{2} = 85$$

EXAMPLE 2.3

An experiment was conducted to measure the effectiveness of a new procedure for pruning grapes. Each of 13 workers was assigned the task of pruning an acre of grapes. The productivity, measured in worker-hours/acre, is recorded for each person.

4.4   4.9   4.2   4.4   4.8   4.9   4.8   4.5   4.3   4.8   4.7   4.4   4.2

Determine the mode and median productivity for the group.

Solution

First arrange the measurements in order of magnitude:

4.2   4.2   4.3   4.4   4.4   4.4   4.5   4.7   4.8   4.8   4.8   4.9   4.9

For these data we have two measurements appearing three times each. Hence the data are bimodal with modes of 4.4 and 4.8. The median for the odd number of measurements is the middle score, 4.5.

grouped data median

The median for **grouped data** is slightly more difficult to compute. Since the actual values of the measurements are unknown, we know that the median occurs in a particular class interval, but we do not know where to locate the median within the interval. If we assume that the measurements are spread evenly throughout the interval, we get the following result. Let

L = lower class limit of the interval that contains the median

n = total frequency

cf_b = the sum of frequencies (cumulative frequency) for all classes before the median class

f_m = frequency of the class interval containing the median

w = interval width

Then for grouped data,

$$\text{median} = L + \frac{w}{f_m}(.5n - cf_b)$$

The next example illustrates how to find the median for grouped data.

EXAMPLE 2.4

Table 2.6 is the frequency table for the chick data of Table 2.4. Compute the median weight gain for these data.

TABLE 2.6 Frequency Table for the Chick Data, Example 2.4

Class Interval	f_i	Cumulative f_i	f_i/n	Cumulative f_i/n
3.55–3.65	1	1	.01	.01
3.65–3.75	1	2	.01	.02
3.75–3.85	6	8	.06	.08
3.85–3.95	6	14	.06	.14
3.95–4.05	10	24	.10	.24
4.05–4.15	10	34	.10	.34
4.15–4.25	13	47	.13	.47
4.25–4.35	11	58	.11	.58
4.35–4.45	13	71	.13	.71
4.45–4.55	7	78	.07	.78
4.55–4.65	6	84	.06	.84
4.65–4.75	7	91	.07	.91
4.75–4.85	5	96	.05	.96
4.85–4.95	4	100	.04	1.00
Totals	$n = 100$		1.00	

Solution

Let the cumulative relative frequency for class j equal the sum of the relative frequencies for class 1 through class j. To determine the interval that contains the median, we must find the first interval for which the cumulative relative frequency exceeds .50. This interval will be the one containing the median. For these data, the interval from 4.25 to 4.35 is the first interval for which the cumulative relative frequency exceeds .50, as shown in Table 2.6, column 5. So this interval contains the median. Then

$$L = 4.25 \qquad f_m = 11$$
$$n = 100 \qquad w = .1$$
$$cf_b = 47$$

and

$$\text{median} = L + \frac{w}{f_m}(.5n - cf_b) = 4.25 + \frac{.1}{11}(50 - 47) = 4.28$$

The third, and last, measure of central tendency we will discuss in this text is the **mean** arithmetic mean, known simply as the **mean**.

DEFINITION 2.3

The **arithmetic mean**, or **mean**, of a set of measurements is defined to be the sum of the measurements divided by the total number of measurements.

Quite often when people talk about an "average," they are referring to the mean. Because of the important role that the mean will play in statistical inference in later chapters, we give special symbols to the population mean and the sample mean. The *popu-*

μ *lation mean* is denoted by the Greek letter μ (read "mu"), and the *sample mean* is denoted

$\bar{y}$ by the symbol $\bar{y}$ *(read "y-bar")*. As indicated in Chapter 1, a population of measurements is the complete set of measurements of interest to us; a sample of measurements is a subset of measurements selected from the population of interest. If we let $y_1, y_2, \ldots, y_n$ denote the measurements observed in a sample of size n, then the sample mean $\bar{y}$ can be written as

$$\bar{y} = \frac{\sum_i y_i}{n}$$

where the symbol appearing in the numerator, $\sum_i y_i$, is the notation used to designate a sum of n measurements, y_i

$$\sum_i y_i = y_1 + y_2 + \cdots + y_n$$

The corresponding population mean is μ.

In most situations we will not know the population mean; the sample will be used to make inferences about the corresponding unknown population mean. Details about how this is done and the inferences we can make will be discussed in Chapter 4.

EXAMPLE 2.5

A representative sample of $n = 15$ overdue accounts in a large department store yields the following amounts due:

$55.20	$ 4.88	$271.95
18.06	180.29	365.29
28.16	399.11	807.80
44.14	97.47	9.98
61.61	56.89	82.73

a. Determine the mean amount due for the 15 accounts sampled.
b. If there are a total of 150 overdue accounts, use the sample mean to predict the total amount overdue for all 150 accounts.

Solution

a. The sample mean is computed as follows:

$$\bar{y} = \frac{\sum_i y_i}{15} = \frac{55.20 + 18.06 + \cdots + 82.73}{15} = \frac{2483.56}{15} = \$165.57$$

b. From part (a) we found that the 15 accounts sampled averaged $165.57 overdue. Using this information, we would predict, or estimate, the total amount overdue for the 150 accounts to be 150(165.57) = $24,835.50.

The sample mean formula for grouped data is only slightly more complicated than the formula just presented for ungrouped data. Since we do not know the individual sample measurements, but only the interval to which a measurement is assigned, this formula will be an approximation to the actual sample mean. Hence when the sample measurements are known, the formula for ungrouped data should be used. We will use the same symbol $\bar{y}$ to designate the sample mean for grouped data. If there are k class intervals and

y_i = midpoint of the ith class interval

f_i = frequency associated with the ith class interval

n = the total number of measurements

then

$$\bar{y} = \frac{\sum_i f_i y_i}{n}$$

EXAMPLE 2.6 The data of Example 2.4 are reproduced in Table 2.7, along with two additional columns, y_i and $f_i y_i$, that will be helpful in computing the mean. Compute the sample mean for this set of grouped data.

Solution Adding the entries in the $f_i y_i$ column and substituting into the formula, we find the sample mean to be

$$\bar{y} = \frac{\sum_i f_i y_i}{100} = \frac{429.2}{100} = 4.29$$

Table 2.7 Chick Data, Example 2.6

Class Interval	f_i	y_i	$f_i y_i$
3.55–3.65	1	3.6	3.6
3.65–3.75	1	3.7	3.7
3.75–3.85	6	3.8	22.8
3.85–3.95	6	3.9	23.4
3.95–4.05	10	4.0	40.0
4.05–4.15	10	4.1	41.0
4.15–4.25	13	4.2	54.6
4.25–4.35	11	4.3	47.3
4.35–4.45	13	4.4	57.2
4.45–4.55	7	4.5	31.5
4.55–4.65	6	4.6	27.6
4.65–4.75	7	4.7	32.9
4.75–4.85	5	4.8	24.0
4.85–4.95	4	4.9	19.6
Totals	100		429.2

outliers

trimmed mean

skewness

The mean is a useful measure of the central value of a set of measurements, but it is subject to distortion due to the presence of one or more extreme values in the set. In these situations the extreme values (called **outliers**) pull the mean in the direction of the outliers, thus distorting the mean as a measure of the central value. A variation of the mean, called a **trimmed mean**, drops the highest and lowest extreme values and averages the rest. For example, a 20% trimmed mean drops the highest 20% and the lowest 20% of the measurements and averages the rest. Similarly, a 10% trimmed mean drops the highest and the lowest 10% of the measurements and averages the rest. By trimming the data, we are able to reduce the impact of very large (or small) values on the mean, and thus get a more reliable measure of the central value of the set. This will be particularly important when the sample mean is used to predict the corresponding population central value.

In this section we discussed the mode, median, mean, and trimmed mean. How are these measures of central tendency related for a given set of measurements? The answer depends on the **skewness** of the data. If the distribution is mound-shaped and symmetrical about a single peak, the mode (M_o), median (M_d), mean (μ), and trimmed mean (TM) will all be the same. This is shown using a smooth curve and population quantities in Figure 2.14(a). If the distribution is skewed, having a long tail in one direction and a single peak, the mean is pulled in the direction of the tail; the median falls between the mode and the mean; and depending on the degree of trimming, the trimmed mean usually falls between the median and the mean. Figure 2.14(b) and (c) illustrates this for distributions skewed to the left and to the right.

The important thing to remember is that we are not restricted to using only one measure of central tendency. For some data sets, it will be necessary to use more than one of these measures to provide an accurate descriptive summary of central tendency for the data.

FIGURE 2.14

Relation Among the Mean μ, the Trimmed Mean TM, the Median M_d, and the Mode M_o

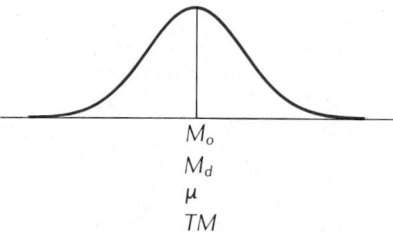

(a) A mound-shaped distribution

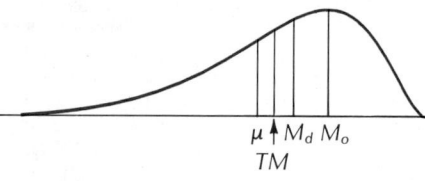

(b) A distribution skewed to the left

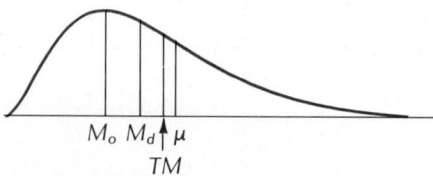

(c) A distribution skewed to the right

MAJOR
CHARACTERISTICS
OF EACH
MEASURE OF
CENTRAL
TENDENCY

Mode

1. It is the most frequent or probable measurement in the data set.
2. There can be more than one mode for a data set.
3. It is not influenced by extreme measurements.
4. Modes of subsets cannot be combined to determine the mode of the complete data set.
5. For grouped data, its value can change depending on the categories used.
6. It is applicable for both qualitative and quantitative data.

Median

1. It is the central value; 50% of the measurements lie above it and 50% fall below it.
2. There is only one median for a data set.
3. It is not influenced by extreme measurements.
4. Medians of subsets cannot be combined to determine the median of the complete data set.
5. For grouped data, its value is rather stable even when the data are organized into different categories.
6. It is applicable to quantitative data only.

Mean

1. It is the arithmetic average of the measurements in a data set.
2. There is only one mean for a data set.
3. Its value is influenced by extreme measurements; trimming can help to reduce the degree of influence.
4. Means of subsets can be combined to determine the mean of the complete data set.
5. It is applicable to quantitative data only.

Measures of central tendency do not provide a complete mental picture of the frequency distribution for a set of measurements. In addition to determining the center of the distribution, we must have some measure of the spread of the data. In the next section we discuss measures of variability, or dispersion.

EXERCISES

2.19 Compute the mean, median, and mode for the following data:

11	17	18	10	22	23	15	17
14	13	10	12	18	18	11	14

2.20 Refer to the data in Exercise 2.19 with the measurements 22 and 23 replaced by 42 and 43. Recompute the mean, median, and mode. Discuss the impact of these extreme measurements on the three measures of central tendency.

2.21 Refer to Exercises 2.19 and 2.20. Compute a 10% trimmed mean for both data sets. Do the extreme values affect the 10% trimmed mean? Would a 5% trimmed mean be affected?

 2.22 Salaries for 40 recent MBA graduates from a major university are summarized here (in $000).

Interval	Frequency
24.9–29.9	6
29.9–34.9	10
34.9–39.9	15
39.9–44.9	7
44.9–49.9	2

Determine the mode, median, and mean for the data shown in this frequency table. What does the relation among the three measures indicate about the shape of the histogram for these data?

2.23 Determine the mode, median, and mean for the following measurements:

10	2	1	5
1	5	7	10
3	4	8	12
5	6	8	9

2.24 Determine the mean, median, and mode for the data presented in the following frequency table:

Class Interval	Frequency
0–2	1
3–5	3
6–8	5
9–11	4
12–14	2

2.25 Exercise capacity (in seconds) was determined for each of 11 patients being treated for chronic heart failure.

906	1320
711	1170
684	1200
837	1056
897	882
1008	

Determine the median and mean.

2.26 Daily crude oil output (in millions of barrels) is shown for the years 1971 to 1986. Compute the mean and median daily output for these years.

Year	Output
1971	9.45
1972	9.40
1973	9.25
1974	8.75
1975	8.30
1976	8.10
1977	8.25
1978	8.70
1979	8.55
1980	8.60
1981	8.55
1982	8.65
1983	8.70
1984	8.70
1985	8.91
1986	8.60

2.27   If additional data and projections are available for the years 1987 to 1990, how would the computations of Exercise 2.26 change?

Year	Output
1987	8.20
1988	7.70
1989	7.20
1990	6.75

2.28   Based on the frequency distribution contained in the data in the accompanying table, what is the class interval width?

Normal Daily Mean Temperatures, Annual Average

Temperature	Frequency f
39–41	3
42–44	2
45–47	8
48–50	10
51–53	9
54–56	10
57–59	8
60–62	7
63–65	3
66–68	3
69–71	2
72–74	0
75–77	2
Total	67

Determine the mode, median, and mean for these data. Would a trimmed mean better describe the center of the distribution than the mean does? Explain.

 2.29 Nitrogen is a limiting factor in the yield of many different plants. In particular, the yield of apple trees is directly related to the nitrogen content of apple tree leaves and must be carefully monitored to protect the trees in an orchard. Research has shown that the nitrogen content should be approximately 2.5% for best yield results. (It should be noted that some researchers report their results in parts per million (ppm); hence 1% would be equivalent to 10,000 ppm.)

To determine the nitrogen content of trees in an orchard, the growing tips of 150 leaves are clipped from trees throughout the orchard. These leaves are ground to form one composite sample, which the researcher assays for percentage of nitrogen. Composite samples obtained from a random sample of 36 orchards throughout the state gave the following nitrogen contents:

2.0968	2.8220	2.1739	1.9928	2.2194	3.0926
2.4685	2.5198	2.7983	2.0961	2.9216	2.1997
1.7486	2.7741	2.8241	2.6691	3.0521	2.9263
2.9367	1.9762	2.3821	2.6456	2.7678	1.8488
1.6850	2.7043	2.6814	2.0596	2.3597	2.2783
2.7507	2.4259	2.3936	2.5464	1.8049	1.9629

a. Round each of these measurements to the nearest hundredth. (Use the convention that 5 is rounded up.)
b. Determine the sample mode for the rounded data.
c. Determine the sample median for the rounded data.
d. Determine the sample mean for the rounded data.

2.30 Refer to the data of Exercise 2.29 rounded to the nearest hundredth. Replace the fourth measurement (2.94) by the value 29.40. Compute the sample mean, median, and mode for these data. Compare these results to those you found in Exercise 2.29.

2.31 Refer to the data of Example 2.5. Since the sample mean is greater than 10 of the 15 observations, suggest a more appropriate measure of central tendency. Compute its value. How does the distribution of amounts for overdue accounts appear to be skewed?

2.32 Effective tax rate (per $1000) on residential property for 10 cities from the south, north, and west are shown here.

South	Rate	North	Rate	West	Rate
Atlanta	23.8	Boston	74.6	Denver	9.1
Baltimore	26.5	Chicago	18.2	Honolulu	8.2
Dallas	24.7	Cleveland	21.4	Kansas City	12.5
Houston	21.8	Columbus	13.1	Los Angeles	12.5
Jacksonville	15.7	Detroit	36.6	Phoenix	16.8
Memphis	19.4	Indianapolis	42.8	St. Louis	21.4
Nashville	11.3	Milwaukee	34.7	San Diego	11.1
New Orleans	8.4	New York	18.4	San Francisco	12.1
San Antonio	19.5	Philadelphia	30.9	San Jose	11.2
Washington, D.C.	16.6	Pittsburgh	37.2	Seattle	15.3

Source: Bureau of the Census, *Statistical Abstract of the United States: 1980* (Washington, D.C.: Government Printing Office, 1980), p. 317.

a. Compute the mean, median, and mode separately for the south, north, and west.

b. Compute the mean, median, and mode for the complete set of 30 measurements.

c. What measure or measures best summarize the center of these distributions? Explain.

2.33   Refer to Exercise 2.32. Average the three group means, the three group medians, and the three group modes, and compare your results to those of part (b). Comment on your findings.

2.4   DESCRIBING DATA ON A SINGLE VARIABLE: MEASURES OF VARIABILITY

The need for some measure of variability is illustrated in the relative frequency histograms of Figure 2.15. All the histograms have the same mean but each has a different spread, **variability**   or **variability**, about the mean. For purposes of illustration, we have shown the histograms as smooth curves.

range   The simplest but least useful measure of data variation is the **range**. Recall that we alluded to the range in Section 2.2. We now present its definition.

DEFINITION 2.4

The **range** of a set of measurements is defined to be the difference between the largest and the smallest measurements of the set.

FIGURE 2.15

Relative Frequency Histograms with Different Variabilities but the Same Mean

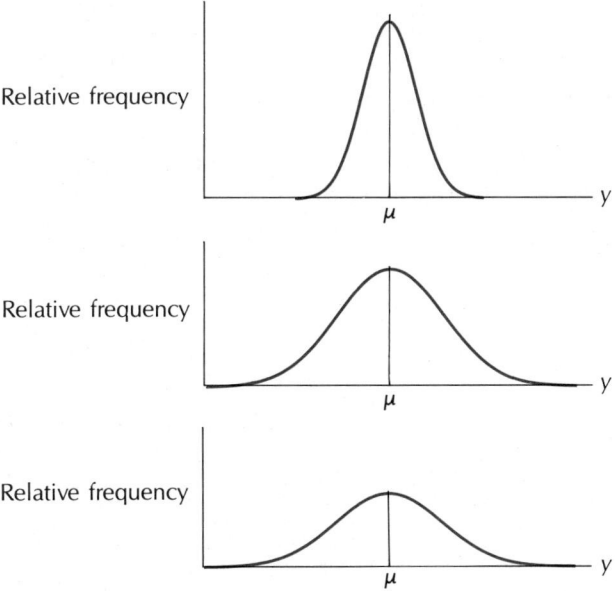

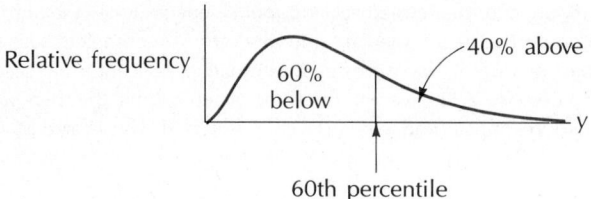

Relative frequency

60% below

40% above

60th percentile

y

EXAMPLE 2.7   Determine the range of the 15 overdue accounts of Example 2.5.

Solution   The smallest measurement is $4.88 and the largest is $807.80. Hence the range is

$$807.80 - 4.88 = \$802.92$$

grouped data   For **group data**, since we do not know the individual measurements, the **range** is
range   taken to be the difference between the upper limit of the last interval and the lower limit
of the first interval.

   Although the range is easy to compute, it is sensitive to outliers since it depends on
the most extreme values. It does not give much information about the pattern of variability.
In Figure 2.15, the middle and the bottom distributions have the same mean and the same
range yet they differ substantially in their variability about the mean. What we seek is a
measure of variability that is more sensitive to the piling up of data about the mean.

percentile   A second measure of variability involves the use of **percentiles**.

DEFINITION 2.5

   The **pth percentile** of a set of n measurements arranged in order of magnitude is
   that value that has at most $p\%$ of the measurements below it and at most
   $(100 - p)\%$ above it.

For example, Figure 2.16 illustrates the 60th percentile of a set of measurements.

   Percentiles are frequently used to describe the results of achievement test scores and
the ranking of a person in comparison to the rest of the people taking an examination.
Specific percentiles of interest are the 25th, 50th, and 75th percentiles, often called the
lower quartile, the middle quartile (median), and the upper quartile, respectively (see
Figure 2.17).

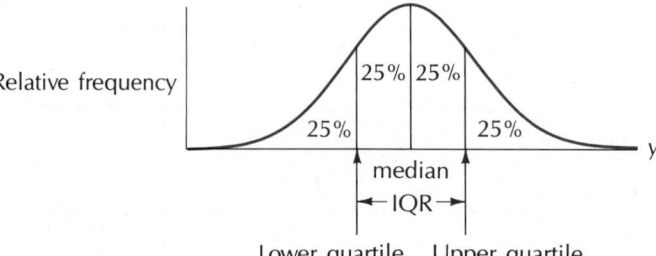

Relative frequency

25% 25%

25%   25%

median

IQR

Lower quartile   Upper quartile

y

**grouped data
percentile**

Now that we have learned how to compute the median (50th percentile) for grouped data, other percentiles follow immediately. Let

P = percentile of interest

L = lower limit of the class interval that includes percentile of interest

n = total frequency

cf_b = cumulative frequency for all class intervals before the percentile class

f_p = frequency of the class interval that includes the percentile of interest

w = interval width

Then, for example, the 65th percentile for a set of grouped data would be computed using the formula

$$P = L + \frac{w}{f_p}(.65n - cf_b)$$

To determine L, f_p, and cf_b, begin with the lowest interval and find the first interval for which the cumulative relative frequency exceeds .65. This interval would contain the 65th percentile.

EXAMPLE 2.8

Refer to the chick data of Table 2.6 Compute the 90th percentile.

Solution

Since the twelfth interval is the first interval for which the cumulative relative frequency exceeds .90, we have

$L = 4.65$

$n = 100$

$cf_b = 84$

$f_{90} = 7$

$w = .1$

Thus the 90th percentile is

$$P_{90} = 4.65 + \frac{.1}{7}[.9(100) - 84] = 4.74$$

This means that 90% of the measurements lie below this value and 10% lie above it.

interquartile range

The second measure of variability, the **interquartile range** is now defined. A slightly different definition of the interquartile range is given along with the box plot (Section 2.6).

DEFINITION 2.6

The **interquartile range (IQR)** of a set of measurements is defined to be the difference between the upper and lower quartiles (see Figure 2.17). That is, IQR = 75th percentile − 25th percentile.

The interquartile range, although more sensitive to data pileup about the midpoint than the range, is still not sufficient for our purposes. In particular, the IQR can be used for comparing the variability of two sets of measurements, but not much useful information is gained from the IQR in interpreting the variability of a single set of measurements.

FIGURE 2.18

Dot Diagram of the
Percentages of
Registered Voters in
Five Cities

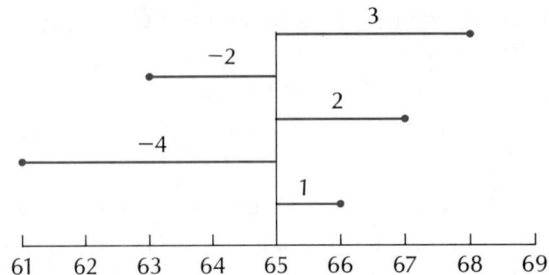

We seek now a sensitive measure of variability, not only for comparing the variabilities of two sets of measurements, but also for interpreting the variability of a single set of mea-

deviation   surements. To do this we work with the **deviation** $y - \bar{y}$ of a measurement y from its mean $\bar{y}$.

To illustrate, suppose we have five sample measurements $y_1 = 68$, $y_2 = 67$, $y_3 = 66$, $y_4 = 63$, and $y_5 = 61$, which represent the percentages of registered voters in five cities who exercised their right to vote at least once during the past year. These measurements are shown in the dot diagram of Figure 2.18. Each measurement is located by a dot above the horizontal axis of the diagram. We use the sample mean

$$\bar{y} = \frac{\sum_i y_i}{n} = \frac{325}{5} = 65$$

to locate the center of the set and we construct horizontal lines in Figure 2.18 to represent the deviations of the sample measurements from their mean. The deviations of the measurements are computed by using the formula $y - \bar{y}$. The five measurements and their deviations are shown in Figure 2.18.

A data set with very little variability will have most of the measurements located near the center of the distribution. Deviations from the mean for a more variable set of measurements would be relatively large.

Many different measures of variability can be constructed by using the deviations $y - \bar{y}$. A first thought would be to use the mean deviation, but this will always equal zero, as it does for our example. A second possibility would be to ignore the minus signs and compute the average of the absolute values. However, a more easily interpreted function of the deviations involves the sum of the squared deviations of the measurements from their

variance   mean. This measure is called the **variance**.

DEFINITION 2.7

The **variance** of a set of n measurements $y_1, y_2, \ldots, y_n$ with mean $\bar{y}$ is the sum of the squared deviations divided by $n - 1$:

$$\frac{\sum_i (y_i - \bar{y})^2}{n - 1}$$

As with the sample and population means, we have special symbols to denote the

s^2 sample and population variances. The symbol s^2 represents the sample variance, and the

σ^2 corresponding population variance is denoted by the symbol σ^2.

The definition for the variance of a set of measurements depends on whether the data are regarded as a sample or population of measurements. The definition we have given here assumes we're working with the sample, since the population measurements usually aren't available. Many statisticians define the sample variance to be the average of the squared deviations, $\sum (y - \bar{y})^2/n$. However, this quantity tends to underestimate the corresponding population variance σ^2. By using the divisor $n - 1$, we get a better estimate of σ^2, and so this is the divisor we will use in our definition of the sample variance.

standard deviation

Another useful measure of variability, the **standard deviation**, involves the square root of the variance.

DEFINITION 2.8

The **standard deviation** of a set of measurements is defined to be the positive square root of the variance.

s

σ

We then have s denoting the sample standard deviation and σ denoting the corresponding population standard deviation.

EXAMPLE 2.9

The time between an electric light stimulus and a bar press to avoid a shock was noted for each of five conditioned rats. Use the data below to compute the sample variance and standard deviation.

shock avoidance times (seconds): 5, 4, 3, 1, 3

Solution

The deviations and the squared deviations are shown below. The sample mean $\bar{y}$ is 3.2.

	y_i	$y_i - \bar{y}$	$(y_i - \bar{y})^2$
	5	1.8	3.24
	4	.8	.64
	3	−.2	.04
	1	−2.2	4.84
	3	−.2	.04
Totals	16	0	8.80

Using the total of the squared deviations column, we find the sample variance to be

$$s^2 = \frac{\sum_i (y_i - \bar{y})^2}{4} = \frac{8.80}{4} = 2.2$$

Computation of the quantities s^2 and s is sometimes simplified using the following algebraic identity:

$$\sum_i (y_i - \bar{y})^2 = \sum_i y_i^2 - \frac{(\sum_i y_i)^2}{n}$$

Hence we have the shortcut formula for s^2 (and s) given here.

SHORTCUT
FORMULA FOR
s^2 AND s

$$s^2 = \frac{1}{n-1}\left[\sum_i y_i^2 - \frac{(\sum_i y_i)^2}{n}\right] \quad \text{and} \quad s = \sqrt{s^2}$$

EXAMPLE 2.10

Use the data of Example 2.9 to compute the sample variance using the shortcut formula.

Solution

It is convenient to construct the following table when we do not have a calculator to perform the calculations:

y_i	y_i^2
5	25
4	16
3	9
1	1
3	9
Totals 16	60

Using the totals from the table, we have

$$s^2 = \frac{1}{4}\left[60 - \frac{(16)^2}{5}\right] = \frac{1}{4}[60 - 51.2] = 2.2$$

which is exactly the result we obtained in Example 2.9.

We can make a simple modification of our shortcut formula to approximate the sample variance if only grouped data are available. Recall that in approximating the sample mean for grouped data, we let y_i and f_i denote the midpoint and frequency, respectively, for the ith class interval. With this notation the sample variance for grouped data is

$$s^2 = \frac{1}{n-1}\left[\sum_i f_i y_i^2 - \frac{(\sum_i f_i y_i)^2}{n}\right]$$

The sample standard deviation is $\sqrt{s^2}$.

EXAMPLE 2.11

Refer to the chick data from Table 2.7 of Example 2.6. Calculate the sample variance and standard deviation for these data.

Solution

In addition to the calculations in Table 2.7, we also need the calculations for $f_i y_i^2$. These calculations, formed by multiplying corresponding elements in the y_i and $f_i y_i$ columns, are shown in the listing below.

$f_i y_i^2$:

12.96	13.69	86.64	91.26	160.00	168.10	229.32
203.39	251.68	141.75	126.96	154.63	115.20	96.04

The sum of the $f_i y_i^2$ calculations is 1851.62. Using this total and the total of $f_i y_i$ in Table 2.7, we can determine s^2 and s.

$$s^2 = \frac{1}{n-1}\left[\sum_i f_i y_i^2 - \frac{\left(\sum_i f_i y_i\right)^2}{n}\right]$$

$$= \frac{1}{99}\left[1851.62 - \frac{(429.2)^2}{100}\right] = \frac{9.49}{99} = .10$$

$$s = \sqrt{.10} = .32$$

We have now discussed several measures of variability, each of which can be used to compare the variabilities of two or more sets of measurements. The standard deviation is particularly appealing for two reasons: (1) we can compare the variabilities of *two or more* sets of data using the standard deviation, and (2) we can also use the results of the rule that follows to interpret the standard deviation of a single set of measurements. This rule applies to data sets with roughly a "mound-shaped" histogram; that is, a histogram that has a single peak, is symmetrical, and tapers off gradually in the tails. Since so many data sets can be classified as mound-shaped, the rule has wide applicability. For this reason it is called the *Empirical Rule*.

EMPIRICAL RULE

Given a set of n measurements possessing a mound-shaped histogram, then

the interval $\bar{y} \pm s$ contains approximately 68% of the measurements
the interval $\bar{y} \pm 2s$ contains approximately 95% of the measurements
the interval $\bar{y} \pm 3s$ contains approximately all the measurements

EXAMPLE 2.12

A sample of 20 days throughout the previous year indicated that the average wholesale price per pound for steers at a particular stockyard was $.61, with a standard deviation of $.07. If the histogram for the measurements is mound-shaped, describe the variability of the data using the Empirical Rule.

Solution

Applying the Empirical Rule, the interval

$$.61 \pm .07 \quad \text{or} \quad \$.54 \text{ to } \$.68$$

contains approximately 68% of the measurements. The interval

$$.61 \pm .14 \quad \text{or} \quad \$.47 \text{ to } \$.75$$

contains approximately 95% of the measurements. The interval

$$.61 \pm .21 \quad \text{or} \quad \$.40 \text{ to } \$.82$$

contains approximately all the measurements.

The results of the Empirical Rule enable us to obtain a check of the calculation of the sample standard deviation s. The Empirical Rule states that approximately 95% of the measurements lie in the interval $\bar{y} \pm 2s$. The length of this interval is therefore $4s$. Since the range approximating s | of the measurements is approximately 4s, we obtain an **approximate value for s** by dividing the range by 4.

> $$\text{Approximate value of } s = \frac{\text{range}}{4}$$

Some people might wonder why we did not equate the range to $6s$ since the interval $\bar{y} \pm 3s$ should contain almost all the measurements. This procedure would yield an approximate value for s that is smaller than the one obtained by the procedure above. If we are going to make an error (as we are bound to do with any approximation), it is better to overestimate the sample standard deviation so that we are not led to believe there is less variability than may be the case.

EXAMPLE 2.13 | The following data represent the percentages of family income allocated to groceries for a sample of 30 shoppers:

26	28	30	37	33	30
29	39	49	31	38	36
33	24	34	40	29	41
40	29	35	44	32	45
35	26	42	36	37	35

For these data $\sum y_i = 1043$ and $\sum y_i^2 = 37{,}331$.

Compute the mean, variance, and standard deviation of the percentage of income spent on food. Check your calculation of s.

Solution | The sample mean is

$$\bar{y} = \frac{\sum_i y_i}{30} = \frac{1043}{30} = 34.77$$

The corresponding sample variance and standard deviation are

$$s^2 = \frac{1}{n-1}\left[\sum_i y_i^2 - \frac{(\sum_i y_i)^2}{n}\right]$$

$$= \frac{1}{29}[37{,}331 - 36{,}261.63] = \frac{1069.37}{29} = 36.87$$

$$s = \sqrt{36.87} = 6.07$$

We can check our calculation of s by using the range approximation. The largest measurement is 49 and the smallest is 24. Hence an approximate value of s is

$$s \approx \frac{\text{range}}{4} = \frac{49-24}{4} = 6.25$$

Note how close the approximation is to our computed value.

While there will not always be such close agreement as found in Example 2.13, the range approximation provides a useful and quick check on the calculation of s.

| 2.5 | CODING TO SIMPLIFY CALCULATIONS (optional) |

Data are frequently coded to simplify the calculations of $\bar{y}$ and s^2 (and s). For example, it is much easier to calculate the mean of the five measurements $-.1, .2, .1, 0,$ and $.2$ than of the five measurements 99.9, 100.2, 100.1, 100.0, and 100.2. The first set of measurements was obtained by subtracting 100 from each measurement in the second set. Similarly, we might wish to simplify a set of measurements by multiplying by a constant. It is easier to work with the set 3, 1, 4, 6, 4, 2 than with the set .003, .001, .004, .006, .004, .002. The first set was obtained by multiplying each element of the second set by 1000.

coding

Data are generally **coded** by performing one or both of the following operations: subtracting (or adding) a constant m from each measurement, or multiplying (or dividing) each measurement by a constant k.

How are the mean ($\bar{y}_c$) and standard deviation (s_c) of the coded measurements related to the mean and standard deviation of the sample measurements $y_1, y_2, \ldots, y_n$? The theorem given next answers this question.

THEOREM 2.1

Let $y_1, y_2, \ldots, y_n$ be n measurements with sample mean $\bar{y}$ and sample standard deviation s. Then we have the following:

1. If we subtract a constant m from each measurement, the mean and standard deviation for the original measurements will be

$$\bar{y} = \bar{y}_c + m \qquad \text{and} \qquad s = s_c$$

2. If we multiply each measurement by a positive constant k, the mean and standard deviation for the original data will be

$$\bar{y} = \frac{\bar{y}_c}{k} \qquad \text{and} \qquad s = \frac{s_c}{k}$$

EXAMPLE 2.14

Use the results of Theorem 2.1 to compute the mean and standard deviation of the measurements .003, .001, .004, .006, .004, .002 by multiplying each by 1000.

Solution

Multiplying each measurement by 1000, we have the coded data 3, 1, 4, 6, 4, 2. For the coded data, then,

$$\bar{y}_c = \frac{\sum y_c}{6} = \frac{20}{6} = 3.33$$

Similarly, with $\sum y_c = 20$ and $\sum y_c^2 = 82$, we have

$$s_c^2 = \frac{1}{5}\left[82 - \frac{(20)^2}{6}\right] = 3.07 \qquad \text{and} \qquad s_c = 1.75$$

Then applying the results of Theorem 2.1, with $k = 1000$,

$$\bar{y} = \frac{3.33}{1000} = .00333 \quad \text{and} \quad s = \frac{1.75}{1000} = .00175$$

coding grouped data   We can also **code grouped data** to simplify the calculations of $\bar{y}$ and s.

RULES FOR CODING GROUPED DATA

1. Based on inspection, select the class interval that you think is likely to contain the mean. Let m denote the midpoint of this interval. (Note: the selection of m is not critical.)
2. Code the interval midpoints as

   $$y_c = \frac{y - m}{w}$$

   where w is the interval width and y is the midpoint (uncoded units).
3. Compute $\bar{y}_c$ and s_c in the usual way for grouped data.
4. $\bar{y} = w\bar{y}_c + m$ and $s = ws_c$.

EXAMPLE 2.15   Use the frequency distribution of crime rates (violent crimes per 100,000 inhabitants) for the data of Table 2.5 shown here to approximate $\bar{y}$ and s.

Class Limits	Midpoint y	f	$y_c = \dfrac{y - m}{w}$	fy_c	fy_c^2
68.5–137.5	103	1	−7	−7	49
137.5–206.5	172	3	−6	−18	108
206.5–275.5	241	5	−5	−25	125
275.5–344.5	310	6	−4	−24	96
344.5–413.5	379	7	−3	−21	63
413.5–482.5	448	11	−2	−22	44
482.5–551.5	517	11	−1	−11	11
551.5–620.5	586	12	0	0	0
620.5–689.5	655	7	1	7	7
689.5–758.5	724	7	2	14	28
758.5–827.5	793	6	3	18	54
827.5–896.5	862	6	4	24	96
896.5–965.5	931	3	5	15	75
965.5–1034.5	1000	3	6	18	108
1034.5–1103.5	1069	1	7	7	49
1103.5–1172.5	1138	1	8	8	64
Totals		90		−17	977

Solution

For the frequency data shown, we selected the eighth interval from the top as the class interval that we thought would contain the mean; then $m = 586$. Step 2 of the coding procedure is to code the midpoints of the intervals using the formula $y_c = (y - m)/w$ where the interval width is $w = 69$. The coded midpoints are shown in column 4 of the table. The remaining two columns provide useful computations for $\bar{y}$ and s.

Using the grouped data formula for $\bar{y}_c$ and the total of column 5 in the table, we have

$$\bar{y}_c = \frac{\sum fy_c}{n} = \frac{-17}{90} = -.189$$

The corresponding mean for the uncoded measurements can be approximated as

$$\bar{y} = w\bar{y}_c + m = 69(-.189) + 586 = 572.96.$$

The computations for s_C^2 and s_C follow in a similar fashion:

$$s_c^2 = \frac{1}{n-1}\left[\sum fy_c^2 - \frac{(\sum fy_c)^2}{n}\right] = \frac{1}{89}\left[977 - \frac{(-17)^2}{90}\right] = \frac{1}{89}(977 - 3.21) = 10.94$$

and $s_c = \sqrt{10.94} = 3.31$. Then multiplying s_c by $w = 69$, we have

$$s = ws_c = 69(3.31) = 228.39$$

You may think that the additional steps required in coding do not really simplify things, especially when most of us have access to a calculator or a computer. However, the mere fact that the numbers are smaller and simpler to manipulate makes the coding worthwhile. More important, by working with smaller numbers, we are probably less prone to make arithmetic errors and are less likely to face serious rounding errors.

EXERCISES

2.34 Three data sets are shown here. Compute $\bar{y}$ and s for each one. Use the short-cut formula for s^2.

Data set 1: 1, 2, 3

Data set 2: .01, .02, .03

Data set 3: 1001, 1002, 1003

2.35 Refer to Exercise 2.34. Compute $\bar{y}$ and s for each data set using the formula

$$s^2 = \frac{\sum_i (y_i - \bar{y})^2}{n-1}$$

Compare your results here to what you obtained in Exercise 2.34. Which formula appears to be more accurate? Why? Do you have any words of caution?

2.36 Based on $\bar{y}$ and s from Exercise 2.35, indicate how the coding steps could be used to compute the sample mean and standard deviation for the three measurements 10,000, 10,001, and 10,002.

| 2.6 | THE BOX PLOT |

box plot

As mentioned earlier in this chapter, a stem-and-leaf plot provides a graphical representation of a set of scores that can be used to examine the shape of the distribution, the range of scores, and where the scores are concentrated. The **box plot**, which builds on the information displayed in a stem-and-leaf plot, is more concerned with the symmetry of the distribution and incorporates numerical measures of central tendency and location in order to study the variability of the scores and the concentration of scores in the tails of the distribution.

Before we show how to construct and interpret a box plot, it is necessary to introduce several new terms that are peculiar to the language of exploratory data analysis (EDA). We are familiar with the definitions for the first, second (median), and third quartiles of a distribution presented earlier in this chapter. The box plot uses the median and **hinges** of a distribution. Hinges are very similar to quartiles of a distribution, but owing to the method by which they are computed for sample data, the lower and upper hinges of a distribution may differ very slightly from the first and third quartiles of a set of scores.

hinges

Having said this, and recognizing the slight distinction, we will compute hinges in this text but refer to them as the lower and upper quartiles of the sample data.

We can now illustrate a *skeletal box plot* by way of an example.

EXAMPLE 2.16

Use the stem-and-leaf plot in Figure 2.19 for the 90 violent crime rates of Table 2.5 to construct a skeletal box plot.

Solution

When the scores are ordered from lowest to highest, the median score and quartile scores are located as follows:

$$\text{median location} = \frac{n + 1}{2}$$

$$\text{quartile location} = \frac{\text{truncated median location} + 1}{2}$$

FIGURE 2.19

Stem-and-Leaf Plot

1	12 71 84
2	01 12 50 65 70 72 89
3	05 06 22 22 36 46 51 70 90
4	04 09 11 36 37 39 41 44 48 51 53 70 80 82 87 94 95 99
5	03 14 21 22 27 48 50 59 60 70 72 74 78 85 92 92
6	07 16 18 21 29 37 38 40 56 68
7	07 09 19 37 39 52 58 66 92 92 94
8	02 18 30 32 43 58 60 69
9	18 25 53 91
10	00 05 68
11	48

where the truncated median location is simply the median location with the decimal .5 omitted where present. For the distribution of $n = 90$ violent crime rates we have

$$\text{median location} = \frac{90 + 1}{2} = 45.5$$

$$\text{truncated median location} = 45$$

and

$$\text{quartile location} = \frac{45 + 1}{2} = 23$$

Since the median location is the 45.5th score in the distribution, we average the 45th and 46th scores to compute the median. For these data the 45th score (counting from the lowest to the highest in Figure 2.19) is 559 and the 46th is 560, hence the median is

$$M = \frac{559 + 560}{2} = 559.5$$

Then to find the lower and upper quartiles for this distribution of scores we determine the 23rd score counting in from the low side of the distribution and from the high side of the distribution, respectively. The 23rd lowest score and 23rd highest score are 436 and 739:

lower quartile, $Q_1 = 436$
upper quartile, $Q_3 = 739$

skeletal box plot

These three descriptive measures and the smallest and largest values in a data set are used to construct a skeletal box plot (see Figure 2.20). The **skeletal box plot** is constructed by drawing a box between the lower and upper quartiles with a solid line drawn across the box to locate the median. A straight line is then drawn connecting the box to the largest value; a second line is drawn from the box to the smallest value. These straight lines are sometimes called whiskers and the entire graph a **box-and-whiskers plot**

box-and-whiskers plot

FIGURE 2.20

Skeletal Box Plot for the Data of Figure 2.19

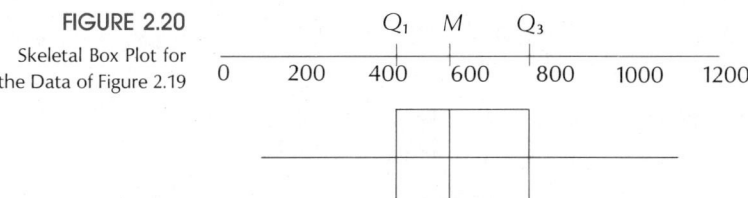

With a quick glance at a skeletal box plot, it is easy to obtain an impression about the following aspects of the data:

1. the lower and upper quartiles, Q_1 and Q_3
2. the interquartile range (IQR), the distance between the lower and upper quartiles
3. the most extreme (lowest and highest) values
4. the symmetry or asymmetry of the distribution of scores

     If we had been presented with Figure 2.20 without having seen the original data, we would have observed that

$$Q_1 \approx 420$$
$$Q_3 \approx 770$$
$$IQR \approx 770 - 420 = 350$$
$$M \approx 570$$

most extreme values: 100 and 1150

Also, because the median is closer to the lower quartile than the upper quartile and because the upper whisker is a little longer than the lower whisker, the distribution is slightly nonsymmetrical. To see that this conclusion is true, construct a frequency histogram for these data (or refer to your results in Exercise 2.18).

The skeletal box plot can be expanded to include more information about extreme values in the tails of the distribution. To do so we need the following additional quantities:

lower inner fence: $Q_1 - 1.5IQR$

upper inner fence: $Q_3 + 1.5IQR$

lower outer fence: $Q_1 - 3IQR$

upper outer fence: $Q_3 + 3IQR$

lower adjacent score: most extreme score in the interval from Q_1 to the lower inner fence

upper adjacent score: most extreme score in the interval from Q_3 to the upper inner fence

Any score beyond an inner fence on either side is called a *mild outlier;* a score beyond an outer fence on either side is called an *extreme outlier.*

EXAMPLE 2.17   Compute the inner and outer fences for the data of Example 2.16. Identify any mild and extreme outliers.

Solution   For these data we found the lower and upper quartiles to be 436 and 739, respectively; the IQR = 739 − 436 = 303. Then

lower inner fence = $436 - 1.5(303) = -18.5$

upper inner fence = $739 + 1.5(303) = 1193.5$

lower outer fence = $436 - 3(303) = -473$

upper outer fence = $739 + 3(303) = 1648$

Also from the stem-and-leaf plot we see that the lower and upper adjacent values are 112 and 1148. Since the upper and lower inner fences are 1193.5 and -18.5 respectively, there are no observations that are beyond the inner fences. Hence there are no mild or extreme outliers.

We now have all the quantities necessary for constructing a box plot.

STEPS IN
CONSTRUCTING
A BOX PLOT

1.   As with a skeletal box plot, mark off a box from the lower quartile to the upper quartile.
2.   Draw a solid line across the box to locate the median.
3.   Mark the location of the upper and lower adjacent values with an x.
4.   Draw a dashed line between each quartile and its adjacent value.
5.   Mark each extreme outlier with the symbol o.

EXAMPLE 2.18

Construct a box plot for the data of Example 2.16.

Solution

The box plot is shown in Figure 2.21.

What information can be drawn from a box plot? First, the center of the distribution of scores is indicated by the median line in the box plot. Second, a measure of the variability of the scores is given by the interquartile range, the length of the box. Recall that the box is constructed between the lower and upper quartiles so it contains the middle 50% of the scores in the distribution, with 25% on either side of the median line inside the box. Third, by examining the relative position of the median line, we can gauge the symmetry of the middle 50% of the scores. For example, if the median line is closer to the lower quartile than the upper, there is a greater concentration of scores on the lower side of the median within the box than on the upper side; a symmetric distribution of scores would have the median line located in the center of the box. Fourth, additional information about skewness is obtained from the lengths of the whiskers; the longer one whisker is relative to the other one, the more skewness there is in the tail with the longer whisker. Fifth, a general assessment can be made about the presence of outliers by examining the number of scores classified as mild outliers and the number classified as extreme outliers.

FIGURE 2.21
The Box Plot for the
Data of Example
2.16

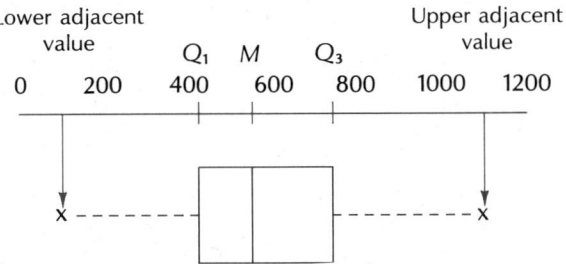

| 2.7 | SUMMARIZING DATA FROM MORE THAN ONE VARIABLE |

In the previous sections we've discussed graphical methods and numerical descriptive methods for summarizing data from a single variable. Frequently, more than one variable is being studied at the same time and, although we might be interested in summarizing the data on each variable separately, we might also be interested in studying relations among the variables. For example, we might be interested in the prime interest rate and in the consumer price index, as well as in the relation between the two. In this section we'll discuss a few techniques for summarizing data from two (or more) variables. Material in this section will provide a brief preview and introduction to chi-square methods (Chapter 6), analysis of variance (Chapters 10, 14, 15, and 16), and regression (Chapters 8, 9, 12, and 13).

contingency table

Consider first the problem of summarizing data from two qualitative variables. Cross-tabulations can be constructed to form a **contingency table**. The rows of the table identify the categories of one variable, and the columns identify the categories of the other variable. The entries in the table are the number of times each value of one variable occurs with each possible value of the other. For example, a television viewing survey was conducted on 1500 individuals. Each individual surveyed was asked to state his or her place of residence and network preference for national news. The results of the survey are shown in Table 2.8. As you can see, 144 urban residents preferred ABC, 135 urban residents preferred CBS, and so on.

The simplest method for looking at relations between variables in a contingency table is to do a percentage comparison based on the row totals, the column totals, or the overall total. If we calculate percentages within each row of Table 2.8, we can compare the distribution of residences within each network preference. A percentage comparison such as this, based on the row totals, is shown in Table 2.9.

Except for ABC, which has the highest urban percent among the networks, the differences among the residence distributions are in the suburban and rural categories. The percent of suburban preferences rises from 43.5% for ABC to 58.1% for NBC. Corresponding shifts downward occur in the rural category. In Chapter 6, we will use chi-square methods to explore further relations between two (or more) qualitative variables.

An extension of the bar graph provides a convenient method for summarizing joint data from a single qualitative and a single quantitative variable. We will discuss this method by way of an example. Suppose that a company wants to investigate the relative effects

TABLE 2.8   Data from a Survey of Television Viewing

| Network Preference | Residence | | | |
	Urban	Suburban	Rural	Total
ABC	144	180	90	414
CBS	135	240	96	471
NBC	108	225	54	387
Other	63	105	60	228
Total	450	750	300	1500

TABLE 2.9   Comparing the Distribution of Residences for Each Network

| | Residence | | | |
Network Preference	Urban	Suburban	Rural	Total
ABC	34.8	43.5	21.7	100 (*n* = 414)
CBS	28.7	50.9	20.4	100 (*n* = 471)
NBC	27.9	58.1	14.0	100 (*n* = 387)
Other	27.6	46.1	26.3	100 (*n* = 228)

of three different employee-incentive systems on productivity. A total of 15 work teams are selected randomly. Of these teams, seven participate in a released-time plan, by which teams achieving certain goals are allowed to take extra time off, with pay; five participate in a bonus-pay plan; and three participate in a profit-sharing plan. The company has a standard productivity measure, and calculates the increased productivity of each work team over a three-month period. Suppose that the results are the following:

released time, R: 16.2   15.6   19.4   18.8   16.9   15.9   17.6

bonus pay, B: 12.4   15.8   14.0   9.8   10.0

profit sharing, P: 4.6   8.0   6.0

Do the data indicate a strong relationship between plan and productivity gain?

The data summarized in Figure 2.22 clearly indicate that productivity gains are generally largest for the released-time plan, R, gains for the bonus-pay plan, B, are in the middle, and gains for the profit-sharing plan, P, are lowest. For now we will use a plot such as that in Figure 2.22. Later, we will use analysis of variance methods (Chapters 10, 14, 15, and 16) to examine the relationships between a quantitative variable and one or more qualitative variables.

Finally, we can construct data plots for summarizing the relation between two quantitative variables. Consider the following example. A manager of a small machine shop

FIGURE 2.22

Relationship Between
Productivity and the
Incentive Plans

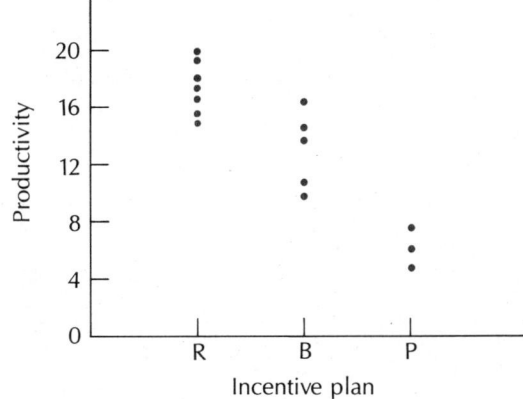

FIGURE 2.23

Scatterplot of Starting
Hourly Wage and
Years Experience

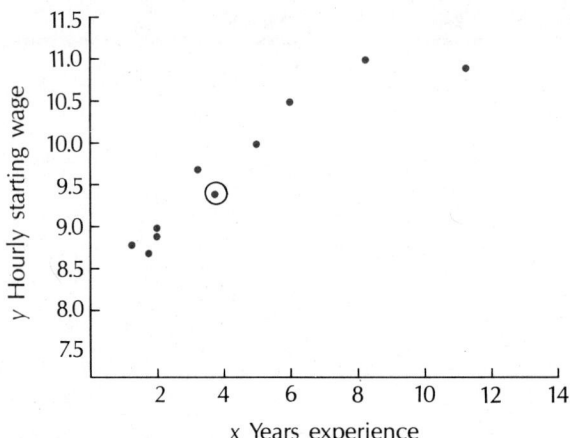

examined the starting hourly wage y offered to machinists with x years of previous experience. The data are shown here:

y (dollars):	8.90	8.70	9.10	9.00	9.79	9.45	10.00	10.65	11.10	11.05
x (years:)	1.25	1.50	2.00	2.00	2.75	4.00	5.00	6.00	8.00	12.00

Is there a relationship between x and y?

scatterplot   One way to summarize these data is to use a **scatterplot**, as shown in Figure 2.23. Each point on the plot represents a machinist with a particular starting wage and years of experience. The point circled corresponds to $y = 9.45$, $x = 4.00$.

In general, the data displayed in Figure 2.23 indicate that, as the years of previous experience x increases, the hourly starting wage y for machinists increases. This basic idea of relating two quantitative variables is discussed and expanded in the chapters on regression (8, 9, 12, and 13).

EXERCISES

2.37   Refer to the television survey data of Table 2.8. Do a percentage comparison based on the column totals. Interpret the data.

2.38   Data on the age at the time of a job turnover and on the reason for the job turnover are displayed here for 250 job changes in a large corporation.

Reason for Turnover	Age (Years)				Total
	≤ 29	30–39	40–49	≥ 50	
Resigned	30	6	4	20	60
Transferred	12	45	4	5	66
Retired/fired	8	9	52	55	124
Total	50	60	60	80	250

Do a percentage comparison based on the row totals and use this to describe the data.

2.39 Refer to Exercise 2.38. What different summary would one get with a percentage comparison based on the column totals? Do this summary and describe your results.

2.40 The lengths of hospital stays were recorded for patients undergoing a particular surgical procedure at each of four hospitals. These data are shown here.

Hospital	Length of Stay (Days)							
A	18	20	22	22	24	26		
B	14	15	17	17	18	19	21	21
C	21	25	27	31				
D	27	33						

a. Compute the mean stay for each hospital.

b. Plot the sample data.

c. Use parts (a) and (b) to describe the data. Which hospital appears to have shorter stays?

2.41 The federal government keeps a close watch on money growth versus targets that have been set for that growth. Below we list two measures of the money supply in the United States, M2 (private checking deposits, cash, and some savings) and M3 (M2 plus some investments), which are given here for 20 consecutive months.

Month	Money Supply (in Trillions of Dollars)	
	M2	M3
1	2.25	2.81
2	2.27	2.84
3	2.28	2.86
4	2.29	2.88
5	2.31	2.90
6	2.32	2.92
7	2.35	2.96
8	2.37	2.99
9	2.40	3.02
10	2.42	3.04
11	2.43	3.05
12	2.42	3.05
13	2.44	3.08
14	2.47	3.10
15	2.49	3.10
16	2.51	3.13
17	2.53	3.17
18	2.53	3.18
19	2.54	3.19
20	2.55	3.20

a. Would a scatterplot describe the relation between M2 and M3?

b. Construct a scatterplot. Is there an obvious relation?

2.42 Refer to Exercise 2.41. What other data plot might be used to describe and summarize these data? Make the plot and interpret your results.

2.8   CALCULATORS, COMPUTERS, AND SOFTWARE SYSTEMS

Electronic calculators can be great aids in performing some of the calculations mentioned in this chapter, especially for small data sets. For example, many calculators have keys for obtaining the sample mean and standard deviation directly after the data are entered. Others can be used with the shortcut formulas of the previous sections to obtain y, s^2, and s. For larger data sets, even hand calculators are of little use because of the time required for entering data and the inability to update an erroneous entry without reentering the entire data set. In these situations, a computer can be of help. Specific programs or more general software systems can be used to perform statistical analyses almost instantaneously even for very large data sets after the data are entered into the computer from a terminal, a magnetic tape, or disk storage. It is not necessary to have knowledge of computer programming to make use of specific programs or software systems for planned analyses. Most have user's manuals that give detailed directions for their use. Others, developed for use at a terminal, provide program prompts that lead the user through the analysis of choice.

There are many statistical software packages available for use on computers. Three of the more commonly used systems are BMDP, SAS, and SPSS. Each is available in a mainframe version, as well as in a personal computer version. Since a software system is a group of programs that work together, it is possible to obtain plots, data descriptions, and complex statistical analyses in a single job. Most people find that they can use any particular system easily, although they are frustrated by minor errors committed on the first few tries. The ability of such packages to perform complicated analyses on large numbers of data more than repays the initial investment of time and irritation.

In general, to use a system you need not learn everything about it. You need to learn about only the programs in which you are interested. Typical steps in a job involve describing your data to the software system, manipulating your data if they are not in the proper format or if you want a subset of your original data set, and then calling the appropriate set of programs or procedures using the key words particular to the software system you are using. The results obtained from calling a program are then displayed at your terminal or sent to your printer.

If you have access to a computer and are interested in using it, find out how to obtain an account, what programs and software systems are available for doing statistical analyses, and where to obtain instruction on data entry for these programs and software systems.

Because computer configurations, operating systems, and text editors vary from site to site, it is best to talk to someone knowledgeable about gaining access to a software system. Once you have mastered the commands to begin executing programs in a software system, you will find that running a job within a given software system is similar from site to site.

Since this isn't a text on computer usage, we won't spend additional time and space on the mechanics, which are best learned by doing. Our main interest is in interpreting the output from these programs. The designers of these programs tend to include in the output everything that a user could conceivably want to know; as a result, in any particular situation, some of the output is irrelevant. When reading computer output look for the values you want; if you don't need or don't understand an output statistic, don't worry. Of course, as you learn more about statistics, more of the output will be meaning-ful. In the meantime, look for what you need and disregard the rest.

There are dangers in using such packages carelessly. A computer is a mindless beast, and will do anything asked of it, no matter how absurd the result might be. For instance, suppose that the data include age, gender (1 = female, 2 = male), religion (1 = Catholic, 2 = Jewish, 3 = Protestant, 4 = other or none), and monthly income of a group of people. If we asked the computer to calculate means we would get means for the variables gender and religion, as well as for age and monthly income, even though these averages are meaningless. Furthermore, it is unlikely that a standard program would warn a user that extreme skewness was distorting a mean value or that the data contained a gross outlier. Used intelligently, these packages are convenient, powerful, and useful—but be sure to examine the output from any computer run to make certain the results make sense. Did anything go wrong? Was something overlooked? In other words, *be skeptical*. One of the important acronyms of computer technology still holds; namely, GIGO: garbage in, garbage out.

Throughout the textbook we will use computer software systems to do some of the more tedious calculations of statistics *after* we have explained how the calculations can be done. Used in this way, computers (and associated graphical and statistical analysis packages) will enable us to spend additional time on interpreting the results of the analyses, rather than on doing the analyses.

2.9   SUMMARY

This chapter was concerned with graphical and numerical description of data. The pie chart, statistical map, and bar graph are particularly appropriate for graphically displaying data obtained from a qualitative variable. The frequency and relative frequency histograms and stem-and-leaf plots are graphical techniques applicable only to quantitative data.

Numerical descriptive measures of data are used to convey a mental image of the distribution of measurements. Measures of central tendency include the mode, the median, and the arithmetic mean. Measures of variability include the range, the interquartile range, the variance, and the standard deviation of a set of measurements. While some disciplines emphasize different measures, we will use the mean and the standard deviation of a set of measurements as the primary numerical measures of central tendency and variability, respectively. One explanation for our choice is that we can not only compare variabilities of *two* sets of measurements using the standard deviation of each, but we can also interpret the variability of a *single* set of measurements using the mean, the standard deviation, and the Empirical Rule.

We extended the concept of data description to summarizing the relations between two qualitative variables. Here cross-tabulations were used to develop percentage com-parisons. We examined plots for summarizing the relations between quantitative and quali-tative variables and between two quantitative variables. Material presented here (namely, summarizing relations among variables) will be discussed and expanded in later chapters on chi-square methods, on the analysis of variance, and on regression.

**KEY
FORMULAS**   Data description

1.   Median, grouped data

$$\text{median} \approx L + \frac{w}{f_m}(.5n - cf_b)$$

2.   Sample mean

$$\bar{y} = \frac{\sum_i y_i}{n}$$

3.   Sample mean, grouped data

$$\bar{y} \approx \frac{\sum_i f_i y_i}{n}$$

4.   Sample variance

$$s^2 = \frac{1}{n-1}\left[\sum_i y_i^2 - \frac{(\sum_i y_i)^2}{n}\right]$$

5.   Sample variance, grouped data

$$s^2 \approx \frac{1}{n-1}\left[\sum_i f_i y_i^2 - \frac{(\sum_i f_i y_i)^2}{n}\right]$$

6.   Sample standard deviation

$$s = \sqrt{s^2}$$

EXERCISES

2.43   To practice using the shortcut method to calculate s^2 and s, consider a small number of sample measurements, say 0, 1, 2, 4, 4.
a. Verify that $\sum_i y_i = 11$ and $\sum_i y_i^2 = 37$.
b. Use the results of part (a) to compute s^2 and s with the shortcut formula.

2.44   The rounded nitrogen contents for the 36 composite apple leaf samples of Exercise 2.29 are presented below.

2.10	2.82	2.17	1.99	2.22	3.09
2.47	2.52	2.80	2.10	2.92	2.20
1.75	2.77	2.82	2.67	3.05	2.93
2.94	1.98	2.38	2.65	2.77	1.85
1.69	2.70	2.68	2.06	2.36	2.28
2.75	2.43	2.39	2.55	1.80	1.96

a. Use the shortcut formula to compute s^2 and s. You can verify that

$$\sum_i y_i = 87.61 \qquad \text{and} \qquad \sum_i y_i^2 = 218.7297$$

b. Use the range approximation to check your calculation of s.

c. To increase your confidence in the Empirical Rule, construct the intervals $\bar{y} \pm s$, $\bar{y} \pm 2s$, and $\bar{y} \pm 3s$. Count the number of rounded nitrogen content readings falling in each of the three intervals. Convert these numbers to percentages and compare your results to the Empirical Rule.

2.45 The College of Dentistry at the University of Florida has made a commitment to develop its entire curriculum around the use of self-paced instructional materials such as video-tapes, slide tapes, syllabi, and so on. It is hoped that each student will proceed at a pace commensurate with his or her ability and that the instructional staff will have more free time for personal consultation in student-faculty interaction. One such instructional module was developed and tested on the first 50 students proceeding through the curriculum. The measurements below represent the number of hours it took these students to complete the required modular material.

16	8	33	21	34	17	12	14	27	6
33	25	16	7	15	18	25	29	19	27
5	12	29	22	14	25	21	17	9	4
12	15	13	11	6	9	26	5	16	5
9	11	5	4	5	23	21	10	17	15

a. Calculate the mode, the median, and the mean for these recorded completion times.

b. Guess the value of s.

c. Compute s by using the shortcut formula and compare your answer to that of part (b).

d. Would you expect the Empirical Rule to describe adequately the variability of these data? Explain.

2.46 Refer to the data of Examples 2.6 and 2.11. We previously computed the sample mean and standard deviation to be 4.29 and .32, respectively. Use the coding procedures of Section 2.5 to compute $\bar{y}$ and s. Proceed assuming that you think the eighth class interval contains the mean.

2.47 Repeat Exercise 2.46 by using $m = 4.4$, the midpoint of the ninth interval.

2.48 Repeat Exercise 2.46 by using $m = 4.5$, the midpoint of the 10th interval. Are your answers to Exercises 2.46, 2.47, and 2.48 identical? If not, check your calculations.

2.49 A study was conducted to determine urine flow of sheep (in milliliters/minute) when in-fused intravenously with the antidiuretic hormone ADH. The urine flows of 10 sheep are recorded here.

0.7	0.5	0.5	0.6	0.5	0.4	0.3	0.9	1.2	0.9

a. Determine the mean, the median, and the mode for these sample data.

b. Suppose that the largest measurement is 6.8 rather than 1.2. How does this affect the mean, the median, and the mode?

2.50 Refer to Exercise 2.49.

a. Compute the range and the sample standard deviation.

b. Check your calculation of s using the range approximation.

c. How are the range and standard deviation affected if the largest measurement is 6.8 rather than 1.2? What about 68?

2.51 Refer to Exercise 2.49. Code the data by multiplying each measurement by 10. Compute the sample mean and standard deviation for the original set using the coded values.

2.52 A stem-and-leaf plot is shown for the telephone data from Exercise 2.15. Compute the mean, median, mode, and standard deviation for the data.

```
        SET DATA IN C1
DATA> 500 350 550 480 610
DATA> 570 620 630 620 570
DATA> 480 550 650 580 570
DATA> 600 480 520 540 610
DATA> 570 580 560 470 570
DATA> 540 590 720 590 650
DATA> 470 530 530 560 550
DATA> 580 560 610 560 510
DATA> 540 540 570 560 520
DATA> 530 570 450 540 580
DATA> END DATA
MTB > PRINT C1
C1
    500     350     550     480     610     570     620     630     620     570     480
    550     650     580     570     600     480     520     540     610     570     580
    560     470     570     540     590     720     590     650     470     530     530
    560     550     580     560     610     560     510     540     540     570     560
    520     530     570     450     540     580

MTB > STEM-AND-LEAF DISPLAY OF C1

Stem-and-leaf of C1        N = 50
Leaf Unit = 10

     1    3 5
     1    4
     7    4 577888
    19    5 012233344444
   (21)   5 5556666677777777888899
    10    6 0111223
     3    6 55
     1    7 2

MTB > STOP
```

2.53 A box plot was constructed for the data of Exercise 2.52 using Minitab. Describe the data using information conveyed by the box plot.

```
        PRINT C1
C1
    500     350     550     480     610     570     620     630     620     570     480
    550     650     580     570     600     480     520     540     610     570     580
    560     470     570     540     590     720     590     650     470     530     530
    560     550     580     560     610     560     510     540     540     570     560
    520     530     570     450     540     580

MTB > BOXPLOT OF C1

                                          ---------
        0                 *  ---------I  +  I-----------          *
                                          --------

        --+---------+---------+---------+---------+---------+-----C1
        350       420       490       560       630       700

MTB >
```

EXERCISES

 2.54   A random sample of 90 standard metropolitan statistical areas (SMSA) was studied to obtain information on murder rates. The murder rate (number of murders per 100,000 people) was recorded, and these data are summarized in the frequency table displayed below.

Class Interval	f_i	Class Interval	f_i
−.5–1.5	2	13.5–15.5	9
1.5–3.5	18	15.5–17.5	4
3.5–5.5	15	17.5–19.5	2
5.5–7.5	13	19.5–21.5	1
7.5–9.5	9	21.5–23.5	1
9.5–11.5	8	23.5–25.5	1
11.5–13.5	7		

Construct a relative frequency histogram for these data.

2.55   Refer to the data of Exercise 2.54.
a. Compute the sample median and the mode.
b. Compute the sample mean.
c. Which measure of central tendency would you use to describe the center of the distribution of murder rates?

2.56   Refer to the data of Exercise 2.54.
a. Compute the interquartile range.
b. Compute the sample standard deviation.

2.57   Refer to the data of Exercise 2.54. If you did not employ coding to compute $\bar{y}$ and s in Exercises 2.55 and 2.56, use the midpoint of the fifth interval as m. Code the sample data to compute $\bar{y}$ and s. Compare your answers to those of Exercises 2.55 and 2.56.

 **2.58**   Every 20 minutes a sample of 10 transistors is drawn from the outgoing product on a production line and tested. The data are summarized below for the first 500 samples of 10 measurements.

y_i	0	1	2	3	4	5	6	7	8	9	10
f_i	170	185	75	25	15	10	8	5	4	2	1

Construct a relative frequency distribution depicting the interquartile range. (Note: y_i in the table is the number of defectives in a sample of 10.)

2.59   Refer to Exercise 2.58.
a. Determine the sample median and the mode.
b. Calculate the sample mean.
c. Based on the mean, the median, and the mode, how is the distribution skewed?

2.60   Refer to Exercise 2.58.
a. Code the sample data to compute the sample standard deviation.
b. Can the Empirical Rule be used to describe this set of measurements?

2.61   The number of persons who volunteered to give a pint of blood at a central donor center was recorded for each of 20 successive Fridays. The data are shown here:

320	370	386	334	325	315	334	301	270	310
274	308	315	368	332	260	295	356	333	250

a. Construct a stem-and-leaf plot.
b. Construct a skeletal box plot and interpret the results.

2.62   Per capita expenditure (dollars) for health and hospital services by state are shown here.

Dollars	f
45–59	1
60–74	4
75–89	9
90–104	9
105–119	12
120–134	6
135–149	4
150–164	1
165–179	3
180–194	0
195–209	1
Total	50

Source: Bureau of the Census, *Statistical Abstract of the United States: 1980* (Washington, D.C.: Government Printing Office, 1980), p. 300.

a. Construct a relative frequency histogram.

b. Compute approximate values for $\bar{y}$ and s from the grouped expenditure data.

2.63   Refer to the data from Exercise 2.32. Eliminate Philadelphia from the north and San Jose and Seattle from the west.

a. Compute $\bar{y}$ for the revised subgroups.

b. Combine the subgroup means ($\bar{y}_i$) to obtain the overall sample mean using the formula

$$\bar{y} = \frac{\sum_i n_i \bar{y}_i}{n}$$

where n_i is the number of observations in subgroup i.

c. Show that the sample mean computed in part (b) is identical to that obtained for the twenty-seven measurements in part (a).

2.64   Refer to Example 2.11. Find the sample mean and the standard deviation by using the coding steps of Section 2.5. Compare your answers here to those of Example 2.11.

2.65   A company revised a long-standing policy to eliminate time clocks and cards for nonexempt employees. Along with this change all employees (exempt and nonexempt) were expected to account for their own time on the job as well as absences due to sickness, vacation, holidays, and so on. The previous policy of allocating a certain number of sick days was eliminated; if an employee was sick he or she was given time off with pay; otherwise, he or she was expected to be working.

In order to see how well the new program was working, the records of a random sample of 15 employees were examined to determine the number of sick days over this year (under the new plan) and the corresponding number for the preceding year. These data are shown here:

Employee	This Year (New Policy)	Preceding Year (Old Policy)
1	0	2
2	0	2
3	0	3
4	0	4
5	2	5
6	1	2
7	1	6
8	3	8
9	1	5
10	0	4
11	5	5
12	6	12
13	1	3
14	2	4
15	12	4

a. Obtain the mean and standard deviation for each column.
b. Based on the sample data, what might you conclude (infer) about the new policies? Explain your reason(s).

2.66 Refer to Exercise 2.65. What happens to $\bar{y}$ and s for each column if we eliminate the two 12s and substitute values of 7? Are the ranges for the old and new policies affected by these substitutions?

2.67 Federal authorities have destroyed considerable amounts of wild and cultivated marijuana plants. The following table shows the number of plants destroyed and the number of arrests for a 12-month period for 15 states.

State	Plants	Arrests
1	110,010	280
2	256,000	460
3	665	6
4	367,000	66
5	4,700,000	15
6	4,500	8
7	247,000	36
8	300,200	300
9	3,100	9
10	1,250	4
11	3,900,200	14
12	68,100	185
13	450	5
14	2,600	4
15	205,844	33

a. Discuss the appropriateness of using the sample mean to describe these two variables.

b. Compute the sample mean, 10% trimmed mean, and 20% trimmed mean. Which trimmed mean seems more appropriate for each variable? Why?

2.68 Refer to Exercise 2.67. Does there appear to be a relation between the number of plants destroyed and the number of arrests? How might you examine this question? What other variable(s) might be related to the number of plants destroyed?

2.69 Monthly readings for the FDC Index, a popular barometer of the health of the pharmaceutical industry, are shown here. As can be seen, the Index has several components—one for pharmaceutical companies, one for diversified companies, one for chain drugstores, and another for drug and medical supply wholesalers.

	Pharmaceuticals	Diversified	Chain	Wholesaler
January	123.1	154.6	393.3	475.5
February	122.4	146.0	407.6	504.1
March	125.2	169.2	405.0	476.6
April	136.1	156.7	415.1	513.3
May	149.3	177.0	418.9	543.5
June	145.7	158.1	443.2	552.6
July	162.4	156.6	419.1	526.2
August	168.0	178.6	404.0	516.3
September	155.6	170.4	391.8	482.1
October	177.0	162.9	410.9	484.0
November	196.6	182.4	459.8	522.6
December	195.2	195.4	431.9	536.8

a. Plot these data on a single graph.

b. Discuss trends within each component and any apparent relations among the separate components of the FDC Index.

2.70 Refer to Exercise 2.69. Compute the percent change for each month of each component of the Index. (Assume that the percent changes in January were 12.3, −.7, 12.1, and 16.1, respectively, for the four components.) Plot these data. Are they more revealing than the original measurements were?

2.71 Closing New York Stock Exchange (NYSE) prices for the components (as of March 1987) of the Dow Jones Industrial Average (DJIA) are shown here:

	Components of Dow Jones Industrial Average	
	Percent of DJIA*	Closing NYSE Stock Price 1/24
Allied-Signal	2.81%	$46.875
Alcoa	2.39	39.875
American Can	3.93	65.500
American Express	3.25	54.125
AT&T	1.35	22.500
Bethlehem Steel	1.06	17.750

EXERCISES

Chevron	2.17	36.250
duPont	3.70	61.750
Eastman Kodak	2.82	47.000
Exxon	3.06	51.000
General Electric	4.12	68.750
General Motors	4.22	70.375
Goodyear	1.90	31.625
Inco	0.85	14.250
IBM	8.99	150.000
Intl. Harvester	0.53	8.875
Intl. Paper	2.96	49.375
McDonald's	4.48	74.750
Merck	8.10	135.125
Minnesota Mining	5.18	86.375
Owens-Illinois	3.35	55.875
Phillip Morris	5.49	91.625
Procter & Gamble	3.94	65.750
Sears, Roebuck	2.22	37.000
Texaco	1.72	28.625
Union Carbide	4.95	82.625
United Technologies	2.73	45.500
U.S. Steel	1.40	23.375
Westinghouse	2.00	44.875
Woolworth	3.61	60.250

*Close on Jan. 24

a. Compute the actual range of the stock prices.
b. The DJIA is actually a weighted average, so only a certain percent of the actual NYSE price is part of the DJIA for each stock. The weighted average can be written as

$$\bar{y}_w = \sum_i \frac{w_i y_i}{n}$$

where y_i is the closing price for stock i, and w_i is the weight attached to stock i. Using the weights (percent of DJIA) listed in the above table, compute the DJIA for this particular day.
c. Refer to part (b). Why might the DJIA be a weighted average, rather than a simple average?

2.72   The number of telephones (per 1000 people) is shown by state in the accompanying table. These data were plotted in Exercise 2.16.

State	Telephones	State	Telephones	State	Telephones
Alabama	500	Louisiana	520	Ohio	550
Alaska	350	Maine	540	Oklahoma	580
Arizona	550	Maryland	610	Oregon	560
Arkansas	480	Massachusetts	570	Pennsylvania	610
California	610	Michigan	580	Rhode Island	560
Colorado	570	Minnesota	560	S. Carolina	510
Connecticut	620	Mississippi	470	S. Dakota	540
Delaware	630	Missouri	570	Tennessee	540
Florida	620	Montana	540	Texas	570
Georgia	570	Nebraska	590	Utah	560
Hawaii	480	Nevada	720	Vermont	520
Idaho	550	New Hampshire	590	Virginia	530
Illinois	650	New Jersey	650	Washington	570
Indiana	580	New Mexico	470	W. Virginia	450
Iowa	570	New York	530	Wisconsin	540
Kansas	600	N. Carolina	530	Wyoming	580
Kentucky	480	N. Dakota	560		

a. Might the Empirical Rule be used to describe the data?

b. Compute $\bar{y}$ and s and count the number (percent) of measurements falling in the intervals $\bar{y} \pm s$, $\bar{y} \pm 2s$, $\bar{y} \pm 3s$.

2.73   Refer to Exercise 2.72. Are there many extreme values affecting $\bar{y}$? Should this have been anticipated based on the data plot in Exercise 2.72? Compute the 10% trimmed mean for these data.

2.74   As one part of a review of middle-manager selection procedures, a study was made of the relation between hiring source (promoted from within, hired from related business, hired from unrelated business) and the three-year job history (additional promotion, same position, resigned, dismissed). The data for 120 middle managers follow.

	Source			
Job History	Within Firm	Related Business	Unrelated Business	Total
Promoted	13	4	10	27
Same position	32	8	18	58
Resigned	9	6	10	25
Dismissed	3	3	4	10
Total	57	21	42	120

a. Calculate job-history percentages within each source.

b. Would you say that there is a strong dependence between source and job history?

EXERCISES

2.75 A survey was taken of 150 residents of major coal-producing states, 200 residents of major oil– and natural gas–producing states, and 450 residents of other states. Each resident chose a most preferred national energy policy. The results are shown in the following SPSS printout.

		STATE		
COUNT ROW PCT COL PCT TOT PCT	COAL	OIL AND GAS	OTHER	ROW TOTAL
OPINION COAL ENCOURAGED	62 32.8 41.3 7.8	25 13.2 12.5 3.1	102 54.0 22.7 12.8	189 23.6
FUSION DEVELOP	3 7.3 2.0 0.4	12 29.3 6.0 1.5	26 63.4 5.8 3.3	41 5.1
NUCLEAR DEVELOP	8 22.2 5.3 1.0	6 16.7 3.0 0.8	22 61.1 4.9 2.8	36 4.5
OIL DEREGULATION	19 12.6 12.7 2.4	79 52.3 39.5 9.9	53 35.1 11.8 6.6	151 18.9
SOLAR DEVELOP	58 15.1 38.7 7.3	78 20.4 39.0 9.8	247 64.5 54.9 30.9	383 47.9
COLUMN TOTAL	150 18.8	200 25.0	450 56.3	800 100.0

CHI SQUARE = 106.19406 WITH 8 DEGREES OF FREEDOM SIGNIFICANCE = 0.0000
CRAMER'S V = 0.25763
CONTINGENCY COEFFICIENT = 0.34233
LAMBDA = 0.01199 WITH OPINION DEPENDENT, = 0.07429 WITH STATE DEPENDENT.

a. Interpret the values 62, 32.8, 41.3, and 7.8 in the upper left cell of the cross-tabulation. Note the labels COUNT, ROW PCT, COL PCT, and TOT PCT at the upper left corner.
b. Which of the percentage calculations seems most meaningful to you?
c. According to the percentage calculations you prefer, does there appear to be a strong dependence between state and opinion?

2.76 A municipal workers' union that represents sanitation workers in many small midwestern cities studied the contracts that were signed in the previous years. The contracts were subdivided into those settled by negotiation without a strike, those settled by arbitration without a strike, and all those settled after a strike. For each contract, the first-year percentage wage increase was determined. Summary figures follow.

Contract type	Negotiation	Arbitration	Poststrike
Mean percentage wage increase	8.20	9.42	8.40
Variance	0.87	1.04	1.47
Standard deviation	0.93	1.02	1.21
Sample size	38	16	6

Does there appear to be a relation between contract type and mean percent wage increase? If you were management rather than union affiliated, which posture would you take in future contract negotiations?

3 PROBABILITY AND PROBABILITY DISTRIBUTIONS

HOW PROBABILITY CAN BE USED IN
MAKING INFERENCES

We stated in Chapter 1 that scientists use inferential statistics to make statements about a population based on information contained in a sample. Because populations are sets of measurements, we need a way to state an inference about them. Graphical and numerical descriptive techniques were presented in Chapter 2. Now let us examine probability, the mechanism for making inferences. This idea is probably best illustrated by means of an example.

Martha Jones, a candidate for Congress, publicly announces that her forthcoming election is a guaranteed success, and she forecasts victory by a substantial margin in all precincts of her district. Somewhat dubious about her claims, a local television station randomly selects 20 names from the voter registration list, calls each voter, and asks the voters whom they will vote for in the upcoming election. Not one of the 20 states that he or she will vote for Jones; all favor her opponent. What do you conclude about Jones's claim to victory in the sampled area?

If Jones were correct in her claim of victory, at least half the voters in the district would favor her and somewhat near this same proportion should be observed in the sample. As it turned out, none of the voters in the sample favored Jones, a result highly contradictory to her claim. Hence we infer that the proportion of voters in the population (the district) favoring Jones is less than 1/2 and that she will lose the district. We conclude that Jones would lose because the sample yielded results highly contradictory to her claim. By "contradictory" we do not mean that it is impossible to select at random 0 voters who favor Jones out of the 20 sampled assuming Jones's claim of victory is correct. We mean, rather, that such a draw is highly *improbable*. Thus we measure the degree of contradiction to Jones's claim of victory in terms of the probability of the observed sample.

To get a better view of the role that probability plays in making this inference, suppose that the sample produced 9 in favor of Jones and 11 in favor of her opponent. Would we consider this result highly improbable and reject Jones's claim? How about 7 in favor and 13 against, or 5 in favor and 15 against? Where do we draw the line? At what point do we decide that the result of the observed sample is so improbable, assuming Jones's claim is correct, that we disagree with her claim? To answer this question we must know how to find the probability of obtaining a particular sample outcome. Knowing this probability we can determine whether we agree or disagree with Jones's claim. Probability is the tool that enables us to make an inference.

Since probability is the tool for making inferences, we might ask: What is probability? In the preceding discussion we used the term *probability* in its everyday sense. Let us examine this idea more closely.

Observations of phenomena can result in many different outcomes, some of which are more likely than others. Numerous attempts have been made to give a precise definition for the probability of an outcome. We will cite a few of these.

classical interpretation   The first interpretation of probability, called the **classical interpretation of probability** arose from games of chance. Typical probability statements of this type are "the probability that a flip of a balanced coin will show 'heads' is 1/2" and "the probability of drawing an ace when a single card is drawn from a standard deck of 52 cards is 4/52." The numerical values for these probabilities arise from the nature of the games. A coin flip has two possible outcomes (a head or tail); the probability of a head should then be 1/2 (1 out of 2). Similarly, there are 4 aces in a standard deck of 52 cards, so the probability of drawing an ace in a single draw is 4/52 or 4 out of 52.

outcome
event   In the classical interpretation of probability, each possible distinct result is called an **outcome**; an **event** is identified as a collection of outcomes. The probability of an event E under the classical interpretation of probability is computed by taking the ratio of the number of outcomes N_e favorable to event E to the total number N of possible outcomes:

$$P(\text{Event } E) = \frac{N_e}{N}$$

The applicability of this interpretation depends on the assumption that all outcomes are equally likely. If this assumption does not hold, the probabilities indicated by the classical interpretation of probability will be in error.

relative frequency interpretation   A second interpretation of probability is called the **relative frequency concept of probability**; this is an empirical approach to probability. If an experiment is repeated a large number of times and event E occurs 30% of the time, then .30 should be a very good approximation to the probability of event E. Symbolically, if an experiment is conducted n different times and if event E occurs on n_e of these trials, then the probability of event E is approximately

$$P(\text{Event } E) \approx \frac{n_e}{n}$$

We say "approximate" because we think of the actual probability P(event E) as the relative frequency of the occurrence of event E over a very large number of observations or repetitions of the phenomenon. The fact that we can check probabilities that have a relative frequency interpretation (by simulating many repetitions of the experiment) makes this interpretation very appealing and practical.

The third interpretation of probability can be used for problems in which it is difficult to imagine a repetition of an experiment. These are "one-shot" situations. For example, the director of a state welfare agency who estimates the probability that a proposed revision in eligibility rules will be passed by the state legislature would not be thinking in terms of a long series of trials. Rather, the director would use a **personal or subjective probability** to make a one-shot statement of belief regarding the likelihood of passage of the proposed legislative revision. The problem with subjective probabilities is that they can vary from person to person and they cannot be checked.

subjective interpretation

Of the three interpretations presented, the relative frequency concept seems to be the most reasonable one since it provides a practical interpretation of the probability for most events of interest. Even though we will never run the necessary repetitions of the experiment to determine the exact probability of an event, the fact that we could check the probability of an event gives meaning to the relative frequency concept. Throughout the remainder of this text we will lean heavily on this interpretation of probability.

EXERCISES

3.1   Indicate which interpretation of the probability statement seems most appropriate.

a. The National Angus Association has stated that there is a 60/40 chance that wholesale beef prices will rise by the summer, that is, a .60 probability of an increase and a .40 probability of a decrease.

b. The quality control section of a large chemical manufacturing company has undertaken an intensive process-validation study. From this study the QC section claims that the probability that the "shelf life" of a newly released batch of chemical will exceed the minimal time specified is .998.

c. A new blend of coffee is being contemplated for release by the marketing division of a large corporation. Preliminary marketing survey results indicate that 550 of a random sample of 1000 potential users rated this new blend better than a brand name competitor. The probability of this happening is approximately .001 assuming that there is actually no difference in consumer preference for the two brands.

d. The probability of receiving a busy signal when attempting to access the company WATS line during the 3:00–5:00 PM time period is .58.

e. The probability that it will rain tomorrow is .30.

f. Within a city the probability of selecting a household at random in which the head of the household is unemployed is .12.

3.2   Give your own personal probability for each of the following situations. It would be instructive to tabulate these probabilities for the entire class. In which cases did you get large disagreements?

a. The federal budget will be balanced in the next fiscal year.

b. You will receive a B or higher in this course.

c. Two or more individuals in the classroom have the same birthday.

d. The New York Giants will win the Super Bowl next year.

e. The total production of Florida oranges next year will exceed this year's production.

3.2   FINDING THE PROBABILITY OF AN EVENT

In the preceding section we discussed three different interpretations of probability. In this section we will use the classical interpretation and the relative frequency concept to illustrate the computation of the probability of an outcome or event. Each possible, distinct result from an experiment is called an *outcome*; an *event* is a collection of one or more outcomes from an experiment. Consider an experiment that consists of tossing two coins, a penny and a dime, and observing the upturned faces. There are four possible outcomes:

TT: tails for both coins

TH: a tail for the penny, a head for the dime

HT: a head for the penny, a tail for the dime

HH: heads for both coins

What is the probability of observing the event, exactly one head from the two coins?

This probability can be obtained easily if we can assume that all four outcomes are equally likely. In this case, that seems quite reasonable. There are $N = 4$ possible outcomes and $N_e = 2$ of these are favorable for the event of interest, observing exactly one head. Hence, by the classical interpretation of probability,

$$P(\text{exactly 1 head}) = \frac{2}{4} = \frac{1}{2}$$

Since the event of interest has a relative frequency interpretation, we could also obtain this same result empirically, using the relative frequency concept. Suppose that a penny and a dime were tossed 2000 times, with the results shown in Table 3.1. Note that this approach yields approximate probabilities that are in agreement with our intuition. That is, intuitively we might expect these outcomes to be equally likely and each to occur with a probability equal to 1/4 or .25. This assumption was made for the classical interpretation.

If we wish to find the probability of tossing two coins and observing exactly one head, we have, from Table 3.1,

$$P(\text{exactly 1 head}) \approx \frac{502 + 496}{2000} = .499$$

This is very close to the theoretical probability, which we have shown to be .5.

TABLE 3.1   Results of 2000 Tossings of a Penny and a Dime

Outcome	Frequency	Relative Frequency
TT	474	474/2000 = .237
TH	502	502/2000 = .251
HT	496	496/2000 = .248
HH	528	528/2000 = .264

The probability of an event, say event A, will always satisfy the property

$$0 \leq P(A) \leq 1$$

That is, the probability of an event lies anywhere in the interval from 0 (the occurrence of the event is impossible) to 1 (the occurrence of an event is a "sure thing").

| 3.3 | BASIC EVENT RELATIONS AND PROBABILITY LAWS |

either *A* or *B* occurs

Suppose that A and B represent two experimental events and that you are interested in a new event, the event that **either *A* or *B* occurs**. For example, suppose that we toss a pair of dice and define the following events:

A: A total of 7 shows

B: A total of 11 shows

Then the event "either A or B occurs" is the event that you toss a total of either 7 or 11 with the pair of dice.

Note that for this example the events A and B are mutually exclusive. That is, if you observe event A (a total of 7), you could not at the same time observe event B (a total of 11). Thus if A occurs, B cannot occur (and vice versa).

DEFINITION 3.1

Two events A and B are said to be **mutually exclusive** if (when the experiment is performed a single time) the occurrence of one of the events excludes the possibility of the occurrence of the other event.

The concept of mutually exclusive events is used to specify a second property that the probabilities of events must satisfy. When two events are mutually exclusive, then the probability that either one of the events will occur is the sum of the event probabilities.

DEFINITION 3.2

If two events, A and B, are mutually exclusive, the **probability** that either event occurs is $P(\text{either } A \text{ or } B) = P(A) + P(B)$

The definition of additivity of probabilities for mutually exclusive events can be extended beyond two events. For example, when we toss a pair of dice, the sum S of the numbers appearing on the dice can assume any one of the values $S = 2, 3, 4, \ldots, 11, 12$. On a single toss of the dice we can observe only one of these values. Therefore, the values $2, 3, \ldots, 12$ represent mutually exclusive events. If we want to find the probability of tossing a sum less than or equal to 4, this probability is

$$P(S \leq 4) = P(2) + P(3) + P(4)$$

For this particular experiment the dice can fall in 36 different equally likely ways. We can observe a 1 on die 1 and a 1 on die 2, denoted by the symbol (1, 1). We can observe a 1 on die 1 and a 2 on die 2, denoted by (1, 2). In other words, for this experiment the possible outcomes are

(1, 1)	(2, 1)	(3, 1)	(4, 1)	(5, 1)	(6, 1)
(1, 2)	(2, 2)	(3, 2)	(4, 2)	(5, 2)	(6, 2)
(1, 3)	(2, 3)	(3, 3)	(4, 3)	(5, 3)	(6, 3)
(1, 4)	(2, 4)	(3, 4)	(4, 4)	(5, 4)	(6, 4)
(1, 5)	(2, 5)	(3, 5)	(4, 5)	(5, 5)	(6, 5)
(1, 6)	(2, 6)	(3, 6)	(4, 6)	(5, 6)	(6, 6)

As you can see, only one of these events, (1, 1), will result in a sum equal to 2. Therefore, we would expect a 2 to occur with a relative frequency of 1/36 in a long series of repetitions of the experiment, and we let $P(2) = 1/36$. The sum $S = 3$ will occur if we observe either of the outcomes (1, 2) or (2, 1). Therefore, $P(3) = 2/36 = 1/18$. Similarly, we find $P(4) = 3/36 = 1/12$. It follows that

$$P(S \leq 4) = P(2) + P(3) + P(4) = \frac{1}{36} + \frac{1}{18} + \frac{1}{12} = \frac{1}{6}$$

In most practical data-collecting situations, we will make observations on a variable. For example, in recording the diastolic blood pressures of hypertensive patients, the variable of interest is diastolic blood pressure and each patient's diastolic blood pressure represents an observation (measurement) on that variable. The fact that these measurements vary in a seemingly random and unpredictable manner leads us to call the variable a

random variable   **random variable**.

Consider a situation in which a single measurement is obtained on a random variable. Since we will observe only one of many possible values of the random variable (for example, only one diastolic blood pressure), it follows that the values of a random variable represent mutually exclusive events (if one value occurs, the others cannot have occurred). We will make particular use of the additive property of the probabilities of mutually exclusive events when we wish to find the probability that a random variable assumes one of two or more values when it is observed in an experiment.

A third property of event probabilities concerns an event and its complement.

DEFINITION 3.3

The **complement** of an event A is the event that A *does not* occur. The complement of A is denoted by the symbol $\bar{A}$.

Thus if we define the complement of an event A as a new event, namely, "A does not occur," it follows that

$$P(A) + P(\bar{A}) = 1$$

For example, refer again to the two-coin-toss experiment. If, in many repetitions of the experiment, the proportion of times you observe event A, "two heads show," is 1/4, then it follows that the proportion of times you observe the event $\bar{A}$, "two heads do not show," is 3/4. Thus $P(A)$ and $P(\bar{A})$ will always sum to 1.

The three properties that the probabilities of events must satisfy can be summarized as follows:

PROPERTIES OF PROBABILITIES

If A and B are any two mutually exclusive events associated with an experiment, then $P(A)$ and $P(B)$ must satisfy the following properties:

1. $0 \leq P(A) \leq 1$ and $0 \leq P(B) \leq 1$
2. $P(\text{either } A \text{ or } B) = P(A) + P(B)$
3. $P(A) + P(\bar{A}) = 1$ and $P(B) + P(\bar{B}) = 1$

DEFINITION 3.4

The **union** of two events A and B is the set of all outcomes that are included in either A or B (or both). The union is denoted as $A \cup B$.

DEFINITION 3.5

The **intersection** of two events A and B is the set of all outcomes that are included in both A and B. The intersection is denoted as $A \cap B$.

These definitions along with the definition of the complement of an event formalize some simple concepts. The event $\bar{A}$ occurs when A *does not*; $A \cup B$ occurs when either A or B occurs; $A \cap B$ occurs when A *and* B occur.

The additivity of probabilities for mutually exclusive events, called the *addition law for mutually exclusive events*, can be extended to give the general addition law.

DEFINITION 3.6

Consider two events A and B; the **probability of the union** of A and B is

$$P(A \cup B) = P(A) + P(B) - P(A \cap B)$$

EXAMPLE 3.1

Events and event probabilities are shown in the following Venn diagram. Use this diagram to determine the probabilities listed.

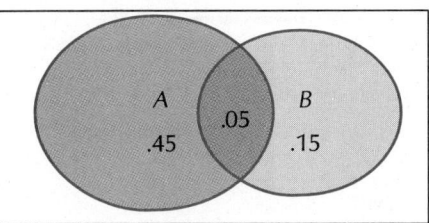

a. $P(A)$, $P(\bar{A})$
b. $P(B)$, $P(\bar{B})$
c. $P(A \cap B)$
d. $P(A \cup B)$

Solution

From the Venn diagram, we are able to determine these probabilities

a. $P(A) = .5$, therefore $P(\bar{A}) = 1 - .5 = .5$
b. $P(B) = .2$, therefore $P(\bar{B}) = 1 - .2 = .8$
c. $P(A \cap B) = .05$
d. $P(A \cup B) = P(A) + P(B) - P(A \cap B) = .5 + .2 - .05 = .65$

3.4   CONDITIONAL PROBABILITY AND INDEPENDENCE

Consider the following situation: the examination of a large number of insurance claims, categorized according to type of insurance and whether or not the claim was fraudulent, produced the results shown in Table 3.2. Suppose you are responsible for checking insurance claims—in particular, for detecting fraudulent claims—and you examine the next claim that is processed. What is the probability of the event F, "the claim is fraudulent"? To answer the question, you examine Table 3.2 and note that 10% of all claims are fraudulent. Thus assuming that the percentages given in the table are reasonable approximations to the true probabilities of receiving specific types of claims, it follows that $P(F) = .10$. Would you say that the risk that you face a fraudulent claim has probability .10? We think not, because you have additional information that may affect the assessment of $P(F)$. For example, you would know the type of policy you were examining (fire, auto, or other).

Suppose that you have the additional information that the claim was associated with a fire policy. Checking Table 3.2, we see that 20% (or .20) of all claims are associated with a fire policy and that 6% (or .06) of all claims are fraudulent fire policy claims. Therefore, it follows that the probability that the claim is fraudulent, given that you know the policy is a fire policy, is

$$P(F \mid \text{fire policy}) = \frac{\text{proportion of claims that are fraudulent fire policy claims}}{\text{proportion of claims that are against fire policies}}$$

$$= \frac{.06}{.20} = .30$$

TABLE 3.2   Categorization of Insurance Claims

Category	Type of policy			Total
	Fire	Auto	Other	
Fraudulent	6%	1%	3%	10%
Nonfraudulent	14%	29%	47%	90%
Total	20%	30%	50%	100%

conditional probability   This probability, $P(F|\text{fire policy})$, is called a **conditional probability** of the event F, that is, the probability of event F given the fact that the event "fire policy" has already occurred. This tells you that 30% of all fire policy claims are fraudulent. The vertical bar in the expression $P(F|\text{fire policy})$ represents the phrase "given that," or simply "given." Thus the expression is read, "the probability of the event F given the event fire policy."

unconditional probability   The probability $P(F) = .10$, called the **unconditional probability** of the event F, gives the proportion of times a claim is fraudulent, that is, the proportion of times event F occurs in a very large (infinitely large) number of repetitions of the experiment (receiving an insurance claim and determining whether the claim is fraudulent). In contrast, the conditional probability of F, given that the claim is for a fire policy, $P(F|\text{fire policy})$, gives the proportion of fire policy claims that are fraudulent. Clearly, the conditional probabilities of F, given the types of policies, will be of much greater assistance in measuring the risk of fraud than the unconditional probability of F.

DEFINITION 3.7

Consider two events A and B with nonzero probabilities, $P(A)$ and $P(B)$. The **conditional probability** of event A given event B is

$$P(A|B) = \frac{P(A \cap B)}{P(B)}$$

The conditional probability of event B given event A is

$$P(B|A) = \frac{P(A \cap B)}{P(A)}$$

This definition for conditional probabilities gives rise to what is referred to as the *multiplication law.*

DEFINITION 3.8

The **probability of the intersection** of two events A and B is

$$P(A \cap B) = P(A)P(B|A)$$
$$= P(B)P(A|B)$$

The only difference between Definitions 3.7 and 3.8, both of which involve conditional probabilities, relates to what probabilities are known and what needs to be calculated. When the intersection probability $P(A \cap B)$ and the individual probability $P(A)$ are known, we can compute $P(B|A)$. When we know $P(A)$ and $P(B|A)$, we can compute $P(A \cap B)$.

EXAMPLE 3.2   Two supervisors are to be selected as safety representatives within the company. Given that there are six supervisors in research and five in development, and each group of two supervisors has the same chance of being selected, find the probability of choosing both supervisors from research.

Solution
|
Let A be the event that the first supervisor selected is from research and let B be the event that the second supervisor is also from research. Clearly we want $P(A \cap B) = P(A)P(B|A)$.

For this example,

$$P(A) = \frac{6}{11} \quad \text{and} \quad P(B|A) = \frac{5}{10}$$

Then

$$P(A \cap B) = \left(\frac{6}{11}\right)\left(\frac{5}{10}\right) = \frac{30}{110} = .27$$

Suppose that the probability of event A is the same regardless of whether event B has or has not occurred. That is, suppose

$$P(A|B) = P(A)$$

Then we say that the occurrence of event A is not dependent on the occurrence of event B or, simply, that A and B are independent events.

DEFINITION 3.9

Two events A and B are **independent events** if

$$P(A|B) = P(A) \quad \text{or if} \quad P(B|A) = P(B)$$

(Note: You can show that if $P(A|B) = P(A)$, then $P(B|A) = P(B)$, and vice versa.)

The concept of independence is of particular importance in sampling. Subsequently we will draw samples from two (or more) populations in order to compare the population means, variances, or some other population parameters. For most of these applications we will select samples in such a way that the observed values in one sample are independent **independent samples** of the values that appear in another sample. We call these **independent samples**.

EXERCISES

3.3   Consider the following outcomes for an experiment:

Outcome	1	2	3	4	5
Probability	.20	.25	.15	.10	.30

Let event A consist of outcomes 1, 3, and 5, and event B consist of outcomes 4 and 5.
a. Find $P(A)$ and $P(B)$.
b. Find P(both A and B occur).
c. Find P(either A or B occurs).

3.4   Refer to Exercise 3.3. Does P(either A or B occurs) $= P(A) + P(B)$? Why or why not?

3.5   A student has to have an accounting course and an economics course the next term. Assuming there are no schedule conflicts, describe the possible outcomes for selecting one section of the accounting course and one of the economics course if there are four possible accounting sections and three possible economics sections.

3.6   An institutional investor is considering a large investment in two of five companies. Suppose that, unknown to the investor, two of the five firms are on shaky grounds with regard to the development of new products.
a. List the possible outcomes for this situation.
b. Determine the probability of choosing two of the three firms that are on better grounds.
c. What's the probability of choosing one of two firms on shaky grounds?
d. What's the probability of choosing the two shakiest firms?

3.7   A survey of workers in two manufacturing sites of a firm included the following question: "How effective is management in responding to legitimate grievances of workers?" The results are shown here.

	Number Surveyed	Number Responding "Poor"
Site 1	192	48
Site 2	248	80

Let A be the event worker comes from site 1 and B be the event the response is "poor." Compute $P(A)$, $P(B)$, and $P(A \cap B)$.

3.8   Refer to Exercise 3.7.
a. Are the events A and B independent?
b. Find $P(B|A)$ and $P(B|\bar{A})$. Are they equal?

3.9   Three coins are tossed and we observe the outcome (heads or tails) for each coin.
a. List the sample space.
b. Define the events:

   A: Observe exactly one head.

   B: Observe one or more heads.

   Find $P(A)$, $P(B)$, $P(A \cap B)$, and $P(A|B)$.
c. Are events A and B independent?

3.10   A large corporation has spent considerable time developing employee performance rating scales to evaluate an employee's job performance on a regular basis, so major adjustments can be made when needed and employees who should be considered for a "fast track" can be isolated. Keys to this latter determination are ratings on the ability of an employee to perform to his or her capabilities and on his or her formal training for the job.

Workload Capacity	Formal Training			
	None	Little	Some	Extensive
Low	.01	.02	.02	.04
Medium	.05	.06	.07	.10
High	.10	.15	.16	.22

The probabilities for being placed on a "fast track" are as indicated for the 12 categories of workload capacity and formal training. The following three events (A, B, and C) are defined:

A: An employee works at the high-capacity level.

B: An employee falls into the highest (Extensive) formal training category.

C: An employee has little or no formal training and works below high capacity.

a. Find $P(A)$, $P(B)$, and $P(C)$.
b. Find $P(A|B)$, $P(A|\bar{B})$, and $P(\bar{B}|C)$.
c. Find $P(A \cup B)$, $P(A \cap C)$, and $P(B \cap C)$.

3.11 The utility company in a large metropolitan area finds that 70% of its customers pay a given monthly bill in full.

a. Suppose two customers are chosen at random from the list of all customers. What is the probability that both customers will pay their monthly bill in full?
b. What's the probability that at least one of them will pay in full?

3.12 Refer to Exercise 3.11. A more detailed examination of the company records indicates that 95% of the customers who pay one monthly bill in full will pay the next monthly bill in full also; only 10% of those who pay less than the full amount one month will pay in full the next month.

a. Find the probability that a customer selected at random will pay two consecutive months in full.
b. Find the probability that a customer selected at random will pay neither of two consecutive months in full.
c. Find the probability that a customer chosen at random will pay exactly one month in full.

3.5   VARIABLES: DISCRETE AND CONTINUOUS

numerical outcomes

Most events of interest result in numerical observations or measurements. If a quantitative variable measured (or observed) in an experiment is denoted by the symbol y, we are interested in the values that y can assume. These values are called **numerical outcomes** The number of students in a class of 50 who earn an A in their biology course is a numerical outcome. The percentage of registered voters who cast ballots in a given election is also a numerical outcome. The quantitative variable y is called a random variable because the value that y assumes in a given experiment is a chance or random outcome.

Random variables are classified as one of two types.

DEFINITION 3.10

> When observations on a quantitative random variable can assume only a countable number of values, the variable is called a **discrete random variable**.

Examples of discrete variables are these:

1. number of bushels of apples per acre for a given orchard this year
2. number of accidents per year at an intersection
3. number of voters in a sample favoring Jones

Note that it is possible to count the number of values that each of these random variables can assume.

DEFINITION 3.11

When observations on a quantitative random variable can assume any one of the uncountable number of values in a line interval, the variable is called a **continuous random variable**.

For example, the daily maximum temperature in Rochester, New York, can assume any of the infinitely many values on a line interval. It could be 89.6, 89.799, or 89.7611114. Typical continuous random variables are temperature, pressure, height, weight, and distance.

The distinction between discrete and continuous random variables is pertinent when we are seeking the probabilities associated with specific values of a random variable. The need for the distinction will be apparent when probability distributions are discussed in later sections of this chapter.

3.6   PROBABILITY DISTRIBUTIONS FOR DISCRETE RANDOM VARIABLES

As previously stated, we need to know the probability of observing a particular sample outcome in order to make an inference about the population from which the sample was drawn. To do this we need to know the probability associated with each value of the variable y. Viewed as relative frequencies, these probabilities generate a distribution of theoretical relative frequencies called the *probability distribution* of y. Probability distributions differ for discrete and continuous variables but the interpretation is essentially the same for both.

The *probability distribution for a discrete random variable* displays the probability $P(y)$ associated with each value of y. This display can be presented as a table, a graph, or a formula. To illustrate, consider the tossing of two coins in Section 3.2 and let y be the number of heads observed. Then y can take the values 0, 1, or 2. From the data of Table 3.1, we can determine the approximate probability for each value of y, as given in Table 3.3. We point out that the relative frequencies in the table are very close to the theoretical relative frequencies (probabilities), which can be shown to be .25, .50, and .25 using the classical interpretation of probability. If we had employed 2,000,000 tosses of the coins instead of 2000, the relative frequencies for $y = 0$, 1, and 2 would be indistinguishable from the theoretical probabilities.

TABLE 3.3   Empirical Sampling Results for y: the Number of Heads in 2000 Tosses of Two Coins

y	Frequency	Relative Frequency
0	474	.237
1	998	.499
2	528	.264

TABLE 3.4   Probability Distribution for the
Number of Heads When
Two Coins Are Tossed

y	$P(y)$
0	.25
1	.50
2	.25

FIGURE 3.1

Probability
Distribution for the
Number of Heads
When Two Coins
Are Tossed

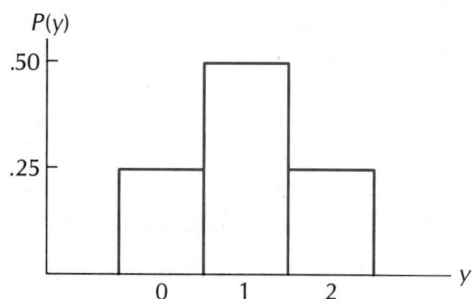

The probability distribution for y, the number of heads in the toss of two coins, is shown in Table 3.4. It is presented graphically as a *probability histogram* in Figure 3.1.

The probability distribution for this simple discrete random variable illustrates three important properties of discrete random variables.

PROPERTIES OF
DISCRETE
RANDOM
VARIABLES

1. The probability associated with every value of y lies between 0 and 1.
2. The sum of the probabilities for all values of y is equal to 1.
3. The probabilities for a discrete random variable are additive. Hence the probability that $y = 1$ or 2 is equal to $P(1) + P(2)$.

The relevance of the probability distribution to statistical inference will be emphasized when we discuss the probability distribution for the binomial random variable.

3.7   A USEFUL DISCRETE RANDOM VARIABLE: THE BINOMIAL

Public opinion polls play an important role in our daily lives. The Gallup and Harris polls, two of the most widely publicized surveys, keep us abreast of national thinking on political and social issues, and to some extent probably make public officials more responsive to the will of the people. Similar polls of a more limited nature are conducted at the state and local levels. Polls are also conducted by business organizations to obtain consumer reaction to new products, sales practices, and company policies.

The public opinion poll and consumer preference poll are examples of a common, frequently conducted sampling situation known as a *binomial experiment*. Because the binomial experiment is conducted in many different disciplines and differs from one situation to another only in the nature of the questions asked, it is useful for us to define its characteristics. We can then apply our knowledge of this one kind of experiment to a wide variety of sampling situations that affect our lives.

For all practical purposes the binomial experiment is identical to the coin-tossing example of previous sections. Here n different coins are tossed and we are interested in the number of heads observed. We assume that the probability of tossing a head on a single trial is π (π may equal .50, as it would for a balanced coin, but in many practical situations π will take some other value between 0 and 1). We also assume that the outcome for any one toss is unaffected by the results of any preceding tosses. These characteristics can be summarized as shown here.

DEFINITION 3.12

A **binomial experiment** is one that has the following properties:

1. The experiment consists of n identical trials.
2. Each trial results in one of two outcomes. We will label one outcome a success and the other a failure.
3. The probability of success on a single trial is equal to π and π remains the same from trial to trial.*
4. The trials are independent; that is, the outcome of one trial does not influence the outcome of any other trial.
5. The random variable y is the number of successes observed during the n trials.

EXAMPLE 3.3   A survey of 500 farmers is conducted to determine the proportion in favor of additional price supports for dairy products. Does this survey satisfy the properties of a binomial experiment?

Solution   To answer this question we check each of the five characteristics of the binomial experiment to determine if they are satisfied.

1. Are there n identical trials? Yes. There are $n = 500$ interviews, all the same.
2. Does each trial result in one of two outcomes? Yes. Each farmer interviewed either favors or does not favor the additional price supports.

---

* Some textbooks and computer programs use the letter p rather than π. We have chosen π to avoid confusion with p-values, discussed in Chapter 4.

3. Is the probability of success the same from trial to trial? Yes. If we let "success" denote a farmer favoring the additional supports, then assuming the list of farmers from which the sample was drawn is large, the probability of success will (for all practical purposes) remain constant from trial to trial.
4. Are the trials independent? Yes. The outcome of one interview is unaffected by the results of the other interviews.
5. Is the random variable of interest to the experimenter the number of successes y in the sample? Yes. We are interested in the number of farmers in the sample of 500 favoring additional price supports for dairy products.

Since all five characteristics are satisfied, the survey represents a binomial experiment.

EXAMPLE 3.4

An economist interviews 75 students in a class of 100 to estimate the proportion of students who expect to obtain a C or better in the course. Is this a binomial experiment?

Solution

Check this experiment against the five characteristics of a binomial.

1. Are there identical trials? Yes. Each of 75 students is interviewed.
2. Does each trial result in one of two outcomes? Yes. Each student either does or does not expect to obtain a grade of C or higher.
3. Is the probability of success the same from trial to trial? No. If we let success denote a student expecting to obtain a C or higher, then the probability of success can change considerably from trial to trial. For example, unknown to the professor, suppose that 75 of the 100 students expect to obtain a grade of C or higher. Then π, the probability of success for the first student interviewed, is $75/100 = .75$. If the student is a failure (does not expect a C or higher), the probability of success for the next student is $75/99 = .76$. Suppose that after 70 students have been interviewed, 60 were successes and 10 were failures. Then the probability of success for the next (71st) student is $15/30 = .50$.

This example shows how the probability of success can change substantially from trial to trial in situations where the sample size is a relatively large portion of the total population size. This experiment does not satisfy the properties of a binomial experiment.

It should be noted that very few real-life situations satisfy perfectly the requirements stated in Definition 3.12, but for many the lack of agreement is so small that the binomial experiment still provides a very good model for reality.

Having defined the binomial experiment and suggested several practical applications, we now examine the probability distribution for the binomial random variable y, the number of successes observed in n trials. Although it would be possible to approximate $P(y)$, the probability associated with a value of y in a binomial experiment, by using a relative frequency approach, it is easier to make use of a general formula for binomial probabilities.

FORMULA FOR COMPUTING $P(y)$ IN A BINOMIAL EXPERIMENT

The probability of observing y successes in n trials of a binomial experiment is

$$P(y) = \frac{n!}{y!(n-y)!} \, \pi^y(1-\pi)^{n-y}$$

where

n = number of trials

π = probability of success on a single trial

$1 - \pi$ = probability of failure on a single trial

y = number of successes in n trials

$n! = n(n-1)(n-2) \cdots (3)(2)(1)$

As indicated above, the notation $n!$ (referred to as n factorial) is used for the product

$$n! = n(n-1)(n-2) \cdots (3)(2)(1)$$

For $n = 3$,

$$n! = 3! = (3)(3-1)(3-2) = (3)(2)(1) = 6$$

Similarly, for $n = 4$,

$$4! = (4)(3)(2)(1) = 24$$

We also note that $0!$ is defined to be equal to 1.

To see how the formula for binomial probabilities can be used to calculate the probability for a specific value of y, consider the following examples.

EXAMPLE 3.5

An experiment consists of tossing a coin two times. If the probability of a head is .5, compute the probability distribution for y, the number of heads, using the binomial formula $P(y)$. Compare your results to those given in Table 3.4.

Solution

Using the formula

$$P(y) = \frac{n!}{y!(n-y)!} \, \pi^y(1-\pi)^{n-y}$$

and substituting for $n = 2$, $\pi = .5$, $y = 0, 1, 2$, we obtain

$$P(y = 0) = \frac{2!}{0!2!} \, (.5)^0(.5)^2 = .25$$

$$P(y = 1) = \frac{2!}{1!1!} \, (.5)(.5) = .50$$

$$P(y = 2) = \frac{2!}{2!0!} \, (.5)^2(.5)^0 = .25$$

Note that these results are identical to those presented in Table 3.4.

EXAMPLE 3.6   A drug company advertises that a new product is effective in the treatment of a particular serious disease. However, the firm also states that an undesirable side effect occurs in 10% of the patients treated. If a physician has four unrelated patients for whom she could prescribe the drug, what is the probability that all four will experience the side effect?

Solution   This example satisfies the characteristics of a binomial experiment, with $n = 4$ trials and π, the probability of observing a side effect, equal to .1. Substituting into the formula $P(y)$ with $y = 4$, $\pi = .1$, $1 - \pi = .9$, and $n = 4$, we have

$$P(y = 4) = \frac{4!}{4!0!} (.1)^4(.9)^0 = (.1)^4 = .0001$$

Thus the probability that all four will experience the side effect is .0001.

EXAMPLE 3.7   Suppose the physician of Example 3.6 treated all four patients and observed the indicated side effect in all four. What might you conclude concerning the drug firm's claim?

Solution   Since the probability of observing a side effect in all four patients is so small (.0001), assuming the side effect appears in only 10% of the patients treated as claimed by the drug firm, we would conclude that the company's claim is incorrect. Indeed, it appears that more than 10% of those treated experience the side effect.

We have discussed probability distributions for discrete random variables and have given an example of a very useful discrete random variable, the binomial. The only other discrete random variable that we will discuss in this text is the Poisson (Chapter 6). The interested reader is referred to Hildebrand and Ott (1987) and Hogg and Craig (1978) for more information about discrete random variables. In the next section we discuss probability distributions with emphasis on the normal distribution.

EXERCISES

3.13   Consider the following class experiment: Toss three coins and observe the number of heads y. Let each student repeat the experiment 10 times, combine the class results, and construct a relative frequency table for y. Note that these frequencies give approximations to the actual probabilities that $y = 0, 1, 2,$ or 3. (Note: Calculate the actual probabilities by using the binomial formula $P(y)$ to compare the approximate results with the actual probabilities.)

3.14   A random sample of 10 members was obtained to ascertain opinions concerning a new wage-package proposal to a large local union by union leaders. If we assume that $\pi = .6$ of all the members have major disagreements with the package, compute the following probabilities using the output shown here:

a. All disagree.
b. Exactly six disagree.
c. Six or more disagree.
d. All agree.

```
        BINOMIAL FOR N = 10,P = .6

     BINOMIAL PROBABILITIES FOR N =   10   AND P = 0.600000

          K          P( X = K )         P(X LESS OR = K)
          0            0.0001              0.0001
          1            0.0016              0.0017
          2            0.0106              0.0123
          3            0.0425              0.0548
          4            0.1115              0.1662
          5            0.2007              0.3669
          6            0.2508              0.6177
          7            0.2150              0.8327
          8            0.1209              0.9536
          9            0.0403              0.9940
         10            0.0060              1.0000
   MTB >
```

3.15   Refer to Exercise 3.14.
     a. Compute probabilities for parts (a) through (d) if $\pi = .3$.
     b. Indicate how you could compute $P(y \le 100)$ if $n = 1000$ for $\pi = .6$.

3.16   Let y be a binomial random variable; compute $P(y)$ for each of the following situations:
     a. $n = 10$, $\pi = .2$, $y = 3$
     b. $n = 4$, $\pi = .4$, $y = 2$
     c. $n = 16$, $\pi = .7$, $y = 12$

3.17   Let y be a binomial random variable with $n = 8$ and $\pi = .4$. Find
     a. $P(y \le 4)$
     b. $P(y > 4)$
     c. $P(y \le 7)$
     d. $P(y > 6)$

3.18   Over a long period of time in a large multinational corporation, 10% of all sales trainees are rated as outstanding, 75% as excellent/good, 10% as satisfactory, and 5% as unsatisfactory. Find the following probabilities for a sample of 10 trainees selected at random.
     a. The probability that two are rated as outstanding.
     b. The probability that two or more are rated as outstanding.
     c. The probability that eight of the 10 are rated either outstanding or excellent/good.
     d. The probability that none of the trainees is rated as unsatisfactory.

3.19   A new technique, balloon angioplasty, is being widely used to open clogged heart valves and vessels. The balloon is inserted via a catheter and is inflated, opening the vessel; thus, no surgery is required. Left untreated, 50% of the people with heart valve disease die within about two years. If experience with this technique suggests that approximately 70% live for more than two years, would the next five patients of the patients treated with balloon angioplasty at a hospital constitute a binomial experiment with $n = 5$, $\pi = .70$? Why or why not?

3.20   A prescription drug firm claims that only 12% of all new drugs shown to be effective in animal tests ever make it through a clinical testing program and onto the market. If a firm has 15 new compounds that have shown effectiveness in animal tests, find the following probabilities:
     a. None reach the market.
     b. One or more reach the market.
     c. Two or more reach the market.

3.21   Does Exercise 3.20 satisfy the properties of a binomial experiment? Why or why not?

3.22   A survey was conducted to investigate the attitudes of nurses working in Veterans Administration hospitals. A random sample of 1000 nurses was contacted using a mailed questionnaire and the number favoring or opposing a particular issue was recorded. If we confine our attention to the nurses' responses to a single question, would this sampling represent a binomial experiment? As with most mail surveys, some of the nurses will not respond. What effect might nonresponders in the sample have on the estimate of the percentage of all Veterans Administration nurses who favor the particular proposition?

3.8   PROBABILITY DISTRIBUTION FOR CONTINUOUS RANDOM VARIABLES

Discrete random variables (such as the binomial) have possible values that are distinct and separate, such as 0 or 1 or 2 or 3. Other random variables are most usefully considered to be *continuous*: their possible values form a whole interval (or range, or continuum). For instance, the one-year return per dollar invested in a common stock could range from 0 to some quite large value. In practice, virtually all random variables assume a discrete set of values; the return per dollar of a million-dollar common-stock investment could be 1.06219423 or 1.06219424 or 1.06219425 or . . . . But, when there are many, many possible values for a random variable, it is sometimes mathematically useful to treat the random variable as continuous.

Theoretically, then, a continuous random variable is one that can assume values associated with infinitely many points in a line interval. We state, without elaboration, that it is impossible to assign a small amount of probability to each value of y (as was done for a discrete random variable) and retain the property that the probabilities sum to 1.

To overcome this difficulty, we revert to the concept of the relative frequency histogram of Chapter 2, where we talked about the probability of y falling in a given interval. Recall that the relative frequency histogram for a population containing a large number of measurements will almost be a smooth curve because the number of class intervals can be made large and the width of the intervals can be decreased. Thus we envision a smooth curve that provides a model for the population relative frequency distribution generated by repeated observation of a continuous random variable. This will be similar to the curve shown in Figure 3.2.

Recall that the histogram relative frequencies are proportional to areas over the class intervals and that these areas possess a probabilistic interpretation. That is, if a measurement is randomly selected from the set, the probability that it will fall in an interval is proportional to the histogram area above the interval. Since a population is the whole (100%, or 1), we want the total area under the probability curve to equal 1. If we let the total area under the curve equal 1, then areas over intervals are exactly equal to the corresponding probabilities.

The graph for the probability distribution for a continuous random variable is shown in Figure 3.2. The ordinate (height of the curve) for a given value of y is denoted by the symbol $f(y)$. Many people are tempted to say that $f(y)$, like $P(y)$ for the binomial random

FIGURE 3.2

Probability
Distribution for a
Continuous Random
Variable

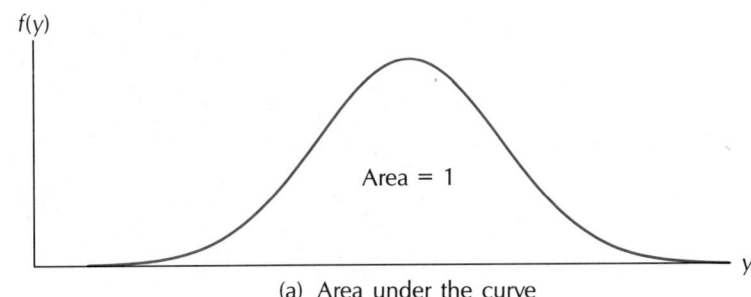

(a) Area under the curve

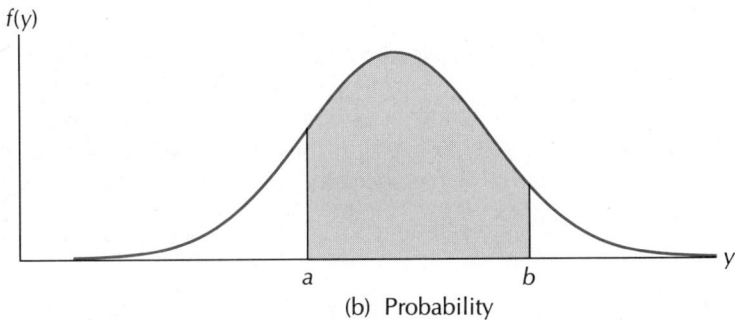

(b) Probability

variable, designates the probability associated with the continuous random variable y. But, as we mentioned before, it is impossible to assign a probability to each of the infinitely many possible values of a continuous random variable. Thus all we can say is that $f(y)$ represents the height of the probability distribution for a given value of y.

The probability that a continuous random variable falls in an interval, say between two points a and b, follows directly from the probabilistic interpretation given to the area over an interval for the relative frequency histogram (Section 2.2) and is equal to the area under the curve over the interval a to b, as shown in Figure 3.2. This probability is written $P(a < y < b)$.

There are curves of many shapes that can be used to represent the population relative frequency distribution for measurements associated with a continuous random variable. Fortunately, the areas under these curves have been tabulated and are ready for use. Thus if we know that student examination scores possess a particular probability distribution, as in Figure 3.3, and if areas under the curve have been tabulated, we could find the probability that a particular student will score more than 80% by looking up the tabulated area, which is shaded in Figure 3.3.

You probably wonder how we know which shape to use for the probability distribution in a given situation. Fortunately, the specific shape chosen for the population frequency distribution (or, equivalently, the probability distribution for the observed variable) will often have little effect on the probability statements associated with our inferences. Thus we can relax in the knowledge that the selection of the exact shape for the probability distribution for a continuous variable is not crucial.

FIGURE 3.3

Hypothetical
Probability
Distribution for
Student Examination
Scores

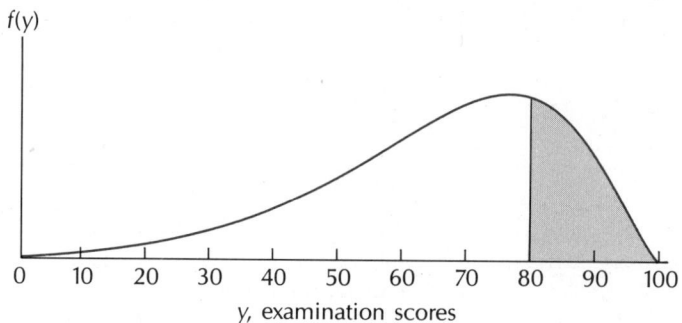

We will find that data collected on continuous variables often possess mound-shaped frequency distributions and that many of these are nearly bell-shaped. A continuous variable (the normal) and its probability distribution (the bell-shaped, normal curve) provide a good model for these types of data. The normally distributed variable also plays a very important role in statistical inference. We will study its bell-shaped probability distribution in detail in the next section.

3.9   A USEFUL CONTINUOUS RANDOM VARIABLE: THE NORMAL DISTRIBUTION

normal curve

Many variables of interest, including several statistics to be discussed in later sections and chapters, have mound-shaped frequency distributions that can be approximated by using a **normal curve**. For example, the distribution of total scores on the Brief Psychiatric Rating Scale for outpatients having a current history of repeated aggressive acts would be mound-shaped. Other practical examples of mound-shaped distributions are social perceptiveness scores of preschool children selected from a particular socioeconomic background, psychomotor retardation scores for patients with circular-type manic-depressive illness, milk yields for cattle of a particular breed, and perceived anxiety scores for residents of a community. Each of these mound-shaped distributions could be approximated with a normal curve.

Since the normal distribution has been well tabulated, areas under a normal curve—which correspond to probabilities—can be used to approximate probabilities associated with the variables of interest in our experimentation. Thus the normal random variable and its associated distribution will play an important role in statistical inference.

The relative frequency histogram for the normal random variable, called the *normal curve* or *normal* probability distribution, is a smooth bell-shaped curve. Figure 3.4 shows a normal curve. If we let *y* represent the normal random variable, then the height of the probability distribution for a specific value of *y* is represented by $f(y)$.* The probabilities associated with a normal curve form the basis for the Empirical Rule.

---

* For the normal distribution $f(y) = (1/\sqrt{2\pi}\sigma)e^{-1/2[(y-\mu)/\sigma]^2}$ where μ and σ are the mean and standard deviation, respectively, of the population of y values.

FIGURE 3.4

A Normal Curve with
Mean μ and Standard
Deviation σ

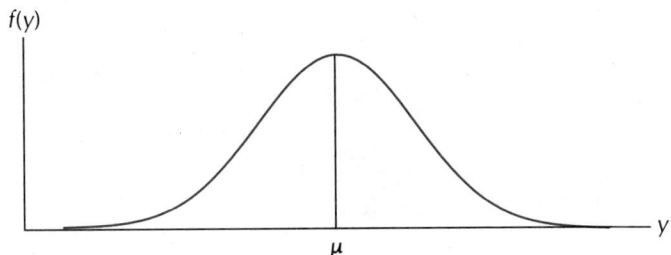

As we see from Figure 3.4, the normal probability distribution is bell-shaped and symmetrical about the mean μ. Although the normal random variable y may theoretically assume values from $-\infty$ to $+\infty$, we know from the Empirical Rule that approximately all the measurements are within 3 standard deviations (3σ) of μ. From the Empirical Rule we also know that if we select a measurement at random from a population of measurements that possesses a mound-shaped distribution, the probability is approximately .68 that the measurement will lie within one standard deviation of its mean (see Figure 3.5). Similarly, we know that the probability is approximately .95 that a value will lie in the interval $\mu \pm 2\sigma$. What we do not know, however, is the probability that the measurement will be within 1.65 standard deviations of its mean, or within 2.58 standard deviations of its mean. The procedure we are going to discuss in this section will enable us to calculate the probability that a measurement falls within any distance of the mean μ for a normal curve.

Since there are many different normal curves (depending on the parameters μ and σ), it might seem to be an impossible task to tabulate areas (probabilities) for all normal curves, especially if each curve requires a separate table. Fortunately, this is not the case. By specifying the probability that a variable y lies within a certain number of standard deviations of its mean (just as we did in using the Empirical Rule), we need only one table of probabilities.

area under a normal curve

z standard deviations

Table 1 in the Appendix gives the **area under a normal curve** from the mean μ to a point z standard deviations ($z\sigma$) to the right of μ (see Figure 3.6). Because of the symmetry of the normal probability distribution, this would be the same as the area between the mean and a point z **standard deviations** to the left of μ.

FIGURE 3.5

Area Under a Normal
Curve Within One
Standard Deviation of
the Mean

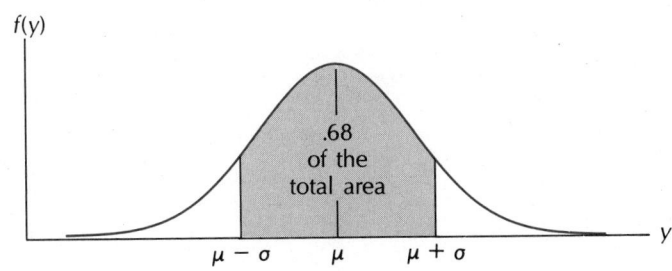

FIGURE 3.6

Area Under a
Normal Curve, As
Given in Table 1
in the Appendix

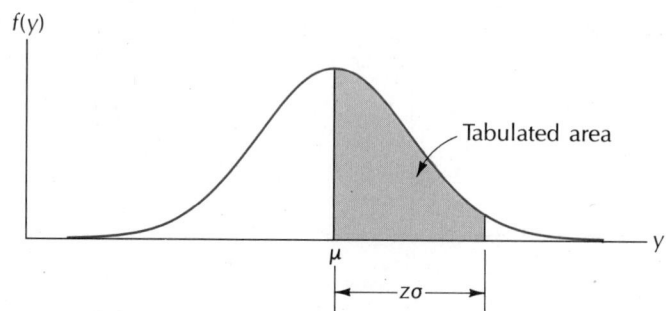

The area shown by the shading in Figure 3.6 is the probability listed in Table 1 in the Appendix. Values of z to the nearest 10th are listed along the left-hand column of Table 1, with z to the nearest 100th along the top of the table. In order to find the probability that a normal random variable will lie in the interval from μ to a point 1.65 standard deviations above the mean, we look up the table entry corresponding to z = 1.65. This probability is .4505 (see Figure 3.7).

To determine the probability that a measurement will fall in the interval from μ to some value y to the right of μ, we first calculate the number of standard deviations that y lies away from the mean by using the formula

formula for z

$$z = \frac{y - \mu}{\sigma}$$

z score   The value of z computed using this formula is sometimes referred to as the **z score** associated with the y-value. Using the computed value of z, we determine the appropriate probability by using Table 1 in the Appendix. Note that we are merely coding the value y by subtracting μ and dividing by σ. (In other words, $y = z\sigma + \mu$.) Figure 3.8 illustrates the values of z corresponding to specific values of y. Thus a value of y two standard deviations below (to the left of) μ corresponds to z = −2.

FIGURE 3.7

Area Under a Normal
Curve from μ to a
Point 1.65 Standard
Deviations Above the
Mean

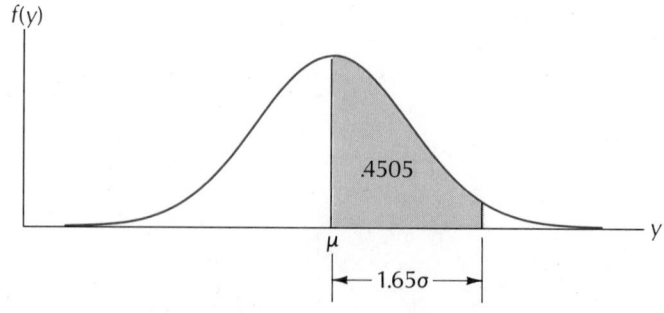

FIGURE 3.8

Relationship Between
Specific Values of y
and $z = (y - \mu)/\sigma$

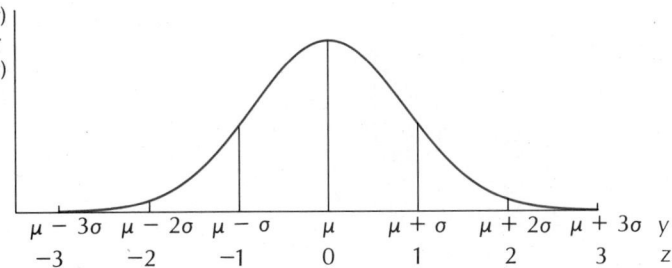

EXAMPLE 3.8

Consider a normal distribution with $\mu = 20$ and $\sigma = 2$. Determine the probability that a measurement will be in the interval from 20 to 23.

Solution

When first working problems of this type, it might be a good idea to draw a picture so that you can see the area in question, as we have in Figure 3.9.

To determine the area under the curve from $\mu = 20$ to the value $y = 23$, we first calculate the number of standard deviations $y = 23$ lies away from the mean.

$$z = \frac{y - \mu}{\sigma} = \frac{23 - 20}{2} = 1.5$$

Thus $y = 23$ lies 1.5 standard deviations above $\mu = 20$. Referring to Table 1 in the Appendix, we find the area corresponding to $z = 1.5$ to be .4332. This is the probability that a measurement falls in the interval from 20 to 23.

Similarly, when finding the probability that a measurement lies in the interval from μ to some value of y to the left of the mean, we again compute z, using

$$z = \frac{y - \mu}{\sigma}$$

negative z   The computed value of z will be negative but we ignore the negative sign and again refer to Table 1 in the Appendix, for the appropriate probability.

FIGURE 3.9

Area Between $\mu = 20$
and $y = 23$,
Example 3.8

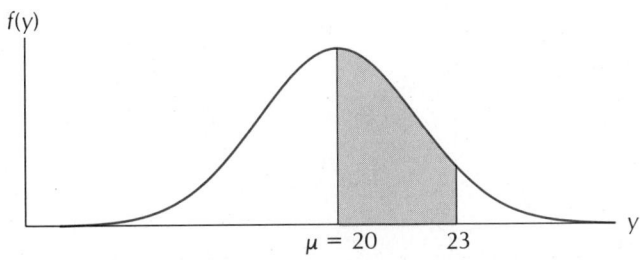

FIGURE 3.10

Area Between $y = 16$
and $\mu = 20$,
Example 3.9

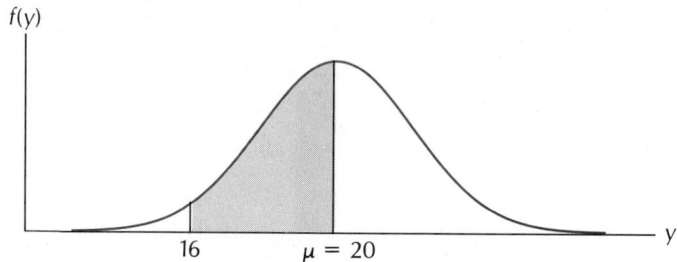

$f(y)$

16        $\mu = 20$ y

EXAMPLE 3.9

For the normal distribution of Example 3.8 with $\mu = 20$ and $\sigma = 2$, find the probability that y will lie in the interval from 16 to 20.

Solution

In determining the area between 16 and 20, we use

$$z = \frac{y - \mu}{\sigma} = \frac{16 - 20}{2} = -2$$

Ignoring the negative sign, we find the appropriate area from Table 1 to be .4772. Thus .4772 is the probability that a measurement falls in the interval from 16 to 20. The area is shown in Figure 3.10.

EXAMPLE 3.10

The mean daily milk production of a herd of Guernsey cows is assumed to be normally distributed with $\mu = 70$ pounds and $\sigma = 13$ pounds.

a.   What is the probability that the milk production for a cow chosen at random will lie in the interval from 60 pounds to 90 pounds?
b.   What is the probability that the milk production for the randomly selected cow will exceed 90 pounds in a given year?

Solution

We begin by drawing a figure to picture the area we are looking for (Figure 3.11). To answer part (a), we must compute two areas, the area between 60 and 70 and the area between 70 and 90. The value $y = 60$ corresponds to a z score of

$$z = \frac{y - \mu}{\sigma} = \frac{60 - 70}{13} = -.77$$

From Table 1, the area between $y = 60$ and $\mu = 70$ is .2794.

FIGURE 3.11

Areas Between 60
and 70 and Between
70 and 90,
Example 3.10

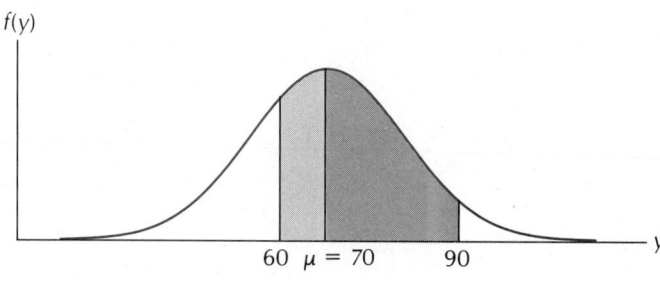

$f(y)$

60 $\mu = 70$ 90 y

FIGURE 3.12

Area Above $y = 90$
for $\mu = 70$ and
$\sigma = 13$

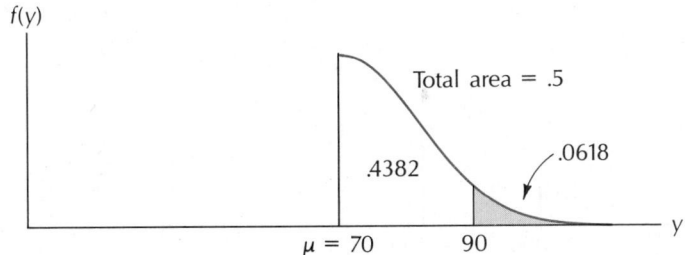

The value $y = 90$ corresponds to a z score of

$$z = \frac{y - \mu}{\sigma} = \frac{90 - 70}{13} = 1.54$$

so the tabulated area between 70 and 90 is .4382. Adding these two areas, we find that the probability that a cow's milk production will lie in the interval from 60 pounds to 90 pounds is .7176.

The probability that the cow's milk production will exceed 90 pounds, part (b), can be computed using the second area computed above. Because of the symmetry of a normal curve, we know that the total area under the curve to the right of μ is .5 (similarly, the total area to the left of μ is .5). We computed the area for the interval from 70 to 90 to be .4382. Subtracting this value from .5, we know that the probability of exceeding 90 is $.5 - .4382 = .0618$. (See Figure 3.12.)

Alternatively, we could use Table 2 in the Appendix (which gives upper-tail probabilities for the normal distribution) to find the answer to part (b) directly. From Table 2, the probability of exceeding a z-value of 1.54 is .0618. This agrees with our previous answer.

In the future, when we need tail areas of a normal curve, it will be easier to use Table 2. But either Table 1 or Table 2 will do; just draw a picture showing the desired area(s).

EXAMPLE 3.11

Annual incomes for career service employees at a large university are approximately normally distributed with a mean of $11,200 and a standard deviation of $900. Find the probability that an employee chosen at random will have an annual income less than $10,000; an income greater than $12,000.

Solution

First we draw a graph showing the areas in question (Figure 3.13). Now we must determine the area between 10,000 and 11,200:

$$z = \frac{y - \mu}{\sigma} = \frac{10,000 - 11,200}{900} = \frac{-1200}{900} = -1.33$$

In a normal distribution, the area to the right of a value 1.33 standard deviations above the mean is, from Table 2 in the Appendix, .0918. Hence, by symmetry, the probability of observing an annual income less than $10,000 is .0918.

FIGURE 3.13
Areas Greater Than
12,000 and Less Than
10,000 for $\mu = 11,200$
and $\sigma = 900$,
Example 3.11

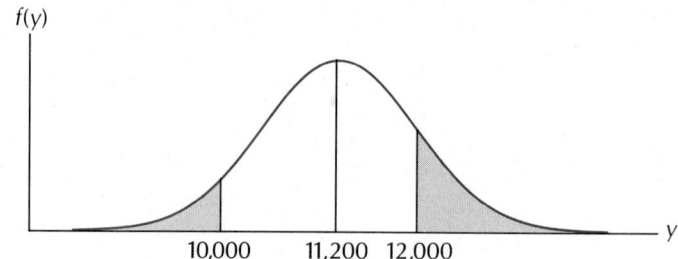

Similarly, to compute the probability of observing a salary above $12,000 we determine the area between 11,200 and 12,000:

$$z = \frac{y - \mu}{\sigma} = \frac{12,000 - 11,200}{900} = .89$$

The area in Table 2 corresponding to $z = .89$ is .1867; this is the desired probability.

EXAMPLE 3.12

From income tax returns in the previous year, it has been found that for a given income classification, the amount of money owed to the government over and above the amount paid in the estimated tax vouchers for the first three payments is approximately normally distributed with a mean of $530 and a standard deviation of $205.

percentile   Find the 75th **percentile** for this distribution of measurements.

Solution

The 75th percentile is, by definition, the value of y such that 75% of the measurements are below it and 25% above it, as shown in Figure 3.14. By referring to Table 2 in the Appendix, we find the value of z corresponding to an area of .25 to the right of $\mu = 530$ is about .67. We find this value of z by first looking for an area in Table 2 that is close to .25. Once we find that, we look across to the z-value and to the top of the table for the hundredths in the z-value. For an area of .25, z is about .67. Substituting into $z = (y - \mu)/\sigma$ and solving for y, we have $y = 667.35$. This value is the 75th percentile of the distribution of owed money for this tax classification.

FIGURE 3.14
Area Under the
Normal Curve for the
75th Percentile,
Example 3.12

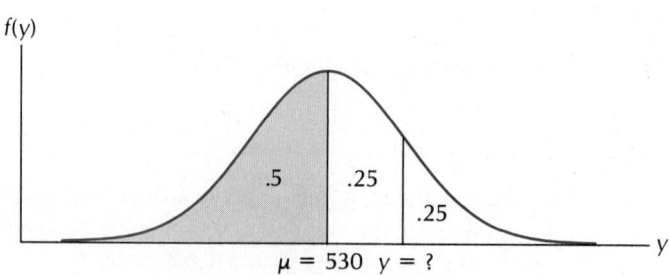

EXERCISES

3.23 Let y be a normal random variable with $\mu = 500$ and $\sigma = 100$. Find the following probabilities.
a. $P(500 < y < 696)$
b. $P(y > 696)$
c. $P(304 < y < 696)$
d. k such that $P(500 - k < y < 500 + k) = .60$

3.24 Suppose that y is a normal random variable with $\mu = 100$ and $\sigma = 15$.
a. Show that $(y < 130)$ is equivalent to $(z < 2)$.
b. Convert $y > 82.5$ to the z score equivalent.
c. Find $P(y < 130)$ and $P(y > 82.5)$.
d. Find $P(y > 106)$, $P(y < 94)$, and $P(94 < y < 106)$.
e. Find $P(y < 70)$, $P(y > 130)$, and $P(70 < y < 130)$.

3.25 Use Table 1 in the Appendix to calculate the area under the curve between these values:
a. $z = 0$ and $z = 1.5$ b. $z = 0$ and $z = 1.8$

3.26 Repeat Exercise 3.25 for these values:
a. $z = -1.96$ and $z = 1.96$ b. $z = -2.33$ and $z = 2.33$

3.27 What is the value of z with an area of .05 to its right? To its left? Hint: Use Table 2 in the Appendix.

3.28 Find the value of z for these areas:
a. an area .01 to the right of z b. an area .10 to the left of z

3.29 Find the probability of observing a value of z greater than these values:
a. 1.96   b. 2.21   c. 2.86   d. 0.73

3.30 Find the probability of observing a value of z less than these values:
a. -1.20 b. -2.62 c. 1.84   d. 2.17

 3.31 Records maintained by the office of budget in a particular state indicate that the amount of time elapsed between the submission of travel vouchers and the final reimbursement of funds has approximately a normal distribution with a mean of 39 days and a standard deviation of 6 days.
a. What is the probability that the elapsed time between submission and reimbursement will exceed 50 days?
b. If you had a travel voucher submitted more than 55 days ago, what might you conclude?

3.10   RANDOM SAMPLING

So far in the text we have discussed "random" samples. What is the importance of random sampling? We must know how the sample was selected so that we can determine probabilities associated with various sample outcomes. The probabilities of samples selected *in a random manner* can be determined and we can use these probabilities to make inferences about the population from which the sample was drawn.

Sample data selected in a nonrandom fashion are frequently distorted by a *selection bias*. A selection bias exists whenever there is a systematic tendency to overrepresent or underrepresent some part of the population. For example, a survey of households conducted during the week entirely between the hours of 9 AM and 5 PM would be severely biased toward households with at least one member at home. Hence any inferences made from the sample data would be biased toward the attributes or opinions of those families with at least one member at home and may not be truly representative of the population of households in the region.

Now we turn to a definition of a *random sample* of n measurements selected from a population containing N measurements $(N > n)$.

DEFINITION 3.13

A sample of n measurements selected from a population is said to be a **random sample** if every different sample of size n from the population has an equal probability of being selected.

EXAMPLE 3.13

Suppose that a population consists of the six measurements. 1, 2, 3, 4, 5, 7. List all possible different samples of two measurements that could be selected from the population. Give the probability associated with each sample in a random sample of $n = 2$ measurements selected from the population.

Solution

All possible samples are listed below.

Sample	Measurements
1	1,2
2	1,3
3	1,4
4	1,5
5	1,7
6	2,3
7	2,4
8	2,5
9	2,7
10	3,4
11	3,5
12	3,7
13	4,5
14	4,7
15	5,7

Now let us suppose that we draw a single sample of $n = 2$ measurements from the 15 possible samples of two measurements. The sample selected is called a random sample if every sample had an equal probability (1/15) of being selected.

It is rather unlikely that we would ever achieve a truly random sample, because the probabilities of selection will not always be exactly equal. But we do the best we can. One of the simplest and most reliable ways to select a random sample of n measurements from a population is to use a table of random numbers (see Table 8 in the Appendix). **random number table**   **Random number tables** are constructed in such a way that, no matter where you start in the table and no matter what direction you move, the digits occur randomly and with equal probability. Thus if we wished to choose a random sample of $n = 10$ measurements from a population containing 100 measurements, we could label the measurements in the population from 0 to 99 (or 1 to 100). Then by referring to Table 8 in the Appendix and choosing a random starting point, the next 10 two-digit numbers going across the page would indicate the labels of the particular measurements to be included in the random sample. Similarly, by moving up or down the page, we would also obtain a random sample.

EXAMPLE 3.14

A small community consists of 850 families. We wish to obtain a random sample of 20 families to ascertain public acceptance of a wage and price freeze. Refer to Table 8 in the Appendix to determine which families should be sampled.

Solution

Assuming that a list of all families in the community is available (such as a telephone directory), we could label the families from 0 to 849 (or, equivalently, from 1 to 850). Then referring to Table 8 in the Appendix, we choose a starting point. Suppose we have decided to start at line 1, column 3. Going down the page we will choose the first 20 three-digit numbers between 000 and 849. From Table 8, we have

015	110	482	333
255	564	526	463
225	054	710	337
062	636	518	224
818	533	524	055

These 20 numbers identify the 20 families that are to be included in our sample.

A telephone directory is not always the best source for names, especially in surveys related to economics or politics. In the 1936 presidential campaign, Franklin Roosevelt was running as the Democratic candidate against the Republican candidate, Governor Alfred Landon of Kansas. This was a difficult time for the nation; the country had not yet recovered from the Great Depression of 1929 to 1933, and there were still 9 million people unemployed.

The *Literary Digest* set out to sample the voting public and predict the winner of the election. Using names and addresses taken from telephone books and club memberships, the *Literary Digest* sent out 10 million questionnaires and got 2.4 million back. Based on the responses to the questionnaire, the *Digest* predicted a Landon victory by 57% to 43%.

At this time George Gallup was starting his survey business. He conducted two surveys. The first one, based on 3000 people, predicted what the results of the *Digest* survey would be long before the *Digest* results were published; the second survey, based on 50,000, was used to forecast *correctly* the Roosevelt victory.

Where did the *Literary Digest* go wrong? The first problem was a severe selection bias. By taking the names and addresses from telephone directories and club memberships, its survey systematically excluded the poor. And, unfortunately for the *Digest*, the vote was split along economic lines; the poor gave Roosevelt a large majority, whereas the rich tended to vote for Landon. A second reason for the error could be due to a *nonresponse bias*. Because only 20% of the 10 million people returned their surveys, and approximately half of those responding favored Landon, one might suspect that maybe the nonrespondents had different preferences than did the respondents. This was, in fact, true.

How, then, does one achieve a random sample? Careful planning and a certain amount of ingenuity are required to have even a decent chance to approximate random sampling. This is especially true when the universe of interest involves people. People can be difficult to work with; they have a tendency to discard mail questionnaires and refuse to participate in personal interviews. Unless we are very careful, the data we obtain may be full of biases having unknown effects on the inferences we are attempting to make.

We do not have sufficient time to explore the topic of random sampling further in this text; entire courses at the undergraduate and graduate levels can be devoted to sample survey research methodology. The important point to remember is that data from a random sample will provide the foundation for making statistical inferences in later chapters. Random samples are not easy to obtain, but with care we can avoid many potential biases that could affect the inferences we make.

EXERCISES

3.32   Define what is meant by a random sample. Is it possible to draw a truly random sample? Comment.

3.33   Suppose that we want to select a random sample of $n = 10$ persons from a population of 800. Use Table 8 in the Appendix to identify the persons to appear in the sample.

3.34   Refer to Exercise 3.33. Identify the elements of a population of $n = 1000$ to be included in a random sample of $n = 15$.

3.35   City officials want to sample the opinions of the homeowners in a community regarding the desirability of increasing local taxes to improve the quality of the public schools. If a random number table is used to identify the homes to be sampled and a home is discarded if the homeowner is not home when visited by the interviewer, is it likely this process will approximate random sampling? Explain.

3.36   A local TV network wants to run an informal survey of individuals who exit from a local voting station to ascertain early results on a proposal to raise funds to move the city-owned historical museum to a new location. How might the network sample voters to approximate random sampling?

3.37   A psychologist was interested in studying women who are in the process of obtaining a divorce to determine whether there are significant attitudinal changes after the divorce has been finalized. Existing records from the geographic area in question show that 798 couples have recently filed for divorce. Assume a sample of 25 women is needed for the study, and use Table 8 in the Appendix to determine which women should be asked to participate in the study. (Hint: Begin in column 2, row 1, and proceed down.)

3.38   Refer to Exercise 3.37. As is the case in most surveys, not all persons chosen for a study will agree to participate. Suppose that five of the 25 women selected refuse to participate. Determine five more women to be included in the study.

3.39   Suppose you have been asked to run a public opinion poll related to an upcoming election. There are 1000 registered voters in a specific precinct and you wish to obtain a random sample of 50 persons. Use a computer program to indicate which individuals are to be included in the sample. A Minitab program is shown below for purposes of illustration. (Note: We assume that there is a list of the 1000 voters, with the numbers 1 to 1000 corresponding to people on the list.)

```
        IRANDOM 50 INTEGERS BETWEEN 1 AND 1000, PUT IN C1
        50 RANDOM INTEGERS BETWEEN     1 AND  1000

MTB > PRINT C1
C1
    715     99    570    914    472    120    985     22    198     53    200
    611    625    497    487    729    380    401    293    872    329    700
    799    939    181    562    356    319    265    213    583    453    150
     80    680   1000    538    749    254    443    595    945    702    767
    645    549    944    548    900    361

MTB >
```

3.11   THE SAMPLING DISTRIBUTION FOR $\bar{y}$

We discussed several different measures of central tendency and variability in Chapter 2, and distinguished between numerical descriptive measures of a population (parameters) and numerical descriptive measures of a sample (statistics). Thus μ and σ are parameters, whereas $\bar{y}$ and s are statistics.

The numerical value that a sample statistic will have cannot be predicted exactly in advance. Even if we knew that a population mean μ was \$216.37 and that the population standard deviation σ was \$32.90—even if we knew the complete population distribution— we could not say that the sample mean $\bar{y}$ would be exactly equal to \$216.37. A sample statistic is a random variable; it is subject to random variation because it is based on a random sample of measurements selected from the population of interest. And, like any other random variable, a sample statistic has a probability distribution. We call the probability distribution of a sample statistic the *sampling distribution* of that statistic. Stated differently, the sampling distribution of a statistic is the population of all values for that statistic.

The actual mathematical derivation of sampling distributions is one of the basic problems of mathematical statistics. We will illustrate how the sampling distribution for $\bar{y}$ can be obtained for a simplified population. Later in the chapter several general results will be presented.

EXAMPLE 3.15   The sample mean $\bar{y}$ is to be calculated from a random sample of size 2 taken from a population consisting of the five values (\$2, \$3, \$4, \$5, \$6). Find the sampling distribution of $\bar{y}$, based on a sample of size 2.

Solution   One way to find the sampling distribution is by counting. There are 10 possible samples of two items from the five items. These are shown here:

Possible Samples of Size 2	Value of $\bar{y}$
2,3	2.5
2,4	3
2,5	3.5
2,6	4
3,4	3.5
3,5	4
3,6	4.5
4,5	4.5
4,6	5
5,6	5.5

Assuming each sample of size 2 is equally likely, it follows that the sampling distribution for $\bar{y}$ based on $n = 2$ observations selected from this population is as indicated here.

$\bar{y}$	$P(\bar{y})$
2.5	1/10
3	1/10
3.5	2/10
4	2/10
4.5	2/10
5	1/10
5.5	1/10

The sampling distribution is shown as a graph in Figure 3.15.

FIGURE 3.15

Sampling Distribution for $\bar{y}$, Example 3.15

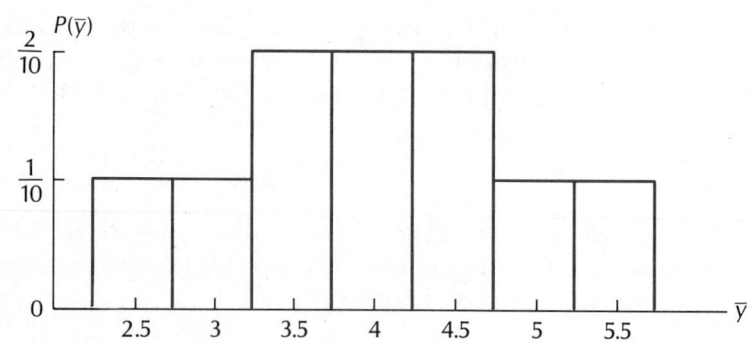

Quite a few of the more common sample statistics have sampling distributions that are normal. A very plausible explanation for this phenomenon is offered by a series of theorems in mathematical statistics called Central Limit Theorems. We will discuss one such theorem (without proof).

CENTRAL LIMIT THEOREM

THEOREM 3.1 ▬▬

If random samples of n measurements are repeatedly drawn from a population with a finite mean μ and a standard deviation σ, then, when n is large, the relative frequency histogram for the sample means (calculated from the repeated samples) will be approximately normal (bell-shaped) with mean μ and standard deviation $\sigma/\sqrt{n}$. (Note: The approximation becomes more precise as n increases.)

Figure 3.16 illustrates Theorem 3.1. Figure 3.16(a) shows the distribution of the original measurement y from which the samples are to be drawn. No specific shape need be assigned to the distribution of y. All we know is that its mean is μ and its variance is **sampling distribution** σ^2. Figure 3.16(b) illustrates the relative frequency histogram, called the **sampling distribution**, for the sample mean $\bar{y}$. Repeated samples of size n are to be drawn from the population illustrated in Figure 3.16(a). For each sample drawn, we compute the sample mean $\bar{y}$. If we were to continue this process over and over again, finally plotting the relative frequency histogram for the sample means, it would appear as in Figure 3.16(b). The mean for the sampling distribution of $\bar{y}$ is μ, the same as that for the original y measurements, and the standard deviation of the sampling distribution is equal to the standard deviation of the y measurements (σ) divided by $\sqrt{n}$. This quantity is designated by $\sigma_{\bar{y}}$ and **standard error of $\bar{y}$** called the **standard error of $\bar{y}$**.

FIGURE 3.16

The Probability Distribution of y and the Sampling Distribution of $\bar{y}$

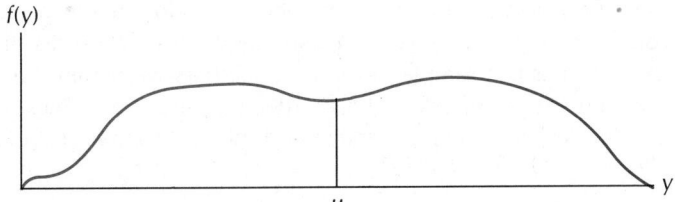

(a) Probability distribution of y, with mean μ and standard deviation σ

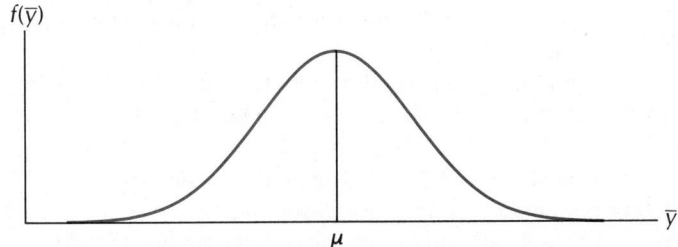

(b) Sampling distribution of $\bar{y}$, with mean μ and standard error $\sigma/\sqrt{n}$

FIGURE 3.17

Characterization
of the Sampling
Distribution for $\bar{y}$

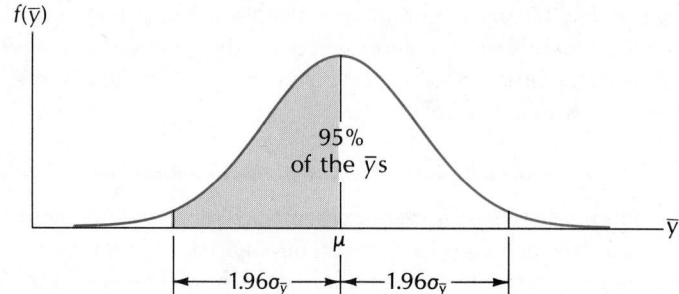

We will be able to use the information we have concerning the distribution of $\bar{y}$ to make inferences about the parameter μ. For example, we know from knowledge of the normal curve that 95% of the $\bar{y}$s will be within 1.96 standard errors ($1.96\sigma_{\bar{y}}$) of their mean. In a given sample where we calculate a single sample mean $\bar{y}$, we would expect our calculated $\bar{y}$ to be within $1.96\sigma_{\bar{y}}$ of μ (see Figure 3.17). So we not only can use $\bar{y}$ to estimate μ but we also can say how close to μ we expect our estimate to be, that is, we can provide a **measure of goodness** for our estimate.

measure of goodness

We can use the table of random numbers (Table 8 in the Appendix) to illustrate empirically how the Central Limit Theorem applies to sample means. Table 3.5 lists the murder rates (number of murders per 100,000 people) associated with 90 metropolitan areas throughout the United States in the north, the south, and the west in 1980. Although it is unrealistic to assume that these 90 murder rates represent a population of measurements, we will, for illustration purposes, take the 90 measurements of Table 3.5 as the population of interest.

To examine the sampling distribution of the sample mean $\bar{y}$—that is, the population of all $\bar{y}$ values—we randomly draw 50 samples of five scores. For example, numbering the cities from 1 to 90 and using the table of random numbers (Table 8 in the Appendix), we could proceed down column 1 and use the first two digits of each five-digit random number. For our first sample of random numbers we obtain the numbers 10, 22, 24, 42, and 37, and we would select the murder rates from the cities assigned these numbers. The result of this first sample is shown in Table 3.6. The sample mean from the first sample is

$$\bar{y} = \frac{\sum y}{n} = \frac{63}{5} = 12.6$$

We repeat this procedure 49 more times to acquire 49 new samples. The 50 sample means are listed in Table 3.7.

To see how the sampling distribution of the sample mean $\bar{y}$ relates to the original distribution for the 90 murder rates, we can construct histograms for both sets of measurements, as shown in Figures 3.18(a) and (b).

As can be seen from Figure 3.18(a), the original population is certainly not mound-shaped or symmetrical. In fact, the distribution is skewed to the right (tails off to the right). But even with a small number of measurements per sample (5), and a small number of samples (50), the sampling distribution of $\bar{y}$ is beginning to look bell-shaped; most of the

TABLE 3.5   Murder Rates (per 100,000 Inhabitants) for 90 Cities Selected from the South, North, and West

South	Rate	North	Rate	West	Rate
Atlanta	14	Albany, N.Y.	2	Bakersfield	21
Augusta, Ga.	13	Allentown, Pa.	2	Boise	3
Baton Rouge	10	Atlantic City	10	Colorado Springs	6
Beaumont, Tex.	12	Canton, Ill.	6	Denver	9
Birmingham	16	Chicago	14	Eugene	2
Charlotte, N.C.	14	Cincinnati	6	Fresno	20
Chattanooga	10	Cleveland	16	Honolulu	8
Columbia, S.C.	10	Detroit	16	Kansas City, Mo.	15
Corpus Christi	16	Evansville	8	Lawton, Okla.	6
Dallas	18	Grand Rapids	6	Los Angeles	23
El Paso	12	Johnstown, Pa.	8	Modesto, Calif.	10
Fort Lauderdale	17	Kankakee, Ill.	6	Oklahoma City	12
Greensboro, N.C.	8	Kenosha, Wis.	4	Oxnard, Calif.	7
Jackson, Miss.	17	Lancaster, Pa.	2	Pueblo, Colo.	4
Knoxville	7	Lansing	3	Sacramento	9
Lexington, Ky.	6	Lima, Ohio	6	St. Louis	15
Lynchburg, Va.	12	Madison, Wis.	2	Salinas, Calif.	8
Macon, Ga.	8	Mansfield, Ohio	3	Salt Lake City	5
Memphis	20	Milwaukee	6	San Diego	10
Monroe, La.	13	Newark, N.J.	11	San Francisco	12
Nashville	14	Paterson, N.J.	9	San Jose	8
Newport News	11	Philadelphia	12	Seattle	7
Orlando, Fla.	10	Pittsfield, Mass.	1	Sioux City	1
Richmond, Va.	12	Racine, Wis.	5	Spokane	4
Roanoke	10	Rockford, Ill.	5	Stockton, Calif.	18
San Antonio	18	South Bend	11	Tacoma	5
Shreveport, La.	20	Springfield, Ill.	6	Topeka	11
Washington, D.C.	11	Syracuse	3	Tucson	9
Wichita Falls, Kans.	13	Vineland, N.J.	9	Vallejo	6
Wilmington, Del.	8	Youngstown	7	Waco, Tex.	15

Source: Department of Justice, *Uniform Crime Reports for the United States: 1980* (Washington, D.C.: Government Printing Office, 1980), pp. 60–86.

TABLE 3.6   Results of Sample 1

City Number	City	Murder Rate
10	Dallas	18
22	Newport News	11
24	Richmond, Va.	12
42	Kankakee	6
37	Cleveland	16
Total		63

TABLE 3.7   Means of 50 Samples of Size 5 Selected from the 90 Murder Rates

12.6	9.5	12.1	6.7	9.6
9.5	13.1	15.6	9.4	9.3
14.9	8.2	12.0	9.1	8.0
11.9	7.1	8.5	10.3	11.0
11.0	9.2	7.9	5.1	9.6
10.7	8.3	6.8	11.3	9.2
9.7	8.7	12.2	12.0	10.3
10.5	9.6	8.3	9.1	7.4
8.2	10.9	14.0	8.3	11.2
7.8	9.4	9.4	5.9	10.3

FIGURE 3.18

Relative Frequency Histograms for the Original Population of Murder Rates and the Sampling Distribution of $\bar{y}$

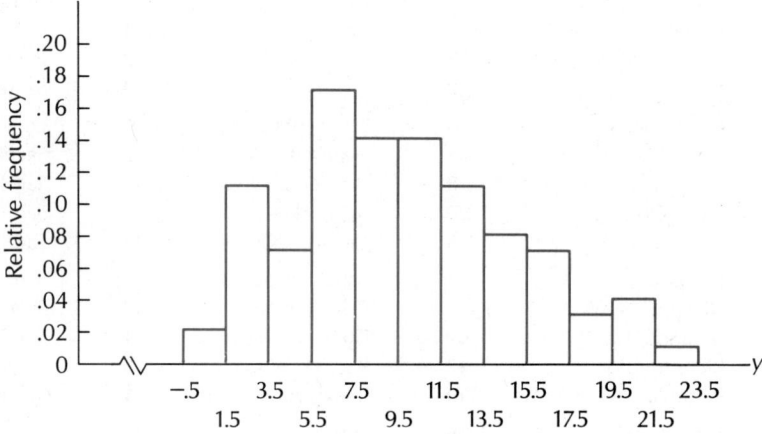

(a) Original population

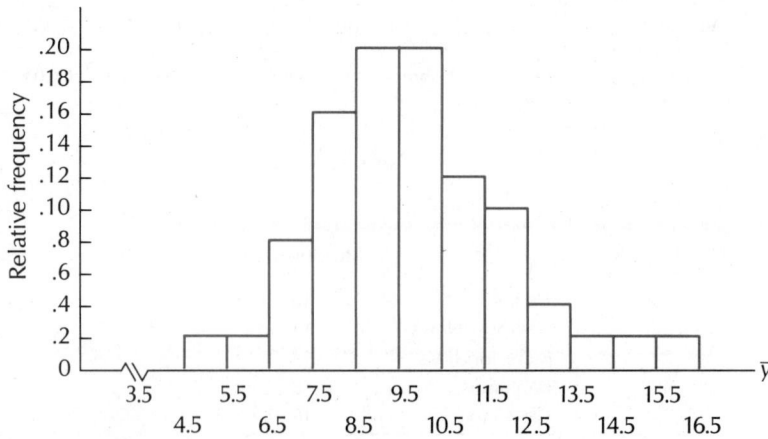

(b) Sampling distribution of $\bar{y}$

sample means are grouped closely about the mean of the population (which in this case can be computed to be $\mu = 9.71$). See Figure 3.18(b).

The illustration we have presented could have been made more convincing by assuming a much larger population and by taking more samples. The important point to remember is that in repeated sampling, $\bar{y}$ will be approximately normally distributed with mean μ and standard error $\sigma/\sqrt{n}$. The approximation will be more precise as n, the sample size for each sample, increases. Thus the frequency histogram for $\bar{y}$ in our example would have been even more bell-shaped if n had been 10 rather than 5, or 15 rather than 10, and so on.

An obvious question is: how large should the sample size be in order for the Central Limit Theorem to hold? Numerous simulation studies have been conducted over the years and the results of these studies suggest that, in general, the Central Limit Theorem holds for $n > 30$. However, one should not apply this rule blindly. If the population is heavily skewed, the sampling distribution for $\bar{y}$ will still be skewed even for $n > 30$. On the other hand if the population is symmetric, the Central Limit Theorem holds for $n < 30$.

So, take a look at the data. If the sample histogram is clearly skewed, then the population will also probably be skewed. Consequently, a value of n much higher than 30 may be required to have one sampling distribution of $\bar{y}$ be approximately normal. Any inference based on the normality of $\bar{y}$ for $n = 30$ under this condition should be examined carefully.

We can use the results of coding, presented in Chapter 2, to extend the Central Limit Theorem to the sample sum $\sum y$. If repeated samples of size n are drawn from a population, and if we compute $\sum y = n\bar{y}$ for each sample drawn, we have, in essence, coded the sample means for each sample by multiplying by n. Applying our coding results from Chapter 2 to the results of the Central Limit Theorem, the mean and the standard error for $\sum y$ will be, respectively, $n\mu$ and $n\sigma/\sqrt{n} = \sqrt{n}\,\sigma$. This result is illustrated in Figure 3.19.

Many of the statistics that we will encounter in later chapters will be either sums or averages of variables. Hence we will employ the Central Limit Theorem discussed in this chapter to specify their sampling distributions.

FIGURE 3.19

Probability Distribution of y and of the Sample Sum $\sum y$

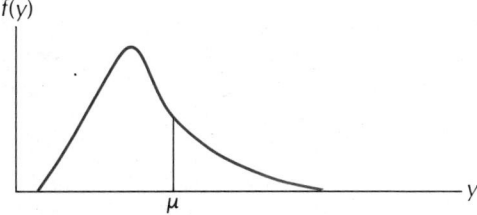

(a) Probability distribution of y, with mean μ and standard deviation σ

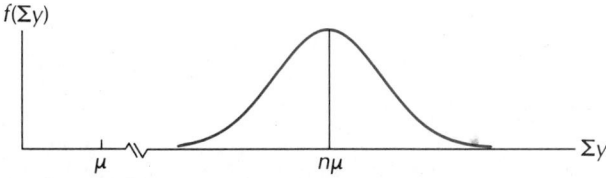

(b) Sampling distribution $\sum y$, with mean $n\mu$ and standard error $\sqrt{n}\sigma$

Usually, a sample statistic is used as an estimate of a population parameter. For example, a sample mean can be used to estimate the corresponding mean μ of the population from which the sample was drawn. The sampling distribution of a sample statistic is then used to determine how accurate the estimate is likely to be. In Example 3.15, the population mean μ is known to be \$4. Obviously, we don't ever know μ in practice. Still, we can use the sampling distribution of $\bar{y}$ to determine the probability that, for example, the computed value of the sample mean will be more than \$.50 away from μ. For Example 3.15, this probability is

$$P(2.5) + P(3) + P(5) + P(5.5) = \frac{4}{10}$$

In general, a sample statistic is used to make inferences about a population parameter. The sampling distribution of the statistic is crucial in determining how good the inference is likely to be.

interpretations of a sampling distribution   **Sampling distributions** can be **interpreted** in at least two ways. One way uses the long-run relative frequency approach. Imagine taking repeated samples of a fixed size from a given population and calculating the value of the sample statistic for each sample. In the long run, the relative frequencies for the possible values of the sample statistic will approach the corresponding sampling distribution probabilities. For example, if one took a large number of samples from the population distribution corresponding to the probabilities of Example 3.15 and, for each sample, computed the sample mean, approximately 20% would have $\bar{y} = 3.5$.

The other way to interpret a sampling distribution makes use of the classical interpretation of probability. Imagine listing all possible samples that could be drawn from a given population. The probability that a sample statistic will have a particular value (say, that $\bar{y} = 3.5$) is then the proportion of all possible samples that yield that value. In Example 3.15, $P(3.5) = 2/10$ corresponds to the fact that two of the 10 samples have a sample mean equal to 3.5. Both the repeated-sampling and the classical approach to finding probabilities for a sample statistic are legitimate.

In practice, though, a sample is taken only once, and only one value of the sample statistic is calculated. A sampling distribution is not something you can see in practice; it is not an empirically observed distribution. Rather it is a theoretical concept, a set of probabilities derived from assumptions about the population and about the sampling method.

There's an unfortunate similarity between the phrase "sampling distribution," meaning the theoretically derived probability distribution of a statistic, and the phrase "sample distribution," which refers to the histogram of individual values actually observed in a particular sample. The two phrases mean very different things. To avoid confusion, we will refer to the distribution of sample values as the **sample histogram** rather than as the sample distribution.

sample histogram

EXERCISES

3.40   A random sample of 16 measurements is drawn from a population with a mean of 60 and a standard deviation of 5. Describe the sampling distribution of $\bar{y}$, the sample mean. Within what interval would you expect $\bar{y}$ to lie approximately 95% of the time?

3.41   Refer to Exercise 3.40. Describe the sampling distribution for the sample sum $\sum y_i$. Is it unlikely (improbable) that $\sum y_i$ would be more than 70 units away from 960? Explain.

3.42   In Exercise 3.40 a random sample of 16 observations was to be selected from a population with $\mu = 60$ and $\sigma = 5$. Assume that the original population of measurements is normal. Use a computer program to simulate the distribution of $\bar{y}$ based on 40 sample means. A Minitab program is shown below in illustration.

```
            NRANDOM 16 OBSN, MU = 60, SIGMA = 5, PUT IN C1
            16 NORMAL OBS. WITH MU =     60.0000 AND SIGMA =      5.0000
    MTB > PRINT C1
    C1
        56.6027    60.6408    61.4819    55.9020    56.6325    66.0543    58.9083
        48.7206    53.5237    56.6236    60.2671    52.8550    62.8772    52.2280
        60.4248    56.0439

    MTB > AVERAGE THE OBSERVATIONS IN C1
        MEAN    =     57.487
    MTB >
```

This set of two instructions would be repeated 39 more times. The complete set of 80 instructions would be entered simultaneously followed by a STOP instruction card.

3.43   Psychomotor retardation scores for a large group of manic-depressive patients were found to be approximately normal with a mean of 930 and a standard deviation of 130.
a. What fraction of the patients scored between 800 and 1100?
b. Less than 800?
c. Greater than 1200?

3.44   Refer to Exercise 3.43.
a. Find the 90th percentile for the distribution of manic-depressive scores. (Hint: Solve for y in the expression $z = (y - \mu)/\sigma$, where z is the number of standard deviations the 90th percentile lies above the mean μ.)
b. Find the interquartile range.

3.45   Federal resources have been tentatively approved for funding the construction of an outpatient clinic. But in order for the designers to present plans for a facility that will handle patient load requirements while still staying within a limited budget, a study of patient demand was made. From studying a similar facility in the area, it was found that the distribution of the number of patients requiring hospitalization during a week could be approximated by a normal distribution with a mean of 125 and a standard deviation of 32.
a. Use the Empirical Rule to describe the distribution of y, the number of patients requesting service in a week.
b. If the facility was built with a 160-patient capacity, what fraction of the weeks might the clinic be unable to handle the demand?

3.46   Refer to Exercise 3.45. What size facility should be built so that the probability of the patient load exceeding the clinic capacity is .05? .01?

3.47   The distribution of the milkfat percentages for Holstein cattle in a particular state during the 1960s was approximately normal with a mean of 3.7 and a standard deviation of .3.
a. What percentage of the Holsteins had a milkfat percentage less than 3?
b. Greater than 4.5?

3.48   Refer to Exercise 3.47.
a. Find the limits within which 90% of the milkfat percentages fell.
b. Compute the 95th percentile for the distribution of milkfat percentages.

3.49   Refer to Exercise 3.47. Suppose a random sample of $n = 25$ Holsteins is selected from the population of Holstein cattle in the state.
a. Describe the distribution of $\bar{y}$, the mean milkfat percentage for the sample of 25 cattle.
b. Compare the distribution of $\bar{y}$ in part (a) to that for a distribution of $\bar{y}$ from a sample of 100 Holsteins.
c. What is the probability that the sample mean milkfat percentage would exceed 4 in part (a)?

3.50   Random samples of size 20 are repeatedly drawn from a normal distribution with a mean of 65 and a standard deviation of 8.
a. Describe the sampling distribution for $\bar{y}$.
b. What fraction of the sample means should be in the interval from 60 to 72?

3.51   Refer to Exercise 3.50.
a. Describe the sampling distribution of the sample sum $\sum_i y_i$.
b. Locate the 25th and 75th percentiles for the sampling distribution of $\sum_i y_i$.

3.12   NORMAL APPROXIMATION TO THE BINOMIAL

The Central Limit Theorem discussed in the previous section will enable us to calculate probabilities for a binomial random variable by approximating the binomial distribution with a normal curve and using normal curve areas as approximations to the desired probabilities. We said in Section 3.7 that probabilities associated with values of y can be computed for a binomial experiment for any values of n or π, but the task becomes more difficult when n gets large. For example, suppose a sample of 1000 voters is polled to determine sentiment toward the consolidation of city and county government. What would be the probability of observing 460 or fewer favoring consolidation if we assume that 50% of the entire population favor the change? Here we have a binomial experiment with $n = 1000$ and π, the probability of selecting a person favoring consolidation, equal to .5. To determine the probability of observing 460 or fewer favoring consolidation in the random sample of 1000 voters, we could compute $P(y)$ using the binomial formula for $y = 460, 459, \ldots, 0$. The desired probability would then be

$$P(y = 460) + P(y = 459) + \cdots + P(y = 0)$$

There would be 461 probabilities to calculate with each one being somewhat difficult due to the factorials. For example, the probability of observing 460 favoring consolidation is

$$P(y = 460) = \frac{1000!}{460!540!}(.5)^{460}(.5)^{540}$$

A similar calculation would be needed for all other values of y.

The normal distribution can be used in many situations to approximate the binomial probability distribution, and areas under the normal curve can be used to *approximate* the actual binomial probabilities. The normal distribution that provides the best approximation to the binomial probability distribution has a mean and a standard deviation given by the formula below.

$$\mu = n\pi \qquad \sigma = \sqrt{n\pi(1 - \pi)}$$

FIGURE 3.20

Approximating
Normal Distribution
for the Binomial
Distribution of
Example 3.16,
$\mu = 500$ and $\sigma = 15.8$

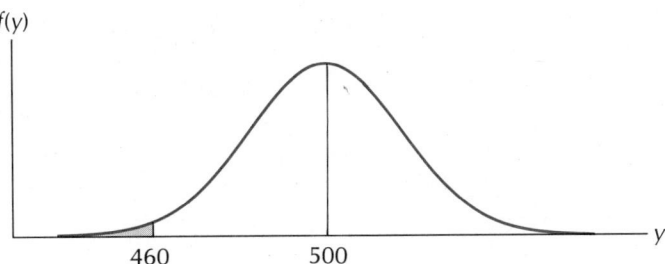

EXAMPLE 3.16

Use the normal approximation to the binomial to compute the probability of observing 460 or fewer in a sample of 1000 favoring consolidation if we assume that 50% of the entire population favor the change.

Solution

The normal distribution used to approximate the binomial distribution will have

$$\mu = n\pi = 1000(.5) = 500$$
$$\sigma = \sqrt{n\pi(1 - \pi)} = \sqrt{1000(.5)(.5)} = 15.8$$

The desired probability is represented by the shaded area shown in Figure 3.20. We calculate the desired area by first computing

$$z = \frac{y - \mu}{\sigma} = \frac{460 - 500}{15.8} = -2.53$$

Referring to Table 1 in the Appendix, we find that the area under the normal curve between 460 and 500 (for $z = 2.53$) is .4943. Thus the probability of observing 460 or fewer favoring consolidation is approximately $.5 - .4943 = .0057$. This probability is shown in Table 2 of the Appendix for $z = 2.53$.

The normal approximation to the binomial distribution can be unsatisfactory if $n\pi < 5$ or $n(1 - \pi) < 5$. If π, the probability of success, is small, and n, the sample size, is modest, the actual binomial distribution is seriously skewed to the right. In such a case, the symmetric normal curve will give a bad approximation. If π is near 1, so $n(1 - \pi) < 5$, the actual binomial will be skewed to the left, and again the normal approximation will not be very good. The normal approximation, as described, is quite good when $n\pi$ and $n(1 - \pi)$ exceed about 20. In the middle zone, $n\pi$ or $n(1 - \pi)$ between 5 and 20, a modification called a

continuity correction

continuity correction makes a substantial contribution to the quality of the approximation.

The point of the continuity correction is that we are using the continuous normal curve to approximate a discrete binomial distribution. A picture of the situation is shown in Figure 3.21.

The binomial probability that $y \leq 5$ is the sum of the areas of the rectangles above 5, 4, 3, 2, 1, and 0. This probability (area) is approximated by the area under the superimposed normal curve to the left of 5. Thus, the normal approximation ignores half of the rectangle above 5. The continuity correction simply includes the area between $y = 5$ and $y = 5.5$. For the binomial distribution with $n = 20$ and $\pi = .30$ (pictured in Figure 3.21), the correction is to take $P(y \leq 5)$ as $P(y \leq 5.5)$. Instead of $P(y \leq 5) = P[z \leq (5 - 20(.3))/\sqrt{20(.3)(.7)}] = P(z \leq -.49) = .3121$, use $P(y \leq 5.5) = P[z \leq (5.5 - 20(.3))/\sqrt{20(.3)(.7)}] = P(z \leq -.24) = .4052$. The actual binomial probability can be shown to be .4164. The general idea of the

FIGURE 3.21

Normal
Approximation
to Binomial

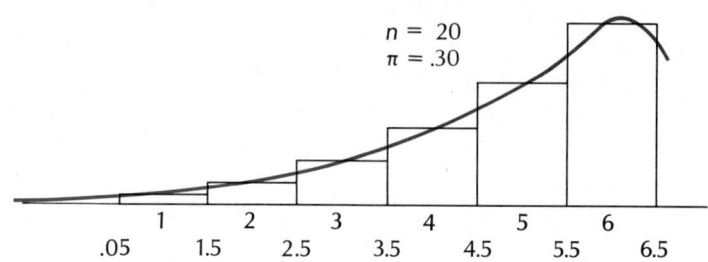

continuity correction is to add or subtract .5 from a binomial value before using normal probabilities. The best way to determine whether to add or subtract is to draw a picture like Figure 3.21.

NORMAL
APPROXIMATION
TO THE BINOMIAL
PROBABILITY
DISTRIBUTION

For large n and π not too near 0 or 1, the distribution of a binomial random variable y may be approximated by a normal distribution with $\mu = n\pi$ and $\sigma = \sqrt{n\pi(1 - \pi)}$. This approximation should be used only if $n\pi \geq 5$ and $n(1 - \pi) \geq 5$. A continuity correction will improve the quality of the approximation in cases where n is not overwhelmingly large.

EXAMPLE 3.17

A large drug company has 100 potential new prescription drugs under clinical test. About 20 percent of all drugs that reach this stage are eventually licensed for sale. What is the probability that at least 15 of the 100 drugs are eventually licensed? Assume that the binomial assumptions are satisfied, and use a normal approximation with continuity correction.

Solution

The mean of y is $\mu = 100(.2) = 20$; the standard deviation is $\sigma = \sqrt{100(.2)(.8)} = 4.0$. The desired probability is that 15 or more drugs are approved. Because $y = 15$ is included, the continuity correction is to take the event as y greater than or equal to 14.5.

$$P(y \geq 14.5) = P\left(z \geq \frac{14.5 - 20}{4.0}\right) = P(z \geq -1.375), \text{ which is about .92.}$$

3.13   SUMMARY

In this chapter we presented an introduction to probability, probability distributions, and sampling distributions. Knowledge of the probabilities of sample outcomes is vital to a statistical inference. Three different interpretations of the probability of an outcome were given: the classical, relative frequency, and subjective interpretations. Although each has a place in statistics, the relative frequency approach has the most intuitive appeal since it can be checked.

Quantitative random variables are classified as either discrete or continuous random variables. The probability distribution for a discrete random variable y is a display of the probability $P(y)$ associated with each value of y. This display may be presented in the form of a histogram, table, or formula.

The binomial is a very important and useful discrete random variable. Many experiments that scientists conduct are similar to a coin-tossing experiment where dichotomous (yes–no) type data are accumulated. The binomial experiment frequently provides an excellent model for computing probabilities of various sample outcomes.

Probabilities associated with a continuous random variable correspond to areas under the probability distribution. Computations of such probabilities were illustrated for areas under the normal curve. The importance of this exercise is borne out by the Central Limit Theorem: any random variable that is expressed as a sum or average will have a normal distribution for sufficiently large sample size. Direct application of the Central Limit Theorem gives the sampling distribution for the sample mean. Since many sample statistics are either sums or averages of random variables, application of the Central Limit Theorem will provide us with information about probabilities of sample outcomes. These probabilities will be vital for the statistical inferences we wish to make.

KEY FORMULAS

1. Binomial probability distribution

$$P(y) = \frac{n!}{y!(n-y)!}\, \pi^y (1-\pi)^{n-y}$$

2. Sampling distribution for $\bar{y}$

    Mean: μ

    Standard error: $\sigma_{\bar{y}} = \sigma/\sqrt{n}$

3. Normal approximation to the binomial

    $\mu = n\pi$

    $\sigma = \sqrt{n\pi(1-\pi)}$

    provided

    $n\pi$ and $n(1-\pi)$ greater than 5.

    Or, equivalently, if

    $$n \geq \frac{5}{\min(\pi,\, 1-\pi)}$$

EXERCISES

3.52 One way to audit expense accounts for a large consulting firm would be to sample all reports dated the last day of each month. Comment on whether or not such a sample would constitute a random sample.

3.53 Critical key-entry errors in the data processing operation of a large district bank occur approximately .1% of the time. If a random sample of 10,000 entries is examined, determine the following:
a. The expected number of errors.
b. The probability of observing less than five errors.
c. The probability of observing less than two errors.

3.54 Use the binomial distribution with $n = 20$, $\pi = .5$ to compare accuracy of the normal approximation to the binomial.
a. Compute the exact probabilities and corresponding normal approximations for $y < 5$.
b. The normal approximation can be improved slightly by taking $P(y \leq 4.5)$. Why should this help? Compare your results.
c. Compute the exact probabilities and corresponding normal approximations with the continuity correction for $P(8 < y < 14)$.

3.55 Let y be a binomial random variable with $n = 10$ and $\pi = .5$.
a. Calculate $P(4 \leq y \leq 6)$.
b. Use a normal approximation without the continuity correction to calculate the same probability. Compare your results. How well did the normal approximation work?

3.56 Refer to Exercise 3.55. Use the continuity correction to compute the probability $P(4 \leq y \leq 6)$. Does the continuity correction help?

3.57 A marketing research firm believes that approximately 25% of all persons mailed a "sweepstakes" offer will respond. If a preliminary mailing of 5000 is conducted in a fixed region,
a. What is the probability that 1000 or fewer will respond?
b. What is the probability that 3000 or more will respond?

3.58 The breaking strengths for 1-foot-square samples of a particular synthetic fabric are approximately normally distributed with a mean of 2250 pounds per square inch (psi) and a standard deviation of 10.2 psi.
a. Find the probability of selecting a 1-foot-square sample of material at random that on testing would have a breaking strength in excess of 2265 psi.
b. Describe the sampling distribution for $\bar{y}$ based on random samples of 15 one-foot sections.

3.59 Refer to Exercise 3.58. Suppose that a new synthetic fabric has been developed that may have a different mean breaking strength. A random sample of 15 one-foot sections is obtained and each section is tested for breaking strength. If we assume that the population standard deviation for the new fabric is identical to that for the old fabric, give the standard deviation for the sampling distribution of $\bar{y}$ using the new fabric.

3.60 Refer to Exercise 3.59. Suppose that the mean breaking strength for the sample of 15 one-foot sections of the new synthetic fabric is 2268. What is the probability of observing a value of $\bar{y}$ equal to or greater than 2268, assuming that the mean breaking strength for the new fabric is 2250, the same as that for the old?

3.61 Refer to Exercise 3.60. Based on your answer in Exercise 3.60, do you believe the new fabric has the same mean breaking strength as the old? (Assume $\sigma = 10.2$.)

3.62 In Figure 3.18 we visually inspected the relative frequency histogram for 50 sample means, each based on $n = 5$ measurements, and noted its bell shape. Another way to determine whether or not a set of measurements is bell-shaped (normal) is to construct a plot of the sample data on *probability paper*. This plot is called a *probability plot*. If the probability plot is approximately a straight line, we say the measurements were selected from a normal population. Use a computer program to construct a probability plot for the 50 sample means. A Minitab program is shown below for illustration. (Note: This program will display the histogram and the probability plot.)

```
        PRINT C6
C6
   12.0    8.8   11.4    5.0    8.8    7.0   12.4   15.0    8.6    8.6   14.2
    7.4   10.2    8.4    7.2   11.2    5.4    7.8    9.6   10.2    9.2    8.8
    6.2    4.4    9.0   10.0    6.4    6.0   10.4    8.4    9.0    7.0   11.0
   11.4    8.6    9.8    8.2    8.2    7.4    5.4    7.4   10.0   13.2    7.4
    8.4    7.0    7.8    7.6    4.2    9.6

MTB > HISTOGRAM OF COL C6

Histogram of C6   N = 50

Midpoint    Count
       4      2    **
       5      3    ***
       6      3    ***
       7      8    ********
       8      8    ********
       9      9    *********
      10      8    ********
      11      4    ****
      12      2    **
      13      1    *
      14      1    *
      15      1    *

MTB > NSCORES OF C6 PUT IN C7
MTB > PLOT C6 VS C7

        15.0+                                                    *
            -                                                *
C6          -
            -                                            *
            -                                          *
        12.0+                                         *
            -                                    * 2
            -                                   *
            -                             2 2*
            -                            2*
         9.0+                         3 2*
            -                   23 3
            -                 *2
            -            3 * 4
            -           *
         6.0+         **
            -        2
            - *     *
            ---------+---------+---------+---------+---------+------C7
                  -1.60     -0.80      0.00      0.80      1.60
```

3.63 A labor union's examining board for the selection of apprentices has a record for admitting 70% of all applicants who satisfy a set of basic requirements. Five members of a minority group recently came before the board, and four out of five were rejected. Find the probability that one or less would be accepted if the record is really .7. Did the board apply a lower probability of acceptance when reviewing the five members of a minority group?

3.64 Suppose that you are regional director of the IRS office and that you are charged with sampling 1% of the returns with gross income levels above $15,000. How might you go about this? Would you use random sampling? How?

3.65 Experts consider high serum cholesterol levels to be associated with an increased incidence of coronary heart disease. Suppose that the logarithm of cholesterol levels for males in a given age bracket are normally distributed with a mean of 2.35 and a standard deviation of 1.2.

a. What percent of the males in this age bracket could be expected to have a serum cholesterol level greater than 250 mg/ml, the upper limit of the clinical normal range?

b. What percent of the males could be expected to have serum cholesterol levels within the clinical normal range of 150–250 mg/ml?

c. If levels above 300 mg/ml are considered very risky, what percent of the adult males in this age bracket could be expected to exceed 300?

3.66 One of the major soft-drink companies changed the "secret" formula for its leading beverage in order to attract new customers. Recently, a marketing research firm interviewed 1000 potential new customers and, after giving them a taste of the newly reformulated beverage, determined the number of these individuals planning to buy the reformulated beverage in the near future.

a. Identify the random variable for the population of $y =$ values of interest.

b. Can you compute the mean and variance? Why or why not?

c. How would you calculate $P(y \leq 250)$?

3.67 Many firms are using or exploring the possibility of using telemarketing techniques—that is, marketing their products via the telephone to supplement the more traditional marketing strategies. Assume a firm finds that approximately one in every 100 calls yields a sale.

a. Find the probability the first sale will occur somewhere in the first five calls.

b. Find the probability the first sale will occur sometime after 10 calls.

3.68 Marketing analysts have determined that a particular advertising campaign should make at least 20% of the adult population aware of the advertised product. After a recent campaign, 25 of 400 adults sampled indicated that they had seen the ad and were aware of the new product.

a. Find the approximate probability of observing $y \leq 25$ given that 20% of the population is aware of the product through the campaign.

b. Based on your answer to part (a), does it appear the ad was successful? Explain.

3.69 One or more specific, minor birth defects occurs with probability .0001 (that is, 1 in 10,000 births). If 20,000 babies are born in a given geographic area in a given year, can we calculate the probability of observing at least one of the minor defects using the binomial or normal approximation to the binomial? Explain.

3.70 The sample mean is to be calculated from a random sample of size $n = 4$ from a population consisting of six values (0, 1, 2, 4, 6, and 8). Find the sampling distribution of $\bar{y}$ based on a sample of size $n = 4$.

3.71 Plot the sampling distribution of Exercise 3.70. Find the mean and median of the sampling distribution of $\bar{y}$.

3.72 Refer to Exercise 3.70. Use the same population to find the sampling distribution for the sample median based on samples of size $n = 4$.

3.73   Plot the sampling distribution of Exercise 3.72. Find the mean and median of the sampling distribution.

3.74   Random samples of size 20, 40, and 80 are drawn from a population with mean $\mu = 100$ and standard deviation $\sigma = 15$. Give the mean and standard error of the distribution of $\bar{y}$ based on samples of size 20, 40, and 80.

3.75   Refer to Exercise 3.74. For each sampling distribution, find the following probabilities:

a. $P(\bar{y} > 105)$

b. $P(\bar{y} < 96)$

c. $P(96 < \bar{y} < 105)$

d. $1 - P(\bar{y} < 94)$

3.76   A random sample of $n = 36$ measurements is selected from a population with mean equal to 40 and standard deviation equal to 12.

a. Describe the sampling distribution of $\bar{y}$.

b. Find $P(\bar{y} > 36)$.

c. Find $P(\bar{y} < 30)$.

d. Find the value of $\bar{y}$ (say k) such that $P(\bar{y} > k) = .05$.

3.77   Refer to Exercise 3.76.

a. Describe the sampling distribution for the sample sum $\sum y_i$.

b. Find $P(\sum y_i > 1440)$.

c. Find $P(\sum y_i > 1540)$.

d. Find the value of $\sum y_i$ (say k) such that $P[k < \sum y_i < k_2] = .95$.

3.78   For each of the following situations, find the expected value and standard error of $\bar{y}$ based on a random sample of size n drawn from a population with mean μ and standard deviation σ.

a. $n = 25$, $\mu = 10$, $\sigma = 10$

b. $n = 100$, $\mu = 10$, $\sigma = 10$

c. $n = 25$, $\mu = 10$, $\sigma = 20$

d. $n = 100$, $\mu = 10$, $\sigma = 20$

3.79   Based on the results of Exercise 3.78, speculate on the effect of increasing the sample size and on the effect of an increase in σ on the standard error of $\bar{y}$.

4 INFERENCES ABOUT μ

INTRODUCTION

We have stated previously that the objective of inferential statistics is to make inferences about population parameters based on information contained in a sample. Our inferences can be phrased in two different ways. Consider the following two situations.

An airline would like to assess the impact of the introduction of a competing flight by another airline. To do this, the airline uses data on passenger loads for 25 days (regarded as a random sample) after the first month of the new competition. The mean passenger count could be used to *estimate* the average daily passenger load for future runs with the new competition.

In the second situation, we are concerned about the effectiveness of a new drug to control a particular species of worms in the stomachs of cattle. Let us suppose that a standard drug exists and that after a fixed period of treatment on the standard drug, the average infestation level is shown to be $\mu = 10.3$. If the new drug is indeed effective, it must have a mean infestation level after treatment at least as low as that for the standard. Thus we might wish to *test* the research hypothesis that the mean infestation level for the test drug is less than 10.3. Note that in this situation we are not asking, "What is the mean infestation level after treatment?" We are asking a more pointed question, "Is the mean infestation level lower than 10.3?"

These two examples illustrate the two different inference-making procedures we can use: **estimation** or **statistical testing**. The two procedures are contrasted by the following questions. In estimating or predicting a population parameter, we are asking the question, "What is the value of the population parameter?" This procedure reflects an experimental situation in which we know very little about the parameter of interest. In a statistical test we are asking the question, "Is the parameter equal to this specific value?" In this situation we do know something about the parameter of interest.

In this chapter we will consider these two inference-making procedures: estimation of a population mean μ and testing a statistical hypothesis about μ.

4.2   ESTIMATION OF μ

The simplest statistical inference problem is point estimation, where we compute a single value (statistic) from the sample data to estimate a population parameter. Suppose that we are interested in estimating a population mean, and that we are willing to assume the underlying population is normal. Then one natural statistic that could be used to estimate the population mean is the sample mean; but we also could use the median and the trimmed mean. Which sample statistic should we use?

A whole branch of mathematical statistics deals with problems related to developing point estimators (the formulas for calculating specific point estimates from sample data) of parameters from various underlying populations and determining whether or not a particular point estimator has certain desirable properties. Fortunately, we will not have to derive these point estimators—they'll be given to us for each parameter. Then, knowing which point estimator (formula) to use for a given parameter, we can develop confidence intervals (interval estimates) for these same parameters.

In this section we deal with point and interval estimation of a population mean μ. Tests of hypotheses about μ will be covered in Section 4.5.

For most problems in this text, the sample mean $\bar{y}$ will be used as a point estimate of μ; it is also used to form an interval estimate for the population mean μ. From the Central Limit Theorem for the sample mean given in Chapter 3, we know that for large n (crudely $n \geq 30$), $\bar{y}$ will be approximately normally distributed, with a mean μ and a standard error $\sigma_{\bar{y}}$. Then from our knowledge of the Empirical Rule and areas under a normal curve, we know that the interval $\mu \pm 2\sigma_{\bar{y}}$, or more precisely, the interval $\mu \pm 1.96\sigma_{\bar{y}}$, includes 95% of the $\bar{y}$s in repeated sampling, as shown in Figure 4.1.

Consider the interval $\bar{y} \pm 1.96\sigma_{\bar{y}}$. Any time $\bar{y}$ lies in the interval $\mu \pm 1.96\sigma_{\bar{y}}$, the interval $\bar{y} \pm 1.96\sigma_{\bar{y}}$ will contain the parameter μ (see Figure 4.2) and this will occur with probability .95. The interval $\bar{y} \pm 1.96\sigma_{\bar{y}}$ represents an interval estimate of μ.

We evaluate the goodness of an interval estimation procedure by examining the fraction of times in repeated sampling that interval estimates would encompass the parameter to be estimated. This fraction, called the **confidence coefficient**, is .95 when using the formula $\bar{y} \pm 1.96\sigma_{\bar{y}}$. That is, 95% of the time in repeated sampling, intervals calculated using the formula $\bar{y} \pm 1.96\sigma_{\bar{y}}$ will contain the mean μ.

confidence coefficient

FIGURE 4.1

Sampling Distribution for $\bar{y}$

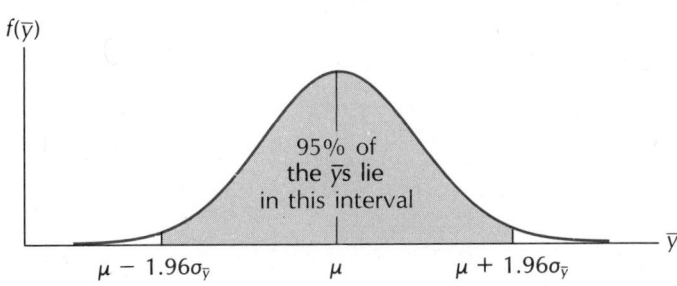

$f(\bar{y})$

95% of the $\bar{y}$s lie in this interval

$\mu - 1.96\sigma_{\bar{y}}$ μ $\mu + 1.96\sigma_{\bar{y}}$

$\bar{y}$

FIGURE 4.2

When the Observed Value of $\bar{y}$ Lies in the Interval $\mu \pm 1.96\sigma_{\bar{y}}$, the Interval $\bar{y} \pm 1.96\sigma_{\bar{y}}$ Contains the Parameter μ

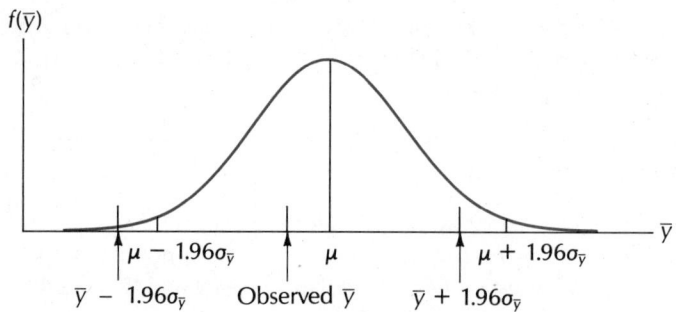

This idea is illustrated in Figure 4.3. Twenty different samples are drawn from a population with mean μ and variance σ^2. For each sample, an interval estimate is computed using the formula $\bar{y} \pm 1.96\sigma_{\bar{y}}$. Note that although the intervals bob about, most of them capture the parameter μ. In fact, if we repeated the process of drawing samples and computing confidence intervals, 95% of the intervals so formed would contain μ.

95% confidence interval

In a given experimental situation, we calculate only one such interval. This interval, called a **95% confidence interval**, represents an interval estimate of μ.

EXAMPLE 4.1

In a random sample of $n = 36$ parochial schools throughout the south, the average number of pupils per school is 379.2, with a standard deviation of 124. Use the sample

FIGURE 4.3

Twenty Interval Estimates Computed by Using $\bar{y} \pm 1.96\sigma_{\bar{y}}$

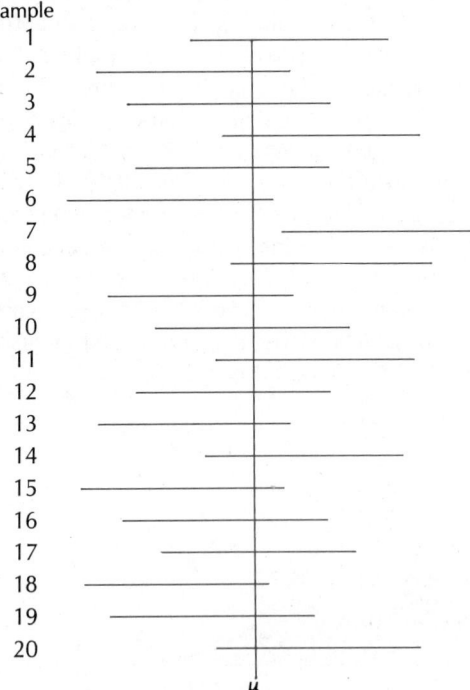

to construct a 95% confidence interval for μ, the mean number of pupils per school for all parochial schools in the south.

Solution

The sample data indicate that $\bar{y} = 379.2$ and $s = 124$. The appropriate 95% confidence interval is then computed by using the formula

$$\bar{y} \pm 1.96\sigma_{\bar{y}}$$

where $\sigma_{\bar{y}} = \sigma/\sqrt{n}$. In Section 4.8 we will present a procedure for obtaining a confidence interval for μ when σ is unknown. However, for all practical purposes, if the sample size is 30 or more, we can estimate the population standard deviation σ with s in the confidence interval formula. With s replacing σ, our interval is

$$379.2 \pm 1.96 \frac{124}{\sqrt{36}} \quad \text{or} \quad 379.2 \pm 40.51$$

The interval from 338.69 to 419.71 forms a 95% confidence interval for μ. In other words, we are 95% sure that the average number of pupils per school for parochial schools throughout the south lies between 338.69 and 419.71.

There are many different confidence intervals for μ, depending on the confidence coefficient we choose. For example, the interval $\mu \pm 2.58\sigma_{\bar{y}}$ includes 99% of the values of $\bar{y}$ in repeated sampling (see Figure 4.4), and the interval $\bar{y} \pm 2.58\sigma_{\bar{y}}$ forms a **99% confidence interval** for μ.

99% confidence interval

$(1 - \alpha)$ = confidence coefficient

We can state a general formula for a confidence interval for μ with a **confidence coefficient of $(1 - \alpha)$**, where α (Greek letter alpha) is between 0 and 1. For a specified value of $(1 - \alpha)$, a $100(1 - \alpha)\%$ confidence interval for μ is given by the following formula. Here we assume that σ is known or that the sample size is large enough to replace σ with s.

CONFIDENCE INTERVAL FOR μ; σ KNOWN

$$\bar{y} \pm z_{\alpha/2}\sigma_y$$

$z_{\alpha/2}$

The quantity $z_{\alpha/2}$ is a value of z having a tail area of $\alpha/2$ to its right. In other words, at a distance of $z_{\alpha/2}$ standard deviations to the right of μ, there is an area of $\alpha/2$ under

FIGURE 4.4

Sampling Distribution of $\bar{y}$

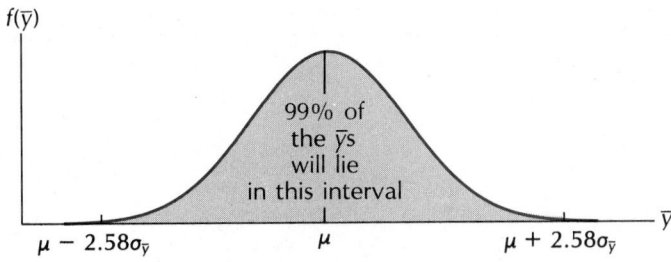

$f(\bar{y})$

99% of the $\bar{y}$s will lie in this interval

$\mu - 2.58\sigma_{\bar{y}}$ μ $\mu + 2.58\sigma_{\bar{y}}$ $\bar{y}$

FIGURE 4.5

Interpretation of $z_{\alpha/2}$ in the Confidence Interval Formula

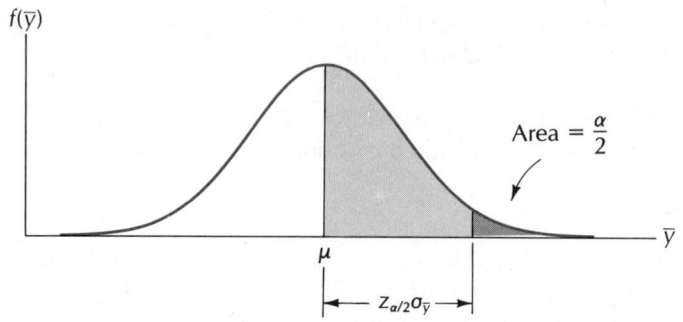

$f(\bar{y})$

Area $= \dfrac{\alpha}{2}$

μ

$z_{\alpha/2}\sigma_{\bar{y}}$

$\bar{y}$

the normal curve. Values of $z_{\alpha/2}$ can be obtained from Table 2 in the Appendix by looking up the z-value corresponding to an area of $\alpha/2$ (see Figure 4.5). Common values of the confidence coefficient $(1 - \alpha)$ and $z_{\alpha/2}$ are given in Table 4.1.

EXAMPLE 4.2

A forester is interested in estimating the average number of "count trees" per acre (trees larger than a specified size) on a 2000-acre plantation. She can then use this information to determine the total timber volume for trees in the plantation. A random sample of $n = 50$ one-acre plots is selected and examined. The average (mean) number of count trees per acre is found to be 27.3, with a standard deviation of 12.1. Use this information to construct a 99% confidence interval for μ, the mean number of count trees per acre for the entire plantation.

Solution

We use the general confidence interval with confidence coefficient equal to .99 and a $z_{\alpha/2}$-value equal to 2.58 (see Table 4.1). Substituting into the formula $\bar{y} \pm 2.58\sigma_{\bar{y}}$ and replacing σ with s in $\sigma_{\bar{y}} = \sigma/\sqrt{n}$, we have

$$27.3 \pm 2.58 \frac{12.1}{\sqrt{50}}$$

This corresponds to the confidence interval 27.3 ± 4.42, that is, the interval from 22.88 to 31.72. Thus we are 99% sure that the average number of count trees per acre is between 22.88 and 31.72.

The discussion in this section has included one rather unrealistic assumption; namely, that the population standard deviation is known. In practice, it's difficult to find situations in which the population mean is unknown, but the standard deviation is known. Usually

TABLE 4.1   Common Values of the Confidence Coefficient $(1 - \alpha)$ and the Corresponding z-Value, $z_{\alpha/2}$

Confidence Coefficient, $(1 - \alpha)$	Area in Table 2, $\alpha/2$	Corresponding z-Value, $z_{\alpha/2}$
.90	.05	1.645
.95	.025	1.96
.98	.01	2.33
.99	.005	2.58

both the mean and the standard deviation must be estimated from the sample. Since σ is estimated by the sample standard deviation s, the actual standard error of the mean, $\sigma/\sqrt{n}$, is naturally estimated by $s/\sqrt{n}$. This estimation introduces another source of random error (s will vary randomly, from sample to sample, around σ) and, strictly speaking, invalidates our confidence interval formula. Fortunately, the formula is still a very good approximation for large sample sizes. As a very rough rule, we can use this formula when n is larger than 30; a better way to handle this issue is described in Section 4.3.

substituting s for σ

Statistical inference-making procedures differ from ordinary procedures in that we not only make an inference, but also provide a measure of how good that inference is. For interval estimation, the width of the confidence interval and the confidence coefficient measure the goodness of the inference. Obviously, for a given confidence coefficient, the smaller the width of the interval, the better the inference. The confidence coefficient, on the other hand, is set by the experimenter to express how much assurance he or she places in whether the interval estimate encompasses the parameter of interest.

EXERCISES

4.1   The sample mean and standard deviation based on a sample of 50 measurements are $\bar{y} = 105$ and $s = 11$.
   a. Calculate a 95% confidence interval for μ.
   b. Calculate a 99% confidence interval for μ.

4.2   Give a careful verbal interpretation of the confidence interval in Exercise 4.1(a).

4.3   Refer to Exercise 4.1.
   a. Discuss the impact of doubling the sample size from $n = 50$ to $n = 100$ on the 95% confidence interval. Assume for discussion purposes that $\bar{y}$ and s are still 105 and 11, respectively.
   b. What impact would quadrupling the sample size have? (Note: Answer this question without doing the calculations.)

 4.4   The caffeine content (in mg) was examined for a random sample of 50 cups of black coffee dispensed by a new machine. The mean and standard deviation were 110 mg and 7.1 mg respectively. Use these data to construct a 98% confidence interval for μ, the mean caffeine content for cups dispensed by the machine.

 4.5   A random sample of the year-end statements of 22 small businesses (under $500,000 in sales) in a city shows the mean gross profit margin to be 5.2% (of sales) with a standard deviation of 3.3%. Use these data to place a 90% confidence interval for μ. Assume $\sigma \approx 3.3$.

 4.6   Recent data from a national survey of 1350 women indicated that the average woman goes to a hair salon once every five weeks and spends on the average $26.40. With a standard deviation of $12.00, use these data to construct a 99% confidence interval for μ.

 4.7   A social worker is interested in estimating the average length of time spent outside of prison for first offenders who later commit a second crime and are sent to prison again. A random sample of $n = 150$ prison records in the county courthouse indicates that the average length of prison-free life between first and second offenses is 3.2 years, with a standard deviation of 1.1 years. Use the sample information to estimate μ, the mean prison-free life between first and second offenses for all prisoners on record in the county courthouse. Construct a 95% confidence interval for μ. Assume that σ can be replaced by s.

 4.8   Refer to Exercise 4.5. What impact would a doubling of the sample size have on the confidence interval?

4.9 The rust mite, a major pest of citrus in Florida, punctures the cells of the leaves and fruit. Damage by rust mites is readily recognizable because the injured fruit will display a brownish (rust) color and be somewhat reduced in size depending on the severity of the attack. If the rust mites are not controlled, the affected groves will have a substantial reduction in both the fruit yield and the fruit quality. In either case the citrus grower suffers financially since the produce will be of a lower grade and sell for less on the fresh fruit market. This year more and more citrus growers have gone to a preventive program of maintenance spraying for rust mites. In evaluating the effectiveness of the program, a random sample of 60 10-acre plots, one plot from each of 60 groves, is selected. These show an average yield of 850 boxes, with a standard deviation of 100 boxes. Give a 95% confidence interval for μ, the average (10-acre) yield for all groves utilizing such a maintenance spraying program. Assume that σ can be replaced by s.

 4.10 An experiment is conducted to examine the susceptibility of root stocks of a variety of lemon trees to a specific larva. Forty of the plants are subjected to the larvae and examined after a fixed period of time. The response of interest is the logarithm of the number of larvae per gram that is counted on each root stock. For these 40 plants the sample mean is 9.02 and the standard deviation is 1.12. Use these data to construct a 90% confidence interval for μ, the mean susceptibility for the population of lemon tree root stocks from which the sample was drawn. Assume that σ can be replaced by s.

 4.11 A mobility study is conducted among a random sample of 900 high school graduates of a particular state over the past 10 years. For each of the persons sampled, the distance between the high school attended and the present permanent address is recorded. For these data, $\bar{y} = 430$ miles and $s = 262$ miles. Using a 95% confidence interval, estimate the average number of miles between a person's high school and present permanent address for high school graduates of the state over the past 10 years. Assume that σ can be replaced by s.

4.3   CHOOSING THE SAMPLE SIZE FOR ESTIMATING μ

How can we determine the number of observations to be included in the sample? The implications of such a question are clear. Data collection costs money. If the sample is too large, time and talent are wasted. Conversely, it is wasteful if the sample is too small, because inadequate information has been purchased for the time and effort expended. Also, it may be impossible to increase the sample size at a later time. Hence the number of observations to be included in the sample will depend on the amount of information the experimenter wants to buy.

Suppose we want to estimate the average amount for accident claims filed against an insurance company. To decide how many claims must be examined, we would have to determine how accurate the company wants to be. For example, the company might indicate that the tolerable error is to be 10 units (± 5 units) or less. Then we would want the confidence interval to be of the form $\bar{y} \pm 5$.

There are two considerations in choosing the appropriate sample size for estimating μ using a confidence interval. The tolerable error establishes the desired width of the confidence interval; the second consideration is the confidence level that should be selected. A wide confidence interval would not be very informative, but the cost of obtaining a narrow confidence interval could be quite large. Similarly, too low a confidence level (say 50%) would mean that the stated confidence interval is likely to be in error, but obtaining a higher level of confidence also would be more expensive.

What constitutes reasonable certainty? In most situations the confidence level is set at 95% or 90%, partly because of tradition and partly because these levels represent (to some people) a reasonable level of certainty. The 95% (or 90%) level translates into a long-run chance of 1 in 20 (or 1 in 10) of not covering the population parameter. This seems reasonable and is comprehensible, whereas 1 chance in 1000 or 1 in 10,000 is just too small.

The tolerable error depends heavily on the context of the problem and only someone who is familiar with the situation can make a reasonable judgment about its magnitude.

When considering a confidence interval for a population mean μ, the plus-or-minus term of the confidence interval is $z_{\alpha/2}\sigma_{\bar{y}}$ where $\sigma_{\bar{y}} = \sigma/\sqrt{n}$. Three quantities determine the value of the plus-or-minus term: the desired confidence level (which determines the z-value used), the standard deviation (σ), and the sample size (which together with σ determines the standard error $\sigma_{\bar{y}}$). Usually a guess must be made about the size of the population standard deviation. (Sometimes an initial sample is taken to estimate the standard deviation; this estimate provides a basis for determining the additional sample size that will be needed.) For a given tolerable error, once the confidence level is specified and an estimate of σ supplied, the required sample size can be calculated using the formula shown here.

If a 95% confidence interval is to be of the form $\bar{y} \pm E$, then solve the expression

$$1.96\sigma_{\bar{y}} = E$$

for n. The width of the interval is $2E$.

In general, if we want to estimate μ using a $100(1-\alpha)\%$ confidence interval of the form $\bar{y} \pm E$, where E is specified, then we solve the equation

$$z_{\alpha/2}\sigma_{\bar{y}} = E$$

for n. This is shown here.

SAMPLE SIZE REQUIRED FOR A $100(1-\alpha)\%$ CONFIDENCE INTERVAL FOR μ OF THE FORM $\bar{y} \pm E$	$$n = \frac{(z_{\alpha/2})^2 \sigma^2}{E^2}$$

Note that determining a sample size to estimate μ requires knowledge of the population variance σ^2 (or standard deviation σ). We can obtain an approximate sample size by estimating σ^2, using one of these two methods:

1. Employ information from a prior experiment to calculate a sample variance s^2. This value is used to approximate σ^2.
2. Use information on the range of the observations to obtain an estimate of σ.

We would then substitute the estimated value of σ^2 in the sample-size equation to determine an approximate sample size n.

We illustrate the procedure for choosing a sample size with two examples.

EXAMPLE 4.3

Union officials are concerned about reports of inferior wages paid to a company's employees under their jurisdiction. It is decided to take a random sample of n wage sheets from the company to estimate the average hourly wage. If it is known that wages in the company have a range of $10 per hour, determine the sample size required to estimate the average hourly wage μ using a 95% confidence interval with width equal to $1.20.

Solution

Since we want a 95% confidence interval with width $1.20, $E = \$.60$. The value that we use to substitute for σ is range/4 = 2.50. Substituting into the formula for n we have

$$n = \frac{(1.96)^2(2.5)^2}{(.60)^2} = 66.69$$

To be on the safe side, we will round this number up to the next integer. A sample size of 67 should give a 95% confidence interval with the desired width of $1.20.

EXAMPLE 4.4

A federal agency has decided to investigate the advertised weight printed on cartons of a certain brand of cereal. The company in question periodically samples cartons of cereal coming off the production line to check their weight. A summary of 1500 of the weights made available to the agency indicates a mean weight of 11.80 ounces per carton and a standard deviation of .75 ounce. Use this information to determine the number of cereal cartons the federal agency must examine to estimate the average weight of cartons being produced now, using a 99% confidence interval of width .50.

Solution

The federal agency has specified that the width of the confidence interval is to be .50, so $E = .25$. Assuming that the weights made available to the agency by the company are accurate, we can take $\sigma = .75$. The required sample size with $z_{\alpha/2} = 2.58$ is

$$n = \frac{(2.58)^2(.75)^2}{(.25)^2} = 59.91$$

That is, the federal agency must obtain a random sample of 60 cereal cartons to estimate the mean weight to within $\pm .25$.

EXERCISES

4.12 Refer to Example 4.3.
   a. How large a sample is needed to obtain a 90% confidence interval with width $0.60? $0.30? $0.15?
   b. In general, for a given confidence level, how much would you increase the sample size to cut the width in half?

 **4.13** The giant size of a new "tough cleaning" laundry detergent has a listed net weight of 42 oz. If the variability in weight has a standard deviation of 2 oz, how many boxes must be sampled to estimate the average fill weight to within $\pm .25$ oz using a 95% confidence interval?

4.14　Refer to Exercise 4.13. Determine the effect of a 90% and a 99% confidence level on the required sample size.

4.15　A biologist would like to estimate the effect of an antibiotic on the growth of a particular bacterium by examining the mean amount of bacteria present per plate of culture when a fixed amount of the antibiotic is applied. Previous experimentation with the antibiotic on this type of bacterium indicates that the standard deviation of the amount of bacteria present is approximately 13 square centimeters. Use this information to determine the number of observations (cultures that must be developed and then tested) to estimate the mean amount of bacteria present, using a 99% confidence interval with a half-width of 3 square centimeters.

4.16　Investigators would like to estimate the average annual taxable income of apartment dwellers in a city to within $500 using a 95% confidence interval. If we assume the annual incomes range from $0 to $40,000, determine the number of observations that should be included in the sample.

4.17　Refer to Exercise 4.16. Determine the required sample size if the desired error in a 95% confidence interval is $E = 250$. Do the same for $E = 1000$. Compare your results to those of Exercise 4.16.

4.18　As part of a much larger study of trends in long-distance telephone usage, a study is to be conducted this month of residential homes with married couples between 25 and 40 years of age. How large a sample should be taken if the mean number of long-distance calls for the month is to be estimated to within one call using a 90% confidence interval? Assume $\sigma \approx 4.0$.

4.4　CONTROL CHARTS FOR A POPULATION MEAN (optional)

control chart　We can extend the notion of a confidence interval for μ to obtain a **control chart**. As consumers, we are vitally interested in product quality. We expect product quality for a particular item to be uniform from one time period to another, and we expect it to live up to the product description advertised by the manufacturer. For example, in buying paint from a paint store, we expect different gallons of the same color to be uniform in color and we expect the color to be identical to that advertised in the paint-sample brochure. Similarly, the Food and Drug Administration (FDA) not only expects but also demands that drug products have uniform potency and meet the standards advertised by the pharmaceutical firm.

Consumers are not the only people interested in product quality. Reputable manufacturers are also concerned that their products meet the standards they have claimed. If the quality of a product falls below the standards advertised by the company, then there is a risk that consumers will reject the product and buy from a competitor. Similarly, if product quality drifts above the standards established by the company, then it would be in the company's interest to upgrade their advertising to reflect the increase in quality.

quality control　**Quality control** techniques have been developed to monitor the ongoing quality of a manufacturing process in order to maintain uniform quality or at least to detect when product quality has shifted. We can monitor product quality of a production process by using a graph called a *control chart*. Thus we could graph the sample mean or sample range for samples collected over a period of time to monitor product quality.

Typically a control chart consists of three lines: a center line, an upper control line, and a lower control line. In a control chart for the mean, successive sample means would

FIGURE 4.6

Control Chart for
Sample Means

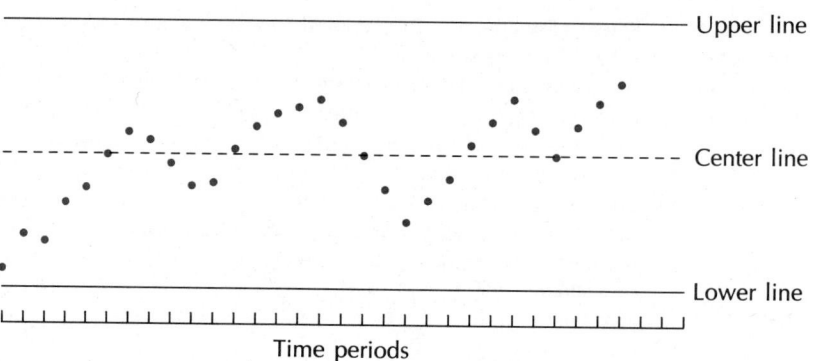

be plotted much as they appear in Figure 4.6. The sample means are shown by the dots in Figure 4.6. If one of the sample means falls outside either the upper or lower control lines, the process is judged to be out of control; that is, it appears that product quality has shifted. At this point company officials and production personnel would try to establish the cause of the shift and would initiate corrective changes in the production process.

center line   Establishing the three control lines is quite simple. The **center line** (denoted by $\bar{y}_c$) represents the average of k sample means, each based on n observations. We generally recommend taking $k \geq 25$ and $n > 3$. These samples should be taken at some time when the process is judged to be under control. Then if we let y_{ij} denote the jth observation in sample i and $\bar{y}_i = \sum_j y_{ij}/n$ denote the mean for sample i, the average of the k sample means is

$$\bar{y}_c = \sum_i \frac{\bar{y}_i}{k} = \frac{\sum_{i,j} y_{ij}}{nk}$$

UCL and LCL for
mean quality   The **upper control line (UCL)** and the **lower control line (LCL)** are computed as follows:

$$\text{UCL} = \bar{y}_c + 3\frac{\sigma}{\sqrt{n}} \qquad \text{and} \qquad \text{LCL} = \bar{y}_c - 3\frac{\sigma}{\sqrt{n}}$$

From knowledge of the Empirical Rule, the inverval $\bar{y}_c \pm (3\sigma/\sqrt{n})$ should contain nearly all the sample means $\bar{y}_i$ in repeated sampling. If a sample mean falls outside this interval, we have either observed an extremely unlikely event or the process quality has changed and $\bar{y}_c$ is no longer an accurate measure of the actual mean product quality. This latter conclusion is more realistic and is used to signal a manufacturing process out of control.

The standard deviation σ in the formulas for the upper and lower control lines can be estimated either by using a pooled sample variance from the k samples or, more quickly, by using the k sample ranges. We will employ the latter procedure. Letting r_i denote the range for the n sample measurements in sample i and $\bar{r}$ the average of the k sample ranges,

estimate of σ for
control chart   we can estimate σ by

$$\hat{\sigma} = \frac{\bar{r}}{d_n}$$

where d_n is obtained from Table 16 of the Appendix. For example, suppose we have $k = 20$ different samples of $n = 7$ observations per sample and $\bar{r} = 5$. Then $d_7 = 2.704$ and $\hat{\sigma} = 5/2.704 = 1.849$.

EXAMPLE 4.5   A company that dyes rugs is interested in establishing a control chart for the mean color when dyeing solid-colored rugs. Although maintaining uniform color is somewhat important for patterned or multicolored rugs, it is much more important for solid-colored rugs, where minor changes in solid colors are readily recognizable. Rug-color quality can be monitored by taking readings on a colorimeter. Twenty-five samples of five measurements each from a rug being dyed red yielded the data listed in Table 4.2. These data were obtained while the manager believed that the process was in control.

Use the data of Table 4.2 to construct a control chart for the mean colorimeter reading.

TABLE 4.2   Colorimeter Readings for the 25 Samples of Example 4.5

Sample	Observation	Sample Sum	Sample Range
1	2.4, 1.8, 0.7, 1.0, 2.5	8.4	1.8
2	2.3, 3.0, 2.5, 1.2, 3.1	12.1	1.9
3	1.3, 1.2, 0.9, 1.2, 3.0	7.6	2.1
4	0.5, 2.2, 2.4, 1.5, 3.0	9.6	2.5
5	2.8, 1.9, 2.6, 1.3, 2.9	11.5	1.6
6	2.4, 3.1, 1.7, 3.3, 2.6	13.1	1.6
7	2.5, 2.9, 1.4, 4.0, 2.1	12.9	2.6
8	1.1, 2.9, 3.0, 1.4, 2.8	11.2	1.9
9	3.3, 2.2, 2.7, 2.8, 2.1	13.1	1.2
10	0.8, 4.2, 2.3, 1.4, 2.1	10.8	3.4
11	0.2, 2.6, 2.3, 0.7, 4.2	10.0	4.0
12	1.8, 1.6, 2.3, 2.1, 1.7	9.5	.7
13	0.1, 3.9, 2.3, 1.4, 1.0	8.7	3.8
14	1.1, 3.1, 1.8, 0.9, 1.8	8.7	2.2
15	0.5, 0.9, 4.0, 2.2, 2.8	10.4	3.5
16	2.9, 3.3, 1.9, 3.1, 2.3	13.5	1.4
17	3.5, 2.0, 2.5, 2.0, 0.3	10.3	3.2
18	2.5, 2.1, 2.7, 1.7, 1.5	10.5	1.2
19	1.1, 3.9, 2.7, 1.2, 1.3	10.2	2.8
20	1.4, 2.0, 2.5, 4.2, 2.4	12.5	2.8
21	2.2, 1.9, 0.7, 1.3, 1.4	7.5	1.5
22	1.5, 1.5, 1.1, 2.3, 2.4	8.8	1.3
23	2.2, 1.3, 2.5, 1.9, 0.7	8.6	1.8
24	1.7, 0.1, 1.8, 0.7, 2.1	6.4	2.0
25	3.7, 1.5, 1.9, 0.6, 4.6	12.3	4.0
Totals		258.2	56.8

Solution

From Table 4.2 we have

$$\bar{y}_c = \frac{\sum_{i,j} y_{ij}}{nk} = \frac{258.2}{5(25)} = 2.07$$

$$\bar{r} = \frac{\sum_i r_i}{k} = \frac{56.8}{25} = 2.27$$

From Table 16 in the Appendix, we have $d_5 = 2.326$ and hence

$$\hat{\sigma} = \frac{\bar{r}}{d_n} = \frac{2.27}{2.326} = .98$$

The center line is then 2.07, with upper and lower control lines given by

$$UCL = \bar{y}_c + 3\frac{\hat{\sigma}}{\sqrt{n}} = 2.07 + \frac{3(.98)}{\sqrt{5}} = 3.38$$

$$LCL = \bar{y}_c - 3\frac{\hat{\sigma}}{\sqrt{n}} = 2.07 - \frac{3(.98)}{\sqrt{5}} = .76$$

As stated previously, an observation falling outside one of the control lines is a signal that something has changed. If σ is known and the control lines are computed by using the known value of σ, a value outside a control line would suggest to us that $\bar{y}_c$ no longer represents the actual mean quality. Unfortunately, when σ is unknown and must be estimated, a value outside one of the control lines could suggest a shift in the mean quality, an increase in σ, or both.

To protect ourselves, we should also keep a control chart on product quality variability. Again, rather than working with the k sample standard deviations, it is simpler to use the k sample ranges. **Upper and lower control lines for product quality variability** are given by

$$UCL = D'_n\bar{r} \qquad \text{and} \qquad LCL = D_n\bar{r}$$

UCL, LCL for product variability

where D'_n and D_n are obtained from Table 17 of the Appendix. For Example 4.5, $\bar{r} = 2.27$, and, based on $n = 5$ observations per sample, $D'_5 = 2.115$ and $D_5 = 0$. Hence the control lines for the sample ranges are

$$UCL = 2.115(2.27) = 4.80 \qquad \text{and} \qquad LCL = 0$$

EXERCISES

4.19 Refer to Example 4.5. Graph the upper and lower control limits and the center line. Plot the sequence of sample means listed below to determine if and when the process is out of control. (Note: Each mean is based on five measurements.)

2.0	1.9	1.6	1.5	1.7	1.8	2.2	2.1	2.0	2.3	2.4	2.7	2.8	2.9

4.20 The labeled amount of ingredient A in a marketed cough drop is 1 in 1500 parts of the total labeled weight (2.2 grams). The assay for the ingredient is based on 10 cough drops that have been dissolved. A sequence of 48 sample means is shown here, expressed as a percent of the total labeled weight. If $\bar{r} = 4.1(\%)$, determine the center line and the upper and lower control limits for μ.

108	110	110
108	107	112
109	107	111
107	108	111
109	108	110
108	109	110
109	108	112
109	110	110
107	109	110
107	108	112
109	107	110
111	110	111
109	111	113
107	111	111
106	110	110
107	111	110

4.5   A STATISTICAL TEST FOR μ

The second type of inference-making procedure is statistical testing (or hypothesis testing). As with estimation procedures, we will make an inference about a population parameter, but here the inference will be of a different sort. In this section we will present a statistical test that will lead to an answer to the question, "Is the population mean equal to a specified value μ_0?" For example, in studying the antipsychotic properties of an experimental compound, we might ask whether the average shock-avoidance response for rats treated with a specific dose is 60, the same value that has been observed after extensive testing using a suitable standard drug.

statistical test   A **statistical test** is based on the concept of proof by contradiction and is composed of the five parts listed here.

FIVE PARTS OF A STATISTICAL TEST

1. Null hypothesis, denoted by H_0
2. Research hypothesis (also called the alternative hypothesis), denoted by H_a
3. Test statistic, denoted by T.S.
4. Rejection region, denoted by R.R.
5. Conclusion

research hypothesis, H_a   For example, in setting up a statistical test concerning the mean yield per acre (in bushels) for a particular variety of soybeans, we may be interested in the **research hypothesis** that the mean yield per acre μ is greater than 520 bushels, the average observed for

FIGURE 4.7

Assuming That H_0 Is
True, Contradictory
Values of $\bar{y}$ Are
in the Upper Tail

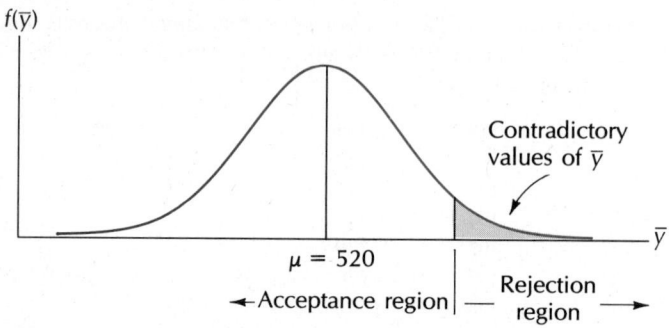

farms throughout a particular state in the past several years. To verify the research hypoth-

null hypothesis, H_0 esis, we try to contradict another hypothesis, called the **null hypothesis**, that $\mu = 520$
(i.e., the highest average yield that is still contradictory to the research hypothesis).

Having stated the null and research hypotheses, we then obtain a random sample of
1-acre yields from farms throughout the state and compute $\bar{y}$ and s, the sample mean and
standard deviation, respectively. The decision to accept the null hypothesis or reject it in

test statistic, T.S. favor of the research hypothesis is based on a **test statistic** or decision maker computed
from the sample data. If the population can be assumed to be more or less mound-shaped,
a logical choice as a decision maker for μ would be $\bar{y}$ or some function of the sample mean.

If we choose $\bar{y}$ as the test statistic, we know that the sampling distribution of $\bar{y}$, assuming the null hypothesis is true, is approximately normal with mean $\mu = 520$. Values of
$\bar{y}$ that are contradictory to the null hypothesis and are in favor of the research hypothesis
will be those that lie in the upper tail of the distribution of $\bar{y}$. See Figure 4.7. These contra-

rejection region, R.R. dictory values form a **rejection region** for our statistical test. If the observed value of $\bar{y}$ falls
in the rejection region of Figure 4.7, we would reject the null hypothesis that the mean
yield per acre is $\mu = 520$ in favor of the research hypothesis that $\mu > 520$. Note that we
are supporting the research hypothesis by contradicting the null hypothesis. If the observed
value of $\bar{y}$ falls in the acceptance region rather than in the rejection region, we do not
reject the null hypothesis. However, this does not mean that we automatically *accept* the
null hypothesis that $H_0: \mu = 520$ (exactly). More will be said on the notion of acceptance
of the H_0 after we have discussed the two types of errors that can be made.

As with any two-way decision process, we can make an error by falsely rejecting the
null hypothesis or by falsely accepting the null hypothesis. We give these errors special
names.

DEFINITION 4.1

A **Type I error** is committed if we reject the null hypothesis when it is true. The
probability of a Type I error is denoted by the symbol α.

DEFINITION 4.2

A **Type II error** is committed if we accept the null hypothesis when it is false and
the research hypothesis is true. The probability of a Type II error is denoted by the
symbol β (Greek letter beta).

Relationships between α and β are shown in Table 4.3.

TABLE 4.3   Relationships Between α and β

	Null Hypothesis	
Decision	True	False
Reject H_0	Type I error α	Correct $1 - \beta$
Accept H_0	Correct $1 - \alpha$	Type II error β

Although it would be desirable to determine the acceptance and rejection regions to simultaneously minimize both α and β, this is not possible. The probabilities associated with Type I and Type II errors are inversely related. For a fixed sample size n, as we change the rejection region to increase α, then β decreases, and vice versa.

To alleviate what appears to be an impossible bind, the experimenter specifies a tolerable probability for a Type I error of the statistical test. Thus the experimenter may choose α to be .01, .05, .10, and so on. Specification of a value for α then locates the rejection region. Determination of the associated probability of a Type II error is more complicated and will be delayed until later in the chapter.

Let us now see how the choice of α locates the rejection region. Returning to our soybean example, we will reject the null hypothesis for large values of the sample mean $\bar{y}$. Suppose we have decided to take a sample of $n = 36$ 1-acre plots, and from these data we compute $\bar{y} = 573$ and $s = 124$. Can we conclude that the mean yield for all farms is above 520?

specifying α Before answering this question we must specify α. If we are willing to take the risk that 1 time in 40 we would incorrectly reject the null hypothesis, then $\alpha = 1/40 = .025$. An appropriate rejection region can be specified for this value of α by referring to the sampling distribution of $\bar{y}$. Assuming that the null hypothesis is true and that σ can be replaced by s, then $\bar{y}$ is normally distributed, with $\mu = 520$ and $\sigma_{\bar{y}} = 124/\sqrt{36} = 20.67$. Since the shaded area of Figure 4.8 corresponds to α, locating a rejection region with an area of .025 in the right tail of the distribution of $\bar{y}$ would be equivalent to determining the value of z that has an area .025 to its right. Referring to Table 2 in the Appendix, this value of z is 1.96. Thus the rejection region for our example would be located 1.96 standard errors $(1.96\sigma_{\bar{y}})$ above the mean $\mu = 520$. If the observed value of $\bar{y}$ is greater than 1.96 standard errors above $\mu = 520$, we reject the null hypothesis, as shown in Figure 4.8.

FIGURE 4.8

Rejection Region for the Soybean Example When $\alpha = .025$

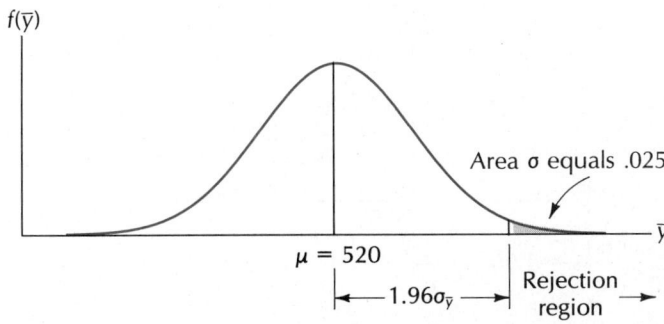

$f(\bar{y})$

Area σ equals .025

$\mu = 520$

$1.96\sigma_{\bar{y}}$

Rejection region

EXAMPLE 4.6   Set up all the parts of a statistical test for the soybean example and use the sample data to reach a decision on whether to accept or reject the null hypothesis. Set $\alpha = .025$. Assume that σ can be replaced by s.

Solution   The five parts of the test are as follows:

H_0: $\mu = 520$

H_a: $\mu > 520$

T.S.: $\bar{y}$

R.R.: For $\alpha = .025$, reject the null hypothesis if $\bar{y}$ lies more than 1.96 standard errors above $\mu = 520$

The computed value of $\bar{y}$ was 573. To determine the number of standard errors that $\bar{y}$ lies above $\mu = 520$, we compute a z score for $\bar{y}$ using the formula

$$z = \frac{\bar{y} - \mu_0}{\sigma_{\bar{y}}}$$

where $\sigma_{\bar{y}} = \sigma/\sqrt{n}$. Substituting into the formula,

$$z = \frac{\bar{y} - \mu_0}{\sigma_{\bar{y}}} = \frac{573 - 520}{124/\sqrt{36}} = 2.56$$

Conclusion: Since the observed value of $\bar{y}$ lies more than 1.96 standard errors above the hypothesized mean $\mu = 520$, we reject the null hypothesis in favor of the research hypothesis and conclude that the average soybean yield per acre is greater than 520.

one-tailed test   The statistical test conducted in Example 4.6 is called a **one-tailed test**, because the rejection region is located in only one tail of the distribution of $\bar{y}$. If our research hypothesis is H_a: $\mu < 520$, small values of $\bar{y}$ would indicate rejection of the null hypothesis. This test would also be one-tailed, but the rejection region would be located in the lower tail of the distribution of $\bar{y}$. Figure 4.9 displays the rejection region for the alternative hypothesis H_a: $\mu < 520$ when $\alpha = .025$.

two-tailed test   We can formulate a **two-tailed test** for the research hypothesis H_a: $\mu \neq 520$, where we are interested in detecting whether the mean yield per acre of soybeans is greater

FIGURE 4.9

Rejection Region for
H_a: $\mu < 520$ When
$\alpha = .025$ for the
Soybean Example

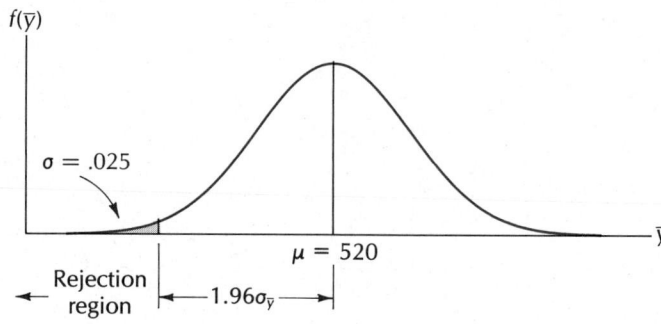

SECTION 4.5   A STATISTICAL TEST FOR μ

FIGURE 4.10 $f(\bar{y})$

Two-Tailed Rejection
Region for $H_a: \mu \neq 520$
When $\alpha = .05$ for the
Soybean Example

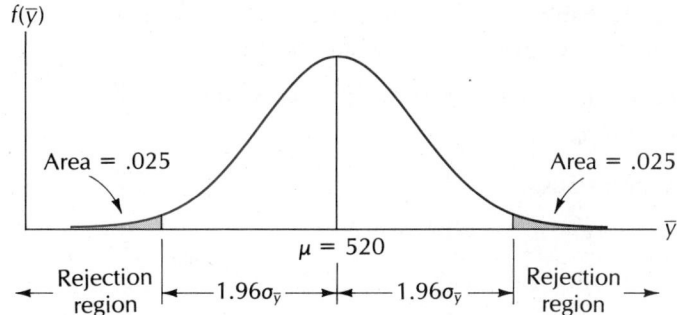

or less than 520. Clearly both large and small values of $\bar{y}$ will contradict the null hypothesis, and we would locate the rejection region in both tails of the distribution of $\bar{y}$. A two-tailed rejection region for $H_a: \mu \neq 520$ and $\alpha = .05$ is shown in Figure 4.10.

EXAMPLE 4.7

A corporation maintains a large fleet of company cars for its salespeople. To check the average number of miles driven per month per car, a random sample of $n = 40$ cars is examined. The mean and standard deviation for the same are 2752 miles and 350 miles, respectively. Records for previous years indicate that the average number of miles driven per car per month was 2600. Use the sample data to test the research hypothesis that the current mean μ differs from 2600. Set $\alpha = .05$ and assume that σ can be replaced by s.

Solution

The research hypothesis for this statistical test is $H_a: \mu \neq 2600$ and the null hypothesis is $H_0: \mu = 2600$. Using $\alpha = .05$, the two-tailed rejection region for this test is located as shown in Figure 4.11.

To determine how many standard errors our test statistic $\bar{y}$ lies away from $\mu = 2600$, we compute

$$z = \frac{\bar{y} - \mu_0}{\sigma/\sqrt{n}} = \frac{2752 - 2600}{350/\sqrt{40}} = 2.75$$

The observed value for $\bar{y}$ lies more than 1.96 standard errors above the mean, so we reject the null hypothesis in favor of the alternative $H_a: \mu \neq 2600$. Since the computed value of $\bar{y}$ is greater than the hypothesized mean $\mu = 2600$, we conclude that the mean number of miles driven is greater than 2600.

FIGURE 4.11 $f(\bar{y})$

Rejection Region for
$H_a: \mu \neq 2600$ When
$\alpha = .05$, Example 4.7

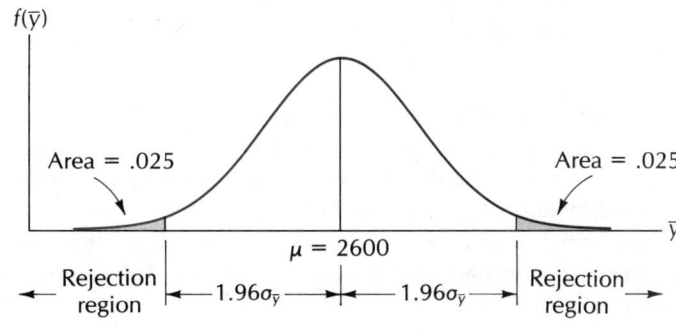

The mechanics of the statistical test for a population mean can be greatly simplified if we use z rather than $\bar{y}$ as a test statistic. Using

H_0: $\mu = \mu_0$ (where μ_0 is some specified value)

H_a: $\mu > \mu_0$

and the test statistic

$$z = \frac{\bar{y} - \mu_0}{\sigma/\sqrt{n}}$$

then for $\alpha = .025$ we would reject the null hypothesis if $z > 1.96$—that is, if $\bar{y}$ lies more than 1.96 standard errors above the mean. Similarly, for the same null hypothesis, $\alpha = .05$, and H_a: $\mu \neq \mu_0$, we would reject the null hypothesis if the computed value of z is greater than 1.96 or less than -1.96, or, equivalently, if $|z| > 1.96$.

test for population mean   The statistical **test for a population mean** is summarized below. For H_0: $\mu = \mu_0$, three different alternatives are given with their corresponding rejection regions. In a given situation you will choose only one of the three alternatives with its associated rejection region.

SUMMARY OF A STATISTICAL TEST FOR μ WITH σ KNOWN

H_0: $\mu = \mu_0$ (μ_0 is specified)

H_a: 1. $\mu > \mu_0$

   2. $\mu < \mu_0$

   3. $\mu \neq \mu_0$

T.S.: $z = \dfrac{\bar{y} - \mu_0}{\sigma/\sqrt{n}}$

R.R.: For a probability α of a Type I error,

1. reject H_0 if $z > z_\alpha$
2. reject H_0 if $z < -z_\alpha$
3. reject H_0 if $|z| > z_{\alpha/2}$

Note: For the time being, if σ is unknown but $n \geq 30$, you may replace σ by s in the standard error $\sigma_{\bar{y}} = \sigma/\sqrt{n}$ and proceed with the test. A more detailed discussion of inferences about μ when σ is unknown is presented later in this chapter.

EXAMPLE 4.8   The average (mean) live weight of a farmer's steers prior to slaughter was 380 pounds in past years. This year his 50 steers were fed on a new diet. Suppose we consider these 50 steers fed on the new diet as a random sample taken from a population of all possible steers that may be fed the diet now or in the future. Use the sample data given below and $\alpha = .01$ to test the research hypothesis that the mean live weight for steers on the new diet is greater than 380. The sample data: $n = 50$; $\bar{y} = 390$; $s = 35.2$.

Solution

Using the sample data with $\alpha = .01$, the five parts of a statistical test are as follows:

H_0: $\mu = 380$

H_a: $\mu > 380$

T.S.: $z = \dfrac{\bar{y} - \mu_0}{\sigma/\sqrt{n}} = \dfrac{390 - 380}{35.2/\sqrt{50}} = \dfrac{10}{35.2/7.07} = 2.01$

R.R.: For $\alpha = .01$ and a one-tailed test, we reject H_0 if $z > z_{.01}$, where $z_{.01} = 2.33$

Conclusion: Since the observed value of z, 2.01, does not exceed 2.33, we might be tempted to accept the null hypothesis that $\mu = 380$. The only problem with this conclusion is that we do not know β, the probability of incorrectly accepting the null hypothesis. To hedge somewhat in situations where z does not fall in the rejection region and β has not been calculated, we recommend stating that there is insufficient evidence to reject the null hypothesis. To reach a conclusion about whether or not to accept H_0, the experimenter would have to compute β. If β is small for reasonable alternative values of μ, then H_0 is accepted. Otherwise, the experimenter should conclude that there is insufficient evidence to reject the null hypothesis.

computing β

We can illustrate the **computation of β**, the probability of a Type II error or equivalently the *power* $1 - \beta$, using the data in Example 4.8. If the null hypothesis is H_0: $\mu = 380$, the probability of incorrectly accepting H_0 will depend on how close the actual mean is to 380. For example, if the actual mean live weight is 400 pounds for steers on the new diet, we would expect β to be much smaller than if the actual mean live weight is 387.

Let us suppose that the actual mean live weight is 395. What is β? With the null and research hypotheses as before,

H_0: $\mu = 380$

H_a: $\mu > 380$

and with $\alpha = .01$, we use Figure 4.12(a) to display β. The shaded portion of Figure 4.12(a) represents β, since this would be the probability of $\bar{y}$ falling in the acceptance region when the null hypothesis is false and μ is actually 395. Similarly, the power of the test for detecting H_a: $\mu = 395$ is $1 - \beta$, the area in the rejection region.

Let us consider two other possible values for μ, namely, 387 and 400. The corresponding values of β are shown as the shaded portions of Figures 4.12(b) and (c), respectively; power is the unshaded portion in the rejection region of Figures 4.12(b) and (c). The three situations illustrated in Figure 4.12 confirm what we alluded to earlier, that is, that the probability of a Type II error β decreases (and hence power increases) the further μ lies away from the hypothesized mean under H_0.

We can readily calculate β for a test concerning μ if we adopt the following notation.

μ_0, μ_a   Let μ_0 denote the hypothesized mean under H_0 and let μ_a denote the actual mean under the research hypothesis. The procedure for calculating β is then as summarized below. Although we never really know the actual mean, we can calculate β for any specified value of μ. The decision whether or not to accept H_0 depends on the magnitude of β for

FIGURE 4.12

The Probability β of a
Type II Error

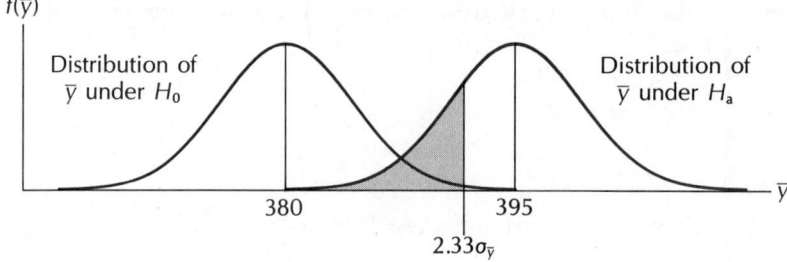

(a) β when H_a is $\mu = 395$

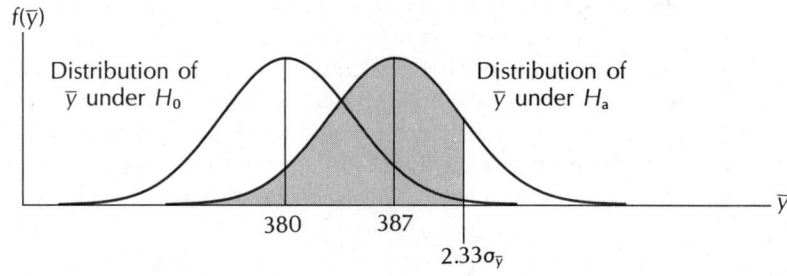

(b) β when H_a is $\mu = 387$

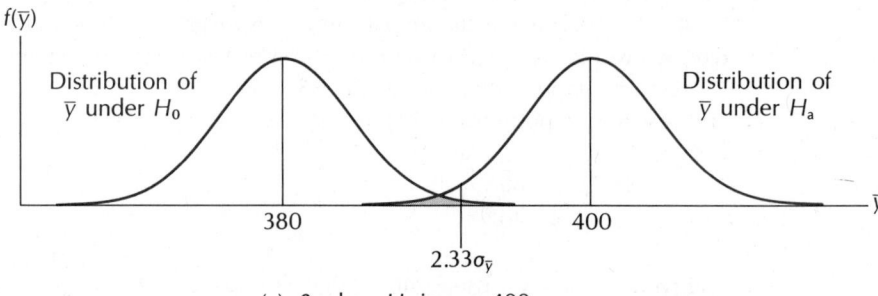

(c) β when H_a is $\mu = 400$

one or more reasonable alternative values. For a one-tailed test of H_0: $\mu = \mu_0$, β is the probability that z is less than

$$z_\alpha - \frac{|\mu_0 - \mu_a|}{\sigma_{\bar{y}}}$$

This probability is written as

$$P\left[z < z_\alpha - \frac{|\mu_0 - \mu_a|}{\sigma_{\bar{y}}}\right]$$

Formulas for β are given here for one- and two-tailed tests. Examples using these formulas follow.

CALCULATION OF
β FOR H_0: $\mu = \mu_0$
WHEN μ_a IS THE
ACTUAL MEAN

1. One-tailed test:

$$\beta = P\left[z < z_\alpha - \frac{|\mu_0 - \mu_a|}{\sigma_{\bar{y}}} \right]; \qquad \text{power} = 1 - \beta$$

2. Two-tailed test:

$$\beta \approx P\left[z < z_{\alpha/2} - \frac{|\mu_0 - \mu_a|}{\sigma_{\bar{y}}} \right]; \qquad \text{power} = 1 - \beta$$

EXAMPLE 4.9

Compute β and the power for the test in Example 4.8 if the actual mean live weight of steers is 395.

Solution

The research hypothesis for Example 4.8 was H_a: $\mu > 380$. Using $\alpha = .01$ and the computing formula for β with $\mu_0 = 380$ and $\mu_a = 395$, we have

$$\beta = P\left[z < z_{.01} - \frac{|\mu_0 - \mu_a|}{\sigma_{\bar{y}}} \right] = P\left[z < 2.33 - \frac{|380 - 395|}{35.2/\sqrt{50}} \right]$$

$$= P[z < 2.33 - 3.01] = P[z < -.68]$$

Referring to Table 2 in the Appendix, the area corresponding to $z = .68$ is .2483. Hence $\beta = .2483$ and power $= 1 - .2483 = .7517$.

Previously, when $\bar{y}$ did not fall in the rejection region, we concluded that there was insufficient evidence to reject H_0 because β was unknown. Now when $\bar{y}$ falls in the acceptance region, we can compute β corresponding to one (or more) alternative values for μ that appear reasonable in light of the experimental setting. Then provided we are willing to tolerate a probability of falsely accepting the null hypothesis equal to the computed value of β for the alternative value(s) of μ considered, our decision is to accept the null hypothesis. Thus in Example 4.9, if we are willing to risk a β error of about .25 of falsely accepting the null hypothesis, we would accept the null hypothesis $\mu = 380$.

EXAMPLE 4.10

Prospective salespeople for an encyclopedia company are now being offered a sales training program. Previous data indicate that the average number of sales per month for those who do not participate in the program is 33. To determine whether the training program is effective or not, a random sample of 35 new employees is given the sales training and then sent out into the field. One month later the mean and standard deviation for the number of sets of encyclopedias sold are 35 and 8.4, respectively. Do the data present sufficient evidence to indicate that the training program enhances sales? Use $\alpha = .05$.

Solution

The five parts to our statistical test are as follows:

H_0: $\mu = 33$

H_a: $\mu > 33$

T.S.: $z = \dfrac{\bar{y} - \mu_0}{\sigma_{\bar{y}}} \approx \dfrac{35 - 33}{8.4/\sqrt{35}} = 1.41$

R.R.: For $\alpha = .05$ we will reject the null hypothesis if $z > z_{.05} = 1.645$

Conclusion: Since the observed value of z does not fall into the rejection region, we reserve judgment on accepting H_0 until we calculate β. That is, we conclude that there is insufficient evidence to reject the null hypothesis that persons on the sales program have the same mean number of sales per month as those not under the program.

EXAMPLE 4.11

Refer to Example 4.10. Suppose that the encyclopedia company thinks that the cost of financing the sales program will be offset by increased sales if those on the program average 38 sales per month. Compute β for $\mu_a = 38$ and, based on the value of β, indicate whether you would accept the null hypothesis.

Solution

Using the computational formula for β with $\mu_0 = 33$, $\mu_a = 38$, and $\alpha = .05$, we have

$$\beta = P\left[z < z_{.05} - \frac{|\mu_0 - \mu_a|}{\sigma_{\bar{y}}} \right] = P\left[z < 1.645 - \frac{|33 - 38|}{8.4/\sqrt{35}} \right] = P[z < -1.88]$$

The area corresponding to $z = 1.88$ in Table 2 of the Appendix is .0301. Hence

$\beta = .0301$; power $= 1 - .0301 = .9699$

Because β is relatively small, we accept the null hypothesis and conclude that the training program has not increased the average sales per month above the point where increased sales would offset the cost of the training program.

In Section 4.2 we discussed how we measure the goodness of interval estimates. The goodness of a statistical test can be measured by the magnitudes of the Type I and Type II errors, α and β. When α is preset at a tolerable level by the experimenter, β is a function of the sample size for a fixed value of μ_a. The larger the sample size, the more information we have concerning μ and hence the smaller the value of β. We will consider now the problem of designing an experiment for testing H_0: $\mu = \mu_0$ when α is specified and β is preset for a fixed actual value μ_a. This problem reduces to determining the sample size needed for the fixed values of α and β.

4.6   CHOOSING THE SAMPLE SIZE FOR TESTING μ

The quantity of information available for a statistical test about μ is measured by the magnitudes of the Type I and II error probabilities, α and β. Suppose that we are interested in testing

H_0: $\mu = \mu_0$

against a one-sided alternative

$$H_a: \mu > \mu_0$$

In addition, suppose that we want the probability of a Type I error to be α and the probability of a Type II error to be β or less when the actual value of μ lies a distance of Δ (delta) or more above μ_0. The sample size necessary to meet these requirements is shown here:

SAMPLE SIZE FOR A ONE-SIDED TEST OF μ

$$n = \sigma^2 \frac{(z_\alpha + z_\beta)^2}{\Delta^2}$$

Note: If σ^2 is unknown, substitute an estimated value to get an approximate sample size.

The same formula applies to the one-sided alternative $H_a: \mu < \mu_0$, with the exception that we want the probability of a Type II error to be of magnitude β or less when the actual value of μ lies a distance of Δ or more below μ_0.

EXAMPLE 4.12

A cereal packager is concerned that one of their machines has a mean fill per package of more than 16 oz, the labeled net weight. While this is not bad from a public relations standpoint, it could cost the packager a great deal of money. Previous experience suggests that the standard deviation of the package fill weights is approximately .225. For

$$H_0: \mu = 16$$
$$H_a: \mu > 16$$

with $\alpha = .05$, determine the sample size required to make $\beta = .01$ or less if the actual mean is 16.1 oz or more. By putting this restriction on β, the packager is saying that she wants a very small probability of falsely accepting $H_0: \mu = 16$, when in fact the actual mean is 16.1 oz or more.

Solution

From previous data the fill weights have a standard deviation approximately equal to .225. The appropriate z-values, $z_{.05}$ and $z_{.01}$, for $\alpha = .05$ and $\beta = .01$ are 1.645 and 2.33, respectively. Using $\Delta = 16.1 - 16 = .1$, the required sample size is

$$n = \frac{(.225)^2(1.645 + 2.33)^2}{(.1)^2} = 79.99 \approx 80$$

That is, the packager must obtain a random sample of $n = 80$ cartons to conduct this test under the specified conditions.

Suppose that after obtaining the sample, the computed value of

$$z = \frac{\bar{y} - 16}{\sigma_{\bar{y}}}$$

does not fall in the rejection region. What is our conclusion? In similar situations in previous sections, our conclusion would have been that there was insufficient evidence to reject H_0. Now, however, knowing that $\beta \leq .01$ when $\mu \geq 16.1$, we would feel safe in our conclusion to accept H_0: $\mu = 16$. No further testing would be required.

With a slight modification of the sample size formula for the one-tailed tests, we can test

H_0: $\mu = \mu_0$
H_a: $\mu \neq \mu_0$

for a specified α and β with $\Delta = |\mu - \mu_0|$. A formula for an approximate sample size when testing μ is presented here.

APPROXIMATE SAMPLE SIZE FOR A TWO-SIDED TEST OF H_0: $\mu = \mu_0$

$$n = \frac{\sigma^2}{\Delta^2} (z_{\alpha/2} + z_{\beta})^2$$

Note: If σ^2 is unknown, substitute an estimated value to get an approximate sample size.

EXERCISES

4.21   Consider the data of Example 4.12 and compute the sample size required for testing H_0: $\mu = 16$ against H_a: $\mu \neq 16$ for $\alpha = .05$ and $\beta \leq .01$ when the actual value of μ lies more than .1 unit away from $\mu_0 = 16$.

4.22   The administrator of a nursing home would like to do a time and motion study of staff time spent per day performing nonemergency-type chores. In particular she would like to test the null hypothesis H_0: $\mu = 16$ (person-hours per day) against H_a: $\mu < 16$. The value of 16 arose from a previous study prior to the introduction of some efficiency measures. How many days must be sampled to test the proposed hypothesis if $\alpha = .05$ and $\beta \leq .10$ when the actual value of μ is 12 hours (a 25% decrease from previous results) or less? Assume $\sigma^2 = 7.64_2$.

4.23   Refer to Exercise 4.22. Determine the sample size for testing H_0: $\mu = 16$ if $\alpha = .05$ and the power of detecting a mean of 13 or less is .80 or more.

4.24   A random sample of 50 measurements from a population yielded $\bar{y} = 40.1$ and $s = 5.6$. Use these data to test the null hypothesis H_0: $\mu = 38$ against the alternative hypothesis H_a: $\mu > 38$. Use $\alpha = .05$ and draw a conclusion. Could you have made a Type II error in this situation? Explain.

4.25   For the same data in Exercise 4.24, determine the power of rejecting H_0: $\mu = 38$ given that the alternative hypothesis is true and $\mu_a = 40$. Do the same for $\mu_a = 42$ and 44 in order to sketch the power of this test for the various alternatives.

4.26   The increase in exercise capacity (in minutes) was recorded for each of 90 adult male patients following treatment for congestive heart failure. Given that the sample results yield $\bar{y} = 2.2$ and $s = 1.05$, use these data to test the null hypothesis H_0: $\mu = 2.0$ versus H_a: $\mu > 2.0$. Use $\alpha = .05$ to draw a conclusion.

4.27   Refer to Exercise 4.26. Sketch a power curve (power versus μ_a) for this test based on $\mu_a = 2.1, 2.2, 2.3,$ and 2.5.

4.7   THE LEVEL OF SIGNIFICANCE OF A STATISTICAL TEST

In Section 4.6 we introduced hypothesis testing along rather traditional lines: we defined the parts of a statistical test along with the two types of errors and their associated probabilities α and β. In recent years many statisticians and other users of statistics have objected to this decision-based approach to hypothesis testing. Rather than running a statistical test with a preset value of α, they argue, we should specify the null and alternative hypotheses, collect the sample data, and determine the weight of the evidence for rejecting the null hypothesis. This weight, given in terms of a probability, is called the

level of significance (or *p*-value) of the statistical test.

level of
significance
We illustrate the calculation of a level of significance with an example.

EXAMPLE 4.13   Refer to Example 4.8.

a.   Rather than specifying a preset value for α, determine the level of significance for the statistical test.

b.   How would the level of significance change if $\bar{y}$ had been 397 rather than 390?

Solution   (a) The null and alternative hypotheses are

$$H_0: \mu = 380$$
$$H_a: \mu > 380$$

From the sample data, the computed value of the test statistic is

$$z = \frac{\bar{y} - 380}{s/\sqrt{n}} = \frac{390 - 380}{35.2/\sqrt{50}} = 2.01$$

The level of significance for this test (i.e., the weight of evidence for rejecting H_0) is the probability of observing a value of $\bar{y}$ greater than 390 assuming the null hypothesis is true. This value can be computed by using the z-value of the test statistic, 2.01, and referring to Table 2 in the Appendix to determine the probability of observing a z-value greater than 2.01. This probability, which is sometimes designated by the

FIGURE 4.13

Level of Significance
for Example 4.13

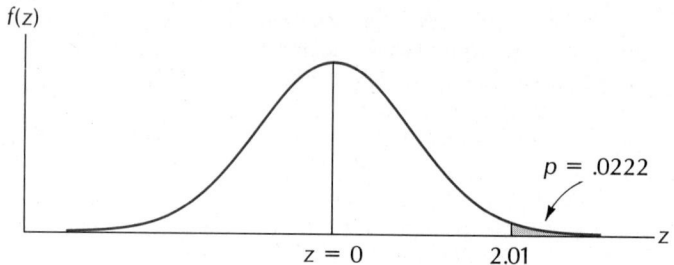

$f(z)$

$p = .0222$

$z = 0$ 2.01

z

p

letter p, is seen to be .0222. This value is shown by the shaded area in Figure 4.13. Thus we would say that the level of significance for this test is .0222.

(b) For $\bar{y} = 397$ the corresponding value of the z statistic is 3.42. Since the largest value of z in Table 2 of the Appendix is 3.09, the p-value is less than .001. We show this as $p < .001$.

As we can see from Example 4.13, the level of significance represents the probability of observing a sample outcome more contradictory to H_0 than the observed sample result. *The smaller the value of this probability, the heavier the weight of the sample evidence for rejecting H_0.* For example, a statistical test with a level of significance of $p = .01$ shows more evidence for the rejection of H_0 than does another statistical test with $p = .20$.

Suppose the null and alternative hypotheses in Example 4.13 had been

H_0: $\mu = 380$
H_a: $\mu < 380$

and the computed value of z had been $z = -2.01$. The level of significance would still be $p = .0222$ (see Figure 4.14).

p-value for one-tailed test

To summarize, the **level of significance for a one-tailed test** can be computed as follows:

For H_a: $\mu > \mu_0$

$$p = P[z > \text{computed } z]$$

FIGURE 4.14

Level of Significance
for H_0: $\mu = 380$,
H_a: $\mu < 380$ and
$z = -2.01$

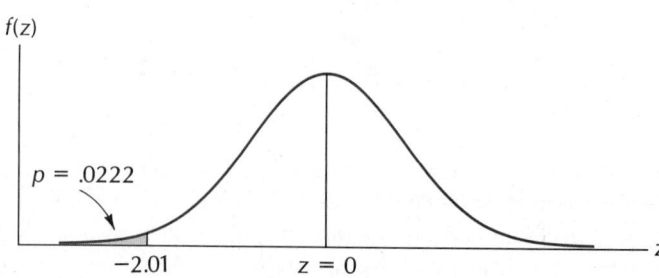

$f(z)$

$p = .0222$

-2.01 $z = 0$

z

For $H_a: \mu < \mu_0$

$$p = P[z < \text{computed } z]$$

p-value for two-tailed test   For two-tailed tests (as determined by the form of H_a), we still compute the probability of obtaining a sample outcome more contradictory to H_0 than the observed result, but the level of significance is commonly taken to be twice this probability. For a two-tailed test, the level of significance can be written as

$$p = 2P[z > |\text{computed } z|]$$

EXAMPLE 4.14   Determine the level of significance for the data of Example 4.8 if the null hypothesis and alternative hypothesis are

$H_0: \mu = 380$

$H_a: \mu \neq 380$

Solution   The computed value of z is 2.01. Since the probability of observing a value of z greater than 2.01 is .0222, the level of significance for the two-tailed statistical test is $p = 2(.0222) = .0444$.

There is much to be said in favor of this approach to hypothesis testing. Rather than reaching a decision directly, the statistician (or person performing the statistical test) presents the experimenter with the weight of evidence for rejecting the null hypothesis. The experimenter can then draw his or her own conclusion. Some experimenters will reject a null hypothesis if $p = .10$, whereas others will require $p < .05$ or $p < .01$ for rejecting the null hypothesis. The experimenter is left to make the decision based on what he or she believes is enough evidence to indicate rejection of the null hypothesis.

Many professional journals have followed this approach by reporting the results of a statistical test in terms of its level of significance. Thus we might read that a particular test was significant at the $p = .05$ level or perhaps the $p < .01$ level. By reporting results this way, the reader is left to draw his or her own conclusion.

One word of warning is needed here. The p-value of .05 has become a magic level, and many seem to feel that a particular null hypothesis should not be rejected unless the test achieves the .05 level or lower. This has resulted in part from the decision-based approach with α preset at .05. Try not to fall into this trap when reading journal articles or reporting the results of your statistical tests. After all, statistical significance at a particular level does not dictate importance or practical significance. Rather, it means that a null hypothesis can be rejected with a specified low risk of error. For example, suppose that a company is interested in determining whether the average number of miles driven per car per month for the sales force has risen above 2600. Sample data from 400 cars show that $\bar{y} = 2640$ and $s = 35$. For these data the z statistic for $H_0: \mu = 2600$ is $z = 22.86$ based on $\sigma = 35$; the level of significance is $p < .0000000001$. Thus even though there has only been a 1.5% increase in the average monthly miles driven for each car, the result is (highly) statistically significant. Is this increase of any practical significance? Probably not. What we have done is proved *conclusively* that the mean μ has increased slightly.

Throughout the text we will conduct statistical tests from both the decision-based approach and from the level-of-significance approach to familiarize you with both avenues of thought. For either approach, remember to consider the practical significance of your findings after drawing conclusions based on the statistical test.

4.8   INFERENCES ABOUT μ, σ UNKNOWN

The estimation and test procedures about μ presented earlier in this chapter were based on the assumption that the population variance was known or that we had enough observations to allow s to be substituted for σ. In this section we will present a test that can be applied when σ is unknown, no matter what the sample size. For example, in determining the average concentration of a drug in the bloodstream one hour after patients suffering from a rare disease are treated with the drug, it might be impossible to obtain a random sample of 30 or more observations at a given time. What test procedure could be used in order to make inferences about μ?

W. S. Gosset faced a similar problem around the turn of the century. As a chemist for Guinness Breweries, he was asked to make judgments on the mean quality of various brews, but was not supplied with large sample sizes to reach his conclusions.

Gosset thought that when he used the test statistic

$$z = \frac{\bar{y} - \mu_0}{\sigma/\sqrt{n}}$$

with σ replaced by s for small sample sizes, he was falsely rejecting the null hypothesis $H_0: \mu = \mu_0$ at a much higher rate than that specified by α. This problem intrigued him, and he set out to derive the distribution and percentage points of the test statistic

$$\frac{\bar{y} - \mu_0}{s/\sqrt{n}}$$

for $n < 30$.

For example, suppose an experimenter sets α at a nominal level, say .05. Then she expects falsely to reject the null hypothesis approximately 1 time in 20. However, Gosset proved that the actual probability of a Type I error for this test was somewhat higher than the nominal level designated by α. He published the results of his study under the pen name Student, because it was against company policy for him to publish his results in his own name at that time. The quantity

$$\frac{\bar{y} - \mu_0}{s/\sqrt{n}}$$

Student's t   is called the t statistic and its distribution is called the *Student's t distribution* or, simply, **Student's t** (see Figure 4.15).

FIGURE 4.15

A t Distribution with
a Normal Distribution
Superimposed

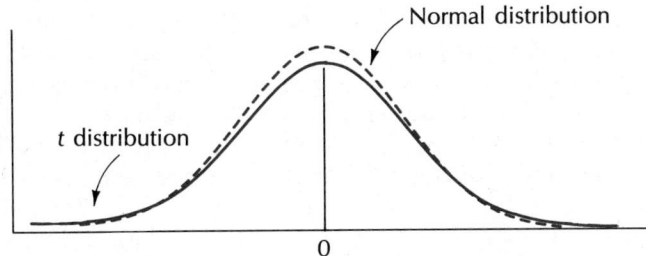

Although the quantity

$$\frac{\bar{y} - \mu_0}{s/\sqrt{n}}$$

will possess a t distribution only when the sample is selected from a normal population, the t distribution provides a reasonable approximation to the distribution of

$$\frac{\bar{y} - \mu_0}{s/\sqrt{n}}$$

when the sample is selected from a population with a mound-shaped distribution. We summarize the properties of t below.

**PROPERTIES OF
STUDENT'S t
DISTRIBUTION**

1. The t distribution, like that of z, is symmetrical about 0.
2. The t distribution is more variable than the z distribution (see Figure 4.15).
3. There are many different t distributions. We specify a particular one by a parameter called the *degrees of freedom* (df). Thus we specify

$$t = \frac{\bar{y} - \mu_0}{s/\sqrt{n}} \qquad df = n - 1$$

4. As n (or equivalently df) increases, the distribution of t approaches the distribution of z.

The phrase "degrees of freedom" sounds awfully mysterious, but the idea will eventually become second nature to you. The technical definition requires advanced mathematics, which we will avoid; on a less technical level the basic idea is that degrees of freedom are pieces of information for estimating σ using s. The standard deviation s for a sample of n measurements is based on the deviations $y_i - \bar{y}$. Because $\sum (y_i - \bar{y}) = 0$ always, if $n - 1$ of the deviations are known, the last (nth) is fixed mathematically to make the sum equal 0. It is therefore noninformative. So, in a sample of n measurements there are $n - 1$ pieces of information (degrees of freedom) about σ.

Because of the symmetry of t, only upper-tail percentage points (probabilities or areas) of the distribution of t have been tabulated; these appear in Table 4 in the Appendix. The degrees of freedom (df) are listed along the left column of the page. An entry in the table **ta** specifies a value of t, say **ta** t_a, such that an area "a" lies to its right. See Figure 4.16. Various values of "a" appear across the top of Table 4 in the Appendix. Thus, for example, with df = 7, the value of t with an area .05 to its right is 1.895 (found in the a = .05 column and df = 7 row).

We can use the t distribution to make inferences about a population mean μ. The sample test concerning μ is summarized next. The only difference between the z test discussed earlier in this chapter and the test given here is that t replaces z. The t test (rather than the z test) should be used any time σ is unknown and the distribution of y-values is mound-shaped.

STATISTICAL TEST ABOUT μ, σ UNKNOWN

H_0: $\mu = \mu_0$

H_a: 1. $\mu > \mu_0$
     2. $\mu < \mu_0$
     3. $\mu \neq \mu_0$

T.S.: $t = \dfrac{\bar{y} - \mu_0}{s/\sqrt{n}}$

R.R.: For a probability α of a Type I error and df = $n - 1$,
   1. reject H_0 if $t > t_\alpha$
   2. reject H_0 if $t < -t_\alpha$
   3. reject H_0 if $|t| > t_{\alpha/2}$

Recall that a denotes the area in the tail of the t distribution. For a one-tailed test with the probability of a Type I error equal to α, we locate the rejection region using the value from Table 4 in the Appendix, for a = α and df = $n - 1$. But for a two-tailed test we would use the t-value from Table 4 corresponding to a = $\alpha/2$ and df = $n - 1$.

Thus for a one-tailed test we reject the null hypothesis if the computed value of t is greater than the t-value from Table 4 in the Appendix, and a = α and df = $n - 1$. Similarly, for a two-tailed test we reject the null hypothesis if $|t|$ is greater than the t-value from Table 4 for a = $\alpha/2$ and df = $n - 1$.

FIGURE 4.16

Illustration of Area Tabulated in Table 4 in the Appendix for the t Distribution

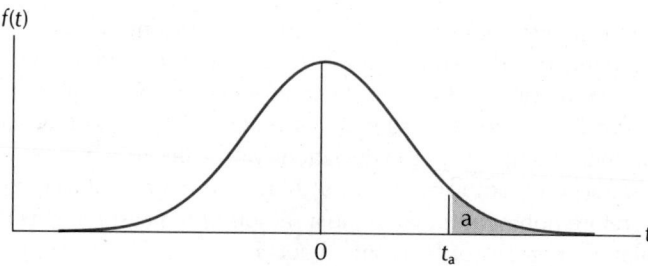

EXAMPLE 4.15

A tire company guarantees that a particular brand of tire has a mean useful lifetime of 42,000 miles or more. A consumer test agency, wishing to verify this claim, observed $n = 10$ tires on a test wheel that simulated normal road conditions. The lifetimes (in thousands of miles) were as follows:

| 42 | 36 | 46 | 43 | 41 | 35 | 43 | 45 | 40 | 39 |

Use these data to determine whether there is sufficient evidence to contradict the manufacturer's claim. Set $\alpha = .05$.

Solution

The null and research hypotheses for this example are

$$H_0: \mu = 42$$

and

$$H_a: \mu < 42$$

Note that we are giving the manufacturer the benefit of the doubt by setting $\mu = 42$ for H_0.

Before setting up the test statistic and rejection region, we must first compute the sample mean and standard deviation. You can verify that

$$\sum_i y_i = 410 \qquad \text{and} \qquad \sum_i y_i^2 = 16,926$$

Then

$$\bar{y} = \frac{\sum_i y_i}{10} = \frac{410}{10} = 41$$

Similarly, substituting into the shortcut formula for s^2, we find

$$s^2 = \frac{1}{9}\left[\sum_i y_i^2 - \frac{(\sum_i y_i)^2}{10}\right] = 12.89$$

$$s = \sqrt{12.89} = 3.59$$

The test statistic, then, is

$$t = \frac{\bar{y} - \mu_0}{s/\sqrt{n}} = \frac{41 - 42}{3.59/\sqrt{10}} = -.88$$

and the rejection region is

R.R.: Reject H_0 if $t < -t_{.05}$

From Table 4 in the Appendix, the critical t-value with a $= .05$ and df $= 9$ is 1.833, so $-t_{.05}$ is -1.833. Since the observed value of t is not less than -1.833, we have insufficient evidence to indicate that the mean lifetime of this brand of tires is less than 42,000 miles.

At this point someone might suggest calculating β, the probability of a Type II error, to see whether we can accept the manufacturer's claim. Unfortunately, this is a much more difficult task for a small-sample test than it is for the large-sample test, and it is beyond the scope of this text. (If you are interested in pursuing the topic, consult *Biometrika Tables for Statisticians*, Volume I.) Our conclusion will be that there is insufficient evidence to reject the company's claim and we should continue sampling.

EXAMPLE 4.16

Refer to Example 4.15. Rather than performing the statistical test with a preset α level, give the level of significance for the test.

Solution

For the one-tailed lower-tail test, the computed t-value is $t = -.88$. If we had an entire table of t areas for each df, this would be no problem. Because of space limitations we show only a few areas (a) for each df. The best we can do for $t = -.88$ and df = 9 is to say that $p > .10$. Based on this probability, the experimenter would probably conclude that there was insufficient evidence to reject the null hypothesis. If you think that the level of significance should be given more precisely, you can refer to more detailed tables of the t distribution in the *Biometrika Tables for Statisticians* or any of several statistical software packages that compute p-values for various test procedures.

In addition to being able to run a statistical test for μ when σ is unknown, we can construct a confidence interval using t. The confidence interval for μ with σ unknown is identical to the corresponding confidence interval for μ when σ is known, with z replaced by t and σ replaced by s.

100(1 − α)% CONFIDENCE INTERVAL FOR μ, σ UNKNOWN

$$\bar{y} \pm t_{\alpha/2} \frac{s}{\sqrt{n}}$$

Note: df = $n - 1$ and the confidence coefficient is $(1 - \alpha)$

EXAMPLE 4.17

In a psychological depth-perception test, a random sample of $n = 14$ airline pilots were asked to judge the distance between two markers at the other end of a laboratory. The sample data (recorded in feet) are listed below.

2.7	2.4	1.9	2.6	2.4	1.9	2.3
2.2	2.5	2.3	1.8	2.5	2.0	2.2

Use the sample data to place a 95% confidence interval on μ, the average recorded distance for this psychological test.

Solution

Before setting up a 95% confidence interval on μ, we must compute $\bar{y}$ and s. You can verify that

$$\sum_i y_i = 31.70 \quad \text{and} \quad \sum_i y_i^2 = 72.79$$

The sample mean, variance, and standard deviation are then

$$\bar{y} = \frac{\sum_i y_i}{14} = \frac{31.70}{14} = 2.26$$

$$s^2 = \frac{1}{13}\left[72.79 - \frac{(31.7)^2}{14}\right] = .078$$

$$s = \sqrt{.078} = .28$$

Referring to Table 4 in the Appendix, the t-value corresponding to a $= .025$ and df $= 13$ is 2.160. Hence the 95% confidence interval is

$$\bar{y} \pm t_{\alpha/2}\frac{s}{\sqrt{n}} \qquad \text{or} \qquad 2.26 \pm \frac{2.160(.28)}{\sqrt{14}}$$

which is the interval $2.26 \pm .16$, or 2.10 to 2.42. Thus we are 95% confident that the interval from 2.10 to 2.42 will encompass the mean μ.

In this section, we have made the formal mathematical assumption that the population is normally distributed. *In practice, no population has exactly a normal distribution.* How does nonnormality of the population distribution affect inferences based on the t distribution?

There are two issues to consider when populations are assumed to be nonnormal. First, what kind of nonnormality is assumed? Second, what possible effects do these specific forms of nonnormality have on the t-distribution procedures? The most important deviations from normality are skewed distributions and heavy-tailed distributions. (Heavy-tailed distributions are roughly symmetric but have outliers.)

skewness and heavy tails

To evaluate the effect of nonnormality as exhibited by skewness or heavy tails, we will consider whether the t-distribution procedures are still approximately correct for these forms of nonnormality and whether there are other more efficient procedures. For example, even if a confidence interval for μ based on t gave nearly correct results for, say, a heavy-tailed population distribution, it might be possible to get a smaller confidence interval width based on a trimmed mean.

The question of approximate correctness of t procedures has been studied extensively. In general, probabilities specified by the t procedures, particularly the confidence level (for confidence intervals) and the Type I error (for statistical tests), have been found to be fairly accurate, even when the population distribution is heavy-tailed (or light-tailed). In contrast, skewness, particularly with small sample sizes, can have a nasty effect on these probabilities, particularly in one-tailed procedures. A t distribution is symmetric, of course. When the population distribution is skewed, the actual sampling distribution of a t statistic is skewed. Although this skewness decreases as the sample size increases, there is no magic sample size that completely deskews the actual sampling distribution.

sensitivity of one-tailed procedures

As a consequence, a nominal 95% confidence interval may actually have 80% or lower confidence if the sample size is in the teens and the population distribution looks like that of Figure 4.17.

FIGURE 4.17 $f_Y(y)$

Skewed Population
Distribution

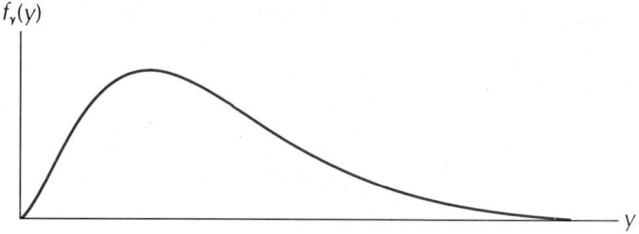

EXAMPLE 4.18   A simulation study takes 1000 samples of size 30 from a symmetric, moderately outlier-prone population. The following results are obtained:

Checking α

Simulation of one-sample t test (1000 samples)
Population shape is moderately outlier-prone

μ	σ	n
50.000	10.0000	30

one-tailed test:

number of times H_0: "mean is 50" is rejected in favor of

α	"mean > 50"	"mean < 50"	total (α doubled)
.100	104	95	199
.050	51	51	102
.025	28	24	52
.010	7	6	13
.005	4	3	7

average t is 0.0077 with variance of 1.086943

Which hypothesis is true in the simulation? Does the outlier-proneness of the population have a serious effect?

Solution   The output indicates that H_0 is $\mu = 50$, and indeed the population mean is 50. Therefore, fractions such as 104/1000 are approximating α, the probability of a Type I error; the fractions are approximations because they are based on 1000 samples, not on an infinite number. Notice that all the fractions are very close to the nominal α values. For example, with a one-tailed α of .025, the observed fractions are .028 and .024.

EXAMPLE 4.19   A simulation study chooses samples of various sizes from a population that is strongly right-skewed. The following results are obtained for a t test:

	μ	σ	n
	50.000	10.0000	10

one-tailed test:

number of times H_0: "mean is 50" is rejected in favor of

α	"mean > 50"	"mean < 50"	total (α doubled)
.100	27	308	335
.050	6	249	255
.025	2	210	212
.010	0	176	176
.005	0	155	155

	μ	σ	n
	50.000	10.0000	30

one-tailed test:

number of times H_0: "mean is 50" is rejected in favor of

α	"mean > 50"	"mean < 50"	total (α doubled)
.100	40	237	277
.050	6	189	195
.025	0	156	156
.010	0	120	120
.005	0	99	99

	μ	σ	n
	50.000	10.0000	60

one-tailed test:

number of times H_0: "mean is 50" is rejected in favor of

α	"mean > 50"	"mean < 50"	total (α doubled)
.100	53	190	243
.050	9	145	154
.025	2	114	116
.010	1	70	71
.005	0	59	59

What effect does the skewness of this population have? What happens as the sample size increases?

Solution

The skewness causes the nominal α probabilities to be seriously wrong, especially for one-tailed tests. The effect is most severe for the smallest n, 10, but it is still severe when n is 60.

The second question, that of the efficiency of t procedures, has only recently been studied seriously. The near-unanimous conclusion from these studies, when the population distribution is symmetric but heavy-tailed, is that **robust methods** such as the Wilcoxon rank sum test discussed in Chapter 5 are more efficient than the corresponding t test for μ. Therefore, the robust procedures tend to have more accurate estimates with smaller

robust methods

standard errors. Unfortunately, less work has been done on the effectiveness of these robust procedures in cases where the population distribution is skewed.

plotting data   So what is a nonexpert to do? First, look at the data. A simple histogram of the data, or some other plotting device, will reveal any gross skewness or extreme outliers. If there's no blatant nonnormality, the nominal t-distribution probabilities should be reasonably correct and the t procedure should be reasonably efficient. If the data values are obviously skewed or heavy-tailed, the t-distribution probabilities and the efficiency of the t procedure are highly suspect. In these situations you may wish to consult other textbooks (such as Hildebrand and Ott, 1987) in which alternative robust procedures are presented. You should at least develop a healthy skepticism toward stated probabilities or confidence levels based on t methods.

Other robust procedures will be mentioned in this text, but we cannot do complete justice to them. We expect that these procedures will be integrated into some of the statistical software systems in the next few years. If such programs are not available, at least a manager can cultivate an alert skepticism about the accuracy of the stated probabilities.

EXERCISES

4.28   A random sample of 10 students in a fourth-grade reading class were thoroughly tested to determine reading speed and reading comprehension. Based on a fixed-length standardized test reading passage, the following speeds (in minutes) and comprehension scores (based on a 100-point scale) were obtained.

Student	1	2	3	4	5	6	7	8	9	10
Reading speed	5	7	15	12	8	7	10	11	13	9
Reading comprehension	60	76	96	100	81	75	85	88	98	83

a. Use the reading speed data to place a 95% confidence interval on μ, the average speed for all fourth-grade students in the large school from which the sample was drawn.
b. Interpret the interval estimate in part (a).
c. How would your inference change by using a 98% confidence interval?

4.29   Refer to Exercise 4.28. Using the reading comprehension data, test the research hypothesis that the mean for all fourth graders on the standardized examination is greater than 80, the statewide average for comparable students the previous year. Give the level of significance for your test. Interpret your findings.

4.30   Refer to Exercise 4.29.

a. Set up all parts for a statistical test of the research hypothesis that the mean score for all fourth graders is different from 80, the statewide average the previous year.
b. Give the level of significance for this test.

4.31   Refer to Exercise 4.28. Use a computer program to construct a 90% confidence interval for the mean total reading score (speed plus comprehension). A Minitab program is shown on the next page for illustrative purposes.

4.32   The amount of sewage and industrial pollutants dumped into a body of water affects the health of the water by reducing the amount of dissolved oxygen available for aquatic life.

```
MTB > READ C1,C2
DATA> 5 60
DATA> 7 76
DATA> 15 96
DATA> 12 100
DATA> 8 81
DATA> 7 75
DATA> 10 85
DATA> 11 88
DATA> 13 98
DATA> 9 83
DATA> END DATA
     10 ROWS READ
MTB > ADD C1 TO C2, PUT IN C3
MTB > TINTERVAL WITH 90 PERCENT CONFIDENCE,DATA IN C3

                N      MEAN    STDEV  SE MEAN   90.0 PERCENT C.I.
C3             10     93.90    15.14    4.79  (   85.12,  102.68)

MTB > PRINT C1,C2,C3
  ROW    C1     C2     C3

    1     5     60     65
    2     7     76     83
    3    15     96    111
    4    12    100    112
    5     8     81     89
    6     7     75     82
    7    10     85     95
    8    11     88     99
    9    13     98    111
   10     9     83     92

MTB >
```

Suppose that weekly readings are taken from the same location in a river over a two-month period. Use the summary data from the computer printout given below to conduct a statistical test of the research hypothesis that the mean dissolved oxygen content is less than 5.0 parts per million, a level some scientists think is marginal for supplying enough dissolved oxygen for fish.

5.100000000
4.900000000
5.600000000
4.200000000
4.800000000
4.500000000
5.300000000
5.200000000

8.000000000   sample size
4.950000000 $\bar{y}$
.2028571428 s^2
.4503966505 s

| 4.9 | SUMMARY |

A population mean can be estimated using point or interval estimation. The goodness of an interval estimate is given by the width of the interval and the confidence coefficient. The general formula for a $100(1 - \alpha)\%$ confidence interval for μ was given along with the sample size formula for planning a study to give a confidence interval of fixed width for μ.

Following the traditional approach to hypothesis testing, a statistical test about μ is composed of five parts: null hypothesis, research hypothesis, test statistic, rejection region, and conclusion. It employs the technique of proof by contradiction. We try to verify the research hypothesis by gathering information to contradict the null hypothesis H_0: $\mu = \mu_0$. As with any two-decision problem, there are two types of errors that can be committed, the rejection of H_0 when H_0 is true—a Type I error—and the acceptance of H_0 when H_0 is false and some alternative is true—a Type II error. The probabilities for these errors, designated by α and β, measure the goodness of the test procedure.

In this chapter we indicated that for a given sample size, α and β are inversely related; as α is increased, β is decreased, and vice versa. If we specify n and α for a given test procedure, we can compute β for alternative values of μ. Sometimes we may wish to specify both α and β *prior* to conducting the investigation. To do this, we determine the sample size required for the specific values of α and β.

We considered an alternative to the traditional decision-based approach for a statistical test of a hypothesis. Rather than relying on a preset level of α, we compute the weight of evidence for rejecting the null hypothesis. This weight, expressed in terms of a probability, is called the level of significance for the test. Most professional journals summarize the results of a statistical test using the level of significance.

The final topic discussed in this chapter concerned inferences about μ when σ is unknown (which is almost always). Through the use of the t distribution we can construct both confidence intervals and a statistical test for μ. Since the t-values of the t distribution approach the z-values of a normal distribution and since σ is almost never known, it is convenient to use t results for all inferences about μ (large or small sample).

KEY FORMULAS   Estimation and tests for μ

1. $100(1 - \alpha)\%$ confidence interval for μ (σ known)

$$\bar{y} \pm z_{\alpha/2}\sigma/\sqrt{n}$$

2. $100(1 - \alpha)\%$ confidence interval for μ (σ unknown)

$$\bar{y} \pm t_{\alpha/2}s/\sqrt{n}, \qquad df = n - 1$$

3. Sample size for estimating μ with a $100(1 - \alpha)\%$ confidence interval, $\bar{y} \pm E$

$$n = \frac{z_{\alpha/2}^2\sigma^2}{E^2}$$

4. Statistical test for μ (σ known)

$H_0: \mu = \mu_0$

T.S.: $z = \dfrac{\bar{y} - \mu_0}{\sigma_{\bar{y}}}$ where $\sigma_{\bar{y}} = \sigma/\sqrt{n}$

5. Statistical test for μ (σ unknown)

$H_0: \mu = \mu_0$

T.S.: $t = \dfrac{\bar{y} - \mu_0}{s/\sqrt{n}}$, df $= n - 1$

6. Calculation of β (and equivalently power) for a test on μ
   a. One-tailed test

$$\beta = P\left(z < z_\alpha - \frac{|\mu_0 - \mu_a|}{\sigma/\sqrt{n}} \right)$$

   b. Two-tailed test

$$\beta \approx P\left(z < z_{\alpha/2} - \frac{|\mu_0 - \mu_a|}{\sigma/\sqrt{n}} \right)$$

7. Sample size for a statistical test on μ
   a. One-tailed test

$$n = \sigma^2 \frac{(z_\alpha + z_\beta)^2}{\Delta^2}$$

   b. Two-tailed test

$$n \approx \sigma^2 \frac{(z_{\alpha/2} + z_\beta)^2}{\Delta^2}$$

EXERCISES

4.33 To test the effectiveness of a new spray for controlling rust mites, we would like to compare the average yield for treated groves with the average yield for untreated groves displayed in previous years. A random sample of 30 one-acre groves is chosen and sprayed according to a recommended schedule. The average yield for the 30-grove sample was 830 boxes, with a standard deviation of 91. Yields from groves in the same area without rust mite maintenance spraying have averaged 760 boxes over previous years. Do these data present sufficient evidence to indicate that the mean yield for groves sprayed with the new preparation is higher than 760 boxes, the average over previous years without spraying? Is this a one-tailed or two-tailed test? Use $\alpha = .05$.

4.34   The board of health of a particular state was called to investigate claims that raw pollutants were being released into the river flowing past a small residential community. By applying financial pressure, the state was able to get the violating company to make major concessions towards the installation of a new water purification system. In the interim, different production systems were to be initiated to help reduce the pollution level of water entering the stream. To monitor the effect of the interim system, a random sample of 50 water specimens was taken throughout the month at a location downstream from the plant. If $\bar{y} = 5.0$ and $s = .70$, use the sample data to determine whether the mean dissolved oxygen count of the water (in ppm) is less than 5.2, the average reading at this location over the past year.

a. List the five parts of the statistical test, using $\alpha = .05$.

b. Conduct the statistical test and state your conclusion.

4.35   Refer to Figure 4.12. Compute β for (b) and (c) by using H_0: $\mu = 380$, H_a: $\mu > 380$. Recall that $n = 50$ and $s = 35.2$.

4.36   An automatic merge system has been installed at the entrance ramp to a major highway. Through the use of a series of pacer lights, drivers are told whether they are traveling at the correct speed to merge successfully onto the highway. Prior to the installation of the system, investigators measured the stress level of many drivers merging onto the highway during rush hours and found the average stress level to be 8.2 on a 10-point scale. Similar testing on a random sample of 50 is to be conducted now that the automatic merge system has been installed. If $\bar{y} = 7.6$ and $s = 1.8$ for the 50 drivers sampled, conduct a statistical test of the research hypothesis that the average stress at peak hours for drivers under the new system is less than 8.2, the average stress level prior to the installation of the automatic merge system. Determine the level of significance of the statistical test. Interpret your findings.

4.37   The search for alternatives to oil as a major source of fuel and energy will inevitably bring about many environmental challenges. These challenges will require solutions to problems in such areas as strip mining and many others. Let us focus on one. If coal is considered as a major source of fuel and energy, we will have to consider ways to keep large amounts of sulfur dioxide (SO_2) and particulates from getting into the air. This is especially important at large government and industrial operations. Here are several possibilities.

1. Build the smokestack extremely high.

2. Remove the SO_2 and particulates from the coal prior to combustion.

3. Remove the SO_2 from the gases after the coal is burned but before the gases are released into the atmosphere. This is accomplished by using a scrubber.

   Several scrubbers have been developed in recent years. Suppose that a new one has been constructed and is set for testing at a given power plant. Fifty samples are obtained at various times from gases emitted from the stack. The mean SO_2 emission is .13 lb per million Btu, with a standard deviation of .05 lb. Use the sample data to construct a statistical test of the null hypothesis H_0: $\mu = .145$, the average emission level for one of the more efficient scrubbers that has been developed. Choose an appropriate alternative hypothesis, with $\alpha = .05$.

4.38   Refer to Exercise 4.11. Construct a 99% confidence interval for μ, the average number of miles between a person's high school and present permanent address for the state sample.

4.39   Refer to Exercise 4.37. Rather than being interested in testing the research hypothesis that $\mu < .145$, the average emission level for one of the more efficient scrubbers, we may wish to estimate the mean emission level for the new scrubber. Use the sample data to construct a 99% confidence interval for μ. Interpret your results.

4.40   As part of an overall evaluation of training methods, an experiment was conducted to determine the average exercise capacity of healthy male army inductees. To do this each

EXERCISES

male in a random sample of 35 healthy army inductees exercised on a bicycle ergometer (a device for measuring work done by the muscles) under a fixed work load until he tired. Blood pressure, pulse rates, and other indicators were carefully monitored to ensure that no one's health was in danger. The exercise capacities (mean time, in minutes) for the 35 inductees are listed below.

23	19	36	12	41	43	19
28	14	44	15	46	36	25
35	25	29	17	51	33	47
42	45	23	29	18	14	48
21	49	27	39	44	18	13

a. Use these data to construct a 95% confidence interval for μ, the average exercise capacity for healthy male inductees. Interpret your findings.

b. How would your interval change using a 99% confidence interval?

4.41 Using the data of Exercise 4.40, determine the number of sample observations that would be required to estimate μ to within one minute, using a 95% confidence interval. (Hint: Substitute $s = 12.36$ for σ in your calculations.)

4.42 A study was conducted to examine the effect of a preparation of mosaic virus on tobacco leaves. In a random sample of $n = 32$ leaves, the mean number of lesions was 22, with a standard deviation of 3. Use these data and a 95% confidence interval to estimate the average number of lesions for leaves affected by a preparation of mosaic virus.

4.43 Refer to Exercise 4.42. Use the sample data to form a 99% confidence interval on μ, the average number of lesions for tobacco leaves affected by a preparation of mosaic virus.

4.44 We all remember being told, "Your fever has subsided and your temperature has returned to normal." What do we mean by the word *normal*? Most people use the benchmark 98.6 Fahrenheit, but this does not apply to all people, only the "average" person. Without putting words into someone's mouth, we might define a person's normal temperature to be his or her average temperature when healthy. But even this definition is cloudy because there is variation in a person's temperature throughout the day. To determine a subject's normal temperature, we recorded it for a random sample of 30 days. On each day selected for inclusion in the sample, the temperature reading was made at 7 AM. The sample mean and standard deviation for these 30 readings were, respectively, 98.4 and .15. Assuming the subject was healthy on the days examined, use these data to estimate the person's 7 AM "normal" temperature using a 90% confidence interval.

4.45 Refer to the data of Exercise 4.40. Suppose that the random sample of 35 inductees was selected from a large group of new army personnel being subjected to a new (and hopefully improved) physical fitness program. Assume previous testing with several thousand personnel over the past several years has shown an average exercise capacity of 29 minutes. Run a statistical test for the research hypothesis that the average exercise capacity is improved for the new fitness program. Give the level of significance for the test. Interpret your findings.

4.46 Refer to Exercise 4.45.

a. How would the research hypothesis change if we were interested in determining whether the new program is better or worse than the physical fitness program for inductees?

b. What is the level of significance for your test?

4.47 In a random sample of 40 hospitals from a list of hospitals with over 100 semiprivate beds, a researcher collected information on the proportion of persons whose bills are covered by a group policy under a major medical insurance carrier. The sample proportions are given in the following chart.

.67	.74	.68	.63	.91	.81	.79	.73
.82	.93	.92	.59	.90	.75	.76	.88
.85	.90	.77	.51	.67	.67	.92	.72
.69	.73	.71	.76	.84	.74	.54	.79
.71	.75	.70	.82	.93	.83	.58	.84

Use the sample data to construct a 90% confidence interval on μ, the average proportion of patients per hospital with group medical insurance coverage.

4.48 Refer to Exercise 4.47. Use the same data to construct a 99% confidence interval.

4.49 Faculty members in a state university system who resign within 10 years of initial employment are entitled to receive the money paid into a retirement system, plus 4% per annum. Unfortunately, experience has shown that the state is extremely slow in returning this money. Concerned about such a practice, a local teachers' organization decides to investigate. From a random sample of 50 employees who resigned from the state university system over the past five years, the average time between the termination date and reimbursement was 75 days, with a standard deviation of 15 days. Use the data to estimate the mean time to reimbursement, using a 95% confidence interval.

4.50 Refer to Exercise 4.49. After a confrontation with the teachers' union, the state promised to make reimbursements within 60 days. Monitoring of the next 40 resignations yields an average of 58 days, with a standard deviation of 10 days. If we assume that these 40 resignations represent a random sample of the state's future performance, estimate the mean reimbursement time, using a 99% confidence interval.

4.51 Refer to Example 4.11. Compute β for $\mu_a = 40$. What would be your conclusion based on the magnitude of β?

4.52 Refer to Exercise 4.51. Using the values of β computed for $\mu_a = 38$ and $\mu_a = 40$, calculate the probability of a Type II error for several other values of μ_a in order to construct a graph of β against μ_a.

4.53 Refer to the data of Exercise 4.40. It can be shown that the sample standard deviation is 12.36. Use a computer program to construct a 90% confidence interval for μ. A Minitab program is shown here for illustration purposes.

```
          SET THE FOLLOWING DATA INTO C6
DATA> 23 19 36 12 41 43 19
DATA> 28 14 44 15 46 36 25
DATA> 35 25 29 17 51 33 47
DATA> 42 45 23 29 18 14 48
DATA> 21 49 27 39 44 18 13
DATA> END DATA
MTB > PRINT C6
C6
    23    19    36    12    41    43    19    28    14    44    15    46    36
    25    35    25    29    17    51    33    47    42    45    23    29    18
    14    48    21    49    27    39    44    18    13

MTB > ZINTERVAL 90 PERCENT CONFIDENCE,SIGMA = 12.36,DATA IN C6

THE ASSUMED SIGMA =12.4

              N    MEAN    STDEV  SE MEAN   90.0 PERCENT C.I.
C6           35    30.51   12.36    2.09  (   27.07,   33.95)

MTB >
```

4.54 Use the data of Exercise 4.47 and the Minitab output shown here to respond to the statements that follow concerning the confidence interval for μ.

```
        SET THE FOLLOWING DATA INTO C4
DATA> .67 .73 .59 .93 .92 .82 .75 .51 .81 .54
DATA> .85 .68 .76 .75 .58 .69 .92 .82 .67 .73
DATA> .71 .77 .91 .74 .88 .74 .71 .9 .83 .72
DATA> .93 .7 .67 .79 .79 .9 .63 .84 .76 .84
DATA> END DATA
MTB > PRINT C4
C4
   0.67   0.73   0.59   0.93   0.92   0.82   0.75   0.51   0.81   0.54   0.85
   0.68   0.76   0.75   0.58   0.69   0.92   0.82   0.67   0.73   0.71   0.77
   0.91   0.74   0.88   0.74   0.71   0.90   0.83   0.72   0.93   0.70   0.67
   0.79   0.79   0.90   0.63   0.84   0.76   0.84

MTB > STANDARD DEVIATION C4
    ST.DEV. =      0.10861
MTB > ZINTERVAL 98 PERCENT CONFIDENCE,SIGMA = .10861,DATA IN C4

THE ASSUMED SIGMA =0.109

                 N      MEAN     STDEV   SE MEAN    98.0 PERCENT C.I.
C4              40     0.7620    0.1086   0.0172   ( 0.7220,  0.8020)

MTB > STOP
```

a. Identify the sample mean for the data shown in the computer output.
b. What is the confidence coefficient for the confidence interval shown?
c. Give the confidence limits and interpret the interval estimate.

4.55 A random sample of birth rates from 40 inner-city areas shows an average of 35 per thousand, with a standard deviation of 6.3. Estimate the mean inner-city birth rate. Use a 95% confidence interval.

4.56 A random sample of 30 standard metropolitan statistical areas (SMSAs) was selected and the ratio (per 1000) of registered voters to the total number of persons 18 years and over was recorded in each area. Use the data below to test the research hypothesis that μ, the average ratio (per 1000), is different from 675, last year's average ratio. Give the level of significance for your test.

802	497	653	600	729	812
751	730	635	605	760	681
807	747	728	561	696	710
641	848	672	740	818	725
694	854	674	683	695	803

4.57 Improperly filled orders are a costly problem for mail-order houses. To estimate the mean loss per incorrectly filled order, a large firm plans to sample n incorrectly filled orders and to determine the added cost associated with each one. It is estimated that the added cost is between $40 and $400. How many incorrectly filled orders must be sampled to estimate the mean additional cost using a 95% confidence interval of width $20?

4.58   Records from a particular hospital were examined to determine the average length of stay for patients being treated for lung cancer. Data from a sample of 100 records showed $\bar{y} = 2.1$ months and $s = 2.6$ months.

a. Would a confidence interval for μ based on t be appropriate? Why or why not?

b. Indicate an alternative procedure for estimating the center of the distribution.

4.59   Refer to Exercise 4.20. Graph the center line and control limits for μ and plot the sequence of sample means shown here to determine if the process is out of control:

109	113	108	108	110	108	107	108	103	107
109	109

4.60   Investigators would like to estimate the average annual taxable income of apartment dwellers in a city to within $500 using a 95% confidence interval. If we assume the annual incomes for apartment dwellers have a range of $40,000, determine the number of observations that should be included in the sample.

4.61   As indicated earlier, the stated weight on the new giant-sized laundry detergent package is 42 oz. Also displayed on the box is the following statement: "Individual boxes of this product may weigh slightly more or less than the marked weight due to normal variations incurred with high-speed packaging machines, but each day's production of detergent average slightly above the marked weight." Discuss how you might attempt to test this claim. Or would it be simpler to modify this claim slightly for testing purposes? State all parts of your test. Would there be any way to determine in advance the sample size required to pick up a specified alternative with power equal to .90, using $\alpha = .05$?

4.62   Congestive heart failure is known to be fatal in a high percentage of cases. A total of 182 patients with chronic left-ventricular failure who were symptomatic in spite of therapy were followed. The length of survival for these patients ranged from 1 to 41 months with a mean of 12 months and a standard deviation of 10. Would a confidence interval for the mean survival of these patients be appropriate? Why or why not?

4.63   After a decade of steadily increasing popularity, the sales of automatic teller machines (ATMs) have been on the decline. In a recent month, a spot check of a random sample of 40 suppliers indicated that shipments averaged 20% lower than those for the corresponding period 1 year ago. Assume the standard deviation is 6.2% and the percentage data appear mound-shaped. Use these data to construct a 99% confidence interval on the mean percentage decrease in shipments of ATMs.

4.64   Suppose the percentage change in shipments of ATMs from the 40 suppliers of Exercise 4.63 ranged from -40% (a 40% decrease) to $+16\%$ (a 16% increase) with a sample mean of -20%, a median decrease of -10%, and a 10% trimmed mean of -12%. Discuss the appropriateness of the t methods for examining the percentage change in shipments of ATMs.

4.65   Doctors have recommended that we try to keep our caffeine intake at 200 mg or less per day. With the following chart, a sample of 35 office workers were asked to record their caffeine intake for a seven-day period.

coffee (6 oz)	100–150 mg
tea (6 oz)	40–110 mg
cola (12 oz)	30 mg
chocolate cake	20–30 mg
cocoa (6 oz)	5–20 mg
milk chocolate (1 oz)	5–10 mg

After the seven-day period, the average daily intake was obtained for each worker. The sample mean and standard deviation of the daily averages were 560 and 160, respectively. Use these data to estimate μ, the average daily intake, using a 90% confidence interval.

4.66 Refer to Exercise 4.65. How many additional observations would be needed to estimate μ to within ± 10 mg with 90% confidence?

4.67 Investigators from the Ohio Department of Agriculture recently selected a junior high school in the area and took samples of the half-pint (8-oz) milk cartons used for student lunches. Based on 25 containers, the investigators found that the cartons were .067 oz short of a full half pint on the average, with a standard deviation of .02.

a. Use these data to test the hypothesis that the average shortfall is zero against a one-sided alternative. Give the p-value for your test.

b. Although .067 oz is only a few drops, predict the annual savings (in pints) for the dairy if it sells 3 million 8-oz cartons of milk each year with this shortweight.

5 INFERENCES ABOUT $\mu_1 - \mu_2$

INFERENCES ABOUT $\mu_1 - \mu_2$: INDEPENDENT SAMPLES

The inferences we have made so far have concerned a parameter from a single population. Quite often we are faced with an inference that concerns a comparison of more than one parameter from different populations. For example, we might wish to compare the mean corn crop yield for two different varieties of corn, the mean annual income for two ethnic groups, the mean nitrogen content of two different lakes, or the mean length of time between administration and eventual relief for two different antivertigo drugs.

In each of these situations, we will assume that we are sampling from two normal populations (1 and 2) with different means μ_1 and μ_2 but identical variances σ^2. We then *draw independent random samples of size n_1 and n_2.* The corresponding sample means and variances are $\bar{y}_i$ and s_i^2 ($i = 1$ or 2). Using the data from the two samples, we would like to make a comparison between the population means μ_1 and μ_2. In particular, we will estimate and test a hypothesis concerning the difference $\mu_1 - \mu_2$.

A logical point estimate for the difference in population means is the sample difference $\bar{y}_1 - \bar{y}_2$. The standard error for the difference in sample means is more complicated than for a single sample mean, but the confidence interval has the same form: point estimate $\pm t$ (standard error). A general confidence interval for $\mu_1 - \mu_2$ with confidence coefficient of $(1 - \alpha)$ is given here.

CONFIDENCE
INTERVAL
FOR $\mu_1 - \mu_2$,
INDEPENDENT
SAMPLES

$$(\bar{y}_1 - \bar{y}_2) \pm t_{\alpha/2} s_p \sqrt{\frac{1}{n_1} + \frac{1}{n_2}}$$

where

$$s_p = \sqrt{\frac{(n_1 - 1)s_1^2 + (n_2 - 1)s_2^2}{n_1 + n_2 - 2}}$$

and

$$df = n_1 + n_2 - 2$$

s_p^2, a
weighted
average

The quantity s_p in the confidence interval is an estimate of the standard deviation σ for the two populations and is formed by combining (pooling) information from the two samples. In fact, s_p^2 is a **weighted average** of the sample variances s_1^2 and s_2^2. For the special case where the sample sizes are the same ($n_1 = n_2$), the formula for s_p^2 reduces to $s_p^2 = (s_1^2 + s_2^2)/2$, the mean of the two sample variances. The degrees of freedom for the confidence interval are a combination of the degrees of freedom for the two samples; that is, $df = (n_1 - 1) + (n_2 - 1) = n_1 + n_2 - 2$.

Recall that we are assuming that the two populations from which we draw the samples have normal distributions with a common variance σ^2. If the confidence interval presented were valid only when these assumptions were met exactly, the estimation procedure would be of limited use. Fortunately, the confidence coefficient remains relatively stable if both distributions are mound-shaped and the sample sizes are approximately equal.

EXAMPLE 5.1

Company officials were concerned about the length of time a particular drug product retained its potency. A random sample, sample 1, of $n_1 = 10$ bottles of the product was drawn from the production line and analyzed for potency. A second sample, sample 2, of $n_2 = 10$ bottles was obtained and stored in a regulated environment for a period of one year.

The readings obtained from each sample are given in Table 5.1.

TABLE 5.1   Potency Reading for Two Samples

Sample 1		Sample 2	
10.2	10.6	9.8	9.7
10.5	10.7	9.6	9.5
10.3	10.2	10.1	9.6
10.8	10.0	10.2	9.8
9.8	10.6	10.1	9.9

Suppose we let μ_1 denote the mean potency for all bottles that might be sampled coming off the production line and μ_2 denote the mean potency for all bottles that may be retained for a period of one year. Estimate $\mu_1 - \mu_2$ by using a 95% confidence interval.

Solution

The necessary sample calculations from the data of Table 5.1 are presented next.

Sample 1	Sample 2
$\sum_j y_{1j} = 103.7$	$\sum_j y_{2j} = 98.3$
$\sum_j y_{1j}^2 = 1076.31$	$\sum_j y_{2j}^2 = 966.81$

Then

$$\bar{y}_1 = \frac{103.7}{10} = 10.37 \qquad\qquad \bar{y}_2 = \frac{98.3}{10} = 9.83$$

$$s_1^2 = \frac{1}{9}\left[1076.31 - \frac{(103.7)^2}{10}\right] = .105 \qquad s_2^2 = \frac{1}{9}\left[966.81 - \frac{(98.3)^2}{10}\right] = .058$$

The estimate of the common standard deviation σ is

$$s_p = \sqrt{\frac{(n_1 - 1)s_1^2 + (n_2 - 1)s_2^2}{n_1 + n_2 - 2}} = \sqrt{\frac{9(.105) + 9(.058)}{18}}$$

which, for $n_1 = n_2 = 9$, reduces to

$$s_p = \sqrt{\frac{.105 + .058}{2}} = .285$$

The t-value based on df $= n_1 + n_2 - 2 = 18$ and a $= .025$ is 2.101. A 95% confidence interval for the difference in mean potencies is

$$(10.37 - 9.83) \pm 2.101(.285)\sqrt{1/10 + 1/10} \text{ or } .54 \pm .268$$

We estimate that the difference in means, $\mu_1 - \mu_2$, lies in the interval .272 to .808.

We can also test a hypothesis about the difference between two population means. As with any test procedure, we begin by specifying a research hypothesis for the difference in population means. Thus we might, for example, specify that the difference $\mu_1 - \mu_2$ is greater than some value D_0. (Note: D_0 will often be zero.) The entire test procedure is summarized here.

A STATISTICAL TEST FOR $\mu_1 - \mu_2$, INDEPENDENT SAMPLES

H_0: $\mu_1 - \mu_2 = D_0$ (D_0 is specified)

H_a: 1. $\mu_1 - \mu_2 > D_0$
    2. $\mu_1 - \mu_2 < D_0$
    3. $\mu_1 - \mu_2 \neq D_0$

T.S.: $t = \dfrac{\bar{y}_1 - \bar{y}_2 - D_0}{s_p\sqrt{1/n_1 + 1/n_2}}$

R.R.: For a Type I error α and df $= n_1 + n_2 - 2$
    1. reject H_0 if $t > t_\alpha$
    2. reject H_0 if $t < -t_\alpha$
    3. reject H_0 if $|t| > t_{\alpha/2}$

EXAMPLE 5.2

An experiment was conducted to compare the mean number of tapeworms in the stomachs of sheep that had been treated for worms against the mean number in those that were untreated. A sample of 14 worm-infected lambs was randomly divided into two groups. Seven were injected with the drug and the remainder were left untreated. After a six-month period, the lambs were slaughtered and the following worm counts were recorded:

Drug-treated sheep	18	43	28	50	16	32	13
Untreated sheep	40	54	26	63	21	37	39

a. Test a hypothesis that there is no difference in the mean number of worms between treated and untreated lambs. Assume that the drug cannot increase the number of worms and hence use the alternative hypothesis that the mean for treated lambs is less than the mean for untreated lambs. Use $\alpha = .05$.
b. Indicate the level of significance for this test.

Solution

(a) The calculations for the samples of treated and untreated sheep are summarized below.

Drug-Treated Sheep	Untreated Sheep
$\sum_j y_{1j} = 200$	$\sum_j y_{2j} = 280$
$\sum_j y_{1j}^2 = 6906$	$\sum_j y_{2j}^2 = 12{,}492$
$\bar{y}_1 = \dfrac{200}{7} = 28.57$	$\bar{y}_2 = \dfrac{280}{7} = 40.0$
$s_1^2 = \dfrac{1}{6}\left[6906 - \dfrac{(200)^2}{7}\right]$	$s_2^2 = \dfrac{1}{6}\left[12{,}492 - \dfrac{(280)^2}{7}\right]$
$\quad = \dfrac{1}{6}[6906 - 5714.29]$	$\quad = \dfrac{1}{6}[12{,}492 - 11{,}200]$
$\quad = 198.62$	$\quad = 215.33$

Under the assumption of equal population variances, the sample variances are combined to form an estimate of the common population standard deviation σ. This assumption appears reasonable based on the sample variances.

$$s_p = \sqrt{\frac{(n_1 - 1)s_1^2 + (n_2 - 1)s_2^2}{n_1 + n_2 - 2}} = \sqrt{\frac{6(198.62) + 6(215.33)}{12}} = 14.39$$

The test procedure for the research hypothesis that the treated sheep will have a mean infestation level (μ_1) less than the mean level (μ_2) for untreated sheep is as follows:

H_0: $\mu_1 - \mu_2 = 0$ (that is, no difference in the mean infestation levels)
H_a: $\mu_1 - \mu_2 < 0$

T.S.: $t = \dfrac{\bar{y}_1 - \bar{y}_2}{s_p\sqrt{1/n_1 + 1/n_2}} = \dfrac{28.57 - 40}{14.39\sqrt{1/7 + 1/7}} = -1.49$

R.R.: For $\alpha = .05$, the critical t-value for a one-tailed test with df $= n_1 + n_2 - 2 = 12$ can be obtained from Table 4 in the Appendix, using a $= .05$. We will reject H_0 if $t < -1.782$.

Conclusion: Since the observed value of t, -1.49, does not fall in the rejection region, we have insufficient evidence to reject the hypothesis that there is no difference in the mean number of worms in treated and untreated lambs.

(b) Using Table 4 in the Appendix with $t = -1.49$ and df $= 12$, we see the level of significance for this test is in the range $.05 < p < .10$.

The test procedures for comparing two population means presented in this section are based on several assumptions. The first and most critical one is that the two samples are independent. Practically, we mean that the two samples are drawn from two different populations and that the elements of one sample are unrelated to those of the second sample. If this assumption is not valid, then the t methods of this section will likely be in error and other methods (such as those presented in Section 5.3) may be appropriate.

The second assumption that we make is that the samples are drawn from normal populations. Fortunately, this assumption is less critical. The reason for this is that for modest-sized samples the Central Limit Theorem of the previous chapter applies and the sampling distributions for $\bar{y}_1$ and $\bar{y}_2$ are approximately normal. With independent samples and the combined sample size $n_1 + n_2 \geq 30$, the t methods of this section should be reasonably accurate even for modest skewness in the two populations. A nonparametric alternative to the t test for independent samples is presented in the next section; this alternative does not require normality.

The third and final assumption is that the two population variances σ_1^2 and σ_2^2 are equal. For now, just examine the sample variances to see that they are approximately equal; later (in Chapter 7), we'll give a test for this assumption. Many efforts have been made to investigate the effect of deviations from the equal variance assumption on the t methods for independent samples. The general conclusion is that for equal sample sizes, the population variances can differ by as much as a factor of 3 (for example, $\sigma_1^2 = 3\sigma_2^2$) and the t methods will still apply. This is remarkable and provides a convincing argument to use equal sample sizes. When the sample sizes are different, the most serious case is when the smaller sample size is associated with the larger variance. In this situation and in others where the sample variances (s_1^2 and s_2^2) suggest that $\sigma_1^2 \neq \sigma_2^2$, there is an approximate t test using the test statistic

$$t' = \frac{\bar{y}_1 - \bar{y}_2}{\sqrt{\dfrac{s_1^2}{n_1} + \dfrac{s_2^2}{n_2}}}$$

Welch (1938) showed that percentage points of a t distribution with modified degrees of freedom can be used to set the rejection region for H_0: $\mu_1 - \mu_2 = D_0$. This t test is summarized here.

APPROXIMATE t TEST FOR INDEPENDENT SAMPLES, UNEQUAL VARIANCE

H_0: $\mu_1 - \mu_2 = D_0$

H_a: 1. $\mu_1 - \mu_2 > D_0$
    2. $\mu_1 - \mu_2 < D_0$
    3. $\mu_1 - \mu_2 \neq D_0$

T.S.: $t' = \dfrac{\bar{y}_1 - \bar{y}_2 - D_0}{\sqrt{\dfrac{s_1^2}{n_1} + \dfrac{s_2^2}{n_2}}}$

R.R.: For a specified value of α,
    1. reject H_0 if $t' > t_\alpha$
    2. reject H_0 if $t' < -t_\alpha$
    3. reject H_0 if $|t'| > t_{\alpha/2}$

where

$$df = \frac{(n_1 - 1)(n_2 - 1)}{(n_2 - 1)c^2 + (1 - c)^2(n_1 - 1)} \qquad \text{where } c = \frac{s_1^2/n_1}{\dfrac{s_1^2}{n_1} + \dfrac{s_2^2}{n_2}}$$

Note: If the computed value of df is not an integer, *round down* to the nearest integer.

The test based on the t' statistic is sometimes referred to as the *separate-variance t test* because we use the separate sample variances s_1^2 and s_2^2, rather than a pooled sample variance.

EXAMPLE 5.3

Refer to the situation explained in Example 5.2. Suppose that only 13 animals were available for analysis at the end of the treatment period. These data are shown here.

Drug-treated sheep	5	13	18	6	4	2	15
Untreated	40	54	26	63	21	37	

Test the research hypothesis H_a: $\mu_1 - \mu_2 < 0$ under the assumption that the two population variances are different. Use $\alpha = .05$.

Solution

It is easy to verify that

$$\bar{y}_1 = 9.00 \qquad \bar{y}_2 = 40.17$$
$$s_1^2 = 38.67 \qquad s_2^2 = 258.17$$

Then the statistical test is set up as follows:

H_0: $\mu_1 - \mu_2 = 0$
H_a: $\mu_1 - \mu_2 < 0$

T.S.: $t' = \dfrac{\bar{y}_1 - \bar{y}_2}{\sqrt{\dfrac{s_1^2}{n_1} + \dfrac{s_2^2}{n_2}}} = \dfrac{9 - 40.17}{\sqrt{\dfrac{38.67}{7} + \dfrac{258.17}{6}}} = -4.47$

In order to compute the rejection region we need

$$c = \frac{s_1^2/n_1}{s_1^2/n_1 + s_2^2/n_2} = \frac{38.67/7}{38.67/7 + 258.17/6} = .114$$

$$c^2 = .013$$

and

$$df = \frac{(n_1 - 1)(n_2 - 1)}{(n_2 - 1)c^2 + (1 - c)^2(n_1 - 1)} = 6.283, \text{ which is rounded to } 6$$

R.R.: For $\alpha = .05$ and $df = 6$ reject H_0 if $t' < -1.943$.

Since $t' = -4.47$ is less than -1.943, we reject H_0 and conclude that μ_1, the mean worm count for treated sheep, is less than that for untreated sheep.

Computer simulations sometimes can help us to understand some of the assumptions underlying our test procedures. One such study was done to compare the pooled t test and separate-variance t test. We'll illustrate this by way of an example.

EXAMPLE 5.4

For the simulation study, it was assumed that we were sampling from the independent normal populations shown here:

Population

1	$\mu_1 = 100$	$\sigma_1 = 15$	$n_1 = 10$
2	$\mu_2 = 100$	$\sigma_2 = 10$	$n_2 = 20$

The study proceeded as follows: A computer program was used to generate a random sample of $n_1 = 10$ observations from Population 1 and a random sample of $n_2 = 20$ from Population 2. The sample statistics ($\bar{y}_1, \bar{y}_2, s_1^2$, and s_2^2) were computed, as were the test statistics t and t'. This process was repeated 999 more times, and for these 1000 samples the program kept track of the number of times t and t' rejected at the upper and lower .05 levels. These results are summarized here:

For H_0: $\mu_1 - \mu_2 = 0$ and $\alpha = .05$

	Pooled t Test	Separate-Variance t Test
H_0 rejected for H_a: $\mu_1 - \mu_2 > 0$	75 (7.5%)	46 (4.6%)
H_0 rejected for H_a: $\mu_1 - \mu_2 < 0$	77 (7.7%)	44 (4.4%)

a. Without running the computer simulations study, which test would you have recommended even without knowing the underlying populations? Explain.
b. What do the computer simulation results tell us about the choice of t or t'? Do they agree with our recommendation in part (a)?

Solution

a. With samples of sizes $n_1 = 10$ and $n_2 = 20$, we don't have much protection against the possibility of unequal variances. Hence the separate-variance t test should be somewhat more reliable.
b. Because we know the underlying populations, one of the assumptions underlying the pooled t test—namely, equal population variances—is violated. Because the population means were equal, H_0 is true and we would have expected to reject H_0 approximately 5% of the time in both the upper and lower tails, due to chance. As can be seen, the pooled t test rejected H_0 more frequently (7.5% above and 7.7% below) than would have been expected. The separate-variance t test, on the other hand, rejects about as often as we would expect. These results agree with our conclusion in part (a).

EXAMPLE 5.5

Another simulation study is done with samples from independent normal populations, with the following results.

Checking α (different sample sizes and σs)

Simulation of two-sample t test (1000 samples)

Popn	μ	σ	n
1	50.000	10.000	10
2	50.000	14.1421	20

Using the pooled-variance t test

One-tailed test:

number of times H_o: "$\mu_1 - \mu_2$ is 0" is rejected in favor of

α	"$\mu_1 - \mu_2 > 0$"	"$\mu_1 - \mu_2 < 0$"	total (α doubled)
.100	71	84	155
.050	32	36	68
.025	13	16	29
.010	4	7	11
.005	1	3	4

using separate variances (t')

One-tailed test:

number of times H_o: "$\mu_1 - \mu_2$ is 0" is rejected in favor of

α	"$\mu_1 - \mu_2 > 0$"	"$\mu_1 - \mu_2 < 0$"	total (α doubled)
.100	103	104	207
.050	43	48	91
.025	21	20	41
.010	7	9	16
.005	1	5	6

average pooled t is -0.0196 with variance of 0.873156
average t' is -0.0226 with variance of 1.088475

What do these results indicate about t versus t'?

Solution

Again, the assumption of equal variances is violated. This time, however, the smaller variance is associated with the smaller sample size; in Example 5.4 the smaller variance was associated with the larger sample size. In this example, we find that the number of false rejections of the null hypothesis is consistently *smaller* than would be indicated by the nominal α. Again, the t' test rejects the null hypothesis just about as often as α indicates.

In this section, we developed pooled-variance t methods based on an assumption of equal population variances. In addition, we introduced the t' statistic for an approximate test when the variances are not equal. Confidence intervals and hypothesis tests based on these different procedures (t or t') need not give identical results. Standard computer packages often report the results of both the pooled-variance and separate-variance t tests. Which should you believe?

If the sample sizes are equal it doesn't matter. The alternative t tests give algebraically identical results when $n_1 = n_2$, and, since the t probabilities are robust to nonnormality and unequal population variances, the test results are quite reliable when $n_1 = n_2$. When n_1 and n_2 are nearly equal, the two results are nearly equal. Only when the sample sizes vary greatly (say 1.5 to 1 or worse) will there be a large difference in results. The evidence in such cases indicates that the separate variance methods contained in computer packages are somewhat more reliable and more conservative.

EXERCISES

5.1   Two different emission-control devices were being tested to determine the average amount of nitric oxide being emitted by an automobile over a one-hour period of time. Twenty cars of the same model and year were selected for the study. Ten cars were randomly selected and equipped with a Type I emission-control device, and the remaining cars were equipped with Type II devices. Each of the 20 cars was then monitored for a one-hour period to determine the amount of nitric oxide emitted.

Use the following data to test the research hypothesis that the mean level of emission for Type I devices (μ_1) is greater than the mean emission level for Type II devices (μ_2). Use $\alpha = .01$.

Type I Device		Type II Device	
1.35	1.28	1.01	0.96
1.16	1.21	0.98	0.99
1.23	1.25	0.95	0.98
1.20	1.17	1.02	1.01
1.32	1.19	1.05	1.02

5.2   It has been estimated that lead poisoning resulting from an unnatural craving (pica) for substances such as paint may affect as many as a quarter of a million children each year, causing them to suffer from severe, irreversible retardation. Explanations for why children voluntarily consume lead range from "improper parental supervision" to "a child's need to mouth objects." Some researchers, however, have been investigating whether the habit of eating such substances has some nutritional explanation. One such study involved a comparison of a regular diet and a calcium-deficient diet on the ingestion of a lead-acetate solution in rats. Each rat in a group of 20 rats was randomly assigned to either an experimental or control group. Those in the control group received a normal diet, while the experimental group received a calcium-deficient diet. Each of the rats occupied a separate cage and was monitored to observe the quantity of a .15% lead-acetate solution consumed during the study period. The sample results are summarized here.

Control group	5.4	6.2	3.1	3.8	6.5	5.8	6.4	4.5	4.9	4.0
Experimental group	8.8	9.5	10.6	9.6	7.5	6.9	7.4	6.5	10.5	8.3

a. Plot the data for the two samples separately. Is there reason to think the assumptions for a *t* test have been violated?

b. Run a test of the research hypothesis that the mean quantity of lead acetate consumed in the experimental group is greater than that consumed in the control group. Use $\alpha = .05$.

5.3   The results of a three-year study to examine the effect of ready-to-eat breakfast cereals on dental caries (tooth decay) in adolescent children were recently reported by Rowe, Anderson, and Wanninger (1974). A sample of 375 adolescent children of both genders from the Ann Arbor, Michigan public schools was enrolled (after parental consent) in the study. Each child was provided with toothpaste and boxes of different varieties of ready-to-eat cereals. Although these were brand-name cereals, each type of cereal was packaged in plain white 7-ounce boxes and labeled as wheat flakes, corn cereal, oat cereal, fruit-flavored corn puffs, corn puffs, cocoa-flavored cereal, and sugared oat cereal. Note that the last four varieties of cereal had been presweetened and the others had not.

Each child received a dental examination at the beginning of the study, twice during the study, and once at the end. The response of interest was the incremental DMF surfaces, that is, the difference between the final (poststudy) and initial (prestudy) number of decayed, missing, and filled (DMF) tooth surfaces. Careful records for each participant were maintained throughout the three years, and at the end of the study, a person was classified as "noneater" if he or she had eaten less than 28 boxes of cereal throughout the study. All others were classified as "eaters." The incremental DMF surface readings for each group are summarized below.

	Sample Size	Sample Mean	Sample Standard Deviation
Noneaters	73	6.41	5.62
Eaters	302	5.20	4.67

Use these data to test the research hypothesis that the mean incremental DMF surface for noneaters is larger than the corresponding mean for eaters. Give the level of significance for your test. Interpret your findings.

5.4   Refer to Exercise 5.3. Although complete details of the original study have not been disclosed, critique the procedure that has been discussed.

5.5   The study of concentrations of atmospheric trace metals in isolated areas of the world has received considerable attention because of the concern that humans might somehow alter the climate of the earth by changing the amount and distribution of trace metals in the atmosphere. Consider a study at the South Pole, where at 10 different sampling periods throughout a two-month period, 10,000 standard cubic meters (scm) of air were obtained and analyzed for metal concentrations. The results associated with magnesium and europium are listed below. (Note: Magnesium results are in units of 10^{-9} g/scm; europium results are in units of 10^{-15} g/scm.)

	Sample Size	Sample Mean	Sample Standard Deviation
Magnesium	10	1.0	2.21
Europium	10	17.0	12.65

Note that $s > \bar{y}$ for the magnesium data. Would you expect the data to be normally distributed? Explain.

5.6   Refer to Exercise 5.5. Could we run a t test comparing the mean metal concentrations for magnesium and europium? Why or why not?

5.7   Refer to Example 5.1.

a. Does it appear from the sample means and variances that any of the underlying assumptions for pooled t methods have been violated?

b. Compute t and t' for these data and draw a conclusion based on each statistic. Do these results agree with your preliminary assessment of the underlying assumptions in part (a)?

 5.8 A firm has a generous but rather complicated policy concerning end-of-year bonuses for its lower-level managerial personnel. The policy's key factor is a subjective judgment of "contribution to corporate goals." A personnel officer took samples of 24 female and 36 male managers to see if there was any difference in bonuses, expressed as a percentage of yearly salary. The data are listed here:

Gender	Bonus Percentage								
F	9.2	7.7	11.9	6.2	9.0	8.4	6.9	7.6	7.4
	8.0	9.9	6.7	8.4	9.3	9.1	8.7	9.2	9.1
	8.4	9.6	7.7	9.0	9.0	8.4			
M	10.4	8.9	11.7	12.0	8.7	9.4	9.8	9.0	9.2
	9.7	9.1	8.8	7.9	9.9	10.0	10.1	9.0	11.4
	8.7	9.6	9.2	9.7	8.9	9.2	9.4	9.7	8.9
	9.3	10.4	11.9	9.0	12.0	9.6	9.2	9.9	9.0

A computer program yielded the output shown here.

```
SAMPLE              1        2

MEAN            8.5333   9.6833

ST. DEV.        1.1890   1.0038

SAMPLE SIZE       24       36

SUM OF RANKS     481.0   1349.0


POOLED-VARIANCE T STATISTIC = −4.037
    2-TAILED P-VALUE = 0.0004
SEPARATE-VARIANCE T STATISTIC = −3.901
    2-TAILED P-VALUE = 0.0008
    APPROX DF = 43

RANK SUM Z STATISTIC = −3.787
    2-TAILED P-VALUE = 0.0002
```

a. Identify the value of the pooled-variance t statistic (the usual t test based on the equal variance assumption).
b. Identify the value of the t' statistic.
c. Use both statistics to test the research hypothesis of unequal means at $\alpha = .05$ and at $\alpha = .01$. Does the conclusion depend on which statistic is used?

 5.9 The costs of major surgery vary substantially from one state to another due to differences in hospital fees, doctors' fees, malpractice insurance cost, and rent. A study of hysterectomy costs was done in California and Montana. Based on a random sample of 20 patient records from each state, the following sample statistics were obtained:

	Sample Mean	Sample Standard Deviation
California	$ 6,458	$520
Montana	$12,690	$305

Construct a 95% confidence interval for $\mu_1 - \mu_2$ (the California minus Montana difference).

5.10 A national educational organization monitors reading proficiency for U.S. students on a regular basis using a scale that ranges from 0 to 500. Sample results based on 500 students per category are shown here. Use these data to make the inferences listed below. Assume the pooled standard deviation for any comparison is 100.

			Sample Mean*
Age	9	Male	210
		Female	216
	13	Male	253
		Female	262
	17	Male	283
		Female	293

* What the scale means: 150—Rudimentary reading skills; can follow basic directions. 200—Basic skills; can identify facts from simple paragraphs. 250—intermediate skills; can organize information in lengthy passages. 300—Adapt skills; can understand and explain complicated information. 350—Advanced skills; can understand and explain specialized materials.

a. Construct a meaningful graph that shows age, gender, and mean proficiency scores.
b. Use the sample data to place a 95% confidence interval on the difference in mean proficiences for females and males age 17 years.
c. Compare the mean scores for females, age 13 and 17 years, using a 90% confidence interval. Does the interval include 0? Why might these means be different?

5.11 The organization alluded to in Exercise 5.10 also examined the effect of television viewing on reading proficiency scores for students of the same age categories. The sample means and sample sizes are shown here.

Hours of TV Viewing per Day		Age (Years)		
		9	13	17
0–2	$\bar{y}$	220	267	295
	n	300	310	305
3–5	$\bar{y}$	220	262	284
	n	280	260	250
6+	$\bar{y}$	202	246	270
	n	210	220	230

     a. Plot the sample means on a graph with age and hours of TV viewing per day. Can you make some general statements about the sample data? What's the effect of TV viewing on reading proficiency within a given age group? What's the influence of age within a given TV viewing category?

     b. Construct 99% confidence intervals for the difference between the 0–2 and the 6$^+$ category in each age category. What's your conclusion?

5.12   Class Exercise. Although we haven't discussed the design of experiments, think about how you might conduct a survey to examine the effects on reading proficiency of age and hours per day spent watching TV. What factor or factors might affect the results of your survey?

5.2 A NONPARAMETRIC ALTERNATIVE: THE WILCOXON RANK SUM TEST

The two-sample t test of the previous section was based on several assumptions: independent samples, normality, and equal variances. When the assumptions of normality and equal variances are not valid but the sample sizes are large, the results using a t (or t') test are approximately correct. There is, however, an alternative test procedure that requires less stringent assumptions. This procedure, called the **Wilcoxon rank sum test**, is discussed here.

Wilcoxon rank sum test

    The assumptions for this test are that we have independent random samples taken from two populations. The Wilcoxon rank sum test provides a procedure for testing that two populations are identical but not necessarily normal. Since the two populations are assumed to be identical under the null hypothesis, independent random samples from the respective populations should be similar. One way to measure the similarity between the samples is to jointly rank (from lowest to highest) the measurements from the combined samples and examine the sum of the ranks for measurements in sample 1 (or, equivalently, sample 2). Under the null hypothesis of identical populations, the sum of the ranks for a sample will be proportional to the sample size. We let T denote the sum of the ranks for sample 1. Intuitively, if T is extremely small (or large), we would have evidence to reject the null hypothesis that the two populations are identical.

    Under the null hypothesis, the statistic T will have a sampling distribution with mean and variance given by

$$\mu_T = \frac{n_1(n_1 + n_2 + 1)}{2}$$

$$\sigma_T^2 = \frac{n_1 n_2}{12}(n_1 + n_2 + 1)$$

If, in addition, both sample sizes are 10 or larger, the sampling distribution of T is approximately normal; this allows us to use a z statistic in the Wilcoxon rank sum test.

    The theory behind the Wilcoxon rank sum test assumes that the population distributions are continuous, so that there is zero probability that any two observations are identical. In practice there will often be ties—two or more observations with the same value. For these situations, each observation in a set of tied values receives a rank score equal to the average of the ranks for the set. For example, if two observations are tied for the ranks 3 and 4, each is given a rank of 3.5; the next higher value receives a rank of 5, and so on. When there are ties, there is a correction for the variance formula. Then σ_T^2 is as shown here.

$$\sigma_T^2 = \frac{n_1 n_2}{12}\left[(n_1 + n_2 + 1) - \frac{\sum_j t_j(t_j^2 - 1)}{(n_1 + n_2)(n_1 + n_2 - 1)}\right]$$

where t_j denotes the number of tied ranks in the jth group. Note that when there are no tied ranks,

$$\sigma_T^2 = \frac{n_1 n_2(n_1 + n_2 + 1)}{12}$$

From a practical standpoint, however, unless there are many ties, the correction will have very little effect on the value of σ_T^2. The Wilcoxon rank sum test is summarized here.

WILCOXON
RANK
SUM TEST*

H_0: The two populations are identical.

H_a: 1. Population 1 is shifted to the right of population 2.
2. Population 1 is shifted to the left of population 2.
3. Populations 1 and 2 have different location parameters.

$(n_1 \leq 10, n_2 \leq 10)$

T.S.: T, the sum of the ranks in sample 1

R.R.: For $\alpha = .05$, use Table 3 in the appendix to find critical values for T_U and T_L)
1. Reject H_0 if $T > T_U$
2. Reject H_0 if $T < T_L$
3. Reject H_0 if $T > T_U$ or $T < T_L$

$(n_1, n_2 > 10)$

T.S.: $z = \dfrac{T - \mu_T}{\sigma_T}$

where T denotes the sum of the ranks in sample 1.

R.R.: For a specified value of α,
1. reject H_0 if $z > z_\alpha$
2. reject H_0 if $z < -z_\alpha$
3. reject H_0 if $|z| > z_{\alpha/2}$

* This test is equivalent to the Mann–Whitney U test, Conover (1980).

TABLE 5.2   Dissolved Oxygen Measurements (in ppm), Example 5.6

Before Cleanup		After Cleanup	
11.0	11.6	10.2	10.8
11.2	11.7	10.3	10.8
11.2	11.8	10.4	10.9
11.2	11.9	10.6	11.1
11.4	11.9	10.6	11.1
11.5	12.1	10.7	11.3

EXAMPLE 5.6

Environmental engineers were interested in determining whether a cleanup project on a nearby lake was effective. Prior to initiation of the project, 12 water samples had been obtained at random from the lake and analyzed for the amount of dissolved oxygen (in ppm). Due to diurnal fluctuations in the dissolved oxygen, all measurements were obtained at the 2 PM peak period. The before and after data are presented in Table 5.2.

a. Use $\alpha = .05$ to test the following hypotheses:

H_0: The distributions of measurements for before cleanup and six months after the cleanup project began are identical.

H_a: The distribution of dissolved oxygen measurements before the cleanup project is shifted to the right of the corresponding distribution of measurements for six months after initiating the cleanup project. (It should be noted that a cleanup project has been effective in one sense if the dissolved oxygen drops over a period of time.)

For convenience, the data have been arranged in ascending order in Table 5.2.

b. Has the correction for ties made much of a difference?

Solution

(a) First we must jointly rank the combined sample of 24 observations by assigning the rank of 1 to the smallest observation, the rank of 2 to the next smallest, and so on. When two or more measurements are the same, we assign all of them a rank equal to the average of the ranks they occupy. The sample measurements and associated ranks (shown in parentheses) are listed in Table 5.3.

Since n_1 and n_2 are both greater than 10, we will use the test statistic z. If we are trying to detect a shift to the left in the distribution after the cleanup, we would expect the sum of the ranks for the observations in sample 1 to be large. Thus we will reject H_0 for large values of $z = (T - \mu_T)/\sigma_T$.

TABLE 5.3   Dissolved Oxygen Measurements and Ranks
for Example 5.6

Before Cleanup		After Cleanup	
11.0	(10)	10.2	(1)
11.2	(14)	10.3	(2)
11.2	(14)	10.4	(3)
11.2	(14)	10.6	(4.5)
11.4	(17)	10.6	(4.5)
11.5	(18)	10.7	(6)
11.6	(19)	10.8	(7.5)
11.7	(20)	10.8	(7.5)
11.8	(21)	10.9	(9)
11.9	(22.5)	11.1	(11.5)
11.9	(22.5)	11.1	(11.5)
12.1	(24)	11.3	(16)
	$T = 216$		

Grouping the measurements with tied ranks, we have 18 groups. These groups are listed below with the corresponding values of t_j, the number of tied ranks in the group.

Rank(s)	Group	t_j
1	1	1
2	2	1
3	3	1
4.5, 4.5	4	2
6	5	1
7.5, 7.5	6	2
9	7	1
10	8	1
11.5, 11.5	9	2
14, 14, 14	10	3
16	11	1
17	12	1
18	13	1
19	14	1
20	15	1
21	16	1
22.5, 22.5	17	2
24	18	1

For all groups with $t_j = 1$, there is no contribution for

$$\frac{\sum_j t_j(t_j^2 - 1)}{(n_1 + n_2)(n_1 + n_2 - 1)}$$

in σ_T^2 since $t_j^2 - 1 = 0$. Thus we will need only $t_j = 2, 3$.

Substituting our data in the formulas, we obtain

$$\mu_T = \frac{n_1(n_1 + n_2 + 1)}{2} = \frac{12(12 + 12 + 1)}{2} = 150$$

$$\sigma_T^2 = \frac{n_1 n_2}{12}\left[(n_1 + n_2 + 1) - \frac{\sum t_j(t_j^2 - 1)}{(n_1 + n_2)(n_1 + n_2 - 1)}\right]$$

$$= \frac{12(12)}{12}\left[25 - \frac{6 + 6 + 6 + 24 + 6}{24(23)}\right] = 12(25 - .0870) = 298.956$$

$$\sigma_T = 17.29$$

The computed value of z is

$$z = \frac{T - \mu_T}{\sigma_T} = \frac{216 - 150}{17.29} = 3.82$$

Since this value exceeds 1.645, we reject H_0 and conclude that the distribution of before-cleanup measurements is shifted to the right of the corresponding distribution of after-cleanup measurements; that is, the after-cleanup measurements on dissolved oxygen tend to be smaller than the corresponding before-cleanup measurements.

(b) The value of σ_T^2 without correcting for ties is

$$\sigma_T^2 = \frac{12(12)(25)}{12} = 300$$

and

$$\sigma_T = 17.32$$

For this value of σ_T, $z = 3.81$ rather than 3.82 found by applying the correction. This should help you understand how little effect the correction has on the final result unless there are very many ties.

The Wilcoxon rank sum test is an alternative to the two-sample t test that requires fewer assumptions. In particular, Wilcoxon's test does not require normality for the two populations, only that they be identical under H_0. When the assumptions underlying a t test hold, the t test will be more likely to declare an existing difference. This only seems logical since the t test uses the magnitudes of observations rather than just their relative magnitudes (ranks). But when the assumptions for a t test are violated, the Wilcoxon rank sum test is the more informative test and is more likely to declare a difference when it exists. This is particularly true when nonnormality of the populations is present in the form of severe skewness or extreme outliers.

EXAMPLE 5.7   To investigate the effect of skewness on the pooled-variance t test as well as the Wilcoxon rank sum test, 1000 samples were drawn from a squared-exponential population; this population was extremely right-skewed. The following results were obtained.

Checking α (different sample sizes; same σs)

Simulation of two-sample t test (1000 samples)

Popn	μ	σ	n
1	50.000	10.0000	5
2	50.000	10.0000	25

using the pooled-variance t test

One-tailed test:

	number of times H_0: "$\mu_1 - \mu_2$ is 0" is rejected in favor of		
α	"$\mu_1 - \mu_2 > 0$"	"$\mu_1 - \mu_2 < 0$"	total (α doubled)
.100	146	35	181
.050	95	3	98
.025	51	0	51
.010	25	0	25
.005	15	0	15

Results of Wilcoxon rank sum test using z as test statistic

	number of times H_0: "two populations are identical" rejected in favor of		
α	Popn1 rt of Popn2	Popn1 left of Popn2	total (α doubled)
.100	102	93	195
.050	37	51	88
.025	15	30	45
.010	5	14	19
.005	4	3	7

What do the results indicate about the effect of skewness on the two tests?

Solution

The null hypothesis is true in this simulation; both means are 50. The actual number of rejections of the null hypothesis by the *t* test is far from what is indicated by the nominal α value for one-tailed probabilities. The Wilcoxon rank sum test, which doesn't assume normal populations, appears to be rejecting the null hypothesis the correct number of times.

EXERCISES

5.13 A plumbing contractor was interested in making her operation more efficient by cutting down on the average distance between service calls while still maintaining at least the same level of business activity. One plumber (Plumber 1) was assigned a dispatcher who monitored all his incoming requests for service and outlined a service strategy for that day. Plumber 2 was to continue as she had in the past, by providing service in roughly sequential order for stacks of service calls received. The total daily mileages for these two plumbers are recorded below for a total of 18 days (three work weeks).

Plumber 1	88.2	94.7	101.8	102.6	89.3	95.7
	78.2	80.1	83.9	86.1	89.4	71.4
	92.4	85.3	87.5	94.6	92.7	84.6

Plumber 2	105.8	117.6	119.5	126.8	108.2	114.7
	90.2	95.6	110.1	115.3	109.6	112.4
	104.6	107.2	109.7	102.9	99.1	111.5

a. Plot the sample data for each plumber and compute $\bar{y}$ and s.

b. Based on your findings in part (a), which procedure appears more appropriate for comparing the distributions?

5.14   Computer output is shown below for the data of Exercise 5.13 for a t test of $H_0: \mu_1 - \mu_2 = 0$ and a Wilcoxon rank sum test (which is equivalent to the Mann–Whitney test shown here).

```
        SET THE DATA FOR THE FIRST PLUMBER INTO COL C1
DATA> 88.2 94.7 101.8 102.6 89.3 95.7 78.2 80.1 83.9 86.1 89.4 71.4
DATA> 92.4 85.3 87.5 94.6 92.7 84.6
DATA> END DATA
MTB > PRINT C1
C1
   88.2    94.7   101.8   102.6    89.3    95.7    78.2    80.1    83.9
   86.1    89.4    71.4    92.4    85.3    87.5    94.6    92.7    84.6

MTB > HISTOGRAM OF C1

Histogram of C1   N = 18

Midpoint   Count
      70       1  *
      75       0
      80       2  **
      85       4  ****
      90       5  *****
      95       4  ****
     100       1  *
     105       1  *

MTB > SET THE DATA FOR THE SECOND PLUMBER INTO COL C2
DATA> 105.8 117.6 119.5 126.8 108.2 114.7 90.2 95.6 110.1 115.3 109.6
DATA> 112.4 104.6 107.2 109.7 102.9 99.1 111.5
DATA> END DATA
MTB > PRINT C2
C2
  105.8   117.6   119.5   126.8   108.2   114.7    90.2    95.6   110.1
  115.3   109.6   112.4   104.6   107.2   109.7   102.9    99.1   111.5

MTB > HISTOGRAM OF C2

Histogram of C2   N = 18

Midpoint   Count
      90       1  *
      95       1  *
     100       1  *
     105       4  ****
     110       6  ******
     115       2  **
     120       2  **
     125       1  *

MTB >
              TWOSAMPLE T WITH 95 PERCENT CONFIDENCE ON DATA IN C1 AND C2

        TWOSAMPLE T FOR C1 VS C2
               N      MEAN     STDEV    SE MEAN
        C1  18     88.81      7.89       1.86
        C2  18    108.93      8.73       2.06

        95 PCT CI FOR MU C1 - MU C2: (-25.77, -14.48)
        TTEST MU C1 = MU C2 (VS NE): T=-7.26 P=0.0000 DF=33.7
```

```
MTB > MANN-WHITNEY TEST WITH 95 PERCENT CONFIDENCE ON DATA IN C1 AND C2
Mann-Whitney Confidence Interval and Test
    C1          N =  18      MEDIAN =          88.75
    C2          N =  18      MEDIAN =         109.65
    POINT ESTIMATE FOR ETA1-ETA2 IS       -20.3002
    95.2  PCT C.I. FOR ETA1-ETA2 IS (     -25.4,      -14.9)
    W =      183.0
    TEST OF ETA1 = ETA2  VS.  ETA1 N.E. ETA2 IS SIGNIFICANT AT   0.0000

MTB >
```

a. Compare the results for these two tests and draw a conclusion about the effectiveness of the dispatcher program.

b. Comment on the appropriateness or inappropriateness of the t test based on your findings in Exercise 5.13(a) and the output shown here.

c. Does it matter which test was used here? Might it be reasonable to run both tests in certain situations? Why?

 5.15 An experiment was conducted to compare the weights of the combs of roosters fed two different vitamin-supplemented diets. Twenty-eight healthy roosters were randomly divided into two groups, with one group receiving diet I and the other receiving diet II. After the study period the comb weight (in milligrams) was recorded for each rooster. These data are given below.

Diet I	73	130	115	144	127	126	112	76	68	101	126	49	110	123

Diet II	80	72	73	60	55	74	67	89	75	66	93	75	68	76

a. Use the Wilcoxon rank sum test to determine whether there is a difference in the distributions of comb weights for the two groups. Use $\alpha = .05$.

b. Can you suggest other statistical procedures that might be appropriate for analyzing the same data? Which would you suggest?

5.16 Refer to Exercise 5.15. Suppose the experimenter was interested in determining whether the comb weights for diet I were selected from a distribution shifted above (to the right of) that for comb weights from diet II. Run an appropriate Wilcoxon rank sum test and give the p-value. Draw a conclusion.

5.17 A computer simulation was done to compare the t test to the Wilcoxon rank sum test when sampling from two identical (nonnormal) populations. Chosen for the study were two identical, right-skewed populations with means and standard deviations as shown

Population 1	$\mu_1 = 100$	$\sigma_1 = 20$	$n_1 = 10$

Population 2	$\mu_2 = 100$	$\sigma_2 = 20$	$n_2 = 20$

The simulation study consisted of 1000 runs where random samples of size $n_1 = 10$ and $n_2 = 20$ were drawn from the two populations; after each run, the pooled t test was run to test $H_0: \mu_1 - \mu_2 = 0$ and the Wilcoxon rank sum test was run to test H_0: The populations were identical. The number (%) of times out of 1000 that H_0 was rejected at the upper and lower .05 levels is recorded for each test.

	Pooled t Test $H_0: \mu_1 - \mu_2 = 0$
H_0 rejected, $H_a: \mu_1 - \mu_2 > 0$	87
H_0 rejected, $H_a: \mu_1 - \mu_2 < 0$	15

	Wilcoxon Rank Sum Test H_0: Populations Are Identical
H_0 rejected, H_a: population 1 is to right of population 2	43
H_0 rejected, H_a: population 1 is to left of population 2	54

What do the simulation results indicate in regards to the effect of skewness in the performance of the pooled t test and the Wilcoxon rank sum test?

5.18   A simulation study evaluates the effect of severe skewness on the relative usefulness of the pooled-variance t test, the t' test, and the rank sum test. Independent samples are taken from the severely skewed distribution. The results are as follows:

Checking α (same sample sizes and σs)

Popn	μ	σ	n
1	50.000	10.0000	10
2	50.000	10.0000	10

using the pooled-variance t test

One-tailed test:

	number of times H_0: "$\mu_1 - \mu_2$ is 0" is rejected in favor of		
α	"$\mu_1 - \mu_2 > 0$"	"$\mu_1 - \mu_2 < 0$"	total (α doubled)
.100	99	108	207
.050	32	37	69
.025	11	12	23
.010	3	3	6
.005	0	0	0

using separate variances (t')

One-tailed test:

	number of times H_0: "$\mu_1 - \mu_2$ is 0" is rejected in favor of		
α	"$\mu_1 - \mu_2 > 0$"	"$\mu_1 - \mu_2 < 0$"	total (α doubled)
.100	94	98	192
.050	27	27	54
.025	6	10	16
.010	1	1	2
.005	0	0	0

Results of Wilcoxon rank sum test using z as test statistic

	number of times H_0: "two populations are identical" rejected in favor of		
α	Popn1 rt of Popn2	Popn1 left of Popn2	total (α doubled)
.100	105	110	215
.050	44	50	94
.025	24	29	53
.010	6	12	18
.005	1	3	4

For any of the tests, are the nominal α values grossly wrong?

5.3   A QUICK PORTABLE STATISTIC: THE TUKEY–DUCKWORTH TEST (optional)

In addition to providing alternative analyses when underlying assumptions are violated, some nonparametric statistical techniques are so easy to remember and use that they can be quickly applied without a desk calculator, a computer, or a reference table for critical values. In short, they can be carried anywhere and applied in many situations to provide a quick preliminary conclusion.

Throughout the text we will insert these portable statistics where appropriate to make you aware of how they may help you in situations where it is not possible to do a formal analysis either because time will not permit it or because appropriate reference material (formulas, critical values) or equipment (computer hardware and software) is not available.

Tukey–Duckworth two-sample test   The first technique we will consider is the **Tukey–Duckworth two-sample test** (Tukey, 1959) to determine if two independent samples were drawn from identical populations. This test can be used for sample sizes satisfying the following inequalities:

$$4 \le n_1 \le n_2 \le 30 \qquad n_2 \le \frac{4n_1}{3} + 3$$

It should be noted that the designation of population 1 and population 2 is completely arbitrary. Modifications of this procedure have been suggested by other authors (see, for example, Neave (1966 and 1975)).

TUKEY–DUCKWORTH TWO-SAMPLE TEST

H_0: The populations are identical.

H_a: The populations are different (a two-tailed test).

Test procedure

1. Determine the largest and smallest measurement in each sample.
2. For the sample that contains the largest value in the combined samples, count all measurements that are larger than the largest measurement in the other sample.
3. For the other sample, count all measurements that are smaller than the smallest measurement of the first sample.
4. Let C denote the sum of the two counts. For $\alpha = .05$, .01, or .001, reject H_0 if $C \ge 7$, 10, or 13, respectively.

C

EXAMPLE 5.8   Thirty different 1-acre plots were randomly divided into two groups, with 15 plots per group. The plots in the first group were fertilized with brand A fertilizer and those in the second with brand B. Each of the 30 one-acre plots was then planted in corn. Yields (in bushels) are presented in Table 5.4 for each plot. Use these data to determine whether there is a difference in yields for the two brands of fertilizer. Use $\alpha = .05$.

TABLE 5.4   Yields of Corn (in Bushels) for Two Different Brands of Fertilizer, Example 5.8

Group 1 (Brand A)		Group 2 (Brand B)	
96	89	98	92
92	94	94	89
98	80	92	95
82	97	84	92
86	84	99	96
87	85	96	101
93	83	98	103
81		96	

Solution   We can proceed immediately with the Tukey–Duckworth two-sample test since our sample sizes satisfy the criteria $4 \le n_1 \le n_2 \le 30$ and $n_2 \le (4n_1/3) + 3$. First we must determine the largest and smallest measurements for each sample, as shown below.

	Group 1	Group 2
Largest	98	103
Smallest	80	84

Group 2 contains the largest measurement (103) for the combined samples. The number of measurements in Group 2 larger than 98, which is the largest measurement in Group 1, is 3. Similarly, the number of measurements in Group 1 less than 84, which is the smallest measurement in Group 2, is 4. Since $C = 3 + 4 = 7$, we reject the null hypothesis that the populations of corn yields corresponding to the two fertilizers are identical, at the $\alpha = .05$ level.

EXERCISES

5.19   a. Refer to the data of Exercise 5.13 and use the Tukey–Duckworth test to reach a conclusion concerning the utility of a dispatcher.

b. Compare your conclusion to the one drawn in Exercise 5.14 after viewing the output for a t test and a Wilcoxon rank sum test.

5.20   Refer to data of Example 5.6.

a. What conclusion would you draw regarding the cleanup program using a Tukey–Duckworth test?

b. Compare your conclusion in part (a) to the results of Example 5.6.

5.21   Comment on the interchangeability of a two-sample t test, the Wilcoxon rank sum test, and the Tukey–Duckworth test. When might you use one over the others? When might you use two or more for the same data set?

| 5.4 | INFERENCES ABOUT $\mu_1 - \mu_2$: PAIRED DATA |

The methods we presented in the preceding three sections were appropriate for situations in which independent random samples are obtained from two populations. These methods are not appropriate for studies or experiments in which each measurement in one sample is *matched* or *paired* with a particular measurement in the other sample. In this section we will deal with methods for analyzing "paired" data. We begin with an example.

EXAMPLE 5.9

Insurance adjusters are concerned about the high estimates they are receiving from garage I for auto repairs compared to garage II. To verify their suspicions each of 15 cars recently involved in an accident was taken to both garages for separate estimates of repair costs. Use a two-sample t test to analyze these data.

Solution

Computer output for these data is shown here.

```
        PRINT C1 C2 C3
    ROW    C1      C2      C3

      1    7.6     7.3     0.3
      2   10.2     9.1     1.1
      3    9.5     8.4     1.1
      4    1.3     1.5    -0.2
      5    3.0     2.7     0.3
      6    6.3     5.8     0.5
      7    5.3     4.9     0.4
      8    6.2     5.3     0.9
      9    2.2     2.0     0.2
     10    4.8     4.2     0.6
     11   11.3    11.0     0.3
     12   12.1    11.0     1.1
     13    6.9     6.1     0.8
     14    7.6     6.7     0.9
     15    8.4     7.5     0.9

MTB > TWOSAMPLE T WITH 95 PERCENT CONFIDENCE FOR DATA IN C1 AND C2

TWOSAMPLE T FOR C1 VS C2
        N      MEAN     STDEV    SE MEAN
C1  15        6.85      3.20     0.827
C2  15        6.23      2.94     0.759

95 PCT CI FOR MU C1 - MU C2: (-1.691, 2.918)
TTEST MU C1 = MU C2 (VS NE): T=0.55 P=0.59 DF=27.8

MTB >
```

From the output we see there is a consistent difference in the sample means ($\bar{y}_1 - \bar{y}_2 = .62$). But this difference is rather small considering the variability of the measurements ($s_1 = 3.20$, $s_2 = 2.94$). In fact the computed t-value (0.55) has a p-value of 0.59, indicating very little evidence of a difference in the average claim estimates for the two garages.

TABLE 5.5   Repair Estimates (in Hundreds of Dollars),
Example 5.9

Car	Garage I	Garage II
1	7.6	7.3
2	10.2	9.1
3	9.5	8.4
4	1.3	1.5
5	3.0	2.7
6	6.3	5.8
7	5.3	4.9
8	6.2	5.3
9	2.2	2.0
10	4.8	4.2
11	11.3	11.0
12	12.1	11.0
13	6.9	6.1
14	7.6	6.7
15	8.4	7.5
Totals	$\bar{y}_1 = 6.85$	$\bar{y}_2 = 6.23$

A closer glance at the data in Table 5.5 indicates there is something about the conclusion that is inconsistent with our intuition. For all but one of the 15 cars, the estimate from garage I was higher than that from garage II. From our knowledge of the binomial distribution, the probability of observing garage I estimates higher in $y = 14$ or more of the $n = 15$ trials assuming no difference ($\pi = .5$) for garages I and II is

$$P(y = 14 \text{ or } 15) = P(y = 14) + P(y = 15)$$

$$= \binom{15}{14}(.5)^{14}(.5) + \binom{15}{15}(.5)^{15}$$

This probability is .000 to three decimal places. Using this binomial probability we would argue that the observed sample results are highly contradictory to the null hypothesis of equality of estimates for the two garages. Where did we go wrong? Why are there such conflicting results?

The explanation of the difference in the conclusion for a t test and the conclusion based on the binomial distribution is that one of the basic assumptions, independent samples, has been violated by the way the experiment was conducted. The adjusters obtained a measurement from both garages for each car rather than having a random sample of 15 cars examined by garage I and a second sample of cars examined by garage II.

As you can see from the data in Table 5.5, the repair estimates for a given car are about the same but vary considerably from car to car. These differences caused large variability among estimates for a given garage and tend to cancel any differences between the two garages. This fact was recognized when the study was planned. By having both garages give an estimate on each car, we can calculate the difference between the two garages for each car, and hence cancel out the car-to-car variability.

A proper analysis of the paired data in Example 5.9 makes use of the 15 difference measurements to test the null hypothesis that the mean difference, μ_d, is D_0. This hypothesis is equivalent to H_0: $\mu_1 - \mu_2 = D_0$. A summary of the test procedure is given here.

PAIRED *t* TEST

H_0: $\mu_d = D_0$

H_a: 1. $\mu_d > D_0$
    2. $\mu_d < D_0$
    3. $\mu_d \neq D_0$

T.S.: $t = \dfrac{\bar{d} - D_0}{s_d/\sqrt{n}}$ where $\bar{d}$ and s_d are the sample mean and standard deviation of the n differences

R.R.: For a specified value of α and df $= n - 1$
    1. reject H_0 if $t > t_\alpha$
    2. reject H_0 if $t < -t_\alpha$
    3. reject H_0 if $|t| > t_{\alpha/2}$

EXAMPLE 5.10

Refer to the data of Example 5.9 and perform a paired t test. Draw a conclusion based on $\alpha = .05$.

Solution

For these data, the parts of the statistical test are

H_0: $\mu_d = \mu_1 - \mu_2 = 0$

H_a: $\mu_d > 0$

T.S.: $t = \dfrac{\bar{d}}{s_d/\sqrt{n}}$

R.R.: For df $= n - 1 = 14$, reject H_0 if $t > t_{.05}$

Before computing t we must first calculate s_d, the sample standard deviation of the differences. We can calculate s_d by using our shortcut formula for a sample variance or by using a calculator:

$$s_d^2 = \frac{1}{n-1}\left[\sum_i d_i^2 - \frac{(\sum_i d_i)^2}{n}\right]$$

For the data of Table 5.5

$$\sum_i d_i = .3 + 1.1 + 1.1 + \cdots + .9 = 9.2, \quad \bar{d} = \frac{9.2}{15} = .61 \text{ and}$$

$$\sum_i d_i^2 = (.3)^2 + (1.1)^2 + (1.1)^2 + \cdots + (.9)^2 = 7.82$$

Hence for $n = 15$ differences,

$$s_d^2 = \frac{1}{14}\left[7.82 - \frac{(9.2)^2}{15}\right] = .156$$

$$s_d = \sqrt{.156} = .394$$

Substituting into the test statistic t, we have

$$t = \frac{\bar{d} - 0}{s_d/\sqrt{n}} = \frac{.61}{.394/\sqrt{15}} = 6.00$$

Indeed $t = 6.00$ is far beyond all tabulated t values for df $= 14$, so the p-value is less than .005; presumably p is much less than .005. We conclude that the mean repair estimate for garage I is greater than that for garage II. This conclusion agrees with our intuitive finding based on the binomial distribution.

The corresponding general $100(1 - \alpha)\%$ confidence interval for μ_d based on paired data is shown here.

$100(1 - \alpha)\%$ CONFIDENCE INTERVAL FOR μ_d BASED ON PAIRED DATA	$\bar{d} \pm t_{\alpha/2}\, s_d/\sqrt{n}$ where n is the number of pairs of observations (and hence the number of differences) and df $= n - 1$.

The use of these t procedures depends on the assumption that the population of *differences* is normally distributed. For small samples, plot the sample differences; if severe skewness or outliers are present, the binomial test or the signed-rank test of Section 5.5 should be used.

EXERCISES

5.22   Consider the paired data shown here.

Pair	y_1	y_2
1	21	29
2	28	30
3	17	21
4	24	25
5	27	33

a. Run a paired t test and give the p-value for the test.
b. What would your conclusion be using an argument related to the binomial distribution? Does it agree with part (a)? When might these two approaches not agree?

5.23   An agricultural experiment station was interested in comparing the yields for two new varieties of corn. Because the investigators thought that there might be a great deal of variability in yield from one farm to another, each variety was randomly assigned to a different 1-acre plot on each of seven farms. The 1-acre plots were planted; the corn was harvested at maturity. The results of the experiment (in bushels of corn) are listed here.

Farm	1	2	3	4	5	6	7
Variety A	48.2	44.6	49.7	40.5	54.6	47.1	51.4
Variety B	41.5	40.1	44.0	41.2	49.8	41.7	46.8

Use these data to test the null hypothesis that there is no difference in mean yields for the two varieties of corn. Use $\alpha = .05$.

5.24   Thirty sets of identical twins were asked to participate in a one-year study designed to measure certain social attitudes. One twin from each set was randomly assigned to live in the home of a minority family, while the other twin stayed at home. After one year each person was asked to respond to a long questionnaire designed to detect and measure well-defined attitudes. Let sample 1 denote the combined questionnaire scores for those persons who lived at home and sample 2 denote the set of scores for those who lived with a family from a minority class.

Set of Twins	Home Environment, y_1	Minority Environment, y_2	Difference
1	78	71	7
2	75	70	5
3	68	66	2
4	92	85	7
5	55	60	−5
6	74	72	2
7	65	57	8
8	80	75	5
9	98	92	6
10	52	56	−4
11	67	63	4
12	55	52	3
13	49	48	1
14	66	67	−1
15	75	70	5
16	90	88	2
17	89	80	9
18	73	65	8
19	61	60	1
20	76	74	2
21	81	76	5
22	89	78	11
23	82	78	4
24	70	62	8
25	68	73	−5
26	74	73	1
27	85	75	10
28	97	88	9
29	95	94	1
30	78	75	3
	$\bar{y}_1 = 75.23$	$\bar{y}_2 = 71.43$	$\bar{d} = \bar{y}_1 - \bar{y}_2 = 3.8$

a. Plot the sample differences. Is there any reason to believe that a t test is inappropriate?

b. Test the null hypothesis

H_0: $\mu_1 - \mu_2 = 0$ (the population mean scores for those not exposed and those exposed to a minority environment are identical)

against the alternative

H_a: $\mu_1 - \mu_2 \neq 0$ (the population mean scores are different for the two environments).

Use $\alpha = .05$.

5.25   Suppose that we wish to estimate the difference between the mean monthly salaries of male and female sales representatives. Since there is a great deal of salary variability from company to company, it was decided to filter out the variability due to companies by making male-female comparisons within each company. One male and one female with the required background and work experience will be selected from each company. If the range of differences in salaries (between males and females) within a company is approximately $300 per month, determine the number of companies that must be examined to estimate the difference in mean monthly salary for males and females. Use a 95% confidence interval with a half width of $5. (Hint: Refer to Section 4.3.)

5.26   Refer to Exercise 5.25. If $n = 35$, $\bar{d} = 120$, and $s_d = 250$, construct a 90% confidence interval for μ_d, the mean difference in salaries for male and female sales representatives.

5.5   A NONPARAMETRIC ALTERNATIVE: WILCOXON SIGNED-RANK TEST

The Wilcoxon signed-rank test, which makes use of the sign and the magnitude of the rank of the differences between pairs of measurements, provides an alternative to the paired t test. The formal null hypothesis for Wilcoxon's signed-rank test is that the population distribution of differences is symmetrical about D_0; the test is sensitive to the distribution of differences being shifted to the right or left of D_0. In most cases D_0 is 0; otherwise we subtract D_0 from every measurement and proceed as if $D_0 = 0$. The test uses the nonzero differences ranked in absolute value from lowest to highest. If two or more measurements have the same nonzero difference (ignoring sign), we assign each difference a rank equal to the average of the occupied ranks. The appropriate sign is then attached to the rank of each difference.

Before summarizing the Wilcoxon signed-rank test, we define the following notation:

n = the number of pairs of observations with a nonzero difference

T_+ = the sum of the positive ranks; if there are no positive ranks, $T_+ = 0$

T_- = the sum of the negative ranks; if there are no negative ranks, $T_- = 0$

T = the smaller of T_+ and T_-, ignoring their signs

μ_T $$\mu_T = \frac{n(n + 1)}{4}$$

σ_T $$\sigma_T = \sqrt{\frac{n(n + 1)(2n + 1)}{24}}$$

g groups   If we group all differences assigned the same rank together, and there are g such groups, the variance of T is

$$\sigma_T^2 = \frac{1}{24}\left[n(n + 1)(2n + 1) - \frac{1}{2}\sum_j t_j(t_j - 1)(t_j + 1)\right]$$

t_j   where t_j is the number of tied ranks in the jth group. Note that if there are no tied ranks, $g = n$ and $t_j = 1$ for all groups. The formula then reduces to

$$\sigma_T^2 = \frac{n(n + 1)(2n + 1)}{24}$$

The Wilcoxon signed-rank test is presented here.

WILCOXON SIGNED-RANK TEST

H_0: The distribution of differences is symmetric around D_0. (D_0 is specified; usually D_0 is 0.)

H_a: 1. The differences tend to be larger than D_0.
2. The differences tend to be smaller than D_0.
3. Either 1. or 2. is true (two-sided H_a).

$(n \leq 50)$
T.S.: 1. $T = |T_-|$
2. $T = T_+$
3. $T = $ smaller of $|T_-|$, T_+

R.R.: For a specified value of α (one-tailed .05, .025, .01, or .005; two-tailed .10, .05, .02, .01) and fixed number of nonzero differences n, reject H_0 if the value of T is less than or equal to the appropriate entry in Table 9 in the Appendix.

$(n > 50)$
T.S.: Compute the test statistic

$$z = \frac{T - \frac{n(n + 1)}{4}}{\sqrt{\frac{n(n + 1)(2n + 1)}{24}}}$$

R.R: For cases 1 and 2, reject H_0 if $z < -z_\alpha$; for case 3 reject H_0 if $z < -z_{\alpha/2}$.

EXAMPLE 5.11   Two different brands of fertilizer (A and B) were compared on each of 10 different 2-acre plots. Each plot was subdivided into 1-acre subplots, with brand A randomly assigned to one subplot and brand B to the other. Sixty pounds per acre of fertilizers were then applied to subplots. The data for barley yields in bushels per acre are listed in Table 5.6 by fertilizer and plot.

TABLE 5.6 Barley Yields (in Bushels) by Plot and by Fertilizer, Example 5.11

| | Barley Yield | | |
| | Fertilizer A | Fertilizer B | Difference |
Plot	y_1	y_2	$y_1 - y_2$
1	312	346	-34
2	333	372	-39
3	356	392	-36
4	316	351	-35
5	310	330	-20
6	352	364	-12
7	389	375	14
8	313	315	-2
9	316	327	-11
10	346	378	-32

Use the Wilcoxon signed-rank test to test the hypothesis that the distributions of barley yields for the two brands of fertilizer are identical against the alternative that they are different. Use $\alpha = .05$.

Solution

First we must rank (from lowest to highest) the absolute values of the $n = 10$ differences. These ranks appear in column 2 of Table 5.7. The appropriate sign is then attached to each rank (see column 3 in Table 5.7). The sum of the positive and negative ranks, are, respectively,

$$T_+ = 4$$
$$T_- = -7 + (-10) + \cdots + (-6) = -51$$

Thus T, the smaller of T_+ and T_-, ignoring the sign, is 4.

TABLE 5.7 Rankings for the Data of Table 5.6

Plot	Rank of Difference $|y_1 - y_2|$	Rank with Appropriate Sign
1	7	-7
2	10	-10
3	9	-9
4	8	-8
5	5	-5
6	3	-3
7	4	4
8	1	-1
9	2	-2
10	6	-6

For a two-tailed test with $n = 10$ and $\alpha = .05$, we see from Table 9 in the Appendix that we will reject H_0 if T is less than or equal to 8. Thus we reject H_0 and conclude that the distributions of barley yields for the two brands of fertilizers are different. Barley yields for fertilizer A tend to be smaller than (to the left of) corresponding yields for fertilizer B.

The choice of an appropriate paired-sample test follows the guidelines mentioned for unpaired data in Section 5.2. If the assumptions of the t test are satisfied—in particular, if the distribution of differences is roughly normal—the t test is more powerful. If the distribution of differences is grossly skewed, the nominal t probabilities may be misleading. If the distribution is roughly symmetric but has heavy tails (as indicated by the presence of outliers), the signed-rank test may be more powerful. Often, the tests will yield essentially the same conclusion.

Even with this discussion you might still be confused as to which statistical test (or confidence interval) to apply in a given situation where there is a choice of two or more methods. When in doubt, do several different tests; computing costs are usually minimal, especially with the availability of many different statistical software packages such as Minitab, SAS, and SPSS. If the results from the different analyses yield different results, you should identify the peculiarities of the data set to understand why the results differ. If the results agree, and if there are no blatant violations of assumptions, you should be very confident in your conclusions.

This particular "hedging" strategy is appropriate not only for paired data, but for many of the situations we have discussed. Since computer software makes it easy to run alternative analyses on the same data, the potential concern about assumptions often can be put to rest when the alternative analyses yield essentially the same results.

EXERCISES

5.27   Refer to Exercise 5.24.
   a. Using the data in the table, run a Wilcoxon signed-rank test. Give the p-value and draw a conclusion.
   b. Compare your conclusions here to those in Exercise 5.24. Does it make a difference which test (t or signed rank) is used?

5.28   Two judges were asked to rate separately each of 22 inmates on his rehabilitative potential. These data appear next.

Inmate	Judge 1	Judge 2	Inmate	Judge 1	Judge 2
1	6	5	12	9	8
2	12	11	13	10	8
3	3	4	14	6	7
4	9	10	15	12	9
5	5	2	16	4	3
6	8	6	17	5	5
7	1	2	18	6	4
8	12	9	19	11	8
9	6	5	20	5	3
10	7	4	21	10	9
11	6	6	22	10	11

Use the computer output shown here to reach a conclusion about the following.

a. H_0: The distribution of differences is symmetrical about 0 versus H_a: The difference tends to be larger than 0. What is your conclusion? What is the *p*-value?

b. How would the results of part (a) compare to those from a paired *t* test? Use the output below to draw conclusions.

OBS	JUDGE-1	JUDGE-2	DIFF
1	6	5	1
2	12	11	1
3	3	4	−1
4	9	10	−1
5	5	2	3
6	8	6	2
7	1	2	−1
8	12	9	3
9	6	5	1
10	7	4	3
11	6	6	0
12	9	8	1
13	10	8	2
14	6	7	−1
15	12	9	3
16	4	3	1
17	5	5	0
18	6	4	2
19	11	8	3
20	5	3	2
21	10	9	1
22	10	11	−1

N = 22

PAIRED T TEST

VARIABLE	LABEL	N	MEAN	STANDARD STD DEVIATION	STD ERROR OF MEAN	T	PR > |T|
DIFF	DIFFERENCE IN RATINGS	22	1.09090909	1.47709789	0.31491833	3.46	0.0023

WILCOXON SIGNED RANK TEST
STATISTICAL ANALYSIS SYSTEM—NPAR 360 INTERFACE

WILCOXON MATCHED-PAIRS SIGNED-RANKS TEST
WITH ONE-TAIL PROBABILITIES OF THIS OR GREATER T

HIGHER GROUP	LOWER GROUP	N(TOTAL)	N(SIGNED)	WILCOXON T	PROBABILITY
JUDGE__1	JUDGE__2	22	20	30.00	0.0026

5.29   The effect of Benzedrine on the heart rate of dogs (in beats per minute) was examined in an experiment on 14 dogs chosen for the study. Each dog was to serve as its own control, with half of the dogs assigned to receive Benzedrine during the first study period and the other half assigned to receive a placebo (saline solution). All dogs were examined to determine the heart rates after two hours on the medication. After two weeks in which no medication was given, the regimens for the dogs were switched for the second study period. The dogs previously on Benzedrine were given the placebo while the others received Benzedrine. Again heart rates were measured after two hours.

The sample data below are not arranged in the order in which they were taken, but have been summarized by regimen. Use these data to test the research hypothesis that the distribution of heart rates for the dogs when receiving Benzedrine is shifted to the right of that for the same animals when on the placebo. Use a one-tailed Wilcoxon signed-rank test with $\alpha = 0.5$.

Dog	Placebo	Benzedrine
1	250	258
2	271	285
3	243	245
4	252	250
5	266	268
6	272	278
7	293	280
8	296	305
9	301	319
10	298	308
11	310	320
12	286	293
13	306	305
14	309	313

5.6   CHOOSING SAMPLE SIZES FOR INFERENCES ABOUT $\mu_1 - \mu_2$

Sections 4.3 and 4.5 were devoted to sample-size calculations to obtain a confidence interval about μ with a fixed width and specified degree of confidence or to conduct a statistical test concerning μ with predefined levels for α and β. Similar calculations can be made for inferences about $\mu_1 - \mu_2$ with either independent samples or with paired data. Determining the sample size for a $100(1 - \alpha)\%$ confidence interval about $\mu_1 - \mu_2$ of width $2E$ based on independent samples is possible by solving the following expression for n. We will assume that both samples are of the same size.

$$z_{\alpha/2}\sigma\sqrt{\frac{1}{n} + \frac{1}{n}} = E$$

Note that in this formula σ is the common population standard deviation and that we have assumed equal sample sizes.

SAMPLE SIZES
FOR A
100(1 − α)%
CONFIDENCE
INTERVAL FOR
$\mu_1 - \mu_2$ OF THE
FORM $\bar{y}_1 - \bar{y}_2 \pm E$,
INDEPENDENT
SAMPLES

$$n = \frac{2z_{\alpha/2}^2 \sigma^2}{E^2}$$

Note: If σ is unknown, substitute an estimated value to get an approximate sample size.

The sample sizes obtained using this formula are usually approximate because we have to substitute an estimated value of σ, the common population standard deviation. This estimate will probably be based on an educated guess from information on a previous study or on the range of population values.

Corresponding sample sizes for one- and two-sided tests of H_0: $\mu_1 - \mu_2 = D_0$ based on specific values of α and β are shown here.

SAMPLE SIZES
FOR TESTING H_0:
$\mu_1 - \mu_2 = D_0$,
INDEPENDENT
SAMPLES

One-sided test $n = 2\sigma^2 \dfrac{(z_\alpha + z_\beta)^2}{\Delta^2}$

Two-sided test $n \approx 2\sigma^2 \dfrac{(z_{\alpha/2} + z_\beta)^2}{\Delta^2}$

where $n_1 = n_2 = n$ and the probability of a Type II error is to be $\leq \beta$ when the true difference $|\mu_1 - \mu_2| \geq \Delta$.

Note: If σ is unknown, substitute an estimated value to obtain an approximate sample size.

EXAMPLE 5.12

An experiment was done to determine the effect on dairy cattle of a diet supplemented with liquid whey. While no differences were noted in milk production measurements among cattle given a standard diet (7.5 kg of grain plus hay by choice) with water and those on the standard diet and liquid whey only, a considerable difference between the groups was noted in the amount of hay ingested. Suppose that one tests the null hypothesis of no difference in mean hay consumption for the two diet groups of dairy cattle. For a two-tailed test with $\alpha = .05$, determine the approximate number of dairy cattle that should be included in each group if we want $\beta \leq .10$ for $|\mu_1 - \mu_2| \geq .5$. Previous experimentation has shown σ to be approximately .8.

Solution

From the description of the problem we have $\alpha = .05$, $\beta \leq .10$ for $\Delta = |\mu_1 - \mu_2| \geq .5$ and $\sigma = .8$. Table 2 in the Appendix gives us $z_{.025} = 1.96$ and $z_{0.10} = 1.28$. Substituting into the formula we have

$$n \approx \frac{2(.8)^2(1.96 + 1.28)^2}{(.5)^2} = 53.75 \text{ or } 54$$

That is, we need 54 cattle per group to run the desired test.

Sample-size calculation can also be done using the formulas shown when $n_1 \neq n_2$. In this situation we let n_2 be some multiple m (e.g., $m = .5$) of n_1; then we substitute $(m + 1)/m$ for 2 in the sample size formulas. After solving for n_1, $n_2 = mn_1$.

Sample sizes for estimating μ_d and conducting a statistical test for μ_d based on paired data (differences) are found using the formulas of Chapter 4 for μ. The only change is that we're working with a single sample of differences rather than a single sample of y values. For convenience the appropriate formulas are shown here.

SAMPLE SIZE REQUIRED FOR A 100(1 − α)% CONFIDENCE INTERVAL FOR μ_d OF THE FORM $\bar{d} \pm E$	$$n = \frac{z_{\alpha/2}^2 \sigma_d^2}{E^2}$$ Note: If σ_d is unknown, substitute an estimated value to obtain approximate sample size.

SAMPLE SIZES FOR ONE- AND TWO-SIDED TESTS OF H_0: $\mu_d = D_0$	One-sided test $$n = \frac{\sigma_d^2(z_\alpha + z_\beta)^2}{\Delta^2}$$ Two-sided test $$n \approx \frac{\sigma_d^2(z_{\alpha/2} + z_\beta)^2}{\Delta^2}$$ where the probability of a Type II error is β or less if the true difference $\mu_d \geq \Delta$. Note: If σ_d is unknown, substitute an estimated value to obtain an approximate sample size.

5.7   SUMMARY

In this chapter we have considered inferences about $\mu_1 - \mu_2$. The first set of methods was based on independent random samples being selected from the populations of interest. We learned how to sample data to run a statistical test or to construct a confidence interval for $\mu_1 - \mu_2$ using t methods. Wilcoxon's rank sum test, which does not require normality of the underlying populations, was presented as an alternative to the t test. Finally, the Tukey–Duckworth test was introduced as a quick, portable test that can be used as a preliminary test when time or circumstance dictates that a formal test cannot be run immediately.

The second major set of procedures can be used to make comparisons between two populations when the sample measurements are paired. In this situation we no longer have independent random samples and hence the procedures of Sections 5.1–5.3 (t methods, Wilcoxon's rank sum, and the Tukey–Duckworth test) are inappropriate. The test and estimation methods for paired data are based on the sample differences for the paired measurements or the ranks of the differences. The paired t test and corresponding confidence interval based on the difference measurements were introduced and found to be identical to the single sample t methods of Chapter 4. The nonparametric alternative to the paired t test is Wilcoxon's signed-rank test.

The material presented in Chapters 4 and 5 lays the foundations of statistical inference (estimation and testing) for the remainder of the text. It would be good to review periodically the material in this chapter as new topics are introduced, so that you retain the basic elements of statistical inference.

KEY FORMULAS

Inferences about $\mu_1 - \mu_2$

1. $100(1 - \alpha)\%$ confidence interval for $\mu_1 - \mu_2$, independent samples; y_1 and y_2 approximately normal; $\sigma_1^2 = \sigma_2^2$

$$\bar{y}_1 - \bar{y}_2 \pm t_{\alpha/2} s_p \sqrt{\frac{1}{n_1} + \frac{1}{n_2}}$$

where

$$s_p = \sqrt{\frac{(n_1 - 1)s_1^2 + (n_2 - 1)s_2^2}{n_1 + n_2 - 2}} \qquad \text{and} \qquad df = n_1 + n_2 - 2$$

2. t test for $\mu_1 - \mu_2$, independent samples; y_1 and y_2 approximately normal; $\sigma_1^2 = \sigma_2^2$

$$H_0: \mu_1 - \mu_2 = D_0$$

$$\text{T.S.: } t = \frac{\bar{y}_1 - \bar{y}_2 - D_0}{s_p \sqrt{\frac{1}{n_1} + \frac{1}{n_2}}} \qquad df = n_1 + n_2 - 2$$

3. t' test for $\mu_1 - \mu_2$, unequal variance, independent samples; y_1 and y_2 approximately normal;

$$H_0: \mu_1 - \mu_2 = D_0$$

$$\text{T.S.: } t' = \frac{\bar{y}_1 - \bar{y}_2 - D_0}{\sqrt{\frac{s_1^2}{n_1} + \frac{s_2^2}{n_2}}} \qquad df = \frac{(n_1 - 1)(n_2 - 1)}{(n_2 - 1)c^2 + (1 - c)^2(n_1 - 1)}$$

where

$$c = \frac{s_1^2/n_1}{\frac{s_1^2}{n_1} + \frac{s_2^2}{n_2}}$$

4. Wilcoxon's rank sum test, independent samples

   H_0: The two populations are identical

   $(n_1 \leq 10, n_2 \leq 10)$

   T.S.: T, the sum of the ranks in sample 1

   $(n_1, n_2 > 10)$

   T.S.: $z = \dfrac{T - \mu_T}{\sigma_T}$ where T denotes the sum of the ranks in sample 1

   $$\mu_T = \frac{n_1(n_1 + n_2 + 1)}{2} \qquad \text{and} \qquad \sigma_T = \sqrt{\frac{n_1 n_2}{12}(n_1 + n_2 + 1)}$$

   provided there are no tied ranks.

5. Paired t test; difference approximately normal

   H_0: $\mu_d = D_0$

   T.S.: $t = \dfrac{\bar{d} - D_0}{s_d/\sqrt{n}}$ df $= n - 1$ where n is the number of differences

6. $100(1 - \alpha)\%$ confidence interval for μ_d, paired data; differences approximately normal

   $$\bar{d} \pm t_{\alpha/2} s_d/\sqrt{n}$$

7. Wilcoxon's signed-rank test, paired data

   H_0: The distribution of differences is symmetrical about D_0

   T.S.: $n > 50$

   $$z = \frac{T - \mu_T}{\sigma_T} \quad \text{where } \mu_T = \frac{n(n + 1)}{4} \quad \text{and} \quad \sigma_T = \sqrt{\frac{n(n + 1)(2n + 1)}{24}}$$

   provided there are no tied ranks.

8. Independent samples: sample sizes for estimating $\mu_1 - \mu_2$ with a $100(1 - \alpha)\%$ confidence interval, $\bar{y}_1 - \bar{y}_2 \pm E$

   $$n = \frac{2z_{\alpha/2}^2 \sigma^2}{E^2}$$

9. Independent samples: sample sizes for a test of H_0: $\mu_1 - \mu_2 = D_0$

   a. One-sided test:

   $$n = \frac{2\sigma^2(z_\alpha + z_\beta)^2}{\Delta^2}$$

   b. Two-sided test:

   $$n \approx \frac{2\sigma^2(z_{\alpha/2} + z_\beta)^2}{\Delta^2}$$

EXERCISES

5.30 A study was conducted to compare the flavor of cheddar cheese made two different ways: with milk and with milkfat plus buttermilk solids. Cheese was processed in large vats, one vat for each of the different types of fat, and stored for six months. A panel of 15 trained judges was employed to compare each of the cheeses. On the day of testing, the cheese was brought to room temperature. Each judge was presented with an unmarked square of each type of cheese and asked to taste the cheeses and judge them for cheddar flavor on a 0-to-8 point scale.

			Pronounced (almost identical to a
None	Slight	Moderate	typical cheddar flavor)
0 1 2	3 4	5 6	7 8

Each judge's score was the difference between the rating on cheese made from milk and the rating on cheese made from milkfat supplemented with buttermilk solids. The sample data are summarized below.

Sample size: 15 (number of differences)

Sample mean: 1.3

Sample standard deviation: 6.04

Use these data to test the null hypothesis that the mean score (mean difference in cheddar flavor ratings for the two cheeses) is zero against a two-sided alternative hypothesis. Give the level of significance of your test and interpret the results of this test.

5.31 Refer to Exercise 5.30. Construct a 95% confidence interval for the mean difference in cheddar ratings for the two cheeses. Are the results from the interval estimate consistent with the statistical test in Exercise 5.30?

5.32 A study was conducted on 16 dairy cattle. Eight cows were randomly assigned to a liquid regimen of water only (group 1); the others received liquid whey only (group 2). In addition, each animal was given 7.5 kg of grain per day and allowed to graze on hay at will. While no significant differences were observed between the groups in the dairy milk production gauges, such as milk production and fat content of the milk, the following data on daily hay consumption (in kg/cow) were of interest:

Group 1	15.1	14.9	14.8	14.2	13.1	12.8	15.5	15.9
Group 2	6.8	7.5	8.6	8.4	8.9	8.1	9.2	9.5

Use these sample data to test the research hypothesis that there is a difference in mean hay consumption for the two diets. Use $\alpha = .05$.

5.33 Refer to Example 5.12. Suppose we only wish to detect $\mu_1 - \mu_2 > 0$. Determine the sample size required when $\alpha = .05$ and $\beta \le .10$ when $\mu_1 - \mu_2 \ge .5$.

5.34 A study is conducted to compare the average number of years of service at the age of retirement for military personnel in 1970 versus 1980. A random sample of career records is to be obtained for each year. Previous information suggests that the common population standard deviation is approximately four years. Determine the number of observations to be collected from records for each year if we wish to estimate the mean difference in age using a 95% confidence interval with a half width of .5 years.

5.35 Refer to Exercise 5.32. Construct a 90% confidence interval for the difference in mean hay consumption for cattle on the two diets.

5.36   Over the past decade or more there has been a steady decrease in the national average Scholastic Aptitude Test (SAT) score. Many parents, teachers, and administrators have been concerned about this trend and have sought ways to halt the decline, at least at the local school level. To this end a group of 50 students (24 males and 26 females), matched as nearly as possible according to socioeconomic background, received parental permission to participate in a study to examine the effect of classroom atmosphere (strict or liberal) on student performance, as measured on a standardized achievement test score at the end of the year. The 50 students were randomly divided into two groups of 25 students each (12 males and 13 females), with group I to study under a strict, closely regulated classroom atmosphere while group II attended classes under a very permissive atmosphere. Since groups rotated to different subject matter classes throughout the day, it was possible to employ the same subject matter teachers for both groups. After nine months under this program, all students were given the same standardized test: the verbal test and the mathematics test. The results are given below.

	Group I			Group II		
	Verbal	Math	Total	Verbal	Math	Total
Sample mean	45.2	46.6	91.8	40.3	43.4	83.7
Sample standard deviation	6.0	4.7	8.1	5.1	5.9	8.4

a. Use the sample verbal data for the two groups to test the research hypothesis that the group with a more regimented classroom atmosphere will score higher, on the average, than the less restricted group. Use $\alpha = .05$.

b. Would your results change using a two-tailed test with $\alpha = .05$?

5.37   Refer to Exercise 5.36. Using the verbal data, construct a 90% confidence interval on the difference in the mean scores for the two groups.

5.38   Using the mathematics test-score data of Exercise 5.36, construct a 99% confidence interval for the difference in mean responses between the two groups.

5.39   Refer to Exercise 5.36. Using the math data, perform the same type of statistical test with $\alpha = .05$.

5.40   Refer to Exercise 5.36. Some people might argue that the best measure of student performance using this standardized test is the sum of the verbal and mathematics scores. Use these data to test the research hypothesis that the two groups have different mean test scores. Use $\alpha = .01$.

 5.41   Refer to Exercise 5.40. Use the sample data in Exercise 5.36 to construct a 98% confidence interval on the difference in mean test scores.

 5.42   An industrial concern has experimented with several different mixtures of the four components magnesium, sodium nitrate, strontium nitrate, and binder, which comprise a rocket propellant. The company has found that two mixtures in particular give higher flare illumination values than the others. Mixture 1 consists of a blend composed of the proportions .40, .10, .42, and .08, respectively, for the four components of the mixture; Mixture 2 consists of a blend using the proportions .60, .27, .10, and .03. Twenty different blends (10 of each mixture) are prepared and tested to obtain the flare illumination values. These data appear below (in units of 1000 candles).

Mixture 1	185	192	201	215	170	190	175	172	198	202
Mixture 2	221	210	215	202	204	196	225	230	214	217

a. Plot the sample data. Which test(s) could be used to compare the mean illumination values for the two mixtures?

EXERCISES

b. Give the level of significance of the test and interpret your findings.

5.43 Refer to Exercise 5.42. Instead of conducting a statistical test, use the sample data to answer the question, "What is the difference in mean flare illumination for the two mixtures?"

5.44 Refer to Example 5.2. Suppose the seventh untreated animal died before the study was completed. Analyze the remaining observations to compare the two population means. Assume that $\sigma_1^2 \neq \sigma_2^2$. Give the level of significance for your test.

5.45 Refer to the computer printout for the statistical test of the data in Exercise 5.2.
a. Any problems in running a t test?
b. Compare these computer results to your calculations for Exercise 5.2.
c. Give the value of the test statistic and the level of significance for a t test of the research hypothesis that the experimental mean is greater than the control mean.

Control group	5.4	6.2	3.1	3.8	6.5	5.8	6.4	4.5	4.9	4.0
Experimental group	8.8	9.5	10.6	9.6	7.5	6.9	7.4	6.5	10.5	8.3

```
                                TWO SAMPLE T-TEST

                                 TTEST PROCEDURE

VARIABLE: LEADPCT

 GROUP      N        MEAN       STD DEV    STD ERROR     MINIMUM       MAXIMUM     VARIANCES      T       DF    PROB > |T|

 CONTROL    10   5.06000000   1.18902388  0.37600236   3.10000000    6.50000000   UNEQUAL    -5.8507    17.2    0.0001
 EXPERMT    10   8.56000000   1.47135614  0.46528366   6.50000000   10.60000000   EQUAL      -5.8507    18.0    0.0001

 FOR HO: VARIANCES ARE EQUAL, F'=    1.53 WITH 9 AND 9 DF     PROB > F'= 0.5356
```

 5.46 A study of anxiety was conducted among residents of a southeastern metropolitan area. Each person selected for the study was asked to check a yes or a no for the presence of each of 12 anxiety symptoms. Anxiety scores ranged from 0 to 12, with higher scores related to higher perceived presence of anxiety symptoms. The results for a random sample of 50 residents, categorized by sex, are summarized below.

	Sample Size	Mean	Standard Deviation
Female	26	5.26	3.2
Male	24	7.02	3.9

Use these data to test the research hypothesis that the mean perceived anxiety score is different for males and females. Give the level of significance for your test.

 5.47 A clinical trial was conducted to determine the effectiveness of drug A in the treatment of symptoms associated with alcohol withdrawal. A total of 30 patients were treated (under blinded conditions) with drug and another 30 with an identical-appearing placebo. The average symptom score for the two groups after one week of therapy was 1.5 and 6.3, respectively. (Note: higher symptom scores indicate more withdrawal "problems.") The corresponding standard deviations were 3.1 and 4.2.
a. Compare the mean total symptom scores for the two groups. Give the p-value for a two sample t test of $H_0: \mu_1 - \mu_2 = 0$ versus $H_a: \mu_1 - \mu_2 < 0$. Draw conclusions.
b. Suppose the average total symptoms scores were 6.8 and 12.2 prior to therapy. How would this affect your conclusions? How could you guard against possible baseline (pretreatment) differences?

5.48   Two analysts, supposedly of identical abilities, each measure the parts per million of a certain type of chemical impurity in drinking water. It is claimed that analyst 1 tends to give higher readings than analyst 2. To test this theory, each of six water samples is divided and then analyzed by both analysts separately. The data are shown in the accompanying table (readings in ppm).

Water Sample	Analyst 1	Analyst 2
1	31.4	28.1
2	37.0	37.1
3	44.0	40.6
4	28.8	27.3
5	59.9	58.4
6	37.6	38.9

a. Is there evidence to indicate that analyst 1 reads higher on the average than analyst 2? Give the level of significance for your test.

b. What would be the conclusion using a Wilcoxon test? Compare your results to part (a).

5.49   A single leaf was taken from each of 11 different tobacco plants. Each was divided in half; one half was chosen at random and treated with preparation 1 and the other half received preparation II. The object of the experiment was to compare the effects of the two preparations of mosaic virus on the number of lesions on the half leaves after a fixed period of time. These data are recorded here.

	Number of Lesions on the Half Leaf	
Tobacco Plant	Preparation I	Preparation II
1	18	14
2	20	15
3	9	6
4	14	12
5	38	32
6	26	30
7	15	9
8	10	2
9	25	18
10	7	3
11	13	6

For $\alpha = .05$, use Wilcoxon's signed-rank test to examine the research hypothesis that the distributions of lesions are different for the two populations.

5.50   An investigator plans to compare the mean number of particles of effluent in water collected at two different locations in a water treatment plant. If the standard deviation for particle counts is expected to be approximately 6 for the counts in samples taken at

EXERCISES

5.51 each of the locations, determine the sample sizes required to estimate the mean difference in particle of effluent using a 99% confidence interval of width 1 (particle).

The weight gains for $n_1 = n_2 = 8$ rats tested on diets 1 and 2 are summarized here. Set up a statistical test for $\mu_1 - \mu_2$, the difference in average weight gained for the two diets. Use $\alpha = .05$ and draw conclusions.

Diet 1		Diet 2
25	Σ_y	26.2
.005	s	.045
8	n	8

5.52 Refer to the computer simulation in Example 5.4. Another computer simulation study of 1000 samples was run using the same sample sizes but different population standard deviations. The independent normal populations were as shown here:

Population

1	$\mu_1 = 100$	$\sigma_1 = 15$	$n_1 = 10$
2	$\mu_2 = 100$	$\sigma_2 = 10$	$n_2 = 10$

For each run of the simulation study, the pooled t test and separate-variance t test were run. The number (%) of times t and t' were rejected at the upper and lower .05 levels is recorded here.

	Pooled t Test	Separate-Variance t-Test
H_0 rejected, H_1: $\mu_1 - \mu_2 < 0$	43 (4.3%)	44 (4.6%)
H_0 rejected, H_1: $\mu_1 - \mu_2 > 0$	48 (4.8%)	46 (4.4%)

a. Before reviewing the results of the simulation study, how well do you think the pooled t test will perform even though one of the assumptions is obviously not satisfied? Explain.

b. Which test performed better? Which would you recommend and why?

5.53 The following memorandum opinion on statistical significance was issued by the judge in a trial involving many scientific issues. The opinion has been stripped of some legal jargon and has been taken out of context. Still, it can give us an understanding of how others deal with the problem of ascertaining the meaning of statistical significance. Read this memorandum and comment on the issues raised regarding statistical significance

Memorandum Opinion

This matter is before the Court upon two evidentiary issues that were raised in anticipation of trial. First, it is essential to determine the appropriate level of statistical significance for the admission of scientific evidence.

With respect to statistical significance, no statistical evidence will be admitted during the course of the trial unless it meets a confidence level of 95%.

Every relevant study before the court has employed a confidence level of at least 95%. In addition, plaintiffs concede that social scientists routinely utilize a 95% confidence level. Finally, all legal authorities agree that statistical evidence is inadmissable unless it meets the 95% confidence level required by statisticians. Therefore, because plaintiffs advance no reasonable basis to alter the accepted approach of mathematicians to the test of statistical significance, no statistical evidence will be admitted at trial unless it satisfies the 95% confidence level.

5.54   Certain baseline determinations were made on 182 patients entered in a study of survival in males suffering from congestive heart failure. At the time these data were summarized, 88 deaths had been observed. This table summarizes the baseline data for the survivors and nonsurvivors. The variables listed below "heart rate" are measures of the severity of the heart failure. The arrows to the left of each variable indicates the direction of improvement.

a. Discuss these baseline findings.

b. What assumptions have the authors made when doing these t tests?

Baseline Characteristics of Patients with Severe Chronic Left-Ventricular Failure Due to Cardiomyopathy Chronic Left-Ventricular Failure Due to Cardiomyopathy			
Variable	Nonsurvivors ($n = 88$)	Survivors ($n = 94$)	t Test p-Value
Age (yr)	57 ± 10	56 ± 8	NS
Duration of symptoms (mo)	45 ± 43	39 ± 27	NS
Heart rate (beats/min)	87 ± 15	83 ± 16	NS
↓Mean arterial pressure (mm Hg)	87 ± 13	94 ± 13	<0.001
↓Left-ventricular filling pressure (mm Hg)	29 ± 7	24 ± 9	<0.001
↑Cardiac index (liters/min/m^2)	2.0 ± 0.7	2.5 ± 0.8	<0.001
↑Stroke volume (ml/beat)	45 ± 16	59 ± 5	<0.001
↓Systemic vascular resistance (units)	25 ± 10	21 ± 8	<0.01
↑Stroke work (g-m)	35 ± 19	56 ± 33	<0.001

Values are listed as mean $\pm$ standard deviation.

5.55   Hospital administrators studied the patterns of length of hospital stays with particular attention paid to those patients having health-maintenance organization (HMO) payment sources versus those with non-HMO payment sources. The graph shown here summarizes the sample data.

EXERCISES

Average length of stay (days)

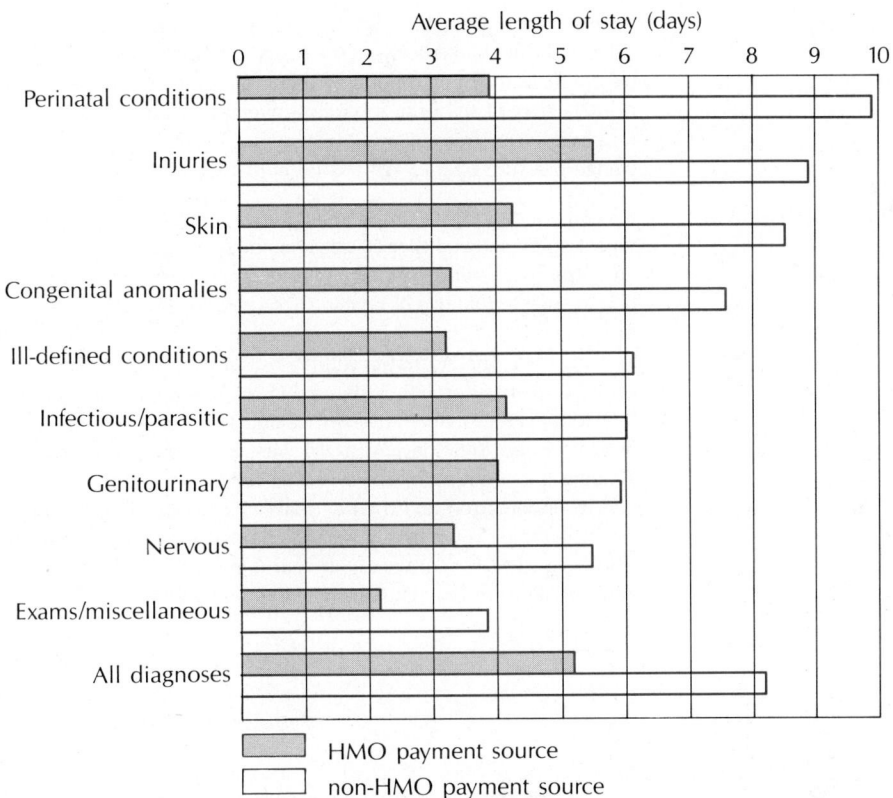

Sources: American Hospital Assn.: Twin Cities Metropolitan Health Board

a. What general observations would you draw from the graph? What additional information would you need to make more definitive statements regarding these results?
b. Suppose that across all diagnoses, the sample statistics were as shown here:

	Sample Mean	Sample Size	Sample Standard Deviation
HMO	5.0 (days)	120	1.3
Non-HMO	8.1	130	1.9

Use these data to test $H_0: \mu_1 - \mu_2 = 0$ versus $H_a: \mu_1 - \mu_2 \neq 0$. Give the p-value for your test.

5.56 Refer to Exercise 5.55. Also run the t' test and compare your results. Which (if any) test is better for these data?

5.57 An abstract for the results of a study of 10 congestive heart-failure patients is given here. Read the abstract and try to interpret the results.

An experimental compound was studied in 10 patients suffering from congestive heart failure. Certain variables were measured at baseline and then four hours after intravenous treatment with the compound. The compound was shown to increase cardiac index from 11.1 to 34.3% from a baseline average of 2.41 ± 0.49 1/min/m^2 ($p < .01$), heart rate by 6–10% from 72 ± 12 beats/min ($p < .02$) and decreased pulmonary capillary wedge pressure by 15.3–24.2% from 18.7 ($p < .001$).

5.58 Several antidepressant drugs have been studied in the treatment of cocaine abusers. One recent study showed that 20 cocaine abusers who were treated with an antidepressant in an outpatient setting experienced decreases in cravings after two weeks and some reduction in their actual use of cocaine. Comment on these results. Are they compelling? Why or why not?

5.59 In April 1986, The *Australian Journal of Statistics* (Vol. 30, No. 1, pp. 23–44) published the results of a study of S. R. Butler and H. W. Marsh on reading and arithmetic achievement for students from non–English-speaking families. All kindergarten students from seven public schools in Sidney, Australia were included in the original sample of 392 children. Reading and arithmetic achievement tests were administered at the start of the study during kindergarten and then at years 1, 2, 3, and 6 of the primary school.

The table shown here gives the characteristics of the 286 of the original 392 students who were available for testing at year 6 ($n = 226$ students from English-speaking families, and $n = 60$ from non–English-speaking families).

	Group	
Characteristics	English-Speaking Family ($n = 226$) $\bar{y}$	Non–English-Speaking Family ($n = 60$) $\bar{y}$
Age (in months)	67.17	67.15
Gender (1 = male, 2 = female)	1.50	1.55
Number of children in family	2.54	2.62
Ordinal position in family (1 = oldest child, etc.)	1.89	1.82
Father's occupation (1 = most skilled, 17 = least skilled)	8.26*	11.50
Peabody Picture Vocabulary IQ	99.26*	74.45

*Statistically significant, $p < .01$

a. Can you suggest better ways to summarize these baseline characteristics?
b. What test(s) may have been used to compare these characteristics?
c. What other characteristics could or should have been examined to make a direct comparison of reading and arithmetic achievement?
d. What effect (if any) might the attrition rate have on the study results? Recall 106 (27%) of the original 392 students were not available for testing at year 6.

6 CATEGORICAL DATA

6.1 INTRODUCTION

Up to this point we have been concerned primarily with sample data measured on a quantitative scale. However, we sometimes encounter situations where levels of the variable of interest are identified by name or rank only and we are interested in the number of observations occurring at each level of the variable. Data obtained from these types of variables are called **categorical** or **count data**. For example, an item coming off an assembly line may be classified into one of three quality classes: acceptable, second, or reject. Similarly, a traffic study might require a count and classification of the type of transportation used by commuters along a major access road into a city. A pollution study might be concerned with the number of different alga species identified in samples from a lake and the number of times each species is identified. A consumer protection group might be interested in the results of a prescription fee survey to compare prices on a checklist of medications in different sections of a large city.

In this chapter we will examine specific inferences that can be made from experiments involving categorical data.

6.2 THE MULTINOMIAL EXPERIMENT AND CHI-SQUARE GOODNESS-OF-FIT TEST

The examples in Section 6.1 all exhibit, to a reasonable degree of approximation, the characteristics of a **multinomial experiment**

THE MULTINOMIAL EXPERIMENT

1. The experiment consists of n identical trials.
2. Each trial results in one of k outcomes.
3. The probability that a single trial will result in outcome i is π_i, $i = 1, 2, \ldots, k$, and remains constant from trial to trial. (Note: $\sum_i \pi_i = 1$.)
4. The trials are independent.
5. We are interested in n_i, the number of trials resulting in outcome i. (Note: $\sum_i n_i = n$.)

multinomial distribution   The probability distribution for the number of observations resulting in each of the k outcomes, called the **multinomial distribution**, is given by the formula

$$P(n_1, n_2, \ldots, n_k) = \frac{n!}{n_1! n_2! \cdots n_k!} \pi_1^{n_1} \pi_2^{n_2} \cdots \pi_k^{n_k}$$

Recall from Chapter 3, where we discussed the binomial probability distribution, that

$$n! = n(n-1) \cdots 1$$

and

$$0! = 1$$

We can use the formula for the multinomial distribution to compute the probability of particular events.

EXAMPLE 6.1   Previous experience with the breeding of a particular herd of cattle suggests that the probability of obtaining one healthy calf from a mating is .83. Similarly, the probabilities of obtaining zero or two healthy calves are, respectively, .15 and .02. If a farmer breeds three dams from the herd, find the probability of obtaining exactly three healthy calves.

Solution   Assuming the three dams are chosen at random, this experiment can be viewed as a multinomial experiment with $n = 3$ trials and $k = 3$ outcomes. These outcomes are listed below with the corresponding probabilities.

Outcome	Number of Progeny	Probability, π_i
1	0	.15
2	1	.83
3	2	.02

Note that outcomes 1, 2, and 3 refer to the events that a dam produces zero, one, or two healthy calves, respectively. Similarly, n_1, n_2, and n_3 refer to the number of dams producing zero, one, or two healthy progeny, respectively. To obtain exactly three healthy progeny, we must observe one of the following possible events.

$$A: \begin{cases} 1 \text{ dam gives birth to no healthy progeny: } n_1 = 1 \\ 1 \text{ dam gives birth to 1 healthy progeny: } n_2 = 1 \\ 1 \text{ dam gives birth to 2 healthy progeny: } n_3 = 1 \end{cases}$$

$$B: \quad 3 \text{ dams give birth to 1 healthy progeny: } \begin{cases} n_1 = 0 \\ n_2 = 3 \\ n_3 = 0 \end{cases}$$

For event A with $n = 3$ and $k = 3$,

$$P(n_1 = 1, n_2 = 1, n_3 = 1) = \frac{3!}{1!1!1!}(.15)^1(.83)^1(.02)^1 \approx .015$$

Similarly, for event B,

$$P(n_1 = 0, n_2 = 3, n_3 = 0) = \frac{3!}{0!3!0!}(.15)^0(.83)^3(.02)^0 = (.83)^3 \approx .572$$

Thus the probability of obtaining exactly three healthy progeny from three dams is the sum of the probabilities for events A and B; namely, $.015 + .572 \approx .59$.

Our primary interest in the multinomial distribution is as a probability model underlying statistical tests about the probabilities $\pi_1, \pi_2, \ldots, \pi_k$. We will hypothesize specific values for the πs and then determine whether the sample data agree with the hypothesized values. One way to test such a hypothesis is to examine the observed number of trials resulting in each outcome and to compare this to the number we would *expect* to result in each outcome. For instance, in our previous example, we gave the probabilities associated with zero, one, and two progeny as .15, .83, and .02. If we were to examine a sample of 100 mated dams, we would expect to observe 15 dams that produce no healthy progeny. Similarly, we would expect to observe 83 dams that produce one healthy calf and two dams that produce two healthy calves.

DEFINITION 6.1

In a multinomial experiment where each trial can result in one of k outcomes, the **expected number of outcomes** of type i in n trials is $n\pi_i$, where π_i is the probability that a single trial results in outcome i.

In 1900 Karl Pearson proposed the following test statistic to test the specified probabilities:

χ^2

$$\chi^2 = \sum_i \left[\frac{(n_i - E_i)^2}{E_i} \right]$$

where n_i represents the number of trials resulting in outcome i and E_i represents the number of trials we would expect to result in outcome i when the hypothesized probabilities represent the actual probabilities assigned to each outcome. (The symbol χ is the Greek

FIGURE 6.1

Chi-Square
Probability
Distribution for
df = 4

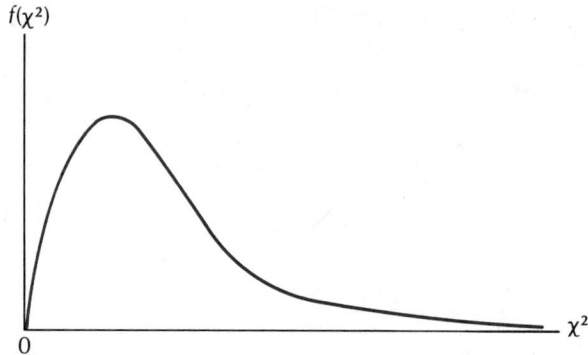

cell probabilities

observed cell counts
expected cell counts

chi-square
distribution

letter chi.) Frequently, we will refer to the probabilities $\pi_1, \pi_2, \ldots, \pi_k$ as **cell probabilities**, one cell corresponding to each of the k outcomes. The observed numbers $n_1, n_2, \ldots, n_k$ corresponding to the k outcomes will be called **observed cell counts**, and the expected numbers $E_1, E_2, \ldots, E_k$ will be referred to as **expected cell counts**.

Suppose that we hypothesize values for the cell probabilities $\pi_1, \pi_2, \ldots, \pi_k$. We can then calculate the expected cell counts by using Definition 6.1 to examine how well the observed data fit, or agree, with what we would expect to observe. Certainly if the hypothesized π-values are correct, the observed cell counts n_i should not deviate greatly from the expected cell counts E_i, and the computed value of χ^2 should be small. Similarly, when one or more of the hypothesized cell probabilities are incorrect, the observed and expected cell counts will differ substantially, making χ^2 large.

The distribution of the quantity χ^2 can be approximated by a **chi-square distribution** provided that the expected cell counts E_i are fairly large. We will not give the mathematical formula for the chi-square probability distribution, but instead will list its properties.

1. The chi-square distribution is a nonsymmetrical distribution (see Figure 6.1).
2. There are many chi-square distributions. We obtain a particular one by specifying the degrees of freedom (df).

The chi-square goodness-of-fit test based on k specified cell probabilities will have $k - 1$ degrees of freedom. Upper-tail values of the test statistic

$$\chi^2 = \sum_i \left[\frac{(n_i - E_i)^2}{E_i} \right]$$

can be found in Table 5 in the Appendix. Entries in the table are values of χ^2 that have an area "a" to the right under the curve. The degrees of freedom are specified in the left column of the table, and values of "a" are listed across the top of the table. Thus for df = 14, the value of chi-square with an area a = .10 to its right under the curve is 21.0642 (see Figure 6.2).

FIGURE 6.2

Critical Value of the Chi-Square Distribution for a = .10 and df = 14

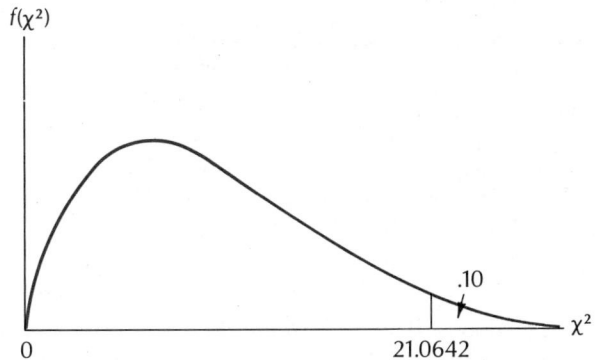

We can now summarize the chi-square goodness-of-fit test concerning k specified cell probabilities.

CHI-SQUARE GOODNESS-OF-FIT TEST

> Null hypothesis: $\pi_i = \pi_{i0}$ for categories $i = 1, \ldots, k$, π_{i0} are specified probabilities or proportions.
>
> Alternative hypothesis: At least one of the cell probabilities differs from the hypothesized value.
>
> Test statistic: $\chi^2 = \sum_i \left[\dfrac{(n_i - E_i)^2}{E_i} \right]$, where n_i is the observed number in category i and $E_i = n\pi_{i0}$ is the expected number under H_0.
>
> Rejection region: Reject H_0 if χ^2 exceeds the tabulated critical value for a $= \alpha$ and df $= k - 1$.

Some researchers (see, for example, Siegel (1956) and Dixon and Massey (1969)) recommend that all the E_is should be 5 or more before performing this test. This requirement is perhaps too stringent. Cochran (1954) indicates that the approximation should be quite good if no E_i is less than 1 and no more than 20% of the E_is are less than 5. We recommend applying Cochran's guidelines for determining whether χ^2 can be approximated with a chi-square distribution. We can combine categories if some of the E_is are too small, but care should be taken so that the combination of categories does not change the nature of the hypothesis to be tested.

EXAMPLE 6.2 ┃ A test drug is to be compared against a standard drug preparation useful in the maintenance of patients suffering from high blood pressure. Over many clinical trials at

TABLE 6.1   Results of Clinical Trials Using the Standard Preparation, Example 6.2

Category	Percentage
Marked decrease in blood pressure	50%
Moderate decrease in blood pressure	25%
Slight decrease in blood pressure	10%
Stationary or slight increase in blood pressure	15%

many different locations, patients suffering from comparable hypertension (as measured by the New York Heart Association (NYHA) Classification) have been administered the standard therapy. Responses to therapy for this large patient group were classified into one of four response categories. Table 6.1 lists the categories and percentages of patients treated on the standard preparation who have been classified in each category.

A clinical trial is conducted with a random sample of 200 patients suffering from high blood pressure. All patients are required to be listed according to the same hypertensive categories of the NYHA Classification as those studied under the standard preparation. Use the sample data in Table 6.2 to test the hypothesis that the cell probabilities associated with the test preparation are identical to those for the standard. Use $\alpha = .05$.

Solution

This experiment possesses the characteristics of a multinomial experiment, with $n = 200$ and $k = 4$ outcomes.

Outcome 1: A person's blood pressure will decrease markedly after treatment on the test drug.

Outcome 2: A person's blood pressure will decrease moderately after treatment on the test drug.

Outcome 3: A person's blood pressure will decrease slightly after treatment on the test drug.

Outcome 4: A person's blood pressure will remain stationary or increase slightly after treatment on the test drug.

The null and alternative hypotheses are then

$$H_0: \pi_1 = .50, \pi_2 = .25, \pi_3 = .10, \pi_4 = .15$$

TABLE 6.2   Sample Data for Example 6.2

Category	Observed Cell Counts
1	120
2	60
3	10
4	10

TABLE 6.3   Observed and Expected Cell Numbers for Example 6.2

Category	Observed Cell Number, n_i	Expected Cell Number, E_i
1	120	200(.50) = 100
2	60	200(.25) = 50
3	10	200(.10) = 20
4	10	200(.15) = 30

and

H_a: At least one of the cell probabilities is different from the hypothesized value.

Before computing the test statistic, we must determine the expected cell numbers. These data are given in Table 6.3.

Since all the expected cell numbers are large, we may calculate the chi-square statistic and compare it to a tabulated value of the chi-square distribution.

$$\chi^2 = \sum_i \left[\frac{(n_i - E_i)^2}{E_i} \right]$$

$$= \frac{(120 - 100)^2}{100} + \frac{(60 - 50)^2}{50} + \frac{(10 - 20)^2}{20} + \frac{(10 - 30)^2}{30}$$

$$= 4 + 2 + 5 + 13.33 = 24.33$$

For the probability of a Type I error set at $\alpha = .05$, we look up the value of the chi-square statistic for a = .05 and df = $k - 1 = 3$. The critical value from Table 5 in the Appendix is 7.81473.

R.R.: Reject H_0 if $\chi^2 > 7.81473$

Conclusion: Since the computed value of χ^2 is greater than 7.81473, we reject the null hypothesis and conclude that at least one of the cell probabilities differs from that specified under H_0. Practically, it appears that a much higher proportion of patients treated with the test preparation fall into the moderate and marked improvement categories.

The assumptions needed for running a chi-square goodness-of-fit test are those associated with a multinomial experiment, of which the key ones are independence of the trials and constant cell probabilities. Independence of the trials would be violated if, for example, several patients from the same family were included in the sample since hypertension has a strong hereditary component. The assumption of constant cell probabilities would be violated if the study were conducted over a period of time during which the standards of medical practice shifted, allowing for other "standard" therapies.

The test statistic for the chi-square goodness-of-fit test is the sum of k terms, which is the reason the degrees of freedom depend on k, the number of categories, rather than on n, the total sample size. However, there are only $k - 1$ degrees of freedom, rather than k of them, because the sum of the $n_i - E_i$ terms must be equal to $n - n = 0$; $k - 1$ of the observed minus expected differences are free to vary, but the last one (kth) is determined by the condition that the sum of the $n_i - E_i$ equals zero.

This goodness-of-fit test has been used extensively over the years to test various scientific theories. Unlike previous statistical tests, however, the hypothesis of interest is the null hypothesis, not the research (or alternative) hypothesis. Unfortunately, the logic behind running a statistical test does not hold. In the standard situation where the research (alternative) hypothesis is the one of interest to the scientist, we formulate a suitable null hypothesis and gather data to reject H_0 in favor of H_a. Thus we "prove" H_a by contradicting H_0.

Not so with the chi-square goodness-of-fit test. If a scientist has a set theory and wants to show that sample data conform to or "fit" that theory, she wants to accept H_0. From our previous work, there is the potential for committing a Type II error in accepting H_0. Here, as with other tests, the calculation of β probabilities is difficult. In general, for a goodness-of-fit test, the potential for committing a Type II error is high if n is small or if k, the number of categories, is large. Even if the expected cell counts E_i conform to our recommendations, the probability a Type II error could be large. So, the results of a chi-square goodness-of-fit test should always be viewed suspiciously. Don't automatically accept the null hypothesis as fact given that H_0 was not rejected.

EXERCISES

6.1   Hypothetical data are presented here. Use these data to run a chi-square goodness-of-fit test with H_0: $\pi_1 = .2$, $\pi_2 = .15$, $\pi_3 = .40$, $\pi_4 = .15$, and $\pi_5 = .10$. Use $\alpha = .05$. Do the data fit the hypothesized probabilities?

Category	Observed Cell Number, n_i
1	60
2	50
3	130
4	40
5	20
Total	300

6.2   Use the data of Exercise 6.1 to run a chi-square goodness-of-fit test with this new null hypothesis—H_0: $\pi_1 = .15$, $\pi_2 = .20$, $\pi_3 = .45$, $\pi_4 = .15$, and $\pi_5 = .05$. Again use $\alpha = .05$. Compare your results to those of Exercise 6.1. How sensitive does this test appear to be for the cell probabilities specified under H_0? What conclusion can be drawn if we do *not* reject H_0?

 6.3   Over the past five years, an insurance company has had a mix of 40% whole life policies, 20% universal life policies, 25% annual renewable-term (ART) policies, and 15% other types of policies. A change in this mix over the long haul could require a change in the commission structure, reserves, and possibly investments. A sample of 1000 policies issued over the last few months gave the following results

Category	Observed Cell Number, n_i
Whole Life	320
Universal Life	280
ART	240
Other	160
Total	1000

Use these data to assess whether there has been a shift from the historical percentages. Give the p-value for your test. Which policies (if any) seem to be more popular?

6.4 A work-study program was developed with a university and several industries in the surrounding community. Students were to work with industrial sociologists during a three-month internship. Equal numbers of students from the university were sent to a chemical, a textile, and a pharmaceutical industry. Students completing the program were classified according to the industry in which they interned. Consider the following data as a random sample of the many students who could have completed the program. Test the null hypothesis that the probability that a finishing student interned in a pharmaceutical, chemical, or textile industry is 1/3. Use $\alpha = .01$ with n_i the number of students in group i finishing the program.

Group	n_i
Pharmaceutical	20
Chemical	13
Textile	30

6.5 An experiment was conducted to determine whether the proportion of mentally ill patients of each social class housed in a county facility agrees with the social class distribution of the county. The observed cell numbers for the 400 patients classified are given below.

Lower: 215   Upper-middle: 60
Lower-middle: 100   Upper: 25

Use these data to test the null hypothesis

$\pi_1 = .25$ $\pi_3 = .20$
$\pi_2 = .48$ $\pi_4 = .07$

where the πs are the hypothesized proportions of persons in the respective social-class categories in the county. Use $\alpha = .05$ and draw conclusions.

6.6 In previous presidential elections in a given locality, 50% of the registered voters were Republicans, 40% were Democrats, and 10% were registered as independents. Prior to the upcoming election, a random sample of 200 registered voters showed that 90 were registered as Republicans, 80 as Democrats, and 30 as independents. Test the research hypothesis that the distribution of registered voters is different from that in previous election years. Give the p-value for your test. Draw conclusions.

6.7   A local doctor suspects that there is a seasonal trend in the occurrence of the common cold. He estimates that 40% of the cases each year occur in the winter, 40% in the spring, 10% in the summer, and 10% in the fall. The following information was collected from a random sample of 1000 cases of patients with the common cold over the past year:

Season	Frequency
Winter	374
Spring	292
Summer	169
Fall	165

Would you agree with the doctor's estimates, based on the sample information? Perform a statistical test using $\alpha = .05$. Draw conclusions.

6.8   Refer to Exercise 6.7. What would the null hypothesis be if the doctor claimed that there are no differences in the percentages of cases over the seasons? Test the hypothesis that there is no seasonal trend in the occurrence of the common cold. Give the level of significance of your test.

    6.9   Previous experimentation with a drug product developed for the relief of depression was conducted with normal adults with no signs of depression. We will assume a large data bank is available from studies conducted with normals and, for all practical purposes, the data bank can represent the population of responses for normals. Each of the adults participating in one of these studies was asked to rate the drug as ineffective, mildly effective, or effective. The percentages of respondents in these categories were 60%, 30%, and 10%, respectively. In a new study of depressed adults, a random sample of 85 adults responded as follows:

Ineffective: 30
Mildly effective: 35
Effective: 20

Is there evidence to indicate a different percentage distribution of responses for depressed adults than for normals? Give the level of significance for your test and draw conclusions.

    6.10   In random sampling, 40 newspaper editors were interviewed to determine their opinions on the degree of future suppression of freedom of the press brought about by recent court decisions. The editors' opinions are summarized here.

Degree of Suppression	Frequency
None	8
Very little	8
Moderate	10
Severe	14

Use these data to test the null hypothesis that each category is equally preferred. Use $\alpha = .05$.

6.3   INFERENCES ABOUT THE BINOMIAL PARAMETER π

The binomial experiment discussed in Chapter 3 is a special case of the multinomial experiment, where each trial results in one of two outcomes, which we labeled as either a success or a failure. Recall that we designated π as the probability of a success and $(1 - \pi)$ as the probability of a failure. Then the probability distribution for y, the number of successes in n identical trials, is

$$P(y) = \frac{n!}{y!(n - y)!} \, \pi^y (1 - \pi)^{n-y}$$

The point estimate of the binomial parameter π is one that we would choose intuitively. In a random sample of n from a population in which the proportion of elements classified as successes is π, the best estimate of the parameter π is the sample proportion of successes. Letting y denote the number of successes in the n sample trials, the sample proportion is

$$\hat{\pi} = \frac{y}{n}$$

We observed in Section 3.12 that y possesses a mound-shaped probability distribution that can be approximated by using a normal curve when

$$n \geq \frac{5}{\min(\pi, \, 1 - \pi)} \qquad \text{(or equivalently, } n\pi \geq 5 \text{ and } n(1 - \pi) \geq 5)$$

In a similar way, the distribution of $\hat{\pi} = y/n$ can be approximated by a normal distribution with a mean and a standard error as given below.

MEAN AND STANDARD ERROR OF $\hat{\pi}$	$\mu_{\hat{\pi}} = \pi$ $\sigma_{\hat{\pi}} = \sqrt{\dfrac{\pi(1 - \pi)}{n}}$

The normal approximation to the distribution of $\hat{\pi}$ can be applied under the same condition as that for approximating y by using a normal distribution. In fact, the approximation for both y and $\hat{\pi}$ becomes more precise for large n. Henceforth in this text we will assume that $\hat{\pi}$ can be adequately approximated by using a normal distribution, and we will base all our inferences on results from our previous study of the normal distribution.

A confidence interval can be obtained for π using the methods of Chapter 4 for μ, by replacing $\bar{y}$ with $\hat{\pi}$ and $\sigma_{\bar{y}}$ with $\sigma_{\hat{\pi}}$. A general $100(1 - \alpha)\%$ confidence interval for the binomial parameter is given here.

CONFIDENCE
INTERVAL
FOR π, WITH
CONFIDENCE
COEFFICIENT
OF $(1 - \alpha)$

$$\hat{\pi} \pm z_{\alpha/2}\sigma_{\hat{\pi}}$$

where

$$\hat{\pi} = \frac{y}{n} \quad \text{and} \quad \sigma_{\pi} = \sqrt{\frac{\pi(1 - \pi)}{n}}$$

Note: Since π is unknown, replace π by $\hat{\pi}$ in $\sigma_{\hat{\pi}}$.

EXAMPLE 6.3

Response to an advertising display was measured by counting the number of people who purchased the product out of the total number exposed to the display. If 330 purchased the product out of a total of 870 exposed, estimate the proportion of all persons exposed who will buy the product. Use a 90% confidence interval.

Solution

For these data,

$$\hat{\pi} = \frac{330}{870} = .38$$

$$\sigma_{\hat{\pi}} = \sqrt{\frac{(.38)(.62)}{870}} = .016$$

The confidence coefficient for our example is .90. Recall from Chapter 4 that we can obtain $z_{\alpha/2}$ by looking up the z-value in Table 2 in the Appendix corresponding to an area of $(\alpha/2)$. For a confidence coefficient of .90, the z-value corresponding to an area of .05 is 1.645. Hence the 90% confidence interval on the proportion of persons who will purchase the product after exposure to this display is

$$.38 \pm 1.645(.016) \qquad \text{or} \qquad .38 \pm .026$$

The confidence interval for π is based on a normal approximation to a binomial, which is appropriate provided n is sufficiently large. The rule we've specified is that both $n\pi$ and $n(1 - \pi)$ should be at least 5, but since π is the unknown parameter, we'll require that $n\hat{\pi}$ and $n(1 - \hat{\pi})$ be at least 5. When the sample size is too small and violates this rule, the confidence interval usually will be too wide to be of any use. For example, with $n = 20$ and $\hat{\pi} = .2$, the rule is not satisfied, since $n\hat{\pi} = 4$. The 95% confidence interval based on these data would be $.025 < \pi < .375$, which is practically useless. Very few product managers would be willing to launch a new product if the expected increase in market share was between .025 and .375.

Keep in mind, however, that a sample size that is sufficiently large to satisfy the rule *does not* guarantee that the interval will be informative. It only judges the adequacy of the normal approximation to the binomial—the basis for the confidence level.

Sample size calculations for estimating π follow very closely the procedures we developed for inferences about μ. In Chapter 4 the required sample size for a $100(1 - \alpha)$%

confidence interval for π of the form $\hat{\pi} \pm E$ (where E is specified) is found by solving the expression

$$z_{\alpha/2}\sigma_{\hat{\pi}} = E$$

for n. This result is shown here.

SAMPLE SIZE REQUIRED FOR A $100(1 - \alpha)\%$ CONFIDENCE INTERVAL FOR π OF THE FORM $\hat{\pi} \pm E$

$$n = \frac{z_{\alpha/2}^2\pi(1 - \pi)}{E^2}$$

Note: Since π is not known, either substitute an educated guess or use $\pi = .5$. Use of $\pi = .5$ will generate the largest possible sample size for the specified confidence interval width, $2E$, and will thus give a conservative answer to the required sample size.

EXAMPLE 6.4

A large public opinion polling agency plans to conduct a national survey to determine the proportion of employed adults who fear losing their job within the next year. How many workers must be polled to estimate to within .02 using a 95% confidence interval?

Solution

By design, the agency wants the interval of the form $\hat{\pi} \pm .02$. The sample size necessary to achieve this accuracy is given by

$$n = \frac{z_{\alpha/2}^2\pi(1 - \pi)}{E^2}$$

where $z_{\alpha/2} = 1.96$ and $E = .02$. If a previous survey has been run recently we could use the sample proportion from that survey to substitute for π; otherwise we could use $\pi = .5$. Using $\pi = .5$ the required sample size is

$$n = \frac{(1.96)^2(.5)(.5)}{(.02)^2} = 2401$$

That is, 2401 workers would have to be surveyed to estimate π to within .02.

A statistical test about a binomial parameter π is very similar to the large-sample test concerning a population mean presented in Chapter 4. These results are summarized next, with three different alternative hypotheses along with their corresponding rejection regions. Recall that only one alternative is chosen for a particular problem.

SUMMARY OF A
STATISTICAL TEST
FOR π

H_0: $\pi = \pi_0$ (π_0 is specified)

H_a: 1. $\pi > \pi_0$
      2. $\pi < \pi_0$
      3. $\pi \neq \pi_0$

T.S.: $z = \dfrac{\hat{\pi} - \pi_0}{\sigma_{\hat{\pi}}}$

R.R.: For a probability α of a Type I error
      1. reject H_0 if $z > z_\alpha$
      2. reject H_0 if $z < -z_\alpha$
      3. reject H_0 if $|z| > z_{\alpha/2}$
      Note: Under H_0,

$$\sigma_{\hat{\pi}} = \sqrt{\frac{\pi_0(1 - \pi_0)}{n}}$$

EXAMPLE 6.5

Sports car owners in a town complain that their cars are judged differently from family-style cars at the state vehicle inspection station. Previous records indicate that 30% of all passenger cars fail the inspection on the first time through. In a random sample of 150 sports cars, 60 failed the inspection on the first time through. Is there sufficient evidence to indicate that the percentage of first failures for sports cars is higher than the percentage for all passenger cars? Use $\alpha = .05$.

Solution

The appropriate statistical test is as follows.

H_0: $\pi = .30$

H_a: $\pi > .30$

T.S.: $z = \dfrac{\hat{\pi} - \pi_0}{\sigma_{\hat{\pi}}}$

R.R.: For $\alpha = .05$, we will reject H_0 if $z > 1.645$.

Using the sample data,

$$\hat{\pi} = \frac{60}{150} = .4 \qquad \text{and} \qquad \sigma_{\hat{\pi}} = \sqrt{\frac{(.3)(.7)}{150}} = .037$$

Also,

$$n\pi_0 = 150(.3) = 45$$

and

$$n(1 - \pi_0) = 150(.7) = 105$$

The test statistic is then

$$z = \frac{.4 - .3}{.037} = 2.7$$

Since the observed value of z exceeds 1.645, we conclude that sports cars at the vehicle inspection station have a first-failure rate greater than .3. However, we must

> be careful not to attribute this difference to a difference in standards for sports cars and family-style cars. Parallel testing of sports cars versus other cars would have to be conducted to eliminate other sources of variability that would perhaps account for the higher first-failure rate for sports cars.

The z test for π, like the confidence interval for π based on z, depends on the adequacy of the normal approximation to the binomial. When can you use the z test for π? Generally speaking, you should view the results of a z test for π skeptically if either $n\pi_0$ or $n(1 - \pi_0)$ is 2 or less. If both $n\pi_0$ and $n(1 - \pi_0)$ are at least 5, the z test should be accurate. But for the same sample size n, z tests based on more extreme values of π_0 are less accurate than are those based on values of π_0 closer to .5. For example for $n = 5000$, a test of H_0: $\pi = .001$ (for which $n\pi_0 = 5$) would be much more suspect than would a test of H_0: $\pi = .01$ (for which $n\pi_0 = 50$).

EXERCISES

6.11 Hypothetical sample results from a binomial experiment with $n = 150$ yielded $\hat{\pi} = .2$.
  a. Does this experiment satisfy the sample-size requirement for a confidence interval for π based on z? What sample sizes would be suspect, given the same sample proportion?
  b. Construct a 90% confidence interval for π.

6.12 Experts have predicted that approximately 1 in 12 tractor-trailer units will be involved in an accident this year. One of the reasons for this is that 1 in 3 tractor-trailer units has an imminently hazardous mechanical condition, probably related to the braking systems on the vehicle.
      A survey of 50 tractor-trailer units passing through a weighing station confirmed that 19 had a potentially serious braking system problem.
  a. Do the binomial assumptions hold?
  b. Can a normal approximation to the binomial be applied here to get a confidence interval for π?
  c. Give a 95% confidence interval for π using these data. Is the interval informative? What could be done to decrease the width of the interval, assuming $\hat{\pi}$ remained the same?

6.13 In a study of self-medication practices, a random sample of 1230 adults completed a survey. Some of the medical conditions that were self-treated are shown here:

Medical Condition	Home Remedy	% Responding
Sore throat—not related to a cold	Salt water or baking soda mouth wash	30
Burns—other than sunburn	Cold water/butter	28
Overindulgence in alcohol	Homebrew	25
Overweight	Diet	22
Pain associated with injury	Hot or cold compress	21

Summarize the results of this part of the survey using a 95% confidence interval for each medical condition.

6.14   In the survey discussed in Exercise 6.13, 441 of the adults reported they had a cough or cold recently and 260 of the respondents said they had treated the condition with an over-the-counter (OTC) remedy. These data are summarized here.

Survey respondents reporting problem	441
Number of patients using any OTC remedy	260
Patients using specific classes of OTC remedies:	
Adult pain relievers	110
Adult cold caps/tabs	57
Cough remedies	44
Allergy/hay fever remedies	9
Liquid cold remedies	35
Sprays/inhalers	4
Children's pain reliever	22
Cough drops	13
Sore-throat lozenges/gum	9
Children's cold caps/tabs	13
Nose drops	9
Chest rubs/ointments	9
Anesthetic throat lozenges	4
Room vaporizers	4
Other product	4

a. How might these data be organized and summarized? Would percentages help? Do the percentages add to 100%? Why or why not?

b. Based on these data, which classes of OTC remedies could be summarized using a 95% confidence interval for π?

 6.15   A national columnist recently reported the results of a survey on marriage and the family. Part of the column has been paraphrased here.

> The Ingredients of Marriage
>
> The Gallup people offered respondents a list of well-known ingredients. Here in the United States, such elements as faithfulness, mutual respect, and under-standing ranked at the top. These were followed by enough money, same back-ground, good housing, and agreement in politics. Seventy-five percent of the respondents voted for "a good sex life," 59% for children, 52% for common interests, 48% for "living away from in-laws," and 43% for "sharing household chores." (In West Germany, by contrast, only 52% voted for a good sex life and only 19% for sharing household chores.)

a. How could you display the results of this survey in a graph or table?

b. Would you use a confidence interval to convey more information about the "true" percentages expressing an opinion on the various ingredients of a good marriage? Why or why not?

c. What qualms might you have about the way this survey has been reported?

 6.16   A substantial part of the U.S. population is "technologically illiterate," according to experts at a National Technological Literacy Conference organized by the National Science Foundation and Pennsylvania State University. At this conference, the results of a national survey of 2000 adults showed that:

•   70% do not understand radiation

- 40% think space-rocket launchings change the weather and that some unidentified flying objects are actually visitors from other planets
- More than 80% do not understand how telephones work
- 75% do not have a clear understanding of what computer software is
- 72% do not understand the gross national product

a. How might you display these data in a graph or table? Construct the display.
b. The problem with many newspaper articles reporting the results of a survey is that conclusions are given without sufficient details about the study for the reader to assess the data and reach a separate conclusion. What details are missing here for you to reach your own conclusion?

6.17 More and more people are dining out—so say the results of national surveys. Compared to 1978, here are some figures

Meal Eaten Away from Home	1978	Now
Breakfast	3%	5%
Lunch	18%	20%
Dinner	16%	16%

a. If these data were based on random samples of 1500 adults in 1978 and at present, what conclusions can be drawn for each meal? Is a normal approximation to the binomial valid here?
b. Can we conclude from the data shown here that more people are eating out? Why or why not?

6.18 The benign mucosal cyst is the most common lesion of a pair of sinuses in the upper jawbone. In a random sample of 800 males, 35 persons were observed to have a benign mucosal cyst.

a. Would it be appropriate to use a normal approximation in conducting a statistical test of the null hypothesis H_0: $\pi = .096$ (the highest incidence in previous studies among males)? Explain.
b. Conduct a statistical test of the research hypothesis H_a: $\pi < .096$. Use $\alpha = .05$.

6.19 National public opinion polls are based on interviews of as few as 1500 persons in a random sampling of public sentiment toward one or more issues. These interviews are commonly done in person, because mail returns are poor and telephone interviews tend to reach older people, thus biasing the results. Suppose that a random sample of 1500 persons is surveyed to determine the proportion of the adult public in agreement with recent energy conservation proposals.

a. If 560 indicate they favor the policies set forth by the current administration, estimate π, the proportion of adults holding a "favor" opinion. Use a 95% confidence interval. What is the half width of the confidence interval?
b. How many persons must be surveyed to have a 95% confidence interval with a half width of .01?

6.20 A sample of 20 crayfish of all sizes was obtained from a large lake to estimate the proportion of crayfish that exhibit more than 9 (ppb) units of mercury. Of those sampled, eight exceeded 9 units. Use these data to estimate π, the proportion of all crayfish in the lake with a mercury level greater than 9, using a 95% confidence interval.

6.21 Simulate the binomial distribution for $n = 20$ and $\pi = .4$, using a computer program. Do this by obtaining y, the number of successes in 20 trials when sampling from a binomial distribution with $\pi = .4$. Repeat this experiment 39 more times, for a total of 40 repetitions of the experiment.

a. Plot the sample data (y-values) in a relative frequency histogram.
b. Compute the sample mean and standard deviation. Compare your answers to the *actual* mean and standard deviation of y. (Hint: A Minitab program is given here for purposes of illustration.)

```
         BRANDOM 40 EXPERIMENTS WITH N = 20, P = .4, PUT IN C1
         40 BINOMIAL EXPERIMENTS WITH N =  20 AND P = 0.4000


         SUMMARY

         VALUE      FREQUENCY
           3           1
           4           1
           5           2
           6           2
           7           4
           8           6
           9           9
          10           7
          11           5
          12           2
          14           1
MTB > HISTOGRAM C1

Histogram of C1   N = 40

Midpoint    Count
       3        1  *
       4        1  *
       5        2  **
       6        2  **
       7        4  ****
       8        6  ******
       9        9  *********
      10        7  *******
      11        5  *****
      12        2  **
      13        0
      14        1  *


MTB > MEAN C1
    MEAN     =       8.7250
MTB > STANDARD DEVIATION C1
    ST.DEV. =       2.2644
MTB >
```

6.22   Refer to Example 6.4. Suppose a recently done survey resulted in $\hat{\pi}$ = .15. Use this guessed value in the computation of an appropriate sample size. Comment on the differences in your answer here and that in Example 6.4.

6.4   OPERATING CHARACTERISTIC CURVES AND CONTROL CHARTS FOR π (optional)

Two techniques are particularly appropriate for monitoring product quality as measured by π, the fraction of items that are defective. One technique, **control charts**, was dis-

control charts

cussed in Chapter 4 as applied to μ; the charts can be used to monitor π as a measure of the ongoing quality of a manufacturing process. We are especially interested in detecting a shift in product quality. The second technique, called **lot acceptance sampling**, provides a means to screen (sample) ingoing raw materials or outgoing production from a plant where the product is shipped in large quantities (often called "lots"). We begin by discussing lot acceptance sampling.

lot acceptance sampling

Manufacturers are interested in minimizing not only the amount or proportion of defective raw material to be used in the production process, but also the proportion of defective finished products shipped from the plant. Thus they would like to sample, or screen, shipments of raw materials entering the plant and reject those shipments (lots) that contain too high a proportion of defectives. Similarly, they must screen the final product to make certain that a shipment does not contain too high a proportion of defectives.

The most obvious type of screen (sampling plan) to employ would be a careful inspection of each item from the lot. Unfortunately, this screen would be both costly and time-consuming. In addition, it would still be subject to errors in reporting brought about by human fatigue.

statistical sampling plan

Another type of screen is called a **statistical sampling plan**. Here we obtain a random sample of n items from the lot. Each item of the sample is inspected, and if y, the number of defectives observed in the sample, is less than or equal to some predetermined number "a," we accept the lot. Thus a statistical sampling plan is designated by n, the sample size, and "a," the acceptance number. If the lot is accepted ($y \leq a$), we conclude that the proportion of defectives π in the lot is small and acceptable. However, if $y > a$, we reject the lot and conclude that π is too large (above an acceptable level of defectives).

operating characteristic (OC) curve

We can characterize the goodness of a particular sampling plan (n, a) by constructing an **operating characteristic (OC) curve**. The OC curve for a sampling plan is a graph displaying the probability of accepting a lot for various values of π, the proportion of defective items in the lot (see Figure 6.3). As you can see, the probability of accepting a lot decreases as the proportion of defectives within the lot increases.

We can construct an OC curve by computing the probability of accepting a lot—namely, $P(y \leq a)$—for several values of π. Consider the sampling plan $n = 4$ and $a = 0$.

FIGURE 6.3

OC Curve for a Sampling Plan

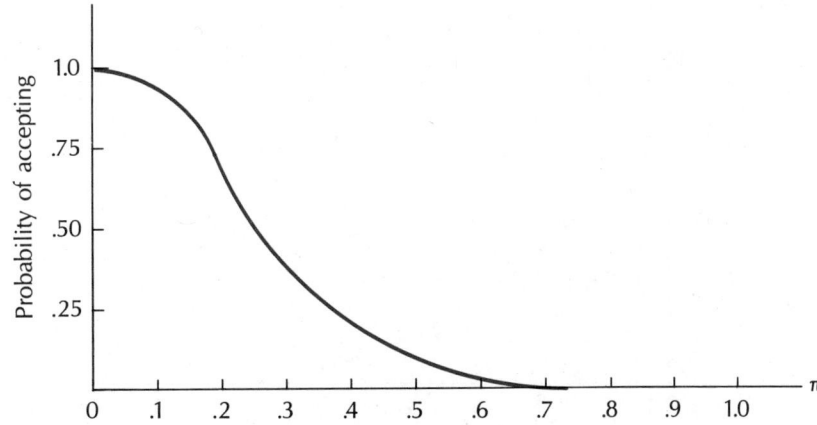

TABLE 6.4   Calculation of $P(y = 0)$ for $n = 4$ and $\pi = .1, .2,$ and .4

Fraction of Defectives, π	Probability of Accepting, $P(y = 0)$
.1	$\dfrac{4!}{0!4!}(.1)^0(.9)^4 = .656$
.2	$\dfrac{4!}{0!4!}(.2)^0(.8)^4 = .410$
.4	$\dfrac{4!}{0!4!}(.4)^0(.6)^4 = .130$

Here we sample 4 items and accept the lot if y, the number of defectives, is zero. Hence we must compute $P(y = 0)$ for $n = 4$ and for different values of π to obtain the OC curve. We use the binomial probability distribution

$$P(y) = \frac{n!}{y!(n - y)!}\,\pi^y(1 - \pi)^{n-y}$$

Table 6.4 shows the results of the calculations for $\pi = .1, .2,$ and .4. Plotting these three points and connecting them, we have the OC curve shown in Figure 6.4.

EXAMPLE 6.6   Construct an operating characteristic curve for the sampling plan ($n = 10, a = 1$).

Solution   The probability of accepting the lot is given by $P(y \leq 1)$. Using the binomial probability distribution, we must calculate $P(y = 0) + P(y = 1)$ for $n = 10$ and various values of π. For $\pi = .1$ and $n = 10$,

$$P(y = 0) = \frac{10!}{0!10!}(.1)^0(.9)^{10} = .349$$

$$P(y = 1) = \frac{10!}{1!9!}(.1)^1(.9)^9 = .387$$

FIGURE 6.4

OC Curve for the Sampling Plan ($n = 4$, $a = 0$)

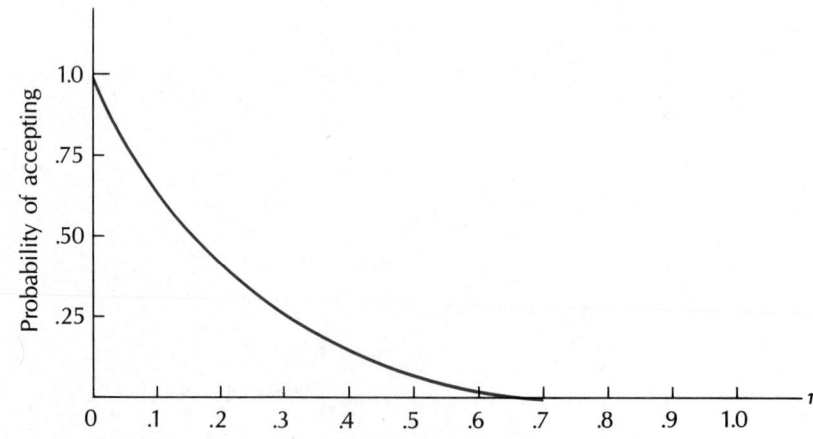

FIGURE 6.5

OC Curve for the Sampling Plan ($n = 10$, $a = 1$), Example 6.6

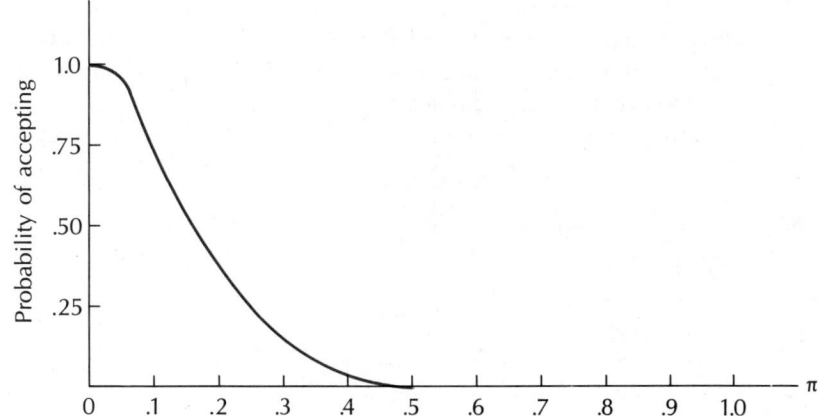

Hence the probability of accepting the lot is $.349 + .387 = .736$. Similarly, for $\pi = .2$ and $.4$, the probabilities of accepting the lot are found to be $.376$ and $.046$, respectively. Graphing our results, we have the OC curve presented in Figure 6.5.

Several comments should be made concerning statistical sampling plans. First, each plan (n, a) is unique, so the inspector must choose a sampling plan that possesses characteristics suitable for his or her particular problem. In general, for a fixed value of "a," an increase in n makes the graph of the probability of acceptance drop sharply as π increases; increasing "a" for a fixed n increases the probability of acceptance for values of π. Second, the OC curve for a particular sampling plan can be thought of as a plot of the probability of a Type II error for the null hypothesis H_0: $\pi = 0$ for various actual values of π, when the rejection region is $y > a$.

control charts   The other method of monitoring product quality makes use of **control charts** for π. Recall from Chapter 4 that a control chart typically consists of three lines. In a control chart for π, successive sample proportions $\hat{\pi}$ would be plotted and might appear as shown in Figure 6.6. If one of the sample proportions falls outside either the upper or lower control line, the process is judged to be out of control; that is, the proportion of defectives π in the production has shifted.

FIGURE 6.6

Control Chart for the Sample Proportion $\hat{\pi}$

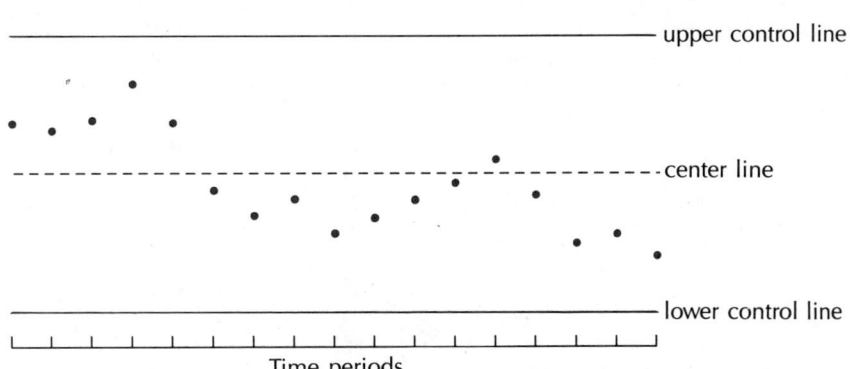

We can compute the three control lines in the following manner. The center line is designated by $\hat{\pi}_c$. If we obtain k different random samples of n observations each, $\hat{\pi}_c$ is the sample proportion of defectives for the entire set of kn measurements, or, equivalently, $\hat{\pi}_c$ is the average of the sample proportions computed for the k samples of n measurements. The upper control line (UCL) and lower control line (LCL) are then

UCL

$$UCL = \hat{\pi}_c + 3\sqrt{\frac{\hat{\pi}_c(1 - \hat{\pi}_c)}{n}}$$

and

LCL

$$LCL = \hat{\pi}_c - 3\sqrt{\frac{\hat{\pi}_c(1 - \hat{\pi}_c)}{n}}$$

EXAMPLE 6.7

A pharmaceutical firm has been investigating the possibility of having hospital personnel supplied with small disposable vials that can be used to perform many of the standard laboratory analyses. For a particular analysis, such as blood sugar, the technician would insert a measured amount of fluid (perhaps blood) in an appropriate vial and observe its color when thoroughly mixed with the fluid already stored in the vial. By comparing the optical density of the combined fluid to a color-coded chart, the technician would have a reading on the blood sugar level of the patient. Quite obviously, the system must be tightly controlled to ensure that the vials are correctly sealed with the proper amount of fluid prior to shipment to the hospital laboratories. The data in Table 6.5 give the proportion of defectives in 30 different samples (taken from 30 different production hours) of 50 vials each. Use these data to construct the three control lines.

TABLE 6.5 Thirty Sample Proportions Each Based on 50 Observations, Example 6.7

Sample	Sample Proportion Defective	Sample	Sample Proportion Defective
1	.18	16	.18
2	.10	17	.14
3	.18	18	.16
4	.16	19	.14
5	.12	20	.12
6	.12	21	.18
7	.16	22	.16
8	.18	23	.18
9	.12	24	.18
10	.12	25	.14
11	.16	26	.16
12	.18	27	.18
13	.12	28	.16
14	.18	29	.18
15	.14	30	.10

Solution

The sum of the 30 sample proportions is 4.58, so

$$\hat{\pi}_c = \frac{4.58}{30} = .15 \qquad \text{and} \qquad 1 - \hat{\pi}_c = .85$$

Substituting into the formulas for the control limits, we have

$$UCL = .15 + 3\sqrt{\frac{(.15)(.85)}{50}} = .15 + 3(.05) = .30$$

and

$$LCL = .15 - 3\sqrt{\frac{(.15)(.85)}{50}} = .15 - 3(.05) = 0$$

We have discussed the binomial probability distribution and various count data problems that utilize the binomial distribution. In the next section we will consider inferences concerning two binomial parameters.

EXERCISES

6.23 Sketch the operating characteristic curve for the sampling plan $(n = 5, a = 0)$.

6.24 Refer to Exercise 6.23. Superimpose the OC curve for the sampling plan with $n = 10$ and $a = 0$ on that for $(n = 5, a = 0)$. What is the effect of increasing the sample size n while holding the acceptance number "a" constant?

6.25 Refer to the control limits for π obtained in Example 6.7. Suppose that you are now in charge of quality control for the vial production line. In the next 15 samples of 50 vials, you observe the following numbers of defectives:

Number	1	2	3	4	5	6	7	8	9	10	11	12	13	14	15
Defective	9	8	6	5	2	4	6	8	9	9	12	13	14	10	12

Determine whether the process has remained in control.

6.5   COMPARING TWO BINOMIAL PROPORTIONS

Many practical problems involve the comparison of two binomial parameters. For example, social scientists may wish to compare the proportions of women who take advantage of prenatal health services for two communities representing different socioeconomic backgrounds. Or, the director of marketing may wish to compare the public awareness of a new product recently launched and that of a competitor's product.

For comparisons of this type we assume that independent random samples are drawn from two binomial populations with unknown parameters designated by π_1 and π_2. If y_1 successes are observed for the random sample of size n_1 from population 1 and y_2 successes are observed for the random sample of size n_2 from population 2, then the point estimates of π_1 and π_2 are the observed sample proportions $\hat{\pi}_1$ and $\hat{\pi}_2$, respectively:

$$\hat{\pi}_1 = \frac{y_1}{n_1} \quad \text{and} \quad \hat{\pi}_2 = \frac{y_2}{n_2}$$

This notation is summarized here.

NOTATION FOR COMPARING TWO BINOMIAL PROPORTIONS

	Population	
	1	2
Population proportion	π_1	π_2
Sample size	n_1	n_2
Number of successes	y_1	y_2
Sample proportion	$\hat{\pi}_1 = \dfrac{y_1}{n_1}$	$\hat{\pi}_2 = \dfrac{y_2}{n_2}$

Inferences about two binomial proportions are usually phrased in terms of their difference $\pi_1 - \pi_2$, and we use the difference in sample proportions $\hat{\pi}_1 - \hat{\pi}_2$ as part of a confidence interval or statistical test. The sampling distribution for $\hat{\pi}_1 - \hat{\pi}_2$ can be approximated by a normal distribution with mean and standard error given by

$$\mu_{\hat{\pi}_1 - \hat{\pi}_2} = \pi_1 - \pi_2$$

and

$$\sigma_{\hat{\pi}_1 - \hat{\pi}_2} = \sqrt{\frac{\pi_1(1 - \pi_1)}{n_1} + \frac{\pi_2(1 - \pi_2)}{n_2}}$$

This approximation is appropriate if we apply the same requirements to both binomial populations that we did in recommending a normal approximation to a binomial (see Chapter 3). Thus the normal approximation to the distribution of $\hat{\pi}_1 - \hat{\pi}_2$ is appropriate if both $n\pi$ and $n(1 - \pi)$ are 5 or more for each population. Since π_1 and π_2 are never known, make your judgment on the validity of the approximation using $n\hat{\pi}$ and $n(1 - \hat{\pi})$ for each sample.

Confidence intervals and statistical tests about $\pi_1 - \pi_2$ are straightforward and follow the format we used for comparisons using $\mu_1 - \mu_2$. Interval estimation is summarized here; it takes the usual form, point estimate $\pm z$ (standard error).

$100(1 - \alpha)\%$ CONFIDENCE INTERVAL FOR $\pi_1 - \pi_2$

$$\hat{\pi}_1 - \hat{\pi}_2 \pm z_{\alpha/2}\sigma_{\hat{\pi}_1 - \hat{\pi}_2}$$

where

$$\sigma_{\hat{\pi}_1 - \hat{\pi}_2} = \sqrt{\frac{\pi_1(1 - \pi_1)}{n_1} + \frac{\pi_2(1 - \pi_2)}{n_2}}$$

Note: Substitute $\hat{\pi}_1$ and $\hat{\pi}_2$ for π_1 and π_2 in the formula for $\sigma_{\hat{\pi}_1 - \hat{\pi}_2}$. When the normal approximation is valid for $\hat{\pi}_1 - \hat{\pi}_2$, very little error will result from this substitution.

EXAMPLE 6.8

In a survey to analyze the funeral expenditures for various social classes, a random sample of 162 families from the working (blue-collar) class was taken to determine the funeral expenses for a recent death in the family. Of the 162 families contacted, 61 spent over \$800 on the funeral. A similar survey was conducted within the middle/upper classes. Of 189 families contacted, 106 spent more than \$800. Estimate $\pi_1 - \pi_2$, the difference in the proportions of families who have spent more than \$800 for a recent family death. Use a 95% confidence interval to interpret your findings.

Solution

The point estimate of $\pi_1 - \pi_2$ is the difference in sample proportions, $\hat{\pi}_1 - \hat{\pi}_2$:

$$\hat{\pi}_1 - \hat{\pi}_2 = \frac{61}{162} - \frac{106}{189} = .376 - .561 = -.185$$

Note also that $n\hat{\pi}$ and $n(1 - \hat{\pi})$ are 5 or more for both samples, implying that the normal approximation to the binomial is appropriate.

The standard error for $\hat{\pi}_1 - \hat{\pi}_2$ is estimated by

$$\sqrt{\frac{\hat{\pi}_1(1 - \hat{\pi}_1)}{n_1} + \frac{\hat{\pi}_2(1 - \hat{\pi}_2)}{n_2}} = \sqrt{\frac{.376(.624)}{162} + \frac{.561(.439)}{189}} = .052$$

A 95% confidence interval for $\pi_1 - \pi_2$ has $z_{\alpha/2} = 1.96$ and is of the form

point estimate $\pm z_{\alpha/2}$ (standard error)

Substituting into this formula we have

$$-.185 \pm 1.96(.052) \quad \text{or} \quad -.185 \pm .102$$

This interval indicates that π_2 is larger than π_1; we are 95% confident that the difference in the proportions of families paying more than \$800 per funeral for the working class (π_1) and the middle/upper class (π_2) lies in the interval $-.287$ to $-.083$.

We can readily formulate a statistical test for the equality of two binomial parameters. The test statistic for testing $H_0: \pi_1 - \pi_2 = 0$ is a z statistic having the familiar form

$$z = \frac{\text{point estimate}}{\text{standard error}} = \frac{\hat{\pi}_1 - \hat{\pi}_2}{\sigma_{\hat{\pi}_1 - \hat{\pi}_2}}$$

The standard error is slightly different from what we used for a confidence interval. When H_0 is true, $\pi_1 = \pi_2$; we'll call the common value π. Then

$$\sigma_{\hat{\pi}_1 - \hat{\pi}_2} = \sqrt{\frac{\pi_1(1 - \pi_1)}{n_1} + \frac{\pi_2(1 - \pi_2)}{n_2}} = \sqrt{\pi(1 - \pi)\left(\frac{1}{n_1} + \frac{1}{n_2}\right)}$$

The best estimate of π, the proportion of successes common to both populations, is

$$\hat{\pi} = \frac{\text{total number of successes}}{\text{total number of trials}} = \frac{y_1 + y_2}{n_1 + n_2}$$

We have summarized the test procedure here.

STATISTICAL TEST FOR COMPARING TWO BINOMIAL PROPORTIONS

H_0: $\pi_1 - \pi_2 = 0$

H_a: 1. $\pi_1 - \pi_2 > 0$

    2. $\pi_1 - \pi_2 < 0$

    3. $\pi_1 - \pi_2 \neq 0$

T.S.: $z = \dfrac{\hat{\pi}_1 - \hat{\pi}_2}{\sigma_{\hat{\pi}_1 - \hat{\pi}_2}}$ where $\sigma_{\hat{\pi}_1 - \hat{\pi}_2} = \sqrt{\pi(1 - \pi)\left(\dfrac{1}{n_1} + \dfrac{1}{n_2}\right)}$

and π is approximated by $\hat{\pi} = \dfrac{y_1 + y_2}{n_1 + n_2}$

R.R.: For a given value of α,

1. reject H_0 if $z > z_\alpha$
2. reject H_0 if $z < -z_\alpha$
3. reject H_0 if $|z| > z_{\alpha/2}$

Note: $n\hat{\pi}$ and $n(1 - \hat{\pi})$ must be greater than or equal to 5 for both populations in order for the normal approximation (and hence for this test) to hold.

EXAMPLE 6.9

In a recent survey of county high school students ($n_1 = 100$ males and $n_2 = 100$ females), 58 of the males and 46 of the females sampled said they consume alcohol on a regular basis. Use the sample data to conduct a test H_0: $\pi_1 - \pi_2 = 0$ against the one-sided alternative H_a: $\pi_1 - \pi_2 > 0$, that a higher proportion of males than females consume alcohol on a regular basis. Use $\alpha = .05$.

Solution

The four parts of the statistical test are shown here:

H_0: $\pi_1 - \pi_2 = 0$

H_a: $\pi_1 - \pi_2 > 0$

T.S.: $z = \dfrac{\hat{\pi}_1 - \hat{\pi}_2}{\sigma_{\hat{\pi}_1 - \hat{\pi}_2}}$ where $\sigma_{\hat{\pi}_1 - \hat{\pi}_2} = \sqrt{\pi(1 - \pi)\left(\dfrac{1}{n_1} + \dfrac{1}{n_2}\right)}$

R.R.: For $\alpha = .05$, reject H_0 if $z > 1.645$.

From the sample data we find

$$\hat{\pi}_1 = \frac{58}{100} = .58, \quad \hat{\pi}_2 = \frac{46}{100} = .46, \quad \text{and} \quad \hat{\pi} = \frac{58 + 46}{100 + 100} = .52$$

Note also that $n\hat{\pi}$ and $n(1 - \hat{\pi})$ are 5 or more for both samples, validating the normal approximations to the binomial.

Substituting into the test statistic we obtain

$$z = \frac{.58 - .46}{\sqrt{.52(.48)\left(\dfrac{1}{100} + \dfrac{1}{100}\right)}} = \frac{.12}{.071} = 1.70$$

Conclusion: Since $z = 1.70$ is greater than 1.645, we reject $H_0: \pi_1 - \pi_2 = 0$; we have shown that a higher proportion of high school males than females in the county studied consumes alcohol on a regular basis.

EXERCISES

6.26 A random sample of $n_1 = 1000$ observations was obtained from a binomial population with $\pi_1 = .4$. Another random sample, independent of the first sample, was selected from a binomial population with $\pi_2 = .2$. Does the normal approximation hold? Describe the sampling distribution for $\hat{\pi}_1 - \hat{\pi}_2$.

6.27 In a study to compare two binomial proportions, $n_1 = 50$, $n_2 = 40$, $y_1 = 20$, and $y_2 = 15$. Use these hypothetical data to construct a 90% confidence interval for $\pi_1 - \pi_2$.

6.28 Refer to Exercise 6.27. How large a sample should we take from each population in order to have a 90% confidence interval of the form $\hat{\pi}_1 - \hat{\pi}_2 \pm .01$? (Hint: Assuming that equal sample sizes will be taken from the two populations, solve the expression

$$z_{\alpha/2}\sigma_{\hat{\pi}_1 - \hat{\pi}_2} = .01$$

for n, the common sample size. Use $\hat{\pi}_1 = .40$ and $\hat{\pi}_2 = .375$ from Exercise 6.27.)

6.29 A law student believes that the proportion of registered Republicans in favor of additional tax incentives is greater than the proportion of registered Democrats in favor of such incentives. The student acquired independent random samples of 200 Republicans and 200 Democrats and found 109 Republicans and 86 Democrats in favor of additional tax incentives. Use these data to test $H_0: \pi_1 - \pi_2 = 0$ versus $H_a: \pi_1 - \pi_2 > 0$. Give the level of significance for your test.

6.30 In a comparison of the incidence of tumor potential in two strains of rats, 100 rats (50 males, 50 females) were selected from each of two strains and were examined for a period of one year. All the rats were approximately the same age and were housed and fed under comparable conditions. Use the accompanying one-year sample data to construct a 95% confidence interval for the difference in the proportions of rats exhibiting tumor potential for the two strains.

One-Year Results		
	Strain A	Strain B
Sample size	100	100
Number exhibiting tumor potential	25	15

6.31 There may be a remedy for male pattern baldness—at least that's what millions of males hope, if the FDA approves Upjohn's minoxidil for such a use. Minoxidil was investigated in a large, 27-center study where patients were randomly assigned to receive topical minoxidil or an identical-appearing "placebo." Ignoring the center-to-center variation, suppose the preliminary results were as follows:

	Sample Size	% with New Hair Growth
Minoxidil group	310	32
Placebo	309	20

a. Use these data to test H_0: $\pi_1 - \pi_2 = 0$ versus H_a: $\pi_1 - \pi_2 \neq 0$. Give the p-value for your test.

b. If you were working for the FDA, what additional information might you want to examine in this study?

6.32 Is cocaine deadlier than heroin? A study reported in the *Journal of the American Medical Association* found that rats with unlimited access to cocaine had poorer health, had more behavior disturbances, and died at a higher rate than did a corresponding group of rats given unlimited access to heroin. The death rates after 30 days on the study were as follows:

	% Dead at 30 Days
Cocaine group	90
Heroin group	36

a. Suppose that 100 rats were used in each group. Conduct a test of H_0: $\pi_1 - \pi_2 = 0$ versus H_a: $\pi_1 - \pi_2 > 0$. Give the p-value for your test.

b. What implications are there for human use of the two drugs?

6.6   THE POISSON DISTRIBUTION

In Chapter 3 (and again in this chapter) we indicated that the normal distribution provides a good approximation to the binomial distribution provided $n \geq 5/\min(\pi, 1 - \pi)$. This requirement was needed to ensure that the binomial distribution was reasonably symmetric. However, there are many instances when the binomial probability distribution is sufficiently skewed so as to render the normal approximation inappropriate. For example, in observing patients administered a new drug product in a properly conducted clinical trial, the number of persons experiencing a particular side effect might be quite small. If π (the probability of observing a person with the side effect) is .001, $\min(\pi, 1 - \pi) = .001$. For this example, in order to approximate the binomial with a normal distribution, the sample size would have to be equal to or greater than $5/.001 = 5000$.

In 1837, S. D. Poisson developed a discrete probability distribution, suitably called the

Poisson distribution   **Poisson distribution**, which provides a good approximation to the binomial when π is small and n is large but $n\pi$ is less than 5. The probability of observing y successes in the n trials is given by the formula

$$P(y) = \frac{\mu^y e^{-\mu}}{y!}$$

$e = 2.71828$ where e is a constant, the number 2.71828, and μ is the average value of y. Table 7 in the Appendix gives Poisson probabilities for various values of the parameter μ. For approximating binomial probabilities using the Poisson distribution, take

$$\mu = n\pi$$

EXAMPLE 6.10   Refer to the clinical trial alluded to at the beginning of this section, where $n = 1000$ patients were treated with a new drug. Compute the probability that none of a sample of $n = 1000$ patients experiences a particular side effect (such as nausea) when $\pi = .001$.

Solution   The mean of the binomial distribution is $\mu = n\pi = 1000(.001) = 1$. Substituting into the Poisson probability distribution with $\mu = 1$, we have

$$P(y = 0) = \frac{(1)^0 e^{-1}}{0!} = e^{-1} = \frac{1}{2.71828} = .367879$$

(Note also from Table 7 in the Appendix that the entry corresponding to $y = 0$ and $\mu = 1$ is .367879.)

EXAMPLE 6.11   Suppose that after a clinical trial of a new medication involving 1000 patients, no patient experienced nausea. Would it be reasonable to infer that less than .001 of the entire population would experience this side effect while taking the drug?

Solution   Certainly not. We computed the probability of observing $y = 0$ in $n = 1000$ trials assuming $\pi = .001$ (i.e., assuming .1% of the population would experience nausea) to be .368. Since this probability is quite large, it would not be wise to infer that $\pi < .001$.

Although the Poisson distribution provides a useful approximation to the binomial under certain conditions, the application of the Poisson distribution is not limited to these situations. In particular, the Poisson distribution has been useful in finding the probability of y occurrences of an event that *occurs randomly* over an interval of time, volume, space, and so on, provided certain assumptions are met.

1. Events occur one at a time; two or more events do not occur precisely at the same time.
2. The occurrence of an event in a given period is independent of the occurrence of the event in a nonoverlapping period; that is, the occurrence (or nonoccurrence) of an event during one period does not change the probability of an event occurring in some later period.

In many discussions of this topic, a third assumption is added:

3.   The expected number of events during any one period is the same as that during any other period.

Although the underlying mathematics is made easier with this third assumption, this appears to be irrelevant; the first two assumptions are sufficient for a Poisson distribution to apply.

While these assumptions seem to be somewhat restrictive, many situations appear to satisfy these conditions. For example, the number of arrivals of customers at a checkout counter, parking lot toll booth, inspection station, or garage repair shop during a specified time interval (such as one minute) could be approximated with a Poisson probability distribution. Similarly, the number of clumps of algae of a particular species observed in a unit volume of lake water visible under a microscope could be approximated by a Poisson probability distribution.

Confronted with a set of measurements, we may now wish to check the assumption that the data follow a Poisson probability distribution. To do this we make use of the **tests using Poisson distribution** goodness-of-fit test of Section 6.2, using the test statistic

$$\chi^2 = \sum_i \left[\frac{(n_i - E_i)^2}{E_i} \right]$$

There are two types of null hypotheses. The first hypothesis is that the data arise from a Poisson distribution with $\mu = \mu_0$; that is, we wish to test H_0: $\mu = \mu_0$ (μ_0 is specified) against the alternative hypothesis H_a: $\mu \neq \mu_0$. The quantity n_i denotes the number of observations in cell i and E_i is the expected number of observations in cell i obtained from the probabilities for a Poisson distribution with mean μ_0. The computed value of the test statistic is then compared to the tabulated chi-square value in Table 5 in the Appendix with a $= \alpha$ and df $= k - 1$, where k is the number of cells.

The second null hypothesis we might be interested in is less specific. We test

H_0: The observed cell counts all come from a common Poisson distribution with mean μ (unspecified).

The alternative is that not all cell counts come from a common Poisson distribution. The test statistic is

$$\chi^2 = \sum_i \left[\frac{(n_i - E_i)^2}{E_i} \right]$$

where for all cells E_i is the expected number of observations in cell i obtained from the probabilities for a Poisson distribution with a mean estimated from the sample data. The rejection region is then located for a $= \alpha$ and df $= k - 2$. Note the difference in the degrees of freedom for the two null hypotheses. In the latter test we lose one degree of freedom because we must estimate the Poisson parameter μ.

EXAMPLE 6.12   Environmental engineers often utilize information contained in the number of different alga species and the number of cell clumps per species to measure the health of a lake. Those lakes exhibiting only a few species but many cell clumps are classified as oligotrophic. In one such investigation a lake sample was analyzed under a microscope to determine the number of clumps of cells per microscopic field. These data are summarized below for 150 fields examined under a microscope. Here y_i denotes the number of cell clumps per field and n_i denotes the number of fields with y_i cell clumps.

y_i	0	1	2	3	4	5	6	≥ 7
n_i	6	23	29	31	27	13	8	13

Use $\alpha = .05$ to test the null hypothesis that the sample data were drawn from a Poisson probability distribution.

Solution

Before we can compute the value of χ^2, first we must estimate the Poisson parameter μ and then compute the expected cell counts. The Poisson mean μ is estimated by using the sample mean $\bar{y}$. For these data,

$$\bar{y} = \frac{\sum_i n_i y_i}{n} = \frac{486}{150} \approx 3.3$$

It should be noted that the sample mean was computed to be 3.3 by using all the sample data before the 13 largest values were collapsed into the final cell. This is why the sample mean computed here was rounded up to 3.3.

The Poisson probabilities for $y = 0, 1, \ldots, 7$ or more can be found in Table 7 in the Appendix with $\mu = 3.3$. These probabilities are shown below.

y_i	0	1	2	3	4	5	6	≥ 7
$P(y_i)$ for $\mu = 3.3$	.037	.121	.201	.221	.182	.120	.066	.051

The expected cell count E_i can be computed for any cell using the formula $E_i = nP(y_i)$. Hence, for our data (with $n = 150$), the expected cell counts are as shown here.

y_i	0	1	2	3	4	5	6	≥ 7
E_i	5.55	18.15	30.15	33.15	27.30	18.00	9.90	7.65

Substituting these values into the test statistic, we have

$$\chi^2 = \sum_i \left[\frac{(n_i - E_i)^2}{E_i} \right]$$

$$= \frac{(6 - 5.55)^2}{5.55} + \frac{(23 - 18.15)^2}{18.15} + \cdots + \frac{(13 - 7.65)^2}{7.65} = 7.01$$

The tabulated value of chi-square for $a = .05$ and df $= k - 2 = 6$ is 12.59. Since the computed value of chi-square does not exceed 12.59, we have insufficient evidence to reject the null hypothesis that the data were collected from a Poisson distribution.

A word of caution is given here for situations in which we are considering this test procedure. As we mentioned previously, when using a chi-square statistic, we should have all expected cell counts fairly large. In particular, we want all $E_i > 1$ and not more than 20% less than 5. In Example 6.12, if values of $y \geq 7$ had been considered individually, the Es would not have satisfied the criteria for the use of χ^2. That is why we combined all values of $y \geq 7$ into one category.

EXERCISES

 6.33 Use a Poisson approximation to the binomial to find $P(y \leq 2)$ for $n = 1500$ and $\pi = .002$; also find for $\pi = .003$.

6.34 Compute the following Poisson probabilities using Table 7 in the Appendix.
a. $P(y = 1)$ given $\mu = .5$, $\mu = 1.0$, and $\mu = 3.0$
b. $P(y > 1)$ given $\mu = 1.7$, $\mu = 2.5$, and $\mu = 4.2$
c. $P(y < 5)$ given $\mu = 0.2$, $\mu = 1.0$, and $\mu = 2.0$

 6.35 Cars arrive at the exit gate for airport long-term parking at a rate of six per minute during rush hour. Find the following probabilities using Table 7 in the Appendix. (y is the number of cars arriving during any given minute in rush hour.)
a. $P(y = 0)$
b. $P(y > 1)$
c. $P(y > .3)$

 6.36 A firm is considering using telemarketing techniques to supplement traditional marketing methods. It is estimated that one of every 100 calls results in a sale. Suppose that 250 calls are made in a single day:
a. Write an expression for the probability that there are less than six sales—do not do the mathematics.
b. What assumptions are you making in part (a)?
c. Use a normal approximation to compute $P(y < 6)$.
d. Compute $P(y < 6)$ using the Poisson distribution.
e. Which approximation (part (c) or part (d)) appears better? Why?

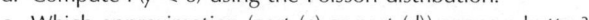 **6.37** A certain birth defect occurs with probability .0001; that is, one of every 10,000 babies has this defect. If 5000 babies are born at a particular hospital in a given year, what approximation should be used? What is the approximate probability that there is at least one baby with the defect?

 6.38 One portion of a study to determine the effectiveness of an exclusive bus lane was directed at examining the number of conflicts (driving situations that could result in an accident) at a major intersection during a specified period of time. A previous study prior to the installation of the exclusive bus lane indicated that the number of conflicts per five minutes during the 7:00 to 9:00 AM peak period could be adequately approximated by a Poisson distribution with $\mu = 2$. The following data were based on a sample of 40 days; y_i denotes the number of conflicts and n_i denotes the number of five-minute periods during which y was observed.

y_i	0	1	2	3	4	5	≥ 6
n_i	90	230	240	130	68	30	12

a. Does the Poisson assumption appear to hold?

b. Use these data to test the research hypothesis that the mean number of conflicts per five minutes differs from 2. Hint: Use a chi-square test based on Poisson probabilities.

 6.39 The number of shutdowns per day caused by a breaking of the thread was noted for a nylon spinning process over a period of 90 days. Use the sample data below to determine if the number of shutdowns per day follows a Poisson distribution. Use $\alpha = .05$. In the listing of the data, y_i denotes the number of shutdowns per day and n_i denotes the number of days with y_i shutdowns.

y_i	0	1	2	3	4	≥ 5
n_i	20	28	15	8	7	12

6.7 r x c CONTINGENCY TABLES: CHI-SQUARE TEST OF INDEPENDENCE

In all our calculations so far in this text, we have assumed that only one measurement is taken on each sampling unit. We might obtain the yield for an acre planted in wheat, a blood pressure reading on a patient who is being administered an anesthetic, or a measurement on the number of potential conflicts at a highway intersection during a one-hour period. However, research problems in the sciences frequently involve more than one variable. If measurements are taken on two (or more) variables for each sam-

bivariate and multivariate data

pling unit, we say that we have **bivariate** (or **multivariate**) **data**.

As with univariate count data, where the data may be summarized in a table, we frequently arrange bivariate data in a two-way table. For example, in a study of the public approval of a proposed high-speed bus lane for commuters, the interviewers might also ask individuals information about their occupations. We could then classify each person by his or her opinion concerning the new lane (favor, do not favor, undecided) and his or her occupation (white-collar worker, blue-collar worker, laborer).

What is the objective of such a classification? In most studies either we wish to determine whether the two variables are related (dependent) or we wish to predict one variable based on knowledge of the other. This section deals with a test of independence for bivariate count data arranged in a two-way table. The two-way tables are sometimes

contingency tables

called **contingency tables** because the alternative hypothesis in our test is that the two variables are dependent (i.e., there is a contingency between the two variables).

Consider the problem in which we would like to determine whether the following two variables are dependent: employee classification (staff, faculty, administrator) at a university and an employee's opinion about whether or not the local chapter of the teachers' union should be the sole collective bargaining agent for employee benefits. A random sample of 200 employees is taken from employee records and each employee is classified according to both variables. Suppose the results of the survey appear as shown in Table 6.6. Is there evidence to indicate that a person's opinion concerning collective

TABLE 6.6   Classification of 200 Employees by Classification and Opinion on Collective Bargaining

| | Opinion on Collective Bargaining by Teachers' Union | | | |
Employee Classification	Favor	Do Not Favor	Undecided	Totals
Staff	30	15	15	60
Faculty	40	50	10	100
Administrator	10	25	5	40
Totals	80	90	30	200

bargaining depends on his or her employment status? That is, can we conclude that the two variables are dependent?

independence   To answer this question we must define the concept of **independence**

DEFINITION 6.2

Two variables that have been categorized in a two-way table are **independent** if the probability that a measurement is classified into a given cell of the table is equal to the probability of being classified into that row times the probability of being classified into that column. This must be true for all cells of the table.

For example, suppose that the probability of selecting a person favoring the teachers' union in the university survey is π_1, the probability of selecting one who does not favor the union for collective bargaining is π_2, and the probability of selecting a person who is undecided is π_3 (note: $\pi_1 + \pi_2 + \pi_3 = 1$). Similarly, suppose that the probabilities of selecting a staff member, a faculty member, or an administrator are, respectively, π_A, π_B, and π_C (where $\pi_A + \pi_B + \pi_C = 1$). Then the two variables, employee classification and opinion concerning the teachers' union, are independent if the probability of classifying a person into a specific cell of the two-way table is obtained by multiplying the respective row and column probabilities. These ideas are illustrated in Table 6.7.

TABLE 6.7   Cell Probabilities Showing Independence for the Collective Bargaining Survey

| | Opinion | | |
Employee Classification	Favor, π_1	Do Not Favor, π_2	Undecided, π_3
Staff, π_A	$\pi_A \pi_1$	$\pi_A \pi_2$	$\pi_A \pi_3$
Faculty, π_B	$\pi_B \pi_1$	$\pi_B \pi_2$	$\pi_B \pi_3$
Administrator, π_C	$\pi_C \pi_1$	$\pi_C \pi_2$	$\pi_C \pi_3$

A test of the independence of two variables arranged in a two-way table makes use of the test statistic

test statistic

$$\chi^2 = \sum_{i,j} \left[\frac{(n_{ij} - E_{ij})^2}{E_{ij}} \right]$$

where n_{ij} and E_{ij} are, respectively, the observed and expected number of measurements falling in the cell for the ith row and the jth column.

DEFINITION 6.3

The expected number of measurements E_{ij} falling in the i, j cell (cell of the ith row and jth column of the table) is taken to be

$$E_{ij} = \frac{(\text{row } i \text{ total})(\text{column } j \text{ total})}{n}$$

when the two variables are independent. (Note: If E_{ij} is not an integer, it should not be rounded to an integer value for the tests that follow.)

EXAMPLE 6.13   Compute the expected number of measurements falling into each cell of Table 6.6.

Solution   The expected number of measurements falling in the 1, 1 cell (first row, first column) is

$$E_{11} = \frac{(\text{row 1 total})(\text{column 1 total})}{n} = \frac{(60)(80)}{200} = 24$$

Similarly, the expected number of measurements in the 3, 2 cell is

$$E_{32} = \frac{(\text{row 3 total})(\text{column 2 total})}{n} = \frac{(40)(90)}{200} = 18$$

These and the remaining cell counts appear in Table 6.8.

Note that the expected counts in a row sum to the same row total as do the observed cell counts. The same applies for columns.

TABLE 6.8   Expected Cell Counts for the Collective Bargaining Survey

Employee Classification	Opinion			
	Favor	Do Not Favor	Undecided	Totals
Staff	24	27	9	60
Faculty	40	45	15	100
Administrator	16	18	6	40
Totals	80	90	30	200

We can now summarize the *chi-square test of independence* for data arranged in a two-way table.

CHI-SQUARE
TEST OF
INDEPENDENCE

Null hypothesis: The two variables are independent.

Alternative hypothesis: The two variables are dependent.

Test statistic: $\chi^2 = \sum_{i,j} \left[\dfrac{(n_{ij} - E_{ij})^2}{E_{ij}} \right]$

Rejection region: Reject H_0 if χ^2 exceeds the tabulated value of chi-square (Table 5 in the Appendix) for a = α and df = $(r - 1)(c - 1)$, where

r = number of rows in the table

c = number of columns in the table

The guidelines that Cochran (1954) proposed for the E_{ij}s (see Section 6.2) are still in effect when we use the chi-square test of independence. While agreeing with Cochran, Conover (1971) goes even further by stating that when the E_{ij}s are all of about the same magnitude and both r and c are large, then even if the E_{ij}s are as small as 1, the approximation by a chi-square distribution will still be good. These guidelines give us a great deal of flexibility in applying the chi-square test without having to collapse some of the categories.

EXAMPLE 6.14

Conduct a chi-square test of independence for the teachers' union data in Table 6.6. Use α = .05.

Solution

Using the observed cell counts of Table 6.6 and the expected cell counts of Table 6.8, we can substitute these values into the test statistic.

$$\chi^2 = \sum_{i,j} \left[\frac{(n_{ij} - E_{ij})^2}{E_{ij}} \right]$$

$$= \frac{(30 - 24)^2}{24} + \frac{(15 - 27)^2}{27} + \frac{(15 - 9)^2}{9} + \frac{(40 - 40)^2}{40} + \frac{(50 - 45)^2}{45}$$

$$+ \frac{(10 - 15)^2}{15} + \frac{(10 - 16)^2}{16} + \frac{(25 - 18)^2}{18} + \frac{(5 - 6)^2}{6}$$

$$= 18.2$$

The critical value of χ^2 for a = .05 and df = $(r - 1)(c - 1) = 2(2) = 4$ is 9.48773. Since the computed value, 18.2, exceeds 9.48773, we reject H_0 and conclude that the two variables are dependent. In particular, we say that the proportion of persons favoring the teachers' union as the collective bargaining agent varies depending on the employee status. From Table 6.6 we see that a much higher proportion of the staff members favor the teachers' union as the collective bargaining agent than of either the faculty or administrators.

The only function of this chi-square test is to determine whether the observed dependence is due to random fluctuations. When H_0: independence is rejected in favor of H_a: dependence, we conclude that the two variables that have been summarized in a contingency table are related; that is, an outcome on one variable affects or is affected by an outcome on the other variable. The rejection of H_0 does not, however, indicate the strength or type of the relation between the two variables. As was discussed in Chapter 2, a percentage comparison can help to identify how the variables are related, and certain "measures of association" that will not be discussed in this text can help to quantify the strength of the relation between the two variables. For more details about measures of association see Hildebrand et al. (1977) or Ott, Larson, and Mendenhall (1987).

The same chi-square test statistic applies to a slightly different sampling procedure. In our discussion of the chi-square goodness-of-fit test, an implicit assumption has been that the data summarized in the contingency table resulted from a single random sample from the population of interest. Often separate random samples are taken from the *sub-populations* defined by the rows (or columns) of the contingency table. For example, the data of Table 6.6 might have resulted from separate random samples of sizes 60 (staff), 100 (faculty), and 40 (administrator) rather than from a single overall sample of 200 individuals. When random samples are taken for categories of the row (or column) variable, the test is called a *test of homogeneity* of the row (or column) distributions. So, when sampling by rows, the percentage distributions across columns is the same from row to row. This would be seen in a percentage comparison by rows for the sample data. Similarly, when sampling by columns, the percentage distribution across rows is the same from column to column.

Since the mechanics and conclusions are the same for the chi-square test of independence and for the chi-square test of homogeneity of the distributions, we will not worry about the distinction between the two tests.

EXERCISES

6.40   A survey of student opinion concerning a proposed tuition increase was taken to determine whether student opinion was independent of gender. The results of 300 interviews are recorded in the accompanying table. Run a chi-square test of independence and give the level of significance for the test results. Draw conclusions.

	Opinion		
Gender of Student	Favor Increase	Oppose	Undecided
Female	91 ( )*	54 ( )*	13
Male	59	69	14

* Note: you need to use the formula for computing the expected cell count for only these cells; the remaining cell counts can be computed by subtraction from the appropriate row or column total.

6.41 A scientist was interested in testing the effectiveness of a new drug product in controlling worms in the small intestine of sheep. A prestudy test was used to select 40 sheep with approximately the same level of infestation. These sheep were then randomly divided into two groups of 20. Those in the first group were given the drug product; those in the second group received no treatment and served as a control group. After two weeks, each of the 40 sheep was examined and classified as either "responder" or "nonresponder" depending on the observed worm count. The sample data are summarized here:

Classification	Group 1 (Drug-Treated)	Group 2 (Control)
Responder	15 ( )	7
Nonresponder	5	13

a. Compute the expected cell counts.
b. Run a chi-square test of independence with $\alpha = .05$. State the null hypothesis for this test and draw appropriate conclusions.

6.42 A carcinogenicity study was conducted to examine the tumor potential of a drug product scheduled for initial testing in humans. A total of 300 rats (150 males and 150 females) were studied for a six-month period. At the beginning of the study 100 rats (50 males, 50 females) were randomly assigned to the control group, 100 to the low-dose group, and the remaining 100 (50 males, 50 females) to the high-dose group. On each day of the six-month period, the rats in the control group received an injection of an inert solution, whereas those in the drug groups received an injection of the solution plus drug. The sample data are shown in the accompanying table.

| | Number of Tumors | |
Rat Group	One or More	None
Control	10	90
Low dose	14	86
High dose	19	81

a. Give the percentages of rats with one or more tumors for each of the three groups.
b. Conduct a chi-square test of independence with $\alpha = .05$.
c. Does there appear to be a drug-related problem regarding tumors for this drug product? That is, as the dose is increased, does there appear to be an increase in the proportion of rats with tumors?

6.43 Computer output for the data of Exercise 6.42 is shown here. Compare the output with your results in Exercise 6.42.

CATEGORICAL ANALYSIS

TABLE OF R_GROUP BY N_TUMORS

R_GROUP   N_TUMORS

FREQUENCY PERCENT ROW PCT COL PCT	> = 1	NONE	TOTAL
CONTROL	10 3.33 10.00 23.26	90 30.00 90.00 35.02	100 33.33
HIGHDOSE	19 6.33 19.00 44.19	81 27.00 81.00 31.52	100 33.33
LOWDOSE	14 4.67 14.00 32.56	86 28.67 86.00 33.46	100 33.33
TOTAL	43 14.33	257 85.67	300 100.00

STATISTICS FOR TABLE OF R_GROUP BY N_TUMORS

STATISTIC	DF	VALUE	PROB
CHI-SQUARE	2	3.312	0.191
LIKELIHOOD RATIO CHI-SQUARE	2	3.327	0.189
MANTEL-HAENSZEL CHI-SQUARE	1	0.649	0.420
PHI		0.105	
CONTINGENCY COEFFICIENT		0.104	
CRAMER'S V		0.105	

SAMPLE SIZE = 300

6.44 A total of 210 emphysema patients entering a clinic over a one-year period were treated with one of two drugs (either the standard drug, A, or an experimental compound, B) for a period of one week. After this period, each patient's condition was rated as greatly improved, improved, or no change. The sample results are shown here.

		Patient's Condition	
Therapy	No Change	Improved	Greatly Improved
Standard, A	20	35	45
Experimental, B	15	45	50

a. Make a percentage comparison for the rows; does there appear to be a difference in the two therapies?

b. Run a chi-square test of independence and draw a conclusion. Does it agree with your speculation in part (a)?

6.45 Refer to Exercise 6.44. The sum of the expected values must equal $n = 210$, and the expected values will add to the appropriate row and column totals. Determine the minimum number of expected values that need to be computed. How is this number related to the degrees of freedom for the test?

 6.46 The market research group of a particular firm conducted a survey in three cities to compare the sales potential of a new soft drink. Each person contacted was asked to try the new drink and to classify it as excellent, satisfactory, or unsatisfactory. The results of the survey are summarized in the accompanying table. Use the Minitab computer output shown here to conduct a chi-square test of independence. Give the approximate level of significance for your test and draw conclusions.

Classification	City 1	City 2	City 3
Excellent	62	51	45
Satisfactory	28	30	35
Unsatisfactory	10	19	20

```
        READ THE TABLE INTO C1,C2,C3
DATA> 62 51 45
DATA> 28 30 35
DATA> 10 19 20
DATA> END DATA
        3 ROWS READ
MTB > PRINT C1,C2,C3
  ROW   C1    C2    C3

   1    62    51    45
   2    28    30    35
   3    10    19    20

MTB > CHISQUARE ANALYSIS ON TABLE IN C1,C2,C3

Expected counts are printed below observed counts

            C1      C2      C3     Total
    1       62      51      45      158
          52.7    52.7    52.7

    2       28      30      35       93
          31.0    31.0    31.0

    3       10      19      20       49
          16.3    16.3    16.3

Total      100     100     100      300

ChiSq =   1.65 +   0.05 +   1.12 +
          0.29 +   0.03 +   0.52 +
          2.46 +   0.44 +   0.82 = 7.38

df = 4

MTB >
```

6.47 A university conducted a self-study to satisfy the requirements for accreditation. One aspect of the self-study concerned faculty evaluations. Through the use of student evaluations of their instructors, each faculty member was classified both by rank and by ability as a teacher. Use the accompanying results to test the null hypothesis of independence of the two classifications. Use $\alpha = .05$, and draw a conclusion.

Teaching Evaluation	Rank			
	Instructor	Assistant Professor	Associate Professor	Professor
Above average	36	62	45	50
Average	48	50	35	43
Below average	30	13	20	35

6.48 A survey of admissions practices at a liberal arts college was conducted to determine whether there appeared to be a difference in the acceptance rates for white and minority (nonwhite) applicants. The results of this survey, which combine information from 4000 applicants, are shown in the accompanying table. Use the accompanying computer printout to conduct a chi-square test of independence. Use Table 5 in the Appendix to obtain an approximate level of significance for the test and draw conclusions.

Applicant Accepted?	Applicant		Total
	Nonwhite	White	
Yes	38	126	164
No	362	3474	3836
Total	400	3600	4000

```
        READ DATA IN C1,C2
DATA> 38 126
DATA> 362 3474
DATA> END DATA
      2 ROWS READ
MTB > CHISQUARE FOR DATA IN C1,C2

Expected counts are printed below observed counts

          C1      C2     Total
    1     38     126      164
        16.4   147.6

    2    362    3474     3836
       383.6  3452.4

Total   400    3600     4000

ChiSq =  28.45 +   3.16 +
          1.22 +   0.14 = 32.96
df = 1

MTB >
```

6.49   Computer output for the data of Exercise 6.42 is shown here. Compare the Minitab output with your results in Exercise 6.42 and the output in Exercise 6.43.

```
         READ THE FOLLOWING TABLE INTO C1,C2
DATA> 10 90
DATA> 14 86
DATA> 19 81
DATA> END DATA
      3 ROWS READ
MTB > PRINT C1,C2
 ROW    C1    C2

  1     10    90
  2     14    86
  3     19    81

MTB > CHISQUARE ANALYSIS ON TABLE IN C1,C2

Expected counts are printed below observed counts

           C1       C2     Total
   1       10       90      100
         14.3     85.7

   2       14       86      100
         14.3     85.7

   3       19       81      100
         14.3     85.7

Total      43      257      300

ChiSq =  1.31 +   0.22 +
         0.01 +   0.00 +
         1.52 +   0.25 = 3.31
df = 2

MTB >
```

6.8   COMBINING SETS OF $r \times c$ CONTINGENCY TABLES (optional)

In the previous section we discussed the chi-square test of independence for examining the dependence of two variables based on data arranged in a contingency table. Suppose a pharmaceutical company is developing a drug product for the treatment of epilepsy. In each of several clinics, patients are assigned at random to either a placebo or the new drug and treated for a period of two months. At the end of the study each patient is rated as either improved or not improved. If 100 patients (50 per treatment group) are to be enrolled in a particular clinic and we observed 40 and 15 patients improved in the

TABLE 6.9   Number (%) of Patients Improved

	Improved	Not Improved	Total
New drug	40 (80%)	10	50
Placebo	15(30%)	35	50

new drug and placebo groups, respectively, the data could be displayed as shown in Table 6.9 and analyzed using the chi-square methods of the previous section. The null hypothesis of independence of the two classifications (treatment group and rating) could be restated in terms of the proportions, π_1 and π_2, of improved patients for the two populations. The new H_0 would be H_0: $\pi_1 - \pi_2 = 0$; namely, that there is no difference in the proportions of improved patients for the drug and placebo groups. Rejection of H_0 using the chi-square statistic from the test of independence test indicates that the population proportions are different for the two treatment groups.

This same scenario can be extended to more than one clinic and we can extend our test procedure to deal with a set of q clinics ($q \geq 2$). For this situation we would observe the sample percentage improved for the drug and placebo groups in each clinic; the data could be summarized using Table 6.10. The test for comparing the drug and placebo proportions combines sample information across the separate contingency tables to answer the question of whether, on the average, the improvement rates are the same for the two treatment groups. Before we do this, however, we need some additional notation, shown in Table 6.11.

Cochran (1954) proposed a test statistic for the hypothesis of no difference (on the average) for the improvement rates for a set of q 2 × 2 contingency tables. This same problem was addressed by Mantel and Haenszel (1959) and also extended to cover a set of q 2 × c contingency tables. For 2 × 2 tables the Mantel–Haenszel statistic for testing the equality of the improvement rates, on the average, can be written as

$$\chi_{MH}^2 = \frac{\left\{\sum_h \left(n_{h11} - \frac{n_{h1.}n_{h.1}}{n_{h..}}\right)\right\}^2}{\sum_h \frac{n_{h1.}n_{h2.}n_{h.1}n_{h.2}}{n_{h..}^2(n_{h..} - 1)}}$$

TABLE 6.10   Summary Table for a Set of 2 × 2 Contingency Tables

	Clinic	Improved	Not Improved
1	Drug		
	Placebo		
2	Drug		
	Placebo		
	$\vdots$		
q	Drug		
	Placebo		

TABLE 6.11   General Notation for a Set of 2 × 2 Contingency Tables

Table	Treatment	Response Category 1	Response Category 2	Total
1	1	n_{111}	n_{112}	$n_{11.}$
	2	n_{121}	n_{122}	$n_{12.}$
	Total	$n_{1.1}$	$n_{1.2}$	$n_{1..}$
2	1	n_{211}	n_{212}	$n_{21.}$
	2	n_{221}	n_{222}	$n_{22.}$
	Total	$n_{2.1}$	$n_{2.2}$	$n_{2..}$
$\vdots$				
h	1	n_{h11}	n_{h12}	$n_{h1.}$
	2	n_{h21}	n_{h22}	$n_{h2.}$
	Total	$n_{h.1}$	$n_{h.2}$	$n_{h..}$
$\vdots$				

which follows a chi-square distribution with df = 1. Let's see how this works for a set of sample data.

EXAMPLE 6.15   The pharmaceutical study discussed previously was extended to three clinics. In each clinic as patients qualified for the study and gave their consent to participate, they were assigned to either the drug or placebo groups according to a predetermined random code. Each clinic was to treat 50 patients per group. The study results are summarized in Table 6.12.

TABLE 6.12   Study Results

	Clinic	Improved	Not Improved	Total
1	Drug	40 (80%)	10	50
	Placebo	15 (30%)	35	50
	Total	55	45	100
2	Drug	35 (70%)	15	50
	Placebo	20 (40%)	30	50
	Total	55	45	100
3	Drug	43 (86%)	7	50
	Placebo	31 (62%)	19	50
	Total	74	26	100
Total		184	116	300

Use these data to test the null hypothesis of no difference in the improvement rates, on the average. Use the Mantel–Haenszel chi-square statistic and give the p-value for the test.

Solution

The necessary row and column totals in each clinic are given in Table 6.12. The numerator of the Mantel–Haenszel statistic is

$$\left\{\sum_h \left(n_{h11} - \frac{n_{h1.}n_{h.1}}{n_{h..}}\right)\right\}^2 = \left\{\left(40 - \frac{50(55)}{100}\right) + \left(35 - \frac{50(55)}{100}\right) + \left(43 - \frac{50(74)}{100}\right)\right\}^2$$

$$= (12.5 + 7.5 + 6)^2 = 676$$

whereas the denominator is

$$\sum_h \frac{n_{h1.}n_{h2.}n_{h.1}n_{h.2}}{n_{h..}^2(n_{h..}-1)} = \frac{50(50)(55)(45)}{(100)^2(99)} + \frac{50(50)(55)(45)}{(100)^2(99)} + \frac{50(50)(74)(26)}{(100)^2(99)}$$

$$= 6.25 + 6.25 + 4.8586 = 17.3586$$

Substituting, we obtain

$$\chi_{MH}^2 = \frac{676}{17.3586} = 38.9432$$

For df $= 1$, this result is significant at the $p < .001$ level. As can be seen from the sample data, the drug-treated groups have consistently higher improvement rates than the placebo groups.

EXAMPLE 6.16

Sample data are not always as obvious and conclusive as those given in Example 6.15. Use the revised sample data shown in Table 6.13 to conduct a Mantel–Haenszel test. Give the p-value for your test and interpret your findings.

TABLE 6.13   Revised Study Results

	Clinic	Improved	Not Improved	Total
1	Drug	35 (70%)	15	50
	Placebo	26 (52%)	24	50
	Total	61	39	100
2	Drug	28 (56%)	22	50
	Placebo	29 (58%)	21	50
	Total	57	43	100
3	Drug	37 (74%)	13	50
	Placebo	24 (48%)	26	50
	Total	61	39	100

Solution

Using the row and column totals of Table 6.13, the numerator and denominator of χ^2_{MH} can be shown to be 110.25 and 18.21, respectively. The Mantel–Haenszel statistic is then

$$\chi^2_{MH} = 6.05$$

Based on df $= 1$, this test result is significant at the $p < .001$ level. We conclude that although the drug product did not have a higher improvement rate in all three clinics, the data combined across clinics indicates that, on the average, the drug improvement rate is higher than the placebo rate ($p < .001$).

Mantel and Haenszel also extended this test procedure to cover the situation in which we want a combined test based on sample data displayed in a set of q $2 \times c$ contingency tables. Returning to our example, suppose rather than having two response categories (e.g., improved, not improved) we have c different categories such as (worse, same, or better) or (none, slight, moderate, completely well). For these situations it is possible to score the categories of the scale and run a Mantel–Haenszel test based on the difference in mean scores for the two treatment groups. Because the formulas become more involved, we will revert to available statistical software.

EXAMPLE 6.17

The data shown in Table 6.14 appeared in Sugiura and Otaka (1974). They summarize by age category and dose of radiation the numbers of deaths from leukemia (LD) observed at the Atomic Bombs Casualty Commission (ABCC) and the corresponding numbers of individuals who did not die from leukemia (NLD) during the period from 1950 to 1970. Run a Mantel–Haenszel mean score test (based on uniform scores) to compare the mean levels of radiation for the two groups (leukemia death and non–death from leukemia) across the five age categories. Interpret the results.

TABLE 6.14   Deaths from Leukemia Observed at ABCC (1950–1970)

Age (Years)	Survival Status	Not in City	Dose (Rads)				
			0–9	10–49	50–99	100–199	200+
0–9	LD	0	7	3	1	4	11
	NLD	5015	10,752	2989	694	418	387
10–19	LD	5	4	6	1	3	6
	NLD	5973	11,811	2620	771	792	820
20–34	LD	2	8	3	1	3	9
	NLD	5669	10,828	2798	797	596	624
35–49	LD	3	19	4	2	1	10
	NLD	6158	12,645	3566	972	694	608
50+	LD	3	7	3	2	2	6
	NLD	3695	9053	2415	655	393	289

***** PARCAT *****

GENERALIZED COCHRAN–MANTEL–HAENSZEL TEST STATISTICS
FOR AVERAGE PARTIAL ASSOCIATION IN THREE-WAY CONTINGENCY TABLES

P.46 DEATHS FROM LEUKEMIA—UNIFORM SCORES FOR DOSE   00060015

COLUMN SCORES: UNIFORM   1.00   2.00
ROW SCORES: UNIFORM   1.00   2.00   3.00   4.00   5.00   6.00

SUMMARY ACROSS TABLES

****

A. SUMMARY OF INDIVIDUAL TABLE STATISTICS

| TABLE | TABLE | SAMPLE | MULTIVARIATE | | | MEAN SCORE | | | CORRELATION | | |
NO.	FREQ.	SIZE	Q	D.F.	P	QMS	D.F.	P	QMA	D.F.	P
1	1	20281	248.05	5	0.0	248.05	5	0.0	127.51	1	0.0
2	1	22812	43.89	5	0.0	43.89	5	0.0	29.32	1	0.0
3	1	21338	100.56	5	0.0	100.56	5	0.0	60.23	1	0.0
4	1	24682	88.85	5	0.0	88.85	5	0.0	38.02	1	0.0
5	1	16523	84.74	5	0.0	84.74	5	0.0	40.56	1	0.0
TOTAL		105636	566.09	25	0.0	566.09	25	0.0	295.65	5	0.0

B. GENERALIZED COCHRAN–MANTEL–HAENSZEL STATISTICS

| SAMPLE | MULTIVARIATE | | | MEAN SCORE | | | CORRELATION | | |
SIZE	Q(CMH)	D.F.	P	Q(CMMS)	D.F.	P	Q(CMMA)	D.F.	P
105636	461.66	5	0.0	461.66	5	0.0	262.62	1	0.0

****

***** PARCAT *****

GENERALIZED COCHRAN–MANTEL–HAENSZEL TEST STATISTICS
FOR AVERAGE PARTIAL ASSOCIATION IN THREE-WAY CONTINGENCY TABLES
P.46 DEATHS FROM LEUKEMIA—MIDPOINT SCORES FOR DOSE   00249515

COLUMN SCORES: UNIFORM   1.00   2.00
ROW SCORES: USER SPECIFIED   0.0   4.50   29.50   74.50   149.50   300.00

SUMMARY ACROSS TABLES

****

A. SUMMARY OF INDIVIDUAL TABLE STATISTICS

| TABLE | TABLE | SAMPLE | MULTIVARIATE | | | MEAN SCORE | | | CORRELATION | | |
NO.	FREQ.	SIZE	Q	D.F.	P	QMS	D.F.	P	QMA	D.F.	P
1	1	20281	248.05	5	0.0	248.05	5	0.0	226.32	1	0.0
2	1	22812	43.89	5	0.0	43.89	5	0.0	39.01	1	0.0
3	1	21338	100.56	5	0.0	100.56	5	0.0	94.29	1	0.0
4	1	24682	88.85	5	0.0	88.85	5	0.0	65.87	1	0.0
5	1	16523	84.74	5	0.0	84.74	5	0.0	76.04	1	0.0
	TOTAL	105636	566.09	25	0.0	566.09	25	0.0	501.52	5	0.0

B. GENERALIZED COCHRAN–MANTEL–HAENSZEL STATISTICS

| SAMPLE | MULTIVARIATE | | | MEAN SCORE | | | CORRELATION | | |
SIZE	Q(CMH)	D.F.	P	Q(CMMS)	D.F.	P	Q(CMMA)	D.F.	P
105636	461.66	5	0.0	461.66	5	0.0	426.60	1	0.0

Solution
The test procedure based on mean scores can be summarized as follows:

H_0: There is no difference, on the average, in mean radiation scores for the LD and NLD groups

H_a: The mean radiation scores are different

T.S.: Generalized Mantel–Haenszel χ^2 statistic

R.R.: Reject H_0 if the observed chi-square value exceeds χ_α^2 based on df $= 1$

The generalized Mantel–Haenszel mean score test can be run using just about any set of scores. One possibility would be to use uniform scores (for example, 0, 1, 2, 3, 4, and 5) to the radiation dose categories listed in Table 6.14. Another possibility would be to use the midpoints of the intervals for the dose categories. This scoring system works fine for all but the first and last categories. Here, in order to compute the chi-square statistic, we will make the arbitrary assignment of 0 to the first (Not in city) category and 300 to the 200+ category of dose. The results of the corresponding mean score tests for the two scoring systems are 262.6 and 426.6, respectively, shown in the preceding computer output under the label of Q(CMMA). Thus we can conclude that for either scoring system, there is a highly significant difference in the mean radiation effect for the LD and NLD groups across the different age categories. In particular, since we are comparing two groups—those who died from leukemia (LD) and those who did not (NLD)—the higher mean associated with the LD group combined across the different categories indicates strong evidence of a connection between the level of radiation exposure and survival status. A more detailed analysis of these data is presented in Landis et al. (1978).

EXERCISES

6.50
A sample of 1200 individuals arrested for driving under the influence of alcohol was obtained from police records. The research recorded the gender, socioeconomic status (from occupation information), and the number of previous alcohol-related arrests. These data are shown here:

Socioeconomic Status	Number of Previous Alcohol-Related Arrests	Male	Female
Low	0	110	130
	1–2	60	50
	>2	30	20
Medium	0	105	101
	1–2	75	55
	>2	20	44
High	0	90	80
	1–2	80	60
	>2	30	60

Use the output shown here to run a generalized Mantel–Haenszel test based on uniform scores to compare the average number of previous arrests for males and females. Give a *p*-value for your test and interpret your results. Would the scores 0, 1.5, and 2.5 give different results?

***** PARCAT *****

GENERALIZED COCHRAN–MANTEL–HAENSZEL TEST STATISTICS
FOR AVERAGE PARTIAL ASSOCIATION IN THREE-WAY CONTINGENCY TABLES

P.48 NO. OF ALC. RELATED ARRESTS—UNIFORM SCORES FOR NO.   00280016

COLUMN SCORES: UNIFORM   1.00   2.00   3.00
ROW SCORES: UNIFORM   1.00   2.00

SUMMARY ACROSS TABLES

****

A. SUMMARY OF INDIVIDUAL TABLE STATISTICS

TABLE NO.	TABLE FREQ.	SAMPLE SIZE	MULTIVARIATE Q	D.F.	P	MEAN SCORE QMS	D.F.	P	CORRELATION QMA	D.F.	P
1	1	400	4.56	2	0.1021	4.49	1	0.0340	4.49	1	0.0340
2	1	400	12.12	2	0.0023	3.56	1	0.0591	3.56	1	0.0591
3	1	400	13.41	2	0.0012	6.54	1	0.0105	6.54	1	0.0105
TOTAL		1200	30.10	6	0.0	14.60	3	0.0022	14.60	3	0.0022

B. GENERALIZED COCHRAN–MANTEL–HAENSZEL STATISTICS

SAMPLE SIZE	MULTIVARIATE Q(CMH)	D.F.	P	MEAN SCORE Q(CMMS)	D.F.	P	CORRELATION Q(CMMA)	D.F.	P
1200	16.11	2	0.0003	2.17	1	0.1406	2.17	1	0.1406

****

***** PARCAT *****

GENERALIZED COCHRAN–MANTEL–HAENSZEL TEST STATISTICS
FOR AVERAGE PARTIAL ASSOCIATION IN THREE-WAY CONTINGENCY TABLES

P.48 NO. OF ALC. RELATED ARRESTS—MID-PT. SCORES FOR NO.   00360016

COLUMN SCORES: USER SPECIFIED   0.0   1.50   2.50
ROW SCORES: UNIFORM   1.00   2.00

SUMMARY ACROSS TABLES

****

A. SUMMARY OF INDIVIDUAL TABLE STATISTICS

TABLE NO.	TABLE FREQ.	SAMPLE SIZE	MULTIVARIATE Q	D.F.	P	MEAN SCORE QMS	D.F.	P	CORRELATION QMA	D.F.	P
1	1	400	4.56	2	0.1021	4.56	1	0.0327	4.56	1	0.0327
2	1	400	12.12	2	0.0023	2.38	1	0.1230	2.38	1	0.1230
3	1	400	13.41	2	0.0012	4.99	1	0.0254	4.99	1	0.0254
TOTAL		1200	30.10	6	0.0	11.94	3	0.0076	11.94	3	0.0076

B. GENERALIZED COCHRAN–MANTEL–HAENSZEL STATISTICS

SAMPLE SIZE	MULTIVARIATE Q(CMH)	D.F.	P	MEAN SCORE Q(CMMS)	D.F.	P	CORRELATION Q(CMMA)	D.F.	P
1200	16.11	2	0.0003	1.08	1	0.2987	1.08	1	0.2987

****

6.51   Refer to Exercise 6.50. Combine the 0 and 1–2 categories and run the Mantel–Haenszel test for sets of 2 × 2 contingency tables. Interpret your findings.

```
***** PARCAT *****

            GENERALIZED COCHRAN–MANTEL–HAENSZEL TEST STATISTICS
        FOR AVERAGE PARTIAL ASSOCIATION IN THREE-WAY CONTINGENCY TABLES

        P.48 NO. OF ALC. RELATED ARRESTS—COMBINING 0 AND 1–2 CATEGS.    00390016

COLUMN SCORES: USER SPECIFIED    1.00    1.00    2.00
ROW SCORES: UNIFORM              1.00    2.00

                        SUMMARY ACROSS TABLES
*************************************************************************************
                A. SUMMARY OF INDIVIDUAL TABLE STATISTICS
 TABLE   TABLE   SAMPLE      MULTIVARIATE           MEAN SCORE            CORRELATION
  NO.    FREQ.    SIZE    Q     D.F.     P      QMS    D.F.    P      QMA    D.F.    P
   1       1      400    4.56     2    0.1021   2.28    1    0.1311   2.28    1    0.1311
   2       1      400   12.12     2    0.0023  10.69    1    0.0011  10.69    1    0.0011
   3       1      400   13.41     2    0.0012  12.87    1    0.0003  12.87    1    0.0003

        TOTAL   1200   30.10     6     0.0     25.84    3     0.0    25.84    3     0.0

                B. GENERALIZED COCHRAN–MANTEL–HAENSZEL STATISTICS
                 SAMPLE      MULTIVARIATE           MEAN SCORE            CORRELATION
                  SIZE   Q(CMH)  D.F.    P     Q(CMMS) D.F.    P     Q(CMMA) D.F.    P
                  1200   16.11     2    0.0003  11.55    1    0.0007  11.55    1    0.0007
*************************************************************************************
```

6.9   SUMMARY

This chapter has dealt with categorical data representing the number (frequency) of observations falling into each possible cell. For a single variable, we're interested in the cell counts of the separate categories of the variable. When observations are made on two or more variables, we're interested in the frequencies associated with the cells of the contingency table formed by cross-classifying observations according to the variables of interest.

Categorical data obtained from a single variable arise in a number of practical situations. We discussed a chi-square goodness-of-fit test that is used to test whether the sample frequencies (and percentages) associated with categories of a variable agree with what would be expected according to hypothesized cell percentages. We also examined estimation and test procedures for a binomial proportion π and for comparing two binomial proportions based on independent samples. Inferences related to a Poisson random variable were also discussed briefly.

Two variable categorical data problems were introduced using a chi-square test of independence for data displayed in an $r \times c$ contingency table. The Mantel–Haenszel test for combining information across sets of 2×2 contingency tables and for providing a mean score test on data combined across sets of $2 \times c$ contingency tables were also presented. Finally, it should be stressed that this chapter gives only a brief introduction to problems in the analysis of categorical data. Sequences of courses can be developed at the undergraduate (but more likely the graduate) level.

KEY FORMULAS

1. Multinomial distribution

$$P(n_1, n_2, \ldots, n_k) = \frac{n!}{n_1! n_2! \ldots n_k!} \pi_1^{n_1} \pi_2^{n_2} \cdots \pi_k^{n_k}$$

2. Chi-square goodness-of-fit test

$H_0: \pi_i = \pi_{i0}$

T.S.: $\chi^2 = \sum_i \left[\frac{(n_i - E_i)^2}{E_i} \right]$

where

$E_i = n\pi_{i0}$

3. Confidence interval for π

$\hat{\pi} \pm z_{\alpha/2} \sigma_{\hat{\pi}}$

where

$\hat{\pi} = \frac{y}{n}$

and

$\sigma_{\hat{\pi}} = \sqrt{\frac{\hat{\pi}(1 - \hat{\pi})}{n}}$

4. Sample size required for a $100(1 - \alpha)\%$ confidence interval of the form $\hat{\pi} \pm E$

$$n = \frac{z_{\alpha/2}^2 \pi(1 - \pi)}{E^2}$$

Hint: Use $\pi = .5$ if no estimate is available.

5. Statistical test for π

H_0: $\pi = \pi_0$

T.S.: $z = \dfrac{\hat{\pi} - \pi_0}{\sigma_{\hat{\pi}}}$

where

$$\sigma_{\hat{\pi}} = \sqrt{\frac{\pi_0(1 - \pi_0)}{n}}$$

6. Confidence interval for $\pi_1 - \pi_2$

$$\hat{\pi}_1 - \hat{\pi}_2 \pm z_{\alpha/2}\sigma_{\hat{\pi}_1 - \hat{\pi}_2}$$

where

$$\sigma_{\hat{\pi}_1 - \hat{\pi}_2} = \sqrt{\frac{\hat{\pi}_1(1 - \hat{\pi}_1)}{n_1} + \frac{\hat{\pi}_2(1 - \hat{\pi}_2)}{n_2}}$$

7. Statistical test for $\pi_1 - \pi_2$

H_0: $\pi_1 - \pi_2 = 0$

T.S.: $z = \dfrac{\hat{\pi}_1 - \hat{\pi}_2}{\sigma_{\hat{\pi}_1 - \hat{\pi}_2}}$

where

$$\sigma_{\hat{\pi}_1 - \hat{\pi}_2} = \sqrt{\hat{\pi}(1 - \hat{\pi})\left(\frac{1}{n_1} + \frac{1}{n_2}\right)}$$

and

$$\hat{\pi} = \frac{y_1 + y_2}{n_1 + n_2}$$

8. Chi-square test of independence

$$\chi^2 = \sum_{i,j}\left[\frac{(n_{ij} - E_{ij})^2}{E_{ij}}\right]$$

where

$$E_{ij} = \frac{(\text{row } i \text{ total})(\text{column } j \text{ total})}{n}$$

EXERCISES

 6.52 A sociologist studied the relationship between male skin color (light, medium, and dark) and job-mobility orientation (high, medium, and low). Use these data to run a chi-square test of independence. Interpret your results.

Job-Mobility Orientation	Male Skin Color		
	Light	Medium	Dark
High	35	84	51
Medium	49	78	23
Low	10	13	6

 6.53 A study was conducted to determine the relationship between annual income and number of children per family. Compute percentages for each of the income categories, then run a chi-square test of independence and draw conclusions.

Number of Children per Family	Annual Income		
	< $20,000	$20,000–$40,000	> $40,000
≤2 children	38	45	22
>2 children	220	95	30

6.54 Refer to Exercise 6.53. Assume that the data were obtained from the east. Data were also obtained from an additional 300 persons in the south and the west. Use these data and the accompanying computer output to determine whether there is a difference in annual income (on the average) for families with two or fewer than two children compared to families with more than two children. Combine data across the east, south, and west using uniform scores. Draw conclusions.

	Number of Children per Family	Annual Income		
		< $20,000	$20,000–$40,000	> $40,000
South	≤2 children	25	38	40
	>2 children	120	50	27
West	≤2 children	36	39	27
	>2 children	95	60	43

```
***** PARCAT *****

              GENERALIZED COCHRAN–MANTEL–HAENSZEL TEST STATISTICS
           FOR AVERAGE PARTIAL ASSOCIATION IN THREE-WAY CONTINGENCY TABLES

                P.50 ANNUAL INCOME—UNIFORM SCORES FOR INCOME      00060000
  COLUMN SCORES: UNIFORM        1.00    2.00    3.00
  ROW SCORES: UNIFORM           1.00    2.00


                               SUMMARY ACROSS TABLES
  *****************************************************************************************
                    A. SUMMARY OF INDIVIDUAL TABLE STATISTICS
  TABLE    TABLE    SAMPLE      MULTIVARIATE            MEAN SCORE              CORRELATION
  NO.      FREQ.    SIZE     Q       D.F.    P      QMS     D.F.    P       QMA     D.F.    P
  1        1        450     27.16    2      0.0    26.59    1      0.0     26.59    1      0.0
  2        1        300     40.83    2      0.0    40.25    1      0.0     40.25    1      0.0
  3        1        300      4.40    2      0.1107  3.25    1      0.0716   3.25    1      0.0716

           TOTAL    1050    72.39    6      0.0    70.08    3      0.0     70.08    3      0.0


                    B. GENERALIZED COCHRAN–MANTEL–HAENSZEL STATISTICS
           SAMPLE        MULTIVARIATE           MEAN SCORE              CORRELATION
           SIZE     Q(CMH)   D.F.   P      Q(CMMS)   D.F.   P      Q(CMMA)   D.F.   P
           1050     61.61    2      0.0    58.83     1      0.0    58.83     1      0.0

  *****************************************************************************************
```

  6.55 A random sample of 145 people of various occupations was taken to investigate the public opinion on police treatment. Each person was asked whether he or she would expect the police to treat him or her as good, better, or worse than a common criminal. The following table summarizes the results:

	Expected Treatment			
Occupation	Better	As Good	Worse	Totals
Unemployed	6	23	11	40
Blue-collar worker	17	30	8	55
White-collar worker	16	28	6	50
Totals	39	81	25	145

Is there sufficient evidence to indicate that the expected treatment is independent of occupation? Use $\alpha = .10$.

6.56 A sociological study was conducted to determine whether there is a relationship between the length of time blue-collar workers remain in their first job and the amount of their education. From union membership records, a random sample of persons was classified. The data are shown below.

a. Use the computer output that follows to identify the expected cell numbers.

b. Test the research hypothesis that the variable "length of time on first job" is related to the variable "amount of education."

Years on First Job	Years of Education			
	0–4.5	4.5–9	9–13.5	13.5
0–2.5	5	21	30	33
2.5–5	15	35	40	30
5–7.5	22	16	15	30
7.5	28	10	8	10

c. Give the level of significance for the test.

d. Draw your conclusions.

CATEGORICAL ANALYSES

TABLE OF YRS—JOB BY YRS—ED

YRS—JOB FREQUENCY PERCENT ROW PCT COL PCT	YRS—ED				
	0–4.5	4.5–9	9–13.5	13.5	TOTAL
0–2.5	5 1.44 5.62 7.14	21 6.03 23.60 25.61	30 8.62 33.71 32.26	33 9.48 37.08 32.04	89 25.57
2.5–5	15 4.31 12.50 21.43	35 10.06 29.17 42.68	40 11.49 33.33 43.01	30 8.62 25.00 29.13	120 34.48
5–7.5	22 6.32 26.51 31.43	16 4.60 19.28 19.51	15 4.31 18.07 16.13	30 8.62 36.14 29.13	83 23.85
7.5	28 8.05 50.00 40.00	10 2.87 17.86 12.20	8 2.30 14.29 8.60	10 2.87 17.86 9.71	56 16.09
TOTAL	70 20.11	82 23.56	93 26.72	103 29.60	348 100.00

(continued)

STATISTICS FOR 2-WAY TABLES

CHI-SQUARE	57.830	DF =	9 PROB = 0.0001
PHI	0.408		
CONTINGENCY COEFFICIENT	0.377		
CRAMER'S V	0.235		
LIKELIHOOD RATIO CHI-SQUARE	55.605	DF =	9 PROB = 0.0001

CATEGORICAL ANALYSES

OBS	YRS—JOB	YRS—ED	FREQ
1	0–2.5	0–4.5	5
2	0–2.5	4.5–9	21
3	0–2.5	9–13.5	30
4	0–2.5	13.5	33
5	2.5–5	0–4.5	15
6	2.5–5	4.5–9	35
7	2.5–5	9–13.5	40
8	2.5–5	13.5	30
9	5–7.5	0–4.5	22
10	5–7.5	4.5–9	16
11	5–7.5	9–13.5	15
12	5–7.5	13.5	30
13	7.5	0–4.5	28
14	7.5	4.5–9	10
15	7.5	9–13.5	8
16	7.5	13.5	10

N = 16

 6.57 The personnel department of a large corporation was interested in determining the relationship between performance ratings of recently hired employees and the employees' college grade-point averages. To do this, a random sample of 90 records was obtained, examined, and classified in the following two-way table:

	College Grade-Point Average		
Performance Rating	A	B	C
Above average	19	8	3
Average	9	12	15
Below average	6	5	13

Is there evidence to indicate a relationship between the two variables "performance rating" and "college grade-point average"? Use $\alpha = .05$.

EXERCISES

6.58   Television research to date suggests that "zipping" and "zapping" of commercials by VCR viewers is uncommon. Zipping is the use of the VCR remote control to fast-forward past commercials; zapping is the use of the remote control to change channels when commercials appear. Based on a random sample of 2000 users of VCRs, 66% said that they did not skip the commercials. Obviously, with more than 30 million households with VCRs, widespread zipping and zapping would have a tremendous impact on the rates charged for TV advertising. Use the data to construct a 95% confidence interval for π, the proportion of VCR users that do not skip commercials.

6.59   Two researchers at Johns Hopkins University have studied the use of drug products in the elderly. Patients in a recent study were asked the extent to which physicians counseled them with regard to their drug therapies. The researchers found the following:

- 25.4% of the patients said their physicians did not explain what the drug was supposed to do
- 91.6% indicated they were not told how the drug might "bother" them
- 47.1% indicated their physicians did not ask how the drug "helped" or "bothered" them after therapy was started
- 87.7% indicated the drug was not changed after discussion on how the therapy was helping or bothering them.

a. Assume that 500 patients were interviewed in this study. Summarize each of these results using a 95% confidence interval.

b. Do you have any comments about the validity of any of these results?

6.60   People over the age of 40 years tend to notice changes in their digestive systems that alter what and how much they can eat. A study was conducted to see whether this observation applies across different ethnic segments of our society. Random samples of Anglo-Saxons, Germans, Latin Americans, Italians, Spaniards, and Blacks were obtained. The data from this survey are summarized here:

Ethnic Group	Sample Size Responding (60 of Each Group Were Contacted)	Number Reporting Altered Digestive System
Anglo-Saxon	55	7
German	58	6
Latin American	52	34
Italian	54	38
Spanish	30	20
Black	49	31

a. Does it appear that there may be a bias due to the response rates?

b. Compare the rates (π_is) for the Anglo-Saxon and German groups using a 95% confidence interval.

6.61   Refer to Exercise 6.60. There seem to be two distinct rates—those around 12% and those around 70%. Combine the sample data for the first two groups and for the last four groups. Use these data to test the hypothesis—H_0: $\pi_1 - \pi_2 = 0$ versus H_a: $\pi_1 - \pi_2 < 0$. Here π_1 corresponds to the population rate for the first combined group, and π_2 is the corresponding proportion for the second combined group. Give the p-value for your test.

6.62   Two sets of 60 ninth graders were taught Algebra I by different methods. The experimental group used self-paced modules developed for use at a computer with a display screen. The control group was given formal lectures by the teacher. At the end of the four-month period, a comprehensive, standardized test was given to both groups. The experimental group had 65% scoring above 80 (out of 100), whereas the control group had just 47% above 80. Use these data to compare the percentages above 80 for the two groups. Give the p-value for your test.

6.63   Refer to Exercise 6.62. How might you have designed this study? What additional data might you want to collect?

6.64   A recent study conducted for a large university compared the effects of intensive behavioral training to change Type A habits on the incidence of a second heart attack. A total of 290 patients who recently had a nonfatal heart attack were studied. Those randomized to Group 1 received the intensive behavioral training in addition to routine medical care. The training was designed to help the patient to slow down and be more relaxed. Those assigned to Group 2 received only the routine medical care. The data from this five-year study are summarized here:

	Sample Size	Number of Second Heart Attacks Fatal and Nonfatal
Group 1	140	17
Group 2	150	29

Use these data to test H_0: $\pi_1 - \pi_2 = 0$ versus H_a: $\pi_1 - \pi_2 < 0$. Give a p-value for your test and draw a conclusion.

6.65   A random sample of faculty members of a state university system was polled and classified by university and by which of the three collective bargaining agents (union 101, union 102, union 103) was preferred. The data appear below, and computer output follows.

	Bargaining Agent		
University	101	102	103
1	42	29	12
2	31	23	6
3	26	28	2
4	8	17	37

CATEGORICAL ANALYSES

TABLE OF UNIV BY B__AGENT

UNIV	B__AGENT			
FREQUENCY PERCENT ROW PCT COL PCT	101	102	103	TOTAL
1	42 16.09 50.60 39.25	29 11.11 34.94 29.90	12 4.60 14.46 21.05	83 31.80
2	31 11.88 51.67 28.97	23 8.81 38.33 23.71	6 2.30 10.00 10.53	60 22.99
3	26 9.96 46.43 24.30	28 10.73 50.00 28.87	2 0.77 3.57 3.51	56 21.46
4	8 3.07 12.90 7.48	17 6.51 27.42 17.53	37 14.18 59.68 64.91	62 23.75
TOTAL	107 41.00	97 37.16	57 21.84	261 100.00

STATISTICS FOR 2-WAY TABLES

CHI-SQUARE	75.197	DF =	6	PROB = 0.0001
PHI	0.537			
CONTINGENCY COEFFICIENT	0.473			
CRAMER'S V	0.380			
LIKELIHOOD RATIO CHI-SQUARE	71.991	DF =	6	PROB = 0.0001

CATEGORICAL ANALYSES

OBS	UNIV	B__AGENT	FREQ
1	1	101	42
2	1	102	29
3	1	103	12
4	2	101	31
5	2	102	23
6	2	103	6
7	3	101	26
8	3	102	28
9	3	103	2
10	4	101	8
11	4	102	17
12	4	103	37

N = 12

a. Identify the expected cell numbers.

b. Use the computer output to determine whether there is evidence to indicate a difference in the distribution of preference across the four state universities.

c. Give the level of significance for the test.

d. Draw your conclusions.

 6.66 An advertising firm selected to conduct a market awareness study for a brand of house paint obtained information from a national survey of 1500 randomly selected homeowners. Each homeowner selected for the survey was asked if he or she was familiar with a newly marketed line of interior latex paints. If 465 responded affirmatively, use the sample data to test the research hypothesis that the company has reached more than 30% of the homeowners with recent advertising. Use $\alpha = .05$.

6.67 Refer to Exercise 6.54. Use the computer output here to determine whether scores of 10, 30, and 60 make a difference in the conclusions. What about scores of 15, 30, and 100? What conclusions can be drawn and what reservations might you have?

```
***** PARCAT *****

        GENERALIZED COCHRAN–MANTEL–HAENSZEL TEST STATISTICS
      FOR AVERAGE PARTIAL ASSOCIATION IN THREE-WAY CONTINGENCY TABLES


       P.50 ANNUAL INCOME—SCORES OF 10,30,60 FOR INCOME      00140000

COLUMN SCORES: USER SPECIFIED     10.00    30.00    60.00
ROW SCORES: UNIFORM                1.00     2.00


                        SUMMARY ACROSS TABLES
************************************************************************************
                A. SUMMARY OF INDIVIDUAL TABLE STATISTICS
```

TABLE NO.	TABLE FREQ.	SAMPLE SIZE	MULTIVARIATE			MEAN SCORE			CORRELATION		
			Q	D.F.	P	QMS	D.F.	P	QMA	D.F.	P
1	1	450	27.16	2	0.0	25.20	1	0.0	25.20	1	0.0
2	1	300	40.83	2	0.0	38.65	1	0.0	38.65	1	0.0
3	1	300	4.40	2	0.1107	2.76	1	0.0966	2.76	1	0.0966
	TOTAL	1050	72.39	6	0.0	66.61	3	0.0	66.61	3	0.0

```
              B. GENERALIZED COCHRAN–MANTEL–HAENSZEL STATISTICS
```

SAMPLE SIZE	MULTIVARIATE			MEAN SCORE			CORRELATION		
	Q(CMH)	D.F.	P	Q(CMMS)	D.F.	P	Q(CMMA)	D.F.	P
1050	61.61	2	0.0	54.97	1	0.0	54.97	1	0.0

```
************************************************************************************
```

(continued)

SECTION 6.9   SUMMARY

```
***** PARCAT *****

            GENERALIZED COCHRAN–MANTEL–HAENSZEL TEST STATISTICS
        FOR AVERAGE PARTIAL ASSOCIATION IN THREE-WAY CONTINGENCY TABLES

        P.50 ANNUAL INCOME—SCORES OF 15,30,100 FOR INCOME      00170000

COLUMN SCORES: USER SPECIFIED      15.00      30.00      100.00
ROW SCORES: UNIFORM                 1.00       2.00

                          SUMMARY ACROSS TABLES
*********************************************************************************
                  A. SUMMARY OF INDIVIDUAL TABLE STATISTICS
```

TABLE NO.	TABLE FREQ.	SAMPLE SIZE	MULTIVARIATE			MEAN SCORE			CORRELATION		
			Q	D.F.	P	QMS	D.F.	P	QMA	D.F.	P
1	1	450	27.16	2	0.0	18.69	1	0.0	18.69	1	0.0
2	1	300	40.83	2	0.0	31.72	1	0.0	31.72	1	0.0
3	1	300	4.40	2	0.1107	1.60	1	0.2057	1.60	1	0.2057
TOTAL		1050	72.39	6	0.0	52.01	3	0.0	52.01	3	0.0

```
              B. GENERALIZED COCHRAN–MANTEL–HAENSZEL STATISTICS
```

SAMPLE SIZE	MULTIVARIATE			MEAN SCORE			CORRELATION		
	Q(CMH)	D.F.	P	Q(CMMS)	D.F.	P	Q(CMMA)	D.F.	P
1050	61.61	2	0.0	41.02	1	0.0	41.02	1	0.0

```
*********************************************************************************
```

6.68   Faculty members at each of three universities were classified according to political ideology (left or right) and according to their academic tolerance (low, medium, or high).

University	Political Ideology	Academic Tolerance		
		Low	Medium	High
1	Left	11	16	44
	Right	15	19	11
2	Left	20	25	36
	Right	50	22	6
3	Left	5	3	4
	Right	30	33	25

a. Use descriptive statistics to characterize the three universities.
b. Conduct a Mantel–Haenszel mean score test based on uniform scores to compare the left and right ideologies across the three universities.

```
***** PARCAT *****
             GENERALIZED COCHRAN–MANTEL–HAENSZEL TEST STATISTICS
           FOR AVERAGE PARTIAL ASSOCIATION IN THREE-WAY CONTINGENCY TABLES

        P.54 POLITICAL IDEOLOGY—UNIFORM SCORES FOR ACADEMIC TOLERANCE     00510000

  COLUMN SCORES: UNIFORM        1.00    2.00    3.00
  ROW SCORES: UNIFORM           1.00    2.00

                              SUMMARY ACROSS TABLES
******************************************************************************************
                      A. SUMMARY OF INDIVIDUAL TABLE STATISTICS
  TABLE   TABLE   SAMPLE      MULTIVARIATE           MEAN SCORE           CORRELATION
  NO.     FREQ.   SIZE     Q      D.F.    P      QMS    D.F.    P      QMA    D.F.    P
   1       1       116    15.50    2    0.0004  13.17    1    0.0003  13.17    1    0.0003
   2       1       159    34.22    2    0.0     34.01    1    0.0     34.01    1    0.0
   3       1       100     0.71    2    0.7003   0.01    1    0.9144   0.01    1    0.9144

           TOTAL   375    50.42    6    0.0     47.19    3    0.0     47.19    3    0.0

                      B. GENERALIZED COCHRAN–MANTEL–HAENSZEL STATISTICS
           SAMPLE       MULTIVARIATE            MEAN SCORE            CORRELATION
           SIZE     Q(CMH)  D.F.    P      Q(CMMS)  D.F.    P     Q(CMMA)  D.F.    P
           375      41.90    2    0.0      39.82     1    0.0     39.82     1    0.0
******************************************************************************************
```

 6.69   A study examining the effectiveness of a drug product for the treatment of arthritis was conducted at four different centers. At each center 100 patients were treated, 50 with the test drug and 50 with an identical-appearing placebo. The data are shown here.

Clinic	Treatment	Global Outcome			
		Worse	Same	Better	Much Better/ Completely Well
1	Placebo	10	15	17	8
	Test drug	12	14	10	14
2	Placebo	6	20	22	2
	Test drug	4	15	10	21
3	Placebo	7	25	12	6
	Test drug	5	22	12	11
4	Placebo	2	14	20	14
	Test drug	1	12	15	22

a. Compare the percentage distributions for the two treatment groups at the four universities.

EXERCISES

    b. One investigator wanted to collapse the global outcome categories into improved and not improved. Comment on this.

    c. Conduct a Mantel–Haenszel test on the collapsed data of part (b) using uniform scores. Draw conclusions.

6.70   Computer output for the full data of Exercise 6.69 is shown here. Compare your findings here to those in Exercise 6.69. Any lessons learned?

```
***** PARCAT *****
          GENERALIZED COCHRAN–MANTEL–HAENSZEL TEST STATISTICS
       FOR AVERAGE PARTIAL ASSOCIATION IN THREE-WAY CONTINGENCY TABLES

          P.55 GLOBAL OUTCOME—UNIFORM SCORES FOR OUTCOME     00610000

  COLUMN SCORES: UNIFORM      1.00    2.00    3.00    4.00
  ROW SCORES: UNIFORM         1.00    2.00

                        SUMMARY ACROSS TABLES
```

```
                    A. SUMMARY OF INDIVIDUAL TABLE STATISTICS
```

TABLE NO.	TABLE FREQ.	SAMPLE SIZE	MULTIVARIATE			MEAN SCORE			CORRELATION		
			Q	D.F.	P	QMS	D.F.	P	QMA	D.F.	P
1	1	100	3.63	3	0.3042	0.08	1	0.7789	0.08	1	0.7789
2	1	100	21.10	3	0.0001	8.84	1	0.0029	8.84	1	0.0029
3	1	100	1.98	3	0.5775	1.72	1	0.1896	1.72	1	0.1896
4	1	100	2.95	3	0.3995	1.93	1	0.1647	1.93	1	0.1647
	TOTAL	400	29.65	12	0.0031	12.57	4	0.0136	12.57	4	0.0136

```
                B. GENERALIZED COCHRAN–MANTEL–HAENSZEL STATISTICS
```

SAMPLE SIZE	MULTIVARIATE			MEAN SCORE			CORRELATION		
	Q(CMH)	D.F.	P	Q(CMMS)	D.F.	P	Q(CMMA)	D.F.	P
400	20.80	3	0.0001	8.38	1	0.0038	8.38	1	0.0038

6.71   Legislators of a particular state were concerned that the enrollment (which affects budget allocations) at a particular university within the state system had been padded by allowing students to overenroll or to enroll for courses that required no academic work. To substantiate their initial findings, a random sample of 200 graduate students (from the 5000 currently enrolled) was interviewed. If 20 students stated they had been allowed to pad their enrollments in the past quarter, use those data to construct a 99% confidence interval for π, the proportion of the entire student body with padded enrollments. Interpret your results.

6.72   The relative sensitivities of two fuses were tested under controlled conditions by firing 40 rounds with Type I and 60 rounds with Type II, the firings being conducted in random order. Each round was classified according to whether the fuse functioned or not, with the following results:

Type of Fuse	Functioned	Did Not Function	Total
I	10	30	40
II	40	20	60
Total	50	50	100

a. Test the hypothesis that there is no difference between the sensitivities of the two fuses. Use $\alpha = .05$.

b. Use the previous data to demonstrate the relationship between the z test and the chi-square test of independence in a 2 × 2 table. Hint: Compare z^2 and x^2 for the two tests.

 6.73 A study was conducted to investigate whether there is any relationship between voting record and education. One hundred and five citizens selected at random were interviewed as to how often they vote and what level of formal education they had achieved. The results were as follows:

Education	How Often Do You Vote?			Totals
	Never	Some Elections	All Elections	
Less than h.s. level	11	12	11	34
High school	7	15	13	35
College	2	20	14	36
Totals	20	47	38	105

Is there evidence to indicate a relationship between level of education and frequency of voting? Use $\alpha = .05$.

6.74 An extension of the traffic study for the implementation of a priority bus lane involved the sampling of public opinion concerning the bus lane during various phases of the study. Three different phases were to be studied. Phase 0 required the bus drivers to use the existing traffic lanes. In Phase 1, bus drivers made use of the exclusive bus lane with no preemption of the traffic signals at the intersection; Phase 2 allowed the bus drivers to extend the "green time" on a traffic signal to allow them to pass through before the light changed. Use the sample data below to determine whether the distribution of persons favoring, not favoring, or undecided changes from phase to phase. Use $\alpha = .05$.

Phase	Opinion			Totals
	Favor	Do Not Favor	Undecided	
0	80	90	30	200
1	60	112	28	200
2	50	125	25	200

EXERCISES

 6.75 The quality assurance unit of a large pharmaceutical company was engaged in comparing two new formulations of a drug product in tablet form. Both tablets contained the same amount of active ingredient, but varied in size, shape, and excipient (an inert substance that acts as a vehicle). A random sample of 100 tablets was obtained from a pilot batch for each formulation. The number of tablets classified as acceptable (or not acceptable) with regard to potency is shown here for each formulation.

Formulation	Number Acceptable	Number Not Acceptable	Sample Size
1	84	16	100
2	96	4	100

Use these data to place a 95% confidence interval about $\pi_1 - \pi_2$, the difference in the proportions of acceptable tablets.

6.76 Refer to Exercise 6.75.
a. Run a two-sided statistical test of H_0: $\pi_1 - \pi_2 = 0$. Draw conclusions.
b. Run a chi-square test of independence for these data. Use $\alpha = .05$. Compare your results to part (a).
c. Suggest a relationship between the two tests (Hint: Compare z^2 and χ^2 for the two tests.

 6.77 A matter of public concern is whether there is an association between the use of saccharin as a sweetener and the development of cancer. Suppose that 25 independent, controlled studies have been conducted at elevated doses (greater than or equal to 100 times the normal yearly human intake) in a particular strain of mice. In two of these studies, the treated group of mice had a higher incidence of cancer than in the corresponding control group of mice; no meaningful differences were observed between the treated and controlled mice in the other studies. What conclusions might you draw from these 25 studies? What additional data would you like to see? How do these results extrapolate to the human situation? What error rate should be controlled more carefully?

 6.78 A survey of drivers was obtained to compare the proportions who use seatbelts regularly for various age categories. These data are shown here:

| Age (Years) | Regularity of Seatbelt Usage | | | |
	Always	Regularly	Sometimes	Never
16–20	1	10	70	19
21–25	4	8	80	8
26–30	8	10	77	5
>30	15	30	49	6

Analyze the data and draw conclusions.

 6.79 Entry-level diastolic blood pressures for a group of 12 hypertensive females are shown in the next table, along with a corresponding diastolic blood pressure following two weeks of treatment with an antihypertensive medication. Note: the goal of the therapy is to reduce the diastolic blood pressure to 90 mm Hg.

Patient	Entry Level Diastolic BP	BP Following 2 weeks
1	105	89
2	110	87
3	115	93
4	107	90
5	108	88
6	116	90
7	114	91
8	110	90
9	112	89
10	106	91
11	109	90
12	111	92

a. What percent of the patients were controlled at ≤ 90 mm Hg?

b. Give a p-value for the test $H_0: \pi = 0$ versus $H_a: \pi > 0$.

6.80 Refer to Exercise 6.79.

a. Consider using a t test for $H_0: \mu_d = 0$ versus $H_a: \mu_d > 0$. Might there be a problem with one (or more) of the assumptions? If so, which ones and why?

b. Suggest and perform a suitable test of location for the difference data. Give the p-value and interpret your findings.

6.81 A recent poll asked the following question of a random sample of 100 people in various countries: From what you have heard or read, which of these statements comes closest to the way you feel about the United States' involvement in the conflict in Nicaragua? (1) The United States should begin to withdraw its support of the contras. (2) The United States should carry on its present level of support. (3) The United States should increase its level of support.

Country	Withdraw Support	Maintain Present Level	Increase Support	Don't Know
Argentina	57	6	6	31
Australia	21	43	24	12
Brazil	76	5	5	14
Canada	41	16	23	20
Finland	81	4	5	10
France	72	8	5	15
Great Britain	45	15	15	25
India	66	4	8	22
Sweden	79	10	4	7
Uruguay	62	10	5	23
United States	31	10	53	6
West Germany	58	11	14	17

a. Would it be meaningful to do a percentage comparison of the data?

b. Suggest a statistical test to compare the responses across countries.

c. Run the test you selected in (b) (use $\alpha = .05$) and draw conclusions.

6.82 A sample of 150 patients suffering from severe acquired immunodeficiency syndrome (AIDS) were treated with an experimental compound. Without treatment, these patients were given little chance for survival over the next six months. Baseline characteristics are shown here.

Baseline Characteristics	
Gender (% male)	92%
Age in years ($\bar{y} \pm SD$)	32 ± 6
Weight in kg ($\bar{y} \pm SD$)	68 ± 7

For the 120 patients who were treated for more than one week, the six-month survival rate was 52%, the survival rate for the 30 not receiving more than one week of therapy was only 17%.

a. Use the survival data to give 95% confidence intervals on six-month survival rates for those patients treated for more than one week and those not.

b. Would it be valid to make a comparison of these two survival rates for the subgroups of $n_1 = 120$ and $n_2 = 30$ patients? Why or why not?

c. Estimate the overall survival rate for the AIDS patients using a 95% confidence interval. Is this a more valid estimate of the survival rate for patients treated with the drug than is the estimate based on those treated for more than one week?

6.83 As part of a research study conducted at a large behavioral treatment program, 160 subjects quitting smoking were randomized to one of four treatment programs.

A: Reading and audiovisual (A/V) material only, no professional contact
B: Reading and A/V material, low professional contact
C: Reading and A/V material, high professional contact
D: High professional contact only

Assessments were made at 4, 12, 26, and 52 weeks, with the results (abstinence rates) shown here:

Program	Time 4	12	26	52
A	50%	45%	35%	30%
B	70%	50%	47%	35%
C	95%	75%	60%	40%
D	90%	68%	56%	38%

a. Plot the sample data and draw some tentative conclusions.

b. At a given timepoint, what statistical test would be appropriate for comparing the abstinence rates?

c. Run these comparisons for each timepoint and draw conclusions. In general, what seems to be the pattern of abstinence?

6.84 A survey was conducted among regular shoppers at three different suburban shopping centers. Subjects were given a questionnaire to be completed in private and deposited in a centrally located, locked collection box. A total of 1000 questionnaires were distributed, and 400 were returned. One of the items on the questionnaire had to do with family income levels.

		Family Income	
Shopping Center	< 50,000	50–100,000	> 100,000
A	60	25	15
B	66	50	9
C	127	40	8
Total	253	115	32

a. Do a percentage comparison of the family income distributions for the three shopping centers. Do the centers appear to have the same distributions?

b. Run a test of significance to compare the income distributions. Give a p-value for your test and draw a conclusion.

c. What effect might the response rate have on the conclusions drawn from this survey?

6.85 Three supermarkets have a policy of advertising specials on certain days of the week to attract customers. A customer count is maintained at these supermarkets from the hours of 11:00 A.M. to 2:00 P.M for the five weekdays. The data collected was the following:

			Day of the Week		
Supermarket	Mon	Tues	Wed	Thurs	Fri
1	605	650	702	663	568
2	696	741	750	827	663
3	540	668	528	572	516

a. Do the data indicate that the choice of a supermarket depends on the day of the week? Use a procedure that has a 0.01 chance of being in error if this conclusion is drawn. If you decide that choice of a supermarket is related to day of the week, briefly describe the nature of that relationship based on the sample data.

b. Do the data support the hypothesis that the number of customers expected (for all three supermarkets together) is the same for the days Monday through Friday? Use a procedure that has a 0.05 chance of being in error if it is concluded that different days have different customer counts.

6.86 A survey was conducted among large corporations in the United States. Each was given a questionnaire to be completed and returned. One of the items of information sought had to do with the frequency of market offerings in the past year through three underwriters. The results compiled from returned questionnaires were as follows:

Underwriter	Frequency of Offerings		
	0	1	2 or More
1	58	26	25
2	56	56	20
3	97	30	42

Use the output provided to determine whether percentage of market offerings differs among underwriters. Give a *p*-value for your test and draw a conclusion.

CATEGORICAL ANALYSIS

TABLE OF WRITER BY OFFERS

WRITER   OFFERS

FREQUENCY PERCENT ROW PCT COL PCT	0	1	2	TOTAL
1	58	26	25	109
	14.15	6.34	6.10	26.59
	53.21	23.85	22.94	
	27.49	23.21	28.74	
2	56	56	20	132
	13.66	13.66	4.88	32.20
	42.42	42.42	15.15	
	26.54	50.00	22.99	
3	97	30	42	169
	23.66	7.32	10.24	41.22
	57.40	17.75	24.85	
	45.97	26.79	48.28	
TOTAL	211	112	87	410
	51.46	27.32	21.22	100.00

STATISTICS FOR TABLE OF WRITER BY OFFERS

STATISTIC	DF	VALUE	PROB
CHI-SQUARE	4	23.977	0.000
LIKELIHOOD RATIO CHI-SQUARE	4	23.488	0.000
MANTEL–HAENSZEL CHI-SQUARE	1	0.087	0.769
PHI		0.242	
CONTINGENCY COEFFICIENT		0.235	
CRAMER'S V		0.171	

SAMPLE SIZE = 410

6.87   The following data give the observed frequencies of errors per page of unread galley proof for a sample of 40 pages from a certain journal publisher.

Errors/Page	Observed Frequencies
0	5
1	9
2	5
3	7
4	4
5	2
6	3
7	2
8	1
9	0
10	2

Conduct a test to determine whether the errors per page follow a Poisson distribution with a mean rate of 3.2. Use $\alpha = .10$.

7 INFERENCES ABOUT POPULATION VARIANCES

7.1 INTRODUCTION

When most people think of statistical inference, they think of inferences concerning population means. However, the population parameter that answers an experimenter's practical questions will vary from one situation to another, and sometimes the **variability** of a population is more important than its mean. For example, the producer of a drug product is certainly concerned with controlling the mean potency of tablets, but he or she must also worry about the variation in potency from one tablet to another. Excessive potency or an underdose could be very harmful to a patient. Hence the manufacturer would like to produce tablets with the desired mean potency and with as little variation in potency (as measured by σ or σ^2) as possible.

Inferential problems about a population variance are similar to those for a population mean. We can estimate or test hypotheses about a single population variance or compare two variances.

7.2 ESTIMATION AND TESTS FOR A POPULATION VARIANCE

The sample variance

$$s^2 = \frac{\sum (y - \bar{y})^2}{n - 1}$$

FIGURE 7.1

Upper-Tail and
Lower-Tail Values of
Chi-Square

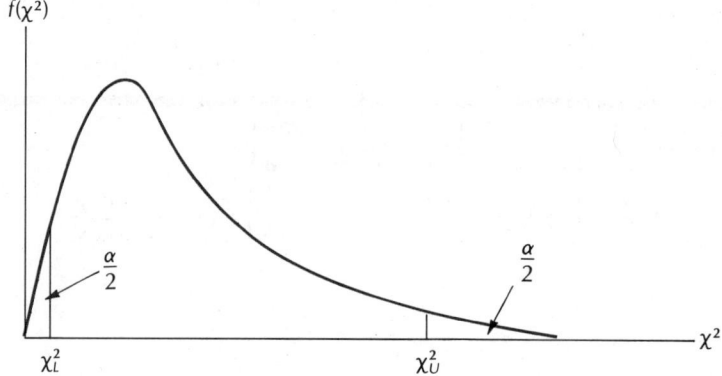

can be used for inferences concerning a population variance σ^2. For a random sample of n measurements drawn from a normal population with mean μ and variance σ^2, s^2 provides a **point estimate for σ^2**. In addition, the quantity $(n-1)s^2/\sigma^2$ follows a chi-square distribution with df $= n - 1$. We can use this information to form a confidence interval for σ^2.

point estimate for σ^2

**GENERAL
CONFIDENCE
INTERVAL FOR
σ^2 (OR σ) WITH
CONFIDENCE
COEFFICIENT
$(1 - \alpha)$**

$$\frac{(n-1)s^2}{\chi_U^2} < \sigma^2 < \frac{(n-1)s^2}{\chi_L^2}$$

where χ_U^2 is the upper-tail value of chi-square for df $= n - 1$ with area $\alpha/2$ to its right, and χ_L^2 is the lower-tail value with area $\alpha/2$ to its left (see Figure 7.1). We can determine χ_U^2 and χ_L^2 for a specific value of df by obtaining the critical value in Table 5 of the Appendix corresponding to a $= \alpha/2$ and a $= 1 - \alpha/2$, respectively.

Note: The confidence interval for σ is found by taking square roots throughout.

EXAMPLE 7.1

The variability in milk production for a 305-day lactation period was observed for a random sample of 15 Holstein cows. Use the milk-yield data in Table 7.1 to estimate σ^2, the population variance of milk yields, using a 95% confidence interval.

Solution

For these data we find

$$\sum y = 188.424 \qquad \sum y^2 = 2431.470$$

TABLE 7.1   Milk Production Data (in 1000 Pounds), Example 7.1

12.928	13.812	11.036
12.120	14.358	9.248
14.972	8.998	9.980
14.044	10.620	11.990
14.788	14.744	14.786

Substituting into the shortcut formula for s^2 (Chapter 2), we have

$$s^2 = \frac{1}{n-1}\left[\sum y^2 - \frac{(\sum y)^2}{n}\right] = \frac{1}{14}\left[2431.470 - \frac{(188.424)^2}{15}\right] = 4.612$$

The confidence coefficient for our example is $1 - \alpha = .95$. The upper-tail chi-square value can be obtained from Table 5 in the Appendix, for $df = n - 1 = 14$ and $a = \alpha/2 = .025$. Similarly, the lower-tail chi-square value is obtained from Table 5 with $a = 1 - \alpha/2 = .975$. Thus

$$\chi_U^2 = 26.1190 \qquad \chi_L^2 = 5.62872$$

The 95% confidence interval is then

$$\frac{14(4.612)}{26.1190} < \sigma^2 < \frac{14(4.612)}{5.62872}$$

or

$$2.472 < \sigma^2 < 11.471$$

In addition to estimating a population variance, we can construct a statistical test of the null hypothesis that σ^2 equals a specified value, σ_0^2. This test procedure is summarized here.

STATISTICAL TEST FOR σ^2 (OR σ)

H_0: $\sigma^2 = \sigma_0^2$ (σ_0^2 is specified)

H_a: 1. $\sigma^2 > \sigma_0^2$
    2. $\sigma^2 < \sigma_0^2$
    3. $\sigma^2 \neq \sigma_0^2$

T.S.: $\chi^2 = \dfrac{(n-1)s^2}{\sigma_0^2}$

R.R.: For a specified value of α,

1. reject H_0 if χ^2 is greater than χ_U^2, the upper-tail value for $a = \alpha$ and $df = n - 1$
2. reject H_0 if χ^2 is less than χ_L^2, the lower-tail value for $a = 1 - \alpha$ and $df = n - 1$
3. reject H_0 if χ^2 is greater than χ_U^2, based on $a = \alpha/2$ and $df = n - 1$, or less than χ_L^2, based on $a = 1 - \alpha/2$ and $df = n - 1$

EXAMPLE 7.2

A manufacturer of a specific pesticide useful in the control of household bugs claims that its product retains most of its potency for a period of at least six months. More specifically, it claims that the drop in potency from zero to six months will vary in the interval from 0% to 8%. To test the manufacturer's claim, a consumer group obtained a random sample of 20 containers of pesticide from the manufacturer. Each can was tested for potency and then stored for a period of six months at room temperature. After the storage period, each can was again tested for potency. The drop in potency was recorded for each can and the sample variance for the drops in potencies was computed to be $s^2 = 6.2$. Use these data to determine whether there is sufficient evidence to indicate that the population of potency drops has more variability than that claimed by the manufacturer. Use $\alpha = .05$.

Solution The manufacturer has claimed that the population of potency reductions has a range of 8%. Dividing the range by 4, we obtain an approximate population standard deviation of $\sigma = 2\%$ (or $\sigma^2 = 4$).

The appropriate null and alternative hypotheses are

H_0: $\sigma^2 = 4$ (i.e., we assume the manufacturer's claim is correct)

H_a: $\sigma^2 > 4$ (i.e., there is more variability than claimed by the manufacturer)

Using the computed sample variance based on 20 observations, the test statistic and rejection region are as follows:

T.S.: $\chi^2 = \dfrac{(n-1)s^2}{\sigma_0^2} = \dfrac{19(6.2)}{4} = 29.45$

R.R.: For $\alpha = .05$, we will reject H_0 if the computed value of chi-square is greater than 30.1435, obtained from Table 5 in the Appendix for a = .05 and df = 19.

Conclusion: Since the computed value of chi-square, 29.45, is less than the critical value, 30.1435, there is insufficient evidence to reject the manufacturer's claim, based on $\alpha = .05$. However, the consumer group is not prepared to accept H_0: $\sigma^2 = 4$. Rather, since $s^2 = 6.2$ and the p-value of the test is $.05 < p < .10$, it would be wise to do additional testing with a larger sample size before reaching a definite conclusion.

normality assumption The inferences we discussed about σ^2 and σ using chi-square distributions are based on the assumption that the underlying population from which the sample measurements are drawn is normal. This **normality assumption** is very important, much more important than it was for the z and t methods of previous chapters. We were fortunate in dealing with t tests about means because the Central Limit Theorem had wide applicability; with tests about variances (or standard deviations) there is no comparable theorem that helps "ensure" normality. So, always plot the data. If there is a suggestion of nonnormality, p-values obtained from the chi-square test or the confidence coefficient for an interval estimate for σ^2 (or σ) could be in serious error. Although there are some nonparametric alternatives that could be used (e.g., jackknife methods) we will not present one here because they are too complicated from a computational standpoint. The interested reader is referred to Hollander and Wolfe (1973) or Miller (1964).

EXERCISES

7.1 In Chapter 4, Exercise 4.6, we discussed several alternatives to alleviating SO_2 pollution from smokestacks in operations using coal as a major source of fuel or energy. One suggestion was to employ a scrubber that cleans the gases after the coal has been burned but before the gases are released into the atmosphere. In 50 samples of gases emitted from a stack equipped with a new scrubber, the sample mean and standard deviation SO_2 emission were .13 and .05 pounds per million Btu, respectively. Use these data to construct a 95% confidence interval for σ^2.

7.2   Refer to Exercise 7.1. Suppose that testing on the leading commercial scrubber on a comparable stack has indicated an SO_2 emission standard deviation of .707 pounds per million Btu. Is there sufficient evidence that the new scruber (Exercise 7.1) has a lower population variance? Use $\alpha = .05$.

7.3   As part of a detailed driver-training program, school officials are requiring teenagers to take a depth-perception test. In one phase of this test, the student is asked to judge the distance between a parked vehicle and a pedestrian stationed a given distance from the student. The recorded distances in feet are listed below for 15 driver-education students.

5	8	7	7	10	6	4	11
6	8	4	9	9	6	5	

Use these data to construct a 99% confidence interval for σ^2, the variance of the depth-perception distances.

| **7.3** | ESTIMATION AND TESTS FOR COMPARING TWO POPULATION VARIANCES |

checking equal variance assumption

One of the major applications of a test for the equality of two population variances is for **checking** the validity of the **equal variance assumption** (that is, $\sigma_1^2 = \sigma_2^2$) for a two-sample t test. First we hypothesize two populations of measurements that are normally distributed. We label these populations as 1 and 2, respectively. We are interested in comparing the variance of population 1, σ_1^2, to the variance of population 2, σ_2^2.

When independent random samples have been drawn from the respective populations, the ratio

$$\frac{s_1^2}{s_2^2} \Big/ \frac{\sigma_1^2}{\sigma_2^2} = \frac{s_1^2}{\sigma_1^2}\left(\frac{\sigma_2^2}{s_2^2}\right)$$

***F* distribution**

possesses a probability distribution in repeated sampling referred to as an ***F* distribution** The formula for the probability distribution is omitted here, but we will specify its properties.

PROPERTIES OF THE *F* DISTRIBUTION

1. Unlike t or z, F can assume only positive values.
2. The F distribution, unlike the normal distribution or the t distribution, is non-symmetrical. (See Figure 7.2.)
3. There are many F distributions and each one has a different shape. We specify a particular one by designating the degrees of freedom associated with s_1^2 and s_2^2. We denote these quantities by df_1 and df_2, respectively.
4. Tail values for the F distribution are tabulated and appear in Table 6 in the Appendix.

FIGURE 7.2

Distribution of s_1^2/s_2^2, the F Distribution

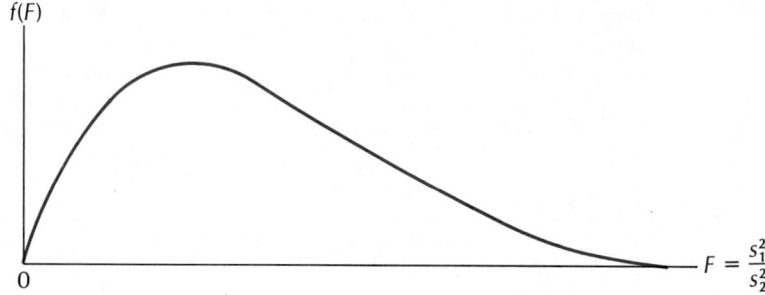

Table 6 in the Appendix records upper-tail values of F corresponding to areas a = .25, .10, .05, .025, .01, .005, and .001. The degrees of freedom for s_1^2, designated by df_1 are indicated across the top of the table; df_2, the degrees of freedom for s_2^2, appear in the first column to the left. Values of a are given in the next column. Thus, for $df_1 = 6$ and $df_2 = 4$, the critical values of F corresponding to a = .25, .10, .05, .025, .01, .005, and .001 are, respectively, 2.08, 4.01, 6.16, 9.20, 15.21, 21.97, and 50.53. It follows that only 5% of the measurements from an F distribution with $df_1 = 6$ and $df_2 = 4$ would exceed 6.16 in repeated sampling. (See Figure 7.3.) Similarly, for $df_1 = 24$ and $df_2 = 10$, the critical values of F corresponding to tail areas of a = .01 and .001 are, respectively, 4.33 and 7.64.

A statistical test of the null hypothesis $\sigma_1^2 = \sigma_2^2$ utilizes the test statistic s_1^2/s_2^2. When H_0 is true, s_1^2/s_2^2 follows an F distribution with $df_1 = n_1 - 1$ and $df_2 = n_2 - 1$. If upper-tail and lower-tail values of F were given in Table 6 in the Appendix, we would have no difficulty in performing the test. Unfortunately, only upper-tail values of F are given. To alleviate this situation, we are at liberty to identify either of the two populations as population 1. For a one-tailed alternative hypothesis, the populations are designated 1 and 2 so that H_a is of the form $\sigma_1^2 > \sigma_2^2$. Then the rejection region is located in the upper-tail of the F distribution. For a two-tailed alternative, we designate the population with the larger sample variance as population 1. By this convention, we again are concerned with only upper-tail rejection regions. The upper-tail F-value for a two-tailed test can then be obtained from Table 6 in the Appendix.

FIGURE 7.3

Critical Value for the F Distribution; $df_1 = 6$ and $df_2 = 4$

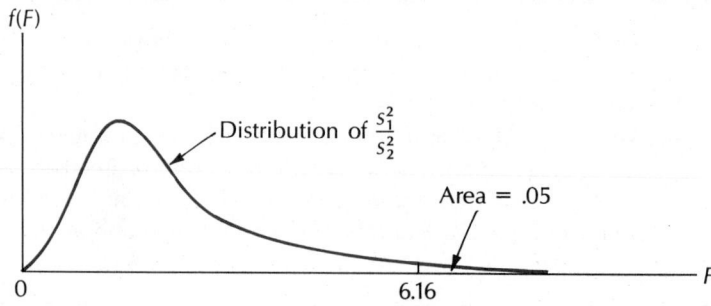

We summarize the test procedure below.

A STATISTICAL TEST COMPARING σ_1^2 AND σ_2^2

H_0: $\sigma_1^2 = \sigma_2^2$

H_a: 1. $\sigma_1^2 > \sigma_2^2$

   2. $\sigma_1^2 \neq \sigma_2^2$

T.S.: $F = \dfrac{s_1^2}{s_2^2}$

R.R.: For a specified value of α,

1. reject H_0 if F exceeds the tabulated value of F for a $= \alpha$, $df_1 = n_1 - 1$, and $df_2 = n_2 - 1$
2. reject H_0 if F exceeds the tabulated value of F for a $= \alpha/2$, $df_1 = n_1 - 1$, and $df_2 = n_2 - 1$

EXAMPLE 7.3

Previously we discussed an experiment in which company officials were concerned about the length of time a particular drug product retained its potency. A random sample of 10 bottles was obtained from the production line and each bottle was analyzed to determine its potency. A second sample of 10 bottles was obtained and stored in a regulated environment for one year. Potency readings were obtained on these bottles at the end of the year. The sample data were then used to place a confidence interval on $\mu_1 - \mu_2$, the difference in mean potencies for the two time periods.

Although we did not stress this at the time, in order to use t in the confidence interval or in a statistical test, we do require that the samples be drawn from normal populations with possibly different means *but* with a common variance. Use the sample data summarized below to test the equality of the population variances. Use $\alpha = .05$. Sample 1 data are the readings taken immediately after production and sample 2 data are the readings taken one year after production. Draw conclusions.

Sample 1: $\bar{y}_1 = 10.37$, $s_1^2 = 0.105$

Sample 2: $\bar{y}_2 = 9.83$, $s_2^2 = 0.058$

Solution

The four parts of the statistical test of H_0: $\sigma_1^2 = \sigma_2^2$ are shown here.

H_0: $\sigma_1^2 = \sigma_2^2$

H_a: $\sigma_1^2 \neq \sigma_2^2$

T.S.: $F = \dfrac{s_1^2}{s_2^2} = \dfrac{0.105}{0.058} = 1.81$

R.R.: For a two-tailed test with $\alpha = .05$, we will reject H_0 if $F > F_{.025,9,9} = 4.03$. Since 1.81 does not fall in the rejection region, we cannot reject H_0: $\sigma_1^2 = \sigma_2^2$.

Also, since 1.81 is not greater than $F_{.10,9,9} = 2.44$, we should feel confident that the assumption of equality of variances holds for the t-methods used with these data.

We can now formulate a confidence interval for the ratio σ_1^2/σ_2^2.

GENERAL CONFIDENCE INTERVAL FOR σ_1^2/σ_2^2 WITH CONFIDENCE COEFFICIENT $(1 - \alpha)$	$$\frac{s_1^2}{s_2^2} F_L < \frac{\sigma_1^2}{\sigma_2^2} < \frac{s_1^2}{s_2^2} F_U$$ If F_{df_1,df_2} represents the $\alpha/2$ upper-tail value of an F distribution with df_1 and df_2 degrees of freedom and F_{df_2,df_1} represents the $\alpha/2$ upper-tail value of an F distribution with the degrees of freedom reversed, then $$F_L = \frac{1}{F_{df_1,df_2}} \quad \text{and} \quad F_U = F_{df_2,df_1}$$ where $df_1 = n_1 - 1$ and $df_2 = n_2 - 1$. Note: The confidence interval for σ_1/σ_2 is found by taking square roots throughout.

It should be noted that although our estimation procedure for σ_1^2/σ_2^2 is appropriate for any confidence coefficient $(1 - \alpha)$, Table 6 allows us to construct confidence intervals for σ_1^2/σ_2^2 with the more commonly used confidence coefficients, such as .90, .95, .98, .99, and so on. For more detailed tables of the F distribution, see Pearson and Hartley (1966).

EXAMPLE 7.4

The life length of an electrical component was studied under two operating voltages, V_1 and V_2. Ten different components were randomly assigned to each of the two operating voltages. Use the data below to find a 90% confidence interval for σ_1^2/σ_2^2, the ratio of the variances in life lengths for the two populations, populations 1 and 2, corresponding to the components studied under V_1 and V_2, respectively.

Voltage V_1: $n_1 = 10$ $s_1^2 = .51$
Voltage V_2: $n_2 = 10$ $s_2^2 = .20$

Solution

Before constructing our confidence interval, we must obtain F_{df_1,df_2} and F_{df_2,df_1}. For $n_1 = n_2 = 10$, $df_1 = df_2 = 9$, and hence F_{df_1,df_2} and F_{df_2,df_1} are the same. For a 90% confidence interval (i.e., $1 - \alpha = .90$), we must look up the .05 F-value based on $df_1 = 9$ and $df_2 = 9$. This value is 3.18. The quantities F_L and F_U are

$$F_L = \frac{1}{3.18} \quad \text{and} \quad F_U = 3.18$$

Substituting into the confidence interval formula, we have

$$\frac{.51}{.20} \left(\frac{1}{3.18} \right) < \frac{\sigma_1^2}{\sigma_2^2} < \frac{.51}{.20} (3.18)$$

$$.80 < \frac{\sigma_1^2}{\sigma_2^2} < 8.11$$

We are 90% confident that the ratio of population variances corresponding to voltages V_1 and V_2 lies in the interval .80 to 8.11.

EXAMPLE 7.5

Refer to Example 7.4. Suppose one of the components on V_1 was damaged by the experimenter midway through the test period and had to be removed from the study. Then with $n_1 = 9$ and $n_2 = 10$, $df_1 = 8$ and $df_2 = 9$. Assuming s_1^2 and s_2^2 are as given in Example 7.4, set up a 90% confidence interval for σ_1^2/σ_2^2.

Solution

The appropriate .05 F-values can be obtained from Table 6 in the Appendix.

$$F_{8,9} = 3.23 \quad \text{and} \quad F_L = \frac{1}{3.23}$$

$$F_{9,8} = 3.39 \quad \text{and} \quad F_U = 3.39$$

We then have the confidence interval

$$\frac{.51}{.20}\left(\frac{1}{3.23}\right) < \frac{\sigma_1^2}{\sigma_2^2} < \frac{.51}{.20}(3.39)$$

or $.79 < \sigma_1^2/\sigma_2^2 < 8.64$

sensitivity of assumptions

The inferences about σ_1^2/σ_2^2 based on the F distribution are *very* sensitive to departures from normality of the underlying distributions. The first precaution that you should take is to plot the data for each sample separately. If there is a hint that one or both of the populations may not be normal, be very careful about the inferences you make on σ_1^2/σ_2^2 using the F distribution; the p-value or confidence coefficient may be substantially different from what you found using the test or confidence interval based on F.

Several alternative procedures are available and are discussed in detail in other textbooks. For example, the Ansari–Bradley test can be used to compare the variances of two populations having the same median (i.e., same location). In contrast, the Moses test and the Miller jackknife procedure are used for comparing the variances of two populations when the populations' medians are unknown and are assumed unequal. The interested reader is referred to Hollander and Wolfe (1973) for details concerning these alternatives to the F-methods of this section.

7.4   SUMMARY

In this chapter we discussed procedures for making inferences concerning a population variance and the ratio of two population variances. Estimation and statistical tests concerning σ^2 make use of the chi-square probability distribution with $df = n - 1$. Inferences concerning the ratio of two population variances utilize an F distribution with $df_1 = n_1 - 1$ and $df_2 = n_2 - 1$.

The need for inferences concerning one or more population variances can be traced to our discussion of numerical descriptive measures of a population in Chapter 2. To describe or make inferences about a population of measurements, we cannot always rely on the mean, a measure of central tendency. Many times in evaluating or comparing the performance of individuals on a psychological test, the consistency of manufactured products emerging from a production line, or the yields of a particular variety of corn, we gain important information by studying the population variance.

In the next chapter we return to a comparison of population means. In particular, we will consider a single-test procedure, called the analysis of variance, for comparing two or more population means.

KEY FORMULAS

1. $100(1 - \alpha)\%$ confidence interval for σ^2 (or σ)

$$\frac{(n-1)s^2}{\chi_U^2} < \sigma^2 < \frac{(n-1)s^2}{\chi_L^2}$$

or

$$\sqrt{\frac{(n-1)s^2}{\chi_U^2}} < \sigma < \sqrt{\frac{(n-1)s^2}{\chi_L^2}}$$

2. Statistical test for σ^2

H_0: $\sigma^2 = \sigma_0^2$ (σ_0^2 is specified)

T.S.: $\chi^2 = \dfrac{(n-1)s^2}{\sigma_0^2}$

3. $100(1 - \alpha)\%$ confidence interval for σ_1^2/σ_2^2 (or σ_1/σ_2)

$$\frac{s_1^2}{s_2^2} F_L < \frac{\sigma_1^2}{\sigma_2^2} < \frac{s_1^2}{s_2^2} F_U$$

where

$$F_L = \frac{1}{F_{df_1,df_2}} \qquad \text{and} \qquad F_U = F_{df_2,df_1}$$

or

$$\sqrt{\frac{s_1^2}{s_2^2} F_L} < \frac{\sigma_1}{\sigma_2} < \sqrt{\frac{s_1^2}{s_2^2} F_U}$$

4. Statistical test for σ_1^2/σ_2^2

H_0: $\sigma_1^2 = \sigma_2^2$

T.S.: $F = \dfrac{s_1^2}{s_2^2}$

EXERCISES

 7.4 Two consumer research groups are vying for a large government contract. Since subjective evaluations of consumer products will be made by judges during the study, government officials prefer to award the contract to a company that utilizes judges with consistent ratings (of course, other qualifications are also evaluated before awarding the contract). One measure of consistency is the variability of judges' scores on the same item.

Before issuing the contract, a test is conducted in which 25 judges from each company are asked to rate a single item. The sample variances are given here:

company A: $s_1^2 = .50$ company B: $s_2^2 = .15$

Use these data to test the hypothesis that the variances of the judges' ratings are the same for the two populations. The alternative hypothesis is that the variances are different. Use $\alpha = .10$.

 7.5 Refer to Exercise 5.15, in which we were interested in comparing the weights of the combs of roosters fed one of two vitamin-supplemented diets. The Wilcoxon rank sum test was suggested as a test of the hypothesis that the two populations were identical. Would it have been appropriate to run a t test comparing the two population means? Explain.

 7.6 A consumer-protection magazine was interested in comparing tires purchased from two different companies, each claiming their tires would last the same number of miles. A sample of five tires of each brand was obtained and tested under simulated road conditions. The number of miles before significant deterioration in tread was recorded for all tires. The data are given below (in 1000 miles).

Brand I	40.6	35.9	48.5	36.4	38.3
Brand II	40.9	40.2	42.5	39.1	42.6

a. Construct a 98% confidence interval for the ratio of the two population variances.
b. How does the confidence interval change if we're interested in σ_1/σ_2 rather than in σ_1^2/σ_2^2?

 7.7 A random sample of 20 patients, each of whom has suffered from depression, was selected from a mental hospital, and each patient was administered the Brief Psychiatric Rating Scale. The scale consists of a series of adjectives that the patient scores according to his or her mood. Extensive testing in the past has shown that ratings in certain mood adjectives tend to be similar and hence are grouped together as jointly measuring one or more components of a person's mood. For example, a group consisting of certain adjectives seems to be measuring depression. Let us suppose that the mean and standard deviation of the 20 patients in the group are 13.2 and 4.6, respectively.

a. Place a 99% confidence interval on σ^2, the variance of the population of patients' scores from which this sample was drawn.
b. What's the critical assumption underlying the inference? Do you know whether this assumption is valid for these data?

7.8 Refer to Exercise 7.7. Suppose that extensive testing in a large number of depressed patients throughout the century has indicated that the population standard deviation of scores for the depression adjectives is 5.9. Use the sample data of Exercise 7.7 to test the research hypothesis that the standard deviation for all patients who might be treated for depression in this hospital is less than 5.9. Give a p-value for these data and draw conclusions.

 7.9 A pharmaceutical company manufactures a particular brand of antihistamine tablets. In the quality control division, certain tests are routinely performed to determine whether the product being manufactured meets specific performance criteria prior to release of the product onto the market. In particular, the company requires that the potencies of the tablets lie in the range of 90% to 110% of the labeled drug amount.

a. If the company is manufacturing 25-mg tablets, within what limits must tablet potencies lie?

b. A random sample of 30 tablets is obtained from a recent batch of antihistamine tablets. The data for the potencies of the tablets are given below. Is the assumption of normality warranted for inferences about the population variances?

c. Translate the company's 90% to 110% specifications on the range of the product potency into a statistical test concerning the population variance for potencies. Draw conclusions based on $\alpha = .05$.

24.1	27.2	26.7	23.6	26.4	25.2
25.8	27.3	23.2	26.9	27.1	26.7
22.7	26.9	24.8	24.0	23.4	25.0
24.5	26.1	25.9	25.4	22.9	24.9
26.4	25.4	23.3	23.0	24.3	23.8

 7.10 A study was conducted to compare the variabilities in strengths of 1-inch-square sections of a synthetic fiber produced under two different procedures. A random sample of nine squares from each process was obtained and tested.

a. Plot the data for each sample separately.

b. Is the assumption of normality warranted?

c. If permissible from part (b), use the data (psi) below to test the research hypothesis that the population variances corresponding to the two procedures are different. Use $\alpha = .10$.

Procedure 1	74	90	103	86	75	102	97	85	69
Procedure 2	59	66	73	68	70	71	82	69	74

7.11 Refer to Example 7.2. Construct a 95% confidence interval for σ^2, and use this interval to help interpret the findings of the consumer group. Does it appear that the test of Example 7.2 had much power to detect an increase in σ^2 of 25% over the claimed value? Explain.

7.12 The risk of an investment is measured in terms of the variance in the return that could be observed. Random samples of 10 yearly returns were obtained from two different portfolios.

	Portfolio	
	1	2
Sample mean Return (000)	132	146
Sample variance	10.9	25.6
Sample size	10	10

Does Portfolio 2 have a higher risk? Give a p-value for your test.

7.13 Refer to Exercise 7.12. Are there any differences in the average returns for the two portfolios? Indicate the method you used in arriving at a conclusion, and explain why you used it.

7.14   Two different modeling techniques for assessing the resale value of houses were considered. A random sample of 12 existing listings was taken and each house was valued using the two techniques. These data are shown here.

Assessed Value of Listing (000)
Technique

Listing	1	2
1	155	138
2	137	128
3	248	230
4	136	146
5	102	95
6	87	82
7	63	67
8	129	134
9	144	149
10	270	292
11	157	150
12	51	48

a. Plot the data. Does it appear that the two modeling techniques give similar results?
b. Give an estimate of the mean and standard error of the difference between estimates for the two methods.

7.15   Refer to Exercise 7.14. Place a 90% confidence interval on the variance of the difference in estimates. Give the corresponding interval for σ.

7.16   Refer to Exercises 7.14 and 7.15. What is the critical assumption concerning the sample data? How would you check this assumption? Do the data suggest the assumption holds? Do you have any cautions about the inferences in Exercise 7.15?

8 Linear regression and correlation

INTRODUCTION

In Chapters 4 and 5 we considered estimation and the test of a hypothesis concerning a population mean or the difference between two population means. The problems we encountered were relatively simple and straightforward. Estimation of a population mean can become more involved, however, if the variable of interest, often called the **dependent variable**, is affected by one or more additional variables, called **independent variables**.

For example, suppose we are interested in estimating the mean weight gain per month for steers fed on a particular variety of feed. The dependent variable, weight gain, could be affected by many variables: initial weight of the steer, amount of feed offered per day, protein content of the feed, water content of the feed, and so on. The problem of estimating the mean weight gain per month must now take into account the **levels**, or **settings**, of the independent variables.

Suppose we want to estimate the mean weight gain (in pounds per day) for steers fed a high-concentrate feed containing 15% protein, 10% water, and the rest carbohydrates. Here 15% is a setting, or level, of the independent variable "protein content of feed" and 10% is a setting of the independent variable, "water content of feed." Similarly, we might wish to estimate, or predict, the mean weight gain for other combinations of settings of the independent variables. Thus estimating a population mean becomes a problem of estimating a population mean for each setting of the independent variables.

This estimation problem can be greatly simplified if we consider models relating a response (dependent variable) to a set of independent variables. In Chapters 8 and 9 we will develop the concepts of linear regression. We will formulate a model and develop estimation and test procedures when the dependent variable is related to one independent variable.

In later chapters we will expand the material of Chapters 8 and 9 to include models and inferences for multiple regression where the dependent variable is related to more than one independent variable. These same models will be useful in many other prediction or estimation problems. For example, a biologist may wish to predict an animal's pulse rate based on the amount of a particular drug administered and the length of time since the drug's administration. A political scientist may wish to predict the outcome of an election (in terms of numbers of votes cast for a particular candidate) based on various socioeconomic factors and previous voting records of the population under study. A sociologist may wish to relate or predict the average number of prison-free years between first and second offenses for persons characterized by various variables, such as age, gender, number of years of schooling, IQ, and so on. A market analyst might want to predict the year-end sales for a company based on various economic indices. Each of these problems bears certain similarities and can be attacked using the models to be proposed in this chapter.

equation of straight line   The simplest type of model relating a response y to a single quantitative independent variable x is given by the **equation of a straight line**

$$y = \beta_0 + \beta_1 x$$

y-intercept β_0
slope β_1
deterministic model

where β_0 is the **y-intercept** (value of y when $x = 0$) and β_1 is the **slope** of the straight line (change in y for a unit change in x). For a given equation, β_0 and β_1 are constants. An equation of this form is called a **deterministic model** because there is no error in reading y. That is, for a given value of independent variable x, we can predict y exactly using the deterministic equation $y = \beta_0 + \beta_1 x$.

Although deterministic models are simple to use, they are unrealistic in many situations, since a dependent variable y cannot always be adequately represented by a deterministic equation in one or more quantitative independent variables. Consider the data of Table 8.1, which gives hospital expenses covered by an insurance carrier and the number of days of confinement for a random sample of $n = 10$ patients. In the table we let y be the

TABLE 8.1   Hospital Expense Data

Expense, y	Number of Days of Confinement, x
$ 50	1
175	3
180	6
200	7
60	2
140	4
420	12
540	15
170	5
300	9

dependent variable (expense) and x the independent variable (number of days of confinement). Suppose the insurance carrier and hospital administrators are interested in estimating the average hospital expense for a given number of days of confinement. Can we use a deterministic model for this problem?

scatter diagram   First, let us draw a picture of the data. The data of Table 8.1 can be plotted by using a **scatter diagram**. In a scatter diagram we draw a vertical axis and a horizontal axis, labeled y and x, respectively. The $n = 10$ data points are then plotted as shown in Figure 8.1.

You can see from Figure 8.1 that a straight line would adequately describe the trend in the data. However, we cannot predict y *exactly* for a given value of x. Thus, for example, we could not predict the average value of y for $x = 8$ days of confinement using a deterministic model of the form $y = \beta_0 + \beta_1 x$.

A model that allows for the possibility that the observations do not all lie on a straight line is the model

$$y = \beta_0 + \beta_1 x + \varepsilon$$

random error ε   where ε is a **random error**. In this model ε represents the difference between a measurement y and a point on the line $\beta_0 + \beta_1 x$. The random error term ε takes into account all unpredictable and unknown factors that are not included in the model. For example, the amount of hospital expenses covered by the insurance carrier could be affected by such factors as the type of operation, complications following the operation, expenses previously submitted for coverage, and whether the insured's spouse has another insurance plan that could cover the expenses. The combined effects of all these and other factors not included in the model contribute to ε.

average value of   One assumption made concerning the random error is that the **average value of ε**
ε is 0 for fixed x   **for a given value of x is 0**. Thus since β_0 and β_1 are constants, the *average* of y (often
expected value of y   called the **expected value of y**) for a fixed value of x is $\beta_0 + \beta_1 x$. This line, denoted by

$$E(y) = \beta_0 + \beta_1 x$$

FIGURE 8.1

Scatter Diagram of the 10 Data Points for the Hospital Expense Data of Table 8.1

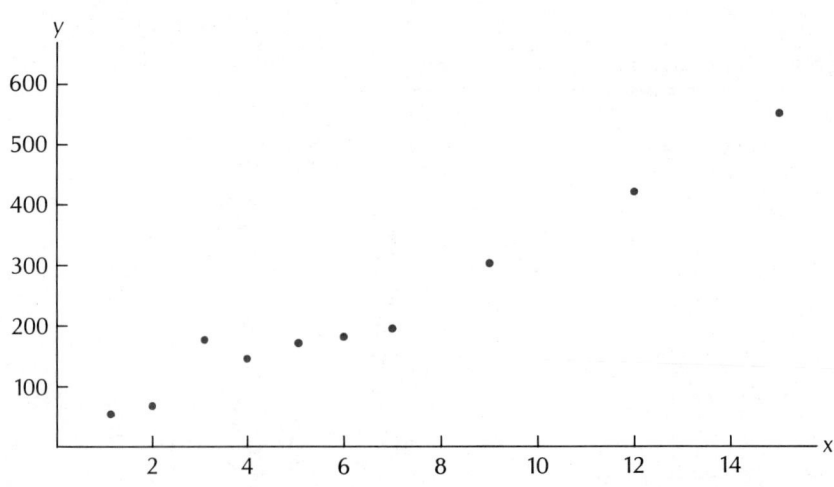

FIGURE 8.2

A Plot of
$E(y) = \beta_0 + \beta_1 x$ for
the Hospital Expense
Data of Table 8.1

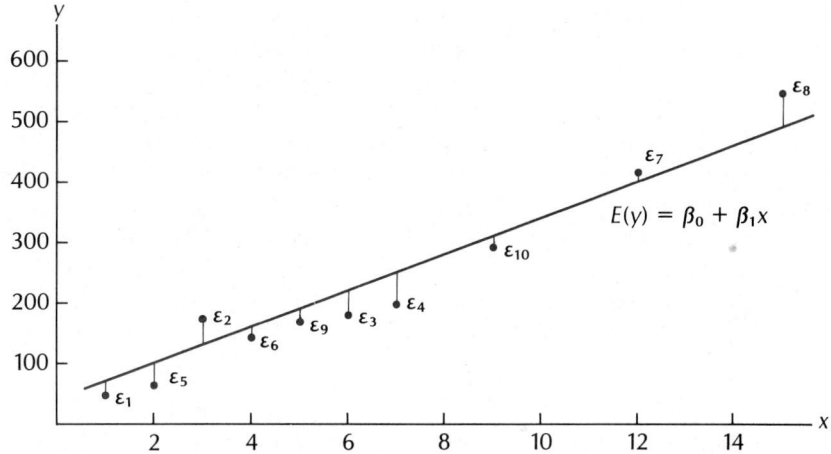

is shown in Figure 8.2. A point on the line denotes the average value of y for the corresponding setting of x. The difference between a sample data point and the expected value of y (a point on the line $E(y) = \beta_0 + \beta_1 x$) is ε. Thus the observed values of y deviate above or below the line by a random amount ε. The random errors associated with the 10 data points listed in Table 8.1 are pictured in Figure 8.2.

Unfortunately, since β_0 and β_1 are unknown parameters, we will never know the precise, location of the line $E(y) = \beta_0 + \beta_1 x$. All we will have in a given experimental situation will be the n data points. In the next section we will show how to use the sample information to construct estimates, $\hat{\beta}_0$ and $\hat{\beta}_1$, of the parameters β_0 and β_1 to be used in formulating an estimate of the line $E(y) = \beta_0 + \beta_1 x$. Appropriate confidence intervals and tests of hypotheses will also be discussed in Chapter 9.

8.2   LINEAR REGRESSION AND THE METHOD OF LEAST SQUARES

Consider the problem of obtaining estimates for parameters in the model

$$y = \beta_0 + \beta_1 x + \varepsilon$$

linear regression   for the **linear regression**

$$E(y) = \beta_0 + \beta_1 x$$

There are many ways to determine an estimate of $E(y)$, which is represented by the equation

$$\hat{y} = \hat{\beta}_0 + \hat{\beta}_1 x$$

eyeball fitting   One procedure, called the **eyeball-fitting** technique, requires that we plot the data on a scatter diagram and then use a ruler to draw what we feel is the straight line that most accurately displays the linear trend of the data. Unfortunately, if each of us was given the same set of data, we might each come up with a different prediction equation.

method of least squares   The **method of least squares** is, in many respects, a formalization of the eyeball-fitting routine just discussed. If we let $\hat{y}$ denote the predicted value of y for a given value of x,

residual, or error of prediction   then the **error of prediction** (often called the **residual**) is $y - \hat{y}$, the difference between the actual value of y and what we predict it to be. *The method of least squares chooses the prediction line $\hat{y} = \hat{\beta}_0 + \hat{\beta}_1 x$ that minimizes the sum of the squared errors of prediction $\sum (y - \hat{y})^2$ for all sample points.* We can denote the sum of the squared errors of prediction for the linear model $y = \beta_0 + \beta_1 x + \varepsilon$ by

$$\sum (y - \hat{y})^2 = \sum (y - \hat{\beta}_0 - \hat{\beta}_1 x)^2$$

Thus the method of least squares consists of finding those estimates $\hat{\beta}_0$ and $\hat{\beta}_1$ that minimize $\sum (y - \hat{y})^2$.

While the procedure for deriving these estimates involves use of the calculus, we

least squares estimates   can summarize the results. The estimates, called **least squares estimates**, that minimize $\sum (y - \hat{y})^2$ are computed as shown here.

LEAST SQUARES ESTIMATES OF β_1 AND β_0	$\hat{\beta}_1 = \dfrac{S_{xy}}{S_{xx}}$ and $\hat{\beta}_0 = \bar{y} - \hat{\beta}_1 \bar{x}$ where S_{xx} $S_{xx} = \sum (x - \bar{x})^2 = \sum x^2 - \dfrac{(\sum x)^2}{n}$ S_{xy} $S_{xy} = \sum (x - \bar{x})(y - \bar{y}) = \sum xy - \dfrac{(\sum x)(\sum y)}{n}$

These ideas can probably be best understood by working an example.

EXAMPLE 8.1   In a random sample of $n = 9$ steers, the live weights and dressed weights were recorded. In Table 8.2 we let y denote the dressed weight (in hundreds of pounds) and x denote the corresponding live weight (in hundreds of pounds). Use the sample data to obtain least squares estimates for the model

$$y = \beta_0 + \beta_1 x + \varepsilon$$

Solution   When we do not have the use of a calculator, the least squares estimates can be computed fairly easily if we construct a summary table, such as that shown in Table 8.3.

TABLE 8.2   Sample Data for Example 8.1; Live Weight (x) and Dressed Weight (y) of Steers

x	y
4.2	2.8
3.8	2.5
4.8	3.1
3.4	2.1
4.5	2.9
4.6	2.8
4.3	2.6
3.7	2.4
3.9	2.5

Using the computational formulas for S_{xx} and S_{xy}, we have, from Table 8.3,

$$S_{xx} = \sum x^2 - \frac{(\sum x)^2}{n} = 155.48 - \frac{(37.2)^2}{9}$$

$$= 155.48 - 153.76 = 1.72$$

$$S_{xy} = \sum xy - \frac{(\sum x)(\sum y)}{n} = 99.02 - \frac{(37.2)(23.7)}{9}$$

$$= 99.02 - 97.96 = 1.06$$

The least squares estimate for β_1 is

$$\hat{\beta}_1 = \frac{S_{xy}}{S_{xx}} = \frac{1.06}{1.72} = .616$$

TABLE 8.3   Summary Table for the Data of Example 8.1

	x	x^2	y	y^2	xy
	4.2	17.64	2.8	7.84	11.76
	3.8	14.44	2.5	6.25	9.50
	4.8	23.04	3.1	9.61	14.88
	3.4	11.56	2.1	4.41	7.14
	4.5	20.25	2.9	8.41	13.05
	4.6	21.16	2.8	7.84	12.88
	4.3	18.49	2.6	6.76	11.18
	3.7	13.69	2.4	5.76	8.88
	3.9	15.21	2.5	6.25	9.75
Totals	37.2	155.48	23.7	63.13	99.02

The sample means $\bar{x}$ and $\bar{y}$ are

$$\bar{x} = \frac{\sum x}{n} = \frac{37.2}{9} = 4.133$$

and

$$\bar{y} = \frac{\sum y}{n} = \frac{23.7}{9} = 2.633$$

Substituting our calculated values into the formula for $\hat{\beta}_0$, we have

$$\hat{\beta}_0 = \bar{y} - \hat{\beta}_1\bar{x} = 2.633 - .616(4.133) = .087$$

The least squares equation for these data is

$$\hat{y} = .087 + .616x$$

which is plotted in Figure 8.3 with the sample data superimposed. The predicted value of y for any value of x is given by a point on the line and can be computed by using the equation

$$\hat{y} = .087 + .616x$$

Algebraically, it can be shown that from a least squares fit of the model $y = \beta_0 + \beta_1 x + \varepsilon$

$$y_i - \bar{y} = (y_i - \hat{y}_i) + (\hat{y}_i - \bar{y})$$

FIGURE 8.3

Plot of the Least Squares Equation for the Data of Example 8.1

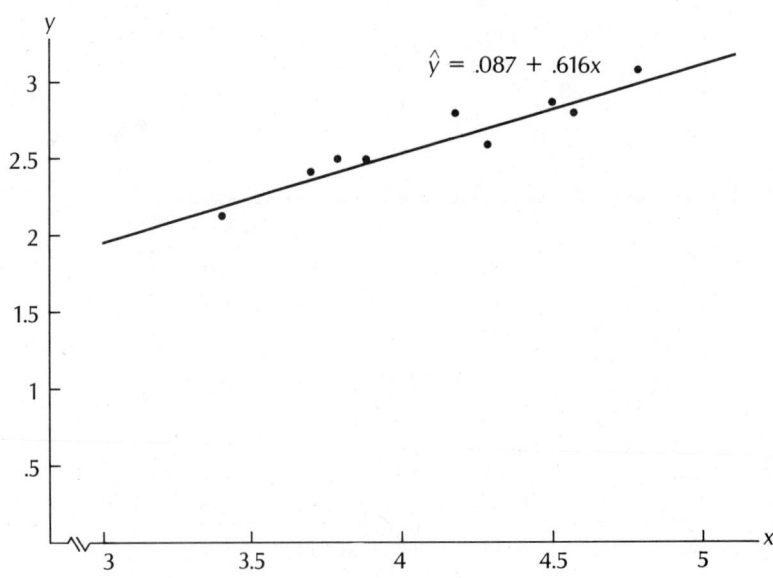

and that

$$\sum_i (y_i - \bar{y})^2 = \sum (y_i - \hat{y}_i)^2 + \sum (\hat{y}_i - \bar{y})^2$$

While the proof of this equality is beyond the scope of this text, we can obtain an intuitive understanding of this relationship by considering the following situation.

Suppose that we wish to use the model

$$y = \beta_0 + \varepsilon$$

In this model β_0 represents the population mean for the variable y, and, intuitively, we would estimate its value using the sample mean $\bar{y}$. (You can confirm this result by using the formula for the estimated intercept $\hat{\beta}_0$ in a linear model.) Since $\hat{y} = \bar{y}$ for this model, the sum of the squared errors of prediction is $\sum (y - \bar{y})^2$.

Now suppose the variable y is related to an independent variable x. From our previous work, we could fit the model $y = \beta_0 + \beta_1 x + \varepsilon$ to obtain

$$\hat{y} = \hat{\beta}_0 + \hat{\beta}_1 x$$

For this model the sum of the squared prediction errors is

$$\sum (y - \hat{y})^2$$

In Figure 8.4 we have presented two prediction equations: $\hat{y} = \bar{y}$ for the model $y = \beta_0 + \varepsilon$ and $\hat{y} = \hat{\beta}_0 + \hat{\beta}_1 x$ for the model $y = \beta_0 + \beta_1 x + \varepsilon$. Note that we can express the distance between an observation y and the sample mean $\bar{y}$ as the sum of two components, $(\hat{y} - \bar{y})$ and $(y - \hat{y})$. The quantity $(\hat{y} - \bar{y})$ represents that portion of the overall distance that can be attributed to the independent variable x (through the prediction equation $\hat{y} = \hat{\beta}_0 + \hat{\beta}_1 x$). The quantity $(y - \hat{y})$ represents that portion of the distance between y and $\bar{y}$ that cannot be accounted for by the independent variable x (and that we attribute to error). Combining this information for all sample observations, we can express the total variability in the sample measurements about the sample mean, $\sum (y - \bar{y})^2$, called the **sum of squares about the mean**, as the sum of the squared deviations of the predicted values

sum of squares about the mean

FIGURE 8.4

Relationship Between $\sum (y - \bar{y})^2$ and $\sum (y - \hat{y})^2$

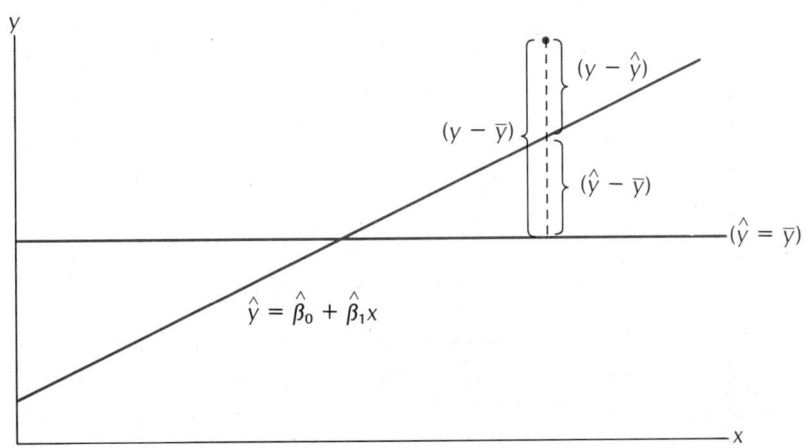

sum of squares due
to regression
sum of squares for
error

from $\bar{y}$ called the **sum of squares due to regression**, $\sum (\hat{y} - \bar{y})^2$, and the sum of the squared errors of prediction, $\sum (y - \hat{y})^2$, called the **sum of squares for error**. Thus we have

Sum of squares about the mean = Sum of squares due to regression
+ Sum of squares for error

There is another way to view this equation. It can be shown that the sample mean $\bar{y}$ is also the average of the fitted values. Therefore the sum of squares due to regression $\sum (\hat{y} - \bar{y})^2$ depicts variability in the fitted values. Similarly $\sum (y - \hat{y})^2$ represents variability in the y-values about the fitted values. As a result, the total variability in the y-values can be written as

$$\underset{\substack{\text{total variability} \\ \text{in y-values}}}{\sum (y - \bar{y})^2} = \underset{\substack{\text{variability} \\ \text{explained by model}}}{\sum (\hat{y} - \bar{y})^2} + \underset{\substack{\text{unexplained} \\ \text{variability}}}{\sum (y - \hat{y})^2}$$

Figure 8.4 shows how the total variability is partitioned into the two components.

Obviously, if we're interested in predicting y based on the independent variable x, the larger the explained variability is relative to the unexplained variability, the better the model "fits" the data, and this should lead to more precise prediction of y based on x.

EXAMPLE 8.2

Consider the five data points listed in columns 1 and 2 of Table 8.4.
a. Fit the model

$$y = \beta_0 + \beta_1 x + \varepsilon$$

b. Verify that

$$\sum (y - \bar{y})^2 = \sum (y - \hat{y})^2 + \sum (\hat{y} - \bar{y})^2$$

Solution

a. For these data, it can be shown that

$$S_{xx} = \sum x^2 - \frac{(\sum x)^2}{n} = 66 - \frac{(16)^2}{5} = 14.8$$

$$S_{xy} = \sum xy - \frac{(\sum x)(\sum y)}{n} = \frac{145 - (16)(38)}{5} = 23.4$$

$$\hat{\beta}_1 = 1.58$$

and

$$\hat{\beta}_0 = \bar{y} - \hat{\beta}_1 \bar{x} = 7.6 - 1.58(3.2) = 2.54$$

b. For each x-value we then compute $\hat{y}$ from the least squares prediction equation. We also compute the quantities $(y - \bar{y})$, $(y - \hat{y})$, and $(\hat{y} - \bar{y})$. These quantities are displayed in Table 8.4.

TABLE 8.4   Data and Computations for Example 8.2

x	y	$\hat{y}$	$y - \bar{y}$	$y - \hat{y}$	$\hat{y} - \bar{y}$
1	4	4.1216	−3.6000	−.1216	−3.4784
2	6	5.7027	−1.6000	.2973	−1.8973
3	7	7.2838	−.6000	−.2838	−.3162
4	9	8.8649	1.4000	.1351	1.2649
6	12	12.0271	4.4000	−.0271	4.4271

From columns, 4, 5, and 6 in the table, we have

$$\sum (y - \bar{y})^2 = 37.2000$$
$$\sum (y - \hat{y})^2 = .2027$$
$$\sum (\hat{y} - \bar{y})^2 = 36.9982$$

Note that, except for rounding errors,

$$\sum (y - \bar{y})^2 = \sum (y - \hat{y})^2 + \sum (\hat{y} - \bar{y})^2$$

Intuitively, since the explained variability $\sum (\hat{y} - \bar{y})^2 = 36.9982$ accounts for almost all of the total variability $\sum (y - \bar{y})^2 = 37.2000$, the model appears to fit the data very well. A scatterplot of y versus x with the prediction equation superimposed shows how tightly the data group about the prediction equation.

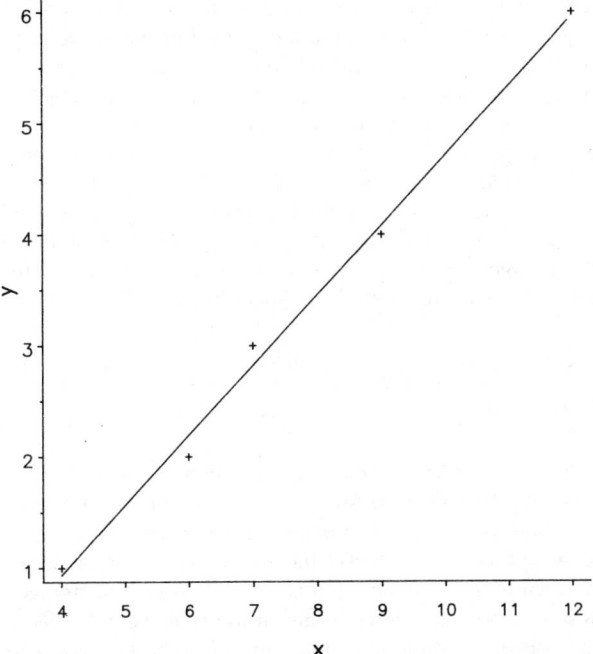

EXERCISES

8.1   The following data were obtained in a study of sales volume (per district) as a fraction of the number of client contacts per month.

Sales Volume ($1,000)	Average Number of Client Contacts per Month
y	x
15	10
26	15
28	17
30	20
32	23
86	46
109	53
95	48
130	59
160	65

a. Plot the data.
b. Eyeball a linear fit to the data and guess the value of the intercepts and slope.
c. Predict sales for $x = 50$.

8.2   Refer to Exercise 8.1
a. Obtain the linear regression equation $\hat{y} = \hat{\beta}_0 + \hat{\beta}_1 x$ using the method of least squares. Compare this line to the one obtained in Exercise 8.1.
b. Predict sales for $x = 50$ and compare to Exercise 8.1(c).

8.3   An experiment was conducted to examine the effect of different concentrations of pectin on the firmness of canned sweet potatoes. Three concentrations were used: 0%, 1.5%, and 3% pectin by weight. Six number 303 × 406 cans were packed with sweet potatoes in a 25% (by weight) sugar solution. Two cans were randomly assigned to each of the pectin concentrations with the appropriate percentage of pectin added to the sugar syrup. The cans were then sealed and placed in a 25°C environment for 30 days. At the end of the storage time, the cans were opened and a firmness determination made for the contents of each can. These data appear below.

Pectin concentration	0%, 0%	1.5%, 1.5%	3.0%, 3.0%
Firmness reading	50.5, 46.8	62.3, 67.7	80.1, 79.2

a. Let x denote the pectin concentration of a can and y denote the firmness reading following the 30 days of storage at 25° C. Plot the sample data in a scatter diagram.
b. Obtain least squares estimates for the parameters in the model $y = \beta_0 + \beta_1 x + \varepsilon$.

8.4   Refer to Exercise 8.3. Predict the firmness for a can of sweet potatoes treated with a 1% concentration of pectin (by weight) after 30 days of storage at 25° C.

8.5   A study was conducted to examine the quality of fish after seven days in ice storage. Ten raw fish of the same kind and approximately the same size were caught and prepared for ice storage. Two of the fish were placed in storage immediately after being caught, two were placed in storage 3 hours after being caught, and two each were placed in storage at 6, 9, and 12 hours after being caught. Let y denote a measurement of fish quality (on a 10-point scale) after the seven days of storage, and x denote the time after being caught that the fish were placed in ice packing. The sample data appear below.

y	8.5	8.4	7.9	8.1	7.8	7.6	7.3	7.0	6.8	6.7
x	0	0	3	3	6	6	9	9	12	12

a. Plot the sample data in a scatter diagram.
b. Use the method of least squares to obtain estimates of the parameters in the model $y = \beta_0 + \beta_1 x + \varepsilon$.

8.6 Refer to Exercise 8.5. Predict the seven-day quality score of a fish placed in ice storage 10 hours after being caught. Would you be willing to predict a quality score for fish placed in storage 18 hours after being caught?

| 8.3 | QUICK PORTABLE STATISTICS |

Sometimes it is necessary to get a quick, approximate fit to the model $y = \beta_0 + \beta_1 x + \varepsilon$. The procedure can be used provided we obtain three or more (x, y) observations. The rules for constructing the approximate linear regression are given here.

PROCEDURE FOR
OBTAINING AN
APPROXIMATE
LINEAR
REGRESSION
EQUATION

1. Plot the data in a scatter diagram.
2. Using two lines parallel to the vertical axis, divide the data into three groups of *roughly* the same number of observations.
3. For the lower-end group, find the median point; that is, the point corresponding to the medians for x and y based on measurements in that group.
4. Repeat (3) for the upper-end group.
5. Connect the two median points using a straight line. This line represents an approximate linear regression equation $\hat{y} = \hat{\beta}_0 + \hat{\beta}_1 x$.

EXAMPLE 8.3   Data were obtained on the reduction in cholesterol count (in mg per 100 ml of blood serum) for a random sample of 15 male volunteers participating in a study involving a low-cholesterol diet. Each volunteer participated in the study for four weeks. A pre-study cholesterol reading was obtained prior to beginning the diet, and the reduction in the count was observed for the four-month period. In addition to cholesterol levels, ages of the volunteers were also recorded. These data are given in Table 8.5. Construct an approximate linear regression line.

TABLE 8.5   Age and Cholesterol Reduction Data, Example 8.3

Age	Reduction in Cholesterol	Age	Reduction in Cholesterol
45	30	31	40
43	52	26	17
46	45	22	28
49	38	58	44
50	62	60	61
37	55	52	58
34	25	27	45
30	30		

Solution

We begin with a scatter diagram of the data (see Figure 8.5). The data are then divided into three approximately equal groups with two vertical lines. These have been inserted onto the diagram in Figure 8.6. The median points for the lower-end group can be found by drawing a vertical line that evenly divides the x-values and a horizontal line that evenly divides the y-values. The intersection of these two lines is the median point for the lower-end group (see Figure 8.6). Similarly, for the upper-end group, we find the median point indicated in Figure 8.6. The straight line joining the median points for the two end groups is the line that approximates the least squares equation for these data.

FIGURE 8.5

Scatter Diagram of the Age and Cholesterol Count Data, Example 8.3

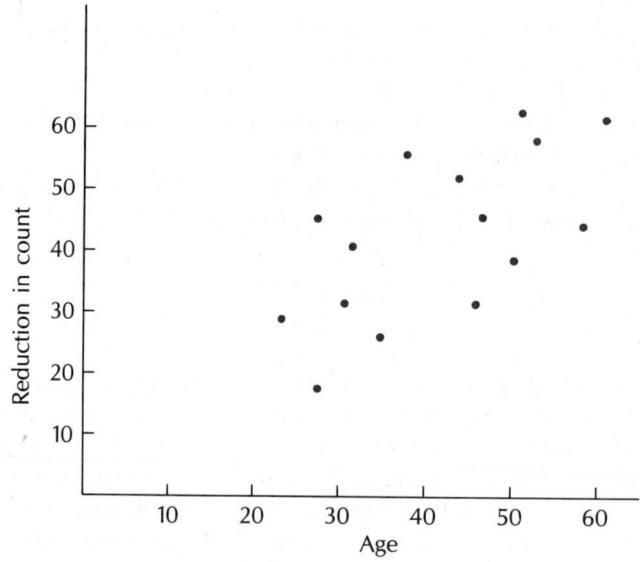

FIGURE 8.6

Approximate Linear
Regression Line

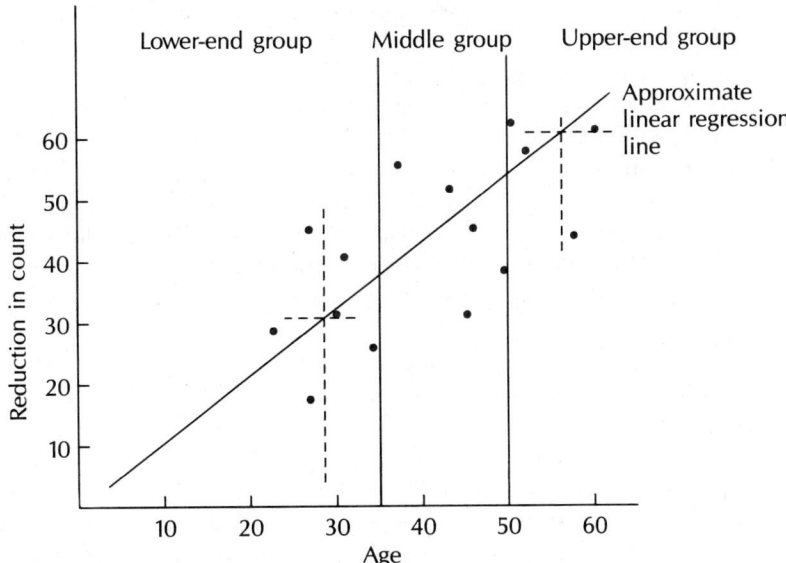

EXAMPLE 8.4   Refer to Example 8.3.

a. Write an equation for the approximate linear regression line shown in Figure 8.6.
b. Predict reduction in cholesterol for a 55-year-old male who uses the low cholesterol diet.

Solution

a. From Figure 8.6 the y intercept appears to be $\hat{\beta}_0 \approx -1$; the slope (change in y for unit change in x) is $\hat{\beta}_1 \approx 1$. Hence $\hat{y} = -1 + x$.
b. For a 55-year-old male, $x = 55$ and $\hat{y} = -1 + 55 = 54$.

Although we've presented a good deal of material on linear regression so far in this chapter, we have not yet developed or discussed any formal estimation or test procedures; all inferences have been of an informal, seat-of-the-pants style. Inferences for linear regression will be developed in detail in Chapter 9. But before we leave our introduction to regression, we will consider several additional topics: ways to linearize data, correlations, and a few extensions beyond the linear regression setting.

8.4   TRANSFORMATIONS TO LINEARIZE DATA

There are many reasons why we might want to re-express (transform) either the independent or dependent variable. One such reason is that we wish to simplify the underlying model. For example, basic books on finance show that if a quantity y grows at a rate

r per unit time then the value y_t at time t is $y_t = y_0 e^{rt}$, a nonlinear model, where y_0 is the value at time 0. One way to deal with the model is to take natural logarithms of both sides to obtain $\log y_t = \log y_0 + rt$; this is a linear model. We will find that there will be many other situations in which we can use a transformation on either the dependent variable, the independent variable, or both variables to eliminate curvature in the data and hence simplify our analysis of the data.

How will we know which transformation will help to linearize the data? In situations that are characterized by the finance model, the appropriate transformation is obvious. In other situations in which we have no known model and only a scatter plot of the data, the choice of transformation to linearize the data is more a matter of trial and error. However, there are a few general guidelines that can be of help.

First, consider transformations of the dependent variable that will help linearize the sample data. For example, suppose a marketing research firm has observed the following trends in sales (y) as a function of mass media advertising expenses (x) for 10 different companies selling a similar product. These data are displayed in Table 8.6. A scatterplot of y versus x is shown in Figure 8.7.

What re-expression of the dependent variable y will help to linearize the data? In order to answer this question we make use of Figure 8.8, which displays the relationship between x and various transformations on the dependent variable y. This figure tells us that when a plot of y versus x is linear, successive plots of x versus y^2 and y^3 will curve upward whereas plots of x versus $\sqrt{y}$, $\log y$, $-1/y$, and $-1/y^2$ will curve downward, with each successive plot having more severe curvature. The order of progression from curvature upward to curvature downward is

transformations on y

$$y^3$$
$$y^2$$
$$y$$
$$\sqrt{y}$$
$$\log y$$
$$-1/y$$
$$-1/y^2$$

TABLE 8.6   Sales and Expenditures

Company	Sales (y)	Expenditure (x)
1	2.5	1.0
2	2.6	1.6
3	2.7	2.5
4	5.0	3.0
5	5.3	4.0
6	9.1	4.6
7	14.8	5.0
8	17.5	5.7
9	23.0	6.0
10	28.0	7.0

FIGURE 8.7

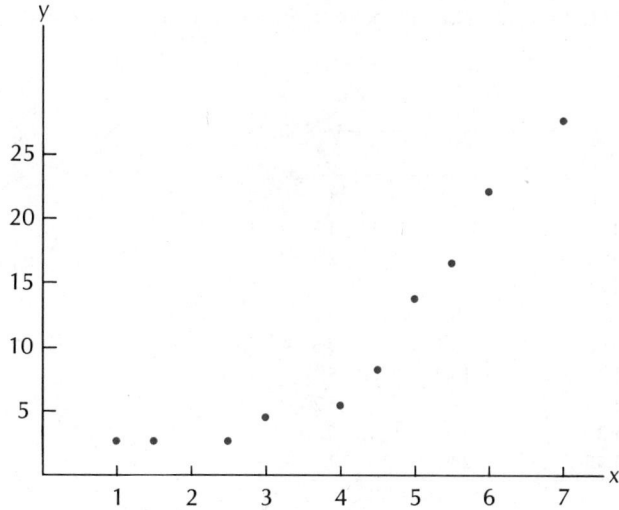

FIGURE 8.8

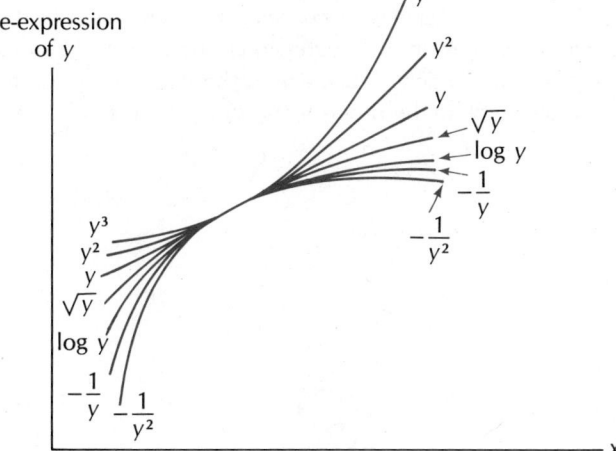

So if, in a scatterplot of the independent variable x versus the dependent variable y, the plot curves upward, we must proceed down on the scale to choose a transformation of the dependent variable that linearizes the data. For our data we might consider either $\sqrt{y}$ or log y initially.

EXAMPLE 8.5   ▌   Re-express the y data from Table 8.6 using a log y transformation to linearize the data.

Solution   The x, y data from Table 8.6 have been reconstructed here with an additional column for $\log_{10} y$.

$\log_{10} y$	y	x
0.40	2.5	1
0.41	2.6	1.6
0.43	2.7	2.5
0.70	5	3
0.72	5.3	4
0.96	9.1	4.6
1.17	14.8	5
1.24	17.5	5.7
1.36	23	6
1.45	28	7

The scatterplot of log y versus x in Figure 8.9 shows that this transformation on the dependent variable does a reasonable job of linearizing the data.

Sometimes it is convenient to consider a re-expression of the independent variable for linearizing the data. A similar progression of transformations exists for the independent variable as for the dependent variable with the direction of the re-expression depending on the shape of the plot. The order of transformation is the same as for transformations

FIGURE 8.9

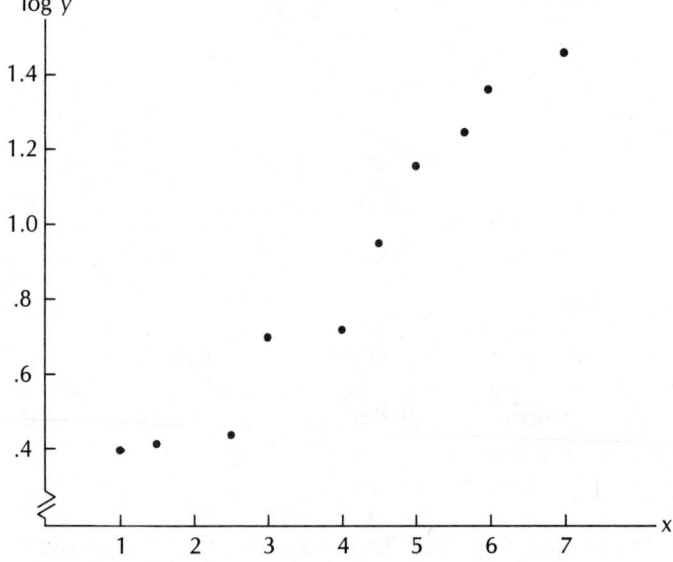

on the dependent variable; namely,

$$x^3$$

$$x^2$$

$$x$$

$$\sqrt{x}$$

transformations on x

$$\log x$$

$$-1/x$$

$$-1/x^2$$

etc.

There are differences, though, in which transformations help to linearize the sample data when a scatterplot has curvature. For example, when the plot of y versus x turns upward, choices of transformations on x to linearize the data are selected by *moving up rather than down* the list of transformations. This is just the opposite of what we would do using a transformation on y.

For the data displayed in Figure 8.7, we could consider an x^2 or x^3 transformation to linearize the data. Table 8.7 and Figure 8.10 display a re-expression of x using an x^2 transformation.

TABLE 8.7

y	x	x^2
2.5	1	1
2.6	1.6	2.56
2.7	2.5	6.25
5	3	9
5.3	4	16
9.1	4.6	21.16
14.8	5	25
17.5	5.7	32.49
23	6	36
28	7	49

FIGURE 8.10

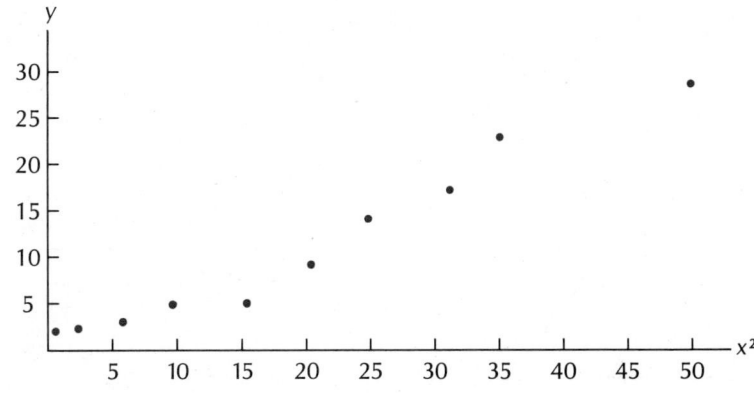

Since this transformation does a better job of linearizing the data than does the log y transformation of Example 8.5, we can use the linear regression model

$$y = \beta_0 + \beta_1 x^2 + \varepsilon$$

to analyze the sample data in Table 8.6. To analyze data for a model such as this, we merely define a new variable, say $z = x^2$, and proceed with a linear regression model using y and z.

To avoid some of the confusion associated with choosing an appropriate re-expression, we have summarized four general shapes of a scatterplot and the choices of transformations on either x and y for linearizing the data in Figure 8.11.

We have indicated that re-expressing the dependent or independent variable can transform a nonlinear model into the framework of a linear regression model for some variables. In addition, transformations on dependent variables can eliminate certain types of non-constant variances. More discussion about this topic will be presented in Chapter 13.

FIGURE 8.11

Transformations for x and y to Linearize Data

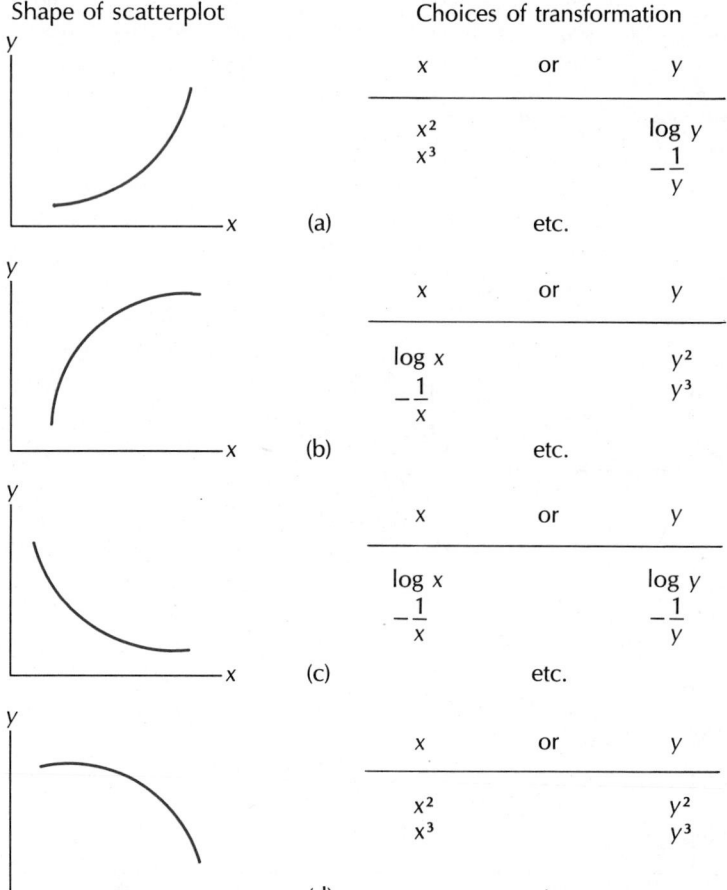

Shape of scatterplot		Choices of transformation		
		x	or	y
(a)		x^2 x^3		$\log y$ $-\frac{1}{y}$
			etc.	
(b)		$\log x$ $-\frac{1}{x}$		y^2 y^3
			etc.	
(c)		$\log x$ $-\frac{1}{x}$		$\log y$ $-\frac{1}{y}$
			etc.	
(d)		x^2 x^3		y^2 y^3
			etc.	

EXERCISES

8.7 The sales, in thousands of units, of a small electronics firm for the last 10 years have been

Year	1	2	3	4	5	6	7	8	9	10
Sales	2.60	2.85	3.02	3.45	3.69	4.26	4.73	5.16	5.91	6.50

a. Plot sales against year.
b. Plot log sales against year. (It doesn't matter whether you use log base 10 or log base e.)
c. Which plot appears more nearly linear?

8.8 Refer to the data of Exercise 8.7.
a. Calculate the regression equation with sales as the dependent variable.
b. Calculate the regression equation with log sales as the dependent variable.
c. Which model appears to provide a better linear fit?

8.9 Use each of the regression equations of Exercise 8.8 to forecast sales in year 11. (You will have to convert log sales back to sales.) Which prediction appears more plausible?

8.10 Data relating advertising (x) to sales for a given product is shown here.

x	15	10	18	16	15	12	10	17	15	13	15	20	18	12
y	365	320	357	375	381	335	312	345	362	349	371	331	340	351

a. Plot the y versus x.
b. Is there evidence of nonlinearity in the plot?

8.5 CORRELATION

In this section we will extend our study of the relationships between two variables. Not only might we like to predict the value of one variable (the dependent variable) based on information on an independent variable, as we have done in previous sections, but we might also wish to provide a measure of the strength of the relationship between these variables. This idea will be the topic of this section.

One measure of the strength of the relationship between two variables x and y is **correlation coefficient** called the coefficient of linear correlation, or, simply, the **correlation coefficient**. Given n **sample correlation** pairs of observations (x_i, y_i), we can compute the **sample correlation coefficient** r as **coefficient r**

$$r = \frac{S_{xy}}{\sqrt{S_{xx}S_{yy}}}$$

where

s_{yy} $$S_{yy} = \sum y^2 - \frac{(\sum y)^2}{n}$$

You will immediately note the similarity between r and the slope of the least squares equation

$$\hat{y} = \hat{\beta}_0 + \hat{\beta}_1 x$$

relating y to x.

$$\hat{\beta}_1 = \frac{S_{xy}}{S_{xx}}$$

$$r = \hat{\beta}_1 \sqrt{\frac{S_{xx}}{S_{yy}}}$$

For experimental situations in which not all x-values and y-values are the same, both S_{xx} and S_{yy} are positive. Then r and $\hat{\beta}_1$ have the same sign. Because of the relationship between r and $\hat{\beta}_1$, the sample correlation coefficient r measures the strength of the linear relationship between x and y and is used to estimate the corresponding population coefficient of linear correlation ρ (Greek letter rho).

PROPERTIES OF
r AND r^2

1. r lies between -1 and $+1$. $r > 0$ indicates a positive linear relationship and $r < 0$ a negative linear relationship between x and y. $r = 0$ indicates no linear relationship between x and y. (See Figure 8.12.)
2. r^2 gives the proportion of the total variability in the y-values that can be accounted for by the independent variable x.

FIGURE 8.12

Interpretation of r

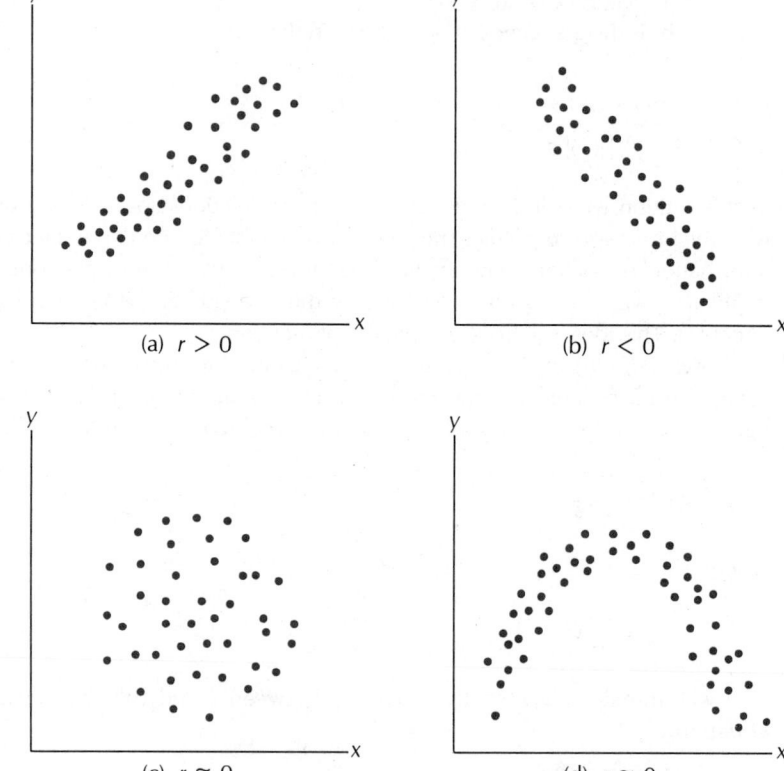

(a) $r > 0$

(b) $r < 0$

(c) $r \approx 0$

(d) $r \approx 0$

TABLE 8.8

Observed Meter Flow Rate, y	Actual Flow Rate, x
1.4	1
2.3	2
3.1	3
4.2	4
5.1	5
5.8	6
6.8	7
7.6	8
8.7	9
9.5	10

EXAMPLE 8.6

An engineer is interested in calibrating a flow meter to be used on a liquid soap production line. For the test 10 different flow rates are fixed and the corresponding meter readings observed. These data are shown in Table 8.8.

a. Plot the sample data. Do the data look linear?

b. Compute the correlation coefficient between the observed and actual flow rates.

Solution

a. A plot of the data is shown here.

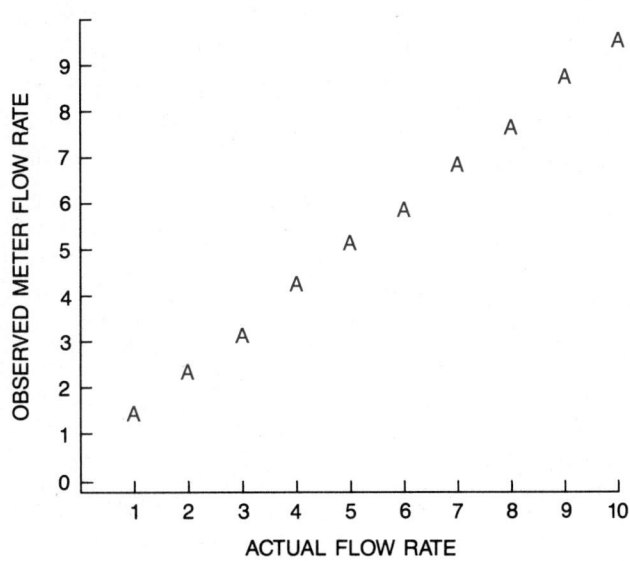

b. For these data, it can be shown that

$$S_{yy} = \sum y^2 - \frac{\left(\sum y\right)^2}{n} = 364.09 - \frac{(54.5)^2}{10} = 67.065$$

$$S_{xx} = \sum x^2 - \frac{\left(\sum x\right)^2}{n} = 385 - \frac{(55)^2}{10} = 82.5$$

and

$$S_{xy} = \sum xy - \frac{(\sum x)(\sum y)}{n} = 374.1 - \frac{55(54.5)}{10} = 74.35$$

Combining, we have

$$r = \frac{S_{xy}}{\sqrt{S_{xx}S_{yy}}} = \frac{74.35}{\sqrt{82.5(67.065)}} = .9996$$

That is, there appears to be a very strong positive linear relationship between the flow rate x and the instrument reading y.

We can illustrate how r^2 measures the proportion of the total y variability accounted for by x for the linear model $y = \beta_0 + \beta_1 x + \varepsilon$. As noted in Section 8.2, the total variability of the y values about their mean $\bar{y}$ can be expressed as

$$\sum (y - \bar{y})^2 = \sum (y - \hat{y})^2 + \sum (\hat{y} - \bar{y})^2$$

where $\sum (\hat{y} - \bar{y})^2$ is that portion of the total variability that can be accounted for by the independent variable x and $\sum (y - \hat{y})^2$ is the sum of squares for error (SSE). Using a computational formula, we can rewrite the total sum of squares S_{yy} as

$$S_{yy} = \sum (y - \bar{y})^2 = \sum y^2 - \frac{(\sum y)^2}{n}$$

Similarly, it can be shown that the sum of squares for error for a linear regression model can be written as

$$\sum (y - \hat{y})^2 = S_{yy} - \frac{S_{xy}^2}{S_{xx}}$$

and by subtraction, the sum of squares due to regression is

$$\sum (\hat{y} - \bar{y})^2 = \frac{S_{xy}^2}{S_{xx}}$$

Then expressing both $\sum (y - \hat{y})^2$ and $\sum (\hat{y} - \bar{y})^2$ as a proportion of $S_{yy} = \sum (y - \bar{y})^2$, we have

$$\frac{\sum (y - \hat{y})^2}{S_{yy}} = 1 - \frac{S_{xy}^2}{S_{xx}S_{yy}} = 1 - r^2$$

and

$$\frac{\sum (\hat{y} - \bar{y})^2}{S_{yy}} = \frac{S_{xy}^2}{S_{xx}S_{yy}} = r^2$$

Thus r^2 represents that proportion of the total variability of the y-values that is accounted for by the independent variable x. Similarly, $1 - r^2$ represents that proportion of the total variability of the y-values that is not accounted for by the variable x.

EXAMPLE 8.7   Refer to Example 8.6. What percent of the variability in flow meter readings (y) is accounted for by the actual flow rate (x)?

Solution   For these data $r^2 = (.9996)^2 = .9992$. Thus 99.92% of the variability in y is accounted for by x.

The ordinary correlation coefficient r assesses the linear association between two variables x and y. In certain situations the variable y may increase (or decrease) with increases in x but not necessarily in a linear fashion. When this happens the correlation coefficient r will not depict the full extent of the relation between x and y. One way to handle this is to consider transforming the data; another way is to consider alternate models. These models will be discussed in multiple regression (Chapters 12 and 13). Another approach is to use the rank correlation coefficient, which measures the *monotonic* association between y and x. That is, the rank correlation coefficient measures whether y increases (or decreases) with x, even when the relation between y and x is not necessarily linear.

The rank order correlation coefficient is easy to calculate. We simply rank all x-values and all y-values separately and then calculate the ordinary correlation coefficient for the ranks. This correlation based on the ranks is called **Spearman's rank order correlation coefficient** r_s.

Spearman's rank order correlation coefficient r_s

EXAMPLE 8.8   A corporation examined the relationship between profits ($1000) and the percent of operating capacity being used for each of 12 plants.

Profits ($1000), y	% of Operating Capacity, x
2.5	50
6.2	57
3.1	61
4.6	68
7.3	77
4.5	80
6.1	82
11.6	85
10.0	89
14.2	91
16.1	95
19.5	99

a.   Plot the data. Are the data linear?
b.   Compute the rank correlation coefficient.

Solution

a.

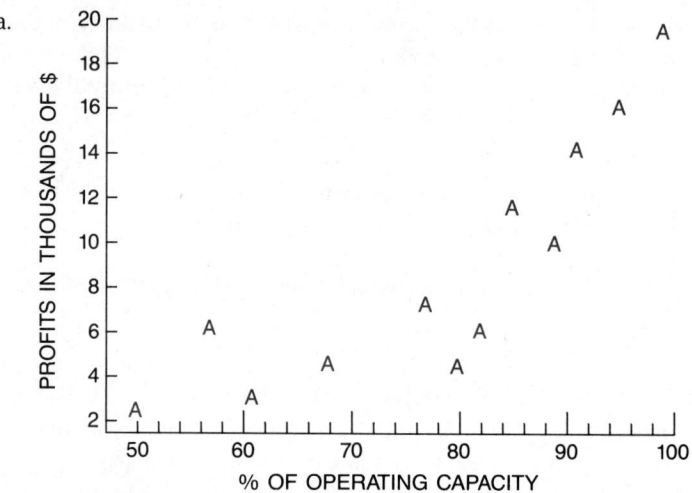

PLOT OF DATA

PLOT OF Y*X   LEGEND: A = 1 OBS, B = 2 OBS, ETC.

b. To compute r_s, first we need the ranks (from low to high) separately for each variable. These are shown here.

Rank on Profits	Rank on % Capacity
1	1
6	2
2	3
4	4
7	5
3	6
5	7
9	8
8	9
10	10
11	11
12	12

Then if we let y denote the ranks on profits and x denote rank on % of operating capacity, we compute r_s as we would r. To do this we need S_{yy}, S_{xx}, and S_{xy}.

For these data, it can be shown that

$$S_{yy} = 143$$

$$S_{xx} = 143$$

and

$$S_{xy} = 125.$$

Hence the rank correlation coefficient is

$$r_s = \frac{125}{\sqrt{143(143)}} = 0.874$$

For these data y increases with x, but as seen in the figure on p. 324, the relation between y and x is not linear.

If there are no ties in ranks for either of the two variables, there is a simpler formula for r_s that makes use of d_i, the difference between the y rank and x rank on observation i:

$$r_s = 1 - \frac{6 \sum d_i^2}{n(n^2 - 1)}$$

where n is the number of x_i, y_i observations.

EXAMPLE 8.9   Compute r_s for the data of Example 8.8 using the simpler computational formula.

Solution   The differences in ranks are shown here:

Rank on Profit	Rank on % Capacity	d_i
1	1	0
6	2	4
2	3	−1
4	4	0
7	5	2
3	6	−3
5	7	−2
9	8	1
8	9	−1
10	10	0
11	11	0
12	12	0

Then $\sum d_i^2 = 36$ and

$$r_s = 1 - \frac{6 \sum d_i^2}{n(n^2 - 1)} = 1 - \frac{6(36)}{12(143)} = .874$$

Note: This agrees with what we obtained in Example 8.8 except for rounding errors.

EXERCISES

8.11   An instructor believes that true–false tests are as effective as problem-type tests in judging a student's proficiency in mathematics. A test consisting of half true–false questions and half problems was given to 10 calculus students selected at random. The test score results were as follows:

Student	1	2	3	4	5	6	7	8	9	10
T–F questions	48	40	25	10	16	21	23	19	35	32
Problems	45	47	20	12	12	15	25	16	30	32

a. Plot the data. Are the data linear?

b. Calculate the correlation coefficient to measure the strength of the linear relation for the two sets of test scores.

8.12 Compute the rank correlation coefficient for the data of Exercise 8.7. Which measure of association seems more appropriate, r or r_s?

8.13 An experiment was conducted to investigate the amplitude of the shock wave recorded on sensors placed at different distances from an explosive charge. The charge is to be detonated underground, with three sensors placed at each of the three different distances from the charge, as illustrated in Figure 8.13. The shock-wave amplitudes are recorded and summarized according to the distance from the explosion. These data are given below.

Distance, x	5	5	5	10	10	10	15	15	15
Amplitude, y	8.6	8.2	8.1	5.8	6.2	6.1	5.2	4.8	4.7

a. Plot the sample data. Are the data linear?
b. Choose between r and r_s as a measure of the strength of the relation between distance and amplitude. Compute its value.

FIGURE 8.13

Location of Sensors from the Charge for the Shock-Wave Experiment of Exercise 8.13

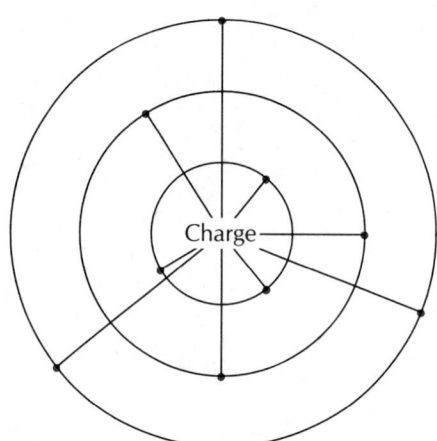

8.14 A forester was interested in training an assistant to estimate the timber volume of a standing tree. Having trained her, the forester calibrated the assistant's estimates against known timber volumes. Perhaps a better way to quantify the assistant's estimate would be to base it on an objective reading, such as the basal area of the tree. If, indeed, volume is related to basal area, the assistant would have an objective way to estimate the timber volume of a tree. A random sample of 12 trees was obtained. For each tree included in the sample, the basal area x was recorded along with the cubic-foot volume after the tree was felled. These data appear below.

Tree	1	2	3	4	5	6	7	8	9	10	11	12
Basal area, x	.3	.5	.4	.9	.7	.2	.6	.5	.8	.4	.8	.6
Volume, y	6	9	7	19	15	5	12	9	20	9	18	13

a. Plot the data.
b. Obtain the linear regression equation $\hat{y} = \hat{\beta}_0 + \hat{\beta}_1 x$.
c. Compute and interpret the correlation coefficient between basal area and timber volume.

8.15   An equal number of families from eight different cities of various sizes were asked how much money they spend for food, clothing, and housing per year. The city sizes and average family responses are summarized below. (City size in 1000s, expenditure in $1000s.)

City Size	30	50	75	100	150	200	175	120
Expenditure	65	77	79	80	82	90	84	81

a. Plot the data.
b. Compute the correlation coefficient r.

8.16   Compute r_s for the data of Exercise 8.15. Is the shortcut formula appropriate? Why or why not?

8.6   A LOOK AHEAD: MULTIPLE REGRESSION

Multiple regression will be discussed in detail in Chapters 12 and 13. In this section we will give you a brief idea of what lies ahead. Linear regression deals with situations in which a dependent variable can be expressed in a model as a linear function of a single quantitative independent variable. Multiple regression models, on the other hand, express the dependent variable using more than one independent variable or higher-degree terms than a single independent variable.* The adjective "multiple" refers to the fact that there is more than one term in the model (excluding the intercept).

Multiple regression models relating a response y to a single quantitative variable could be of the form

multiple regression: single x

$$y = \beta_0 + \beta_1 x_1 + \beta_2 x_1^2 + \varepsilon$$
$$y = \beta_0 + \beta_1 x_1 + \beta_2 x_1^2 + \beta_3 x_1^3 + \varepsilon$$

Plots of data for some of the previous examples and exercises of this chapter have indicated that a linear regression model may not be appropriate. For some of these situations a model with higher-degree terms could represent a viable alternative.

For other situations it is quite possible that more than one independent variable is related to the response and should be included in the model. For example, in studying the yield of a tomato crop, several independent variables—amount of fertilizer (x_1), amount of water (x_2), hours of sunlight on clear days (x_3)—could all have an effect on the yield. Hence to formulate a model that adequately represents the yield of a tomato crop, we should include all these terms in a model. Typical multiple regression models relating a response y to two quantitative independent variables, x_1 and x_2, are shown here:

multiple regression: two xs

$$y = \beta_0 + \beta_1 x_1 + \beta_2 x_2 + \varepsilon$$
$$y = \beta_0 + \beta_1 x_1 + \beta_2 x_2 + \beta_3 x_1 x_2 + \varepsilon$$
$$y = \beta_0 + \beta_1 x_1 + \beta_2 x_1^2 + \beta_3 x_2 + \beta_4 x_2^2 + \beta_5 x_1 x_2 + \beta_6 x_1^2 x_2$$
$$+ \beta_7 x_1 x_2^2 + \beta_8 x_1^2 x_2^2 + \varepsilon$$

Several of these models are illustrated in Figure 8.14.

* Some textbooks would not classify these models as multiple regression models.

FIGURE 8.14

Graphical Illustrations
of Several Models
Relating a Response y
to Two Independent
Variables x_1 and x_2

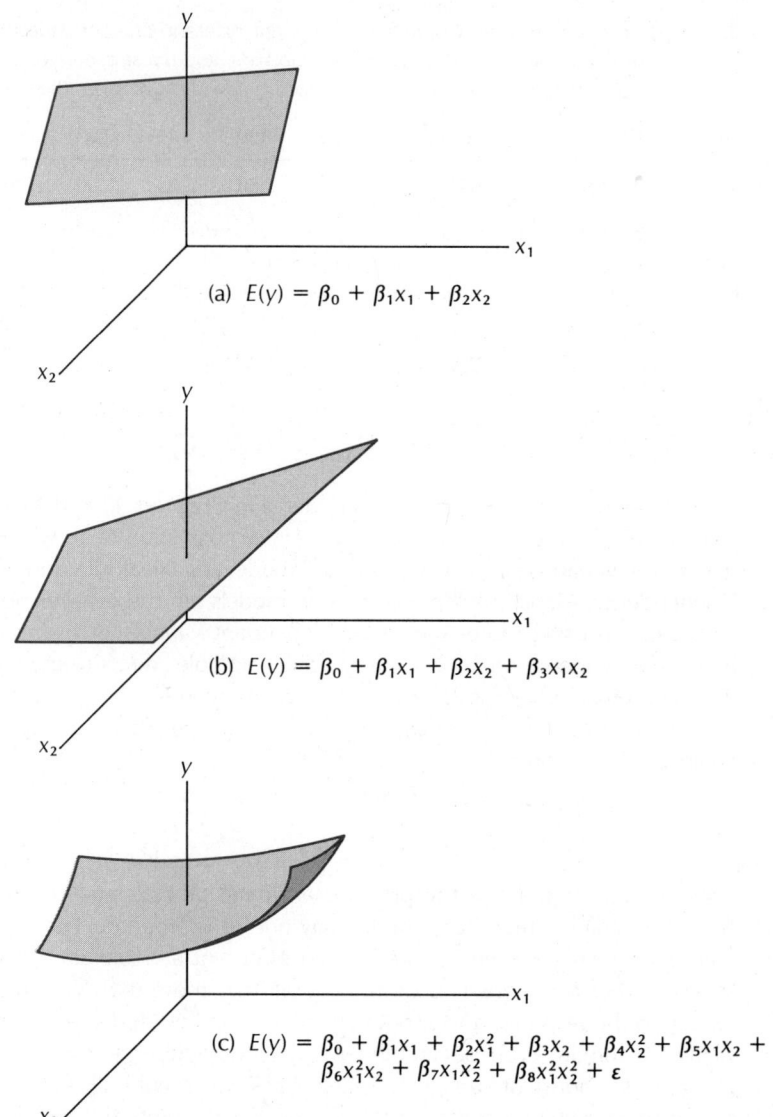

(a) $E(y) = \beta_0 + \beta_1 x_1 + \beta_2 x_2$

(b) $E(y) = \beta_0 + \beta_1 x_1 + \beta_2 x_2 + \beta_3 x_1 x_2$

(c) $E(y) = \beta_0 + \beta_1 x_1 + \beta_2 x_1^2 + \beta_3 x_2 + \beta_4 x_2^2 + \beta_5 x_1 x_2 + \beta_6 x_1^2 x_2 + \beta_7 x_1 x_2^2 + \beta_8 x_1^2 x_2^2 + \varepsilon$

Similarly, typical multiple regression models relating y to a set of three independent variables are as follows:

**multiple regression:
three xs**

$$y = \beta_0 + \beta_1 x_1 + \beta_2 x_2 + \beta_3 x_3 + \varepsilon$$

$$y = \beta_0 + \beta_1 x_1 + \beta_2 x_2 + \beta_3 x_3 + \beta_4 x_1 x_2 + \beta_5 x_1 x_3 + \beta_6 x_2 x_3 + \varepsilon$$

$$y = \beta_0 + \beta_1 x_1 + \beta_2 x_1^2 + \beta_3 x_2 + \beta_4 x_2^2 + \beta_5 x_3 + \beta_6 x_3^2 + \beta_7 x_1 x_2 + \beta_8 x_1 x_3 + \beta_9 x_2 x_3 + \varepsilon$$

Again, the particular choice of a model will depend on the actual experimental situation under study and the number of variables that affect the response. It suffices to say here that we will try to choose a model that best describes the relationship between the independent variables of interest and the response y.

degree

pth-degree term

Individual terms in a multiple regression model are classified by their exponents. The **degree** of a term is given by the sum of the exponents for the independent variables appearing in the term. Thus any independent variable x_i that appears in the regression model as x_i^p is called a **pth-degree term**. Hence x_i, x_i^2, x_i^3 are called first-, second-, and third-degree terms, respectively. Similarly, if two independent variables x_i and x_j appear together as $x_i^p x_j^q$, the term is called a $(p + q)$th-degree term. We would classify the terms $x_i x_j$, $x_i^2 x_j$, $x_i x_j^2$, $x_i^2 x_j^2$ as second-, third-, third-, and fourth-degree terms, respectively.

EXAMPLE 8.10

An experimenter believes that the multiple regression model

$$y = \beta_0 + \beta_1 x_1 + \beta_2 x_1^2 + \beta_3 x_2 + \beta_4 x_1 x_2 + \varepsilon$$

adequately represents the relationship between a dependent variable y and two independent variables x_1 and x_2. Identify the degree of all terms in the model containing one or more independent variables.

Solution

Following the procedures just discussed for assigning degrees to terms, we have the following:

Term	Degree
$\beta_1 x_1$	first
$\beta_2 x_1^2$	second
$\beta_3 x_2$	first
$\beta_4 x_1 x_2$	second

first-order model

We just indicated that individual terms of a multiple regression model are classified by their exponents. We can also identify specific models by the types of terms that appear in the model. A **first-order model** is a multiple regression model that contains all possible first-degree terms in the independent variables.*

EXAMPLE 8.11

Write a first-order multiple regression model relating a response y to the independent variables x_1, x_2, and x_3.

Solution

A first-order model in x_1, x_2, and x_3 includes all first-degree terms in these variables. The appropriate model is

$$y = \beta_0 + \beta_1 x_1 + \beta_2 x_2 + \beta_3 x_3 + \varepsilon$$

second-order model

A **second-order model** is a multiple regression model that includes all possible first- and second-degree terms in the independent variables.

EXAMPLE 8.12

Write a second-order model relating a dependent variable y to the independent variables x_1, x_2, and x_3.

---

* Some textbooks refer to these models as multiple linear regression models because they have only first-degree (linear) terms in the independent variables.

Solution | The appropriate second-order model would include the terms of the corresponding first-order model (Example 8.11) and all possible second-degree terms.

$$y = \beta_0 + \beta_1 x_1 + \beta_2 x_2 + \beta_3 x_3 + \beta_4 x_1^2 + \beta_5 x_2^2 + \beta_6 x_3^2 + \beta_7 x_1 x_2 + \beta_8 x_1 x_3 + \beta_9 x_2 x_3 + \varepsilon$$

Higher-order models represent obvious extensions of the first- and second-order models. Several exercises at the end of this chapter offer you additional practice in formulating these models.

8.7   SUMMARY

This chapter begins to lay the foundation for linear and multiple regression where a dependent variable y can be represented as a function of one or more independent variables. For the linear regression model $y = \beta_0 + \beta_1 x + \varepsilon$, we discussed a procedure for obtaining estimates of the parameters β_0 and β_1. The method of least squares chooses values for $\hat{\beta}_0$ and $\hat{\beta}_1$ in the linear regression line

$$\hat{y} = \hat{\beta}_0 + \hat{\beta}_1 x$$

so that the quantity $\sum (y - \hat{y})^2$ is minimized. For a given data set, these values are computed as

$$\hat{\beta}_1 = \frac{S_{xy}}{S_{xx}} \quad \text{and} \quad \hat{\beta}_0 = \bar{y} - \hat{\beta}_1 \bar{x}$$

where

$$S_{xx} = \sum x^2 - \frac{(\sum x)^2}{n}$$

and

$$S_{xy} = \sum xy - \frac{(\sum x)(\sum y)}{n}$$

We also introduced a quick, portable way to approximate a linear regression equation and transformations for the independent and dependent variables, which may help to linearize data.

The correlation coefficient, which is closely related to the estimated slope $\hat{\beta}_1$, was presented as a measure of the strength of the linear relationship. When two variables are related in a positive or negative fashion, but not necessarily linearly related, the rank correlation coefficient r_s can be used as a measure of the strength of the monotonic relation between y and x. For r_s, ranks rather than original values for x and y are used in the calculations.

Finally we presented a brief look at what's ahead beyond linear regression. After learning how to make statistical inferences (estimation and statistical tests) for linear regression in Chapter 9 we will be in a position to expand the model to include more than one independent variable. Model relations and methods of inferences for multiple regression are presented in Chapters 12 and 13.

KEY FORMULAS: LINEAR REGRESSION AND CORRELATION	

KEY FORMULAS: LINEAR REGRESSION AND CORRELATION

1. Formulas for least squares estimates, $\hat{\beta}_0$ and $\hat{\beta}_1$

$$\hat{\beta}_1 = \frac{S_{xy}}{S_{xx}} \quad \text{and} \quad \hat{\beta}_0 = \bar{y} - \hat{\beta}_1 \bar{x}$$

where

$$S_{xx} = \sum x^2 - \frac{(\sum x)^2}{n}$$

and

$$S_{xy} = \sum xy - \frac{(\sum x)(\sum y)}{n}$$

2. Correlation coefficient

$$r = \frac{S_{xy}}{\sqrt{S_{xx}S_{yy}}} \quad \text{or} \quad r = \hat{\beta}_1 \sqrt{\frac{S_{xx}}{S_{yy}}}$$

3. Rank order correlation coefficient (Spearman's)

$$r_s = 1 - \frac{6 \sum d_i^2}{n(n^2 - 1)}$$

when there are no tied ranks for either the x or y variables.

EXERCISES

8.17 Suppose that a response y is related to two independent variables x_1 and x_2.
a. Write a first-order regression model.
b. Write three different regression models relating y to x_1 and x_2 using first- and second-degree terms.

8.18 Identify the degree for all terms in the model

$$y = \beta_0 + \beta_1 x_1 + \beta_2 x_1^2 + \beta_3 x_2 + \beta_4 x_1 x_2 + \beta_5 x_1^2 x_2 + \varepsilon$$

8.19 Sketch the deterministic model

$$y = 1.5 + 1.0x^2$$

for x in the range $-3 \leq x \leq 3$. (Hint: Substitute different values for x into the model to determine corresponding values for y. Plot the x and y points.)

8.20 Sketch the deterministic model

$$y = 1.5 + 2.5x + 1.0x^2$$

for x in the range $-3 \leq x \leq 3$.

8.21 Sketch the deterministic model

$$y = 1.5 - 2.5x + 1.0x^2$$

for x in the range $-3 \leq x \leq 3$. Compare this sketch to that for Exercise 8.20.

8.22   Refer to the data of Exercise 8.13.
   a. Determine an approximate linear regression line using the method of Section 8.3.
   b. Determine the linear regression line using least squares estimates for β_0 and β_1. Compare your results to part (a).

8.23   Refer to Exercise 8.14. Use the method of Section 8.3 to obtain an approximate linear regression line. How does it compare to the linear regression line shown for the same data in the Minitab output displayed here?

```
        SET AREA INTO C1
DATA> .3 .5 .4 .9 .7 .2 .6 .5 .8 .4 .8 .6
DATA> END DATA
MTB > SET VOLUME INTO C2
DATA> 6 9 7 19 15 5 12 9 20 9 18 13
DATA> END DATA
MTB > PRINT C1 C2
  ROW    C1     C2

    1    0.3     6
    2    0.5     9
    3    0.4     7
    4    0.9    19
    5    0.7    15
    6    0.2     5
    7    0.6    12
    8    0.5     9
    9    0.8    20
   10    0.4     9
   11    0.8    18
   12    0.6    13


MTB > BRIEF 2
MTB > REGRESS Y IN C2 ON 1 PREDICTOR IN C1

The regression equation is
C2 = - 1.23 + 23.4 C1


Predictor       Coef       Stdev      t-ratio
Constant       -1.234      1.080       -1.14
C1             23.404      1.815       12.90


s = 1.295      R-sq = 94.3%     R-sq(adj) = 93.8%

Analysis of Variance

SOURCE        DF         SS          MS
Regression     1       278.90      278.90
Error         10        16.77        1.68
Total         11       295.67

Unusual Observations
Obs.     C1        C2      Fit Stdev.Fit  Residual   St.Resid
   9   0.800    20.000    17.489   0.576     2.511      2.17R

R denotes an obs. with a large st. resid.


MTB > STOP
*** Minitab Release 5.1.3 *** Minitab, Inc. ***
```

EXERCISES

8.24 Refer to Example 8.8. Compute the correlation coefficient r for these data. Distinguish between the interpretation of r and r_s for these data.

8.25 Consider the multiple regression model

$$y = \beta_0 + \beta_1 x_1 + \beta_2 x_1^2 + \beta_3 x_1^3 + \beta_4 x_1^4 + \varepsilon$$

a. Specify the degree of each term in the model.
b. What is the order of the model?

8.26 Write a second-order model relating a response y to the independent variables x_1, x_2, and x_3.

8.27 Consider the multiple regression model

$$y = \beta_0 + \beta_1 x_1 + \beta_2 x_1^2 + \beta_3 x_2 + \beta_4 x_1 x_2 + \varepsilon$$

a. Specify the degree of each term in the general linear model.
b. Is this a first- or second-order model? Explain.

8.28 Write a second-order model relating a response y to four independent variables (x_1, x_2, x_3, and x_4).

8.29 Sketch plots for these deterministic models for $-1 \le x \le 1$:
a. $y = 1.2 + x$
b. $y = 1.2 + x + .4x^2$
c. $y = 1.2 + x + .4x^2 + .6x^3$

8.30 Sketch a graph using the five data points given below.

x	1	2	3	4	5
y	5	10	14	21	26

8.31 Refer to the sketch in Exercise 8.30. Write a regression model relating y to x.

8.32 Sketch a graph of the response y as a function of the independent variable x using the five data points given below.

x	0	-1	2	3	-2
y	15	12	31	50	11

8.33 Using the sketch drawn in Exercise 8.32, indicate the form (without values for the βs) of a multiple regression model relating the response y to the independent variable x.

8.34 If a second-order model relating y to x has one peak (or, equivalently, one valley) when sketched, and a third-order model has one peak and one valley when sketched, how many peaks and valleys do you think a fourth-order model has when sketched? Sketch a typical fourth-order model relating a response y to an independent variable x.

8.35 Earnings from a particular stock are listed below for the past seven years.

Year	1987	1986	1985	1984	1983	1982	1981
Earnings per share	2.30	1.80	1.50	1.20	1.05	1.10	1.20

Sketch these seven data points.

8.36 Refer to the sketch in Exercise 8.35.
a. Suggest an appropriate multiple regression model relating earnings per share to the independent variable "year."
b. Compute the rank order correlation coefficient r_s.

8.37 Yields in bushels of tomatoes are shown here for 12 equal-sized plots.
a. Plot the data.
b. Based on the plot, specify a linear or multiple regression model that may be appropriate.
c. Compute either r or r_s depending on the model selected in (b).

Yield of 12 Equal-Sized Plots of Tomato Plantings for
Different Amounts of Fertilizer

Plot	Yield, y (in Bushels)	Amount of Fertilizer, x (in Pounds per Plot)
1	24	12
2	18	5
3	31	15
4	33	17
5	26	20
6	30	14
7	20	6
8	25	23
9	25	11
10	27	13
11	21	8
12	29	18

8.38   An experiment was conducted to examine the weight gain for chickens treated with various doses of a growth promotant. From a group of 15 chickens of approximately the same age and weight, a random assignment of three chickens was made to each dose group. The specified dose x (mg/kg) of growth promotant was added to the feed daily for a fixed period of time. Weight gains (in lb) are shown here.

Dose x	Weight Gain y		
0 control	1.5	1.8	1.7
0.2	2.3	2.0	1.8
0.4	4.3	3.7	4.1
0.8	5.7	5.9	6.2
1.6	7.9	7.7	7.5

a. Plot the sample data.
b. Fit the data to obtain the linear regression equation $\hat{y} = \hat{\beta}_0 + \hat{\beta}_1 x$.

8.39   Refer to Exercise 8.38. Compute the correlation coefficient r as a measure of the strength of the linear relation between dose and weight gain. What economic implications do you see for this promotant?

8.40   The correlation coefficient can sometimes be used to measure the reliability (consistency or reproducibility) of a test. This can be done by examining the same individuals with the same test on two separate occasions or by examining the same individuals on two different, equivalent tests. Discuss the pros and cons of these two methods for measuring the reliability of a test. Is one method to be preferred to the other?

8.41   Each student in a group of 100 second-grade students was administered two equivalent forms of a word fluency test. The test required that a student write as many words as possible beginning with a given letter during a five-minute period. The second test was the same as the first except that the letter was changed. The sample correlation coefficient was .75; interpret this result. What factors might contribute to the less-than-complete agreement between the two test results?

8.42   A sociologist working for the government of a large city collected data on the number of nonviolent crimes (in 1000s) reported and the increase (or decrease) of all crimes over the previous reporting period. Quarterly data are shown here:

Quarter	Nonviolent Crimes	Increase (or Decrease) in All Crimes
1	7.2	14.1
2	6.4	14.5
3	6.6	13.3
4	7.3	13.6
5	7.5	15.2
6	6.9	15.7
7	7.1	15.3
8	7.4	14.8
9	7.6	16.1
10	7.3	16.6
11	7.1	16.2
12	7.0	15.9

a. Plot the nonviolent crime data versus quarter. Also plot the increase versus quarter on the same graph.

b. Does there appear to be a relationship between the two crime variables?

c. Compute the correlation coefficient between nonviolent crimes and the increase in all crimes.

8.43 Refer to Exercise 8.42.

a. Lag the increase in crimes by one quarter and superimpose these data on the plot of nonviolent crimes by quarter.

b. Compute the correlation coefficient of the nonviolent crimes and lagged increase in crimes. (Note: You will only have 11 data points corresponding to quarters 1–11.)

 8.44 A study was conducted to examine the efficiencies of various manufacturing sites of a large corporation. At each site, the average number of acceptable cartons of manufactured goods per month was recorded, as was the average number of hours of assembly-line operation per month. These data are shown here.

Location	Average Number of Acceptable Cartons (000) y	Average Number of Hours of Line Operation x
1	12	20
2	11	38
3	15	40
4	16	45
5	20	57
6	18	68
7	22	74
8	26	79
9	20	81
10	21	86
11	27	93
12	32	104
13	33	110
14	34	120
15	31	138

a. Plot the sample data.
b. Obtain a least squares fit to these data, using a linear regression model.
c. Plot the least squares prediction equation on the graph of part (a). Does this model appear to fit the data adequately?

8.45   Refer to the data of Exercise 8.44. Suppose the last four data points corresponding to locations 12, 13, 14, and 15 were as shown here, rather than as indicated in the previous exercise.

Location	y	x
12	25	104
13	23	110
14	20	120
15	15	138

a. Plot the entire new data set for the 15 locations.
b. Would a linear regression model fit these data well? If not, can you suggest a transformation of the x variable to linearize the plot? Plot the transformed data.
c. Refer to part (b). Fit the transformed data using a linear regression model.

8.46   Refer to Exercise 8.45. What other way might you have linearized the plot using a transformation? Plot these transformed data as well. Which transformation seems more suitable for these data?

8.47   Refer to the data of Exercise 8.44. Obtain an appropriate linear regression equation using the median approach of Section 8.3. How close is this approximation to the least squares fit of Exercise 8.44?

 8.48   The fuel of a new four-cylinder diesel engine was studied under various external, controlled operating temperatures. For each setting, two different engines were studied and the fuel consumption recorded.

Observation	External Temperature (°F) x	Fuel Consumption (Gallons) y
1	20	25
2	20	26
3	30	28
4	30	27
5	40	32
6	40	35
7	50	42
8	50	46
9	60	55
10	60	53
11	70	55
12	70	57
13	80	60
14	80	58
15	90	61
16	90	58

a. Plot the sample data. Do you think a linear regression line will be an adequate model?

b. Fit the least squares regression model $y = \beta_0 + \beta_1 x + \varepsilon$, and draw the prediction equation on the graph of part (a).

c. What's the sample correlation coefficient for these data?

8.49 Refer to Exercise 8.48. Since there does not appear to be a simple, suitable transformation on one or both of the variables to linearize the data, rank the data separately and compute the rank order correlation coefficient measure.

8.50 The maximum volume of oxygen uptake (VO_2 max) has been used as a measure of cardiac status in healthy individuals as well as in persons suffering from cardiac-related illnesses (such as congestive heart failure). The VO_2 max readings for 12 healthy adult males following strenuous exercise are recorded here. In general, VO_2 max decreases with any increase in activity level.

Individual	VO_2 Max y	Duration of Exercise (Minutes) x
1	82	10.0
2	73	9.5
3	68	10.2
4	74	10.5
5	66	11.0
6	63	11.3
7	58	11.6
8	54	12.0
9	56	12.1
10	51	12.5
11	55	12.8
12	44	13.0

a. Plot the data.

b. Does a linear regression equation seem to be appropriate for these data?

c. Fit the data using the model $y = \beta_0 + \beta_1 x + \varepsilon$.

8.51 Refer to Exercise 8.50.

a. Obtain the least squares prediction for the same data, except that the final observation is $x = 13$, $y = 30$.

b. How well does a linear regression equation fit these data?

c. What effect does the observation from part (a) have on the prediction equation, compared to the corresponding prediction equation of Exercise 8.50? (Hint: Plot both the regression equations.)

9

INFERENCES RELATED TO LINEAR REGRESSION AND CORRELATION

INTRODUCTION

In Chapter 8 we gave formulas for finding the least squares estimates for β_0 and β_1 in the linear regression model

$$y_i = \beta_0 + \beta_1 x_i + \varepsilon_i \qquad i = 1, 2, \ldots, n$$

We would now like to use these estimates to make inferences about the relationship between y and x. For example, suppose that x represents the amount of force applied to a one-foot section of steel and y denotes the corresponding increase in width of the steel sample. Applying the results of Chapter 8, we would obtain a random sample of n observations and compute the least squares estimates for β_0 and β_1 as

$$\hat{\beta}_1 = \frac{S_{xy}}{S_{xx}} \qquad \text{and} \qquad \hat{\beta}_0 = \bar{y} - \hat{\beta}_1 \bar{x}$$

It might be of interest in this problem to test to see if there is a positive linear relationship between x and y. To do this we could conduct a statistical test of the null hypothesis H_0: $\beta_1 = 0$ against the alternative hypothesis H_a: $\beta_1 > 0$. The estimate $\hat{\beta}_1$ will be used in this test.

9.2 INFERENCES ABOUT β_0 AND β_1

Before we can make any inferences about parameters in the linear regression model, we need to expand on the assumptions we have for the model. Previously, we assumed that the random error term ε_i associated with observation y_i has expectation zero. In addition, we will assume the following:

1. $\varepsilon_1, \varepsilon_2, \ldots, \varepsilon_n$ are independent of each other.
2. ε for a given setting of the independent variable x is normally distributed with mean 0 and variance σ_ε^2. The variance σ_ε^2 is constant for all settings of x.

These two assumptions imply that y_i is normally distributed with mean $\beta_0 + \beta_1 x_i$ and constant variance σ_ε^2 and that y_i and y_j are independent. See Figure 9.1.

expected values   Under these assumptions, both $\hat{\beta}_0$ and $\hat{\beta}_1$ have sampling distributions that are normal with means (called **expected values**) and standard errors as shown here.

EXPECTED VALUES AND STANDARD ERRORS FOR $\hat{\beta}_0$ AND $\hat{\beta}_1$ IN LINEAR REGRESSION

$$\mu_{\hat{\beta}_0} = \beta_0 \qquad \mu_{\hat{\beta}_1} = \beta_1$$

$$\sigma_{\hat{\beta}_0} = \sigma_\varepsilon \sqrt{\frac{\sum x^2}{n S_{xx}}} \qquad \sigma_{\hat{\beta}_1} = \frac{\sigma_\varepsilon}{\sqrt{S_{xx}}}$$

FIGURE 9.1

Illustration of the Two Assumptions for Linear Regression

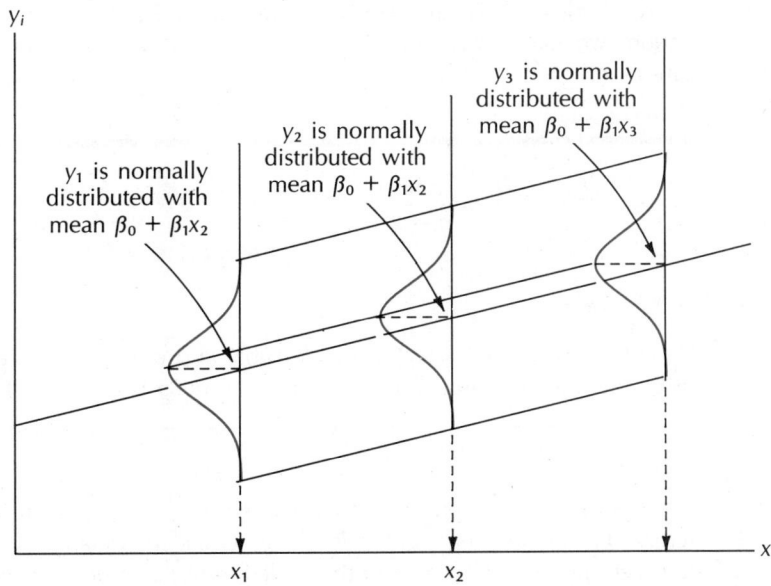

y_i

y_1 is normally distributed with mean $\beta_0 + \beta_1 x_2$

y_2 is normally distributed with mean $\beta_0 + \beta_1 x_2$

y_3 is normally distributed with mean $\beta_0 + \beta_1 x_3$

x_1 x_2 x

EXAMPLE 9.1

In examining the weight loss of a compound, a chemist hypothesized that weight loss y (in pounds) is linearly related to the relative humidity x of the room in which the process operates. From a sample of 12 observations, we find $\bar{y} = 4.8$, $\bar{x} = 6.5$, $S_{xy} = 138.2$, $S_{xx} = 101$, and $\sum x^2 = 608.1$. Compute $\hat{\beta}_0$, $\hat{\beta}_1$, and their standard errors.

Solution

For these data

$$\hat{\beta}_1 = \frac{S_{xy}}{S_{xx}} = 1.37$$

$$\sigma_{\hat{\beta}_1} = \frac{\sigma_\varepsilon}{\sqrt{S_{xx}}} = \frac{\sigma_\varepsilon}{10.05}$$

and

$$\hat{\beta}_0 = \bar{y} - \hat{\beta}_1\bar{x} = 4.8 - 1.37(6.5) = -4.11$$

$$\sigma_{\hat{\beta}_0} = \sigma_\varepsilon\sqrt{\frac{\sum x^2}{nS_{xx}}} = .71\sigma_\varepsilon$$

The problem with our solution to Example 9.1 is that we still don't have the standard errors for $\hat{\beta}_0$ and $\hat{\beta}_1$ because we don't know σ_ε, the standard deviation of the εs, or we don't have an estimate of its value. In Chapter 8 we showed that

$$\underbrace{\sum (y - \bar{y})^2}_{\substack{\text{sum of squares} \\ \text{about the mean}}} = \underbrace{\sum (\hat{y} - \bar{y})^2}_{\substack{\text{sum of squares} \\ \text{for regression}}} + \underbrace{\sum (y - \hat{y})^2}_{\substack{\text{sum of squares} \\ \text{for error}}}$$

This last quantity is called by many different names: residual sum of squares, sum of squares about the regression, and sum of squares for error (SSE), to name just a few. For the remainder of this text, we will use the term SSE. Dividing SSE by $n - 2$ degrees of freedom we can obtain an estimate of σ_ε^2 in linear regression. We designate this estimate as s_ε^2.

ESTIMATE OF σ_ε^2 AND σ_ε FOR LINEAR REGRESSION

$$s_\varepsilon^2 = \frac{\sum (y - \hat{y})^2}{n - 2} = \frac{SSE}{n - 2}$$

$$s_\varepsilon = \sqrt{\frac{SSE}{n - 2}}$$

Note: If computer software is not available and calculations are done by hand or by using a calculator, use the shortcut formula $SSE = S_{yy} - \hat{\beta}_1 S_{xy}$.

EXAMPLE 9.2

The yield per plot in bushels of corn was observed on $n = 10$ plots that had been fertilized in varying degrees. We let the independent variable x denote the amount of fertilizer applied. The data and the coded fertilizer values are recorded in Table 9.1. Use the sample data of Table 9.1 to obtain the least squares prediction equation $\hat{y}$

TABLE 9.1 Corn Yield Data for Example 9.2

Yield (in Bushels)	Fertilizer (in Pounds per Plot)
12	2
13	2
13	3
14	3
15	4
15	4
14	5
16	5
17	6
18	6

for the linear regression model $y = \beta_0 + \beta_1 x + \varepsilon$. Also, calculate estimates of σ_ε^2 and σ_ε.

Solution For these data (using a computer or hand computations) it is easily seen that

$$n = 10 \qquad \sum x = 40$$
$$\sum y = 147 \qquad \bar{x} = 4.0$$
$$\bar{y} = 14.7 \qquad \sum x^2 = 180$$
$$\sum y^2 = 2193 \qquad \sum xy = 611$$

Substituting into appropriate linear regression formulas we find the least square estimates for β_1 and β_0 to be, respectively,

$$\hat{\beta}_1 = \frac{S_{xy}}{S_{xx}} = \frac{611 - \dfrac{40(147)}{10}}{180 - \dfrac{(40)^2}{10}} = \frac{23}{20} = 1.15$$

and

$$\hat{\beta}_0 = \bar{y} - \hat{\beta}_1 \bar{x} = 14.7 - 1.15(4.0) = 10.10$$

The estimate for σ_ε^2 (and σ_ε) requires that we find

$$S_{yy} = \sum y^2 - \frac{(\sum y)^2}{n} = 2193 - \frac{(147)^2}{10} = 32.10$$

and

$$SSE = S_{yy} - \hat{\beta}_1 S_{xy} = 32.10 - 1.15(23) = 5.65$$

The estimate for σ_ε^2 from these data is

$$s_\varepsilon^2 = \frac{SSE}{n-2} = \frac{5.65}{8} = 0.71$$

and hence $s_\varepsilon = 0.84$.

EXAMPLE 9.3

Refer to Example 9.2 to compute the *estimated* standard errors for $\hat{\beta}_0$ and $\hat{\beta}_1$.

Solution

The formulas for the standard errors are, respectively,

$$\sigma_{\hat{\beta}_0} = \sigma_\varepsilon \sqrt{\frac{\sum x^2}{nS_{xx}}} \quad \text{and} \quad \sigma_{\hat{\beta}_1} = \frac{\sigma_\varepsilon}{\sqrt{S_{xx}}}$$

Using $s_\varepsilon = 0.84$ as the estimate of σ_ε and substituting into the above formulas, we obtain the estimated standard errors for $\hat{\beta}_0$ and $\hat{\beta}_1$

$$\hat{\sigma}_{\hat{\beta}_0} = s_\varepsilon \sqrt{\frac{\sum x^2}{nS_{xx}}} = 0.84 \sqrt{\frac{180}{10(20)}} = 0.80$$

and

$$\hat{\sigma}_{\hat{\beta}_1} = \frac{s_\varepsilon}{\sqrt{S_{xx}}} = \frac{0.84}{4.47} = 0.19$$

estimated standard errors

Note that by substituting s_ε for σ_ε in the formulas for $\sigma_{\hat{\beta}_1}$ and $\sigma_{\hat{\beta}_0}$ we have obtained the **estimated standard errors** for $\hat{\beta}_1$ and $\hat{\beta}_0$. In practice, since we will never know σ_ε and hence will always substitute an estimate of its value in the standard error formulas, we will drop the word "estimated" and simply call $\hat{\sigma}_{\hat{\beta}_1}$ and $\hat{\sigma}_{\hat{\beta}_0}$ the standard errors for $\hat{\beta}_1$ and $\hat{\beta}_0$, respectively.

An additional assumption for the ε_i will allow us to construct confidence intervals and statistical tests for β_1 and β_0. If we assume that the ε_i from the linear regression model

$$y_i = \beta_0 + \beta_1 x_i + \varepsilon_i$$

are normally distributed, we can specify confidence intervals for β_0 and β_1 using the formula, estimate $\pm t$ standard error. Fortunately, for large sample sizes, the confidence intervals are robust against modest departures from normality; other assumptions will be discussed in Section 9.6.

$100(1 - \alpha)\%$ CONFIDENCE INTERVALS FOR β_0 AND β_1 IN LINEAR REGRESSION

$$\hat{\beta}_0 \pm t_{\alpha/2} s_\varepsilon \sqrt{\frac{\sum x^2}{nS_{xx}}}$$

and

$$\hat{\beta}_1 \pm t_{\alpha/2} \frac{s_\varepsilon}{\sqrt{S_{xx}}}$$

where

$$s_\varepsilon = \sqrt{\frac{SSE}{n - 2}}$$

EXAMPLE 9.4

Use the data from Example 9.2 to develop 95% confidence intervals for β_0 and β_1.

Solution

The calculations from Example 9.2 yielded the linear regression equation $\hat{y} = 10.10 + 1.15x$. The $t_{.025}$ value for df = 8 is 2.306, $s_\varepsilon = 0.84$, $S_{xx} = 20$, and $\sum x^2 = 180$. Substituting these values into the appropriate formulas we obtain the 95% confidence intervals shown here.

$$\beta_0: \quad 10.10 \pm 2.306(0.84)\sqrt{\frac{180}{10(20)}} \quad \text{or} \quad 10.10 \pm 1.84$$

$$\beta_1: \quad 1.15 \pm 2.306\frac{(0.84)}{\sqrt{20}} \quad \text{or} \quad 1.15 \pm 0.43$$

In other words we are 95% confident that the true value of the intercept β_0 lies somewhere in the interval $8.26 \le \beta_0 \le 11.94$. Similarly, we are 95% confident that the true slope β_1 lies somewhere in the interval $0.72 \le \beta_1 \le 1.58$.

EXAMPLE 9.5

A restaurant operating on a "reservations only" basis would like to use the number of advance reservations x to predict the number of dinners y to be prepared. Data on reservations and number of dinners served for one day chosen at random from each week in a 100-week period gave the following results:

$$\bar{x} = 150 \qquad\qquad \bar{y} = 120$$
$$\sum (x - \bar{x})^2 = 90{,}000 \qquad \sum (y - \bar{y})^2 = 70{,}000$$
$$\sum (x - \bar{x})(y - \bar{y}) = 60{,}000$$

a. Find the least squares estimates $\hat{\beta}_0$ and $\hat{\beta}_1$ for the linear regression line $\hat{y} = \hat{\beta}_0 + \hat{\beta}_1 x$.
b. Predict the number of meals to be prepared if the number of reservations is 135.
c. Construct a 90% confidence interval for the slope. Does information on x (number of advance reservations) help in predicting y (number of dinners prepared)?

Solution

a. The least squares estimates are given by

$$\hat{\beta}_1 = \frac{S_{xy}}{S_{xx}} = \frac{60{,}000}{90{,}000} = 0.67$$

and

$$\hat{\beta}_0 = \bar{y} - \hat{\beta}_1 \bar{x} = 120 - 0.67(150) = 19.50$$

b. The predicted number of meals required for the number of advance reservations equal to 135 is
$$\hat{y} = 19.50 + 0.67(135) = 109.95 \qquad \text{or} \qquad 110$$

c. The 90% confidence interval for β_1 uses the formula

$$\hat{\beta}_1 \pm t \text{ standard error}$$

where the standard error is $s_\varepsilon/\sqrt{S_{xx}}$.

Although Table 4 in the Appendix does not list a t-value for $a = .05$ and df = 98, we'll use the t-value for the next higher df (df = 120); this value is 1.658.

The standard deviation s_ε can be computed using the summary sample data

$$s_\varepsilon^2 = \frac{SSE}{n-2}$$

where

$$\begin{aligned} SSE &= S_{yy} - \hat{\beta}_1 S_{xy} \\ &= 70,000 - 0.67(60,000) \\ &= 29,800 \end{aligned}$$

Thus

$$s_\varepsilon = \sqrt{\frac{29,800}{98}} = \sqrt{304.08} = 17.44$$

and the 90% confidence interval for β_1 is

$$0.67 \pm 1.658 \frac{(17.44)}{\sqrt{90,000}}$$

or

$$0.67 \pm .10$$

Since we are 90% confident that the true value of β_1 lies somewhere in the interval $.57 \le \beta_1 \le .77$ and since $\beta_1 = 0$ does not lie in this interval, it appears that the number of advance reservations is a useful predictor of the number of meals to be prepared in the context of a linear regression model, $y = \beta_0 + \beta_1 x + \varepsilon$.

Statistical tests for β_0 and β_1 use a t statistic of the form $t = $ estimate/standard error. Three different research hypotheses are presented along with the corresponding rejection regions. For a particular experimental situation, we must choose one of the specific alternatives shown here.

STATISTICAL TESTS FOR β_0 AND β_1

Intercept, β_0	Slope, β_1
H_0: $\beta_0 = 0$	H_0: $\beta_1 = 0$
H_a: 1. $\beta_0 > 0$	H_a: 1. $\beta_1 > 0$
2. $\beta_0 < 0$	2. $\beta_1 < 0$
3. $\beta_0 \ne 0$	3. $\beta_1 \ne 0$
T.S.: $t = \dfrac{\hat{\beta}_0}{s_\varepsilon \sqrt{\dfrac{\sum x^2}{nS_{xx}}}}$	T.S.: $t = \dfrac{\hat{\beta}_1}{\dfrac{s_\varepsilon}{\sqrt{S_{xx}}}}$
R.R.: For a given value of α and df $= n - 2$ 1. reject H_0 if $t > t_\alpha$ 2. reject H_0 if $t < -t_\alpha$ 3. reject H_0 if $|t| > t_{\alpha/2}$	R.R.: Same as those shown for β_0

EXAMPLE 9.6

Refer to the data of Example 9.5. Confirm the conclusion we reached concerning β_1 by conducting a test of $H_0: \beta_1 = 0$ versus $H_a: \beta_1 \neq 0$. Use $\alpha = .10$.

Solution

The parts of the statistical test are given here

$H_0: \beta_1 = 0$

$H_a: \beta_1 \neq 0$

T.S.: $t = \dfrac{\hat{\beta}_1}{\dfrac{s_\varepsilon}{\sqrt{S_{xx}}}} = \dfrac{0.67}{\dfrac{17.44}{\sqrt{90,000}}} = 11.53$

R.R.: For a two-tailed test with $\alpha = .10$ and df = 98, we will reject H_0 if $|t| > 1.645$.

Conclusion: Since $t = 11.53$ is greater than 1.645, we have sufficient evidence to reject H_0. It does appear that x is useful in predicting y.

EXAMPLE 9.7

The laboratory of a hospital participating in the clinical trial of an antibiotic drug had to be validated to see that the laboratory personnel could accurately assay blood samples "spiked" with fixed amounts of the antibiotic. The validation consisted of the following experiment. Thirteen spiked samples (with amounts known only to the study investigator) were sent to the laboratory to be assayed for the amount of the antibiotic present. The results of the validation experiment are shown here. (Note: The spiked samples with known amounts added were supplied in a blinded fashion to the laboratory.) The amounts found are the assay results found by the hospital laboratory.

Amount Added ($\mu g/ml$) x	Amount Found ($\mu g/ml$) y
0	0
5	4.5
5	5.0
5	4.8
10	8.9
10	8.9
10	8.9
20	17.0
20	18.2
20	15.4
40	32.6
40	36.1
40	31.5

a. Plot the sample data.
b. Fit these data to a linear regression model.

c. Test the null hypothesis $H_0: \beta_1 = 1$ versus $H_a: \beta_1 \neq 1$. Give the p-value for your test.

Solution

a. A plot of the sample data is shown here.

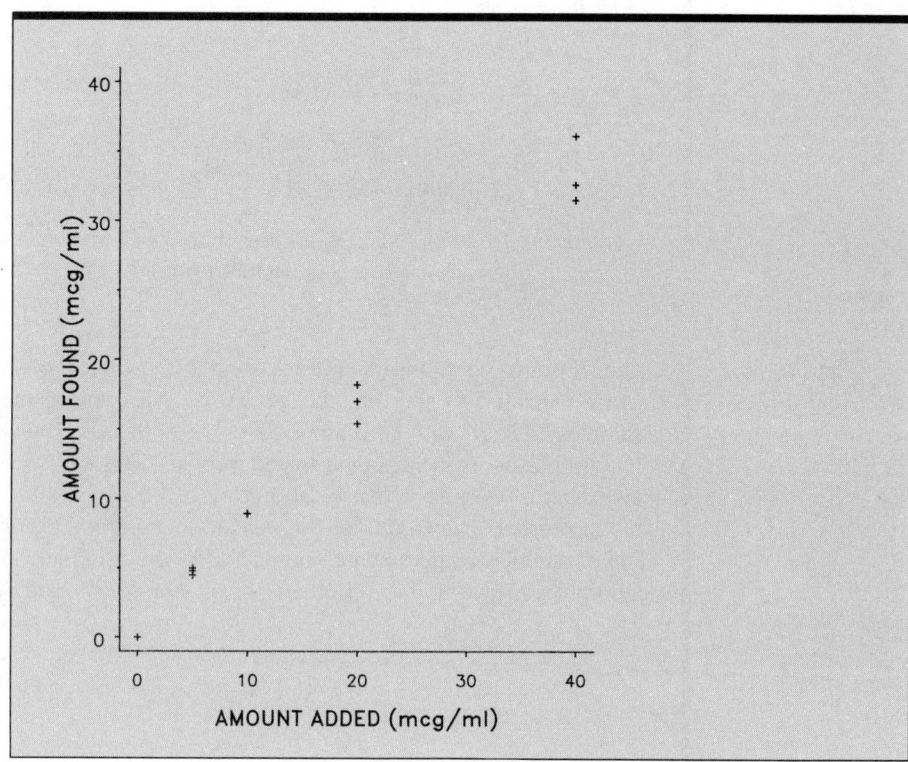

b. For the data, it can be shown that $\hat{\beta}_1 = .822$ with a standard error of $s_\varepsilon / \sqrt{S_{xx}} = .024$; the estimate of β_0 is $\hat{\beta}_0 = .529$ with a standard error of .537. The linear regression equation is then

$$\hat{y} = .529 + .822x$$

c. The statistical test for $H_0: \beta_1 = 1$ is a slight variation of the test for $H_0: \beta_1 = 0$. The test statistic for this variation is

$$t = \frac{\hat{\beta}_1 - \beta_{1,0}}{s_\varepsilon / \sqrt{S_{xx}}}$$

where $\beta_{1,0}$ is the hypothesized value of β_1 under H_0. The test of $H_0: \beta_1 = 1$ is shown here.

$H_0: \beta_1 = 1$
$H_a: \beta_1 \neq 1$

T.S.: $t = \dfrac{.822 - 1}{.024} = -7.42$

R.R.: Based on df $= 11$ and Table 4 of the Appendix, the p-value for the test result is $p < .001$.

A statistical test about β_1 can be restated in terms of an F statistic and put into the format of an "analysis of variance." Recall the total variability can be partitioned into two components: the sum of squares due to regression (SSREG) and the sum of squares for error (SSE). As indicated previously, the larger SSREG is relative to SSE, the better the model "fits" the data. It should come as no surprise, then, that we can form a test statistic based on SSREG and SSE. In particular, for

$$H_0: \beta_1 = 0$$

and

$$H_a: \beta_1 \neq 0$$

we can form an F statistic as

$$F = \frac{\text{SSREG}/1}{\text{SSE}/(n-2)} = \frac{\text{SSREG}}{s_\varepsilon^2}$$

which, under H_0, follows an F distribution with $df_1 = 1$ and $df_2 = n - 2$.

The F statistic shown here is the ratio of the explained variation to the unexplained variation divided by the respective degrees of freedom. So large values of F would provide evidence for rejection of H_0 in favor of H_a.

The resulting F test is sometimes summarized in an analysis of variance table, which lists the sources of variability, the sums of squares, their degrees of freedom, the "mean squares" (i.e., sums of squares divided by degrees of freedom), and the F statistic. An analysis of variance table for a linear regression problem is shown in Table 9.2.

Note that the F statistic is written as the ratio of the mean square regression (MSREG) to the mean square for error. This is equivalent to $F = \text{SSREG}/s_\varepsilon^2$, since $df_1 = 1$ and $s_\varepsilon^2 = \text{MSE}$. We will, however, use the more standard notation involving mean squares, to conform with notation to be used when we study analysis of variance techniques in more detail in Chapters 10, 14, 15, and 16.

TABLE 9.2   Analysis of Variance Table for Linear Regression

Source Due to	Sum of Squares (SS)	df	Mean Square (MS)	F
Regression	SSREG	1	MSREG $=$ SSREG/1	$F = $ MSREG/MSE
Error	SSE	$n - 2$	MSE $=$ SSE/$(n - 2)$	
Total	SSTotal	$n - 1$		

EXAMPLE 9.8   Refer to the SAS output shown here for the data of Example 9.7.

a. Show that the square of the t statistic in Example 9.7 is equal to the computed F statistic shown in the output.

b. Locate the analysis of variance table for these data in the output and reconstruct it in the format shown in this section.

```
DEP VARIABLE: FOUND
                                                    ANALYSIS OF VARIANCE

                                        SUM OF          MEAN
                     SOURCE     DF      SQUARES         SQUARE        F VALUE      PROB>F

                     MODEL       1    1675.71027     1675.71027      1149.032      0.0001
                     ERROR      11    16.04203876    1.45836716
                     C TOTAL    12    1691.75231

                     ROOT MSE         1.207629       R-SQUARE        0.9905
                     DEP MEAN        14.75385        ADJ R-SQ        0.9897
                     C.V.             8.185179

                                            PARAMETER ESTIMATES

                                        PARAMETER       STANDARD       T FOR H0:
                     VARIABLE   DF      ESTIMATE         ERROR         PARAMETER=0    PROB > |T|

                     INTERCEP    1      0.52906977     0.53691888        0.985        0.3456
                     ADDED       1      0.82187597     0.02424601       33.897        0.0001


TEST: B1EQ1        NUMERATOR:    78.7103   DF:    1    F VALUE:   53.9715
                   DENOMINATOR:   1.45837  DF:   11    PROB >F :   0.0001
```

Solution   a. The computed value of t from Example 9.7 was $t = -7.42$ and hence $t^2 = 55.06$. Except for rounding errors, in the hand computation of t, this is equal to the value of the F statistic, 53.9715.

b. The only difference between the analysis of variance table shown in the output and the one that we have constructed here using the format of this section is that the regression source of variability is designated by MODEL and the degrees of freedom column precedes the SS column in the output. The reconstructed analysis of variance table is shown here.

Source	SS	df	MS	F	p
Regression	1675.71027	1	1675.71027	1149.032	.0001
Error	16.04203876	11	1.45836716		
Total	1691.75231	12			

A variation on the inferences discussed so far arises when the experimenter knows the value of the intercept β_0. For example in the calibration of an instrument, if known sample values x_i yield corresponding instrument readings y_i, then at least theoretically a value of $x_i = 0$ should give rise to an instrument reading of $y_i = 0$. In this situation, it might be logical to assume that the fitted regression goes through the origin $(0, 0)$; that is to say $\beta_0 = 0$.

When β_0 is known, we have the following changes in our estimation and test procedure for β_1. The least squares estimate of β_1 is

$$\hat{\beta}_1 = \frac{\sum x_i y_i - \beta_0 \sum x_i}{\sum x_i^2} \qquad \left(\text{Note: if } \beta_0 = 0,\ \hat{\beta}_1 = \frac{\sum x_i y_i}{\sum x_i^2}\right)$$

and the standard error of $\hat{\beta}_1$ is $\dfrac{s_\varepsilon}{\sqrt{\sum x_i^2}}$ where s_ε^2 is now computed as

$$s_\varepsilon^2 = \frac{\sum (y_i - \hat{y}_i)^2}{n - 1}$$

The test procedure for β_1 when β_0 is known is as shown here.

STATISTICAL TEST FOR β_1, WHERE INTERCEPT (β_0) IS KNOWN	H_0: $\beta_1 = 0$
	
	H_a: 1. $\beta_1 > 0$
	     2. $\beta_1 < 0$
	     3. $\beta_1 \neq 0$
	
	T.S.: $t = \dfrac{\hat{\beta}_1}{s_\varepsilon^2 / \sqrt{\sum x_i^2}}$
	
	where
	
	$\hat{\beta}_1 = \dfrac{\sum x_i y_i - \beta_0 \sum x_i}{\sum x_i^2}$
	
	and
	
	$s_\varepsilon^2 = \dfrac{\sum (y_i - \hat{y}_i)^2}{n - 1}$
	
	R.R.: For a given value of α and df $= n - 1$,
	    1. reject H_0 if $t > t_\alpha$
	    2. reject H_0 if $t < -t_\alpha$
	    3. reject H_0 if $|t| > t_{\alpha/2}$

In this section we have developed inferences about β_0 and β_1 in parallel. However, in most situations the slope is the more important parameter. Some statistical software packages (e.g., SPSS) even omit the standard error for the intercept. Also we will not be presenting inferences concerning the correlation coefficient since r is a multiple of $\hat{\beta}_1$. (Recall that $r = \hat{\beta}_1 \sqrt{S_{xx}/S_{yy}}$.) Confidence limits for ρ can be obtained by multiplying the limits for β_1 by $\sqrt{S_{xx}/S_{yy}}$. Similarly, a test of H_0: $\rho = 0$ (no linear correlation between x and y) is equivalent to a test of H_0: $\beta_1 = 0$ (no linear relation between x and y).

EXERCISES

9.1   Recall in Exercise 8.2 that we were interested in examining the relationship between the different concentrations of pectin (0%, 1.5%, and 3% by weight) on the firmness of canned sweet potatoes after storage in a controlled 25° C environment. The sample data for 6 cans are repeated below.

y (firmness)	50.5	46.8	62.3	67.7	80.1	79.2
x (concentration of pectin)	0	0	1.5	1.5	3.0	3.0

a. Obtain the least squares estimates for the parameters in the model $y = \beta_0 + \beta_1 x + \varepsilon$.
b. Obtain an estimate of σ_ε^2.
c. Give the standard error of $\hat{\beta}_1$.

9.2   Refer to Exercise 9.1. Perform a statistical test of the null hypothesis that there is no linear relationship between the concentration of pectin and the firmness of canned sweet potatoes after thirty days of storage at 25° C. Give the p-value for this test and draw conclusions.

9.3   The extent of disease transmission can be affected greatly by the viability of infectious organisms suspended in the air. Because of the infectious nature of the disease under study, the viability of these organisms must be studied in an airtight chamber. One way to do this is to disperse an aerosol cloud, prepared from a solution containing the organisms, into the chamber. The biological recovery at any particular time is the percentage of the total number of organisms suspended in the aerosol that are viable. The data below are the biological recovery percentages computed from 13 different aerosol clouds. For each of the clouds, recovery percentages were determined at different times.

Cloud	Time, x (in Minutes)	Biological Recovery (%)
1	0	70.6
2	5	52.0
3	10	33.4
4	15	22.0
5	20	18.3
6	25	15.1
7	30	13.0
8	35	10.0
9	40	9.1
10	45	8.3
11	50	7.9
12	55	7.7
13	60	7.7

a. Plot the data.
b. Since there is some curvature, try to linearize the data using the log of the biological recovery.

9.4   Refer to Exercise 9.3.
a. Fit the linear regression model, $y = \beta_0 + \beta_1 x + \varepsilon$, where y is the log biological recovery.
b. Compute an estimate of σ_ε.
c. Identify the standard errors of $\hat{\beta}_0$ and $\hat{\beta}_1$.

9.5   Refer to Exercise 9.3. Conduct a test of the null hypothesis that $\beta_1 = 0$. Use $\alpha = .05$.

9.6 Refer to Exercise 9.3. Place a 95% confidence interval on β_0, the mean log biological recovery percentage at time zero. Interpret your findings. (Note: $E(y) = \beta_0$ when $x = 0$.)

9.7 An experiment was conducted to examine the relationship between the weight gain of chickens whose diets had been supplemented by different amounts of the amino acid lysine and the amount of lysine ingested. Since the percentage of lysine is known, and we can monitor the amount of feed consumed, we can determine the amount of lysine eaten. A random sample of 12 two-week-old chickens was selected for the study. Each was caged separately and was allowed to eat at will from feed composed of a base supplemented with lysine. The sample data summarizing weight gains and amounts of lysine eaten over the test period are given here. (In the data, y represents weight gain in grams, and x represents the amount of lysine ingested in grams.)

a. Plot the data in a scatter diagram. Does a linear model seem appropriate?

b. Fit the linear regression model $y = \beta_0 + \beta_1 x + \varepsilon$

Chick	y	x	Chick	y	x
1	14.7	.09	7	17.2	.11
2	17.8	.14	8	18.7	.19
3	19.6	.18	9	20.2	.23
4	18.4	.15	10	16.0	.13
5	20.5	.16	11	17.8	.17
6	21.1	.23	12	19.4	.21

9.8 Refer to Exercise 9.7.

a. Compute an estimate of σ_ε^2.

b. Identify the standard error of $\hat{\beta}_1$.

c. Conduct a statistical test of the research hypothesis that for this diet preparation and length of study, there is a direct (positive) linear relationship between weight gain and the amount of lysine eaten.

9.9 Refer to Exercises 9.7 and 9.8.

a. For this example would it make sense to give any physical interpretation to β_0? (Hint: The lysine was mixed in the feed.)

b. Consider an alternative model relating weight gain to amount of lysine ingested:

$$y = \beta_1 x + \varepsilon.$$

Distinguish between this model and the model $y = \beta_0 + \beta_1 x + \varepsilon$.

9.10 a. Refer to part (b) of Exercise 9.9. Compute $\hat{\beta}_1$ for the model $y = \beta_1 x + \varepsilon$,

$$\hat{\beta}_1 = \frac{\sum xy}{\sum x^2}$$

b. Which of the two models, $y = \beta_0 + \beta_1 x + \varepsilon$ or $y = \beta_1 x + \varepsilon$, appears to give a better fit to the sample data? (Plot the two prediction equations on a graph of the sample observations.)

9.11 Refer to Example 9.7. Use the sample data to construct a 95% confidence interval for the intercept β_0. Does the interval include zero as a possible value? Does the validation experiment conform to theory in regard to the intercept?

9.12 Refer to Example 9.7. Compute the mean percent (of the added amount) received for each level of x. Plot these data. If there is a problem with the use of the assay at this laboratory, at what ranges of x does this occur?

9.13   Interest rates charged for home mortgages have, in general, declined over recent months. With the apparent favorable influence for new home building, the data shown here are the prevailing mortgage interest rates and the number of housing starts in a midwestern city over a period of 18 months.

Month	Interest Rate x	Number of Housing Starts y
1	10.5	360
2	10.3	340
3	10.6	370
4	11.4	360
5	11.8	330
6	11.3	300
7	11.0	290
8	10.5	340
9	10.2	360
10	10.0	370
11	9.8	380
12	9.8	390
13	9.9	375
14	10.0	350
15	10.0	345
16	9.9	360
17	9.8	380
18	9.7	395

a. Plot the data.
b. Use these data to obtain a linear regression equation.
c. Is the slope significantly different from zero?

9.14   Refer to Exercise 9.13. Predict the number of housing starts for interest rates of 10.2% and 9.5%. Do you predict that the prevailing interest rate will increase or decrease next month (month 19)?

FIGURE 9.2

Locating the Median
Lines and Quadrants
for the Quadrant Sum
Test

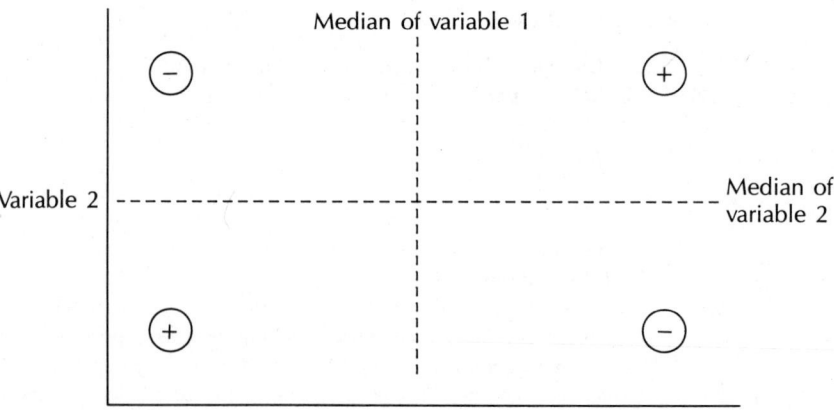

| 9.3 | QUICK PORTABLE STATISTICS |

One portable statistic for approximating a linear regression equation was discussed in Chapter 8. There is a portable statistical test that is useful for ascertaining whether or not there is an association between two variables.

QUADRANT SUM
TEST FOR
ASSOCIATION

H_0: The two variables are not correlated.

H_a: The two variables are correlated (a two-tailed test).

Test Procedure

1. Plot the data using a scatter diagram.
2. Draw a median line parallel to each axis. The median line parallel to the vertical axis will designate the median (midpoint) value of the variable plotted along the horizontal axis. Similarly, the median line parallel to the horizontal axis will correspond to the median of the variable plotted along the vertical axis (see Figure 9.2).
3. Beginning in the upper-right quadrant and moving counterclockwise, label the quadrants $+$, $-$, $+$, $-$, respectively (see Figure 9.2).
4. Obtain the following four counts:
   a. Beginning from the right side of the scatter diagram and moving toward the left along the horizontal median line, count all observations (dots) that are on the same side of the horizontal median line. Stop counting when you encounter the first observation on the other side of the horizontal median line. Attach the sign of the quadrant to this count.
   b. Beginning from the top of the scatter diagram and moving down along the vertical median line, count all observations that are on the same side of the median line. Stop counting when you encounter the first observation on the other side of the vertical median line. Attach the sign of the quadrant to this count.
   c. Repeat step a, moving from left to right.
   d. Repeat step b, moving from bottom to top.
5. Let C denote the sum of the counts (with their appropriate signs) obtained in part (4).
6. For $\alpha = .10$, $.05$, or $.01$, reject H_0 if $|C| \geq 9$, 11, or 13, respectively.
7. If C is positive, the two variables are positively correlated; if C is negative, the two variables are negatively correlated.

EXAMPLE 9.9

Recall that in Example 8.3 we had data on the age of volunteers and their reductions in cholesterol counts after four weeks on a low-cholesterol diet. For convenience, the data are shown in Table 9.3.

TABLE 9.3   Age and Cholesterol Reduction Data

Age (Years)	Reduction in Cholesterol	Age (Years)	Reduction in Cholesterol
45	30	31	40
43	52	26	17
46	45	22	28
49	38	58	44
50	62	60	61
37	55	52	58
34	25	27	45
30	30		

Use the quadrant sum test with $\alpha = .05$ to determine whether there is a relationship between the age of a volunteer and the four-week cholesterol reduction.

Solution   We must first construct a scatter diagram. This is shown in Figure 9.3. The four quadrants are labeled $+$, $-$, $+$, $-$ going counterclockwise from the upper right. Using the age and cholesterol count data in Table 9.3, we find the median age to be 43 and the median cholesterol count to be 44. The median lines have been drawn on the scatter diagram by using dotted lines.

1. Beginning at the extreme right, we count observations until we must cross the horizontal median to count the next observation (dot). There are four observations (see Figure 9.3). We attach a plus sign to this count since the four observations were in the upper-right quadrant.
2. Beginning from the top, we count three observations before we must cross the vertical median line. Again we assign a plus sign to these three measurements.
3. Beginning from the extreme left, we count two observations before we must cross the horizontal median line. Since this observation was in the lower-left quadrant, it receives a plus sign.

FIGURE 9.3

Scatter Diagram for Example 9.9

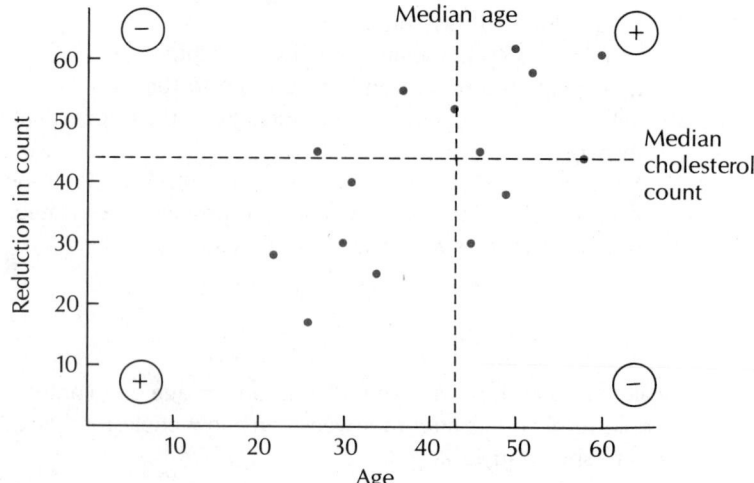

4. Similarly, from the bottom we count four observations before we must cross the vertical median line. This count is assigned a plus sign because the measurements are in the lower-left quadrant.

5. The combined count (with appropriate sign) is

$$C = 4 + 3 + 2 + 4 = 13$$

Since $|C| > 11$, we reject the null hypothesis, at the $\alpha = .05$ level, that the variables are uncorrelated. The sign of C indicates that there is a positive correlation between the age and cholesterol count reduction observed for persons treated with this diet.

EXERCISES

9.15 An experiment was conducted to examine the relationship between the weight of plaque scraped from teeth and the DNA concentration of the plaque. To do this a random sample of 10 males (ages 18 to 20 years) was obtained. Each person was given a normal diet plus 14 2-mg tablets of sucrose per day. No toothbrushing was allowed. At the end of four days, plaque samples were obtained from the 10 subjects. These data are shown below.

Person	1	2	3	4	5	6	7	8	9	10
Plaque weight (mg)	42.7	52.3	24.6	33.4	41.8	36.7	27.0	47.3	31.4	33.9
DNA (μg)	206	303	175	214	226	246	181	251	154	247

Use the quadrant sum test to determine if there is a significant association between plaque weight and DNA concentration.

9.16 Refer to Exercise 9.15. Do the data appear to be linear? How might you examine the linear relationship?

9.17 The thermal pollution of automobiles was studied for 1987 and 1988 models. The data relating automobile weight to BTU (in 1000s) per vehicle mile are shown in the table below.
a. Plot the data in a scatter diagram.
b. Is there evidence for an association based on the quadrant sum test?

Weight, x (in 1000 lb)	BTU per Vehicle Mile, y (in 1000s)
1.8	4
2.6	5.2
4.2	8.5
5.0	11.6
4.8	10.1
3.4	6.3

9.18 Refer to Exercise 9.17.
a. Use the approximate regression equation procedure of Section 8.3 to obtain an approximate fit to the data.

    b. Compare the approximate prediction equation to the actual least squares prediction equation.

    c. Plot both equations [the approximate prediction equation and the least squares line on your graph of part (a) of Exercise 9.17].

9.4   INFERENCES CONCERNING $E(y)$

The methods of previous sections can be expanded to include inferences concerning the average (expected) value of y for a given setting of the independent variable. For example, in evaluating the effects of different levels of advertising expenditure x on sales y, it may be of interest to estimate the average sales per month for a given level of expenditure x. The estimate of $E(y)$ for a specific setting of x can be obtained by evaluating the prediction equation

$$\hat{y} = \hat{\beta}_0 + \hat{\beta}_1 x$$

at that setting. It can be shown that in repeated sampling at a particular setting of x, the sampling distribution of $\hat{y}$ has a mean

$$E(y) = \beta_0 + \beta_1 x$$

and a variance given by

$$V(\hat{y}) = \sigma_\varepsilon^2 \left(\frac{1}{n} + \frac{(x - \bar{x})^2}{S_{xx}} \right)$$

Again assuming that the εs are normally distributed, a $100(1 - \alpha)\%$ confidence interval for $E(y)$ is given by the formula below.

$100(1 - \alpha)\%$ CONFIDENCE INTERVAL FOR $E(y)$

$$\hat{y} \pm t_{\alpha/2} s_\varepsilon \sqrt{\frac{1}{n} + \frac{(x - \bar{x})^2}{S_{xx}}}$$

where

$$s_\varepsilon^2 = \frac{SSE}{n - 2}$$

and the t-value is based on df $= n - 2$.

EXAMPLE 9.10

Use the data of Example 9.2 to give a 90% confidence interval for the mean corn yield when 5 pounds of fertilizer are applied to a plot.

Solution

The prediction equation in Example 9.2 was

$$\hat{y} = 10.10 + 1.15x$$

where x = fertilizer applied. For our example we need $x = 5$ so that

$$\hat{y} = 10.10 + 1.15(5) = 15.85$$

The variance of $\hat{y}$ can be computed by using $S_{xx} = 20$, $s_\varepsilon = 0.84$, $\bar{x} = 4$, and $n = 10$. The t-value in Table 4 in the Appendix for $a = .05$ and df $= n - 2 = 8$ is 1.86. Hence the appropriate confidence interval for the average corn yield per plot when 5 pounds of fertilizer are applied is

$$15.85 \pm 1.86(.84) \sqrt{\frac{1}{10} + \frac{(5-4)^2}{20}} \qquad \text{or} \qquad 15.85 \pm .61$$

that is, 15.24 to 16.46.

EXAMPLE 9.11

In Example 9.10 we constructed a 90% confidence interval for the mean corn yield when 5 pounds of fertilizer are applied. Use the same sample data to construct a 90% confidence interval on $E(y)$ for any specific value of fertilizer in the range from 2 to 6. Graph your results.

Solution

Using the results from Example 9.10, $\hat{y} = 10.10 + 1.15x$, $s_\varepsilon = .84$, and

$$\sqrt{\frac{1}{n} + \frac{(x - \bar{x})^2}{S_{xx}}} = \sqrt{.1 + \frac{(x - 4)^2}{20}}$$

Our 90% confidence interval for $E(y)$, then, is of the form

$$\hat{y} \pm 1.86(.84) \sqrt{.1 + \frac{(x - 4)^2}{20}}$$

All we need to do is substitute a specific value of x in this form to determine a confidence interval. For fertilizer settings of 2, 3, 4, 5, and 6 the 90% confidence limits are given below.

x	90% Confidence Interval
2	11.544 to 13.256
3	12.945 to 14.155
4	14.206 to 15.194
5	15.245 to 16.455
6	16.144 to 17.856

confidence bands

Plotting the endpoints of the confidence intervals and connecting the points, we get the general 90% confidence interval on $E(y)$ for any value of x between 2 and 6. The graph in Figure 9.4 displays 90% **confidence bands** for $E(y)$. Notice how the width of the confidence interval (the vertical distance between the two curves of the graph) varies for different values of x. $E(y)$ can be estimated more precisely for values of x in the center of the experimental region. The widening of the gap between the bands at the extremities of the experimental region indicates that it would be unwise to extrapolate (try to estimate $E(y)$) beyond the region of experimentation.

FIGURE 9.4

90% Confidence Band
for E(y) Using the Data
of Example 9.10

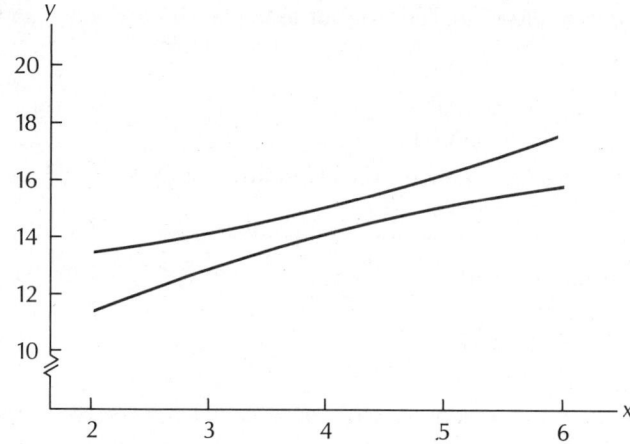

A statistical test concerning $E(y)$ for a given setting of the independent variable in linear regression can also be formulated using the test setup shown below. Thus, for example, we might wish to test that the mean corn yield per plot is 16 when 6 pounds of fertilizer are applied.

TEST OF A HYPOTHESIS CONCERNING $E(y)$

H_0: $E(y) = \mu_0$

H_a: 1. $E(y) > \mu_0$
       2. $E(y) < \mu_0$
       3. $E(y) \neq \mu_0$

T.S.: $t = \dfrac{\hat{y} - \mu_0}{s_\varepsilon \sqrt{\dfrac{1}{n} + \dfrac{(x - \bar{x})^2}{S_{xx}}}}$

R.R.: For a general value of α and df $= n - 2$
      1. reject H_0 if $t > t_\alpha$
      2. reject H_0 if $t < -t_\alpha$
      3. reject H_0 if $|t| > t_{\alpha/2}$

EXAMPLE 9.12   An experiment was run to examine the rate of growth of a particular type of bacteria. The growth y was determined for two different cultures at five equally spaced time intervals (1, 2, 3, 4, and 5 hours past culture seeding).

	Time (hrs)				
Growth Rate, y	1	2	3	4	5
Culture 1	8.0	9.0	9.1	10.2	10.4
Culture 2	8.5	9.2	9.3	9.8	10.1

Use the computer output shown below to answer the following questions.

a. Determine the least squares fit to the linear regression model $y = \beta_0 + \beta_1 x + \varepsilon$.
b. Conduct a test of $H_0: \beta_1 = 0$. Give a p-value for the test.
c. Conduct a test of $E(y) = 9.5$ where $x = 3.5$. Use $\alpha = .05$.

Solution

a. The linear regression equation is $\hat{y} = 7.89 + 0.49x$.
b. The computed value of t for $H_0: \beta_1 = 0$ is shown as 8.42; the corresponding two-sided p-value is $p < .0001$.
c. Although the results of a statistical test of $H_0: E(y) = 9.5$ is not given directly, we can use the 95% confidence interval for $E(y)$ shown under observation eleven ($x = 3.5$) to reach a conclusion. Since the confidence limits $9.40 - 9.81$ include the value 9.5, we have insufficient evidence to reject $H_0: E(y) = 9.5$.

```
LISTING OF DATA

OBS    TIME    RATE

  1     1.0     8.0
  2     1.0     8.5
  3     2.0     9.0
  4     2.0     9.2
  5     3.0     9.1
  6     3.0     9.3
  7     4.0    10.2
  8     4.0     9.8
  9     5.0    10.4
 10     5.0    10.1
 11     3.5

N=    11
```

LINEAR REGRESSION

GENERAL LINEAR MODELS PROCEDURE

DEPENDENT VARIABLE: RATE   GROWTH RATE

SOURCE	DF	SUM OF SQUARES	MEAN SQUARE	F VALUE	PR > F	R-SQUARE	C.V.
MODEL	1	4.80200000	4.80200000	70.88	0.0001	0.898578	2.7809
ERROR	8	0.54200000	0.06775000		ROOT MSE		RATE MEAN
CORRECTED TOTAL	9	5.34400000			0.26028830		9.36000000

SOURCE	DF	TYPE I SS	F VALUE	PR > F	DF	TYPE III SS	F VALUE	PR > F
TIME	1	4.80200000	70.88	0.0001	1	4.80200000	70.88	0.0001

PARAMETER	ESTIMATE	T FOR HO: PARAMETER=0	PR > |T|	STD ERROR OF ESTIMATE
INTERCEPT	7.89000000	40.87	0.0001	0.19303497
TIME	0.49000000	8.42	0.0001	0.05820223

OBSERVATION	OBSERVED VALUE	PREDICTED VALUE	RESIDUAL	LOWER 95% CL FOR MEAN	UPPER 95% CL FOR MEAN
1	8.00000000	8.38000000	-0.38000000	8.05123956	8.70876044
2	8.50000000	8.38000000	0.12000000	8.05123956	8.70876044
3	9.00000000	8.87000000	0.13000000	8.63753127	9.10246873
4	9.20000000	8.87000000	0.33000000	8.63753127	9.10246873
5	9.10000000	9.36000000	-0.26000000	9.17019007	9.54980993
6	9.30000000	9.36000000	-0.06000000	9.17019007	9.54980993
7	10.20000000	9.85000000	0.35000000	9.61753127	10.08246873
8	9.80000000	9.85000000	-0.05000000	9.61753127	10.08246873
9	10.40000000	10.34000000	0.06000000	10.01123956	10.66876044
10	10.10000000	10.34000000	-0.24000000	10.01123956	10.66876044
11 *		9.60500000		9.40367617	9.80632383

* OBSERVATION WAS NOT USED IN THIS ANALYSIS

```
     SUM OF RESIDUALS                       0.00000000
     SUM OF SQUARED RESIDUALS               0.54200000
     SUM OF SQUARED RESIDUALS - ERROR SS   -0.00000000
     PRESS STATISTIC                        0.88400885
     FIRST ORDER AUTOCORRELATION           -0.20885609
     DURBIN-WATSON D                        2.04501845
```

EXAMPLE 9.13

Refer to Example 9.12. Suppose we wanted to give a confidence interval for $E(y)$ when $x = 8.5$. What problem(s) might we encounter?

Solution

Since $x = 8.5$ is well outside the range of experimentation $(1 \leq x \leq 5)$, one might have cause for concern due to extrapolation. The assumed linear model seems to fit the data well in the experimental region; it might be completely inappropriate near $x = 8.5$. Hence any inferences based on the confidence interval for $E(y)$ at $x = 8.5$ would have to be viewed skeptically.

9.5   PREDICTING y FOR A GIVEN VALUE OF x

In Section 9.4, we were concerned with estimating the expected value of y for a given value of x. Suppose, however, that after obtaining a least squares prediction equation for **predict y**   the general linear model, an investigator would like to **predict** the actual value of y (say the next measurement) for a given value of the independent variable x. Note that this problem differs from the problem discussed in the previous section in that we do not want to estimate the average value of y for a given value of x; rather, we wish to predict what a particular observation will be for that same setting of x.

We still use the least squares equation $\hat{y}$ as our predictor, but the corresponding interval about the observation y is called a *prediction interval*. (Prediction intervals are constructed about variables, whereas confidence intervals are constructed about parameters.)

GENERAL
100$(1 - \alpha)$%
PREDICTION
INTERVAL

$$\hat{y} \pm t_{\alpha/2} s_\varepsilon \sqrt{1 + \frac{1}{n} + \frac{(x - \bar{x})^2}{S_{xx}}}$$

where

$$s_\varepsilon^2 = \frac{SSE}{n - 2}$$

and $t_{\alpha/2}$ is based on df $= n - 2$

Note the similarity between the confidence interval for $E(y)$ and the prediction interval for the variable y. The only difference is that the above prediction interval has a 1 added to the quantity under the square root sign. This makes the interval wider to account for the fact that we're predicting a variable (future value of y) rather than a constant $E(y)$.

EXAMPLE 9.14

Use the data of Example 9.2 to predict the actual crop yield for a plot fertilized with 5 pounds of fertilizer. Place a 90% prediction interval about the actual value of y.

Solution

Using our previous work from Example 9.10, the predicted value of y (using $\hat{y}$) at $x = 5$ is $\hat{y} = 15.85$. Also, for $x = 5$, $\bar{x} = 4$, $n = 10$, and $S_{xx} = 20$,

$$\frac{1}{n} + \frac{(x - \bar{x})^2}{S_{xx}} = .15$$

The corresponding t-value for a = .05 and df = $n - 2 = 8$ is 1.86. Hence the 90% prediction interval is

$$15.85 \pm 1.86(.84)\sqrt{1 + .15} \qquad \text{or} \qquad 15.85 \pm 1.68$$

that is, 14.17 to 17.53.

Note that the above interval is almost three times wider than the corresponding interval for $E(y)$ of Example 9.10. This is to be expected since here we are placing an interval about a quantity that may vary, while in Example 9.10 we were placing an interval about $E(y)$, which cannot vary. Since both intervals are called 90% intervals, the prediction interval must be wider to have the same fraction of intervals (.90) covering y in repeated sampling.

EXAMPLE 9.15

a. Refer to the data of Example 9.2. Construct a general 90% prediction interval for y when x takes the values 2, 3, 4, 5, and 6.
b. Graph your results to show a 90% prediction band for y when $2 \le x \le 6$.
c. On the graph of part (b) superimpose the graph of the 90% confidence band for $E(y)$.

Solution

a. From previous calculations, $\hat{y} = 10.10 + 1.15x$, $s_\varepsilon = .84$, and

$$\sqrt{1 + \frac{1}{n} + \frac{(x - \bar{x})^2}{S_{xx}}} = \sqrt{1 + \frac{1}{10} + \frac{(x - 4)^2}{20}}$$

Hence a general 90% prediction interval is of the form

$$\hat{y} \pm 1.86(.84)\sqrt{1.1 + \frac{(x - 4)^2}{20}}$$

Substituting the values $x = 2, 3, 4, 5,$ and 6 into this form we have the intervals given here.

x	90% Prediction Interval
2	10.618 to 14.182
3	11.874 to 15.226
4	13.061 to 16.339
5	14.174 to 17.526
6	15.218 to 18.782

b, c. Plotting the endpoints of the prediction intervals and connecting the dots, we obtain the 90% prediction bands, shown by the solid lines in Figure 9.5. The dotted lines indicate the corresponding 90% confidence bands for $E(y)$. Notice that the prediction bands are wider than the corresponding confidence bands to allow for the fact that we are predicting the value of a random variable rather than estimating a parameter.

FIGURE 9.5

90% Prediction and
Confidence Bands for
y and $E(y)$,
Example 9.15

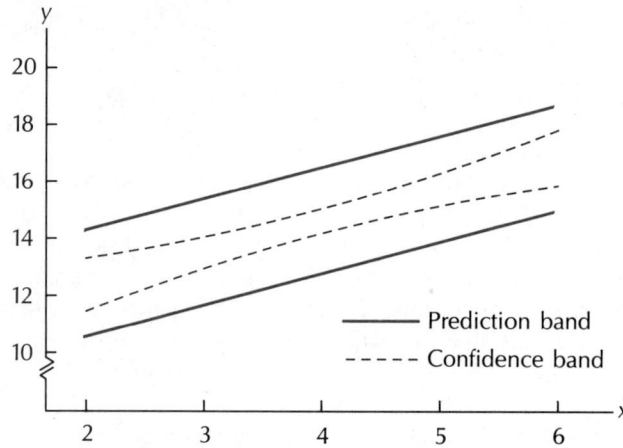

A study of Figure 9.5 and the formulas for these confidence and prediction intervals should suggest factors that may influence the precision of our confidence intervals for $E(y)$ and prediction intervals for y. First, the plus or minus terms in the confidence interval and prediction interval involve $t_{\alpha/2}$, s_ε, n, $(x - \bar{x})^2$, and S_{xx}. Forgetting about $t_{\alpha/2}$ and s_ε for the moment, the width of these intervals will decrease as n increases, $(x - \bar{x})^2$ decreases, and S_{xx} increases. Obviously, if we take more observations, n increases. The quantity $(x - \bar{x})^2$ can be made smaller by making predictions at values of x closer to the mean of the x-values $(\bar{x})$. This is seen in Figure 9.5, where the confidence (and prediction) interval is wider for values of x farther away from the center of the region $\bar{x} = 4$. Finally, we can increase $S_{xx} = \sum (x_i - \bar{x})^2$ and hence improve the confidence and prediction intervals based on our model by increasing the spread of the x-values in our sample. This is certainly an important point when the experimenter has control of the x-values. However, there is a point of diminishing returns. The width of the confidence interval for $E(y)$ and the prediction interval for y are adequate measures of precision *assuming* the model adequately fits the data. If the x-values are spread too far to make S_{xx} large (say, the values $x \neq 1$, 5, 6, 7, and 11 in Example 9.15), a linear regression model may no longer adequately describe the relation between x and y, thus rendering invalid the confidence interval for $E(y)$ and prediction interval for y discussed in this section.

The point is that you should understand the factors affecting the precision of the confidence and prediction intervals based on a linear regression model. However, you should not apply these methods blindly. Plot the data; see how well the model fits the data; plot the confidence and prediction intervals to see the penalty associated with predictions away from $\bar{x}$, and so on.

EXERCISES

9.19   Refer to Exercise 9.3. For the least squares equation

$$\hat{y} = \hat{\beta}_0 + \hat{\beta}_1 x$$

estimate the mean log biological recovery percentage at 30 minutes, using a 95% confidence interval.

9.20   Refer to Exercise 9.7. Estimate the mean weight gain for chickens fed on a diet supplemented with lysine if .19 grams of lysine were ingested over a study period of the same duration. Use a 95% confidence interval.

9.21   Refer to Exercise 9.20. Construct a 95% prediction interval for the weight gain of a chick chosen at random and observed to ingest .19 grams of lysine. Compare your results to the confidence interval of Exercise 9.20.

9.22   Using the data of Exercise 9.19, construct a 95% prediction interval for the log biological recovery percentage at 30 minutes. Compare your result to the confidence interval on $E(y)$ of Exercise 9.19.

 9.23   A chemist is interested in determining the weight loss y of a particular compound as a function of the amount of time the compound is exposed to the air. The data in the table below give the weight losses associated with $n = 12$ settings of the independent variable, exposure time.

Weight Loss and Exposure Time Data

Weight Loss, y (in Pounds)	Exposure Time (in Hours)
4.3	4
5.5	5
6.8	6
8.0	7
4.0	4
5.2	5
6.6	6
7.5	7
2.0	4
4.0	5
5.7	6
6.5	7

a. Find the least squares prediction equation for the model

$$y = \beta_0 + \beta_1 x + \varepsilon$$

b. Test $H_0: \beta_1 = 0$; give the p-value for $H_a: \beta_1 > 0$ and draw conclusions.

9.24   Refer to Exercise 9.23 and the SAS computer output shown here.

```
        LISTING OF DATA

    OBS     WT_LOSS     TIME

     1        4.3        4
     2        5.5        5
     3        6.8        6
     4        8.0        7
     5        4.0        4
     6        5.2        5
     7        6.6        6
     8        7.5        7
     9        2.0        4
    10        4.0        5
    11        5.7        6
    12        6.5        7

    N=        12
```

LINEAR REGRESSION

GENERAL LINEAR MODELS PROCEDURE

DEPENDENT VARIABLE: WT_LOSS   WEIGHT LOSS (LBS)

SOURCE	DF	SUM OF SQUARES	MEAN SQUARE	F VALUE	PR > F	R-SQUARE	C.V.
MODEL	1	26.00416667	26.00416667	40.22	0.0001	0.800888	14.5970
ERROR	10	6.46500000	0.64650000		ROOT MSE		WT_LOSS MEAN
CORRECTED TOTAL	11	32.46916667			0.80405224		5.50833333

SOURCE	DF	TYPE I SS	F VALUE	PR > F	DF	TYPE III SS	F VALUE	PR > F
TIME	1	26.00416667	40.22	0.0001	1	26.00416667	40.22	0.0001

PARAMETER	ESTIMATE	T FOR H0: PARAMETER=0	PR > |T|	STD ERROR OF ESTIMATE
INTERCEPT	-1.73333333	-1.49	0.1677	1.16518239
TIME	1.31666667	6.34	0.0001	0.20760539

OBSERVATION	OBSERVED VALUE	PREDICTED VALUE	RESIDUAL	LOWER 95% CL INDIVIDUAL	UPPER 95% CL INDIVIDUAL
1	4.30000000	3.53333333	0.76666667	1.54371634	5.52295033
2	5.50000000	4.85000000	0.65000000	2.97100515	6.72899485
3	6.80000000	6.16666667	0.63333333	4.28767181	8.04566152
4	8.00000000	7.48333333	0.51666667	5.49371634	9.47295033
5	4.00000000	3.53333333	0.46666667	1.54371634	5.52295033
6	5.20000000	4.85000000	0.35000000	2.97100515	6.72899485
7	6.60000000	6.16666667	0.43333333	4.28767181	8.04566152
8	7.50000000	7.48333333	0.01666667	5.49371634	9.47295033
9	2.00000000	3.53333333	-1.53333333	1.54371634	5.52295033
10	4.00000000	4.85000000	-0.85000000	2.97100515	6.72899485
11	5.70000000	6.16666667	-0.46666667	4.28767181	8.04566152
12	6.50000000	7.48333333	-0.98333333	5.49371634	9.47295033

SUM OF RESIDUALS	0.00000000
SUM OF SQUARED RESIDUALS	6.46500000
SUM OF SQUARED RESIDUALS - ERROR SS	-0.00000000
PRESS STATISTIC	10.03092643
FIRST ORDER AUTOCORRELATION	0.60849016
DURBIN-WATSON D	0.54253674

LINEAR REGRESSION

GENERAL LINEAR MODELS PROCEDURE

DEPENDENT VARIABLE: WT_LOSS   WEIGHT LOSS (LBS)

SOURCE	DF	SUM OF SQUARES	MEAN SQUARE	F VALUE	PR > F	R-SQUARE	C.V.
MODEL	1	26.00416667	26.00416667	40.22	0.0001	0.800888	14.5970
ERROR	10	6.46500000	0.64650000		ROOT MSE		WT_LOSS MEAN
CORRECTED TOTAL	11	32.46916667			0.80405224		5.50833333

SOURCE	DF	TYPE I SS	F VALUE	PR > F	DF	TYPE III SS	F VALUE	PR > F
TIME	1	26.00416667	40.22	0.0001	1	26.00416667	40.22	0.0001

PARAMETER	ESTIMATE	T FOR H0: PARAMETER=0	PR > |T|	STD ERROR OF ESTIMATE
INTERCEPT	-1.73333333	-1.49	0.1677	1.16518239
TIME	1.31666667	6.34	0.0001	0.20760539

OBSERVATION	OBSERVED VALUE	PREDICTED VALUE	RESIDUAL	LOWER 95% CL FOR MEAN	UPPER 95% CL FOR MEAN
1	4.30000000	3.53333333	0.76666667	2.66793184	4.39873483
2	5.50000000	4.85000000	0.65000000	4.28346174	5.41653826
3	6.80000000	6.16666667	0.63333333	5.60012840	6.73320493
4	8.00000000	7.48333333	0.51666667	6.61793184	8.34873483
5	4.00000000	3.53333333	0.46666667	2.66793184	4.39873483
6	5.20000000	4.85000000	0.35000000	4.28346174	5.41653826
7	6.60000000	6.16666667	0.43333333	5.60012840	6.73320493
8	7.50000000	7.48333333	0.01666667	6.61793184	8.34873483
9	2.00000000	3.53333333	-1.53333333	2.66793184	4.39873483
10	4.00000000	4.85000000	-0.85000000	4.28346174	5.41653826
11	5.70000000	6.16666667	-0.46666667	5.60012840	6.73320493
12	6.50000000	7.48333333	-0.98333333	6.61793184	8.34873483

SUM OF RESIDUALS	0.00000000
SUM OF SQUARED RESIDUALS	6.46500000
SUM OF SQUARED RESIDUALS - ERROR SS	-0.00000000
PRESS STATISTIC	10.03092643
FIRST ORDER AUTOCORRELATION	0.60849016
DURBIN-WATSON D	0.54253674

a. Identify the 95% confidence bands for $E(y)$ when $4 \leq x \leq 7$.
b. Identify the 95% prediction bands for y, $4 \leq x \leq 7$.
c. Distinguish between the meaning of the confidence bands and prediction bands in parts (a) and (b).

9.6   EXAMINING LACK OF FIT IN LINEAR REGRESSION

In our study of linear regression we have been concerned with how well a linear regression model $y = \beta_0 + \beta_1 x + \varepsilon$ fits but only from an intuitive standpoint. We could examine a scatterplot of the data to see if it looked linear and we could test whether the slope differed from zero; however, we had no way of testing to see if a higher-order model would be a more appropriate model for the relationship between y and x. This section will outline situations in which we can test for the validity of a linear regression model.

Pictures (or graphs) are always a good starting point for examining lack of fit. First use a scatterplot of y versus x. Second, a plot of residuals $y_i - \hat{y}_i$ versus predicted values $\hat{y}_i$ may give an indication of the following problems:

1. Outliers or erroneous observations. In examining the residual plot, your eye will naturally be drawn to data points with unusually high (in absolute value) residuals.
2. Violation of the assumptions. For the model $y = \beta_0 + \beta_1 x + \varepsilon$, we've assumed a linear relation between y and the dependent variable x, and independent, normally distributed errors with a constant variance.

The residual plot for a model and data set that has none of these apparent problems would look much like the plot in Figure 9.6. Note from this plot there are no extremely large residuals (and hence no apparent outliers) and that there is no trend in the residuals to indicate that the linear model is inappropriate. When a higher-order model is more appropriate, a residual plot more like that shown in Figure 9.7 would be observed.

FIGURE 9.6

Residual Plot with No Apparent Pattern

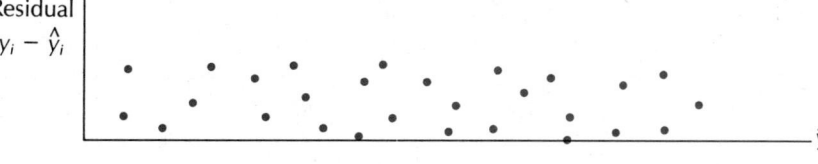

FIGURE 9.7

Residual Plot Showing the Need for a Higher-Order Model

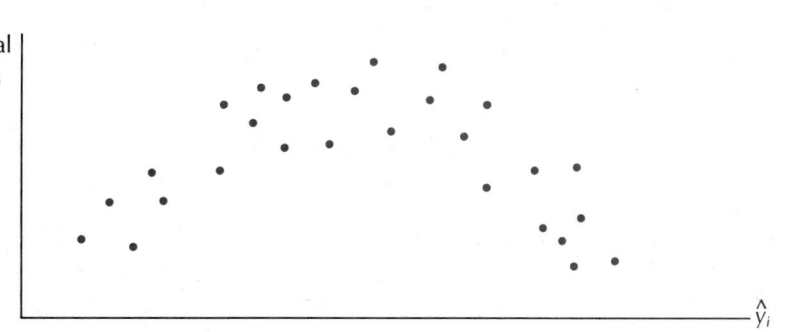

FIGURE 9.8

Residual Plot Showing Homogeneous Error Variances

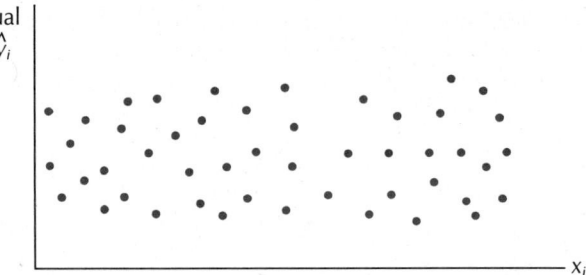

FIGURE 9.9

Residual Plot Showing Error Variances Increasing with x

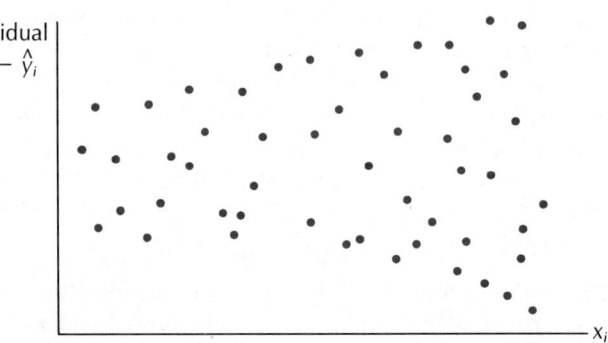

A check of the constant variance assumption can be addressed in the y versus x scatterplot or with a plot of the residuals $(y_i - \hat{y}_i)$ versus x_i. For example a pattern of residuals as shown in Figure 9.8 indicates homogeneous error variances across values of x; Figure 9.9 indicates the error variances increase with increasing values of x.

The question of independence of the errors and normality of the errors will be addressed later in Chapter 13. We will illustrate some of the points we've learned so far about residuals by way of an example.

EXAMPLE 9.16

The amount of heat loss was examined for a new brand of thermal panes. Random assignment of three different panes was made to each of the three outdoor temperature settings being considered. For each trial the window temperature was controlled at 68 °F and 50% relative humidity.

Outdoor Temperature (°F)	Heat Loss
20	86, 80, 77
40	78, 84, 75
60	33, 38, 43

a. Plot the data.
b. Fit the linear regression model $y = \beta_0 + \beta_1 x + \varepsilon$ and test $H_0: \beta_1 = 0$ (give the p-value for your test).
c. Compute $\hat{y}_i$ and $y_i - \hat{y}_i$ for the nine observations. Plot $y_i - \hat{y}_i$ versus $\hat{y}_i$.
d. Does the constant variance assumption seem reasonable?

Solution

The computer output shown here can be used to address the four parts of this example.

a. The scatterplot of y versus x certainly shows a downward linear trend, and there may be evidence of curvature as well.

b. The linear regression model seems to fit the data well, and the test of $H_0: \beta_1 = 0$ is significant at the $p = .0023$ level. But is this the best model for the data?

c. The plot of residuals $(y_i - \hat{y}_i)$ against the predicted values $\hat{y}_i$ is similar to Figure 9.7, suggesting that we may need additional terms in our model.

d. Since residuals associated with $x = 20$ (the first three), $x = 40$ (the second three), and $x = 60$ (the third three) are easily located, we really do not need a separate plot of residuals versus x to examine the constant variance assumption. It is clear from the original scatterplot and the residual plot shown that we do not have a problem.

```
          SAS

OBS    TEMP    H_LOSS

 1      20       86
 2      20       80
 3      20       77
 4      40       78
 5      40       84
 6      40       75
 7      60       33
 8      60       38
 9      60       43

N=       9
```

REGRESSION ANALYSIS OF HEAT LOSS VS. TEMPERATURE

GENERAL LINEAR MODELS PROCEDURE

DEPENDENT VARIABLE: H_LOSS   HEAT LOSS

SOURCE	DF	SUM OF SQUARES	MEAN SQUARE	F VALUE	PR > F	R-SQUARE	C.V.
MODEL	1	2773.50000000	2773.50000000	21.70	0.0023	0.756134	17.1276
ERROR	7	894.50000000	127.78571429		ROOT MSE		H LOSS MEAN
CORRECTED TOTAL	8	3668.00000000			11.30423435		66.00000000

SOURCE	DF	TYPE I SS	F VALUE	PR > F	DF	TYPE III SS	F VALUE	PR > F
TEMP	1	2773.50000000	21.70	0.0023	1	2773.50000000	21.70	0.0023

PARAMETER	ESTIMATE	T FOR H0: PARAMETER=0	PR > |T|	STD ERROR OF ESTIMATE
INTERCEPT	109.00000000	10.93	0.0001	9.96939762
TEMP	-1.07500000	-4.66	0.0023	0.23074672

OBSERVATION	OBSERVED VALUE	PREDICTED VALUE	RESIDUAL
1	86.00000000	87.50000000	-1.50000000
2	80.00000000	87.50000000	-7.50000000
3	77.00000000	87.50000000	-10.50000000
4	78.00000000	66.00000000	12.00000000
5	84.00000000	66.00000000	18.00000000
6	75.00000000	66.00000000	9.00000000
7	33.00000000	44.50000000	-11.50000000
8	38.00000000	44.50000000	-6.50000000
9	43.00000000	44.50000000	-1.50000000

```
     SUM OF RESIDUALS                          0.00000000
     SUM OF SQUARED RESIDUALS                894.50000000
     SUM OF SQUARED RESIDUALS - ERROR SS       0.00000000
     FIRST ORDER AUTOCORRELATION               0.36109558
     DURBIN-WATSON D                           1.27277809
```

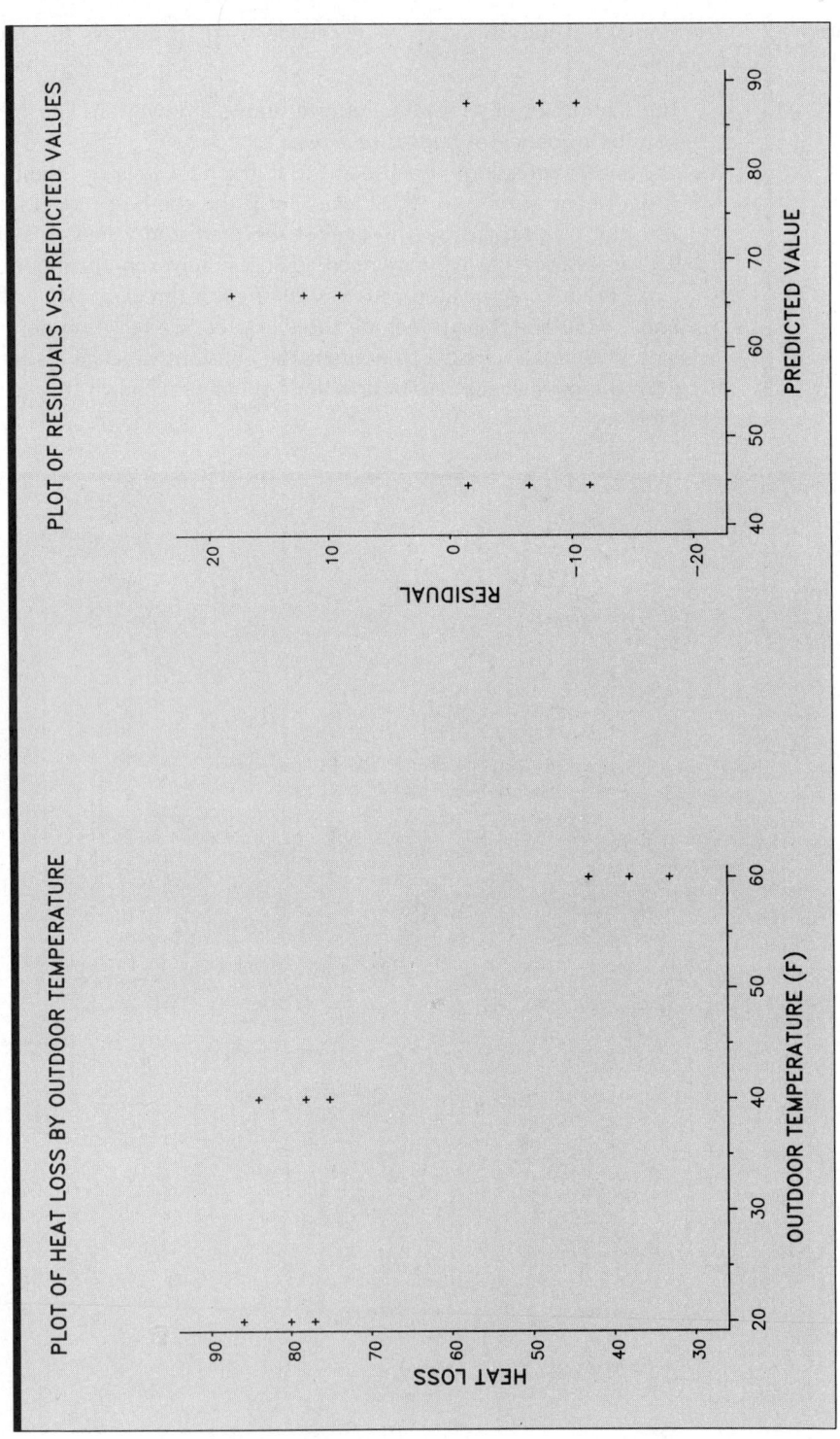

How can we test for the apparent lack of fit of the linear regression model in Example 9.16? When there is more than one observation per level of the independent variable, we can conduct a test for lack of fit of the fitted model by partitioning SSE into two parts, one **pure experimental error** and the other **lack of fit**. Let y_{ij} denote the response for the jth observation at the ith level of the independent variable. Then if there are n_i observations at the ith level of the independent variable, the quantity

pure experimental error

lack of fit

$$\sum_j (y_{ij} - \bar{y}_i)^2$$

provides a measure of what we will call pure experimental error. This sum of squares has $n_i - 1$ degrees of freedom.

Similarly, for each of the other levels of x, we can compute a sum of squares due to pure experimental error. The pooled sum of squares

$$SSP_{exp} = \sum_{ij} (y_{ij} - \bar{y}_i)^2$$

called the sum of squares for pure experimental error, has $\sum_i (n_i - 1)$ degrees of freedom. With SS_{Lack} representing the remaining portion of SSE, we have

$$\begin{array}{ccccc} SSE = & SSP_{exp} & + & SS_{Lack} \\ & \text{due to pure} & & \text{due to lack} \\ & \text{experimental} & & \text{to fit} \\ & \text{error} & & \end{array}$$

If SSE is based on $n - 2$ degrees of freedom in the linear regression model then SS_{Lack} will have df $= n - 2 - \sum_i (n_i - 1)$.

Under the null hypothesis that our model is correct, we can form independent estimates of σ_ε^2, the model error variance, by dividing SSP_{exp} and SS_{Lack} by their respective degrees of freedom; these estimates are called **mean squares** and denoted by MSP_{exp} and MS_{Lack}, respectively.

mean squares

The test for lack of fit is summarized here.

A TEST FOR LACK OF FIT IN LINEAR REGRESSION

H_0: A linear regression model is appropriate.

H_a: A linear regression model is not appropriate.

T.S.: $F = \dfrac{MS_{Lack}}{MSP_{exp}}$

R.R.: For specified value of α, reject H_0 (the adequacy of the model) if the computed value of F exceeds the table value for $df_1 = n - 2 - \sum_i (n_i - 1)$ and $df_2 = \sum_i (n_i - 1)$.

Conclusion: If the F test is significant, this indicates that the linear regression model is inadequate. A nonsignificant result indicates that there is insufficient evidence to suggest that the linear regression model is inappropriate.

EXAMPLE 9.17

Refer to the data of Example 9.16. Conduct a test for lack of fit of the linear regression model.

Solution

Using a calculator it is easy to show that the contributions to experimental error for the differential levels of x are as shown here.

Level of x	$\bar{y}_i$	Contribution to Pure Experimental Error $\sum_j (y_{ij} - \bar{y}_i)^2$	$n_i - 1$
20	81	42	2
40	79	42	2
60	38	50	2
Total		134	6

Summarizing these results, we have

$$SSP_{exp} = \sum_{ij} (y_{ij} - y_i)^2 = 134$$

The output shown for Example 9.16 gives SSE = 894.5, hence by subtraction

$$SS_{Lack} = SSE - SSP_{exp} = 894.5 - 134 = 760.5$$

The sum of squares due to pure experimental error has $\sum_i (n_i - 1) = 6$ degrees of freedom; it follows that with $n = 9$, SS_{Lack} has $n - 2 - \sum_i (n_i - 1) = 1$ degree of freedom. We find that

$$MSP_{exp} = \frac{SSP_{exp}}{6} = \frac{134}{6} = 22.33$$

and

$$MS_{Lack} = \frac{SS_{Lack}}{1} = 760.5$$

The F statistic for the test of lack of fit is

$$F = \frac{MS_{Lack}}{MSP_{exp}} = \frac{760.5}{22.33} = 34.06$$

Using $df_1 = 1$, $df_2 = 6$, and $\alpha = .05$ we will reject H_0 if $F \geq 5.99$.

Since the computed value of F exceeds 5.99, we reject H_0 and conclude that there is significant lack of fit for a linear regression model. The scatterplot shown in Example 9.16 confirms this nonlinearity.

To summarize: in situations where there is more than one y-value at one or more levels of x, it is possible to conduct a formal test for lack of fit of the linear regression model. This test should precede any inferences made using the fitted linear regression line. If the test for lack of fit is significant, some higher-order polynomial in x may be more appropriate. A scatterplot of the data and a residual plot from the linear regression line should help in selecting the appropriate model. More information on the selection of an appropriate model will be discussed along with multiple regression (Chapters 12, 13).

If the F test for lack of fit is not significant, proceed with inferences based on the fitted linear regression line.

EXERCISES

 9.25 A manufacturer of laundry detergent was interested in testing a new product prior to market release. One area of concern was the relationship between the height of the detergent suds in a washing machine as a function of the amount of detergent added in the wash cycle. For a standard size washing machine tub filled to the full level, random assignments of amounts of detergent were made and tested on the washing machine. The data appear below.

Height, y	Amount, x
28.1, 27.6	6
32.3, 33.2	7
34.8, 35.0	8
38.2, 39.4	9
43.5, 46.8	10

a. Plot the data.
b. Fit a linear regression model.
c. Use a residual plot to investigate possible lack of fit.

9.26 Refer to Exercise 9.25.
a. Conduct a test for lack of fit of the linear regression model.
b. If the model is appropriate, give a 95% prediction band for y.

9.27 Refer to Exercise 8.10. Conduct a test for lack of fit and draw conclusions.

9.7 THE CALIBRATION PROBLEM: PREDICTING x FOR A GIVEN VALUE OF y

In experimental situations we are often interested in estimating the value of the independent variable corresponding to a measured value of the dependent variable. This problem will be illustrated for the case in which the dependent variable y is linearly related to an independent variable x.

Consider the calibration of an instrument that measures the flow rate of a chemical process. Let x denote the actual flow rate and y denote a reading on the calibrating instrument. In the calibration experiment the flow rate is controlled at n levels x_i, and the corresponding instrument readings y_i are observed. Suppose we assume a model of the form

$$y_i = \beta_0 + \beta_1 x_i + \varepsilon_i$$

where the ε_is are independent identically distributed normal random variables with mean zero and variance σ_ε^2. Then using the n data points (x_i, y_i), we can obtain the least squares estimates $\hat{\beta}_0$ and $\hat{\beta}_1$. Sometime in the future the experimenter will be interested in estimating the flow rate x from a particular instrument reading y.

The most commonly used estimate is found by replacing $\hat{y}$ by y and solving the least squares equation $\hat{y} = \hat{\beta}_0 + \hat{\beta}_1 x$ for x:

$$\hat{x} = \frac{y - \hat{\beta}_0}{\hat{\beta}_1}$$

Two different inverse prediction problems will be discussed here. The first is for predicting x corresponding to an *observed* value of y; the second is for predicting x corresponding to the mean of $m > 1$ values of y that were obtained independent of the regression data. The solution to the first inverse problem is shown here.

CASE 1: PREDICTING x BASED ON AN OBSERVED y-VALUE

Predictor of x: $\hat{x} = \dfrac{y - \hat{\beta}_0}{\hat{\beta}_1}$

$100(1 - \alpha)\%$ prediction limits for x:

$$\hat{x}_U, \hat{x}_L = \bar{x} + \frac{1}{1 - c^2}\left[(\hat{x} - \bar{x}) \pm \frac{t_{\alpha/2} s_\varepsilon}{\hat{\beta}_1} \sqrt{\frac{n+1}{n}(1 - c^2) + \frac{(\hat{x} - \bar{x})^2}{S_{xx}}} \right]$$

where

$$s_\varepsilon^2 = \frac{SSE}{n - 2}, \qquad c^2 = \frac{t_{\alpha/2}^2 s_\varepsilon^2}{\hat{\beta}_1^2 S_{xx}}$$

and $t_{\alpha/2}$ is based on df $= n - 2$.

It should be noted that since

$$t = \frac{\hat{\beta}_1}{s_\varepsilon / \sqrt{S_{xx}}}$$

is the test statistic for H_0: $\beta_1 = 0$, $c = t_{\alpha/2}/t$. We will require that $|t| > t_{\alpha/2}$; that is, β_1 must be significantly different from zero. Then $c^2 < 1$ and $0 < (1 - c^2) < 1$. The greater the strength of the linear relationship between x and y, the larger the quantity $(1 - c^2)$, making the width of the prediction interval narrower. Note also that we will get a better prediction of x when $\hat{x}$ is closer to the center of the experimental region, as measured by $\bar{x}$. Combining a prediction at an endpoint of the experimental region with a weak linear relationship between x and y ($t \approx t_{\alpha/2}$ and $c^2 < 1$) can create extremely wide limits for the prediction of x.

EXAMPLE 9.18   In Example 8.6 an engineer was interested in calibrating a flow rate meter. The data are shown in Table 9.4.

TABLE 9.4   Data for the Calibration Problem of
Example 9.18

Flow Rate, x	Instrument Reading, y
1	1.4
2	2.3
3	3.1
4	4.2
5	5.1
6	5.8
7	6.8
8	7.6
9	8.7
10	9.5

Use these data to place a 95% prediction interval on x, the actual flow rate corresponding to an instrument reading of 4.0.

Solution

For these data, we found that $S_{xy} = 74.35$, $S_{xx} = 82.5$, and $S_{yy} = 67.065$. It follows that $\hat{\beta}_1 = 74.35/82.5 = .9012$, $\hat{\beta}_0 = \bar{y} - \hat{\beta}_1\bar{x} = 5.45 - (.9012)(5.5) = .4934$ and $SSE = S_{yy} - \hat{\beta}_1 S_{xy} = 67.065 - (.9012)(74.35) = .0608$. The estimate of σ_ε^2 is based on $n - 2 = 8$ degrees of freedom.

$$s_\varepsilon^2 = \frac{SSE}{n-2} = \frac{.0608}{8} = .0076$$

$$s_\varepsilon = .0872$$

For $\alpha = .05$, the t-value for df $= 8$ and a $= .025$ is 2.306.

$$c^2 = \frac{t_{\alpha/2}^2 s_\varepsilon^2}{\hat{\beta}_1^2 S_{xx}} = \frac{(2.306)^2(.0076)}{(.9012)^2(82.5)} = .0006$$

and $1 - c^2 = .9994$. Using $\hat{x} = 3.8910$, the upper and lower prediction limits for x when $y = 4.0$ are as follows:

$$\hat{x}_U = 5.5 + \frac{1}{.9994}\left[-1.6090 + \frac{2.306(.0872)}{.9012}\sqrt{\frac{11}{10}(.9994) + \frac{(-1.6090)^2}{82.5}}\right]$$

$$= 5.5 + \frac{1}{.9994}(-1.6090 + .2373) = 4.1274$$

$$\hat{x}_L = 5.5 + \frac{1}{.9994}(-1.6090 - .2373) = 3.6526$$

Thus the 95% prediction limits for x are 3.65 to 4.13. These limits are shown in Figure 9.10.

The solution to the second inverse prediction problem is summarized next.

FIGURE 9.10

95% Prediction
Interval for x when
y = 4.0, Example 9.18

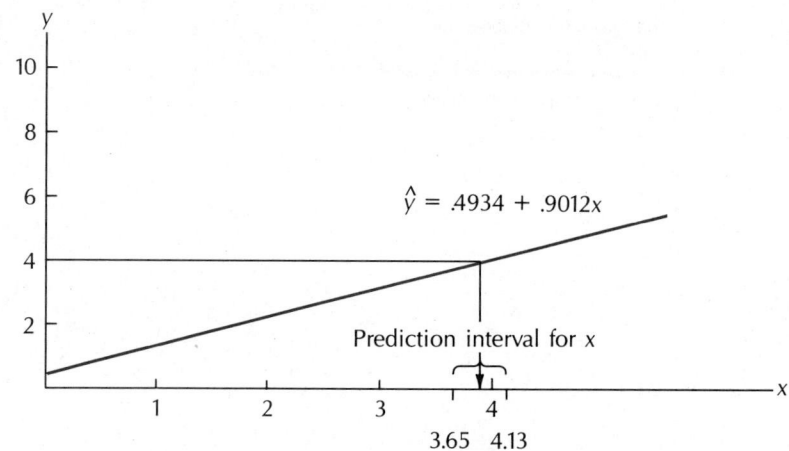

CASE 2:
PREDICTING x
BASED ON m
y-VALUES

Predicting the value of x corresponding to $100P\%$ of the mean of m independent y-values. For $0 \leq P \leq 1$,

Predictor of x: $\hat{x} = \dfrac{P\bar{y}_m - \hat{\beta}_0}{\hat{\beta}_1}$

$$\hat{x}_L, \hat{x}_U = \bar{x} + \frac{1}{1-c^2}\left[(\hat{x} - \bar{x}) \pm \frac{t_{\alpha/2}}{\hat{\beta}_1}\sqrt{\left(s_y^2 P^2 + \frac{s_\varepsilon^2}{n}\right)(1 - c^2) + \frac{(\hat{x} - \bar{x})^2 s_\varepsilon^2}{S_{xx}}}\right]$$

where $\bar{y}_m$ and $s_{\bar{y}}$ are the mean and standard error, respectively, of m independent y-values.

EXERCISES

9.28   A particular forester has become adept at estimating the volume (in cubic feet) of trees on a particular site prior to a timber sale. Since his operation has now expanded, he would like to train another person to assist in estimating the cubic-foot volume of trees. He decides to calibrate his assistant's estimations of actual tree volume. The forester selects a random sample of trees soon to be felled. For each tree, the assistant is to guess the cubic-foot volume y. In addition, the forester obtains the actual cubic-foot volume x after the tree has been chopped down. From these data the forester obtains the calibration curve for the model

$$y = \beta_0 + \beta_1 x + \varepsilon$$

Then in the near future he can use the calibration curve to correct the assistant's estimates of tree volumes. The sample data are summarized below.

Tree	1	2	3	4	5	6	7	8	9	10
Estimated volume, y	12	14	8	12	17	16	14	14	15	17
Actual volume, x	13	14	9	15	19	20	16	15	17	18

Fit the calibration curve using the method of least squares. Does the evidence indicate that the slope is significantly greater than zero? Use $\alpha = .05$.

9.29 Refer to Exercise 9.28.

a. Predict the actual tree volume for a tree the assistant estimates to have a cubic-foot volume of 13.

b. Place a 95% prediction interval on x, the actual tree volume in part (a).

 9.30 Data from 24 patients were obtained to examine the relationship between dose (amount of drug) and cumulative urine volume for a drug product being studied as a diuretic. These data are shown here in the computer output.

a. Locate the linear regression equation. Identify the independent and dependent variables.

b. Use the output to predict dose based on individual y-values of 10, 14, and 19cc. What are the corresponding 99% prediction limits for each of those cases?

```
************************************************** V01-B **************************************************
LINEAR REGRESSION

  FILES
    INDEPENDENT: DOSE
    DEPENDENT: CUMVOL
```

	DOSE		CUMVOL	
ORIGINAL	NATURAL LOG		ORIGINAL	ARCSIN
6.00000	1.79176		7.10000	0.26972
6.00000	1.79176		11.50000	0.34598
6.00000	1.79176		8.40000	0.29405
6.00000	1.79176		8.00000	0.28676
6.00000	1.79176		9.40000	0.31161
6.00000	1.79176		12.00000	0.35374
9.00000	2.19722		13.20000	0.37183
9.00000	2.19722		14.70000	0.39348
9.00000	2.19722		12.70000	0.36438
9.00000	2.19722		15.50000	0.40465
9.00000	2.19722		18.40000	0.44333
9.00000	2.19722		14.40000	0.38923
13.50000	2.60269		12.10000	0.35528
13.50000	2.60269		15.80000	0.40878
13.50000	2.60269		13.80000	0.38061
13.50000	2.60269		20.40000	0.46863
13.50000	2.60269		22.70000	0.49661
13.50000	2.60269		17.00000	0.42499
20.25000	3.00815		19.80000	0.46114
20.25000	3.00815		15.60000	0.40603
20.25000	3.00815		25.30000	0.52706
20.25000	3.00815		13.50000	0.37624
20.25000	3.00815		24.80000	0.52129
20.25000	3.00815		20.90000	0.47481

		Y	Y	Y	Y	Y
	X	MEAN	VARIANCE	#	MINIMUM	MAXIMUM
	1.79176	0.31031	0.00113	6	0.26972	0.35374
	2.19722	0.39448	0.00079	6	0.36438	0.44333
	2.60269	0.42248	0.00282	6	0.35528	0.49661
	3.00815	0.46109	0.00368	6	0.37624	0.52706
TOTAL		0.39709	0.00210	24		

ESTIMATED LINE

	ESTIMATE	95% CONFIDENCE LIMITS
SLOPE:	0.11847	[0.07553, 0.16140]
INTERCEPT:	0.11277	[0.00791, 0.21763]

INDEX OF SIGNIFICANCE OF SLOPE: 0.131

CORRELATION COEFFICIENT: 0.77342

TABLE OF RESIDUALS

X VALUE	OBSERVED Y	PREDICTED Y	STANDARDIZED RESIDUAL
1.79176	0.26972	0.32504	−1.20333
1.79176	0.34598	0.32504	0.45544
1.79176	0.29405	0.32504	−0.67412
1.79176	0.28676	0.32504	−0.83269
1.79176	0.31161	0.32504	−0.29204
1.79176	0.35374	0.32504	0.62432
2.19722	0.37183	0.37307	−0.02713
2.19722	0.39348	0.37307	0.44388
2.19722	0.36438	0.37307	−0.18910
2.19722	0.40465	0.37307	0.68689
2.19722	0.44333	0.37307	1.52821
2.19722	0.38923	0.37307	0.35134
2.60269	0.35528	0.42111	−1.43193
2.60269	0.40878	0.42111	−0.26814
2.60269	0.38061	0.42111	−0.88100
2.60269	0.46863	0.42111	1.03361
2.60269	0.49661	0.42111	1.64217
2.60269	0.42499	0.42111	0.08438
3.00815	0.46114	0.46914	−0.17405
3.00815	0.40603	0.46914	−1.37276
3.00815	0.52706	0.46914	1.25964
3.00815	0.37624	0.46914	−2.02086*
3.00815	0.52129	0.46914	1.13415
3.00815	0.47481	0.46914	0.12312

* RESIDUAL IS GREATER THAN OR EQUAL TO 2 TIMES STANDARD DEVIATION

PREDICTION OF X
  99% PREDICTION LIMITS FOR INDIVIDUAL X VALUE

INPUT VALUE	PREDICTION	PREDICTION LIMITS
10.00000	5.83575*	[2.92557, 7.74469]
14.00000	9.82768	[7.25149, 12.37486]
19.00000	17.37828	[13.58904, 29.75220]

* X VALUE LIES OUTSIDE RANGE OF VALUES IN THE INDEPENDENT FILE

PREDICTION OF X
  % OF CONTROL MEAN
  CONTROL VALUES:

10.00000
20.00000
30.00000
12.00000

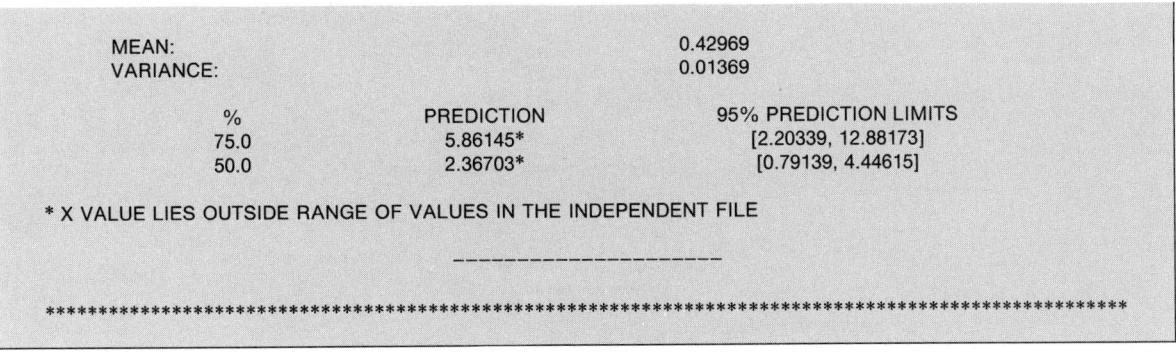

%	PREDICTION	95% PREDICTION LIMITS
75.0	5.86145*	[2.20339, 12.88173]
50.0	2.36703*	[0.79139, 4.44615]

MEAN: 0.42969
VARIANCE: 0.01369

* X VALUE LIES OUTSIDE RANGE OF VALUES IN THE INDEPENDENT FILE

9.31   Refer to the output of Exercise 9.30. Suppose the investigator wanted to predict the dose of the diuretic that would produce a response equivalent to 50% (and 75%) of the response obtained from four patients treated with a known diuretic. Predict x and give appropriate limits for each of these situations.

9.8   SUMMARY

This chapter followed closely the material presented in Chapter 8 for constructing a linear regression equation. Here we showed how to make inferences (using confidence intervals and statistical tests) related to a linear regression model. First we considered confidence intervals and tests related to β_0 and β_1, the intercept and slope in the linear regression model. Second, we developed a confidence interval and statistical test for $E(y)$, the average (expected) value of y for a given setting of the independent variable x. By considering different values of x, we showed how to generate a confidence band for $E(y)$. Next we developed a prediction interval and corresponding prediction bands for predicting the actual value (perhaps the next value) of y for a given value of x.

Finally, after these sections on inferences related to linear regression, we presented ways to examine lack of fit of a linear regression model to a data set, ways to linearize the relationship between y and x using transformations, and several quick, portable statistics.

This chapter has provided an important foundation for the further discussions of regression in Chapters 12 and 13.

KEY FORMULAS: LINEAR REGRESSION AND CORRELATION

1. Means and standard errors for $\hat{\beta}_0$ and $\hat{\beta}_1$

$$\mu_{\hat{\beta}_0} = \beta_0 \qquad\qquad \mu_{\hat{\beta}_1} = \beta_1$$

$$\sigma_{\hat{\beta}_0} = \sigma_\varepsilon \sqrt{\frac{\sum x^2}{nS_{xx}}} \qquad \sigma_{\hat{\beta}_1} = \frac{\sigma_\varepsilon}{\sqrt{S_{xx}}}$$

Note: For all practical problems, σ_ε is unknown and must be estimated from the sample data.

2. Calculation of s_ε, an estimate of σ_ε for linear regression

$$s_\varepsilon = \sqrt{\frac{SSE}{n-2}} \text{ where } SSE = S_{yy} - \hat{\beta}_1 S_{xy}$$

3. $100(1 - \alpha)\%$ confidence intervals for β_0 and β_1

$$\beta_0: \hat{\beta}_0 \pm t_{\alpha/2} s_\varepsilon \sqrt{\frac{\sum x^2}{n S_{xx}}}$$

$$\beta_1: \hat{\beta}_1 \pm t_{\alpha/2} \frac{s_\varepsilon}{\sqrt{S_{xx}}}$$

4. Statistical tests for β_0 and β_1

$H_0: \beta_0 = 0$ $\qquad\qquad\qquad\qquad\qquad$ $H_0: \beta_1 = 0$

$\text{T.S.: } t = \dfrac{\hat{\beta}_0}{s_\varepsilon \sqrt{\dfrac{\sum x^2}{n S_{xx}}}}, \quad df = n-2 \qquad \text{T.S.: } t = \dfrac{\hat{\beta}_1}{s_\varepsilon / \sqrt{S_{xx}}}, \quad df = n-2$

5. Alternative test for β_1

$H_0: \beta_1 = 0$

$\text{T.S.: } F = SSR_{EC} / s_\varepsilon^2$

$\qquad df_1 = 1, \qquad df_2 = n-2$

6. Test for β_1 (intercept known)

$H_0: \beta_1 = 0$

$\text{T.S.: } t = \dfrac{\hat{\beta}_1}{s_\varepsilon / \sum x_i^2}, \qquad df = n-1$

where

$$\hat{\beta}_1 = \frac{\sum x_i y_i - \beta_0 \sum x_i}{\sum x_i^2}$$

and

$$s_\varepsilon^2 = \sum (\hat{y}_i - y_i)^2 / (n-1)$$

7. $100(1 - \alpha)\%$ confidence interval for $E(y)$

$$\hat{y} \pm t_{\alpha/2} s_\varepsilon \sqrt{\frac{1}{n} + \frac{(x - \bar{x})^2}{S_{xx}}}$$

8. Statistical test for $E(y)$

$H_0: E(y) = \mu_0$

$$\text{T.S.: } t = \frac{\hat{y} - \mu_0}{s_\varepsilon \sqrt{\dfrac{1}{n} + \dfrac{(x - \bar{x})^2}{S_{xx}}}}, \qquad df = n - 2$$

9. $100(1 - \alpha)\%$ prediction interval for y

$$\hat{y} \pm t_{\alpha/2} s_\varepsilon \sqrt{1 + \frac{1}{n} + \frac{(x - \bar{x})^2}{S_{xx}}}$$

10. Test for lack of fit

H_0: A linear regression model is appropriate

$$\text{T.S.: } F = \frac{MS_{\text{Lack}}}{MSP_{\text{exp}}}$$

where

$$MSP_{\text{exp}} = \frac{SSP_{\text{exp}}}{\sum (n_i - 1)} = \frac{\sum_{i,j} (y_{ij} - \bar{y}_i)^2}{\sum_i (n_i - 1)}$$

and

$$MS_{\text{Lack}} = \frac{SSE - SSP_{\text{exp}}}{n - 2 - \sum (n_i - 1)}$$

11. $100(1 - \alpha)\%$ prediction interval for x

$$\hat{x}_u = \bar{x} + \frac{1}{1 - c^2} [(\hat{x} - \bar{x}) + d]$$

$$\hat{x}_L = \bar{x} + \frac{1}{1 - c^2} [(\hat{x} - \bar{x}) - d]$$

where

$$c^2 = \frac{t_{\alpha/2}^2 s_\varepsilon^2}{\hat{\beta}_1^2 S_{xx}}$$

and

$$d = \frac{t_{\alpha/2}}{\hat{\beta}_1} s_\varepsilon \sqrt{\frac{n + 1}{n} (1 - c^2) + \frac{(\hat{x} - \bar{x})^2}{S_{xx}}}$$

12. $100(1 - \alpha)\%$ prediction interval for x based on m y-values

$$\hat{x}_u = \bar{x} + \frac{1}{1 - c^2} [(\hat{x} - \bar{x}) + g]$$

$$\hat{x}_L = \bar{x} + \frac{1}{1 - c^2} [(\hat{x} - \bar{x}) - g]$$

where

$$\hat{x} = \frac{P\bar{y}_m - \hat{\beta}_0}{\hat{\beta}_1}$$

and

$$g = \frac{t_{\alpha/2}}{\hat{\beta}_1} \sqrt{\left(s_y^2 P^2 + \frac{s_\varepsilon^2}{n}\right)(1 - c^2) + \frac{(\hat{x} - \bar{x})^2 s_\varepsilon^2}{S_{xx}}}$$

EXERCISES

9.32 In Example 8.1, we presented the live and dressed weights for a sample of nine steers. The data are shown here.

Live weight, x	Dressed weight, y
4.2	2.8
3.8	2.5
4.8	3.1
3.4	2.1
4.5	2.9
4.6	2.8
4.3	2.6
3.7	2.4
3.9	2.5

a. Plot the data.
b. In Example 8.1, we found $S_{xx} = 1.72$, $S_{xy} = 1.06$, and $\hat{y} = .087 + .616x$. Is there a positive linear relationship between x and y? Give a p-value associated with your conclusion.

9.33 Refer to Exercise 9.32. Use the data to predict the live weight for $y = 3.0$. Use a 95% prediction interval.

9.34 The thermal pollution data of Exercise 9.17 are shown here.

Weight, x (in 1000 lb)	Btu per Vehicle Mile, y (in 1000s)
1.8	4.0
2.6	5.2
4.2	8.5
5.0	11.6
4.8	10.1
3.4	6.3

a. Compute s_ε^2.
b. Give a 95% confidence interval for β_1 and discuss the meaning of your finding.

9.35 Labor data (in terms of manhours) are presented here for the number of orders processed per month by a large manufacturing center.

Month	Orders Processed, x	Manhours Required to Process Orders, y
1	3000	8000
2	3400	9200
3	4000	10000
4	2800	7500
5	2000	5800
6	1700	5000
7	1400	4400
8	1300	3700
9	1000	3100
10	600	2220
11	1500	4100
12	2200	5500
13	3300	8100
14	3600	9400
15	4100	10600
16	3200	7900

a. Examine the plot of orders versus months. Are there cyclical patterns?

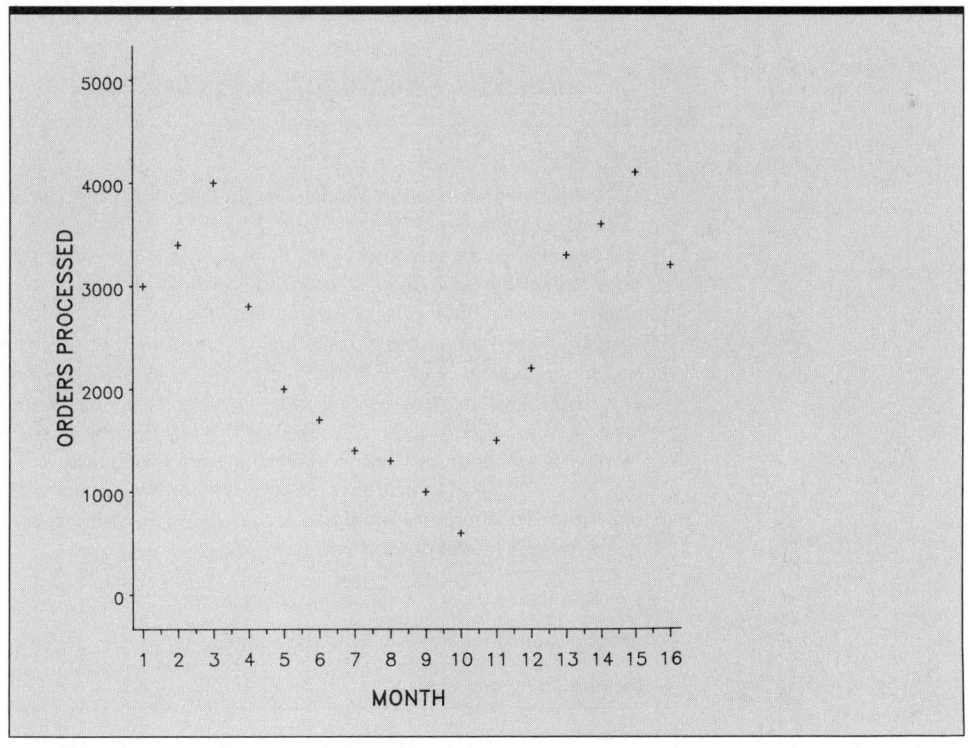

b. From the plot of orders (x) versus manhours (y) ignoring the apparent cyclical effect of part (a), what regression model might adequately describe the data?

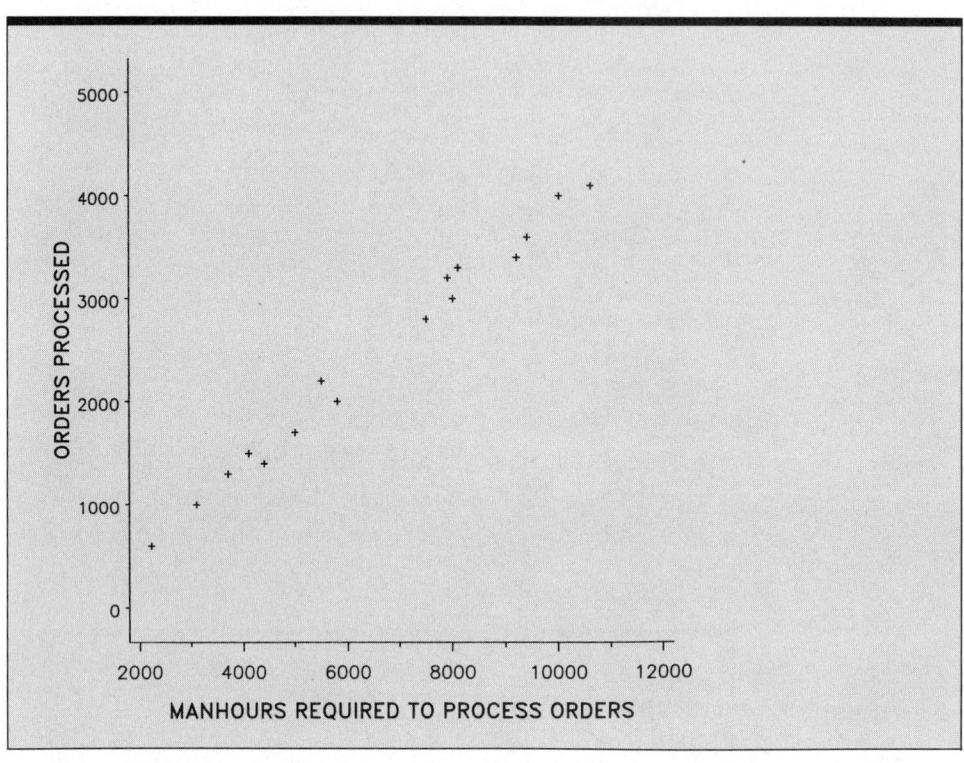

9.36 Refer to Exercise 9.35.

a. Fit the linear regression model $y = \beta_0 + \beta_1 x + \varepsilon$, and draw conclusions about the slope and intercept.

b. Show the results of your test for β_1 in an analysis of variance table.

9.37 Refer to Exercise 9.35. Redo your analysis without an intercept. Does this model provide a better fit to the data? Why or why not? Does it make practical sense to use such a model?

9.38 A company was interested in calibrating an instrument to measure the consistency of the liquid in a paper mill. In the making of paper, small particles of wood fiber are conveyed in a liquid flow to the screening process, where the pulp is separated from the water, dried, and made into paper. Controlling the consistency of the fibers in the liquid flow represents an important step in maintaining the final quality of the paper. Let us assume that for the calibration of the consistency meter, we are able to control the consistency of a simulated laboratory liquid flow at various levels x while monitoring the meter reading y. Assuming a model of the form

$$y = \beta_0 + \beta_1 x + \varepsilon,$$

use the sample data below to fit the model.

Meter reading, y	2.16	2.15	2.17	2.26	2.35	2.39	2.42	2.51
Actual consistency, x	2.16	2.17	2.18	2.20	2.22	2.24	2.26	2.28

Is there sufficient evidence to indicate that the slope is greater than zero? Use $\alpha = .05$.

9.39 Computer output for the data of Exercise 9.38 is shown here.

a. Compare the least squares estimates to those obtained in Exercise 9.38.

b. Give the level of significance for a test of $H_0: \beta_1 = 0$.

c. Identify 95% prediction limits for the independent variable (actual consistency of the flow) when the observed meter reading is 2.20.

```
      2.16, 2.16
   ? 2.17, 2.15
   ? 2.18, 2.17
   ? 2.20, 2.26
   ? 2.22, 2.35
   ? 2.24, 2.39
   ? 2.26, 2.42
   ? 2.28, 2.51
   WANT A LISTING OF THIS FILE (Y OR N)? Y
   WANT TRANSFORMATION OF INDEPENDENT VARIABLE
   (0 = NO, 1 = NATURAL LOG, 2 = COMMON LOG) ? 0
   WANT TRANSFORMATION ON DEPENDENT VARIABLE
   (0 = NO, 1 = NATURAL LOG, 2 = COMMON LOG, 3 = ARCSIN, 4 = SQUARE ROOT) ?
   0

   SIMPLE LINEAR REGRESSION
```

	FILES:	X	Y
	OBSERVATIONS:	INDEPENDENT	DEPENDENT
		2.1600	2.1600
		2.1700	2.1500
		2.1800	2.1700
		2.2000	2.2600
		2.2200	2.3500
		2.2400	2.3900
		2.2600	2.4200
		2.2800	2.5100

	INDEPENDENT	DEPENDENT
NO. OBSERVATIONS	8.0000	8.0000
ARITHMETIC MEANS	2.2138	2.3013
STANDARD DEVIATION	0.0437	0.1361

	Y INTERCEPT	SLOPE
ESTIMATE	−4.5054	3.0747
STANDARD DEVIATION	0.4367	0.1973
T-STATISTIC-SIGNIFICANCE	−10.3159	15.5877
DEGREES OF FREEDOM	6.0000	6.0000
PROBABILITY	0.0000	0.0000
RESIDUAL SUM OF SQUARES	0.0031	
RESIDUAL MEAN SQUARE	0.0005	

```
   WANT TO ESTIMATE EFFECTIVE DOSE (Y OR N)? Y
   ENTER NO. OF ESTIMATES? 1
   ENTER Y VALUES
   ? 2.20
```

ESTIMATE OF X

Y VALUE	ESTIMATE	95% CONFIDENCE INTERVAL
2.2000	2.1808	[2.1597, 2.2002]

9.40   Consider the following data.

y	x
1.0	−1
2.0	−1
1.0	−1
6.0	1
7.0	1
6.5	1
2.0	3
3.0	3

a. Fit the linear regression model, $y = \beta_0 + \beta_1 x + \varepsilon$.

b. Plot the residuals versus $\hat{y}$. Is there evidence of lack of fit?

9.41   Refer to Exercise 9.40 and test for lack of fit. Draw conclusions and make recommendations for a model relating y to x.

9.42   An airline was interested in comparing the estimated time of arrival (ETA) and the actual time of arrival (ATA) for domestic flights. To do this a random sample of 20 flights was taken over the past month for flights with ETAs in the time frame from noon until 6:00 PM. These data are shown here in order of ETA.

ETA	ATA
12:15	12:16
12:30	12:52
12:30	12:40
1:00	1:28
1:30	1:42
2:10	2:20
2:35	2:52
3:05	3:11
3:20	3:55
3:50	4:14
4:15	4:55
4:40	4:46
4:50	5:16
5:00	5:33
5:00	5:20
5:10	5:28
5:15	6:30
5:30	6:05
6:00	7:10
6:00	6:30

a. Plot the data.

b. Would a linear regression model appear appropriate for characterizing the relationship between ETA and ATA? If so, why? If not, why not?

c. Guess the value of the correlation coefficient.

9.43   a. Obtain the linear regression line for the data of Exercise 9.42.

b. Plot the residuals versus $\hat{y}$. Does any recognizable pattern emerge?

c. Can you test for lack of fit? If so, do it and interpret your results.

9.44 Refer to Exercises 9.42 and 9.43. Suppose that you were given a computer printout of the data for Exercise 9.42 and the eleventh observation was shown as (ETA = 4:15, ATA = 9.55).

a. What circumstances might have given rise to such a data point? Might such a value be an error? If so, how might it have occurred?

b. How might you check for errors and odd-values in this or other data sets?

9.45 A study was conducted to examine the effect of different levels of nitrogen on the yield of lettuce plants. Use the data shown here to fit a linear regression model. Test for possible lack of fit of the model.

Coded Nitrogen	Yield (Emergent Stalks per Plot)
1	21, 18, 17
2	24, 22, 26
3	34, 29, 32

9.46 The specific activity of the enzyme sucrase was measured by extracting a portion of the intestines of 24 patients who underwent an intestinal bypass. After the sections were extracted they were homogenized and analyzed for enzyme activity (Carter (1981)). Two different methods can be used to measure the activity of sucrase: the homogenate method and the pellet method. Data for the 24 patients are shown here for the two methods:

Sucrase Activity As Measured by the
Homogenate and Pellet Methods

Patient	Homogenate Method, y	Pellet Method, x
1	18.88	70.00
2	7.26	55.43
3	6.50	18.87
4	9.83	40.41
5	46.05	57.43
6	20.10	31.14
7	35.78	70.10
8	59.42	137.56
9	58.43	221.20
10	62.32	276.43
11	88.53	316.00
12	19.50	75.56
13	60.78	277.30
14	77.92	331.50
15	51.29	133.74
16	77.91	221.50
17	36.65	132.93
18	31.17	85.38
19	66.09	142.34
20	115.15	294.63
21	95.88	262.52
22	64.61	183.56
23	37.71	86.12
24	100.82	226.55

a. Examine the scatterplot of the data. Might a linear model adequately describe the relationship between the two methods?

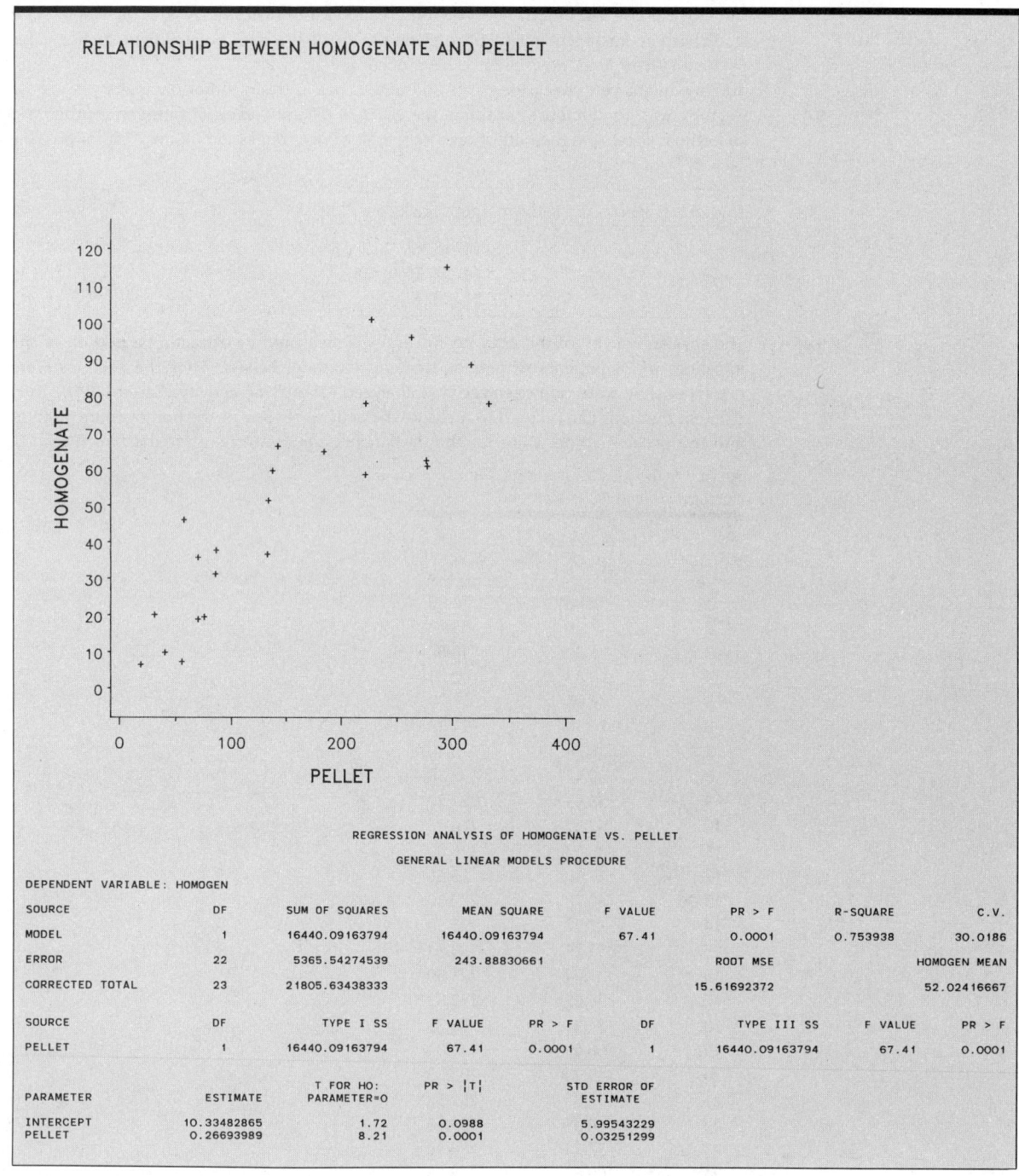

RELATIONSHIP BETWEEN HOMOGENATE AND PELLET

REGRESSION ANALYSIS OF HOMOGENATE VS. PELLET

GENERAL LINEAR MODELS PROCEDURE

DEPENDENT VARIABLE: HOMOGEN

SOURCE	DF	SUM OF SQUARES	MEAN SQUARE	F VALUE	PR > F	R-SQUARE	C.V.
MODEL	1	16440.09163794	16440.09163794	67.41	0.0001	0.753938	30.0186
ERROR	22	5365.54274539	243.88830661		ROOT MSE		HOMOGEN MEAN
CORRECTED TOTAL	23	21805.63438333			15.61692372		52.02416667

SOURCE	DF	TYPE I SS	F VALUE	PR > F	DF	TYPE III SS	F VALUE	PR > F
PELLET	1	16440.09163794	67.41	0.0001	1	16440.09163794	67.41	0.0001

PARAMETER	ESTIMATE	T FOR H0: PARAMETER=0	PR > |T|	STD ERROR OF ESTIMATE
INTERCEPT	10.33482865	1.72	0.0988	5.99543229
PELLET	0.26693989	8.21	0.0001	0.03251299

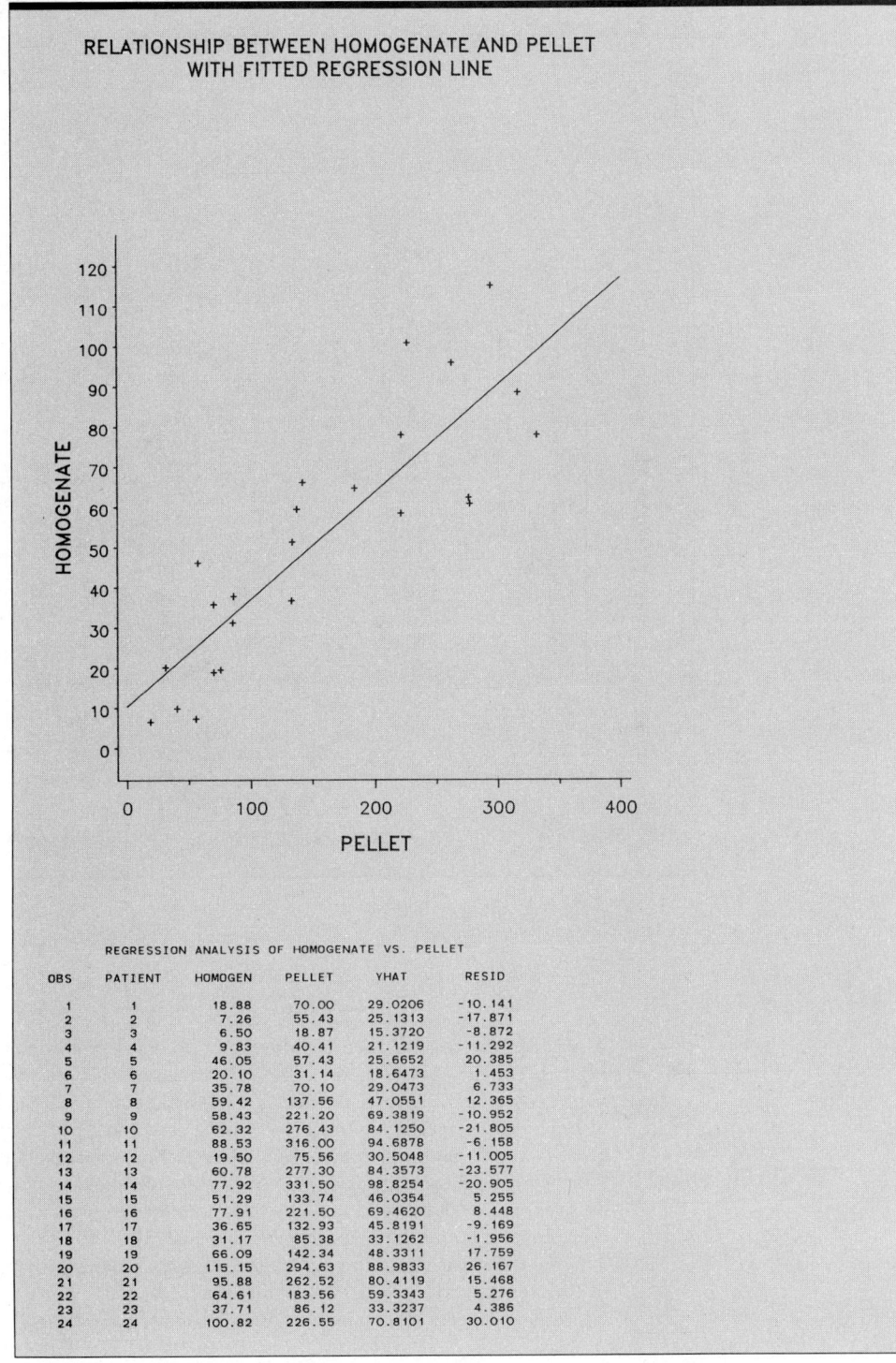

RELATIONSHIP BETWEEN HOMOGENATE AND PELLET
WITH FITTED REGRESSION LINE

REGRESSION ANALYSIS OF HOMOGENATE VS. PELLET

OBS	PATIENT	HOMOGEN	PELLET	YHAT	RESID
1	1	18.88	70.00	29.0206	-10.141
2	2	7.26	55.43	25.1313	-17.871
3	3	6.50	18.87	15.3720	-8.872
4	4	9.83	40.41	21.1219	-11.292
5	5	46.05	57.43	25.6652	20.385
6	6	20.10	31.14	18.6473	1.453
7	7	35.78	70.10	29.0473	6.733
8	8	59.42	137.56	47.0551	12.365
9	9	58.43	221.20	69.3819	-10.952
10	10	62.32	276.43	84.1250	-21.805
11	11	88.53	316.00	94.6878	-6.158
12	12	19.50	75.56	30.5048	-11.005
13	13	60.78	277.30	84.3573	-23.577
14	14	77.92	331.50	98.8254	-20.905
15	15	51.29	133.74	46.0354	5.255
16	16	77.91	221.50	69.4620	8.448
17	17	36.65	132.93	45.8191	-9.169
18	18	31.17	85.38	33.1262	-1.956
19	19	66.09	142.34	48.3311	17.759
20	20	115.15	294.63	88.9833	26.167
21	21	95.88	262.52	80.4119	15.468
22	22	64.61	183.56	59.3343	5.276
23	23	37.71	86.12	33.3237	4.386
24	24	100.82	226.55	70.8101	30.010

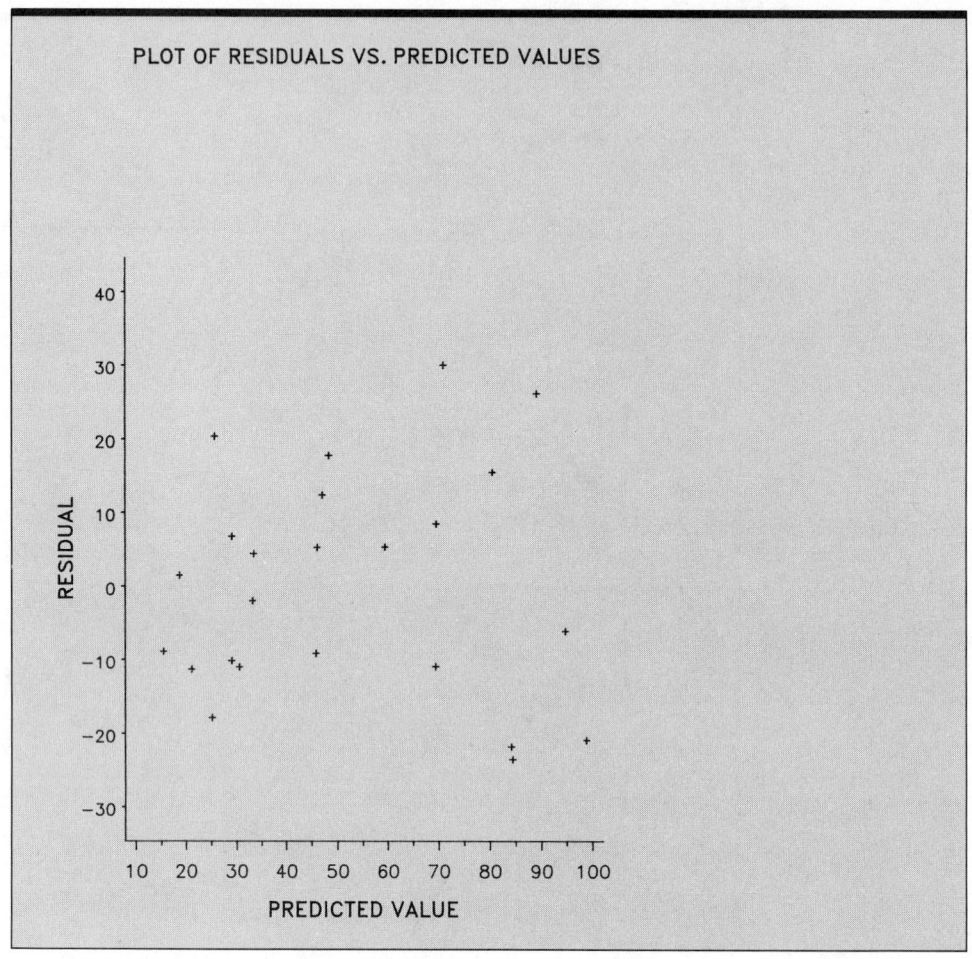

PLOT OF RESIDUALS VS. PREDICTED VALUES

b. Examine the residual plot; are there any potential problems uncovered by the plot?

9.47   Refer to Exercise 9.46. In general the pellet method is more time consuming than the homogenate method, yet it provides a more accurate measure of sucrase activity.

a. How might you estimate the pellet reading based on a particular homogenate reading?

b. How would you develop a confidence (prediction) interval about your point estimate?

9.48   A large chemical company examined the impact of immediate development expenditures on sales ($100,000) by collecting quarterly data for the preceding six years. A portion of the computer output for a linear regression analysis is shown here.

a. Write the linear regression model and the least squares prediction equation.

b. Give results and a conclusion for the test $H_0: \beta_1 = 0$.

c. Can you suggest other variable costs that may be important?

EXERCISES

DEPENDENT VARIABLE: SALES				
SOURCE	DF	SUM OF SQUARES	MEAN SQUARE	F VALUE
MODEL	1	35139.08936255	35139.0893255	24.5
ERROR	22	31582.24397078	1435.55654413	
CORRECTED TOTAL	23	66721.33333333		

SOURCE	DF	TYPE I SS	F VALUE	PR > F
DEVEL	1	35139.08936255	24.48	0.0001

PARAMETER	ESTIMATE	T FOR HO: PARAMETER = 0	PR > |T|	STD ERROR OF ESTIMATE
INTERCEPT	−663.89962895	−3.37	0.0028	197.0663320
DEVEL	144.48242659	4.95	0.0001	29.2031655

PR > F	R-SQUARE	C.V.
0.0001	0.526654	12.2090

STD DEV	SALES MEAN
37.88873901	310.33333333

9.49 A random sample of 14 pharmacies was used to examine the relation between sales volume and the profit before tax (PBT). These data are shown here:

Pharmacy	Sales Volume, x ($1000)	PBT, y ($1000)
1	38	1.3
2	20	2.1
3	48	2.2
4	44	2.6
5	56	3.3
6	39	4.0
7	65	4.1
8	84	4.2
9	82	5.5
10	105	5.7
11	126	7.0
12	52	7.5
13	80	7.7
14	101	7.9

a. Plot the sample data.
b. Calculate the sample correlation coefficient.
c. Is there a significant linear trend between sales (x) and PBT (y)?

9.50 Refer to Exercise 9.49. Refer to a residual plot for these data. Are there any obvious outliers? Do the assumptions for a linear regression model seem to hold?

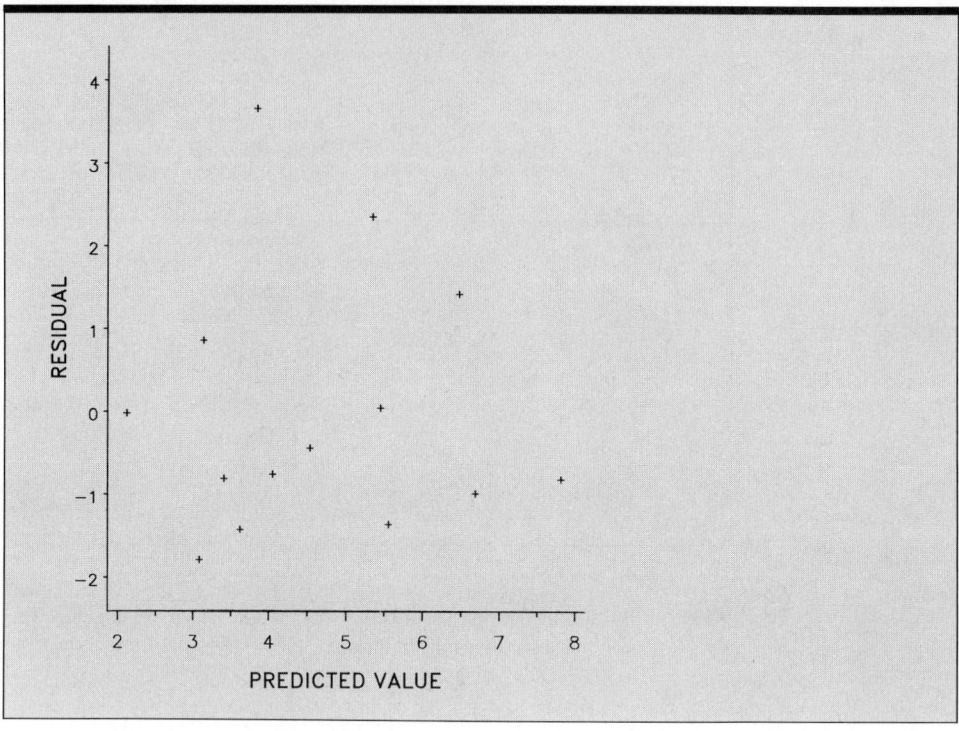

9.51   Refer to the following data:

x	y
1	13.5
1	15.4
1	16.1
2	18.2
2	19.6
2	20.2
4	21.8
4	22.2
4	23.1
8	23.6
8	24.7
8	24.9

a. Plot y versus x.

b. Compute $\log_{10} x$ and plot y versus $\log_{10} x$.

c. Which plot (parts (a) and (b)) looks more linear?

9.52   Refer to the data of Exercise 9.51.

a. Fit a linear regression model for y and x.

b. Compute s_ε^2 for this model.

9.53   Refer to the data of Exercise 9.51.

a. Fit a linear regression model for y and $\log_{10} x$.

EXERCISES

b. Compute s_ε^2 for this model.

c. Compare the fit of this model to that for y versus x in Exercise 9.52 using s_ε^2. Which model provides the better fit? Does this agree with your opinion in Exercise 9.51, based on plots of the data?

9.54 Use the data in Exercise 9.51 and the linear regression of 9.53 to construct a 95% confidence band for $E(y)$. Within what limits would we expect $E(y)$ to be when $x = 5$? When $x = 9$?

9.55 Refer to Exercise 9.54. Construct a 95% prediction band on future values of y. Give the prediction limits for y when $x = 5$ and $x = 9$. Compare these to the confidence limits for $E(y)$ at these same values of x.

9.56 In screening for compounds useful in treating hypertension (high blood pressure), six rats are assigned to each of three groups. The rats in Group 1 receive .1 mg/kg of a test compound; those in Groups 2 and 3 receive .2 and .4 mg/kg, respectively. The response of interest is the decrease in blood pressure two hours postdose, compared to the corresponding predose blood pressure. The data are shown here:

	Dose, x	Blood Pressure Drop (mm Hg), y					
Group 1	.1 mg/kg	10	12	15	16	13	11
Group 2	.2 mg/kg	25	22	26	19	18	24
Group 3	.4 mg/kg	30	32	35	27	26	29

a. Use a software package to fit the model

$$y = \beta_0 + \beta_1 \log_{10} x + \varepsilon$$

b. Use residual plots to examine the fit to the model in part (a).

c. Conduct a statistical test of $H_0: \beta_1 = 0$ vs. $H_a: \beta_1 > 0$. Give the p-value for your test.

9.57 Population and area data are listed here by state. Plot the data. Compute the correlation coefficient. Can inferences be made based on these data?

State	Population (000)	Area (000 sq. mi.)
Maine	1125	33.3
New Hampshire	921	9.3
Vermont	511	9.6
Massachusetts	5737	8.3
Rhode Island	947	1.2
Connecticut	3108	5.0
New York	17558	49.1
New Jersey	7365	7.8
Pennsylvania	11864	45.3
Ohio	10798	41.3
Indiana	5490	36.2
Illinois	11427	56.3
Michigan	9262	58.5
Wisconsin	4706	56.2
Minnesota	4076	84.4
Iowa	2914	56.3

State	Population (000)	Area (000 sq. mi.)
Missouri	4917	69.7
North Dakota	653	70.7
South Dakota	691	77.1
Nebraska	1570	77.4
Kansas	2364	82.2
Delaware	594	2.0
Maryland	4217	10.5
Virginia	5347	40.8
West Virginia	1950	24.2
North Carolina	5882	52.7
South Carolina	3122	31.1
Georgia	5463	58.9
Florida	9746	58.7
Kentucky	3661	40.4
Tennessee	4591	42.1
Alabama	3894	51.8
Mississippi	2521	47.7
Arkansas	2286	53.1
Louisiana	4206	47.8
Oklahoma	3025	70.0
Texas	14229	266.8
Montana	787	147.0
Idaho	944	83.6
Wyoming	470	97.8
Colorado	2890	104.0
New Mexico	1303	121.6
Arizona	2718	114.0
Utah	1461	84.9
Nevada	800	110.6
Washington	4132	68.1
Oregon	2633	97.1
California	23668	158.8
Alaska	402	591.0
Hawaii	965	6.5

U.S. Bureau of Census 1980 Census of Population

 9.58   The following data give the profit ($000) per trip of the Space Shuttle (y) and the dollar amount of items per payload for each trip (x).

Trip:	1	2	3	4	5	6
Items:	7500	12500	15200	9900	8700	15100
Profit:	5	8	9	8	7	10

a. Plot the data.
b. Using the output below, give the least squares prediction equation for profit.
c. Is there a significant linear relationship between the number of items and the profit per trip? Give the value of the test statistic and the p-value.

```
                    REGRESSION ANALYSIS OF PROFIT BY PAYLOAD ITEMS
                         GENERAL LINEAR MODELS PROCEDURE

DEPENDENT VARIABLE: PROFIT

SOURCE                 DF      SUM OF SQUARES      MEAN SQUARE    F VALUE       PR > F      R-SQUARE          C.V.

MODEL                   1        12.20451345      12.20451345      18.57       0.0126      0.822776        10.3491

ERROR                   4         2.62881988       0.65720497                  ROOT MSE                  PROFIT MEAN

CORRECTED TOTAL         5        14.83333333                                0.81068179                  7.83333333


SOURCE                 DF          TYPE I SS      F VALUE      PR > F      DF       TYPE III SS      F VALUE      PR > F

ITEMS                   1        12.20451345       18.57       0.0126       1      12.20451345       18.57       0.0126


                                  T FOR H0:      PR > |T|         STD ERROR OF
PARAMETER             ESTIMATE    PARAMETER=0                       ESTIMATE

INTERCEPT           2.37654568        1.82        0.1436         1.30880887
ITEMS               0.00047519        4.31        0.0126         0.00011027
```

9.59 Refer to Exercise 9.58. Use the output shown here to find the Spearman's rank order correlation. Is there a significant positive relationship between x and y? What's the value of the test statistic and the p-value for the test that lead you to this conclusion?

```
                              SPEARMANS RANK ORDER CORRELATION

VARIABLE      N              MEAN            STD DEV            MEDIAN            MINIMUM            MAXIMUM

PROFIT        6         7.83333333        1.72240142        8.00000000        5.00000000       10.00000000
ITEMS         6     11483.33333333     3287.80575258    11200.00000000     7500.00000000    15200.00000000

            SPEARMAN CORRELATION COEFFICIENTS / PROB > |R| UNDER H0:RHO=0 / N = 6
                                      PROFIT     ITEMS

                          PROFIT     1.00000    0.92763
                                     0.0000     0.0077

                          ITEMS      0.92763    1.00000
                                     0.0077     0.0000
```

9.60 A study was conducted to examine the list price of residential properties and the actual sale price as listed in the records of the County Court House. A random sample of 20 was taken from the list of residential sales recorded over the past six months. These data are shown here.

Sale	Sale Price ($000)	List Price ($000)
1	45.0	49.9
2	58.0	59.0
3	66.5	69.0
4	67.5	75.0
5	69.0	74.0
6	72.5	79.5
7	74.0	80.0
8	74.9	78.0
9	82.0	89.9

Sale	Sale Price ($000)	List Price ($000)
10	84.5	88.1
11	83.5	91.0
12	89.0	94.9
13	90.0	93.9
14	93.9	99.9
15	92.0	98.5
16	102.0	105.9
17	106.9	115.0
18	120.5	139.9
19	147.0	165.0
20	206.5	229.9

a. Examine a plot of the data.
b. Use the computer output shown here to give the least squares prediction equation relating sale price (y) to list price (x). Give the p-value for a test of $H_0: \beta_1 = 0$ versus $H_a: \beta_1 \neq 0$.

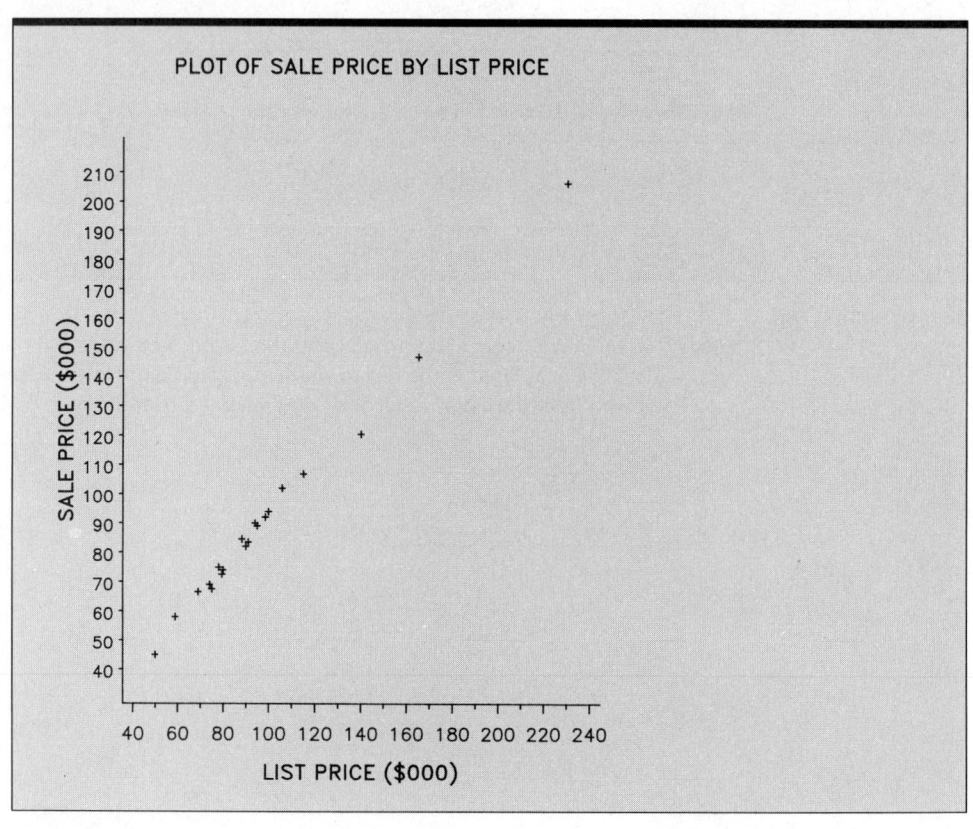

EXERCISES

```
      LIST OF DATA

 OBS    SALE      LIST

  1     45.0      49.9
  2     58.0      59.0
  3     66.5      69.0
  4     67.5      75.0
  5     69.0      74.0
  6     72.5      79.5
  7     74.0      80.0
  8     74.9      78.0
  9     82.0      89.9
 10     84.5      88.1
 11     83.5      91.0
 12     89.0      94.9
 13     90.0      93.9
 14     93.9      99.9
 15     92.0      98.5
 16    102.0     105.9
 17    106.9     115.0
 18    120.5     139.9
 19    147.0     165.0
 20    206.5     229.9
 21              150.0

 N=      21
```

REGRESSION ANALYSIS OF SALE PRICE BY LIST PRICE

GENERAL LINEAR MODELS PROCEDURE

DEPENDENT VARIABLE: SALE

SOURCE	DF	SUM OF SQUARES	MEAN SQUARE	F VALUE	PR > F	R-SQUARE	C.V.
MODEL	1	23515.88443941	23515.88443941	3282.72	0.0001	0.994547	2.9328
ERROR	18	128.94356059	7.16353114		ROOT MSE		SALE MEAN
CORRECTED TOTAL	19	23644.82800000			2.67647738		91.26000000

SOURCE	DF	TYPE I SS	F VALUE	PR > F	DF	TYPE III SS	F VALUE	PR > F
LIST	1	23515.88443941	3282.72	0.0001	1	23515.88443941	3282.72	0.0001

PARAMETER	ESTIMATE	T FOR HO: PARAMETER=0	PR > |T|	STD ERROR OF ESTIMATE
INTERCEPT	5.29292116	3.28	0.0042	1.61538230
LIST	0.86998005	57.30	0.0001	0.01518421

OBSERVATION	OBSERVED VALUE	PREDICTED VALUE	RESIDUAL	LOWER 95% CL INDIVIDUAL	UPPER 95% CL INDIVIDUAL
1	45.00000000	48.70492575	-3.70492575	42.73547556	54.67437595
2	58.00000000	56.62174423	1.37825577	50.72152042	62.52196804
3	66.50000000	65.32154475	1.17845525	59.48167736	71.16141214
4	67.50000000	70.54142506	-3.04142506	64.72966185	76.35318827
5	69.00000000	69.67144501	-0.67144501	63.85542576	75.48746426
6	72.50000000	74.45633529	-1.95633529	68.66158932	80.25108127
7	74.00000000	74.89132532	-0.89132532	69.09825364	80.68439700
8	74.90000000	73.15136522	1.74863478	67.35133604	78.95139439
9	82.00000000	83.50412784	-1.50412784	77.73521919	89.27303648
10	84.50000000	81.93816374	2.56183626	76.16613943	87.71018805
11	83.50000000	84.46110589	-0.96110589	78.69382064	90.22839114
12	89.00000000	87.85402810	1.14597190	82.09078034	93.61727585
13	90.00000000	86.98404804	3.01595196	81.22002076	92.74807533
14	93.90000000	92.20392836	1.69607164	86.44193001	97.96592670
15	92.00000000	90.98595628	1.01404372	85.22405313	96.74785943
16	102.00000000	97.42380867	4.57619133	91.65748313	103.19013420
17	106.90000000	105.34062714	1.55937286	99.55564618	111.12560811
18	120.50000000	127.00313044	-6.50313044	121.09405289	132.91220799
19	147.00000000	148.83962974	-1.83962974	142.70308369	154.97617580
20	206.50000000	205.30133512	1.19866488	198.18192546	212.42074479
21 *	.	135.78992896	.	129.80114167	141.77871626

* OBSERVATION WAS NOT USED IN THIS ANALYSIS

REGRESSION ANALYSIS OF SALE PRICE BY LIST PRICE

GENERAL LINEAR MODELS PROCEDURE

DEPENDENT VARIABLE: SALE

```
      SUM OF RESIDUALS                        -0.00000000
      SUM OF SQUARED RESIDUALS               128.94356059
      SUM OF SQUARED RESIDUALS - ERROR SS     -0.00000000
      PRESS STATISTIC                        161.70259371
      FIRST ORDER AUTOCORRELATION              0.06291667
      DURBIN-WATSON D                          1.75657045
```

c. How well does the linear regression equation fit the data? Explain.

9.61   Refer to Exercise 9.60. Predict the sale price of a house to be listed at 150,000. Within what limits should the sale price be? Hint: Use 95% limits.

9.62   A supermarket chain conducted an experiment to investigate the effect of price (P) on the weekly demand y (in pounds) for a house brand of coffee. Eight supermarket stores that had nearly equal past records of demand for the product were used in the experiment. Eight prices were randomly assigned to the stores and were advertised using the same procedures. The number of pounds of coffee sold during the following week was recorded for each of the stores and is shown below:

Demand (y, Pounds)	Price (P, Dollars)
1120	3.00
999	3.10
932	3.20
884	3.30
807	3.40
760	3.50
701	3.60
688	3.70

The data were analyzed using SAS with the results shown in the computer output below and on the following page:

a. Suppose that a supermarket that had been selling coffee for $3.70/pound is considering a raise in price to $3.80/pound. Give a 90% confidence interval for the expected change in sales per week.

b. The supermarket in (a) that is considering a raise in price has decided that the price increase will be desirable as long as average sales will not drop below 550 pounds/week. Compute a point estimate for the estimated average sales in subsequent weeks if the price is raised to $3.80/pound. Can the supermarket be 95% certain that sales will be at least 550 pounds? (In answering these questions, assume that a linear regression model is a correct one.)

c. Comment on the assumption of linearity for this model. Are there any (1) theoretical or (2) empirical reasons to question this assumption?

d. What is the interpretation of the intercept for this analysis? Is the intercept a meaningful number here? Comment.

```
        LISTING OF DATA

    OBS     DEMAND    PRICE

     1       1120      3.0
     2        999      3.1
     3        932      3.2
     4        884      3.3
     5        807      3.4
     6        760      3.5
     7        701      3.6
     8        688      3.7
     9                 3.8
```

EXERCISES

```
                           REGRESSION ANALYSIS OF DEMAND BY PRICE

                              GENERAL LINEAR MODELS PROCEDURE

DEPENDENT VARIABLE: DEMAND
```

SOURCE	DF	SUM OF SQUARES	MEAN SQUARE	F VALUE	PR > F	R-SQUARE	C.V.
MODEL	1	155246.72023808	155246.72023808	182.89	0.0001	0.968235	3.3824
ERROR	6	5093.15476192	848.85912699		ROOT MSE		DEMAND MEAN
CORRECTED TOTAL	7	160339.87500000			29.13518709		861.37500000

SOURCE	DF	TYPE I SS	F VALUE	PR > F	DF	TYPE III SS	F VALUE	PR > F
PRICE	1	155246.72023808	182.89	0.0001	1	155246.72023808	182.89	0.0001

PARAMETER	ESTIMATE	T FOR H0: PARAMETER=0	PR > |T|	STD ERROR OF ESTIMATE
INTERCEPT	2898.09523810	19.20	0.0001	150.95636910
PRICE	-607.97619048	-13.52	0.0001	44.95656970

OBSERVATION	OBSERVED VALUE	PREDICTED VALUE	RESIDUAL	LOWER 95% CL FOR MEAN	UPPER 95% CL FOR MEAN
1	1120.00000000	1074.16666667	45.83333333	1028.14833941	1120.18499393
2	999.00000000	1013.36904762	-14.36904762	976.06459051	1050.67350473
3	932.00000000	952.57142857	-20.57142857	922.44536171	982.69749543
4	884.00000000	891.77380952	-7.77380952	865.97538716	917.57223189
5	807.00000000	830.97619048	-23.97619048	805.17776811	856.77461284
6	760.00000000	770.17857143	-10.17857143	740.05250457	800.30463829
7	701.00000000	709.38095238	-8.38095238	672.07649527	746.68540949
8	688.00000000	648.58333333	39.41666667	602.56500607	694.60166059
9 *	.	587.78571429	.	532.23599174	643.33543684

```
* OBSERVATION WAS NOT USED IN THIS ANALYSIS

        SUM OF RESIDUALS                         0.00000000
        SUM OF SQUARED RESIDUALS                 5093.15476190
        SUM OF SQUARED RESIDUALS - ERROR SS      -0.00000002
        PRESS STATISTIC                          12885.97459378
        FIRST ORDER AUTOCORRELATION              -0.00347221
        DURBIN-WATSON D                          1.28943863
```

 9.63 A computer-equipment outlet sells an imported personal computer (PC) on a franchise basis and performs preventive maintenance and repair service on this PC. The data below have been collected from 16 recent calls on users to perform routine preventive maintenance service. For each call, x is the number of machines serviced and y is the total number of minutes spent by the service person.

x	y	x	y
6	86	2	33
5	95	8	102
1	18	5	65
5	69	2	25
4	62	7	105
7	101	1	17
4	39	4	55
4	53	5	68

a. Based on the data plot and a look at the residuals, does the linear regression equation provide a good fit to the data? Are there any apparent outliers? What could be done to evaluate this further?

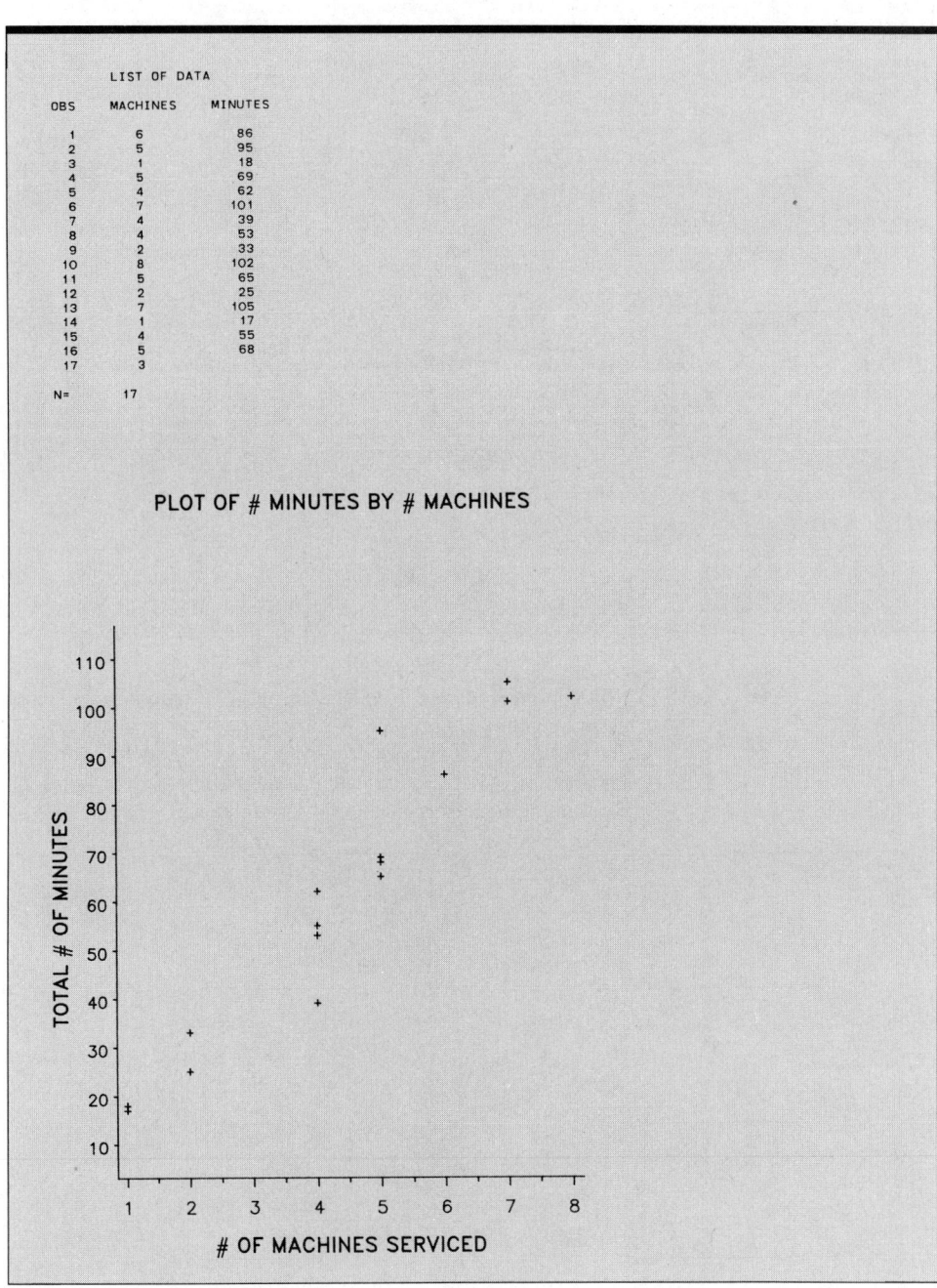

LIST OF DATA

OBS	MACHINES	MINUTES
1	6	86
2	5	95
3	1	18
4	5	69
5	4	62
6	7	101
7	4	39
8	4	53
9	2	33
10	8	102
11	5	65
12	2	25
13	7	105
14	1	17
15	4	55
16	5	68
17	3	

N= 17

PLOT OF # MINUTES BY # MACHINES

```
                            GENERAL LINEAR MODELS PROCEDURE

DEPENDENT VARIABLE: MINUTES     TOTAL NUMBER OF MINUTES

SOURCE              DF        SUM OF SQUARES       MEAN SQUARE      F VALUE     PR > F      R-SQUARE          C.V.

MODEL                1        12363.91848859      12363.91848859    144.85     0.0001      0.911865       14.8866

ERROR               14         1195.01901141         85.35850081               ROOT MSE               MINUTES MEAN

CORRECTED TOTAL     15        13558.93750000                                  9.23896644             62.06250000


SOURCE              DF           TYPE I SS      F VALUE    PR > F      DF        TYPE III SS     F VALUE     PR > F

MACHINES             1        12363.91848859    144.85     0.0001       1      12363.91848859    144.85      0.0001


                                    T FOR HO:    PR > |T|        STD ERROR OF
PARAMETER           ESTIMATE     PARAMETER=0                      ESTIMATE

INTERCEPT           2.06844106       0.38         0.7122          5.49397893
MACHINES           13.71292776      12.04         0.0001          1.13939815


OBSERVATION          OBSERVED         PREDICTED          RESIDUAL       LOWER 95% CL       UPPER 95% CL
                      VALUE            VALUE                            INDIVIDUAL          INDIVIDUAL

     1            86.00000000       84.34600760       1.65399240      63.53813898        105.15387622
     2            95.00000000       70.63307985      24.36692015      50.15063670         91.11552300
     3            18.00000000       15.78136882       2.21863118      -6.24638827         37.80912591
     4            69.00000000       70.63307985      -1.63307985      50.15063670         91.11552300
     5            62.00000000       56.92015209       5.07984791      36.47418705         77.36611713
     6           101.00000000       98.05893536       2.94106464      76.64986616        119.46800456
     7            39.00000000       56.92015209     -17.92015209      36.47418705         77.36611713
     8            53.00000000       56.92015209      -3.92015209      36.47418705         77.36611713
     9            33.00000000       29.49429658       3.50570342       8.26028403         50.72830912
    10           102.00000000      111.77186312      -9.77186312      89.50814781        134.03557843
    11            65.00000000       70.63307985      -5.63307985      50.15063670         91.11552300
    12            25.00000000       29.49429658      -4.49429658       8.26028403         50.72830912
    13           105.00000000       98.05893536       6.94106464      76.64986616        119.46800456
    14            17.00000000       15.78136882       1.21863118      -6.24638827         37.80912591
    15            55.00000000       56.92015209      -1.92015209      36.47418705         77.36611713
    16            68.00000000       70.63307985      -2.63307985      50.15063670         91.11552300
    17 *                           43.20722433                       22.50726242         63.90718625

* OBSERVATION WAS NOT USED IN THIS ANALYSIS

        SUM OF RESIDUALS                         -0.00000000
        SUM OF SQUARED RESIDUALS               1195.01901141
        SUM OF SQUARED RESIDUALS - ERROR SS      -0.00000000
        PRESS STATISTIC                        1466.75264743
        FIRST ORDER AUTOCORRELATION               0.10650215
        DURBIN-WATSON D                           1.77890479
```

b. Based on the original prediction equation, if you were to send a service person on a call to perform preventive maintenance on three PCs, is it likely that she will spend less than 45 minutes? Explain how you reached your conclusion.

9.64   The following data give the number of pounds of meat sold in a week by a grocery store (y) and the number of meat items advertised (x) that week.

Week:	1	2	3	4	5	6
Pounds Sold:	7700	11025	8150	8500	8750	7920
Number of Items:	5	8	9	8	7	10

Give a measure of the strength of relationship between x and y, based on the computer output shown here. Interpret this finding.

SPEARMANS RANK CORRELATION

VARIABLE	N	MEAN	STD DEV	MEDIAN	MINIMUM	MAXIMUM
ITEMS	6	7.83333333	1.72240142	8.00000000	5.00000000	10.00000000
POUNDS	6	8674.16666667	1212.84960596	8325.00000000	7700.00000000	11025.00000000

SPEARMAN CORRELATION COEFFICIENTS / PROB > $|R|$ UNDER HO:RHO=0 / N = 6

	ITEMS	POUNDS
ITEMS	1.00000	-0.02899
	0.0000	0.9565
POUNDS	-0.02899	1.00000
	0.9565	0.0000

10 INTRODUCTION TO THE ANALYSIS OF VARIANCE

10.1 INTRODUCTION

In Chapter 5 we presented methods for comparing two population means, based on independent random samples. Very often the two-sample problem is a simplification of what we encounter in practical situations. For example, suppose we wish to compare the mean hourly wage for nonunion farm laborers from three different ethnic groups (black, white, and Hispanic) employed by a large produce company. Independent random samples of farm laborers would be selected from each of the three ethnic groups (populations). Then using the information from the three sample means, we would try to make an inference about the corresponding population mean hourly wages. Most likely, the sample means would differ, but this does not necessarily imply a difference among the population means for the three ethnic groups. How do you decide whether the differences among the sample means are large enough to imply that the corresponding population means are different? We will answer this question using a statistical testing procedure called an *analysis of variance*.

10.2 THE LOGIC BEHIND AN ANALYSIS OF VARIANCE

The reason we call the testing procedure an analysis of variance can be seen by using the example in Section 10.1. Assume that we wish to compare the three ethnic mean hourly wages based on samples of five workers selected from each of the ethnic groups.

TABLE 10.1   A Comparison of Three Sample Means (Small Amount of Within-Sample Variation)

Sample from Populations		
1	2	3
5.90	5.51	5.01
5.92	5.50	5.00
5.91	5.50	4.99
5.89	5.49	4.98
5.88	5.50	5.02
$\bar{y}_1 = 5.90$	$\bar{y}_2 = 5.50$	$\bar{y}_3 = 5.00$

Although a sample of size five from each of the populations seems pitifully small, it illustrates the basic ideas.

Suppose the sample data (hourly wages, in dollars) are as shown in Table 10.1. Do these data present sufficient evidence to indicate differences among the three population means? A brief visual inspection of the data indicates very little variation with a sample, whereas the variability among the sample means is much larger. Since the variability among the sample means is so large *in comparison to the* **within-sample variation**, we might conclude intuitively that the corresponding population means are different.

within-sample variation

Table 10.2 illustrates a situation in which the sample means are the same as given in Table 10.1 but the variability within a sample is much larger. In contrast to the data in Table 10.1, the **between-sample variation** is small relative to the within-sample variability. We would be less likely to conclude that the corresponding population means differ based on these data.

between-sample variation

The variations in the two sets of data, Tables 10.1 and 10.2, are shown graphically in Figure 10.1. The strong evidence to indicate a difference in population means for the data of Table 10.1 is apparent in Figure 10.1(a). The lack of evidence to indicate a difference in population means for the data of Table 10.2 is indicated by the overlapping of data points for the samples in Figure 10.1(b).

TABLE 10.2   A Comparison of Three Sample Means (Large Amount of Within-Sample Variation)

Sample from population		
1	2	3
5.90	6.31	4.52
4.42	3.54	6.93
7.51	4.73	4.48
7.89	7.20	5.55
3.78	5.72	3.52
$\bar{y}_1 = 5.90$	$\bar{y}_2 = 5.50$	$\bar{y}_3 = 5.00$

FIGURE 10.1

Dot Diagrams for the
Data of Table 10.1
and Table 10.2:
○, Measurement
from Sample 1;
●, Measurement
from Sample 2;
□, Measurement
from Sample 3

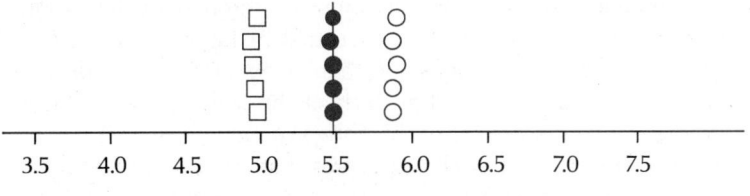

(a) Data from Table 10.1

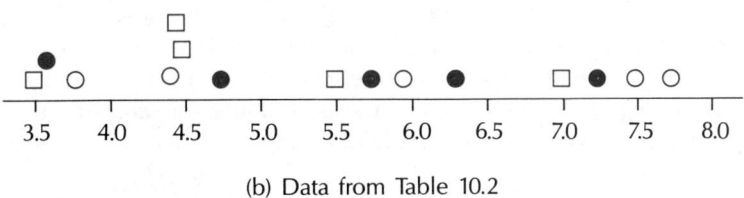

(b) Data from Table 10.2

The preceding discussion, with the aid of Figure 10.1, should indicate what we mean by an **analysis of variance**. All differences in sample means are judged statistically significant (or not) by comparing them to the variation within samples. The details of the testing procedure will be presented in the next section.

analysis of
variance

10.3   A STATISTICAL TEST ABOUT MORE THAN TWO POPULATION MEANS: AN ANALYSIS OF VARIANCE

In Chapter 5, we presented a method for testing the equality of two population means. We hypothesized two normal populations (1 and 2) with means denoted by μ_1 and μ_2, respectively, and a common variance σ^2. To test the null hypothesis that $\mu_1 = \mu_2$, independent random samples of sizes n_1 and n_2 were drawn from the two populations. The sample data were then used to compute the value of the test statistic

$$t = \frac{\bar{y}_1 - \bar{y}_2}{s_p\sqrt{(1/n_1) + (1/n_2)}}$$

where

$$s_p^2 = \frac{(n_1 - 1)s_1^2 + (n_2 - 1)s_2^2}{(n_1 - 1) + (n_2 - 1)} = \frac{(n_1 - 1)s_1^2 + (n_2 - 1)s_2^2}{n_1 + n_2 - 2}$$

pooled estimate
of σ^2

is a pooled estimate of the common population variance σ^2. The rejection region for a specified value of α, the probability of a Type I error, was then found using Table 4 in the Appendix.

Suppose now that we wish to extend this method to test the equality of more than two population means. The test procedure described above applies to only two means and therefore is inappropriate. Hence we will employ a more general method of data analysis, the analysis of variance. We illustrate its use with the following example.

Students from five different campuses throughout the country were surveyed to determine their attitudes toward industrial pollution. Each student sampled was asked a specific number of questions and then given a total score for the interview. Suppose that nine students are surveyed at each of the five campuses and we wish to examine the average student score for each of the five campuses.

We label the set of all test scores that could have been obtained from campus I as population I, and we will assume that this population possesses a mean μ_1. A random sample of $n_1 = 9$ measurements (scores) is obtained from this population to monitor student attitudes toward pollution. The set of all scores that could have been obtained from students on campus II is labeled population II (which has a mean μ_2). The data from a random sample of $n_2 = 9$ scores are obtained from this population. Similarly μ_3, μ_4, and μ_5 represent the means of the populations for scores from campuses III, IV, and V, respectively. We also obtain random samples of nine student scores from each of these populations.

From each of these five samples, we calculate a sample mean and variance. The sample results can then be summarized as shown in Table 10.3.

If we are interested in testing the equality of the population means (i.e., $\mu_1 = \mu_2 = \mu_3 = \mu_4 = \mu_5$), we might be tempted to run all possible pairwise comparisons of two population means. Hence if we assume that the five distributions are approximately normal with the same variance σ^2, we could run 10 t tests comparing two means, as listed here (see Section 5.1).

multiple t tests

Null hypotheses:

$\mu_1 = \mu_2$	$\mu_1 = \mu_4$	$\mu_2 = \mu_3$	$\mu_2 = \mu_5$	$\mu_3 = \mu_5$
$\mu_1 = \mu_3$	$\mu_1 = \mu_5$	$\mu_2 = \mu_4$	$\mu_3 = \mu_4$	$\mu_4 = \mu_5$

One obvious disadvantage to this test procedure is that it is tedious and time-consuming. But a more important and less apparent disadvantage of running multiple t tests to compare means is that the probability of falsely rejecting at least one of the hypotheses increases as the number of t tests increases. Thus, although we may have the probability of a Type I error fixed at $\alpha = .05$ for each individual test, the probability of falsely rejecting *at least one* of those tests is larger than .05. In other words, the combined probability of a Type I error for the set of 10 hypotheses would be larger than the value .05 set for each individual test. Indeed, it can be proved that the combined probability could be as large as .40.

What we need is a single test of the hypothesis "all five population means are equal," which will be less tedious than the individual t tests and can be performed with a specified probability of a Type I error (say, .05). This test is the analysis of variance.

TABLE 10.3   Summary of the Sample Results for Five Populations

	Population				
	I	II	III	IV	V
Sample mean	$\bar{y}_1$	$\bar{y}_2$	$\bar{y}_3$	$\bar{y}_4$	$\bar{y}_5$
Sample variance	s_1^2	s_2^2	s_3^2	s_4^2	s_5^2

First we assume that the five sets of measurements are normally distributed, with means given by $\mu_1, \mu_2, \mu_3, \mu_4,$ and μ_5 and with a common variance σ^2. Next we consider the quantity

s_W^2

$$s_W^2 = \frac{(n_1 - 1)s_1^2 + (n_2 - 1)s_2^2 + (n_3 - 1)s_3^2 + (n_4 - 1)s_4^2 + (n_5 - 1)s_5^2}{(n_1 - 1) + (n_2 - 1) + (n_3 - 1) + (n_4 - 1) + (n_5 - 1)}$$

$$= \frac{(n_1 - 1)s_1^2 + (n_2 - 1)s_2^2 + (n_3 - 1)s_3^2 + (n_4 - 1)s_4^2 + (n_5 - 1)s_5^2}{n_1 + n_2 + n_3 + n_4 + n_5 - 5}$$

Note that this quantity is merely an extension of

$$s_p^2 = \frac{(n_1 - 1)s_1^2 + (n_2 - 1)s_2^2}{n_1 + n_2 - 2}$$

which is used as an estimate of the common variance for two populations for a test of the hypothesis $\mu_1 = \mu_2$ (Section 5.1). Thus s_W^2 represents a combined estimate of the common variance σ^2, and it measures the variability of the observations within the five populations. (The subscript W refers to the within-population variability.)

Next we consider a quantity that measures the variability between or among the population means. If the null hypothesis $\mu_1 = \mu_2 = \mu_3 = \mu_4 = \mu_5$ is true, then the populations are identical, with mean μ and variance σ^2. Drawing single samples from the five populations is then equivalent to drawing five different samples from the same population. What kind of variation might be expected for these sample means? If the variation is too great, we would reject the hypothesis that $\mu_1 = \mu_2 = \mu_3 = \mu_4 = \mu_5$.

To discuss the variation from sample mean to sample mean, we need to know the distribution of the mean of a sample of nine observations in repeated sampling. From Chapter 3 we know that the sampling distribution for $\bar{y}$ based on $n = 9$ measurements will have the same mean μ and variance $\sigma^2/9$. Since we have drawn five samples of nine observations each, we can estimate the variance of the distribution of sample means, $\sigma^2/9$, using the formula

$$\text{sample variance (of the means)} = \frac{\sum \bar{y}^2 - [(\sum \bar{y})^2/5]}{5 - 1}$$

Note that we merely consider the $\bar{y}$s as a sample of five observations and calculate the "sample variance." This quantity estimates $\sigma^2/9$ and hence 9 × (sample variance of the means) estimates σ^2. We designate this quantity as s_B^2; the subscript B designates a measure

s_B^2

of the variability among the sample means for the five populations. For this problem $s_B^2 = (9$ times the sample variance of the means).

Under the null hypothesis that all five population means are identical, we have two estimates of σ^2, namely, s_W^2 and s_B^2. Suppose the ratio

$$\frac{s_B^2}{s_W^2}$$

is used as the test statistic to test the hypothesis that $\mu_1 = \mu_2 = \mu_3 = \mu_4 = \mu_5$. What is the distribution of this quantity if we were to repeat the experiment over and over again, each time calculating s_B^2 and s_W^2?

For our example s_B^2/s_W^2 follows an F distribution, with degrees of freedom that can be shown to be $df_1 = 4$ for s_B^2 and $df_2 = 40$ for s_W^2. The proof of these remarks is beyond the

FIGURE 10.2

Critical Value of F for
$\alpha = .05$, $df_1 = 4$, and
$df_2 = 40$

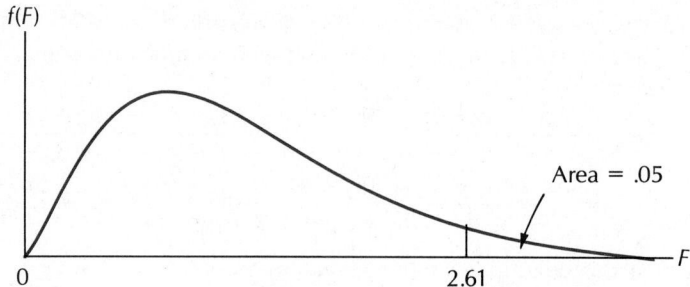

FIGURE 10.2

Critical Value of F for $\alpha = .05$, $df_1 = 4$, and $df_2 = 40$

scope of this text. However, we make use of this result for testing the null hypothesis $\mu_1 = \mu_2 = \mu_3 = \mu_4 = \mu_5$.

The test statistic used to test equality of the population means is

test statistic

$$F = \frac{s_B^2}{s_W^2}$$

When the null hypothesis is true, both s_B^2 and s_W^2 estimate σ^2, and F would be expected to assume a value near $F = 1$. When the hypothesis of equality is false, s_B^2 will tend to be larger than s_W^2 due to the differences among the population means. Hence we will reject the null hypothesis in the upper tail of the distribution of $F = s_B^2/s_W^2$. For α, the probability of a Type I error, equal to .10, .05, .025, .01, .005, or .001, we can locate the rejection region for this one-tailed test using Table 6 in the Appendix, with $df_1 = 4$ and $df_2 = 40$. Thus for $\alpha = .05$ the critical value of $F = s_B^2/s_W^2$ is 2.61. (See Figure 10.2.) If the calculated value of F falls in the rejection region, we conclude that not all five population means are identical.

This procedure can be generalized (and simplified) with only slight modifications in the formulas to test the equality of t (where t is an integer equal to or greater than 2) population means from normal populations with a common variance σ^2. Random samples of sizes $n_1, n_2, \ldots, n_t$ are drawn from the respective populations. We then compute the sample means and variances. The null hypothesis $\mu_1 = \mu_2 = \cdots = \mu_t$ is tested against the alternative that at least one of the population means is different from the others.

Before presenting the generalized test procedure, it is convenient to introduce the notation to be used in the shortcut computational formulas for s_B^2 and s_W^2.

**completely
randomized design**

The experimental setting where a random sample of observations is taken from each of t different populations is called a **completely randomized design**. Consider a completely randomized design where four observations are obtained from each of five populations.

TABLE 10.4   Summary of Sample Data for a Completely Randomized Design

Population	Data				Total	Mean
1	y_{11}	y_{12}	y_{13}	y_{14}	T_1	$\bar{y}_1$
2	y_{21}	y_{22}	y_{23}	y_{24}	T_2	$\bar{y}_2$
3	y_{31}	y_{32}	y_{33}	y_{34}	T_3	$\bar{y}_3$
4	y_{41}	y_{42}	y_{43}	y_{44}	T_4	$\bar{y}_4$
5	y_{51}	y_{52}	y_{53}	y_{54}	T_5	$\bar{y}_5$

If we let y_{ij} denote the jth observation from population i, we could display the sample data for this completely randomized design as shown in Table 10.4. Using Table 10.4, we can introduce notation that is helpful when performing an **analysis of variance (AOV)** for a one-way classification.

AOV

NOTATION NEEDED FOR THE AOV OF A COMPLETELY RANDOMIZED DESIGN

y_{ij}: The jth sample observation selected from population i. For example, y_{23} denotes the third sample observation drawn from population 2.

n_i: The number of sample observations selected from population i. In our data set, n_1, the number of observations obtained from population 1, is 4. Similarly, $n_2 = n_3 = n_4 = n_5 = 4$. However, it should be noted that the sample sizes need not be the same. Thus we might have $n_1 = 12$, $n_2 = 3$, $n_3 = 6$, $n_4 = 10$, and so forth.

n: The total sample size; $n = \sum n_i$. For the data given in Table 10.4, $n = n_1 + n_2 + n_3 + n_4 + n_5 = 20$

T_i: The sum (total) of the sample measurements obtained from population i

G: The sum (grand total) of *all* sample observations; $G = \sum T_i$

$\bar{y}_i$: The average of the n_i sample observations drawn from population i, $\bar{y}_i = T_i/n_i$

$\bar{y}$: The average of all sample observations; $\bar{y} = G/n$

With this notation it is possible to establish the following algebraic identities. (Although we will use these results in later calculations for s_W^2 and s_B^2, the proofs of these identities are beyond the scope of this text.) The variability of the n sample measurements about their mean $\bar{y}$ can be measured using the sum of the squared deviations $(y_{ij} - \bar{y})^2$. This quantity,

$$\text{TSS} = \sum_{i,j} (y_{ij} - \bar{y})^2 = (n - 1)s^2$$

total sum of squares

is called the **total sum of squares** of the measurements about their mean. The double summation in TSS means that we must sum the squared deviations for all rows (i) and columns (j) of the one-way classification.

It is possible to partition the total sum of squares as follows:

$$\sum_{i,j} (y_{ij} - \bar{y})^2 = \sum_{i,j} (y_{ij} - \bar{y}_i)^2 + \sum_{i} n_i(\bar{y}_i - \bar{y})^2$$

The first quantity on the right side of the equation measures the variability of an observation y_{ij} about its sample mean $\bar{y}_i$. Thus

$$\text{SSW} = \sum_{i,j} (y_{ij} - \bar{y}_i)^2 = (n_1 - 1)s_1^2 + (n_2 - 1)s_2^2 + \cdots + (n_t - 1)s_t^2$$

within-sample sum of squares

is a measure of the *within*-sample variability. SSW is referred to as the **within-sample sum of squares** and will be used to compute s_w^2.

The second expression in the total sum of squares equation measures the variability of the sample means $\bar{y}_i$ about the overall mean $\bar{y}$. This quantity, which measures the variability *between* (or among) the sample means, is referred to as the **sum of squares between samples** (SSB) and will be used to compute s_B^2.

between-sample sum of squares

$$SSB = \sum_i n_i(\bar{y}_i - \bar{y})^2$$

Although the formulas for TSS, SSW, and SSB are easily interpreted, they are not easy to use for calculations. Instead, we use the shortcut formulas shown here.

SHORTCUT SUM OF SQUARES FORMULAS FOR A COMPLETELY RANDOMIZED DESIGN

$$TSS = \sum_{i,j} y_{ij}^2 - \frac{G^2}{n}$$

$$SSB = \sum \frac{T_i^2}{n_i} - \frac{G^2}{n}$$

$$SSW = TSS - SSB$$

An analysis of variance for a completely randomized design with t populations has the following null and alternative hypotheses:

H_0: $\mu_1 = \mu_2 = \mu_3 = \cdots = \mu_t$ (i.e., the t population means are equal).

H_a: At least one of the t population means differs from the rest.

The quantities s_B^2 and s_W^2 can be computed using the shortcut formulas

$$s_B^2 = \frac{SSB}{t-1} \qquad s_W^2 = \frac{SSW}{n-t}$$

where $t-1$ and $n-t$ are the degrees of freedom for s_B^2 and s_W^2, respectively.

Historically, people have referred to a sum of squares divided by its degrees of freedom as a **mean square**. Hence s_B^2 is often called the *mean square between samples* and s_W^2 the *mean square within samples*.

mean square

The null hypothesis of equality of the t population means is rejected if

$$F = \frac{s_B^2}{s_W^2}$$

exceeds the tabulated value of F for a $= \alpha$, $df_1 = t - 1$, and $df_2 = n - t$.

After completing the F test, the results of a study are then summarized in an *analysis of variance table*. The format of an **AOV table** is shown in Table 10.5. The AOV table lists the sources of variability in the first column. The second column lists the sums of squares associated with each source of variability. Since we showed that the total sum of squares (TSS) can be partitioned into two parts, then SSB and SSW must add up to TSS in the AOV table. The third column of the table gives the degrees of freedom associated with the sources of variability. Again we have a check; $(t-1) + (n-t)$ must add to $n-1$.

AOV table

TABLE 10.5   An Example of an AOV Table for a Completely Randomized Design

Source	Sum of Squares	Degrees of Freedom	Mean Square	F Test
Between samples	SSB	$t - 1$	$s_B^2 = SSB/(t - 1)$	s_B^2/s_W^2
Within samples	SSW	$n - t$	$s_W^2 = SSW/(n - t)$	
Totals	TSS	$n - 1$		

The mean squares are found in the fourth column, and the F test for the equality of the t population means is given in the fifth column.

EXAMPLE 10.1   A horticulturist was investigating the phosphorus content of tree leaves from three different varieties of apple trees (1, 2, and 3). Random samples of five leaves from each of the three varieties were analyzed for phosphorus content. The data are given in Table 10.6. Use these data to test the hypothesis of equality of the mean phosphorus levels for the three varieties. Use $\alpha = .05$.

Solution   The null and alternative hypotheses for this example are

H_0: $\mu_1 = \mu_2 = \mu_3$

H_a: At least one of the population means differs from the rest.

The sample sizes are $n_1 = n_2 = n_3 = 5$, for which $n = 15$. From the sample data we see that the total (sum) for all observations on variety 1 is $T_1 = 2.30$. Similarly, the totals for varieties 2 and 3 are $T_2 = 3.88$, and $T_3 = 3.54$. The sum of all sample measurements is then

$$G = 9.72$$

Using the sample measurements, the total sum of squares, TSS, is

$$TSS = \sum_{i,j} y_{ij}^2 - \frac{G^2}{n} = (.35)^2 + (.40)^2 + \cdots + (.66)^2 - \frac{(9.72)^2}{15}$$

$$= 6.673 - 6.299 = .374$$

The sample totals can then be used to compute the sum of squares between samples, SSB.

$$SSB = \frac{\sum T_i^2}{n_i} - \frac{G^2}{n} = \frac{(2.30)^2 + (3.88)^2 + (3.54)^2}{5} - \frac{(9.72)^2}{15}$$

$$= 6.575 - 6.299 = .276$$

TABLE 10.6   Phosphorus Content of Leaves from Three Different Trees, Example 10.1

Variety	Phosphorus Content					Totals	Means
1	.35	.40	.58	.50	.47	2.30	0.46
2	.65	.70	.90	.84	.79	3.88	0.78
3	.60	.80	.75	.73	.66	3.54	0.71
Total						9.72	0.65

TABLE 10.7   AOV Table for the Data for Example 10.1

Source	Sum of squares	Degrees of freedom	Mean square	F test
Between samples	.276	2	.276/2 = .138	.138/.008 = 17.25
Within samples	.098	12	.098/12 = .008	
Totals	.374	14		

Then the sum of squares within samples, SSW, is

$$SSW = TSS - SSB = .374 - .276 = .098$$

The AOV table for these data is shown in Table 10.7. The critical value of $F = s_B^2/s_W^2$ is 3.89, which is obtained from Table 6 in the Appendix for a = .05, $df_1 = 2$, and $df_2 = 12$. Since the computed value of F, 17.25, exceeds 3.89, we reject the null hypothesis of equality of the mean phosphorus content for the three varieties. It appears from the data that the mean for variety 1 is smaller than the means for varieties 2 and 3.

EXAMPLE 10.2   A clinical psychologist wished to compare three methods for reducing hostility levels in university students. A certain test (HLT) was used to measure the degree of hostility. A high score on this test indicated great hostility. Eleven students obtaining high and nearly equal scores were used in the experiment. Four were selected at random from among the 11 problem cases and treated with method 1. Four of the remaining seven students were selected at random and treated with method 2. The remaining three students were treated with method 3. All treatments were continued for a one-semester period. Each student was given the HLT test at the end of the semester, with the results shown in Table 10.8. Use these data to perform an analysis of variance to determine if there are differences among mean scores for the three methods. Use $\alpha = .05$.

Solution   The null and alternative hypotheses are

H_0: $\mu_1 = \mu_2 = \mu_3$
H_a: At least one of the population means differs from the rest.

TABLE 10.8   HLT Test Scores, Example 10.2

Method	Test Scores				Total, T_i
1	80	92	87	83	342
2	70	81	78	74	303
3	63	76	70		209
Total					854

TABLE 10.9   AOV Table for the Data of Example 10.2

Source	SS	df	MS	F
Between samples	452.13	2	226.07	$226.07/29.3 = 7.72$
Within samples	234.42	8	29.30	
Totals	686.55	10		

For $n_1 = 4$, $n_2 = 4$, and $n_3 = 3$, we have a total sample size of $n = 11$. The totals from Table 10.8 are

$$T_1 = 342, \qquad T_2 = 303, \qquad T_3 = 209$$

and

$$G = 854$$

Substituting into the computational formulas for TSS and SSB, we have

$$TSS = \sum_{i,j} y_{ij}^2 - \frac{G^2}{n} = (80)^2 + (92)^2 + \cdots + (70)^2 - \frac{(854)^2}{11}$$

$$= 66{,}988 - 66{,}301.45 = 686.55$$

$$SSB = \sum \frac{T_i^2}{n_i} - \frac{G^2}{n} = \frac{(342)^2}{4} + \frac{(303)^2}{4} + \frac{(209)^2}{3} - 66{,}301.45$$

$$= 66{,}753.58 - 66{,}301.45 = 452.13$$

Then

$$SSW = 686.55 - 452.13 = 234.42$$

The AOV table for these data is shown in Table 10.9.

The critical value of F is obtained from Table 6 in the Appendix for a = .05, $df_1 = 2$, and $df_2 = 8$; this value is 4.46. Since the computed value of F, 7.72, exceeds the tabulated value, 4.46, we reject the null hypothesis of equality of the mean scores for the three groups. Computer output shown here verifies the results we obtained by hand.

```
        LISTING OF DATA

  OBS    METHOD    SCORE

   1        1        80
   2        1        92
   3        1        87
   4        1        83
   5        2        70
   6        2        81
   7        2        78
   8        2        74
   9        3        63
  10        3        76
  11        3        70

  N=      11
```

ANALYSIS OF VARIANCE PROCEDURE

GENERAL LINEAR MODELS PROCEDURE

DEPENDENT VARIABLE: SCORE

SOURCE	DF	SUM OF SQUARES	MEAN SQUARE	F VALUE	PR > F	R-SQUARE	C.V.
MODEL	2	452.12878788	226.06439394	7.71	0.0136	0.658556	6.9724
ERROR	8	234.41666667	29.30208333		ROOT MSE		SCORE MEAN
CORRECTED TOTAL	10	686.54545455			5.41313988		77.63636364

SOURCE	DF	TYPE I SS	F VALUE	PR > F	DF	TYPE III SS	F VALUE	PR > F
METHOD	2	452.12878788	7.71	0.0136	2	452.12878788	7.71	0.0136

EXERCISES

 10.1 Sample data from an experiment aimed at comparing the tar content of five different brands of cigarettes gave the following results:

Brand	$\bar{y}$ (mg)	s	n_i
1	9.6	1.3	10
2	10.2	1.4	10
3	10.8	1.1	10
4	11.5	1.2	10
5	13.6	1.5	10

a. Based on your intuition, is there evidence to indicate any differences among the mean contents of the five brands?

b. Run an analysis of variance to confirm or reject your conclusion of part (a).

 10.2 The number of units of production was recorded for a random sample of 10 hourly periods from the three bottling assembly lines of a plant. These data are shown here:

Assembly Line		
1	2	3
290	258	249
265	276	257
286	277	264
275	243	266
288	248	278
250	259	273
279	265	281
294	282	254
285	275	261
293	268	265

a. Plot the data separately for each line. Are there any obvious differences?
b. Identify the means and standard deviations for the three lines using the output shown here.

```
LISTING OF DATA

OBS     LINE     UNITS

 1       1        290
 2       1        265
 3       1        286
 4       1        275
 5       1        288
 6       1        250
 7       1        279
 8       1        294
 9       1        285
10       1        293
11       2        258
12       2        276
13       2        277
14       2        243
15       2        248
16       2        259
17       2        265
18       2        282
19       2        275
20       2        268
21       3        249
22       3        257
23       3        264
24       3        266
25       3        278
26       3        273
27       3        281
28       3        254
29       3        261
30       3        265

N=      30
```

```
           MEANS BY LINE

VARIABLE       MEAN        STANDARD
                           DEVIATION

-------------- LINE=1 ---------------

UNITS     280.50000000   13.89844116

-------------- LINE=2 ---------------

UNITS     265.10000000   12.99957264

-------------- LINE=3 ---------------

UNITS     264.80000000   10.26103742
```

ANALYSIS OF VARIANCE PROCEDURE

GENERAL LINEAR MODELS PROCEDURE

DEPENDENT VARIABLE: UNITS

SOURCE	DF	SUM OF SQUARES	MEAN SQUARE	F VALUE	PR > F	R-SQUARE	C.V.
MODEL	2	1612.46666667	806.23333333	5.17	0.0125	0.277082	4.6209
ERROR	27	4207.00000000	155.81481481		ROOT MSE		UNITS MEAN
CORRECTED TOTAL	29	5819.46666667			12.48258045		270.13333333

SOURCE	DF	TYPE I SS	F VALUE	PR > F	DF	TYPE III SS	F VALUE	PR > F
LINE	2	1612.46666667	5.17	0.0125	2	1612.46666667	5.17	0.0125

10.4   THE MODEL FOR OBSERVATIONS IN A
ONE-WAY CLASSIFICATION

We formulated a model (equation) to relate a response y to a set of quantitative independent variables in Chapter 8. In this section we will consider a model for the one-way classification. While the model at first may appear to be quite different from those of Chapter 8, we will see later that it is very similar.

assumptions   We make the following **assumptions** concerning the sample measurements and the populations from which they were drawn:

1. The samples are independent random samples. Results from one sample in no way affect the measurements observed in another sample.
2. Each sample is selected from a normal population.
3. The mean and variance for population i are, respectively, μ_i and σ^2 $(i = 1, 2, \ldots, t)$.
4. To summarize, we assume that the t populations are normally distributed with different means but a common variance σ^2.

We can now formulate a model (equation) that encompasses the assumptions listed above. Recall that we previously let y_{ij} denote the jth sample observation from population i.

model $$y_{ij} = \mu + \alpha_i + \varepsilon_{ij}$$

This model states that y_{ij}, the jth sample measurement selected from population i, is the
terms   sum of three **terms**. The term μ denotes an overall mean that is an unknown constant. The term α_i denotes an effect due to population i; α_i is an unknown constant. The term ε_{ij} denotes a random error associated with the jth observation from population i. We assume that ε_{ij} is normally distributed, with a mean of 0 and a variance σ_ε^2. In addition, the errors are independent; that is, the error associated with one observation in no way affects the error associated with another observation.

Since the εs are normally distributed with mean 0, the mean or expected value of y_{ij}, denoted by $E(y_{ij})$, is

$$E(y_{ij}) = \mu + \alpha_i$$

That is, y_{ij} has been selected from a population with mean $\mu + \alpha_i$. Since α_i may assume a positive, zero, or negative value, the mean for population i can be greater than, equal to, or less than μ, the overall mean. The variance for each of the t populations can be shown to be σ_ε^2. Finally, because the εs are normally distributed, each of the t populations is normal. A summary of the assumptions for a one-way classification is shown in Table 10.10.

The null hypothesis for a one-way analysis of variance is that $\mu_1 = \mu_2 = \cdots = \mu_t$. Using our model, this would be equivalent to the null hypothesis

$$H_0: \ \alpha_1 = \alpha_2 = \cdots = \alpha_t = 0$$

If H_0 is true, then all populations have the same unknown mean μ. Indeed, many textbooks use this latter null hypothesis for the analysis of variance in a completely randomized design. The corresponding alternative hypothesis is

$$H_a: \text{At least one of the } \alpha_i\text{s differs from zero.}$$

TABLE 10.10   Summary of Some of the Assumptions for a Completely
Randomized Design

Population	Population Mean	Population Variance	Sample Measurements
1	$\mu + \alpha_1$	σ_ε^2	$y_{11}, y_{12}, \ldots, y_{1n_1}$
2	$\mu + \alpha_2$	σ_ε^2	$y_{21}, y_{22}, \ldots, y_{2n_2}$
$\vdots$	$\vdots$	$\vdots$	$\vdots$
t	$\mu + \alpha_t$	σ_ε^2	$y_{t1}, y_{t2}, \ldots, y_{tn_t}$

In this section we have presented a brief discussion of the model associated with the analysis of variance for a completely randomized design. Although some authors bypass an examination of the model, we believe it is a necessary part of an analysis of variance discussion.

You may be concerned with checking the validity of the underlying assumptions in an analysis of variance. In practice, you should always make at least a rough check before proceeding. In the next section we will discuss how to test the "equality of variance" assumption. The assumption of normality is not too critical since we are basing the analysis of variance test on means (and hence the Central Limit Theorem applies). Non-normality in the form of skewed distributions will not affect conclusions drawn from an analysis of variance unless the skewness is severe and the sample sizes are small. However, to guard against gross violations of the normality assumption, the data for each sample should be plotted separately. If the data for one or more of the samples appear non-normal, the Kruskal–Wallis test of Section 10.6 (also referred to as the Kruskal–Wallis one-way analysis of variance by ranks) can be used. The null hypothesis for the Kruskal–Wallis test is that the t populations are identical.

10.5   CHECKING ON THE EQUAL VARIANCE ASSUMPTION

The assumption of equal population variances, like the assumption of normality of the populations, has been made in several places in the text, such as for the t test when comparing two population means and now for the analysis of variance F test in a completely randomized design.

Let us consider first an experiment where we wish to compare t population means based on independent random samples from each of the populations. Recall that we assume we are dealing with normal populations with a common variance σ_ε^2 and possibly different means. If there were just two populations of interest, we could verify the assumption of equality of the two population variances using the F test of Chapter 7. However, with $t > 2$, rather than making all pairwise F tests, we seek a single test that can be used to verify the assumption of equality of the population variances.

The one test we will use in this text for the null hypothesis

Hartley's test $$H_0: \ \sigma_1^2 = \sigma_2^2 = \cdots = \sigma_t^2$$

was proposed by H. O. Hartley (1940 and 1950) and represents a logical extension to the F test for $t = 2$. If s_i^2 denotes the sample variance computed from the ith sample, the test statistic is

F_{max}

$$F_{max} = \frac{s^2_{max}}{s^2_{min}}$$

where s^2_{max} and s^2_{min} are the largest and smallest of the s^2_is, respectively. The test procedure is summarized here.

HARTLEY'S TEST FOR HOMOGENEITY OF POPULATION VARIANCES

H_0: $\sigma^2_1 = \sigma^2_2 = \cdots = \sigma^2_t$, i.e., homogeneity of variances

H_a: Not all population variances are the same.

T.S.: $F_{max} = \dfrac{s^2_{max}}{s^2_{min}}$

R.R.: For a specified value of α, reject H_0 if F_{max} exceeds the tabulated F value (Table 14) for a = α, t, and $df_2 = n - 1$, where n is the number of observations in each sample.

It should be noted that, theoretically, we required the sample sizes to all be the same. In practice, if the sample sizes are nearly equal, the largest n_i can be used for running the test of homogeneity. This procedure will result in the probability of a Type I error being slightly more than the nominal value α.

Several comments should be made. Most practitioners do not routinely run Hartley's test. One reason is that the test is extremely sensitive to departures from normality. So, in checking one assumption (constant variance), the practitioner would have to be very careful about departures from another analysis of variance assumption (normality of the populations). Fortunately, as we mentioned in Chapter 5, the assumption of homogeneity (equality) of population variances is less critical when the sample sizes are substantially different. When the sample sizes are nearly equal, the variances can be markedly different and the p-values for an analysis of variance will still be only mildly distorted. Thus we recommend that Hartley's test be used only for the more extreme cases. In these extreme situations where homogeneity of the population variances is a problem, a transformation of the data may help to stabilize the variances. Then inferences can be made from an analysis of variance.

transformation of data

A **transformation of the sample data** is defined to be a process in which the measurements on the original scale are systematically converted to a new scale of measurement. For example, if the original variable is y and the variances associated with the variable across the treatments are not equal (heterogeneous), it may be necessary to work with a new variable such as $\sqrt{y}$, log y, or some other transformed variable.

How can we select the appropriate transformation? This is no easy task and often takes a great deal of experience in the experimenter's area of application. In spite of these difficulties, we can consider several guidelines for choosing an appropriate transformation.

guidelines for selecting y_T

Many times the variances across the populations of interest are heterogeneous and seem to vary with the magnitude of the population mean. For example, it may be that the larger the population mean, the larger the population variance. When we are able to identify how the variance varies with the population mean, we can define a suitable

SECTION 10.5   CHECKING ON THE EQUAL VARIANCE ASSUMPTION

TABLE 10.11   Transformation to Achieve Uniform Variance

Relationship Between μ and σ^2	y_T	Variance of y_T (for a Given k)
$\sigma^2 = k\mu$ (when $k = 1$, y is a Poisson variable)	$y_T = \sqrt{y}$ or $\sqrt{y + .375}$	$1/4$; $(k = 1)$
$\sigma^2 = k\mu^2$	$y_T = \log y$ or $\log (y + 1)$	1; $(k = 1)$
$\sigma^2 = k\pi(1 - \pi)$ (when $k = 1/n$, y is a binomial variable)	$y_T = \arcsin \sqrt{y}$	$1/4n$; $(k = 1/n)$

TABLE 10.12   Mean Dissolved Oxygen Contents (in ppm) of Three Lakes, Example 10.3

| | Lake | |
1	2	3
0	1	14
2	3	26
1	4	25
3	6	18
1	8	19
2	7	22
3	5	21
4	3	16
1	4	20
5	5	30
$\bar{y} = 2.2$	$\bar{y} = 4.6$	$\bar{y} = 21.1$
$s = 1.55$	$s = 2.07$	$s = 4.84$

transformation from the variable y to a new variable y_T. Three specific situations are presented in Table 10.11.

The first row of Table 10.11 suggests that if y is a Poisson* random variable the variance of y is equal to the mean of y. Thus if the different populations correspond to different Poisson populations, the variances will be heterogeneous provided the means are different. The transformation that will stabilize the variances is $y_T = \sqrt{y}$; or, if the Poisson means are small (under 5), the transformation $y_T = \sqrt{y + .375}$ is better.

EXAMPLE 10.3   The mean dissolved oxygen contents (in ppm) of three different lakes were to be compared based on independent random samples of 10 observations taken from the center of each lake at a depth of 1 foot. The sample data are given in Table 10.12.

a.   Run a test of the equality of the population variances. Use Hartley's test, with $\alpha = .05$.

---

* The Poisson random variable is a useful discrete random variable with applications as an approximation for the binomial (when n is large but $n\pi$ is small) and as a model for events occurring randomly in time. For additional information see Hildebrand and Ott (1987) and Mendenhall (1987).

TABLE 10.13   Square Root Transformations $(\sqrt{y + .375})$ of the Data of Table 10.12

	Lake	
1	2	3
0.612	1.173	3.791
1.541	1.837	5.136
1.173	2.092	5.037
1.837	2.525	4.287
1.173	2.894	4.402
1.541	2.716	4.730
1.837	2.318	4.623
2.092	1.837	4.047
1.173	2.092	4.514
2.318	2.318	5.511

b. Transform the data using $y_T = \sqrt{y + .375}$.

c. Compute the sample means and sample standard deviations for the transformed data.

Solution

a. The F test for the equality of population variances has

$$F_{max} = \frac{(4.84)^2}{(1.55)^2} = 9.75$$

The critical value of F_{max} for a $= .05$, $t = 3$, and $df_2 = 9$ is 5.34. Since F_{max} is greater than 5.34, we reject the hypothesis of homogeneity of the population variances.

b. The most convenient way to transform the data is to make use of a calculator with a square root key or a table of squares and square roots. The square root data appear in Table 10.13.

c. The sample means and standard deviations for the transformed data are shown in Table 10.14. Although the original data had heterogeneous variances, the sample variances are all approximately .25, as indicated in Table 10.11.

$y_T = \log y$ The second transformation indicated in Table 10.11 is for an experimental situation where the population variance is approximately equal to the square of the population mean,

TABLE 10.14   Sample Means and Standard Deviations for the Data in Table 10.13

	Lake		
	1	2	3
Sample mean	1.53	2.18	4.61
Sample standard deviation	.51	.50	.52

TABLE 10.15   Data for Hours of Relief While on
Therapy, Example 10.4

	Treatment	
A	B	C
4.2	4.1	38.7
2.3	10.7	26.3
6.6	14.3	5.4
6.1	10.4	10.3
10.2	15.3	16.9
11.7	11.5	43.1
7.0	19.8	48.6
3.6	12.6	29.5
$\bar{y} = 6.46$	$\bar{y} = 12.34$	$\bar{y} = 27.35$
$s = 3.22$	$s = 4.53$	$s = 15.66$

or, equivalently, where $\sigma = \mu$. Actually, the logarithmic transformation is appropriate any time the **coefficient of variation** σ_i/μ_i is constant across the populations of interest.

coefficient of variation

EXAMPLE 10.4   Irritable bowel syndrome (IBS) is a nonspecific intestinal disorder characterized by abdominal pain and irregular bowel habits. Each person in a random sample of 24 patients having periodic attacks of IBS was randomly assigned to one of three treatment groups, A, B, and C. The number of hours of relief while on therapy is recorded for each patient in Table 10.15.

a. Test for differences among the population variances. Use $\alpha = .05$.
b. Since there are no zero y values, use the transformation $y_T = \ln y$ ("ln" denotes logarithms to the base e) to try to stabilize the variances.
c. Compute the sample means and the sample standard deviations for the transformed data.

Solution   a. The F test for a test of the null hypothesis $H_0: \sigma_1^2 = \sigma_2^2 = \sigma_3^2$ is

$$F_{max} = \frac{(15.66)^2}{(3.22)^2} = \frac{245.24}{10.37} = 23.65$$

Since the computed value of F_{max} exceeds 6.94, the tabulated value (Table 14) for a = .05, t = 3, and $df_2 = 7$, we reject H_0 and conclude that the population variances are different.

b. We can obtain the natural logarithms (log to the base e) for the sample data using either a calculator with an "ln" key or by referring to a table of natural logs (see, for example, the CRC *Standard Mathematical Tables*, 1961). The transformed data are shown in Table 10.16.

TABLE 10.16   Natural Logarithms of the Data in Table 10.15

	Treatment	
A	B	C
1.435	1.411	3.656
0.833	2.370	3.270
1.887	2.660	1.686
1.808	2.342	2.332
2.322	2.728	2.827
2.460	2.442	3.764
1.946	2.986	3.884
1.281	2.534	3.384

c. The sample means and standard deviations for the transformed data are given in Table 10.17. Although the sample variances are not exactly the same, they certainly do not indicate that the corresponding population variances are different.

$y_T = \arcsin \sqrt{y}$

The third transformation listed in Table 10.11 is particularly appropriate for data recorded as percentages or proportions. You will recall that in Chapter 3 we introduced the binomial distribution, where y designates the number of successes in n identical trials and $\hat{\pi} = y/n$ provides an estimate of π, the proportion of experimental units in the population possessing the characteristic. Although we may not have mentioned this while studying the binomial, the variance of $\hat{\pi}$ is given by $\pi(1 - \pi)/n$. Thus if the response variable is $\hat{\pi}$, the proportion of successes in a random sample of n observations, then the variance of $\hat{\pi}$ will vary, depending on the values of π for the populations from which the samples were drawn. See Table 10.18.

Since the variance of $\hat{\pi}$ is symmetrical about $\pi = .5$, the variance of $\hat{\pi}$ for $\pi = .7$ and $n = 20$ would be .0105. Similarly, we can determine $\pi(1 - \pi)/n$ for other values of $\pi > .5$. The important thing to note is that if the populations have values of π in the vicinity of approximately .3 to .5, there is very little difference in the variances for $\hat{\pi}$. However, the variance of $\hat{\pi}$ is quite variable for either large or small values of π, and for these situations we should consider the possibility of transforming the sample proportions to stabilize the variances.

The transformation we recommend is $\arcsin \sqrt{\hat{\pi}}$ (sometimes written as $\sin^{-1} \sqrt{\hat{\pi}}$). In words, we are transforming the sample proportion into the angle whose sine is $\sqrt{\hat{\pi}}$. Some experimenters express these angles in degrees, others in radians. For consistency we will

TABLE 10.17   Sample Means and Standard Deviations for the Data of Table 10.16

	Treatment		
	A	B	C
Sample mean	1.75	2.43	3.10
Sample standard deviation	.54	.46	.77

TABLE 10.18   Variance of $\hat{\pi}$, the Sample Proportion, for Several Values of π and $n = 20$

Values of π	$\pi(1 - \pi)/n$
.01	.0005
.05	.0024
.1	.0045
.2	.0080
.3	.0105
.4	.0120
.5	.0125

always express our angles in radians. Many calculators include a key for the arcsin transformation; Table 15* of the Appendix provides arcsin computations for various values of $\hat{\pi}$.

EXAMPLE 10.5

In a national public opinion poll, a random sample of 30 registered voters was obtained from each of 24 different standard metropolitan statistical areas (SMSA). Each of the 30 voters in a sample was asked whether he or she favored limiting the FBI director to a fixed term in office (such as 10 years). The data below are the sample proportions for the 24 SMSAs. Transform the data by using $y_T = \arcsin \sqrt{\hat{\pi}}$. Calculate the sample mean and standard deviation for the transformed data.

.13	.60	.33	.03	.43	.43
.17	.70	.47	.10	.60	.60
.30	.10	.57	.20	.20	.67
.53	.20	.70	.33	.30	.77

Solution

Using a calculator or Table 15 in the Appendix, the transformed data are

.37	.89	.61	.17	.72	.72
.42	.99	.76	.32	.89	.89
.58	.32	.86	.46	.46	.96
.82	.46	.99	.61	.58	1.07

The sample mean and standard deviation are, respectively, .66 and .25.

when $\hat{\pi} = 0, 1$

One comment should be made concerning the situation in which a sample proportion of 0 or 1 is observed. For these cases we recommend substituting $1/4n$ and $1 - (1/4n)$, respectively, as the corresponding sample proportions to be used in the calculations.

In this section we discussed Hartley's test for checking the equality of variance assumption and the analysis of variance, and transformations of data that can alleviate the problem of nonconstant variances. As an added benefit, the transformations presented in this section also (sometimes) decrease the nonnormality of the data. Still there will be

* Table 15 in the Appendix gives $2 \arcsin \sqrt{\hat{\pi}}$.

times when the presence of severe skewness or outliers causes nonnormality that could not be eliminated by a transformation. Wilcoxon's rank sum test (Chapter 5) can be used for comparing two populations in the presence of nonnormality when working with two independent samples. For data based on more than two independent samples we can address nonnormality using the Krusal–Wallis test (Section 10.6).

EXERCISES

10.3   The data of Example 10.5 are shown below. Suppose that the four columns represent four geographic locations of the country (NE, SE, NW, SW) and that a random sample of 100 voters was obtained from six selected SMSAs within each geographic location. Analyze the sample data by using the arcsin transformation to determine if there are differences among the four geographic locations. Use $\alpha = .05$.

NE	SE	NW	SW
.13	.10	.03	.20
.17	.20	.10	.30
.30	.33	.20	.43
.53	.47	.33	.60
.60	.57	.43	.67
.70	.70	.60	.77

10.4   Refer to Exercise 10.3. Suppose that the rows correspond to different socioeconomic levels, so that one SMSA was selected from each socioeconomic level in each of the four geographic locations. The sample data then represent the proportion of favorable responses based on independent samples of 100 people for each socioeconomic–geographic location combination.

a. Analyze the transformed data and draw conclusions.

b. Comment on the proposal to take two random samples of size 50 for each socioeconomic–geographic location combination, rather than taking one sample of 100 voters.

10.6   A NONPARAMETRIC ALTERNATIVE: THE KRUSKAL–WALLIS TEST

The concept of a rank sum test can be extended to a comparison of more than two populations. In particular, suppose that n_1 observations are drawn at random from population 1, n_2 from population 2, ..., and n_k from population k. We may wish to test the hypothesis that the k samples were drawn from identical distributions. The following test procedure, sometimes called the Kruskal–Wallis test, is then appropriate.

EXTENSION OF
THE RANK SUM
TEST FOR MORE
THAN TWO
POPULATIONS

H_0: The k distributions are identical.

H_a: Not all the distributions are the same.

T.S.: $H = \dfrac{12}{n(n+1)} \sum_i \dfrac{T_i^2}{n_i} - 3(n+1)$

where n_i is the number of observations from sample i ($i = 1, 2, \ldots, k$), n is the combined sample size; that is, $n = \sum_i n_i$ and T_i denotes the sum of the ranks for the measurements in sample i after the combined sample measurements have been ranked.

R.R.: For a specified value of α, reject H_0 if H exceeds the critical value of χ^2 for $a = \alpha$ and df $= k - 1$.

Note: When there are a large number of ties in the ranks of the sample measurements, use

H'

$H' = \dfrac{H}{1 - \left[\sum_j (t_j^3 - t_j)/(n^3 - n) \right]}$

where t_j is the number of observations in the jth group of tied ranks.

EXAMPLE 10.6

Three random samples of clergymen were drawn, one containing 10 Methodist ministers, the second containing 10 Catholic priests, and the third containing 10 Pentecostal ministers. Each of the clergymen was then examined, using a test to measure his knowledge about causes of mental illness. These test scores are listed in Table 10.19.

Use the data to determine if the three groups of clergymen differ with respect to their knowledge about the causes of mental illness. Use $\alpha = .05$.

Solution

The research and null hypotheses for this example can be stated as follows:

H_a: At least one of the three groups of clergymen differs from the others with respect to knowledge about causes of mental illness.

TABLE 10.19   Scores for Knowledge of Mental Illness for the Clergymen, Example 10.6

Methodist	Catholic	Pentecostal
32	32	28
30	32	21
30	26	15
29	26	15
26	22	14
23	20	14
20	19	14
19	16	11
18	14	9
12	14	8

H_0: There is no difference among the three groups with respect to knowledge about the causes of mental illness (i.e., the samples of scores were drawn from identical populations).

Before computing H we must first jointly rank the 30 test scores from lowest to highest. From Table 10.19 we see that 8 is the lowest test score, and this clergyman is assigned the rank of 1. Similarly, the scores 9, 11, and 12 receive the ranks 2, 3, and 4, respectively. Five clergymen have a test score of 14, and since these 5 scores occupy the ranks 5, 6, 7, 8, and 9, we assign each one a rank of 7, the average of the occupied ranks. In a similar way we can assign the remaining ranks to test scores. Table 10.20 lists the 30 test scores and associated ranks (in parentheses).

Note from Table 10.20 that the sums of the ranks for the three groups are 197, 178, and 90. Hence the computed value of H is

$$H = \frac{12}{30(30 + 1)} \left[\frac{(197)^2}{10} + \frac{(178)^2}{10} + \frac{(90)^2}{10} \right] - 3(30 + 1)$$

$$= \frac{12}{930} (3880.9 + 3168.4 + 810) - 93 = 8.4$$

Since there are groups of tied ranks, we will use H' and compare its value to H. To do this we form the g groups composed of identical ranks, shown in the accompanying table.

From this information we calculate the quantity

$$\frac{\sum_j (t_j^3 - t_j)}{n^3 - n} = \frac{1}{30^3 - 30} [(5^3 - 5) + (2^3 - 2) + (2^3 - 2) + (2^3 - 2) + (3^3 - 3)$$

$$+ (2^3 - 2) + (3^3 - 3)]$$

$$= \frac{192}{26,970} = .0071$$

TABLE 10.20   Test Scores and Ranks for the Clergymen Study

Methodist		Catholic		Pentecostal	
32	(29)	32	(29)	28	(24)
30	(26.5)	32	(29)	21	(18)
30	(26.5)	26	(22)	15	(10.5)
29	(25)	26	(22)	15	(10.5)
26	(22)	22	(19)	14	(7)
23	(20)	20	(16.5)	14	(7)
20	(16.5)	19	(14.5)	14	(7)
19	(14.5)	16	(12)	11	(3)
18	(13)	14	(7)	9	(2)
12	(4)	14	(7)	8	(1)

Rank	Group	t_i
1	1	1
2	2	1
3	3	1
4	4	1
7, 7, 7, 7, 7	5	5
10.5, 10.5	6	2
12	7	1
13	8	1
14.5, 14.5	9	2
16.5, 16.5	10	2
18	11	1
19	12	1
20	13	1
22, 22, 22	14	3
24	15	1
25	16	1
26.5, 26.5	17	2
29, 29, 29	18	3

Substituting this value into the formula for H', we have

$$H' = \frac{H}{1 - .0071} = \frac{8.4}{.9929} = 8.46$$

So, even with more than half of the measurements involved in ties, H' and H are nearly the same. The critical value of chi-square with $a = .05$ and $df = k - 1 = 2$ can be found using Table 5 in the Appendix. This value is 5.99. Since the observed value of H' is greater than 5.99, we reject the null hypothesis and conclude that at least one of the clergy groups has more knowledge about the causes of mental illness than the other two groups.

EXERCISES

10.5   The yields (in pounds) of five different varieties (A, B, C, D, E) of four-year-old orange trees in one orchard were to be compared. A random sample of seven trees of each variety was obtained from the orchard. The yields for these trees are presented below.

A	B	C	D	E
13	27	40	17	36
19	31	44	28	32
39	36	41	41	34
38	29	37	45	29
22	45	36	15	25
25	32	38	13	31
10	44	35	20	30

Conduct a test of the null hypothesis that the five varieties have the same yield distributions. Use $\alpha = .01$. Draw conclusions.

 10.6 In Exercise 5.50, we discussed an experiment to compare the weights of the combs of roosters fed two different vitamin-supplemented diets. Twenty-eight healthy roosters were randomly divided into two groups, with one group receiving diet I and the other receiving diet II. After the study period the comb weight (in mg) was recorded for each rooster. These data are given below.

Diet I	73	130	115	144	127	126	112	76	68	101	126	49	110	123
Diet II	80	72	73	60	55	74	67	89	75	66	93	75	68	76

a. Plot the sample data separately for the two diets.
b. Compute the sample means and variances.
c. Based on parts (a) and (b), is there reason to suspect nonconstant variances or non-normality?
d. Run an analysis of variance (if appropriate) and draw conclusions.

10.7 Refer to Exercise 10.6. Suggest a nonparametric alternative and conduct the test. Compare your results to those obtained from Exercise 10.6, and reach a conclusion based on the results of both tests.

10.7   SUMMARY

In this chapter we presented methods for extending the results of Chapter 5 to include a comparison among t population means. An independent random sample is drawn from each of the t populations. A measure of the within-sample variability is computed as $s_W^2 = SSW/(n - t)$. Similarly, a measure of the between-sample variability is obtained as $s_B^2 = SSB/(t - 1)$.

The decision to accept or reject the null hypothesis of equality of the t population means depends on the computed value of $F = s_B^2/s_W^2$. Under H_0, both s_B^2 and s_W^2 estimate σ_ε^2, the variance common to all t populations. Under the alternative hypothesis, s_B^2 estimates $\sigma_\varepsilon^2 + \theta$, where θ is a positive quantity, whereas s_W^2 still estimates σ_ε^2. Thus large values of F indicate a rejection of H_0. Critical values for F are obtained from Table 6 in the Appendix for $df_1 = t - 1$ and $df_2 = n - t$. This test procedure, called an analysis of variance, is usually summarized in an analysis of variance (AOV) table.

You might be puzzled at this point by the following question: Suppose we reject H_0 and conclude that at least one of the means differs from the rest; which ones differ from the others? This chapter has not answered this question; Chapter 11 attacks this problem through procedures based on multiple comparisons.

In this chapter we also discussed the assumptions underlying an analysis of variance for a completely randomized design. Independent random samples are absolutely necessary. The assumption of normality is least critical because we are dealing with means and the Central Limit Theorem holds for reasonable sample sizes. The equal variance assumption is critical only when the sample sizes are markedly different; this is a good argument for equal (or nearly equal) sample sizes. A test for equality of variances makes use of the F_{max} statistic, s_{max}^2/s_{min}^2.

Sometimes the sample data indicate that the population variances are different. Then, when the relationship between the population mean and the population standard deviation is either known or suspected, it is convenient to transform the sample measurements y to new values y_T in order to stabilize the population variances, using the transformation suggested in Table 10.11. These transformations include the square root, logarithmic, arcsin, and many others.

The topics in this chapter are certainly not covered in exhaustive detail. However, the material is sufficient for training the beginning researcher to be aware of the assumptions underlying his or her project and to consider either running an alternative analysis (such as using a nonparametric statistical method, the Kruskal–Wallis test) when appropriate or applying a transformation to the sample data.

KEY FORMULAS

1. Analysis of variance for a completely randomized design

$$TSS = \sum_{i,j} (y_{ij} - \bar{y})^2$$

$$SSB = \sum_{i} n_i(\bar{y}_i - \bar{y})^2$$

$$SSW = \sum_{i,j} (y_{ij} - \bar{y}_i)^2$$

2. Analysis of variance for a completely randomized design (shortcut formulas)

$$TSS = \sum_{i,j} y_{ij}^2 - \frac{G^2}{n}$$

$$SSB = \sum \frac{T_i^2}{n_i} - \frac{G^2}{n}$$

$$SSW = TSS - SSB$$

3. Model for a completely randomized design

$$y_{ij} = \mu + \alpha_i + \varepsilon_{ij}$$

4. Hartley's test for homogeneity of variances

$$H_0:\ \sigma_1^2 = \sigma_2^2 = \cdots = \sigma_t^2$$

$$\text{T.S.:}\ F_{max} = \frac{s_{max}^2}{s_{min}^2}$$

5. Kruskal–Wallis test

H_0: The k distributions are identical

$$\text{T.S.:}\ H = \frac{12}{n(n+1)} \sum \frac{T_i^2}{n_i} - 3(n+1)$$

EXERCISES

10.8 An experiment was conducted to compare the number of major defectives observed along each of five production lines in which changes were being instituted. Production was monitored continuously during the period of changes, and the number of major defectives was recorded per day for each line. These data are shown here.

	Production line			
1	2	3	4	5
34	54	75	44	80
44	41	62	43	52
32	38	45	30	41
36	32	10	32	35
51	56	68	55	58

a. Compute $\bar{y}$ and s^2 for each sample. Does there appear to be a problem with nonconstant variances?

b. Use a square root transformation on the data and conduct an analysis on the transformed data.

c. Draw your conclusions concerning differences among production lines.

10.9 Do a Kruskal–Wallis test on the data represented in Exercise 10.8. Does this test confirm the conclusions drawn in Exercise 10.8? If the results differ, which analysis do you believe?

10.10 The Agricultural Experiment Station of a university tested two different herbicides and their effects on crop yield. From 90 acres set aside for the experiment, herbicide 1 was used on a random sample of 30 acres, herbicide 2 was used on a second random sample of 30 acres, and the remaining 30 acres were used as a control. At the end of the growing season the yields (in bushels per acre) were

	Sample Mean	Sample Standard Deviation
Herbicide 1	90.2	6.5
Herbicide 2	89.3	7.8
Control 3	85.0	7.4

Use these data to conduct a one-way analysis of variance. Use $\alpha = .05$. Are any of the yields different? If so, which ones?

10.11 Research from the Department of Fruit Crops at a university compared four different preservatives to be used in freezing strawberries. The yield from a strawberry patch was prepared for freezing and randomly divided into four equal groups. Within each group the strawberries were treated with the appropriate preservative and packaged into 8 small plastic bags for freezing at 0° C. Those in group I served as a control group, while those in groups II, III, and IV were assigned one of three newly developed preservatives. After all 32 bags of strawberries were prepared, they were stored at 0° C for a period of six months.

At the end of this time, the contents of each bag were allowed to thaw and then rated on a scale of 1 to 10 points for discoloration. (Note that a low score indicates little discoloration.) These ratings are given below.

Group I	10	8	7.5	8	9.5	9	7.5	7
Group II	6	7.5	8	7	6.5	6	5	5.5
Group III	3	5.5	4	4.5	3	3.5	4	4.5
Group IV	2	1	2.5	3	4	3.5	2	2

a. We might be concerned with the normality of the data. To avoid any problems, refer to Section 10.6 to run a Kruskal–Wallis one-way analysis of variance by ranks. Use $\alpha = .05$.

b. Use the output here to compare the results of a one-way analysis of variance to the results for the Kruskal–Wallis test of part (a). Draw conclusions based on the results of both tests.

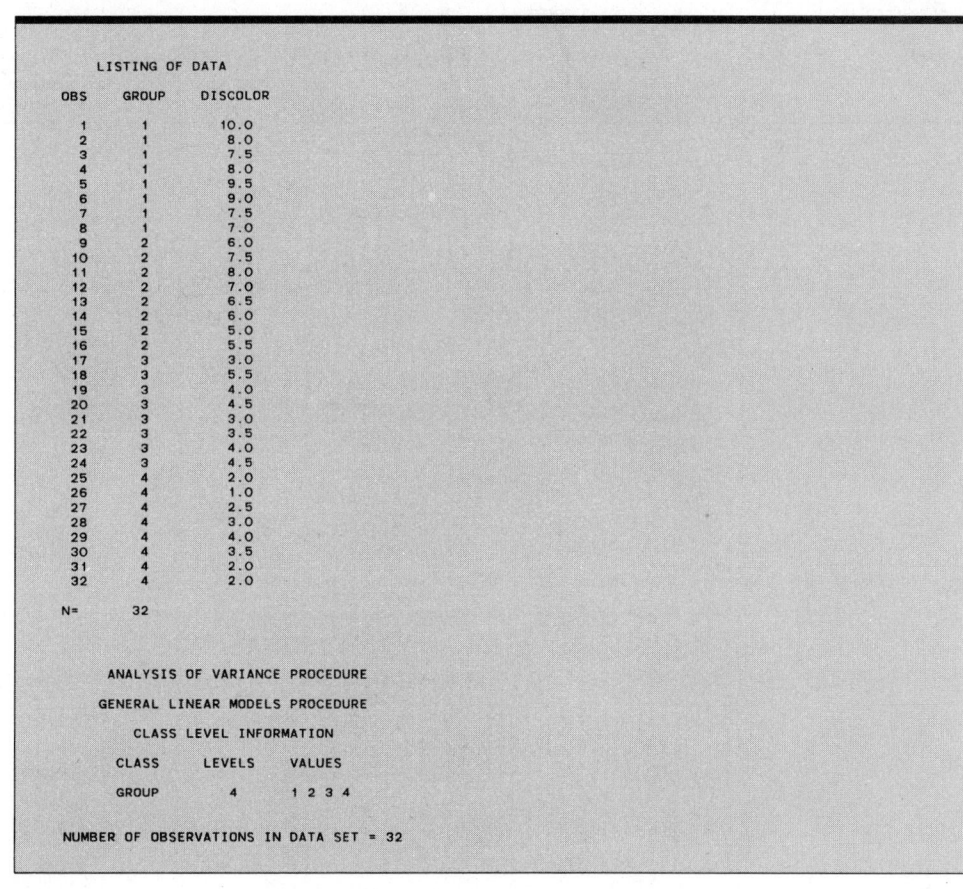

```
        LISTING OF DATA

   OBS      GROUP      DISCOLOR

    1         1          10.0
    2         1           8.0
    3         1           7.5
    4         1           8.0
    5         1           9.5
    6         1           9.0
    7         1           7.5
    8         1           7.0
    9         2           6.0
   10         2           7.5
   11         2           8.0
   12         2           7.0
   13         2           6.5
   14         2           6.0
   15         2           5.0
   16         2           5.5
   17         3           3.0
   18         3           5.5
   19         3           4.0
   20         3           4.5
   21         3           3.0
   22         3           3.5
   23         3           4.0
   24         3           4.5
   25         4           2.0
   26         4           1.0
   27         4           2.5
   28         4           3.0
   29         4           4.0
   30         4           3.5
   31         4           2.0
   32         4           2.0

   N=       32


        ANALYSIS OF VARIANCE PROCEDURE

     GENERAL LINEAR MODELS PROCEDURE

         CLASS LEVEL INFORMATION

     CLASS     LEVELS     VALUES

     GROUP       4        1 2 3 4

   NUMBER OF OBSERVATIONS IN DATA SET = 32
```

ANALYSIS OF VARIANCE PROCEDURE

GENERAL LINEAR MODELS PROCEDURE

DEPENDENT VARIABLE: RANK1   RANK FOR VARIABLE DISCOLOR

SOURCE	DF	SUM OF SQUARES	MEAN SQUARE	F VALUE	PR > F	R-SQUARE	C.V.
MODEL	3	2331.93750000	777.31250000	56.74	0.0001	0.858751	22.4313
ERROR	28	383.56250000	13.69866071		ROOT MSE		RANK1 MEAN
CORRECTED TOTAL	31	2715.50000000			3.70117018		16.50000000

SOURCE	DF	TYPE I SS	F VALUE	PR > F	DF	TYPE III SS	F VALUE	PR > F
GROUP	3	2331.93750000	56.74	0.0001	3	2331.93750000	56.74	0.0001

NUMBER OF OBSERVATIONS IN DATA SET = 32

ANALYSIS OF VARIANCE PROCEDURE

GENERAL LINEAR MODELS PROCEDURE

DEPENDENT VARIABLE: DISCOLOR

SOURCE	DF	SUM OF SQUARES	MEAN SQUARE	F VALUE	PR > F	R-SQUARE	C.V.
MODEL	3	159.18750000	53.06250000	55.67	0.0001	0.856422	18.3771
ERROR	28	26.68750000	0.95312500		ROOT MSE		DISCOLOR MEAN
CORRECTED TOTAL	31	185.87500000			0.97628121		5.31250000

SOURCE	DF	TYPE I SS	F VALUE	PR > F	DF	TYPE III SS	F VALUE	PR > F
GROUP	3	159.18750000	55.67	0.0001	3	159.18750000	55.67	0.0001

10.12 Refer to Exercise 10.11. Suppose that some of the 32 bags were stored improperly and could not be analyzed at the end of the six-month period. In particular, the sample sizes in the four groups were, respectively, 8, 6, 5, and 7. These data appear below.

Group I	10	8	7.5	8	9.5	9	7.5	7
Group II	6	7.5	8	7	6.5	6		
Group III	3	5.5	4	4.5	3			
Group IV	2	1	2.5	3	4	3.5	2	

Rerun an analysis of variance and/or a Kruskal–Wallis test and draw conclusions. Use $\alpha = .05$.

10.13 An experiment was conducted to compare the starch content of tomato plants grown in sandy soil supplemented by one of three different nutrients, A, B, or C. Eighteen tomato seedlings of one particular variety were selected for the study, with six assigned to each of the nutrient groups. All seedlings were planted in a sand culture and maintained under a controlled environment. Those seedlings assigned to nutrient A served as the control group (receiving distilled water only). Plants assigned to nutrient B were fed a weak concentration of Hoagland nutrient, while those assigned to nutrient C received the Hoagland nutrient at full strength. The stem starch contents were determined 25 days after planting and are recorded below, in micrograms per milligram.

Nutrient A	22	20	21	18	16	14
Nutrient B	12	14	15	10	9	6
Nutrient C	7	9	7	6	5	3

a. Run an analysis of variance to test for differences in starch content for the three nutrient groups. Use $\alpha = .05$.
b. Draw your conclusions.

10.14 Although we often have well-planned experiments with equal numbers of observations per treatment, we still end up with unequal numbers at the end of a study. Suppose that although six plants were allocated to each of the nutrient groups of Exercise 10.13, only five survived in group B and four in group C. The data for the stem starch contents are given below.

Nutrient A	22	20	21	18	16	14
Nutrient B	12	14	15	10	9	
Nutrient C	7	9	7	6		

a. Write an appropriate model for this experimental situation. Define all terms.
b. Assuming that nutrients B and C did not cause the plants to die, perform an analysis of variance to compare the treatment means. Use $\alpha = .05$.

10.15 Salary disputes and their eventual resolutions often leave both employers and employees embittered by the entire ordeal. To assess employee reactions to a recently devised salary and fringe benefits plan, the personnel department obtained random samples of 15 employees from each of three divisions: manufacturing, marketing, and research. Each employee sampled was asked to respond (in confidence) to a series of questions. Several employees refused to cooperate, as reflected in the unequal sample sizes. The data are given below.

	Manufacturing	Marketing	Research
Sample size	12	14	11
Sample mean	25.2	32.6	28.1
Sample variance	3.6	4.8	5.3

a. Write a model for this experimental situation.
b. Use the summary of the scored responses to compare the means for the three divisions (the higher a score, the higher the employee acceptance). Use $\alpha = .01$.

10.16 The yields of corn, in bushels per plot, were recorded for four different varieties of corn, A, B, C, and D. In a controlled greenhouse environment, each variety was randomly assigned to eight of 32 plots available for the study. The yields are listed here.

A	2.5	3.6	2.8	2.7	3.1	3.4	2.9	3.5
B	3.6	3.9	4.1	4.3	2.9	3.5	3.8	3.7
C	4.3	4.4	4.5	4.1	3.5	3.4	3.2	4.6
D	2.8	2.9	3.1	2.4	3.2	2.5	3.6	2.7

a. Write an appropriate statistical model.
b. Perform an analysis of variance on these data and draw your conclusions. Use $\alpha = .05$.

10.17 Refer to Exercise 10.16. Perform a Kruskal–Wallis analysis of variance by ranks (with $\alpha = .05$) and compare your results to those in Exercise 10.16.

10.18 Many corporations make use of the Wide Area Telephone System (WATS), where, for a fixed rent per month, the corporation can make as many long distance calls as it likes. Depending on the area of the country in which the corporation is located, it can rent a WATS line for certain geographic bands. For example, in Ohio these bands might include the following states:

Band I:	Ohio	
Band II:	Indiana	Pennsylvania
	Kentucky	Tennessee
	Maryland	Virginia
	Michigan	West Virginia
	North Carolina	Washington, D.C.
Band III:	32 states shown in the map, plus Washington, D.C.	

To monitor the use of the WATS lines, a corporation selected a random sample of 12 calls from each of the following areas in a given month. The length of the conversation (in minutes) was recorded for each call. (Band III excludes states in Band II and Ohio.)

Ohio	2	3	5	8	4	6	18	19	9	6	7	5
Band II	6	8	10	15	19	21	10	12	13	2	5	7
Band III	12	14	13	20	25	30	5	6	21	22	28	11

EXERCISES

FIGURE 10.3   Band II
WATS Line
Coverage

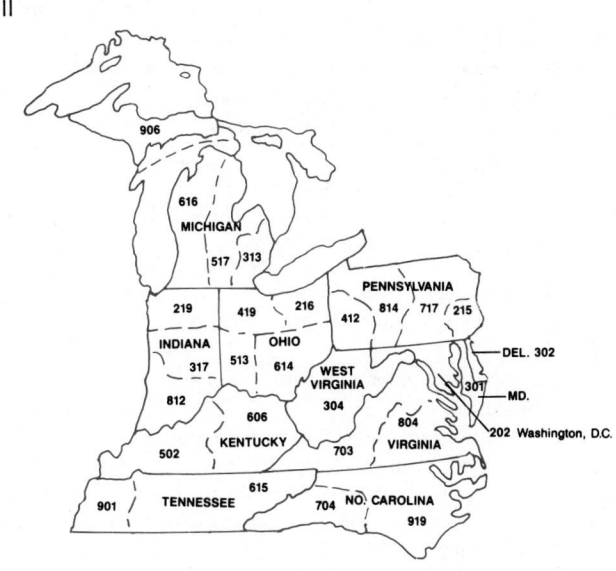

Band III

Perform an analysis of variance to compare the mean lengths of calls for the three areas. Use $\alpha = .05$.

10.19   Refer to Exercise 10.18. Suppose that rather than 12 calls from each area, we obtained a random sample of 15 calls. Use the additional three measurements recorded below for the total of 15 observations per area.

Ohio	10	11	4
Band II	8	28	31
Band III	29	50	120

Analyze the sample data to compare the mean number of call times for the three areas. Use $\alpha = .05$.

10.20   Refer to Exercise 10.18. Some researchers would argue that durations of telephone calls may not be normally distributed. Perform a Kruskal-Wallis one-way analysis of variance by ranks and compare your results to those of Exercise 10.18. Use $\alpha = .05$.

10.21   Examine the SAS computer output that follows for the analysis of variance performed in Example 10.2.
a. Identify the sums of squares and degrees of freedom for methods, error, and total.
b. Give the mean squares and F test for the equality of method means.
c. Give the level of significance for the test in part (b).
d. What is the coefficient of determination?

ANALYSIS OF VARIANCE PROCEDURE

GENERAL LINEAR MODELS PROCEDURE

DEPENDENT VARIABLE: SCORE

SOURCE	DF	SUM OF SQUARES	MEAN SQUARE	F VALUE	PR > F	R-SQUARE	C.V.
MODEL	2	452.12878788	226.06439394	7.71	0.0136	0.658556	6.9724
ERROR	8	234.41666667	29.30208333		ROOT MSE		SCORE MEAN
CORRECTED TOTAL	10	686.54545455			5.41313988		77.63636364

SOURCE	DF	TYPE I SS	F VALUE	PR > F	DF	TYPE III SS	F VALUE	PR > F
METHOD	2	452.12878788	7.71	0.0136	2	452.12878788	7.71	0.0136

10.22   Four different types of pillows were tested by a panel of consumers. Each panelist examined only one pillow and rated it on a scale from 1 (inferior) to 7 (superior). The data are summarized here.

Pillow Type	Rating	n_i
1	1, 3, 5, 7, 2, 3, 4	7
2	7, 6, 7, 7, 6	5
3	1, 2, 3, 2, 3, 2, 1	7
4	4, 3, 4, 1, 5	5

a. Perform a one-way analysis of variance using $\alpha = .05$.

b. Draw conclusions.

10.23 a. Do the data of Exercise 10.22 suggest violations of the AOV assumptions?

b. Suggest an alternate analysis and compare your results to those of Exercise 10.22.

c. Draw final conclusions.

10.24 Use data from either a laboratory course or from some other source to make comparisons of three or more methods, varieties, or plans. Use an analysis of variance with $\alpha = .05$.

10.25 An experiment was conducted to test the effects of five different diets in turkeys. Six turkeys were randomly asssigned to each of the five diet groups and were fed for a fixed period of time.

Group	Weight Gained (lbs)
Control diet	4.1, 3.3, 3.1, 4.2, 3.6, 4.4
Control diet + level 1 of additive A	5.2, 4.8, 4.5, 6.8, 5.5, 6.2
Control diet + level 2 of additive A	6.3, 6.5, 7.2,7.4, 7.8, 6.7
Control diet + level 1 of additive B	6.5, 6.8, 7.3, 7.5, 6.9, 7.0
Control diet + level 2 of additive B	9.5, 9.6, 9.2, 9.1, 9.8, 9.1

a. Plot the data separately for each sample.

b. Compute $\bar{y}_i$ and s_i for each sample.

c. Is there any evidence of unequal variances or nonnormality?

d. Assuming that the five groups were comparable with respect to initial weights of the turkeys, use the weight-gained data to draw conclusions concerning the different diets. Use $\alpha = .05$.

10.26 Run a Kruskal–Wallis test for the data of Exercise 10.25. Do these results confirm what you concluded from an analysis of variance? What overall conclusions can be drawn?

10.27 Some researchers have conjectured that stem-pitting disease in peach tree seedlings might be related to the presence or absence of nematodes in the soil. Hence weed and soil treatment using herbicides might be effective in promoting seedling growth. An experiment was conducted to compare peach tree seedling growth with soil and weeds treated with one of three herbicides:

A: control (no herbicide)

B: herbicide with Nemagone

C: herbicide without Nemagone

Of the 18 seedlings chosen for the study, six were randomly assigned to each treatment group. Soil and weeds in the growing areas for the three groups were treated with the appropriate herbicide. At the end of the study period, the height (in cm) was recorded for each seedling. Use the sample data below to run an analysis of variance for detecting differences among the seedling heights for the three groups. Use $\alpha = .05$. Draw your conclusions.

Herbicide A	66	67	74	73	75	64
Herbicide B	85	84	76	82	79	86
Herbicide C	91	93	88	87	90	86

10.28 Suppose that the data of Exercise 10.8 represented data from five different weeks by day of the week.

Day of Week	Data
Monday	34, 54, 75, 44, 80
Tuesday	44, 41, 62, 43, 52
Wednesday	32, 38, 45, 30, 41
Thursday	36, 32, 10, 32, 35
Friday	51, 56, 68, 55, 58

a. Do a one-way analysis of variance to assess the variability due to days.
b. Draw conclusions.

10.29 Refer to the data of Exercise 10.25. To illustrate the effect that an extreme value can have on conclusions from an analysis of variance, suppose that the weight gained by the fifth turkey in the level 2, additive B group was 12.8 (or 15.8) rather than 9.8.
a. What effect does this have on the assumptions for an analysis of variance?
b. With 9.8 replaced by 15.8, if someone unknowingly ran an analysis of variance, what conclusions would he or she draw?

10.30 Refer to Exercise 10.29. What happens to the Kruskal–Wallis test if the value 9.8 is replaced by 12.8 (or 15.8)? Might there be a reason to run both a Kruskal–Wallis test and an analysis of variance? Why?

10.31 Is the Kruskal–Wallis test more powerful than an analysis of variance, in certain situations, for detecting differences among treatment means? Explain.

10.32 A small corporation makes wire coating (for insulation) on three different machines. The machines are identical with respect to make, model, and age, but seem to differ with respect to variation of the inside diameter dimension (mm). Management wishes to test this variation and collected data from each machine.

Machine A	Machine B	Machine C
105	56	183
3	43	144
90	1	219
217	37	86
22	14	39

Conduct a test for the homogeneity of the population variances. Would it be appropriate to proceed with an analysis of variance based on the results of this test? Explain.

11 | MULTIPLE COMPARISONS

11.1 | INTRODUCTION

In Chapter 10 we introduced a procedure for testing the equality of t population means. The test statistic $F = s_B^2/s_W^2$ was used to determine whether the between-sample variability was large relative to the within-sample variability. If the computed value of F for the sample data exceeded the critical value obtained from Table 6 in the Appendix, the null hypothesis $H_0: \mu_1 = \mu_2 = \cdots = \mu_t$ was rejected in favor of the alternative hypothesis

H_a: At least one of the t population means differs from the rest

While rejection of the null hypothesis does give us some information concerning the population means, we do not know which means differ from each other. For example, does μ_1 differ from μ_2 or μ_3? Does μ_3 differ from μ_4, μ_5, and μ_6? **Multiple-comparison procedures** have been developed to answer questions such as these. While many multiple-comparison procedures have been proposed, we will focus on just a few of the more common methods. After studying these few procedures, you should be able to evaluate the results of most published material using multiple comparisons or to suggest an appropriate multiple-comparison procedure in an experimental situation.

11.2 | NOTATION AND DEFINITIONS

Before developing several different multiple-comparison procedures, we need the following notation and definitions. Consider a one-way classification where we wish to make comparisons among the t population means $\mu_1, \mu_2, \ldots, \mu_t$. These comparisons among t population means can be written in the form

$$l = a_1\mu_1 + a_2\mu_2 + \cdots + a_t\mu_t = \sum_{i=1}^{t} a_i\mu_i$$

where the a_is are constants satisfying the property that $\sum a_i = 0$. For example, if we wanted to compare μ_1 to μ_2, we could write the linear form

$$l = \mu_1 - \mu_2$$

Note here that $a_1 = 1$, $a_2 = -1$, $a_3 = a_4 = \cdots = a_t = 0$, and $\sum_i a_i = 0$. Similarly, we could compare the mean for population 1 to the average of the means for populations 2 and 3. Then l would be of the form

$$l = \mu_1 - \frac{(\mu_2 + \mu_3)}{2}$$

where $a_1 = 1$, $a_2 = a_3 = -\frac{1}{2}$, $a_4 = a_5 = \cdots = a_t = 0$, and $\sum_i a_i = 0$

$\hat{l}$ An estimate of the linear form l, designated by $\hat{l}$, is formed by replacing the μs in l with

linear contrast   their corresponding sample means $\bar{y}_i$. The estimate $\hat{l}$ is called a **linear contrast**

DEFINITION 11.1

$\hat{l} = a_1 \bar{y}_1 + a_2 \bar{y}_2 + \cdots + a_t \bar{y}_t = \sum_i a_i \bar{y}_i$ is called a **linear contrast** among the t sample means and can be used to estimate $l = \sum_i a_i \mu_i$. The a_is are constants satisfying the constraint $\sum_i a_i = 0$.

The variance of the linear contrast $\hat{l}$ can be estimated as follows:

$\hat{V}(\hat{l})$ $$\hat{V}(\hat{l}) = s_W^2 \left[\frac{a_1^2}{n_1} + \frac{a_2^2}{n_2} + \cdots + \frac{a_t^2}{n_t} \right] = s_W^2 \sum_i \frac{a_i^2}{n_i}$$

where n_i is the number of sample observations selected from population i and s_W^2 is the mean square within samples obtained from the analysis of variance table for the one-way classification. If all the sample sizes are the same (i.e., all $n_i = n$), then

$$\hat{V}(\hat{l}) = \frac{s_W^2}{n} \sum_i a_i^2$$

There are many different contrasts that can be formed among the t sample means. However, if each of the sample means is based on the same number of observations (i.e., $n_i = n$), we have the following definition.

DEFINITION 11.2

Two contrasts l_1 and l_2, where

$$\hat{l}_1 = \sum_i a_i \bar{y}_i \qquad \text{and} \qquad \hat{l}_2 = \sum_i b_i \bar{y}_i$$

are said to be **orthogonal** if

$$a_1 b_1 + a_2 b_2 + \cdots + a_t b_t = \sum_i a_i b_i = 0$$

Note: The sample sizes must be the same.

mutually orthogonal   A set of contrasts is said to be **mutually orthogonal** if all pairs of contrasts in the set are orthogonal.

EXAMPLE 11.1   Consider a one-way classification for comparing $t = 4$ population means. Are the following contrasts orthogonal?

$$\hat{l}_1 = \bar{y}_1 - \bar{y}_2 \qquad \hat{l}_2 = \bar{y}_3 - \bar{y}_4$$

Solution   We can rewrite the contrasts in the following form:

$$\hat{l}_1 = \bar{y}_1 - \bar{y}_2 + 0(\bar{y}_3) + 0(\bar{y}_4)$$
$$\hat{l}_2 = 0(\bar{y}_1) + 0(\bar{y}_2) + \bar{y}_3 - \bar{y}_4$$

where we see that $a_1 = 1$, $a_2 = -1$, $a_3 = 0$, $a_4 = 0$, and $b_1 = 0$, $b_2 = 0$, $b_3 = 1$, $b_4 = -1$. It is then apparent that

$$\sum_i a_i b_i = a_1 b_1 + a_2 b_2 + a_3 b_3 + a_4 b_4 = 0$$

and hence the contrasts are orthogonal.

EXAMPLE 11.2   Refer to Example 11.1. Are the contrasts given below orthogonal?

$$\hat{l}_1 = \bar{y}_1 - \bar{y}_2 \qquad \text{and} \qquad \hat{l}_2 = \bar{y}_1 - \bar{y}_3$$

Solution   Rewriting the contrasts as

$$\hat{l}_1 = \bar{y}_1 - \bar{y}_2 + 0(\bar{y}_3) + 0(\bar{y}_4)$$
$$\hat{l}_2 = \bar{y}_1 + 0(\bar{y}_2) - \bar{y}_3 + 0(\bar{y}_4)$$

we have that

$$\sum_i a_i b_i = (1)(1) + (-1)(0) + (0)(-1) + (0)(0) = 1$$

which indicates that the two contrasts are not orthogonal.

The concept of orthogonality between linear contrasts is important in the study of multiple-comparison procedures. Recall that prior to running an analysis of variance among the t population means in a one-way classification, we assumed that

1. The t populations were normally distributed with a common variance σ_ε^2 but different means (under H_0 we assume that the means are equal).
2. Independent random samples were obtained from the t populations.

If we assume that each of the sample means is based on the same number of observations, then it can be shown that $t - 1$ orthogonal contrasts can be formed using the t **t − 1 contrasts**   sample means. These **$t - 1$ contrasts** form a set of mutually orthogonal contrasts. (An easy way to remember $t - 1$ is to refer to the number of degrees of freedom associated with the between-sample source of variability in the AOV table.) In addition, it can be shown that the sums of squares for the $t - 1$ contrasts will add up to the treatment sum of squares.

Mutual orthogonality is desirable because it leads to independence of the $t-1$ sums of squares associated with the orthogonal contrasts. As we will see later in the chapter, methods have also been developed that use nonorthogonal contrasts among the sample means. These methods will be particularly appropriate when the experimenter is making all pairwise comparisons among t treatment means.

EXERCISES

11.1   Consider the expressions

$$\hat{l}_1 = \bar{y}_1 + \bar{y}_2 - 2\bar{y}_3$$
$$\hat{l}_2 = \bar{y}_1 + \bar{y}_2 - 2\bar{y}_4$$

a. Are $\hat{l}_1$ and $\hat{l}_2$ linear contrasts?
b. Are $\hat{l}_1$ and $\hat{l}_2$ orthogonal?

11.2   Refer to Exercise 11.1. Construct a set of three mutually orthogonal contrasts for comparing four population means.

11.3   Write down a set of four mutually orthogonal contrasts to be used in comparing five population means.

11.3   WHICH ERROR RATE IS CONTROLLED?

Let us suppose that an experimenter wishes to compare t population means using c independent (orthogonal) contrasts. Each comparison among the t population means can be tested using a t test of the following form:

t test

H_0: $l = 0$

H_a: $l > 0$ (for a one-tailed test)

T.S.: $t = \dfrac{\hat{l}}{\sqrt{\hat{V}(\hat{l})}} = \dfrac{\hat{l}}{\sqrt{s_W^2 \sum_i a_i^2/n_i}}$

The rejection region for the computed value of the test statistic can be obtained from Table 4 in the Appendix with $a = \alpha/2$ and $df = n - t$.

   If each of the comparisons is tested with the same value of α, and if we assume that s_W^2 has an infinite number of degrees of freedom (so the tests are independent), then when all the null hypotheses are true, the probability of falsely rejecting H_0 on at least one of the t tests can be shown to be $1 - (1 - \alpha)^c$. This quantity is sometimes called an **overall error rate** for the c comparisons. We can see from Table 11.1 that as c increases for a given value of α, the probability of falsely rejecting H_0 on at least one of the t tests becomes quite large. Hence if an experimenter wished to compare $t = 20$ population means by using $c = 10$ orthogonal contrasts, the probability of falsely rejecting H_0 on at least one of the t tests could be as high as .401 when each individual test was performed with $\alpha = .05$.

$1 - (1 - \alpha)^c$ overall error rate

   The results of Table 11.1 are disturbing and may lead us to question significant results when they appear. The problem can be alleviated somewhat by **controlling the overall error rate** rather than the error rate (Type I error) for the individual t test. Suppose, for

controlling overall error rate

TABLE 11.1   A Comparison of the Overall Error Rate for c Independent Contrasts Among $t (t > c)$ Sample Means

c, Number of Contrasts	α, Probability of a Type I Error on an Individual Test		
	.10	.05	.01
1	.100	.050	.010
2	.190	.097	.020
3	.271	.143	.030
4	.344	.185	.039
5	.410	.226	.049
⋮	⋮	⋮	⋮
10	.651	.401	.096

example, that we wished the overall error rate for $c = 10$ orthogonal contrasts among $t = 20$ population means to be .10. What value of α must we use on the individual t tests to achieve an overall error rate of .10? Assuming s_W^2 is based on a large number of degrees of freedom, this problem can be solved by determining the value of α for which

$$1 - (1 - \alpha)^{10} = .10$$

The method of solution for this equation is not important now; we can see from Table 11.1 that by using $\alpha = .01$ for all 10 tests, the overall error rate would be approximately .10.

error rate for nonorthogonal contrasts While controlling the overall error rate for comparisons using orthogonal contrasts is fairly simple, it is difficult to obtain an expression equivalent to $1 - (1 - \alpha)^c$ for comparisons made with nonorthogonal contrasts. For example, suppose we wish to make all pairwise comparisons among $t = 4$ population means. Previous results indicate that we could make $t - 1 = 3$ orthogonal (independent) contrasts, but there are six possible pairwise comparisons among the population means (1 and 2, 1 and 3, 1 and 4, 2 and 3, 2 and 4, and 3 and 4). If each of these six comparisons is made using a t test with $\alpha = .05$, what is the overall error rate? Pearson and Hartley (1942, 1943) and Harter (1957) attacked the problem of determining the probability of falsely rejecting H_0 on at least one of the t tests for nonindependent contrasts. The solution is not easy, however, and is beyond the scope of this text. One alternative is to redefine the overall error rate and to determine a testing procedure that controls the overall error rate at a desired level. Indeed, a major difference among the multiple-comparison procedures we will discuss in the following sections is the error rate that each procedure controls.

11.4   FISHER'S LEAST SIGNIFICANT DIFFERENCE

Recall that we are interested in determining which population means differ after we have rejected the hypothesis of equality of t population means in an analysis of variance. R. A. Fisher (1949) developed a procedure for making pairwise comparisons among a set of t population means. The procedure is called Fisher's least significant difference (LSD).

The α-level of Fisher's LSD is valid for a given comparison only if the LSD is used for independent (orthogonal) comparisons or for preplanned comparisons. However, since many people find Fisher's LSD easy to compute and hence use it for making all possible pairwise comparisons (particularly those that look "interesting" following the completion of the experiment), researchers recommend applying Fisher's LSD only after the F test for treatments has been shown to be significant. This revised approach is sometimes referred to as **Fisher's protected LSD**. Simulation studies (Cramer and Swanson (1973)) suggest that the error rate for the protected LSD is controlled on an experimentwise basis at a level approximately equal to the α-level for the F test.

Fisher's protected LSD

We'll illustrate Fisher's protected procedure, but continue to call it Fisher's LSD. This procedure is summarized below.

FISHER'S LEAST SIGNIFICANT DIFFERENCE PROCEDURE

1. Perform an analysis of variance to test $H_0: \mu_1 = \mu_2 = \cdots = \mu_t$ against the alternative hypothesis that at least one of the means differs from the rest.
2. If there is insufficient evidence to reject H_0 using $F = s_B^2/s_W^2$, we proceed no further.

LSD

3. If H_0 is rejected, define the **least significant difference (LSD)** to be the observed difference between two sample means necessary to declare the corresponding population means different.
4. For a specified value of α, the least significant difference for comparing μ_i to μ_j is

$$LSD = t_{\alpha/2} \sqrt{s_W^2 \left(\frac{1}{n_i} + \frac{1}{n_j} \right)}$$

where n_i and n_j are the respective sample sizes from population i and j and t is the critical t value (Table 4 of the Appendix) for $a = \alpha/2$ and df denoting the degrees of freedom for s_W^2. Note that for $n_i = n_j = n$,

$$LSD = t_{\alpha/2} \sqrt{\frac{2s_W^2}{n}}$$

5. All pairs of sample means are then compared. If $|\bar{y}_i - \bar{y}_j| \geq LSD$, we declare the corresponding population means μ_i and μ_j different.
6. For each pairwise comparison of population means, the probability of a Type I error is fixed at a specified value of α.

EXAMPLE 11.3

Hydrochloric acid (HCL) is used in the preparation of certain dyes. Six different batches of HCL were used to produce a particular dye. Five measurements on the yield (in grams of dye) were obtained from each batch. A summary of the sample data is given in Tables 11.2 and 11.3. Use Fisher's LSD procedure to make all pairwise comparisons among the six population (batch) mean yields. Use $\alpha = .05$.

TABLE 11.2   Summary of the Dye Yields for Example 11.3

Batch	Sample Mean
1	505
2	528
3	564
4	498
5	600
6	470

TABLE 11.3   AOV Table for the Data of Example 11.3

Source	df	SS	MS	F
Between batches	5	56,360	11,272	4.60
Within batches	24	58,824	2,451	
Total	29			

Solution

steps for LSD procedure

We can solve this problem by following the five steps listed for the LSD procedure.

Step 1. We use the AOV table in Table 11.3. The F test of H_0: $\mu_1 = \mu_2 = \cdots = \mu_6$ is based on

$$F = \frac{s_B^2}{s_W^2} = 4.60$$

For $\alpha = .05$ with $df_1 = 5$ and $df_2 = 24$, we reject H_0 if F exceeds 2.62 (see Table 6 in the Appendix).

Steps 2, 3. Since 4.60 is greater than 2.62, we reject H_0 and conclude that at least one of the population means differs from the rest.

Step 4. The least significant difference for comparing two means based on samples of size 5 is then

$$LSD = t_{\alpha/2} \sqrt{\frac{2s_W^2}{5}} = 2.064 \sqrt{\frac{2(2451)}{5}} = 64.63$$

Note that the appropriate t value (2.064) was obtained from Table 4 with $a = \alpha/2 = .025$ and $df = 24$.

Step 5. When we have equal sample sizes, it is convenient to use the following procedure rather than make all pairwise comparisons among the sample means, because the same LSD is to be used for all comparisons.

a. We rank the sample means from lowest to highest.

Population	6	4	1	2	3	5
Sample Mean	470	498	505	528	564	600

b. We compute the sample difference

$$\bar{y}_{largest} - \bar{y}_{smallest}$$

If this difference is greater than the LSD, we declare the corresponding population means significantly different from each other. Next we compute the sample difference

$$\bar{y}_{\text{2nd largest}} - \bar{y}_{\text{smallest}}$$

and compare the result to the LSD. We continue to make comparisons with $\bar{y}_{\text{smallest}}$:

$$\bar{y}_{\text{3rd largest}} - \bar{y}_{\text{smallest}}$$

and so on, until we find either that all sample differences involving $\bar{y}_{\text{smallest}}$ exceed the LSD (and hence the corresponding population means are different) or that a sample difference involving $\bar{y}_{\text{smallest}}$ is less than the LSD. In the latter case we stop and make no further comparisons with $\bar{y}_{\text{smallest}}$. For our data, comparisons with $\bar{y}_{\text{smallest}}$, $\bar{y}_6$, give the following results:

Comparison	Conclusion
$\bar{y}_{\text{largest}} - \bar{y}_{\text{smallest}} = \bar{y}_5 - \bar{y}_6 = 130$	> LSD; proceed
$\bar{y}_{\text{2nd largest}} - \bar{y}_{\text{smallest}} = \bar{y}_3 - \bar{y}_6 = 94$	> LSD; proceed
$\bar{y}_{\text{3rd largest}} - \bar{y}_{\text{smallest}} = \bar{y}_2 - \bar{y}_6 = 58$	< LSD; stop

summary diagram

To summarize our results we make the following diagram:

population   <u> 6   4   1   2 </u>   3   5

Those populations joined by the underline have means that are not significantly different from $\bar{y}_6$. Note that populations 3 and 5 have sample differences with population 6 that exceed the LSD and hence are not underlined.

c. We now make similar comparisons with $\bar{y}_{\text{2nd smallest}}$, $\bar{y}_4$ in this case, using the procedures of part (b).

Comparison	Conclusion
$\bar{y}_5 - \bar{y}_4 = 102$	> LSD; proceed
$\bar{y}_3 - \bar{y}_4 = 66$	> LSD; proceed
$\bar{y}_2 - \bar{y}_4 = 30$	< LSD; stop

population   6   <u> 4   1   2 </u>   3   5

d. Continue with $\bar{y}_{\text{3rd smallest}}$, or $\bar{y}_1$ in our example.

Comparison	Conclusion
$\bar{y}_5 - \bar{y}_1 = 95$	> LSD; proceed
$\bar{y}_3 - \bar{y}_1 = 59$	< LSD; stop

population   6   4   <u> 1   2   3 </u>   5

e. Continue with $\bar{y}_{4\text{th smallest}}$, or $\bar{y}_2$ in our example.

Comparison	Conclusion
$\bar{y}_5 - \bar{y}_2 = 72$	$>$ LSD; proceed
$\bar{y}_3 - \bar{y}_2 = 36$	$<$ LSD; stop

population   6   4   1   <u>2   3</u>   5

f. Continue with $\bar{y}_{5\text{th smallest}}$, or $\bar{y}_3$ in our example.

Comparison	Conclusion
$\bar{y}_5 - \bar{y}_3 = 36$	$<$ LSD; stop

population   6   4   1   2   <u>3   5</u>

g. We can summarize steps (a) through (f) as follows:

population   <u>6   4   1   2</u>   3   5
                <u>4   1   2   3</u>
                      <u>1   2   3</u>
                            <u>2   3   5</u>

Those populations not underlined by a common line are declared to have means that are significantly different according to the least significant difference criterion. Note that we can eliminate the second and fourth lines from the top of part (g) since they are part of the first and third lines, respectively. The revised summary of significant and nonsignificant results is

population   <u>6   4   1   2</u>   3   5
                        <u>1   2   3</u>
                            <u>2   3   5</u>

In conclusion, we have μ_6, μ_4, μ_1, and μ_2 significantly less than μ_5. Also, μ_6 and μ_4 are significantly less than μ_3.

While the LSD procedure described in Example 11.3 may seem quite laborious, its application is quite simple. First we run an analysis of variance. If we reject the null hypothesis of equality of the population means, we compute the LSD for all pairs of sample means. When the sample sizes are the same, this difference is a single number for all pairs. We can then use the stepwise procedure described in steps 5(a) through 5(g) of Example 11.3. We need not write all those steps down, only the summary lines. The final summary (as given in step 5(g)) gives a handy visual display of the pairwise comparisons using Fisher's LSD.

Several remarks should be made concerning the LSD method for pairwise comparisons.

First, there is the possibility that the overall F test in our analysis of variance is significant but that no pairwise differences are significant using the LSD procedure. This apparent anomaly can occur because the null hypothesis $H_0: \mu_1 = \mu_2 = \cdots = \mu_t$ for the F test is equivalent to the hypothesis that all possible comparisons (paired or otherwise) among the population means are zero. For a given set of data, the comparisons that are significant might not be of the form $\mu_i - \mu_j$, the form we are using in our paired comparisons.

Fisher confidence interval

Second, Fisher's LSD procedure can also be used to form a confidence interval for $\mu_i - \mu_j$. A $100(1 - \alpha)\%$ confidence interval has the form

$$(\bar{y}_i - \bar{y}_j) \pm \text{LSD}$$

LSD for equal sample sizes

Third, when all the sample sizes are the same, the LSD for all pairs is

$$t_{\alpha/2} \sqrt{\frac{2s_W^2}{n}}$$

<hr>

11.5   TUKEY'S W PROCEDURE

We are aware of the major drawback of a multiple-comparison procedure with a controlled per-comparison error rate. Even when $\mu_1 = \mu_2 = \cdots = \mu_t$, unless α, the per-comparison error rate (such as with Fisher's unprotected LSD) is quite small, there is a high probability of declaring at least one pair of means significantly different when running multiple comparisons. To avoid this, other multiple-comparison procedures have been developed that control different error rates.

Studentized range distribution

Tukey (1953) proposed a procedure that makes use of the **Studentized range distribution**. When more than two sample means are being compared, to test the largest and smallest sample means, we could use the test statistic

$$\frac{\bar{y}_{\text{largest}} - \bar{y}_{\text{smallest}}}{s_p \sqrt{1/n}}$$

where n is the number of observations in each sample and s_p is a pooled estimate of the common population standard deviation σ. This test statistic is very similar to that for comparing two means (Section 5.1), but it does not possess a t distribution. One reason it does not is that we have waited to determine which two sample means (and hence population means) we would compare until we observed the largest and smallest sample means. This procedure is quite different from that of specifying $H_0: \mu_1 - \mu_2 = 0$, observing $\bar{y}_1$ and $\bar{y}_2$, and forming a t statistic.

The quantity

$$\frac{\bar{y}_{\text{largest}} - \bar{y}_{\text{smallest}}}{s_p \sqrt{1/n}}$$

follows a Studentized range distribution. We will not discuss the properties of this distribution, but will illustrate its use in Tukey's multiple-comparison procedure.

<table>
<tr><td>

TUKEY'S *W*
PROCEDURE

W

$q_\alpha(t, v)$

experimentwise error
rate

</td><td>

1. Rank the t sample means.
2. Two population means μ_i and μ_j are declared different if

$$|\bar{y}_i - \bar{y}_j| \geq W$$

where

$$W = q_\alpha(t, v) \sqrt{\frac{s_W^2}{n}}$$

s_W^2 is the mean square within samples based on v degrees of freedom, $q_\alpha(t, v)$ is the upper-tail critical value of the Studentized range (with a = α) for comparing t different populations, and n is the number of observations in each sample. A discussion follows showing how to obtain values of $q_\alpha(t, v)$ from Table 10 in the Appendix.

3. The error rate that is controlled is an **experimentwise error rate**. Thus the probability of observing an experiment with one or more pairwise comparisons falsely declared significant is specified at α.

</td></tr>
</table>

We can obtain values of $q_\alpha(t, v)$ from Table 10 in the Appendix. Values of v are listed along the left column of the table with values of t across the top row. Upper-tail values for the Studentized range are then presented for a = .05 and .01. For example, in comparing 10 population means based on 9 degrees of freedom for s_W^2, the .05 upper-tail critical value for the Studentized range is $q_{.05}(10, 9) = 5.74$.

EXAMPLE 11.4

Refer to the data of Example 11.3. Use Tukey's *W* procedure with $\alpha = .05$ to make pairwise comparisons among the six population means.

Solution

We can eliminate step 1 since the sample means were ranked in Example 11.3. For

$t = 6$ (we are making pairwise comparisons among six means)

$v = 24$ (s_W^2 had 24 degrees of freedom in the AOV)

$\alpha = .05$ (we specified the experimentwise error rate at .05)

$n = 5$ (there were five sample observations selected from each population)

we find

$q_{.05}(6, 24) = 4.37$

The absolute value of each difference in sample means must then be compared to

$$W = q_\alpha(t, v) \sqrt{\frac{s_W^2}{n}} = 4.37 \sqrt{\frac{2451}{5}} = 96.75$$

By substituting *W* for the LSD, we can use the same stepwise procedure for comparing sample means that we used in step 5 of the solution to Example 11.3.

Having ranked the sample means from low to high, comparisons against $\bar{y}_{smallest}$, which is $\bar{y}_6$, yield

population   <u>6   4   1   2   3</u>   5

Comparisons with $\bar{y}_{2nd\,smallest}$ ($\bar{y}_4$) yield

population   6   <u>4   1   2   3</u>   5

Similarly, comparisons with $\bar{y}_1$, $\bar{y}_2$, and $\bar{y}_3$ yield

population   6   4   <u>1   2   3   5</u>

                                   <u>   </u>

                                         <u>   </u>

Combining our results we obtain

population   <u>6   4   1   2   3</u>   5

                          <u>   </u>

                                    <u>   </u>

                                       <u>   </u>

                                         <u>   </u>

which simplifies to

population   <u>6   4   1   2   3</u>   5

                          <u>   </u>

All populations not underlined by a common line have population means that are significantly different from each other. That is, μ_6 and μ_4 are significantly less than μ_5.

By examining the multiple-comparison summaries using the least significant difference (Example 11.3) and Tukey's W procedure (Example 11.4), we see that Tukey's procedure is more conservative (declares fewer significant differences) than the LSD procedure. For example, applying Tukey's procedure to the data of Table 11.2 shows that μ_3 is no longer significantly larger than μ_6 and μ_4. Similarly, μ_5 is no longer significantly larger than μ_1 and μ_2. The explanation for this is that although both procedures have an experiment-wise error rate, the per-comparison error rate of the protected LSD method has been shown to be larger than that for Tukey's W procedure.

Tukey's procedure can be modified to account for unequal samples using a method proposed originally by Tukey in 1953 and then independently by Kramer in 1956. The procedure, sometimes referred to as the Tukey–Kramer procedure, makes use of the quantity

$$W_{ij} = q_\alpha(t, v) \sqrt{\frac{s_W^2}{2}\left(\frac{1}{n_i} + \frac{1}{n_j}\right)}$$

for determining whether or not μ_i and μ_j are different. Note that W_{ij} becomes the familiar W from Tukey's procedure when $n_i = n_j$. The rest of the procedure remains the same.

Tukey confidence interval   Tukey's procedure can also be used to construct confidence intervals for comparing two means. However, unlike the confidence intervals that can be formed from Fisher's LSD, Tukey's procedure enables us to construct simultaneous confidence intervals for all pairs of treatment differences. For a specified α level from which we compute W, the overall probability is $1 - \alpha$ that all differences $\mu_i - \mu_j$ will be included in an interval of the form

$$(\bar{y}_i - \bar{y}_j) \pm W$$

That is, the probability is $1 - \alpha$ that all the intervals $(\bar{y}_i - \bar{y}_j) \pm W$ include the corresponding population differences $\mu_i - \mu_j$. For unequal sample sizes, W_{ij} replaces W in the confidence interval.

11.6   STUDENT–NEWMAN–KEULS PROCEDURE

The Student–Newman–Keuls (SNK) procedure provides a modification of the Tukey W procedure. Although the SNK procedure also makes use of the Studentized range statistic, different critical values are used depending on the number of steps separating the means being tested. To compare the two procedures, let's refer to Example 11.3. Ranked in order from lowest to highest, the sample means are

Sample mean	470	498	505	528	564	600
Sample i	6	4	1	2	3	5

and the critical value of the Studentized range for Tukey's W procedure is

$$q_\alpha(t, v) = q_{.05}(6, 24) = 4.37$$

This same value of q is used for all pairwise comparisons of the six treatment means.

The SNK procedure makes use of a critical value

$$W_r = q_\alpha(r, v) \sqrt{\frac{s_W^2}{n}}$$

for means that are r steps apart when the t sample means are ranked from lowest to highest. For our example, $\bar{y}_{largest}$ and $\bar{y}_{smallest}$ are 6 "steps" apart and they would be compared using

$$W_6 = q_\alpha(6, v) \sqrt{\frac{s_W^2}{n}}$$

$$= 4.37 \sqrt{\frac{2451}{5}} = 96.75$$

(Note: This is W for Tukey's W procedure.) However, $\bar{y}_{largest}$ and $\bar{y}_{2nd\,smallest}$ are 5 "steps" apart and they would be compared to

$$W_5 = q_\alpha(5, v) \sqrt{\frac{s_W^2}{n}}$$

$$= 4.17 \sqrt{\frac{2451}{5}} = 92.33$$

TABLE 11.4   Values of r, $q_\alpha(r, v)$ and W_r for Example 11.3

r	2	3	4	5	6
$q_\alpha(r, v)$	2.92	3.53	3.90	4.17	4.37
W_r	64.65	78.16	86.35	92.33	96.75

The complete set of critical values W_r needed for the data of Example 11.3 is shown in Table 11.4. Values of $q_\alpha(r, v)$ are obtained from Table 10 in the Appendix by replacing t by r.

The Student–Newman–Keuls procedure, which relies on the number of ordered steps between two sample means when determining the significance of an observed sample difference, has neither an experimentwise nor a per-comparison error rate. Rather, the error rate is defined for means the same number of ordered steps apart. Since the critical value W_r decreases as the number of steps between the means being compared decreases, the SNK procedure is less conservative and hence will generally declare more significant differences than will Tukey's W procedure, which utilizes the largest value for W no matter how many steps separate the means being compared.

The SNK procedure is summarized here.

SNK PROCEDURE

1. Rank the t sample means from lowest to highest.
2. For two means $\bar{y}_i$ and $\bar{y}_j$ that are r steps apart, we declare μ_i and μ_j different if

    $$|\bar{y}_i - \bar{y}_j| \geq W_r$$

    where $W_r = q_\alpha(r, v)\sqrt{s_W^2/n}$, n is the number of observations per sample, s_W^2 is the mean square within samples from the AOV table, v is the degrees of freedom for s_W^2, and $q_\alpha(r, v)$ is the critical value of the Studentized range. Values of $q_\alpha(r, v)$ are given in Table 10 in the Appendix for $\alpha = .05$ and .01.

    (Note: Use the column labeled t to locate the desired value for r.)

EXAMPLE 11.5   Refer to the data of Example 11.3. Run the SNK procedure to make all pairwise comparisons based on $\alpha = .05$.

Solution   The critical values of W_r to be used were shown previously.

1. Beginning with $\bar{y}_{largest}(\bar{y}_6)$, every sample mean is compared to $\bar{y}_{smallest}(\bar{y}_6)$ using the appropriate value of W_r. These results are summarized here.

Comparison	W_r	Conclusion
$\bar{y}_5 - \bar{y}_6 = 130$	96.75	> 96.75; proceed
$\bar{y}_3 - \bar{y}_6 = 94$	92.33	> 92.33; proceed
$\bar{y}_2 - \bar{y}_6 = 58$	86.35	< 86.35; stop

2. Similarly, we can make comparisons with $\bar{y}_{2\text{nd smallest}}(\bar{y}_4)$, $\bar{y}_1$, $\bar{y}_2$, and $\bar{y}_3$;

Comparison	W_r	Conclusion
$\bar{y}_5 - \bar{y}_4 = 102$	92.33	>92.33; proceed
$\bar{y}_3 - \bar{y}_4 = 66$	86.35	<86.35; stop
$\bar{y}_5 - \bar{y}_1 = 95$	86.35	>86.35; proceed
$\bar{y}_3 - \bar{y}_1 = 59$	78.16	<78.16; stop
$\bar{y}_5 - \bar{y}_2 = 72$	78.16	<78.16; stop
$\bar{y}_5 - \bar{y}_3 = 36$	64.65	<64.65; stop

The results of these multiple comparisons made using the SNK procedure are shown here:

population   6   4   1   2   3   5
            _____
                            _____

All populations not underlined by a common line have population means that are different from each other. These results using the SNK procedure are slightly different from those shown for Tukey's W procedure. In particular, μ_6 and μ_3 are declared different as well as μ_5 and μ_1 for the SNK procedure. This example illustrates the fact that the SNK procedure will tend to declare more differences (and hence be less conservative) than Tukey's W procedure.

The SNK procedure can be modified to account for different sample sizes. When $n_i \neq n_j$, the value of W_r for means r steps apart is modified in the same way as Tukey's procedure. Here we would use

$$q_\alpha(r, v) \sqrt{\frac{s_W^2}{2}\left(\frac{1}{n_i} + \frac{1}{n_j}\right)}$$

to assess whether μ_i differs from μ_j.

11.7   DUNCAN'S NEW MULTIPLE RANGE TEST

Duncan (1955) developed a multiple-comparison procedure for obtaining all pairwise comparisons among t sample means. Although his procedure makes use of the Studentized range, his error rate is neither on an experimentwise basis (as with Tukey's) nor on a per-comparison basis. When the sample means have been ranked from lowest to highest, the error rate is designated in the following way. In general, if two sample means are r steps

TABLE 11.5   Duncan Protection Level Using $\alpha = .05$ When the Sample Means Are r Steps Apart

Number of Steps Apart, r	Protection Level, $(1 - .05)^{r-1}$	Probability of Falsely Rejecting H_0, $1 - (1 - .05)^{r-1}$
2	.950	.050
3	.903	.097
4	.857	.143
5	.815	.185
6	.774	.226
7	.735	.265

protection level   apart, Duncan defines the **protection level** as

$$(1 - \alpha)^{r-1}$$

The probability of falsely rejecting the equality of two population means when the sample means are r steps apart is then taken to be

error rate $$1 - (1 - \alpha)^{r-1}$$

For $\alpha = .05$, we illustrate the concept of a protection level in Table 11.5.

Duncan's reasons for allowing the protection level to decrease for increasing values of r have their basis in results presented in Section 11.2. As we indicated there, it is possible to form $t - 1$ orthogonal contrasts for comparing t treatment means. Using those contrasts, we can partition the treatment sum of squares into $t - 1$ single degree-of-freedom sums of squares. (We will discuss this partitioning later.) If we assume the degrees of freedom for s_W^2 are quite large, then the $(t - 1)$ F statistics are nearly independent. Then when each F test is conducted at a preset α value, and we assume $\mu_1 = \mu_2 = \cdots = \mu_t$, the probability of rejecting H_0 for one or more contrasts is approximately

$$1 - (1 - \alpha)^{t-1}$$

Duncan argued that since experimenters have little or no reservations in performing these multiple F tests for orthogonal contrasts even though the overall α level increases with t, it is reasonable to construct a multiple-comparison test for which the protection level decreases with the number of sample means included in a comparison. Thus Duncan uses a α-value equal to the quantity $1 - (1 - \alpha)^{r-1}$ when a pair of sample means are r steps apart $(r = 2, \ldots, t)$.

Because the protection level decreases with increasing r, *Duncan's multiple range test is very powerful*. That is, there is a high probability of declaring a difference when there is actually a difference between the population means. This has been one of the reasons Duncan's procedure has been extremely popular among researchers.

We summarize Duncan's new multiple range test for pairwise comparisons of t population means.

DUNCAN'S NEW MULTIPLE RANGE TEST

1. Rank the t sample means.
2. Two population means are declared significantly different if the absolute value of their sample differences exceeds

w'_r

$$W'_r = q'_\alpha(r, v)\sqrt{\frac{s^2_W}{n}}$$

where n is the number of observations in each sample mean, s^2_W is the mean square within samples obtained from the analysis of variance table, v is the number of degrees of freedom for s^2_W, and $q'_\alpha(r, v)$ is the critical value of the Studentized range required for Duncan's procedure when the means being compared are r steps apart. Values of $q'_\alpha(r, v)$ are given in Table 11 in the Appendix for $\alpha = .05$ or $.01$ and various combinations of r and v.

$q'_\alpha(r, v)$

We illustrate the use of Duncan's procedure with the data of Example 11.3.

EXAMPLE 11.6

Refer to the data of Example 11.3. Run Duncan's multiple range test with $\alpha = .05$ to make all pairwise comparisons among the six population means.

Solution

Recall from Example 11.3 that $n = 5$, $s^2_W = 2451$, and $v = 24$. Using this information we can set up the following table for r, $q'_\alpha(r, v)$, and W'_r.

r	2	3	4	5	6
$q'_\alpha(r, v)$	2.92	3.07	3.15	3.22	3.28
W'_r	64.65	67.97	69.74	71.29	72.62

For example, when two means are $r = 2$ steps apart, $q'_{.05}(2, 24)$ is 2.92. Then

$$W'_2 = q'_{.05}(2, 24)\sqrt{\frac{s^2_W}{n}} = 2.92\sqrt{\frac{2451}{5}} = 64.65$$

Thus two sample means $r = 2$ steps apart will be declared significantly different if the absolute value of their difference exceeds 64.7. The remainder of the entries in the table for different values of r were computed in a similar manner.

The sample means, ranked in order from lowest to highest, are

Population	6	4	1	2	3	5
Sample mean	470	498	505	528	564	600

Beginning with the largest mean, $\bar{y}_5$, each sample mean is compared to the smallest mean, $\bar{y}_6$, using the appropriate value of W'_r. For example, $\bar{y}_5$ and $\bar{y}_6$ are $r = 6$ steps apart and their difference must be compared to $W'_6 = 72.6$. The comparisons with $\bar{y}_{smallest}$ are shown in the following table.

Comparison	Conclusion
$\bar{y}_5 - \bar{y}_6 = 130$	>72.62; proceed
$\bar{y}_3 - \bar{y}_6 = 94$	>71.29; proceed
$\bar{y}_2 - \bar{y}_6 = 58$	<69.74; stop

Similarly, comparisons are made with $\bar{y}_4$, $\bar{y}_1$, $\bar{y}_2$, and $\bar{y}_3$ as follows:

Comparison	Conclusion
$\bar{y}_5 - \bar{y}_4 = 102$	>71.29; proceed
$\bar{y}_3 - \bar{y}_4 = 66$	<69.74; stop
$\bar{y}_5 - \bar{y}_1 = 95$	>69.74; proceed
$\bar{y}_3 - \bar{y}_1 = 59$	<67.97; stop
$\bar{y}_5 - \bar{y}_2 = 72$	>67.97; proceed
$\bar{y}_3 - \bar{y}_2 = 36$	<64.65; stop
$\bar{y}_5 - \bar{y}_3 = 36$	<64.65; stop

The results of these pairwise comparisons can be summarized as usual.

population	6	4	1	2	3	5

In conclusion, μ_6, μ_4, μ_1, and μ_2 are significantly less than μ_5; μ_6 is significantly less than μ_3.

We can compare the results obtained by using Duncan's new multiple range test to those obtained for Fisher's LSD, Tukey's procedure, and the SNK procedure. Based on the summary lines for the four procedures (pages 445, 448, and 451), we see that Duncan's procedure for the data of Table 11.2 results in conclusions more like the LSD and SNK procedures than Tukey's. The only difference in the conclusions for Fisher's LSD and Duncan's procedure is that in using the LSD, the means for populations 3 and 4 were found to be different—for Duncan's, they were not. The difference between the SNK and Duncan's procedures is with μ_2 and μ_5.

Unlike the other multiple-comparison procedures discussed, Duncan's new multiple range test cannot be used to form confidence intervals for the pairwise differences $\mu_i - \mu_j$. We can, however, adopt Duncan's procedure for unequal sample sizes if the n_is are nearly equal by replacing n by

$$\tilde{n} = \frac{t}{\dfrac{1}{n_1} + \dfrac{1}{n_2} + \cdots + \dfrac{1}{n_t}}$$

in the formula for W'_r. Duncan (1957) and Kramer (1957) have also presented extensions for situations where the population variances are different and where the sample means are correlated.

11.8   THE k RATIO RULE FOR MAKING PAIRWISE COMPARISONS AMONG TREATMENT MEANS

In this chapter we have spent considerable time discussing the merits of different multiple-comparison procedures. Fisher's LSD procedure can be used to make all pairwise comparisons among t population means while the controlled error rate for each comparison is at a specified level α. We found that even when $\mu_1 = \mu_2 = \cdots = \mu_t$, unless the per-comparison error rate α is quite small, the probability of falsely declaring at least one pair of means significantly different is quite large for Fisher's LSD.

To avoid this difficulty, Tukey proposed a multiple-comparison procedure utilizing an experimentwise error rate. Numerous other authors have offered solutions to this same problem, but none of these has adequately answered the question of which error rate to control.

The dilemma (see Duncan (1975)) of multiple comparisons can be illustrated with the following simplistic example. Suppose we wish to make pairwise comparisons among t (t very large) population means. In addition, suppose the null and alternative hypotheses for each of the pairwise comparisons is of the form

$$H_0: \ \mu_i - \mu_j = 0$$
$$H_a: \ \mu_i - \mu_j \neq 0$$

and that for each test the LSD, based on $\alpha = .05$, is used to determine whether two population means are significantly different.

Case I. Suppose that only 5% of all the differences in population means were declared significant. Intuitively, since we set $\alpha = .05$, we would be inclined to think we erred by declaring 5% of the differences significant when in fact all the population means are identical. To guard against declaring too many differences significant, we would certainly want to increase the magnitude of the absolute difference (LSD) in sample means required for significance. This is precisely the approach taken by Tukey (and others, such as Scheffé), who required that the absolute difference in sample means exceed W (where $W > \text{LSD}$) before declaring significance. Certainly the use of Tukey's procedure would be preferred to Fisher's LSD, to protect against a situation as described in case I.

Case II. Suppose, however, that only 5% of the population differences were declared to be nonsignificant. Then, intuitively, we might think we had erred on the opposite side and declared 5% of the population differences nonsignificant when in fact all t population means are different. To protect against such an occurrence, we would decrease the magnitude of the absolute difference in sample means required to declare significance. Clearly in situations such as this, Fisher's LSD procedure would be preferable to Tukey's W procedure.

Waller and Duncan (1969) proposed a multiple-comparison procedure that uses the sample data to help in determining whether we need to use a conservative rule (much like Tukey's) or a nonconservative rule (much like Fisher's). The procedure makes use of the computed value of $F = s_B^2/s_W^2$, the test statistic for the null hypothesis $H_0: \mu_1 = \mu_2 = \cdots = \mu_t$. If the computed value of F is small, then the sample data tend to indicate that the population means are rather homogeneous. For this situation, to protect against case I, we would require a large absolute difference in sample means to declare significance. In contrast, if the computed value of F is large, the sample data would tend to indicate or confirm that the population means are heterogeneous. For this situation, to protect against case II, we would require a small absolute difference in sample means to declare significance.

While a complete explanation of this procedure is well beyond the scope of this text, adequate tables have been developed to enable us to apply it to multiple comparisons resulting from the analysis of variance for a completely randomized design (Chapter 10), and from the other standard experimental designs (to be discussed in Chapter 14). There are two restrictions that are placed on the Waller–Duncan procedure. First, we require all sample sizes to be the same, and, second, the procedure should not be used when a priori we would expect certain of the population means to differ more than others.

k error weight ratio   Fisher's LSD required us to choose the comparisonwise error rate α. Now we must specify the **error weight ratio k**, which designates the seriousness of a Type I error relative to a Type II error. While we cannot go into a detailed discussion of the rationale for choosing k, we can provide several guidelines. Whereas typical values of α for Fisher's LSD may be .10, .05, or .01, corresponding values of k, the error weight ratio, are 50, 100, or 500.

The Waller–Duncan test procedure is summarized here.

WALLER–DUNCAN
k RATIO
PROCEDURE

1. Choose k, the error weight ratio.
2. Perform an analysis of variance to obtain the computed value of F for

$$H_0: \mu_1 = \mu_2 = \cdots = \mu_t$$

3. Compute

$$\text{LSD} = t_c \sqrt{s_W^2 \left(\frac{2}{n}\right)}$$

t_c where t_c is obtained from Tables 12 or 13 in the Appendix and is based on k, df_1, df_2, and F; s_W^2 is the mean square error from the AOV table; and n is the number of observations selected from each population. (Note: This procedure requires that we have the same number of observations from each population.)

4. Follow a stepwise procedure similar to that for Fisher's LSD to declare

$$\mu_i - \mu_j \neq 0 \quad \text{if} \quad |\bar{y}_i - \bar{y}_j| > \text{LSD}$$

Before we proceed with an example, let us consider the format of Tables 12 and 13. Because of the extensiveness of the tables, only two values of k (100 and 500) have been included. (Tables for $k = 50$ are presented in Waller and Duncan (1972).) The table entry is the t_c-value corresponding to specific values of k, df_1 (df for MST), df_2 (df for MSE), and $F = \text{MST}/\text{MSE}$. Thus for $k = 100$, $df_1 = 6$, $df_2 = 8$, and $F = 3.0$, the t-value for Waller–Duncan's least significant difference is 2.61.

TABLE 11.6 Situations in Which to Use a or b for Interpolation in the Waller–Duncan Procedure

F	df_1	df_2	
		≤ 100	> 100
≤ 2.4	≤ 60	a	a
	> 60	a	b

F	df_1	df_2	
		≤ 20	> 20
> 2.4	≤ 20	a	b
	> 20	b	b

Because of the complexity of the tables, not all values of df_1, df_2, and F can be given for a fixed error weight ratio. Fortunately, it is possible to interpolate to obtain the appropriate value of t_c.

interpolation   **Interpolation** will often be required for values of F. To obtain the appropriate value of t_c when the computed value of F falls between two tabulated F values, interpolate linearly using one of two quantities

$$a = \frac{1}{\sqrt{F}} \qquad \text{or} \qquad b = \sqrt{\frac{F}{F-1}}$$

Table 11.6 indicates under what situations we use a and under what conditions we interpolate using b.

EXAMPLE 11.7   Refer to Example 11.3. We were interested in comparisons among 6 batches used to produce a particular dye. The computed value of F was 4.60, based on $df_1 = 5$ and $df_2 = 24$. Find the value of t_c to be used in a k ratio procedure for pairwise comparisons among the 6 batch means. Use an error weight ratio of 100.

Solution   From Table 12 in the Appendix we note that there are no t_c-values listed for $F = 4.60$; we must interpolate between the values listed for $F = 4.0$ and $F = 6.0$. Note also that there are no t_c-values for $df_1 = 5$. However, very little difference is observed whether we use $df_1 = 4$ or 6. For $k = 100$, $df_1 = 4$, $df_2 = 24$, and $F = 4.0$, the table entry is 2.18. For the same settings except $df_1 = 6$, $t_c = 2.19$. There is no change in t_c for $k = 100$, $df_2 = 24$, and $F = 6.0$ when $df_1 = 4$ or 6. In both cases $t_c = 2.06$.

Because df_1 has very little effect on t_c in the range of 4 to 6 for this problem, we will work with $df_1 = 4$. To obtain that t_c-value corresponding to 4.60, we see from Table 11.6 that we must interpolate using

$$b = \sqrt{\frac{F}{F-1}}$$

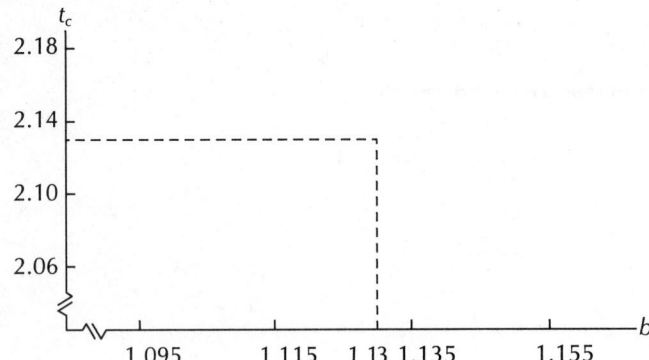

From Table 12 in the Appendix, for $F = 4.0$, then $b = 1.155$, and for $F = 6.0$, then $b = 1.095$. The distance between 1.155 and 1.095 is .060. Computing b for $F = 4.60$, we have

$$b = \sqrt{\frac{4.60}{3.60}} = 1.13$$

We now interpolate linearly with respect to b. First we draw a graph with the horizontal axis labeled with b and the vertical axis with t_c. We plot the values of t_c corresponding to $b = 1.155$ and $b = 1.095$. Then we connect these two points with a straight line. We use the straight line to determine the interpolated value of t_c corresponding to $b = 1.13$. (See Figure 11.1.) As can be seen from Figure 11.1, the interpolated value of t_c corresponding to $b = 1.13$ is $t_c = 2.13$.

EXAMPLE 11.8

Refer to Example 11.3 and perform all pairwise comparisons between population means using the k ratio procedure with $k = 100$.

Solution

In Example 11.7 we found the appropriate value of t_c to be 2.13. Recalling that $s_W^2 = 2451$ and that there were five observations per batch, the least significant difference using the Waller–Duncan approach is

$$\text{LSD} = t_c \sqrt{s_W^2 \left(\frac{2}{n}\right)} = 2.13 \sqrt{2451\left(\frac{2}{5}\right)} = 66.69$$

The stepwise procedure of Fisher's LSD can be employed to summarize our sample findings. The lines summarizing our results are seen to be

population	6	4	1	2	3	5

It is interesting to compare our results in Example 11.8 with those of previous multiple-comparison procedures for the same set of data. We see that our results are identical to the results from Duncan's multiple range test and very similar to those from Fisher's LSD and the SNK procedure. The reason that the k ratio results are similar to the results for the less conservative multiple-comparison procedures is due to the fact that the value of $F = s_B^2 / s_W^2$ is fairly large, indicating that the population means are more heterogeneous. This situation requires a nonconservative rule. If the computed value of F was small, we would have found the results of the k ratio procedure similar to one of the more conservative rules.

11.9   SCHEFFÉ'S *S* METHOD

The five multiple-comparison procedures discussed so far have been developed for pairwise comparisons among t population means. A more general procedure, proposed by Scheffé (1953), can be used to make all possible comparisons among the t population means. Although Scheffé's procedure can be applied to pairwise comparisons among the t population means, it is more conservative (less sensitive) than any of the other three multiple comparison procedures for detecting significant differences among pairs of population means.

SCHEFFÉ'S *S* METHOD FOR MULTIPLE COMPARISONS

1. Consider any linear comparison among the t population means of the form

    $$l = a_1 \mu_1 + a_2 \mu_2 + \cdots + a_t \mu_t$$

    We wish to test the null hypothesis

    $$H_0: l = 0$$

    against the alternative

    $$H_a: l \neq 0$$

2. The test statistic is

    $$\hat{l} = a_1 \bar{y}_1 + a_2 \bar{y}_2 + \cdots + a_t \bar{y}_t$$

3. Let

    S

    $$S = \sqrt{\hat{V}(\hat{l})} \sqrt{(t-1) F_{\alpha, df_1, df_2}}$$

    where, from Section 11.2,

    $$\hat{V}(\hat{l}) = s_W^2 \sum_i \frac{a_i^2}{n_i}$$

    t is the total number of population means, F_{α, df_1, df_2} is the upper-tail critical value of the F distribution with $a = \alpha$, $df_1 = t - 1$, and df_2 is the degrees of freedom for s_W^2.

4. For a specified value of α, we reject H_0 if $|\hat{l}| > S$.

5. The error rate that is controlled is an *experimentwise error rate*. If we consider all imaginable contrasts, the probability of observing an experiment with one or more contrasts falsely declared significant is designated by α.

EXAMPLE 11.9

Refer to the data of Table 11.2. Suppose that three of the batches (6, 4, and 2) were prepared from one concentration of HCl, and the other three batches (1, 3, and 5) were prepared from another concentration of HCl. Use the sample data and Scheffé's procedure to compare the mean dye yields for the two different concentrations of HCl. Let $\alpha = .05$.

Solution

Assume the even-numbered batches are from concentration I of HCl and the odd-numbered batches from concentration II. A contrast of particular importance is

$$\hat{I} = \bar{y}_1 + \bar{y}_3 + \bar{y}_5 - \bar{y}_2 - \bar{y}_4 - \bar{y}_6$$

which compares the means for batches from concentration II to those for concentration I. In particular, we would like to test

$H_0: I = 0$

$H_a: I \neq 0$

The estimated value of I is

$$\hat{I} = \bar{y}_1 + \bar{y}_3 + \bar{y}_5 - \bar{y}_2 - \bar{y}_4 - \bar{y}_6$$
$$= 505 + 564 + 600 - 528 - 498 - 470 = 173$$

To compute

$$S = \sqrt{\hat{V}(\hat{I})}\sqrt{(t-1)F_{\alpha,df_1,df_2}}$$

we must first calculate $\hat{V}(\hat{I})$. Using the formula

$$\hat{V}(\hat{I}) = s_W^2 \sum_i \frac{a_i^2}{n_i}$$

with all sample sizes equal to 5 and $s_W^2 = 2451$, we have

$$\hat{V}(\hat{I}) = 2451 \left[\frac{1}{5} + \frac{1}{5} + \frac{1}{5} + \frac{1}{5} + \frac{1}{5} + \frac{1}{5} \right] = 2941.2$$

From Table 6 for $\alpha = .05$, $df_1 = t - 1 = 5$, and $df_2 = 24$ (the degrees of freedom for s_W^2),

$$F_{.05,5,24} = 2.62$$

The computed value of S is then

$$S = \sqrt{2941.2}\sqrt{5(2.62)} = (54.23)(3.62) = 196.31$$

Since the absolute value of $\hat{I}$, 173, does not exceed $S = 196.31$, we have insufficient evidence to indicate that the means for batches from concentration II differ from those for concentration I.

Scheffé confidence interval

Scheffé's method can also be used for constructing a simultaneous **confidence interval** for all possible (not necessarily pairwise) contrasts using the t treatment means. In particular, there is a probability equal to $1 - \alpha$ that all possible comparisons of the form $I = \sum a_i\mu_i$, where $\sum a_i = 0$, will be encompassed by intervals of the form

$$\hat{I} - S < I < \hat{I} + S$$

11.10   SUMMARY

Four different multiple-comparison procedures (Fisher's, Tukey's, SNK, and Duncan's) were presented for making pairwise comparisons of t population means. Another procedure, Scheffé's, can be applied to any linear combination (including pairwise comparisons) of the means. For each procedure we have tried to indicate which error rate is controlled and how conservative the procedure is relative to the others presented. Since all the pairwise, multiple-comparison procedures compute the magnitude of the difference $|\bar{y}_i - \bar{y}_j|$ that is needed to declare μ_i and μ_j different, we can get some feel for how conservative one procedure is relative to another by comparing the magnitudes of the differences required for significance using the data of Example 11.3. This information is shown in Table 11.7.

As can be seen, Scheffé's procedure is *very* conservative and should not be used for pairwise comparisons.

A fifth procedure for pairwise multiple comparisons was the Waller–Duncan k ratio test. Since critical differences depend on the computed value of the F test for treatments, it can be either a conservative or a nonconservative rule, depending on the weight of sample evidence for rejection of the null hypothesis; the smaller the value of F, the more conservative the rule.

Which procedure should you use? We generally prefer the SNK procedure for efficacy (effectiveness) comparisons. But our reasons for this choice have a great deal to do with our work setting and the regulations surrounding our decision. Since our environment may be entirely different from yours, the decision regarding which procedure to use, and when to use it, is up to the individual. For a given problem, determine whether your decisions regarding differences should, in general, be more (or less) conservative. Then choose a procedure that exhibits the desired characteristic.

TABLE 11.7   Critical Difference $|\bar{y}_i - \bar{y}_j|$ for Sample Means r Steps Apart

Multiple-Comparison Procedure	Number of Steps Separating Means				
	2	3	4	5	6
Fisher's	64.63	64.63	64.63	64.63	64.63
SNK	64.65	78.16	86.35	92.33	96.75
Tukey's	96.75	96.75	96.75	96.75	96.75
Duncan's	64.65	67.97	69.74	71.29	72.62
Scheffé's	113.33	113.33	113.33	113.33	113.33

KEY
FORMULAS:
MULTIPLE
COMPARI-
SONS

1. Fisher's LSD procedure

$$LSD = t_{\alpha/2} \sqrt{s_W^2 \left(\frac{1}{n_i} + \frac{1}{n_j} \right)}$$

2. Tukey's W procedure

$$W = q_\alpha(t, v) \sqrt{\frac{s_W^2}{2} \left(\frac{1}{n_i} + \frac{1}{n_j} \right)}$$

3. SNK procedure

$$W_r = q_\alpha(r, v) \sqrt{\frac{s_W^2}{2} \left(\frac{1}{n_i} + \frac{1}{n_j} \right)}$$

4. Duncan's multiple range test

$$W_r' = q_\alpha'(r, v) \sqrt{\frac{s_W^2}{n}} \qquad (\text{Note: } n_i = n_j)$$

5. Scheffé's method

$$S = \sqrt{\hat{V}(\hat{l})} \sqrt{(t - 1)F_{\alpha, df_1, df_2}}$$

where

$$\hat{V}(\hat{l}) = s_W^2 \sum \frac{a_i^2}{n_i}$$

EXERCISES

11.4   Refer to the data of Example 10.1. There a horticulturist was investigating the phosphorus content of the leaves from three different varieties of apple trees.
a. Perform an analysis of variance.
b. Use Duncan's multiple range test procedure to run all pairwise comparisons. Use $\alpha = .05$.
c. Compare your conclusions in part (b) to those in the SAS computer output shown here.

```
          LISTING OF DATA

   OBS     VARIETY     PHOSPHOR

    1         1          0.35
    2         1          0.40
    3         1          0.58
    4         1          0.50
    5         1          0.47
    6         2          0.65
    7         2          0.70
    8         2          0.90
    9         2          0.84
   10         2          0.79
   11         3          0.60
   12         3          0.80
   13         3          0.75
   14         3          0.73
   15         3          0.66

   N=        15
```

EXERCISES

```
                          ANALYSIS OF VARIANCE PROCEDURE
                          GENERAL LINEAR MODELS PROCEDURE

DEPENDENT VARIABLE: PHOSPHOR

SOURCE              DF      SUM OF SQUARES    MEAN SQUARE    F VALUE      PR > F      R-SQUARE          C.V.

MODEL               2         0.27664000      0.13832000      16.97      0.0003      0.738810       13.9317

ERROR               12        0.09780000      0.00815000                 ROOT MSE              PHOSPHOR MEAN

CORRECTED TOTAL     14        0.37444000                                0.09027735             0.64800000


SOURCE              DF         TYPE I SS    F VALUE    PR > F      DF       TYPE III SS    F VALUE    PR > F

VARIETY             2         0.27664000     16.97     0.0003      2       0.27664000      16.97     0.0003




                  ANALYSIS OF VARIANCE PROCEDURE
                  GENERAL LINEAR MODELS PROCEDURE


     DUNCAN'S MULTIPLE RANGE TEST FOR VARIABLE: PHOSPHOR
     NOTE: THIS TEST CONTROLS THE TYPE I COMPARISONWISE ERROR RATE,
           NOT THE EXPERIMENTWISE ERROR RATE


            ALPHA=0.05  DF=12  MSE=0.00815


        NUMBER OF MEANS        2        3
        CRITICAL RANGE    0.124165  0:130067


  MEANS WITH THE SAME LETTER ARE NOT SIGNIFICANTLY DIFFERENT.

       DUNCAN   GROUPING         MEAN     N  VARIETY

                    A          0.77600    5  2
                    A
                    A          0.70800    5  3

                    B          0.46000    5  1
```

 11.5 An experiment was conducted to compare the effectiveness of five different weight-reducing agents. A random sample of 50 males was randomly divided into five equal groups, with preparation A assigned to the first group, B to the second group, and so on. Each person in the experiment was given a prestudy physical and told how many pounds overweight he was. A comparison of the mean number of pounds overweight for the groups showed no significant differences. The study program was then begun, with each group taking the prescribed preparation for a fixed period of time. At the end of the study period, weight losses were recorded. The data are given here.

A	12.4	10.7	11.9	11.0	12.4	12.3	13.0	12.5	11.2	13.1
B	9.1	11.5	11.3	9.7	13.2	10.7	10.6	11.3	11.1	11.7
C	8.5	11.6	10.2	10.9	9.0	9.6	9.9	11.3	10.5	11.2
D	8.7	9.3	8.2	8.3	9.0	9.4	9.2	12.2	8.5	9.9
E	12.7	13.2	11.8	11.9	12.2	11.2	13.7	11.8	11.5	11.7

Run an analysis of variance to determine if there are any significant differences among the weight losses for the five diet preparations. Use $\alpha = .05$.

11.6   Refer to Exercise 11.5. Run all pairwise comparisons of population means using the following procedures.

   a. Fisher's LSD, $\alpha = .05$
   b. Tukey's W, $\alpha = .05$
   c. Duncan's multiple range test, $\alpha = .05$
   d. Compare the conclusions for the three procedures.

11.7   Use a computer program to run an analysis of variance for the data of Exercise 11.5. Draw conclusions from the output. Compare your results to the conclusions for Exercise 11.5.

11.8   Refer to Exercise 11.5. Examine the widths of 95% confidence intervals for pairwise comparisons of diets A, B, . . . , E, using Fisher's LSD, Tukey's W, and Scheffé's S method.

11.9   Refer to Exercise 11.5. Suppose that preparation D was a placebo. Use Scheffé's procedure to make the following comparisons. Set $\alpha = .05$.

   a. $\mu_D - \frac{1}{4}(\mu_A + \mu_B + \mu_C + \mu_E)$
   b. $\mu_A - \mu_E$
   c. $\mu_A - \frac{1}{2}(\mu_B + \mu_E)$
   d. $\mu_A - \frac{1}{3}(\mu_B + \mu_C + \mu_E)$

11.10 Refer to Exercise 11.9. Give the corresponding Scheffé 95% confidence intervals for these four comparisons.

11.11 Refer to Exercise 10.27.

   a. Use Fisher's LSD procedure with $\alpha = .05$ to declare significant differences.
   b. Use Tukey's W method with $\alpha = .05$ to make all pairwise comparisons of population means for the data of Exercise 10.27. Compare your conclusions to those of part (a).
   c. How would your conclusions differ using the SNK procedure?

11.12 Refer to Exercise 10.27.

   a. Use the Waller–Duncan method, with $k = 100$.
   b. Compare your results for parts (a) and (b) of Exercise 11.11 and with part (a) of this exercise.

11.13 Refer to Exercise 10.13. Use Duncan's multiple range procedure to declare significant differences among the population means. Set $\alpha = .05$.

11.14 Refer to Exercise 11.5. Determine the appropriate value of t_c for the Waller–Duncan k ratio procedure when $k = 100$ and $k = 500$.

11.15 Make all pairwise comparisons among the 5 diet preparations of Exercise 11.5 using the Waller–Duncan method. Use an error weight ratio of $k = 100$. Compare your results to those of Exercise 11.6.

11.16 Use Duncan's multiple range procedure to compare the four population means of Exercise 10.16.

11.17 Refer to Exercise 10.12. Perform all pairwise comparisons of the four population means. Be sure to identify the procedure you use and the error rate controlled. Use $\alpha = .05$.

11.18 Refer to Exercise 10.12. Set up confidence intervals for all pairwise comparisons of treatment means in Exercise 11.4, using Tukey's W procedure. Interpret your findings.

11.19 Refer to the data of Exercise 10.19. Make all pairwise comparisons of the three areas using the Waller–Duncan procedure with $k = 100$.

 11.20 The nitrogen contents of red clover plants inoculated with three strains of *Rhizobium* are given below.

3DOK1	3DOK5	3DOK7
19.4	18.2	20.7
32.6	24.6	21.0
27.0	25.5	20.5
32.1	19.4	18.8
33.0	21.7	18.6
	20.8	20.1
		21.3

Is there evidence of a difference in the effects of the three treatments on nitrogen content? Analyze the data completely and draw conclusions.

11.21 An experiment was conducted to compare three methods of grass-seed preparation: mechanical scarification (ms), acid dip (ad), and hot water dip (hw). One hundred grass seeds were placed in each of 150 petri dishes. Among the 150 dishes, 50 were randomly assigned ms, 50 ad, and 50 hw. After a period of two weeks, the germination rates were checked for each dish.

Method	Mean Germination	Standard Deviation
ms	65.3	7.2
ad	82.1	5.4
hw	73.8	6.5

Analyze these data using a one-way analysis and draw conclusions.

11.22 Refer to Exercise 11.21. Use the SNK procedure to identify method differences. Summarize your results.

11.23 An experiment was conducted to assess the relative merits of four different gasoline blends. Twenty automobiles of the same type, model, and engine size were used with five randomly assigned to each of the blends. Summary test data for the blends are shown here.

Blend	Mean (mpg)	Standard Deviation
1 (control)	26.2	4.3
2 (control — additive x)	28.1	5.6
3 (control — additive y)	29.6	5.1
4 (control — additives x and y)	38.2	7.3

Run an analysis of variance and draw conclusions. Use $\alpha = .05$.

11.24 Refer to Exercise 11.23 and consider the following linear contrasts. Describe what each contrast is measuring.

a. $\hat{l}_1 = \bar{y}_1 + \bar{y}_2 - \bar{y}_3 - \bar{y}_4$

b. $\hat{l}_2 = \bar{y}_1 + \bar{y}_3 - \bar{y}_2 - \bar{y}_4$

c. $\hat{l}_3 = \bar{y}_1 - \bar{y}_2 - \bar{y}_3 + \bar{y}_4$

11.25 Use Scheffé's method to conduct a test of significance on $\hat{l}_3$ of Exercise 11.24, based on $\alpha = .05$. What do you conclude?

11.26 Construct a confidence interval for l_1 and l_2 of Exercise 11.24 using Scheffé's method. Draw conclusions.

11.27 Refer to Exercise 10.25. Use Tukey's procedure to run all pairwise comparisons with $\alpha = .05$. Summarize your results.

11.28 How would the SNK procedure change your conclusions to Exercise 11.27?

11.29 Refer to Exercise 11.27. Suppose that levels 1 and 2 for additives A and B are 1.0 and 3.0 mg/kg, respectively.

a. Plot the mean weight gained versus $x = $ mg/kg for the two additives (include the control data, $x = 0$).

b. Fit separate linear regression models for additives A and B.

11.30 Refer to Exercise 11.29. Estimate the slopes for the two lines using 95% confidence intervals. Does it appear that additives A and B have different regression lines? (A method for comparing slopes will be presented in Chapter 12.)

12 MULTIPLE REGRESSION AND THE GENERAL LINEAR MODEL

The simplest type of regression model relating the dependent variable y to a quantitative independent variable x is the one discussed in Chapter 8:

$$y = \beta_0 + \beta_1 x + \varepsilon$$

Under the assumption that the average value of ε (also called the **expected value of ε**) for a given value of x is $E(\varepsilon) = 0$, this model indicates that the expected value of y for a given value of x is described by the straight line

$$E(y) = \beta_0 + \beta_1 x$$

Not all data sets are adequately described by a model the expectation of which is a straight line. For example, consider the data of Table 12.1, which gives the yields (in bushels) for 14 equal-sized plots planted in tomatoes for different levels of fertilization. It is evident from the scatterplot in Figure 12.1 that a linear equation will not adequately represent the relationship between yield and the amount of fertilizer applied to the plot. The reason for this is that, whereas a modest amount of fertilizer may well enhance the crop yield, too much fertilizer can be destructive.

A model for this physical situation might be

$$y = \beta_0 + \beta_1 x + \beta_2 x^2 + \varepsilon$$

TABLE 12.1   Yield of 14 Equal-Sized Plots of Tomato
Plantings for Different Amounts
of Fertilizer

Plot	Yield, y (in Bushels)	Amount of Fertilizer, x (in lbs per Plot)
1	24	12
2	18	5
3	31	15
4	33	17
5	26	20
6	30	14
7	20	6
8	25	23
9	25	11
10	27	13
11	21	8
12	29	18
13	29	22
14	26	25

FIGURE 12.1

Scatterplot of the
Yield Versus Fertilizer
Data in Table 12.1

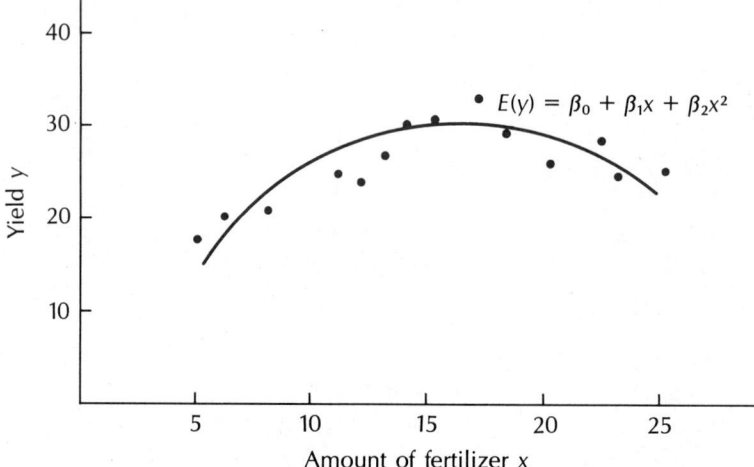

Again with the assumption that $E(\varepsilon) = 0$, the expected value of y for a given value of x is

$$E(y) = \beta_0 + \beta_1 x + \beta_2 x^2$$

One such line is plotted in Figure 12.1, superimposed on the data of Table 12.1.

A general polynomial regression model relating a dependent variable y to a single quantitative independent variable x is given by

$$y = \beta_0 + \beta_1 x + \beta_2 x^2 + \cdots + \beta_p x^p + \varepsilon$$

with

$$E(y) = \beta_0 + \beta_1 x + \beta_2 x^2 + \cdots + \beta_p x^p$$

The choice of p and hence the choice of an appropriate regression model will depend on the experimental situation.

multiple regression model The **multiple regression model** that relates a dependent variable y to a set of quantitative independent variables is a direct extension of a polynomial regression model in one independent variable. We write the multiple regression model as

$$y = \beta_0 + \beta_1 x_1 + \beta_2 x_2 + \cdots + \beta_k x_k + \varepsilon$$

cross-product term Any of the independent variables may be powers of other independent variables; for example, x_2 might be x_1^2. In fact, there are many other possibilities; x_3 might be a **cross-product term** equal to $x_1 x_2$, x_4 might be log x_1, etc. The only restriction is that no x is a perfect linear function of other xs.

The simplest type of multiple regression equation is a first-order model, in which each of the independent variables appears but there are no cross-product terms or terms in powers of the independent variables. For example, when three quantitative independent variables are involved, the first-order multiple regression model would be

$$y = \beta_0 + \beta_1 x_1 + \beta_2 x_2 + \beta_3 x_3 + \varepsilon$$

For these first-order models we can attach some meaning to the βs. The parameter β_0 is the y-intercept, which represents the expected value of y when all xs are zero. For cases in which it does not make sense to have all the xs be zero, β_0 (or its estimate) should be used only as part of the prediction equation, and not given an interpretation by itself.

partial slopes The other parameters $(\beta_1, \beta_2, \ldots, \beta_k)$ in the multiple regression equation are sometimes called **partial slopes**. In linear regression, the parameter β_1 is the slope of the regression line and it represents the expected change in y for a unit increase in x. In a first-order multiple regression model, β_1 represents the expected change in y for a unit increase in x_1 *when all other xs are held constant*. In general then, $\beta_j (j \neq 0)$ represents the expected change in y for a unit increase in x_j while holding all other xs constant.

additive effects Besides the usual assumptions for a multiple regression model (see Chapter 8) there is an additional assumption that is implied when we use a first-order multiple regression model. Since the expected change in y for a unit change in x_j is constant and does not depend on values of the other xs, we are in fact assuming that the effects of the independent variables are **additive**.

When might this additional assumption of additivity be warranted? Figure 12.2(a) shows a scatterplot of y versus x_1; Figure 12.2(b) shows the same plot with an ID attached to the different levels of a second independent variable x_2 (x_2 takes on the values of 1, 2, or 3). From Figure 12.2(a) we see that y is approximately linear in x_1. The parallel lines of Figure 12.2(b) corresponding to the three levels of the independent variable x_2 indicate that the expected change in y for a unit change in x_1 remains the same no matter which level of x_2 is used. These data suggest that the effects of x_1 and x_2 are additive; hence, a first-order model of the form $y = \beta_0 + \beta_1 x_1 + \beta_2 x_2 + \varepsilon$ is appropriate.

interaction Figure 12.3 displays a situation in which **interaction** is present between the variables x_1 and x_2. Even though a scatterplot of y versus x_1 is as shown in Figure 12.2(a), the

FIGURE 12.2

(a) Scatterplot of y
Versus x_1;
(b) Scatterplot of y
Versus x_1, Indicating
Additivity of Effects
for x_1 and x_2

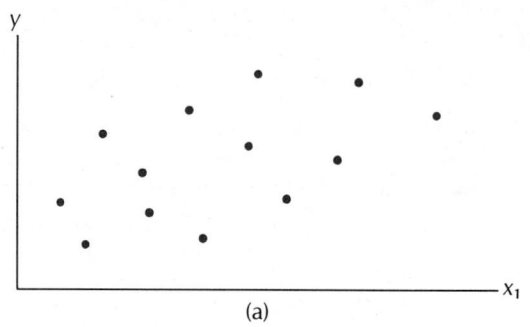

(a)

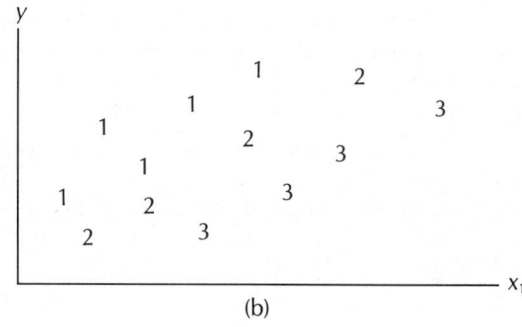

(b)

FIGURE 12.3

Scatterplot of y
Versus x_1, Indicating
Nonadditivity
(Interaction) of
Effects Between
x_1 and x_2

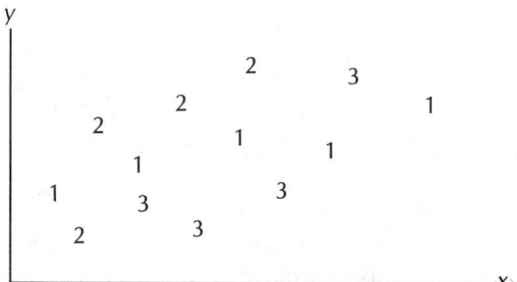

nonparallel lines of Figure 12.3 indicate that the expected change in y for a unit change in x_1 now depends on the level of x_2. When this occurs, the independent variables x_1 and x_2 are said to interact. A first-order model, which assumes additivity of the effects, would not be appropriate here. At the very least, we would include a cross-product term (x_1x_2) in the model.

The simplest new model allowing for interaction between x_1 and x_2 is

$$y = \beta_0 + \beta_1 x_1 + \beta_2 x_2 + \beta_3 x_1 x_2 + \varepsilon$$

Note that for a given value of x_2 (say $x_2 = 2$), the expected value of y is

$$E(y) = \beta_0 + \beta_1 x_1 + \beta_2(2) + \beta_3 x_1(2)$$
$$= (\beta_0 + 2\beta_2) + (\beta_1 + 2\beta_3)x_1$$

Here the intercept and slope are $(\beta_0 + 2\beta_2)$ and $(\beta_1 + 2\beta_3)$, respectively. The corresponding intercept and slope for $x_2 = 3$ can be shown to be $(\beta_0 + 3\beta_2)$ and $(\beta_1 + 3\beta_3)$. Clearly, the slopes of the two regression lines are not the same, and hence we have nonparallel lines.

Not all experiments can be modeled using a first-order multiple regression model. For these situations where a higher-order multiple regression model may be appropriate, it will be more difficult to assign a literal interpretation to the βs because of the presence of terms that contain cross-products or powers of the independent variables. Our focus will be on finding a multiple regression model that provides a good fit to the sample data, not on interpreting individual βs, except as they relate to the overall model.

The models that we have described briefly have been for regression problems where the experimenter is interested in developing a model to relate a response to one or more *quantitative* independent variables. The problem of modeling an experimental situation is not restricted to the quantitative independent-variable case.

Consider the problem of writing a model for an experimental situation in which a response y is related to a set of *qualitative* independent variables or to both quantitative and qualitative independent variables. For the first situation (relating y to one or more qualitative independent variables), let's suppose that we want to compare the average number of lightning discharges per minute for a storm, as measured from two different tracking posts located 30 miles apart. If we let y denote the number of discharges recorded on an oscilloscope during a one-minute period, we could write the following two models:

for tracking post 1: $y = \mu_1 + \varepsilon$

for tracking post 2: $y = \mu_2 + \varepsilon$

Thus we assume that observations at tracking post 1 randomly "bob" about a population mean μ_1. Similarly, at tracking post 2, observations differ from a population mean μ_2 by a random amount ε. These two models are not new and could have been used to describe observations when comparing two population means in Chapter 5. What is new is that we can combine these two models into a single model of the form

$$y = \beta_0 + \beta_1 x_1 + \varepsilon$$

dummy variable   where β_0 and β_1 are unknown parameters, ε is a random error term, and x_1 is a **dummy variable** with the following interpretation. We let

$x_1 = 1$ if an observation is obtained from tracking post 2

$x_1 = 0$ if an observation is obtained from tracking post 1

For observations obtained from tracking post 1, we substitute $x_1 = 0$ into our model to obtain

$$y = \beta_0 + \beta_1(0) + \varepsilon = \beta_0 + \varepsilon$$

Hence $\beta_0 = \mu_1$, the population mean for observations from tracking post 1. Similarly, by substituting $x_1 = 1$ into our model, the equation for observations from tracking post 2 is

$$y = \beta_0 + \beta_1(1) + \varepsilon = \beta_0 + \beta_1 + \varepsilon$$

Since $\beta_0 = \mu_1$ and $\beta_0 + \beta_1$ must equal μ_2, we have $\beta_1 = \mu_2 - \mu_1$, the difference in means between observations from tracking posts 2 and 1.

This model, $y = \beta_0 + \beta_1 x_1 + \varepsilon$, which relates y to the qualitative independent variable tracking post, can be extended to a situation in which the qualitative variable has more than two levels. We do this by using more than one dummy variable. Consider a one-way classification in which we're interested in four treatments. We could write the model

$$y = \beta_0 + \beta_1 x_1 + \beta_2 x_2 + \beta_3 x_3 + \varepsilon$$

where

$$x_1 = 1 \text{ if treatment 2,} \qquad x_1 = 0 \text{ otherwise}$$
$$x_2 = 1 \text{ if treatment 3,} \qquad x_2 = 0 \text{ otherwise}$$
$$x_3 = 1 \text{ if treatment 4,} \qquad x_3 = 0 \text{ otherwise}$$

To interpret the βs in this equation, it is convenient to construct a table of the expected values. Since ε has expectation zero, the general expression for the expected value of y is

$$E(y) = \beta_0 + \beta_1 x_1 + \beta_2 x_2 + \beta_3 x_3$$

The expected value for observations on treatment 1 is found by substituting $x_1 = 0$, $x_2 = 0$, and $x_3 = 0$; after this substitution we find $E(y) = \beta_0$. The expected value for observations on treatment 2 is found by substituting $x_1 = 1$, $x_2 = 0$, and $x_3 = 0$ into the $E(y)$ formula; this substitution yields $E(y) = \beta_0 + \beta_1$. Substitutions of $x_1 = 0$, $x_2 = 1$, $x_3 = 0$ and $x_1 = 0$, $x_2 = 0$, $x_3 = 1$ yield expected values for treatments 3 and 4, respectively. These expected values are summarized in Table 12.2.

If we identify the mean of treatment 1 as μ_1, the mean of treatment 2 as μ_2, and so on, then from Table 12.2 we have

$$\mu_1 = \beta_0, \qquad \mu_2 = \beta_0 + \beta_1, \qquad \mu_3 = \beta_0 + \beta_2, \qquad \text{and} \qquad \mu_4 = \beta_0 + \beta_3$$

Solving these equations for the βs, we have

$$\beta_0 = \mu_1 \qquad\qquad \beta_2 = \mu_3 - \mu_1$$
$$\beta_1 = \mu_2 - \mu_1 \qquad \beta_3 = \mu_4 - \mu_1$$

Any comparison among the treatment means can be phrased in terms of the βs. For example, the comparison $\mu_4 - \mu_3$ could be written as $\beta_3 - \beta_2$. Likewise, $\mu_3 - \mu_2$ could be written as $\beta_2 - \beta_1$.

EXAMPLE 12.1

Consider a hypothetical situation for a one-way classification (with $t = 4$) in which we know the means for the four treatments. If $\mu_1 = 7$, $\mu_2 = 9$, $\mu_3 = 6$, and $\mu_4 = 15$, determine values for β_0, β_1, β_2, and β_3 in the model

$$y = \beta_0 + \beta_1 x_1 + \beta_2 x_2 + \beta_3 x_3 + \varepsilon$$

TABLE 12.2 Expected Values for the One-way Classification with Four Treatments

	Treatment		
1	2	3	4
$E(y) = \beta_0$	$E(y) = \beta_0 + \beta_1$	$E(y) = \beta_0 + \beta_2$	$E(y) = \beta_0 + \beta_3$

where

$$x_1 = 1 \text{ if treatment 2,} \qquad x_1 = 0 \text{ otherwise}$$
$$x_2 = 1 \text{ if treatment 3,} \qquad x_2 = 0 \text{ otherwise}$$
$$x_3 = 1 \text{ if treatment 4,} \qquad x_3 = 0 \text{ otherwise}$$

Solution   Based on what was presented in Table 12.2, we know that

$$\beta_0 = \mu_1$$
$$\beta_1 = \mu_2 - \mu_1$$
$$\beta_2 = \mu_3 - \mu_1$$

and

$$\beta_3 = \mu_4 - \mu_1$$

Using the known values for μ_1, μ_2, μ_3, and μ_4, it follows that

$$\beta_0 = 7$$
$$\beta_1 = 9 - 7 = 2$$
$$\beta_2 = 6 - 7 = -1$$

and

$$\beta_3 = 15 - 7 = 8$$

EXAMPLE 12.2   Refer to Example 12.1. Express $\mu_3 - \mu_2$ and $\mu_3 - \mu_4$ in terms of the βs. Check your findings by substituting values for the βs.

Solution   Using the relationship between the βs and the μs, it can be seen that

$$\beta_2 - \beta_1 = (\mu_3 - \mu_1) - (\mu_2 - \mu_1) = \mu_3 - \mu_2$$

and

$$\beta_2 - \beta_3 = (\mu_3 - \mu_1) - (\mu_4 - \mu_1) = \mu_3 - \mu_4$$

Substituting computed values for the βs we have

$$\beta_2 - \beta_1 = -1 - (2) = -3$$

and

$$\beta_2 - \beta_3 = -1 - (8) = -9$$

These computed values are identical to the "known" differences for $\mu_3 - \mu_2$ and $\mu_3 - \mu_4$, respectively.

EXAMPLE 12.3   Use dummy variables to write the model for a general one-way classification with t treatments. Identify the βs.

Solution   The model could be written in the form

$$y = \beta_0 + \beta_1 x_1 + \beta_2 x_2 + \cdots + \beta_{t-1} x_{t-1} + \varepsilon$$

where

$$x_1 = 1 \text{ if treatment 2,} \qquad x_1 = 0 \text{ otherwise}$$
$$x_2 = 1 \text{ if treatment 3,} \qquad x_2 = 0 \text{ otherwise}$$
$$\vdots \qquad\qquad\qquad \vdots$$
$$x_{t-1} = 1 \text{ if treatment } t, \qquad x_{t-1} = 0 \text{ otherwise}$$

The table of expected values would be

	Treatments		
1	2	$\cdots$	t
$E(y) = \beta_0$	$E(y) = \beta_0 + \beta_1$	$\cdots$	$E(y) = \beta_0 + \beta_{t-1}$

from which we obtain

$$\beta_0 = \mu_1$$
$$\beta_1 = \mu_2 - \mu_1$$
$$\vdots$$
$$\beta_{t-1} = \mu_t - \mu_1$$

The procedure just described for a one-way classification can be applied to any experimental situation in which a response y is related to one or more qualitative independent variables. In the one-way classification we have a response related to the variable "treatments," and for t levels of the treatments, we enter $(t - 1)$ βs into our model, using dummy variables. More will be said about the use of the models for more than one qualitative independent variable in Chapters 14 and 15, where we consider the analysis of variance for several different experimental designs.

12.2   THE GENERAL LINEAR MODEL

It is important at this point to recognize that a single general model can be used for multiple regression models in which a response is related to a set of quantitative independent variables, and for models that relate y to a set of qualitative independent vari-

general linear model   ables. This model, called the **general linear model**, has the form

$$y = \beta_0 + \beta_1 x_1 + \beta_2 x_2 + \cdots + \beta_k x_k + \varepsilon$$

For multiple regression models, the xs represent independent variables (such as weight, amount of water, etc.), independent variables raised to powers, and cross-product terms involving the independent variables. A few regression models were discussed in Section 12.1; more about the use of the general linear model in regression will be discussed in the remainder of this chapter and in Chapter 13.

The xs of the general linear model represent dummy variables (coded 0 and 1) or products of dummy variables when y is related to a set of qualitative independent variables. We discussed how to use dummy variables for representing y in terms of a single qualitative variable in Section 12.1; the same approach can be used to relate y to more than one qualitative independent variable. This will be discussed in Chapter 14, where we present more analysis of variance techniques.

The general linear model can also be used for the case in which y is related to both qualitative and quantitative independent variables. A particular example of this is discussed in Section 13.5 and other applications are presented in Chapter 18.

Why is this model called the general *linear* model, especially since it can be used for polynomial models? The word "linear" in the general linear model refers to how the βs are entered in the model, not to how the independent variables appear in the model. A general linear model is linear (used in the usual algebraic sense) in the βs.

Why are we discussing the general linear model now? The techniques that we will develop in this chapter for making inferences about a single β, a set of βs, and $E(y)$ in multiple regression are those that apply to any general linear model. Thus, using general linear model techniques, we have a common thread to inferences about multiple regression (Chapters 12 and 13) and the analysis of variance (Chapters 14, 15, 16, and 18). As you study these six chapters, try whenever possible to make the connection back to a general linear model; we'll help you with this connection.

EXERCISES

12.1   a. Write a first-order multiple regression model relating a response y to three qualitative independent variables.
    b. Show how this model can be written as a general linear model.

12.2   Write a second-order multiple regression model relating a response y to three quantitative independent variables. Include all possible terms. (Hint: A first-order model contains terms in the x_j; a second-order model includes these terms as well as squares and cross-products.)

12.3   Refer to Exercise 12.2. Show that the model you wrote can be written in the form of a general linear model. Identify the terms.

12.4   Consider the model

$$y = \beta_0 + \beta_1 x_1 + \beta_2 x_2 + \varepsilon$$

where

$$x_1 = \begin{cases} 1 & \text{if treatment 2} \\ 0 & \text{otherwise} \end{cases}$$

$$x_2 = \begin{cases} 1 & \text{if treatment 3} \\ 0 & \text{otherwise} \end{cases}$$

    a. Interpret the βs in the model.
    b. Identify the difference in mean responses for treatments 2 and 3 using the model.

12.5   (Optional) Refer to Exercise 12.4. Suppose that the model is expanded to include the term $\beta_3 x_3$, where x_3 is a dummy variable for the qualitative variable "location."

$$x_3 = \begin{cases} 1 & \text{if location 2} \\ 0 & \text{otherwise} \end{cases}$$

a. Interpret the βs for this model. (Hint: Consider all combinations of the three treatments and two locations.)

b. Write the difference in mean response for treatments 2 and 3 for location 2. Is it the same for location 1?

c. Identify an experimental situation in which this model might be a reasonable approximation.

12.6   (Optional) A study was done to examine the effect of a quantitative independent variable (age) on reaction time (as measured by braking time). The experiment included males and females. Two models were proposed:

$$y = \beta_0 + \beta_1 x_1 + \beta_2 x_2 + \varepsilon$$

and

$$y = \beta_0 + \beta_1 x_1 + \beta_2 x_2 + \beta_3 x_1 x_2$$

where

x_1 = age (in years)

and

$$x_2 = \begin{cases} 1 & \text{female} \\ 0 & \text{if male} \end{cases}$$

Interpret the βs for the two models and explain a practical difference between the two models.

12.3   LEAST SQUARES SOLUTION TO THE GENERAL LINEAR MODEL

The general linear model relates a response y to a set of independent variables (qualitative or quantitative). For a random sample of n measurements we can write the ith observation as

$$y_i = \beta_0 + \beta_1 x_{i1} + \beta_2 x_{i2} + \cdots + \beta_k x_{ik} + \varepsilon_i \qquad (i = 1, 2, \ldots, n; \, n > k)$$

where $x_{i1}, x_{i2}, \ldots, x_{ik}$ are the settings of the independent variables corresponding to the observation y_i.

In order to find least squares estimates for $\beta_0, \beta_1, \ldots$, and β_k in a general linear model we follow the same procedure that we did for a linear regression model in Chapter 8. A random sample of n observations is obtained; the least squares prediction equation

$$\hat{y} = \hat{\beta}_0 + \hat{\beta}_1 x_1 + \cdots + \hat{\beta}_k x_k$$

is found by choosing $\hat{\beta}_0, \hat{\beta}_1, \ldots, \hat{\beta}_k$ to minimize the expression $SSE = \sum_i (y_i - \hat{y}_i)^2$. But, although it was easy to write down the solutions to $\hat{\beta}_0$ and $\hat{\beta}_1$ for the linear regression model,

$$y = \beta_0 + \beta_1 x + \varepsilon$$

the estimates for $\beta_0, \beta_1, \ldots, \beta_k$ must be found by solving a set of simultaneous equations, called the *normal equations*, shown here.

	y_i	$\hat{\beta}_0$	$x_{i1}\hat{\beta}_1$	$\cdots$	$x_{ik}\hat{\beta}_k$
1	$\sum y_i$ $=$	$n\hat{\beta}_0$ $+$	$\sum x_{i1}\hat{\beta}_1$ $+\cdots+$		$\sum x_{ik}\hat{\beta}_k$
x_{i1}	$\sum x_{i1}y_i$ $=$	$\sum x_{i1}\hat{\beta}_0$ $+$	$\sum x_{i1}^2\hat{\beta}_1$ $+\cdots+$		$\sum x_{i1}x_{ik}\hat{\beta}_k$
$\vdots$	$\vdots$				
x_{ik}	$\sum x_{ik}y_i$ $=$	$\sum x_{ik}\hat{\beta}_0$ $+$	$\sum x_{ik}x_{i1}\hat{\beta}_1$ $+\cdots+$		$\sum x_{ik}^2\hat{\beta}_k$

Note the pattern associated with these equations. By labeling the rows and columns as we have done, we can obtain any term in the normal equations by multiplying the row and column elements and summing. For example, the last term in the second equation is found by multiplying the row element (x_{i1}) by the column element $(x_{ik}\hat{\beta}_k)$ and summing; the resulting term is $\sum x_{i1}x_{ik}\hat{\beta}_k$. Because all terms in the normal equations can be formed in this way, it is fairly simple to write down the equations to be solved in order to obtain the least squares estimates $\hat{\beta}_0, \hat{\beta}_1, \ldots, \hat{\beta}_k$. The solution to these equations is not necessarily trivial; that's why we'll enlist the help of various statistical software packages for their solution.

EXAMPLE 12.4   In Exercise 9.23 we presented data for the weight loss of a compound for different amounts of time the compound was exposed to the air. Additional information was also available on the humidity of the environment during exposure. The complete data are presented in Table 12.3.

TABLE 12.3   Weight Loss, Exposure Time, and Relative Humidity Data for Example 12.4

Weight Loss, y (lbs)	Exposure Time, x_1 (hrs)	Relative humidity, x_2
4.3	4	.20
5.5	5	.20
6.8	6	.20
8.0	7	.20
4.0	4	.30
5.2	5	.30
6.6	6	.30
7.5	7	.30
2.0	4	.40
4.0	5	.40
5.7	6	.40
6.5	7	.40

a. Set up the normal equations for this regression problem if the assumed model is

$$y = \beta_0 + \beta_1 x_1 + \beta_2 x_2 + \varepsilon$$

where

x_1 is exposure time and x_2 is relative humidity

b. Use the computer output shown here to determine the least squares estimates of β_0, β_1, and β_2. Predict weight loss for 6.5 hours of exposure and a relative humidity of .35.

LISTING OF DATA

OBS	WT_LOSS	TIME	HUMID
1	4.3	4.0	0.20
2	5.5	5.0	0.20
3	6.8	6.0	0.20
4	8.0	7.0	0.20
5	4.0	4.0	0.30
6	5.2	5.0	0.30
7	6.6	6.0	0.30
8	7.5	7.0	0.30
9	2.0	4.0	0.40
10	4.0	5.0	0.40
11	5.7	6.0	0.40
12	6.5	7.0	0.40
13		6.5	0.35

N = 13

LEAST SQUARES ANALYSIS

GENERAL LINEAR MODELS PROCEDURE

DEPENDENT VARIABLE: WT_LOSS WEIGHT LOSS (POUNDS)

SOURCE	DF	SUM OF SQUARES	MEAN SQUARE	F VALUE	PR > F	R-SQUARE	C.V.
MODEL	2	31.12416667	15.56208333	104.13	0.0001	0.958576	7.0181
ERROR	9	1.34500000	0.14944444				WT_LOSS
						STD DEV	MEAN
CORRECTED TOTAL	11	32.46916667				0.38658045	5.50833333

SOURCE	DF	TYPE I SS	F VALUE	PR > F	DF	TYPE IV SS	F VALUE	PR > F
TIME	1	26.00416667	174.01	0.0001	1	26.00416667	174.01	0.0001
HUMID	1	5.12000000	34.26	0.0002	1	5.12000000	34.26	0.0002

PARAMETER	ESTIMATE	T FOR H0: PARAMETER = 0	PR > |T|	STD ERROR OF ESTIMATE
INTERCEPT	0.66666667	0.96	0.3620	0.69423219
TIME	1.31666667	13.19	0.0001	0.09981464
HUMID	−8.00000000	−5.85	0.0002	1.36676829

OBSERVATION	OBSERVED VALUE	PREDICTED VALUE	RESIDUAL	LOWER 95% CL FOR MEAN	UPPER 95% CL FOR MEAN
1	4.30000000	4.33333333	−0.03333333	3.80984168	4.85682499
2	5.50000000	5.65000000	−0.15000000	5.23518217	6.06481783
3	6.80000000	6.96666667	−0.16666667	6.55184883	7.38148450
4	8.00000000	8.28333333	−0.28333333	7.75984168	8.80682499
5	4.00000000	3.53333333	0.46666667	3.11090353	3.95576314
6	5.20000000	4.85000000	0.35000000	4.57345478	5.12654522
7	6.60000000	6.16666667	0.43333333	5.89012144	6.44321189
8	7.50000000	7.48333333	0.01666667	7.06090353	7.90576314
9	2.00000000	2.73333333	−0.73333333	2.20984168	3.25682499
10	4.00000000	4.05000000	−0.05000000	3.63518217	4.46481783
11	5.70000000	5.36666667	0.33333333	4.95184883	5.78148450
12	6.50000000	6.68333333	−0.18333333	6.15984168	7.20682499
13*		6.42500000		6.05268960	6.79731040

*OBSERVATION WAS NOT USED IN THIS ANALYSIS

SUM OF RESIDUALS	0.00000000
SUM OF SQUARED RESIDUALS	1.34500000
SUM OF SQUARED RESIDUALS − ERROR SS	−0.00000000
PRESS STATISTIC	2.61233345
FIRST ORDER AUTOCORRELATION	0.15902520
DURBIN–WATSON D	1.65613383

Solution

a. The three normal equations for this model are shown here.

	y_i	$\hat{\beta}_0$	$x_{i1}\hat{\beta}_1$	$x_{i2}\hat{\beta}_2$
1	$\sum y_i =$	$n\hat{\beta}_0 +$	$\sum x_{i1}\hat{\beta}_1 +$	$\sum x_{i2}\hat{\beta}_2$
x_{i1}	$\sum x_{i1}y_i =$	$\sum x_{i1}\hat{\beta}_0 +$	$\sum x_{i1}^2\hat{\beta}_1 +$	$\sum x_{i1}x_{i2}\hat{\beta}_2$
x_{i2}	$\sum x_{i2}y_i =$	$\sum x_{i2}\hat{\beta}_0 +$	$\sum x_{i2}x_{i1}\hat{\beta}_1 +$	$\sum x_{i2}^2\hat{\beta}_2$

For these data, we have

$$\sum y_i = 66.10, \qquad \sum x_{i1} = 66, \qquad \sum x_{i2} = 3.60,$$
$$\sum x_{i1}y_i = 383.3, \qquad \sum x_{i2}y_i = 19.19,$$
$$\sum x_{i1}^2 = 378, \qquad \sum x_{i2}^2 = 1.16, \quad \text{and} \quad \sum x_{i1}x_{i2} = 19.8$$

Substituting these values into the normal equation yields the result shown here:

$$66.1 = 12\hat{\beta}_0 + 66\hat{\beta}_1 + 3.6\hat{\beta}_2$$
$$383.3 = 66\hat{\beta}_0 + 378\hat{\beta}_1 + 19.8\hat{\beta}_2$$
$$19.19 = 3.6\hat{\beta}_0 + 19.8\hat{\beta}_1 + 1.16\hat{\beta}_2$$

b. The normal equations of part (a) could be solved to determine $\hat{\beta}_0$, $\hat{\beta}_1$, and $\hat{\beta}_2$. The solution would agree with that shown here in the output. The least squares prediction equation is

$$\hat{y} = 0.667 + 1.317x_1 - 8.000x_2$$

where x_1 is exposure time and x_2 is relative humidity. Substituting $x_1 = 6.5$ and $x_2 = .35$, we have

$$\hat{y} = 0.667 + 1.317(6.5) - 8.000(.35) = 6.428$$

This value agrees with the predicted value shown as observation 13 in the output, except for rounding errors.

EXERCISES

12.7   A pharmaceutical firm would like to obtain information on the relationship between the dose level and potency of a drug product. To do this, each of 15 test tubes is inoculated with a virus culture and incubated for five days at 30° C. Three test tubes are randomly assigned to each of the five different dose levels to be investigated (2, 4, 8, 16, and 32 mg). Each tube is injected with only one dose level and the response of interest (a measure of the protective strength of the product against the virus culture) is obtained. The data are given below.

Dose Level	Response
2	5, 7, 3
4	10, 12, 14
8	15, 17, 18
16	20, 21, 19
32	23, 24, 29

a. Plot the data.
b. Fit a linear regression model to these data.
c. What other regression model might be appropriate?
d. SAS computer output is shown for both a linear and quadratic regression line. Which regression equation appears to fit the data better? Why?

OBS	DOSE	RESPONSE
1	2	5
2	2	7
3	2	3
4	4	10
5	4	12
6	4	14
7	8	15
8	8	17
9	8	18
10	16	20
11	16	21
12	16	19
13	32	23
14	32	24
15	32	29
N=	15	

DEPENDENT VARIABLE: RESPONSE

SOURCE	DF	SUM OF SQUARES	MEAN SQUARE	F VALUE	PR > F	R-SQUARE	C.V.
MODEL	1	590.91612903	590.91612903	44.28	0.0001	0.773046	23.1207
ERROR	13	173.48387097	13.34491315		ROOT MSE		RESPONSE MEAN
CORRECTED TOTAL	14	764.40000000			3.65306900		15.80000000

SOURCE	DF	TYPE I SS	F VALUE	PR > F	DF	TYPE III SS	F VALUE	PR > F
DOSE	1	590.91612903	44.28	0.0001	1	590.91612903	44.28	0.0001

PARAMETER	ESTIMATE	T FOR HO: PARAMETER=0	PR > |T|	STD ERROR OF ESTIMATE
INTERCEPT	8.66666667	6.07	0.0001	1.42786770
DOSE	0.57526882	6.65	0.0001	0.08645016

OBSERVATION	OBSERVED VALUE	PREDICTED VALUE	RESIDUAL
1	5.00000000	9.81720430	-4.81720430
2	7.00000000	9.81720430	-2.81720430
3	3.00000000	9.81720430	-6.81720430
4	10.00000000	10.96774194	-0.96774194
5	12.00000000	10.96774194	1.03225806
6	14.00000000	10.96774194	3.03225806
7	15.00000000	13.26881720	1.73118280
8	17.00000000	13.26881720	3.73118280
9	18.00000000	13.26881720	4.73118280
10	20.00000000	17.87096774	2.12903226
11	21.00000000	17.87096774	3.12903226
12	19.00000000	17.87096774	1.12903226
13	23.00000000	27.07526882	-4.07526882
14	24.00000000	27.07526882	-3.07526882
15	29.00000000	27.07526882	1.92473118

SUM OF RESIDUALS	0.00000000
SUM OF SQUARED RESIDUALS	173.48387097
SUM OF SQUARED RESIDUALS - ERROR SS	-0.00000000
FIRST ORDER AUTOCORRELATION	0.53691663
DURBIN-WATSON D	0.77105118

LINEAR REGRESSION ANALYSIS
PLOT OF RESIDUALS VS. PREDICTED VALUES

DEPENDENT VARIABLE: RESPONSE

SOURCE	DF	SUM OF SQUARES	MEAN SQUARE	F VALUE	PR > F	R-SQUARE	C.V.
MODEL	2	673.82061986	336.91030993	44.63	0.0001	0.881503	17.3887
ERROR	12	90.57938014	7.54828168		ROOT MSE		RESPONSE MEAN
CORRECTED TOTAL	14	764.40000000			2.74741363		15.80000000

SOURCE	DF	TYPE I SS	F VALUE	PR > F	DF	TYPE III SS	F VALUE	PR > F
DOSE	1	590.91612903	78.28	0.0001	1	205.97013882	27.29	0.0002
DOSE*DOSE	1	82.90449083	10.98	0.0062	1	82.90449083	10.98	0.0062

PARAMETER	ESTIMATE	T FOR H0: PARAMETER=0	PR > |T|	STD ERROR OF ESTIMATE
INTERCEPT	4.48366013	2.71	0.0191	1.65720388
DOSE	1.50632511	5.22	0.0002	0.28836373
DOSE*DOSE	-0.02698714	-3.31	0.0062	0.00814314

OBSERVATION	OBSERVED VALUE	PREDICTED VALUE	RESIDUAL
1	5.00000000	7.38836180	-2.38836180
2	7.00000000	7.38836180	-0.38836180
3	3.00000000	7.38836180	-4.38836180
4	10.00000000	10.07716635	-0.07716635
5	12.00000000	10.07716635	1.92283365
6	14.00000000	10.07716635	3.92283365
7	15.00000000	14.80708412	0.19291588
8	17.00000000	14.80708412	2.19291588
9	18.00000000	14.80708412	3.19291588
10	20.00000000	21.67615433	-1.67615433
11	21.00000000	21.67615433	-0.67615433
12	19.00000000	21.67615433	-2.67615433
13	23.00000000	25.05123340	-2.05123340
14	24.00000000	25.05123340	-1.05123340
15	29.00000000	25.05123340	3.94876660

SUM OF RESIDUALS	0.00000000
SUM OF SQUARED RESIDUALS	90.57938014
SUM OF SQUARED RESIDUALS - ERROR SS	0.00000000
FIRST ORDER AUTOCORRELATION	0.21674175
DURBIN-WATSON D	1.33139643

QUADRATIC REGRESSION ANALYSIS
PLOT OF RESIDUALS VS. PREDICTED VALUES

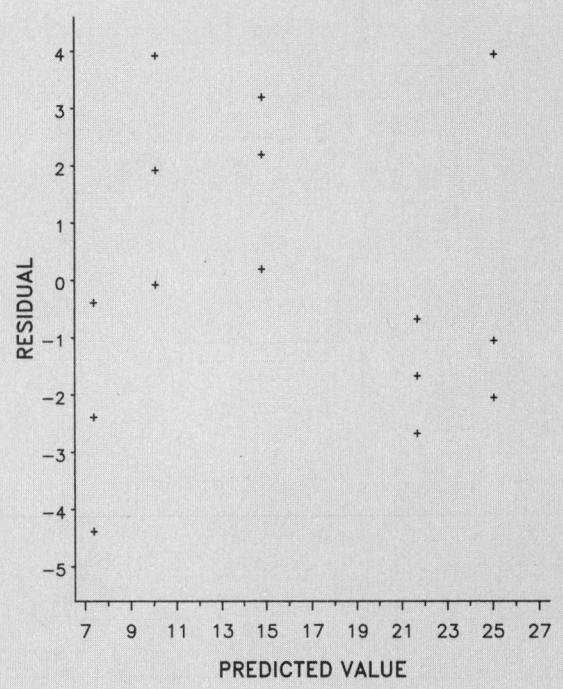

12.8   Refer to the data of Exercise 12.7. Many times a logarithmic transformation can be used on the dose levels to linearize the response with respect to the independent variable.

a. Refer to a set of log tables (see, for example, the Chemical Rubber Company tables) or use a calculator to obtain the logarithms of the five dose levels.

b. If x_1 denotes the log dose, fit the model

$$y = \beta_0 + \beta_1 x_1 + \varepsilon$$

A residual plot is shown here in the output.

```
                          REGRESSION ANALYSIS USING LOG DOSE

                            GENERAL LINEAR MODELS PROCEDURE

DEPENDENT VARIABLE: RESPONSE
```

SOURCE	DF	SUM OF SQUARES	MEAN SQUARE	F VALUE	PR > F	R-SQUARE	C.V.
MODEL	1	710.53333333	710.53333333	171.48	0.0001	0.929531	12.8834
ERROR	13	53.86666667	4.14358974		ROOT MSE		RESPONSE MEAN
CORRECTED TOTAL	14	764.40000000			2.03558094		15.80000000

SOURCE	DF	TYPE I SS	F VALUE	PR > F	DF	TYPE III SS	F VALUE	PR > F
LOGDOSE	1	710.53333333	171.48	0.0001	1	710.53333333	171.48	0.0001

PARAMETER	ESTIMATE	T FOR H0: PARAMETER=0	PR > |T|	STD ERROR OF ESTIMATE
INTERCEPT	1.20000000	0.97	0.3480	1.23260547
LOGDOSE	16.16671673	13.09	0.0001	1.23457641

REGRESSION ANALYSIS USING LOG DOSES
PLOT OF RESIDUALS VS. PREDICTED VALUES

c. Compare your results in part (b) to those for Exercise 12.7. Does the logarithmic transformation provide a better linear fit than that in Exercise 12.7?

12.9   The abrasive effect of a wear tester for experimental fabrics was tested on a particular fabric while run at six different machine speeds. Forty-eight identical 5-inch-square pieces of fabric were cut, with eight squares randomly assigned to each of the six machine speeds 100, 120, 140, 160, 180, and 200 revolutions per minute (rpm). The order of assignment of the squares to the machine was random, with each square tested for a three-minute period at the appropriate machine setting. The amount of wear was measured and recorded for each square. The data appear below.

Machine Speed (in rpm)	Wear
100	23.0, 23.5, 24.4, 25.2, 25.6, 26.1, 24.8, 25.6
120	26.7, 26.1, 25.8, 26.3, 27.2, 27.9, 28.3, 27.4
140	28.0, 28.4, 27.0, 28.8, 29.8, 29.4, 28.7, 29.3
160	32.7, 32.1, 31.9, 33.0, 33.5, 33.7, 34.0, 32.5
180	43.1, 41.7, 42.4, 42.1, 43.5, 43.8, 44.2, 43.6
200	54.2, 43.7, 53.1, 53.8, 55.6, 55.9, 54.7, 54.5

a. Generate a graph of the data (since the variability within a speed is about the same for all speeds, you can save time while still maintaining the trend by plotting the sample mean for each speed).

b. What type of regression model appears appropriate?

c. Output for linear, quadratic, and cubic regression models is shown on the following pages. Which regression equation gives a better fit? Why?

d. Is there anything peculiar about the data? What might have happened?

```
LISTING OF DATA

OBS     SPEED    WEAR

  1      100     23.0
  2      100     23.5
  3      100     24.4
  4      100     25.2
  5      100     25.6
  6      100     26.1
  7      100     24.8
  8      100     25.6
  9      120     26.7
 10      120     26.1
 11      120     25.8
 12      120     26.3
 13      120     27.2
 14      120     27.9
 15      120     28.3
 16      120     27.4
 17      140     28.0
 18      140     28.4
 19      140     27.0
 20      140     28.8
 21      140     29.8
 22      140     29.4
 23      140     28.7
 24      140     29.3
 25      160     32.7
 26      160     32.1
 27      160     31.9
 28      160     33.0
 29      160     33.5
 30      160     33.7
 31      160     34.0
```

```
   LISTING OF DATA

  OBS    SPEED    WEAR

   32     160     32.5
   33     180     43.1
   34     180     41.7
   35     180     42.4
   36     180     42.1
   37     180     43.5
   38     180     43.8
   39     180     44.2
   40     180     43.6
   41     200     54.2
   42     200     43.7
   43     200     53.1
   44     200     53.8
   45     200     55.6
   46     200     55.9
   47     200     54.7
   48     200     54.5

  N=      48
```

LINEAR REGRESSION ANALYSIS

GENERAL LINEAR MODELS PROCEDURE

DEPENDENT VARIABLE: WEAR

SOURCE	DF	SUM OF SQUARES	MEAN SQUARE	F VALUE	PR > F	R-SQUARE	C.V.
MODEL	1	4326.79207143	4326.79207143	291.47	0.0001	0.863693	11.0305
ERROR	46	682.84709524	14.84450207		ROOT MSE		WEAR MEAN
CORRECTED TOTAL	47	5009.63916667			3.85285635		34.92916667

SOURCE	DF	TYPE I SS	F VALUE	PR > F	DF	TYPE III SS	F VALUE	PR > F
SPEED	1	4326.79207143	291.47	0.0001	1	4326.79207143	291.47	0.0001

PARAMETER	ESTIMATE	T FOR HO: PARAMETER=0	PR > $\lvert T \rvert$	STD ERROR OF ESTIMATE
INTERCEPT	-6.76547619	-2.70	0.0096	2.50470943
SPEED	0.27796429	17.07	0.0001	0.01628129

OBSERVATION	OBSERVED VALUE	PREDICTED VALUE	RESIDUAL
1	23.00000000	21.03095238	1.96904762
2	23.50000000	21.03095238	2.46904762
3	24.40000000	21.03095238	3.36904762
4	25.20000000	21.03095238	4.16904762
5	25.60000000	21.03095238	4.56904762
6	26.10000000	21.03095238	5.06904762
7	24.80000000	21.03095238	3.76904762
8	25.60000000	21.03095238	4.56904762
9	26.70000000	26.59023810	0.10976190
10	26.10000000	26.59023810	-0.49023810
11	25.80000000	26.59023810	-0.79023810
12	26.30000000	26.59023810	-0.29023810
13	27.20000000	26.59023810	0.60976190
14	27.90000000	26.59023810	1.30976190
15	28.30000000	26.59023810	1.70976190
16	27.40000000	26.59023810	0.80976190
17	28.00000000	32.14952381	-4.14952381
18	28.40000000	32.14952381	-3.74952381
19	27.00000000	32.14952381	-5.14952381
20	28.80000000	32.14952381	-3.34952381
21	29.80000000	32.14952381	-2.34952381
22	29.40000000	32.14952381	-2.74952381
23	28.70000000	32.14952381	-3.44952381
24	29.30000000	32.14952381	-2.84952381
25	32.70000000	37.70880952	-5.00880952
26	32.10000000	37.70880952	-5.60880952
27	31.90000000	37.70880952	-5.80880952
28	33.00000000	37.70880952	-4.70880952
29	33.50000000	37.70880952	-4.20880952
30	33.70000000	37.70880952	-4.00880952

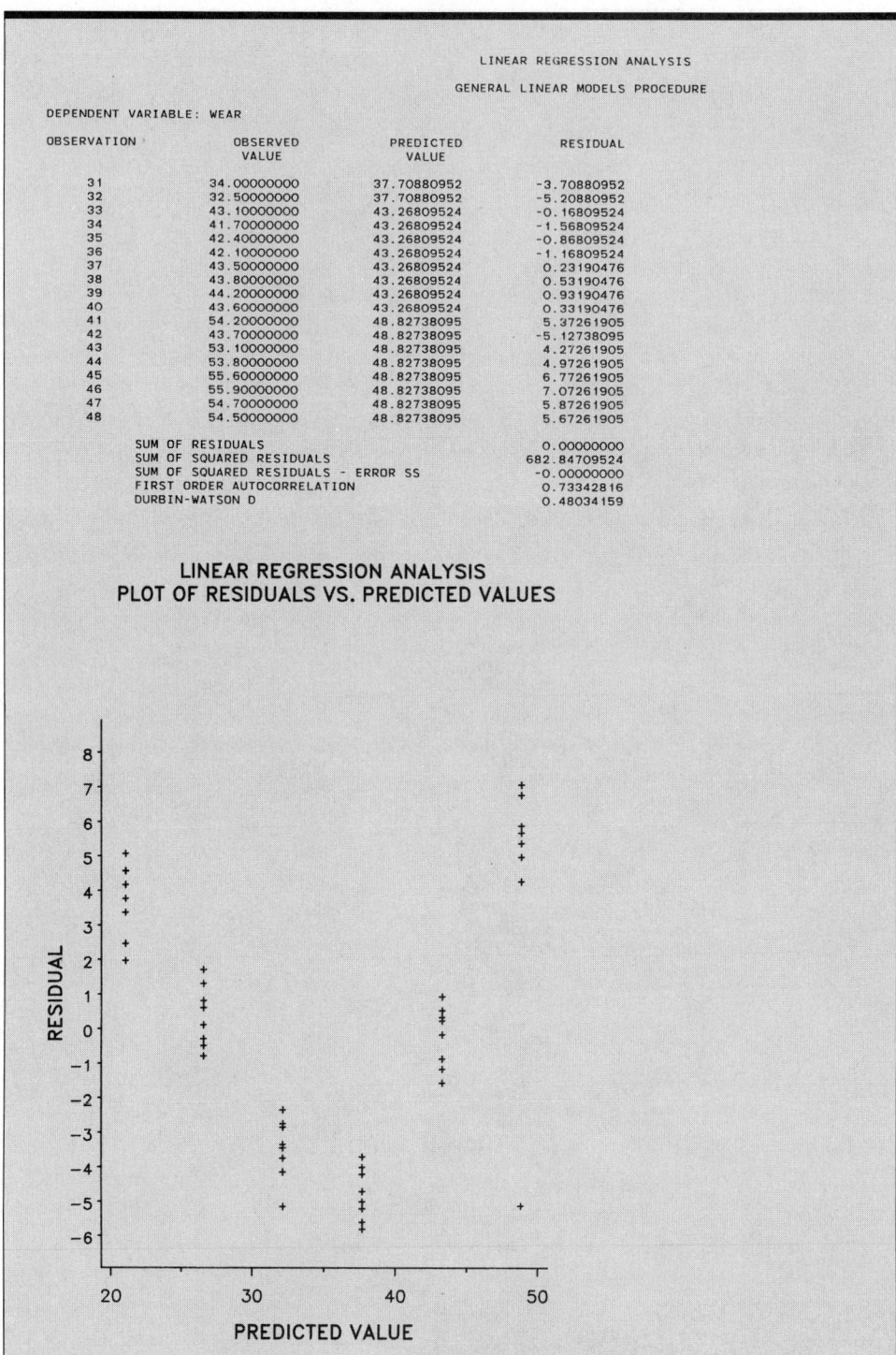

LINEAR REGRESSION ANALYSIS

GENERAL LINEAR MODELS PROCEDURE

DEPENDENT VARIABLE: WEAR

OBSERVATION	OBSERVED VALUE	PREDICTED VALUE	RESIDUAL
31	34.00000000	37.70880952	-3.70880952
32	32.50000000	37.70880952	-5.20880952
33	43.10000000	43.26809524	-0.16809524
34	41.70000000	43.26809524	-1.56809524
35	42.40000000	43.26809524	-0.86809524
36	42.10000000	43.26809524	-1.16809524
37	43.50000000	43.26809524	0.23190476
38	43.80000000	43.26809524	0.53190476
39	44.20000000	43.26809524	0.93190476
40	43.60000000	43.26809524	0.33190476
41	54.20000000	48.82738095	5.37261905
42	43.70000000	48.82738095	-5.12738095
43	53.10000000	48.82738095	4.27261905
44	53.80000000	48.82738095	4.97261905
45	55.60000000	48.82738095	6.77261905
46	55.90000000	48.82738095	7.07261905
47	54.70000000	48.82738095	5.87261905
48	54.50000000	48.82738095	5.67261905

SUM OF RESIDUALS	0.00000000
SUM OF SQUARED RESIDUALS	682.84709524
SUM OF SQUARED RESIDUALS - ERROR SS	-0.00000000
FIRST ORDER AUTOCORRELATION	0.73342816
DURBIN-WATSON D	0.48034159

LINEAR REGRESSION ANALYSIS
PLOT OF RESIDUALS VS. PREDICTED VALUES

```
                              QUADRATIC REGRESSION ANALYSIS

                             GENERAL LINEAR MODELS PROCEDURE

DEPENDENT VARIABLE: WEAR
```

SOURCE	DF	SUM OF SQUARES	MEAN SQUARE	F VALUE	PR > F	R-SQUARE	C.V.
MODEL	2	4839.89302381	2419.94651190	641.53	0.0001	0.966116	5.5604
ERROR	45	169.74614286	3.77213651		ROOT MSE		WEAR MEAN
CORRECTED TOTAL	47	5009.63916667			1.94219888		34.92916667

SOURCE	DF	TYPE I SS	F VALUE	PR > F	DF	TYPE III SS	F VALUE	PR > F
SPEED	1	4326.79207143	1147.04	0.0001	1	261.47616353	69.32	0.0001
SPEED*SPEED	1	513.10095238	136.02	0.0001	1	513.10095238	136.02	0.0001

PARAMETER	ESTIMATE	T FOR HO: PARAMETER=0	PR > |T|	STD ERROR OF ESTIMATE
INTERCEPT	63.13928571	10.31	0.0001	6.12529888
SPEED	-0.70507143	-8.33	0.0001	0.08468583
SPEED*SPEED	0.00327679	11.66	0.0001	0.00028096

OBSERVATION	OBSERVED VALUE	PREDICTED VALUE	RESIDUAL
1	23.00000000	25.40000000	-2.40000000
2	23.50000000	25.40000000	-1.90000000
3	24.40000000	25.40000000	-1.00000000
4	25.20000000	25.40000000	-0.20000000
5	25.60000000	25.40000000	0.20000000
6	26.10000000	25.40000000	0.70000000
7	24.80000000	25.40000000	-0.60000000
8	25.60000000	25.40000000	0.20000000
9	26.70000000	25.71642857	0.98357143
10	26.10000000	25.71642857	0.38357143
11	25.80000000	25.71642857	0.08357143
12	26.30000000	25.71642857	0.58357143
13	27.20000000	25.71642857	1.48357143
14	27.90000000	25.71642857	2.18357143
15	28.30000000	25.71642857	2.58357143
16	27.40000000	25.71642857	1.68357143
17	28.00000000	28.65428571	-0.65428571
18	28.40000000	28.65428571	-0.25428571
19	27.00000000	28.65428571	-1.65428571
20	28.80000000	28.65428571	0.14571429
21	29.80000000	28.65428571	1.14571429
22	29.40000000	28.65428571	0.74571429
23	28.70000000	28.65428571	0.04571429
24	29.30000000	28.65428571	0.64571429
25	32.70000000	34.21357143	-1.51357143
26	32.10000000	34.21357143	-2.11357143
27	31.90000000	34.21357143	-2.31357143
28	33.00000000	34.21357143	-1.21357143
29	33.50000000	34.21357143	-0.71357143
30	33.70000000	34.21357143	-0.51357143
31	34.00000000	34.21357143	-0.21357143
32	32.50000000	34.21357143	-1.71357143
33	43.10000000	42.39428571	0.70571429
34	41.70000000	42.39428571	-0.69428571
35	42.40000000	42.39428571	0.00571429
36	42.10000000	42.39428571	-0.29428571
37	43.50000000	42.39428571	1.10571429
38	43.80000000	42.39428571	1.40571429
39	44.20000000	42.39428571	1.80571429
40	43.60000000	42.39428571	1.20571429
41	54.20000000	53.19642857	1.00357143
42	43.70000000	53.19642857	-9.49642857
43	53.10000000	53.19642857	-0.09642857
44	53.80000000	53.19642857	0.60357143
45	55.60000000	53.19642857	2.40357143
46	55.90000000	53.19642857	2.70357143
47	54.70000000	53.19642857	1.50357143
48	54.50000000	53.19642857	1.30357143

```
        SUM OF RESIDUALS                        0.00000000
        SUM OF SQUARED RESIDUALS              169.74614286
        SUM OF SQUARED RESIDUALS - ERROR SS    0.00000000
        FIRST ORDER AUTOCORRELATION            0.25731691
        DURBIN-WATSON D                        1.44142233
```

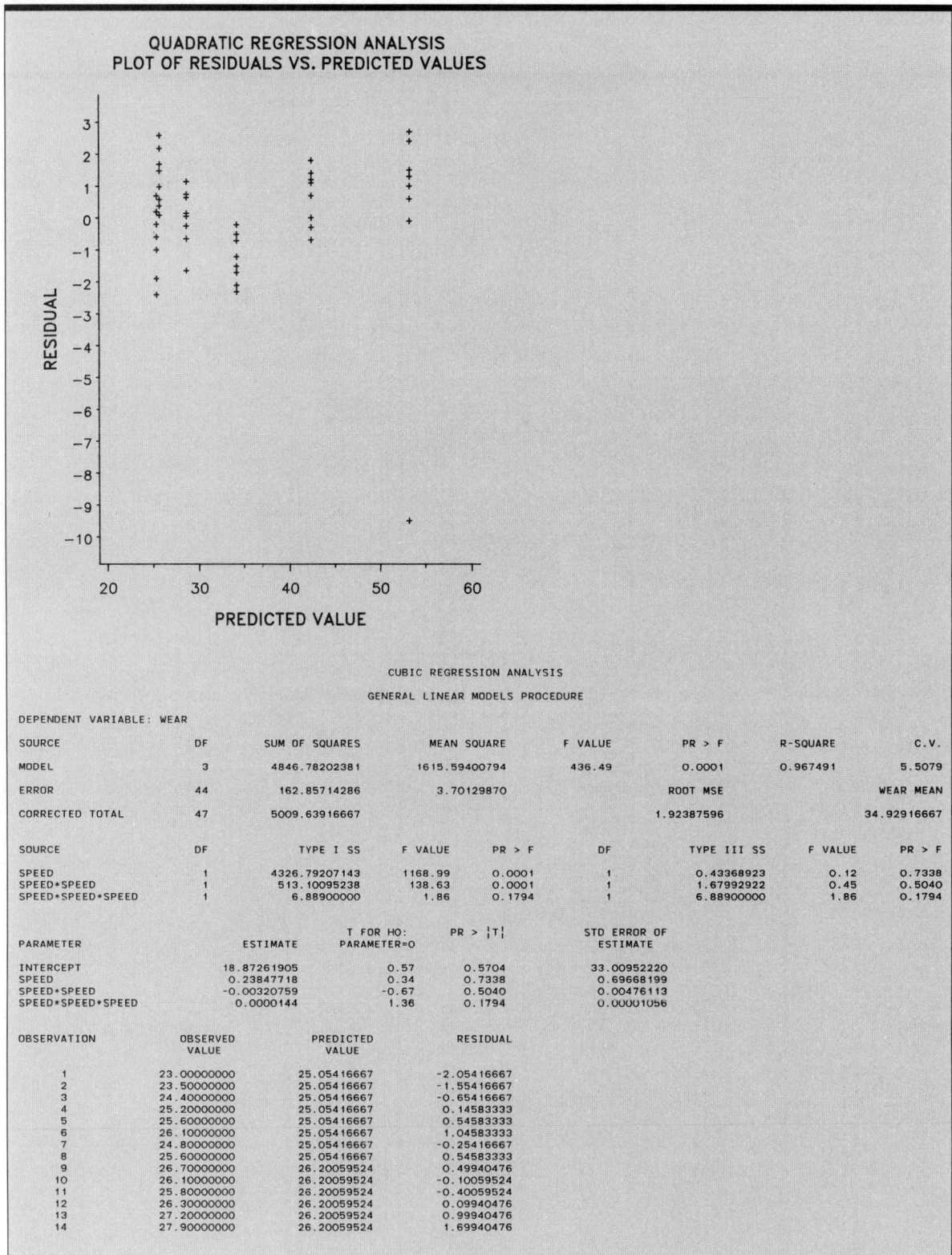

QUADRATIC REGRESSION ANALYSIS
PLOT OF RESIDUALS VS. PREDICTED VALUES

CUBIC REGRESSION ANALYSIS

GENERAL LINEAR MODELS PROCEDURE

DEPENDENT VARIABLE: WEAR

SOURCE	DF	SUM OF SQUARES	MEAN SQUARE	F VALUE	PR > F	R-SQUARE	C.V.
MODEL	3	4846.78202381	1615.59400794	436.49	0.0001	0.967491	5.5079
ERROR	44	162.85714286	3.70129870		ROOT MSE		WEAR MEAN
CORRECTED TOTAL	47	5009.63916667			1.92387596		34.92916667

SOURCE	DF	TYPE I SS	F VALUE	PR > F	DF	TYPE III SS	F VALUE	PR > F
SPEED	1	4326.79207143	1168.99	0.0001	1	0.43368923	0.12	0.7338
SPEED*SPEED	1	513.10095238	138.63	0.0001	1	1.67992922	0.45	0.5040
SPEED*SPEED*SPEED	1	6.88900000	1.86	0.1794	1	6.88900000	1.86	0.1794

PARAMETER	ESTIMATE	T FOR HO: PARAMETER=0	PR > |T|	STD ERROR OF ESTIMATE
INTERCEPT	18.87261905	0.57	0.5704	33.00952220
SPEED	0.23847718	0.34	0.7338	0.69668199
SPEED*SPEED	-0.00320759	-0.67	0.5040	0.00476113
SPEED*SPEED*SPEED	0.0000144	1.36	0.1794	0.00001056

OBSERVATION	OBSERVED VALUE	PREDICTED VALUE	RESIDUAL
1	23.00000000	25.05416667	-2.05416667
2	23.50000000	25.05416667	-1.55416667
3	24.40000000	25.05416667	-0.65416667
4	25.20000000	25.05416667	0.14583333
5	25.60000000	25.05416667	0.54583333
6	26.10000000	25.05416667	1.04583333
7	24.80000000	25.05416667	-0.25416667
8	25.60000000	25.05416667	0.54583333
9	26.70000000	26.20059524	0.49940476
10	26.10000000	26.20059524	-0.10059524
11	25.80000000	26.20059524	-0.40059524
12	26.30000000	26.20059524	0.09940476
13	27.20000000	26.20059524	0.99940476
14	27.90000000	26.20059524	1.69940476

DEPENDENT VARIABLE: WEAR

OBSERVATION	OBSERVED VALUE	PREDICTED VALUE	RESIDUAL
15	28.30000000	26.20059524	2.09940476
16	27.40000000	26.20059524	1.19940476
17	28.00000000	28.93095238	-0.93095238
18	28.40000000	28.93095238	-0.53095238
19	27.00000000	28.93095238	-1.93095238
20	28.80000000	28.93095238	-0.13095238
21	29.80000000	28.93095238	0.86904762
22	29.40000000	28.93095238	0.46904762
23	28.70000000	28.93095238	-0.23095238
24	29.30000000	28.93095238	0.36904762
25	32.70000000	33.93690476	-1.23690476
26	32.10000000	33.93690476	-1.83690476
27	31.90000000	33.93690476	-2.03690476
28	33.00000000	33.93690476	-0.93690476
29	33.50000000	33.93690476	-0.43690476
30	33.70000000	33.93690476	-0.23690476
31	34.00000000	33.93690476	0.06309524
32	32.50000000	33.93690476	-1.43690476
33	43.10000000	41.91011905	1.18988095
34	41.70000000	41.91011905	-0.21011905
35	42.40000000	41.91011905	0.48988095
36	42.10000000	41.91011905	0.18988095
37	43.50000000	41.91011905	1.58988095
38	43.80000000	41.91011905	1.88988095
39	44.20000000	41.91011905	2.28988095
40	43.60000000	41.91011905	1.68988095
41	54.20000000	53.54226190	0.65773810
42	43.70000000	53.54226190	-9.84226190
43	53.10000000	53.54226190	-0.44226190
44	53.80000000	53.54226190	0.25773810
45	55.60000000	53.54226190	2.05773810
46	55.90000000	53.54226190	2.35773810
47	54.70000000	53.54226190	1.15773810
48	54.50000000	53.54226190	0.95773810

SUM OF RESIDUALS	0.00000000
SUM OF SQUARED RESIDUALS	162.85714286
SUM OF SQUARED RESIDUALS - ERROR SS	-0.00000000
FIRST ORDER AUTOCORRELATION	0.23779258
DURBIN-WATSON D	1.49287270

CUBIC REGRESSION ANALYSIS
PLOT OF RESIDUALS VS. PREDICTED VALUES

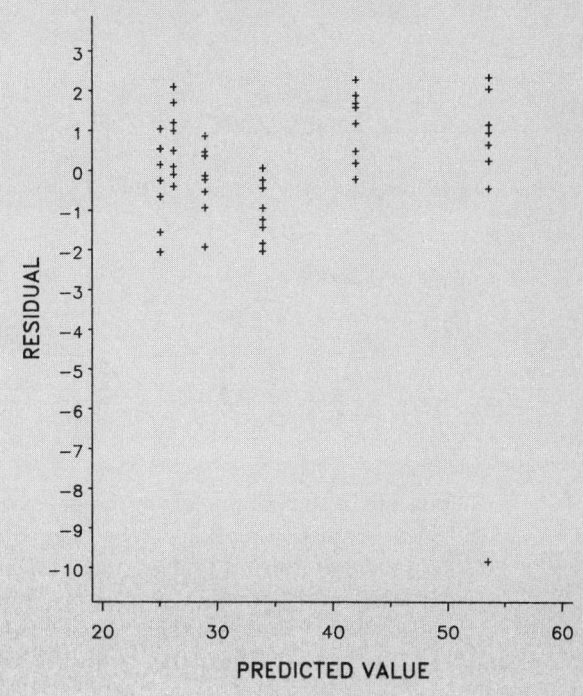

489

12.10 Refer to the data of Exercise 12.9. Suppose that another variable was controlled and that the first four squares at each speed were treated with a .2 concentration of protective coating, whereas the second four squares were treated with a .4 concentration of the same coating. x_1 denotes the machine speed and x_2 denotes the concentration of the protective coating. Fit these models using available statistical software. Which model seems to provide a better fit to the data? Why?

$$y = \beta_0 + \beta_1 x_1 + \beta_2 x_1^2 + \beta_3 x_2 + \varepsilon$$
$$y = \beta_0 + \beta_1 x_1 + \beta_2 x_1^2 + \beta_3 x_2 + \beta_4 x_1 x_2 + \beta_5 x_1^2 x_2 + \varepsilon$$

12.11 In Exercise 9.25 we compared the height of detergent suds to the amount of detergent added. Another variable that could affect the height of the suds is the degree of agitation in the wash cycle (measured in minutes). The complete data are presented here.

Height, y	Agitation, x_1	Amount, x_2
28.1	1	6
32.3	1	7
34.8	1	8
38.2	1	9
43.5	1	10
60.3	2	6
63.7	2	7
65.4	2	8
69.2	2	9
72.9	2	10
88.2	3	6
89.3	3	7
94.1	3	8
95.7	3	9
100.6	3	10

Fit the following two models using available statistical software.

$$y = \beta_0 + \beta_1 x_1 + \beta_2 x_2 + \varepsilon$$
$$y = \beta_0 + \beta_1 x_1 + \beta_2 x_2 + \beta_3 x_1 x_2 + \varepsilon$$

12.4   INFERENCES ABOUT A SINGLE PARAMETER IN THE GENERAL LINEAR MODEL

We make inferences about any of the parameters in the general linear model just as we did for β_0 and β_1 in the linear regression model, $y = \beta_0 + \beta_1 x + \varepsilon$.

Before we do this, however, we must introduce the *coefficient of determination*, R^2. The coefficient of determination for the model $y = \beta_0 + \beta_1 x_1 + \cdots + \beta_k x_k + \varepsilon$ is defined as the proportion of the variability in the dependent variable y that is accounted for by the independent variables $x_1, x_2, \ldots, x_k$ of the model. When there is only one independent variable in the regression equation, the coefficient of determination is r^2, the square of the simple correlation coefficient between y and the independent variable, presented in Chapter 8. The coefficient of determination is designated by the symbol $R^2_{y \cdot x_1 x_2 \cdots x_k}$ and is computed as follows:

$$R^2_{y \cdot x_1 x_2 \cdots x_k} = \frac{S_{yy} - SSE}{S_{yy}}, \qquad 0 \le R^2 \le 1$$

where

$$S_{yy} = \sum y_i^2 - \frac{(\sum y_i)^2}{n}$$

and

$$SSE = \sum (y_i - \hat{y}_i)^2$$
$$= \sum y_i^2 - \hat{\beta}_0 \sum y_i - \hat{\beta}_1 \sum x_{i1} y_i - \cdots - \hat{\beta}_k \sum x_{ik} y_i$$

EXAMPLE 12.5

Refer to the data for Example 12.4 and compute $R^2_{y \cdot x_1 x_2}$. Compare this to the value shown in the computer output for that.

Solution

a. For these data we have

$$S_{yy} = \sum y_i^2 - (\sum y_i)^2 / n$$
$$= 396.57 - (66.1)^2 / 12$$
$$= 396.57 - 364.10 = 32.47$$

and

$$SSE = \sum y_i^2 - \hat{\beta}_0 \sum y_i - \hat{\beta}_1 \sum x_{i1} y_i - \hat{\beta}_2 \sum x_{i2} y_i$$
$$= 396.57 - 0.667(66.1) - 1.317(383.3) - (-8.000)(19.19)$$
$$= 1.195^*$$

Substituting into the formula for R^2 we find

$$R^2_{y \cdot x_1 x_2} = \frac{32.47 - 1.195}{32.47} = 0.963$$

b. The output for Example 12.4 lists the coefficient of determination as 0.958576. Our result in part (a) agrees with this value except for rounding error.

There is no general relationship between the coefficient of determination R^2 and squares of the individual correlation coefficients r_{yx_j}. If the independent variables are uncorrelated, then

$$R^2 = r^2_{yx_1} + r^2_{yx_2} + \cdots + r^2_{yx_k}$$

---

* It is important to carry as many decimals as possible to avoid rounding errors. Had we used the eight places to the right of the decimal for the $\hat{\beta}$s, the value of SSE would be 1.345, as shown in the output.

But when the independent variables are themselves correlated, it is difficult to separate R^2 into the predictive contribution of each independent variable. Most problems that have a model with more than one independent variable are affected (to a greater or lesser degree) by *collinearity* or **multicollinearity**, where the independent variables are themselves correlated. For these situations, where the *x*s account for overlapping pieces of the variability in the *y*-values, we often find that

$$R^2_{y \cdot x_1 x_2 \cdots x_k} < r^2_{yx_1} + r^2_{yx_2} + \cdots + r^2_{yx_k}$$

We can now write down the expression for $s_{\hat{\beta}_j}$, the *estimated standard error for* $\hat{\beta}_j$ in the general linear model.

margin: multicollinearity

---

margin heading: ESTIMATED STANDARD ERROR FOR $\hat{\beta}_j$ IN THE GENERAL LINEAR MODEL

$$s_{\hat{\beta}_j} = s_\varepsilon \sqrt{\frac{1}{S_{x_j x_j}(1 - R^2_{x_j \cdot x_1 \cdots x_{j-1} x_{j+1} \cdots x_k})}}$$

where

$$s_\varepsilon = \sqrt{\frac{SSE}{n - (k + 1)}}$$

is the standard deviation of the fitted line,

$$S_{x_j x_j} = \sum_i x_{ij}^2 - \frac{\left(\sum_i x_{ij}\right)^2}{n}$$

is the sum of squares,

$$\sum_i (x_{ij} - \bar{x}_j)^2$$

for the variable x_j and $R^2_{x_j \cdot x_1 \cdots x_{j-1} x_{j+1} \cdots x_k}$ is the coefficient of determination for the model with x_j as the *dependent* variable and all other *x*s in the model.

This formula for the estimated standard error of $\hat{\beta}_j$ is *not* recommended for use in computing $s_{\hat{\beta}_j}$; we will rely on software packages for that. The reason we choose to present this formula is for its interpretive value, especially in the presence of multicollinearity. If the independent variable x_j is highly correlated with one or more of the other independent variables, $R^2_{x_j \cdot x_1 \cdots x_{j-1} x_{j+1} \cdots x_k}$ will be large and $1 - R^2$ will be small, making $s_{\hat{\beta}_j}$ large. This makes sense. By definition $\hat{\beta}_j$ is the estimated change in *y* for a one-unit increase in x_j, *while holding the other xs constant.* In the presence of correlation between x_j and the other *x*s, it will be difficult to estimate the change in *y* for a unit increase x_j while the other *x*s remain constant. This difficulty is reflected in the large value of $s_{\hat{\beta}_j}$.

EXAMPLE 12.6
a. Compute estimated standard errors for $\hat{\beta}_1$ and $\hat{\beta}_2$ in the data of Example 12.4.
b. Compare these standard errors to those shown in the output for Example 12.4.

Solution

a. The quantities we need for $s_{\hat{\beta}_1}$ and $s_{\hat{\beta}_2}$ are $S_{x_1x_1}$, $S_{x_2x_2}$, s_ε, $R^2_{x_1 \cdot x_2}$ and $R^2_{x_2 \cdot x_1}$. From Example 12.4 we have

$$S_{x_1x_1} = \sum x_{i1}^2 - \left(\sum x_{i1}\right)^2/n = 378 - (66)^2/12 = 15$$

and

$$S_{x_2x_2} = \sum x_{i2}^2 - \left(\sum x_{i2}\right)^2/n = 1.16 - (3.6)^2/12 = .08$$

Using SSE = 1.345, the value obtained in Example 12.5 after carrying sufficient decimal places, it follows that

$$s_\varepsilon = \sqrt{\frac{\text{SSE}}{n - (k + 1)}} = \sqrt{\frac{1.345}{9}} = .387$$

The quantity $R^2_{x_1 \cdot x_2}$ obtained by fitting the linear regression model

$$x_1 = \beta_0 + \beta_1 x_2 + \varepsilon$$

is simply $r^2_{x_1x_2}$, the square of the correlation coefficient for x_1 and x_2

$$r_{x_1x_2} = \frac{S_{x_1x_2}}{\sqrt{S_{x_1x_1}S_{x_2x_2}}}$$

For these data

$$S_{x_1x_2} = \sum x_{i1}x_{i2} - \frac{\sum x_{i1}\sum x_{i2}}{n} = 19.8 - \frac{66(3.6)}{12} = 0$$

Hence $r_{x_1x_2} = 0$; x_1 and x_2 are uncorrelated. It follows that

$$s_{\hat{\beta}_1} = .387\sqrt{\frac{1}{15(1 - 0)}} = .387\sqrt{\frac{1}{15}} = 0.100$$

and since $R^2_{x_1 \cdot x_2} = R^2_{x_2 \cdot x_1} = 0$,

$$s_{\hat{\beta}_2} = .387\sqrt{\frac{1}{.08}} = 1.368$$

b. Except for slight rounding errors, the estimated standard errors computed for $\hat{\beta}_1$ and $\hat{\beta}_2$ agree with those shown in the output of Example 12.4.

Knowing the estimated standard error for $\hat{\beta}_j$, we can write down formulas for interval estimation of β_j and a statistical test for β_j. These are summarized here.

100 $(1 - \alpha)$% CONFIDENCE INTERVAL FOR β_j

$$\hat{\beta}_j \pm t_{\alpha/2}s_{\hat{\beta}_j}$$

where $t_{\alpha/2}$ is the tabulated t-value for df $= n - (k + 1)$ and $a = \alpha/2$.

STATISTICAL TEST FOR $H_0: \beta_j = 0$

$H_0: \beta_j = 0$

$H_a:$ 1. $\beta_j > 0$

     2. $\beta_j < 0$

     3. $\beta_j \neq 0$

T.S.: $t = \dfrac{\hat{\beta}_j}{s_{\hat{\beta}_j}}$

R.R.: For df $= n - (k + 1)$ and specified value α, reject H_0 if

     1. $t > t_\alpha$

     2. $t < -t_\alpha$

     3. $|t| > t_{\alpha/2}$

EXAMPLE 12.7

An experiment was conducted to examine the potential for deterioration of a new commercial paint when exposed to the atmosphere. Under controlled conditions (settings of the independent variables, temperature x_1 and exposure time x_2) the deterioration y of the paint was measured. These data are shown here.

y	120	101	110	105	92	130
x_1 (°C)	−10	−10	0	0	10	10
x_2 (months)	1	3	2	2	1	3

a. Consider the model $y = \beta_0 + \beta_1 x_1 + \beta_2 x_2 + \varepsilon$. Give an interpretation to the test $H_0: \beta_2 = 0$.

b. SAS was used to fit the model of part (a). A portion of the output is shown below. Test $H_0: \beta_2 = 0$ and draw a conclusion.

c. Distinguish between the model shown in part (a) and the model $y = \beta_0 + \beta_1 x_1 + \beta_2 x_2 + \beta_3 x_1 x_2 + \varepsilon$. Which model is more appropriate?

```
        LISTING OF DATA

 OBS     Y      X1     X2

  1     120    -10     1
  2     101    -10     3
  3     110      0     2
  4     105      0     2
  5      92     10     1
  6     130     10     3

 N=       6
```

MULTIPLE LINEAR REGRESSION ANALYSIS

GENERAL LINEAR MODELS PROCEDURE

DEPENDENT VARIABLE: Y

SOURCE	DF	SUM OF SQUARES	MEAN SQUARE	F VALUE	PR > F	R-SQUARE	C.V.
MODEL	2	90.50000000	45.25000000	0.16	0.8575	0.097382	15.2476
ERROR	3	838.83333333	279.61111111		ROOT MSE		Y MEAN
CORRECTED TOTAL	5	929.33333333			16.72157621		109.66666667

SOURCE	DF	TYPE I SS	F VALUE	PR > F	DF	TYPE III SS	F VALUE	PR > F
X1	1	0.25000000	0.00	0.9780	1	0.25000000	0.00	0.9780
X2	1	90.25000000	0.32	0.6097	1	90.25000000	0.32	0.6097

PARAMETER	ESTIMATE	T FOR HO: PARAMETER=0	PR > |T|	STD ERROR OF ESTIMATE
INTERCEPT	100.16666667	5.55	0.0116	18.06136659
X1	0.02500000	0.03	0.9780	0.83607881
X2	4.75000000	0.57	0.6097	8.36078811

Solution

a. In general, $H_0: \beta_j = 0$ states that x_j has no predictive power over and above that provided by the other independent variables. It does not say that x_j has no predictive power by itself. For our model, $y = \beta_0 + \beta_1 x_1 + \beta_2 x_2 + \varepsilon$, $H_0: \beta_2 = 0$ states that x_2 has no predictive power over and above x_1.

b. The t-value for $H_0: \beta_2 = 0$ is $t = 0.57$; since this value is less than $t_{.05,3} = 2.353$ (and $t_{.10,3} = 1.638$) we have insufficient evidence to reject H_0. The variable x_2 does not appear to have any predictive power over and above x_1. In fact, since $R^2 = .097$ (i.e. 9.7%) neither x_1 nor x_2 seems to provide much predictive power for y.

c. The model $y = \beta_0 + \beta_1 x_1 + \beta_2 x_2 + \beta_3 x_1 x_2 + \varepsilon$ allows for an "interaction" between the independent variables x_1 and x_2; that is, it allows for the fact that the expected change in y for a unit change in x_1 *does* depend on the level of x_2. A scatterplot of y vs. x_1 with levels of x_2 marked is shown in Figure 12.4.

FIGURE 12.4

Scatterplot of y Versus x_1 with the Levels of x_2 (1, 2, or 3) Marked

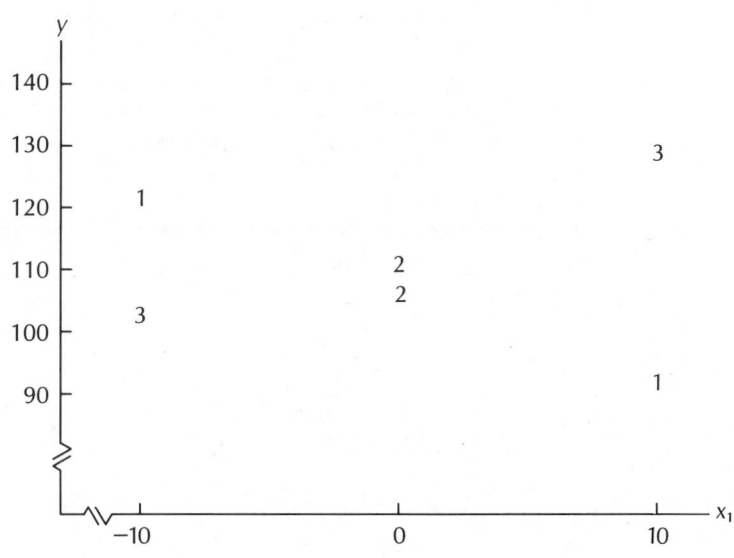

Figure 12.4 shows that the expected change in y for unit change in x_1 depends on the level of x_2 being considered. For example, when $x_2 = 1$, y decreases from 120 to 92 when x_1 changes from -10 to $+10$. However, when $x_2 = 3$, y increases from 101 to 130 when x_1 changes from -10 to $+10$. It is clear then that a model such as

$$y = \beta_0 + \beta_1 x_1 + \beta_2 x_2 + \beta_3 x_1 x_2 + \varepsilon$$

which allows for nonadditivity (interaction) may be more appropriate.

EXERCISES

12.12   In Exercise 9.3 we presented data on the biological recovery of organisms suspended in aerosol clouds formed in an airtight chamber and the times at which the recoveries were determined. These data are shown here for your convenience.

Cloud	Time, x (in Minutes)	Biological Recovery (%)
1	0	70.6
2	5	52.0
3	10	33.4
4	15	22.0
5	20	18.3
6	25	15.1
7	30	13.0
8	35	10.0
9	40	9.1
10	45	8.3
11	50	7.9
12	55	7.7
13	60	7.7

a. Because the assumption of equal variances at different settings of x (time) would probably not be satisfied, logarithms (to the base 10) of the biological recoveries were used. Give logs of the biological recoveries.

b. Plot the transformed biological recoveries versus time and suggest a possible model.

12.13   Refer to the output shown here for fitting the regression model $y = \beta_0 + \beta_1 x + \beta_2 x^2 + \varepsilon$ to the data of Exercise 12.12.

```
                 LISTING OF DATA

   OBS    TIME    RECOVERY    LOG_REC    TIME_2

    1       0       70.6      1.84880       0
    2       5       52.0      1.71600      25
    3      10       33.4      1.52375     100
    4      15       22.0      1.34242     225
    5      20       18.3      1.26245     400
    6      25       15.1      1.17898     625
    7      30       13.0      1.11394     900
    8      35       10.0      1.00000    1225
    9      40        9.1      0.95904    1600
   10      45        8.3      0.91908    2025
   11      50        7.9      0.89763    2500
   12      55        7.7      0.88649    3025
   13      60        7.7      0.88649    3600

   N=      13
```

```
                                  REGRESSION ANALYSIS
                            GENERAL LINEAR MODELS PROCEDURE

DEPENDENT VARIABLE: LOG_REC    LOG BASE 10 OF RECOVERY

SOURCE              DF      SUM OF SQUARES      MEAN SQUARE     F VALUE       PR > F      R-SQUARE          C.V.

MODEL               2         1.28344124        0.64172062     1217.91       0.0001      0.995911        1.9209

ERROR              10         0.00526904        0.00052690                 ROOT MSE               LOG_REC MEAN

CORRECTED TOTAL    12         1.28871028                                   0.02295439               1.19500589


SOURCE              DF       TYPE I SS       F VALUE     PR > F      DF      TYPE III SS     F VALUE     PR > F
TIME                1        1.15231437      2186.95     0.0001       1      0.40321970      765.26      0.0001
TIME_2              1        0.13112687       248.86     0.0001       1      0.13112687      248.86      0.0001


                               T FOR H0:      PR > |T|       STD ERROR OF
PARAMETER          ESTIMATE    PARAMETER=0                    ESTIMATE

INTERCEPT         1.85047412      112.17       0.0001        0.01649659
TIME             -0.03533741      -27.66       0.0001        0.00127741
TIME_2            0.00032372       15.78       0.0001        0.00002052
```

    a. Determine the prediction equation.

    b. Determine s_ε and the standard errors for $\hat{\beta}_1$ and $\hat{\beta}_2$.

12.14   Refer to Exercise 12.13. Conduct a test of $H_0: \beta_2 = 0$ and interpret your findings.

12.15   Refer to Exercise 12.12.

    a. Give an interpretation to β_0 for the model
$$y = \beta_0 + \beta_1 x + \varepsilon$$

    b. Would the model $y = \beta_1 x + \varepsilon$ be an alternate model? Why?

12.16   Refer to Example 12.7. Use the output here to test $H_0: \beta_3 = 0$ for the model $y = \beta_0 + \beta_1 x_1 + \beta_2 x_2 + \beta_3 x_1 x_2 + \varepsilon$. Give a p-value for this test and draw a conclusion. Does the test of $H_0: \beta_3 = 0$ reflect what was seen in the scatterplot of Example 12.7?

```
        LISTING OF DATA

 OBS    Y     X1    X2    X1X2

  1    120   -10    1    -10
  2    101   -10    3    -30
  3    110     0    2      0
  4    105     0    2      0
  5     92    10    1     10
  6    130    10    3     30

 N=     6
                                  REGRESSION ANALYSIS
                            GENERAL LINEAR MODELS PROCEDURE

DEPENDENT VARIABLE: Y

SOURCE              DF      SUM OF SQUARES      MEAN SQUARE     F VALUE       PR > F      R-SQUARE          C.V.

MODEL               3        902.75000000     300.91666667      22.64       0.0426      0.971395        3.3244

ERROR               2         26.58333333      13.29166667                 ROOT MSE                    Y MEAN

CORRECTED TOTAL     5        929.33333333                                  3.64577381              109.66666667


SOURCE              DF       TYPE I SS       F VALUE     PR > F      DF      TYPE III SS     F VALUE     PR > F
X1                  1        0.25000000        0.02      0.9035       1      638.45000000     48.03      0.0202
X2                  1       90.25000000        6.79      0.1211       1       90.25000000      6.79      0.1211
X1X2                1      812.25000000       61.11      0.0160       1      812.25000000     61.11      0.0160


                               T FOR H0:      PR > |T|       STD ERROR OF
PARAMETER          ESTIMATE    PARAMETER=0                    ESTIMATE

INTERCEPT        100.16666667     25.44       0.0015        3.93788578
X1                -2.82500000     -6.93       0.0202        0.40760990
X2                 4.75000000      2.61       0.1211        1.82288690
X1X2               1.42500000      7.82       0.0160        0.18228869
```

12.17 Consider the data shown here.

y	x_1	x_2
3	-2	-1
10	-1	0
12	0	0
12	1	0
13	2	1

a. Compute r_{yx_1}, r_{yx_2} and $r_{x_1x_2}$.

b. Does $R^2_{y \cdot x_1x_2} = r^2_{yx_1} + r^2_{yx_2}$? Interpret your finding.

12.18 An experiment was conducted to examine the effect of temperature (x_1) and humidity (x_2) on the yield of a production process. The sample data are shown here.

Yield, y	Temperature, x_1	Humidity, x_2
65	70	50
78	100	50
52	130	50
70	70	80
77	100	80
83	130	80

a. Use the computer output shown here to determine the least squares fit to the model

$$y = \beta_0 + \beta_1 x_1 + \beta_2 x_2 + \beta_3 x_1^2 + \beta_4 x_1 x_2 + \varepsilon$$

b. Identify and interpret R^2.

```
            LISTING OF DATA

    OBS    YIELD    TEMP    HUMID

     1      65       70      50
     2      78      100      50
     3      52      130      50
     4      70       70      80
     5      77      100      80
     6      83      130      80

    N=      6
```

REGRESSION ANALYSIS

GENERAL LINEAR MODELS PROCEDURE

DEPENDENT VARIABLE: YIELD

SOURCE	DF	SUM OF SQUARES	MEAN SQUARE	F VALUE	PR > F	R-SQUARE	C.V.
MODEL	4	506.50000000	126.62500000	1.05	0.6152	0.808030	15.4866
ERROR	1	120.33333333	120.33333333		ROOT MSE		YIELD MEAN
CORRECTED TOTAL	5	626.83333333			10.96965511		70.83333333

SOURCE	DF	TYPE I SS	F VALUE	PR > F	DF	TYPE III SS	F VALUE	PR > F
TEMP	1	0.00000000	0.00	1.0000	1	38.72351234	0.32	0.6715
HUMID	1	204.16666667	1.70	0.4168	1	85.14150943	0.71	0.5548
TEMP*TEMP	1	133.33333333	1.11	0.4837	1	133.33333333	1.11	0.4837
TEMP*HUMID	1	169.00000000	1.40	0.4462	1	169.00000000	1.40	0.4462

PARAMETER	ESTIMATE	T FOR H0: PARAMETER=0	PR > |T|	STD ERROR OF ESTIMATE
INTERCEPT	35.00000000	0.27	0.8323	129.81759917
TEMP	1.28333333	0.57	0.6715	2.26227334
HUMID	-1.05555556	-0.84	0.5548	1.25488346
TEMP*TEMP	-0.01111111	-1.05	0.4837	0.01055556
TEMP*HUMID	0.01444444	1.19	0.4462	0.01218851

12.19   In Exercise 8.11 we discussed how well true–false tests and problem-type tests predict a student's proficiency in mathematics. Use the sample data (shown here) to see how well results on a true–false test predict results on a problem-type test.

Student	1	2	3	4	5	6	7	8	9	10
T–F	48	40	25	10	16	21	23	19	35	32
Problems	45	47	20	12	12	15	25	16	30	32

a. Fit the linear regression model

$$y = \beta_0 + \beta_1 x + \varepsilon$$

where y is the problem test score and x is the corresponding true–false test score for a student.

b. Test for the significance of β_1; give the p-value for your test and draw a conclusion. What does this say about p?

c. Compute R and R^2; interpret your findings. What is the value of r?

12.5   INFERENCES CONCERNING $E(y)$ AND y

The methods of previous sections can be expanded to include inferences concerning the average value of y for a given setting of the independent variables. For example, suppose that in a regression study we obtain the prediction equation $\hat{y} = 22.6 + 1.2x$. What does it mean to predict y when $x = 10$? It could mean predicting the average (expected) y-value for all cases having $x = 10$; we would predict $\hat{y} = 22.6 + 1.2(10) = 34.6$. Or it could mean predicting the y-value for one particular case having $x = 10$; again we would use $\hat{y} = 22.6 + 1.2(10) = 34.6$. The difference in the two solutions is in the standard errors. Because the standard errors are more complicated to write down, we will make use of software packages to develop inferences about $E(y)$ and y. Those readers interested in the actual formulas are referred to Section 9.5 for the single independent variable case and to Section 12.7 for a discussion of the matrix formulas for the k independent variable case.

We will illustrate the output for data analyzed by hand in Chapter 9 so that the output can be compared to previous work when necessary. However, keep in mind that the software packages are completely general and can be used for any number of independent variables.

EXAMPLE 12.8   In Example 9.2, we recorded the corn yield (in bushels) for 10 plots with varying degrees of fertilization. These data are repeated here for your convenience.

Yield (Bushels)	Fertilizer (lbs per Plot)
12	2
13	2
13	3
14	3
15	4
15	4
14	5
16	5
17	6
18	6

a. Use the output shown here to obtain a fit to the linear regression equation $y = \beta_0 + \beta_1 x + \varepsilon$.

LISTING OF DATA

OBS	YIELD	FERTILIZ
1	12	2
2	13	2
3	13	3
4	14	3
5	15	4
6	15	4
7	14	5
8	16	5
9	17	6
10	18	6

N = 10

LINEAR REGRESSION ANALYSIS

GENERAL LINEAR MODELS PROCEDURE

DEPENDENT VARIABLE: YIELD   YIELD (BUSHELS)

SOURCE	DF	SUM OF SQUARES	MEAN SQUARE	F VALUE	PR > F	R-SQUARE	C.V.
MODEL	1	26.45000000	26.45000000	37.45	0.0003	0.823988	5.7169
ERROR	8	5.65000000	0.70625000		STD DEV		YIELD MEAN
CORRECTED TOTAL	9	32.10000000			0.84038682		14.70000000

SOURCE	DF	TYPE I SS	F VALUE	PR > F	DF	TYPE IV SS	F VALUE	PR > F
FERTILIZ	1	26.45000000	37.45	0.0003	1	26.45000000	37.45	0.0003

PARAMETER	ESTIMATE	T FOR H0: PARAMETER = 0	PR > |T|	STD ERROR OF ESTIMATE
INTERCEPT	10.10000000	12.67	0.0001	0.79726094
FERTILIZ	1.15000000	6.12	0.0003	0.18791620

OBSERVATION	OBSERVED VALUE	PREDICTED VALUE	RESIDUAL	LOWER 90% CL FOR MEAN	UPPER 90% CL FOR MEAN
1	12.00000000	12.40000000	−0.40000000	11.54404094	13.25595906
2	13.00000000	12.40000000	0.60000000	11.54404094	13.25595906
3	13.00000000	13.55000000	−0.55000000	12.94474555	14.15525445
4	14.00000000	13.55000000	0.45000000	12.94474555	14.15525445
5	15.00000000	14.70000000	0.30000000	14.20581181	15.19418819
6	15.00000000	14.70000000	0.30000000	14.20581181	15.19418819
7	14.00000000	15.85000000	−1.85000000	15.24474555	16.45525445
8	16.00000000	15.85000000	0.15000000	15.24474555	16.45525445
9	17.00000000	17.00000000	−0.00000000	16.14404094	17.85595906
10	18.00000000	17.00000000	1.00000000	16.14404094	17.85595906

SUM OF RESIDUALS	0.00000000
SUM OF SQUARED RESIDUALS	5.65000000
SUM OF SQUARED RESIDUALS—ERROR SS	0.00000000
PRESS STATISTIC	8.79139107
FIRST ORDER AUTOCORRELATION	−0.25221239
DURBIN–WATSON D	2.29911504

b. Find a 90% confidence interval for $E(y)$ when $x = 5$.

c. Construct 90% confidence bands for $E(y)$ when $2 \le x \le 6$.

Solution

a. The output gives $\hat{y} = 10.1 + 1.15x$.

b. The 90% confidence interval for $E(y)$ when $x = 5$ is 15.245 to 16.455. Our hand calculations of Example 9.10 agree with this result except for rounding errors.

c. The 90% confidence bands for $E(y)$ can be constructed using the 90% confidence intervals shown in the output for $x = 2$, 3, 4, 5, and 6. These results agree with our hand calculations in Example 9.11. First plot the endpoints of the confidence intervals for $x = 2$, 3, 4, 5, and 6. Then if we connect all the lower endpoints and all the upper endpoints, we obtain the 90% confidence bands for $E(y)$. (See Figure 12.5.)

FIGURE 12.5

90% Confidence Bands for $E(y)$, Example 12.8

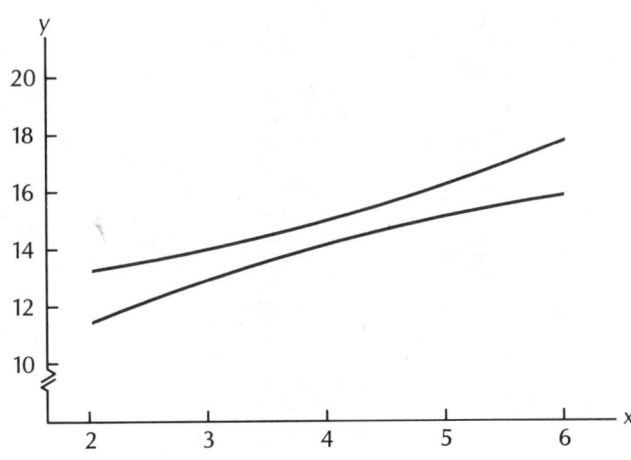

There is another problem closely related to that of estimating $E(y)$ for a given setting of the independent variables. Suppose that after obtaining least squares estimates for the βs of a multiple regression equation, we want to *predict* the actual value of y (say the next observation) for a given setting of the independent variables. Note that this problem differs from the problem discussed previously in that we do not want to estimate the average value of y for a given value of x, but rather we wish to predict what a particular observation will be for that same setting of x.

prediction interval   We still use the least squares equation y as our predictor, but the corresponding interval about the observation y is called a **prediction interval**. (Prediction intervals are constructed about variables, whereas confidence intervals are constructed about parameters.) Again we will use computer output to illustrate the solution to such problems since the algebraic formulas can be quite involved without the use of matrices.

EXAMPLE 12.9 ▌ Refer to the data for Example 12.8 and the computer output shown here.

LINEAR REGRESSION ANALYSIS

GENERAL LINEAR MODELS PROCEDURE

DEPENDENT VARIABLE: YIELD   YIELD (BUSHELS)

SOURCE	DF	SUM OF SQUARES	MEAN SQUARE	F VALUE	PR > F	R-SQUARE	C.V.
MODEL	1	26.45000000	26.45000000	37.45	0.0003	0.823988	5.7169
ERROR	8	5.65000000	0.70625000		STD DEV		YIELD MEAN
CORRECTED TOTAL	9	32.10000000			0.84038682		14.70000000

SOURCE	DF	TYPE I SS	F VALUE	PR > F	DF	TYPE IV SS	F VALUE	PR > F
FERTILIZ	1	26.45000000	37.45	0.0003	1	26.45000000	37.45	0.0003

PARAMETER	ESTIMATE	T FOR H0: PARAMETER = 0	PR > |T|	STD ERROR OF ESTIMATE
INTERCEPT	10.10000000	12.67	0.0001	0.79726094
FERTILIZ	1.15000000	6.12	0.0003	0.18791620

OBSERVATION	OBSERVED VALUE	PREDICTED VALUE	RESIDUAL	LOWER 90% CL FOR MEAN	UPPER 90% CL FOR MEAN
1	12.00000000	12.40000000	−0.40000000	10.61817913	14.18182087
2	13.00000000	12.40000000	0.60000000	10.61817913	14.18182087
3	13.00000000	13.55000000	−0.55000000	11.87412630	15.22587370
4	14.00000000	13.55000000	0.45000000	11.87412630	15.22587370
5	15.00000000	14.70000000	0.30000000	13.06096319	16.33903681
6	15.00000000	14.70000000	0.30000000	13.06096319	16.33903681
7	14.00000000	15.85000000	−1.85000000	14.17412630	17.52587370
8	16.00000000	15.85000000	0.15000000	14.17412630	17.52587370
9	17.00000000	17.00000000	−0.00000000	15.21817913	18.78182087
10	18.00000000	17.00000000	1.00000000	15.21817913	18.78182087

SUM OF RESIDUALS	0.00000000
SUM OF SQUARED RESIDUALS	5.65000000
SUM OF SQUARED RESIDUALS—ERROR SS	0.00000000
PRESS STATISTIC	8.79139107
FIRST ORDER AUTOCORRELATION	−0.25221239
DURBIN–WATSON D	2.29911504

FIGURE 12.6

90% Prediction and Confidence Bands for y and $E(y)$, Example 12.9

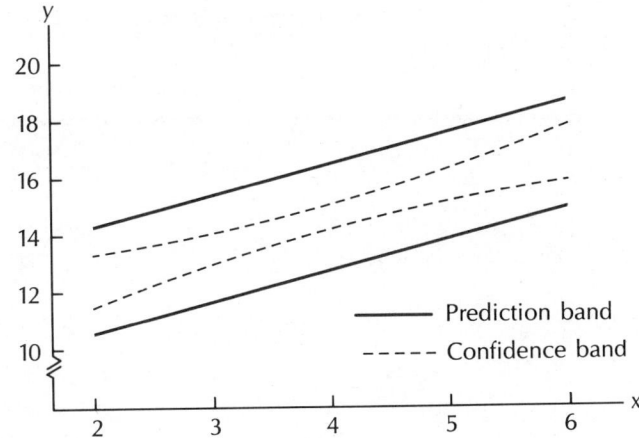

a. Place a 90% prediction interval about the value of y when $x = 5$.
b. Develop 90% prediction bands for y when $2 \leq x \leq 6$.
c. Graph both the 90% confidence bands of Example 12.8 and the 90% prediction bands of part (b). Comment on your findings.

Solution

a. From the output shown here, the 90% prediction interval corresponding to $x = 5$ is 14.174 to 17.526. This result agrees with our hand calculations in Example 9.14.

b. The additional intervals necessary to construct a 90% prediction band for y when $2 \leq x \leq 6$ can be extracted from the output. These intervals agree with those obtained by hand in Example 9.14.

c. Plotting the endpoints of the intervals shown for $x = 2, 3, 4, 5$, and 6 and then connecting all the lower endpoints and all the upper endpoints, we obtain the 90% prediction bands shown in Figure 12.6.

As we have seen previously, the prediction bands are wider than the corresponding confidence bands to allow for the fact that we are predicting the value of a random variable rather than estimating a parameter.

EXERCISES

12.20   Refer to Exercise 12.12. For the regression equation

$$\hat{y} = \hat{\beta}_0 + \hat{\beta}_1 x + \hat{\beta}_2 x^2$$

estimate the mean log biological recovery percentage at 30 minutes using a 95% confidence interval.

12.21   Refer to Example 12.9. In general, how would the prediction bands change if you used a higher degree of certainty (say 99%)? What if you use a lower degree of certainty (say 80%)?

12.22   A portion of a computer output for the weight loss data of Example 12.4 is shown here. Locate 95% confidence bands for $E(y)$.

```
        LISTING OF DATA

  OBS    WT_LOSS    TIME    HUMIDITY

   1       4.3        4        0.2
   2       5.5        5        0.2
   3       6.8        6        0.2
   4       8.0        7        0.2
   5       4.0        4        0.3
   6       5.2        5        0.3
   7       6.6        6        0.3
   8       7.5        7        0.3
   9       2.0        4        0.4
  10       4.0        5        0.4
  11       5.7        6        0.4
  12       6.5        7        0.4

  N=      12
```

```
                          REGRESSION ANALYSIS

                    GENERAL LINEAR MODELS PROCEDURE

DEPENDENT VARIABLE: WT_LOSS
```

SOURCE	DF	SUM OF SQUARES	MEAN SQUARE	F VALUE	PR > F	R-SQUARE	C.V.
MODEL	2	31.12416667	15.56208333	104.13	0.0001	0.958576	7.0181
ERROR	9	1.34500000	0.14944444		ROOT MSE		WT_LOSS MEAN
CORRECTED TOTAL	11	32.46916667			0.38658045		5.50833333

SOURCE	DF	TYPE I SS	F VALUE	PR > F	DF	TYPE III SS	F VALUE	PR > F
TIME	1	26.00416667	174.01	0.0001	1	26.00416667	174.01	0.0001
HUMIDITY	1	5.12000000	34.26	0.0002	1	5.12000000	34.26	0.0002

PARAMETER	ESTIMATE	T FOR HO: PARAMETER=0	PR > |T|	STD ERROR OF ESTIMATE
INTERCEPT	0.66666667	0.96	0.3620	0.69423219
TIME	1.31666667	13.19	0.0001	0.09981464
HUMIDITY	-8.00000000	-5.85	0.0002	1.36676829

OBSERVATION	OBSERVED VALUE	PREDICTED VALUE	RESIDUAL	LOWER 95% CL FOR MEAN	UPPER 95% CL FOR MEAN
1	4.30000000	4.33333333	-0.03333333	3.80984144	4.85682523
2	5.50000000	5.65000000	-0.15000000	5.23518198	6.06481802
3	6.80000000	6.96666667	-0.16666667	6.55184865	7.38148469
4	8.00000000	8.28333333	-0.28333333	7.75984144	8.80682523
5	4.00000000	3.53333333	0.46666667	3.11090334	3.95576333
6	5.20000000	4.85000000	0.35000000	4.57345465	5.12654535
7	6.60000000	6.16666667	0.43333333	5.89012132	6.44321201
8	7.50000000	7.48333333	0.01666667	7.06090334	7.90576333
9	2.00000000	2.73333333	-0.73333333	2.20984144	3.25682523
10	4.00000000	4.05000000	-0.05000000	3.63518198	4.46481802
11	5.70000000	5.36666667	0.33333333	4.95184865	5.78148469
12	6.50000000	6.68333333	-0.18333333	6.15984144	7.20682523

```
     SUM OF RESIDUALS                           0.00000000
     SUM OF SQUARED RESIDUALS                   1.34500000
     SUM OF SQUARED RESIDUALS - ERROR SS       -0.00000000
     PRESS STATISTIC                            2.61233345
     FIRST ORDER AUTOCORRELATION                0.16902520
     DURBIN-WATSON D                            1.65613383
```

12.23   Refer to Exercise 9.7. Construct a 95% prediction interval for the weight gain of a chick chosen at random and observed to ingest .19 grams of lysine. Compare your results to the confidence interval of Exercise 9.20.

12.24   Using the data of Exercise 12.20, construct a 95% prediction interval for the log biological recovery percentage at 30 minutes. Compare your result to the confidence interval on $E(y)$ of Exercise 12.20.

INFERENCES CONCERNING A SET OF βs IN A GENERAL LINEAR MODEL

The general linear model that we have discussed has the form

$$y = \beta_0 + \beta_1 x_1 + \cdots + \beta_k x_k + \varepsilon$$

where the xs associated with qualitative independent variables are dummy variables and the xs associated with quantitative independent variables have a quantitative interpretation (such as initial weight, amount of water added, and stirring time). The ability to write both qualitative and quantitative independent variables in the framework of a general linear model will enable us to develop one procedure for simultaneously testing the significance of one or more βs of the model.

Consider a model for relating the sales volume y to advertising expenditure x_1, with the form

$$y = \beta_0 + \beta_1 x_1 + \beta_2 x_1^2 + \varepsilon$$

In Section 12.4 we considered the problem of testing the significance of a single β in the general linear model. Thus a test of $H_0: \beta_2 = 0$ would be a test of the null hypothesis that y is linearly related to x_1 (sales expenditure). However, an experimenter might wish to test

$$H_0: \beta_1 = \beta_2 = 0$$

which hypothesizes that y is *not related* in a linear or quadratic way to the independent variable x_1.

The procedure for simultaneously testing that a set of βs is equal to zero in the general linear model will have the following null and research hypotheses:

$$H_0: \beta_{g+1} = \beta_{g+2} = \cdots = \beta_k = 0 \qquad (k > g)$$
$$H_a: \text{At least one of the } \beta\text{s is nonzero.}$$

To formulate a test statistic, we must specify two models, model 1—often referred to as the **complete model**—and model 2—referred to as the **reduced model**

complete model,
reduced model

Model 1: $y = \beta_0 + \beta_1 x_1 + \beta_2 x_2 + \cdots + \beta_g x_g + \beta_{g+1} x_{g+1} \cdots + \beta_k x_k + \varepsilon \qquad (k > g)$
Model 2: $y = \beta_0 + \beta_1 x_1 + \beta_2 x_2 + \cdots + \beta_g x_g + \varepsilon$

You will note that model 1 represents the general linear model and model 2 is a general linear model under the assumption that H_0 is true. Thus model 1 contains k independent variables, model 2 contains g independent variables, and we are testing that a set of $(k - g)\beta$s is equal to zero. The xs in the two models represent either quantitative independent variables or dummy variables associated with qualitative independent variables.

Using the method of least squares, we fit both models separately, and for each model we calculate the sum of squares for error. Letting SSE_1 and SSE_2 denote the sum of squares for error for models 1 and 2, respectively, we can examine the difference $\mathbf{SSE_2 - SSE_1}$. For n sample observations, the degrees of freedom for a sum of squares for error can be computed as $n -$ (the number of parameters in the model). Hence SSE_1 and SSE_2 have,

$SSE_2 - SSE_1$

df

respectively, $n - (k + 1)$ and $n - (g + 1)$ degrees of freedom. When H_0 is true, SSE_2 and SSE_1 will be of approximately the same magnitude, although SSE_1 will be less than SSE_2 owing to the fact that the complete model has more terms in it. When H_a is true and at least one of the βs under test is different from zero, SSE_2 will be much larger than SSE_1. Since $SSE_2 - SSE_1$ will be greater than zero under either H_0 or H_a, we call the difference **SS$_{drop}$** $SSE_2 - SSE_1$ the **drop** *in the sum of squares for error attributable to the variables* x_{g+1}, $x_{g+2}, \ldots, x_k$. If the sum of squares drop is large, this implies that the sum of squares for error has been greatly reduced by including the variables x_{g+1}, $x_{g+2}, \ldots, x_k$ in the model, and intuitively we would reject H_0.

How large must $SSE_2 - SSE_1$ be in order to reject the hypothesis that x_{g+1}, $x_{g+2}, \ldots, x_k$ are unrelated to the response y? When H_0 is true, the quantity $SS_{drop} = SSE_2 - SSE_1$ divided by the number of parameters under test in H_0 $(k - g)$ provides an unbiased estimate of σ_ε^2, the variance associated with an observation in the general linear model. We designate this estimate as the mean square drop:

MS$_{drop}$
$$MS_{drop} = \frac{SSE_2 - SSE_1}{k - g}$$

The mean square error for model 1,

$$MSE_1 = \frac{SSE_1}{n - (k + 1)}$$

also provides an unbiased estimate of σ_ε^2. It can be shown that when H_0 is true, the ratio

F statistic
$$\frac{MS_{drop}}{MSE_1}$$

follows an F distribution, with $k - g$ and $n - (k + 1)$ degrees of freedom, respectively.

When H_a is true, MSE_1 is still an unbiased estimate of σ_ε^2, whereas MS_{drop} is an unbiased estimate of $\sigma_\varepsilon^2 +$ (a positive function of β_{g+1}, $\beta_{g+2}, \ldots, \beta_k$). Thus large values of $F = MS_{drop}/MSE_1$ will indicate rejection of the null hypothesis.

EXAMPLE 12.10

In Example 12.4 we considered a situation in which a chemist was interested in determining the weight loss y of a compound as a function of the length of time the compound was exposed to the air and the relative humidity of the environment during exposure. An experiment was performed using three relative humidities and four exposure times, with weight loss (in lbs) recorded for each factor–level combination. We obtained a least squares fit to a first-order model, relating the response to the two variables.

Now let us consider a model of the form

$$y = \beta_0 + \beta_1 x_1 + \beta_2 x_2 + \beta_3 x_2^2 + \beta_4 x_1 x_2 + \beta_5 x_1 x_2^2 + \varepsilon$$

where $x_1 =$ exposure time and $x_2 =$ relative humidity. The computer output shows the least squares fit for the complete and reduced model corresponding to H_0: $\beta_4 = \beta_5 = 0$. Give all parts of the F test of H_0: $\beta_4 = \beta_5 = 0$ and interpret the results of your test.

```
        LISTING OF DATA

  OBS    WT_LOSS    TIME    HUMIDITY

   1       4.3        4       0.2
   2       5.5        5       0.2
   3       6.8        6       0.2
   4       8.0        7       0.2
   5       4.0        4       0.3
   6       5.2        5       0.3
   7       6.6        6       0.3
   8       7.5        7       0.3
   9       2.0        4       0.4
  10       4.0        5       0.4
  11       5.7        6       0.4
  12       6.5        7       0.4

  N=      12
```

COMPLETE MODEL

GENERAL LINEAR MODELS PROCEDURE

DEPENDENT VARIABLE: WT_LOSS

SOURCE	DF	SUM OF SQUARES	MEAN SQUARE	F VALUE	PR > F	R-SQUARE	C.V.
MODEL	5	32.04216667	6.40843333	90.05	0.0001	0.986849	4.8430
ERROR	6	0.42700000	0.07116667		ROOT MSE		WT_LOSS MEAN
CORRECTED TOTAL	11	32.46916667			0.26677081		5.50833333

SOURCE	DF	TYPE I SS	F VALUE	PR > F	DF	TYPE III SS	F VALUE	PR > F
TIME	1	26.00416667	365.40	0.0001	1	0.28212844	3.96	0.0936
HUMIDITY	1	5.12000000	71.94	0.0001	1	0.16601864	2.33	0.1775
HUMIDITY*HUMIDITY	1	0.60166667	8.45	0.0271	1	0.24448677	3.44	0.1132
TIME*HUMIDITY	1	0.19600000	2.75	0.1481	1	0.09174312	1.29	0.2995
TIME*HUMIDI*HUMIDI	1	0.12033333	1.69	0.2412	1	0.12033333	1.69	0.2412

PARAMETER	ESTIMATE	T FOR H0: PARAMETER=0	PR > |T|	STD ERROR OF ESTIMATE
INTERCEPT	-9.69000000	-1.39	0.2150	6.99071885
TIME	2.48000000	1.99	0.0936	1.24556547
HUMIDITY	75.50000000	1.53	0.1775	49.43184702
HUMIDITY*HUMIDITY	-152.00000000	-1.85	0.1132	82.00762160
TIME*HUMIDITY	-10.00000000	-1.14	0.2995	8.80747788
TIME*HUMIDI*HUMIDI	19.00000000	1.30	0.2412	14.61163920

REDUCED MODEL

GENERAL LINEAR MODELS PROCEDURE

DEPENDENT VARIABLE: WT_LOSS

SOURCE	DF	SUM OF SQUARES	MEAN SQUARE	F VALUE	PR > F	R-SQUARE	C.V.
MODEL	3	31.72583333	10.57527778	113.81	0.0001	0.977106	5.5338
ERROR	8	0.74333333	0.09291667		ROOT MSE		WT_LOSS MEAN
CORRECTED TOTAL	11	32.46916667			0.30482235		5.50833333

SOURCE	DF	TYPE I SS	F VALUE	PR > F	DF	TYPE III SS	F VALUE	PR > F
TIME	1	26.00416667	279.87	0.0001	1	26.00416667	279.87	0.0001
HUMIDITY	1	5.12000000	55.10	0.0001	1	0.30844037	3.32	0.1059
HUMIDITY*HUMIDITY	1	0.60166667	6.48	0.0345	1	0.60166667	6.48	0.0345

PARAMETER	ESTIMATE	T FOR H0: PARAMETER=0	PR > |T|	STD ERROR OF ESTIMATE
INTERCEPT	-3.29166667	-2.00	0.0810	1.64904855
TIME	1.31666667	16.73	0.0001	0.07870479
HUMIDITY	20.50000000	1.82	0.1059	11.25162025
HUMIDITY*HUMIDITY	-47.50000000	-2.54	0.0345	18.66648065

Solution   The null hypothesis $H_0: \beta_4 = \beta_5 = 0$ states there is no interaction (linear or quadratic) between the independent variables exposure time x_1 and relative humidity x_2. The complete and reduced models for this null hypothesis are

$$\text{complete model: } y = \beta_0 + \beta_1 x_1 + \beta_2 x_2 + \beta_3 x_2^2 + \beta_4 x_1 x_2 + \beta_5 x_1 x_2^2 + \varepsilon$$
$$\text{reduced model: } y = \beta_0 + \beta_1 x_1 + \beta_2 x_2 + \beta_3 x_2^2 + \varepsilon$$

From the computer output we have

$$SSE_1 = .427, \qquad df_1 = 6$$

and from the reduced model

$$SSE_2 = .743, \qquad df_2 = 8$$

It follows that

$$SS_{drop} = SSE_2 - SSE_1 = .743 - .427 = .316$$

$$MS_{drop} = \frac{SSE_2 - SSE_1}{2} = .158$$

and

$$MSE_1 = \frac{SSE_1}{6} = .071$$

The parts of the test are summarized here.

$H_0: \beta_4 = \beta_5 = 0$

$H_a:$ At least one different from zero

T.S.: $F = \dfrac{MS_{drop}}{MSE_1} = \dfrac{.158}{.071} = 2.23$

R.R.: For $\alpha = .05$ the critical F value based on 2 and 6 degrees of freedom is 5.14. Since the observed value of F does not exceed the critical value, we have insufficient evidence to reject H_0. Practically, we are unable to detect any linear or quadratic interaction between time of exposure and relative humidity.

EXAMPLE 12.11   Consider an experiment to investigate the effects of four different pesticides on the yield of fruit from three different varieties of citrus trees. Write a model that contains main effects for both qualitative variables but no interaction. Use the data, reproduced in Table 12.4, to test for no difference among mean yields for pesticides by fitting complete and reduced models. Use $\alpha = .05$.

a.   Write a model that contains dummy variables for pesticides and for varieties.
b.   Interpret the βs.
c.   Fit complete and reduced models to test the hypothesis of no differences among mean yields for pesticides. Use $\alpha = .05$.

TABLE 12.4   Data for the Fruit Yield Experiment of Example 12.11

	Presticide			
Variety	1	2	3	4
1	29	50	43	53
2	41	58	42	73
3	66	85	69	85

Solution

We could use the model

$$y = \beta_0 + \overbrace{\beta_1 x_1 + \beta_2 x_2}^{\text{varieties}} + \overbrace{\beta_3 x_3 + \beta_4 x_4 + \beta_5 x_5}^{\text{pesticides}} + \varepsilon$$

where $x_1, x_2, \ldots, x_5$ are dummy variables defined in the following way:

a.

$x_1 = 1$ if variety 2, $x_1 = 0$ otherwise

$x_2 = 1$ if variety 3, $x_2 = 0$ otherwise

$x_3 = 1$ if pesticide 2, $x_3 = 0$ otherwise

$x_4 = 1$ if pesticide 3, $x_4 = 0$ otherwise

and

$x_5 = 1$ if pesticide 4, $x_5 = 0$ otherwise

b. We can form a table of expected values by substituting appropriate values for the dummy variables. For example an observation on variety 1, pesticide 1 has $x_1 = x_2 = \cdots = x_5 = 0$; hence $E(y) = \beta_0$. Similarly, an observation on variety 2, pesticide 3 has $x_1 = 1$, $x_2 = 0$, $x_3 = 0$, $x_4 = 1$, and $x_5 = 0$; so $E(y) = \beta_0 + \beta_1 + \beta_4$. The remaining entries for the expected value table shown here were computed in the same way.

	Pesticide			
Variety	1	2	3	4
1	β_0	$\beta_0 + \beta_3$	$\beta_0 + \beta_4$	$\beta_0 + \beta_5$
2	$\beta_0 + \beta_1$	$\beta_0 + \beta_1 + \beta_3$	$\beta_0 + \beta_1 + \beta_4$	$\beta_0 + \beta_1 + \beta_5$
3	$\beta_0 + \beta_2$	$\beta_0 + \beta_2 + \beta_3$	$\beta_0 + \beta_2 + \beta_4$	$\beta_0 + \beta_2 + \beta_5$

It is clear from this table that

β_0 is the expected value (mean yield) for an observation on variety 1, pesticide 1.

β_1 is the difference in expected values (mean yields) for varieties 2 and 1 (for a given pesticide).

β_2 is the difference in expected values for varieties 3 and 1 (for a given pesticide).

β_3 is the difference in expected values for pesticides 2 and 1 (for a given variety).

$\vdots$

β_5 is the difference in expected values for pesticides 4 and 1 (for a given variety).

c. The null hypothesis for testing no difference among mean yields for pesticides is

H_0: $\beta_3 = \beta_4 = \beta_5 = 0$

For this test, the complete and reduced models are

complete model: $y = \beta_0 + \beta_1 x_1 + \beta_2 x_2 + \cdots + \beta_5 x_5 + \varepsilon$

reduced model: $y = \beta_0 + \beta_1 x_1 + \beta_2 x_2 + \varepsilon$

From the computer output shown here,

$$SSE_1 = 151.50, \quad df_1 = 6, \quad MSE_1 = 25.25$$
$$SSE_2 = 1342.50, \quad df_2 = 9, \quad MSE_2 = 149.17$$

Hence $SS_{drop} = 1342.50 - 151.50 = 1191.00$ and $MS_{drop} = SS_{drop}/3 = 397.00$. The entire test is shown here.

```
                    LISTING OF DATA

  OBS    VARIETY    PEST_CDE    YIELD    X1    X2    X3    X4    X5

   1        1          1         29      0     0     0     0     0
   2        1          2         50      0     0     1     0     0
   3        1          3         43      0     0     0     1     0
   4        1          4         53      0     0     0     0     1
   5        2          1         41      1     0     0     0     0
   6        2          2         58      1     0     1     0     0
   7        2          3         42      1     0     0     1     0
   8        2          4         73      1     0     0     0     1
   9        3          1         66      0     1     0     0     0
  10        3          2         85      0     1     1     0     0
  11        3          3         69      0     1     0     1     0
  12        3          4         85      0     1     0     0     1

  N=       12
```

COMPLETE MODEL

GENERAL LINEAR MODELS PROCEDURE

DEPENDENT VARIABLE: YIELD

SOURCE	DF	SUM OF SQUARES	MEAN SQUARE	F VALUE	PR > F	R-SQUARE	C.V.
MODEL	5	3416.16666667	683.23333333	27.06	0.0005	0.957535	8.6887
ERROR	6	151.50000000	25.25000000		ROOT MSE		YIELD MEAN
CORRECTED TOTAL	11	3567.66666667			5.02493781		57.83333333

SOURCE	DF	TYPE I SS	F VALUE	PR > F	DF	TYPE III SS	F VALUE	PR > F
X1	1	112.66666667	4.46	0.0791	1	190.12500000	7.53	0.0336
X2	1	2112.50000000	83.66	0.0001	1	2112.50000000	83.66	0.0001
X3	1	169.00000000	6.69	0.0414	1	541.50000000	21.45	0.0036
X4	1	84.50000000	3.35	0.1171	1	54.00000000	2.14	0.1940
X5	1	937.50000000	37.13	0.0009	1	937.50000000	37.13	0.0009

PARAMETER	ESTIMATE	T FOR H0: PARAMETER=0	PR > |T|	STD ERROR OF ESTIMATE
INTERCEPT	31.25000000	8.79	0.0001	3.55316760
X1	9.75000000	2.74	0.0336	3.55316760
X2	32.50000000	9.15	0.0001	3.55316760
X3	19.00000000	4.63	0.0036	4.10284454
X4	6.00000000	1.46	0.1940	4.10284454
X5	25.00000000	6.09	0.0009	4.10284454

REDUCED MODEL

GENERAL LINEAR MODELS PROCEDURE

DEPENDENT VARIABLE: YIELD

SOURCE	DF	SUM OF SQUARES	MEAN SQUARE	F VALUE	PR > F	R-SQUARE	C.V.
MODEL	2	2225.16666667	1112.58333333	7.46	0.0123	0.623704	21.1182
ERROR	9	1342.50000000	149.16666667		ROOT MSE		YIELD MEAN
CORRECTED TOTAL	11	3567.66666667			12.21338064		57.83333333

SOURCE	DF	TYPE I SS	F VALUE	PR > F	DF	TYPE III SS	F VALUE	PR > F
X1	1	112.66666667	0.76	0.4074	1	190.12500000	1.27	0.2881
X2	1	2112.50000000	14.16	0.0045	1	2112.50000000	14.16	0.0045

PARAMETER	ESTIMATE	T FOR H0: PARAMETER=0	PR > |T|	STD ERROR OF ESTIMATE
INTERCEPT	43.75000000	7.16	0.0001	6.10669032
X1	9.75000000	1.13	0.2881	8.63616427
X2	32.50000000	3.76	0.0045	8.63616427

H_0: $\beta_3 = \beta_4 = \beta_5 = 0$

H_a: At least 1 of the βs is different from zero.

T.S.: $F = \dfrac{MS_{drop}}{MSE_1} = \dfrac{397.00}{25.25} = 15.72$

R.R.: For $\alpha = .05$, reject H_0 if $F > F_{.05}$ with 3 and 6 degrees of freedom. This value is 4.76. Since $F > 4.76$, we reject H_0 and conclude that the pesticide means are not all equal.

EXERCISES

12.25   Refer to Example 12.10. Under the assumption that there is no interaction between the variables "exposure time" and "relative humidity," start with the complete model

$$y = \beta_0 + \beta_1 x_1 + \beta_2 x_2 + \beta_3 x_2^2 + \varepsilon$$

Use an appropriate reduced model to test the hypothesis of no linear or quadratic effect due to x_2. Use $\alpha = .05$.

12.26   Refer to Exercise 12.25.
a. Fit a complete and reduced model to test for no linear effect due to x_1. Use $\alpha = .05$.
b. How else might you test this same hypothesis without fitting complete and reduced models?

12.7   MATRIX NOTATION FOR THE GENERAL LINEAR MODEL (optional)

Recall that a model relating a response y to a set of k independent variables of the form

$$y = \beta_0 + \beta_1 x_1 + \beta_2 x_2 + \cdots + \beta_k x_k + \varepsilon$$

has been called the general linear model. If a sample of n ($n > k$) measurements is obtained for n settings of the independent variables $x_1, x_2, \ldots, x_k$, we can write an individual observation as

$$y_i = \beta_0 + \beta_1 x_{i1} + \beta_2 x_{i2} + \cdots + \beta_k x_{ik} + \varepsilon_i \qquad (i = 1, 2, \ldots, n)$$

where $x_{i1}, x_{i2}, \ldots, x_{ik}$ are the settings of the independent variables for the response y_i and ε_i is the random error for the ith response.

The entire set of n observations can be expressed in the general linear model using matrix notation. Let the $n \times 1$ matrix $\mathbf{Y}$

$$\mathbf{Y} = \begin{bmatrix} y_1 \\ y_2 \\ \vdots \\ y_n \end{bmatrix}$$

be the matrix of observations, and let the $n \times (k + 1)$ matrix $\mathbf{X}$

$$\mathbf{X} = \begin{bmatrix} 1 & x_{11} & x_{12} & \cdots & x_{1k} \\ 1 & x_{21} & x_{22} & \cdots & x_{2k} \\ \vdots & \vdots & \vdots & & \vdots \\ 1 & x_{n1} & x_{n2} & \cdots & x_{nk} \end{bmatrix}$$

be a matrix of settings for the independent variables augmented with a column of 1s. The first row of $\mathbf{X}$ contains a 1 and the settings on the k independent variables for the first observation. Row 2 contains a 1 and corresponding settings on the independent variables for y_2. Similarly, the other rows contain settings for the remaining observations. Let

$$\boldsymbol{\beta} = \begin{bmatrix} \beta_0 \\ \beta_1 \\ \beta_2 \\ \vdots \\ \beta_k \end{bmatrix}$$

be a $(k + 1) \times 1$ matrix containing the unknown parameters for the general linear model and let

$$\boldsymbol{\varepsilon} = \begin{bmatrix} \varepsilon_1 \\ \varepsilon_2 \\ \vdots \\ \varepsilon_n \end{bmatrix}$$

general linear model, matrix notation   be an $n \times 1$ matrix of errors associated with the n observations. Then we can write the **general linear model** in **matrix notation** as

$$\begin{bmatrix} y_1 \\ y_2 \\ \vdots \\ y_n \end{bmatrix} = \begin{bmatrix} 1 & x_{11} & x_{12} & \cdots & x_{1k} \\ 1 & x_{21} & x_{22} & \cdots & x_{2k} \\ \vdots & \vdots & \vdots & & \vdots \\ 1 & x_{n1} & x_{n2} & \cdots & x_{nk} \end{bmatrix} \begin{bmatrix} \beta_0 \\ \beta_1 \\ \vdots \\ \beta_k \end{bmatrix} + \begin{bmatrix} \varepsilon_1 \\ \varepsilon_2 \\ \vdots \\ \varepsilon_n \end{bmatrix}$$

or simply

$$\mathbf{Y} = \mathbf{X}\boldsymbol{\beta} + \boldsymbol{\varepsilon}$$

Note that to obtain the equation for y_1, we multiply row one of $\mathbf{X}$ times the matrix of βs and then add ε_1 to obtain

$$\begin{bmatrix} y_1 \\ \\ \\ \\ \\ \end{bmatrix} = \begin{bmatrix} 1 & x_{11} & x_{12} & \cdots & x_{1k} \\ \\ \\ \\ \\ \end{bmatrix} \begin{bmatrix} \beta_0 \\ \beta_1 \\ \vdots \\ \beta_k \end{bmatrix} + \begin{bmatrix} \varepsilon_1 \\ \\ \\ \\ \end{bmatrix}$$

or

$$y_1 = \beta_0 + \beta_1 x_{11} + \beta_2 x_{12} + \cdots + \beta_k x_{1k} + \varepsilon_1$$

which is precisely as it was defined previously. In fact, any observation y_i is obtained by multiplying the ith row of $\mathbf{X}$ times the matrix of βs and then adding ε_i.

If we let the matrix

$$\hat{\boldsymbol{\beta}} = \begin{bmatrix} \hat{\beta}_0 \\ \hat{\beta}_1 \\ \hat{\beta}_2 \\ \vdots \\ \hat{\beta}_k \end{bmatrix}$$

represent the matrix of least squares estimates for the parameters of the general linear model, then, provided the matrix $\mathbf{X'X}$ has an inverse, we can find these estimates using the matrix equation given here.

$$\hat{\boldsymbol{\beta}} = (\mathbf{X'X})^{-1}\mathbf{X'Y}$$

This matrix solution gives the set of parameter estimates $\hat{\beta}_0, \hat{\beta}_1, \ldots, \hat{\beta}_k$ in the general linear model that minimizes $\sum_i (y_i - \hat{y}_i)^2$ for the data collected.

EXAMPLE 12.12

Refer to the data of Example 8.1, where we related the dressed weight y of a steer to its live weight x using the linear regression model

$$y = \beta_0 + \beta_1 x + \varepsilon$$

Use matrices to find the least squares estimates of β_0 and β_1. Compare your results to those obtained in Example 8.1.

Solution

The model $y = \beta_0 + \beta_1 x + \varepsilon$ can be considered a general linear model with $k = 1$ independent variable. Using the $n = 9$ sample observations, we can specify the following matrices:

$$\mathbf{Y} = \begin{bmatrix} 2.8 \\ 2.5 \\ 3.1 \\ 2.1 \\ 2.9 \\ 2.8 \\ 2.6 \\ 2.4 \\ 2.5 \end{bmatrix} \qquad \mathbf{X} = \begin{bmatrix} 1 & 4.2 \\ 1 & 3.8 \\ 1 & 4.8 \\ 1 & 3.4 \\ 1 & 4.5 \\ 1 & 4.6 \\ 1 & 4.3 \\ 1 & 3.7 \\ 1 & 3.9 \end{bmatrix}$$

Note that the second column of **X** gives the settings (live weights) corresponding to the observed dressed weights (y).

The transpose of **X** is

$$\mathbf{X}' = \begin{bmatrix} 1 & 1 & 1 & 1 & 1 & 1 & 1 & 1 & 1 \\ 4.2 & 3.8 & 4.8 & 3.4 & 4.5 & 4.6 & 4.3 & 3.7 & 3.9 \end{bmatrix}$$

Thus

$$\mathbf{X}'\mathbf{X} = \begin{bmatrix} 1 & 1 & 1 & 1 & 1 & 1 & 1 & 1 & 1 \\ 4.2 & 3.8 & 4.8 & 3.4 & 4.5 & 4.6 & 4.3 & 3.7 & 3.9 \end{bmatrix} \begin{bmatrix} 1 & 4.2 \\ 1 & 3.8 \\ 1 & 4.8 \\ 1 & 3.4 \\ 1 & 4.5 \\ 1 & 4.6 \\ 1 & 4.3 \\ 1 & 3.7 \\ 1 & 3.9 \end{bmatrix}$$

$$= \begin{bmatrix} 9 & 37.2 \\ 37.2 & 155.48 \end{bmatrix}$$

and

$$\mathbf{X}'\mathbf{Y} = \begin{bmatrix} 1 & 1 & 1 & 1 & 1 & 1 & 1 & 1 & 1 \\ 4.2 & 3.8 & 4.8 & 3.4 & 4.5 & 4.6 & 4.3 & 3.7 & 3.9 \end{bmatrix} \begin{bmatrix} 2.8 \\ 2.5 \\ 3.1 \\ 2.1 \\ 2.9 \\ 2.8 \\ 2.6 \\ 2.4 \\ 2.5 \end{bmatrix}$$

$$= \begin{bmatrix} 23.7 \\ 99.02 \end{bmatrix}$$

The inverse of $\mathbf{X'X}$ can be shown to be

$$(\mathbf{X'X})^{-1} = \begin{bmatrix} 10.0439 & -2.4031 \\ -2.4031 & .5814 \end{bmatrix}$$

so the least squares estimates of β_0 and β_1 are

$$\hat{\beta} = (\mathbf{X'X})^{-1}\mathbf{X'Y}$$

$$= \begin{bmatrix} 10.0439 & -2.4031 \\ -2.4031 & .5814 \end{bmatrix} \begin{bmatrix} 23.7 \\ 99.02 \end{bmatrix} = \begin{bmatrix} .085 \leftarrow \\ .617 \leftarrow \end{bmatrix} \begin{matrix} \hat{\beta}_0 \\ \hat{\beta}_1 \end{matrix}$$

and the linear regression line is

$$\hat{y} = .085 + .617x$$

This result is identical to that obtained for Example 8.1, except for rounding errors.

EXAMPLE 12.13

Refer to the data of Example 12.4, in which a chemist was interested in determining the weight loss y of a compound as a function of the exposure time and relative humidity. Set up all the matrices for fitting the regression model $y = \beta_0 + \beta_1 x_1 + \beta_2 x_2 + \varepsilon$.

Solution

The matrices required to obtain $\hat{\beta}$ are

$$\mathbf{Y} = \begin{bmatrix} 4.3 \\ 5.5 \\ \vdots \\ 6.5 \end{bmatrix} \qquad \mathbf{X} = \begin{bmatrix} 1 & 4 & .20 \\ 1 & 5 & .20 \\ \vdots & \vdots & \vdots \\ 1 & 7 & .40 \end{bmatrix}$$

$$\mathbf{X'X} = \begin{bmatrix} 12 & 66 & 3.6 \\ 66 & 378 & 19.8 \\ 3.6 & 19.8 & 1.16 \end{bmatrix} \qquad \text{and} \qquad \mathbf{X'Y} = \begin{bmatrix} 66.1 \\ 383.3 \\ 19.19 \end{bmatrix}$$

From these matrices one could compute

$$\hat{\beta} = (\mathbf{X'X})^{-1}\mathbf{X'Y}$$

The important thing is not the computation, but the understanding that all general model problems (multiple regression and others) have a common format and are solved via matrices using the same formulas.

In a similar way, all the inferential procedures about general linear models that were discussed earlier in this chapter can be set up using matrices and solved simply using available statistical software packages.

We begin as follows. Under the assumptions that the ε_i are independent and each normally distributed with mean 0 and variance σ_ε^2, the least squares estimator $\hat{\beta}_i$ has mean β_i (the parameter estimated) and standard error $\sqrt{v_{ii}}\sigma_\varepsilon$, where v_{ii} is the ith diagonal element of $(\mathbf{X'X})^{-1}$

$$(\mathbf{X'X})^{-1} = \begin{bmatrix} v_{00} & v_{01} & \cdots & v_{0k} \\ v_{10} & v_{11} & \cdots & v_{1k} \\ \vdots & \vdots & & \vdots \\ v_{k0} & v_{k1} & \cdots & v_{kk} \end{bmatrix}$$

Note that the first diagonal element is labeled v_{00} to correspond to $\hat{\beta}_0$, so the standard error of $\hat{\beta}_0$ is $\sqrt{v_{00}}\sigma_\varepsilon$.

EXAMPLE 12.14

Refer to Example 12.4. For the regression model $y = \beta_0 + \beta_1 x_1 + \beta_2 x_2 + \varepsilon$, it can be shown that

$$(\mathbf{X'X})^{-1} = \begin{bmatrix} 3.2250 & -0.3667 & -3.7500 \\ -0.3667 & 0.0667 & 0.0000 \\ -3.7500 & 0.0000 & 12.5000 \end{bmatrix}$$

a. Determine the standard errors for $\hat{\beta}_0$, $\hat{\beta}_1$, and $\hat{\beta}_2$.
b. If MSE = .149, estimate the standard errors for $\hat{\beta}_0$, $\hat{\beta}_1$, and $\hat{\beta}_2$. Compare your values to those given in the output for Example 12.4.

Solution

a. Using the standard error formula based on matrices

$$\sigma_{\hat{\beta}_0} = \sqrt{v_{00}}\sigma_\varepsilon = \sqrt{3.2250}\,\sigma_\varepsilon$$
$$\sigma_{\hat{\beta}_1} = \sqrt{v_{11}}\sigma_\varepsilon = \sqrt{0.0667}\,\sigma_\varepsilon$$

and

$$\sigma_{\hat{\beta}_2} = \sqrt{v_{22}}\sigma_\varepsilon = \sqrt{12.5000}\,\sigma_\varepsilon$$

b. Substituting $\sqrt{\text{MSE}} = \sqrt{.149} = .386$ for σ_ε, the (estimated) standard errors for $\hat{\beta}_0$, $\hat{\beta}_1$, and $\hat{\beta}_2$ are, respectively, .693, .100, and 1.365. These agree (except for rounding errors) with the standard errors shown in the output for Example 12.4.

Using the previous result on standard errors and the fact that $\text{SSE} = \mathbf{Y'Y} - \hat{\boldsymbol{\beta}}'\mathbf{X'Y}$, it is possible to make inferences about a single β_i or a set of βs.

INFERENCES ABOUT A SINGLE β_i AND A SET OF βs

100$(1 - \alpha)$% confidence interval for β_i: $\hat{\beta}_i \pm t_{\alpha/2}s_\varepsilon\sqrt{v_{ii}}$

Statistical test for H_0: $\beta_i = 0$ T.S. $t = \dfrac{\hat{\beta}_i}{s_\varepsilon\sqrt{v_{ii}}}$

where

$$s_\varepsilon = \sqrt{\text{MSE}} = \sqrt{\frac{SSE}{n - (k + 1)}}$$

Test for H_0: $\beta_{g+1} = \beta_{g+2} = \cdots = \beta_k = 0$ makes use of $\text{SSE} = \mathbf{Y'Y} - \hat{\boldsymbol{\beta}}'\mathbf{X'Y}$ computed for both the complete and reduced models.

A confidence interval for $E(y)$ and the corresponding prediction interval for y are as follows:

$100(1 − \alpha)\%$ confidence interval for $E(y)$

$$\hat{y} \pm t_{\alpha/2}s_{\varepsilon}\sqrt{\ell'(\mathbf{X'X})^{-1}\ell}$$

$100(1 − \alpha)\%$ prediction interval for y

$$\hat{y} \pm t_{\alpha/2}s_{\varepsilon}\sqrt{1 + \ell'(\mathbf{X'X})^{-1}\ell}$$

where the matrix

$$\ell = \begin{bmatrix} 1 \\ x_1 \\ x_2 \\ \vdots \\ x_k \end{bmatrix}$$

displays the desired settings of the independent variable and

$$s_{\varepsilon} = \sqrt{MSE} = \sqrt{\frac{SSE}{n − (k + 1)}}$$

The results of this section indicate that with a basic understanding of a few matrix operations, it is possible to use a common format for all the estimation and test procedures discussed to date for models in the form of a general linear model. It is this common format that has enabled computer software vendors to develop statistical software packages with broad applicability rather than creating separate programs for each type of problem.

EXERCISES

12.27   The data from Exercise 8.3 related to the firmness of canned sweet potatoes treated with various concentrations of pectin and stored at 25°C are shown here.

y (firmness)	50.5	46.8	62.3	67.7	80.1	79.2
x (concentration of pectin)	0	0	1.5	1.5	3.0	3.0

a. Obtain the least squares estimates for the parameters in the model $y = \beta_0 + \beta_1 x + \varepsilon$ by using the matrix approach.
b. Give the standard error of $\hat{\beta}_1$.
c. Obtain an estimate of σ_{ε}.

12.28   Refer to Exercise 12.27.
a. Give an estimate of the standard error of $\hat{\beta}_1$.
b. Perform a statistical test of the null hypothesis that there is no linear relationship between the concentration of pectin and the firmness of canned sweet potatoes after thirty days of storage at 25° C. Use $\alpha = .05$.

12.29   Refer to Exercise 12.12.

a. Using a matrix approach, fit the general linear model

$$y = \beta_0 + \beta_1 x + \beta_2 x^2 + \varepsilon$$

to obtain $\hat{y}$.

b. Compute an estimate of σ_ε^2.

c. Identify the standard errors of $\hat{\beta}_0$, $\hat{\beta}_1$, and $\hat{\beta}_2$.

12.30   Refer to Exercise 12.12. Conduct a test of the null hypothesis that $\beta_2 = 0$, that is, the log of the biological recovery percentage (y) is linearly related to time (x). Use $\alpha = .05$.

12.31   Refer to Exercise 12.12. Place a 95% confidence interval on β_0, the mean log biological recovery percentage at time zero. (Note: $E(y) = \beta_0$ when $x = 0$.)

12.32   Refer to Example 12.8. Set up all matrices for developing a 90% confidence interval on $E(Y)$, the mean corn yield, when 5 pounds of fertilizer are applied.

| 12.8 | SUMMARY |

This chapter consolidates the material for expressing a response y as a function of one or more independent variables. Multiple regression models (where all the independent variables are quantitative) and models that incorporate information on qualitative variables were discussed and can be represented in the form of a general linear model

$$y = \beta_0 + \beta_1 x_1 + \beta_2 x_2 + \cdots + \beta_k x_k + \varepsilon$$

After discussing various models and the interpretation of βs in these models, we presented the normal equations used in obtaining the least squares estimates $\hat{\beta}$.

A confidence interval and statistical test about an individual parameter β_j were developed using $\hat{\beta}_j$ and the standard error of $\hat{\beta}_j$. We also considered a statistical test about a set of βs, a confidence interval for $E(y)$ based on a set of xs, and a prediction interval for a given set of xs.

All of these inferences involve a fair to moderate amount of numerical calculation unless statistical software programs or packages are available. Sometimes these calculations can be done by hand if one is familiar with matrix operations (see Section 12.7 and the Appendix to this chapter). However, even these methods become unmanageable as the number of independent variables increases. So, the message should be very clear. Inferences about general linear models should be done using available computer software to facilitate the analysis and to minimize computational errors. Our job in these situations is to review and interpret the output.

Aside from a few exercises that will probe your understanding of the mechanics involved with these calculations, most of the exercises in the remainder of this chapter and in the regression problems of the next chapter will make extensive use of computer output.

EXERCISES

12.33   The data presented here illustrate the effects of correlated and uncorrelated independent variables.

y	7	10	16	12	16	15	19	22	20	23	22	24
x_1	1	2	3	4	1	2	3	4	1	2	3	4
x_2	1	1	1	1	2	2	2	2	3	3	3	3
x_3	1	3	5	7	9	11	13	15	17	19	21	23

a. Plot x_1 versus x_2, x_1 versus x_3, and x_2 versus x_3.

b. Which of the plots in part (a) indicate uncorrelated independent variables?

12.34 Computer output for the data from Exercise 12.33 is shown for the following models:

1. $y = \beta_0 + \beta_1 x_1 + \beta_2 x_2 + \varepsilon$
2. $y = \beta_0 + \beta_1 x_1 + \beta_3 x_3 + \varepsilon$
3. $y = \beta_0 + \beta_2 x_2 + \beta_3 x_3 + \varepsilon$

and

4. $y = \beta_0 + \beta_1 x_1 + \beta_2 x_2 + \beta_3 x_3 + \varepsilon$

What do you observe based on the output for these four models? Does it confirm what you saw from the plots in Exercise 12.33?

```
            LISTING OF DATA

    OBS     Y     X1    X2    X3

     1      7     1     1     1
     2     10     2     1     3
     3     16     3     1     5
     4     12     4     1     7
     5     16     1     2     9
     6     15     2     2    11
     7     19     3     2    13
     8     22     4     2    15
     9     20     1     3    17
    10     23     2     3    19
    11     22     3     3    21
    12     24     4     3    23

   N=      12
```

REGRESSION ANALYSIS, MODEL 1

GENERAL LINEAR MODELS PROCEDURE

DEPENDENT VARIABLE: Y

SOURCE	DF	SUM OF SQUARES	MEAN SQUARE	F VALUE	PR > F	R-SQUARE	C.V.
MODEL	2	290.60000000	145.30000000	35.28	0.0001	0.886877	11.8218
ERROR	9	37.06666667	4.11851852		ROOT MSE		Y MEAN
CORRECTED TOTAL	11	327.66666667			2.02941334		17.16666667

SOURCE	DF	TYPE I SS	F VALUE	PR > F	DF	TYPE III SS	F VALUE	PR > F
X1	1	48.60000000	11.80	0.0074	1	48.60000000	11.80	0.0074
X2	1	242.00000000	58.76	0.0001	1	242.00000000	58.76	0.0001

PARAMETER	ESTIMATE	T FOR H0: PARAMETER=0	PR > |T|	STD ERROR OF ESTIMATE
INTERCEPT	1.66666667	0.82	0.4327	2.02941334
X1	1.80000000	3.44	0.0074	0.52399227
X2	5.50000000	7.67	0.0001	0.71750597

REGRESSION ANALYSIS, MODEL 2

GENERAL LINEAR MODELS PROCEDURE

DEPENDENT VARIABLE: Y

SOURCE	DF	SUM OF SQUARES	MEAN SQUARE	F VALUE	PR > F	R-SQUARE	C.V.
MODEL	2	290.60000000	145.30000000	35.28	0.0001	0.886877	11.8218
ERROR	9	37.06666667	4.11851852		ROOT MSE		Y MEAN
CORRECTED TOTAL	11	327.66666667			2.02941334		17.16666667

SOURCE	DF	TYPE I SS	F VALUE	PR > F	DF	TYPE III SS	F VALUE	PR > F
X1	1	48.60000000	11.80	0.0074	1	2.42517483	0.59	0.4625
X3	1	242.00000000	58.76	0.0001	1	242.00000000	58.76	0.0001

PARAMETER	ESTIMATE	T FOR HO: PARAMETER=0	PR > |T|	STD ERROR OF ESTIMATE
INTERCEPT	7.85416667	5.01	0.0007	1.56633788
X1	0.42500000	0.77	0.4625	0.55384459
X3	0.68750000	7.67	0.0001	0.08968825

REGRESSION ANALYSIS, MODEL 3

GENERAL LINEAR MODELS PROCEDURE

DEPENDENT VARIABLE: Y

SOURCE	DF	SUM OF SQUARES	MEAN SQUARE	F VALUE	PR > F	R-SQUARE	C.V.
MODEL	2	290.60000000	145.30000000	35.28	0.0001	0.886877	11.8218
ERROR	9	37.06666667	4.11851852		ROOT MSE		Y MEAN
CORRECTED TOTAL	11	327.66666667			2.02941334		17.16666667

SOURCE	DF	TYPE I SS	F VALUE	PR > F	DF	TYPE III SS	F VALUE	PR > F
X2	1	242.00000000	58.76	0.0001	1	2.42517483	0.59	0.4625
X3	1	48.60000000	11.80	0.0074	1	48.60000000	11.80	0.0074

PARAMETER	ESTIMATE	T FOR HO: PARAMETER=0	PR > |T|	STD ERROR OF ESTIMATE
INTERCEPT	9.76666667	5.22	0.0005	1.87102665
X2	-1.70000000	-0.77	0.4625	2.21537835
X3	0.90000000	3.44	0.0074	0.26199614

REGRESSION ANALYSIS, MODEL 4

GENERAL LINEAR MODELS PROCEDURE

DEPENDENT VARIABLE: Y

SOURCE	DF	SUM OF SQUARES	MEAN SQUARE	F VALUE	PR > F	R-SQUARE	C.V.
MODEL	2	290.60000000	145.30000000	35.28	0.0001	0.886877	11.8218
ERROR	9	37.06666667	4.11851852		ROOT MSE		Y MEAN
CORRECTED TOTAL	11	327.66666667			2.02941334		17.16666667

SOURCE	DF	TYPE I SS	F VALUE	PR > F	DF	TYPE III SS	F VALUE	PR > F
X1	1	48.60000000	11.80	0.0074	0	0.00000000	.	.
X2	1	242.00000000	58.76	0.0001	0	0.00000000	.	.
X3	0	0.00000000			0	0.00000000	.	.

PARAMETER	ESTIMATE	T FOR HO: PARAMETER=0	PR > |T|	STD ERROR OF ESTIMATE
INTERCEPT	1.66666667 B	0.82	0.4327	2.02941334
X1	1.80000000 B	3.44	0.0074	0.52399227
X2	5.50000000 B	7.67	0.0001	0.71750597
X3	0.00000000 B			

NOTE: THE X'X MATRIX HAS BEEN DEEMED SINGULAR AND A GENERALIZED INVERSE HAS BEEN EMPLOYED TO SOLVE THE NORMAL EQUATIONS.
THE ABOVE ESTIMATES REPRESENT ONLY ONE OF MANY POSSIBLE SOLUTIONS TO THE NORMAL EQUATIONS. ESTIMATES FOLLOWED BY
THE LETTER B ARE BIASED AND DO NOT ESTIMATE THE PARAMETER BUT ARE BLUE FOR SOME LINEAR COMBINATION OF PARAMETERS
(OR ARE ZERO). THE EXPECTED VALUE OF THE BIASED ESTIMATORS MAY BE OBTAINED FROM THE GENERAL FORM OF ESTIMABLE
FUNCTIONS. FOR THE BIASED ESTIMATORS, THE STD ERR IS THAT OF THE BIASED ESTIMATOR AND THE T VALUE TESTS
HO: E(BIASED ESTIMATOR) = 0. ESTIMATES NOT FOLLOWED BY THE LETTER B ARE BLUE FOR THE PARAMETER.

12.35   Refer to Exercise 12.34. Might there be a problem with correlated independent variables for one or more of these models? How would this be detected?

12.36   An experiment was conducted to determine the effects of two bonding agents (starch and calcium phosphate) on the dissolution properties of the tablet form of a drug product. Sample data were collected and then used to determine the prediction equation $\hat{y} = 77.5 + 0.2x_1 + 10x_1^2 + 5.8x_2 + 6.5x_1x_2$. If $n = b$, $\sum y = 425$, $\sum y^2 = 30{,}731$, and SSE = 120.45, determine R^2.

12.37   In Exercise 8.13 we presented an experiment to investigate the shock wave amplitudes recorded from sensors placed at fixed distances from an explosive charge. These data are presented here for your convenience.

Distance, x	5	5	5	10	10	10	15	15	15
Amplitude, y	8.6	8.2	8.1	5.8	6.2	6.1	5.2	4.8	4.7

a. Plot the data.

b. Determine the quadratic regression equation $y = \beta_0 + \beta_1 x + \beta_2 x^2$ from the output shown.

c. Compute and interpret the coefficient of determination.

```
       LISTING OF DATA

   OBS     DISTANCE     AMPTUDE

    1          5          8.6
    2          5          8.2
    3          5          8.1
    4         10          5.8
    5         10          6.2
    6         10          6.1
    7         15          5.2
    8         15          4.8
    9         15          4.7

   N=     9
```

COMPLETE MODEL

GENERAL LINEAR MODELS PROCEDURE

DEPENDENT VARIABLE: AMPTUDE

SOURCE	DF	SUM OF SQUARES	MEAN SQUARE	F VALUE	PR > F	R-SQUARE	C.V.
MODEL	2	17.98222222	8.99111111	147.13	0.0001	0.980017	3.8559
ERROR	6	0.36666667	0.06111111		ROOT MSE		AMPTUDE MEAN
CORRECTED TOTAL	8	18.34888889			0.24720662		6.41111111

SOURCE	DF	TYPE I SS	F VALUE	PR > F	DF	TYPE III SS	F VALUE	PR > F
DISTANCE	1	17.34000000	283.75	0.0001	1	1.92666667	31.53	0.0014
DISTANCE*DISTANCE	1	0.64222222	10.51	0.0176	1	0.64222222	10.51	0.0176

PARAMETER	ESTIMATE	T FOR H0: PARAMETER=0	PR > |T|	STD ERROR OF ESTIMATE
INTERCEPT	11.70000000	18.81	0.0001	0.62212301
DISTANCE	-0.79333333	-5.61	0.0014	0.14129035
DISTANCE*DISTANCE	0.02266667	3.24	0.0176	0.00699206

12.38   Refer to Exercise 12.37. Give the p-value for a test of H_0: $\beta_2 = 0$. Interpret your findings. Does this agree with what you saw in the data for Exercise 12.37?

12.39   Refer to Example 12.4. Determine 95% prediction bands for the data using the output shown here.

LEAST SQUARES ANALYSIS

GENERAL LINEAR MODELS PROCEDURE

DEPENDENT VARIABLE: WT—LOSS   WEIGHT LOSS (POUNDS)

SOURCE	DF	SUM OF SQUARES	MEAN SQUARE	F VALUE	PR > F	R-SQUARE	C.V.
MODEL	2	31.12416667	15.56208333	104.13	0.0001	0.958576	7.0181
ERROR	9	1.34500000	0.14944444		STD DEV		WT—LOSS MEAN
CORRECTED TOTAL	11	32.46916667			0.38658045		5.50833333

SOURCE	DF	TYPE 1 SS	F VALUE	PR > F	DF	TYPE IV SS	F VALUE	PR > F
TIME	1	26.00416667	174.01	0.0001	1	26.00416667	174.01	0.0001
HUMID	1	5.12000000	34.26	0.0002	1	5.12000000	34.26	0.0002

PARAMETER	ESTIMATE	T FOR H0: PARAMETER = 0	PR > T	STD ERROR OF ESTIMATE
INTERCEPT	0.66666667	0.96	0.3620	0.69423219
TIME	1.31666667	13.19	0.0001	0.09981464
HUMID	−8.00000000	−5.85	0.0002	1.36676829

OBSERVATION	OBSERVED VALUE	PREDICTED VALUE	RESIDUAL	LOWER 95% CL INDIVIDUAL	UPPER 95% CL INDIVIDUAL
1	4.30000000	4.33333333	−0.03333333	3.31411004	5.35255662
2	5.50000000	5.65000000	−0.15000000	4.68209172	6.61790828
3	6.80000000	6.96666667	−0.16666667	5.99875839	7.93457495
4	8.00000000	8.28333333	−0.28333333	7.26411004	9.30255662
5	4.00000000	3.53333333	0.46666667	2.56213843	4.50452824
6	5.20000000	4.85000000	0.35000000	3.93280326	5.76719674
7	6.60000000	6.16666667	0.43333333	5.24946993	7.08386341
8	7.50000000	7.48333333	0.01666667	6.51213843	8.45452824
9	2.00000000	2.73333333	−0.73333333	1.71411004	3.75255662
10	4.00000000	4.05000000	−0.05000000	3.08209172	5.01790828
11	5.70000000	5.36666667	0.33333333	4.39875839	6.33457495
12	6.50000000	6.68333333	−0.18333333	5.66411004	7.70255662
13*		6.42500000		5.47453294	7.37546706

* OBSERVATION WAS NOT USED IN THIS ANALYSIS

SUM OF RESIDUALS	0.00000000
SUM OF SQUARED RESIDUALS	1.34500000
SUM OF SQUARED RESIDUALS—ERROR SS	−0.00000000
PRESS STATISTIC	2.61233345
FIRST ORDER AUTOCORRELATION	0.15902520
DURBIN–WATSON D	1.65613383

12.40   A study of demand for imported subcompact cars consisted of data from 12 metropolitan areas. The variables were:

DEMAND: imported subcompact car sales as a percentage of total sales

EDUC: average number of years of schooling completed by adults

INCOME: per capita income

POPN: area population

FAMSIZE: average size of intact families

BMDP output is shown below.
a. Write down the regression equation. Place the standard error of each coefficient below the coefficient, perhaps in parentheses.
b. Locate R^2 and the residual standard deviation.

MULTIPLE R   0.9813   STD. ERROR OF EST.   1.7777
MULTIPLE R-SQUARE   0.9629

ANALYSIS OF VARIANCE

	SUM OF SQUARES	DF	MEAN SQUARE	F RATIO	P(TAIL)
REGRESSION	574.073	4	143.518	45.412	0.00004
RESIDUAL	22.123	7	3.160		

VARIABLE		COEFFICIENT	STD. ERROR	STD. REG COEFF	T	P(2 TAIL)
INTERCEPT		−44.01318				
EDUC	2	5.08694	1.970	0.471	2.582	0.036
INCOME	3	3.01868	2.204	0.351	1.370	0.213
POPN	4	0.20527	0.373	0.045	0.550	0.600
FAMSIZE	5	−5.40869	3.850	−0.228	−1.405	0.203

12.41   Refer to Exercise 12.40.
a. Draw conclusions from the statistical tests summarized in the output.
b. Would you consider an alternate model? If so, what would it be?

12.42   A study was conducted to examine the effects of temperature and lighting intensity on office productivity. Let y denote a measure of productivity, x_1 denote office temperature in °F, and x_2 denote lighting intensity.
a. Write a first-order regression model.
b. Write a second-order regression model allowing for all possible terms in x_1 and x_2.

12.43   Refer to Exercise 12.42. Here are the data:

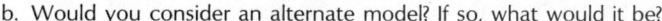

Y	45	49	47	57	48	53	51	54	56	64
x_1	64	64	66	66	68	68	70	70	72	72
x_2	60	65	60	65	60	65	60	65	60	65

a. Suggest plots that may help to formulate a regression model.
b. Fit the model that you think is appropriate and draw some conclusions.
c. Do you have any recommendations for further experimentation?

12.44   The following data were recorded for a regression study:

y	15	12	14	18	19	16	17	26	20	22	24
x_1	6	7	7	8	8	9	9	10	10	11	12
x_2	10	12	13	14	15	15	16	17	18	19	19

a. Write a first-order regression model.

b. Use the sample data to fit the model.

c. Calculate the residual standard deviation.

12.45   Refer to Exercise 12.44.

a. Calculate the estimated standard errors for $\hat{\beta}_1$ and $\hat{\beta}_2$.

b. Compute 90% confidence intervals for β_1 and β_2.

 **12.46**   Computer output is shown here for a study of consumer product sales for a company as a function of advertising expense.

```
CORRELATIONS

            ADV  ADVSQ  SALES
             1     3      2
ADV      1  1.000
ADVSQ    3  0.994  1.000
SALES    2  0.363  0.274  1.000


SQUARED MULTIPLE CORRELATIONS OF EACH INDEPENDENT VARIABLE WITH ALL OTHER
INDEPENDENT VARIABLES
(MEASURES OF MULTICOLLINEARITY OF PREDICTOR VARIABLES)
AND TESTS OF SIGNIFICANCE OF MULTIPLE REGRESSION


DEGREES OF FREEDOM FOR F-STATISTICS ARE     1 AND        12

    VARIABLE                           SIGNIFICANCE
NO.    NAME      SMC     F-STATISTIC   (P LESS THAN)
 1     ADV      0.98895    1074.10       0.00000
 3     ADVSQ    0.98895    1074.10       0.00000

SQUARED MULTIPLE CORRELATIONS (SMC) OF EACH DEPENDENT VARIABLE WITH THE
INDEPENDENT VARIABLES AND TESTS OF SIGNIFICANCE OF MULTIPLE REGRESSION


DEGREES OF FREEDOM FOR F-STATISTICS ARE     2 AND        11

    VARIABLE                           SIGNIFICANCE
NO.    NAME      SMC     F-STATISTIC   (P LESS THAN)
 2     SALES    0.80721    23.03         0.00012
```

a. Identify the regression equation.

b. Give the p-value for a test of H_0: $\beta_2 = 0$. Does this suggest that a quadratic regression model is to be preferred to a linear regression model?

SECTION 12.8   SUMMARY

12.47   The manager of documentation for a computer software firm wants to forecast the time required to document moderate-sized computer programs. Records are available for 26 programs. The variables are y = number of writer-days needed, x_1 = number of subprograms, x_2 = average number of lines per subprogram, $x_3 = x_1 x_2$, $x_4 = x_2^2$, and $x_5 = x_1 x_2^2$. A portion of the output from a regression analysis (SAS) of the data is shown here:

DEPENDENT VARIABLE: Y

SOURCE	DF	SUM OF SQUARES	MEAN SQUARE	F VALUE	PR > F	R SQUARE	C.V.
MODEL	5	2546.02735209	509.20547042	44.31	0.0001	0.917195	11.9597
ERROR	20	229.85726330	11.49286316		ROOT MSE		Y MEAN
CORRECTED TOTAL	25	2775.88461538			3.39011256		28.34615385

PARAMETER	ESTIMATE	T FOR HO: PARAMETER = 0	PR > |T|	STD. ERROR OF ESTIMATE
INTERCEPT	−16.81979712	−1.45	0.1636	11.63104920
X1	1.47018752	4.02	0.0007	0.36594367
X2	0.99477822	1.63	0.1194	0.61144114
X1X2	−0.02400705	−1.01	0.3243	0.02375645
X2SQ	−0.01031004	−1.40	0.1774	0.00737400
X1X2SQ	0.00024957	0.71	0.4862	0.00035178

a. Write the multiple regression model and locate the residual standard deviation.
b. What does the variable x_3 represent in terms of the problem?
c. Does x_3 have a statistically significant predictive value as "last predictor in"?

12.48   The model $y = \beta_0 + \beta_1 x_1 + \beta_2 x_2 + \varepsilon$ was fit to the data of Exercise 12.47. Selected output is shown here:

DEPENDENT VARIABLE: Y

SOURCE	DF	SUM OF SQUARES	MEAN SQUARE	F VALUE	PR > F	R SQUARE	C.V.
MODEL	2	2516.12160091	1258.06080045	111.39	0.0001	0.906422	11.8558
ERROR	23	259.76301448	11.29404411		ROOT MSE		Y MEAN
CORRECTED TOTAL	25	2775.88461538			3.36066126		28.34615385

PARAMETER	ESTIMATE	T FOR HO: PARAMETER = 0	PR > |T|	STD. ERROR OF ESTIMATE
INTERCEPT	0.84008527	0.24	0.8089	3.43374955
X1	1.01583472	12.81	0.0001	0.07929252
X2	0.05582624	1.08	0.2897	0.05150660

a. Write the complete and reduced-form estimated models.
b. Is the improvement in R^2 obtained by adding x_3, x_4, and x_5 statistically significant at $\alpha = .05$? What is the p-value for this test?

12.49 A producer of various feed additives for cattle conducted a study of the number of days of feedlot time required to bring beef cattle to market weight. Eighteen steers of essentially identical age and weight were purchased and brought to a feedlot. Each steer was fed a diet with a specific combination of protein content, antibiotic concentration, and percentage of feed supplement. The data were:

STEER:	1	2	3	4	5	6	7	8	9
PROTEIN:	10	10	10	10	10	10	15	15	15
ANTIBIO:	1	1	1	2	2	2	1	1	1
SUPPLEM:	3	5	7	3	5	7	3	5	7
TIME:	88	82	81	82	83	75	80	80	75
STEER:	10	11	12	13	14	15	16	17	18
PROTEIN:	15	15	15	20	20	20	20	20	20
ANTIBIO:	2	2	2	1	1	1	2	2	2
SUPPLEM:	3	5	7	3	5	7	3	5	7
TIME:	77	76	72	79	74	75	74	70	69

Computer output from a regression analysis follows:

MULTIPLE R	0.9490		STD. ERROR OF EST.		1.7096
MULTIPLE R-SQUARE	0.9007				

ANALYSIS OF VARIANCE

	SUM OF SQUARES	DF	MEAN SQUARE	F RATIO	P(TAIL)
REGRESSION	371.0832	3	123.6944	42.323	0.0000
RESIDUAL	40.9166	14	2.9226		

VARIABLE		COEFFICIENT	STD. ERROR	STD. REG COEFF	T	P(2 TAIL)	TOLERANCE
INTERCEPT		102.70834					
PROTEIN	2	−0.83333	0.09870	−0.711	−8.443	0.0000	1.00000
ANTIBIO	3	−4.00000	0.80590	−0.418	−4.963	0.0002	1.00000
SUPPLEM	4	−1.37500	0.24675	−0.469	−5.572	0.0001	1.00000

a. Write down the regression equation.
b. Find the standard deviation.
c. Find the R^2 value.

12.50 Refer to Exercise 12.49.

a. Predict the feedlot time required for a steer fed 15% protein, 1.5% antibiotic concentration, and 5% supplement.
b. Do these values of the independent variables represent a major extrapolation from the data?
c. Give a 95% confidence interval for the mean time predicted in part (a).

Note: This requires access to computer software for multiple regression.

12.51 The data of Exercise 12.49 were also analyzed by a regression model using only protein content as an independent variable, yielding the following output:

MULTIPLE R	0.7111			STD. ERROR OF EST.		3.5678
MULTIPLE R-SQUARE	0.5057					

ANALYSIS OF VARIANCE

	SUM OF SQUARES	DF	MEAN SQUARE	F RATIO	P(TAIL)
REGRESSION	208.3332	1	208.3332	16.367	0.0009
RESIDUAL	203.6667	16	12.7292		

VARIABLE		COEFFICIENT	STD. ERROR	STD. REG COEFF	T	P(2 TAIL)	TOLERANCE
INTERCEPT		89.83334					
PROTEIN	2	−0.83333	0.20599	−0.711	−4.046	0.0009	1.00000

a. Write the regression equation.

b. Find the R^2 value.

c. Test the null hypothesis that the coefficients of ANTIBIO and SUPPLEM are zero at $\alpha = .05$.

12.52 The accompanying table gives demographic data for 12 male patients with congestive heart failure enrolled in a study of an experimental compound.

a. Summarize these data using a box plot for each variable.

b. Construct scatterplots to display (1) age by cardiac index (CI) and by pulmonary capillary wedge pressure (PCWP) and (2) disease duration by CI and by PCWP.

Is there evidence of a correlation between age and CI or PCWP? What about correlation between duration of disease and CI or PCWP?

Demographic Data for Patients with Heart Failure (NYHA Class III or IV)

Patient	Age (yrs)	Disease Duration	Height (cm)	Weight (kg)	Baseline Cardiac Index (L/Min/M²)	Baseline Pulmonary Capillary Wedge Pressure (mm Hg)
01	67	5 yr	172.0	57.0	1.6	40
02	45	2 yr	170.0	67.0	2.4	25
03	59	8 yr	172.7	102.0	2.2	39
04	63	1 yr	175.3	74.9	1.7	39
05	55	1 yr	172.7	92.0	2.3	34
06	65	1 yr	178.0	90.0	1.6	36
07	62	2 yr	163.0	67.0	1.4	36
08	60	1 yr	182.5	72.0	2.2	17
09	72	2 yr	168.0	71.0	1.3	37
10	44	3 mo	163.0	68.0	2.4	28
11	63	5 yr	172.0	82.0	2.1	38
12	63	1 yr	163.0	64.0	1.1	36

12.53 The data of Exercise 12.52 were used to fit several multiple regression models. $y_1 = $ CI, $y_2 = $ PCWP, $x_1 = $ age, $x_2 = $ disease duration.

a. $y_1 = \beta_0 + \beta_1 x_1 + \beta_2 x_2 + \varepsilon$

b. $y_1 = \beta_0 + \beta_1 x_1 + \beta_2 x_2 + \beta_3 x_1 x_2 + \varepsilon$

c. $y_2 = \beta_0 + \beta_1 x_1 + \beta_2 x_2 + \varepsilon$

d. $y_2 = \beta_0 + \beta_1 x_1 + \beta_2 x_2 + \beta_3 x_1 x_2 + \varepsilon$

Which model provides the best fit to the cardiac index data? To the pulmonary capillary wedge pressure data? Do these analyses confirm what you concluded in Exercise 12.52? Explain.

```
           LISTING OF DATA

  OBS    PATIENT    AGE    DURATION    CI    PCWP

    1        1       67      5.00      1.6    40
    2        2       45      2.00      2.4    25
    3        3       59      8.00      2.2    39
    4        4       63      1.00      1.7    39
    5        5       55      1.00      2.3    34
    6        6       65      1.00      1.6    36
    7        7       62      2.00      1.4    36
    8        8       60      1.00      2.2    17
    9        9       72      2.00      1.3    37
   10       10       44      0.25      2.4    28
   11       11       63      5.00      2.1    38
   12       12       63      1.00      1.1    36

  N=       12
```

REGRESSION ANALYSIS, MODEL 1

GENERAL LINEAR MODELS PROCEDURE

DEPENDENT VARIABLE: CI

SOURCE	DF	SUM OF SQUARES	MEAN SQUARE	F VALUE	PR > F	R-SQUARE	C.V.
MODEL	2	1.56955279	0.78477639	9.30	0.0065	0.673869	15.6333
ERROR	9	0.75961388	0.08440154		ROOT MSE		CI MEAN
CORRECTED TOTAL	11	2.32916667			0.29051943		1.85833333

SOURCE	DF	TYPE I SS	F VALUE	PR > F	DF	TYPE III SS	F VALUE	PR > F
AGE	1	1.36216181	16.14	0.0030	1	1.53467632	18.18	0.0021
DURATION	1	0.20739097	2.46	0.1514	1	0.20739097	2.46	0.1514

PARAMETER	ESTIMATE	T FOR HO: PARAMETER=O	PR > |T|	STD. ERROR OF ESTIMATE
INTERCEPT	4.47562208	7.00	0.0001	0.63976685
AGE	-0.04620336	-4.26	0.0021	0.01083529
DURATION	0.06039479	1.57	0.1514	0.03852829

REGRESSION ANALYSIS, MODEL 2

GENERAL LINEAR MODELS PROCEDURE

DEPENDENT VARIABLE: CI

SOURCE	DF	SUM OF SQUARES	MEAN SQUARE	F VALUE	PR > F	R-SQUARE	C.V.
MODEL	3	1.57161289	0.52387096	5.53	0.0237	0.674753	16.5592
ERROR	8	0.75755378	0.09469422		ROOT MSE		CI MEAN
CORRECTED TOTAL	11	2.32916667			0.30772426		1.85833333

SOURCE	DF	TYPE I SS	F VALUE	PR > F	DF	TYPE III SS	F VALUE	PR > F
AGE	1	1.36216181	14.38	0.0053	1	0.64786628	6.84	0.0309
DURATION	1	0.20739097	2.19	0.1772	1	0.00015009	0.00	0.9692
AGE*DURATION	1	0.00206010	0.02	0.8864	1	0.00206010	0.02	0.8864

PARAMETER	ESTIMATE	T FOR HO: PARAMETER=O	PR > |T|	STD. ERROR OF ESTIMATE
INTERCEPT	4.59930709	4.27	0.0027	1.07814691
AGE	-0.04833990	-2.62	0.0309	0.01848097
DURATION	-0.02240952	-0.04	0.9692	0.56287924
AGE*DURATION	0.00137554	0.15	0.8864	0.00932590

APPENDIX   MATRIX OPERATIONS

```
                          REGRESSION ANALYSIS, MODEL 3
                        GENERAL LINEAR MODELS PROCEDURE
```

DEPENDENT VARIABLE: PCWP

SOURCE	DF	SUM OF SQUARES	MEAN SQUARE	F VALUE	PR > F	R-SQUARE	C.V.
MODEL	2	221.88101159	110.94050579	3.26	0.0862	0.420030	17.2873
ERROR	9	306.36898841	34.04099871		ROOT MSE		PCWP MEAN
CORRECTED TOTAL	11	528.25000000			5.83446645		33.75000000

SOURCE	DF	TYPE I SS	F VALUE	PR > F	DF	TYPE III SS	F VALUE	PR > F
AGE	1	162.57201147	4.78	0.0567	1	115.29741682	3.39	0.0989
DURATION	1	59.30900012	1.74	0.2194	1	59.30900012	1.74	0.2194

PARAMETER	ESTIMATE	T FOR H0: PARAMETER=0	PR > |T|	STD. ERROR OF ESTIMATE
INTERCEPT	7.29878609	0.57	0.5839	12.84835977
AGE	0.40047458	1.84	0.0989	0.21760372
DURATION	1.02132713	1.32	0.2194	0.77375900

```
                          REGRESSION ANALYSIS, MODEL 4
                        GENERAL LINEAR MODELS PROCEDURE
```

DEPENDENT VARIABLE: PCWP

SOURCE	DF	SUM OF SQUARES	MEAN SQUARE	F VALUE	PR > F	R-SQUARE	C.V.
MODEL	3	228.56514750	76.18838250	2.03	0.1878	0.432684	18.1348
ERROR	8	299.68485250	37.46060656		ROOT MSE		PCWP MEAN
CORRECTED TOTAL	11	528.25000000			6.12050705		33.75000000

SOURCE	DF	TYPE I SS	F VALUE	PR > F	DF	TYPE III SS	F VALUE	PR > F
AGE	1	162.57201147	4.34	0.0708	1	21.54668440	0.58	0.4700
DURATION	1	59.30900012	1.58	0.2438	1	4.08124565	0.11	0.7498
AGE*DURATION	1	6.68413591	0.18	0.6838	1	6.68413591	0.18	0.6838

PARAMETER	ESTIMATE	T FOR H0: PARAMETER=0	PR > |T|	STD. ERROR OF ESTIMATE
INTERCEPT	14.34402569	0.67	0.5224	21.44389171
AGE	0.27877467	0.76	0.4700	0.36757883
DURATION	-3.69530133	-0.33	0.7498	11.19543293
AGE*DURATION	0.07835228	0.42	0.6838	0.18548824

APPENDIX: MATRIX OPERATIONS

A *matrix* is defined to be a rectangular array of numbers. We will indicate a particular matrix by a boldface capital letter. The numbers of a matrix, called *elements*, appear in rows and columns, as indicated in Figure 12.7.

dimension of matrix   Note that in addition to identifying a matrix by a capital boldface letter, we can also indicate the **dimension of a matrix** by specifying the number of rows and columns in the matrix. Thus a 3 × 3 (read "3 by 3") matrix contains 3 rows and 3 columns, a 2 × 3 matrix contains 2 rows and 3 columns, and a 1 × 4 matrix contains 1 row and 4 columns.

identity matrix   One important type of matrix is the **identity matrix**. An identity matrix, denoted by **I**, is a square matrix (same number of rows and columns) the diagonal elements of which, proceeding from the upper left to the lower right of the matrix, are 1, with all off-diagonal elements 0. Three identity matrices are shown in Figure 12.7.

FIGURE 12.7

2 × 2, 3 × 3, and 4 × 4 Identity Matrices

$$I = \begin{bmatrix} 1 & 0 \\ 0 & 1 \end{bmatrix}$$

$$I = \begin{bmatrix} 1 & 0 & 0 \\ 0 & 1 & 0 \\ 0 & 0 & 1 \end{bmatrix}$$

$$I = \begin{bmatrix} 1 & 0 & 0 & 0 \\ 0 & 1 & 0 & 0 \\ 0 & 0 & 1 & 0 \\ 0 & 0 & 0 & 1 \end{bmatrix}$$

As with other quantities used in statistics and mathematics, we will want to perform operations with matrices, such as addition, multiplication, and so on. Thus in the following discussions, we define the matrix operations we will need in our statistical work.

addition of matrices

*Two matrices, **A** and **B**, can be added only if they have the same dimensions.* For example, we could add two 3 × 3 matrices because both matrices have the same dimensions, but we could not add a 2 × 3 matrix and a 3 × 3 matrix. *The sum of two matrices, **A** and **B**, the dimensions of which are the same forms a new matrix by adding the corresponding elements of **A** and **B**.* This new matrix, **A** + **B**, has the same dimensions as **A** and **B**.

We illustrate addition with some examples.

EXAMPLE 12.15

Suppose a 2 × 2 matrix **A** and a 2 × 2 matrix **B** are as shown below.

$$A = \begin{bmatrix} 1 & 3 \\ 2 & 5 \end{bmatrix} \qquad B = \begin{bmatrix} 4 & 1 \\ 3 & 7 \end{bmatrix}$$

Find the sum of the two matrices.

Solution

Since the two matrices **A** and **B** have the same dimensions we can add them. The sum of the two matrices, denoted by **A** + **B**, is

$$A + B = \begin{bmatrix} 1 & 3 \\ 2 & 5 \end{bmatrix} + \begin{bmatrix} 4 & 1 \\ 3 & 7 \end{bmatrix} = \begin{bmatrix} (1+4) & (3+1) \\ (2+3) & (5+7) \end{bmatrix} = \begin{bmatrix} 5 & 4 \\ 5 & 12 \end{bmatrix}$$

EXAMPLE 12.16

For the 2 × 3 matrix **A** and 2 × 3 matrix **B** shown below, find **A** + **B**.

$$A = \begin{bmatrix} 3 & 4 & -6 \\ 9 & 1 & 1 \end{bmatrix} \qquad B = \begin{bmatrix} 1 & 7 & 4 \\ 0 & -9 & 5 \end{bmatrix}$$

Solution

The sum of these two matrices is

$$\mathbf{A} + \mathbf{B} = \begin{bmatrix} (3 + 1) & (4 + 7) & (-6 + 4) \\ (9 + 0) & (1 - 9) & (1 + 5) \end{bmatrix} = \begin{bmatrix} 4 & 11 & -2 \\ 9 & -8 & 6 \end{bmatrix}$$

Using Example 12.16, you can easily verify that the addition of two matrices $\mathbf{A}$ and $\mathbf{B}$ is commutative; that is,

$$\mathbf{A} + \mathbf{B} = \mathbf{B} + \mathbf{A}$$

**multiplication
of matrices**

We can multiply a matrix $\mathbf{A}$ *times a matrix* $\mathbf{B}$ *if the number of columns in* $\mathbf{A}$ *is equal to the number of rows in* $\mathbf{B}$. For example, we could multiply a 2×3 matrix $\mathbf{A}$ times a 3×2 matrix $\mathbf{B}$ because the number of columns in $\mathbf{A}$ and the number of· rows in $\mathbf{B}$ is the same, namely, 3. Similarly, we could multiply a 1×5 matrix $\mathbf{A}$ times a 5×4 matrix $\mathbf{B}$.

The multiplication of matrices is somewhat complex. Basically, an element in the product matrix $\mathbf{AB}$ is found by multiplying each element in a row in $\mathbf{A}$ times each element in the corresponding column in $\mathbf{B}$ and adding the results. This procedure is best illustrated and understood by working some examples.

EXAMPLE 12.17

Let the 2×2 matrix $\mathbf{A}$ and the 2×1 matrix $\mathbf{B}$ be given as follows:

$$\mathbf{A} = \begin{bmatrix} 3 & 1 \\ 2 & 4 \end{bmatrix} \qquad \mathbf{B} = \begin{bmatrix} 1 \\ 2 \end{bmatrix}$$

Find the product $\mathbf{AB}$.

Solution

The first thing to note in the multiplication of two matrices $\mathbf{A}$ and $\mathbf{B}$ is that the resulting product $\mathbf{AB}$ will be a new matrix with dimensions given by

number of rows of $\mathbf{AB}$ = number of rows of $\mathbf{A}$

number of columns of $\mathbf{AB}$ = number of columns of $\mathbf{B}$.

In our example

$$\underset{2 \times 2}{\mathbf{A}} \quad \underset{2 \times \underline{1}}{\mathbf{B}}$$

will be a new 2×1 matrix. There will be two elements in the new matrix, one in the first row and first column and one in the second row and first column.

The element in the first row, first column, is found by multiplying the elements of the first *row* of $\mathbf{A}$ times the corresponding elements of the first *column* of $\mathbf{B}$ and adding the result:

$$\begin{bmatrix} 3 & 1 \\ 2 & 4 \end{bmatrix}\begin{bmatrix} 1 \\ 2 \end{bmatrix} = \begin{bmatrix} 3 \cdot 1 + 1 \cdot 2 \end{bmatrix} = \begin{bmatrix} 5 \\ \end{bmatrix}$$

The element in the second row, first column, of the new matrix is found by multiplying the elements of the *second* row of **A** times the elements of the *first* column of **B** and adding the result.

$$\begin{bmatrix} 3 & 1 \\ 2 & 4 \end{bmatrix}\begin{bmatrix} 1 \\ 2 \end{bmatrix} = \begin{bmatrix} \\ 2 \cdot 1 + 4 \cdot 2 \end{bmatrix} = \begin{bmatrix} \\ 10 \end{bmatrix}$$

The product **AB** is then

$$\mathbf{AB} = \begin{bmatrix} 3 & 1 \\ 2 & 4 \end{bmatrix}\begin{bmatrix} 1 \\ 2 \end{bmatrix} = \begin{bmatrix} 5 \\ 10 \end{bmatrix}$$

EXAMPLE 12.18

Let a 2×2 matrix **A** and a 2×3 matrix **B** be given as follows:

$$\mathbf{A} = \begin{bmatrix} 1 & 2 \\ 0 & 3 \end{bmatrix} \qquad \mathbf{B} = \begin{bmatrix} 4 & 4 & 0 \\ 8 & 3 & 2 \end{bmatrix}$$

a. Find **AB**.
b. Find **BA**.

Solution

a. Again we will illustrate this matrix multiplication in separate parts. First we know that the new matrix will be a 2×3 matrix.

$$\begin{matrix} \mathbf{A} & \mathbf{B} \\ {\scriptstyle 2 \times 2} & {\scriptstyle 2 \times \underline{3}} \end{matrix}$$

The element in the first row, first column, of the new matrix (called the (1, 1) element) is obtained by multiplying the elements in the first row of **A** times corresponding elements in the first column of **B** and adding:

$$(1, 1) \text{ element} = \begin{bmatrix} 1 & 2 \\ 0 & 3 \end{bmatrix}\begin{bmatrix} 4 & 4 & 0 \\ 8 & 3 & 2 \end{bmatrix} = \begin{bmatrix} 1 \cdot 4 + 2 \cdot 8 & & \\ & & \end{bmatrix}$$

The element in the first row, second column (called the (1, 2) element), of the new matrix is formed by multiplying the elements of the first row of **A** and the second column of **B** and adding.

$$(1, 2) \text{ element} = \begin{bmatrix} 1 & 2 \\ 0 & 3 \end{bmatrix}\begin{bmatrix} 4 & 4 & 0 \\ 8 & 3 & 2 \end{bmatrix} = \begin{bmatrix} & 1 \cdot 4 + 2 \cdot 3 & \\ & & \end{bmatrix}$$

The remaining elements are found in a similar manner.

$$(1, 3) \text{ element} = \begin{bmatrix} 1 & 2 \\ 0 & 3 \end{bmatrix}\begin{bmatrix} 4 & 4 & 0 \\ 8 & 3 & 2 \end{bmatrix} = \begin{bmatrix} & & 1 \cdot 0 + 2 \cdot 2 \\ & & \end{bmatrix}$$

$$(2, 1) \text{ element} = \begin{bmatrix} 1 & 2 \\ 0 & 3 \end{bmatrix}\begin{bmatrix} 4 & 4 & 0 \\ 8 & 3 & 2 \end{bmatrix} = \begin{bmatrix} & & \\ 0 \cdot 4 + 3 \cdot 8 & & \end{bmatrix}$$

$$(2, 2) \text{ element} = \begin{bmatrix} 1 & 2 \\ 0 & 3 \end{bmatrix}\begin{bmatrix} 4 & 4 & 0 \\ 8 & 3 & 2 \end{bmatrix} = \begin{bmatrix} & & \\ & 0 \cdot 4 + 3 \cdot 3 & \end{bmatrix}$$

$$(2, 3) \text{ element} = \begin{bmatrix} 1 & 2 \\ 0 & 3 \end{bmatrix}\begin{bmatrix} 4 & 4 & 0 \\ 8 & 3 & 2 \end{bmatrix} = \begin{bmatrix} & & \\ & & 0 \cdot 0 + 3 \cdot 2 \end{bmatrix}$$

Combining out results we have

$$\mathbf{AB} = \begin{bmatrix} 20 & 10 & 4 \\ 24 & 9 & 6 \end{bmatrix}$$

b. To multiply

$$\underset{2 \times 3}{\mathbf{B}} \qquad \text{times} \qquad \underset{2 \times 2}{\mathbf{A}}$$

the number of columns in **B** must equal the number of rows in **A**. Since this does not hold in this example, we cannot multiply **B** times **A**. *Note: This also implies that* **BA** *is not necessarily equal to* **AB**.

transpose A′   The **transpose** of a matrix **A** is a new matrix, denoted by **A′** (called "A prime"), formed by interchanging the corresponding rows and columns of the original matrix **A**. Thus the first row of **A** becomes the first column of **A′**, and so on.

EXAMPLE 12.19   Determine the transpose of the matrix

$$\mathbf{A} = \begin{bmatrix} 1 & 5 & 6 \\ 3 & 2 & 4 \end{bmatrix}$$

Solution   We obtain the transpose of **A** by forming a new matrix, where the first and second rows of **A** become the first and second columns of the new matrix.

$$\mathbf{A'} = \begin{bmatrix} 1 & 3 \\ 5 & 2 \\ 6 & 4 \end{bmatrix}$$

Note that **A** had dimensions 2×3, whereas **A′** has the reverse dimensions, 3×2.

determinant   We can associate a number with every square matrix (that is, the number of rows equals the number of columns), which we call the **determinant** of a matrix. The determinant of a 1×1 matrix, that is, a matrix with only one element, is the value of that element. Thus the determinant of the matrix $\mathbf{A} = [4]$ is 4. The computation of determinants will be illustrated for 2×2 and 3×3 matrices.

Let **A** be a 2×2 matrix with elements

$$\mathbf{A} = \begin{bmatrix} a_{11} & a_{12} \\ a_{21} & a_{22} \end{bmatrix}$$

The determinant of the matrix **A**, denoted by $|\mathbf{A}|$, is the quantity

$$|\mathbf{A}| = a_{11}a_{22} - a_{21}a_{12}$$

Note that the determinant of a 2×2 matrix is formed by multiplying elements along the diagonal of the matrix from upper left to lower right and then subtracting the product of the diagonal elements from lower left to upper right (see Figure 12.8).

FIGURE 12.8

Calculations for the
Determinant of a
2×2 Matrix

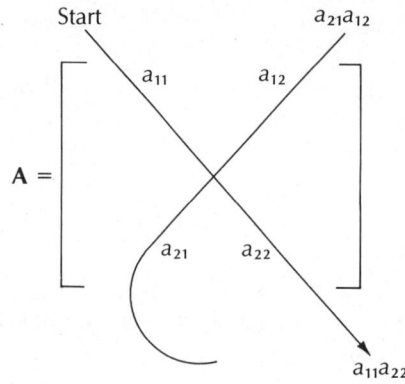

EXAMPLE 12.20

Find the determinant of the matrix

$$A = \begin{bmatrix} 2 & 3 \\ 1 & 5 \end{bmatrix}$$

Solution

Using the computational formula, we have

$$|A| = 2(5) - 1(3) = 10 - 3 = 7$$

The determinant of a 3×3 matrix can be computed in a similar way. Let

$$A = \begin{bmatrix} a_{11} & a_{12} & a_{13} \\ a_{21} & a_{22} & a_{23} \\ a_{31} & a_{32} & a_{33} \end{bmatrix}$$

Then the determinant of this matrix is

$$|A| = a_{11}a_{22}a_{33} + a_{21}a_{32}a_{13} + a_{31}a_{23}a_{12} - a_{13}a_{22}a_{31} - a_{23}a_{32}a_{11} - a_{33}a_{21}a_{12}$$

Although this computation is more difficult than that for the 2×2 matrix, it can be remembered easily using the procedure shown in Figure 12.9.

EXAMPLE 12.21

Find the determinant of the matrix

$$A = \begin{bmatrix} 1 & 3 & 1 \\ 2 & 1 & 1 \\ 1 & 3 & 3 \end{bmatrix}$$

Solution

The first three terms of $|A|$, starting in the upper-left corner, are

$$1(1)(3) = 3 \qquad 2(3)(1) = 6 \qquad 1(1)(3) = 3$$

The last three terms are

$$1(1)(1) = 1 \qquad 1(3)(1) = 3 \qquad 3(2)(3) = 18$$

Combining, we have

$$|A| = 3 + 6 + 3 - 1 - 3 - 18 = -10$$

FIGURE 12.9

Computation of $|\mathbf{A}|$ for a 3 × 3 Matrix

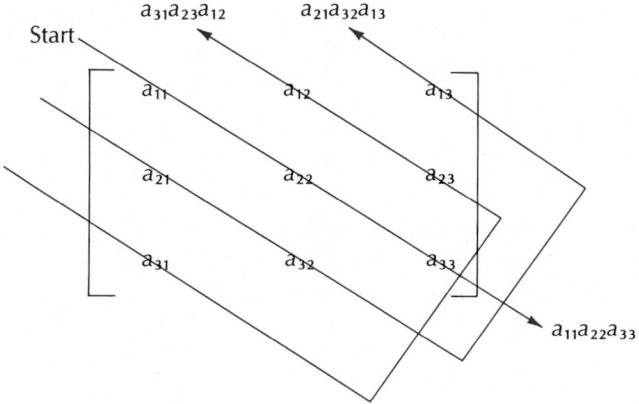

(a) Computation of first three terms of $|\mathbf{A}|$

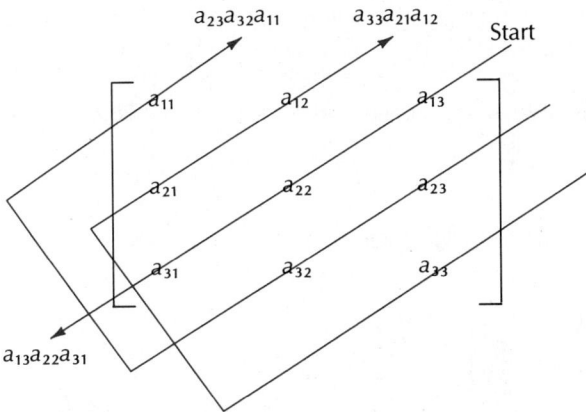

(b) Computation of last three terms of $|\mathbf{A}|$

Although there are computational formulas for calculating the determinant of a square matrix with more than three rows (and columns), the computational tricks for 2 × 2 and 3 × 3 matrices do not extend to higher-dimensional matrices. *When working with larger matrices, we will use a software package to obtain a solution.*

cofactor of element   We can now use the definition of a determinant to define a new concept. The **cofactor** associated with the (i, j) element of a square matrix $\mathbf{A}$ is defined to be $(-1)^{i+j}$ times the determinant of the matrix formed by deleting all the elements in row i and column j of the matrix $\mathbf{A}$. We illustrate this idea with an example.

EXAMPLE 12.22   Find the cofactors of the (1, 2) element and the (2, 2) element of the matrix

$$\mathbf{A} = \begin{bmatrix} 1 & 3 & 1 \\ 2 & 1 & 1 \\ 1 & 3 & 3 \end{bmatrix}$$

Solution

To find the cofactor of the (1, 2) element of **A**, we delete the first row and second column of **A**.

$$\begin{bmatrix} 1 & 3 & 1 \\ 2 & 1 & 1 \\ 1 & 3 & 3 \end{bmatrix}$$

The determinant of the matrix that remains is $2(3) - 1(1) = 6 - 1 = 5$, so the cofactor of the (1, 2) element is $(-1)^{1+2}5 = -5$.

To find the cofactor of the (2, 2) element of **A**, we first delete the second row and second column of the matrix.

$$\begin{bmatrix} 1 & 3 & 1 \\ 2 & 1 & 1 \\ 1 & 3 & 3 \end{bmatrix}$$

The determinant of the remaining matrix is $1(3) - 1(1) = 2$. Hence the cofactor of the (2, 2) element is $(-1)^{2+2}2 = 2$.

cofactor matrix

A **cofactor matrix** associated with a square matrix **A** is a new matrix formed by replacing each element by its cofactor.

EXAMPLE 12.23

Find the cofactor matrix associated with the matrix

$$\mathbf{A} = \begin{bmatrix} 1 & 3 & 1 \\ 2 & 1 & 1 \\ 1 & 3 & 3 \end{bmatrix}$$

Solution

We have already found the cofactors associated with the (1, 2) and (2, 2) elements in Example 12.22. In a similar way we can find the cofactors of the other elements. Table 12.5 summarizes these computations.

TABLE 12.5   Computations for the Cofactor Matrix **A** in Example 12.23

Element (i, j)	Determinant after Deleting ith Row and jth Column	Cofactor of Element (i, j)
(1, 1)	$1(3) - 3(1) = 0$	0
(1, 2)	see Example 12.22	-5
(1, 3)	$2(3) - 1(1) = 5$	$(-1)^{1+3}5 = 5$
(2, 1)	$3(3) - 3(1) = 6$	$(-1)^{2+1}6 = -6$
(2, 2)	see Example 12.22	2
(2, 3)	$1(3) - 1(3) = 0$	0
(3, 1)	$3(1) - 1(1) = 2$	$(-1)^{3+1}2 = 2$
(3, 2)	$1(1) - 2(1) = -1$	$(-1)^{3+2}(-1) = 1$
(3, 3)	$1(1) - 2(3) = -5$	$(-1)^{3+3}(-5) = -5$

Hence the cofactor matrix is

$$\begin{bmatrix} 0 & -5 & 5 \\ -6 & 2 & 0 \\ 2 & 1 & -5 \end{bmatrix}.$$

inverse A^{-1}

The final matrix concept that we present makes use of many of the previous operations and results we have discussed. The **inverse** of a square matrix **A**, denoted by $\mathbf{A}^{-1}$ (read "A inverse"), is a new matrix with the property that both

$$\mathbf{AA}^{-1} = \mathbf{I} \qquad \text{and} \qquad \mathbf{A}^{-1}\mathbf{A} = \mathbf{I}$$

It should be noted that not all square matrices have an inverse; only those with a nonzero determinant have inverses.

Now we will show how we can apply our previous results to obtain the inverse of a 2×2 matrix and a 3×3 matrix.

A PROCEDURE FOR OBTAINING THE INVERSE OF A 2×2 OR 3×3 SQUARE MATRIX A

1. Find the determinant of **A**. If the determinant is nonzero, proceed; otherwise, the inverse does not exist.
2. Find the cofactor matrix associated with **A**.
3. Find the transpose of the cofactor matrix.
4. Divide each element of this transposed matrix by $|\mathbf{A}|$. The resulting matrix is $\mathbf{A}^{-1}$.

Note: For matrices larger than 3×3, we will find the inverse of a matrix by using a computer software package.

EXAMPLE 12.24

Find the inverse of the matrix

$$\mathbf{A} = \begin{bmatrix} 1 & 3 & 1 \\ 2 & 1 & 1 \\ 1 & 3 & 3 \end{bmatrix}$$

Solution

Referring to Examples 12.21 and 12.23 and following the steps for obtaining $\mathbf{A}^{-1}$, we have the following.

1. $|\mathbf{A}| = -10$.
2. The cofactor matrix of **A** is

$$\begin{bmatrix} 0 & -5 & 5 \\ -6 & 2 & 0 \\ 2 & 1 & -5 \end{bmatrix}$$

3. The transpose of the cofactor matrix is

$$\begin{bmatrix} 0 & -6 & 2 \\ -5 & 2 & 1 \\ 5 & 0 & -5 \end{bmatrix}$$

4. Dividing each element by $|\mathbf{A}| = -10$, we find the inverse of $\mathbf{A}$ to be

$$\mathbf{A}^{-1} = \begin{bmatrix} 0 & .6 & -.2 \\ .5 & -.2 & -.1 \\ -.5 & 0 & .5 \end{bmatrix}$$

Note that

$$\mathbf{A}\mathbf{A}^{-1} = \begin{bmatrix} 1 & 3 & 1 \\ 2 & 1 & 1 \\ 1 & 3 & 3 \end{bmatrix} \begin{bmatrix} 0 & .6 & -.2 \\ .5 & -.2 & -.1 \\ -.5 & 0 & .5 \end{bmatrix} = \begin{bmatrix} 1 & 0 & 0 \\ 0 & 1 & 0 \\ 0 & 0 & 1 \end{bmatrix}$$

$$\mathbf{A}^{-1}\mathbf{A} = \begin{bmatrix} 0 & .6 & -.2 \\ .5 & -.2 & -.1 \\ -.5 & 0 & .5 \end{bmatrix} \begin{bmatrix} 1 & 3 & 1 \\ 2 & 1 & 1 \\ 1 & 3 & 3 \end{bmatrix} = \begin{bmatrix} 1 & 0 & 0 \\ 0 & 1 & 0 \\ 0 & 0 & 1 \end{bmatrix}$$

Hence our calculations are correct.

EXAMPLE 12.25 Find the inverse of the matrix

$$\mathbf{A} = \begin{bmatrix} 5 & 0 & 0 \\ 0 & 6 & 0 \\ 0 & 0 & 2 \end{bmatrix}$$

Solution

diagonal matrix

Although we could follow the general procedure for the inverse of a matrix, we can simplify our work when we have a **diagonal matrix**, that is, a matrix with nonzero elements on the diagonal from the upper left to the lower right and with all other elements being 0. The inverse of any diagonal matrix $\mathbf{A}$ is a matrix with each diagonal element equal to the inverse of the corresponding diagonal element of $\mathbf{A}$. All other elements are zero. Thus we can immediately write $\mathbf{A}^{-1}$ as

$$\mathbf{A}^{-1} = \begin{bmatrix} \frac{1}{5} & 0 & 0 \\ 0 & \frac{1}{6} & 0 \\ 0 & 0 & \frac{1}{2} \end{bmatrix}$$

EXERCISES

(optional) **12.54** Consider the two matrices

$$\mathbf{A} = \begin{bmatrix} 2 & 1 \\ 3 & 2 \end{bmatrix} \qquad \mathbf{B} = \begin{bmatrix} 2 & 1 \\ 1 & 1 \end{bmatrix}$$

a. Compute $\mathbf{A} + \mathbf{B}$.
b. Verify that $\mathbf{A} + \mathbf{B} = \mathbf{B} + \mathbf{A}$.
c. Compute $\mathbf{A}\mathbf{B}$.

12.55 Refer to Exercise 12.54.
a. Find the matrices $\mathbf{A}'$ and $\mathbf{B}'$.
b. Compute $\mathbf{A}^{-1}$.

12.56 Refer to Exercise 12.54. Compute $(\mathbf{A}\mathbf{B})^{-1}$.

12.57 Consider the 3×3 matrix

$$\mathbf{A} = \begin{bmatrix} 3 & 0 & 2 \\ 0 & 2 & 0 \\ 2 & 0 & 2 \end{bmatrix}$$

a. Find $\mathbf{A}'$.
b. Compute $|\mathbf{A}|$.

12.58 Refer to Exercise 12.57. Find $\mathbf{A}^{-1}$. Verify that $\mathbf{A}\mathbf{A}^{-1} = \mathbf{A}^{-1}\mathbf{A} = \mathbf{I}$.

12.59 Consider the matrix

$$\mathbf{A} = \begin{bmatrix} 1 & 1 \\ 1 & 2 \\ 1 & 3 \end{bmatrix}$$

a. Find $\mathbf{A}'$.
b. Compute $\mathbf{A}'\mathbf{A}$.

12.60 Refer to Exercise 12.59. Compute $(\mathbf{A}'\mathbf{A})^{-1}$.

12.61 Refer to Exercises 12.59 and 12.60. Let

$$\mathbf{B} = \begin{bmatrix} 10 \\ 22 \end{bmatrix}$$

Compute $(\mathbf{A}'\mathbf{A})^{-1}\mathbf{B}$.

12.62 Consider the two matrices

$$\mathbf{A} = \begin{bmatrix} 1 & -1 & 1 \\ 2 & 0 & 1 \\ 3 & 1 & 3 \end{bmatrix} \qquad \mathbf{B} = \begin{bmatrix} 1 & 0 & 0 \\ 0 & 1 & 0 \\ 0 & 0 & 1 \end{bmatrix}$$

a. Compute $\mathbf{A} + \mathbf{B}$.
b. Compute $\mathbf{A}\mathbf{B}$.

12.63 Refer to Exercise 12.62.
a. Find $|\mathbf{A}|$.
b. Compute $\mathbf{A}^{-1}$.

13 MORE ON MULTIPLE REGRESSION

13.1 INTRODUCTION

In Chapter 12 we presented the background information needed in order to use multiple regression. We discussed the general linear model and its use in multiple regression and introduced the normal equations, a set of simultaneous equations used in obtaining least squares estimates for the βs of a multiple regression equation. Next we presented standard errors associated with the $\hat{\beta}_j$ and their use in inferences about a single parameter β_j, a set of βs, $E(y)$ and a future value of y. Finally we condensed all of these inferential techniques using matrices (in an optional appendix).

This chapter is devoted to putting multiple regression into practice. How does one begin to develop an appropriate multiple regression for a given problem? While there are no hard and fast rules, we can offer a few hints to enable you to put multiple regression into practice.

First, for each problem you must decide upon the dependent variable and candidate independent variables for the regression equation. This selection process will be discussed in Section 13.2. Next, in Section 13.3, we consider how one selects the form of the multiple regression equation. The final step in the process of developing a multiple regression is to check for violation of the underlying assumptions. Tools for assessing the validity of the assumptions will be discussed in Section 13.4.

Following these steps *once* for a given problem will not ensure that you have an appropriate model. Rather, the regression equation seems to evolve as these steps are applied repeatedly, depending on the particular problem. For example, having considered candidate independent variables (step 1) and selected the form for a regression model involving some of these variables (step 2), we may find that certain assumptions have been violated (step 3). This will mean that we may have to return to either step 1 or step 2, but, hopefully, we have learned from our previous deliberations and can modify the variables under consideration and/or the model(s) selected for consideration. Eventually, a regression model will emerge that meets the needs of the experimenter. Then the analysis techniques of Chapter 12 can be used to draw inferences about model parameters $E(y)$ and y.

13.2 SELECTING THE VARIABLES (Step 1)

The problem of trying to select reasonable candidate independent variables for inclusion in a multiple regression model can be a difficult task, but the time and attention paid to this selection process will reap benefits later in the form of better predictions. The input of a person knowledgeable in the subject matter field is a valuable source of advice on reasonable (independent) variables that could influence the response (dependent variable) of interest.

For example, in trying to predict annual sales for a recently released personal computer, it would be wise to consult an electrical engineer and perhaps other technical people to learn about the unique software and hardware features of the computer relative to the competition. Is it a new generation computer or merely a minor modification of what's presently available? In addition, it would be wise to consult with a marketing person who has more than passing knowledge of factors that have affected the marketability of other microcomputers. With information obtained from these two sources, one could piece together a list of candidate independent variables. Ideally, this list would contain variables that are closely related to the dependent variable, but not to one another.

EXAMPLE 13.1

Construct a list of independent variables that might be useful in developing a multiple regression model for predicting sales of a new microcomputer.

Solution

The list of candidate variables is quite long. For convenience we have divided the list into those variables related to the attributes of the microprocessor and those related to the competition and the general market. The list, however, is not exhaustive.

Microprocessor	Competition/Market
1. Purchase price	1. Number of competitors
2. Ease of use (Is it user friendly?)	2. Purchase price of market leader
3. Specificity	3. Maintenance fee of market leader
4. Annual maintenance fee	4. Volume of the market
5. Number of possible applications	5. GNP for last three quarters
6. Number of unique features	
7. Can it be copied?	

One way to sort out which independent variables should be included in the regression model from the list of variables generated from discussions with experts is to resort to any one of a number of selection procedures. We will consider several of these in this text; for further details, the reader can consult with Draper and Smith (1981).

The first selection procedure involves performing *all possible regressions* with the dependent variable and one or more of the independent variables from the list of candidate variables. Obviously this approach should not be attempted unless the analyst has access to a computer with suitable software and sufficient core to run a large number of regression models relatively efficiently.

For purposes of illustration, we will use hypothetical data on prescription sales data (volume per month) obtained for a random sample of 20 independent pharmacies. These data along with data on the total floor space, percent of floor space allocated to the prescription department, the number of parking spaces available for the store, whether or not the pharmacy is in a shopping center, and the per capita income for the surrounding community are recorded in Table 13.1.

Before running all possible regressions for the data of Table 13.1, we need to consider what criterion should be used to select the best-fitting equation from all possible regressions. The first and perhaps simplest criterion for selecting the best regression equation from the set of all possible regression equations is to compute an estimate of the error variance σ_ε^2 using $s_\varepsilon^2 = \text{MSE} = \text{SSE}/[n - (k + 1)]$. Since this quantity is used in most inferences (statistical tests and confidence intervals) about model parameters and $E(y)$, it would seem reasonable to choose the model that has the smallest value of s_ε^2. A second criterion makes use of the *coefficient of determination R^2* for each model; by examining in detail the models that have the highest R^2 values, we can see whether there is some consistent pattern that suggests the number and identity of the variables to include in the model.

EXAMPLE 13.2   Refer to the data of Table 13.1. Use the R^2 criterion to determine the best-fitting regression equation for 1, 2, 3, and 4 independent variables.

TABLE 13.1   Data on 20 Independent Pharmacies

OBS	PHARMACY	VOLUME	FLOOR__SP	PRESC__RX	PARKING	SHOPCNTR	INCOME
1	1	22	4900	9	40	1	18
2	2	19	5800	10	50	1	20
3	3	24	5000	11	55	1	17
4	4	28	4400	12	30	0	19
5	5	18	3850	13	42	0	10
6	6	21	5300	15	20	1	22
7	7	29	4100	20	25	0	8
8	8	15	4700	22	60	1	15
9	9	12	5600	24	45	1	16
10	10	14	4900	27	82	1	14
11	11	18	3700	28	56	0	12
12	12	19	3800	31	38	0	8
13	13	15	2400	36	35	0	6
14	14	22	1800	37	28	0	4
15	15	13	3100	40	43	0	6
16	16	16	2300	41	20	0	5
17	17	8	4400	42	46	1	7
18	18	6	3300	42	15	0	4
19	19	7	2900	45	30	1	9
20	20	17	2400	46	16	0	3

N = 20

TABLE 13.2   Best-Fitting Models, R^2 Criterion, Example 13.2

Number of Independent Variables	R^2	Variables
1	0.439	Prescription sales
2	0.666	Floor space, prescription sales
3	0.691	Floor space, prescription sales, shopping center
4	0.694	All except per capita income

Solution | SAS output is provided here, and the best-fitting equations are summarized in Table 13.2.

REGRESSION ANALYSES
PROC RSQUARE—ALL POSSIBLE SUBSETS ANALYSIS

N = 20   REGRESSION MODELS FOR DEPENDENT VARIABLE VOLUME

NUMBER IN MODEL	R-SQUARE	C(P)	VARIABLES IN MODEL
1	0.00480421	30.45388047	PARKING
1	0.03353172	29.11293360	FLOOR__SP
1	0.04105340	28.76183600	SHOPCNTR
1	0.14798995	23.77023759	INCOME
1	0.43933184	10.17094219	PRESC__RX
2	0.04210776	30.71262010	PARKING SHOPCNTR
2	0.06855667	29.47803470	FLOOR__SP PARKING
2	0.20543099	23.08899693	PARKING INCOME
2	0.23487329	21.71468547	FLOOR__SP INCOME
2	0.25653635	20.70349407	FLOOR__SP SHOPCNTR
2	0.49576794	9.53661080	SHOPCNTR INCOME
2	0.53142435	7.87223587	PRESC__RX PARKING
2	0.54748785	7.12242198	PRESC__RX INCOME
2	0.64706473	2.47435928	PRESC__RX SHOPCNTR
2	0.66566267	1.60624219	FLOOR__SP PRESC__RX
3	0.25569607	22.74271718	FLOOR__SP PARKING INCOME
3	0.26507110	22.30510820	FLOOR__SP PARKING SHOPCNTR
3	0.49828073	11.41931841	PARKING SHOPCNTR INCOME
3	0.50012580	11.33319388	FLOOR__SP SHOPCNTR INCOME
3	0.60243233	6.55771633	PRESC__RX PARKING INCOME
3	0.64711563	4.47198330	PRESC__RX SHOPCNTR INCOME
3	0.66259120	3.74961255	PRESC__RX PARKING SHOPCNTR
3	0.66641145	3.57129027	FLOOR__SP PRESC__RX INCOME
3	0.67943313	2.96346249	FLOOR__SP PRESC__RX PARKING
3	0.69072432	2.43641080	FLOOR__SP PRESC__RX SHOPCNTR
4	0.50128901	13.27889728	FLOOR__SP PARKING SHOPCNTR INCOME
4	0.66300855	5.73013127	PRESC__RX PARKING SCHOPCNTR INCOME
4	0.68058567	4.90966443	FLOOR__SP PRESC__RX PARKING INCOME
4	0.69326657	4.31774327	FLOOR__SP PRESC__RX SHOPCNTR INCOME
4	0.69873953	4.06227626	FLOOR__SP PRESC__RX PARKING SHOPCNTR
5	0.70007369	6.00000000	FLOOR__SP PRESC__RX PARKING SHOPCNTR INCOME

Although there is a good jump in R^2 going from one to two independent variables, very little improvement is seen thereafter. Hence the best-fitting model based on the R^2 criterion involves the independent variables floor space and prescription sales.

One problem with using R^2 as a criterion for the best-fitting regression equation is that R^2 increases for each independent variable, even when the new x has very little predictive power. Other possible criteria for selecting the best regression that do not increase with the addition of each are presented here.

We should keep in mind that the object of our search is to choose the subset of independent variables that generates the best prediction equation for *future* values of y; unfortunately, however, since we do not know these future values, we focus on criteria that choose the best fitting regression equations to the known sample y-values. One possible bridge between this emphasis on the best fit to the known sample y-values and that on choosing the best predictor of future y-values is to split the sample data into two parts—one part used for fitting the various regression equations and the other part for validating how well the prediction equations can predict "future" values. Although there is no universally accepted rule for deciding how many of the data should be included in the "fitting" portion of the sample and how many go into the "validating" portion of the sample, it's reasonable to split the total sample in half, provided the total sample size n is greater than $2p + 20$ where p is the number of parameters in the largest potential regression model. A possible criterion for the best prediction equation would be to minimize $\sum (y_i - \hat{y}_i)^2$ for the validating portion of the total sample.

Once the regression model is selected from the data splitting approach, the entire set of sample data is used to obtain the final prediction equation. So, even though it appears we would only use part of the data, the entire data set is used to obtain the final prediction equation.

Observations do cost money, however, and it may be impractical to obtain enough observations to apply the data splitting approach for choosing the best-fitting regression equation. In these situations, a form of validation can be accomplished using the PRESS statistic. For a sample of y-values and a proposed regression model relating y to a set of xs, we first remove the first observation and fit the model using the remaining $n - 1$ observation. Based on the fitted equation we estimate the first observation (denoted by $\hat{y}_1^*$) and compute the residual $y_1 - \hat{y}_1^*$. This process is repeated $n - 1$ times, successively removing the second, third, . . . , nth observation, each time computing the residual for the removed observation. The PRESS statistic is defined as

$$PRESS = \sum_{i=1}^{n} (y_i - \hat{y}_i^*)^2$$

The model that gives the smallest value for the PRESS statistic is chosen as the best-fitting model.

EXAMPLE 13.3   Compute the PRESS statistic for the data of Table 13.1 to determine the best-fitting regression equation.

Solution   SAS output is provided here. The best-fitting model based on the lowest value of the PRESS statistic involves the independent variables floor space and prescription sales.

```
                              LISTING OF DATA

    OBS      PHARMACY    VOLUME     FLOOR_SP    PRESC_RX     PARKING    SHOPCNTR    INCOME

     1          1          22        4900          9           40         1          18
     2          2          19        5800         10           50         1          20
     3          3          24        5000         11           55         1          17
     4          4          28        4400         12           30         0          19
     5          5          18        3850         13           42         0          10
     6          6          21        5300         15           20         1          22
     7          7          29        4100         20           25         0           8
     8          8          15        4700         22           60         1          15
     9          9          12        5600         24           45         1          16
    10         10          14        4900         27           82         1          14
    11         11          18        3700         28           56         0          12
    12         12          19        3800         31           38         0           8
    13         13          15        2400         36           35         0           6
    14         14          22        1800         37           28         0           4
    15         15          13        3100         40           43         0           6
    16         16          16        2300         41           20         0           5
    17         17           8        4400         42           46         1           7
    18         18           6        3300         42           15         0           4
    19         19           7        2900         45           30         1           9
    20         20          17        2400         46           16         0           3

   N=         20
```

```
                         REGRESSION ANALYSIS

PRESS STATISTIC FOR REGRESSION MODELS. DEPENDENT VARIABLE VOLUME

  NUMBER     PRESS STATISTIC      VARIABLES
    IN                            IN MODEL
  MODEL

     1          907.636          PARKING
                887.545          SHOPCNTR
                869.668          FLOOR_SP
                772.163          INCOME
                516.391          PRESC_RX

     2          975.912          PARKING SHOPCNTR
                916.644          PARKING FLOOR_SP
                797.404          FLOOR_SP INCOME
                787.578          PARKING INCOME
                762.507          FLOOR_SP SHOPCNTR
                547.150          INCOME PRESC_RX
                485.820          SHOPCNTR INCOME
                479.976          PARKING PRESC_RX
                368.757          SHOPCNTR PRESC_RX
                347.007          FLOOR_SP PRESC_RX

     3          890.550          PARKING FLOOR_SP INCOME
                819.792          PARKING FLOOR_SP SHOPCNTR
                602.214          FLOOR_SP SHOPCNTR INCOME
                523.006          PARKING SHOPCNTR INCOME
                513.246          PARKING INCOME PRESC_RX
                482.387          FLOOR_SP INCOME PRESC_RX
                455.424          SHOPCNTR INCOME PRESC_RX
                378.166          PARKING SHOPCNTR PRESC_RX
                371.671          PARKING FLOOR_SP PRESC_RX
                370.843          FLOOR_SP SHOPCNTR PRESC_RX

     4          684.190          PARKING FLOOR_SP SHOPCNTR INCOME
                513.468          PARKING FLOOR_SP INCOME PRESC_RX
                471.086          FLOOR_SP SHOPCNTR INCOME PRESC_RX
                458.014          PARKING SHOPCNTR INCOME PRESC_RX
                405.832          PARKING FLOOR_SP SHOPCNTR PRESC_RX

     5          513.915          PARKING FLOOR_SP SHOPCNTR INCOME PRESC_RX
```

To this point we've considered criteria for selecting the best-fitting regression model from a subset of independent variables. In general, if we choose a model that leaves out one or more "important" predictor variables, our model is *underspecified* and the additional variability in the y-values that would be accounted for with these variables becomes part of the estimated error variance. At the other end of the spectrum, if we choose a model that contains one or more "extraneous" predictor variables, our model is *overspecified* and we stand the chance of having a *multicollinearity* problem. This problem will be dealt with later. The point is that a final criterion, based on the C_p statistic, seems to balance some pros and cons of previously presented selection criteria, along with the problems of over- and underspecification, to arrive at a choice of the best-fitting subset regression equation. The C_p statistic (see Mallows, 1973) is

$$C_p = \frac{SSE_p}{s_\varepsilon^2} - (n - 2p)$$

where SSE_p is the sum of squares for error from a model with p parameters (including β_0) and s_ε^2 is the mean square error from the regression equation with the largest number of independent variables. For a given selection problem compute C_p for every regression equation that is fit. Theory suggests that the best-fitting model should have $C_p \approx p$.

EXAMPLE 13.4

Refer to the output of Example 13.2. Determine the value of C_p for all possible regressions with 1, 2, 3, 4, and 5 independent variables. Select the best-fitting equation for 1, 2, 3, and 4 independent variables. Which regression equation seems to give the best overall fit, based on the C_p statistic?

Solution

The best-fitting models are summarized in Table 13.3. Based on the C_p criterion, there would be very little difference between the best-fitting models for 2, 3, or 4 independent variables in the model. The most "important" predictive variables seem to be floor space and prescription sales since they appear in the best-fitting models for 2, 3, and 4 independent variables. Note that these are the same important independent variables found in Example 13.2.

Best subset regression provides another procedure for finding the best-fitting regression equation from a set of candidate independent variables. The beauty of this procedure is that it uses an algorithm to avoid running all possible regressions. The user indicates the number (k) of best subset regressions desired. Some programs also allow the user to specify the criterion (for example, C_p or maximum R^2) but others fix the criterion. For instance in the Biomedical System of programs, BMDP9R allows for the best subset regression using the C_p statistic. Having indicated a value for k, the program computes the best k subset regressions on the basis of the C_p statistic and then identifies the best of the best. We will illustrate this with the data of Table 13.1.

TABLE 13.3   Best-Fitting Models, C_p Criterion

Number of Independent Variables	p	C_p	Variables
1	2	10.17	Prescription sales
2	3	1.61	Floor space, prescription sales
		2.47	Prescription sales, shopping center
3	4	2.96	Floor space, prescription sales, parking spaces
4	5	4.06	Floor space, prescription sales, parking spaces, shopping center

EXAMPLE 13.5 | Refer to the BDMP output shown here to find the best subset regression equation based on the C_p criterion for the data of Table 13.1.

NUMBER OF CASES READ 20
SUMMARY STATISTICS FOR EACH VARIABLE

VARIABLE	MEAN	STANDARD DEVIATION	COEFFICIENT OF VARIATION	SMALLEST VALUE	LARGEST VALUE	SMALLEST STANDARD SCORE	LARGEST STANDARD SCORE	SKEWNESS	KURTOSIS
3 FLOORSP	3932.50000	1177.67333	0.299472	1800.00000	5800.00000	−1.81	1.59	−0.18	−1.25
4 PRESC__RX	27.55000	12.99585	0.471719	9.00000	46.00000	−1.43	1.42	−0.05	−1.62
5 PARKING	38.80000	16.84168	0.434064	15.00000	82.00000	−1.41	2.57	0.60	−0.04
6 SHOPCNTR	0.45000	0.51042	1.134262	0.0	1.00000	−0.88	1.08	0.19	−2.06
7 INCOME	11.15000	6.01992	0.539903	3.00000	22.00000	−1.35	1.80	0.29	−1.44
2 VOLUME	17.15000	6.28511	0.366479	6.00000	29.00000	−1.77	1.89	0.02	−0.71

VALUES FOR KURTOSIS GREATER THAN ZERO INDICATE DISTRIBUTIONS
WITH HEAVIER TAILS THAN THE NORMAL DISTRIBUTION.

Solution | The output is shown here. The best-fitting equation involves floor space and the prescription sales as independent variables as we obtained in Example 13.2. The C_p value is 1.61 and the prediction equation is

$$\hat{y} = 48.291 - 0.004(\text{floor space}) - 0.582(\text{prescription sales})$$

STATISTICS FOR "BEST" SUBSET

MALLOWS' CP	1.61
SQUARED MULTIPLE CORRELATION	0.66566
MULTIPLE CORRELATION	0.81588
ADJUSTED SQUARED MULT. CORR.	0.62633
RESIDUAL MEAN SQUARE	14.760993
STANDARD ERROR OF EST.	3.842004
F-STATISTIC	16.92
NUMERATOR DEGREES OF FREEDOM	2
DENOMINATOR DEGREES OF FREEDOM	17
SIGNIFICANCE (TAIL PROB.)	0.0001

NOTE THAT THE ABOVE F-STATISTIC AND
ASSOCIATED SIGNIFICANCE TEND TO BE
LIBERAL WHENEVER A SUBSET OF VARIABLES
IS SELECTED BY THE CP OR ADJUSTED
R-SQUARE CRITERIA.

NO.	VARIABLE NAME	REGRESSION COEFFICIENT	STANDARD ERROR	STAND. COEF.	T-STAT.	2TAIL SIG.	TOLERANCE	CONTRIBUTION TO R-SQ
	INTERCEPT	48.2909	6.89043	7.683	7.01	0.000		
3	FLOOR__SP	−0.00384228	0.00113262	−0.720	−3.39	0.003	0.436658	0.22633
4	PRESC__RX	−0.581890	0.102637	−1.203	−5.67	0.000	0.436658	0.63213

THE CONTRIBUTION TO R-SQUARED FOR EACH VARIABLE IS THE AMOUNT
BY WHICH R-SQUARED WOULD BE REDUCED IF THAT VARIABLE WERE
REMOVED FROM THE REGRESSION EQUATION.

There are a number of other procedures that can be used to select the best regression and, although we won't spend a great deal more time on this subject, we will mention briefly the *backward elimination method* and *stepwise regression procedure*.

The backward elimination method begins with fitting the regression model, which contains all the candidate independent variables. For each independent variable x_j we compute

$$F_j = SSdrop_j/MSE \qquad j = 1, 2, \ldots$$

where $SSdrop_j$ is the drop in the sum of squares error obtained for the complete model, which contains all xs except x_j. MSE is the mean square error for the complete model. Let min F_j denote the smallest F_j value. If min $F_j < F_\alpha$ where α is the preselected significance level, remove the independent variable corresponding to min F_j from the regression equation. The backward elimination process then begins all over again with one variable removed from the list of candidate independent variables.

Backward elimination starts with the complete model with all independent variables entered and eliminates variables one at a time until a reasonable candidate regression model is found. This occurs when, in a particular step, min $F_j > F_\alpha$; the resulting complete model is the best-fitting regression equation. Stepwise regression, on the other hand, works in the other direction starting with the model $y = \beta_0 + \varepsilon$ and adding variables one at a time until a stopping criterion is satisfied. At the initial stage of the process, the first variable entered into the equation is the one with the largest F test for regression. At the second stage the two variables to be included in the model are the variables with the largest F test for regression of two variables. Note that the variable entered in the first step may not be included in the second step; that is, the best single variable may not be one of the best two variables. Because of this some people use a simplified stepwise regression (sometimes called *forward selection*) whereby once a variable is entered, it cannot be eliminated from the regression equation at a later stage.

EXAMPLE 13.6   Use the data of Example 13.2 to find the variables to be included in a regression equation based on backward elimination. Comment on your findings.

```
                    REGRESSION ANALYSIS, USING BACKWARD ELIMINATION

                BACKWARD ELIMINATION PROCEDURE FOR DEPENDENT VARIABLE VOLUME

STEP 0    ALL VARIABLES ENTERED      R SQUARE = 0.70007369      C(P) =   6.00000000

                    DF          SUM OF SQUARES       MEAN SQUARE          F      PROB>F

REGRESSION          5           525.44030541        105.08806108        6.54     0.0025
ERROR              14           225.10969459         16.07926390
TOTAL              19           750.55000000

                  B VALUE          STD ERROR         TYPE II SS          F      PROB>F

INTERCEPT       42.08710826
FLOOR_SP        -0.00241878        0.00183889        27.81923726        1.73     0.2095
PRESC_RX        -0.50046955        0.16429694       149.19783807        9.28     0.0087
PARKING         -0.03690284        0.06546687         5.10907792        0.32     0.5819
SHOPCNTR        -3.09957355        3.24983522        14.62673442        0.91     0.3564
INCOME           0.10666360        0.42742012         1.00135642        0.06     0.8066

BOUNDS ON CONDITION NUMBER:       7.823107,       117.1991
----------------------------------------------------------------------------------------
```

```
STEP 1     VARIABLE INCOME REMOVED      R SQUARE = 0.69873952      C(P) =    4.06227626

                      DF           SUM OF SQUARES          MEAN SQUARE             F       PROB>F

REGRESSION            4             524.43894899          131.10973725           8.70     0.0008
ERROR                15             226.11105101           15.07407007
TOTAL                19             750.55000000

                   B VALUE             STD ERROR          TYPE II SS              F       PROB>F

INTERCEPT        43.46782063
FLOOR_SP         -0.00228513           0.00170330          27.13112543           1.80     0.1997
PRESC_RX         -0.52910174           0.11386382         325.48983690          21.59     0.0003
PARKING          -0.03952477           0.06256589           6.01580808           0.40     0.5371
SHOPCNTR         -2.71387948           2.76799605          14.49041122           0.96     0.3424

BOUNDS ON CONDITION NUMBER:     5.071729,      46.98862
----------------------------------------------------------------------------------------------

STEP 2     VARIABLE PARKING REMOVED     R SQUARE = 0.69072432      C(P) =    2.43641080

                      DF           SUM OF SQUARES          MEAN SQUARE             F       PROB>F

REGRESSION            3             518.42314091          172.80771364          11.91     0.0002
ERROR                16             232.12685909           14.50792869
TOTAL                19             750.55000000

                   B VALUE             STD ERROR          TYPE II SS              F       PROB>F

INTERCEPT        42.82702645
FLOOR_SP         -0.00247284           0.00164539          32.76871130           2.26     0.1523
PRESC_RX         -0.52941361           0.11170410         325.87978038          22.46     0.0002
SHOPCNTR         -3.03834296           2.66836223          18.81002755           1.30     0.2716

BOUNDS ON CONDITION NUMBER:     4.917388,      30.31995
----------------------------------------------------------------------------------------------

                    REGRESSION ANALYSIS, USING BACKWARD ELIMINATION

               BACKWARD ELIMINATION PROCEDURE FOR DEPENDENT VARIABLE VOLUME

STEP 3     VARIABLE SHOPCNTR REMOVED    R SQUARE = 0.66566267      C(P) =    1.60624219

                      DF           SUM OF SQUARES          MEAN SQUARE             F       PROB>F

REGRESSION            2             499.61311336          249.80655668          16.92     0.0001
ERROR                17             250.93688664           14.76099333
TOTAL                19             750.55000000

                   B VALUE             STD ERROR          TYPE II SS              F       PROB>F

INTERCEPT        48.29085530
FLOOR_SP         -0.00384228           0.00113262         169.87259933          11.51     0.0035
PRESC_RX         -0.58189034           0.10263739         474.44587802          32.14     0.0001

BOUNDS ON CONDITION NUMBER:     2.290122,       9.160487
----------------------------------------------------------------------------------------------

ALL VARIABLES IN THE MODEL ARE SIGNIFICANT AT THE 0.1000 LEVEL.

               SUMMARY OF BACKWARD ELIMINATION PROCEDURE FOR DEPENDENT VARIABLE VOLUME

                      VARIABLE    NUMBER    PARTIAL      MODEL
               STEP   REMOVED      IN       R**2         R**2        C(P)          F       PROB>F

                1     INCOME        4       0.0013      0.6987      4.06228      0.0623     0.8066
                2     PARKING       3       0.0080      0.6907      2.43641      0.3991     0.5371
                3     SHOPCNTR      2       0.0251      0.6657      1.60624      1.2965     0.2716
```

Solution   SAS output is shown for a backward elimination procedure applied to the data of Table 13.1. As indicated, backward elimination begins with all (five) candidate variables in the regression equation. This is designated as step 0 in the backward elimination process. Then one by one, independent variables are eliminated until min $F_j > F_\alpha$. Note that in step 1, the variable income is removed and in step 2, the variable parking is removed from the regression equation. Step 3 is the final step in the process for this example; the variable shopping center is removed. As indicated in the output, the remaining variables comprise the best-fitting regression equation based on backward elimination. That equation is

$$\hat{y} = 48.291 - 0.004(\text{floor space}) - 0.582(\text{prescription sales})$$

This is identical to the result we obtained from the other variable selection procedures.

EXAMPLE 13.7   Note the results of stepwise regression applied to the data of Table 13.1.

Solution   The SAS output for the data of Table 13.1 is shown here. Stepwise regression begins with the model $y = \beta_0 + \varepsilon$ and adds variables one at a time. For these data, the variable prescription sales was entered in step 1 of the stepwise procedure, the variable floor space was added to the regression model in step 2, and the variable shopping center was added in step 3. No other variables met the entrance criterion of $p = .5$ for inclusion in the model. If the criterion was more selective, requiring a relatively small p-value (say .15 or less) for each new independent variable, the stepwise regression procedure would not include the variable shopping center in step 3 (with a p-value of .2716) and we would arrive at the same best-fitting regression equation as we obtained previously with other methods.

```
                        REGRESSION ANALYSIS, USING FORWARD SELECTION

                     FORWARD SELECTION PROCEDURE FOR DEPENDENT VARIABLE VOLUME

STEP 1    VARIABLE PRESC_RX ENTERED      R SQUARE = 0.43933184      C(P) =    10.17094219

                    DF          SUM OF SQUARES        MEAN SQUARE           F        PROB>F

REGRESSION          1           329.74051403        329.74051403        14.10      0.0014
ERROR              18           420.80948597         23.37830478
TOTAL              19           750.55000000

                  B VALUE         STD ERROR         TYPE II SS            F        PROB>F

INTERCEPT        25.98133346
PRESC_RX         -0.32055657       0.08535423       329.74051403        14.10      0.0014

BOUNDS ON CONDITION NUMBER:          1,                1
---------------------------------------------------------------------------------------------

STEP 2    VARIABLE FLOOR_SP ENTERED      R SQUARE = 0.66566267      C(P) =    1.60624219

                    DF          SUM OF SQUARES        MEAN SQUARE           F        PROB>F

REGRESSION          2           499.61311336        249.80655668        16.92      0.0001
ERROR              17           250.93688664         14.76099333
TOTAL              19           750.55000000

                  B VALUE         STD ERROR         TYPE II SS            F        PROB>F

INTERCEPT        48.29085530
FLOOR_SP         -0.00384228       0.00113262       169.87259933        11.51      0.0035
PRESC_RX         -0.58189034       0.10263739       474.44587802        32.14      0.0001

BOUNDS ON CONDITION NUMBER:      2.290122,        9.160487
---------------------------------------------------------------------------------------------

STEP 3    VARIABLE SHOPCNTR ENTERED      R SQUARE = 0.69072432      C(P) =    2.43641080

                    DF          SUM OF SQUARES        MEAN SQUARE           F        PROB>F

REGRESSION          3           518.42314091        172.80771364        11.91      0.0002
ERROR              16           232.12685909         14.50792869
TOTAL              19           750.55000000

                  B VALUE         STD ERROR         TYPE II SS            F        PROB>F

INTERCEPT        42.82702645
FLOOR_SP         -0.00247284       0.00164539        32.76871130         2.26      0.1523
PRESC_RX         -0.52941361       0.11170410       325.87978038        22.46      0.0002
SHOPCNTR         -3.03834296       2.66836223        18.81002755         1.30      0.2716

BOUNDS ON CONDITION NUMBER:      4.917388,        30.31995
---------------------------------------------------------------------------------------------

NO OTHER VARIABLES MET THE 0.5000 SIGNIFICANCE LEVEL FOR ENTRY INTO THE MODEL.

                        REGRESSION ANALYSIS, USING FORWARD SELECTION

             SUMMARY OF FORWARD SELECTION PROCEDURE FOR DEPENDENT VARIABLE VOLUME

          VARIABLE    NUMBER    PARTIAL     MODEL
   STEP   ENTERED       IN       R**2       R**2       C(P)         F        PROB>F

     1    PRESC_RX      1        0.4393     0.4393     10.1709     14.1046    0.0014
     2    FLOOR_SP      2        0.2263     0.6657      1.6062     11.5082    0.0035
     3    SHOPCNTR      3        0.0251     0.6907      2.4364      1.2965    0.2716
```

In a typical regression problem, you ascertain which variables are potential candidates for inclusion in a regression model (step 1) by discussions with experts and/or by using any one of a number of possible selection procedures. For example, we could run all possible regressions, apply a best-subset regression approach, or follow a stepwise regression (a backward elimination) procedure. This list is by no means exhaustive. Sometimes the various criteria do single out the same model as best (or near best, as seen with the data of Table 13.1). Other times you may get different models from the different criteria. Which approach is best? Which one should we believe and use?

The most important response to these questions is that with the availability and accessibility of large-scale computer and major software systems, it is possible to work effectively with any of these selection procedures; no one procedure is universally accepted as better than the others. Hence rather than attempting to use some or all of the procedures, you should begin to use one method (perhaps because of the availability of particular software in your computer facility) and learn as much as you can about it by continued use. Then you will be well equipped to solve almost any regression problem to which you are exposed.

EXERCISES

 13.1   (Class Project) The director of admissions at your college or university is interested in developing a regression model that will be useful in predicting a student's end-of-the-year grade point average (GPA) based on his or her high school record. Discuss this project among yourselves and seek out additional experts in order to develop a list of candidate independent variables for inclusion in the regression model. Should only one model be developed or should you consider more than one regression model? Might dummy variables be useful?

13.2   (Class Project) See Exercise 13.1. Obtain data from the admissions office and apply one of the selection procedures to identify a possible regression model.

 13.3   A random sample of 45 students at a state university was asked to decide whether each of the following acts should be considered a crime. The acts presented were aggravated assault, armed robbery, arson, atheism, auto theft, burglary, civil disobedience, communism, drug addiction, embezzlement, forcible rape, gambling, homosexuality, land fraud, nazism, payola, price fixing, prostitution, sexual abuse of children, sexual discrimination, shoplifting, strikes, strip mining, treason, and vandalism. For each student the interviewer determined the number of acts considered a crime and other information concerning the interviewee (years of college, age, income of parents, and gender). The data are shown here. Use the output of a best subset regression program to ascertain which variables should be included in the model. Can you suggest other variables that should have been addressed in the interview?

LISTING OF DATA

OBS	CRIME	AGE	COLLEGE	INCOME	SEX
1	23	16	2	63	1
2	25	18	2	72	1
3	22	18	2	75	1
4	16	18	2	61	0
5	19	19	2	65	1
6	19	19	2	70	1
7	18	20	2	78	1
8	16	19	2	76	0
9	12	18	2	53	0
10	13	19	2	56	0

11	16	19	2	59	1
12	13	20	2	55	0
13	13	21	2	60	0
14	14	20	2	52	0
15	14	24	3	54	0
16	13	25	3	55	0
17	16	25	3	55	0
18	16	27	4	56	1
19	14	28	4	52	1
20	20	38	4	59	0
21	25	29	4	63	1
22	19	30	4	55	1
23	23	31	4	59	0
24	25	32	4	52	1
25	22	32	4	55	1
26	25	31	4	57	0
27	17	30	4	46	1
28	14	29	4	35	0
29	12	29	4	32	0
30	10	28	4	30	0
31	8	27	4	29	0
32	7	26	4	28	0
33	5	25	4	25	0
34	9	24	3	33	0
35	7	23	3	26	0
36	9	23	3	28	1
37	10	22	3	38	0
38	4	22	3	24	0
39	6	22	3	28	0
40	8	21	3	29	1
41	11	21	2	35	1
42	10	20	2	33	0
43	6	19	2	27	0
44	7	21	3	24	0
45	15	21	2	53	1

N=   45

REGRESSION ANALYSIS
BACKWARD ELIMINATION PROCEDURE FOR INDEPENDENT VARIABLE CRIMES

BACKWARD ELIMINATION PROCEDURE FOR DEPENDENT VARIABLE CRIME

STEP 0   ALL VARIABLES ENTERED   R SQUARE = 0.82783940   C(P) =   5.00000000

	DF	SUM OF SQUARES	MEAN SQUARE	F	PROB>F
REGRESSION	4	1301.62108953	325.40527238	48.09	0.0001
ERROR	40	270.69002158	6.76725054		
TOTAL	44	1572.31111111			

	B VALUE	STD ERROR	TYPE II SS	F	PROB>F
INTERCEPT	-10.82338752				
INCOME	0.29025487	0.03141812	577.57817022	85.35	0.0001
AGE	0.43238152	0.20236447	30.89427247	4.57	0.0388
COLLEGE	-0.02399594	1.22148794	0.00261162	0.00	0.9844
SEX	2.45416550	0.87466592	53.27648156	7.87	0.0077

BOUNDS ON CONDITION NUMBER: 7.476669,   68.21544

STEP 1   VARIABLE COLLEGE REMOVED   R SQUARE = 0.82783774   C(P) =   3.00038592

	DF	SUM OF SQUARES	MEAN SQUARE	F	PROB>F
REGRESSION	3	1301.61847791	433.87282597	65.72	0.0001
ERROR	41	270.69263320	6.60225935		
TOTAL	44	1572.31111111			

	B VALUE	STD ERROR	TYPE II SS	F	PROB>F
INTERCEPT	-10.82193315				
INCOME	0.29058236	0.02630415	805.71727230	122.04	0.0001
AGE	0.42872187	0.07806990	199.10244384	30.16	0.0001
SEX	2.45108843	0.84997169	54.90370062	8.32	0.0062

BOUNDS ON CONDITION NUMBER: 1.202437,   10.21103

ALL VARIABLES IN THE MODEL ARE SIGNIFICANT AT THE 0.1000 LEVEL.

SUMMARY OF BACKWARD ELIMINATION PROCEDURE FOR DEPENDENT VARIABLE CRIME

STEP	VARIABLE REMOVED	NUMBER IN	PARTIAL R**2	MODEL R**2	C(P)	F	PROB>F
1	COLLEGE	3	0.0000	0.8278	3.00039	0.0004	0.9844

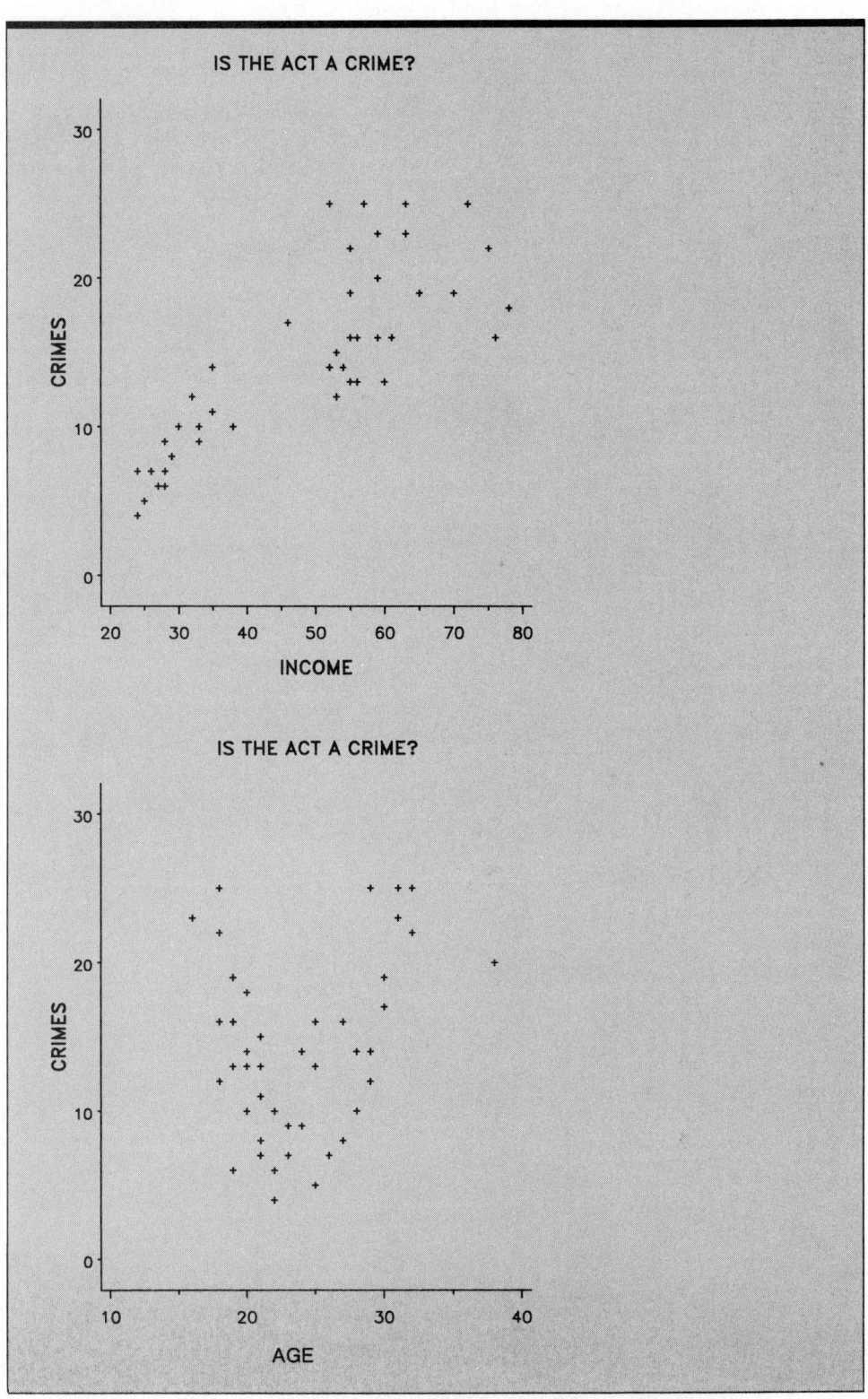

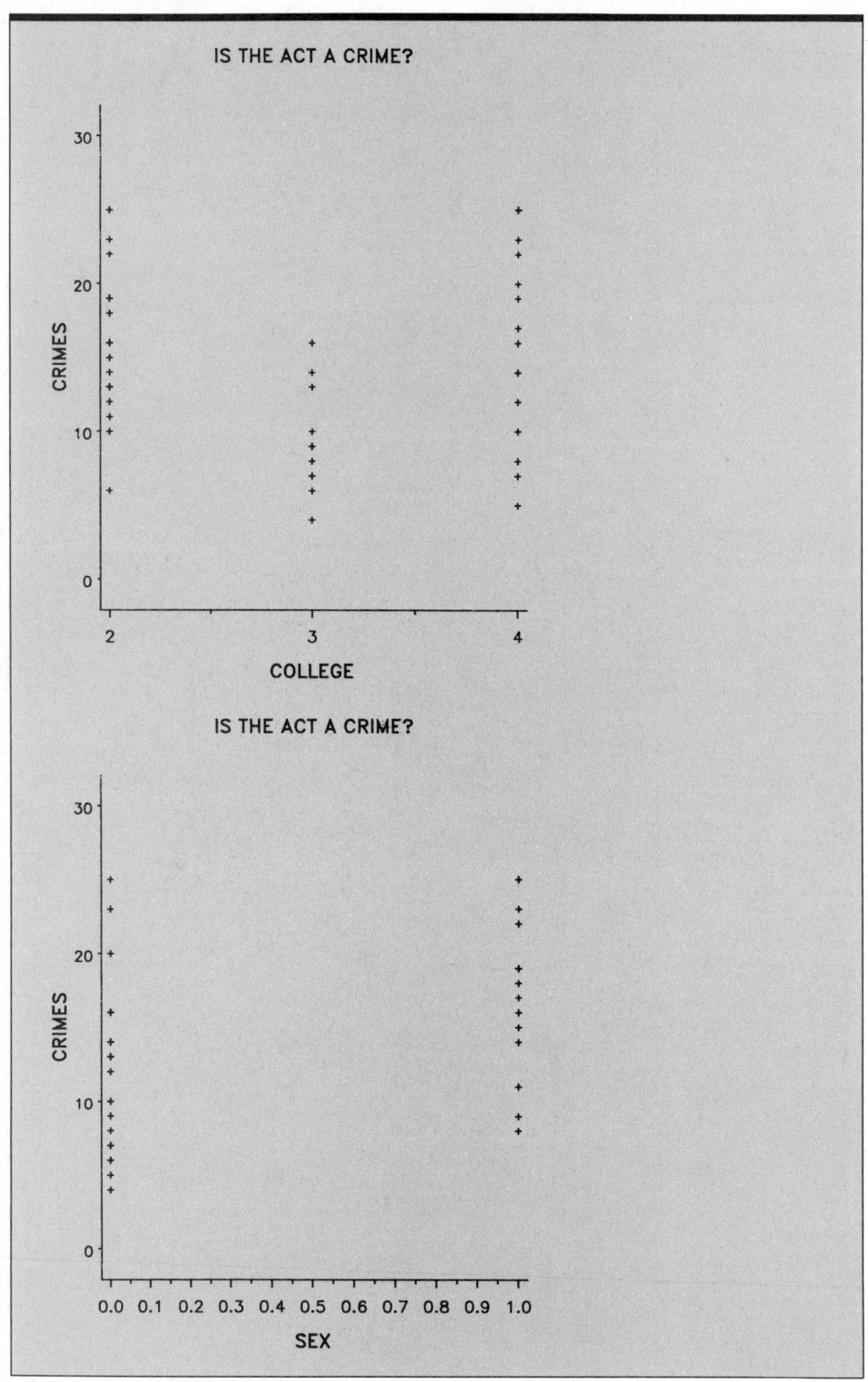

13.4   Refer to Exercise 13.3. Computer output from a stepwise regression program is shown here. Comment on the results of this analysis compared to that done in Exercise 13.3.

```
                              REGRESSION ANALYSIS
                  FORWARD SELECTION PROCEDURE FOR INDEPENDENT VARIABLE CRIMES

                    FORWARD SELECTION PROCEDURE FOR DEPENDENT VARIABLE CRIME

STEP 1    VARIABLE INCOME ENTERED      R SQUARE = 0.66453936      C(P) =    36.94132731

                    DF           SUM OF SQUARES        MEAN SQUARE          F      PROB>F

REGRESSION          1            1044.86262180        1044.86262180      85.18    0.0001
ERROR               43            527.44848931          12.26624394
TOTAL               44           1572.31111111

                   B VALUE           STD ERROR         TYPE II SS          F      PROB>F

INTERCEPT        -0.19647505
INCOME            0.30177022         0.03269660        1044.86262180      85.18    0.0001

BOUNDS ON CONDITION NUMBER:            1,              1
------------------------------------------------------------------------------------------

STEP 2    VARIABLE AGE ENTERED         R SQUARE = 0.79291863      C(P) =     9.11353325

                    DF           SUM OF SQUARES        MEAN SQUARE          F      PROB>F

REGRESSION          2            1246.71477730         623.35738865      80.41    0.0001
ERROR               42            325.59633381           7.75229366
TOTAL               44           1572.31111111

                   B VALUE           STD ERROR         TYPE II SS          F      PROB>F

INTERCEPT       -11.33832496
INCOME            0.32018698         0.02624270        1154.03879316     148.86    0.0001
AGE               0.43163600         0.08458942         201.85215549      26.04    0.0001

BOUNDS ON CONDITION NUMBER:        1.01928,        4.077119
------------------------------------------------------------------------------------------

STEP 3    VARIABLE SEX ENTERED         R SQUARE = 0.82783774      C(P) =     3.00038592

                    DF           SUM OF SQUARES        MEAN SQUARE          F      PROB>F

REGRESSION          3            1301.61847791         433.87282597      65.72    0.0001
ERROR               41            270.69263320           6.60225935
TOTAL               44           1572.31111111

                   B VALUE           STD ERROR         TYPE II SS          F      PROB>F

INTERCEPT       -10.82193315
INCOME            0.29058236         0.02630415         805.71727230     122.04    0.0001
AGE               0.42872187         0.07806990         199.10244384      30.16    0.0001
SEX               2.45108843         0.84997169          54.90370062       8.32    0.0062

BOUNDS ON CONDITION NUMBER:        1.202437,       10.21103
------------------------------------------------------------------------------------------

NO OTHER VARIABLES MET THE 0.5000 SIGNIFICANCE LEVEL FOR ENTRY INTO THE MODEL.


                              REGRESSION ANALYSIS
                  FORWARD SELECTION PROCEDURE FOR INDEPENDENT VARIABLE CRIMES

             SUMMARY OF FORWARD SELECTION PROCEDURE FOR DEPENDENT VARIABLE CRIME

          VARIABLE   NUMBER   PARTIAL    MODEL
  STEP    ENTERED      IN      R**2       R**2       C(P)          F       PROB>F

   1      INCOME       1       0.6645     0.6645     36.9413     85.1820    0.0001
   2      AGE          2       0.1284     0.7929      9.1135     26.0377    0.0001
   3      SEX          3       0.0349     0.8278      3.0004      8.3159    0.0062
```

    13.5   A company is interested in the effects of various food additives (protein and antibiotics) on the amount of time it takes to bring cattle to a desired market weight. Discuss what variable should be examined in arriving at a multiple regression equation for predicting the time to market weight.

| 13.3 | MODEL FORMATION (Step 2) |

In Section 13.2 we suggested several ways to develop a list of candidate independent variables for a given regression problem. We can and should seek the advice of experts in the subject matter area to provide a starting point and we can employ any one of several selection procedures to come up with a possible regression model. This section involves refining the information gleaned from step 1 in order to develop a useful multiple regression model.

Having chosen a subset of k independent variables to be candidates for inclusion in the multiple regression and the dependent variable y, we still may not know the actual relationship between the dependent and independent variables. Suppose the assumed regression model is of a lower order than is the actual model relating y to $x_1, x_2, \ldots, x_k$. Then provided there is more than one observation per factor-level combination of the independent variables, we can conduct a test of the inadequacy of a fitted polynomial model using the equation $F = MS_{Lack}/MSE$ as discussed in Chapter 9.

Another way to examine an assumed (fitted) model for lack of fit is to examine scatterplots of residuals $(y_i - \hat{y}_i)$ versus x_j. For example, suppose that step 1 has indicated that the variables x_1, x_2, and x_3 constitute a reasonable subset of independent variables to be related to a response y using a multiple regression equation. Not knowing which polynomial function of the independent variables to use, we could start by fitting the multiple linear regression model

$$y = \beta_0 + \beta_1 x_1 + \beta_2 x_2 + \beta_3 x_3 + \varepsilon$$

to obtain the least squares prediction equation $y = \hat{\beta}_0 + \hat{\beta}_1 x_1 + \hat{\beta}_2 x_2 + \hat{\beta}_3 x_3$. A plot of the residuals $(y_i - \hat{y}_i)$ versus each one of the xs would shed some light as to which higher-degree terms may be appropriate.

EXAMPLE 13.8

In a radioimmunoassay a hormone with a radioactive trace is added to a test tube containing an antibody that is specific to that hormone. The two will combine to form an antigen–antibody complex. In order to measure the extent of the reaction of the hormone with the antibody, we measure the amount of hormone that is bound to the antibody relative to the amount remaining free. Typically, experimenters measure the ratio of the bound/free radioactive count (y) for each dose of hormone (x) added to a test tube. Frequently, the relation between y and x is nearly linear. Data from 11 test tubes in a radioimmunoassay experiment are shown in Table 13.4.

a. Plot the sample data and fit the linear regression model

$$y = \beta_0 + \beta_1 x + \varepsilon$$

b. Plot the residuals versus count and versus $\hat{y}$. Does a linear model adequately fit the data?

c. Suggest an alternative (if appropriate).

SECTION 13.3 MODEL FORMATION (Step 2)

TABLE 13.4 Radioimmunoassay Data, Example 13.8

Bound/Free Count	Dose (Concentration)
9.900	0.00
10.465	0.25
10.312	0.50
13.633	0.75
20.784	1.00
36.164	1.25
62.045	1.50
78.327	1.75
90.307	2.00
97.348	2.25
102.686	2.50

Solution ▌ Computer output is shown here.

```
    LISTING OF DATA

  OBS      COUNT    DOSE

   1       9.900    0.00
   2      10.465    0.25
   3      10.312    0.50
   4      13.633    0.75
   5      20.784    1.00
   6      36.164    1.25
   7      62.045    1.50
   8      78.327    1.75
   9      90.307    2.00
  10      97.348    2.25
  11     102.686    2.50

  N=      11
```

SIMPLE LINEAR REGRESSION

GENERAL LINEAR MODELS PROCEDURE

DEPENDENT VARIABLE: COUNT

SOURCE	DF	SUM OF SQUARES	MEAN SQUARE	F VALUE	PR > F	R-SQUARE	C.V.
MODEL	1	13577.44212015	13577.44212015	111.44	0.0001	0.925273	22.8242
ERROR	9	1096.54306185	121.83811798		ROOT MSE		COUNT MEAN
CORRECTED TOTAL	10	14673.98518200			11.03803053		48.36100000

SOURCE	DF	TYPE I SS	F VALUE	PR > F	DF	TYPE III SS	F VALUE	PR > F
DOSE	1	13577.44212015	111.44	0.0001	1	13577.44212015	111.44	0.0001

PARAMETER	ESTIMATE	T FOR HO: PARAMETER=0	PR > |T|	STD ERROR OF ESTIMATE
INTERCEPT	-7.18881818	-1.15	0.2780	6.22628894
DOSE	44.43985455	10.56	0.0001	4.20973967

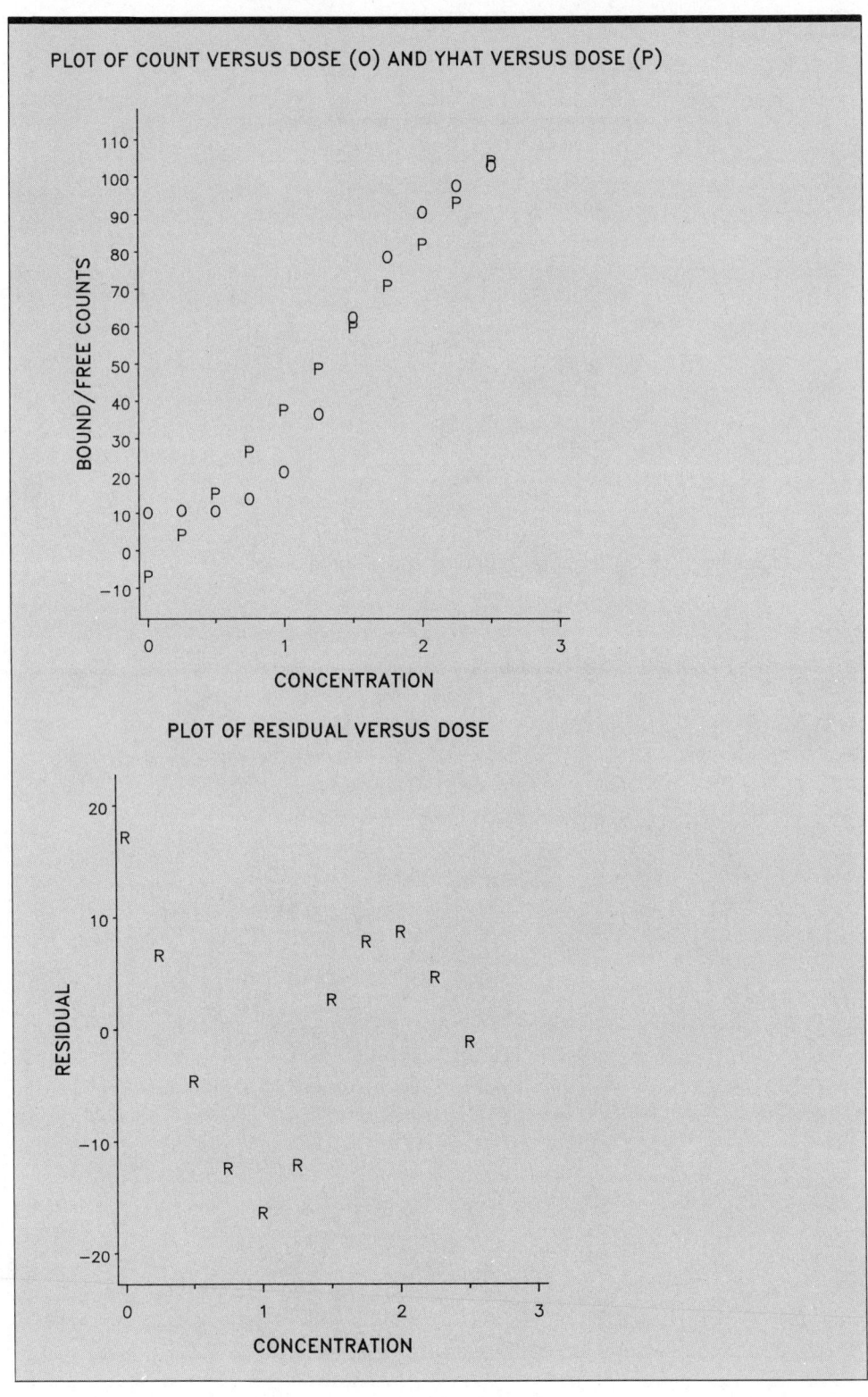

558

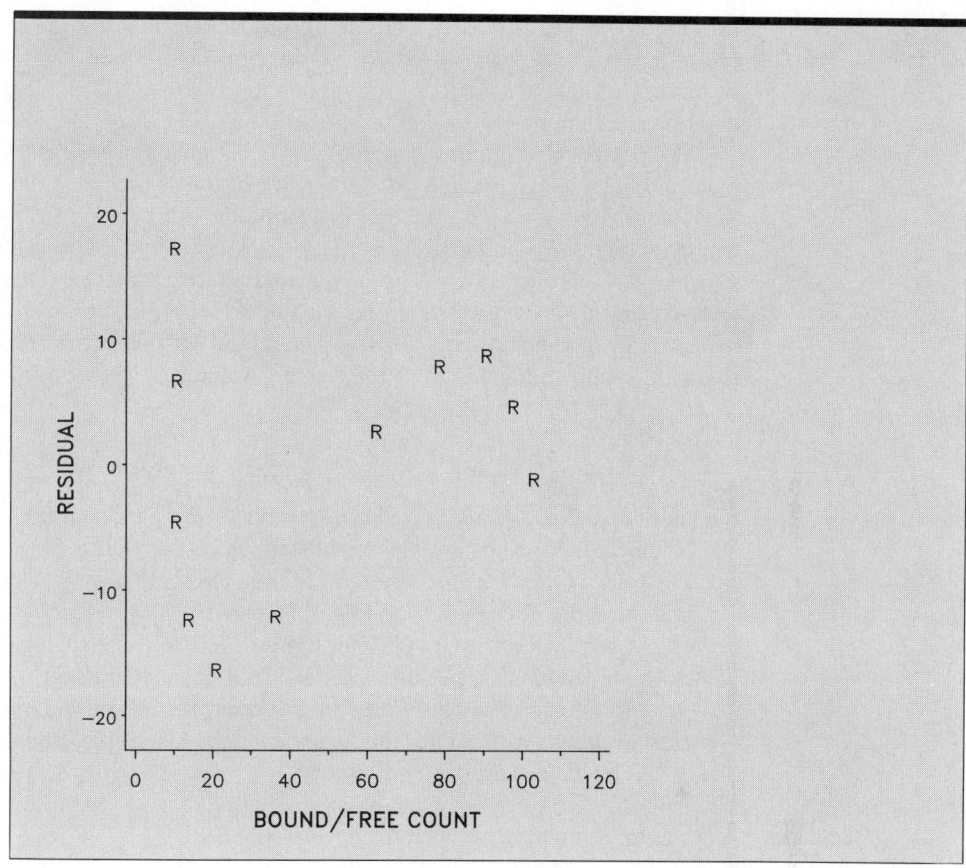

a. and b. The linear fit is

$$\hat{y} = -7.189 + 44.440x$$

The plot of y (count) versus x (concentration) clearly shows a lack of fit of the linear regression model; the residual plots confirm this same lack of fit. The linear regression underestimates counts at the lower and upper ends of the concentration scale and overestimates at the middle concentrations.

c. A possible alternative model would be a quadratic model in concentration,

$$y = \beta_0 + \beta_1 x + \beta_2 x^2 + \varepsilon$$

More will be said about this later in the chapter.

Scatterplots are not very helpful in detecting interactions among the independent variables, other than for the two independent variable case. The reason is that there are just too many variables for most practical problems and it is difficult to present the inter-relationships among independent variables and their joint effects on the response y using two-dimensional scatterplots. Perhaps the most reasonable suggestion is to use one of the best subset regression methods of the previous section, some trial-and-error fitting of models using the candidate independent variables, and a bit of common sense to determine which interaction terms should be used in the multiple regression model.

The presence of dummy variables (for qualitative independent variables) presents no major problem for ascertaining the adequacy of the fit of a polynomial model. The important thing to remember is that when quantitative and dummy variables are included in the same regression model, by substituting any combination of 1s and 0s for the dummy variables, we obtain a regression in the quantitative variables. Hence plotting methods for detecting an inadequate fit should be applied separately for each setting of the dummy variables. By examining plots carefully, we could also detect potential differences in the forms of the polynomial models for different settings of the dummy variables.

EXAMPLE 13.9

A company analyst is interested in developing a regression model for predicting automobile sales for standard and luxury models of a particular make in a given territory. Empirical discussions and some substantive research into previous sales patterns for the company in that territory tend to indicate that the prevailing interest rate for car loans and the price per gallon of gasoline are the key predictive variables. The number of cars sold per month (in 000) for the previous 18 months is shown here for gasoline-powered standard and luxury models. Fit a linear regression model and use residual plots to determine what (if any) higher-order terms are required. Do the same conclusions hold for standard and luxury models? Make suggestions for additional terms in the multiple regression equation.

Solution

a. A multiple regression model of the form

$$y = \beta_0 + \beta_1 x_1 + \beta_2 x_2 + \beta_3 x_3 + \varepsilon$$

where

y = number of sales per month (in 000)

x_1 = price per gallon

x_2 = interest rate

$$x_3 = \begin{cases} 1 & \text{if standard} \\ 0 & \text{if luxury} \end{cases}$$

was fit to the data. From the output the regression equation is

$$\hat{y} = 56.074 - 16.144 x_1 - 2.332 x_2 + 14.422 x_3$$

Substituting $x_3 = 0$ and 1 into this equation, we obtain the separate regression equations for the luxury and standard cars, respectively:

$x_3 = 0$ (luxury cars)

$$\hat{y} = 56.074 - 16.144 x_1 - 2.332 x_2$$

$x_3 = 1$ (standard cars)

$$\hat{y} = 56.074 - 16.144x_1 - 2.332x_2 + 14.422$$
$$= 70.496 - 16.144x_1 - 2.332x_2$$

Plots of y versus x_1 and x_2 for the two model types show clear negative linear relationships between sales and price per gallon of gasoline or interest rates. However, the slopes appear to be greater for the standard model than for the luxury model. This is borne out in the residual plots for the two models.

```
                LISTING OF DATA

  OBS     MONTH      Y       X1      X2     X3

    1       1      22.1     1.39    12.1     1
    2       1       7.2     1.39    12.1     0
    3       2      15.4     1.44    12.2     1
    4       2       5.4     1.44    12.2     0
    5       3      11.7     1.45    12.3     1
    6       3       7.6     1.45    12.3     0
    7       4      10.3     1.32    14.2     1
    8       4       2.5     1.32    14.2     0
    9       5      11.4     1.35    15.8     1
   10       5       2.4     1.35    15.8     0
   11       6       7.5     1.28    16.3     1
   12       6       1.7     1.28    16.3     0
   13       7      13.0     1.26    16.5     1
   14       7       4.3     1.26    16.5     0
   15       8      12.8     1.26    14.7     1
   16       8       3.7     1.26    14.7     0
   17       9      14.6     1.25    13.4     1
   18       9       3.9     1.25    13.4     0
   19      10      18.9     1.24    12.9     1
   20      10       7.0     1.24    12.9     0
   21      11      19.3     1.20    11.2     1
   22      11       6.8     1.20    11.2     0
   23      12      30.1     1.20    10.9     1
   24      12      10.1     1.20    10.9     0
   25      13      28.2     1.18    10.3     1
   26      13       9.4     1.18    10.3     0
   27      14      25.6     1.10     9.7     1
   28      14       7.9     1.10     9.7     0
   29      15      37.5     1.11     9.6     1
   30      15      14.1     1.11     9.6     0
   31      16      36.1     1.14     9.1     1
   32      16      14.5     1.14     9.1     0
   33      17      39.8     1.17     7.8     1
   34      17      14.9     1.17     7.8     0
   35      18      44.3     1.18     8.3     1
   36      18      15.6     1.18     8.3     0

  N=      36
```

LINEAR REGRESSION

GENERAL LINEAR MODELS PROCEDURE

DEPENDENT VARIABLE: Y

SOURCE	DF	SUM OF SQUARES	MEAN SQUARE	F VALUE	PR > F	R-SQUARE	C.V.
MODEL	3	3713.14622454	1237.71540818	54.61	0.0001	0.836596	31.8796
ERROR	32	725.25377546	22.66418048		ROOT MSE		Y MEAN
CORRECTED TOTAL	35	4438.40000000			4.76069118		14.93333333

SOURCE	DF	TYPE I SS	F VALUE	PR > F	DF	TYPE III SS	F VALUE	PR > F
X1	1	902.74323281	39.83	0.0001	1	68.06149557	3.00	0.0927
X2	1	938.39854728	41.40	0.0001	1	938.39854728	41.40	0.0001
X3	1	1872.00444444	82.60	0.0001	1	1872.00444444	82.60	0.0001

PARAMETER	ESTIMATE	T FOR H0: PARAMETER=0	PR > |T|	STD ERROR OF ESTIMATE
INTERCEPT	56.07441586	5.60	0.0001	10.00797958
X1	-16.14364096	-1.73	0.0927	9.31581287
X2	-2.33218910	-6.43	0.0001	0.36244333
X3	14.42222222	9.09	0.0001	1.58689706

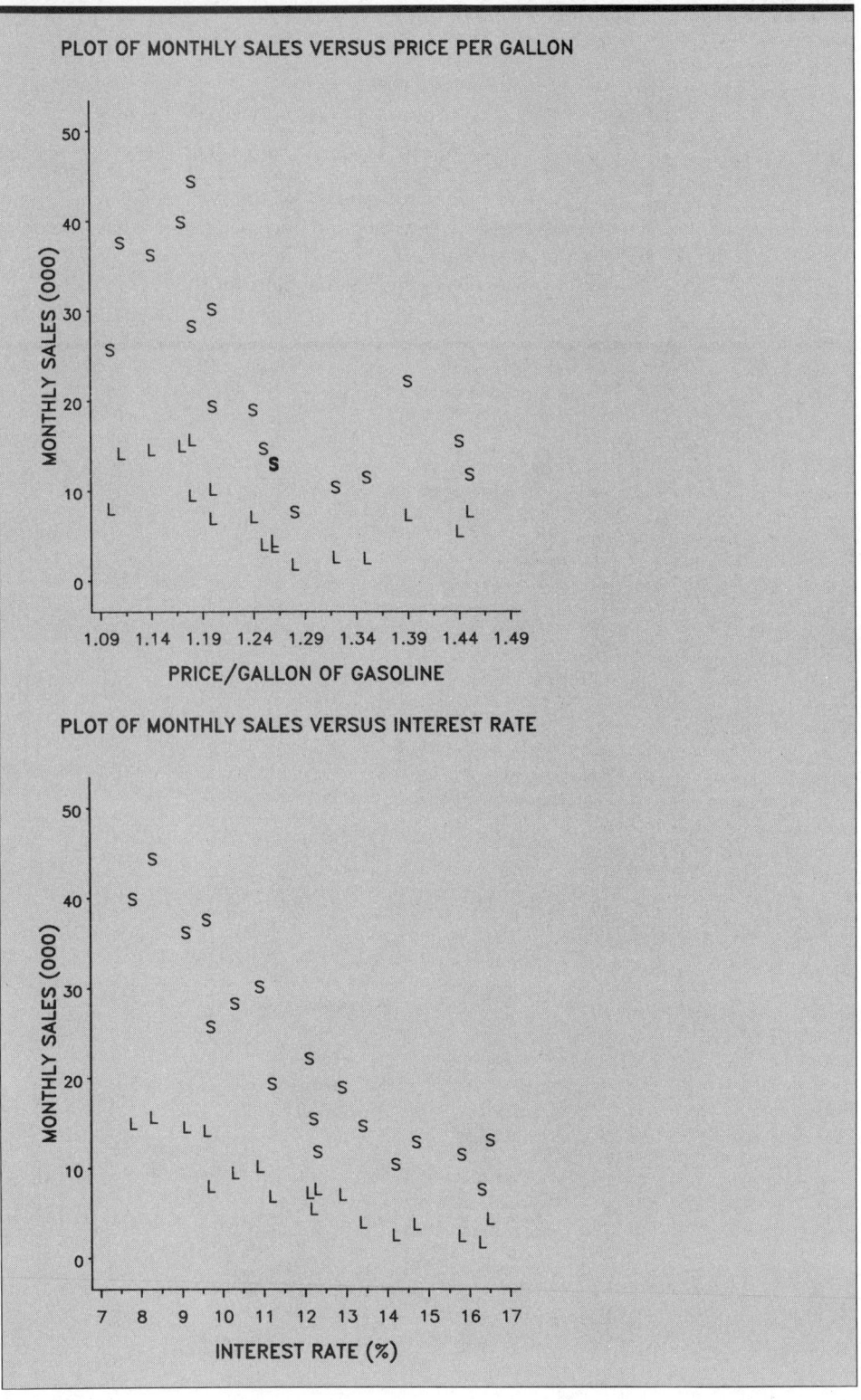

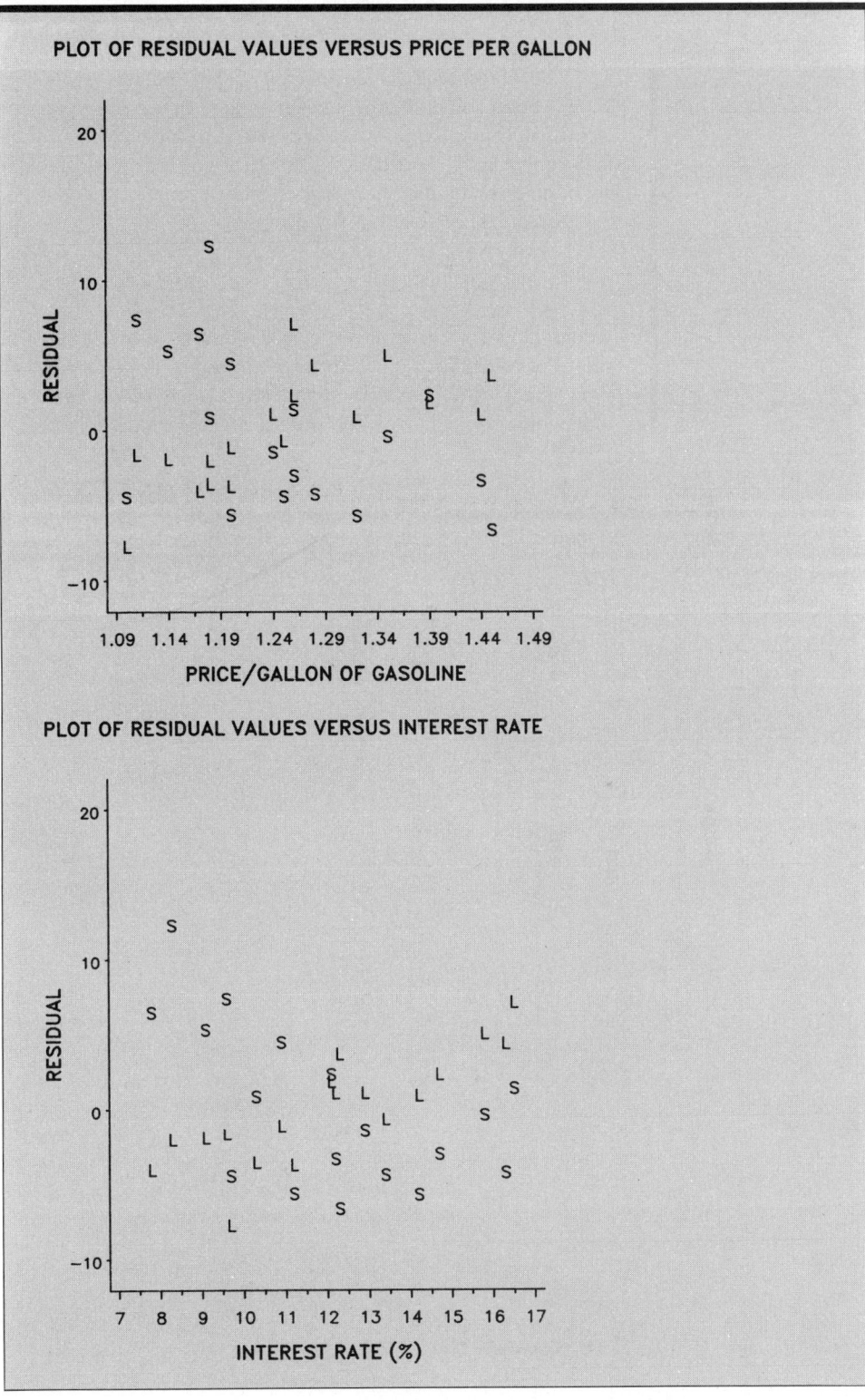

PLOT OF RESIDUAL VALUES VERSUS PRICE PER GALLON

PLOT OF RESIDUAL VALUES VERSUS INTEREST RATE

Plots of residuals versus price per gallon and versus interest rates for luxury models show underestimation for smaller values of x_1 and x_2 and overestimation for the larger values of x_1 and x_2. Corresponding residual plots for the standard models show fairly good fits to the data, although there may be some curvature that could be accounted for by including higher-order terms in x_1 and x_2 in the regression model.

A regression model of the form

$$y = \beta_0 + \beta_1 x_1 + \beta_2 x_1^2 + \beta_3 x_2 + \beta_4 x_2^2 + \beta_5 x_3 + \beta_6 x_1 x_3 + \beta_7 x_1^2 x_3 + \beta_8 x_2 x_3 + \beta_9 x_2^2 x_3 + \varepsilon$$

would allow for curvature in y (sales) due to x_1 (price per gallon) and to x_2 (interest rate); the model also allows for different regression coefficients for the two car models. One might also consider adding interaction terms between the two quantitative independent variables. Some output for this model follows.

LISTING OF DATA

OBS	MONTH	Y	X1	X2	X3	X1_2	X2_2
1	1	22.1	1.39	12.1	1	1.9321	146.41
2	1	7.2	1.39	12.1	0	1.9321	146.41
3	2	15.4	1.44	12.2	1	2.0736	148.84
4	2	5.4	1.44	12.2	0	2.0736	148.84
5	3	11.7	1.45	12.3	1	2.1025	151.29
6	3	7.6	1.45	12.3	0	2.1025	151.29
7	4	10.3	1.32	14.2	1	1.7424	201.64
8	4	2.5	1.32	14.2	0	1.7424	201.64
9	5	11.4	1.35	15.8	1	1.8225	249.64
10	5	2.4	1.35	15.8	0	1.8225	249.64
11	6	7.5	1.28	16.3	1	1.6384	265.69
12	6	1.7	1.28	16.3	0	1.6384	265.69
13	7	13.0	1.26	16.5	1	1.5876	272.25
14	7	4.3	1.26	16.5	0	1.5876	272.25
15	8	12.8	1.26	14.7	1	1.5876	216.09
16	8	3.7	1.26	14.7	0	1.5876	216.09
17	9	14.6	1.25	13.4	1	1.5625	179.56
18	9	3.9	1.25	13.4	0	1.5625	179.56
19	10	18.9	1.24	12.9	1	1.5376	166.41
20	10	7.0	1.24	12.9	0	1.5376	166.41
21	11	19.3	1.20	11.2	1	1.4400	125.44
22	11	6.8	1.20	11.2	0	1.4400	125.44
23	12	30.1	1.20	10.9	1	1.4400	118.81
24	12	10.1	1.20	10.9	0	1.4400	118.81
25	13	28.2	1.18	10.3	1	1.3924	106.09
26	13	9.4	1.18	10.3	0	1.3924	106.09
27	14	25.6	1.10	9.7	1	1.2100	94.09
28	14	7.9	1.10	9.7	0	1.2100	94.09
29	15	37.5	1.11	9.6	1	1.2321	92.16
30	15	14.1	1.11	9.6	0	1.2321	92.16
31	16	36.1	1.14	9.1	1	1.2996	82.81
32	16	14.5	1.14	9.1	0	1.2996	82.81
33	17	39.8	1.17	7.8	1	1.3689	60.84
34	17	14.9	1.17	7.8	0	1.3689	60.84
35	18	44.3	1.18	8.3	1	1.3924	68.89
36	18	15.6	1.18	8.3	0	1.3924	68.89

N= 36

LINEAR REGRESSION

GENERAL LINEAR MODELS PROCEDURE

DEPENDENT VARIABLE: Y

SOURCE	DF	SUM OF SQUARES	MEAN SQUARE	F VALUE	PR > F	R-SQUARE	C.V.
MODEL	9	4203.08834550	467.00981617	51.60	0.0001	0.946983	20.1455
ERROR	26	235.31165450	9.05044825		ROOT MSE		Y MEAN
CORRECTED TOTAL	35	4438.40000000			3.00839629		14.93333333

SOURCE	DF	TYPE I SS	F VALUE	PR > F	DF	TYPE III SS	F VALUE	PR > F
X1	1	902.74323281	99.75	0.0001	1	0.02708386	0.00	0.9568
X1*X1	1	270.27042800	29.86	0.0001	1	0.02587043	0.00	0.9578
X2	1	684.25426357	75.60	0.0001	1	29.10270106	3.22	0.0846
X2*X2	1	90.22789329	9.97	0.0040	1	15.45933401	1.71	0.2027
X3	1	1872.00444444	206.84	0.0001	1	2.24286224	0.25	0.6228
X1*X3	1	218.90085296	24.19	0.0001	1	7.39894543	0.82	0.3742
X1*X1*X3	1	24.95517426	2.76	0.1088	1	8.24522060	0.91	0.3486
X2*X3	1	124.22121378	13.73	0.0010	1	29.98826625	3.31	0.0802
X2*X2*X3	1	15.51084238	1.71	0.2019	1	15.51084238	1.71	0.2019

PARAMETER	ESTIMATE	T FOR H0: PARAMETER=0	PR > |T|	STD ERROR OF ESTIMATE
INTERCEPT	40.56734222	0.28	0.7785	142.71008115
X1	12.20097631	0.05	0.9568	223.03567864
X1*X1	-4.59508961	-0.05	0.9578	85.94630524
X2	-5.32756126	-1.79	0.0846	2.97096029
X2*X2	0.15363504	1.31	0.2027	0.11755197
X3	-100.46988523	-0.50	0.6228	201.82253225
X1*X3	285.19322700	0.90	0.3742	315.42008163
X1*X1*X3	-116.01343994	-0.95	0.3486	121.54643051
X2*X3	-7.64808102	-1.82	0.0802	4.20157234
X2*X2*X3	0.21763441	1.31	0.2019	0.16624360

PLOT OF RESIDUAL VALUES VERSUS (INTEREST RATE)^2

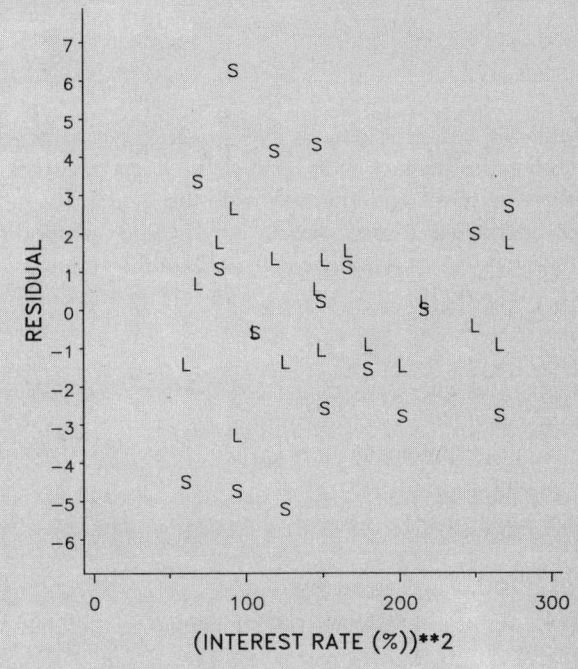

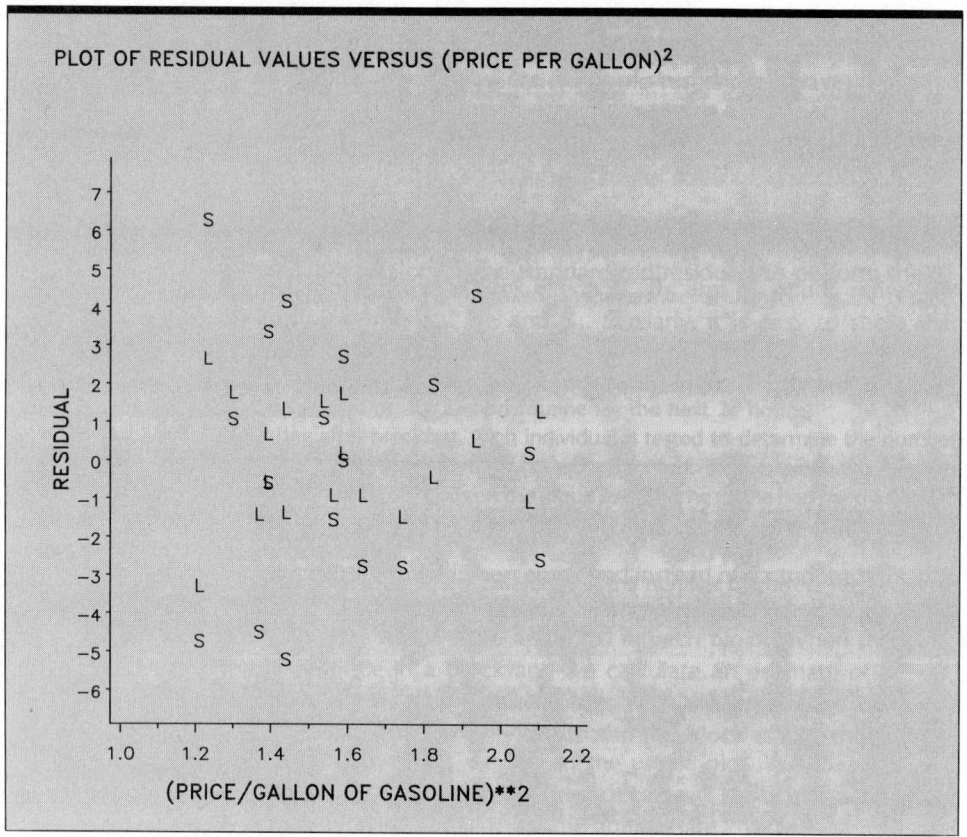

PLOT OF RESIDUAL VALUES VERSUS (PRICE PER GALLON)²

So far in this section we have considered lack of fit only as it relates to polynomial terms and interaction terms. However, sometimes the lack of fit is unrelated to the fact that we have not included enough higher-degree terms and interactions in the model, but rather is related to the fact that y is not adequately represented by any polynomial model in the subset of independent variables. A model that is *nonlinear* in the βs may be appropriate.

The Cobbs–Douglas production equation

$$y = \alpha_1 l^{\alpha_2} c^{\alpha_3}$$

is an example of a nonlinear equation in the constants, α_1, α_2, and α_3. Here y is production, l is the labor input, and c is the capital input. However a logarithmic transformation enables us to treat the equation as if we had a general linear model.

$$\log y = \log \alpha_1 + \alpha_2 \log l + \alpha_3 \log c$$
$$= \beta_0 + \beta_1 x_1 + \beta_2 x_2$$

where x_1 is the log labor input and x_2 is the log capital input.

Not all nonlinear models can be transformed as we did this one. For example, the model

$$y = \alpha_0 e^{-\alpha_1 x_1} + \alpha_2 e^{-\alpha_3 x_2}$$

cannot be transformed to a general linear model. Even so, we would like to obtain estimates for the parameters (αs) in the nonlinear equation.

The remaining material in this section should be considered optional. We'll use computer software and output to illustrate the fitting of nonlinear models. The logic behind what we are doing is the same used in the least squares method for the general linear model; in fact the procedure is sometimes called **nonlinear least squares**. The sum of squares for error is defined as before,

nonlinear least squares

$$SSE = \sum_i (y_i - \hat{y}_i)^2$$

The problem is to find a method for obtaining estimates $\hat{\alpha}_1, \hat{\alpha}_2, \ldots$ that will minimize SSE. The set of simultaneous equations used for finding these estimates is again called the set of normal equations, but unlike least squares for the general linear model, the form of the normal equations depends on the form of the nonlinear model being used. Also, since the normal equations involve nonlinear functions of the parameters, their solutions can be quite complicated. Because of this technical difficulty, a number of iterative methods have been developed for obtaining a solution to the normal equations.

For those of you with a background in the calculus, the normal equations for a nonlinear model involve partial derivatives of the nonlinear function with respect to each of the parameters α_i. Fortunately most of the computer software packages currently marketed (for example, SAS, NONLIN, BMDP) approximate the derivative and do not require one to give the form of the normal equations; only the form of the nonlinear equation is needed. We will illustrate this with the data from a previous example.

Recall that in Example 13.8 we fit a linear regression model to the radioimmunoassay data; a residual plot for that model suggested that a quadratic model might be more appropriate

$$y = \beta_0 + \beta_1 x + \beta_2 x^2 + \varepsilon$$

Computer output for the revised model is shown here. Note that the cyclical pattern is still apparent in the residual plot and hence the quadratic model is still inadequate.

```
                              QUADRATIC REGRESSION

                         GENERAL LINEAR MODELS PROCEDURE

DEPENDENT VARIABLE: COUNT

SOURCE              DF      SUM OF SQUARES      MEAN SQUARE      F VALUE      PR > F       R-SQUARE           C.V.

MODEL               2       13964.37989560      6982.18994780    78.72        0.0001       0.951642         19.4746

ERROR               8         709.60528640        88.70066080                 ROOT MSE                    COUNT MEAN

CORRECTED TOTAL    10       14673.98518200                                    9.41810282                  48.36100000


SOURCE              DF         TYPE I SS       F VALUE     PR > F       DF       TYPE III SS      F VALUE      PR > F

DOSE                1       13577.44212015     153.07      0.0001        1      153.70381752       1.73       0.2245
DOSE*DOSE           1         386.93777546       4.36      0.0702        1      386.93777546       4.36       0.0702


                                    T FOR HO:      PR > |T|      STD ERROR OF
PARAMETER          ESTIMATE       PARAMETER=0                      ESTIMATE

INTERCEPT          2.88439860         0.40         0.6982         7.17520734
DOSE              17.57794312         1.32         0.2245        13.35331689
DOSE*DOSE         10.74476457         2.09         0.0702         5.14445966
```

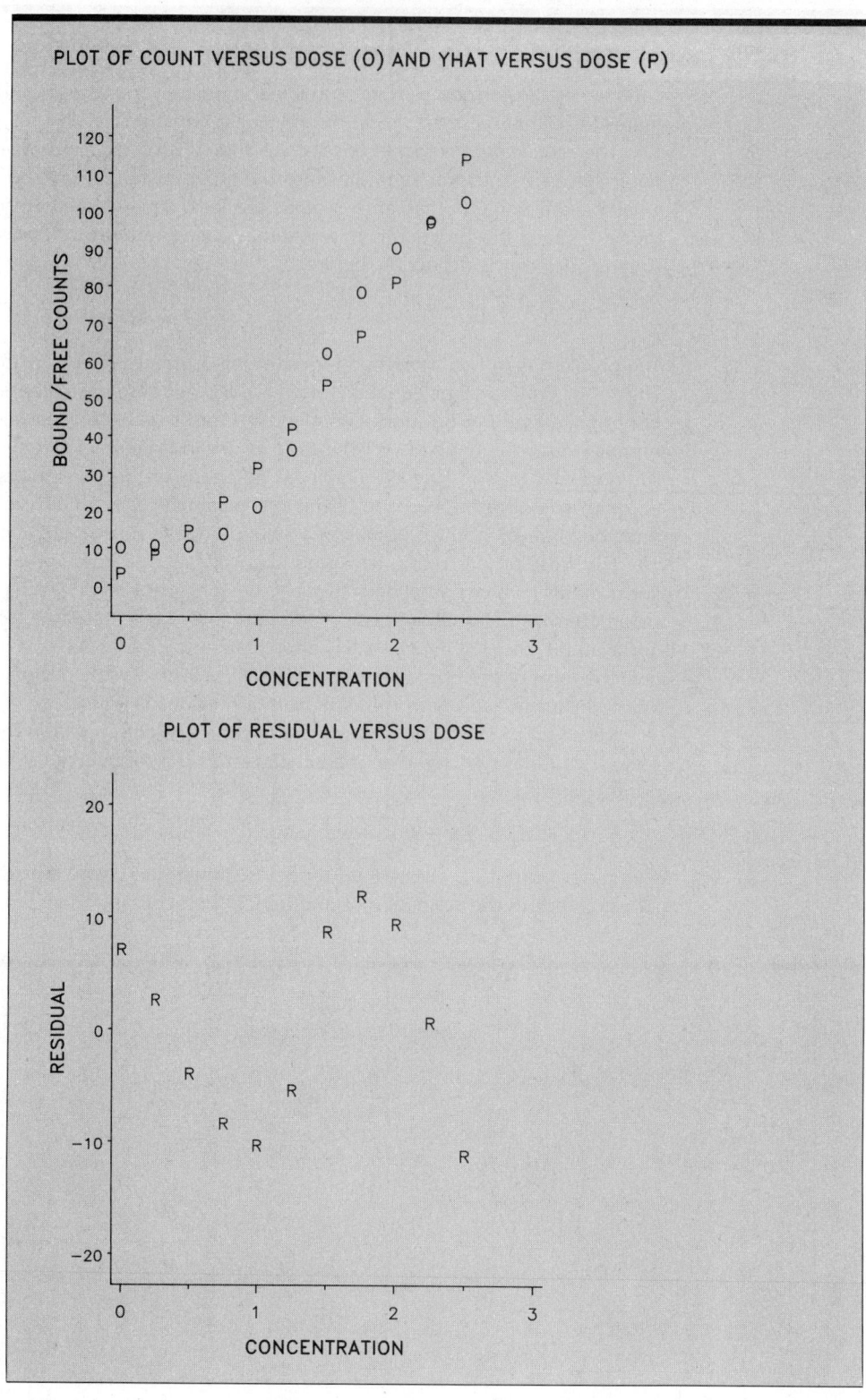

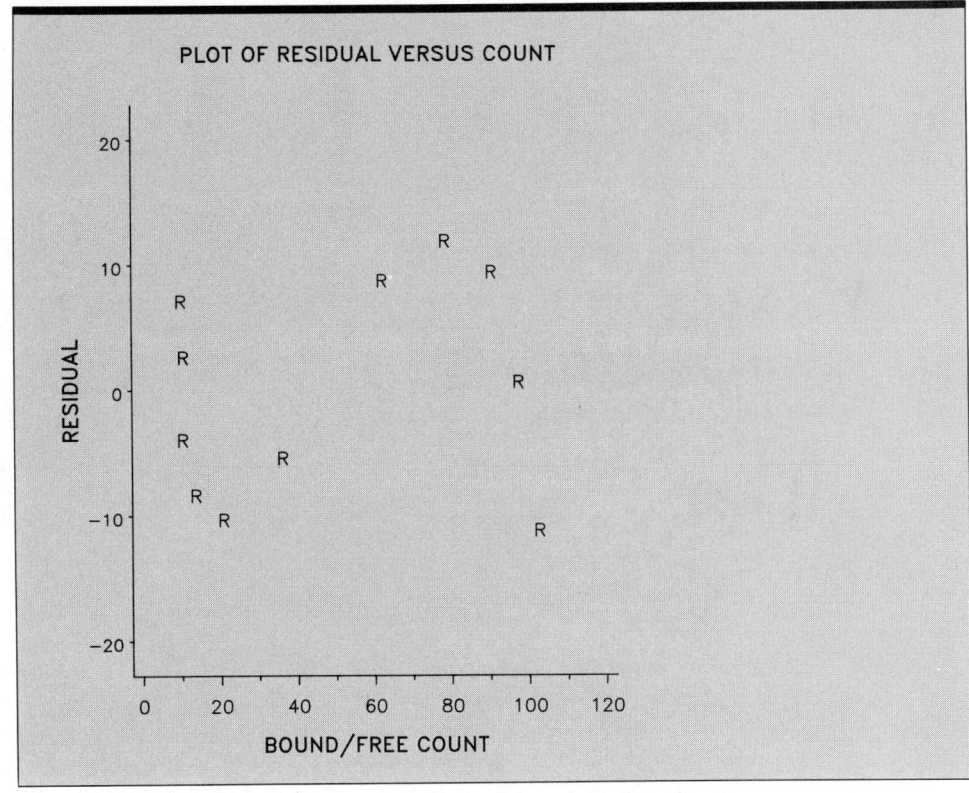

PLOT OF RESIDUAL VERSUS COUNT

A nonlinear model that may help to flatten the **S**-shape of the data plot shown in the output has the following form:

$$y = \left[(\beta_0 - \beta_3) \bigg/ \left(1 + \left(\frac{x}{\beta_2} \right)^{\beta_1} \right) \right] + \beta_3$$

EXAMPLE 13.10   Use a nonlinear estimation program to fit the radioimmunoassay data to the above model.

Solution   SAS was used to fit this model to the sample data. As can be seen from the residual plot, the nonlinear model provides a much better fit to the sample data than either the linear or quadratic model.

By way of explanation the parameters have the following interpretations. These can be seen by the data

β_0: Value of y at the lower end of the curve

β_3: Value of y at the upper end of the curve

β_1: Concentration (x) corresponding to the value of y midway between β_0 and β_3

β_2: A measure of the slope

We can also use the fitted equation to predict y (count ratio) based on concentration.

```
      LISTING OF DATA

   OBS      COUNT     DOSE

    1        9.900    0.00
    2       10.465    0.25
    3       10.312    0.50
    4       13.633    0.75
    5       20.784    1.00
    6       36.164    1.25
    7       62.045    1.50
    8       78.327    1.75
    9       90.307    2.00
   10       97.348    2.25
   11      102.686    2.50

N=      11
                        NON-LINEAR REGRESSION

   NON-LINEAR LEAST SQUARES SUMMARY STATISTICS    DEPENDENT VARIABLE COUNT

      SOURCE                  DF SUM OF SQUARES    MEAN SQUARE

      REGRESSION               4   40390.959650   10097.739913
      RESIDUAL                 7       9.675063       1.382152
      UNCORRECTED TOTAL       11   40400.634713

      (CORRECTED TOTAL)       10   14673.985182


      PARAMETER     ESTIMATE      ASYMPTOTIC           ASYMPTOTIC 95 %
                                  STD. ERROR         CONFIDENCE INTERVAL
                                                     LOWER         UPPER
         BO        10.3172000   0.6302686688     8.82683942    11.80756051
         B1         5.3701025   0.2558487401     4.76511153     5.97509351
         B2         1.4863325   0.0154082328     1.44989755     1.52276749
         B3       107.3776302   1.7275763169   103.29252858   111.46273188


         ASYMPTOTIC CORRELATION MATRIX OF THE PARAMETERS

   CORR         BO              B1              B2              B3

   BO        1.0000          0.4316          0.1141         -0.2553
   B1        0.4316          1.0000         -0.5150         -0.8088
   B2        0.1141         -0.5150          1.0000          0.7938
   B3       -0.2553         -0.8088          0.7938          1.0000
```

PLOT OF COUNT VERSUS DOSE (O) AND YHAT VERSUS DOSE (P)

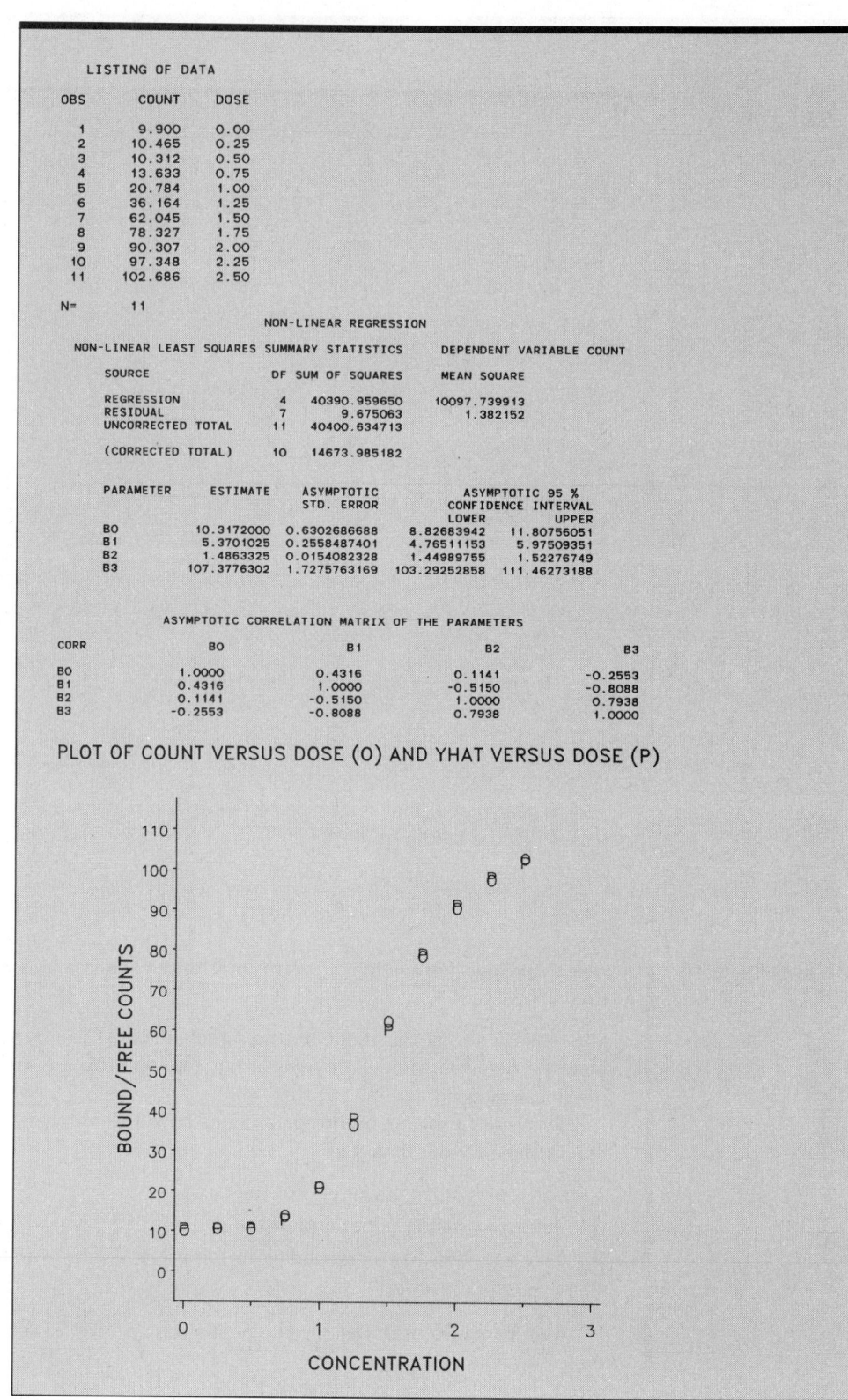

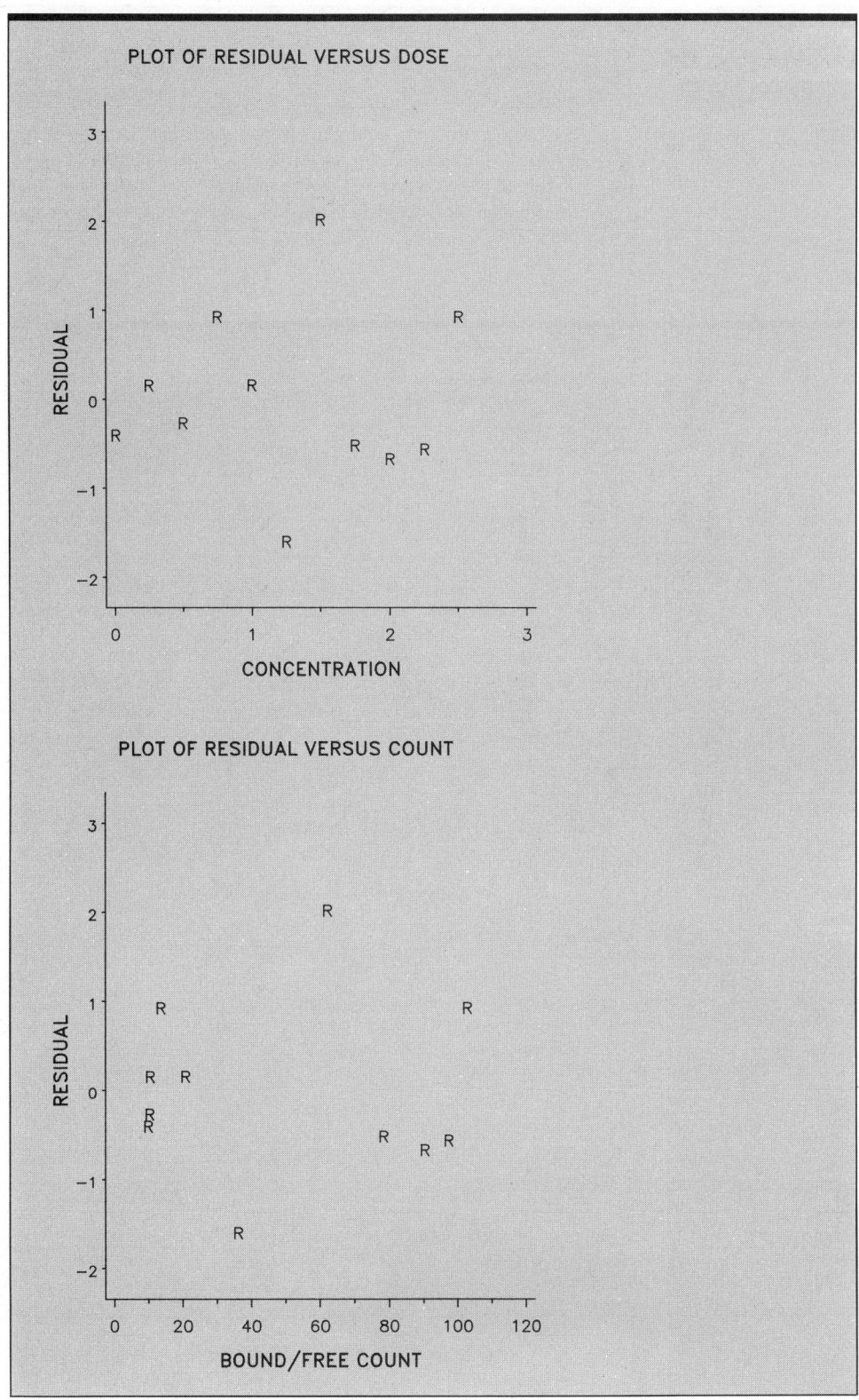

EXERCISES

 13.6 Peak blood level data (in mg/ml) were obtained for 20 patients for a single dose of a drug product. In addition to the peak blood level, the patient's weight (lb) and the amount of drug (mg) were recorded. Use the output shown here to fit a linear regression line and use residual plots to identify possible additional terms to be included in the regression model.

OBS	BLOOD	DOSE	WEIGHT
1	300	1	120
2	250	1	135
3	210	1	150
4	150	1	128
5	210	2	150
6	230	2	160
7	350	2	135
8	270	2	180
9	380	4	132
10	330	4	148
11	270	4	190
12	240	4	195
13	340	8	150
14	330	8	160
15	180	8	200
16	320	8	140
17	270	16	195
18	290	16	170
19	315	16	161
20	350	16	145

GENERAL LINEAR MODELS PROCEDURE

DEPENDENT VARIABLE: BLOOD   PEAK BLOOD LEVEL

SOURCE	DF	SUM OF SQUARES	MEAN SQUARE	F VALUE	PR > F	R-SQUARE	C.V.
MODEL	2	22290.44079456	11145.22039728	3.68	0.0468	0.302392	19.6953
ERROR	17	51423.30920544	3024.90054150		STD DEV		BLOOD MEAN
CORRECTED TOTAL	19	73713.75000000			54.99909582		279.25000000

SOURCE	DF	TYPE I SS	F VALUE	PR > F	DF	TYPE IV SS	F VALUE	PR > F
DOSE	1	8452.70329301	2.79	0.1129	1	16119.33338339	5.33	0.0338
WEIGHT	1	13837.73750155	4.57	0.0473	1	13837.73750155	4.57	0.0473

PARAMETER	ESTIMATE	T FOR H0: PARAMETER = 0	PR > |T|	STD ERROR OF ESTIMATE
INTERCEPT	432.60229380	5.11	0.0001	84.69454320
DOSE	5.54666579	2.31	0.0338	2.40278001
WEIGHT	−1.19428513	−2.14	0.0473	0.55838151

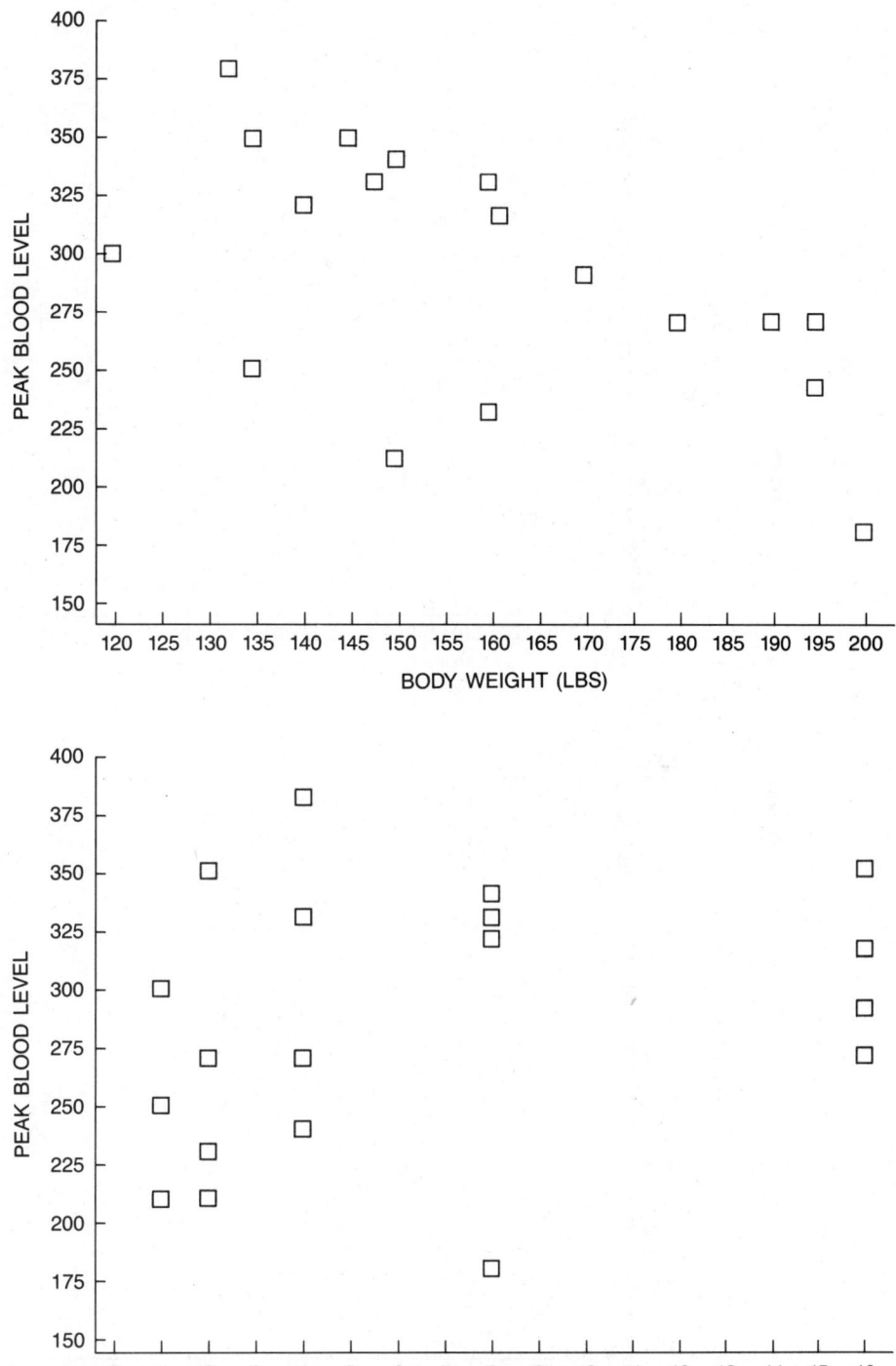

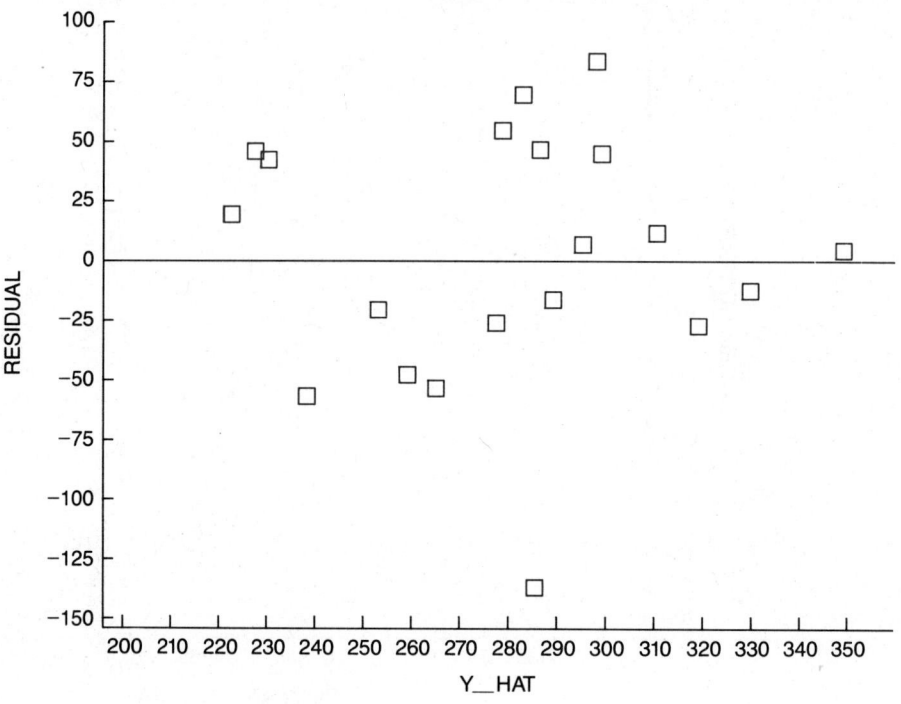

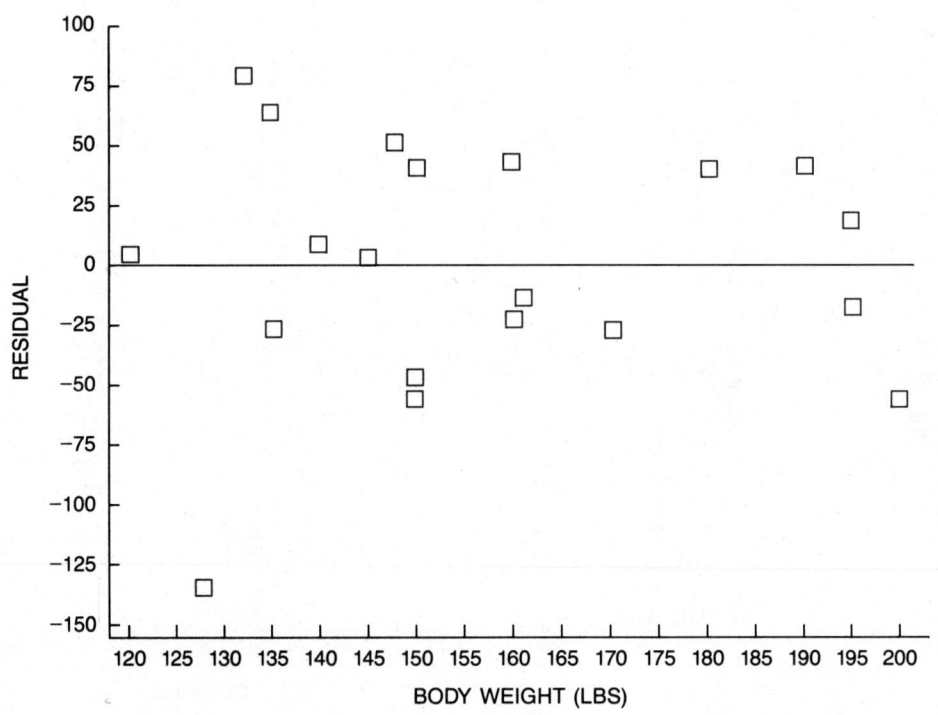

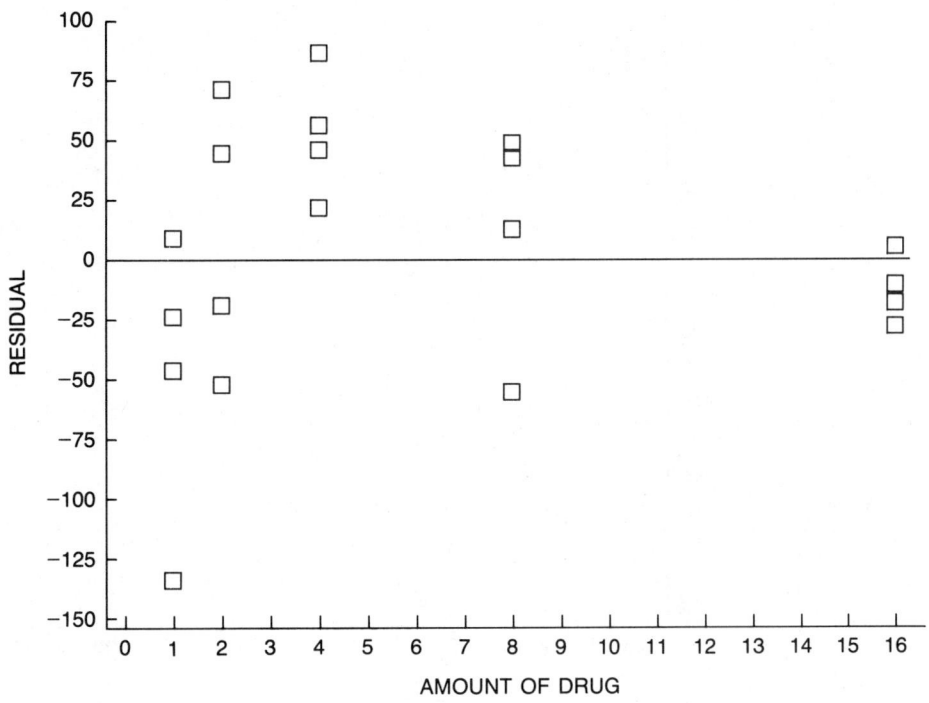

13.7   Refer to Exercise 13.6. Identify and discuss the fit of the model for the output shown here.

GENERAL LINEAR MODELS PROCEDURE

DEPENDENT VARIABLE: BLOOD   PEAK BLOOD LEVEL

SOURCE	DF	SUM OF SQUARES	MEAN SQUARE	F VALUE	PR > F	R-SQUARE	C.V.
MODEL	3	41167.19623371	13722.39874457	6.75	0.0038	0.558474	16.1510
ERROR	16	32546.55376629	2034.15961039		STD DEV		BLOOD MEAN
CORRECTED TOTAL	19	73713.75000000			45.10165862		279.25000000

SOURCE	DF	TYPE I SS	F VALUE	PR > F	DF	TYPE IV SS	F VALUE	PR > F
LOG__DOSE	1	13690.00000000	6.73	0.0196	1	9899.32954861	4.87	0.0423
WEIGHT	1	21798.10707872	10.72	0.0048	1	309.92533215	0.15	0.7014
LOG__DOSE*WEIGHT	1	5679.08915499	2.79	0.1142	1	5679.08915499	2.79	0.1142

PARAMETER	ESTIMATE	T FOR H0: PARAMETER = 0	PR > |T|	STD ERROR OF ESTIMATE
INTERCEPT	288.06239408	2.25	0.0390	128.09498236
LOG__DOSE	402.52746916	2.21	0.0423	182.46734172
WEIGHT	−0.34416249	−0.39	0.7014	0.88171355
LOG__DOSE*WEIGHT	−1.98696306	−1.67	0.1142	1.18916731

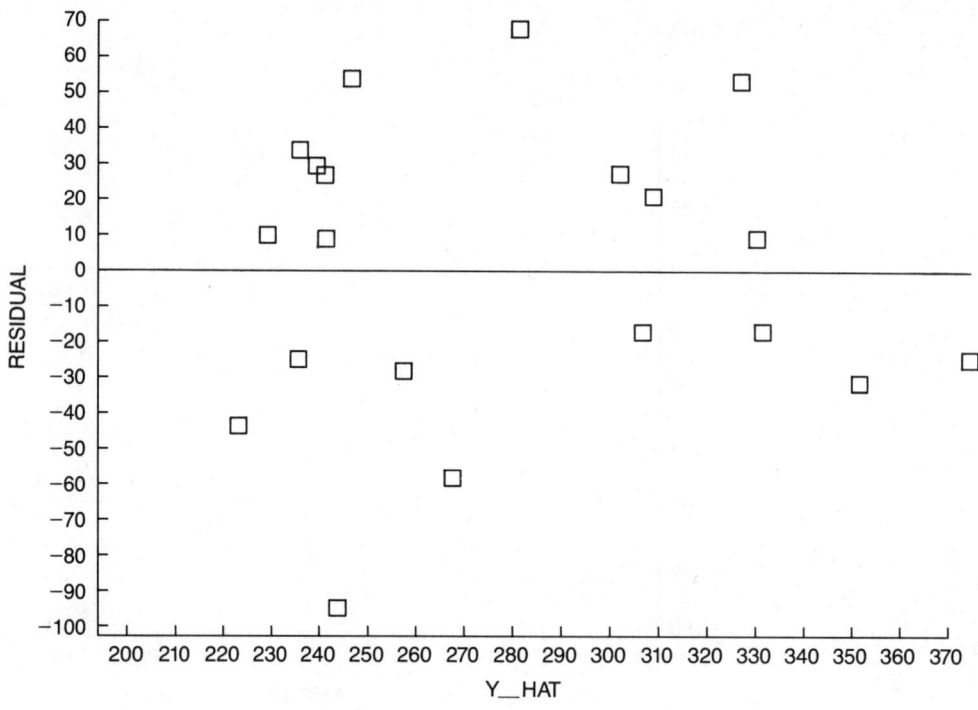

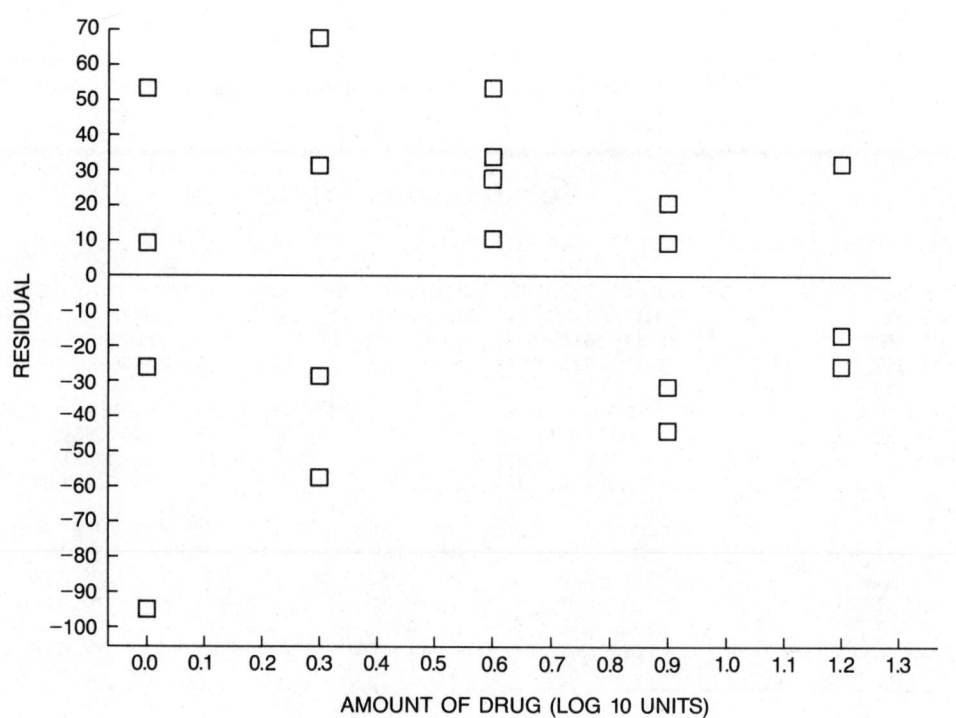

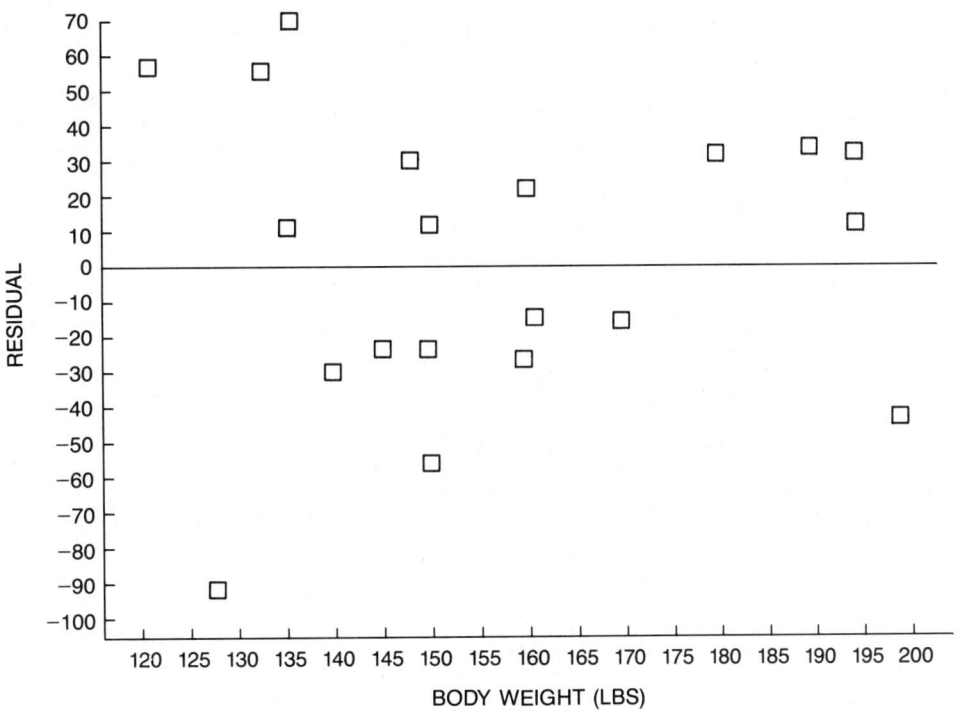

13.8   (Optional) Several different forms of nonlinear models are useful in predicting the plasma concentration from single or multiple doses of a drug product. The simplest model is

$$C(t) = \frac{k_a D}{V(k_a - k_e)} (e^{-k_e t} - e^{-k_a t})$$

where $C(t)$ is the concentration of drug in the body at time t, D/V is the ratio of the amount of drug absorbed to the volume of blood, k_a is the absorption rate of the drug, and k_e is the elimination rate. The data for an experiment consist of concentrations obtained from blood samples at various points in time. The problem is to fit the data to the model to obtain estimates of k_a, k_e, and D/V. The concentration data shown here were fit using NONLIN. Give estimates of the model parameters. Predict concentration when $t = 3.0$. Does this model seem to fit the data? Note: Estimates of the model parameter are given in the order $\hat{k}_a$, $\hat{k}_e$, and $\widehat{D/V}$.

FUNCTION 1			
X	OBSERVED Y	X	OBSERVED Y
0.170000	5.10000	5.00000	18.5000
0.250000	12.1000	6.00000	14.5000
0.500000	133.000	7.00000	10.8000
0.750000	127.000	8.00000	10.1000
1.00000	109.600	10.0000	7.30000
1.50000	77.0000	12.0000	3.60000
2.00000	59.1000	16.0000	1.80000
3.00000	41.3000	24.0000	1.00000
4.00000	25.2000		

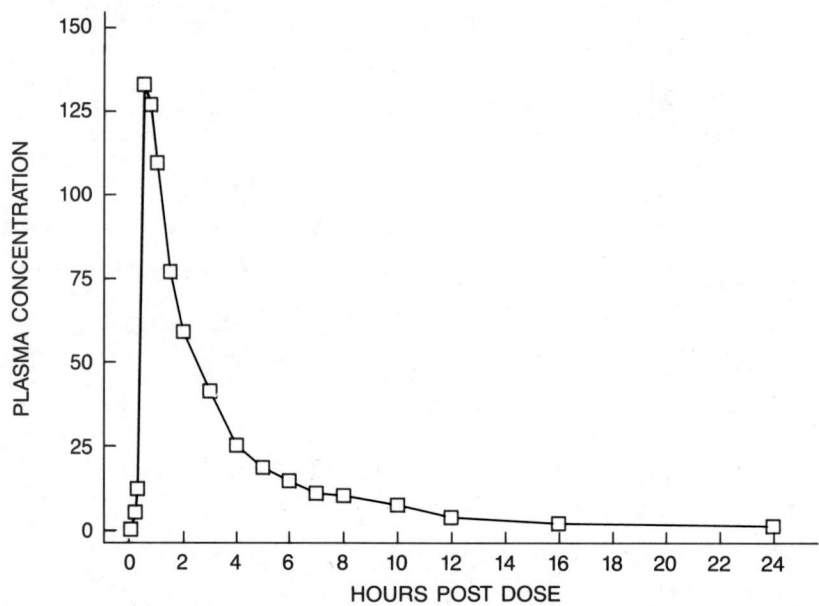

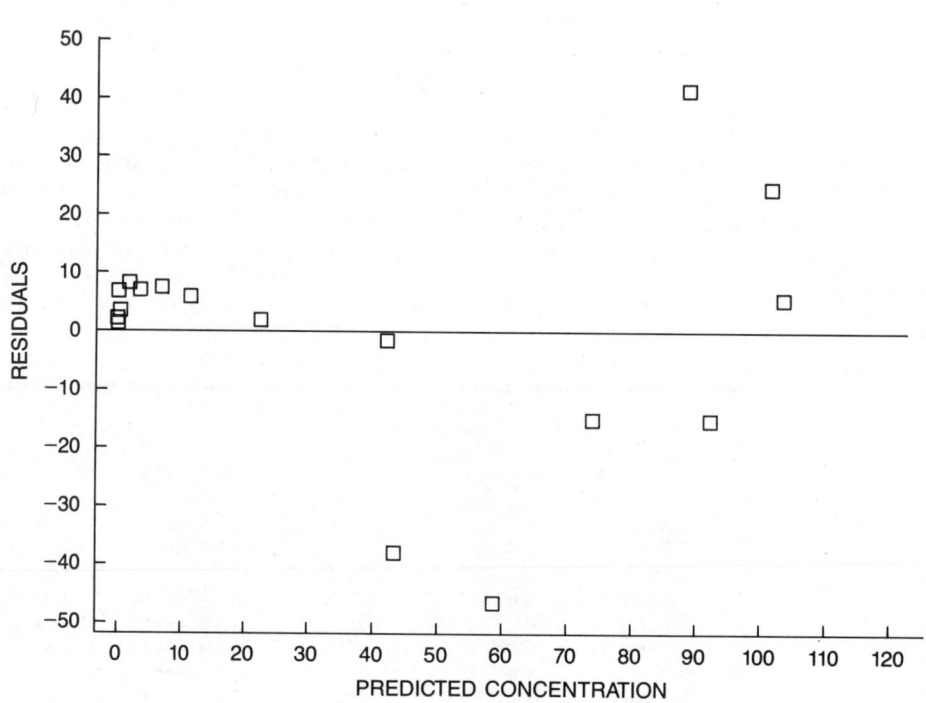

SUMMARY OF NON-LINEAR ESTIMATION

AFTER 25 ITERATIONS THE ESTIMATES AND THEIR VARIABILITY ARE:

	NO.	ESTIMATE	STD. DEV.	95% CONFIDENCE LIMITS		
$\hat{K}A$	1	1.60860	1.26013	−1.09411	4.31131	UNIVAR
				−2.44040	5.65760	S PLANE
$\hat{K}E$	2	0.663720	0.465020	−0.333646	1.66109	UNIVAR
				−0.830461	2.15790	S PLANE
$\widehat{D/V}$	3	194.207	109.909	−41.5237	429.938	UNIVAR
				−158.947	547.362	S PLANE

PREDICTED CONCENTRATION AT T = 9 IS 0.8414
HALF-LIFE KA = 0.4309
HALF-LIFE K = 1.0443

FUNCTION 1   ... ARE PREDICTED POINTS,   *** ARE OBSERVED POINTS

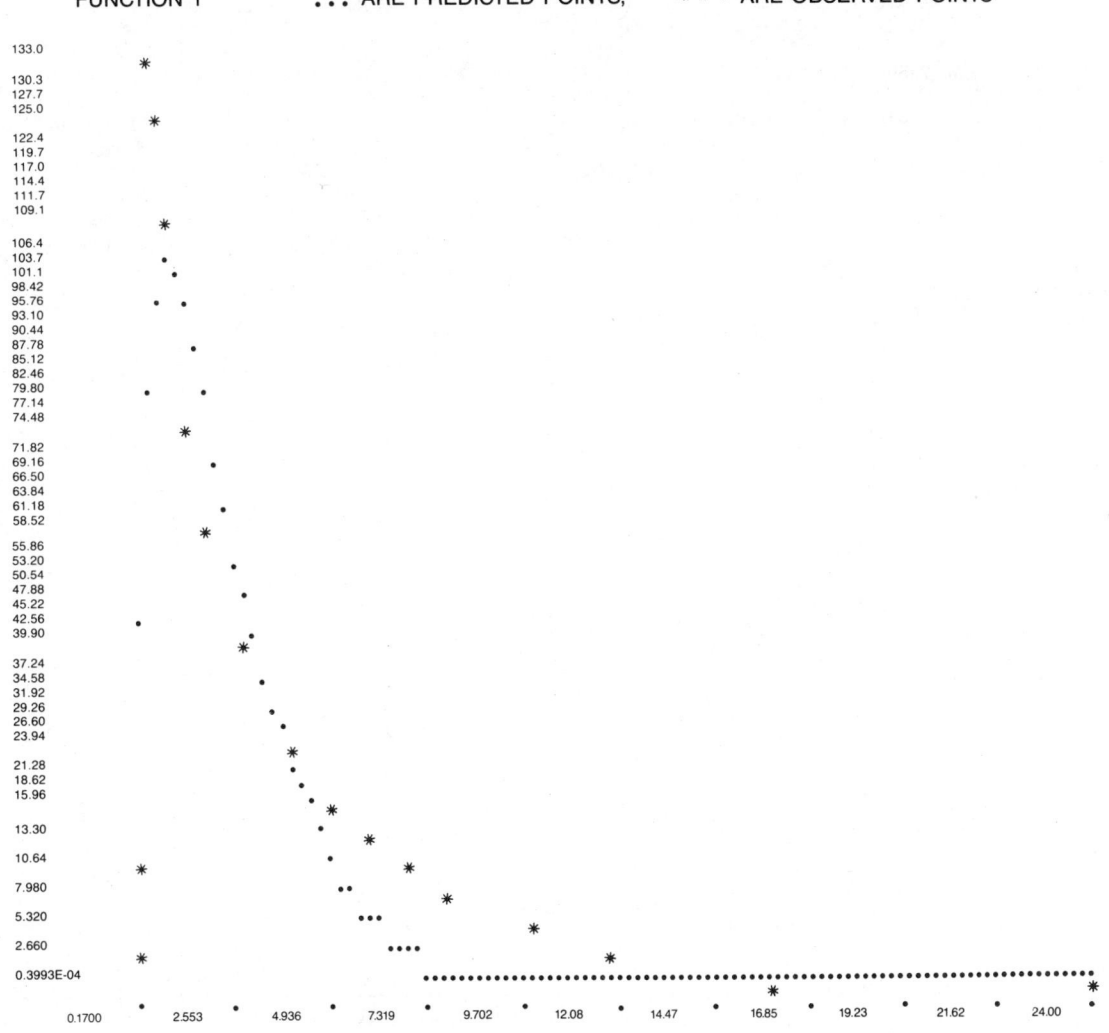

13.9   (Optional) Refer to Exercise 13.8. A second, more complicated model may be appropriate for the data of the previous exercise. The model is

$$C(t) = k_a \frac{D}{V} \left[\frac{\alpha - k_{21}}{(\alpha - \beta)(k_a - \alpha)} e^{-\alpha t} + \frac{k_{21} - \beta}{(\alpha - \beta)(k_a - \beta)} e^{-\beta t} + \frac{k_a - k_{21}}{(k_a - \alpha)(k_a - \beta)} e^{-k_a t} \right]$$

where $\alpha + \beta = k_e + k_{12} + k_{21}$ and $\alpha\beta = k_e k_{21}$. Use the output shown here to determine whether this model provides a better fit to the sample concentration data. Note: The estimates are shown in the order $\hat{k}_a$, $\hat{\alpha}$, $\hat{\beta}$, $\hat{k}_{21}$, and $\widehat{D/V}$.

SUMMARY OF NONLINEAR ESTIMATION

AFTER 13 ITERATIONS THE ESTIMATES AND THEIR VARIABILITY ARE:

NO.	ESTIMATE	STD. DEV.	95% CONFIDENCE LIMITS		
1	6.83567	26.5429	−50.9965	64.6679	UNIVAR
			−98.5367	112.208	S PLANE
2	4.19859	3.06471	−2.47885	10.8760	UNIVAR
			−7.96796	16.3651	S PLANE
3	0.556948	0.140483	0.250861	0.863036	UNIVAR
			−0.754321E − 03	1.11465	S PLANE
4	11.5141	150.713	−316.862	339.890	UNIVAR
			−586.799	609.827	S PLANE
5	62.8093	878.265	−1850.77	1976.39	UNIVAR
			−3423.80	3549.42	S PLANE

PREDICTED AT T = 9 IS 1.3690
HALF-LIFE KA = 0.1014
HALF-LIFE B = 1.2445
K1 = −0.283 K2 = −0.762 K3 = 0.479

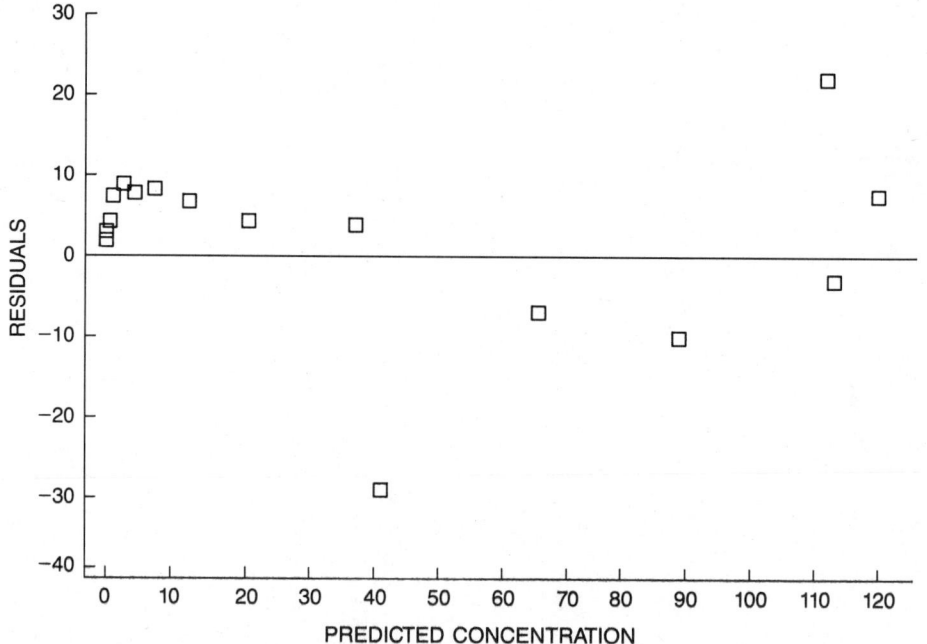

FUNCTION 1   ARE PREDICTED POINTS,   ARE OBSERVED POINTS

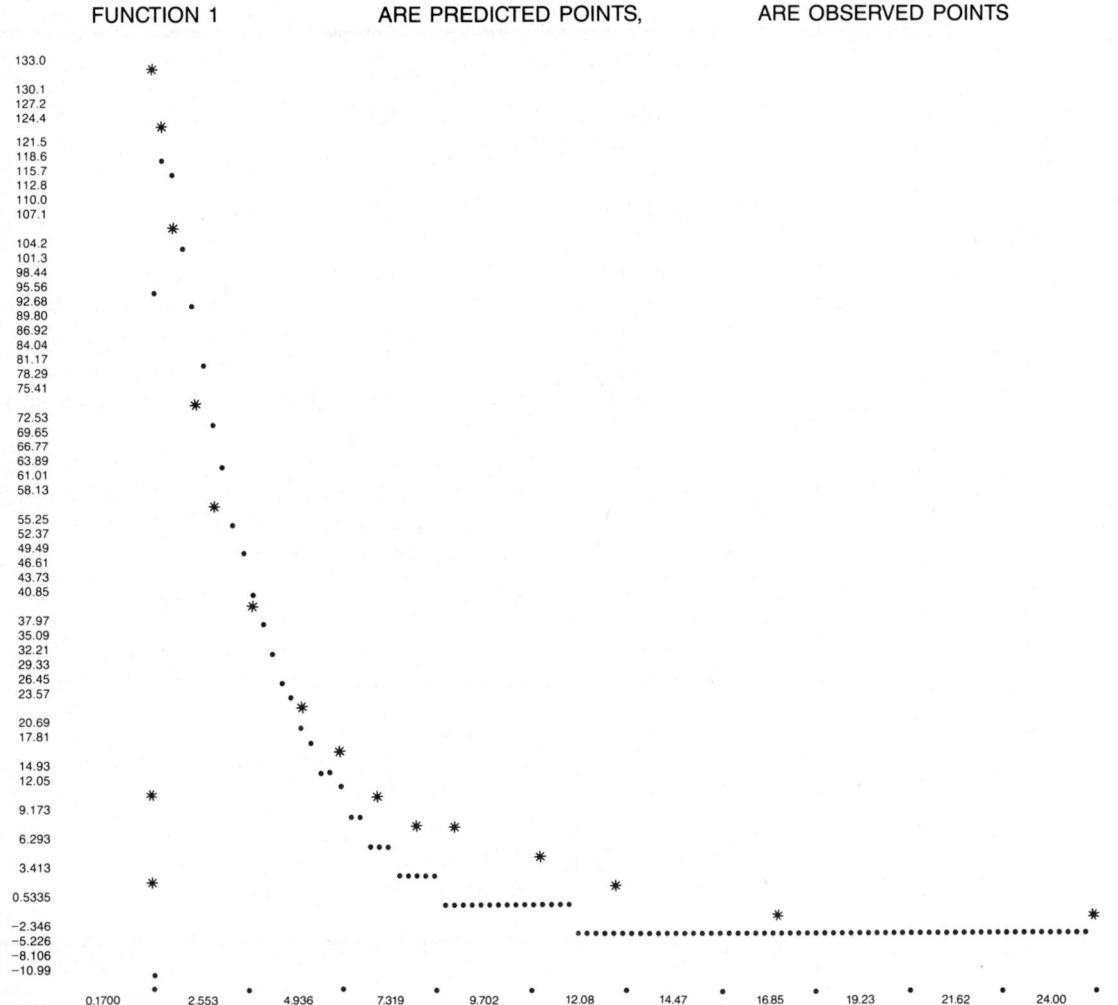

13.4 RESIDUAL ANALYSIS: CHECKING MODEL ASSUMPTIONS (Step 3)

Now that we have identified possible independent variables (step 1) and considered the form of the multiple regression model (step 2), we should check whether the assumptions underlying the chosen model are valid. Recall that in Chapter 12 we indicated that the basic assumptions for a regression model of the form

$$y_i = \beta_0 + \beta_1 x_{i1} + \beta_2 x_{i2} + \cdots + \beta_k x_{ik} + \varepsilon_i$$

are

1. Zero expectation: $E(\varepsilon_i) = 0$ for all i.
2. Constant variance: $V(\varepsilon_i) = \sigma_\varepsilon^2$ for all i.
3. Normality: ε_i is normally distributed.
4. Independence: The ε_i are independent.

TABLE 13.5   Data for Example 13.11

y	11	10	2	14	22	10	20	19	32	23	40	37
x	.5	1	1.2	1.4	1.7	1.8	2	2.3	2.5	2.8	3	3.1
y	30	43	55	29	45	60	53	30	42	25	63	51
x	3.5	3.6	3.8	4.2	4.4	5.1	5.2	5.4	5.5	6	6.2	6.3

Note that since the assumptions for multiple regression are written in terms of the random errors ε_i, it would seem reasonable to check the assumptions by using the residuals $y_i - \hat{y}_i$, which are *estimates* of the ε_i.

The first assumption, zero expectation, deals with model selection and whether or not additional independent variables need to be included in the model. If we have done our job in steps 1 and 2, Assumption 1 should hold. The use of residual plots to check for inadequacy (lack of fit) of the model was discussed briefly in Chapter 8 and again in Section 13.3.

The assumptions of constant variance can be examined using residual plots. One of the simplest residual plots for detecting nonconstant variance is a plot of the residuals versus the predicted values, $\hat{y}_i$. Most of the available statistical software systems can provide these plots as part of the regression analysis.

EXAMPLE 13.11

The data that are shown in Table 13.5 were fit to the model $y = \beta_0 + \beta_1 x + \beta_2 x^2 + \varepsilon$ using SAS. Examine the plot residuals versus $\hat{y}_i$ to detect possible nonconstant variance. Can you identify a pattern of nonconstant variance?

Solution

As can be seen from the SAS residual plot, the magnitudes of the residuals are generally increasing with the magnitudes of the predicted values of y, suggesting possible nonconstant variance. And, since y_i is directly related to x via the regression model (i.e., y increases with x), the residuals are increasing with the magnitude of the xs. This pattern in the residuals suggests that the variance of ε_i (and hence $V(y_i)$) is increasing with x. The accompanying plot of y vs. x tends to bear this out.

```
LISTING OF DATA

OBS    Y      X

  1    11    0.5
  2    10    1.0
  3     2    1.2
  4    14    1.4
  5    22    1.7
  6    10    1.8
  7    20    2.0
  8    19    2.3
  9    32    2.5
 10    23    2.8
 11    40    3.0
 12    37    3.1
 13    30    3.5
 14    43    3.6
 15    55    3.8
 16    29    4.2
 17    45    4.4
 18    60    5.1
 19    53    5.2
 20    30    5.4
 21    42    5.5
 22    25    6.0
 23    63    6.2
 24    51    6.3

N=     24
```

QUADRATIC REGRESSION

GENERAL LINEAR MODELS PROCEDURE

DEPENDENT VARIABLE: Y

SOURCE	DF	SUM OF SQUARES	MEAN SQUARE	F VALUE	PR > F	R-SQUARE	C.V.
MODEL	2	4458.40551726	2229.20275863	20.45	0.0001	0.660717	32.7142
ERROR	21	2289.42781607	109.02037219		ROOT MSE		Y MEAN
CORRECTED TOTAL	23	6747.83333333			10.44128211		31.91666667

SOURCE	DF	TYPE I SS	F VALUE	PR > F	DF	TYPE III SS	F VALUE	PR > F
X	1	4141.99721230	37.99	0.0001	1	970.14118949	8.90	0.0071
X*X	1	316.40830496	2.90	0.1032	1	316.40830496	2.90	0.1032

PARAMETER	ESTIMATE	T FOR HO: PARAMETER=0	PR > |T|	STD ERROR OF ESTIMATE
INTERCEPT	-6.87174737	-0.77	0.4472	8.87156858
X	17.10536096	2.98	0.0071	5.73414408
X*X	-1.34903610	-1.70	0.1032	0.79186923

PLOT OF Y*X

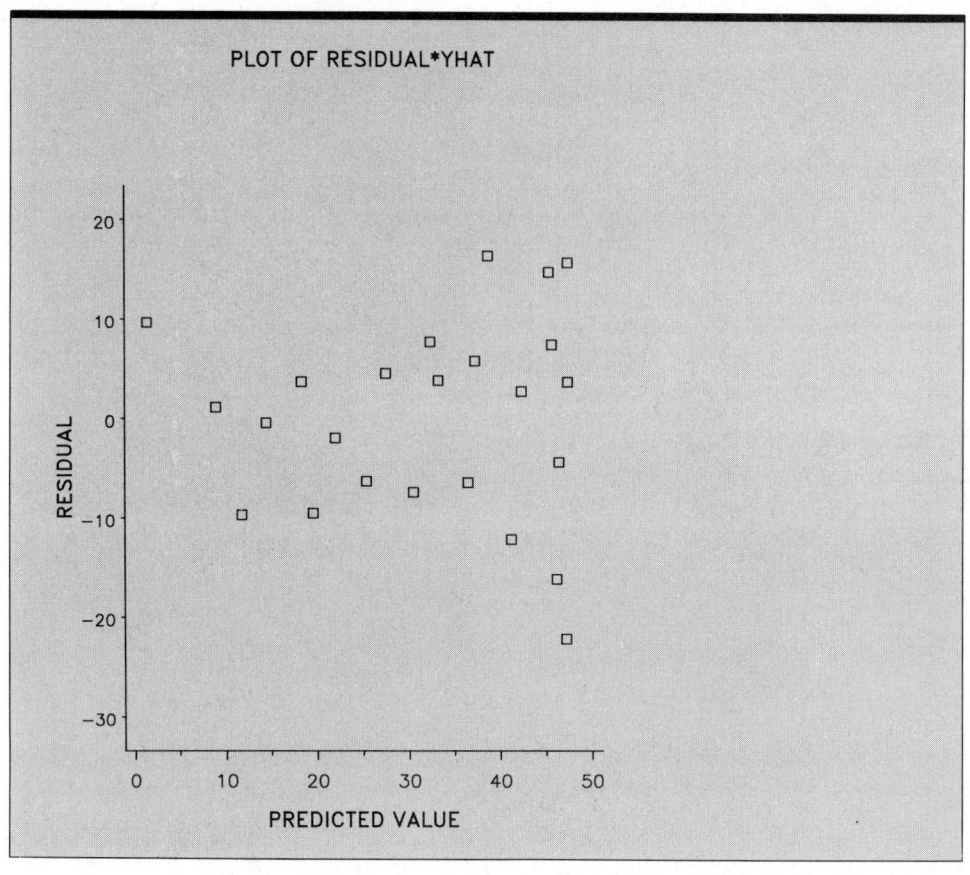

PLOT OF RESIDUAL*YHAT

What are the consequences of having a nonconstant variance problem in a regression model? First, if the variance about the regression line is not constant, the least squares estimates may not be as accurate as possible. A technique called **weighted least squares** (see Draper and Smith (1981)) will give more accuracy. But, perhaps more important, the weighted least squares technique improves the statistical tests (F and t tests) on model parameters and the interval estimates for parameter because they are, in general, based on smaller standard errors.

weighted least squares

The more serious pitfall involved with inferences in the presence of nonconstant variance seems to be for estimates $E(y)$ and predictions of y. For these inferences, the point estimate y is sound but the width of the interval may be too large or too small depending on whether we're predicting in a low or high variance section of the experimental region.

The remedy for nonconstant variance seems to be weighted least squares. We will not cover this technique in the text. However, when the nonconstant variance possesses a pattern related to y, a reexpression (transformation) of y may resolve the problem, accomplishing the same end as weighted least squares. Several transformations for y were discussed in Chapter 8; ones that help to stabilize the variance when there is a pattern to the nonconstant variance were discussed in Chapter 10 for analysis of variance. They can also be applied in certain regression situations as well.

EXAMPLE 13.12   Refer to the data of Example 13.11, where we detected a nonconstant variance problem. Since the variance about the regression line seemed to increase with x, a square root transformation on y was tried to stabilize the variance. Examine the computer output and residual plot shown here to determine whether the nonconstant variance problem has been eliminated.

```
        LISTING OF DATA

  OBS     Y     X     SQRT_Y

   1     11    0.5    3.31662
   2     10    1.0    3.16228
   3      2    1.2    1.41421
   4     14    1.4    3.74166
   5     22    1.7    4.69042
   6     10    1.8    3.16228
   7     20    2.0    4.47214
   8     19    2.3    4.35890
   9     32    2.5    5.65685
  10     23    2.8    4.79583
  11     40    3.0    6.32456
  12     37    3.1    6.08276
  13     30    3.5    5.47723
  14     43    3.6    6.55744
  15     55    3.8    7.41620
  16     29    4.2    5.38516
  17     45    4.4    6.70820
  18     60    5.1    7.74597
  19     53    5.2    7.28011
  20     30    5.4    5.47723
  21     42    5.5    6.48074
  22     25    6.0    5.00000
  23     63    6.2    7.93725
  24     51    6.3    7.14143

   N=      24
```

QUADRATIC REGRESSION

GENERAL LINEAR MODELS PROCEDURE

DEPENDENT VARIABLE: SQRT_Y

SOURCE	DF	SUM OF SQUARES	MEAN SQUARE	F VALUE	PR > F	R-SQUARE	C.V.
MODEL	2	45.00905315	22.50452657	24.68	0.0001	0.701561	17.6571
ERROR	21	19.14652533	0.91173930		ROOT MSE		SQRT_Y MEAN
CORRECTED TOTAL	23	64.15557847			0.95485041		5.40772758

SOURCE	DF	TYPE I SS	F VALUE	PR > F	DF	TYPE III SS	F VALUE	PR > F
X	1	39.57102314	43.40	0.0001	1	13.13323564	14.40	0.0011
X*X	1	5.43803000	5.96	0.0235	1	5.43803000	5.96	0.0235

PARAMETER	ESTIMATE	T FOR HO: PARAMETER=0	PR > |T|	STD ERROR OF ESTIMATE
INTERCEPT	1.18979490	1.47	0.1573	0.81130083
X	1.99021795	3.80	0.0011	0.52438482
X*X	-0.17685626	-2.44	0.0235	0.07241607

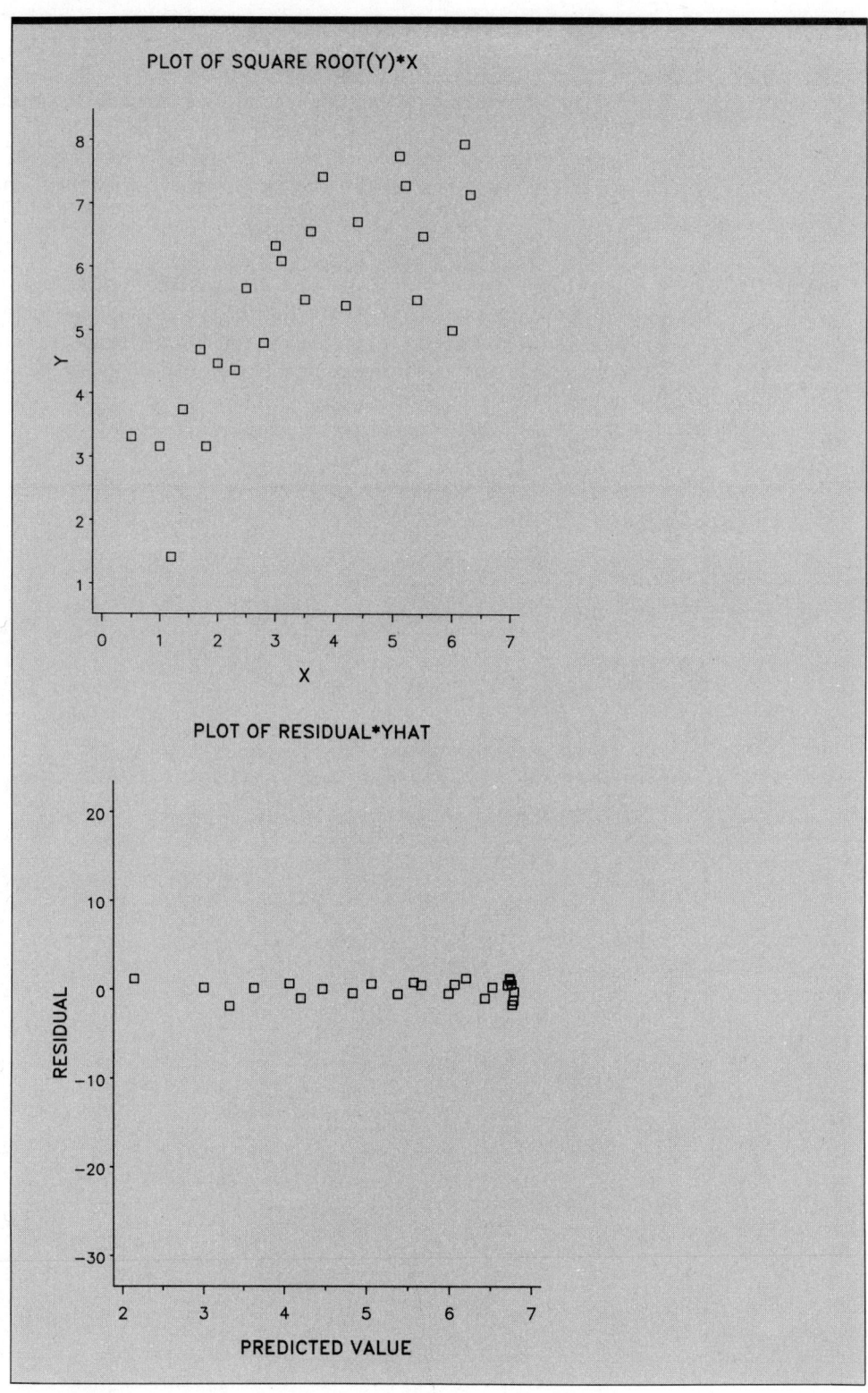

Solution ▌ The output shown here documents that this model provides a much better fit to the sample data; note, especially, the residual plot.

The third assumption for multiple regression is that of normality of the ε_i. Skewness and/or outliers are examples of forms of nonnormality that may be detected through the use of certain scatterplots and residual plots.

A plot of the residuals in the form of a histogram or a stem-and-leaf plot will help to detect skewness. By assumption, the ε_i are normally distributed with mean zero. If a histogram of the residuals is not symmetrical about zero, some skewness is present. For example, the residual plot in Figure 13.1(a) is symmetrical on zero and suggests no skewness. In contrast, the residual plot in Figure 13.1(b) is skewed to the right.

probability plot   Another way to detect nonnormality is through the use of a normal **probability plot** of the residuals. The idea behind the plot is that if the residuals are normally distributed, the normal probability plot will be approximately a straight line.

The normal probability plot of the residuals is simple to construct using specially designed graph paper called *normal probability paper* (see Figure 13.2). The horizontal axis is scaled in uniform units while the vertical axis goes from 0.01 to 0.99 corresponding to

FIGURE 13.1

Top: Residuals Centered on Zero; Bottom: Residuals Skewed to Right

Middle of
interval   Number of observations

−2.0	3	× × ×
−1.5	10	× × × × × × × × × ×
−1.0	16	× × × × × × × × × × × × × × × ×
−0.5	15	× × × × × × × × × × × × × × ×
0.0	20	× × × × × × × × × × × × × × × × × × × ×
0.5	15	× × × × × × × × × × × × × × ×
1.0	11	× × × × × × × × × × ×
1.5	6	× × × × × ×
2.0	3	× × ×
2.5	0	
3.0	1	×

Middle of
interval   Number of observations

−2.0	3	× × ×
−1.5	10	× × × × × × × × × ×
−1.0	16	× × × × × × × × × × × × × × × ×
−0.5	15	× × × × × × × × × × × × × × ×
0.0	18	× × × × × × × × × × × × × × × × × ×
0.5	12	× × × × × × × × × × × ×
1.0	7	× × × × × × ×
1.5	5	× × × × ×
2.0	3	× × ×
2.5	3	× × ×
3.0	2	× ×
3.5	0	
4.0	1	×

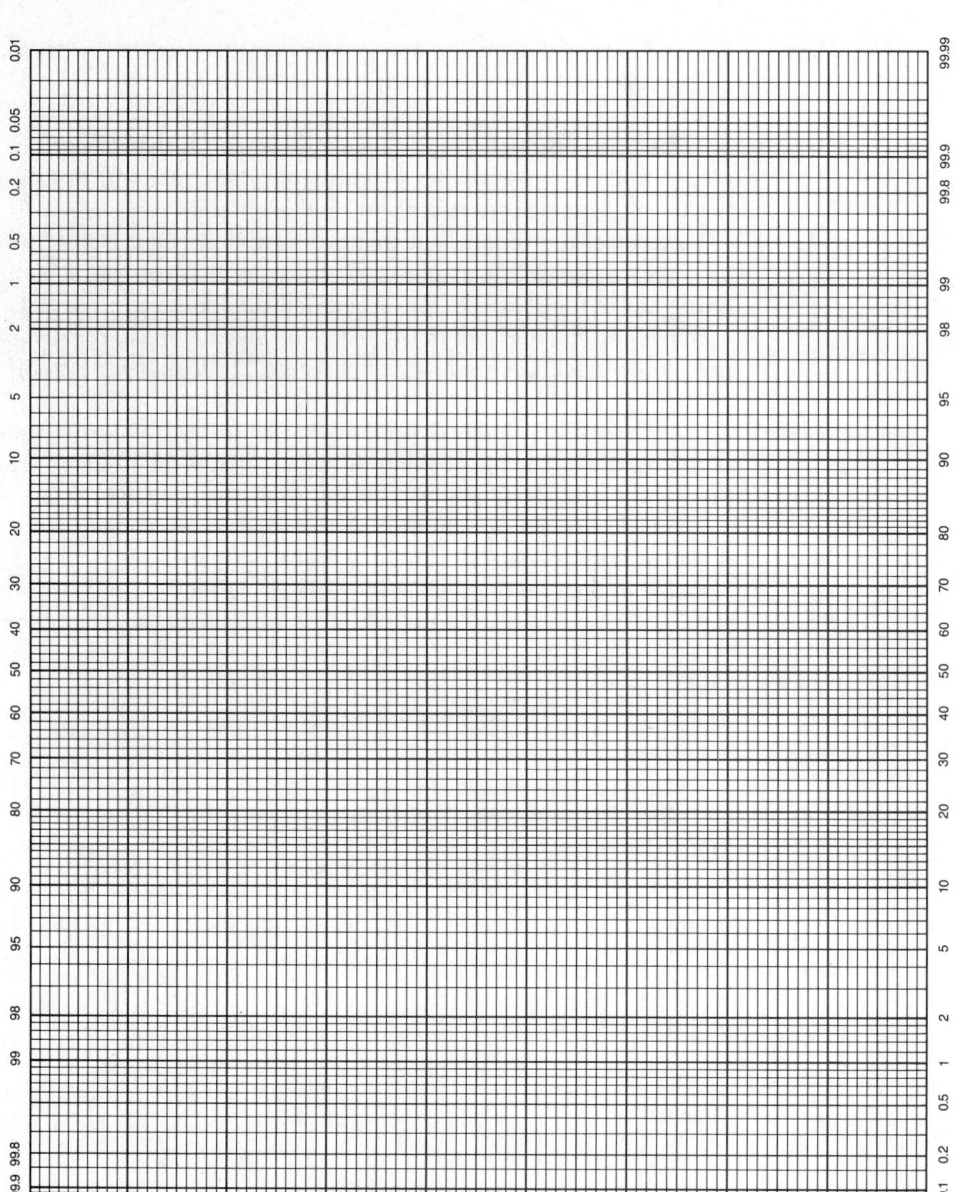

FIGURE 13.2

Normal Probability
Paper

FIGURE 13.3

S-Shape of the
Normal cdf

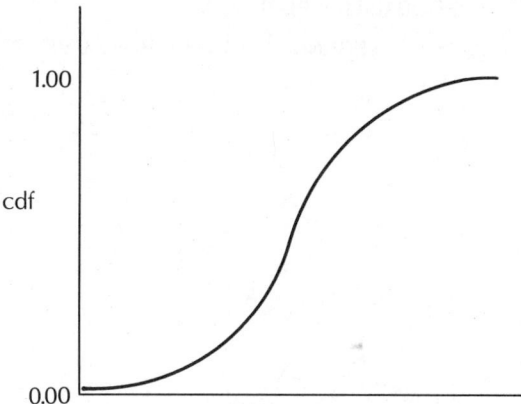

the cumulative distribution function (cdf) of a normal distribution. The cdf of a normal distribution plot using uniformly scaled units along the vertical and horizontal axes would be S-shaped, as shown in Figure 13.3. The scaling used for the vertical axis of normal probability paper "stretches" the vertical axis to straighten out the S-shape of the normal cdf.

The steps we follow in constructing a normal probability plot are listed below.

CONSTRUCTING A NORMAL PROBABILITY PLOT

1. Compute the residuals; divide each one by $s_\varepsilon = \sqrt{MSE}$ to standardize it. (Note: The standardized residuals from a normal distribution should have mean 0 and standard deviation 1; hence almost all of the standardized residuals should lie between -3 and $+3$.)
2. Number the standardized residuals from the largest negative (1) to the largest positive (n).
3. Label the horizontal axis of a piece of normal probability paper from -5 to $+5$ corresponding to values of the standardized residuals.
4. Plot the value of the ith ordered standardized residual on the x axis and $(i - .5)/n$ along the y axis.
5. If the plot is nearly linear, it suggests the normality assumption is not violated.

Most regression software packages offer an option for generating probability plots. This is illustrated in Example 13.13.

EXAMPLE 13.13

Refer to the data of Example 13.11. Use the computer output shown here to determine whether there is evidence of nonnormality.

Solution

The fact that the probability plot is nearly linear suggests that the normality assumption has not been violated for the data of Example 13.11.

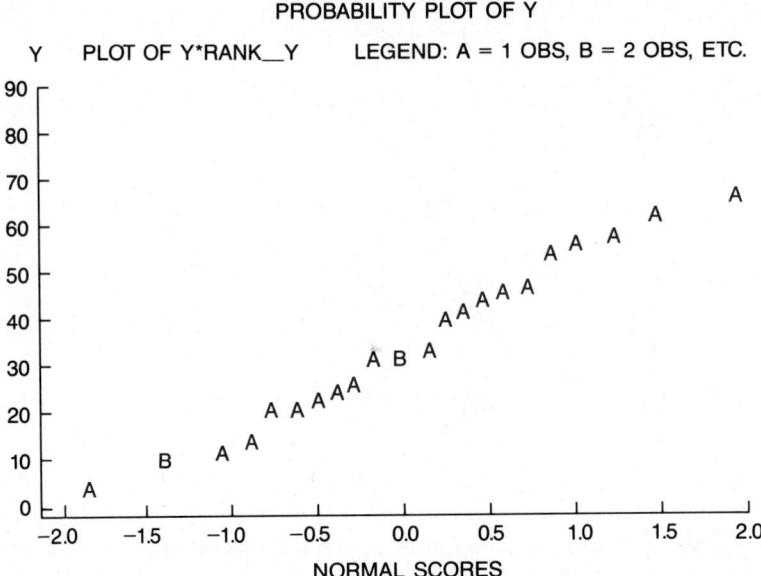

PROBABILITY PLOT OF Y

Y   PLOT OF Y*RANK__Y   LEGEND: A = 1 OBS, B = 2 OBS, ETC.

NORMAL SCORES

The presence of one or more outliers is perhaps a more subtle form of nonnormality that may be detected by using a scatterplot and one or more residual plots. For the linear regression model $y = \beta_0 + \beta_1 x + \varepsilon$ a scatterplot of y versus x will help detect the presence of an outlier. This is shown in Table 13.6 and Figure 13.4 (on the next page). It certainly appears that the circled data point is an outlier.

TABLE 13.6   Listing of Data

Obs	x	y
1	10	120
2	20	115
3	21	250
4	27	210
5	29	300
6	33	330
7	40	295
8	44	400
9	52	380
10	56	460
11	62	125
12	68	510
N = 12		

Computer output for a linear fit to the data of Table 13.6 is shown here, along with a residual plot and a normal probability plot. Again the data point corresponding to the suspected outlier (62, 125) is circled in each plot.

FIGURE 13.4   PLOT OF Y*X

Scatterplot of the
Data in Table 13.6

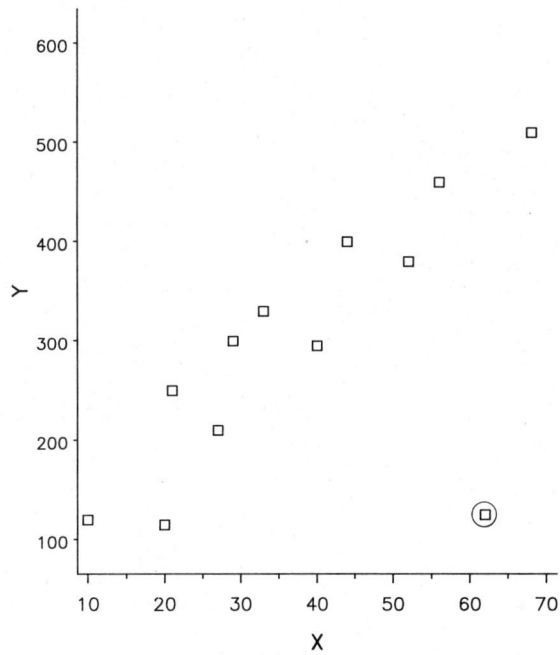

LISTING OF DATA

OBS	X	Y
1	10	120
2	20	115
3	21	250
4	27	210
5	29	300
6	33	330
7	40	295
8	44	400
9	52	380
10	56	460
11	62	125
12	68	510

N=   12

LINEAR REGRESSION

GENERAL LINEAR MODELS PROCEDURE

DEPENDENT VARIABLE: Y

SOURCE	DF	SUM OF SQUARES	MEAN SQUARE	F VALUE	PR > F	R-SQUARE	C.V.
MODEL	1	77200.98612250	77200.98612250	6.61	0.0278	0.398033	37.0999
ERROR	10	116755.26387750	11675.52638775		ROOT MSE		Y MEAN
CORRECTED TOTAL	11	193956.25000000			108.05334973		291.25000000

SOURCE	DF	TYPE I SS	F VALUE	PR > F	DF	TYPE III SS	F VALUE	PR > F
X	1	77200.98612250	6.61	0.0278	1	77200.98612250	6.61	0.0278

PARAMETER	ESTIMATE	T FOR H0: PARAMETER=0	PR > |T|	STD ERROR OF ESTIMATE
INTERCEPT	114.35739677	1.51	0.1610	75.53322553
X	4.59461307	2.57	0.0278	1.78679922

OBSERVATION	OBSERVED VALUE	PREDICTED VALUE	RESIDUAL
1	120.00000000	160.30352748	-40.30352748
2	115.00000000	206.24965819	-91.24965819
3	250.00000000	210.84427126	39.15572874
4	210.00000000	238.41194969	-28.41194969
5	300.00000000	247.60117583	52.39882417
6	330.00000000	265.97962811	64.02037189
7	295.00000000	298.14191961	-3.14191961
8	400.00000000	316.52037189	83.47962811
9	380.00000000	353.27272646	26.72272354
10	460.00000000	371.65572874	88.34427126
11	125.00000000	399.22340716	-274.22340716
12	510.00000000	426.79108559	83.20891441

SUM OF RESIDUALS	0.00000000
SUM OF SQUARED RESIDUALS	116755.26387750
SUM OF SQUARED RESIDUALS - ERROR SS	-0.00000000
FIRST ORDER AUTOCORRELATION	-0.36022054
DURBIN-WATSON D	2.64722728

PLOT OF RESIDUAL*YHAT

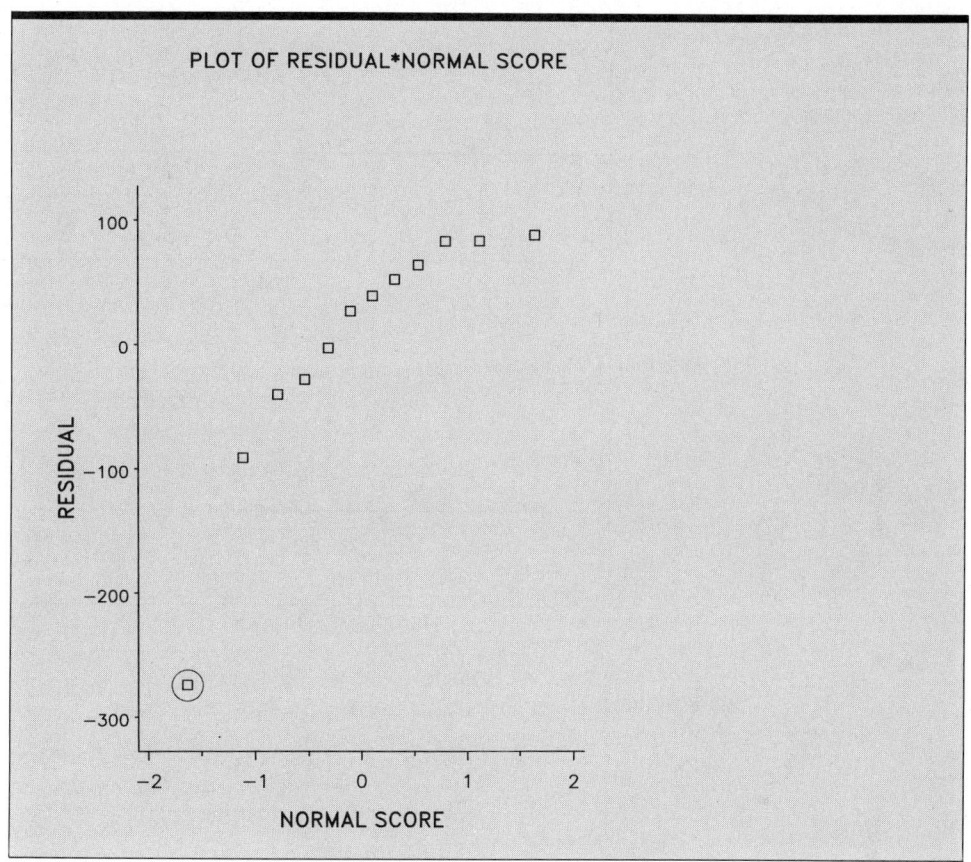

This data set helps to illustrate one of the problems in trying to identify outliers. Sometimes a single plot is not sufficient. For this example, the scatterplot and the probability plot clearly identify the outlier, whereas the residual plot is less conclusive since the outlier adversely affects the linear fit to the data by pulling the fitted line toward the outlier. This makes some of the other residuals larger than they should be. The message is clear: *don't jump to conclusions without examining the data in several different ways.* The problem becomes even more difficult with multiple regression where simple scatterplots are not possible.

  The final assumption is that the ε_i are statistically independent, and hence uncorrelated. When the time sequence of the observations is known, as is the case with **time series** *data* where observations are taken at successive points in time, it is possible to construct a plot of the residuals versus time to observe where the residuals are **serially correlated**. If, for example, there is a positive serial correlation, adjacent residuals (in time) tend to be similar; negative serial correlation implies that adjacent residuals are dissimilar. These patterns of positive and negative serial correlation are displayed in Figures 13.5(a) and 13.5(b), respectively. Figure 13.5(c) shows a residual plot with no apparent serial correlation.

  A formal statistical test for serial correlation is based on the *Durbin–Watson statistic.* Let $\hat{\varepsilon}_t$ denote the residual at time t and n the total number of time points. Then the

time series

serial correlation

FIGURE 13.5

(a) Positive Serial
Correlation;
(b) Negative Serial
Correlation; (c) No
Apparent Serial
Correlation

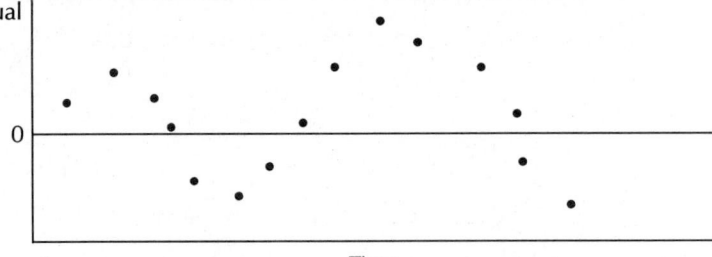

(a) Residual

Time

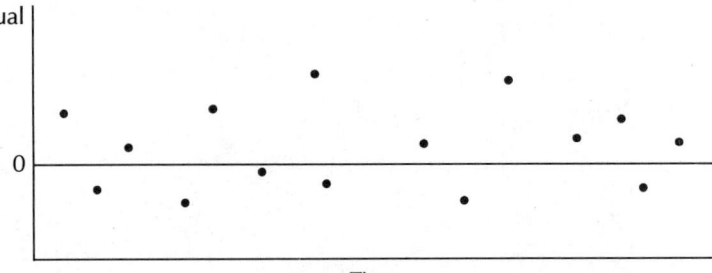

(b) Residual

Time

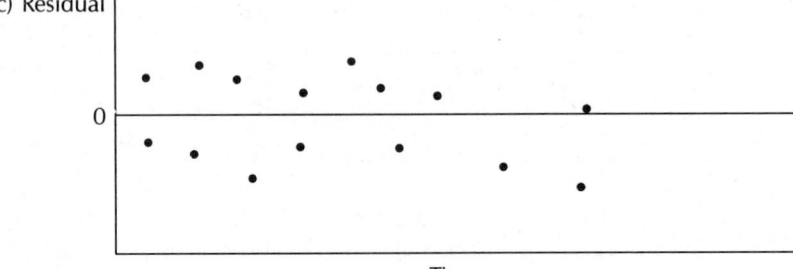

(c) Residual

Time

Durbin–Watson test statistic is

**Durbin–Watson
statistic**

$$d = \frac{\sum_{t=1}^{n-1} (\hat{\varepsilon}_{t+1} - \hat{\varepsilon}_t)^2}{\sum_t \hat{\varepsilon}_t^2}$$

The logic behind this statistic is as follows: If there is a positive serial correlation, then successive residuals will be similar and their squared difference $(\hat{\varepsilon}_{t+1} - \hat{\varepsilon}_t)^2$ will tend to be smaller than it would be if the residuals were uncorrelated. Similarly, if there is a negative serial correlation among the residuals, the squared difference of successive residuals will tend to be larger than when no correlation exists.

**positive and
negative serial
correlation**
When there is no serial correlation, the expected value of the Durbin–Watson test statistic d is approximately 2.0; **positive serial correlation** makes $d < 2.0$ and **negative serial correlation** makes $d > 2.0$. Although critical values of d have been tabulated by J. Durbin and G. S. Watson (1951), values of d less than approximately 1.5 (or greater than approximately 2.5) lead one to suspect positive (or negative) serial correlation.

EXAMPLE 13.14

Sample data corresponding to retail sales for a particular line of personalized computers by month are shown here.

Month, x	Sales (Millions of Dollars), y
1	6.0
2	6.3
3	6.1
4	6.8
5	7.5
6	8.0
7	8.1
8	8.5
9	9.0
10	8.7
11	7.9
12	8.2
13	8.4
14	9.0

Plot the data. Also plot the residuals by time based on a linear regression equation. Does there appear to be serial correlation?

Solution

It is clear from the scatterplot of the sample data and from the residual plot of the linear regression that there is serial correlation present in the data.

```
        LISTING OF DATA

    OBS     MONTH     SALES

     1        1        6.0
     2        2        6.3
     3        3        6.1
     4        4        6.8
     5        5        7.5
     6        6        8.0
     7        7        8.1
     8        8        8.5
     9        9        9.0
    10       10        8.7
    11       11        7.9
    12       12        8.2
    13       13        8.4
    14       14        9.0

    N=       14
```

LINEAR REGRESSION

GENERAL LINEAR MODELS PROCEDURE

DEPENDENT VARIABLE: SALES

SOURCE	DF	SUM OF SQUARES	MEAN SQUARE	F VALUE	PR > F	R-SQUARE	C.V.
MODEL	1	10.57539560	10.57539560	34.30	0.0001	0.740833	7.1645
ERROR	12	3.69960440	0.30830037		ROOT MSE		SALES MEAN
CORRECTED TOTAL	13	14.27500000			0.55524802		7.75000000

SOURCE	DF	TYPE I SS	F VALUE	PR > F	DF	TYPE III SS	F VALUE	PR > F
MONTH	1	10.57539560	34.30	0.0001	1	10.57539560	34.30	0.0001

PARAMETER	ESTIMATE	T FOR HO: PARAMETER=0	PR > |T|	STD ERROR OF ESTIMATE
INTERCEPT	6.13296703	19.57	0.0001	0.31344787
MONTH	0.21560440	5.86	0.0001	0.03681259

OBSERVATION	OBSERVED VALUE	PREDICTED VALUE	RESIDUAL
1	6.00000000	6.34857143	-0.34857143
2	6.30000000	6.56417582	-0.26417582
3	6.10000000	6.77978022	-0.67978022
4	6.80000000	6.99538462	-0.19538462
5	7.50000000	7.21098901	0.28901099
6	8.00000000	7.42659341	0.57340659
7	8.10000000	7.64219780	0.45780220
8	8.50000000	7.85780220	0.64219780
9	9.00000000	8.07340659	0.92659341
10	8.70000000	8.28901099	0.41098901
11	7.90000000	8.50461538	-0.60461538
12	8.20000000	8.72021978	-0.52021978
13	8.40000000	8.93582418	-0.53582418
14	9.00000000	9.15142857	-0.15142857

```
SUM OF RESIDUALS                         0.00000000
SUM OF SQUARED RESIDUALS                 3.69960440
SUM OF SQUARED RESIDUALS - ERROR SS     -0.00000000
FIRST ORDER AUTOCORRELATION              0.66819228
DURBIN-WATSON D                          0.62457541
```

PLOT OF SALES VERSUS MONTH

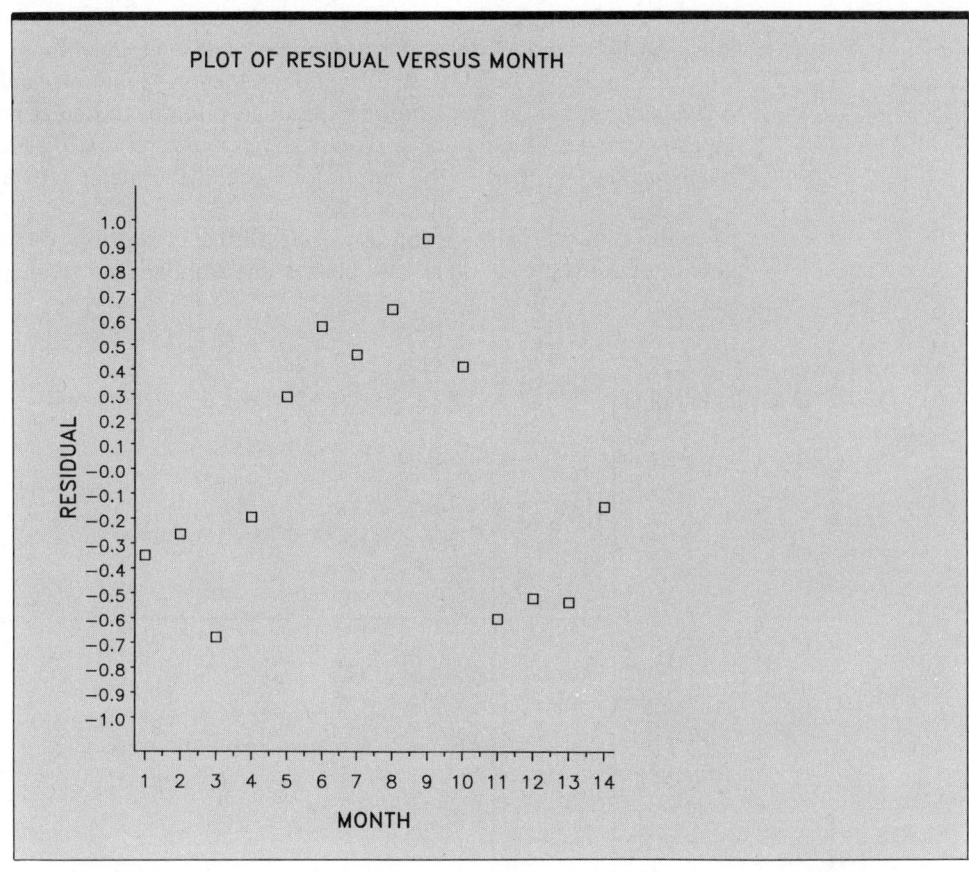

PLOT OF RESIDUAL VERSUS MONTH

EXAMPLE 13.15   Determine the value of the Durbin–Watson statistic for the data of Example 13.14. Does it confirm the impressions you obtained from the plots?

Solution   Based on the output of Example 13.14, we find $d = 0.62457541$. Since this value is much less than 1.5, we have evidence of positive serial correlation; the residual plot bears this out.

If serial correlation is suspected then the proposed multiple regression model is inappropriate and some alternative must be sought. A study of the many approaches to analyzing time series data where the errors are not independent can consume many years; hence we cannot expect to solve many of these problems within the confines of this text. We will, however, suggest a simplified regression approach, based on *first differences*, which may alleviate the problem.

Regression based on first differences is simple to use and, as might be expected, is only a crude approach to the problem of serial correlation. For a simple linear regression of y on x, we compute the differences $y_t - y_{t-1}$ and $x_t - x_{t-1}$. A regression of the $n - 1$ y differences on the corresponding $n - 1$ x differences may eliminate the serial correlation. If not, you should consult someone more familiar with analyzing time series data.

The residual plots that we have discussed can be useful in diagnosing problems in fitting regression models to data. Unfortunately however, they, too, can be misleading because the residuals are subject to random variation. Some researchers have suggested that it is better to use "standardized" residuals in order to detect problems with a fitted regression model. One particular type of standardized residual, called the Studentized residual, has become part of the output for some of the major software packages such as SAS.

If the software package you use works with standardized residuals, you can replace plots of the ordinary residuals with plots of the standardized residuals to perform the di-

FIGURE 13.6

Residual Plots for
Exercise 13.10

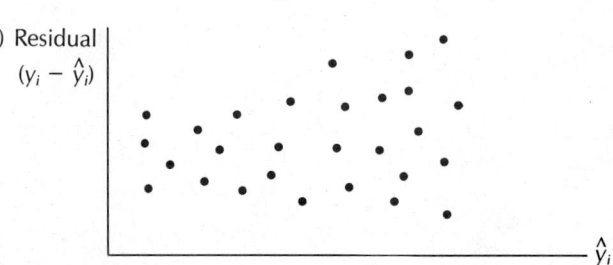

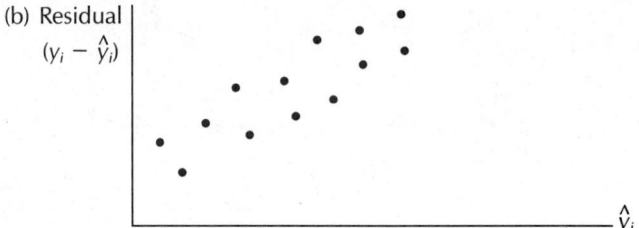

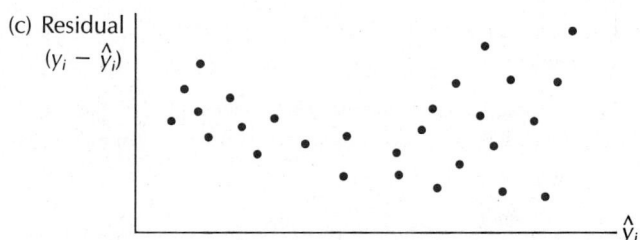

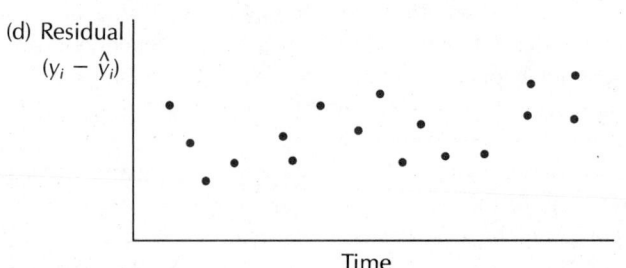

agnostic evaluation of the fit of a regression model. In theory, these standardized residuals have a mean of 0 and a standard deviation of 1. So large residuals would be ones with an absolute value of, say, 3 or more.

EXERCISES

13.10   Several different patterns of residuals are shown in the plots of Figure 13.6. Indicate whether or not the plot suggests a problem, and, if so, indicate the potential problem and a possible solution.

13.11   Refer to the data of Example 13.14. Form first differences and regress the y differences on the x differences. Is there evidence of serial correlation for the difference model? What plot(s) did you use to reach a conclusion?

13.12   What is the value of the Durbin–Watson statistic for the data of Exercise 13.11? Does it agree with your previous conclusion?

13.13   A researcher in the social sciences examined the relationship between the rate (per 1000) of nonviolent crimes y based on the rate 5 years ago x_1, the present unemployment rate x_2 for cities. Data from 20 different cities are shown here.

OBS	RATE	RATE 5	UNEMPLOY
1	20	14	14.3
2	10	10	7.4
3	14	16	12.0
4	15	10	10.7
5	13	16	8.1
6	4	12	6.5
7	11	8	8.6
8	16	7	10.4
9	13	12	10.5
10	13	20	12.2
11	15	14	8.3
12	8	10	8.0
13	15	10	12.2
14	15	20	14.1
15	6	13	8.3
16	3	2	8.7
17	5	10	9.5
18	13	14	14.4
19	14	16	13.8
20	7	8	10.0

N = 20

Use the output shown here to:

a. Determine the fit to the model

$$y = \beta_0 + \beta_1 x_1 + \beta_2 x_2 + \beta_3 x_1 x_2 + \varepsilon.$$

b. Examine the assumptions underlying the regression model. Discuss whether the assumptions appear to hold. If they don't, suggest possible remedies.

```
            LISTING OF DATA

   OBS    RATE    RATE_5    UNEMPLOY

    1      20       14       14.3
    2      10       10        7.4
    3      14       16       12.0
    4      15       10       10.7
    5      13       16        8.1
    6       4       12        6.5
    7      11        8        8.6
    8      16        7       10.4
    9      13       12       10.5
   10      13       20       12.2
   11      15       14        8.3
   12       8       10        8.0
   13      15       10       12.2
   14      15       20       14.1
   15       6       13        8.3
   16       3        2        8.7
   17       5       10        9.5
   18      13       14       14.4
   19      14       16       13.8
   20       7        8       10.0

   N=      20
```

MULTIPLE REGRESSION

GENERAL LINEAR MODELS PROCEDURE

DEPENDENT VARIABLE: RATE

SOURCE	DF	SUM OF SQUARES	MEAN SQUARE	F VALUE	PR > F	R-SQUARE	C.V.
MODEL	3	227.86512995	75.95504332	7.10	0.0030	0.571091	28.4388
ERROR	16	171.13487005	10.69592938		ROOT MSE		RATE MEAN
CORRECTED TOTAL	19	399.00000000			3.27046317		11.50000000

SOURCE	DF	TYPE I SS	F VALUE	PR > F	DF	TYPE III SS	F VALUE	PR > F
RATE_5	1	87.59158010	8.19	0.0113	1	53.18992084	4.97	0.0404
UNEMPLOY	1	94.79946979	8.86	0.0089	1	79.79956840	7.46	0.0148
RATE_5*UNEMPLOY	1	45.47408006	4.25	0.0558	1	45.47408006	4.25	0.0558

PARAMETER	ESTIMATE	T FOR H0: PARAMETER=0	PR > |T|	STD ERROR OF ESTIMATE
INTERCEPT	-29.87534146	-2.13	0.0492	14.03734719
RATE_5	2.33165740	2.23	0.0404	1.04558496
UNEMPLOY	3.81909377	2.73	0.0148	1.39820005
RATE_5*UNEMPLOY	-0.20308407	-2.06	0.0558	0.09849250

OBSERVATION	OBSERVED VALUE	PREDICTED VALUE	RESIDUAL
1	20.00000000	16.72347242	3.27652758
2	10.00000000	6.67430534	3.32569466
3	14.00000000	14.26816091	-0.26816091
4	15.00000000	12.57554050	2.42445950
5	13.00000000	12.04614112	0.95385888
6	4.00000000	7.08809946	-3.08809946
7	11.00000000	7.64994023	3.35005977
8	16.00000000	11.38031534	4.61968466
9	13.00000000	12.61643923	0.38356077
10	13.00000000	13.79823765	-0.79823765
11	15.00000000	10.86797161	4.13202839
12	8.00000000	7.74725719	0.25274281
13	15.00000000	15.25792012	-0.25792012
14	15.00000000	13.33732118	1.66267882
15	6.00000000	10.22191198	-4.22191198
16	3.00000000	4.48042637	-1.48042637
17	5.00000000	10.42963681	-5.42963681
18	13.00000000	16.82106410	-3.82106410
19	14.00000000	15.29370850	-1.29370850
20	7.00000000	10.72212993	-3.72212993

MULTIPLE REGRESSION

GENERAL LINEAR MODELS PROCEDURE

DEPENDENT VARIABLE: RATE

```
        SUM OF RESIDUALS                      -0.00000000
        SUM OF SQUARED RESIDUALS             171.13487005
        SUM OF SQUARED RESIDUALS - ERROR SS   -0.00000000
        FIRST ORDER AUTOCORRELATION            0.29418193
        DURBIN-WATSON D                        1.26794898
```

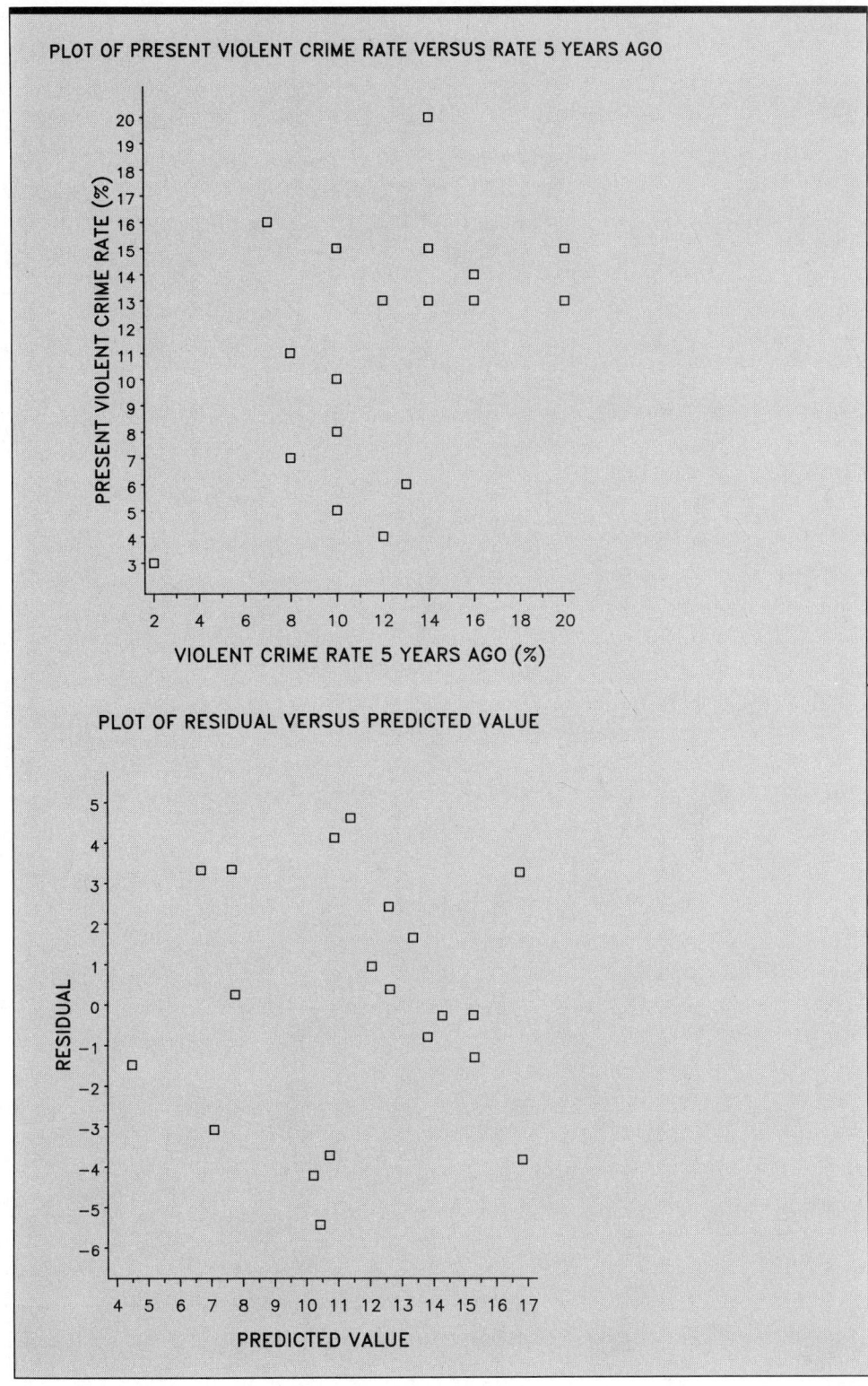

PLOT OF PRESENT VIOLENT CRIME RATE VERSUS RATE 5 YEARS AGO

PLOT OF RESIDUAL VERSUS PREDICTED VALUE

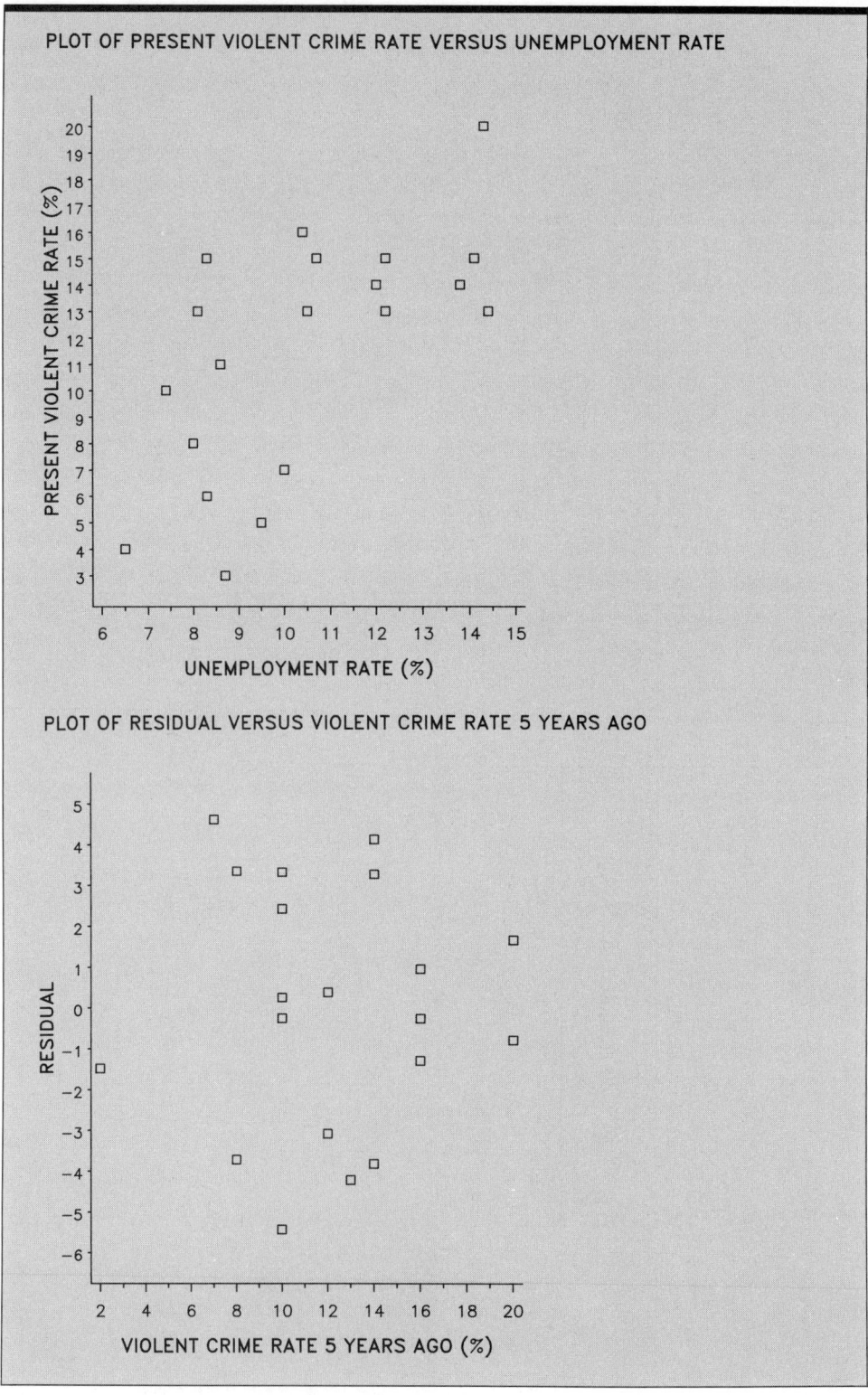

PLOT OF PRESENT VIOLENT CRIME RATE VERSUS UNEMPLOYMENT RATE

PLOT OF RESIDUAL VERSUS VIOLENT CRIME RATE 5 YEARS AGO

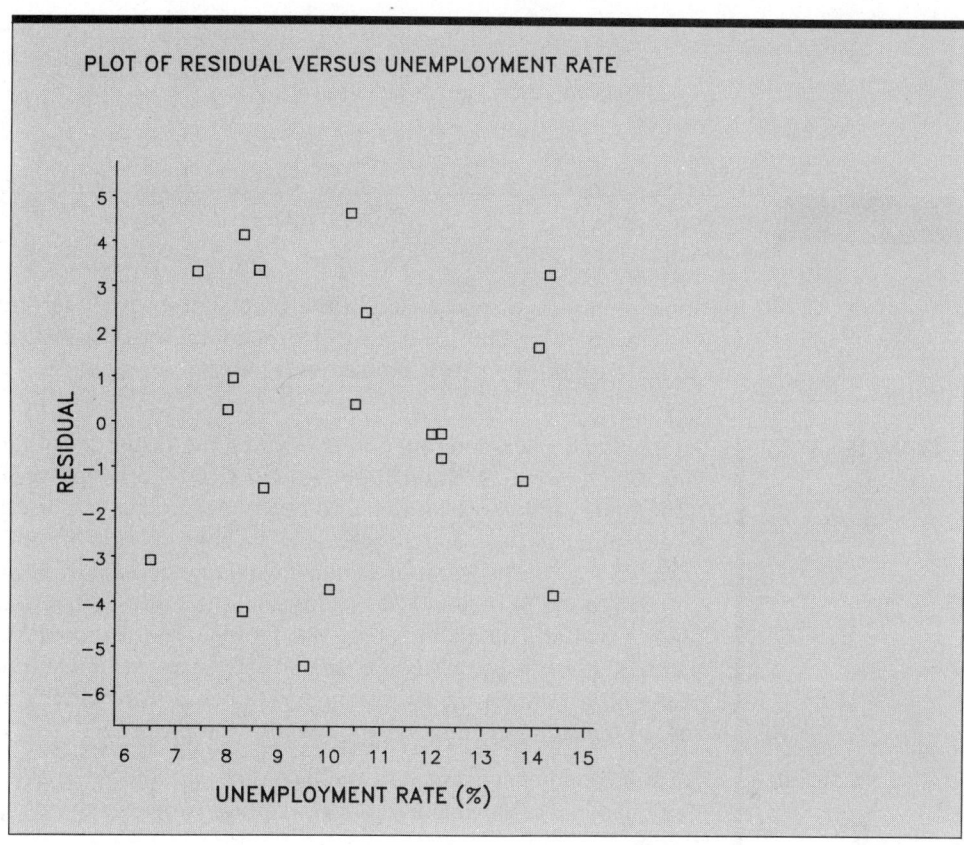

PLOT OF RESIDUAL VERSUS UNEMPLOYMENT RATE

13.14   Refer to Exercise 13.13. Predict y for $x_1 = 9$ and $x_2 = 16$. Might there be a problem with this prediction? If so, why?

13.15   Estimates ($\hat{y}$s) and residuals from a securities firm's regression model for the prediction of earnings per share (per quarter) are shown here for 25 different high-technology companies. Is there any evidence that the assumptions have been violated? Are any additional tests or plots warranted?

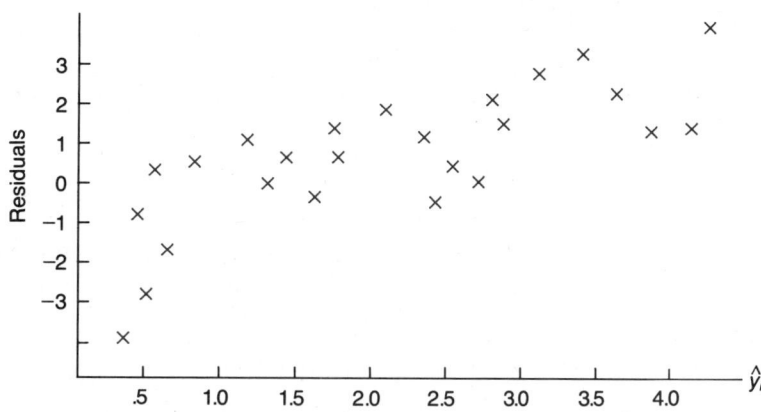

13.16   Refer to Exercise 13.15. Suppose that these data represent estimates and residuals for the earnings per share for the past 25 quarters of a single company. Assess the possibility of serial correlation. Are any adjustments required? If so, which?

13.5   ODDS AND ENDS: COMPARING THE SLOPES OF TWO OR MORE REGRESSION LINES

This topic represents a special case of the general problem of constructing a multiple regression equation for both qualitative and quantitative independent variables. The best way to illustrate this particular problem is by way of an example.

EXAMPLE 13.16

An investigator was interested in comparing the responses of rats to different doses of two drug products (A and B). The study called for a sample of 60 rats of a particular strain to be randomly allocated into two equal groups. The first group of rats was to receive Drug A, with 10 rats randomly assigned to each of three doses (5, 10, and 20 mg). Similarly, the 30 rats in Group 2 were to receive Drug B, with 10 rats randomly assigned to the 5-, 10-, and 20-mg doses. In the study each rate received its assigned dose, and after a 30-minute observation period, it was scored for signs of anxiety on a 0-to-30-point scale. Assume a rat's anxiety score is a linear function of the dosage of the drug. Write a model relating a rat's scores to the two independent variables "drug product" and "drug dose." Interpret the βs.

Solution

For this experimental situation, we have one qualitative variable (drug product) and one quantitative variable (drug dose). Letting x_1 denote the drug dose, we have the model

$$y = \beta_0 + \beta_1 x_1 + \beta_2 x_2 + \beta_3 x_1 x_2 + \varepsilon$$

where

$x_1 =$ drug dose
$x_2 = 1$ if product B $x_2 = 0$ otherwise

The expected value for y in our model is

$$E(y) = \beta_0 + \beta_1 x_1 + \beta_2 x_2 + \beta_3 x_1 x_2$$

Substituting $x_2 = 0$ and $x_2 = 1$, respectively, for Drugs A and B, we obtain the expected rat anxiety score for a given dose:

Drug A: $E(y) = \beta_0 + \beta_1 x_1$
Drug B: $E(y) = \beta_0 + \beta_1 x_1 + \beta_2 + \beta_3 x_1 = (\beta_0 + \beta_2) + (\beta_1 + \beta_3) x_1$

linear regression lines

These two expected values represent **linear regression lines**. The parameters in the model can be interpreted in terms of the slopes and intercepts associated with these regression lines. In particular,

y-intercept β_0: y-intercept for product A regression line
slope β_1: slope of product A regression line

FIGURE 13.7

Comparing Two
Regression Lines

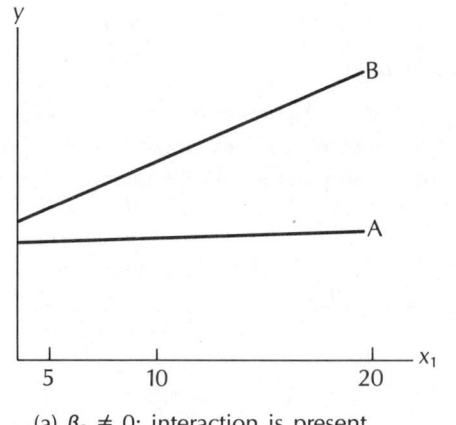

(a) $\beta_3 \neq 0$; interaction is present

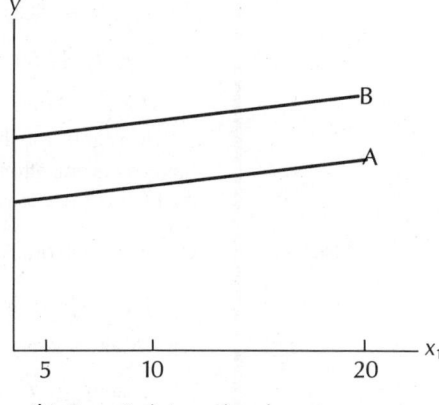

(b) $\beta_3 = 0$; interaction is not present

β_2: difference in the y-intercepts of the regression lines for products B and A

β_3: difference in the slopes of the regression lines for products B and A

intersecting lines

Figure 13.7(a) indicates a situation where $\beta_3 \neq 0$ (that is, there is an interaction between the two variables "drug product" and "drug dose"). Thus the regression lines are not parallel. Figure 13.7(b) indicates a case in which $\beta_3 = 0$ (no interaction), which results in parallel regression lines.

parallel lines

Indeed, many other experimental situations are possible, depending on the signs and magnitudes of the parameters β_0, β_1, β_2, and β_3.

EXAMPLE 13.17

Sample data for the experiment discussed in Example 13.16 are listed in Table 13.7. The response of interest is an anxiety score obtained from trained investigators. Use

TABLE 13.7   Rat Anxiety Scores, Example 13.17

Drug Product	Drug Dose (mg)					
	5		10		20	
A	15	16	18	16	20	17
	16	15	17	15	19	18
	18	16	18	19	21	21
	13	17	19	18	18	20
	19	15	20	16	19	17
	av = 16		av = 17.6		av = 19.0	
B	16	15	19	18	24	23
	17	15	21	20	25	24
	18	18	22	21	23	22
	17	17	23	22	25	26
	15	16	20	19	25	24
	av = 16.4		av = 20.5		av = 24.1	

these data to fit the general linear model

$$y = \beta_0 + \beta_1 x_1 + \beta_2 x_2 + \beta_3 x_1 x_2 + \varepsilon$$

Of particular interest to the experimenter is a comparison between the slopes of the regression lines. A difference in slopes would indicate that the drug products have different effects on the anxiety of the rats. Conduct a statistical test of the equality of the two slopes. Use $\alpha = .05$.

Solution   Using the complete model

$$y = \beta_0 + \beta_1 x_1 + \beta_2 x_2 + \beta_3 x_1 x_2 + \varepsilon$$

we obtain a least squares fit of

$$\hat{y} = 15.30 + .19 x_1 - .70 x_2 + .30 x_1 x_2$$

with $SSE_1 = 133.63$ (see the computer output that follows).

```
          LISTING OF DATA

  OBS    PRODUCT    DOSE    SCORE

   1        0         5      15
   2        0         5      16
   3        0         5      16
   4        0         5      15
   5        0         5      18
   6        0         5      16
   7        0         5      13
   8        0         5      17
   9        0         5      19
  10        0         5      15
  11        0        10      18
  12        0        10      16
  13        0        10      17
  14        0        10      15
  15        0        10      18
  16        0        10      19
  17        0        10      19
  18        0        10      18
  19        0        10      20
  20        0        10      16
  21        0        20      20
  22        0        20      17
  23        0        20      19
  24        0        20      18
  25        0        20      21
  26        0        20      21
  27        0        20      18
  28        0        20      20
  29        0        20      19
  30        0        20      17
  31        1         5      16
  32        1         5      15
  33        1         5      17
  34        1         5      15
  35        1         5      18
  36        1         5      18
  37        1         5      17
  38        1         5      17
  39        1         5      15
  40        1         5      16
  41        1        10      19
  42        1        10      18
  43        1        10      21
  44        1        10      20
  45        1        10      22
  46        1        10      21
  47        1        10      23
  48        1        10      22
  49        1        10      20
  50        1        10      19
  51        1        20      24
  52        1        20      23
  53        1        20      25
  54        1        20      24
  55        1        20      23
```

```
            LISTING OF DATA

   OBS    PRODUCT    DOSE    SCORE

   56        1        20      22
   57        1        20      25
   58        1        20      26
   59        1        20      25
   60        1        20      24

   N=       60
```

FULL MODEL

GENERAL LINEAR MODELS PROCEDURE

DEPENDENT VARIABLE: SCORE

SOURCE	DF	SUM OF SQUARES	MEAN SQUARE	F VALUE	PR > F	R-SQUARE	C.V.
MODEL	3	442.10476190	147.36825397	61.76	0.0001	0.767898	8.1588
ERROR	56	133.62857143	2.38622449		ROOT MSE		SCORE MEAN
CORRECTED TOTAL	59	575.73333333			1.54474091		18.93333333

SOURCE	DF	TYPE I SS	F VALUE	PR > F	DF	TYPE III SS	F VALUE	PR > F
DOSE	1	272.00476190	113.99	0.0001	1	42.75238095	17.92	0.0001
PRODUCT	1	117.60000000	49.28	0.0001	1	1.63333333	0.68	0.4116
DOSE*PRODUCT	1	52.50000000	22.00	0.0001	1	52.50000000	22.00	0.0001

PARAMETER	ESTIMATE	T FOR H0: PARAMETER=0	PR > |T|	STD ERROR OF ESTIMATE
INTERCEPT	15.30000000	25.57	0.0001	0.59827558
DOSE	0.19142857	4.23	0.0001	0.04522538
PRODUCT	-0.70000000	-0.83	0.4116	0.84608944
DOSE*PRODUCT	0.30000000	4.69	0.0001	0.06395835

REDUCED MODEL

GENERAL LINEAR MODELS PROCEDURE

DEPENDENT VARIABLE: SCORE

SOURCE	DF	SUM OF SQUARES	MEAN SQUARE	F VALUE	PR > F	R-SQUARE	C.V.
MODEL	2	389.60476190	194.80238095	59.66	0.0001	0.676710	9.5443
ERROR	57	186.12857143	3.26541353		ROOT MSE		SCORE MEAN
CORRECTED TOTAL	59	575.73333333			1.80704553		18.93333333

SOURCE	DF	TYPE I SS	F VALUE	PR > F	DF	TYPE III SS	F VALUE	PR > F
DOSE	1	272.00476190	83.30	0.0001	1	272.00476190	83.30	0.0001
PRODUCT	1	117.60000000	36.01	0.0001	1	117.60000000	36.01	0.0001

PARAMETER	ESTIMATE	T FOR H0: PARAMETER=0	PR > |T|	STD ERROR OF ESTIMATE
INTERCEPT	13.55000000	24.77	0.0001	0.54711020
DOSE	0.34142857	9.13	0.0001	0.03740940
PRODUCT	2.80000000	6.00	0.0001	0.46657715

The reduced model corresponding to the null hypothesis H_0: $\beta_3 = 0$ (that is, the slopes are the same) is

$$y = \beta_0 + \beta_1 x_1 + \beta_2 x_2 + \varepsilon$$

for which we obtain

$$\hat{y} = 13.55 + .34x_1 + 2.80x_2$$

and $SSE_2 = 186.13$. The reduction in the sum of squares for error attributed to $x_1 x_2$ is

$$SS_{drop} = SSE_2 - SSE_1 = 186.13 - 133.63 = 52.50$$

Using $MSE_1 = SSE_1/56 = 133.63/56 = 2.39$ and $MS_{drop} = 52.50$ (since we are testing only one β),

$$F = \frac{MS_{drop}}{MSE_1} = \frac{52.50}{2.39} = 22.00$$

Since the observed value of F exceeds 4.00, the table value for $df_1 = 1$, $df_2 = 56$ (actually, 60), and a $= .05$, we reject H_0 and conclude that the slopes for the two groups are different.

The results presented here for comparing the slope of two regression lines can be readily extended to the comparison of three or more regression lines by including additional dummy variables and all possible interaction terms between the quantitative variable x_1 and the dummy variables. Thus, for example, in comparing the slopes of three regression lines, the model would contain the quantitative variable x_1, two dummy variables x_2 and x_3, and two interaction terms x_1x_2 and x_1x_3.

EXERCISES

 13.17   An experimenter wished to compare the potencies of three different drug products. To do this, 12 test tubes were inoculated with a culture of the virus under study and incubated for two days at 35° C. Four dosage levels (.2, .4, .8, and .16 μg per tube) were to be used from each of the three drug products, with only one dose–drug product combination for each of the 12 test tube cultures. One means of comparing the drug products would be to examine their slopes (with respect to dose).

a. Write a general linear model relating the response y to the independent variables "dose" and "drug product." Make the expected response a linear function of log dose (x_1). Identify the parameters in the model.

b. It would seem reasonable to assume that the three separate response lines have a common intercept β_0, since this would correspond to a zero dosage level of any of the drug products. Change the model of part (a) to reflect this change.

 13.18   Refer to Exercise 13.17.

a. Use the data below to make a comparison among the three slopes. Fit a complete and a reduced model for your test. Use $\alpha = .05$.

| | Drug Product | | |
Dose	A	B	C
.2	2.0	1.8	1.3
.4	4.3	4.1	2.0
.8	6.5	4.9	2.8
1.6	8.9	5.7	3.4

b. Is there evidence to indicate that the slopes are equal?

c. Suggest how you might test the null hypothesis that the intercepts are all equal to zero.

13.6 ODDS AND ENDS—GENCAT: A GENERAL LINEAR MODEL PROGRAM FOR CATEGORICAL DATA (optional)

In this section we will present an example of a procedure that can be used to compare s different multinomial populations, where each population has r possible responses. The procedure represents an extension of our study of two-dimensional contingency tables (Chapter 11). We present the material in this chapter to take advantage of what we have learned about writing models for experimental situations. The example we will discuss makes use of methodology developed by Grizzle, Starmer, and Koch (1969), with the format explained in more detail by Forthofer, Starmer, and Grizzle (1971).

The subject of analyses for multidimensional contingency tables has been widely studied, and we hope only to scratch the surface in this section. However, by following the example, you can see the utility of the approach and then refer to the references for more details of additional applications.

Assume there are s different multinomial populations, each with r categories of re-

π_{ij} sponse. Let π_{ij} (Greek letter pi) denote the cell probability for the ith row and jth column. Then the cell probabilities for the s populations are as listed in Table 13.8. Note that for any row we require

$$\sum_{j} \pi_{ij} = 1$$

The procedure we will consider for analyzing categorical data is similar to the general linear model approach of Chapters 12 and 13. In particular, we construct u different functions of the multinomial population probabilities by fitting the model

model $A\pi = X\beta$

(Note that this model is similar to the general linear model $Y = X\beta$.) The matrix A in our model is a matrix of constants that defines the u different functions of the πs and is used to specify the response. The matrix π is given by

$$\pi = \begin{bmatrix} \pi_{11} \\ \pi_{12} \\ \vdots \\ \pi_{sr} \end{bmatrix}$$

TABLE 13.8 Cell Probabilities for the s Populations

Population	\multicolumn{5}{c}{Category}				
	1	2	3	$\cdots$	r
1	π_{11}	π_{12}	π_{13}	$\cdots$	π_{1r}
2	π_{21}	π_{22}	π_{23}	$\cdots$	π_{2r}
3	π_{31}	π_{32}	π_{33}	$\cdots$	π_{3r}
$\vdots$	$\vdots$	$\vdots$	$\vdots$		$\vdots$
s	π_{s1}	π_{s2}	π_{s3}	$\cdots$	π_{sr}

The design matrix **X** is similar to those used in the general linear model, and $\boldsymbol{\beta}$ is a matrix of v unknown parameters:

$$\boldsymbol{\beta} = \begin{bmatrix} \beta_1 \\ \beta_2 \\ \vdots \\ \beta_v \end{bmatrix} \qquad v \leq u$$

To bring this procedure to life, consider the following example. Suppose that a multi-clinic investigation was conducted to compare four different treatments (nasal sprays) for their effectiveness in relieving the symptoms of ragweed allergy. Each investigator was to obtain a random sample of patients who responded positively to a skin test for ragweed allergy and who volunteered to participate in the study. A double-blind procedure was used, in which each physician was supplied with a random code to assign one of the four sprays to patients but neither the investigator nor the patient knew which medication would be received. On the appointed day, volunteers were asked to report to the physician's

TABLE 13.9   Yes and No Responses for the Multiclinic Investigation

Investigator	Treatment	Response No	Response Yes	Sample Size
1	A	1	2	3
	B	0	1	1
	C	2	1	3
	D	1	2	3
2	A	2	3	5
	B	1	2	3
	C	0	3	3
	D	0	4	4
3	A	10	11	21
	B	13	5	18
	C	12	7	19
	D	7	11	18
4	A	1	0	1
	B	3	0	3
	C	0	1	1
	D	2	0	2
5	A	2	4	6
	B	4	2	6
	C	7	6	13
	D	4	4	8
6	A	15	4	19
	B	14	4	18
	C	13	5	18
	D	14	8	22
				218

office, where they were examined for allergic symptoms due to ragweed. Then each person was shown how to self-administer one dose of the assigned nasal spray. After a four-hour period, each patient was rated as to whether he or she had improved while on the drug (yes or no).

The data appear in Table 13.9. An entry is the number of yes or no responses for a treatment–investigator combination.

Note that this example differs from the two-dimensional contingency tables of Section 6.7. In particular, we have $s = 24$ (six investigators times four treatments) different multinomial populations with each population having $r = 2$ possible categories (yes or no).

The π matrix of unknown probabilities associated with our data is shown below. Several of the individual πs have been identified to show you how the matrix has been constructed using a single subscript.

π matrix

$$\pi = \begin{bmatrix} \pi_1 \\ \pi_2 \\ \pi_3 \\ \vdots \\ \pi_8 \\ \pi_9 \\ \vdots \\ \pi_{48} \end{bmatrix}$$

probability of a no response on treatment A, investigator 1
probability of a yes response on treatment A, investigator 1
probability of a no response on treatment B, investigator 1

probability of a yes response on treatment D, investigator 1
probability of a no response on treatment A, investigator 2

probability of a yes response on treatment D, investigator 6

tests   There are a number of different hypotheses that we might wish to pose concerning the cell probabilities for the 24 multinomial populations.

For example, we might hypothesize that the proportion of no responses is linearly related to a treatment and an investigator effect. To do this, we define the following parameters:

μ: overall mean effect
α_1: effect due to investigator 1
α_2: effect due to investigator 2
$\vdots$
α_5: effect due to investigator 5
β_1: effect due to treatment A
β_2: effect due to treatment B
β_3: effect due to treatment C

Note that we have not defined an effect for treatment D or for investigator 6 since these effects can be stated in terms of the previous effects. In particular, we take

Investigator 6 effect: $\alpha_6 = -\alpha_1 - \alpha_2 - \alpha_3 - \alpha_4 - \alpha_5$
Treatment D effect: $\beta_4 = -\beta_1 - \beta_2 - \beta_3$

We now define the following matrix model relating the theoretical cell probabilities (the πs) to a linear function of the αs, βs, and μ. This is given by our general model

$\mathbf{A\pi} = \mathbf{X\beta}$, where $\mathbf{A}$ is a 24×48 matrix of the form

A matrix

$$\mathbf{A} = \begin{bmatrix} 1 & 0 & & & & \\ & & 1 & 0 & & \\ & & & & 1 & 0 & \\ & & & & & \ddots & \\ & & & & & & 1 & 0 \end{bmatrix}$$

In general, we can write the matrix $\mathbf{A}$ as

A*

$$\mathbf{A} = \begin{bmatrix} \mathbf{A^*} & & & \\ & \mathbf{A^*} & & \\ & & \ddots & \\ & & & \mathbf{A^*} \end{bmatrix}$$

where, for our case, $\mathbf{A^*} = [1 \quad 0]$.

The $\mathbf{X}$ matrix is the design matrix corresponding to the $\boldsymbol{\pi}$ matrix. For our example,

X matrix

	μ	α_1	α_2	α_3	α_4	α_5	β_1	β_2	β_3
$\mathbf{X} =$	1	1	0	0	0	0	1	0	0
	1	1	0	0	0	0	0	1	0
	1	1	0	0	0	0	0	0	1
	1	1	0	0	0	0	−1	−1	−1
	1	0	1	0	0	0	1	0	0
	1	0	1	0	0	0	0	1	0
	1	0	1	0	0	0	0	0	1
	1	0	1	0	0	0	−1	−1	−1
	1	0	0	1	0	0	1	0	0
	1	0	0	1	0	0	0	1	0
	1	0	0	1	0	0	0	0	1
	1	0	0	1	0	0	−1	−1	−1
	1	0	0	0	1	0	1	0	0
	1	0	0	0	1	0	0	1	0
	1	0	0	0	1	0	0	0	1
	1	0	0	0	1	0	−1	−1	−1
	1	0	0	0	0	1	1	0	0
	1	0	0	0	0	1	0	1	0
	1	0	0	0	0	1	0	0	1
	1	0	0	0	0	1	−1	−1	−1
	1	−1	−1	−1	−1	−1	1	0	0
	1	−1	−1	−1	−1	−1	0	1	0
	1	−1	−1	−1	−1	−1	0	0	1
	1	−1	−1	−1	−1	−1	−1	−1	−1

For our data, the matrix of parameters is

β matrix

$$\boldsymbol{\beta} = \begin{bmatrix} \mu \\ \alpha_1 \\ \alpha_2 \\ \alpha_3 \\ \alpha_4 \\ \alpha_5 \\ \beta_1 \\ \beta_2 \\ \beta_3 \end{bmatrix}$$

With these three matrices, we see that the model

$$\mathbf{A}\boldsymbol{\pi} = \mathbf{X}\boldsymbol{\beta}$$

indicates that the probability of a no response for a particular multinomial population is linearly related to the treatment and investigator effect for that cell. For example, from the first row of both sides of the equation, we have π_1, the probability of a no response on investigator 1 and treatment A, equal to $\mu + \alpha_1 + \beta_1$.

test for fit Having specified a model, we can perform a number of different tests similar to those indicated by an analysis of variance table for the general linear model. The first test we consider is a test of the adequacy of the linear model $\mathbf{A}\boldsymbol{\pi} = \mathbf{X}\boldsymbol{\beta}$. Some people refer to this as a test for the treatment-by-investigator interaction. Since this test will be performed automatically by the computer program we will use, we will not go into any details here. Then, provided there is no significant interaction, we proceed to examine the main effects due to treatments (sprays) and investigators.

Unlike the analysis of variance tests for the general linear model, our tests will utilize chi-square rather than F statistics. The degrees of freedom for the chi-square test will be the same as the numerator degrees of freedom in a comparable F test. The details of the calculations are left to the computer solution (Gencat).

As with the general linear model, we wish to test that sets of parameters equal zero. Main effects null hypotheses can be stated in matrix notation as

$$H_0: \mathbf{C}\boldsymbol{\beta} = \mathbf{0}$$

investigator effects For testing that there is no effect due to investigators ($\alpha_1 = \alpha_2 = \cdots = \alpha_5 = 0$), we can use the $\mathbf{C}$ matrix

$$\mathbf{C} = \begin{bmatrix} 0 & 1 & 0 & 0 & 0 & 0 & 0 & 0 & 0 \\ 0 & 0 & 1 & 0 & 0 & 0 & 0 & 0 & 0 \\ 0 & 0 & 0 & 1 & 0 & 0 & 0 & 0 & 0 \\ 0 & 0 & 0 & 0 & 1 & 0 & 0 & 0 & 0 \\ 0 & 0 & 0 & 0 & 0 & 1 & 0 & 0 & 0 \end{bmatrix}$$

treatment effects Similarly, for testing $H_0: \beta_1 = \beta_2 = \beta_3 = 0$ (no treatment effects), we use

$$\mathbf{C} = \begin{bmatrix} 0 & 0 & 0 & 0 & 0 & 0 & 1 & 0 & 0 \\ 0 & 0 & 0 & 0 & 0 & 0 & 0 & 1 & 0 \\ 0 & 0 & 0 & 0 & 0 & 0 & 0 & 0 & 1 \end{bmatrix}$$

No $\mathbf{C}$ matrix need be specified for the interaction test; it is performed automatically.

EXAMPLE 13.18

Use the previous data to fit a model relating the probability of a no response to a linear function of a treatment and an investigator effect. Test for interaction and main effects. Use $\alpha = .05$ for each test.

Solution

The computer program we will use (Gencat) is an improved version of the one described in detail in Forthofer, Starmer, and Grizzle (1971). Similar analyses can be done using SAS.

We will identify the control fields (without the appropriate job-control language that is required by your computer center) necessary to run this job.

Columns	Description
1−5	number of categories of response (enter r)
6−10	number of multinomial populations (enter s)
11−15	1
16−20	1
21−25	0
26−30	number of **C** matrices
31−35	0
36−40	0

Parameter specifications

For our data, the parameter specifications are

$$2 \quad 24 \quad 1 \quad 1 \quad 0 \quad 2 \quad 0 \quad 0$$

Note that we are not showing the entire data card here but only the numbers punched on the card.

Data. The data are entered by populations, beginning with a new line for each population. The number of responses falling into the first category of a population is recorded anywhere in the first 10 columns of the card, with the decimal recorded. The first 10 columns are called a *10-column field*. The number of responses falling into the second category of a population is recorded in the next 10-column field (columns 11−20), and so on. For cells with no responses, we enter $1/r$ (in our case, 0.5) rather than a zero. Our data are presented here.

1.0	2.0
0.5	1.0
2.0	1.0
1.0	2.0
2.0	3.0
1.0	2.0
0.5	3.0
0.5	4.0
10.0	11.0
13.0	5.0
12.0	7.0
7.0	11.0

1.0	0.5
3.0	0.5
0.5	1.0
2.0	0.5
2.0	4.0
4.0	2.0
7.0	6.0
4.0	4.0
15.0	4.0
14.0	4.0
13.0	5.0
14.0	8.0

A* *matrix.* The **A*** matrix is entered by rows in 10-column fields with the decimals entered

    1.0   0.0

X *matrix.* The first line for the **X** matrix contains the number of parameters in our model entered on the right side (right-justified) of the first five-column field (columns 1–5). For example, if the **X** matrix contains six parameters, the number 6 would be entered in column 5. For an **X** matrix containing 12 parameters, the number 12 would be entered in columns 4 and 5. Then with the second line, we enter the **X** matrix *by columns* in fields of five, with the decimal entered. Each column starts a new line. These lines are shown for our example.

```
9
1.0  1.0  1.0  1.0  1.0  1.0  1.0  1.0  1.0  1.0  1.0  1.0  1.0  1.0  1.0  1.0
1.0  1.0  1.0  1.0  1.0  1.0  1.0  1.0
1.0  1.0  1.0  1.0  0.0  0.0  0.0  0.0  0.0  0.0  0.0  0.0  0.0  0.0  0.0  0.0
0.0  0.0  0.0  0.0 -1.0 -1.0 -1.0 -1.0
0.0  0.0  0.0  0.0  1.0  1.0  1.0  1.0  0.0  0.0  0.0  0.0  0.0  0.0  0.0  0.0
0.0  0.0  0.0  0.0 -1.0 -1.0 -1.0 -1.0
0.0  0.0  0.0  0.0  0.0  0.0  0.0  0.0  1.0  1.0  1.0  1.0  0.0  0.0  0.0  0.0
0.0  0.0  0.0  0.0 -1.0 -1.0 -1.0 -1.0
0.0  0.0  0.0  0.0  0.0  0.0  0.0  0.0  0.0  0.0  0.0  0.0  1.0  1.0  1.0  1.0
0.0  0.0  0.0  0.0 -1.0 -1.0 -1.0 -1.0
0.0  0.0  0.0  0.0  0.0  0.0  0.0  0.0  0.0  0.0  0.0  0.0  0.0  0.0  0.0  0.0
1.0  1.0  1.0  1.0 -1.0 -1.0 -1.0 -1.0
1.0  0.0  0.0 -1.0  1.0  0.0  0.0 -1.0  1.0  0.0  0.0 -1.0  1.0  0.0  0.0 -1.0
1.0  0.0  0.0 -1.0  1.0  0.0  0.0 -1.0
0.0  1.0  0.0 -1.0  0.0  1.0  0.0 -1.0  0.0  1.0  0.0 -1.0  0.0  1.0  0.0 -1.0
0.0  1.0  0.0 -1.0  0.0  1.0  0.0 -1.0
0.0  0.0  1.0 -1.0  0.0  0.0  1.0 -1.0  0.0  0.0  1.0 -1.0  0.0  0.0  1.0 -1.0
0.0  0.0  1.0 -1.0  0.0  0.0  1.0 -1.0
```

C *matrices.* For each **C** matrix needed for a specific null hypothesis, we first enter a line giving the number of rows of the **C** matrix in a five-column field, right-justified.

Succeeding lines contain the **C** matrix entered *by rows* in five-column fields, with the decimal entered. Each row of **C** starts a new line.

```
5
0.0   1.0   0.0   0.0   0.0   0.0   0.0   0.0   0.0
0.0   0.0   1.0   0.0   0.0   0.0   0.0   0.0   0.0
0.0   0.0   0.0   1.0   0.0   0.0   0.0   0.0   0.0
0.0   0.0   0.0   0.0   1.0   0.0   0.0   0.0   0.0
0.0   0.0   0.0   0.0   0.0   1.0   0.0   0.0   0.0
3
0.0   0.0   0.0   0.0   0.0   0.0   1.0   0.0   0.0
0.0   0.0   0.0   0.0   0.0   0.0   0.0   1.0   0.0
0.0   0.0   0.0   0.0   0.0   0.0   0.0   0.0   1.0
```

A copy of the output follows. Comments have been made within the output to explain the parts of the output in which we are interested.

GENERALIZED CHI-SQUARE ANALYSIS

R =	2	R IS THE NUMBER OF CATEGORIES OF RESPONSE
S =	24	S IS THE NUMBER OF POPULATIONS
U =	1	U IS THE RANK OF THE A MATRIX
MM =	1	MM = 1 IF LEAST SQUARES ANALYSIS IS USED AND ZERO OTHERWISE
ML =	0	ML = 0 IF TESTING A LINEAR HYPOTHESIS AND ML = 1 IF TESTING A LOGARITHMIC HYPOTHESIS
NC =	2	NC IS THE NUMBER OF SETS OF CONTRASTS TO BE TESTED BY LEAST SQUARES ANALYSIS
IK =	0	IK = 1 IF K IS THE IDENTITY MATRIX AND ZERO OTHERWISE
ISW =	0	ISW = 1 IF YOU WISH TO REANALYZE THE DATA ENTERED IN THE PRECEDING PROBLEM AND ZERO OTHERWISE

FREQUENCY TABLE (S × R)

1.	2.
1.	1.
2.	1.
1.	2.
2.	3.
1.	2.
1.	3.
1.	4.
10.	11.
13.	5.
12.	7.
7.	11.
1.	1.
3.	1.
1.	1.
2.	1.
2.	4.
4.	2.
7.	6.
4.	4.
15.	4.
14.	4.
13.	5.
14.	8.

This table gives the number of responses falling into the no and yes categories of each population. (Note: When a zero response was observed for a particular cell, we entered 0.5, which the computer then rounded to 1. for this frequency table.)

PROBABILITY TABLE (S × R)

0.33333	0.66667
0.33333	0.66667
0.66667	0.33333
0.33333	0.66667
0.40000	0.60000
0.33333	0.66667
0.14286	0.85714
0.11111	0.88889
0.47619	0.52381
0.72222	0.27778
0.63158	0.36842
0.38889	0.61111
0.66667	0.33333
0.85714	0.14286
0.33333	0.66667
0.80000	0.20000
0.33333	0.66667
0.66667	0.33333
0.53846	0.46154
0.50000	0.50000
0.78947	0.21053
0.77778	0.22222
0.72222	0.27778
0.63636	0.36364

This table gives the sample proportions of no and yes for each population, based on the frequency table entered into the computer.

CHI-SQUARE = 225.1973   DF = 24   P = 0.0

DESIGN MATRIX   This is our **X** matrix

1.	1.	0.	0.	0.	0.	1.	0.	0.
1.	1.	0.	0.	0.	0.	0.	1.	0.
1.	1.	0.	0.	0.	0.	0.	0.	1.
1.	1.	0.	0.	0.	0.	−1.	−1.	−1.
1.	0.	1.	0.	0.	0.	1.	0.	0.
1.	0.	1.	0.	0.	0.	0.	1.	0.
1.	0.	1.	0.	0.	0.	0.	0.	1.
1.	0.	1.	0.	0.	0.	−1.	−1.	−1.
1.	0.	0.	1.	0.	0.	1.	0.	0.
1.	0.	0.	1.	0.	0.	0.	1.	0.
1.	0.	0.	1.	0.	0.	0.	0.	1.
1.	0.	0.	1.	0.	0.	−1.	−1.	−1.
1.	0.	0.	0.	1.	0.	1.	0.	0.
1.	0.	0.	0.	1.	0.	0.	1.	0.
1.	0.	0.	0.	1.	0.	0.	0.	1.
1.	0.	0.	0.	1.	0.	−1.	−1.	−1.
1.	0.	0.	0.	0.	1.	1.	0.	0.
1.	0.	0.	0.	0.	1.	0.	1.	0.
1.	0.	0.	0.	0.	1.	0.	0.	1.
1.	0.	0.	0.	0.	1.	−1.	−1.	−1.
1.	−1.	−1.	−1.	−1.	−1.	1.	0.	0.
1.	−1.	−1.	−1.	−1.	−1.	0.	1.	0.
1.	−1.	−1.	−1.	−1.	−1.	0.	0.	1.
1.	−1.	−1.	−1.	−1.	−1.	−1.	−1.	−1.

ESTIMATED MODEL 3 PARAMETERS   These are least squares estimates of the model parameters.

0.53613D 00   −0.92943D-01   −0.30425D 00   0.20820D-01   0.19908D 00   −0.20553D-01   −0.14319D-01   0.10228D 00
0.14669D 01

VARIANCE COVARIANCE MATRIX OF THE ESTIMATED MODEL PARAMETERS

0.16851D 02	0.18434D-02	−0.12320D-03	−0.11849D-02	0.12164D-02	−0.47443D-03	0.10812D-03	0.78970D-05
−0.96186D-05							
0.18434D-02	0.15851D-01	−0.33372D-02	−0.23548D-02	−0.48790D-02	−0.30238D-02	−0.24452D-03	0.40121D-03
−0.12907D-03							
−0.12320D-03	−0.33372D-02	0.79601D-02	−0.40146D-03	−0.28617D-02	−0.10533D-02	0.10747D-03	0.46891D-03
−0.52128D-04							
−0.11849D-02	−0.23548D-02	−0.40146D-03	0.37066D-02	−0.16988D-02	−0.31286D-04	−0.12321D-03	−0.87075D-04
0.24444D-04							
0.12164D-02	−0.48790D-02	−0.28617D-02	−0.16988D-02	0.13577D-01	−0.25169D-02	0.39683D-03	−0.94708D-03
0.51481D-03							
−0.47443D-03	−0.30238D-02	−0.10533D-02	−0.31286D-04	−0.25169D-02	0.65797D-02	0.97526D-04	0.20367D-03
−0.47156D-03							
0.10812D-03	−0.24452D-03	0.10747D-03	−0.12321D-03	0.39683D-03	0.97526D-04	0.28285D-02	−0.95478D-03
−0.93921D-03							
0.78970D-05	0.40121D-03	0.46891D-03	−0.87075D-04	−0.94708D-03	0.20367D-03	−0.95478D-03	0.28637D-02
−0.97081D-03							
−0.96186D-05	−0.12907D-03	−0.52128D-04	0.24444D-04	0.51481D-03	−0.47156D-03	−0.93921D-03	−0.97081D-03
0.28066D-02							

CHI-SQUARE DUE TO ERROR = 7.0677   DF = 15   P = 0.9557   This is our chi-square test for interaction. The p-value is the probability of χ^2 being greater than the observed value.

F(P) LINEAR MODEL

0.33333D 00	0.33333D 00	0.66667D 00	0.33333D 00	0.40000D 00	0.33333D 00	0.14286D 00	0.11111D 00
0.47619D 00	0.72222D 00	0.63158D 00	0.38889D 00	0.66667D 00	0.85714D 00	0.33333D 00	0.80000D 00
0.33333D 00	0.66667D 00	0.53846D 00	0.50000D 00	0.78947D 00	0.77778D 00	0.72222D 00	0.63636D 00

F(P) PREDICTED FROM FITTED MODEL

0.42886D 00	0.54546D 00	0.45785D 00	0.34055D 00	0.21755D 00	0.33415D 00	0.24654D 00	0.12924D 00
0.54263D 00	0.65922D 00	0.57161D 00	0.45432D 00	0.72089D 00	0.83749D 00	0.74988D 00	0.63258D 00
0.50125D 00	0.61785D 00	0.53024D 00	0.41294D 00	0.71965D 00	0.83625D 00	0.74864D 00	0.63134D 00

F(P)-F(P) PREDICTED = RESIDUAL

−0.95531D-01	−0.21213D 00	0.20882D 00	−0.72208D-02	0.18245D 00	−0.81888D-03	−0.10368D 00	−0.18134D-01
−0.66436D-01	0.62998D-01	0.59965D-01	−0.65428D-01	−0.54225D-01	0.19654D-01	−0.41655D 00	0.16742D 00
−0.16792D 00	0.48816D-01	0.82207D-02	0.87056D-01	0.69823D-01	−0.58471D-01	−0.26416D-01	0.50226D-02

C MATRIX   The **C** matrix for investigators

0.	1.	0.	0.	0.	0.	0.	0.	0.
0.	0.	1.	0.	0.	0.	0.	0.	0.
0.	0.	0.	1.	0.	0.	0.	0.	0.
0.	0.	0.	0.	1.	0.	0.	0.	0.
0.	0.	0.	0.	0.	1.	0.	0.	0.

ESTIMATED MODEL CONTRASTS

−0.92943D-01	−0.30425D 00	0.20820D-01	0.19908D 00	−0.20553D-01

STANDARD DEVIATIONS OF THE ESTIMATED MODEL CONTRASTS

0.12590D 00	0.89219D-01	0.60882D-01	0.11652D 00	0.81115D-01

CHI-SQUARE = 25.0614   DF = 5   P = 0.0001   This is our chi-square test for investigators.

```
                            C MATRIX
  0.     0.     0.     0.     0.     0.     1.     0.     0.
  0.     0.     0.     0.     0.     0.     0.     1.     0.              The C matrix for treatments
  0.     0.     0.     0.     0.     0.     0.     0.     1.

                 ESTIMATED MODEL CONTRASTS
        −0.14319D-01      0.10228D 00      0.14669D-01

     STANDARD DEVIATIONS OF THE ESTIMATED MODEL CONTRASTS
          0.53184D-01     0.53514D-01     0.52978D-01

          CHI-SQUARE = 5.7173     DF = 3     P = 0.1262        This is our chi-square test for treatments.
```

> To summarize the results of this analysis, the test for interaction between investigators and treatments (sprays) was nonsignificant ($p = .9557$). The tests for main effects showed a highly significant effect ($p = .0001$) due to differences among investigators, but the treatment effect achieved a level of significance of only .1262. Thus, although there were significant differences in the proportions of no responses for different investigators, there did not appear to be differences in the proportions of no responses for the different sprays.

13.7   SUMMARY

This chapter is one of the key chapters in the text because it presents some of the practical problems associated with multiple regression problems. Step 1 of the process is to decide upon the dependent variable and a set of candidate independent variables for inclusion in the model. We discussed the invaluable nature of information from an expert in the subject matter field and the utility of some of the best-subset regression techniques for choosing which variables to include in the model.

Step 2 is involved with the actual polynomial form of the particular multiple regression equation. In particular, attention should be paid to lack of fit of a proposed model to data collected on the dependent and independent variables of interest. A formal test for lack of fit of a polynomial model is possible where there are repetitions of observations at one or more than one settings of the independent variables. Lack of fit can also be examined using residual plots.

Following steps 1 and 2 as we've discussed them can sometimes be a problem depending on the data that are available. For example, if data are available on many variables at the time when the multiple regression model is being formulated, then consultation with experts and application of one (or more) of the best-subset regression can be useful in culling the list of potential candidate independent variables (step 1). The regression model is then modified in step 2 based on the discussions and analyses of step 1. Sometimes, however, data are not available on many possible independent variables. For these situations, step 1 consists of discussions with experts to determine which variables may be important predictors; data are then gathered on these variables. After the data are ob-

620

tained on these candidate independent variables, the subset regression techniques and the model formulation techniques of step 2 can be applied to refine the model.

The final step of the multiple regression problem is to check the underlying assumptions of multiple regression: zero expectation, constant variance, normality, and independence. Although some formal tests were presented, violation of the assumption is checked best by closely examining the data using scatterplots and various residual plots. The more experience one gains in examining and interpreting data with these plots, the better will be the resulting regression equations.

EXERCISES

13.19 Use the data below to fit a model. Plot the data and suggest a polynomial model.

y	7	8	6	12	15	13	7	10	11	14	16	17
x	10	10	10	15	15	15	20	20	20	25	25	25

13.20 Refer to the data of Exercise 13.19.
a. Fit the model $y = \beta_0 + \beta_1 x + \beta_2 x^2 + \beta_3 x^3 + \varepsilon$.
b. Test for lack of fit using $\alpha = .05$.
c. Examine a residual plot for violation of the regression assumptions.

13.21 Refer to Exercise 13.19. Suppose that the third, fifth, sixth, and tenth observations are missing.
a. Fit a cubic model.
b. Examine the residuals and compare the fits for the models of Exercises 13.20 and 13.21.

 13.22 A pharmaceutical firm wanted to obtain information on the relationship between the dose level of a drug product and its potency. To do this, each of 15 test tubes were inoculated with a virus culture and incubated for five days at 30° C. Three test tubes were randomly assigned to each of the five different dose levels to be investigated (2, 4, 8, 16, and 32 mg). Each tube was injected with only one dose level and the response of interest (a measure of the protective strength of the product against the virus culture) was obtained. The data are given below.

Dose Level	Response
2	5, 7, 3
4	10, 12, 14
8	15, 17, 18
16	20, 21, 19
32	23, 24, 29

a. Plot the data.
b. Fit both a linear and a quadratic model to these data.
c. Which model seems more appropriate?
d. Compare your results in part (b) to those obtained in the SPSS computer output that follows. (*Note:* VARO1 = dose, VARO2 = response, VARO3 = dose^2.)

EXAMPLE OF REGRESSION MODELS

FILE   NONAME

********************************* MULTIPLE REGRESSION **************************** VARIABLE LIST 1
                                                                                  REGRESSION LIST 1

DEPENDENT VARIABLE. .   VAR02

VARIABLE(S) ENTERED ON STEP NUMBER 1. .   VAR01

		ANALYSIS OF VARIANCE	DF	SUM OF SQUARES	MEAN SQUARE	F
MULTIPLE R	0.87923					
R SQUARE	0.77305	REGRESSION	1	590.91613	590.91613	44.28025
ADJUSTED R SQUARE	0.75559	RESIDUAL	13	173.48387	13.34491	
STANDARD ERROR	3.65307					

------------VARIABLES IN THE EQUATION------------   --------VARIABLES NOT IN THE EQUATION--------

VARIABLE	B	BETA	STD ERROR B	F	VARIABLE	BETA IN	PARTIAL	TOLERANCE	F
VARO1	0.57527	0.87923	0.08645	44.280					
(CONSTANT)	8.66667								

MAXIMUM STEP REACHED

EXAMPLE OF REGRESSION MODELS

FILE   NONAME

********************************* MULTIPLE REGRESSION **

DEPENDENT VARIABLE: VAR02   FROM   VARIABLE LIST 1
                                        REGRESSION LIST 1

SEQNUM	OBSERVED VAR02	PREDICTED VAR02	RESIDUAL	PLOT OF STANDARDIZED RESIDUAL
1	5.000000	9.817204	−4.817204	
2	7.000000	9.817204	−2.817204	
3	3.000000	9.817204	−6.817204	
4	10.00000	10.86774	−.9677419	
5	12.00000	10.96774	1.032258	
6	14.00000	10.96774	3.032258	
7	15.00000	13.26882	1.731182	
8	17.00000	13.26882	3.731182	
9	18.00000	13.26882	4.731182	
10	20.00000	17.87096	2.129032	
11	21.00000	17.87096	3.129032	
12	19.00000	17.87096	1.129032	
13	23.00000	27.07526	−4.075269	
14	24.00000	27.07526	−3.075269	
15	29.00000	27.07526	1.924730	

(PLOT axis: −2.0   −1.0   0.0   1.0   2.0)

DURBIN–WATSON TEST OF RESIDUAL DIFFERENCES COMPARED BY CASE ORDER (SEQNUM).

VARIABLE LIST 1.   REGRESSION LIST 1.   DURBIN–WATSON TEST   0.77105

```
****************************************** MULTIPLE REGRESSION **************************** VARIABLE LIST 1
                                                                                         REGRESSION LIST 1


DEPENDENT VARIABLE. .    VAR02
VARIABLE(S) ENTERED ON STEP NUMBER 1. .      VAR01
                                             VAR03

MULTIPLE R              0.93888    ANALYSIS OF VARIANCE    DF    SUM OF SQUARES    MEAN SQUARE        F
R SQUARE                0.88150    REGRESSION               2.      673.82062       336.91031    44.63404
ADJUSTED R SQUARE       0.86175    RESIDUAL                12.       90.57938         7.54828
STANDARD ERROR          2.74741
```

------------VARIABLES IN THE EQUATION------------					------VARIABLES NOT IN THE EQUATION------			
VARIABLE	B	BETA	STD ERROR B	F	VARIABLE	BETA IN	PARTIAL	TOLERANCE	F
VAR01	1.50633	2.30224	0.28836	27.287					
VAR03	−0.02699	−1.46062	0.00814	10.983					
(CONSTANT)	4.48366								

```
ALL VARIABLES ARE IN THE EQUATION

EXAMPLE OF REGRESSION MODELS

FILE    NONAME

****************************************** MULTIPLE REGRESSION *************************************************

DEPENDENT VARIABLE:    VAR02    FROM        VARIABLE LIST 1
                                            REGRESSION LIST 1

                 OBSERVED      PREDICTED                      PLOT OF STANDARDIZED RESIDUAL
    SEQNUM         VAR02          VAR02        RESIDUAL      −2.0      −1.0      0.0      1.0      2.0
       1         5.000000       7.388367     −2.388361                          *
       2         7.000000       7.388367     −.3883618                           *    |
       3         3.000000       7.388367     −4.388361                 *          |
       4         10.00000      10.07717      −.7716632E-01                        |
       5         12.00000      10.07717       1.922833                            |   *
       6         14.00000      10.07717       3.922833                            |     *
       7         15.00000      14.80708       .1929159                           *|
       8         17.00000      14.80708       2.192915                           *|
       9         18.00000      14.80708       3.192915                            |   *
      10         20.00000      21.67615      −1.676154                        *    |
      11         21.00000      21.67615      −.6761543                          *  |
      12         19.00000      21.67615      −2.676154                       *     |
      13         23.00000      25.05122      −2.051233                       *     |
      14         24.00000      25.05122      −1.051233                         *   |
      15         29.00000      25.05122       3.948766                            |  *

DURBIN–WATSON TEST OF RESIDUAL DIFFERENCES COMPARED BY CASE ORDER (SEQNUM).

VARIABLE LIST 1.    REGRESSION LIST 1.    DURBIN–WATSON TEST 1.33140
```

13.23 Refer to the data of Exercise 13.22. Many times a logarithmic transformation can be used on the dose levels to linearize the response with respect to the independent variable.

a. Refer to a set of log tables or an electronic calculator to obtain the logarithms of the five dose levels.

b. Where x_1 denotes the log dose, fit the model

$$y = \beta_0 + \beta_1 x_1 + \varepsilon$$

c. Compare your results in part (b) to those shown in the computer printout that follows. (*Note:* VAR04 = log dose.)

d. Which of the three models seems more appropriate? Why?

EXAMPLE OF REGRESSION MODELS

FILE   NONAME

** MULTIPLE REGRESSION ***************************** VARIABLE LIST 1
                                                                                           REGRESSION LIST 1

DEPENDENT VARIABLE. .   VAR02

VARIABLE(S) ENTERED ON STEP NUMBER 1   VAR04

		ANALYSIS OF VARIANCE	DF	SUM OF SQUARES	MEAN SQUARE	F
MULTIPLE R	0.96412					
R SQUARE	0.92953	REGRESSION	1.	710.53334	710.53334	171.47775
ADJUSTED R SQUARE	0.92411	RESIDUAL	13.	53.86666	4.14359	
STANDARD ERROR	2.03558					

------------ VARIABLES IN THE EQUATION ------------   ------ VARIABLES NOT IN THE EQUATION ------

VARIABLE	B	BETA	STD ERROR E	F	VARIABLE	BETA IN	PARTIAL	TOLERANCE	F
VAR04	16.16672	0.96412	1.23458	171.478					
(CONSTANT)	1.20000								

MAXIMUM STEP REACHED

EXAMPLE OF REGRESSION MODELS

FILE   NONAME

** MULTIPLE REGRESSION **

DEPENDENT VARIABLE: VAR02   FROM   VARIABLE LIST 1
                                        REGRESSION LIST 1

SEQNUM	OBSERVED VAR02	PREDICTED VAR02	RESIDUAL	PLOT OF STANDARDIZED RESIDUAL −2.0 −1.0 0.0 1.0 2.0
1	5.000000	6.066669	− 1.066667	* |
2	7.000000	6.066669	.9333332	|*
3	3.000000	6.066669	−3.066667	* |
4	10.00000	10.93333	− .9333344	* |
5	12.00000	10.93333	1.066665	| *
6	14.00000	10.93333	3.066665	| *
7	15.00000	15.80000	− .7999990	* |
8	17.00000	15.80000	1.200001	|*
9	18.00000	15.80000	2.200001	| *
10	20.00000	20.66666	− .6666647	* |
11	21.00000	20.66666	.3333353	|†
12	19.00000	20.66666	−1.666664	* |
13	23.00000	25.53333	−2.533335	* |
14	24.00000	25.53333	−1.533335	* |
15	29.00000	25.53333	3.466664	| *

DURBIN–WATSON TEST OF RESIDUAL DIFFERENCES COMPARED BY CASE ORDER (SEQNUM).

VARIABLE LIST 1.   REGRESSION LIST 1.   DURBIN–WATSON TEST 1.71667

13.24   An experiment was conducted to examine the weather resistance of a new commercial paint as a function of two independent variables, temperature x_1 and exposure time x_2. The sample data are listed below.

y	120	101	110	105	92	130
$x_1(°C)$	-10	-10	0	0	10	10
x_2(months)	1	3	2	2	1	3

a. Fit the model

$$y = \beta_0 + \beta_1 x_1 + \beta_2 x_2 + \beta_3 x_1 x_2 + \varepsilon$$

b. Examine the residuals and comment on your findings.

13.25   Refer to Exercise 13.24.

a. Could we fit the following model?

$$y = \beta_0 + \beta_1 x_1 + \beta_2 x_1^2 + \beta_3 x_2 + \beta_4 x_2^2 + \beta_5 x_1 x_2 + \beta_6 x_1 x_2^2 + \beta_7 x_1^2 x_2 + \beta_8 x_1^2 x_2^2 + \varepsilon$$

b. Test for lack of fit of the model in Exercise 13.24. Make a recommendation.

13.26   (Optional) Refer to Example 13.18. If we wish to make pairwise comparisons among the four treatments, we can do so by identifying different **C** matrices for a null hypothesis of the form $\mathbf{C}\boldsymbol{\beta} = \mathbf{0}$. For example, to test for no significant difference between treatments A and B, we would use the **C** matrix.

$$\mathbf{C} = [0 \quad 0 \quad 0 \quad 0 \quad 0 \quad 0 \quad 1 \quad -1 \quad 0]$$

Identify the **C** matrices for the following comparisons. (Hint: For comparisons with treatment D, recall that $\beta_4 = -\beta_1 - \beta_2 - \beta_3$.)

a. A versus C   b. A versus D   c. B versus D   d. C versus D

13.27   (Optional) Refer to Exercise 13.26. Run the Gencat procedure to make the five pairwise treatment comparisons of the nasal sprays:

A versus B, A versus C, A versus D, B versus D, C versus D

Note that the number of **C** matrices to be used will change the parameter card of Example 13.18.

13.28   The abrasive effect of a wear tester for experimental fabrics was tested on a particular fabric while run at six different machine speeds. Forty-eight identical 5-inch-square pieces of fabric were cut, with eight squares randomly assigned to each of the six machine speeds 100, 120, 140, 160, 180, and 200 revolutions per minute (rpm). The order of assignment of the squares to the machine was random, with each square tested for a three-minute period at the appropriate machine setting. The amount of wear was measured and recorded for each square. The data appear in the accompanying table.

a. Plot the mean data per rpm level and suggest a model.

b. Fit the suggested model to the data.

c. Suggest which residual plots might be useful in checking the assumptions underlying the model.

Machine Speed (in rpm)	Wear
100	23.0, 23.5, 24.4, 25.2, 25.6, 26.1, 24.8, 25.6
120	26.7, 26.1, 25.8, 26.3, 27.2, 27.9, 28.3, 27.4
140	28.0, 28.4, 27.0, 28.8, 29.8, 29.4, 28.7, 29.3
160	32.7, 32.1, 31.9, 33.0, 33.5, 33.7, 34.0, 32.5
180	43.1, 41.7, 42.4, 42.1, 43.5, 43.8, 44.2, 43.6
200	54.2, 43.7, 53.1, 53.8, 55.6, 55.9, 54.7, 54.5

13.29 Refer to the data of Exercise 13.28. Suppose that another variable was controlled and that the first four squares at each speed were treated with a .2 concentration of protective coating, while the second four squares were treated with a .4 concentration of the same coating. Given that x_1 denotes the machine speed and x_2 denotes the concentration of the protective coating, fit these models:

$$y = \beta_0 + \beta_1 x_1 + \beta_2 x_1^2 + \beta_3 x_2 + \varepsilon$$
$$y = \beta_0 + \beta_1 x_1 + \beta_2 x_1^2 + \beta_3 x_2 + \beta_4 x_1 x_2 + \beta_5 x_1^2 x_2 + \varepsilon$$

13.30 A laundry detergent manufacturer was interested in testing a new product prior to market release. One area of concern was the relationship between the height of the detergent suds in a washing machine as a function of the amount of detergent added and the degree of agitation in the wash cycle. For a standard size washing machine tub filled to the full level, random assignments of different agitation levels (measured in minutes) and amounts of detergent were made and tested on the washing machine. The data are as shown in the accompanying table.
a. Plot the data and suggest a model.
b. Does the assumption of normality appear to hold?
c. Fit an appropriate model.
d. Use residual plots to detect possible violations of the assumptions.

Height, y	Agitation, x_1	Amount, x_2
28.1	1	6
32.3	1	7
34.8	1	8
38.2	1	9
43.5	1	10
60.3	2	6
63.7	2	7
65.4	2	8
69.2	2	9
72.9	2	10
88.2	3	6
89.3	3	7
94.1	3	8
95.7	3	9
100.6	3	10

13.31   Refer to Exercise 13.30. Would the following model be more appropriate? Why or why not?

$$y = \beta_0 + \beta_1 x_1 + \beta_2 x_1^2 + \beta_3 x_2 + \beta_4 x_2^2 + \beta_5 x_1 x_2 + \beta_6 x_1 x_2^2 + \beta_7 x_1^2 x_2 + \beta_8 x_1^2 x_2^2 + \varepsilon$$

13.32   Refer to the data of Exercise 13.30.

a. Can we test for lack of fit for the following model?

$$y = \beta_0 + \beta_1 x_1 + \beta_2 x_1^2 + \beta_3 x_2 + \beta_4 x_2^2 + \beta_5 x_1 x_2 + \beta_6 x_1 x_2^2 + \beta_7 x_1^2 x_2 + \beta_8 x_1^2 x_2^2 + \varepsilon$$

b. Write the complete model for the sample data. Note that if there were replication at one or more design points, the number of degrees of freedom for SS_{Lack} would be identical to the difference between the number of parameters in the complete model and the number of parameters in the model of part (a).

13.33   Refer to Example 13.9.

a. Identify the parameters in the model.

b. Fit the "complete" model.

c. Draw conclusions relative to the standard and luxury models.

13.34   (Optional) The time to failure of a particular jet engine component has a probability distribution given by

$$f(t) = \frac{\beta t^{\beta-1}}{\theta^\beta} e^{-(t/\theta)^\beta} \qquad t, \theta, \beta > 0$$

For this same component type we can represent the rate of failure for a population of components as

$$h(t) = \frac{\beta t^{\beta-1}}{\theta^\beta}$$

Data were collected for the first 24 months of engine service, and component failure rates were obtained for each month. SAS was used to fit the nonlinear $h(t)$ function, sometimes referred to as the *hazard* function.

a. Identify estimates of the scale parameter (θ) and the shape parameter (β).

```
        LISTING OF DATA

  OBS     MONTH      RATE

    1       1       0.17977
    2       2       0.00000
    3       3       0.00000
    4       4       0.32022
    5       5       0.00000
    6       6       0.00000
    7       7       0.00000
    8       8       0.00000
    9       9       0.48333
   10      10       0.35694
   11      11       1.59485
   12      12       1.10666
   13      13       1.23159
   14      14       1.48853
   15      15       0.32607
   16      16       2.87241
   17      17       2.68547
   18      18       0.50282
   19      19       0.00000
   20      20       2.49903
   21      21       2.80140
   22      22       3.85508
   23      23       2.54602
   24      24       2.43984

  N=      24
```

```
              NONLINEAR REGRESSION FIT OF WEIBULL FAILURE RATE

NONLINEAR LEAST SQUARES SUMMARY STATISTICS      DEPENDENT VARIABLE RATE

    SOURCE              DF SUM OF SQUARES      MEAN SQUARE

    REGRESSION           2   51.749061649     25.874530824
    RESIDUAL            22   13.458898572      0.611768117
    UNCORRECTED TOTAL   24   65.207960220

    (CORRECTED TOTAL)   23   34.176887829


    PARAMETER    ESTIMATE      ASYMPTOTIC              ASYMPTOTIC 95 %
                              STD. ERROR          CONFIDENCE INTERVAL
                                                  LOWER          UPPER
    SCALE       7.226386345   1.7737555131     3.5478718589   10.904900832
    SHAPE       2.717670156   0.4829055984     1.7161932182    3.719147093


         ASYMPTOTIC CORRELATION MATRIX OF THE PARAMETERS

         CORR          SCALE             SHAPE

         SCALE         1.0000            0.9866
         SHAPE         0.9866            1.0000
```

b. Comment on the fit of the data based on the plots.

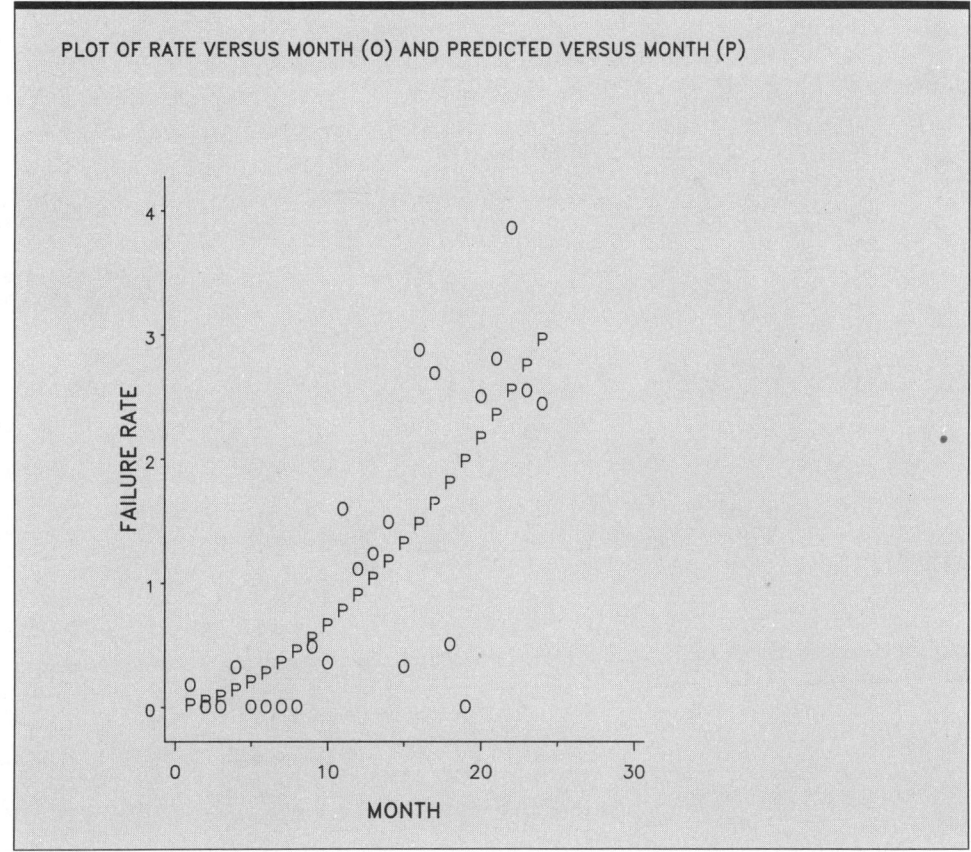

PLOT OF RATE VERSUS MONTH (O) AND PREDICTED VERSUS MONTH (P)

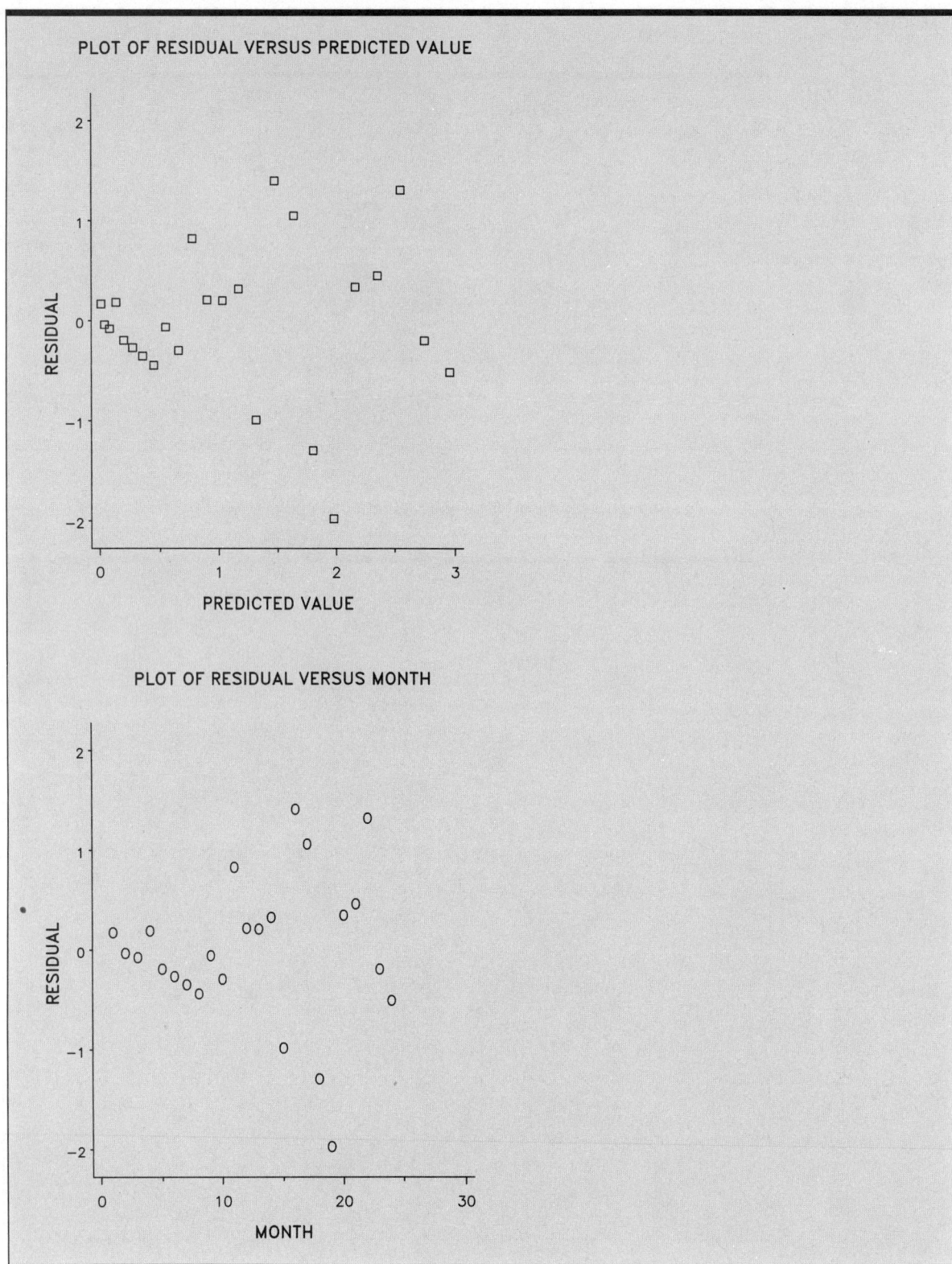

PLOT OF RESIDUAL VERSUS PREDICTED VALUE

PLOT OF RESIDUAL VERSUS MONTH

13.35 Refer to Exercise 13.34. What other technique might we use to analyze these data? Comment on any potential pitfalls.

13.36 The solubility of a solution was examined for six different temperature settings, shown in the accompanying table.

a. Plot the data, and fit as appropriate.

y, Solubility by Weight	x, Temperature (in °C)
43, 45, 42	0
32, 33, 37	25
21, 28, 29	50
15, 14, 9	75
12, 10, 8	100
7, 6, 2	125

b. Test for lack of fit if possible. Use $\alpha = .05$.

c. Examine the residuals and draw conclusions.

13.37 Refer to Exercise 13.36. Suppose we are missing observations 5, 8, and 14.

a. Fit the model $y = \beta_0 + \beta_1 x + \beta_2 x^2 + \varepsilon$.

b. Test for lack of fit, using $\alpha = .05$.

c. Again examine the residuals.

13.38 Refer to the data of Exercise 13.29.

a. Test for lack of fit of the model

$$y = \beta_0 + \beta_1 x_1 + \beta_2 x_1^2 + \beta_3 x_2 + \beta_4 x_1 x_2 + \beta_5 x_1^2 x_2 + \varepsilon$$

b. Write the complete model for this experimental situation.

13.39 Refer to the data of Exercise 13.22. Test for lack of fit of a quadratic model.

13.40 (Optional) In a random sample of 498 motorists on a major toll road, each driver was classified according to the following categories.

1. Transportation status (number of passengers including the driver):

 1, 2, 3, 4 or more

2. Destination area of the city for the driver: NE, SE, NW, SW.

3. Whether or not the driver planned to use the proposed public transportation system: yes, no.

	Yes					No			
Number of Passengers	Destination Area				Number of Passengers	Destination Area			
	NE	SE	NW	SW		NE	SE	NW	SW
1	7	10	15	22	1	31	27	15	7
2	4	8	13	14	2	18	16	9	4
3	13	10	21	29	3	12	13	4	3
4 or more	28	17	32	35	4 or more	15	18	16	12

These data, the number of motorists in each classification, are summarized in two two-way tables. Analyze these data using Gencat or some other multidimensional contingency table analysis program. If you use Gencat, follow the format of Example 13.18. Draw your conclusions.

13.41   Refer to Example 13.17. Suggest another way to test for parallelism. Use the computer output given in Example 13.17.

13.42   A psychologist is interested in examining the effects of sleep deprivation on a person's ability to perform simple arithmetic tasks. To do this, prospective subjects are screened to obtain individuals whose daily sleep patterns were closely matched. From this group, 20 subjects are chosen. Each individual selected is randomly assigned to one of five groups, four individuals per group.

Group 1: 0 hours of sleep

Group 2: 2 hours of sleep

Group 3: 4 hours of sleep

Group 4: 6 hours of sleep

Group 5: 8 hours of sleep

All subjects are then placed on a standard routine for the next 24 hours.

The following day after breakfast, each individual is tested to determine the number of arithmetic additions done correctly in a 10-minute period. That evening the amount of sleep each person is allowed depends on the group to which he or she had been assigned. The following morning after breakfast, each person is again tested using a different, but equally difficult, set of additions.

Let the response of interest be the difference in the number of correct responses on the first test day minus the number correct on the second test day. The data are presented here.

Group	Response, y
1	39, 33, 41, 40
2	25, 29, 34, 26
3	10, 18, 14, 17
4	4, 6, −1, 9
5	−5, 0, −3, −8

a. Plot the sample data and use the plot to suggest a model.

b. Fit the suggested model.

c. Examine the fitted model for possible violation of assumptions.

13.43   An experiment was conducted to determine the relationship between the amount of warping y for a particular alloy and the temperature (in °C) under which the experiment was conducted. The sample data appear below. Note that three observations were taken at each temperature setting.

Amount of Warping	Temperature (°C)
10, 13, 12	15
14, 12, 11	20
14, 12, 16	25
18, 19, 22	30
25, 21, 20	35
23, 25, 26	40
30, 31, 34	45
35, 33, 38	50

EXERCISES

Use the computer output that follows to complete parts (a) through (d).

a. Plot the data to determine whether a linear or quadratic model appears more appropriate.

b. If a linear model is fit, indicate the prediction equation. Superimpose the prediction equation over the scatter diagram of y versus x.

c. If a quadratic model is fit, identify the prediction equation. Superimpose the quadratic prediction equation on the scatter diagram. Which fit looks better, the linear or the quadratic?

d. Predict the amount of warping at a temperature of 27° C, using both the linear and the quadratic prediction equations.

```
        LISTING OF DATA

   OBS     WARPING    TEMP

    1        10        15
    2        13        15
    3        12        15
    4        14        20
    5        12        20
    6        11        20
    7        14        25
    8        12        25
    9        16        25
   10        18        30
   11        19        30
   12        22        30
   13        25        35
   14        21        35
   15        20        35
   16        23        40
   17        25        40
   18        26        40
   19        30        45
   20        31        45
   21        34        45
   22        35        50
   23        33        50
   24        38        50

   N=       24
```

QUADRATIC REGRESSION

GENERAL LINEAR MODELS PROCEDURE

DEPENDENT VARIABLE: WARPING

SOURCE	DF	SUM OF SQUARES	MEAN SQUARE	F VALUE	PR > F	R-SQUARE	C.V.
MODEL	1	1571.62698413	1571.62698413	265.55	0.0001	0.923491	11.3593
ERROR	22	130.20634921	5.91847042		ROOT MSE		WARPING MEAN
CORRECTED TOTAL	23	1701.83333333			2.43279066		21.41666667

SOURCE	DF	TYPE I SS	F VALUE	PR > F	DF	TYPE III SS	F VALUE	PR > F
TEMP	1	1571.62698413	265.55	0.0001	1	1571.62698413	265.55	0.0001

PARAMETER	ESTIMATE	T FOR HO: PARAMETER=0	PR > |T|	STD ERROR OF ESTIMATE
INTERCEPT	-1.53968254	-1.03	0.3138	1.49370995
TEMP	0.70634921	16.30	0.0001	0.04334604

OBSERVATION	OBSERVED VALUE	PREDICTED VALUE	RESIDUAL
1	10.00000000	9.05555556	0.94444444
2	13.00000000	9.05555556	3.94444444
3	12.00000000	9.05555556	2.94444444
4	14.00000000	12.58730159	1.41269841
5	12.00000000	12.58730159	-0.58730159
6	11.00000000	12.58730159	-1.58730159
7	14.00000000	16.11904762	-2.11904762
8	12.00000000	16.11904762	-4.11904762

QUADRATIC REGRESSION

GENERAL LINEAR MODELS PROCEDURE

DEPENDENT VARIABLE: WARPING

9	16.00000000	16.11904762	-0.11904762
10	18.00000000	19.65079365	-1.65079365
11	19.00000000	19.65079365	-0.65079365
12	22.00000000	19.65079365	2.34920635
13	25.00000000	23.18253968	1.81746032
14	21.00000000	23.18253968	-2.18253968
15	20.00000000	23.18253968	-3.18253968
16	23.00000000	26.71428571	-3.71428571
17	25.00000000	26.71428571	-1.71428571
18	26.00000000	26.71428571	-0.71428571
19	30.00000000	30.24603175	-0.24603175
20	31.00000000	30.24603175	0.75396825
21	34.00000000	30.24603175	3.75396825
22	35.00000000	33.77777778	1.22222222
23	33.00000000	33.77777778	-0.77777778
24	38.00000000	33.77777778	4.22222222

SUM OF RESIDUALS	0.00000000
SUM OF SQUARED RESIDUALS	130.20634921
SUM OF SQUARED RESIDUALS - ERROR SS	0.00000000
FIRST ORDER AUTOCORRELATION	0.47433866
DURBIN-WATSON D	0.90755753

QUADRATIC REGRESSION

GENERAL LINEAR MODELS PROCEDURE

DEPENDENT VARIABLE: WARPING

SOURCE	DF	SUM OF SQUARES	MEAN SQUARE	F VALUE	PR > F	R-SQUARE	C.V.
MODEL	2	1613.92063492	806.96031746	192.76	0.0001	0.948342	9.5535
ERROR	21	87.91269841	4.18631897		ROOT MSE		WARPING MEAN
CORRECTED TOTAL	23	1701.83333333			2.04604960		21.41666667

SOURCE	DF	TYPE I SS	F VALUE	PR > F	DF	TYPE III SS	F VALUE	PR > F
TEMP	1	1571.62698413	375.42	0.0001	1	0.15969355	0.04	0.8470
TEMP*TEMP	1	42.29365079	10.10	0.0045	1	42.29365079	10.10	0.0045

PARAMETER	ESTIMATE	T FOR H0: PARAMETER=0	PR > |T|	STD ERROR OF ESTIMATE
INTERCEPT	9.17857143	2.55	0.0186	3.59852022
TEMP	-0.04682540	-0.20	0.8470	0.23974742
TEMP*TEMP	0.01158730	3.18	0.0045	0.00364553

13.44 One use of multiple regression is in the setting of performance standards. In other words, a regression equation can be used to predict how well an individual ought to perform when certain conditions are met. In a study of this type, designed to identify an equation that could be used to predict the sales of individual salespeople, data from a random sample of 50 sales territories from four sections of the country (northeast, southeast, midwest, and west) were collected. Data on individual sales performances, as well as on several potential predictor variables, were collected. The variables were as follows.

y = sales-territory performance measured by aggregate sales, in units, credited to territory salesperson

x_1 = time with company (months)

x_2 = advertising, or company effort (dollar expenditures in ads in territory)

x_3 = market share (the weighted average of past market share magnitudes for four previous years)

x_4 = indicator variable for section of country (1 = northeast, 0 = otherwise)

x_5 = indicator variable for section of country (1 = southeast, 0 = otherwise)

x_6 = indicator variable for section of country (1 = midwest, 0 = otherwise)

x_7 = indicator variable (1 = male salesperson, 0 = female salesperson)

These data were analyzed using Minitab, with the following results:

```
MTB > DESCRIBE C1–C10

         N      MEAN    MEDIAN    TRMEAN    STDEV    SEMEAN
Y       50      3335      3396      3277     1579       223
X1      50     96.62     85.00     93.86    66.33      9.38
X2      50      5002      5069      4915     2370       335
X3      50     7.335     7.305     7.297    1.668     0.236
C5      50     2.460     2.000     2.455    1.129     0.160
X4      50    0.8200    1.0000    0.8636   0.3881    0.0549
X5      50    0.2600    0.0000    0.2273   0.4431    0.0627
X6      50    0.2600    0.0000    0.2273   0.4431    0.0627
X7      50    0.2400    0.0000    0.2045   0.4314    0.0610

         MIN      MAX        Q1        Q3
Y        131     7205      2033      4367
X1       000   237.00     40.00    144.25
X2       222    10832      3038      6564
X3     4.131   11.205     5.987     8.569
C5     1.000    4.000     1.000     3.250
X4    0.0000   1.0000    1.0000    1.0000
X5    0.0000   1.0000    0.0000    1.0000
X6    0.0000   1.0000    0.0000    1.0000
X7    0.0000   1.0000    0.0000    0.2500
MTB > REGRESS 'Y' ON 7 'X1' 'X2' 'X3' 'X4' 'X5' 'X6' 'X7'
```

The regression equation is
$$Y = 16.4 - 0.000546X1 + 0.667X2 + 0.0302X3 - 0.116X4 - 0.041X5 - 33.3X6 - 33.6X7$$

```
Predictor       Coef        Stdev      t-ratio
Constant      16.3944      0.2931       55.94
   X1        -0.0005463   0.0007607     -0.72
   X2         0.666689    0.000047    14315.65
   X3         0.03024     0.06467        0.47
   X4        -0.1163      0.1128        -1.03
   X5        -0.0412      0.1201        -0.34
   X6       -33.3155      0.1204      -276.81
   X7       -33.6118      0.1185      -283.70
S = 0.2864    R-sq = 100.0%    R-Sq(adj) = 100.0%

Analysis of Variance
SOURCE        DF        SS          MS
Regression     7    122189056    17455576
Error         42            3           0
Total         49    122189056
```

```
SOURCE      DF      SEQ SS
  X1         1     33243924
  X2         1     88931584
  X3         1            1
  X4         1           80
  X5         1         4972
  X6         1         1880
  X7         1         6602
```

Obs.	X1	Y	Fit	Stdev. Fit	Residual	St. Resid.
1	62	3407.00	3406.54	0.09	0.46	1.68
2	70	131.00	131.17	0.14	−0.17	−0.69
3	186	4650.00	4649.93	0.09	0.07	0.27
4	13	1971.00	1970.91	0.11	0.09	0.35
5	20	4168.00	4167.94	0.11	0.06	0.21
6	0	3047.00	3047.28	0.10	−0.28	−1.03
7	31	1196.00	1195.91	0.13	0.09	0.36
8	61	2415.00	2414.91	0.10	0.09	0.34
9	48	1987.00	1987.12	0.09	−0.12	−0.46
10	101	2214.00	2213.84	0.10	0.16	0.61
11	145	4333.00	4333.14	0.27	−0.14	−1.36X
12	200	6253.00	6253.08	0.12	−0.08	−0.29
13	81	1714.00	1713.87	0.12	0.13	0.49
14	124	5146.00	5146.01	0.09	−0.01	−0.04
15	24	3469.00	3469.27	0.11	−0.27	−1.04
16	216	4124.00	4123.60	0.11	0.40	1.53
17	232	3851.00	3851.17	0.14	−0.17	−0.69
18	109	2172.00	2171.83	0.10	0.17	0.64
19	75	1743.00	1743.25	0.12	−0.25	−0.97
20	5	2269.00	2268.93	0.11	0.07	0.27
21	12	3429.00	3429.24	0.10	−0.24	−0.88
22	90	1986.00	1985.83	0.10	0.17	0.64
23	209	3623.00	3623.21	0.12	−0.21	−0.82
24	167	5429.00	5429.16	0.15	−0.16	−0.64
25	170	4511.00	4511.22	0.10	−0.22	−0.81
26	42	1478.00	1477.94	0.12	0.06	0.24
27	167	3385.00	3385.22	0.11	−0.22	−0.84
28	98	1660.00	1660.84	0.11	−0.84	−3.16R
29	144	1212.00	1211.69	0.12	0.31	1.20
30	78	4592.00	4592.00	0.09	0.00	0.00
31	116	2876.00	2875.85	0.09	0.15	0.55
32	89	4349.00	4349.02	0.09	−0.02	−0.06
33	37	2096.00	2095.80	0.09	0.20	0.72
34	34	5308.00	5308.07	0.11	−0.07	−0.26
35	165	5731.00	5730.01	0.10	0.99	3.70R
36	41	1121.00	1120.84	0.11	0.16	0.62
37	80	2356.00	2355.91	0.12	0.09	0.34
38	140	7205.00	7204.80	0.13	0.20	0.79
39	48	3562.00	3561.96	0.13	0.04	0.15
40	203	4133.00	4132.94	0.11	0.06	0.23
41	71	2049.00	2049.12	0.09	−0.12	−0.42
42	13	2512.00	2511.90	0.09	0.10	0.36
43	144	3722.00	3721.89	0.09	0.11	0.40
44	11	2806.00	2805.74	0.13	0.26	1.01
45	34	1477.00	1477.10	0.09	−0.10	−0.37
46	94	4040.00	4039.96	0.08	0.04	0.16
47	237	6633.00	6633.36	0.12	−0.36	−1.37
48	115	3203.00	3203.04	0.12	−0.04	−0.17
49	66	4423.00	4423.27	0.10	−0.27	−1.00
50	113	5563.00	5563.38	0.10	−0.38	−1.40

R denotes an obs. with a large st. resid.
X denotes an obs. whose X value gives it large influence.

Conduct a test to determine whether salespersons in the west make more (other things being equal) than salespersons in the northeast. Give the null and alternative hypotheses, the computed and the critical values of the test statistic, and your conclusion.

13.45 Refer to Exercise 13.44. What is the estimated average increase in sales territory performance of a salesperson when advertising in the territory increases by $1000?

13.46 Refer to Exercise 13.44. Conduct a test to determine whether or not males sell (on average) 200 units more than females (other things being equal). Use $\alpha = .05$.

13.47 Refer to Exercise 13.44. A particular concern of one company sales manager is that different regional attitudes may well affect the performance of males and females unequally.
   a. Suggest a new regression model that allows for the possibility of an interaction effect between the four regions of the country and the sex of the salesperson.
   b. Interpret the "new" βs in this model.

13.48 A random sample of 22 residential properties was used in a regression of price on nine different independent variables. The variables used in this study were:

Price = selling price (dollars)

Baths = number of baths (powder room = 1/2 bath)

Beda = dummy variable for number of bedrooms (1 = 2 bedrooms, 0 = otherwise)

Bedb = dummy variable for number of bedrooms (1 = 3 bedrooms, 0 = otherwise)

Bedc = dummy variable for number of bedrooms (1 = 4 bedrooms, 0 = otherwise)

Cara = dummy variable for type of garage (1 = no garage, 0 = otherwise)

Carb = dummy variable for type of garage (1 = one-car garage, 0 = otherwise)

Age = age in years

Lot = lot size in square yards

Dom = days on the market

In this study, homes had two, three, four, or five bedrooms and either no garage or one- or two-car garages. Hence, we are using two dummy variables to code for the three categories of garage.

The data were analyzed using Minitab, with the following results.

```
        PRINT C1-C10
ROW     PRICE  BATHS  BEDA  BEDB  BEDC  CARA  CARB  AGE   LOT   DOM

  1     25750   1.0     1     0     0     1     0    23   9680   164
  2     37950   1.0     0     1     0     0     1     7   1889    67
  3     46450   2.5     0     1     0     0     0     9   1941   315
  4     46550   2.5     0     0     1     1     0    18   1813    61
  5     47950   1.5     1     0     0     0     1     2   1583   234
  6     49950   1.5     0     1     0     0     0    10   1533   116
  7     52450   2.5     0     0     1     0     0     4   1667   162
  8     54050   2.0     0     1     0     0     1     5   3450    80
  9     54850   2.0     0     1     0     0     0     5   1733    63
 10     52050   2.5     0     1     0     0     0     5   3727   102
 11     54392   2.5     0     1     0     0     0     7   1725    48
 12     53450   2.5     0     1     0     0     0     3   2811   423
 13     59510   2.5     0     1     0     0     1    11   5653   130
 14     60102   2.5     0     1     0     0     0     7   2333   159
 15     63850   2.5     0     0     1     0     0     6   2022   314
 16     62050   2.5     0     0     0     0     0     5   2166   135
```

```
   17     69450      2.0      0        1        0        0        0       15    1836      71
   18     82304      2.5      0        0        1        0        0        8    5066     338
   19     81850      2.0      0        1        0        0        0        0    2333     147
   20     70050      2.0      0        1        0        0        0        4    2904     115
   21    112450      2.5      0        0        1        0        0        1    2930      11
Continue? YES
   22    127050      3.0      0        0        1        0        0        9    2904      36

MTB > DESCRIBE C1-C10

                 N      MEAN    MEDIAN    TRMEAN     STDEV    SEMEAN
PRICE           22     62023     54621     60585     22749      4850
BATHS           22     2.182     2.500     2.200     0.524     0.112
BEDA            22    0.0909    0.0000    0.0500    0.2942    0.0627
BEDB            22     0.591     1.000     0.600     0.503     0.107
BEDC            22    0.2727    0.0000    0.2500    0.4558    0.0972
CARA            22    0.0909    0.0000    0.0500    0.2942    0.0627
CARB            22    0.1818    0.0000    0.1500    0.3948    0.0842
AGE             22      7.45      6.50      7.05      5.48      1.17
LOT             22      2895      2250      2624      1868       398
DOM             22     149.6     123.0     142.9     109.8      23.4

               MIN       MAX        Q1        Q3
PRICE        25750    127050     49450     69600
BATHS        1.000     3.000     2.000     2.500
BEDA        0.0000    1.0000    0.0000    0.0000
BEDB         0.000     1.000     0.000     1.000
BEDC        0.0000    1.0000    0.0000    1.0000
CARA        0.0000    1.0000    0.0000    0.0000
CARB        0.0000    1.0000    0.0000    0.0000
AGE          0.00     23.00      4.00      9.25
LOT          1533      9680      1793      3060
DOM          11.0     423.0      66.0     181.5
MTB >
```

REGRESS C1 ON 9 PREDICTORS IN C2-C10

The regression equation is
PRICE = 39617 + 11686 BATHS + 15128 BEDA + 2477 BEDB + 26114 BEDC - 44023 CARA
        - 12375 CARB - 506 AGE + 3.40 LOT - 86.0 DOM

```
Predictor        Coef       Stdev      t-ratio
Constant        39617       30942         1.28
BATHS           11686       10428         1.12
BEDA            15128       26254         0.58
BEDB             2477       17783         0.14
BEDC            26114       18118         1.44
CARA           -44023       22775        -1.93
CARB           -12375       10759        -1.15
AGE              -506        1111        -0.46
LOT             3.399       2.504         1.36
DOM            -86.05       35.72        -2.41
```

s = 16531   R-sq = 69.8%   R-sq(adj) = 47.2%

Analysis of Variance

```
SOURCE       DF          SS             MS
Regression    9   7588195840      843132864
Error        12   3279394048      273282848
Continue? YES
Total        21  10867590144
```

```
SOURCE      DF        SEQ SS
BATHS        1     3352323072
BEDA         1       24291496
BEDB         1      668205888
BEDC         1      261898224
CARA         1     1261090304
CARB         1      133807624
AGE          1           5848
LOT          1      300736096
DOM          1     1585837312

Unusual Observations
Obs.   BATHS      PRICE       Fit Stdev.Fit  Residual   St.Resid
  7     2.50      52450     84651      7506    -32201      -2.19R
 16     2.50      62050     62050     16531         0         * X

R denotes an obs. with a large st. resid.
X denotes an obs. whose X value gives it large influence.

MTB >




        REGRESS REGRESS C1 ON 7 PREDICTORS IN C2,C3,C5,C6,C7,C9,C10

The regression equation is
PRICE = 39091 + 11712 BATHS + 14183 BEDA + 24531 BEDC - 50962 CARA - 12121 CARB
        + 3.08 LOT - 84.8 DOM

Predictor      Coef        Stdev     t-ratio
Constant      39091        21446       1.82
BATHS         11712         9531       1.23
BEDA          14183        16759       0.85
BEDC          24531         9021       2.72
CARA         -50962        15878      -3.21
CARB         -12121        10010      -1.21
LOT           3.082        2.231       1.38
DOM          -84.81        33.24      -2.55

s = 15443      R-sq = 69.3%     R-sq(adj) = 53.9%

Analysis of Variance

SOURCE      DF          SS            MS
Regression   7    7528777728    1075539712
Error       14    3338812416     238486608
Total       21   10867590144

Continue? YES
SOURCE      DF        SEQ SS
BATHS        1     3352323072
BEDA         1       24291496
BEDC         1      929454592
CARA         1     1261501440
CARB         1      133856232
LOT          1      274448000
DOM          1     1552902528
```

```
Unusual Observations
Obs.   BATHS     PRICE       Fit  Stdev.Fit  Residual   St.Resid
  7     2.50     52450      84299      6973    -31849     -2.31R

R denotes an obs. with a large st. resid.

MTB > REGRESS C1 ON 6 PREDICTORS IN C2,C5,C6,C7,C9,C10

The regression equation is
PRICE = 44534 + 8336 BATHS + 24649 BEDC - 47007 CARA - 10588 CARB + 3.54 LOT
        - 76.7 DOM

Predictor      Coef        Stdev     t-ratio
Constant      44534        20264        2.20

BATHS          8336         8574        0.97
BEDC          24649         8934        2.76
CARA         -47007        15030       -3.13
CARB         -10588         9751       -1.09
LOT           3.539         2.144       1.65
DOM          -76.67        31.51       -2.43

s = 15296      R-sq = 67.7%     R-sq(adj) = 54.8%

Analysis of Variance

SOURCE        DF           SS           MS
Regression     6    7357974528   1226329088
Error         15    3509615104    233974336
Total         21   10867590144

Continue? YES
SOURCE        DF        SEQ SS
BATHS          1    3352323072
BEDC           1     883193344
CARA           1    1307168128
CARB           1     111305152
LOT            1     318872864
DOM            1    1385112064

Unusual Observations
Obs.   BATHS     PRICE       Fit  Stdev.Fit  Residual   St.Resid
  7     2.50     52450      83502      6843    -31052     -2.27R

R denotes an obs. with a large st. resid.

MTB > REGRESS C1 ON 5 PREDICTORS IN C5,C6,C7,C9,C10

The regression equation is
PRICE = 62606 + 28939 BEDC - 52659 CARA - 14153 CARB + 3.52 LOT - 75.6 DOM

Predictor      Coef        Stdev     t-ratio
Constant      62606         8056        7.77
BEDC          28939         7755        3.73
CARA         -52659        13837       -3.81
CARB         -14153         9019       -1.57
LOT           3.523         2.140       1.65
DOM          -75.64        31.44       -2.41
```

```
s = 15270        R-sq = 65.7%    R-sq(adj) = 54.9%

Analysis of Variance

SOURCE        DF           SS           MS
Regression     5    7136792576    1427358464
Error         16    3730797312     233174832

Total         21   10867590144

Continue? YES
SOURCE        DF        SEQ SS
BEDC           1     2901187584
CARA           1     2274636288
CARB           1      292810432
LOT            1      318495200
DOM            1     1349662976


Unusual Observations
Obs.    BEDC      PRICE      Fit Stdev.Fit  Residual   St.Resid
  1     0.00      25750    31641    13849      -5891     -0.92 X
  4     1.00      46550    40659    13849       5891      0.92 X
  7     1.00      52450    85164     6614     -32714     -2.38R
 22     1.00     127050    99052     7948      27998      2.15R

R denotes an obs. with a large st. resid.
X denotes an obs. whose X value gives it large influence.


MTB > REGRESS C1 ON 4 PREDICTORS IN C5,C6,C9,C10


The regression equation is
PRICE = 59313 + 31921 BEDC - 48742 CARA + 3.02 LOT - 69.0 DOM

Predictor      Coef        Stdev      t-ratio
Constant      59313         8105         7.32
BEDC          31921         7836         4.07
CARA         -48742        14183        -3.44
LOT           3.025        2.206         1.37
DOM          -69.00        32.46        -2.13

s = 15913        R-sq = 60.4%    R-sq(adj) = 51.1%


Analysis of Variance

SOURCE        DF           SS           MS
Regression     4    6562672128    1640668032
Error         17    4304917504     253230448
Total         21   10867590144

Continue? YES
SOURCE        DF        SEQ SS
BEDC           1     2901187584
CARA           1     2274636288
LOT            1      242949280
DOM            1     1143899008
```

```
Unusual Observations
Obs.     BEDC      PRICE        Fit Stdev.Fit  Residual   St.Resid
  1      0.00      25750      28533    14284     -2783      -0.40 X
  4      1.00      46550      43767    14284      2783       0.40 X
  7      1.00      52450      85098     6893    -32648      -2.28R
 22      1.00     127050      97533     8221     29517       2.17R

R denotes an obs. with a large st. resid.
X denotes an obs. whose X value gives it large influence.

MTB > REGRESS C1 ON 3 PREDICTORS IN C5,C6,C10

The regression equation is
PRICE = 66338 + 30129 BEDC - 38457 CARA - 60.4 DOM

Predictor        Coef        Stdev      t-ratio
Constant        66338         6433        10.31
BEDC            30129         7913         3.81
CARA           -38457        12329        -3.12
DOM             -60.41        32.62       -1.85

s = 16298      R-sq = 56.0%     R-sq(adj) = 48.7%

Analysis of Variance

SOURCE       DF            SS           MS
Regression    3    6086432256   2028810752
Error        18    4781157888    265619888
Total        21   10867590144

SOURCE       DF         SEQ SS
BEDC          1     2901187584
CARA          1     2274636288
DOM           1      910608192

Continue? YES
Unusual Observations
Obs.     BEDC      PRICE        Fit Stdev.Fit  Residual   St.Resid
  1      0.00      25750      17975    12322      7775       0.73 X
  4      1.00      46550      54326    12322     -7776      -0.73 X
  7      1.00      52450      86682     6960    -34232      -2.32R
 22      1.00     127050      94293     8065     32757       2.31R

R denotes an obs. with a large st. resid.
X denotes an obs. whose X value gives it large influence.

MTB > REGRESS C1 ON 2 PREDICTORS IN C5,C6


The regression equation is
PRICE = 57231 + 29518 BEDC - 35840 CARA

Predictor        Coef        Stdev      t-ratio
Constant        57231         4403        13.00
BEDC            29518         8396         3.52
CARA           -35840        13006        -2.76

s = 17308      R-sq = 47.6%     R-sq(adj) = 42.1%
```

```
Analysis of Variance

SOURCE       DF           SS            MS
Regression    2    5175823872    2587911936
Error        19    5691765760     299566624
Total        21   10867590144

SOURCE       DF       SEQ SS
BEDC          1    2901187584
CARA          1    2274636288

Continue? YES
Unusual Observations
Obs.    BEDC    PRICE      Fit Stdev.Fit  Residual  St.Resid
  1     0.00    25750    21391    12939      4359      0.38 X
  4     1.00    46550    50909    12939     -4359     -0.38 X
  7     1.00    52450    86749     7391    -34299     -2.19R
 22     1.00   127050    86749     7391     40301      2.58R

R denotes an obs. with a large st. resid.
X denotes an obs. whose X value gives it large influence.

MTB > REGRESS C1 ON 1 PREDICTOR IN C5

The regression equation is
PRICE = 54991 + 25785 BEDC

Predictor      Coef       Stdev    t-ratio
Constant      54991        4989      11.02
BEDC          25785        9554       2.70

s = 19958     R-sq = 26.7%    R-sq(adj) = 23.0%

Analysis of Variance

SOURCE       DF           SS            MS
Regression    1    2901187584    2901187584
Error        20    7966402048     398320096
Total        21   10867590144

Unusual Observations
Obs.    BEDC    PRICE      Fit Stdev.Fit  Residual  St.Resid
 22     1.00   127050    80776     8148     46274      2.54R

R denotes an obs. with a large st. resid.

MTB >
```

Using the full regression model (nine independent variables), estimate the average difference in selling price between:

a. Properties with no garage and properties with a one-car garage.

b. Properties with a one-car and properties with a two-car garage.

c. Properties with no garage and properties with a two-car garage.

13.49 Refer to Exercise 13.48. Conduct a test using the full regression model to determine whether the depreciation (decrease) in house price per year of age is less than $2500. Give the null hypothesis for your test and the *p*-value. Draw a conclusion.

13.50   Refer to Exercise 13.48. Suppose that we wished to modify our nine-variable model to allow for the possibility that the relationship between "price" and "age" differs depending on the number of bedrooms.

a. Formulate such a model.

b. What combination of model parameters represents the difference between a five-bedroom, one-garage home and a two-bedroom, two-garage home?

13.51   Refer to Exercise 13.48. What is your choice of a "best" model (at the .05 level) from the original set of nine variables? Why did you choose this model?

13.52   Refer to Exercise 13.48. In another study involving the same 22 properties, "price" was regressed on a single independent variable, "list," which was the listing price of the property in thousands of dollars.

```
      PRINT C1,C2
  ROW     PRICE        LIST

    1     25.750      29.900
    2     37.950      39.900
    3     46.450      44.900
    4     46.550      47.500
    5     47.950      49.900
    6     49.950      49.900
    7     52.450      53.000
    8     54.050      54.900
    9     54.850      54.900
   10     52.050      55.900
   11     54.392      55.900
   12     53.450      56.000
   13     59.510      62.000
   14     60.102      62.500
   15     63.850      63.900
   16     62.050      66.900
   17     69.450      72.500
   18     82.304      82.254
   19     81.850      82.900
   20     70.050      99.900
   21    112.450     117.000
Continue? YES
   22    127.050     139.000

MTB > DESCRIBE C1,C2

                N      MEAN    MEDIAN    TRMEAN     STDEV    SEMEAN
  PRICE        22     62.02     54.62     60.59     22.75      4.85
  LIST         22     65.52     55.95     63.63     25.55      5.45

               MIN       MAX        Q1        Q3
  PRICE      25.75    127.05     49.45     69.60
  LIST       29.90    139.00     49.90     74.94

MTB > REGRESS C1 ON 1 PREDICTOR IN C2

The regression equation is
PRICE = 5.41 + 0.864 LIST
```

```
Predictor        Coef        Stdev       t-ratio
Constant        5.406        3.363          1.61
LIST          0.86411      0.04797         18.01

s = 5.616       R-sq = 94.2%     R-sq(adj) = 93.9%

Analysis of Variance


        PRINT C1,C2
  ROW      PRICE        LIST

   1      25.750      29.900
   2      37.950      39.900
   3      46.450      44.900
   4      46.550      47.500
   5      47.950      49.900
   6      49.950      49.900
   7      52.450      53.000
   8      54.050      54.900
   9      54.850      54.900
  10      52.050      55.900
  11      54.392      55.900
  12      53.450      56.000
  13      59.510      62.000
  14      60.102      62.500
  15      63.850      63.900
  16      62.050      66.900
  17      69.450      72.500
  18      82.304      82.254
  19      81.850      82.900
  20      70.050      99.900
  21     112.450     117.000
Continue? YES
  22     127.050     139.000

MTB > DESCRIBE C1,C2

            N      MEAN    MEDIAN    TRMEAN    STDEV    SEMEAN
PRICE      22     62.02     54.62     60.59    22.75      4.85
LIST       22     65.52     55.95     63.63    25.55      5.45

          MIN       MAX        Q1        Q3
PRICE    25.75    127.05     49.45     69.60
LIST     29.90    139.00     49.90     74.94

MTB > REGRESS C1 ON 1 PREDICR

* ERROR * 2 IS TOO FEW ARGUMENTS

MTB >


SOURCE      DF         SS         MS
Regression   1      10237      10237
Error       20        631         32
Total       21      10868
```

```
Unusual Observations
Obs.    LIST     PRICE      Fit Stdev.Fit  Residual   St.Resid
 20      100     70.05    91.73     2.04    -21.68     -4.14R
 22      139    127.05   125.52     3.72      1.53      0.36 X

R denotes an obs. with a large st. resid.
X denotes an obs. whose X value gives it large influence.
MTB > STOP
*** Minitab Release 5.1.3 *** Minitab, Inc. ***
Storage available 896302
```

a. Using the regression results, predict the selling price of a home that is listed at $70,000.
b. What is the chance that your prediction is off by more than $3000?

13.53 A study was conducted involving the relationship between the selling price (in thousands of dollars) of a home and two independent variables, the number of rooms and the number of square feet. The following data were collected on 22 properties sold in a particular residential area.

Row	Price	Rooms	Sq ft
1	25.75	5	986
2	37.95	5	998
3	46.45	7	1690
4	46.55	8	1829
5	47.95	6	1186
6	49.95	6	1734
7	52.45	7	1684
8	54.05	7	1846
9	54.85	7	1690
10	52.05	7	1910
11	54.39	7	1784
12	53.45	6	1690
13	59.51	7	1590
14	60.10	8	1855
15	63.85	8	2212
16	62.05	10	2784
17	69.45	7	2190
18	82.30	8	2259
19	81.85	7	1919
20	70.05	7	1685
21	112.45	10	2654
22	127.05	10	2756

Use the computer output shown here to address parts a, b, and c.
a. Conduct a test to see whether the two variables, "rooms" and "sq ft," taken together contain information about "price." Use $\alpha = .05$.
b. Conduct a test to see whether the coefficient of "rooms" is equal to zero. Use $\alpha = .05$.
c. Conduct a test to see whether the coefficient of "sq ft" is equal to zero. Use $\alpha = .05$.

```
          LISTING OF DATA

   OBS     PRICE    ROOMS    SQ_FT

    1      25.75      5        986
    2      37.95      5        998
    3      46.45      7       1690
    4      46.55      8       1829
    5      47.95      6       1186
    6      49.95      6       1734
    7      52.45      7       1684
    8      54.05      7       1846
    9      54.85      7       1690
   10      52.05      7       1910
   11      54.39      7       1784
   12      53.45      6       1690
   13      59.51      7       1590
   14      60.10      8       1855
   15      63.85      8       2212
   16      62.05     10       2784
   17      69.45      7       2190
   18      82.30      8       2259
   19      81.85      7       1919
   20      70.05      7       1685
   21     112.45     10       2654
   22     127.05     10       2756

   N=     22
```

MULTIPLE REGRESSION

GENERAL LINEAR MODELS PROCEDURE

DEPENDENT VARIABLE: PRICE

SOURCE	DF	SUM OF SQUARES	MEAN SQUARE	F VALUE	PR > F	R-SQUARE	C.V.
MODEL	2	6816.77693315	3408.38846658	15.99	0.0001	0.627265	23.5416
ERROR	19	4050.68890321	213.19415280		ROOT MSE		PRICE MEAN
CORRECTED TOTAL	21	10867.46583636			14.60116957		62.02272727

SOURCE	DF	TYPE I SS	F VALUE	PR > F	DF	TYPE III SS	F VALUE	PR > F
ROOMS	1	6357.37363636	29.82	0.0001	1	109.54214216	0.51	0.4822
SQ_FT	1	459.40329679	2.15	0.1585	1	459.40329679	2.15	0.1585

PARAMETER	ESTIMATE	T FOR H0: PARAMETER=0	PR > |T|	STD ERROR OF ESTIMATE
INTERCEPT	-16.97597859	-0.90	0.3815	18.94658431
ROOMS	4.33606202	0.72	0.4822	6.04912439
SQ_FT	0.02551127	1.47	0.1585	0.01737891

OBSERVATION	OBSERVED VALUE	PREDICTED VALUE	RESIDUAL
1	25.75000000	29.85843923	-4.10843923
2	37.95000000	30.16457441	7.78542559
3	46.45000000	56.49049414	-10.04049414
4	46.55000000	64.37262206	-17.82262206
5	47.95000000	39.29675434	8.65324566
6	49.95000000	53.27692781	-3.32692781
7	52.45000000	56.33742655	-3.88742655
8	54.05000000	60.47025156	-6.42025156
9	54.85000000	56.49049414	-1.64049414
10	52.05000000	62.10297254	-10.05297254
11	54.39000000	58.88855310	-4.49855310
12	53.45000000	52.15443213	1.29556787
13	59.51000000	53.93936760	5.57063240
14	60.10000000	65.03591496	-4.93591496
15	63.85000000	74.14343673	-10.29343673
16	62.05000000	97.40800460	-35.35800460
17	69.45000000	69.24612687	0.20387313
18	82.30000000	75.34246621	6.95753379
19	81.85000000	62.33257393	19.51742607
20	70.05000000	56.36293782	13.68706218
21	112.45000000	94.09154009	18.35845991
22	127.05000000	96.69368917	30.35631083

MULTIPLE REGRESSION

GENERAL LINEAR MODELS PROCEDURE

DEPENDENT VARIABLE: PRICE

```
       SUM OF RESIDUALS                        0.00000000
       SUM OF SQUARED RESIDUALS             4050.68890321
       SUM OF SQUARED RESIDUALS - ERROR SS    -0.00000000
       FIRST ORDER AUTOCORRELATION             0.39259574
       DURBIN-WATSON D                         0.98314796
```

13.54   Refer to Exercise 13.53.

   a. Explain the apparent inconsistency between the result of part (a) and the results of parts (b) and (c).

   b. What do you think would happen to the T-ratio of "sq ft" if "rooms" were dropped from the model?

13.55   A study was conducted to determine whether infection surveillance and control programs have reduced the rates of hospital-acquired infection in U.S. hospitals. This data set consists of a random sample of 28 hospitals selected from 338 hospitals participating in a larger study.

   Each line of the data set provides information on variables for a single hospital. The variables are as follows:

Risk = Output variable, average estimated probability of acquiring infection in hospital (in percent)

Stay = Input variable, average length of stay of all patients in hospital (in days)

Age = Input variable, average age of patients (in years)

RCR = Input variable, ratio of number of cultures performed to number of patients without signs or symptoms of hospital-acquired infection (times 100)

School = Dummy input variable for medical school affiliation, 1 = yes, 0 = no

DV_1 = Dummy input variable for region of country, 1 = northeast, 0 = other

DV_2 = Dummy input variable for region of country, 1 = north central, 0 = other

DV_3 = Dummy input variable for region of country, 1 = south, 0 = other

---

(Note that there are four geographic regions of the country—northeast, north central, south, and west. These four regions of the country require only three dummy variables to code for them.)

   The data were analyzed using SAS with the following results.

```
                          LISTING OF DATA

   OBS    RISK    STAY    AGE    RCR    SCHOOL    DV1    DV2    DV3

    1     4.1     7.13    55.7    9.0      0        0      0      1
    2     1.6     8.82    58.2    3.8      0        1      0      0
    3     2.7     8.34    56.9    8.1      0        0      1      0
    4     5.6     8.95    53.7   18.9      0        0      0      1
    5     5.7    11.20    56.5   34.5      0        0      0      0
    6     5.1     9.76    50.9   21.9      0        1      0      0
    7     4.6     9.68    57.8   16.7      0        0      1      0
    8     5.4    11.18    45.7   60.5      1        1      0      0
    9     4.3     8.67    48.2   24.4      0        0      1      0
   10     6.3     8.84    56.3   29.6      0        0      0      0
   11     4.9    11.07    53.2   28.5      1        0      0      0
   12     4.3     8.30    57.2    6.8      0        0      1      0
   13     7.7    12.78    56.8   46.0      1        0      0      0
   14     3.7     7.58    56.7   20.8      0        1      0      0
   15     4.2     9.00    56.3   14.6      0        0      1      0
   16     5.6    10.12    51.7   14.9      1        0      1      0
   17     5.5     8.37    50.7   15.1      0        1      0      0
   18     4.6    10.16    54.2    8.4      1        0      0      1
   19     6.5    19.56    59.9   17.2      0        0      0      0
   20     5.5    10.90    57.2   10.6      0        1      0      0
   21     1.8     7.67    51.7    2.5      0        0      1      0
   22     4.2     8.88    51.5   10.1      0        0      1      0
   23     5.6    11.48    57.6   20.3      0        0      0      0
   24     4.3     9.23    51.6   11.6      0        1      0      0
   25     7.6    11.41    61.1   16.6      0        0      0      0
   26     7.8    12.07    43.7   52.4      0        1      0      0
   27     3.1     8.63    54.0    8.4      0        0      0      0
   28     3.9    11.15    56.5    7.7      0        0      0      0

   N=     28
```

EXERCISES

CORRELATION

VARIABLE	N	MEAN	STD DEV	SUM	MINIMUM	MAXIMUM
STAY	28	10.03321429	2.37286052	280.93000000	7.13000000	19.56000000
AGE	28	54.33928571	4.08024503	1521.50000000	43.70000000	61.10000000
RCR	28	19.28214286	14.32881204	539.90000000	2.50000000	60.50000000
SCHOOL	28	0.17857143	0.39002103	5.00000000	0.00000000	1.00000000
DV1	28	0.28571429	0.46004371	8.00000000	0.00000000	1.00000000
DV2	28	0.28571429	0.46004371	8.00000000	0.00000000	1.00000000
DV3	28	0.10714286	0.31497039	3.00000000	0.00000000	1.00000000

PEARSON CORRELATION COEFFICIENTS / PROB > |R| UNDER HO:RHO=0 / N = 28

	STAY	AGE	RCR	SCHOOL	DV1	DV2	DV3
STAY	1.00000	0.18019	0.35014	0.20586	-0.07993	-0.32591	-0.19127
	0.0000	0.3589	0.0678	0.2933	0.6860	0.0906	0.3296
AGE	0.18019	1.00000	-0.47243	-0.23498	-0.39490	-0.06737	0.01678
	0.3589	0.0000	0.0111	0.2287	0.0375	0.7334	0.9325
RCR	0.35014	-0.47243	1.00000	0.41016	0.23847	-0.31552	-0.17682
	0.0678	0.0111	0.0000	0.0302	0.2217	0.1019	0.3681
SCHOOL	0.20586	-0.23498	0.41016	1.00000	-0.08847	-0.08847	0.13998
	0.2933	0.2287	0.0302	0.0000	0.6544	0.6544	0.4774
DV1	-0.07993	-0.39490	0.23847	-0.08847	1.00000	-0.40000	-0.21909
	0.6860	0.0375	0.2217	0.6544	0.0000	0.0349	0.2627
DV2	-0.32591	-0.06737	-0.31552	-0.08847	-0.40000	1.00000	-0.21909
	0.0906	0.7334	0.1019	0.6544	0.0349	0.0000	0.2627
DV3	-0.19127	0.01678	-0.17682	0.13998	-0.21909	-0.21909	1.00000
	0.3296	0.9325	0.3681	0.4774	0.2627	0.2627	0.0000

STEPWISE REGRESSION
BACKWARD ELIMINATION, DEPENDENT VARIABLE RISK

BACKWARD ELIMINATION PROCEDURE FOR DEPENDENT VARIABLE RISK

STEP 0   ALL VARIABLES ENTERED   R SQUARE = 0.60724861   C(P) =   8.00000000

	DF	SUM OF SQUARES	MEAN SQUARE	F	PROB>F
REGRESSION	7	39.49805177	5.64257882	4.42	0.0041
ERROR	20	25.54623394	1.27731170		
TOTAL	27	65.04428571			

	B VALUE	STD ERROR	TYPE II SS	F	PROB>F
INTERCEPT	-1.07800774				
STAY	0.23613428	0.11569116	5.32126218	4.17	0.0547
AGE	0.04359681	0.07810854	0.39793239	0.31	0.5829
RCR	0.06923673	0.02278287	11.79650358	9.24	0.0065
SCHOOL	-0.41516871	0.64822732	0.52395194	0.41	0.5291
DV1	-0.26955673	0.68941266	0.19527144	0.15	0.6999
DV2	-0.19268071	0.71943459	0.09162010	0.07	0.7916
DV3	0.70243224	0.88962481	0.79632801	0.62	0.4390

BOUNDS ON CONDITION NUMBER: 2.315515,   94.11721

STEP 1   VARIABLE DV2 REMOVED   R SQUARE = 0.60584002   C(P) =   6.07172885

	DF	SUM OF SQUARES	MEAN SQUARE	F	PROB>F
REGRESSION	6	39.40643167	6.56773861	5.38	0.0017
ERROR	21	25.63785404	1.22085019		
TOTAL	27	65.04428571			

	B VALUE	STD ERROR	TYPE II SS	F	PROB>F
INTERCEPT	-1.81224950				
STAY	0.24597088	0.10725430	6.42096620	5.26	0.0322
AGE	0.05262498	0.06888511	0.71251762	0.58	0.4534
RCR	0.07154787	0.02061408	14.70713325	12.05	0.0023
SCHOOL	-0.42280540	0.63312506	0.54445805	0.45	0.5115
DV1	-0.15497958	0.52853481	0.10496975	0.09	0.7722
DV3	0.83288104	0.72780215	1.59882767	1.31	0.2653

BOUNDS ON CONDITION NUMBER: 1.929521,   53.56369

BACKWARD ELIMINATION PROCEDURE FOR DEPENDENT VARIABLE RISK

STEP 2   VARIABLE DV1 REMOVED   R SQUARE = 0.60422621   C(P) =   4.15390906

	DF	SUM OF SQUARES	MEAN SQUARE	F	PROB>F
REGRESSION	5	39.30146193	7.86029239	6.72	0.0006
ERROR	22	25.74282379	1.17012835		
TOTAL	27	65.04428571			

	B VALUE	STD ERROR	TYPE II SS	F	PROB>F
INTERCEPT	-2.21637907				
STAY	0.24760767	0.10486035	6.52437780	5.58	0.0275
AGE	0.05898907	0.06400415	0.99394033	0.85	0.3667
RCR	0.07087867	0.02005725	14.61240661	12.49	0.0019
SCHOOL	-0.38736862	0.60843670	0.47429829	0.41	0.5309
DV3	0.87192445	0.70049715	1.81291925	1.55	0.2263

BOUNDS ON CONDITION NUMBER: 1.905871,   36.65382

STEP 3   VARIABLE SCHOOL REMOVED   R SQUARE = 0.59693428   C(P) =   2.52523447

	DF	SUM OF SQUARES	MEAN SQUARE	F	PROB>F
REGRESSION	4	38.82716364	9.70679091	8.52	0.0002
ERROR	23	26.21712207	1.13987487		
TOTAL	27	65.04428571			

	B VALUE	STD ERROR	TYPE II SS	F	PROB>F
INTERCEPT	-2.30479519				
STAY	0.23848508	0.10252510	6.16764346	5.41	0.0292
AGE	0.06257589	0.06292612	1.12722159	0.99	0.3304
RCR	0.06713326	0.01892561	14.34276871	12.58	0.0017
DV3	0.76072793	0.66954727	1.47147677	1.29	0.2676

BOUNDS ON CONDITION NUMBER: 1.741914,   23.03492

STEP 4   VARIABLE AGE REMOVED   R SQUARE = 0.57960421   C(P) =   1.40772979

	DF	SUM OF SQUARES	MEAN SQUARE	F	PROB>F
REGRESSION	3	37.69994205	12.56664735	11.03	0.0001
ERROR	24	27.34434367	1.13934765		
TOTAL	27	65.04428571			

	B VALUE	STD ERROR	TYPE II SS	F	PROB>F
INTERCEPT	0.88480344				
STAY	0.28060533	0.09334523	10.29588785	9.04	0.0061
RCR	0.05622030	0.01541554	15.15391450	13.30	0.0013
DV3	0.74723631	0.66925498	1.42032908	1.25	0.2753

BOUNDS ON CONDITION NUMBER: 1.162616,   10.11556

STEPWISE REGRESSION
BACKWARD ELIMINATION, DEPENDENT VARIABLE RISK

BACKWARD ELIMINATION PROCEDURE FOR DEPENDENT VARIABLE RISK

STEP 5   VARIABLE DV3 REMOVED   R SQUARE = 0.55776787   C(P) =   0.51969728

	DF	SUM OF SQUARES	MEAN SQUARE	F	PROB>F
REGRESSION	2	36.27961297	18.13980648	15.77	0.0001
ERROR	25	28.76467275	1.15058691		
TOTAL	27	65.04428571			

	B VALUE	STD ERROR	TYPE II SS	F	PROB>F
INTERCEPT	1.15123509				
STAY	0.26598212	0.09287658	9.43651980	8.20	0.0084
RCR	0.05416385	0.01538042	14.26927648	12.40	0.0017

BOUNDS ON CONDITION NUMBER: 1.139728,   4.558912

ALL VARIABLES IN THE MODEL ARE SIGNIFICANT AT THE 0.1000 LEVEL.

SUMMARY OF BACKWARD ELIMINATION PROCEDURE FOR DEPENDENT VARIABLE RISK

STEP	VARIABLE REMOVED	NUMBER IN	PARTIAL R**2	MODEL R**2	C(P)	F	PROB>F
1	DV2	6	0.0014	0.6058	6.07173	0.0717	0.7916
2	DV1	5	0.0016	0.6042	4.15391	0.0860	0.7722
3	SCHOOL	4	0.0073	0.5969	2.52523	0.4053	0.5309
4	AGE	3	0.0173	0.5796	1.40773	0.9889	0.3304
5	DV3	2	0.0218	0.5578	0.51970	1.2466	0.2753

    a. Does the set of seven input variables contain information about the output variable, "risk"? Give a p-value for your test.

    b. Based on the full regression model (seven input variables), can we be at least 95% certain that hospitals in the south have at least .5% higher risk of infection than hospitals in the west, all other things being equal?

13.56   Refer to Exercise 13.55.

    a. Consider the following two statements:

       1. There is multicollinearity between region of the country and whether or not a hospital has a medical school.

       2. There is an interaction effect between region of the country and whether or not a hospital has a medical school.

      What is the difference between these two statements? What evidence is needed to ascertain the truth or falsity of the statements? Is this evidence present in the accompanying output? If it is, do you think the statements are true or false?

    b. Construct a model that allows for the possibility of an interaction effect between region of the country and medical school affiliation. For this model, what is the difference in intercept between a hospital in the northeast affiliated with a medical school and a hospital in the west not affiliated with one?

13.57   Refer to Exercise 13.55. Suppose that we decide to eliminate from the full model some variables that we think contribute little to explaining the output variable. What would your final choice of a model be? Why would you choose this model?

13.58   Refer to Exercise 13.55. Predict the infection risk of a patient in a medical school–affiliated hospital in the northeast, where the average stay of patients is 10 days, the average age is 64, and the routine culturing ratio is 20%. Is this prediction an interpolation or an extrapolation? How do you know?

13.59   Thirty volunteers participated in the following experiment. The subjects took their own pulse rates (which is easiest to do by holding the thumb and forefinger of one hand on the pair of arteries on the side of the neck). They were then asked to flip a coin. If their coin came up heads, they ran in place for one minute. Then everyone took their own pulse rates again. The difference in the before and after pulse rates was recorded, as well as other data on student characteristics. A regression was run to "explain" the pulse rate differences using the other variables as independent variables. The variables were:

    Pulse = Difference between the before and after pulse rates

    Run = Dummy variable, 1 = did not run in place, 0 = ran in place

    Smoke = Dummy variable, 1 = does not smoke, 0 = smokes

    Height = Height in inches

    Weight = Weight in pounds

    Phys1 = Dummy variable, 1 = a lot of physical exercise, 0 = otherwise

    Phys2 = Dummy variable, 1 = moderate physical exercise, 0 = otherwise

---

    a. Perform an appropriate test to determine whether the entire set of independent variables explains a significant amount of the variability of "pulse." Draw a conclusion based on $\alpha = .01$.

    b. Does multicollinearity seem to be a problem here? What is your evidence? What effect does multicollinearity have on your ability to make predictions using regression?

```
                          LISTING OF DATA

     OBS    PULSE    RUN    SMOKE    HEIGHT    WEIGHT    PHYS1    PHYS2

      1      -29      0       1        66       140        0        1
      2      -17      0       1        72       145        0        1
      3      -14      0       0        73       160        1        0
      4      -22      0       0        73       190        0        0
      5      -21      0       1        69       155        0        1
      6      -25      0       1        73       165        0        0
      7       -5      0       1        72       150        1        0
      8       -9      0       1        74       190        0        1
      9      -18      0       1        72       195        0        1
     10      -23      0       1        71       138        0        1
     11      -14      0       0        74       160        0        0
     12      -21      0       1        72       155        0        1
     13        8      0       0        70       153        1        0
     14      -13      0       1        67       145        0        1
     15      -21      0       1        71       170        1        0
     16       -1      0       1        72       175        1        0
     17      -16      0       0        69       175        0        1
     18      -15      1       1        68       145        0        0
     19        4      1       0        75       190        0        1
     20       -3      1       1        72       180        1        0
     21        2      1       0        67       140        0        1
     22       -5      1       1        70       150        0        1
     23       -1      1       1        73       155        0        1
     24       -5      1       1        74       148        1        0
     25       -6      1       0        68       150        0        1
     26       -6      1       0        73       155        0        1
     27        8      1       0        66       130        0        1
     28       -1      1       1        69       160        0        1
     29       -5      1       1        66       135        1        0
     30       -3      1       1        75       160        1        0

   N=     30
```

```
                                         CORRELATION

VARIABLE     N            MEAN              STD DEV              SUM            MINIMUM            MAXIMUM

RUN          30       0.43333333         0.50400693         13.00000000      0.00000000         1.00000000
SMOKE        30       0.66666667         0.47946330         20.00000000      0.00000000         1.00000000
HEIGHT       30      70.86666667         2.77592275       2126.00000000     66.00000000        75.00000000
WEIGHT       30     158.63333333        17.53908607       4759.00000000    130.00000000       195.00000000
PHYS1        30       0.30000000         0.46609160          9.00000000      0.00000000         1.00000000
PHYS2        30       0.56666667         0.50400693         17.00000000      0.00000000         1.00000000
```

PEARSON CORRELATION COEFFICIENTS / PROB > $|R|$ UNDER HO:RHO=0 / N = 30

```
                   RUN       SMOKE      HEIGHT     WEIGHT     PHYS1      PHYS2

   RUN          1.00000    -0.09513   -0.12981   -0.25056    0.01468    0.08597
                0.0000      0.6170     0.4942     0.1817     0.9386     0.6515

   SMOKE       -0.09513     1.00000    0.01727   -0.06834    0.15430   -0.04757
                0.6170      0.0000     0.9278     0.7197     0.4156     0.8029

   HEIGHT      -0.12981     0.01727    1.00000    0.59885    0.19189   -0.28919
                0.4942      0.9278     0.0000     0.0005     0.3097     0.1211

   WEIGHT      -0.25056    -0.06834    0.59885    1.00000    0.01392   -0.11221
                0.1817      0.7197     0.0005     0.0000     0.9418     0.5549

   PHYS1        0.01468     0.15430    0.19189    0.01392    1.00000   -0.74863
                0.9386      0.4156     0.3097     0.9418     0.0000     0.0001

   PHYS2        0.08597    -0.04757   -0.28919   -0.11221   -0.74863    1.00000
                0.6515      0.8029     0.1211     0.5549     0.0001     0.0000
```

```
                          STEPWISE REGRESSION
            BACKWARD ELIMINATION, DEPENDENT VARIABLE PULSE

        BACKWARD ELIMINATION PROCEDURE FOR DEPENDENT VARIABLE PULSE

STEP 0    ALL VARIABLES ENTERED      R SQUARE = 0.62973045     C(P) =    7.00000000

                  DF         SUM OF SQUARES        MEAN SQUARE          F       PROB>F

REGRESSION         6          1850.58887109       308.43147852        6.52      0.0004
ERROR             23          1088.11112891        47.30917952
TOTAL             29          2938.70000000
```

	B VALUE	STD ERROR	TYPE II SS	F	PROB>F
INTERCEPT	-31.68830679				
RUN	11.40166481	2.66171908	868.07553823	18.35	0.0003
SMOKE	-6.89029281	2.74454278	298.18154585	6.30	0.0195
HEIGHT	0.13169561	0.60021947	2.27754970	0.05	0.8283
WEIGHT	0.02303608	0.09440380	2.81697901	0.06	0.8094
PHYS1	13.43465041	4.25117641	472.47616161	9.99	0.0044
PHYS2	.7.80635269	3.97815470	182.17065424	3.85	0.0619

BOUNDS ON CONDITION NUMBER: 2.464274,   62.50691

STEP 1   VARIABLE HEIGHT REMOVED   R SQUARE = 0.62895543   C(P) =   5.04814181

	DF	SUM OF SQUARES	MEAN SQUARE	F	PROB>F
REGRESSION	5	1848.31132139	369.66226428	8.14	0.0001
ERROR	24	1090.38867861	45.43286161		
TOTAL	29	2938.70000000			

	B VALUE	STD ERROR	TYPE II SS	F	PROB>F
INTERCEPT	-24.25519127				
RUN	11.43076116	2.60516294	874.68284765	19.25	0.0002
SMOKE	-6.85327902	2.68448142	296.10525519	6.52	0.0175
WEIGHT	0.03529782	0.07456145	10.18209732	0.22	0.6402
PHYS1	13.44838310	4.16556957	473.54521380	10.42	0.0036
PHYS2	7.65315557	3.83795325	180.65576063	3.98	0.0576

BOUNDS ON CONDITION NUMBER: 2.406131,   40.22006

STEPWISE REGRESSION
BACKWARD ELIMINATION, DEPENDENT VARIABLE PULSE

BACKWARD ELIMINATION PROCEDURE FOR DEPENDENT VARIABLE PULSE

STEP 2   VARIABLE WEIGHT REMOVED   R SQUARE = 0.62549060   C(P) =   3.26336637

	DF	SUM OF SQUARES	MEAN SQUARE	F	PROB>F
REGRESSION	4	1838.12922407	459.53230602	10.44	0.0001
ERROR	25	1100.57077593	44.02283104		
TOTAL	29	2938.70000000			

	B VALUE	STD ERROR	TYPE II SS	F	PROB>F
INTERCEPT	-18.30152045				
RUN	11.13212935	2.48810400	881.24648295	20.02	0.0001
SMOKE	-6.96302377	2.63262467	307.96107626	7.00	0.0139
PHYS1	13.32514812	4.09240540	466.72897076	10.60	0.0032
PHYS2	7.45071026	3.75440264	173.37705597	3.94	0.0583

BOUNDS ON CONDITION NUMBER: 2.396734,   27.36375

ALL VARIABLES IN THE MODEL ARE SIGNIFICANT AT THE 0.1000 LEVEL.

SUMMARY OF BACKWARD ELIMINATION PROCEDURE FOR DEPENDENT VARIABLE PULSE

STEP	VARIABLE REMOVED	NUMBER IN	PARTIAL R**2	MODEL R**2	C(P)	F	PROB>F
1	HEIGHT	5	0.0008	0.6290	5.04814	0.0481	0.8283
2	WEIGHT	4	0.0035	0.6255	3.26337	0.2241	0.6402

c. Based on the full regression model (six dependent variables), compute a point estimate of the average increase in "pulse" for individuals who engaged in a lot of physical activity compared to those who engaged in little physical activity. Can we be 95% certain that the actual average increase is greater than zero?

13.60 Refer to Exercise 13.59.

a. Give the implied regression line of pulse-rate difference on height and weight for a smoker who did not run in place and who has engaged in little physical activity.

b. Consider the following two statements:
   1. There is multicollinearity between the "smoke" variable and the physical activity dummy variables.
   2. There is an interaction effect between the "smoke" variable and the physical activity dummy variables.

   Is there any difference between these two statements? Explain the relationships that would exist in the data set if each of these two statements were correct.

13.61 Refer to Exercise 13.59.
   a. What is your choice of a good predictive equation? Why did you choose that particular equation?
   b. The model as constructed does not contain any interaction effects. Construct a model that allows for the possibility of an interaction effect between each pair of qualitative variables.

 13.62 The data for this exercise were taken from a chemical assay of calcium discussed in Brown, Healy, and Kearns (1981). A set of standard solutions is prepared and these and the unknowns are read on a spectrophotometer in arbitrary units (y). A linear regression model is fit to the standards and the values of the unknowns (x) are read off from this. The preparation of the standard and unknown solutions involves a fair amount of laboratory manipulation, and the actual concentrations of the standards may differ slightly from their target values, the very precise instrumentation being capable of detecting this. The target values are 2.0, 2.0, 2.5, 3.0, 3.0 mmol per liter; the "duplicates" are made up independently. The sequence of reading the standards and unknowns is repeated four times. Two specimens of each unknown are included in each assay and the four sequences of readings are done twice, first with the flame conditions in the instrument optimized, and then with a slightly weaker flame.

The data in the following table relate to assays on the above pattern of a set of six unknowns performed by four laboratories.

The standards are identified as 2.0A, 2.0B, 2.5, 3.0A, 3.0B; the unknowns are identified as U1, U2, W1, W2, Y1, Y2.

y: spectrophotometer reading

x: actual mmol per liter.

Laboratory/Solution		Measurements			
1	W1	1206	1202	1202	1201
1	2.0A	1068	1071	1067	1066
1	W2	1194	1193	1189	1185
1	2.0B	1072	1068	1064	1067
1	U1	1387	1387	1384	1380
1	2.5	1333	1321	1326	1317
1	U2	1394	1390	1383	1376
1	3.0A	1579	1576	1578	1572
1	Y1	1478	1480	1473	1466
1	3.0B	1579	1571	1579	1567
1	Y2	1483	1477	1482	1472
2	W1	1017	1017	1012	1020
2	2.0A	910	916	915	921

(continued)

Laboratory/Solution		Measurements			
2	W2	1012	1018	1015	1023
2	2.0B	913	923	914	921
2	U1	1188	1199	1197	1202
2	2.5	1129	1148	1136	1147
2	U2	1186	1196	1193	1199
2	3.0A	1359	1378	1370	1373
2	Y1	1263	1280	1280	1279
2	3.0B	1349	1361	1359	1363
2	Y2	1259	1269	1259	1265
3	W1	1090	1098	1090	1100
3	2.0A	969	975	969	972
3	W2	1088	1092	1087	1085
3	2.0B	969	960	960	966
3	U1	1270	1261	1261	1269
3	2.5	1196	1196	1209	1200
3	U2	1261	1268	1270	1273
3	3.0A	1451	1440	1439	1449
3	Y1	1352	1349	1353	1343
3	3.0B	1439	1433	1433	1445
3	Y2	1349	1353	1349	1355
4	2.0A	1122	1117	1119	1120
4	W2	1256	1254	1256	1263
4	W1	1260	1251	1252	1264
4	2.0B	1122	1110	1111	1116
4	U2	1453	1447	1451	1455
4	2.5	1386	1381	1381	1387
4	U1	1450	1446	1448	1457
4	3.0A	1656	1663	1659	1665
4	Y2	1543	1548	1543	1545
4	3.0B	1658	1658	1661	1660
4	Y1	1545	1546	1548	1544

a. Plot y versus x for the standards, one graph for each laboratory.

b. Fit the linear regression equation $y = \beta_0 + \beta_1 x + \varepsilon$ for each laboratory and predict the value of x corresponding to the y for each of the unknowns. Compute the standard deviation of predicted values of x based on the four predicted x-values for each of the unknowns.

c. Which laboratory appears to make better predictions of x, mmol of calcium per liter? Why?

13.63 Refer to Exercise 13.62. Suppose you average the y-values for each of the unknowns and fit the ys in the linear regression model of Exercise 13.62.

a. Do your linear regression lines change for each of the laboratories?

b. Will predictions of x change based on these new regression lines for the four laboratories? Explain.

13.64   Refer to Exercise 13.62. Using the independent variable x, suggest a single general linear model that could be used to fit the data from all four laboratories. Identify the parameters in this general linear model.

13.65   Refer to Exercise 13.64.

a. Fit the data to the model of Exercise 13.64.

b. Give separate regression models for each of the laboratories.

c. How do these regression models compare to the previous regression equations for the laboratories?

d. What advantage(s) might there be to fitting a single model rather than separate models for the laboratories?

14 ANALYSIS OF VARIANCE FOR SOME STANDARD EXPERIMENTAL DESIGNS

14.1   INTRODUCTION

The design of an experiment is the process of planning an experiment. A large part of scientific reasoning consists of drawing conclusions from experiments (studies) that have been carefully designed, appropriately conducted, and properly analyzed. In this chapter we will present a brief preview of some standard experimental designs and their analyses.

Section 14.2 provides a brief review of the single-factor analysis of variance discussed in Chapter 10. The design associated with this analysis of variance is called the completely randomized design, where the focus of interest is the comparison of treatment means for different levels of the single factor. Sections 14.3 and 14.4 deal with extensions of the completely randomized design where the focus remains the same—namely, treatment comparisons for the factor of interest—but where other "nuisance" factors must be controlled. For each of these designs, we will consider the arrangement of treatments, the advantages and disadvantages of the design, a model, and an analysis of variance for data from such a design. Section 14.5 introduces factorial experiments that focus on the evaluation of the effects of two or more independent variables (factors) on a response rather than on comparisons of treatment means as in the designs of Sections 14.2 through 14.4. Particular attention is given to measuring the effects of each factor alone or in combination with the other factors. Not all designs focus on either comparison of treatment means or examination of the effects of factors on a response. In Section 14.6 we discuss designs that combine the attributes of the "block" designs of Section 14.3 and 14.4 with those of factorial experiments in Section 14.5. The remaining sections of this chapter deal with multiple-comparison procedures for the different experimental designs and a useful unifying thread showing the relationship between regression and the analysis of variance.

14.2   COMPLETELY RANDOMIZED DESIGN

Recall that the completely randomized design is concerned with the comparison of t population (treatment) means $\mu_1, \mu_2, \ldots, \mu_t$. We assume that there are t different populations from which we are to draw independent random samples of sizes $n_1, n_2, \ldots, n_t$, respectively. Or, in the terminology of the design of experiments, we assume that there are $n_1 + n_2 + \cdots + n_t$ homogeneous *experimental units* (people or objects on which a measurement is made). Then treatments are randomly allocated to the experimental units in such a way that n_1 units receive treatment 1, n_2 receive treatment 2, and so on. The objective of the experiment is to make inferences about the corresponding treatment means.

Consider the following example. A horticultural laboratory is interested in examining the leaves of apple trees to detect nutritional deficiencies using three different laboratory procedures. In particular, the laboratory would like to determine whether there is a difference in mean assay readings for apple leaves utilizing three different laboratory procedures (A, B, C). The experimental units in this investigation are apple tree leaves and the treatments are the three levels of the qualitative variable "laboratory procedure." If a single analyst takes a random sample of nine leaves from the same tree, randomly assigns three leaves to each of the three procedures, and assays the leaves using the assigned treatment, we could use the three sample means to estimate the corresponding mean leaf nutritional deficiency for the three laboratory test procedures, or use the analysis of variance methods of Chapter 10 to run a statistical test of the hypothesis that all three treatment means are identical. The design used for this investigation is a completely randomized design with three observations for each treatment.

The completely randomized design has several advantages and disadvantages when used as an experimental design for comparing t treatment means.

ADVANTAGES AND DISADVANTAGES OF A COMPLETELY RANDOMIZED DESIGN

Advantages
1. It is extremely easy to construct the design.
2. The design is easy to analyze even though the sample sizes might not be the same for each treatment.
3. The design can be used for any number of treatments.

Disadvantages

1. Although the completely randomized design can be used for any number of treatments, it is best suited for situations in which there are relatively few treatments.
2. The experimental units to which treatments are applied must be as homogeneous as possible. Any extraneous sources of variability will tend to inflate the error term, making it more difficult to detect differences among the treatment means.

14.3   RANDOMIZED BLOCK DESIGN

Let us now change the horticultural problem slightly and see how well the completely randomized design suits our needs. Suppose that, rather than relying upon one analyst, we use three analysts for the leaf assays. If we randomly assigned three apple leaves to each of the analysts, we might end up with a randomization scheme like the one listed in Table 14.1.

Even though we still have three observations for each treatment in this scheme, any differences that we may observe among the leaf determinations for the three laboratory procedures may be due entirely to differences among the analysts who assayed the leaves. For example, if we tested the hypothesis $H_0: \mu_A - \mu_B = 0$ against $H_a: \mu_A - \mu_B \neq 0$ and were led to reject H_0, we would not be able to tell whether μ_A differs from μ_B because assays from analyst 1 are different from those for analyst 2 or because the properties of determinations by procedure A differ markedly from those for procedure B. This example illustrates a situation in which the nine experimental units (tree leaves) are affected by an extraneous source of variability: the analyst. In this case the units differ markedly and would not be a homogeneous set on which we could base an evaluation of the effects of the three treatments.

The completely randomized design just described can be modified to gain additional information concerning the means μ_A, μ_B, and μ_C. We can block out the undesirable variability among analysts by using the following experimental design. We restrict our randomization of treatments to experimental units to ensure that each analyst performs a determination using each of the three procedures. The order of these determinations for each analyst is randomized. One such randomization is listed in Table 14.2. Note that each analyst will assay three leaves, one leaf for each of the three procedures. Hence pairwise comparisons among the laboratory procedures that utilize the sample means will

TABLE 14.1 Random Assignment of the Nine Leaves to the Three Analysts

	Analyst	
1	2	3
A	B	C
A	B	C
A	B	C

TABLE 14.2 A Different Assignment of Leaves to Analysts

	Analyst	
1	2	3
A	B	A
C	A	B
B	C	C

be free of any variability among analysts. For example, if we ran the test

$$H_0: \mu_A - \mu_B = 0$$
$$H_a: \mu_A - \mu_B \neq 0$$

and rejected H_0, the difference between μ_A and μ_B would be due to a difference between the nutritional deficiencies detected by procedures A and B and not due to a difference among the analysts, since each analyst would have assayed one leaf for each of the three procedures.

randomized block design, blocks

This design, which represents an extension to the completely randomized design, is called a **randomized block design**; the analysts in our experiment are called **blocks**. By using this design, we have effectively filtered out any variability among the analysts, enabling us to make more precise comparisons among the *treatment* means μ_A, μ_B, and μ_C.

In general, we can use a randomized block design to compare t different treatment means when an extraneous source of variability (blocks) is present. If there are b different blocks, we would run each of the t treatments in each block to filter out the block-to-block variability. In our example we had $t = 3$ treatment means (laboratory procedures) and $b = 3$ blocks (analysts).

The randomized block design has certain advantages and disadvantages, as shown here.

ADVANTAGES AND DISADVANTAGES OF THE RANDOMIZED BLOCK DESIGN

Advantages

1. It is a useful design for comparing t treatment means in the presence of a single extraneous source of variability.
2. The statistical analysis is simple.
3. The design is easy to construct.
4. It can be used to accommodate any number of treatments in any number of blocks.

Disadvantages

1. Since the experimental units within a block must be homogeneous, the design is best suited for a relatively small number of treatments.
2. This design controls for only one extraneous source of variability (due to blocks). Additional extraneous sources of variability tend to increase the error term, making it more difficult to detect treatment differences.
3. The effect of each treatment on the response must be approximately the same from block to block.

The definition of a randomized block design is given next.

DEFINITION 14.1

A **randomized block design** is an experimental design for comparing t treatments in b blocks. Treatments are randomly assigned to experimental units within a block, with each treatment appearing exactly once in every block.

TABLE 14.3   A Randomized Block Design

Treatment	Block 1	2	$\cdots$	b	Totals	Mean
1	y_{11}	y_{12}	$\cdots$	y_{1b}	T_1	$\bar{y}_1$
2	y_{21}	y_{22}	$\cdots$	y_{2b}	T_2	$\bar{y}_2$
$\vdots$	$\vdots$	$\vdots$		$\vdots$	$\vdots$	$\vdots$
t	y_{t1}	y_{t2}	$\cdots$	y_{tb}	T_t	$\bar{y}_t$
Totals	B_1	B_2	$\cdots$	B_b	G	
Mean	$\bar{B}_1$	$\bar{B}_2$	$\cdots$	$\bar{B}_b$		$\bar{y}$

Consider the data for a randomized block design as arranged in Table 14.3. Using Table 14.3 we can introduce notation that is helpful in performing an analysis of variance. This notation is presented here.

NOTATION NEEDED FOR THE AOV OF A RANDOMIZED BLOCK DESIGN

y_{ij}: observation for treatment i in block j

t: number of treatments

b: number of blocks

n: total number of sample measurements; $n = bt$

T_i: total for all observations receiving treatment i

B_j: total for all observations in block j

G: total for all sample observations

$\bar{y}_i$: sample mean for treatment i; $\bar{y}_i = T_i/b$

$\bar{B}_j$: sample mean for block j; $\bar{B}_j = B_j/t$

$\bar{y}$: overall sample mean; $\bar{y} = G/n$

total sum of squares

The **total sum of squares** of the measurements about their mean $\bar{y}$ is defined as before:

$$\text{TSS} = \sum_{ij} (y_{ij} - \bar{y})^2$$

It is possible to partition the total sum of squares into three separate sources of variability: one due to the variability among treatments, one due to the variability among blocks, and one due to the variability among the y_{ij}s that is not accounted for by either **partition of TSS** treatments or blocks. We call this final source of variability "error." The **partition of TSS** can be shown to take the following form:

$$\sum_{ij} (y_{ij} - \bar{y})^2 = b \sum_i (\bar{y}_i - \bar{y})^2 + t \sum_j (\bar{B}_j - \bar{y})^2 + \sum_{ij} (y_{ij} - \bar{y}_i - \bar{B}_j + \bar{y})^2$$

The first quantity on the right side of the equation measures the variability of the treatment means $\bar{y}_i$ from the overall mean. Thus

$$\text{SST} = b \sum_i (\bar{y}_i - \bar{y})^2$$

between-treatment sum of squares

called the **between-treatment sum of squares**, is a measure of the between-treatment variability. Similarly, the second quantity,

$$SSB = t \sum_j (\bar{B}_j - \bar{y})^2$$

between-block sum of squares

sum of squares for error

measures the variability between the block means $\bar{B}_j$ and the overall mean. It is called the **between-block sum of squares**. The third source of variability, referred to as the **sum of squares for error** (*SSE*), represents the variability in the y_{ij}s not accounted for by the block and treatment sources.

Although the sum of squares formulas just discussed are instructive, they are not convenient to use in calculations. The shortcut formulas given below are more convenient tools for calculations.

SHORTCUT SUMS OF SQUARES FORMULAS FOR A RANDOMIZED BLOCK DESIGN

$$TSS = \sum_{ij} y_{ij}^2 - \frac{G^2}{n}$$

$$SST = \sum_i \frac{T_i^2}{b} - \frac{G^2}{n}$$

$$SSB = \sum_j \frac{B_j^2}{t} - \frac{G^2}{n}$$

$$SSE = TSS - SST - SSB$$

The model for an observation in a randomized block design can be written in the form

model

$$y_{ij} = \mu + \alpha_i + \beta_j + \varepsilon_{ij}$$

where the terms of the model are defined as follows:

μ: an overall mean, which is an unknown constant

α_i: an effect due to treatment i; α_i is an unknown constant

β_j: an effect due to block j; β_j is an unknown constant

ε_{ij}: a random error associated with the response on treatment i, block j. We assume that the ε_{ij}s are normally distributed with mean 0 and unknown variance σ_ε^2. In addition, the errors are assumed to be independent. (Technically speaking, we need not assume that the ε_{ij}s are normally distributed at this point, but we must make this assumption prior to running an analysis of variance.)

The assumptions given above for our model can be shown to imply that y_{ij}, the response on treatment i in block j, is normally distributed with mean $\mu + \alpha_i + \beta_j$ and variance σ_ε^2. A table of population means (expected values) for the data of Table 14.3 is shown in Table 14.4.

Several comments should be made concerning the table of expected values. First, two observations that receive the same treatment (appear in the same row of Table 14.4) have

TABLE 14.4   Expected Values for the y_{ij}s in a Randomized Block Design

Treatment	Block			
	1	2	$\cdots$	b
1	$E(y_{11}) = \mu + \alpha_1 + \beta_1$	$E(y_{12}) = \mu + \alpha + \beta_2$	$\cdots$	$E(y_{1b}) = \mu + \alpha_1 + \beta_b$
2	$E(y_{21}) = \mu + \alpha_2 + \beta_1$	$E(y_{22}) = \mu + \alpha_2 + \beta_2$	$\cdots$	$E(y_{2b}) = \mu + \alpha_2 + \beta_b$
$\vdots$	$\vdots$	$\vdots$		$\vdots$
t	$E(y_{t1}) = \mu + \alpha_t + \beta_1$	$E(y_{t2}) = \mu + \alpha_t + \beta_2$	$\cdots$	$E(y_{tb}) = \mu + \alpha_t + \beta_b$

population means that differ by block effects only. For example, the expected values associated with y_{11} and y_{12} (two observations receiving treatment 1) are, respectively, $\mu + \alpha_1 + \beta_1$ and $\mu + \alpha_1 + \beta_2$. Thus the difference in their means is $\beta_1 - \beta_2$, which accounts for the fact that y_{11} appeared in block 1 and y_{12} in block 2. Second, two observations appearing in the same block (in the same column of Table 14.4) have means that differ by a treatment effect only. For example, y_{11} and y_{21} both appear in block 1. The difference in their means is, from Table 14.4,

$$(\mu + \alpha_1 + \beta_1) - (\mu + \alpha_2 + \beta_1) = \alpha_1 - \alpha_2$$

which accounts for the fact that the observations received different treatments. Finally, when two observations receive a different treatment *and* appear in different blocks, their expected values differ by effects due to treatments and to blocks. Thus y_{11} and y_{22} have expectations that differ by

$$(\mu + \alpha_1 + \beta_1) - (\mu + \alpha_2 + \beta_2) = (\alpha_1 - \alpha_2) + (\beta_1 - \beta_2)$$

filtering   Using the information we have learned concerning the model for a randomized block design, we can illustrate the concept of **filtering** and show how the randomized block design filters out the variability due to blocks. Consider a randomized block design with $t = 3$ treatments (I, II, and III) laid out in $b = 3$ blocks as shown in Table 14.5.

The model for this randomized block design is

$$y_{ij} = \mu + \alpha_i + \beta_j + \varepsilon_{ij} \qquad (i = 1, 2, 3; j = 1, 2, 3)$$

Suppose we wish to estimate the difference in mean response for treatments II and I, namely, $\alpha_2 - \alpha_1$. The difference in sample means, $\bar{y}_{II} - \bar{y}_I$, would represent a point estimate

TABLE 14.5   Randomized Block Design with $t = 3$ Treatments and $b = 3$ Blocks

Block	Treatment		
1	I	II	III
2	I	III	II
3	III	I	II

of $\alpha_2 - \alpha_1$. By substituting into our model, we have

$$\bar{y}_{\mathrm{I}} = \sum_j \frac{y_{ij}}{3}$$

$$= \frac{1}{3}[(\mu + \alpha_1 + \beta_1 + \varepsilon_{11}) + (\mu + \alpha_1 + \beta_2 + \varepsilon_{12}) + (\mu + \alpha_1 + \beta_3 + \varepsilon_{13})]$$

$$= \mu + \alpha_1 + \bar{\beta} + \bar{\varepsilon}_1$$

where $\bar{\beta}$ represents the mean of the three block effects β_1, β_2, and β_3, and $\bar{\varepsilon}_1$ represents the mean of the three random errors ε_{11}, ε_{12}, and ε_{13}. Similarly, it is easy to show that

$$\bar{y}_{\mathrm{II}} = \mu + \alpha_2 + \bar{\beta} + \bar{\varepsilon}_2$$

and hence

$$\bar{y}_{\mathrm{II}} - \bar{y}_{\mathrm{I}} = (\alpha_2 - \alpha_1) + (\bar{\varepsilon}_2 - \bar{\varepsilon}_1)$$

Note how the block effects cancel, leaving the quantity $(\bar{\varepsilon}_2 - \bar{\varepsilon}_1)$ as the error of estimation when using $\bar{y}_{\mathrm{II}} - \bar{y}_{\mathrm{I}}$ to estimate $\alpha_2 - \alpha_1$.

If a completely randomized design had been employed instead of a randomized block design, treatments would have been assigned to experimental units at random and it is quite unlikely that each treatment would have appeared in each block. When the same treatment appears more than once in a block and we calculate an estimate of $\alpha_2 - \alpha_1$ using $\bar{y}_{\mathrm{II}} - \bar{y}_{\mathrm{I}}$, all block effects would not cancel out as they did previously. Then the error of estimation would include not only $\bar{\varepsilon}_2 - \bar{\varepsilon}_1$ but also the block effects that do not cancel; that is,

$$\bar{y}_{\mathrm{II}} - \bar{y}_{\mathrm{I}} = \alpha_2 - \alpha_1 + (\bar{\varepsilon}_2 - \bar{\varepsilon}_1) + (\text{block effects that do not cancel})$$

Hence the randomized block design filters out variability due to blocks by decreasing the error of estimation for a comparison of treatment means.

Having indicated the shortcut formulas for calculating sums of squares, and having specified the model associated with a randomized block design, we can now formulate **statistical test** the analysis of variance. The null hypothesis of no difference among the treatment means is equivalent to testing

$$H_0: \alpha_1 = \alpha_2 = \cdots = \alpha_t = 0$$

As we observed in Table 14.4, any time we compare the mean response of two treatments (say i and i') in the same block, the difference in their mean response is $\alpha_i - \alpha_{i'}$. Thus under H_0 we are assuming that treatments have the same mean response within a block.

The alternative hypothesis is

H_a: At least one α_i is different from zero (i.e., at least one of the treatment means differs from the rest).

The test statistic is

$$F = \frac{\mathrm{MST}}{\mathrm{MSE}}$$

TABLE 14.6   Analysis of Variance Table for a Randomized Block Design

Source	SS	df	MS	F
Treatments	SST	$t-1$	$\text{MST} = \text{SST}/(t-1)$	MST/MSE
Blocks	SSB	$b-1$	$\text{MSB} = \text{SSB}/(b-1)$	MSB/MSE
Error	SSE	$(b-1)(t-1)$	$\text{MSE} = \text{SSE}/(b-1)(t-1)$	
Totals	TSS	$bt-1$		

where MST and MSE are mean squares computed from the appropriate sums of squares in the AOV table of Table 14.6.

unbiased estimates   When H_0: $\alpha_1 = \alpha_2 = \cdots = \alpha_t = 0$ is true, both MST and MSE are **unbiased estimates** of σ_ε^2, the variance associated with the observations in our model. That is, when H_0 is true, both MST and MSE have a mean value in repeated sampling equal to σ_ε^2, and we would **expected mean** expect $F = \text{MST}/\text{MSE}$ to be near 1. When H_a is true, the mean of MSE, called the **expected square for error** **mean square for error**, is still

$$E(\text{MSE}) = \sigma_\varepsilon^2$$

However, MST is no longer unbiased for σ_ε^2. In fact, the expected mean square for treatments can be shown to be

$$E(\text{MST}) = \sigma_\varepsilon^2 + b\theta_T$$

where θ_T is a positive function of the α_is. Because MST will tend to overestimate σ_ε^2 under H_a, the ratio $F = \text{MST}/\text{MSE}$ will be greater than 1, and we will reject H_0 in the upper tail of the distribution of F.

For a specified probability of a Type 1 error, the F test for H_0: $\alpha_1 = \alpha_2 = \cdots = \alpha_t = 0$ will reject H_0 if the computed value of F exceeds the critical value of F for a $= \alpha$, $\text{df}_1 = t-1$, and $\text{df}_2 = (b-1)(t-1)$. Note that df_1 and df_2 correspond to the degrees of freedom for MST and MSE, respectively, in the AOV table.

test for effects   We may also be interested in testing whether it was advantageous to block. That is, is **of blocks** there an **effect due to blocks** that we have effectively filtered out? Recall that the expected values for two observations from different blocks receiving the same treatment differed by an effect due to blocks only (see Table 14.4). The hypothesis of no effect due to blocks can be written in the form

$$H_0: \beta_1 = \beta_2 = \cdots = \beta_b = 0$$

The alternative hypothesis and test statistic are then

H_a: At least one of the βs differs from zero.

$$\text{T.S.:} \ F = \frac{\text{MSB}}{\text{MSE}}$$

Under H_0, both MSB and MSE are unbiased estimates of σ_ε^2 (i.e., $E(\text{MSB}) = E(\text{MSE}) = \sigma_\varepsilon^2$). But under H_a, the expected mean squares are

$$E(\text{MSB}) = \sigma_\varepsilon^2 + t\theta_B \qquad \text{and} \qquad E(\text{MSE}) = \sigma_\varepsilon^2$$

TABLE 14.7   Number of Seedlings by Insecticide and Plot, Example 14.1

	Plot			
Insecticide	1	2	3	4
1	56	49	65	60
2	84	78	94	93
3	80	72	83	85

where θ_B is a positive function of the βs. Since MSB will tend to overestimate σ_ε^2 under H_a, we will reject H_0 for large values of F. For a specified value of α, we can locate the rejection region in the F table of the Appendix for $a = \alpha$, $df_1 = b - 1$, and $df_2 = (b - 1)(t - 1)$.

EXAMPLE 14.1

An experiment was conducted to compare the effect of three different insecticides on a particular variety of string beans. Four different plots were prepared, with each plot subdivided into three rows. A suitable distance was maintained between the rows within a plot. Each row was planted with 100 seeds and then maintained under the insecticide assigned to the row. The insecticides were randomly assigned to the rows within a plot so that each insecticide appeared in one row in all four plots. The response of interest was the number of seedlings that emerged per row. These data are given in Table 14.7.

a. Set up an appropriate statistical model for this experimental situation.
b. Run an analysis of variance to compare the three insecticides.
c. Summarize your results in an AOV table. Use $\alpha = .05$.

Solution

We recognize this experimental design as a randomized block design with $b = 4$ blocks (plots) and $t = 3$ treatments (insecticides) per block. The appropriate statistical model is

$$y_{ij} = \mu + \alpha_i + \beta_j + \varepsilon_{ij} \qquad (i = 1, 2, 3; j = 1, 2, 3, \ldots, 4)$$

From the sample data, the treatment and block totals can be shown to be

Insecticides:	$T_1 = 230$	Plots:	$B_1 = 220$
	$T_2 = 349$		$B_2 = 199$
	$T_3 = \underline{320}$		$B_3 = 242$
	$G = 899$		$B_4 = \underline{238}$
			$G = 899$

Substituting into the corresponding shortcut formulas for the sums of squares, we have

$$SST = \sum_i \frac{T_i^2}{b} - \frac{G^2}{n} = \frac{(230)^2 + (349)^2 + (320)^2}{4} - \frac{(899)^2}{12}$$

$$= 69{,}275.25 - 67{,}350.08 = 1925.17$$

TABLE 14.8   AOV Table for the Data of Example 14.1

Source	SS	df	MS	F
Treatments	1925.17	2	962.59	962.59/3.92 = 245.56
Blocks	386.25	3	128.75	128.75/3.92 = 32.84
Error	23.50	6	3.92	
Totals	2334.92	11		

$$SSB = \sum_j \frac{B_j^2}{t} - \frac{G^2}{n} = \frac{(220)^2 + (199)^2 + (242)^2 + (238)^2}{3} - \frac{(899)^2}{12}$$

$$= 67{,}736.33 - 67{,}350.08 = 386.25$$

Then

$$TSS = \sum_{ij} y_{ij}^2 - \frac{G^2}{n} = (56)^2 + (49)^2 + \cdots + (85)^2 - 67{,}350.08$$

$$= 2334.92$$

By subtraction,

$$SSE = TSS - SST - SSB = 2334.92 - 1925.17 - 386.25 = 23.50$$

The analysis of variance table in Table 14.8 summarizes our results. Note that the mean square for a source in the AOV table is computed by dividing the sum of squares for that source by its degrees of freedom.

The F test for treatments, namely

H_0: $\alpha_1 = \alpha_2 = \alpha_3 = 0$ (no differences among treatment means)

makes use of the F statistic MST/MSE. Since the computed value of F, 245.56, is greater than the tabulated F-value, 5.14, based on $df_1 = 2$, $df_2 = 6$, and a = .05, we reject H_0 and conclude that there are differences among the three treatment (insecticide) means.

The test for blocks

H_0: $\beta_1 = \beta_2 = \beta_3 = \beta_4 = 0$ (no differences among plot means)

utilizes the computed value of F = MSB/MSE as the test statistic. Since the computed value of F, 32.84, is greater than the tabulated value, 4.76, based on $df_1 = 3$, $df_2 = 6$, and a = .05, we reject the null hypothesis and conclude that there are differences among the plot means. Further, since the $F_{.01}$ value (9.78) is also greatly exceeded, the level of significance for the test is much less than .01.

EXAMPLE 14.2   Compare the computer output shown here to the hand calculations for the data of Example 14.1.

ANALYSIS OF VARIANCE PROCEDURE

DEPENDENT VARIABLE: NO__SEED

SOURCE	DF	SUM OF SQUARES	MEAN SQUARE	F VALUE	PR > F	R-SQUARE	C.V.
MODEL	5	2311.41666667	462.28333333	118.03	0.0001	0.989935	2.6417
ERROR	6	23.50000000	3.91666667		ROOT MSE		NO__SEED MEAN
CORRECTED TOTAL	11	2334.91666667			1.97905701		74.91666667

SOURCE	DF	ANOVA SS	F VALUE	PR > F
INSECT	2	1925.16666667	245.77	0.0001
PLOT	3	386.25000000	32.87	0.0004

Solution   Except for minor rounding errors in our hand calculations, the results from the output agree with the analysis of variance of Example 14.1. Note that the p-value for treatment (insecticide) differences is .0001.

We have discussed randomized block designs briefly in Chapter 10 and in more detail here. But we might still ask whether blocking has increased our precision for comparing treatment means in a given experiment. Let MSE_{RB} and MSE_{CR} denote the mean square errors for a randomized block design and a completely randomized design, respectively. One measure of precision for the two designs is the variance of $\bar{y}_i$, the ith treatment mean ($i = 1, 2, \ldots, t$). For a randomized block design, the estimated variance for $\bar{y}_i$, is MSE_{RB}/b. A similar expression for a completely randomized design is MSE_{CR}/r, where r is the number of observations (replications) of each treatment required to satisfy the relationship

$$\frac{MSE_{CR}}{r} = \frac{MSE_{RB}}{b} \quad \text{or} \quad \frac{MSE_{CR}}{MSE_{RB}} = \frac{r}{b}$$

relative efficiency   The quantity r/b is called the **relative efficiency** of the randomized block design. The larger the value of MSE_{CR} to MSE_{RB}, the larger r must be to obtain the same precision for a treatment mean as obtained with the randomized block design.

Although we never perform an analysis of variance for both the completely randomized design and the randomized block design every time we employ a randomized block design, we can use the mean squares from the analysis of variance for the randomized block design to obtain the relative efficiency by using the formula

$$\frac{MSE_{CR}}{MSE_{RB}} = \frac{(b-1)MSB + b(t-1)MSE}{(bt-1)MSE}$$

EXAMPLE 14.3   Refer to Example 14.1. Compute the relative efficiency of the randomized block design.

Solution   From the AOV table in Example 14.1, $MSB = 128.75$ and $MSE = 3.92$. Hence the relative efficiency of this randomized block design relative to a completely randomized design is

$$\frac{MSE_{CR}}{MSE_{RB}} = \frac{3(128.75) + 4(2)(3.92)}{11(3.92)} = 9.68$$

That is, approximately 10 times as many observations of each treatment would be required in a completely randomized design to obtain the same precision for treatment comparisons as with this randomized block design.

EXERCISES

14.1   a. Analyze the hypothetical data shown here for a randomized block design with $b = 6$ blocks and $t = 2$ treatments. Use the analysis of variance methods of this section.

Block	Treatment	
	A	B
1	58	47
2	324	331
3	206	163
4	94	75
5	39	30
6	418	397

b. Give the efficiency of the randomized block design relative to the completely randomized design. Explain.

14.2   Refer to Exercise 14.1. Analyze these same data using the paired t-methods of Chapter 5. Compare your results. (Hint: $t^2 = F$ for the test on treatments.)

14.3   An experiment is conducted to compare four different mixtures of the components oxidizer, binder, and fuel used in the manufacturing of rocket propellant. The four mixtures under test, corresponding to settings of the mixture proportions for oxide, are shown here.

Mixture	Oxidizer	Binder	Fuel
1	0.4	0.4	0.2
2	0.4	0.2	0.4
3	0.6	0.2	0.2
4	0.5	0.3	0.2

To compare the four mixtures, five different samples of propellant are prepared from each mixture and readied for testing. Each of five investigators is randomly assigned one sample of each of the four mixtures and asked to measure the propellant thrust. These data are summarized below.

Mixture	Investigator				
	1	2	3	4	5
1	2340	2355	2362	2350	2348
2	2658	2650	2665	2640	2653
3	2449	2458	2432	2437	2445
4	2403	2410	2418	2397	2405

a. Identify the blocks and treatments for this experimental design.
b. Indicate the method of randomization.
c. Why would this design be preferable to a completely randomized design?

14.4   Refer to Exercise 14.3.

a. Use the computer output shown here to conduct an analysis of variance. Use $\alpha = .05$.

b. What conclusions can you draw concerning the best mixture from the four tested? (Note: The higher the response value, the better the rocket propellant thrust.)

c. Compute the relative efficiency of the randomized block design.

ANALYSIS OF VARIANCE PROCEDURE

DEPENDENT VARIABLE: THRUST

SOURCE	DF	SUM OF SQUARES	MEAN SQUARE	F VALUE	PR > F	R-SQUARE	C.V.
MODEL	7	261713.45000000	37387.63571429	542.96	0.0001	0.996853	0.3368
ERROR	12	826.30000000	68.85833333		ROOT MSE		THRUST MEAN
CORRECTED TOTAL	19	262539.75000000			8.29809215		2463.75000000
SOURCE	DF	ANOVA SS	F VALUE	PR > F			
MIXTURE	3	261260.95000000	1264.73	0.0001			
INVEST	4	452.50000000	1.64	0.2273			

14.5   A study was undertaken to compare the starting salaries of bachelor's degree candidates at a particular university for the academic years 1985–1986 and 1986–1987. Since there could be a great deal of variability in salaries from one discipline to another, the investigator blocked on curricula within the university. Then the median salary for bachelor's candidates for 1985–1986 and for 1986–1987 was obtained from the most recent lists of graduates to ascertain the starting salary. It should be noted that only those students who had accepted a job were considered in this study. Use the following sample data of starting salaries (in $000) to conduct an analysis of variance, testing separately for an effect due to treatments and blocks. Use $\alpha = .05$ for both tests.

Curriculum	1986–1987	1985–1986
Accounting	22.0	21.5
Agricultural sciences	20.6	20.3
Biological sciences	20.0	19.8
Business (general)	24.8	24.6
Chemistry	22.4	22.2
Computer science	26.9	26.5
Engineering (civil)	24.3	24.9
Engineering (chemical)	23.0	22.6
Humanities	16.5	16.6
Mathematics	20.7	21.1
Social sciences	16.9	16.8

14.6   Examine the computer output shown here for the data of Example 14.2.

a. Identify the sums of squares and degrees of freedom for plots, insecticides, and error.

b. Give the F test and p-values for the null hypothesis of no difference in insecticide means. Draw a conclusion.

c. Give the F test and p-value for the null hypothesis of no difference in plot means. Draw a conclusion.

d. Compare your results to those of Example 14.2.

ANALYSIS OF VARIANCE FOR 1-ST DEPENDENT VARIABLE

SOURCE	SUM OF SQUARES	DEGREES OF FREEDOM	MEAN SQUARE	F	PROB. F EXCEEDED
MEAN	67350.06250	1	67350.06250	17195.78991	0.000
GROUP	1925.16382	2	962.58179	246.39508	0.000
LEVEL	386.24829	3	128.74942	32.87225	0.000
ERROR	23.49994	6	3.91666		

14.7 The computer output shown here gives the analysis of variance for a taste-test experiment where each of 12 persons was asked to sample three new formulations of a widely used bulk laxative.

MTB > TWOWAY ANOVA OF 'RATING' BY 'FORM' AND 'PERSON'

ANALYSIS OF VARIANCE RATING

SOURCE	DF	SS	MS
FORM	2	767.4	383.7
PERSON	11	5301.2	481.9
ERROR	22	1532.6	69.7
TOTAL	35	7601.2	

MTB > TABLE BY 'FORM';
SUBC > MEANS OF 'RATING'.

ROWS: FORM

	RATING MEAN
1	57.667
2	46.917
3	49.250
ALL	51.278

a. Run an F test to compare the formulation means. Give the p-value for your test.
b. Also run an F test comparing the people. Do you think it was appropriate to block on people? Explain.

14.8 Refer to Exercise 14.7.
a. Which formulation means appear to be different?
b. Calculate 95% confidence intervals for all pairwise differences in formulation means. Which means appear different based on these intervals? Explain.

14.4 LATIN SQUARE DESIGN

The apple leaf problem of Sections 14.2 and 14.3 can be complicated further in the following way. Suppose that each leaf assay takes a long time and only one can be done by each analyst per day. If we used the randomized block design of Table 14.2, letting the first

TABLE 14.9   A Randomized Block Design for the
Leaf Assay in the Presence of a
Day Effect

Day	Analyst		
	1	2	3
1	A	B	A
2	C	A	B
3	B	C	C

row denote day 1, the second row denote day 2, and the third row denote day 3, the design could be listed as shown in Table 14.9.

Suppose now that we tested $H_0: \mu_A - \mu_B = 0$ against $H_a: \mu_A - \mu_B \neq 0$. Two procedure A determinations were done on day 1 and one on day 2, whereas procedure B was used on each of the three days. Thus if we reject H_0, we would not be certain whether μ_A differed from μ_B because of a difference in the laboratory procedures or because of a difference among the three days. Sometimes laboratory equipment must be calibrated daily and new chemical solutions must be prepared. Differences in determinations from day to day could be due to differences among the solutions or to differences in calibration accuracy.

This example illustrates a situation in which the experimental units (leaves) are affected by a second extraneous source of variability, days. We can modify the randomized block design to filter out this second source of variability, the variability among days, in addition to filtering out the first source, variability among analysts. To do this we restrict our randomization to ensure that each treatment appears in each row (day) and in each column (analyst). One such randomization is shown in Table 14.10. Note that the test procedures have been assigned to analysts and to days so that each procedure is performed once a day and once by each analyst. Hence pairwise comparisons among treatment procedures that involve the sample means are free of variability among days and analysts.

Latin square design   This experimental design is called a **Latin square design**. In general, a Latin square design can be used to compare t treatment means in the presence of two extraneous sources of variability, which we block off into t rows and t columns. The t treatments are then randomly assigned to the rows and columns so that each treatment appears in every row and every column of the design (see Table 14.10).

The advantages and disadvantages of the Latin square design are listed here.

TABLE 14.10   Assignment of Leaves to Analysts
and Days

Day	Analyst		
	1	2	3
1	A	B	C
2	B	C	A
3	C	A	B

ADVANTAGES AND DISADVANTAGES OF THE LATIN SQUARE DESIGN

Advantages

1. The design is particularly appropriate for comparing t treatment means in the presence of two sources of extraneous variation, each measured at t levels.
2. The analysis is still quite simple.

Disadvantages

1. Although a Latin square can be constructed for any value of t, it is best suited for comparing t treatments when $5 \leq t \leq 10$.
2. Any additional extraneous sources of variability tend to inflate the error term, making it more difficult to detect differences among the treatment means.
3. The effect of each treatment on the response must be approximately the same across rows and columns.

The definition of a Latin square design is given here.

DEFINITION 14.2

A $t \times t$ **Latin square design** contains t rows and t columns. The t treatments are randomly assigned to experimental units within the rows and columns so that each treatment appears in every row and in every column.

A typical randomization scheme for a 4×4 Latin square comparing the treatments I, II, III, and IV is shown in Table 14.11. Note that each treatment appears in all four rows and all four columns.

The notation for a Latin square design is only slightly more complicated than that for a randomized block design.

NOTATION NEEDED FOR THE AOV OF A $t \times t$ LATIN SQUARE DESIGN

y_{ijk}: response on treatment i in row j and column k

t: number of treatments; also the number of rows and the number of columns

n: total number of sample measurements; $n = t^2$

T_i: total for all observations receiving treatment i

$\bar{y}_i$: sample mean for treatment i; $\bar{y}_i = T_i/t$

R_j: total for all observations in row j

$\bar{R}_j$: sample mean for row j; $\bar{R}_j = R_j/t$

C_k: total for all observations in column k

$\bar{C}_k$: sample mean for column k; $\bar{C}_k = C_k/t$

G: total for all sample measurements

$\bar{y}$: overall sample mean; $\bar{y} = G/n$

TABLE 14.11   A 4 × 4 Latin Square Design

		Column		
Row	1	2	3	4
1	I	II	III	IV
2	II	III	IV	I
3	III	IV	I	II
4	IV	I	II	III

partitioning TSS   With this notation we can show a partition of the total sum of squares into four components. The first three components measure variability among the treatments, rows, and columns, respectively. The other source is due to random error.

$$\sum_{\substack{\text{over} \\ \text{all } ys}} (y_{ijk} - \bar{y})^2 = t \sum_i (\bar{y}_i - \bar{y})^2 + t \sum_j (\bar{R}_j - \bar{y})^2 + t \sum_k (\bar{C}_k - \bar{y})^2$$

$$+ \sum_{\substack{\text{over} \\ \text{all } ys}} (y_{ijk} - \bar{y}_i - \bar{R}_j - \bar{C}_k + 2\bar{y})^2$$

Note that the total sum of squares is obtained by summing over only two of the three subscripts. Even though observations are identified by treatment, row, and column, by summing over treatments (i) and rows (j), we have also summed over columns (k).

The algebraic verification of the partitioning of TSS into the four components is not important here and is beyond the scope of this text. We will concentrate instead on the interpretation of the partitioning.

The first quantity on the right side of the equation for TSS measures the variability of the treatment means $\bar{y}_i$ about the overall mean $\bar{y}$. As before, we call this source of variability the sum of squares between treatments:

$$SST = t \sum_i (\bar{y}_i - \bar{y})^2$$

Similarly, the second and third terms of the equation measure, respectively, the variability between rows and the variability between columns. These are designated by

$$SSR = t \sum_j (\bar{R}_j - \bar{y})^2$$

$$SSC = t \sum_k (\bar{C}_k - \bar{y})^2$$

The final source of variability, designated as the sum of squares for error (SSE), represents all additional variability in the measurements not accounted for by rows, columns, or treatments.

The simplified computational formulas useful in the AOV of a Latin square are given here.

SHORTCUT
SUMS OF
SQUARES
FORMULAS
FOR A
LATIN SQUARE
DESIGN

$$\text{TSS} = \sum_{\substack{\text{over} \\ \text{all ys}}} y_{ijk}^2 - \frac{G^2}{n}$$

$$\text{SST} = \sum_i \frac{T_i^2}{t} - \frac{G^2}{n}$$

$$\text{SSR} = \sum_j \frac{R_j^2}{t} - \frac{G^2}{n}$$

$$\text{SSC} = \sum_k \frac{C_k^2}{t} - \frac{G^2}{n}$$

$$\text{SSE} = \text{TSS} - \text{SST} - \text{SSR} - \text{SSC}$$

model   The **model** for a response in a Latin square design is the same as that for a randomized block design, with the addition of one more term to account for the second blocking variable. Thus

$$y_{ijk} = \mu + \alpha_i + \beta_j + \gamma_k + \varepsilon_{ijk}$$

where the terms are defined as follows:

y_{ijk}: The response on treatment i in row j and column k.

μ: An overall mean; μ is a constant.

α_i: An effect due to treatment i; α_i is a constant.

β_j: An effect due to row j; β_j is a constant.

γ_k: An effect due to column k; γ_k is a constant.

ε_{ijk}: A random error associated with the response for treatment i in row j and column k. We assume the ε_{ijk}s are normally distributed with mean 0 and unknown variance σ_ε^2. As before, the εs are assumed to be independent.

These assumptions for this model imply that y_{ijk}, the response for treatment i in row j and column k, is normally distributed with mean

$$E(y_{ijk}) = \mu + \alpha_i + \beta_j + \gamma_k$$

and variance σ_ε^2.

filtering   We can use the model to illustrate how a Latin square design **filters** out extraneous variability due to rows and columns. For purposes of illustration we will consider the Latin square design shown in Table 14.11. If we wish to estimate $\alpha_3 - \alpha_1$, the difference in mean response for treatments III and I, using the sample difference $\bar{y}_{III} - \bar{y}_I$, we can substitute into our model to obtain expressions for $\bar{y}_{III}$ and $\bar{y}_I$. If y_{ijk} denotes the observation in treatment i in row j and column k, we have from Table 14.11

$$\bar{y}_I = \frac{1}{4}(y_{111} + y_{142} + y_{133} + y_{124})$$

$$= \mu + \alpha_1 + \frac{1}{4}(\beta_1 + \beta_2 + \beta_3 + \beta_4) + \frac{1}{4}(\gamma_1 + \gamma_2 + \gamma_3 + \gamma_4) + \bar{\varepsilon}_1$$

where $\bar{\varepsilon}_1$ is the mean of the random errors for the 4 observations on treatment 1. Similarly,

$$\bar{y}_{III} = \frac{1}{4}(y_{331} + y_{322} + y_{313} + y_{344})$$

$$= \mu + \alpha_3 + \frac{1}{4}(\beta_1 + \beta_2 + \beta_3 + \beta_4) + \frac{1}{4}(\gamma_1 + \gamma_2 + \gamma_3 + \gamma_4) + \bar{\varepsilon}_3$$

Then the sample difference is

$$\bar{y}_{III} - \bar{y}_I = \alpha_3 - \alpha_1 + (\bar{\varepsilon}_3 - \bar{\varepsilon}_1)$$

and the error of estimation for $\alpha_3 - \alpha_1$ is $\bar{\varepsilon}_3 - \bar{\varepsilon}_1$

If a randomized block design had been used with blocks representing rows, treatments would be randomized within the rows only. It is quite possible for the same treatment to appear more than once in the same column. Then the sample difference would be

$$\bar{y}_{III} - \bar{y}_I = \alpha_3 - \alpha_1 + (\bar{\varepsilon}_3 - \bar{\varepsilon}_1) + \text{(column effects that do not cancel)}$$

Thus the error of estimation would be inflated by the column effects that do not cancel out. Following the same reasoning, if a completely randomized design was used when a Latin square design was appropriate, the error of estimation would be inflated by both row and column effects that do not cancel out.

statistical tests   We can test specific hypotheses concerning the parameters in our model. In particular, we may wish to test the hypothesis of no difference among the t treatment means. This hypothesis can be stated in the form

H_0: $\alpha_1 = \alpha_2 = \cdots = \alpha_t = 0$ (i.e., the t treatment means are identical).

The alternative hypothesis would be

H_a: At least one of the α_is differs from the rest (i.e., at least one treatment mean is different from the others).

TABLE 14.12   AOV Table for a $t \times t$ Latin Square Design

Source	SS	df	MS	F
Treatments	SST	$t - 1$	MST = SST/$(t - 1)$	MST/MSE
Rows	SSR	$t - 1$	MSR = SSR/$(t - 1)$	MSR/MSE
Columns	SSC	$t - 1$	MSC = SSC/$(t - 1)$	MSC/MSE
Error	SSE	$t^2 - 3t + 2$	MSE = SSE/$(t^2 - 3t + 2)$	
Totals	TSS	$t^2 - 1$		

The test statistic for our test would be

$$F = \frac{MST}{MSE}$$

For our model,

$$E(MSE) = \sigma_\varepsilon^2 \qquad \text{and} \qquad E(MST) = \sigma_\varepsilon^2 + t\theta_T$$

Since it can be shown that θ_T is zero under H_0 and is a positive function of the αs under H_a, we will reject H_0 in the upper tail of the F distribution. The appropriate degrees of freedom are obtained from the AOV table shown in Table 14.12.

SSE
df for SSE
We should note that we can compute **SSE** and the degrees of freedom for SSE by subtraction. Thus knowing the degrees of freedom for treatments, rows, and columns, we can subtract this sum from $t^2 - 1$ to obtain the degrees of freedom for error.

Tests similar to that for treatments can be formulated for rows and columns.

test for rows   The **test for rows** is as follows:

H_0: $\beta_1 = \beta_2 = \cdots = \beta_t = 0$ (i.e., no effect due to rows)

H_a: At least one of the βs differs from zero.

T.S.: $F = \dfrac{MSR}{MSE}$

test for columns   The **test for columns** is as follows:

H_0: $\gamma_1 = \gamma_2 = \cdots = \gamma_t = 0$ (i.e., no effect due to columns)

H_a: At least one of the γs differs from zero.

T.S.: $F = \dfrac{MSC}{MSE}$

EXAMPLE 14.4   A petroleum company was interested in comparing the miles per gallon achieved by four different gasoline blends (I, II, III, and IV). Because there can be considerable variability due to drivers and due to car models, these two extraneous sources of variability were included as "blocking" variables in the following Latin square design.

Each operator drove each car model over a standard course with the assigned gasoline blend by the Latin square design of Table 14.13. The miles per gallon data

TABLE 14.13   4 × 4 Latin Square Assignment for the Gasoline Blend Experiment of Example 14.4

Driver	Car model			
	1	2	3	4
1	IV	II	III	I
2	II	III	I	IV
3	III	I	IV	II
4	I	IV	II	III

TABLE 14.14　　Sample Data for the Gasoline Blend Study of Example 14.4

Driver	Car Model								Totals
	1		2		3		4		
1	IV	15.5	II	33.9	III	13.2	I	29.1	91.7
2	II	16.3	III	26.6	I	19.4	IV	22.8	85.1
3	III	10.8	I	31.1	IV	17.1	II	30.3	89.3
4	I	14.7	IV	34.0	II	19.7	III	21.6	90.0
Totals		57.3		125.6		69.4		103.8	356.1

are shown in Table 14.14. Use the sample data in Table 14.14 to perform an analysis of variance. Make all appropriate tests using $\alpha = .05$.

Solution

Using Table 14.14, we find the treatment totals to be

$$T_I = 94.3 \qquad T_{II} = 100.2 \qquad T_{III} = 72.2 \qquad T_{IV} = 89.4$$

Thus

$$SST = \sum_i \frac{T_i^2}{t} - \frac{G^2}{n} = \frac{(94.3)^2 + (100.2)^2 + (72.2)^2 + (89.4)^2}{4} - \frac{(356.1)^2}{16}$$

$$= \frac{32{,}137.73}{4} - 7925.45 = 108.98$$

Similarly,

$$SSR = \sum_j \frac{R_j^2}{t} - \frac{G^2}{n} = \frac{(91.7)^2 + \cdots + (90.0)^2}{4} - 7925.45$$

$$= 7931.35 - 7925.45 = 5.9$$

$$SSC = \sum_k \frac{C_k^2}{t} - \frac{G^2}{n} = \frac{(57.3)^2 + \cdots + (103.8)^2}{4} - 7925.45$$

$$= 8662.36 - 7925.45 = 736.91$$

$$TSS = \sum y_{ijk}^2 - \frac{G^2}{n} = (15.5)^2 + (33.9)^2 + \cdots + (21.6)^2 - 7925.45$$

$$= 8801.05 - 7925.45 = 875.6$$

The sum of squares for error can be found by subtraction:

$$SSE = TSS - SST - SSR - SSC = 875.6 - 108.98 - 5.9 - 736.91 = 23.81$$

The results of these calculations and F tests for treatments, rows, and columns can be summarized in an analysis of variance table, as shown in Table 14.15.

The F-value for $a = .05$, $df_1 = 3$, and $df_2 = 6$ is 4.76. Since the computed value of F for treatments and for columns exceeds 4.76, we conclude that there are significant differences among the blends and among the car models.

TABLE 14.15   AOV Table for the Data of Example 14.4

Source	SS	df	MS	F
Treatments (blends)	108.98	3	36.33	9.15
Rows (drivers)	5.90	3	1.97	0.50
Columns (car models)	736.91	3	245.64	61.87
Error	23.81	6	3.97	
Totals	875.6	15		

As with the randomized block design, we can compare the efficiency of the Latin square design to that of the completely randomized design. Let MSE_{LS} and MSE_{CR} denote the mean square errors, respectively, for a Latin square design and a completely randomized design. The **relative efficiency** is

relative efficiency

$$\frac{MSE_{CR}}{MSE_{LS}} = \frac{MSR + MSC + (t-1)MSE}{(t+1)MSE}$$

EXAMPLE 14.5   Computer output is shown for the analysis of variance of Example 14.4. Compare the output to our hand calculations.

```
                      ANALYSIS OF VARIANCE PROCEDURE

DEPENDENT
VARIABLE:         MILES
SOURCE            DF   SUM OF SQUARES  MEAN SQUARE  F VALUE    PR > F    R-SQUARE      C.V.
MODEL              9     851.79062500   94.64340278   23.85    0.0005   0.972809     8.9504
ERROR              6      23.80875000    3.96812500            ROOT MSE             MILES MEAN
CORRECTED TOTAL   15     875.599937500                        1.99201531           22.25625000

SOURCE     DF   ANOVA SS      F. VALUE   PR > F
DRIVER      3     5.89687500     0.50     0.6987
MODEL       3   736.91187500    61.90     0.0001
BLEND       3   108.98187500     9.15     0.0117
```

Solution   Except for minor rounding errors that we may have made in our hand calculations, the results of Example 14.4 agree with the output.

EXAMPLE 14.6   Refer to the data of Example 14.4. Compute the efficiency of the Latin square design relative to a completely randomized design.

Solution   For these data, $t = 4$, $MSR = 1.97$, $MSC = 245.64$, and $MSE = 3.97$. Substituting into the formula for relative efficiency, we have

$$\frac{MSE_{CR}}{MSE_{LS}} = \frac{1.97 + 245.64 + 3(3.97)}{5(3.97)} = 13.07$$

That is, it would take approximately 13 times as many observations in using a completely randomized design to gather the same amount of information on the treatment means as it would take when using the Latin square design.

EXERCISES

 14.9 An experiment was planned to compare two different fertilizer placements (broadcast band) and two different rates of fertilizer flow on watermelon yields. Recent research has shown that broadcast application (scattering over the outer area) of fertilizer is superior to bands of fertilizer applied near the seed for watermelon yields. For this experiment the investigators wished to compare two nitrogen–phosphorus–potassium (broadcast and band) fertilizers applied at a rate of 160–70–135 lb/acre and two brands of micronutrients (A and B). These four combinations were to be studied in a Latin square field plot.

The treatments were randomly assigned according to a Latin square design conducted over a large farm plot, which was divided into rows and columns. A watermelon plant dry weight was obtained for each row–column combination 30 days after the emergence of the plants. These data are shown below.

Row	Column								
	1		**2**		**3**		**4**	
1	I	1.75	III	1.43	IV	1.28	II	1.66
2	II	1.70	I	1.78	III	1.40	IV	1.31
3	IV	1.35	II	1.73	I	1.69	III	1.41
4	III	1.45	IV	1.36	II	1.65	I	1.73

Treatment I: broadcast, A   Treatment III: band, a
Treatment II: broadcast, B   Treatment IV: band, B

a. Write an appropriate statistical model for this experiment.
b. Use the data to run an analysis of variance. Give the p-value for each test and draw conclusions.

 14.10 Refer to Exercise 14.9. In addition to obtaining 30-day-emergence dry weights, the watermelon yields (in tons per acre) were also recorded after the growing season. Use the data below to conduct an analysis of variance ($\alpha = .05$). Draw conclusions.

Row	Column			
	1	**2**	**3**	**4**
1	9.5	6.8	4.9	7.1
2	7.9	9.1	6.6	5.3
3	5.6	7.6	8.7	6.7
4	7.1	5.4	6.9	8.8

14.11 Refer to the data of Exercise 14.10.

ROW	COL	TRT	YIELD
1	1	1	9.5
1	2	3	6.8
1	3	4	4.9
1	4	2	7.1
2	1	2	7.9
2	2	1	9.1
2	3	3	6.6
2	4	4	5.3
3	1	4	5.6
3	2	2	7.6
3	3	1	8.7
3	4	3	6.7
4	1	3	7.1
4	2	4	5.4
4	3	2	6.9
4	4	1	8.8

ANALYSIS OF VARIANCE PROCEDURE

DEPENDENT VARIABLE: YIELD

SOURCE	DF	SUM OF SQUARES	MEAN SQUARE	F VALUE	PR > F	R-SQUARE	C.V.
MODEL	9	29.77000000	3.30777778	248.08	0.0001	0.997320	1.6206
ERROR	6	0.08000000	0.01333333		ROOT MSE		YIELD MEAN
CORRECTED TOTAL	15	29.85000000			0.11547005		7.12500000

SOURCE	DF	ANOVA SS	F VALUE	PR > F
ROW	3	0.07500000	1.88	0.2347
COL	3	1.26000000	31.50	0.0005
TRT	3	28.43500000	710.88	0.0001

a. Locate the analysis of variance in the output shown here.

b. Compare your results to those of Exercise 14.10.

14.12   Refer to Table 14.13. Derive the error of estimation for estimating $\alpha_4 - \alpha_2$ using $\bar{y}_{IV} - \bar{y}_{II}$.

14.5   FACTORIAL EXPERIMENTS

In Section 14.2 we reviewed the completely randomized design for comparing $t \geq 2$ treatment means. Sections 14.3 and 14.4 were devoted to a discussion of the randomized block design and the Latin square design. Both designs are extensions of the completely randomized design; the randomized block design allows one to compare t treatment means in the presence of one extraneous (nuisance) source of variability, whereas the Latin square design allows the experimenter to control two such sources in order to compare treatment means. Suppose now that, rather than comparing t treatment means, we wish to examine the effects of two or more independent variables on a response y. For example, suppose that we want to examine the effects of temperature x_1 and pressure x_2 on the bond strength y of a new adhesive product. Two major problems arise. First, we must consider the number of levels and the actual settings of these levels for each independent variable (factor). Second, having chosen the levels for each factor, we must choose the factor-level combinations (treatments) that will be applied to the experimental units.

The ability to choose appropriate settings for the independent variables depends a great deal on the experimenter's knowledge of the physical situation under study. Then, assuming the experimenter has chosen the levels of each independent variable, he or she faces the task of deciding which factor-level combinations should be assigned to the experimental units. For purposes of illustration, suppose that an experimenter is interested in examining the effects of two independent variables, nitrogen and phosphorus, on the yield of a crop. For simplicity we will assume that two levels have been selected for the study of each factor: 40 and 60 pounds per plot for nitrogen, 10 and 20 pounds per plot for phosphorus. For this study the experimental units are small, relatively homogeneous plots that have been partitioned from the acreage of a farm.

one-at-a-time approach   One approach suggested for examining the effects of two or more factors on a response is called the **one-at-a-time approach**. To examine the effect of a single variable, an experimenter varies the levels of this variable while holding the levels of the other independent variables fixed. This process is continued until the effect of each variable on the response has been examined while holding the other independent variables constant. For our experiment the factor-level combinations chosen might be as shown in Table 14.16. These factor-level combinations are illustrated in Figure 14.1.

From the graph in Figure 14.1 we see that there is one difference that can be used to measure the effects of nitrogen and phosphorus separately. The difference in response for combinations 1 and 2 would estimate the effect of nitrogen; the difference in response for combinations 2 and 3 would estimate the effect of phosphorus.

TABLE 14.16   Factor-Level Combinations for a One-at-a-Time Approach

Combination	Nitrogen	Phosphorus
1	60	10
2	40	10
3	40	20

FIGURE 14.1

Factor-Level Combinations for a One-at-a-Time Approach

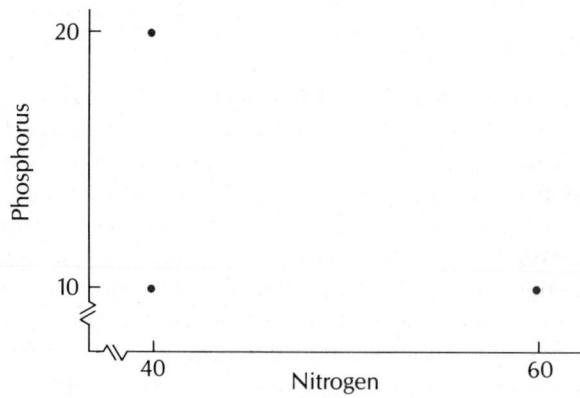

TABLE 14.17   Yields for the Three Factor-Level Combinations

Observation (yield)	Nitrogen	Phosphorus
145	60	10
125	40	10
160	40	20
?	60	20

Hypothetical yields corresponding to the three factor-level combinations of our experiment are given in Table 14.17. Suppose the experimenter is interested in using the sample information to determine the factor-level combination that will give the maximum yield. From the table we see that crop yield increases when the nitrogen application is increased from 40 to 60 (holding phosphorus at 10). Yield also increases when the phosphorus setting is changed from 10 to 20 (at a fixed nitrogen setting of 40). Thus it might seem logical to predict that increasing both the nitrogen and phosphorus applications to the soil will result in a larger crop yield. The fallacy in this argument is that our prediction is based on the assumption that the effect of one factor is the same for both levels of the other factor.

We know from our investigation what happens to yield when the nitrogen application is increased from 40 to 60 for a phosphorus setting of 10. But will the yield also increase by approximately 20 units when the nitrogen application is changed from 40 to 60 at a setting of 20 for phosphorus?

To answer this question we could apply the factor-level combination of 60 nitrogen–20 phosphorus to another experimental plot and observe the crop yield. If the yield is 180, then the information obtained from the three factor-level combinations would be correct and would have been useful in predicting the factor-level combination that produces the greatest yield. However, suppose the yield obtained from the high settings of nitrogen and phosphorus turns out to be 110. If this happens, the two factors nitrogen and phosphorus **interaction**   are said to **interact**. That is, the effect of one factor on the response does not remain the same for different levels of the second factor, and the information obtained from the one-at-a-time approach would lead to a faulty prediction.

The two outcomes just discussed for the crop yield at 60–20 setting is displayed in Figure 14.2, along with the yields at the three initial design points. Figure 14.2 (top) illustrates a situation with no interaction between the two factors. The effect of nitrogen on yield is the same for both levels of phosphorus. In contrast, Figure 14.2 (bottom) illustrates a case in which the two factors nitrogen and phosphorus do interact.

We have seen that the one-at-a-time approach to investigating the effect of two factors on a response is suitable only for situations in which the two factors do not interact. Although this was illustrated for the simple case in which two factors were to be investigated at each of two levels, the inadequacies of a one-at-a-time approach are even more salient when trying to investigate the effects of more than two factors on a response.

factorial experiment   **Factorial experiments** are useful for examining the effects of two or more factors on a response y, whether or not interaction exists. As before, the choice of the number of levels of each variable and the actual settings of these variables is important. But assuming

FIGURE 14.2

Yields of the
Three Design Points
and Possible Yield at
a Fourth Design Point

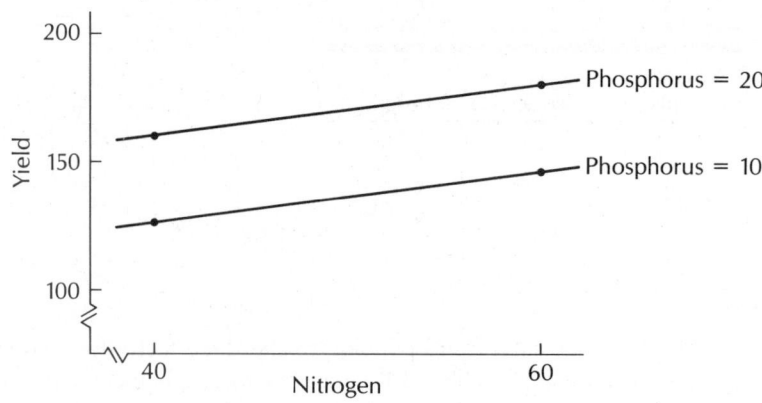

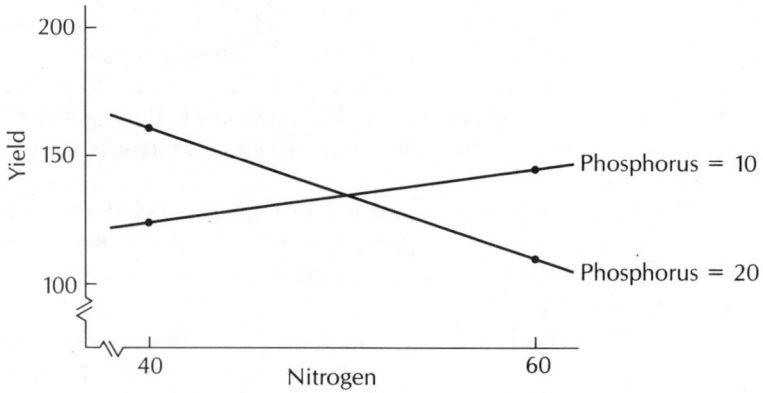

we have made these selections with help from an investigator knowledgeable in the area being examined, we must decide at what factor-level combinations we will observe y.

Classically, factorial experiments have not been referred to as designs, because they deal with the choice of levels and the selection of factor-level combinations (treatments) rather than with how the treatments are assigned to experimental units. Unless otherwise specified, we will assume that treatments are assigned to experimental units at random, and we will refer to a factorial experiment as an experimental design.

DEFINITION 14.3

A **factorial experiment** is an experiment in which the response y is observed at all factor-level combinations of the independent variables.

Using our previous example, if we are interested in examining the effect of two levels of nitrogen x_1 at 40 and 60 pounds per plot and two levels of phosphorus x_2 at 10 and 20 pounds per plot on the yield of a crop, we must decide how to prepare plots to observe yield. A 2×2 factorial experiment for this example is shown in Table 14.18. The four factor-level combinations are assigned at random to the experimental units.

TABLE 14.18 2 × 2 Factorial Experiment for Crop Yield

Factor-Level x_1	Combinations x_2
40	10
40	20
60	10
60	20

Similarly, if we wished to examine x_1 at the two levels 40 and 60 and x_2 at the three levels 10, 15, and 20, the 2 × 3 factorial experiment would have the factor-level combinations shown in Table 14.19.

EXAMPLE 14.7

An auto manufacturer is interested in examining the effect of engine speed x_1, measured in revolutions per minute, and ground speed x_2, measured in miles per hour, on gasoline mileage. The investigators, in consultation with company mechanics and other personnel, decided to consider settings of x_1 at 800, 1000, and 1200 and settings of x_2 at 30, 50, and 70. Give the factor-level combinations to be used in a 3 × 3 factorial experiment.

Solution

Using the definition of factorial experiment, we would observe gasoline mileage at the following settings of x_1 and x_2:

x_1	800	800	800	1000	1000	1000	1200	1200	1200
x_2	30	50	70	30	50	70	30	50	70

The examples of factorial experiments presented in this section have concerned two independent variables. However, the procedure applies to any number of factors and levels per factor. Thus if we had four different factors x_1, x_2, x_3, and x_4 at two, three, three, and

TABLE 14.19 2 × 3 Factorial Experiment for Crop Yield

Factor-Level x_1	Combinations x_2
40	10
40	15
40	20
60	10
60	15
60	20

four levels, respectively, we could formulate a $2 \times 3 \times 3 \times 4$ factorial experiment by considering all $2 \cdot 3 \cdot 3 \cdot 4 = 72$ factor-level combinations.

One final comparison should be made between the one-at-a-time approach and a factorial experiment. Not only do we get information concerning factor interactions using a factorial experiment, but also, when there are no interactions, we get the same amount of information about the effects of each individual factor using fewer observations. To illustrate this idea let us consider the 2×2 factorial experiment with nitrogen and phosphorus. If there is no interaction between the two factors, the data appear as shown in Figure 14.3(a). For convenience, the data are reproduced in Table 14.20, with the four sample

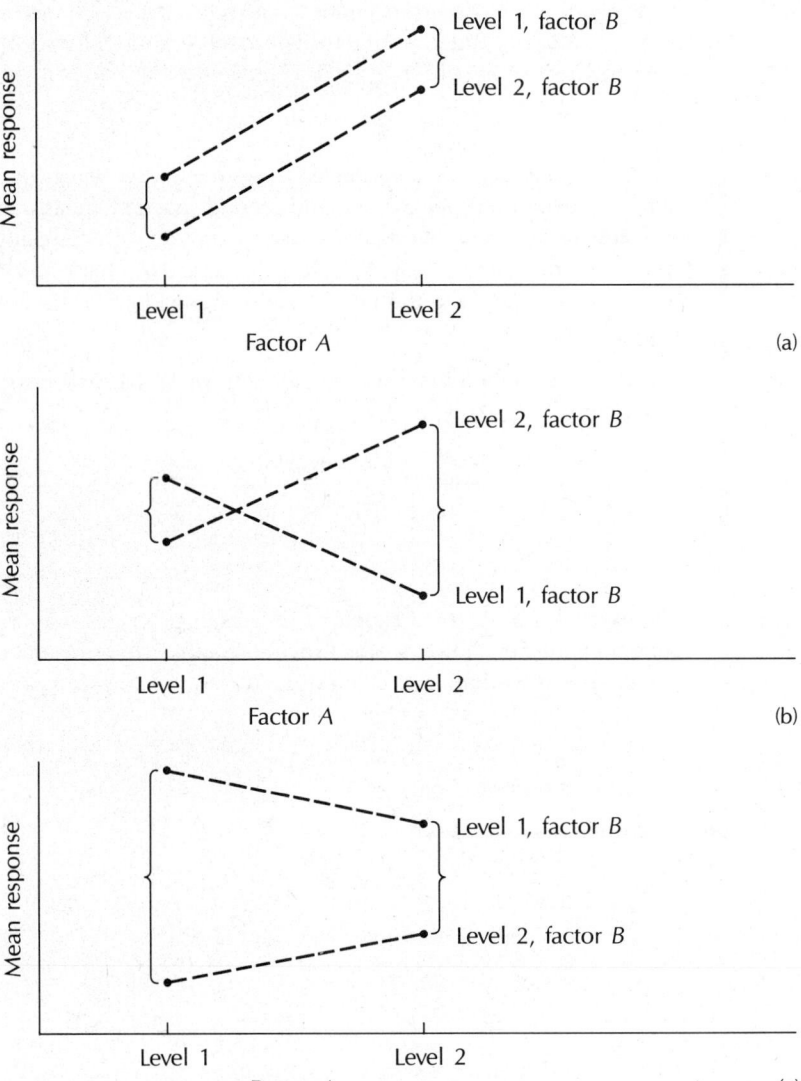

FIGURE 14.3

Illustrations of the Absence and Presence of Interaction in a 2×2 Factorial Experiment; (a) factors A and B do not interact; (b) factors A and B interact; (c) factors A and B interact.

TABLE 14.20   Factor-Level Combinations for a 2 × 2 Factorial Experiment

Combination	Yield	Nitrogen	Phosphorus
1	145	60	10
2	125	40	10
3	165	40	20
4	180	60	20

combinations designated by the numbers 1 through 4. If a 2 × 2 factorial experiment is used and no interaction exists between the two factors, we can obtain two independent differences to examine the effects of each of the factors on the response. Thus from Table 14.20, the differences between observations 1 and 4 and the difference between observations 2 and 3 would be used to measure the effect of phosphorus. Similarly, the difference between observations 4 and 3 and the difference between observations 2 and 1 would be used to measure the effect of the two levels of nitrogen on plot yield.

If we employed a one-at-a-time approach for the same experimental situation, it would take six observations (two observations at each of the three initial factor-level combinations shown in Table 14.20) to obtain the same number of independent differences for examining the separate effects of nitrogen and phosphorus when no interaction is present.

The model for a two-factor factorial experiment, such as the 2 × 2 factorial experiment with nitrogen and phosphorus, is

$$y_{ij} = \mu + \alpha_i + \beta_j + \varepsilon_{ij}$$

with the usual assumptions for the constants α_i and β_j and the random error ε_{ij}. Here y_{ij} denotes an observation taken at the ith level of factor A and the jth level of factor B. Since terms included in the model are added to one another, the model is sometimes **additive model**   referred to as an **additive model**. Expected values for a 2 × 2 factorial experiment are shown in Table 14.21

This model assumes that difference in population means (expected values) for any two levels of factor A is the same no matter what level of B we're considering. The same property holds when comparing two levels of factor B. For example, the difference in mean response for levels 1 and 2 of factor A is the same value, $\alpha_1 - \alpha_2$, no matter what level of factor B we are considering. Thus a test for no differences among the two levels of factor A would be of the form $H_0: \alpha_1 - \alpha_2 = 0$. Similarly, the difference between levels of

TABLE 14.21   Expected Values for a 2 × 2 Factorial Experiment

	Factor B	
Factor A	Level 1	Level 2
Level 1	$\mu + \alpha_1 + \beta_1$	$\mu + \alpha_1 + \beta_2$
Level 2	$\mu + \alpha_2 + \beta_1$	$\mu + \alpha_2 + \beta_2$

TABLE 14.22   Expected Values for a 2 × 2 Factorial Experiment, with
Replications*

	Factor B	
Factor A	Level 1	Level 2
Level 1	$\mu + \alpha_1 + \beta_1 + \alpha\beta_{11}$	$\mu + \alpha_1 + \beta_2 + \alpha\beta_{12}$
Level 2	$\mu + \alpha_2 + \beta_1 + \alpha\beta_{21}$	$\mu + \alpha_2 + \beta_2 + \alpha\beta_{22}$

factor B is $\beta_1 - \beta_2$ for either level of factor A, and a test of no difference between the factor B means is $H_0: \beta_1 - \beta_2 = 0$. This phenomenon was also noted for the randomized block design.

interaction   If the assumption of additivity of terms in the model does not hold, then we may need another model that employs terms to account for **interaction**. Consider a two-factor factorial experiment with r observations (sometimes called r replications) per cell. For this experimental situation we can write the model

$$y_{ijk} = \mu + \alpha_i + \beta_j + \alpha\beta_{ij} + \varepsilon_{ijk}$$

where y_{ijk} is the response obtained for the kth observation at the jth level of factor A and jth level of factor B and $\alpha\beta_{ij}^*$ is an effect due to the ith level of A and ith level of B. The expected values for a 2 × 2 factorial experiment with r observations per cell are presented in Table 14.22.

As can be seen from Table 14.22, the difference in mean response for levels 1 and 2 of factor A on level 1 of factor B is

$$(\alpha_1 - \alpha_2) + (\alpha\beta_{11} - \alpha\beta_{21}),$$

but for level 2 of factor B this difference is

$$(\alpha_1 - \alpha_2) + (\alpha\beta_{12} - \alpha\beta_{22})$$

Since the difference in mean response for levels 1 and 2 of factor A is *not* the same for different levels of factor B, the model is no longer additive, and we say that the two factors **interaction** **interact**.

DEFINITION 14.4

Two factors A and B are said to **interact** if the difference in mean responses for two levels of one factor is not constant across levels of the second factor.

In measuring the octane rating of gasoline, interaction can occur when two components of the blend are combined to form a gasoline mixture. The octane properties of the

---

* It is convenient to use the notation $\alpha\beta_{ij}$ to denote a new term for the model. It does not, however, represent the multiplication of two terms α_i and β_j.

blended mixture may be quite different than would be expected by examining each component of the mixture. Interaction in this situation could have a positive or negative effect on the performance of the blend, in which case the components are said to potentiate or antagonize one another.

profile plot   We can amplify the notion of an interaction with the **profile plots** shown in Figure 14.3. As we see from Figure 14.3(a), when no interaction is present, the difference in the mean response between levels 1 and 2 of factor B (as indicated by the braces) is the same for both levels of factor A. However, for the two illustrations in Figures 14.3(b) and (c), we see that the difference between the levels of factor B changes from level 1 to level 2 of factor A. For these cases we have an interaction between the two factors.

It should be noted that an interaction is not restricted to two factors. With three factors A, B, and C, we might have an interaction between factors A and B, A and C, and B and C, and the two-factor interactions would have interpretations that follow immediately from Definition 14.4. Thus the presence of an AC interaction indicates that the difference in mean response for levels of factor A varies across levels of factor C. A three-way interaction between factors A, B, and C might indicate that the difference in mean response for levels of C changes across combinations of levels for factors A and B.

In general, the analysis of variance table for a k-factor factorial experiment depends on whether or not we have $r > 1$ replications per cell. Before presenting these tables, we need the notation defined here.

NOTATION NEEDED FOR THE AOV OF A k-FACTOR FACTORIAL EXPERIMENT WITH $r > 1$ REPLICATIONS PER CELL

n_A: number of observations at each level of factor A

A_i: sum of the n_A observations receiving the ith level of factor A ($i = 1, 2, \ldots, a$)

B_j: sum of the n_B observations receiving the jth level of factor B ($j = 1, 2, \ldots, b$)

C_k: sum of the n_C observations receiving the kth level of factor C ($k = 1, 2, \ldots c$)

$\vdots$

n_{AB}: number of observations at each combination of levels of factors A and B

$(AB)_{ij}$: sum of the n_{AB} observations receiving the ith level of A and jth level of B

$(AC)_{ik}$: sum of the n_{AC} observations receiving the ith level of A and kth level of C

$(BC)_{jk}$: sum of the n_{BC} observations receiving the jth level of B and kth level of C

$\vdots$

n_{ABC}: number of observations at each combination of levels of the three factors A, B, and C

$(ABC)_{ijk}$: sum of the n_{ABC} observations receiving the ith level of A, jth level of B, and kth level of C

$\vdots$

Partitioning sums of squares　　The appropriate AOV formulas for a k-factor factorial experiment with r observations per cell can be subdivided into sums of squares for main effects (variability between levels of a single factor), two-way interactions, three-way interactions, and so on.

main effects　　The sums of squares for **main effects** are

$$\text{SSA} = n_A \sum_i (\bar{A}_i - \bar{y})^2 = \sum_i \frac{A_i^2}{n_A} - \frac{G^2}{n}$$

$$\text{SSB} = n_B \sum_j (\bar{B}_j - \bar{y})^2 = \sum_j \frac{B_j^2}{n_B} - \frac{G^2}{n}$$

$$\text{SSC} = n_C \sum_k (\bar{C}_k - \bar{y})^2 = \sum_k \frac{C_k^2}{n_C} - \frac{G^2}{n}$$

and so on.

two-way interactions　　The sums of squares for **two-way interactions** are

$$\text{SSAB} = n_{AB} \sum_{ij} (\overline{AB}_{ij} - \bar{A}_i - \bar{B}_j + \bar{y})^2 = \sum_{ij} \frac{(AB)_{ij}^2}{n_{AB}} - \text{SSA} - \text{SSB} - \frac{G^2}{n}$$

$$\text{SSAC} = n_{AC} \sum_{ik} (\overline{AC}_{ik} - \bar{A}_i - \bar{C}_k + \bar{y})^2 = \sum_{ik} \frac{(AC)_{ik}^2}{n_{AC}} - \text{SSA} - \text{SSC} - \frac{G^2}{n}$$

$$\text{SSBC} = n_{BC} \sum_{jk} (\overline{BC}_{jk} - \bar{B}_j - \bar{C}_k + \bar{y})^2 = \sum_{jk} \frac{(BC)_{jk}^2}{n_{BC}} - \text{SSB} - \text{SSC} - \frac{G^2}{n}$$

and so on.

three-way interactions　　The sums of squares for **three-way interactions** are

$$\text{SSABC} = n_{ABC} \sum_{ijk} (\overline{ABC}_{ijk} - \overline{AB}_{ij} - \overline{AC}_{ik} - \overline{BC}_{jk} + \bar{A}_i + \bar{B}_j + \bar{C}_k - \bar{y})^2$$

$$= \sum_{ijk} \frac{(ABC)_{ijk}^2}{n_{ABC}} - \text{SSAB} - \text{SSAC} - \text{SSBC} - \text{SSA} - \text{SSB} - \text{SSC} - \frac{G^2}{n}$$

and so on.

You will note that all the main effects formulas are identical except for interchanging of the appropriate letters. Similar substitutions apply to the interaction sums of squares. And, in general, if we wished to compute the sums of squares associated with $ABC \ldots$ interaction, we would use the formula

$$\text{SSABC} \ldots = \sum_{ijk\ldots} \frac{(ABC \ldots)_{ijk}^2}{n_{ABC\ldots}} - \begin{bmatrix} \text{sums of squares associated} \\ \text{with \textit{all} subcombinations} \\ \text{of the factors } A, B, C, \ldots \end{bmatrix} - \frac{G^2}{n}$$

We illustrate the AOV table for a three-factor factorial experiment (i.e., $k = 3$), with a levels of A, b levels of B, c levels of C, and $r = 1$ observation per cell. The AOV table is given in Table 14.23.

The total sum of squares formula is as before:

$$\text{TSS} = \sum y^2 - \frac{G^2}{n}$$

TABLE 14.23   AOV for a Three-Factor Factorial Experiment Without Replication

Source	SS	df	MS
Main effects			
A	SSA	$a - 1$	$MSA = SSA/(a - 1)$
B	SSB	$b - 1$	$MSB = SSB/(b - 1)$
C	SSC	$c - 1$	$MSC = SSC/(c - 1)$
Interactions			
AB	SSAB	$(a - 1)(b - 1)$	$MSAB = SSAB/(a - 1)(b - 1)$
AC	SSAC	$(a - 1)(c - 1)$	$MSAC = SSAC/(a - 1)(c - 1)$
BC	SSBC	$(b - 1)(c - 1)$	$MSBC = SSBC/(b - 1)(c - 1)$
ABC	SSABC	$(a - 1)(b - 1)(c - 1)$	$MSABC = SSABC/(a - 1)(b - 1)(c - 1)$
Totals	TSS	$abc - 1$	

You will note, however, that there is no source of variability designated as error since there are no more degrees of freedom. This is true for any factorial with $r = 1$ observation per cell.

F tests   To construct **F tests** for the sources of variability listed in the AOV table, we must do one of two things:

1. Assume that one or more of the higher-order interactions are negligible. The affected sums of squares are then combined to form an error sum of squares. (Note that this is why we illustrated the 2×2 factorial [with $r = 1$] using an additive model.)

2. Replicate the experiment to generate additional degrees of freedom for error.

EXAMPLE 14.8   An experiment was conducted to determine the effects of four different pesticides on the yield of fruit from three different varieties (B_1, B_2, B_3) of a citrus tree. Four trees from each variety were randomly selected from an orchard. The four pesticides were then randomly assigned to trees of a particular variety and applications were made according to recommended levels. Yields of fruit, in bushels per tree, were obtained after the test period. These data appear in Table 14.24. Set up an analysis of variance table, computing all the sums of squares and the mean squares. Use $\alpha = .05$ for all F tests.

TABLE 14.24   Data for Example 14.8; Yield of Fruit (in Bushels per Tree)

Variety, B	Pesticide, A				
	1	2	3	4	Totals
1	29	50	43	53	175
2	41	58	42	73	214
3	66	85	69	85	305
Totals	136	193	154	211	694

Solution   The experiment just described is a 3×4 factorial with one observation per cell. Factor A, pesticides, is investigated at $a = 4$ levels and factor B, varieties, at $b = 3$ levels. The sources of variability and the corresponding degrees of freedom are shown below.

Source	df
Pesticide, A	3
Variety, B	2
AB	6
Total	11

Using the AOV formulas, we have, from Table 14.24

$$n = ab = 12 \qquad n_A = 3 \qquad n_B = 4$$

$$\frac{G^2}{n} = \frac{(694)^2}{12} = 40{,}136.33$$

$$SSA = \sum_i \frac{A_i^2}{n_A} - \frac{G^2}{n}$$

$$= \frac{(136)^2 + (193)^2 + (154)^2 + (211)^2}{3} - 40{,}136.33$$

$$= 41{,}327.33 - 40{,}136.33 = 1191$$

$$SSB = \sum_j \frac{B_j^2}{n_B} - \frac{G^2}{n} = \frac{(175)^2 + (214)^2 + (305)^2}{4} - 40{,}136.33$$

$$= 42{,}361.5 - 40{,}136.33 = 2225.17$$

$$TSS = \sum y^2 - \frac{G^2}{n} = (29)^2 + (50)^2 + \cdots + (85)^2 - 40{,}136.33$$

$$= 43{,}704 - 40{,}136.33 = 3567.67.$$

The formula for the AB interaction is

$$SSAB = \sum_{ij} \frac{(AB)_{ij}^2}{n_{AB}} - SSA - SSB - \frac{G^2}{n}$$

For $n_{AB} = 1$ observation at each AB combination, we have

$$\sum_{ij} \frac{(AB)_{ij}^2}{n_{AB}} = (29)^2 + (50)^2 + \cdots + (85)^2 = 43{,}704$$

Then

$$SSAB = 43{,}704 - 1191 - 2225.17 - 40{,}136.33 = 151.50$$

This same result could be obtained by subtraction as

$$SSAB = TSS - SSA - SSB = 3567.67 - 1191 - 2225.17 = 151.50$$

The analysis of variance table is then as shown in Table 14.25.

TABLE 14.25   AOV Table for the Fruit Yield Experiment of
Example 14.8

Source	SS	df	MS
Pesticide, A	1191.00	3	397
Varieties, B	2225.17	2	1112.59
AB	151.50	6	25.25
Totals	3567.67	11	

As can be seen from this AOV table, we have no degrees of freedom remaining for error.

If we are willing to assume that the AB interaction is negligible, we can designate the AB interaction source as error and use SSAB as SSE. For our data, if we are willing to assume that the difference in mean yield for the two pesticides remains constant for all varieties, the AOV table would be as shown in Table 14.26.

To test for no difference among pesticides, with $\alpha = .05$, we use the test statistic

$$F = \frac{MSA}{MSE} = \frac{397.00}{25.25} = 15.72$$

Since the computed value of F exceeds 4.76, the tabulated value for $a = .05$, $df_1 = 3$, and $df_2 = 6$, we reject H_0 and conclude that there are differences among the mean yields for the four pesticides.

Similarly, to test for no difference among varieties, we use

$$F = \frac{MSB}{MSE} = 44.06$$

Based on $\alpha = .05$, the computed value of F exceeds the tabulated value of 5.14 for $a = .05$, $df_1 = 2$, and $df_2 = 6$, and we conclude that there are differences among the mean yields for varieties.

EXAMPLE 14.9   Refer to Example 14.8. Suppose that before beginning the study, the experimenter realized that it cannot be assumed that the AB interaction was negligible in order to perform F tests in an analysis of variance; so the experimenter decided to use 24 citrus trees (eight of each variety) and randomly assigned two trees to each factor-level combination.

TABLE 14.26   AOV Table for the Fruit Yield Experiment of Example 14.8, with
the AB Interaction Designated As Error

Source	SS	df	MS	F
A	1191.00	3	397.00	$397/25.25 = 15.72$
B	2225.17	2	1112.59	$1112.59/25.25 = 44.06$
Error	151.50	6	25.25	
Totals	3567.67	11		

TABLE 14.27   Data for the 3 × 4 Factorial Experiment of Fruit Yield, Example 14.9 with $r = 2$ Observations per Cell

	Pesticide, A				
Variety, B	1	2	3	4	Totals
1	49	50	43	53	375
	39	55	38	48	
2	55	67	53	85	474
	41	58	42	73	
3	66	85	69	85	626
	68	92	62	99	
Totals	318	407	307	443	1475

a. Construct a profile plot of the data.
b. Write an appropriate model.
c. Use the yield data of Table 14.27 to conduct an analysis of variance. Use $\alpha = .05$ for all F tests.

Solution

a. To construct a profile plot we need the $\bar{y}_{ij}s$, the sample means for the factor-level combinations of A and B. These are given in Table 14.28.

TABLE 14.28   Sample Means for Factor-Level Combinations of A and B

	Pesticide, A			
Variety, B	1	2	3	4
1	44	52.5	40.5	50.5
2	48	62.5	47.5	79
3	67	88.5	65.5	92

FIGURE 14.4

Profile Plot for Example 14.9

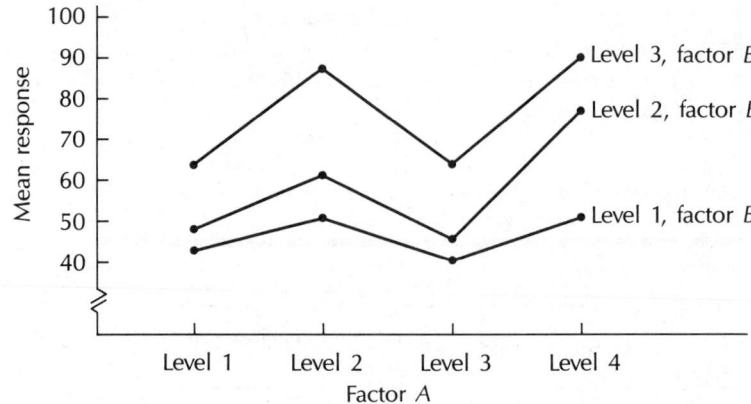

TABLE 14.29 $(AB)_{ij}$ Totals for the Data of Table 14.27

	Pesticide, A				
Variety, B	1	2	3	4	Totals
1	88	105	81	101	375
2	96	125	95	158	474
3	134	177	131	184	626
Totals	318	407	307	443	1475

Using these sample means it is easy to form the profile plot shown in Figure 14.4.

b. The profile plot of part (a) does not indicate much evidence of an interaction between factors A and B since the lines are essentially parallel. However, since we have two observations per factor-level combination, we'll include interaction terms in the model and run the test for interaction as part of the analysis of variance. The model is

$$y_{ijk} = \mu + \alpha_i + \beta_j + \alpha\beta_{ij} + \varepsilon_{ijk}$$

c. The sums of squares for an analysis of variance make use of the data totals shown in Tables 14.27 and 14.29.

$$SSA = \sum_i \frac{A_i^2}{n_A} - \frac{G^2}{n} = \frac{(318)^2 + (407)^2 + (307)^2 + (443)^2}{6} - 90{,}651.04$$

$$= 92{,}878.5 - 90{,}651.04 = 2227.46$$

$$SSB = \sum_j \frac{B_j^2}{n_B} - \frac{G^2}{n} = \frac{(375)^2 + (474)^2 + (626)^2}{8} - 90{,}651.04$$

$$= 94{,}647.13 - 90{,}651.04 = 3996.09$$

To compute the AB interaction, we must calculate the totals $(AB)_{ij}$ for all cells of the table. These are given in Table 14.29.

Using the data of Table 14.29 we compute SSAB.

$$SSAB = \sum_{ij} \frac{(AB)_{ij}^2}{n_{AB}} - SSA - SSB - \frac{G^2}{n}$$

$$= \frac{(88)^2 + (105)^2 + \cdots + (184)^2}{2} - 2227.46 - 3996.09 - 90{,}651.04$$

$$= 97{,}331.5 - 2227.46 - 3996.09 - 90{,}651.04 = 456.91$$

From Table 14.27, the total sum of squares is computed as

$$TSS = \sum y^2 - \frac{G^2}{n} = (49)^2 + (39)^2 + (50)^2 + \cdots + (85)^2 + (99)^2 - 90{,}651.04$$

$$= 97{,}839 - 90{,}651.04 = 7187.96$$

TABLE 14.30   AOV Table for the Fruit Yield Experiment of Example 14.9

Source	SS	df	MS	F
A	2227.46	3	742.49	$742.49/42.29 = 17.56$
B	3996.09	2	1998.05	$1998.05/42.29 = 47.25$
AB	456.91	6	76.15	$76.15/42.29 = 1.80$
Error	507.50	12	42.29	

The sum of squares and degrees of freedom for error can be obtained by subtraction:

$$SSE = TSS - SSA - SSB - SSAB$$
$$= 7187.96 - 2227.46 - 3996.09 - 456.91 = 507.50$$
$$df = 23 - 3 - 2 - 6 = 12$$

The analysis of variance table for this 3×4 factorial experiment with two observations per cell is given in Table 14.30.

The first test of significance would be to test for no interaction between factors A and B. The F statistic is

$$F = \frac{MSAB}{MSE} = \frac{76.15}{42.29} = 1.80$$

The computed value of F does not exceed the tabulated value of 3.00 for $a = .05$, $df_1 = 6$, and $df_2 = 12$. Hence we have insufficient evidence to indicate an interaction between A and B; this confirms what we saw in the profile plot. We then proceed as in the AOV table and test separately for differences among pesticides and among varieties. Comparing the computed F-values to the appropriate critical value from the tables, we find a significant effect due to both factors. The patterns of responses are displayed in the profile plot.

The results of an F test for main effects for A or B must be interpreted very care-

significant interaction   fully in the presence of a **significant interaction** The first thing we would do is to con-

FIGURE 14.5

Significant Interaction
Present; Tests on
Main Effects Are
Appropriate

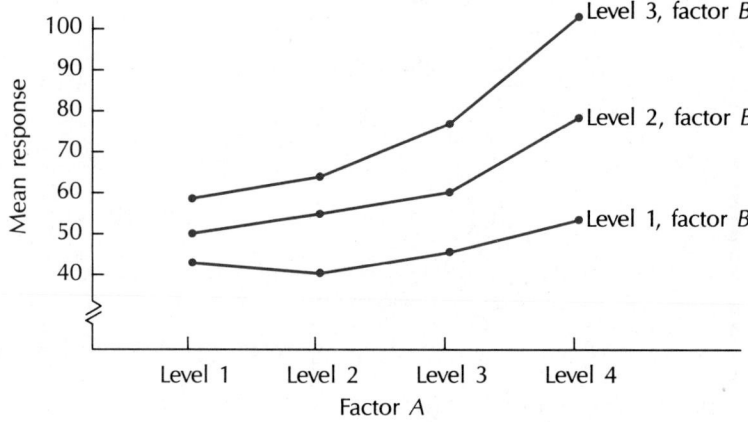

FIGURE 14.6

Significant Interaction
Present; Tests on
Main Effects Are
Inappropriate

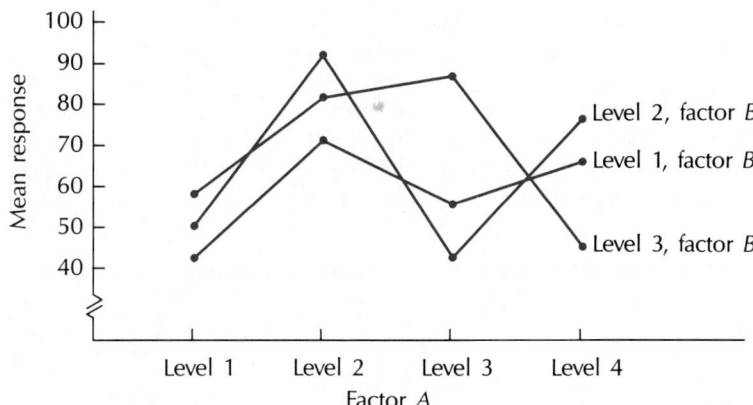

struct a profile plot using the sample means. If the profile plot for Example 14.9 had appeared as shown in Figure 14.5 there would have been an indication of an interaction between factors A and B (pesticides and varieties). Indeed, the F test for interaction would undoubtedly have been significant.

Would F tests for main effects have been appropriate for the profile plot of Figure 14.5? The answer is yes because this is an orderly interaction; the *order* of the means for levels of factor B is always the same even though the *magnitude* of the differences between levels of factor B may change from level to level of factor A. Clearly, the profile plot in Figure 14.5 shows that the level 3 mean of factor B is always larger than the means for levels 2 and 1. Similarly, the level 2 mean for B is always larger than the mean for level 1, factor B.

When the interaction is orderly, a test on main effects can be meaningful; however, not all interactions are so well behaved. The profile in Figure 14.6 shows a situation in which a test of main effects in the presence of interaction would be inappropriate.

EXERCISES

 14.13   A national trade association asked relocation experts from four different companies to estimate the total transfer costs for the same 10 hypothetical employees in order to generate data for input into proposed revisions of corporate tax structures. These data ($000) are shown here:

Company	Employee									
	1	2	3	4	5	6	7	8	9	10
1	76	93	109	74	54	95	64	92	141	87
2	75	95	102	68	52	90	65	96	120	76
3	80	98	110	82	67	88	80	101	133	91
4	76	95	108	75	64	98	76	97	136	93

a. Identify the design.
b. Write the model, identifying terms.
c. Give the analysis of variance table without performing the calculations.

14.14   Compute the sums of squares for the analysis of variance table of Exercise 14.13. Perform appropriate F tests and draw conclusions. Can a test be made for the interaction effect?

14.15   SAS output for the data for Exercise 14.13 is shown here. Compare the AOV from SAS to the one you obtained in Exercise 14.14.

RANDOMIZED BLOCK DESIGN

ANALYSIS OF VARIANCE PROCEDURE

DEPENDENT VARIABLE: DOLLAR

SOURCE	DF	SUM OF SQUARES	MEAN SQUARE	F VALUE	PR > F	R-SQUARE	C.V.
MODEL	12	16065.80000000	1338.81666667	67.11	0.0001	0.967563	5.0015
ERROR	27	538.60000000	19.94814815		ROOT MSE		DOLLAR MEAN
CORRECTED TOTAL	39	16604.40000000			4.46633498		89.30000000

SOURCE	DF	ANOVA SS	F VALUE	PR > F
COMPANY	3	497.40000000	8.31	0.0004
EMPLOYEE	9	15568.40000000	86.72	0.0001

14.6   FACTORIAL EXPERIMENTS COMBINED WITH BLOCKING DESIGNS

In Section 14.5, we defined a factorial experiment to be an experiment in which the response y is observed at all factor-level combinations of the independent variables. The major difference between the block designs (randomized block and Latin square) and factorial designs lies in their design objectives. With the randomized block design, an experimenter wishes to block out (or filter) variability due to a nuisance variable so that precise comparisons may be made among the treatments. Similarly, a Latin square design filters out variability due to two nuisance variables in order to obtain precise comparisons among the treatments of interest. In contrast, the factorial experiment is used to investigate the effect(s) of k factors (independent variables) on the response.

The differences in design objectives lead to different methods of randomization of treatments to experimental units. In the randomized block design, treatments are applied at random to experimental units within a block, with the stipulation that each treatment appears on time in a block. Treatments for a Latin square can be assigned at random across columns within a row, subject to the restriction that each treatment appears in every row and every column. Treatments for a factorial experiment correspond to factor-level combinations. Treatments are then assigned at random to the experimental units, with the only restriction being that each factor-level combination must appear the same number of times.

Not all designs can be classified as either a block design or a factorial experiment. Sometimes the objectives of a study are such that we wish to investigate the effects of certain factors on a response while blocking out certain other extraneous sources of variability. Such experiments may require an experimental design that is a combination of a block design and a factorial experiment. This can be illustrated with the following example.

EXAMPLE 14.10

An investigator wants to examine the effects of two factors (each measured at three levels) on a response y. It is determined that $r = 2$ observations are desired at each factor-level combination, but only nine observations can be done each day. Propose an appropriate experimental design.

Solution

Since nine observations can be obtained each day, it would be possible to run a complete replication of the 3×3 factorial experiment on two different days to get the desired number of observations. The design is shown here

Day 1				Day 2			
Factor B				Factor B			
Factor A	1	2	3	Factor A	1	2	3
1				1			
2				2			
3				3			

It should be noted that this design is really a randomized block design where the blocks are days and the treatments are the nine factor-level combinations of the 3×3 factorial experiment. So, with the randomized block design, we are able to block or filter out the variability due to the nuisance variable days while comparing the treatments. Since the treatments are factor-level combinations from a factorial experiment, we can examine the effects of the two factors (A and B) on the response while filtering out the day-to-day variability.

The analysis of variance for this design follows from our discussions in Sections 14.2 and 14.4.

EXAMPLE 14.11

Construct an analysis of variance table identifying the sources of variability and the degrees of freedom for the 3×3 factorial experiment laid off in a randomized block design with $b = 2$ discussed in Example 14.10.

Solution

The analysis of variance table for a randomized block design with $t = 9$ and $b = 2$ is shown here:

Source	SS	df
Treatments	SST	8
Blocks	SSB	1
Error	SSE	8
Total	TSS	17

Since the treatments of this randomized block are the nine factor-level combinations of a 3×3 factorial experiment, we can subdivide the sum of squares treatment (SST) into the sources of variability for a 3×3 factorial experiment from Section 14.5. The revised AOV table is shown here.

Source	SS	df
Treatments	SST	8
A	SSA	2
B	SSB	2
AB	SSAB	4
Blocks	SSB	1
Error	SSE	8
Total	TSS	17

So, rather than running an overall test comparing the treatement means using $F = MST/MSE$, we could conduct the analysis of variance for a factorial experiment in order to examine the interaction and mean effects. These F tests would use the appropriate numerator mean squares (MSAB, MSA, and MSB) and MSE from this analysis.

EXERCISES

14.16   Diagram a design that has a 3×5 factorial experiment laid off in a randomized block design with $b = 3$ blocks. Give the complete analysis of variance table (sources, SSs, dfs).

14.17   Diagram a design which has a $2 \times 4 \times 3$ factorial experiment laid off in a randomized block design with $b = 2$ blocks. Give the complete analysis of variance for this experimental design.

14.7   THE ESTIMATION OF TREATMENT DIFFERENCES AND MULTIPLE COMPARISONS

We have emphasized the analysis of variance associated with a randomized block design, a Latin square design, and factorial experiments. There are times, however, when we might be more interested in estimating the difference in mean response for two treatments (different levels of the same factor or different combinations of levels). For example, an environmental engineer might be more interested in estimating the difference in the mean dissolved oxygen content for a lake before and after rehabilitative work than in testing to see whether there is a difference. Thus the engineer is asking the question, "What is the difference in mean dissolved oxygen content?" instead of the question, "Is there a difference between the mean content before and after the cleanup project?"

The same formula can be used to estimate the difference in treatment means for a randomized block design, a Latin square design, and k-factor factorial experiment. Let $\bar{y}_i$ denote the mean response for treatment i, $\bar{y}_{i'}$, denote the mean response for treatment i', and n_t denote the number of observations in each treatment. A $100(1 - \alpha)\%$ confidence interval on $\mu_i - \mu_{i'}$, the difference in mean response for the two treatments, is defined as shown here.

100(1 − α)% CONFIDENCE INTERVAL FOR THE DIFFERENCE IN TREATMENT MEANS	$$(\bar{y}_i - \bar{y}_{i'}) \pm t_{\alpha/2} s_\varepsilon \sqrt{\frac{2}{n_t}}$$ where s_ε is the square root of MSE in the AOV table and $t_{\alpha/2}$ can be obtained from Table 4 in the Appendix for a = $\alpha/2$ and the degrees of freedom for MSE.

EXAMPLE 14.12

A company was interested in comparing three different display panels for use by air traffic controllers. Each display panel was to be examined under five different simulated emergency conditions. Thirty highly trained air traffic controllers with similar work experience were enlisted for the study. A random assignment of controllers to display panel–emergency conditions was made, with two controllers assigned to each factor-level combination. The time (in seconds) required to stabilize the emergency situation was recorded for each controller at a panel-emergency condition. These data appear in Table 14.31.

a. Construct a profile plot.

b. Run an analysis of variance that includes a test for interaction.

Solution

a. The sample means are shown in Figure 14.7.

	Emergency Condition, A				
Display Panel, B	1	2	3	4	5
1	17	33	24.5	37.5	13.5
2	14	31.5	22.5	36.5	13
3	26	44	38.5	54.5	26

TABLE 14.31   Display Panel Data for Example 14.12 (Time in Seconds)

Display Panel, B	Emergency Condition, A					Totals
	1	2	3	4	5	
1	18 16	31 35	22 27	39 36	15 12	251
2	13 15	33 30	24 21	35 38	10 16	235
3	24 28	42 46	40 37	52 57	28 24	378
Totals	114	217	171	257	105	864

FIGURE 14.7

Plot of Panel
Means for Each
Emergency Condition

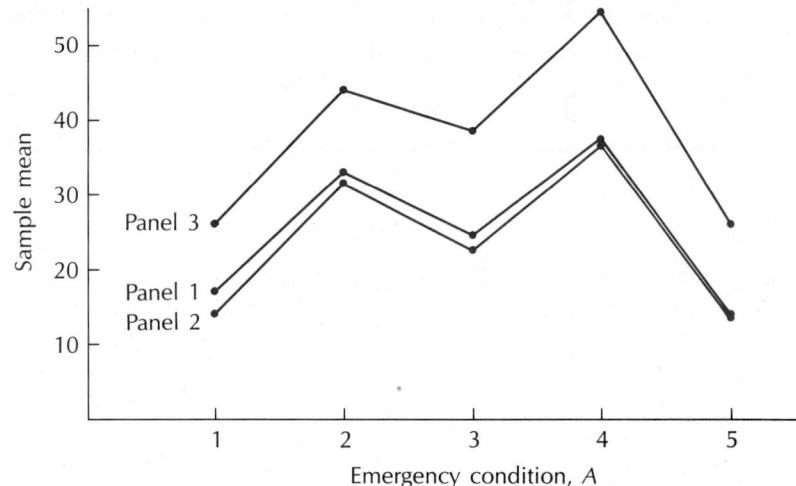

b. Substituting into the appropriate sums of squares formulas and using the data from Table 14.31, we have

$$\frac{G^2}{n} = \frac{(864)^2}{30} = 24{,}883.2$$

$$SSA = \sum_i \frac{A_i^2}{n_A} - \frac{G^2}{n} = \frac{(114)^2 + (217)^2 + \cdots + (105)^2}{6} - 24{,}883.2$$

$$= 27{,}733.33 - 24{,}883.2 = 2850.13$$

$$SSB = \sum_j \frac{B_j^2}{n_B} - \frac{G^2}{n} = \frac{(251)^2 + (235)^2 + (378)^2}{10} - 24{,}883.2$$

$$= 26{,}111 - 24{,}883.2 = 1227.8$$

Combining the two observations in each cell, we obtain a table of $(AB)_{ij}$ totals, as shown in Table 14.32.

TABLE 14.32 $(AB)_{ij}$ Totals for the Data of Table 14.31

Display Panel, B	Emergency Condition, A					Totals
	1	2	3	4	5	
1	34	66	49	75	27	251
2	28	63	45	73	26	235
3	52	88	77	109	52	378
Totals	114	217	171	257	105	864

Substituting into the formula for SSAB, we have

$$SSAB = \sum_{ij} \frac{(AB)_{ij}^2}{n_{AB}} - SSA - SSB - \frac{G^2}{n}$$

$$= \frac{(34)^2 + (66)^2 + \cdots + (52)^2}{2} - 2850.13 - 1227.8 - 24{,}883.2$$

$$= 29{,}006 - 2850.13 - 1227.8 - 24{,}883.2 = 44.87$$

The total sum of squares can be computed using the original observations in Table 14.31.

$$TSS = \sum y^2 - \frac{G^2}{n}$$

$$= (18)^2 + (16)^2 + (31)^2 + \cdots + (28)^2 + (24)^2 - 24{,}883.2$$

$$= 29{,}112 - 24{,}883.2 = 4228.8$$

By subtraction we find SSE to be

$$SSE = TSS - SSA - SSB - SSAB$$

$$= 4228.8 - 2850.13 - 1227.8 - 44.87 = 106$$

We can summarize our computations in an AOV table, Table 14.33.

The first test of significance should be a test for interaction between factors A and B. Since the computed value of F, .79, is less than the critical value of F, 2.64, for $a = .05$, $df_1 = 8$, and $df_2 = 15$, we have insufficient evidence to indicate an interaction between emergency conditions and display panels. As indicated by the profile plot, the difference in mean reaction time for controllers on two different display panels appears constant for all five emergency conditions.

EXAMPLE 14.13   Refer to Example 14.12. Since there is no interaction, estimate the difference in mean response (reaction time) for display panels 2 and 3 using a 95% confidence interval.

Solution   For these data

$$\bar{y}_3 = \frac{378}{10} = 37.8 \quad \text{and} \quad \bar{y}_2 = \frac{235}{10} = 23.5$$

TABLE 14.33   AOV Table for the Data of Table 14.31

Source	SS	df	MS	F
Emergency conditions A	2850.13	4	712.53	712.53/7.07 = 100.78
Display panel B	1227.8	2	613.9	613.9/7.07 = 86.83
AB	44.87	8	5.61	5.61/7.07 = .79
Error	106	15	7.07	
Totals	4228.8	29		

The t-value for a $= .025$ and df $= 23$ is 2.069; the estimate of σ_ε is

$$s_\varepsilon = \sqrt{\text{MSE}} = 2.66$$

The appropriate confidence interval is

$$\bar{y}_3 - \bar{y}_2 \pm t_{\alpha/2} s_\varepsilon \sqrt{\frac{2}{n_t}}$$

$$37.8 - 23.5 \pm 2.069(2.66) \sqrt{\frac{2}{10}}$$

$$14.3 \pm 2.46$$

or 11.84 to 16.76.

Assume now that we have chosen analysis of variance rather than estimation as the inferential technique for answering our practical questions. We proceed to compute the appropriate sums of squares and perform F tests on the sources of variability. Having rejected a hypothesis of the form "no difference in mean response for levels of factor A," do we stop and draw no further conclusion? No, we do not, because the F test is only a preliminary test to determine if there are any differences among the treatment means. We must then try to determine which levels of the factor differ from the rest.

As mentioned in Chapter 11, we could perform multiple t tests among the factor-level means, but the overall error rate could be quite large. Presumably we would proceed with one of the **multiple comparison procedures**, such as Tukey's, Duncan's, or Scheffé's, with a controlled error rate. All these procedures can be performed following an analysis of variance for a randomized block design, a Latin square design, or a k-factorial experiment. The quantity s_w^2 is replaced by MSE in the various formulas, and the sample size n refers to the number of observations per mean in the comparison. The degrees of freedom for MSE are obtained from the AOV table.

Using multiple comparison procedure

EXAMPLE 14.14

Refer to Example 14.12 and the data in Table 14.31. Use Tukey's W procedure to locate significant differences among display panels.

Solution

For Tukey's W procedure we must compute

$$W = q_\alpha(t, v) \sqrt{\frac{s_w^2}{n}}$$

where s_w^2 is MSE from the AOV table, based on $v = 15$ degrees of freedom, and $q_\alpha(t, v)$ is the upper-tail critical value of the Studentized range (with a $= \alpha$) for comparing t different population means. The value of $q_\alpha(t, v)$ from Table 10 in the Appendix for comparing the three display panel means is

$$q_{.05}(3, 15) = 3.67$$

For $n = 10$ observations per mean,

$$W = q_\alpha(t, v) \sqrt{\frac{s_w^2}{n}} = 3.67 \sqrt{\frac{7.07}{10}} = 3.09$$

The display panel means are, from Table 14.31,

$$\bar{B}_1 = \frac{251}{10} = 25.1 \qquad \bar{B}_2 = \frac{235}{10} = 23.5 \qquad \bar{B}_3 = \frac{378}{10} = 37.8$$

First we rank the sample means from lowest to highest:

Display panel	2	1	3
Means	23.5	25.1	37.8

For two means that differ (in absolute value) by more than $W = 3.09$, we declare them to be significantly different from each other. The results of our multiple comparison procedure are summarized below.

Display panel   <u>2   1</u>   3

Thus display panels 1 and 2 both have a mean reaction time significantly lower than display panel 3, but we are unable to detect any difference between panels 1 and 2.

14.8   RELATIONSHIP BETWEEN REGRESSION AND ANALYSIS OF VARIANCE (optional)

The link between regression methods and analysis of variance procedures is provided by the general linear model. In Section 12.1 we showed how dummy variables could be used to describe a response y as a function of a qualitative independent variable using a general linear model. Recall that the model for a completely randomized design (with t treatments) could be written as

$$y = \beta_0 + \beta_1 x_1 + \beta_2 x_2 + \cdots + \beta_{t-1} x_{t-1} + \varepsilon$$

where

$$x_1 = 1 \text{ if treatment 2, } x_1 = 0 \text{ otherwise}$$
$$x_2 = 1 \text{ if treatment 3, } x_2 = 0 \text{ otherwise}$$
$$\vdots$$
$$x_{t-1} = 1 \text{ if treatment } t, x_{t-1} = 0 \text{ otherwise}$$

For this model we have

$$\beta_0 = \mu_1$$
$$\beta_1 = \mu_2 - \mu_1$$
$$\beta_2 = \mu_3 - \mu_1$$
$$\vdots$$
$$\beta_{t-1} = \mu_t - \mu_1$$

We also indicated in Chapter 12 that the general linear model could be used to relate y to more than one qualitative independent variable. For a randomized block design, we have a response related to two qualitative independent variables: blocks and treatments.

TABLE 14.34   Design for $b = 3$ Blocks and $t = 4$
Treatments, Example 14.15

Blocks	Treatments			
1	I	III	II	IV
2	II	IV	III	I
3	IV	I	II	III

Applying our previous results, if there are b blocks and t treatments, we will enter $(b - 1)$ βs for blocks and $(t - 1)$ βs for treatments into our model.

EXAMPLE 14.15   Write a model for a randomized block design to compare $t = 4$ treatments in $b = 3$ blocks. The design is shown in Table 14.34.

Solution   The model can be written in the form

$$y = \beta_0 + \overbrace{\beta_1 x_1 + \beta_2 x_2}^{\text{blocks}} + \overbrace{\beta_3 x_3 + \beta_4 x_4 + \beta_5 x_5}^{\text{treatments}} + \varepsilon$$

where

$x_1 = 1$ if block 2 $x_1 = 0$ otherwise

$x_2 = 1$ if block 3 $x_2 = 0$ otherwise

$x_3 = 1$ if treatment II $x_3 = 0$ otherwise

$x_4 = 1$ if treatment III $x_4 = 0$ otherwise

$x_5 = 1$ if treatment IV $x_5 = 0$ otherwise

We can easily interpret the βs using a table of expected values. To obtain the expected value of an observation for a given block-treatment combination, we substitute appropriate values for $x_1, x_2, \ldots, x_5$ in the formula

$$E(y) = \beta_0 + \beta_1 x_1 + \beta_2 x_2 + \cdots + \beta_5 x_5$$

For example, the expected value for an observation in block 2 on treatment II would have $x_1 = 1$, $x_2 = 0$, $x_3 = 1$, $x_4 = 0$, and $x_5 = 0$, giving

$$E(y) = \beta_0 + \beta_1(1) + \beta_2(0) + \beta_3(1) + \beta_4(0) + \beta_5(0) = \beta_0 + \beta_1 + \beta_3$$

The table of expected values is given in Table 14.35.

TABLE 14.35   Table of Expected Values for a Randomized Block Design, with $b = 3$ and $t = 4$, Example 14.15

Block	Treatment			
	I	II	III	IV
1	β_0	$\beta_0 + \beta_3$	$\beta_0 + \beta_4$	$\beta_0 + \beta_5$
2	$\beta_0 + \beta_1$	$\beta_0 + \beta_1 + \beta_3$	$\beta_0 + \beta_1 + \beta_4$	$\beta_0 + \beta_1 + \beta_5$
3	$\beta_0 + \beta_2$	$\beta_0 + \beta_2 + \beta_3$	$\beta_0 + \beta_2 + \beta_4$	$\beta_0 + \beta_2 + \beta_5$

The mean response for block 1 and treatment I is β_0. If we consider any treatment and compare blocks 2 and 1, the difference in the mean response is β_1. For example, when using treatment II, the difference in the mean response for blocks 2 and 1 is, from Table 14.35,

$$(\beta_0 + \beta_1 + \beta_3) - (\beta_0 + \beta_3) = \beta_1$$

Note that this is true for any treatment. Hence $\beta_1 = \mu_2 - \mu_1$, the difference in the mean response for blocks 2 and 1. Similarly, in comparing blocks 3 and 1 for a given treatment, the difference is $\beta_2 = \mu_3 - \mu_1$.

In the same way, we can compare two treatments for a given block. For example it follows immediately that $\beta_3 = \mu_{II} - \mu_I$, the difference in mean response for treatments II and I. Similarly, in block 3 the difference in mean response for treatments II and I is

$$(\beta_0 + \beta_2 + \beta_3) - (\beta_0 + \beta_2) = \beta_3$$

In the same fashion, we can show that

$$\beta_4 = \mu_{III} - \mu_I$$
$$\beta_5 = \mu_{IV} - \mu_I$$

These results can easily be extended to a Latin square design.

EXAMPLE 14.16

Nylon is spun on a series of machines. When a break in the nylon thread occurs during the spinning process, the machine operator must stop the machine and rethread the nylon prior to continuing. Investigators would like to compare the output of three different spinning machines (I, II, III) using three different operators (A, B, C). To control day-to-day variability in addition to operator variability, it was decided to use a 3 × 3 Latin square design, as shown in Table 14.36. Write an appropriate model for this Latin square design, using dummy variables. Interpret the βs for the model.

Solution

The model for this Latin square design must relate a response to three qualitative independent variables, rows (operators), columns (days), and treatments (machines). For each qualitative variable, we will include two βs.

$$y = \beta_0 + \overbrace{\beta_1 x_1 + \beta_2 x_2}^{\text{operators}} + \overbrace{\beta_3 x_3 + \beta_4 x_4}^{\text{days}} + \overbrace{\beta_5 x_5 + \beta_6 x_6}^{\text{machines}} + \varepsilon$$

TABLE 14.36   Latin Square Design for the Spinning Machines Experiment, Example 14.16

		Day	
Operator	1	2	3
A	I	II	III
B	II	III	I
C	III	I	II

where

$x_1 = 1$ if operator B $x_1 = 0$ otherwise

$x_2 = 1$ if operator C $x_2 = 0$ otherwise

$x_3 = 1$ if day 2 $x_3 = 0$ otherwise

$x_4 = 1$ if day 3 $x_4 = 0$ otherwise

$x_5 = 1$ if machine II $x_5 = 0$ otherwise

$x_6 = 1$ if machine III $x_6 = 0$ otherwise

For this model, with $E(\varepsilon) = 0$, we can write the expected value of y as

$$E(y) = \beta_0 + \beta_1 x_1 + \beta_2 x_2 + \beta_3 x_3 + \beta_4 x_4 + \beta_5 x_5 + \beta_6 x_6$$

By substituting appropriate values for the dummy variables $x_1, x_2, \ldots, x_6$, we obtain the table of expected values, Table 14.37.

Although it is readily apparent that β_0 represents the mean response for machine I run by operator A during day 1, it is more difficult to identify the other βs. Since each machine appears in each row (operator) and each column (day), the βs due to rows and columns will be eliminated when comparing two machines. If we average the three expected values for observations on machine I and the three expected values from Table 14.37 for machine II, we have

$$\text{average for machine I} = \frac{1}{3} [\beta_0 + (\beta_0 + \beta_2 + \beta_3) + (\beta_0 + \beta_1 + \beta_4)]$$

$$= \beta_0 + \frac{1}{3}(\beta_1 + \beta_2 + \beta_3 + \beta_4)$$

$$\text{average for machine II} = \frac{1}{3} [(\beta_0 + \beta_1 + \beta_5) + (\beta_0 + \beta_3 + \beta_5)$$

$$+ (\beta_0 + \beta_2 + \beta_4 + \beta_5)]$$

$$= \beta_0 + \beta_5 + \frac{1}{3}(\beta_1 + \beta_2 + \beta_3 + \beta_4)$$

The difference in these averages represents $\mu_{II} - \mu_{I}$. Subtracting, we have

$$\beta_5 = \mu_{II} - \mu_{I}$$

TABLE 14.37 Expected Values for the 3 × 3 Latin Square Design of the Spinning Machines Experiment, Example 14.16

		Day		
Operator		1	2	3
A	I β_0		II $\beta_0 + \beta_3 + \beta_5$	III $\beta_0 + \beta_4 + \beta_6$
B	II $\beta_0 + \beta_1 + \beta_5$		III $\beta_0 + \beta_1 + \beta_3 + \beta_6$	I $\beta_0 + \beta_1 + \beta_4$
C	III $\beta_0 + \beta_2 + \beta_6$		I $\beta_0 + \beta_2 + \beta_3$	II $\beta_0 + \beta_2 + \beta_4 + \beta_5$

Similarly, by comparing machine III to machine I, we can show that

$$\beta_6 = \mu_{III} - \mu_I$$

The same reasoning can be used to obtain

$$\beta_1 = \mu_B - \mu_A$$
$$\beta_2 = \mu_C - \mu_A$$
$$\beta_3 = \mu_2 - \mu_1$$
$$\beta_4 = \mu_3 - \mu_1$$

Now that the βs have been identified, these interpretations are exactly what we might have imagined, having examined the completely randomized design and a randomized block design.

main effects terms   The models that we have written in this section have had only **main effects terms** (terms involving only one x) for each of the qualitative independent variables. But it is also **interaction terms**   possible to write models containing **interaction terms**: terms involving the product of xs between two or more variables.

factorial experiment   Consider a 2×3 **factorial experiment** in which an experimenter would like to compare three different diet preparations (A, B, C) under two different diet plans. A fixed number of overweight persons would be assigned to each of the six factor-level combinations. Since it is possible that the difference in mean response (weight loss) for two different diet preparations is not the same for each diet plan, we must include interaction terms as well as main effects in our models. An appropriate model is given by

model

$$y = \beta_0 + \overbrace{\beta_1 x_1 + \beta_2 x_2 + \beta_3 x_3}^{\text{main effects}} + \overbrace{\beta_4 x_1 x_2 + \beta_5 x_1 x_3}^{\text{interaction}} + \varepsilon$$

where

$$x_1 = 1 \text{ if plan } 2 \qquad x_1 = 0 \text{ otherwise}$$
$$x_2 = 1 \text{ if diet B} \qquad x_2 = 0 \text{ otherwise}$$
$$x_3 = 1 \text{ if diet C} \qquad x_3 = 0 \text{ otherwise}$$

The main effects terms in our model are

$\beta_1 x_1$ for diet plans
$\beta_2 x_2$ and $\beta_3 x_3$ for diet preparations

Interaction terms are formed from cross-products of the xs involved in main effects. Thus from the product of x_1 and x_2, we have the interaction term $\beta_4 x_1 x_2$. Similarly, we have $\beta_5 x_1 x_3$ from the product of x_1 and x_3. It might also appear that we should have an interaction term involving the product of x_2 with x_3. However, *interaction terms are always formed by products of x values from different variables*. Intuitively this makes sense, since an interaction measures how two (or more) variables react when combined. Hence no term involving levels of the same variable could contribute towards measuring this effect.

To interpret the βs associated with our model, we again form a table of expected values by substituting appropriate combinations for x_1, x_2, and x_3 into the general formula

$$E(y) = \beta_0 + \beta_1 x_1 + \beta_2 x_2 + \beta_3 x_3 + \beta_4 x_1 x_2 + \beta_5 x_1 x_3$$

For example, the expected response on plan 2 and diet B can be found by substituting $x_1 = 1$, $x_2 = 1$, and $x_3 = 0$ into $E(y)$.

$$E(y) = \beta_0 + \beta_1(1) + \beta_2(1) + \beta_3(0) + \beta_4(1)(1) + \beta_5(1)(0)$$
$$= \beta_0 + \beta_1 + \beta_2 + \beta_4$$

The table of expected values is given in Table 14.38.

From Table 14.38 we have the following interpretations for main effects:

β_0: mean response for plan 1, diet A

β_1: difference in mean response for plans 2 and 1 on diet A

β_2: difference in mean response for diets B and A on plan 1

β_3: difference in mean response for diets C and A on plan 1

The interaction βs are slightly more complicated to interpret since they measure the failure of diet preparations to have the same effect across different diet plans. Using the first two columns of Table 14.38, we can find β_4 as the sum of the expected values in the diagonal left-to-right direction minus the sum of the expected values in the diagonal right-to-left direction.

	A	B
1	β_0	$\beta_0 + \beta_2$
2	$\beta_0 + \beta_1$	$\beta_0 + \beta_1 + \beta_2 + \beta_4$

$$[\beta_0 + (\beta_0 + \beta_1 + \beta_2 + \beta_4)] - [(\beta_0 + \beta_2) + (\beta_0 + \beta_1)] = \beta_4$$

Similarly, taking the first and third columns of Table 14.38 and subtracting the right-to-left sum from the left-to-right sum, we have

$$[\beta_0 + (\beta_0 + \beta_1 + \beta_3 + \beta_5)] - [(\beta_0 + \beta_3) + (\beta_0 + \beta_1)] = \beta_5$$

no interaction   We should note that if there were **no interaction** between the variables "diet plan" and "diet preparation," the parameters β_4 and β_5 would both equal zero. Using Table 14.38, with $\beta_4 = \beta_5 = 0$, we have the following interpretations for the parameters in the

TABLE 14.38   Expected Values of a 2 × 3 Factorial Experiment

Diet Plan	Diet Preparation		
	A	B	C
1	β_0	$\beta_0 + \beta_2$	$\beta_0 + \beta_3$
2	$\beta_0 + \beta_1$	$\beta_0 + \beta_1 + \beta_2 + \beta_4$	$\beta_0 + \beta_1 + \beta_3 + \beta_5$

reduced model:

$$y = \beta_0 + \beta_1 x_1 + \beta_2 x_2 + \beta_3 x_3 + \varepsilon$$

β_0: mean response for diet A, plan 1

$\beta_1 = \mu_2 - \mu_1$: the difference in mean response for plans 2 and 1

$\beta_2 = \mu_B - \mu_A$: the difference in mean response for diets B and A

$\beta_3 = \mu_C - \mu_A$: the difference in mean response for diets C and A

Note that in the absence of interaction, the interpretation of a main effect term for one variable does not depend on the level of another variable. For example, with interaction present,

$$\beta_1 = \mu_2 - \mu_1 \text{ on diet A (that is, the difference in mean response for plans 2 and}$$
1 while on diet A)

and without interaction,

$$\beta_1 = \mu_2 - \mu_1 \text{ (that is, the difference in mean response for plans 2 and 1 for any}$$
diet preparation)

EXAMPLE 14.17   Refer to the experiment discussed above, which involved the investigation of diet plans and diet preparations. Write a complete model (indicate main effects and interactions) for an experiment to compare four diet preparations using three diet plans.

Solution   The model must relate a response to two qualitative independent variables: diet plans and diet preparations. The variable "diet plan" will have $3 - 1 = 2$ main effects terms; the variable "diet preparation" will have $4 - 1 = 3$ main effects terms, and there will be $2(3) = 6$ interaction terms formed from products of x values, one from each of the two variables. The model can be written as follows:

$$y = \beta_0 + \overbrace{\beta_1 x_1 + \beta_2 x_2}^{\text{diet plans}} + \overbrace{\beta_3 x_3 + \beta_4 x_4 + \beta_5 x_5}^{\text{diet preparations}}$$

$$+ \overbrace{\beta_6 x_1 x_3 + \beta_7 x_1 x_4 + \beta_8 x_1 x_5 + \beta_9 x_2 x_3 + \beta_{10} x_2 x_4 + \beta_{11} x_2 x_5}^{\text{interaction terms}} + \varepsilon$$

where

$x_1 = 1$ if plan 2 $x_1 = 0$ otherwise

$x_2 = 1$ if plan 3 $x_2 = 0$ otherwise

$x_3 = 1$ if diet B $x_3 = 0$ otherwise

$x_4 = 1$ if diet C $x_4 = 0$ otherwise

$x_5 = 1$ if diet D $x_5 = 0$ otherwise

Being able to write models for regression problems *and* models for analysis of variance problems using a general linear model means that there is a link between regression and analysis of variance methods. We have already seen in previous chapters that we can make tests concerning a set of βs in a regression model by fitting complete and reduced models (see Chapter 12). The same procedure can also be done to make the required F tests for an analysis of variance if the model is written as a general linear model. The study described in Example 14.8 can be used to illustrate this fact.

The study design described in Example 14.8 is a 3×4 factorial involving three varieties and four pesticides. The no-interaction model, written in the form of a general linear model, is

$$y = \beta_0 + \beta_1 x_1 + \beta_2 x_2 + \beta_3 x_3 + \beta_4 x_4 + \beta_5 x_5 + \varepsilon$$

where

$$\text{varieties} \begin{cases} x_1 = 1 \text{ if variety 2} \\ \quad = 0 \text{ otherwise} \\ x_2 = 1 \text{ if variety 3} \\ \quad = 0 \text{ otherwise} \end{cases} \quad \text{pesticides} \begin{cases} x_3 = 1 \text{ if pesticide 2} \\ \quad = 0 \text{ otherwise} \\ x_4 = 1 \text{ if pesticide 3} \\ \quad = 0 \text{ otherwise} \\ x_5 = 1 \text{ if pesticide 4} \\ \quad = 0 \text{ otherwise} \end{cases}$$

The hypothesis of no difference among the pesticide means, $H_0: \beta_3 = \beta_4 = \beta_5 = 0$, is the same as that specified for an analysis of variance F test on pesticides. In addition the sum of squares drop (SS_{drop}) obtained by fitting models 1 and 2 is identical to the sum of squares due to pesticides of Example 14.8.

$$\text{model 1:} \quad y = \beta_0 + \overbrace{\beta_1 x_1 + \beta_2 x_2}^{\text{varieties}} + \overbrace{\beta_3 x_3 + \beta_4 x_4 + \beta_5 x_5}^{\text{pesticides}} + \varepsilon$$
$$SSE_1 = 151.50$$

$$\text{model 2:} \quad y = \beta_0 + \beta_1 x_1 + \beta_2 x_2 + \varepsilon$$
$$SSE_2 = 1342.50$$

$$SS_{drop}(\text{due to pesticides}) = SSE_2 - SSE_1 = 1191.00$$

In the same way, the sum of squares drop obtained for testing no difference among the variety means $(H_0: \beta_1 = \beta_2 = 0)$ is identical to the sum of squares due to varieties in the analysis of variance of Example 14.8. Here

$$\text{model 1:} \quad y = \beta_0 + \overbrace{\beta_1 x_1 + \beta_2 x_2}^{\text{varieties}} + \overbrace{\beta_3 x_3 + \beta_4 x_4 + \beta_5 x_5}^{\text{pesticides}} + \varepsilon$$
$$SSE_1 = 151.50$$

$$\text{model 2:} \quad y = \beta_0 + \beta_3 x_3 + \beta_4 x_4 + \beta_5 x_5 + \varepsilon$$
$$SSE_2 = 2376.67$$

Although we have not fit model 2, it can be shown that the sum of squares for error is as given here. Also,

$$SS_{drop}(\text{due to varieties}) = SSE_2 - SSE_1 = 2225.17$$

Note that SSE_1, the sum of squares for error for the complete model, is SSE from the AOV table of Example 14.8.

The method of obtaining sums of squares (SS_{drop}) due to various sources of variability by fitting complete and reduced models can be used for *any* experimental design. All we need to do is to specify an appropriate general linear model. Then by hypothesizing that various parameters in the model are zero, we can obtain SS_{drop} and run an appropriate *F* test. The *F* test is identical to that used in a standard analysis of variance. *In fact, as we have developed this section, we see that an analysis of variance consists of testing hypotheses concerning parameters in the general linear model.*

Why don't we fit complete and reduced models every time we conduct an analysis of variance? While this would certainly be possible, it is not practical. The reason is that many designs are **balanced designs**.

balanced design

DEFINITION 14.5 ▪

> A **balanced design** has each level of one independent variable appearing the same number of times with each level of another independent variable, and this is true for all pairs of independent variables.

For many balanced designs it is possible to obtain shortcut computational formulas for calculating the sum of squares drop for a particular source of variability. For example, by Definition 14.5, it is clear that the factorial experiment, Latin square, randomized block, and completely randomized designs are all balanced designs. As indicated in this chapter, we have shortcut formulas for calculating the sum of squares associated with the various sources of variability in an AOV table. These formulas are simplified expressions for calculating SS_{drop} by fitting a complete and reduced model.

At this point you might be concerned with identifying an unbalanced design. An example of an unbalanced design would be any balanced design with one or more missing observations. Suppose an experimenter was interested in comparing the drop in potency for three different concentrations of a drug product stored at three different temperatures. Assume further that two bottles were to be stored for six weeks at each factor-level combination. Since final potency determinations must be made after the six-week period in order to compute the drop in potency, it is possible that one or more bottles would be broken. If such an accident did occur, we would have an unequal number of observations at the different factor-level combinations, making the design *unbalanced*. The shortcut formulas developed earlier in this chapter would no longer be appropriate for computing sums of squares for sources in the AOV table. While the general procedure for fitting complete and reduced models can be used, we will delay any further discussion of analyzing unbalanced designs until Chapter 15.

14.9   SUMMARY

In this chapter we have extended our discussion of the analysis of variance presented in Chapter 10. We did this by considering several basic experimental designs. The completely randomized design, the randomized block design, and the Latin square design illustrate how we can block out undesirable background variability to obtain more precise comparisons among treatment means. In contrast, the factorial experiment is useful in investigating the effect of one or more factors on a response.

Not all designs can be classified either as a block design or as a factorial experiment. Some designs represent combinations of block designs with factorial experiments. Thus an experimenter may wish to examine the effects of two or more factors on a response while blocking out one or more extraneous sources of variability.

For each design discussed in this chapter we presented a description of the design layout (including arrangement of treatments), potential advantages and disadvantages, a model, and the analysis of variance. Finally, we discussed how one could make comparisons between pairs of treatment means for these designs.

The reader should recognize that the designs presented are only the most basic ones and so far we have only dealt with the situation in which we had a *balanced design*, that is, an equal number of observations per treatment or factor-level combination. Chapter 15 extends the results of this chapter to some unbalanced (and perhaps more practical) situations.

KEY FORMULAS: ANALYSIS OF VARIANCE

1. Randomized block design

$$\text{TSS} = \sum_{ij} (y_{ij} - \bar{y})^2 = \sum_{ij} y_{ij}^2 - \frac{G^2}{n}$$

$$\text{SST} = b \sum_i (\bar{y}_i - \bar{y})^2 = \sum_i \frac{T_i^2}{b} - \frac{G^2}{n}$$

$$\text{SSB} = t \sum_j (\bar{B}_j - \bar{y})^2 = \sum_j \frac{B_j^2}{t} - \frac{G^2}{n}$$

$$\text{SSE} = \text{TSS} - \text{SST} - \text{SSB}$$

2. Latin square design

$$\text{TSS} = \sum_{\substack{\text{over all} \\ ys}} (y_{ijk} - \bar{y})^2 = \sum_{\substack{\text{over all} \\ ys}} y_{ijk}^2 - \frac{G^2}{n}$$

$$\text{SST} = t \sum_i (\bar{y}_i - \bar{y})^2 = \sum_i \frac{T_i^2}{t} - \frac{G^2}{n}$$

$$\text{SSR} = t \sum_j (\bar{R}_j - \bar{y})^2 = \sum_j \frac{R_j^2}{t} - \frac{G^2}{n}$$

$$\text{SSC} = t \sum_k (\bar{C}_k - \bar{y})^2 = \sum_k \frac{C_k^2}{t} - \frac{G^2}{n}$$

$$\text{SSE} = \text{TSS} - \text{SST} - \text{SSR} - \text{SSC}$$

3. *k*-factor factorial experiment

$$SSA = \sum_i \frac{A_i^2}{n_A} - \frac{G^2}{n}$$

$$SSB = \sum_j \frac{B_j^2}{n_B} - \frac{G^2}{n}$$

$$SSC = \sum_k \frac{C_k^2}{n_C} - \frac{G^2}{n}$$

$$\vdots$$

$$SSAB = \sum_{ij} \frac{(AB)_{ij}^2}{n_{AB}} - SSA - SSB - \frac{G^2}{n}$$

$$SSAC = \sum_{ik} \frac{(AC)_{ik}^2}{n_{AC}} - SSA - SSC - \frac{G^2}{n}$$

$$\vdots$$

$$SSABC = \sum_{ijk} \frac{(ABC)_{ijk}^2}{n_{ABC}} - SSAB - SSAC - SSBC - SSA - SSB - SSC - \frac{G^2}{n}$$

$$\vdots$$

4. $100(1 - \alpha)\%$ confidence interval for difference in treatment means

$$\bar{y}_i - \bar{y}_{i'} \pm t_{\alpha/2} s_\varepsilon \sqrt{\frac{2}{n_t}}$$

EXERCISES

 14.18 An experimenter was interested in examining the bond strength of a new adhesive product prepared under three different temperature settings (280° F, 300° F, and 320° F) and four different pressure settings (100, 150, 200, and 250 psi). A single fixed amount of adhesive is to be prepared and tested at each temperature–pressure setting combination. Identify the design.

 14.19 An oil company has been experimenting with a new gasoline additive. As part of the testing program, the company examined the effect on miles per gallon (mpg) of four additive concentrations and five different octane levels for the gasoline. If one gasoline mixture is to be made and tested at each concentration–octane combination, identify the experimental design.

 14.20 A company executive was interested in comparing the cost per mile for all cars of a particular brand with V-8 engines (351 horsepower) and all cars of the same make with six-cylinder engines (250 horsepower). Since the entire fleet of company cars for salespeople consisted of approximately 600 automobiles, it was decided to obtain a random sample of data from 16 cars, eight from each engine type. To avoid geographic variability, a random sample of one car of each engine type was selected from each of eight geographic areas throughout the country. Cost per mile was determined for each car sampled during the period of January 1986 to March 1987. These data appear below.

Geographic area	1	2	3	4	5	6	7	8
351 hp, V-8	.0837	.0564	.0703	.0502	.0638	.0483	.0746	.0694
250 hp, 6-cylinder	.0523	.0371	.0464	.0481	.0535	.0335	.0444	.0528

a. Do the data suggest a difference in the mean cost per mile for the two engine types?

b. Give the level of significance for your test. Draw some conclusions.

c. Can you identify additional sources of variability that could be blocked?

14.21 Write the model and complete an analysis of variance table for a 3 × 5 factorial experiment.

a. How many degrees of freedom do you have for the error term when the two-way interaction is included?

b. How many degrees of freedom do you have for the error term if you assume there is no two-way interaction?

14.22 Examine the analysis of variance for the data of Example 14.8 shown in the BMDP computer output that follows.

ANALYSIS OF VARIANCE FOR 1-ST DEPENDENT VARIABLE

SOURCE	SUM OF SQUARES	DEGREES OF FREEDOM	MEAN SQUARE	F	PROB. F EXCEEDED
MEAN	90650.93750	1	90650.93750	2143.45068	0.000
VARIETY	3996.07349	2	1998.03662	47.24377	0.000
PESTICIDE	2227.44238	3	742.48071	17.55603	0.000
VAR × PEST	456.91479	6	76.15247	1.80063	0.182
ERROR	507.50464	12	42.29205		

a. Construct an analysis of variance table from the computer output.

b. Give the p-value and conclusion for the F test on the variety-by-treatment interaction. Is it appropriate to proceed with tests for main effects?

c. Draw conclusions from the p-values listed for varieties and pesticides.

d. Compare your results to those of Example 14.9.

14.23 A study was conducted to compare the yield of soybeans in a factorial experiment consisting of four manganese rates (from $MnSO_4$) and four copper rates (from $CuSO_4 5H_2O$). A large plot was subdivided into 16 separate subplots, to which the 16 factor-level combinations were applied. Soybeans were then planted over the entire plot in rows 3 feet apart. The sample data appear below (in kilograms/hectare).

		Mn		
Cu	20	50	80	110
1	1558	2003	2490	2830
3	1590	2020	2620	2860
5	1550	2010	2490	2750
7	1328	1760	2280	2630

Treating this 4×4 factorial as a two-way classification, construct a profile plot and write an appropriate statisical model.

14.24 Refer to Exercise 14.23.

a. Perform an analysis of variance. Can you test for an interaction? Use $\alpha = .05$.

b. Assuming that there is no interaction between the two variables Cu and Mn, write a regression model expressing the soybean yield in terms of two quantitative variables (Cu and Mn).

14.25 Write the model and complete an analysis of variance table for a $4 \times 3 \times 6$ factorial experiment. (Assume that the three-way interaction is nonsignficant.)

14.26 An experiment was set up to compare the effect of different soil pH and calcium additives on the increase in trunk diameters for orange trees. Annual applications of elemental sulfur, gypsum, soda ash, and other ingredients were applied to provide pH value levels of 4, 5, 6, and 7. Three levels of a calcium supplement (100, 200, and 300 pounds per acre) were also applied. All factor-level combinations of these two variables were used in the experiment. At the end of a two-year period, three diameters were examined at each factor-level combination. These data appear below.

pH Value	Calcium		
	100	200	300
4.0	5.2	7.4	6.3
	5.9	7.0	6.7
	6.3	7.6	6.1
5.0	7.1	7.4	7.3
	7.4	7.3	7.5
	7.5	7.1	7.2
6.0	7.6	7.6	7.2
	7.2	7.5	7.3
	7.4	7.8	7.0
7.0	7.2	7.4	6.8
	7.5	7.0	6.6
	7.2	6.9	6.4

a. Construct a profile plot. What do the data suggest?

b. Write an appropriate statistical model.

c. Treat the data as a two-way classification and perform an analysis of variance. Use $\alpha = .05$.

14.27 Refer to Exercise 14.9. Use Fisher's LSD to determine significant differences among the four treatments (broadcast and band methods of fertilizer applications).

14.28 Refer to Exercise 14.23. Use Tukey's W procedure to determine differences among the four manganese rates. Use $\alpha = .05$.

14.29 Refer to Exercise 14.26. Use Duncan's multiple range test (for $\alpha = .05$) to declare significant differences.

14.30 An experiment was conducted to compare the average oral body temperature for persons taking one of nine different medications often prescribed for a specific disorder. To do this, each of three investigators with prior clinical experience of this nature was to obtain a random sample of patients from his or her practice who satisfy the study entrance criteria. Then the investigator was to allocate the medications randomly, one to each person. Each patient in the study was given the assigned medication at 6:00 AM of the assigned study day. Temperatures were taken at hourly intervals beginning at 8:00 AM and continuing for 10 hours. During this time the patients were not allowed to do any physical activity and had to lie in bed. To eliminate the variability of temperature readings within a day, the average of the hourly determinations was the response of interest. These data are given in the accompanying table.

a. Write an appropriate statistical model and identify the parameters of the model.

b. Perform an analysis of variance to test for a difference in mean temperatures for medications and then for investigators. Summarize your results in an AOV table. Use $\alpha = .05$. (Hint: To simplify some of your calculations, subtract a constant, such as 97, from each of the sample measurements.)

					Medication				
Investigator	A	B	C	D	E	F	G	H	I
	97.8	98.1	98.0	97.3	97.9	97.9	97.1	98.0	97.8
	97.2	98.1	97.8	97.3	97.8	97.9	97.6	97.8	98.0
1	97.6	98.0	98.1	97.5	97.8	97.8	97.3	98.0	97.7
	97.2	97.7	97.8	97.5	97.7	97.8	97.7	97.9	97.9
	97.6	97.7	97.9	97.6	97.8	97.6	97.5	98.0	97.8
	97.6	97.8	97.9	97.5	97.8	98.0	97.6	97.9	98.0
	97.4	97.7	98.1	97.4	97.8	97.7	97.5	98.0	97.6
2	97.3	97.6	97.8	97.5	97.7	97.8	97.6	97.9	98.0
	97.5	97.7	97.8	97.6	97.7	97.9	97.5	97.9	97.9
	97.5	97.7	97.6	97.7	97.8	97.8	97.3	97.8	97.9
	97.5	97.6	98.0	97.9	97.7	97.9	97.4	97.8	98.0
	97.9	97.7	97.8	97.8	97.8	98.0	97.8	97.8	98.1
3	97.6	97.9	98.1	97.8	97.9	97.7	97.4	98.0	97.9
	97.6	97.9	97.7	97.8	98.0	97.9	97.6	97.9	98.1
	97.7	97.8	98.7	97.6	98.1	97.9	97.6	97.8	97.9

14.31 Refer to Exercise 14.30. Use a computer program to run an analysis of variance. Compare your results to those of Exercise 14.30, in which we coded the measurements for ease of calculations.

14.32 A physician was interested in examining the relationship between work performed by individuals in an exercise tolerance test and the excess weight (as determined by standard weight–height tables) they carried. To do this, a random sample of 28 healthy adult females, ranging in age from 25 to 40, was selected from the community clinic during routine visits for physical examinations. The selection process was restricted so that seven persons were selected from each of the following weight classifications.

Normal weight (less than 10% underweight)
1%–10% overweight
11%–20% overweight
More than 20% overweight

As part of the physical examination, each person was required to exercise on a bicycle ergometer until the onset of fatigue. The time to fatigue (in minutes) was recorded for each person. These data are given below.

Classification	Fatigue Time
Normal	25, 28, 19, 27, 23, 30, 35
1%–10% overweight	24, 26, 18, 16, 14, 12, 17
11%–20% overweight	15, 18, 17, 25, 12, 10, 23
More than 20% overweight	10, 9, 18, 14, 6, 4, 15

a. Identify the experimental design and write an appropriate statistical model.

b. Use $\alpha = .05$ and perform an analysis of variance.

14.33 Refer to Exercise 14.32.

a. How would you design an experiment to investigate the effects of age, gender, and excess weight on fatigue time?

b. Suppose the physician wanted to investigate the relationship among the quantitative variables percentage overweight, age, and fatigue time. Write a possible model.

14.34 An experiment was conducted to compare the heat loss for five different designs for commercial thermal panes. To do this, a sample of five panes of each design was obtained. The panes were then randomly assigned to the five exterior temperature settings (in °F) listed below so that each design appeared in each temperature setting. The interior temperature of the test was controlled at 68° F. Use the sample data below to compare the heat loss associated with the five pane designs.

	Pane Design				
Exterior Temperature Setting (°F)	A	B	C	D	E
80	8.4	8.6	9.2	9.1	10.3
60	8.4	8.7	9.3	9.4	10.7
40	8.9	9.1	9.7	9.9	10.9
20	10.4	10.7	10.6	10.5	11.3
0	10.8	11.2	11.1	11.3	11.6

a. Identify the experimental design and write an appropriate statistical model.

b. Use $\alpha = .05$ to run an analysis of variance.

14.35 Refer to Exercise 14.34. Use Tukey's W procedure to compare the treatment means. Set $\alpha = .05$.

14.36 Refer to Exercise 14.26. Suppose that rather than running three observations in each cell of the 4×3 factorial experiment at the same orange grove, a separate factorial experiment with one observation per cell is run at each of three different orange groves, as shown below.

Grove 1	Grove 2	Grove 3
4×3 factorial	4×3 factorial	4×3 factorial

Assume that the first, second, and third observations at each factor-level combination of Exercise 14.26 represent observations from groves 1, 2, and 3, respectively.

a. Identify the experimental design.
b. Write an appropriate linear statistical model, assuming there is no three-way interaction.
c. Perform an analysis of variance and give the level of significance for each test.

 14.37 An experiment was conducted to examine the effects of different levels of reinforcement and different levels of isolation on children's ability to recall. A single analyst was to work with a random sample of 36 children selected from a relatively homogeneous group of fourth-grade students. Two levels of reinforcement (none and verbal) and three levels of isolation (20, 40, and 60 minutes) were to be used. Students were randomly assigned to the six treatment groups, with a total of six students being assigned to each group.

Each student was to spend a 30-minute session with the analyst. During this time the student was to memorize a specific passage, with reinforcement provided as dictated by the group to which the student was assigned. Following the 30-minute session, the student was isolated for the time specified for his or her group and then tested for recall of the memorized passage. These data appear in the accompanying table.

Level of	Time of Isolation (Minutes)					
Reinforcement	20		40		60	
None	26	19	30	36	6	10
	23	18	25	28	11	14
	28	25	27	24	17	19
Verbal	15	16	24	26	31	38
	24	22	29	27	29	34
	25	21	23	21	35	30

Use the computer output shown here to draw your conclusions.

```
EXAMPLE OF TWO-WAY ANOVA WITH INTERACTION
FILE    NONAME
**************************** ANALYSIS OF VARIANCE ****************************
      VARO3
      VARO1
      VARO2
****************************************************************************

     SOURCE OF          SUM OF                   MEAN                  SIGNIF
     VARIATION          SQUARES        DF        SQUARE        F        OF F

MAIN EFFECTS            352.223        3         117.408      7.441      0.001
   VARO1               196.000        1         196.000     12.423      0.001
   VARO2               156.223        2          78.112      4.951      0.014
2-WAY INTERACTIONS    1058.665        2         529.333     33.549      0.001
   VARO1   VARO2      1058.665        2         529.333     33.549      0.001
EXPLAINED             1410.888        5         282.177     17.885      0.001
RESIDUAL               473.330       30          15.778
TOTAL                 1884.218       35          53.835


   36 CASES WERE PROCESSED.
    0 CASES (0.0 PCT) WERE MISSING.
```

14.38   A food-processing plant has tested several different formulations of a new breakfast drink. Each of six panels rated the 12 different formulations obtained from combining one of three levels of sweetness, one of two levels of caloric content, and one of two colors.
a. Identify the design.
b. Write an appropriate model.
c. Give the analysis of variance table for this design.

14.39   Critique the SAS output shown here for the design described in Exercise 14.38. Has anything been omitted?

SOURCE	SUM OF SQUARES	DEGREES OF FREEDOM	MEAN SQUARE	F	TAIL PROBABILITY
MEAN	171307.55556	1	171307.55556	6234.40	0.0000
A	4149.52778	2	2074.76389	75.51	0.0000
B	624.22222	1	624.22222	22.72	0.0000
C	3200.00000	1	3200.00000	116.46	0.0000
AB	488.52778	2	244.26389	8.89	0.0004
AC	203.08333	2	101.54167	3.70	0.0307
BC	80.22222	1	80.22222	2.92	0.0927
ABC	24.19444	2	12.09722	0.44	0.6459
ERROR	1648.66667	60	27.47778		

14.40   Refer to Exercise 14.39. Assume there was no panel-to-panel variability (and hence MSE was an appropriate measure of error), and draw conclusions about the formulations. Based on the cell means shown here, which ones appear different? Would a series of profile plots help to explain what is happening? Explain.

Sweetness Level	1 Color		2	
	Caloric Level		Caloric Level	
	1	2	1	2
1	59.5	42.5	54.5	40.1
2	66.8	49.6	64.7	50.1
3	52.0	39.3	35.1	30.2

 14.41   Job performance reviews were based on a numerical rating scale for random samples of 12, 9, and 18 employees from three divisions of a corporation.
Summary data are shown here:

Division	n	$\bar{y}$	s
Research	12	21.2	8.3
Development	9	15.4	7.3
Commercial	18	27.4	8.2

a. Identify the design.

b. Write an appropriate model.

14.42 Refer to Exercise 14.41. Perform an analysis of variance and draw conclusions. (Note: A high score is good.)

14.43 Stability data were generated for 2-ml vials manufactured with 30 mg/ml of active ingredient of a drug product. These data are shown here. Note that triplicate measurements were taken at each laboratory and time point.

Time (in Months at 30° C)	Laboratory	Mg/Ml of Active Ingredient	pH
1	1	30.03	3.61
1	1	30.10	3.60
1	1	30.14	3.57
3	1	30.10	3.50
3	1	30.18	3.45
3	1	30.23	3.48
6	1	30.03	3.56
6	1	30.03	3.74
6	1	29.96	3.81
9	1	29.81	3.60
9	1	29.79	3.55
9	1	29.82	3.59
1	2	30.12	3.87
1	2	30.10	3.80
1	2	30.02	3.84
3	2	29.90	3.70
3	2	29.95	3.80
3	2	29.85	3.75
6	2	29.75	3.90
6	2	29.85	3.90
6	2	29.80	3.90
9	2	29.75	3.77
9	2	29.85	3.74
9	2	29.80	3.76

a. Write a model that could be used to relate y (either mg/ml of active ingredient or pH) to the independent variables, time (in months) and laboratory. (Hint: Use a dummy variable and allow for interaction.)

b. Interpret the βs in the model for part (a).

c. Give an analysis of variance table for the model of part (a) without computing the necessary sums of squares.

14.44 Refer to Exercise 14.43. Computer output is shown for an analysis of variance for both dependent variables (i.e., $y_1 = $ mg/ml of active ingredient and $y_2 = $ pH). Draw conclusions about the stability of these 2-ml vials based on these analyses.

EXERCISES

```
                  ANALYSIS OF VARIANCE FOR DEPENDENT VARIABLE MG_ML OF ACTIVE

                          GENERAL LINEAR MODELS PROCEDURE

DEPENDENT VARIABLE: MG_ML
```

SOURCE	DF	SUM OF SQUARES	MEAN SQUARE	F VALUE	PR > F	R-SQUARE	C.V.
MODEL	3	0.38500499	0.12833500	21.12	0.0001	0.760078	0.2602
ERROR	20	0.12152834	0.00607642		ROOT MSE		MG_ML MEAN
CORRECTED TOTAL	23	0.50653333			0.07795138		29.95666667

SOURCE	DF	TYPE I SS	F VALUE	PR > F	DF	TYPE III SS	F VALUE	PR > F
TIME	1	0.29170249	48.01	0.0001	1	0.17123832	28.18	0.0001
LAB	1	0.09126667	15.02	0.0009	1	0.04021903	6.62	0.0182
TIME*LAB	1	0.00203583	0.34	0.5692	1	0.00203583	0.34	0.5692

PARAMETER	ESTIMATE	T FOR HO: PARAMETER=0	PR > |T|	STD ERROR OF ESTIMATE
INTERCEPT	30.20553288	722.07	0.0001	0.04183178
TIME	-0.03941043	-5.31	0.0001	0.00742394
LAB	-0.15219955	-2.57	0.0182	0.05915907
TIME*LAB	0.00607710	0.58	0.5692	0.01049904

```
                  ANALYSIS OF VARIANCE FOR DEPENDENT VARIABLE PH

                          GENERAL LINEAR MODELS PROCEDURE

DEPENDENT VARIABLE: PH
```

SOURCE	DF	SUM OF SQUARES	MEAN SQUARE	F VALUE	PR > F	R-SQUARE	C.V.
MODEL	3	0.30464096	0.10154699	12.67	0.0001	0.655289	2.4196
ERROR	20	0.16025488	0.00801274		ROOT MSE		PH MEAN
CORRECTED TOTAL	23	0.46489583			0.08951393		3.69958333

SOURCE	DF	TYPE I SS	F VALUE	PR > F	DF	TYPE III SS	F VALUE	PR > F
TIME	1	0.00126193	0.16	0.6957	1	0.00663061	0.83	0.3738
LAB	1	0.29703750	37.07	0.0001	1	0.12982259	16.20	0.0007
TIME*LAB	1	0.00634152	0.79	0.3842	1	0.00634152	0.79	0.3842

PARAMETER	ESTIMATE	T FOR HO: PARAMETER=0	PR > |T|	STD ERROR OF ESTIMATE
INTERCEPT	3.55149660	73.93	0.0001	0.04803670
TIME	0.00775510	0.91	0.3738	0.00852514
LAB	0.27344671	4.03	0.0007	0.06793416
TIME*LAB	-0.01072562	-0.89	0.3842	0.01205636

14.45 The following models were fit to the data of Exercise 14.43.

$$y_i = \beta_0 + \beta_1 x_1 + \beta_2 x_2 + \beta_3 x_1 x_2 + \varepsilon$$

where y_1 = mg/ml and y_2 = pH

x_1 = time in months

x_2 = 1 if laboratory 2
     0 otherwise

a. How might you graph these data and the fitted lines? Does a linear relationship between y_i and x_1 seem to fit these data? For both laboratories?

b. For both y_1 and y_2, estimate the difference in the slopes of the two regression lines using a 95% confidence interval.

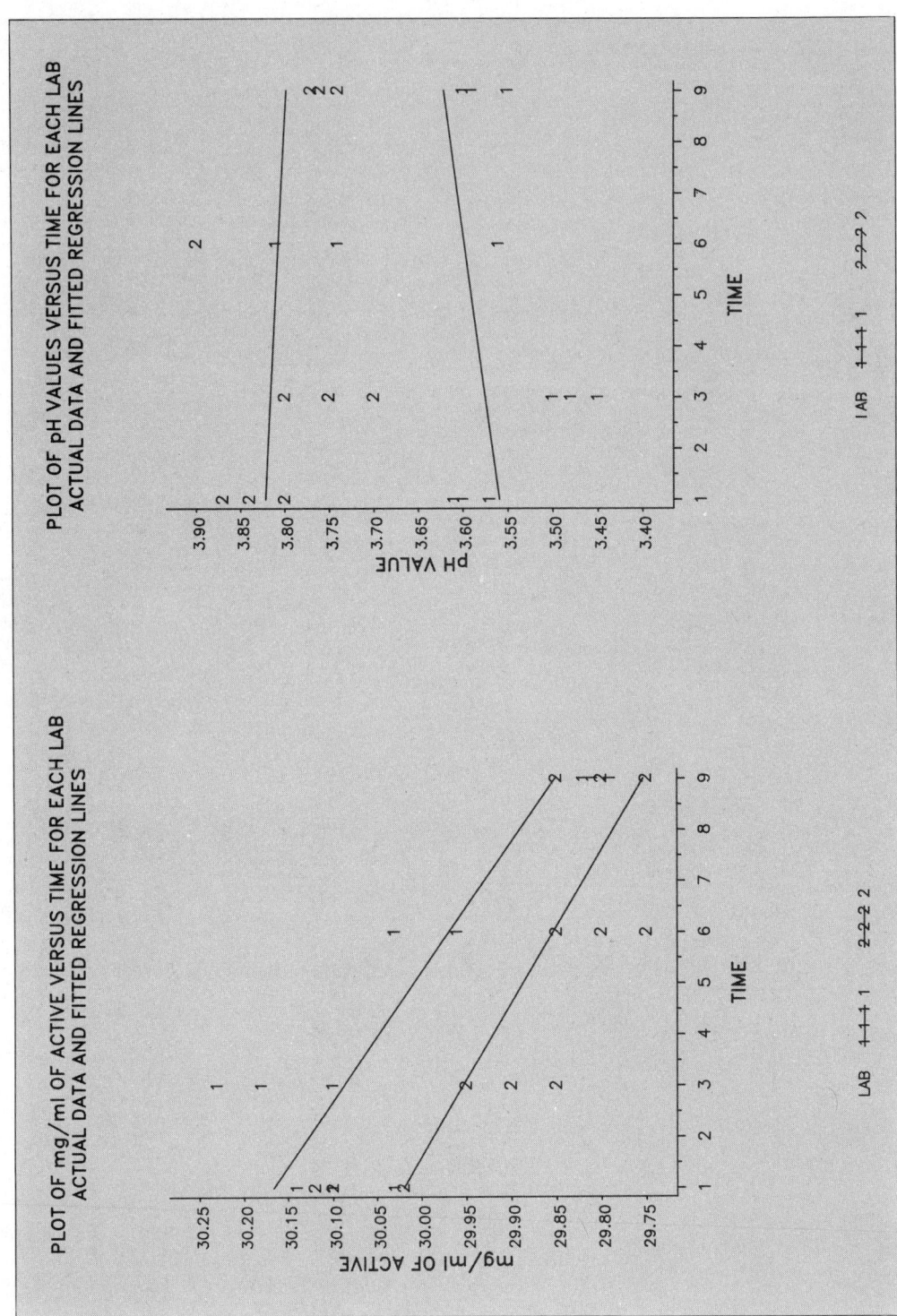

EXERCISES

14.46 Refer to Exercise 14.43. The same type of data on mg/ml and pH were generated at 40° C as were obtained at 30° C in Exercise 14.43. These data are shown here.

Time (in Months at 40° C)	Laboratory	Mg/Ml of Active Ingredient	pH
1	1	30.08	3.61
1	1	30.10	3.60
1	1	30.14	3.59
3	1	30.03	3.39
3	1	30.18	3.45
3	1	30.26	3.29
6	1	29.90	3.63
6	1	29.90	3.71
6	1	29.96	3.65
9	1	29.81	3.51
9	1	29.85	3.38
9	1	29.72	3.32
1	2	30.12	3.80
1	2	30.10	3.70
1	2	30.02	3.81
3	2	29.90	3.70
3	2	29.85	3.80
3	2	29.80	3.75
6	2	29.75	3.80
6	2	29.70	3.70
6	2	29.75	3.70
9	2	29.65	3.64
9	2	29.75	3.68
9	2	29.70	3.60

a. Without computing the necessary sums of squares, give an analysis of variance table for this three-factor experiment. Include all possible interactions.

b. Computer output is shown here for the analysis of variance of part (a) for both dependent variables (i.e., $y_1 = $ mg/ml and $y_2 = $ pH). Interpret the results of these analyses.

ANALYSIS OF VARIANCE FOR DEPENDENT VARIABLE MG_ML OF ACTIVE

GENERAL LINEAR MODELS PROCEDURE

DEPENDENT VARIABLE: MG_ML

SOURCE	DF	SUM OF SQUARES	MEAN SQUARE	F VALUE	PR > F	R-SQUARE	C.V.
MODEL	7	0.98852273	0.14121753	22.08	0.0001	0.794382	0.2672
ERROR	40	0.25586893	0.00639672		ROOT MSE		MG_ML MEAN
CORRECTED TOTAL	47	1.24439167			0.07997952		29.93708333

SOURCE	DF	TYPE I SS	F VALUE	PR > F	DF	TYPE III SS	F VALUE	PR > F
TIME	1	0.72777149	113.77	0.0001	1	0.39245581	61.35	0.0001
LAB	1	0.22963333	35.90	0.0001	1	0.08149613	12.74	0.0009
TIME*LAB	1	0.00107959	0.17	0.6834	1	0.00107959	0.17	0.6834
TEMP	1	0.01840833	2.88	0.0976	1	0.00000857	0.00	0.9710
TIME*TEMP	1	0.00797194	1.25	0.2709	1	0.00170139	0.27	0.6089
LAB*TEMP	1	0.00270000	0.42	0.5196	1	0.00000346	0.00	0.9816
TIME*LAB*TEMP	1	0.00095805	0.15	0.7008	1	0.00095805	0.15	0.7008

ANALYSIS OF VARIANCE FOR DEPENDENT VARIABLE PH

GENERAL LINEAR MODELS PROCEDURE

DEPENDENT VARIABLE: PH

SOURCE	DF	SUM OF SQUARES	MEAN SQUARE	F VALUE	PR > F	R-SQUARE	C.V.
MODEL	7	0.69408435	0.09915491	10.08	0.0001	0.638258	2.7108
ERROR	40	0.39338231	0.00983456		ROOT MSE		PH MEAN
CORRECTED TOTAL	47	1.08746667			0.09916934		3.65833333

SOURCE	DF	TYPE I SS	F VALUE	PR > F	DF	TYPE III SS	F VALUE	PR > F
TIME	1	0.01158186	1.18	0.2843	1	0.00026670	0.03	0.8700
LAB	1	0.56767500	57.72	0.0001	1	0.22710248	23.09	0.0001
TIME*LAB	1	0.00714427	0.73	0.3991	1	0.00714427	0.73	0.3991
TEMP	1	0.08167500	8.30	0.0063	1	0.00007612	0.01	0.9303
TIME*TEMP	1	0.02491888	2.53	0.1193	1	0.01728914	1.76	0.1924
LAB*TEMP	1	0.00030000	0.03	0.8622	1	0.00108908	0.11	0.7410
TIME*LAB*TEMP	1	0.00078934	0.08	0.7784	1	0.00078934	0.08	0.7784

 14.47   Shelf life data for a given product (% of theory) are shown here for each of eight batches of product tested at 0, 3, 6, 12, and 24 weeks.

LISTING OF DATA

BATCH	MONTH	THEORY
1	0	100.7
1	3	100.7
1	6	101.3
1	12	101.0
1	24	102.0
2	0	98.7
2	3	102.3
2	6	100.0
2	12	101.7
2	24	99.3
3	0	102.0
3	3	101.3
3	6	101.7
3	12	101.3
3	24	101.2
4	0	99.7
4	3	100.3
4	6	100.7
4	12	100.7
4	24	100.2
5	0	100.0
5	3	100.7
5	6	101.3
5	12	100.3
5	24	100.7
6	0	98.5
6	3	99.4
6	6	99.3
6	12	100.3
6	24	100.0
7	0	99.7
7	3	100.2
7	6	100.7
7	12	99.3
7	24	100.2
8	0	100.3
8	3	99.0
8	6	101.7
8	12	101.0
8	24	100.0

Three different analyses are shown here. Identify the objective(s) of each analysis and draw conclusions.

```
                              ANALYSIS OF VARIANCE
                                   FULL MODEL

                          GENERAL LINEAR MODELS PROCEDURE
```

DEPENDENT VARIABLE: THEORY

SOURCE	DF	SUM OF SQUARES	MEAN SQUARE	F VALUE	PR > F	R-SQUARE	C.V.
MODEL	15	16.04425000	1.06961667	1.46	0.1977	0.477351	0.8514
ERROR	24	17.56675000	0.73194792		ROOT MSE		THEORY MEAN
CORRECTED TOTAL	39	33.61100000			0.85553955		100.48500000

SOURCE	DF	TYPE I SS	F VALUE	PR > F	DF	TYPE III SS	F VALUE	PR > F
MONTH	1	0.33153125	0.45	0.5074	1	0.33153125	0.45	0.5074
BATCH	7	13.50300000	2.64	0.0360	7	9.47932169	1.85	0.1234
MONTH*BATCH	7	2.20971875	0.43	0.8729	7	2.20971875	0.43	0.8729

```
                              ANALYSIS OF VARIANCE
                                  REDUCED MODEL

                          GENERAL LINEAR MODELS PROCEDURE
```

DEPENDENT VARIABLE: THEORY

SOURCE	DF	SUM OF SQUARES	MEAN SQUARE	F VALUE	PR > F	R-SQUARE	C.V.
MODEL	8	13.83453125	1.72931641	2.71	0.0218	0.411607	0.7949
ERROR	31	19.77646875	0.63795060		ROOT MSE		THEORY MEAN
CORRECTED TOTAL	39	33.61100000			0.79871810		100.48500000

SOURCE	DF	TYPE I SS	F VALUE	PR > F	DF	TYPE III SS	F VALUE	PR > F
MONTH	1	0.33153125	0.52	0.4764	1	0.33153125	0.52	0.4764
BATCH	7	13.50300000	3.02	0.0153	7	13.50300000	3.02	0.0153

```
                              ANALYSIS OF VARIANCE
                                SIMPLE REGRESSION

                          GENERAL LINEAR MODELS PROCEDURE
```

DEPENDENT VARIABLE: THEORY

SOURCE	DF	SUM OF SQUARES	MEAN SQUARE	F VALUE	PR > F	R-SQUARE	C.V.
MODEL	1	0.33153125	0.33153125	0.38	0.5420	0.009864	0.9313
ERROR	38	33.27946875	0.87577549		ROOT MSE		THEORY MEAN
CORRECTED TOTAL	39	33.61100000			0.93582877		100.48500000

SOURCE	DF	TYPE I SS	F VALUE	PR > F	DF	TYPE III SS	F VALUE	PR > F
MONTH	1	0.33153125	0.38	0.5420	1	0.33153125	0.38	0.5420

PARAMETER	ESTIMATE	T FOR H0: PARAMETER=0	PR > |T|	STD ERROR OF ESTIMATE
INTERCEPT	100.38843750	465.41	0.0001	0.21569787
MONTH	0.01072917	0.62	0.5420	0.01743814

 14.48 A manufacturer whose daily supply of raw materials is variable and limited can use the material to produce two different products in various proportions. The profit per unit of raw material obtained by producing each of the two products depends on the length of a product's manufacturing run and hence on the amount of raw material assigned to it. Other factors—such as worker productivity, machine breakdown, and so on—can affect the profit per unit as well, but their net effect on profit is random and uncontrollable. The manufacturer has conducted an experiment to investigate the effect of the level of supply of raw material, "S," and the ratio of its assignment, "R," to the two product manufacturing lines on the profit per unit of raw material. The ultimate goal was to be able to choose the best ratio "R" to match each day's supply of raw materials, "S." The levels of supply of the raw material chosen for the experiment were 15, 18, and 21 tons. The levels of the ratio of allocation to the two product lines were 1/2, 1, and 2. The response was the profit (in cents) per unit of raw material supply obtained from a single day's production. Three replications of each combination were conducted in a random sequence. The data for the 27 days are shown in the following table.

		Raw Material Supply (Tons)		
		15	18	21
Ratio of	1/2	22, 20, 21	21, 19, 20	19, 18, 20
raw material	1	21, 20, 19	23, 24, 22	20, 19, 21
allocation (R)	2	17, 18, 16	21, 11, 20	20, 22, 24

a. Draw conclusions based on the analysis of variance shown here.

```
             LISTING OF DATA

    RATIO     SUPPLY     PROFIT

     0.5        15         22
     0.5        15         20
     0.5        15         21
     0.5        18         21
     0.5        18         19
     0.5        18         20
     0.5        21         19
     0.5        21         18
     0.5        21         20
     1.0        15         21
     1.0        15         20
     1.0        15         19
     1.0        18         23
     1.0        18         24
     1.0        18         22
     1.0        21         20
     1.0        21         19
     1.0        21         21
     2.0        15         17
     2.0        15         18
     2.0        15         16
     2.0        18         21
     2.0        18         11
     2.0        18         20
     2.0        21         20
     2.0        21         22
     2.0        21         24
```

ANALYSIS OF VARIANCE
2 FACTOR FACTORIAL

GENERAL LINEAR MODELS PROCEDURE

DEPENDENT VARIABLE: PROFIT

SOURCE	DF	SUM OF SQUARES	MEAN SQUARE	F VALUE	PR > F	R-SQUARE	C.V.
MODEL	8	93.18518519	11.64814815	2.54	0.0482	0.529907	10.7550
ERROR	18	82.66666667	4.59259259		ROOT MSE		PROFIT MEAN
CORRECTED TOTAL	26	175.85185185			2.14303350		19.92592593

SOURCE	DF	TYPE I SS	F VALUE	PR > F	DF	TYPE III SS	F VALUE	PR > F
RATIO	2	22.29629630	2.43	0.1166	2	22.29629630	2.43	0.1166
SUPPLY	2	4.96296296	0.54	0.5917	2	4.96296296	0.54	0.5917
RATIO*SUPPLY	4	65.92592593	3.59	0.0255	4	65.92592593	3.59	0.0255

```
                    ANALYSIS OF VARIANCE
                     2 FACTOR FACTORIAL

            GENERAL LINEAR MODELS PROCEDURE

                          MEANS

    RATIO   SUPPLY         N       PROFIT

      1       15           3     20.0000000
      1       18           3     23.0000000
      1       21           3     20.0000000
      2       15           3     17.0000000
      2       18           3     17.3333333
      2       21           3     22.0000000
     0.5      15           3     21.0000000
     0.5      18           3     20.0000000
     0.5      21           3     19.0000000
```

b. Identify the two best combinations of "R" and "S." Are these two combinations significantly different? Use a procedure that limits the error rate of all pairwise comparisons of combinations to be no more than 0.05.

 14.49 A manufacturer frequently sends small packages to a customer in another city via air freight, and, in many cases, it is important for a package to reach the customer as soon as possible. Three different firms offer air freight service, including pickup and delivery, on a 24-hour basis. The head of the manufacturer's shipping department would like to know whether the firms differ in speed of service and whether the time of day makes any difference. An experiment is designed to investigate these issues. Packages are sent at random times, and the air freight firm used for each package also is randomly chosen. The customer records the time that each package arrives, so that the time elapsed during shipment can be determined. These times are rounded to the nearest hour. The experimental results for a total of 54 packages are shown in the following table.

	Firm		
Time	1	2	3
Morning	8, 6, 6, 12, 7, 8	11, 11, 9, 10, 8, 11	7, 4, 6, 4, 9, 7
Afternoon	7, 10, 8, 11, 9, 11	10, 13, 10, 12, 11, 10	10, 8, 6, 5, 8, 6
Night	13, 11, 14, 11, 9, 12	12, 6, 9, 9, 10, 6	8, 11, 9, 9, 10, 12

a. Suppose that the above analysis were to be done using the dummy variable approach instead of the ANOVA approach. How many dummy variables would be needed to include both main effects and interaction effects in the model? What would the R^2 for this regression be?

b. What evidence is relevant for deciding whether the choice of best firm will be different at different times of the day? What conclusion would you draw using a 5% level of significance? Construct a graph that depicts the nature of any differences in firm as a function of time of day.

c. Does any firm appear to be better than the other two firms? How could you compare the best firm and the second-best firm using a confidence interval?

```
    LISTING OF DATA

TIME       FIRM    SPEED

MORNING     1        8
MORNING     1        6
MORNING     1        6
MORNING     1       12
MORNING     1        7
MORNING     1        8
MORNING     2       11
MORNING     2       11
MORNING     2        9
MORNING     2       10
MORNING     2        8
MORNING     2       11
MORNING     3        7
MORNING     3        4
MORNING     3        6
MORNING     3        4
MORNING     3        9
MORNING     3        7
AFTRNOON    1        7
AFTRNOON    1       10
AFTRNOON    1        8
AFTRNOON    1       11
AFTRNOON    1        9
AFTRNOON    1       11
AFTRNOON    2       10
AFTRNOON    2       13
AFTRNOON    2       10
AFTRNOON    2       12
AFTRNOON    2       11
AFTRNOON    2       10
AFTRNOON    3       10
AFTRNOON    3        8
AFTRNOON    3        6
AFTRNOON    3        5
AFTRNOON    3        8
AFTRNOON    3        6
NIGHT       1       13
NIGHT       1       11
NIGHT       1       14
NIGHT       1       11
NIGHT       1        9
NIGHT       1       12
NIGHT       2       12
NIGHT       2        6
NIGHT       2        9
NIGHT       2        9
NIGHT       2       10
NIGHT       2        6
NIGHT       3        8
NIGHT       3       11
NIGHT       3        9
NIGHT       3        9
NIGHT       3       10
NIGHT       3       12
```

ANALYSIS OF VARIANCE
2 FACTOR FACTORIAL

GENERAL LINEAR MODELS PROCEDURE

DEPENDENT VARIABLE: SPEED

SOURCE	DF	SUM OF SQUARES	MEAN SQUARE	F VALUE	PR > F	R-SQUARE	C.V.
MODEL	8	154.37037037	19.29629630	6.06	0.0001	0.518537	19.6682
ERROR	45	143.33333333	3.18518519		ROOT MSE		SPEED MEAN
CORRECTED TOTAL	53	297.70370370			1.78470871		9.07407407

SOURCE	DF	TYPE I SS	F VALUE	PR > F	DF	TYPE III SS	F VALUE	PR > F
TIME	2	38.25925926	6.01	0.0049	2	38.25925926	6.01	0.0049
FIRM	2	50.03703704	7.85	0.0012	2	50.03703704	7.85	0.0012
TIME*FIRM	4	66.07407407	5.19	0.0016	4	66.07407407	5.19	0.0016

 14.50   Three dye formulas for a certain synthetic fiber are under consideration by a textile manufacturer who wishes to know whether the three are in fact different in quality. To aid in this decision, the manufacturer conducts an experiment in which five specimens of fabric are cut into thirds, and one third is randomly assigned to be dyed by each of the three dyes. Each piece of fabric is later graded and assigned a score measuring the quality of the dye. The results are as follows.

EXERCISES

Dye	Fabric Specimen				
	1	2	3	4	5
A	74	78	76	82	77
B	81	86	90	93	73
C	95	99	90	87	93

a. Identify the design.

b. Run an analysis of variance and draw conclusions about the dyes.

c. Give a measure of the efficiency of this design to one not blocking on fabric specimens.

```
          LISTING OF DATA

 DYE     SPECIMEN     QUALITY

  A          1          74
  A          2          78
  A          3          76
  A          4          82
  A          5          77
  B          1          81
  B          2          86
  B          3          90
  B          4          93
  B          5          73
  C          1          95
  C          2          99
  C          3          90
  C          4          87
  C          5          93

                     ANALYSIS OF VARIANCE
                  RANDOMIZED BLOCK DESIGN

            GENERAL LINEAR MODELS PROCEDURE

DEPENDENT VARIABLE: QUALITY

SOURCE            DF      SUM OF SQUARES      MEAN SQUARE      F VALUE      PR > F       R-SQUARE         C.V.

MODEL              6      688.00000000      114.66666667        3.34       0.0596       0.714484        6.9022

ERROR              8      274.93333333       34.36666667                   ROOT MSE                  QUALITY MEAN

CORRECTED TOTAL   14      962.93333333                                     5.86230899                84.93333333


SOURCE            DF        TYPE I SS     F VALUE      PR > F         DF        TYPE III SS    F VALUE      PR > F
DYE                2      593.73333333      8.64       0.0100          2      593.73333333      8.64       0.0100
SPECIMEN           4       94.26666667      0.69       0.6216          4       94.26666667      0.69       0.6216
```

14.51 An experiment tested the effect of factory music on workers' production. Four music programs (A, B, C, D) were compared with no music (E). Each program was played for one entire day, and five replications for each program were desired. The length of the experiment was thus five weeks. To control for variation in week and day of week, a Latin square design was adopted for the 25 days of the experiment. Each program was played once on each day of the week and once each week.

Week	Monday		Tuesday		Wednesday		Thursday		Friday	
1	A	133	B	139	C	140	D	140	E	145
2	B	136	C	141	D	143	E	146	A	139
3	C	140	A	138	E	142	B	139	D	139
4	D	129	E	132	A	137	C	136	B	140
5	E	132	D	144	B	143	A	142	C	142

Use the output shown here to analyze these data, and draw conclusions.

```
           LISTING OF DATA

WEEK   DAY       MUSIC   OUTPUT

  1    MONDAY      A      133
  1    TUESDAY     B      139
  1    WEDNDAY     C      140
  1    THURSDAY    D      140
  1    FRIDAY      E      145
  2    MONDAY      B      136
  2    TUESDAY     C      141
  2    WEDNDAY     D      143
  2    THURSDAY    E      146
  2    FRIDAY      A      139
  3    MONDAY      C      140
  3    TUESDAY     A      138
  3    WEDNDAY     E      142
  3    THURSDAY    B      139
  3    FRIDAY      D      139
  4    MONDAY      D      129
  4    TUESDAY     E      132
  4    WEDNDAY     A      137
  4    THURSDAY    C      136
  4    FRIDAY      B      140
  5    MONDAY      E      132
  5    TUESDAY     D      144
  5    WEDNDAY     B      143
  5    THURSDAY    A      142
  5    FRIDAY      C      142
```

ANALYSIS OF VARIANCE
LATIN SQUARE DESIGN

GENERAL LINEAR MODELS PROCEDURE

DEPENDENT VARIABLE: OUTPUT

SOURCE	DF	SUM OF SQUARES	MEAN SQUARE	F VALUE	PR > F	R-SQUARE	C.V.
MODEL	12	313.12000000	26.09333333	2.59	0.0561	0.721741	2.2805
ERROR	12	120.72000000	10.06000000		ROOT MSE		OUTPUT MEAN
CORRECTED TOTAL	24	433.84000000			3.17175031		139.08000000

SOURCE	DF	TYPE I SS	F VALUE	PR > F	DF	TYPE III SS	F VALUE	PR > F
MUSIC	4	11.84000000	0.29	0.8761	4	11.84000000	0.29	0.8761
WEEK	4	123.44000000	3.07	0.0589	4	123.44000000	3.07	0.0589
DAY	4	177.84000000	4.42	0.0200	4	177.84000000	4.42	0.0200

 14.52 The yields of wheat (in lb) are shown here for 25 plots (five farms, five plots per farm). The treatments (fertilizers) applied to each plot are shown in parentheses.

	Plot				
Farm	1	2	3	4	5
1	(D) 10.3	(E) 8.6	(A) 6.7	(C) 7.6	(B) 5.8
2	(E) 8.8	(B) 6.7	(C) 6.7	(A) 4.8	(D) 6.0
3	(A) 6.3	(C) 8.3	(B) 6.8	(D) 8.0	(E) 8.8
4	(C) 8.9	(D) 7.4	(E) 8.2	(B) 6.2	(A) 4.4
5	(B) 7.3	(A) 4.4	(D) 7.7	(E) 6.8	(C) 6.7

a. Identify the designs.

b. Do an analysis of variance and draw conclusions concerning the five fertilizers.

14.53 Refer to Exercise 14.52. Run a multiple-comparison procedure to make all pairwise comparisons of the treatment means. Identify which error rate was controlled.

15 ANALYSIS OF VARIANCE FOR SOME UNBALANCED DESIGNS

INTRODUCTION

We examined the analysis of variance for balanced designs in Chapters 10 and 14, where we used appropriate shortcut formulas (and corresponding computer solutions) to construct AOV tables and set up hypothesis tests. In Chapter 14 we also considered another way of performing an analysis of variance. We found that the null hypothesis under test in an analysis of variance can be expressed in terms of one or more βs in the general linear model. We also saw that the sum of squares associated with a source of variability in the analysis of variance table can be found as the drop in the sum of squares for error obtained from fitting reduced and complete models. While we did not advocate the use of complete and reduced models for obtaining the sums of squares for sources of variability in balanced designs, we did indicate that the procedure was completely general and could be used for any experimental design. In particular, in this chapter we will make use of complete and reduced models for obtaining the sums of squares in the analysis for *unbalanced designs*, where shortcut formulas are no longer readily available and easy to apply.

You might ask why an experimenter would run a study using an unbalanced design, especially since unbalanced designs seem to be more difficult to analyze. In point of fact, most studies do begin by using a balanced design, but for any one of many different reasons, the experimenter is unable to obtain the same number of observations per cell as dictated by the balanced design being employed. Consider a study of three different weight-reducing agents in which five different clinics (blocks) are employed and patients are to be randomly assigned to the three treatment groups according to a randomized

731

block design. Even if the experimenter plans to have five overweight persons assigned to each treatment at each clinic, the final count will almost certainly show an imbalance of persons assigned to each treatment group. Almost every clinic could be expected to have a few people who would not complete the study. Some people might move from the community, others might drop out due to a lack of efficacy in the program, and so on. In addition, the experimenter might find it impossible to locate 15 overweight people at each clinic who are willing to participate in the study. Because an unbalanced design at the end of a study occurs quite often, we must learn now to analyze data arising from unbalanced designs.

| 15.2 | A RANDOMIZED BLOCK DESIGN WITH ONE OR MORE MISSING OBSERVATIONS |

unbalanced design

Any time the number of observations is not the same for all factor-level combinations, we call the design **unbalanced**. Thus a randomized block design or a Latin square design with one or more missing observations is an unbalanced design. We will begin our examination by considering a simple case, a randomized block design with one missing observation.

value of missing observation

The analysis of variance for a randomized block design with one missing observation can be performed rather easily by using the shortcut formulas for a balanced design, after we have estimated the **value of the missing observation**. The formula for the missing observation M is given by

$$M = \frac{tT + bB - G}{(t - 1)(b - 1)}$$

where t is the number of treatments, b is the number of blocks, T is the sum of all the observations on the treatment assigned to the missing observation, B is the sum of all measurements in the block with the missing observation, and G is the sum of all the measurements.

We illustrate the analysis of variance for this design with an example.

EXAMPLE 15.1

An experiment was conducted to determine the nutritional value of diets for cows that are supplemented by whey. Five dairies were involved in the study. Each cow in a sample of four cows from a dairy was randomly assigned to one of the four treatment groups, so that a total of five cows were in each treatment group.

Treatment 1: water only

Treatment 2: whey plus 30.2 liters of water/day

Treatment 3: whey plus 15.1 liters of water/day

Treatment 4: whey only

In addition to the liquid portion of the diet listed for each treatment group, each cow was fed 7.5 kg of grain per day.

TABLE 15.1   Consumption of Hay for Cows, Example 15.1

	Treatment			
Dairy	1	2	3	4
1	15.4	9.6	9.5	8.4
2	14.8	9.3	9.4	—
3	15.9	9.8	9.7	9.3
4	15.5	9.4	9.2	8.1
5	14.7	9.2	9.0	7.9

One response of interest was the amount of hay consumed per day. These data (in kg per animal) are listed in Table 15.1. Unfortunately, as can be seen from the data, the cow on diet 4 from dairy 2 was dropped from the study and no replacement was made. The cow developed an infection (unrelated to the treatment) and was dropped from the study for safety reasons.

Estimate the missing value and then perform an analysis of variance. Use $\alpha = .01$.

Solution

For this randomized block design with $b = 5$ and $t = 4$, the quantities T, B, and G are defined as follows:

$T =$ sum of all observations on treatment 4

$\quad = 8.4 + 9.3 + 8.1 + 7.9 = 33.7$

$B =$ sum of all observations in block 2

$\quad = 14.8 + 9.3 + 9.4 = 33.5$

$G =$ sum of all measurements

$\quad = 15.4 + 9.6 + \cdots + 7.9 = 204.1$

The estimate of the missing value is

$$M = \frac{tT + bB - G}{(t - 1)(b - 1)} = \frac{4(33.7) + 5(33.5) - 204.1}{3(4)}$$

$$= \frac{98.2}{12} = 8.2$$

Having estimated the missing value, we can compute sums of squares for our analysis of variance by using the shortcut formulas of Chapter 14. The treatment and block totals are given by

$T_1 =$	76.3	$B_1 =$	42.9
$T_2 =$	47.3	$B_2 =$	41.7
$T_3 =$	46.8	$B_3 =$	44.7
$T_4 =$	41.9	$B_4 =$	42.2
		$B_5 =$	40.8
Totals	212.3		212.3

TABLE 15.2   AOV Table for the Data of Example 15.1

Source	SS	df	MS	F
Treatments	147.41	3	49.14	982.80
Blocks	2.16	4	.54	10.80
Error	.56	11	.05	
Totals	150.13	18		

Note that the new totals for treatment 4 and for block 2 incorporate the estimated missing observation. Similarly, the sum of all measurements includes the estimated missing value.

$$SST = \frac{(76.3)^2 + \cdots + (41.9)^2}{5} - \frac{(212.3)^2}{20} = 2400.97 - 2253.56 = 147.41$$

$$SSB = \frac{(42.9)^2 + \cdots + (40.8)^2}{4} - 2253.56 = 2.16$$

$$TSS = \sum_{ij} y_{ij}^2 - \frac{G^2}{20} = (15.4)^2 + (9.6)^2 + \cdots + (7.9)^2 - 2253.56$$

$$= 2403.69 - 2253.56 = 150.13$$

By subtraction, SSE = .56.

The only difference in the analysis of variance table for unbalanced and balanced randomized block designs is that since n refers to the number of actual observations, the error for an unbalanced design loses one degree of freedom for each missing observation when compared to the corresponding balanced design. The AOV table for our example is shown in Table 15.2.

The F tests for treatments and blocks are both significant, using $\alpha = .01$ (the critical values of F are 6.22 and 5.67, respectively). As can be seen from the data, those cows on treatment 1 (water only) consumed much more hay than cows on any of the diets supplemented with whey.

comparisons among treatment means   Having seen an analysis of variance, we may wish to make certain **comparisons among the treatment means**. We'll run pairwise comparisons using Fisher's least significant difference. The least significant difference between the treatment with a missing observation and any other treatment mean is

LSD

$$LSD = t_{\alpha/2} \sqrt{MSE\left(\frac{2}{b} + \frac{t}{b(b-1)(t-1)}\right)}$$

For any pair of treatments with no missing value, the least significant difference is as before; namely,

$$LSD = t_{\alpha/2} \sqrt{\frac{2MSE}{b}}$$

fitting complete and reduced models

The formulas for estimating missing observations in a randomized block design become more complicated with more missing data, as do the formulas for least significant differences. Because of this, we will consider **fitting complete and reduced models** to analyze unbalanced designs. We will illustrate the procedure first by examining an unbalanced randomized block design.

Because it would require more data input for a computer solution using the general linear model format presented in Chapters 12 and 13, we will represent the complete and reduced models for testing treatments as follows:

models

complete model: $y_{ij} = \mu + \beta_j + \alpha_i + \varepsilon_{ij}$
(model 1)
reduced model: $y_{ij} = \mu + \beta_j + \varepsilon_{ij}$
(model 2)

where β_j is the jth block effect and α_i is the ith treatment effect.

By fitting model 1 (using SAS or other computer software), we obtain SSE_1. Similarly, a fit of model 2 yields SSE_2. The difference in the two sums of squares for error, $SSE_2 - SSE_1$, gives the drop in the sum of squares due to treatments. Since this is an unbalanced design, the block effects do not cancel out when comparing treatment means as they do in a balanced randomized block design (see Chapter 14). The difference in the sums of squares, $SSE_2 - SSE_1$, has been adjusted for any effects due to blocks caused by the imbalance in the design. This difference is called the sum of squares due to treatments *adjusted for blocks*.

SST_{adj}

$$SSE_2 - SSE_1 = SST_{adj}$$

The sum of squares due to blocks **unadjusted for any treatment differences** is obtained by subtraction:

$$SSB = TSS - SST_{adj} - SSE$$

where SSE and TSS are sums of squares from the complete model. (Note: We could also obtain SSB, the uncorrected sum of squares for blocks, using the shortcut formula of Section 14.3.)

AOV table, treatments

The analysis of variance table for testing the effect of treatments is shown in Table 15.3. In the table n is the number of actual observations.

The corresponding sum of squares for testing the effect of blocks has the same complete model (model 1) as before, and

$$y_{ij} = \mu + \alpha_i + \varepsilon_{ij}$$

TABLE 15.3   AOV Table for Testing the Effects of Treatments, Unbalanced Randomized Block Design

Source	SS	df	MS	F
Blocks	SSB	$b - 1$	—	—
Treatments$_{adj}$	SST_{adj}	$t - 1$	MST_{adj}	MST_{adj}/MSE
Error	SSE	by subtraction	MSE	
Totals	TSS	$n - 1$		

TABLE 15.4   AOV Table for Testing Effects of Blocks, Unbalanced Randomized
Block Design

Source	SS	df	MS	F
Blocks$_{adj}$	SSB$_{adj}$	$b - 1$	MSB$_{adj}$	MSB$_{adj}$/MSE
Treatments	SST	$t - 1$	—	—
Error	SSE	by subtraction	MSE	—
Totals	TSS	$n - 1$		

SSB$_{adj}$ is the reduced model (model 2). The sum of squares drop, $SSE_2 - SSE_1$, is the sum of squares due to blocks after adjusting for the effects of treatments. By subtraction, we obtain

$$SST = TSS - SSB_{adj} - SSE$$

AOV table, blocks   The AOV table is shown in Table 15.4.

Note that SST and SST$_{adj}$ are not the same quantity in an unbalanced design; they will be the same only for a balanced design. Similarly, SSB and SSB$_{adj}$ are different quantities in an unbalanced design. For an unbalanced design we have the following identities:

$$TSS = SST_{adj} + SSB + SSE$$
$$= SST + SSB_{adj} + SSE$$

but

$$TSS \neq SST_{adj} + SSB_{adj} + SSE$$

EXERCISES

15.1   Refer to the data of Example 15.1 and the SAS computer output shown here. Some items are notated to help you identify quantities in the output.

a. Indicate the complete and reduced models for testing treatments.

b. Construct an analysis of variance table for testing treatments. Give the level of significance for your test and draw conclusions.

c. Indicate the complete and reduced models for testing blocks.

d. Construct an analysis of variance table for testing blocks. Give the levels of significance for your test.

```
SAS
DATA COWS;
INPUT Y BLOCKS TREATS;
CARDS;

  19 OBSERVATIONS IN DATA SET COWS      3 VARIABLES
PROC PRINT;
```

OBS	Y	BLOCKS	TREATS
1	15.4	1	1
2	9.6	1	2
3	9.5	1	3
4	8.4	1	4
5	14.8	2	1
6	9.3	2	2
7	9.4	2	3
8	15.9	3	1
9	9.8	3	2
10	9.7	3	3
11	9.3	3	4
12	15.5	4	1
13	9.4	4	2
14	9.2	4	3
15	8.1	4	4
16	14.7	5	1
17	9.2	5	2
18	9.0	5	3
19	7.9	5	4

```
PROC GLM;
CLASSES BLOCKS TREATS;
MODEL Y = BLOCKS TREATS;
TITLE "EXERCISE 15.1";
```

GENERAL LINEAR MODELS PROCEDURE

SOURCE	DF	SUM OF SQUARES	MEAN SQUARE	F VALUE	PROB > F	R-SQUARE	C.V.
REGRESSION	7	143.41548246	20.48792607	394.80383	0.0001	0.99603550	2.12065%
ERROR	11	0.57083333	0.05189394			STD DEV	Y MEAN
CORRECTED TOTAL	18	143.98631579				0.22780241	10.74211

SOURCE	DF	TYPE I SS	F VALUE	PROB > F	TYPE III SS	F VALUE	PROB > F
BLOCKS	4	2.61464912	12.59612	0.0007	2.11266667	10.17781	0.0014
TREATS	3	140.80083333	904.41411	0.0001	140.80083333	904.41411	0.0001

SSB — SST$_{adj}$ — *F* test for treatments (adj) — SSB$_{adj}$ — *F* test for blocks (adj)

15.2 Refer to Example 15.1. Use the least significant difference criterion for identifying which treatments differ from the others. Use $\alpha = .05$.

15.3 Refer to Example 14.1. Suppose that the first observation in block 1 (plot 1) is missing. Analyze the data by estimating the missing value and then performing an analysis of variance. Use $\alpha = .05$.

15.4 Refer to Exercise 15.3. Perform the corresponding analysis by fitting complete and reduced models. Compare your conclusions to those in Exercise 15.3.

15.5 Refer to Exercise 15.1. Fit the reduced model $y_{ij} = \mu + \alpha_i + \varepsilon_{ij}$ to obtain SSE_2. The sum of squares drop will be the sum of squares due to blocks, adjusted for treatments. Verify that this computer value for SSB_{adj} is the same as that shown in TYPE III SS column of the computer output in Exercise 15.1.

15.6 Refer to the data of Exercise 14.3. Suppose that in the rocket propellant test for the second mixture to be analyzed by investigator 3, a piece of equipment malfunctioned. Instead of going back to the laboratories to prepare a duplicate mixture, the investigators proceeded to obtain the remaining propellant thrust data.

> a. Estimate the missing value.
> b. Perform an analysis of variance, using $\alpha = .05$.

15.7   Refer to Exercise 15.6.

> a. Use complete and reduced models to obtain an analysis of variance. Compare your results to those in Exercise 15.6.
> b. How would you analyze the data if the response for mixture 4 and investigator 1 was also missing?

| **15.3** | A LATIN SQUARE DESIGN WITH MISSING DATA |

Recall that a $t \times t$ Latin square design can be used to compare t treatment means while filtering out two additional sources of variability (rows and columns). The treatments are randomly assigned in such a way that each treatment appears in every row and in every column. In this section we will illustrate the method for performing an analysis of variance in a Latin square design when one observation is missing. Then we will use the general method of fitting complete and reduced models with missing observations, described for the randomized block design in Section 15.2, for more complicated designs.

estimating missing value   The formula for **estimating a single missing value** in a Latin square design is

$$M = \frac{t(T + R + C) - 2G}{(t - 1)(t - 2)}$$

where T, R, and C represent the treatment, row, and column totals, respectively, corresponding to the missing observation, and t is the number of treatments in the Latin square design.

EXAMPLE 15.2

A company has considered the properties (such as strength, elongation, and so on) of many different variations of nylon stocking in trying to select the experimental stockings to be placed in extensive consumer acceptance surveys.

Five versions (A, B, C, D, and E) of the stockings have passed the preliminary screening and are scheduled for more extensive testing. As part of the testing, five samples of each type are to be examined for elongation under constant stress by each of five investigators on five separate days. The analyses are to be performed following the random assignment of a Latin square. The elongation data (in centimeters) are displayed in Table 15.5.

TABLE 15.5   Elongation Data for Example 15.2

Investigator	\multicolumn{10}{c}{Day}

Investigator	1		2		3		4		5	
1	B	22.1	A	18.6	C	23.0	E	24.3	D	17.1
2	C	23.5	D	16.5	A	18.7	B	22.0	E	—
3	D	17.4	E	23.8	B	22.8	C	23.9	A	20.0
4	A	20.3	B	23.4	E	25.9	D	18.7	C	24.2
5	E	25.7	C	24.8	D	18.9	A	20.6	B	24.6

Note that the measurement on variety E stockings for investigator 2 is missing and that the experiment was not rerun to obtain an observation. Use the methods of this section to estimate the missing value.

Solution

For our data the treatment, row, and column totals corresponding to the missing observations are

$$T_E = 99.70 \qquad R_2 = 80.70 \qquad C_5 = 85.90$$

Then with $t = r = c = 5$ and $G = 520.80$, we find

$$M = \frac{5(99.7 + 80.7 + 85.9) - 2(520.8)}{4(3)} = 24.2$$

The analysis could now proceed as for a balanced Latin square design, using the shortcut formulas.

Having located a significant effect due to treatments, we can make pairwise treatment comparisons using the formulas below. The least significant difference between the treatment with the missing value and any other treatment is

LSD

$$\text{LSD} = t_{\alpha/2} \sqrt{\text{MSE}\left(\frac{2}{t} + \frac{1}{(t-1)(t-2)}\right)}$$

For any other pair of treatments, the LSD is as before:

$$\text{LSD} = t_{\alpha/2} \sqrt{\frac{2\text{MSE}}{t}}$$

fitting complete and reduced models

For Latin squares with more than one missing observation, it might be easier to use the method of **fitting complete and reduced models** to adjust for imbalances caused by the missing values. In general, using the complete model

$$y_{ijk} = \mu + \alpha_i + \beta_j + \gamma_k + \varepsilon_{ijk}$$

and a computer solution, we would get the analysis of variance table shown in Table 15.6. Note that the sum of squares due to rows is unadjusted, the sum of squares for columns is adjusted for rows, and the sum of squares for treatments is adjusted for both rows and columns.

TABLE 15.6   AOV Table for an Unbalanced Latin Square Design

Source	SS	df	MS	F
Rows	SSR	$t-1$	—	—
Columns (adjusted for rows)	$\text{SSC}_{\text{adj}}\text{R}$	$t-1$	—	—
Treatments (adjusted for rows, columns)	$\text{SST}_{\text{adj}}\text{R, C}$	$t-1$	$\text{MST}_{\text{adj}}\text{R, C}$	$\text{MST}_{\text{adj}}\text{R, C/MSE}$
Error	SSE	by subtraction	MSE	
Totals	TSS	$n-1$		

Even though we do not have all the information for the analysis of variance tables, the corresponding tests for either rows or columns can be obtained from the computer output for this same model by using the TYPE III sums of squares column. Here we obtain $SSR_{adj}C$, T and $SSC_{adj}R$, T. In the computer output the F test and level of significance for these tests are given in the adjacent columns to the right of the partial sums of squares.

EXERCISES

15.8   Refer to Example 15.2. Perform an analysis of variance, using the estimated value 24.2. Use $\alpha = .05$ to draw your conclusions.

15.9   Use the SAS computer output shown here to give an analysis of variance table for testing the effect of treatments adjusted for rows (investigators) and columns (days). Indicate the results of testing separately for effects of rows and columns. Use $\alpha = .05$. Compare your results to those of Exercise 15.8.

```
SAS

DATA NYLON;
INPUT Y INV DAY TRT;
CARDS;

24 OBSERVATIONS IN DATA SET NYLON     4 VARIABLES
PROC PRINT;

OBS      Y       INV     DAY     TRT
  1     22.1      1       1       2
  2     23.5      2       1       3
  3     17.4      3       1       4
  4     20.3      4       1       1
  5     25.7      5       1       5
  6     18.6      1       2       1
  7     16.5      2       2       4
  8     23.8      3       2       5
  9     23.4      4       2       2
 10     24.8      5       2       3
 11     23.0      1       3       3
 12     18.7      2       3       1
 13     22.8      3       3       2
 14     25.9      4       3       5
 15     18.9      5       3       4
 16     24.3      1       4       5
 17     22.0      2       4       2
 18     23.9      3       4       3
 19     18.7      4       4       4
 20     20.6      5       4       1
 21     17.1      1       5       4
 22     20.0      3       5       1
 23     24.2      4       5       3
 24     24.6      5       5       2

PROC GLM;
CLASSES INV DAY TRT;
MODEL Y = INV DAY TRT;
TITLE "EXERCISE 15.9";
```

GENERAL LINEAR MODELS PROCEDURE

SOURCE	DF	SUM OF SQUARES	MEAN SQUARE	F VALUE	PROB > F	R-SQUARE	C.V.
REGRESSION	12	189.95683333	15.82973611	120.65626	0.0001	0.99245994	1.66918%
ERROR	11	1.44316667	0.13119697				
						STD DEV	Y MEAN
CORRECTED TOTAL	23	191.40000000					
						0.36221122	21.70000

SOURCE	DF	TYPE I SS SSR	F VALUE	PROB > F	$SSR_{adj}C$, T TYPE III SS	F VALUE	PROB > F
INV	4	22.32850000	42.54767	0.0001	14.36883333	27.38027	0.0001
DAY	4	2.13400000	4.06640	0.0289	0.94283333	1.79660	0.1994
TRT	4	165.49433333	315.35472	0.0001	165.49433333	315.35472	0.0001
	$SSC_{adj}R$	$SST_{adj}R$, C	F test for treatments (adj)		$SSC_{adj}R$, T	F tests for rows and columns	

15.10 The data of Example 14.4 have been reproduced below. Suppose that a car malfunction invalidated the data for driver 3, model 4, and blend 2. Rather than rent the speedway for another day, the investigators decided to analyze the data without replacing the missing value.

Driver		Car model							
		1		2		3		4	
1	IV	15.5	II	33.9	III	13.2	I	29.1	
2	II	16.3	III	26.6	I	19.4	IV	22.8	
3	III	10.8	I	31.1	IV	17.1	II	—	
4	I	14.7	IV	34.0	II	19.7	III	21.6	

a. Run an analysis of variance by estimating the missing value. Use $\alpha = .05$.
b. Make treatment comparisons by using Fisher's least significant difference, with $\alpha = .05$.

15.11 Use the method of fitting complete and reduced models to obtain an analysis of variance for the data in Exercise 15.10.

15.4   INCOMPLETE BLOCK DESIGNS

So far we have discussed the analysis for unbalanced designs where the imbalance was not planned but rather was caused by some accident during the collection of the sample data. Sometimes, however, we may be forced to design an experiment in which we must sacrifice some balance to perform the experiment. To illustrate, suppose that a regulatory agency would like to compare the mean potencies for three different batches (A, B, C) of the same drug product. Assume for the sake of simplicity that each analyst can do just two analyses per day and that there are three analysts available on a given day. Thus it would be possible to complete a comparison of the three batches on a single day if each analyst examines just two of the three possible batches. One possible design would be the arrangement listed here.

Analyst		
1	2	3
A	C	B
B	A	C

incomplete block design

balanced incomplete block design

We can think of this design as a partial (incomplete) randomized block design, where the number of treatments (batches) per block (analyst) is less than *t*, the number of treatments. In fact, a design such as this has become known as an **incomplete block design**.

There are many different types of incomplete block designs. The one we have constructed belongs to a class of designs that statisticians have called **balanced incomplete block designs**. Although these designs are not balanced as we have defined the term in Definition 14.5, the designs do retain some balance. For example, even though all treatments do not appear in the same block, each pair of treatments appears together in a block the same number of times. The pairs of treatments AB, AC, and BC in our design appear together once in a block. We acheived this "balance" by taking all possible combinations of two of the three treatments for blocks.

Consider now an extension to the balanced incomplete block design, with three treatments per block. Suppose we want to compare nine different batches in sets of three each. If each analyst can run three analyses, how many analysts would be required to maintain a balance similar to that obtained with our previous design?

Following similar logic, we could consider all possible combinations of three treatments with one analyst assigned to each different combination. Without too much trouble, we can show that this would require 84 analysts. Obviously, it would be prohibitive for most companies to employ 84 different analysts to compare the nine treatments in sets of three. Fortunately, there are other balanced incomplete designs that accomplish the experimental objective. One such design is shown in Table 15.7. By employing 12 analysts (blocks), we can compare the nine treatments.

TABLE 15.7   A Balanced Incomplete Block Design for Comparing Nine Treatments with Three Treatments per Block

Block	Treatments			Block	Treatments		
1	A	B	C	7	A	E	I
2	D	E	F	8	B	F	G
3	G	H	I	9	C	D	H
4	A	D	G	10	A	F	H
5	B	E	H	11	B	D	I
6	C	F	I	12	C	E	G

EXAMPLE 15.3   Identify (by quantity) the parameters listed below for the balanced incomplete block design described in the previous discussion.

t: number of treatments
k: number of treatments per block
b: number of blocks
r: number of repetitions of each treatment
λ: number of times each pair of treatments appears together in a block

Solution   From Table 15.7 we see that

$$t = 9 \qquad k = 3 \qquad b = 12 \qquad r = 4$$

In addition, after a cursory check we see that $\lambda = 1$ (i.e., each pair of treatments appears once in a block).

Before considering the analysis of variance for balanced incomplete blocks, we should note that a balanced incomplete block design does not exist for all possible values of t, k, b, and r. While some researchers in statistics have been concerned with methods for constructing balanced incomplete block designs and other more complicated incomplete block designs, we encourage you to refer to tables of incomplete block designs (see Cochran and Cox (1957)) when searching for a design to satisfy certain experimental objectives.

The analysis of variance for a balanced incomplete block design can be performed either by using specifically developed shortcut formulas or by using the method of fitting complete and reduced models as discussed for unbalanced designs. We will present the shortcut formulas for the analysis of variance table shown in Table 15.8.

shortcut formulas   The quantities SSB (the sum of squares unadjusted for treatments) and the total sum of squares are computed as we have done previously:

$$TSS = \sum y^2 - \frac{G^2}{n}$$

and

$$SSB = \sum_j \frac{B_j^2}{k} - \frac{G^2}{n}$$

TABLE 15.8   Analysis of Variance Table for a Balanced Incomplete Block Design

Source	SS	df	MS	F
Blocks	SSB	$b - 1$	—	—
Treatments$_{adj}$	SST$_{adj}$	$t - 1$	MST$_{adj}$	MST$_{adj}$/MSE
Error	SSE	by subtraction	MSE	
Totals	TSS	$n - 1$		

where B_j is the sum of all observations in block j. Then if we define

T_i = sum of all observations on treatment i

$B_{(i)}$ = sum of all measurements for blocks that contain treatment i

the sum of squares for treatments adjusted for blocks is

$$SST_{adj} = \frac{t - 1}{nk(k - 1)} \sum_i (kT_i - B_{(i)})^2$$

The sum of squares for error is found by subtraction:

$$SSE = TSS - SSB - SST_{adj}$$

As indicated in the analysis of variance table, the test statistic for testing the hypothesis of no difference among the treatment means is MST_{adj}/MSE.

EXAMPLE 15.4

A large company enlisted the help of a random sample of 20 potential consumers in a given geographical location to compare the physical characteristics (such as firmness and rebound) of eight experimental pillows and one presently marketed pillow. Since it was assumed that people would have a difficult time in distinguishing differences among the pillows when presented with all nine at once, it was decided to employ the balanced incomplete block design shown in Table 15.7.

After the pillow types were randomly assigned the letters from A to I, tables were prepared with the appropriate pillow types assigned to each table. For example, pillows G, H, and I were placed on Table 3. Each pillow was sealed in an identical white pillowcase and hence could not be distinguished from the others by color. The only marking on the pillowcase was a four-digit number, which provided the investigators with an identification code. With all tables in place, the 20 potential consumers were asked to proceed one at a time through the design from Table 1 to Table 12, stopping at each table to compare the three pillows. All persons were to record a firmness score for each pillow at each table, based on a 1-to-5-point scale (higher score indicates more firmness). The sums of scores for each pillow at the 12 tables are recorded in Table 15.9 (with letters identifying the pillow type).

Use the shortcut formulas of this section to perform an analysis of variance. Use $\alpha = .05$ to test the null hypothesis of no differences among treatment (pillow) means.

Solution

For an analysis using the shortcut formulas, it is convenient to construct a table of totals, as shown in Table 15.10.

From Table 15.10 we find that

$$SST_{adj} = \frac{(t - 1) \sum_i (kT_i - B_{(i)})^2}{nk(k - 1)} = \frac{8(322,112.00)}{36(3)(2)} = 11,930.07$$

Similarly, using the block totals from the raw data table, we obtain

$$SSB = \sum_j \frac{B_j^2}{k} - \frac{G^2}{n} = \frac{368,991.00}{3} - \frac{(2065)^2}{36}$$

$$= 122,997.00 - 118,450.69 = 4546.31$$

TABLE 15.9   Sum of Firmness Scores, Example 15.4

Block (Table)	Treatment (Pillow)						Block Totals
1	A	59	B	26	C	38	123
2	D	85	E	92	F	69	246
3	G	74	H	52	I	27	153
4	A	62	D	70	G	68	200
5	B	27	E	98	H	59	184
6	C	31	F	60	I	35	126
7	A	63	E	85	I	30	178
8	B	22	F	73	G	75	170
9	C	45	D	74	H	51	170
10	A	52	F	76	H	43	171
11	B	18	D	79	I	41	138
12	C	41	E	84	G	81	206
							2065

TABLE 15.10   Totals for the Data of Table 15.9

Treatment	T_i	$B_{(i)}$	$kT_i - B_{(i)}$
A	236	672	36
B	93	615	−336
C	155	625	−160
D	308	754	170
E	359	814	263
F	278	713	121
G	298	729	165
H	205	678	−63
I	133	595	−196
Total	2065		

The total sum of squares is

$$TSS = \sum y^2 - \frac{G^2}{n} = 135,435.00 - 118,450.69 = 16,984.31$$

The analysis of variance table for testing the hypothesis of no differences among the treatment means is shown in Table 15.11.

TABLE 15.11   AOV Table for the Data of Example 15.4

Source	SS	df	MS	F
Blocks	4,546.31	11	—	—
Treatments$_{adj}$	11,930.07	8	1491.26	46.97
Error	507.93	16	31.75	
Totals	16,984.31	35		

Since the compute value of F, 46.97, exceeds the table value, 2.59, for $df_1 = 8$, $df_2 = 16$, and $a = .05$, we conclude that there are differences among the nine treatment means.

comparison among treatment means

Following the observation of a significant F test concerning differences among treatment means, we naturally might like to determine which treatment means are significantly different from the others. To do this, we make use of the following notation: $\hat{\mu}_i$, an estimate of the mean for treatment i, given by

$$\hat{\mu}_i = \bar{y} + \frac{kT_i - B_{(i)}}{t\lambda}$$

where $\bar{y}$ is the overall sample mean. An estimate of the difference between two treatment means i and j is then

$$\hat{\mu}_i - \hat{\mu}_j = \frac{[kT_i - B_{(i)}] - [kT_j - B_{(j)}]}{t\lambda}$$

The least significant difference between any pair of treatment means is

LSD

$$LSD = t_{\alpha/2}\sqrt{\frac{2kMSE}{t\lambda}}$$

EXAMPLE 15.5

Compute the least significant difference for the nine treatment means of Example 15.4. Determine all pairwise differences, using $\alpha = .05$.

Solution

For these data the overall sample mean is $\bar{y} = 2065/36 = 57.36$. Then using the column $kT_i - B_{(i)}$, we have the following estimated treatment means:

Treatment	$\bar{y} + \dfrac{kT_i - B_{(i)}}{t\lambda}$
A	61.36
B	20.03
C	39.58
D	76.25
E	86.58
F	70.80
G	75.69
H	50.36
I	35.58

Using MSE $= 31.75$, based on df $= 16$, we obtain

$$LSD = t_{\alpha/2}\sqrt{\frac{2kMSE}{t\lambda}} = 2.12\sqrt{\frac{2(3)31.75}{9(1)}} = 9.75$$

The nine treatment means are arranged below in ascending order, with a summary of the significant results. Those treatments underlined by a common line are

not significantly different from each other, using the least significant difference criterion (see Chapter 11).

B	I	C	H	A	F	G	D	E
20.03	35.58	39.58	50.36	61.36	70.80	75.69	76.25	86.58

15.5   SUMMARY

In this chapter we have discussed the analysis of variance for some unbalanced designs
beginning with a discussion of the analysis for a randomized block design with one missing
observation. Two possible analyses were proposed. The first required that we estimate the
missing value and then proceed with the usual shortcut formulas developed in Chapter
14. While estimating a single missing value is quite easy to do, the procedure becomes
more difficult when there is more than one missing value. The second procedure, that of
fitting complete and reduced models to obtain adjusted sums of squares, can be used
for one or more missing observations.

With the Latin square design, we again showed how to estimate a single missing observation and proceed with the usual analysis. But, as with the randomized block design,
the method of analysis by fitting complete and reduced models is more appropriate when
there is more than one missing value.

Finally, we considered another class of unbalanced designs, incomplete block designs.
The particular designs we discussed were incomplete randomized block designs in which
not all treatments appear in each block. These incomplete block designs retain a certain
amount of balance, since all pairs of treatments appear together in a block the same
number of times. The analysis for balanced incomplete block designs was illustrated using
appropriate shortcut formulas. While no example was given in the chapter to show a
computer solution for a balanced incomplete block design, we can obtain the analysis of
variance for testing treatment differences following the procedure outlined for a randomized block design with missing observations.

KEY FORMULAS

1. Missing observation, randomized block design

$$M = \frac{tT + bB - G}{(t - 1)(b - 1)}$$

2. Fisher's LSD for a randomized block design
   a. For any pair of treatments with no missing value

$$LSD = t_{\alpha/2} \sqrt{\frac{2MSE}{b}}$$

b. Between the treatment with a missing value and any other treatment

$$LSD = t_{\alpha/2} \sqrt{MSE\left(\frac{2}{b} + \frac{t}{b(b-1)(t-1)}\right)}$$

3. Equalities for randomized block design

$$SSB = TSS - SST_{adj} - SSE$$
$$SST = TSS - SSB_{adj} - SSE$$

4. Missing observation, Latin square design

$$M = \frac{t(T + R + C) - 2G}{(t-1)(t-2)}$$

5. Fisher's LSD for a Latin square design
   a. For any pair of treatments with no missing value

$$LSD = t_{\alpha/2} \sqrt{\frac{2MSE}{t}}$$

   b. Between the treatment with the missing value and any other treatment

$$LSD = t_{\alpha/2} \sqrt{MSE\left(\frac{2}{t} + \frac{1}{(t-1)(t-2)}\right)}$$

6. Sums of squares for an incomplete block design

$$SST_{adj} = \frac{t-1}{nk(k-1)} \sum_i (kT_i - B_{(i)})^2$$

$$SSE = TSS - SSB - SST_{adj}$$

7. Pairwise comparisons of treatment means, incomplete block design

$$\hat{\mu}_i - \hat{\mu}_j = \frac{(kT_i - B_{(i)}) - (kT_j - B_{(j)})}{t\lambda}$$

$$LSD = t_{\alpha/2} \sqrt{\frac{2kMSE}{t\lambda}}$$

EXERCISES

 15.12   A physician was interested in comparing the effects of six different antihistamines in persons extremely sensitive to a ragweed skin allergy test. To do this, a random sample of 10 allergy patients was selected from the physician's private practice, with treatments (antihistamines) assigned to each patient according to the experimental design shown in the following table. Each person then received injections of the assigned antihistamines in different sections of the right arm. The area of redness surrounding the point of injection was measured after a fixed period of time. The data are shown in the table.

Person			Treatments			
1	B	25	A	41	F	40
2	E	37	B	46	A	42
3	C	45	D	33	B	37
4	E	34	D	35	A	46
5	B	31	F	42	D	34
6	C	56	E	36	F	65
7	D	33	A	42	C	67
8	F	49	D	37	E	30
9	C	59	A	40	F	55
10	B	36	C	57	E	34

    a. Identify the design.

    b. Identify the characteristics of the design.

15.13 Refer to the data of Exercise 15.12. Do the data indicate differences among the treatment means? Use $\alpha = .05$.

15.14 Refer to Exercise 15.13. Use the least significant difference criterion for determining treatment differences, with $\alpha = .05$.

15.15 Use a computer program to perform the same analysis as in Exercise 15.13. Compare the results of both exercises.

15.16 Refer to Example 15.4. Use a computer program to perform an analysis of variance. Are your results the same as those found in the example?

15.17 Indicate how you would test for a significant effect due to blocks in a balanced incomplete block design.

 15.18 The marketing research group of a corporation examined the public response to the introduction of a new TV game module by comparing weekly sales ($000) volumes for three different store chains in each of four geographic locations.

Geographic Area		Chain 1	2	3
N	W1	35	17	7
	W2	30	22	12
S	W1	42	30	22
	W2	48	28	19
E	W1	35	35	15
	W2	38	40	20
W	W1	22	43	28
	W2	26	48	23

    a. Write an appropriate model (including an effect for weeks) and the sources of variability in an analysis of variance table.

    b. How would your model change if we analyze the total two-week sales data?

    c. Run an analysis of variance on the two-week sales data using shortcut formulas from Chapter 14.

15.19 Refer to Exercise 15.18. Use Tukey's procedure to compare the different geographic areas by chain means.

15.20 Refer to Exercise 15.18. Suppose that the week 1 data were not available in the north and east for chain 1, due to logistics problems that slowed the introduction of the product by a week.

a. Write an appropriate model.

b. Suggest a method for analyzing the data using available software.

c. Write model(s) for the procedure described in (b).

15.21 A foreign automobile manufacturer is spending hundreds of millions of dollars to construct a large manufacturing plant (about 70 acres under one roof) here in the United States. One of their objectives is to produce cars of high quality in the United States using U.S. workers. One part of the massive orientation program for new employees is to send about 20% of them to the home country for additional training. One measure of the worth of this additional training is whether the product quality is better on assembly lines where 20% of the employees have had the homeland orientation and been able to share it with their fellow employees. Data from six assembly lines (three with the additional orientation) are shown here. Two different inspectors examined each of two cars chosen at random for defects from the assembly lines. Use these data to answer the following questions.

	Assembly line	Inspector 1	Inspector 2		Assembly Line	Inspector 1	Inspector 2
Additional training	1	6	6	No additional training	4	8	7
		3	4			5	5
	2	4	3		5	10	9
		2	2			4	4
	3	2	3		6	15	13
		1	1			7	6

a. Suggest an appropriate dependent variable.

b. Write a model for this experimental situation and identify all terms.

c. Fill out the sources and degrees of freedom for an AOV table.

15.22 Refer to the conditions of Exercise 15.21.

a. Suggests a means to analyze these data.

b. Use the output shown here to draw conclusions.

c. Can you suggest any plots that might be helpful in interpreting the data?

```
                LISTING OF DATA

 OBS    DEFECTS    LINE    INSPECT    TRAIN
  1        6         1        1         1
  2        6         1        2         1
  3        3         1        1         1
  4        4         1        2         1
  5        4         2        1         1
  6        3         2        2         1
  7        2         2        1         1
  8        2         2        2         1
  9        2         3        1         1
 10        3         3        2         1
 11        1         3        1         1
```

12	1	3	2	1
13	8	4	1	0
14	7	4	2	0
15	5	4	1	0
16	5	4	2	0
17	10	5	1	0
18	9	5	2	0
19	4	5	1	0
20	4	5	2	0
21	15	6	1	0
22	13	6	2	0
23	7	6	1	0
24	6	6	2	0

N = 24

FIRST MODEL

GENERAL LINEAR MODELS PROCEDURE
CLASS LEVEL INFORMATION

CLASS	LEVELS	VALUES
INSPECT	2	1 2
LINE	6	1 2 3 4 5 6
TRAIN	2	0 1

NUMBER OF OBSERVATIONS IN DATA SET = 24

FIRST MODEL

GENERAL LINEAR MODELS PROCEDURE

DEPENDENT VARIABLE: DEFECTS   NUMBER OF DEFECTS

SOURCE	DF	SUM OF SQUARES	MEAN SQUARE	F VALUE	PR > F	R-SQUARE	C.V.
MODEL	11	190.83333333	17.34848485	1.98	0.1275	0.645070	54.6100
ERROR	12	105.00000000	8.75000000		STD DEV		DEFECTS MEAN
CORRECTED TOTAL	23	295.83333333			2.95803989		5.41666667

SOURCE	DF	TYPE I SS	F VALUE	PR > F	DF	TYPE III SS	F VALUE	PR > F
TRAIN	1	130.66666667	14.93	0.0023	1	130.66666667	14.93	0.0023
LINE (TRAIN)	4	56.66666667	1.62	0.2329	4	56.66666667	1.62	0.2329
INSPECT	1	0.66666667	0.08	0.7872	1	0.66666667	0.08	0.7872
INSPECT*TRAIN	1	1.50000000	0.17	0.6861	1	1.50000000	0.17	0.6861
INSPECT*LINE(TRAIN)	4	1.33333333	0.04	0.9968	4	1.33333333	0.04	0.9968

SECOND MODEL

GENERAL LINEAR MODELS PROCEDURE

DEPENDENT VARIABLE: DEFECTS   NUMBER OF DEFECTS

SOURCE	DF	SUM OF SQUARES	MEAN SQUARE	F VALUE	PR > F	R-SQUARE	C.V.
MODEL	7	189.50000000	27.07142857	4.07	0.0095	0.640563	47.5929
ERROR	16	106.33333333	6.64583333		STD DEV		DEFECTS MEAN
CORRECTED TOTAL	23	295.83333333			2.57795138		5.41666667

SOURCE	DF	TYPE I SS	F VALUE	PR > F	DF	TYPE III SS	F VALUE	PR > F
TRAIN	1	130.66666667	19.66	0.0004	1	130.66666667	19.66	0.0004
LINE (TRAIN)	4	56.66666667	2.13	0.1240	4	56.66666667	2.13	0.1240
INSPECT	1	0.66666667	0.10	0.7555	1	0.66666667	0.10	0.7555
INSPECT*TRAIN	1	1.50000000	0.23	0.6411	1	1.50000000	0.23	0.6411

THIRD MODEL

GENERAL LINEAR MODELS PROCEDURE

DEPENDENT VARIABLE: DEFECTS   NUMBER OF DEFECTS

SOURCE	DF	SUM OF SQUARES	MEAN SQUARE	F VALUE	PR > F	R-SQUARE	C.V.
MODEL	6	188.00000000	31.33333333	4.94	0.0043	0.635493	46.4965
ERROR	17	107.83333333	6.34313725		STD DEV		DEFECTS MEAN
CORRECTED TOTAL	23	295.83333333			2.51855857		5.41666667

SOURCE	DF	TYPE I SS	F VALUE	PR > F	DF	TYPE III SS	F VALUE	PR > F
TRAIN	1	130.66666667	20.60	0.0003	1	130.66666667	20.60	0.0003
LINE (TRAIN)	4	56.66666667	2.23	0.1084	4	56.66666667	2.23	0.1084
INSPECT	1	0.66666667	0.11	0.7497	1	0.66666667	0.11	0.7497

15.23   Refer to Exercise 15.21. Suppose that inspector 2 was unable to evaluate the second car from assembly line 4 and that inspector 1 missed car 1 from assembly line 3.

a. Does the model change? Suggests a method for analyzing the data.

b. Use the computer output shown here to draw conclusions.

LISTING OF ADJUSTED DATA

OBS	DEFECTS	LINE	INSPECT	TRAIN
1	6	1	1	1
2	6	1	2	1
3	3	1	1	1
4	4	1	2	1
5	4	2	1	1
6	3	2	2	1
7	2	2	1	1
8	2	2	2	1
9	3	3	2	1
10	1	3	1	1
11	1	3	2	1
12	8	4	1	0
13	7	4	2	0
14	5	4	1	0
15	10	5	1	0
16	9	5	2	0
17	4	5	1	0
18	4	5	2	0
19	15	6	1	0
20	13	6	2	0
21	7	6	1	0
22	6	6	2	0

N = 22

FIRST MODEL

GENERAL LINEAR MODELS PROCEDURE
CLASS LEVEL INFORMATION

CLASS	LEVELS	VALUES
INSPECT	2	1 2
LINE	6	1 2 3 4 5 6
TRAIN	2	0 1

NUMBER OF OBSERVATIONS IN DATA SET = 22

FIRST MODEL
GENERAL LINEAR MODELS PROCEDURE
DEPENDENT VARIABLE: DEFECTS   NUMBER OF DEFECTS

SOURCE	DF	SUM OF SQUARES	MEAN SQUARE	F VALUE	PR > F	R-SQUARE	C.V.
MODEL	11	180.81818182	16.43801653	1.60	0.2326	0.638216	57.2637
ERROR	10	102.50000000	10.25000000		STD DEV		DEFECTS MEAN
CORRECTED TOTAL	21	283.31818182			3.20156212		5.59090909

SOURCE	DF	TYPE I SS	F VALUE	PR > F	DF	TYPE III SS	F VALUE	PR > F
TRAIN	1	127.68181818	12.46	0.0055	1	124.32142857	12.13	0.0059
LINE (TRAIN)	4	49.30303030	1.20	0.3683	4	48.75000000	1.19	0.3733
INSPECT	1	0.18750000	0.02	0.8951	1	0.03571429	0.00	0.9541
INSPECT*TRAIN	1	1.02083333	0.10	0.7588	1	0.89285714	0.09	0.7739
INSPECT*LINE (TRAIN)	4	2.62500000	0.06	0.9913	4	2.62500000	0.06	0.9913

SECOND MODEL

GENERAL LINEAR MODELS PROCEDURE
DEPENDENT VARIABLE: DEFECTS   NUMBER OF DEFECTS

SOURCE	DF	SUM OF SQUARES	MEAN SQUARE	F VALUE	PR > F	R-SQUARE	C.V.
MODEL	7	178.19318182	25.45616883	3.39	0.0247	0.628951	49.0125
ERROR	14	105.12500000	7.50892857		STD DEV		DEFECTS MEAN
CORRECTED TOTAL	21	283.31818182			2.74024243		5.59090909

SOURCE	DF	TYPE I SS	F VALUE	PR > F	DF	TYPE III SS	F VALUE	PR > F
TRAIN	1	127.68181818	17.00	0.0010	1	123.52083333	16.45	0.0012
LINE (TRAIN)	4	49.30303030	1.64	0.2191	4	50.14166667	1.67	0.2127
INSPECT	1	0.18750000	0.02	0.8767	1	0.18750000	0.02	0.8767
INSPECT*TRAIN	1	1.02083333	0.14	0.7179	1	1.02083333	0.14	0.7179

THIRD MODEL

GENERAL LINEAR MODELS PROCEDURE
DEPENDENT VARIABLE: DEFECTS   NUMBER OF DEFECTS

SOURCE	DF	SUM OF SQUARES	MEAN SQUARE	F VALUE	PR > F	R-SQUARE	C.V.
MODEL	6	177.17234848	29.52872475	4.17	0.0115	0.625348	47.5799
ERROR	15	106.14583333	7.07638889		STD DEV		DEFECTS MEAN
CORRECTED TOTAL	21	283.31818182			2.66014828		5.59090909

SOURCE	DF	TYPE I SS	F VALUE	PR > F	DF	TYPE III SS	F VALUE	PR > F
TRAIN	1	127.68181818	18.04	0.0007	1	123.52083333	17.46	0.0008
LINE (TRAIN)	4	49.30303030	1.74	0.1933	4	49.27234848	1.74	0.1935
INSPECT	1	0.18750000	0.03	0.8729	1	0.18750000	0.03	0.8729

15.24   Refer to Exercise 15.21. In addition to examining the total number of defects, each defect was classified as major or minor. These data are shown here.

Assembly Line	Inspector			
	1		2	
	Major	Minor	Major	Minor
1	1	5	1	5
	0	3	0	4
2	0	4	0	3
	0	2	0	2
3	0	2	0	3
	0	1	0	1
4	3	7	2	5
	0	5	0	5
5	3	7	3	6
	0	4	0	4
6	5	10	5	8
	1	6	1	5

a. How consistent are the inspectors at evaluating major defects? Minor defects?
b. Choose a dependent variable for analyzing these data.
c. Write an appropriate model, identifying all terms.
d. Fill out the sources and degrees of freedom for an AOV table.

15.25   Refer to Exercise 15.24. Suggest a dependent variable that accounts for both major and minor defects but does not obscure their identity. Would any transformations of the data be appropriate?

16

ANALYSIS OF VARIANCE FOR SOME FIXED-, RANDOM-, AND MIXED-EFFECTS MODELS

16.1 INTRODUCTION

In previous chapters we have been able to write the model for a response in terms of k independent variables, using the **general linear model**

$$y_i = \beta_0 + \beta_1 x_{i1} + \beta_2 x_{i2} + \cdots + \beta_k x_{ik} + \varepsilon_i$$

Initially we assumed $x_1, x_2, \ldots, x_k$ to be independent variables measured without error; $\beta_0, \beta_1, \ldots, \beta_k$ to be unknown parameters; and the random error ε_i associated with observation i to have $E(\varepsilon_i) = 0$. Then we expanded these assumptions to include the following:

1. $\varepsilon_1, \varepsilon_2, \ldots, \varepsilon_n$ are independent of one another.
2. For a given setting of the independent variables $x_1, x_2, \ldots, x_k$, the variance of ε_i is σ_ε^2.

Thus although we had many terms in the model, there was only one source of random variation.

DEFINITION 16.1

A model that can be written in the form of a general linear model with $k > 0$ independent variables and one random component is called a **fixed-effects model**.

All models discussed so far in this text relating a response y to one or more independent variables have fallen into the category of fixed-effects models. Inferences for fixed-effects models are stated in terms of one or more parameters in the general linear model.

Sometimes, however, we must account for more than one source of random variation in an experimental situation, and the model must be expanded to accommodate **components of variation** these additional **components of variation**. For example, the blocks in a randomized block design might represent a random sample of b blocks taken from a population of all possible blocks. Then the effects due to blocks are considered to be random effects rather than fixed effects. Appropriate changes would be made in the model to reflect this difference in interpretation.

Before we give examples of situations that might warrant the inclusion of more than one random source of variability in a model, we define two different types of models.

DEFINITION 16.2

A model that can be written in the form of a general linear model with $k = 0$ independent variables and more than one random component is called a **random-effects model**.

DEFINITION 16.3

A model that can be written in the form of a general linear model with $k > 0$ independent variables and more than one random component is called a **mixed-effects** (or simply **mixed**) model.

In this chapter we will consider various random-effects and mixed models. For each model we will indicate the appropriate analysis of variance and also relate the analysis discussed to those that we would obtain with the corresponding fixed-effects model.

16.2   A ONE-FACTOR EXPERIMENT WITH TREATMENT EFFECTS RANDOM: A RANDOM-EFFECTS MODEL

The best way to illustrate the difference between the fixed- and random-effects models **example** for a one-factor experiment is by an **example**. Suppose we want to compare readings made on the intensities of the electrostatic discharges of lightning at three different tracking stations within a 20-mile radius of the central computing facilities of a university. If these three tracking stations are the only feasible tracking stations for such an operation and inferences are to be about these stations only, then we could write the fixed-effects model in the form of a general linear model (see Section 12.2):

fixed-effects model
$$y = \beta_0 + \beta_1 x_1 + \beta_2 x_2 + \varepsilon$$

where

$x_1 = 1$ if tracking station 2 $x_1 = 0$ otherwise

$x_2 = 1$ if tracking station 3 $x_2 = 0$ otherwise

Also, β_0, β_1, and β_2 are unknown parameters representing mean intensities or differences in mean intensities. Equivalently, using the results of Section 10.4, we could write the fixed-effects model as

$$y_{ij} = \mu + \alpha_i + \varepsilon_{ij}$$

where y_{ij} is the jth observation at tracking station i ($i = 1, 2, 3$), μ is an overall mean, and α_i is a fixed effect due to tracking station i. For both of these models, ε is assumed to be normally distributed, with mean 0 and variance σ^2.

Suppose, however, that rather than being concerned about only these three tracking stations, we consider these stations as a random sample of three taken from the many possible locations for tracking stations. Inferences would now relate not just to what happened at the sampled locations but also to what might happen at other possible locations for tracking stations. A model that can account for this difference in interpretation is

random-effects model   the **random-effects model**

$$y_{ij} = \mu + \alpha_i + \varepsilon_{ij}$$

While the model looks the same as the previous fixed-effects model, some of the

assumptions   **assumptions** are different.

1. μ is still an overall mean, which is an unknown constant.
2. α_i is a random effect due to the ith tracking station. We assume that α_i is normally distributed, with mean 0 and variance σ_α^2.
3. The α_is are independent.
4. As before, ε_{ij} is normally distributed, with mean 0 and variance σ_ε^2.
5. The ε_{ij}s are independent.
6. The random components α_i and ε_{ij} are independent.

The difference between the fixed-effects model and the random-effects model can be illustrated by supposing we were to repeat the experiment. For the fixed-effects model, we would use the same three tracking stations, so it would make sense to make inferences about the mean intensities or differences in mean intensities at these three locations. However, for the random-effects model, we would take another random sample of three tracking stations (i.e., take another sample of three αs). Now rather than concentrating on the effect of a particular group of three αs from one experiment, we would examine the variability of the population of all possible α values. This will be illustrated using the analysis of variance table given in Table 16.1.

The analysis of variance table is the same for a fixed- or random-effects model, with

EMS   the exception that the **expected mean squares (EMS)** columns are different. You will recall

TABLE 16.1   An AOV Table for a One-Factor Experiment: Fixed or Random Model

Source	SS	df	MS	EMS Fixed Effects	EMS Random Effects
Treatments	SST	$t - 1$	MST	$\sigma_\varepsilon^2 + r\theta_T$	$\sigma_\varepsilon^2 + r\sigma_\alpha^2$
Error	SSE	$t(r - 1)$	MSE	σ_ε^2	σ_ε^2
Totals	TSS	$rt - 1$			

that this column was not used in our tables in Chapters 10 and 14, since all mean squares except MSE had an expectation under the alternative hypothesis equal to σ_ε^2 plus a positive constant, which depended on the parameters under test. In general, with t treatments (tracking stations) and r observations per treatment, the **AOV table** would appear as shown in Table 16.1. For the fixed-effects model, θ_T is a positive function of the constants α_i, while σ_α^2 represents the variance of the population of α_i values for the random-effects model. Referring to our example, a test for the equality of the mean intensities at the three tracking stations in the fixed-effects model is (from Chapter 10)

H_0: $\alpha_1 = \alpha_2 = \alpha_3 = 0$ (i.e., the 3 means are identical)

H_a: At least one α is different from zero.

T.S.: $F = \text{MST}/\text{MSE}$, based on $\text{df}_1 = t - 1$ and $\text{df}_2 = t(r - 1)$

A test concerning the variability for the population of α values in the random-effects model makes use of the same test statistic. The null hypothesis and alternative hypothesis are

H_0: $\sigma_\alpha^2 = 0$

H_a: $\sigma_\alpha^2 > 0$

Since we assumed that the αs sampled were selected from a normal population with mean 0 and variance σ_α^2, the null hypothesis states that the αs were drawn from a normal population with mean 0 and variance 0; that is, all α values in the population are equal to zero.

Thus, although the forms of the null hypotheses are different for the two models, the meanings attached to them are very similar. For the fixed-effects model, we are assuming that the sampled αs (which are the only αs) are identically zero, whereas in the random-effects model, the null hypothesis leads us to assume that the sample αs, as well as all other αs in the population, are zero.

The alternative hypotheses are also similar. In the fixed-effects model, we are assuming that at least one of the αs is different from the rest; that is, there is some variability among the set of αs. For the random-effects model, the alternative hypothesis is that $\sigma_\alpha^2 > 0$; that is, not all α values are the same in the population.

EXAMPLE 16.1

Consider the problem we have used to illustrate a one-factor experiment with random treatment effects. Two graduate students working for a professor in electrical engineering have been funded to record lightning discharge intensities (intensities of the electrical field) at three tracking stations. Because of the high frequency of thunderstorms in the summer months (in Florida, storms occur on 80 or more days per year), the graduate students were to choose a point at random on a map of the 20-mile-radius region and assemble their tracking equipment (provided they could get permission of the property owners). Each day during the hours from 8 AM to 5 PM, they were to monitor their instruments until the maximum intensity had been recorded for five separate storms. The process was then repeated separately at the two other locations chosen at random. The sample data (in volts per meter) appear in Table 16.2.

AOV table

test for means

test for σ_α^2

TABLE 16.2   Lightning Discharge Intensities (in Volts per Meter), Example 16.1

Tracking Station	Intensities					Totals
1	20	1050	3200	5600	50	9,920
2	4300	70	2560	3650	80	10,660
3	100	7700	8500	2960	3340	22,600
Total						43,180

a.   Write an appropriate statistical model, defining all terms.
b.   Perform an analysis of variance and interpret your results. Use $\alpha = .05$.

Solution   A model appropriate for this one-factor experiment, with tracking stations chosen at random, is

$$y_{ij} = \mu + \alpha_i + \varepsilon_{ij} \qquad (i = 1, 2, 3; j = 1, 2, \ldots, 5)$$

where the terms in the model were as defined in this section.

The computational formulas for sums of squares are identical to those for a fixed-effects model.

$$\frac{G^2}{n} = \frac{(43,180)^2}{15} = 124,300,826.7$$

$$SST = \frac{\sum T_i^2}{5} - \frac{G^2}{n}$$

$$= 144,560,400 - 124,300,826.7 = 20,259,573.3$$

$$TSS = \sum y_{ij}^2 - \frac{G^2}{n}$$

$$= 232,550,000 - 124,300,826.7 = 108,249,173.3$$

By subtraction,

$$SSE = TSS - SST = 87,989,600$$

We can use these calculations to construct an AOV table, as shown in Table 16.3. The F test for $H_0: \sigma_\alpha^2 = 0$ is based on $df_1 = 2$ and $df_2 = 12$ degrees of freedom. Since the computed value of F, 1.38, does not exceed 3.89, the value in Table 6 for a = .05, $df_1 = 2$, and $df_2 = 12$, we have insufficient evidence to indicate that there is a significant random component due to variability in intensities from tracking station

TABLE 16.3   AOV Table for the Data of Example 16.1

Source	SS	df	MS	EMS	F
Tracking stations	20,259,573.3	2	10,129,786.65	$\sigma_\varepsilon^2 + 5\sigma_\alpha^2$	1.38
Error	87,989,600.0	12	7,332,466.67	σ_ε^2	
Totals	108,249,173.3	14			

to tracking station. Rather, as an electrical engineer postulated, it is probably best to work with a single tracking station, since most of the variability in intensities is related to the distance of the tracking station from the point of discharge, and we have no control of this source.

EXERCISES

 16.1   A pharmaceutical company would like to examine the potency of a liquid medication mixed in large vats. To do this, a random sample of five vats from a month's production was obtained, and four separate samples were selected from each vat.

a. Write a random-effects model for this experimental situation, identifying all terms in the model.

b. Run an analysis of variance for the sample data given below. Use $\alpha = .05$.

Vat 1	Vat 2	Vat 3	Vat 4	Vat 5
3.2	2.6	3.4	4.2	1.8
3.8	2.9	3.9	4.4	2.3
3.5	2.8	3.3	4.3	1.9
3.0	2.0	3.1	4.2	2.1

 16.2   Suppose that the pharmaceutical company of Exercise 16.1 wishes to estimate μ, the expected potency for a measurement made on a vat selected at random. We have not discussed this topic in this section, but as might be expected, we can estimate μ using $\bar{y}$, the mean of the sample data. However, it is not so obvious, but nonetheless true, that the variance of $\bar{y}$ is

$$\frac{\sigma_\varepsilon^2}{20} + \frac{\sigma_\alpha^2}{5}$$

where σ_α^2 is the variance of the random vat effect α_i. Since we do not know σ_ε^2 or σ_α^2, we can form estimates by using the MS and EMS columns of the AOV table shown below.

Source	MS	EMS
Vats	MSA	$\sigma_\varepsilon^2 + 4\sigma_\alpha^2$
Error	MSE	σ_ε^2

From the abbreviated table, we see that MSA has expected mean square of $\sigma_\varepsilon^2 + 4\sigma_\alpha^2$, and hence MSA/20 provides an estimate of the variance of $\bar{y}$, $(\sigma_\varepsilon^2/20) + (\sigma_\alpha^2/5)$. In general, for a random-effects model in a one-factor experiment with the same number of observations per treatment, $\bar{y}$ provides an estimate of μ and $\sqrt{MSA/n}$ provides an estimate of the standard error of $\bar{y}$.

a. Using the sample data of Exercise 16.1, form a point estimate of μ, the average potency for a measurement made on a vat selected at random.

b. Give a 95% confidence interval for μ of part (a). (Hint: The formula is: point estimate ± 2.131 standard error.)

| 16.3 | A TWO-FACTOR EXPERIMENT WITH BOTH FACTORS RANDOM: A RANDOM-EFFECTS MODEL |

The ideas presented for a random-effects model in a one-factor experiment can be extended to any of the block and factorial experimental designs covered in Chapters 10 and 14. Although we will not have time to cover all such situations, we will consider a two-factor experiment in which both factors are random.

Consider an experiment to examine the effects of different analysts and subjects in chemical analyses for the DNA content of plaque. Three female subjects (ages 18–20 years) were chosen for the study. Each subject was allowed to maintain her usual diet, supplemented with 30 mg (15 tablets) of sucrose per day. No toothbrushing or mouthwashing was allowed during the study. At the end of the week, plaque was scraped from the entire dentition of each subject and divided into three samples. Each of three analysts chosen at random was then given an unmarked sample of plaque from each of the subjects and asked to perform an analysis for the DNA content (in micrograms). The two-factor experiment of sample data could then be organized as shown in Table 16.4.

randomized block design   This experimental design is recognized as a **randomized block design**, with subjects representing blocks and analysts being the treatments. The experimental units are samples of plaque scraped from the dentition of subjects. If we assume that the three subjects represent a random sample from a large population of possible subjects, and, similarly, that the three analysts represent a random sample from a large population of possible analysts, we can write the following random-effects model relating DNA concentration to the two factors "analysts" and "subjects":

random-effects model

$$y_{ij} = \mu + \alpha_i + \beta_j + \varepsilon_{ij}$$

We assume the following:

assumptions

1. μ is an overall unknown concentration mean.
2. α_i is a random effect due to the ith analyst; α_i is normally distributed, with mean 0 and variance σ_α^2.
3. The α_is are independent.
4. β_j is a random effect due to the jth subject. β_j is a normally distributed random variable, with mean 0 and variance σ_β^2.
5. The β_js are independent.
6. The α_is and the β_js are independent.

TABLE 16.4   DNA Concentrations for Samples of Plaque

		Subject		
Analyst	1	2	3	Totals
1				
2				
3				
Totals				

TABLE 16.5   AOV Table for a Two-Factor Experiment, a Levels of Factor A and b Levels of Factor B

				EMS	
Source	SS	df	MS	Fixed Effects	Random Effects
A	SSA	$a - 1$	MSA	$\sigma_\varepsilon^2 + b\theta_A$	$\sigma_\varepsilon^2 + b\sigma_\alpha^2$
B	SSB	$b - 1$	MSB	$\sigma_\varepsilon^2 + a\theta_B$	$\sigma_\varepsilon^2 + a\sigma_\beta^2$
Error	SSE	$(a - 1)(b - 1)$	MSE	σ_ε^2	σ_ε^2
Totals	TSS	$ab - 1$			

Again note the difference between assuming that the treatments and blocks are random rather than fixed effects. If, for example, the three analysts chosen for the study were the only analysts of interest, we would be concerned with differences in mean DNA concentrations for these specific analysts. Now, however, treating the effect due to an analyst as a random variable, our inference will be about the population of analysts' effects. Since the mean of this normal population is assumed to be zero, we want to determine whether or not the variance σ_α^2 is greater than zero.

Once again the analysis of variance tables are the same for fixed- or random-effects models in a two-factor experiment, with the exception that the expected mean squares columns are different. The **AOV table** for a general two-factor experiment of a levels of factor A and b levels of factor B and no replication is shown in Table 16.5. Note the difference in the expected mean squares columns.

The computation of sums of squares and mean squares would proceed exactly as shown in Chapter 14, using the appropriate shortcut formulas. The difference in test procedures is illustrated in Table 16.6 for factor A. Similar results would also apply to factor B.

Rather than proceed with an example at this point, we will discuss a random-effects model for a factorial experiment with $r > 1$ observations at each factor-level combination. Then we will illustrate the test procedure.

In Chapter 14 we considered the fixed-effects model for a 2×2 factorial experiment with $r > 1$ observations per cell. The random-effects model for an **$a \times b$ factorial** experiment would be of the same form as the corresponding fixed-effects experiment, but with different assumptions.

$$y_{ijk} = \mu + \alpha_i + \beta_j + \alpha\beta_{ij} + \varepsilon_{ijk}$$

where y_{ijk} is the response for the kth observation at the ith level of factor A and the jth level of factor B; μ, α_i, β_j, and ε_{ijk} are defined as before for the random-effects model without

TABLE 16.6   Difference in Test Procedures for Factor A

Fixed-Effects Model	Random-Effects Model
H_0: $\alpha_1 = \alpha_2 = \cdots = \alpha_a = 0$	H_0: $\sigma_\alpha^2 = 0$
H_a: At least one of the αs differs from the rest.	H_a: $\sigma_\alpha^2 > 0$
T.S.: $F = \dfrac{MSA}{MSE}$	T.S.: $F = \dfrac{MSA}{MSE}$
R.R.: Based on $df_1 = a - 1$, $df_2 = (a - 1)(b - 1)$	R.R.: Same

TABLE 16.7   AOV Table for an $a \times b$ Factorial Experiment with r Observations per Cell

				EMS	
Source	SS	df	MS	Fixed Effects	Random Effects
A	SSA	$a - 1$	MSA	$\sigma_\varepsilon^2 + rb\theta_A$	$\sigma_\varepsilon^2 + r\sigma_{\alpha\beta}^2 + br\sigma_\alpha^2$
B	SSB	$b - 1$	MSB	$\sigma_\varepsilon^2 + ra\theta_B$	$\sigma_\varepsilon^2 + r\sigma_{\alpha\beta}^2 + ar\sigma_\beta^2$
AB	SSAB	$(a - 1)(b - 1)$	MSAB	$\sigma_\varepsilon^2 + r\theta_{AB}$	$\sigma_\varepsilon^2 + r\sigma_{\alpha\beta}^2$
Error	SSE	$ab(r - 1)$	MSE	σ_ε^2	σ_ε^2
Totals	TSS	$abr - 1$			

replication. In addition, we assume the following:

assumptions
1. $\alpha\beta_{ij}$ is a random effect due to the ith level of factor A and the jth level of factor B. $\alpha\beta_{ij}$ is normally distributed, with mean 0 and variance $\sigma_{\alpha\beta}^2$.
2. The $\alpha\beta_{ij}$s are independent.
3. The α_is, β_js, and $\alpha\beta_{ij}$s are independent.

AOV table   The appropriate **AOV tables** for fixed- and random-effects models are shown in Table 16.7.

The appropriate tests using the AB interaction sum of squares are illustrated in Table 16.8 for the two models.

Now, unlike the one-factor experiment and the two-factor experiment without replication, the test statistics for main effects are different for the fixed- and random-effects models. In addition, for the random-effects model, the tests for σ_α^2 and σ_β^2 can proceed even when the test on the AB interaction ($\sigma_{\alpha\beta}^2$) is significant. We've seen previously that for fixed-effects models, a test for main effects in the presence of a significant interaction only seems to make sense when the profile plot suggests that the interaction is "orderly." For random-effects models, we're interested in identifying the various sources of variability (e.g., $\sigma_{\alpha\beta}^2$, σ_α^2, and σ_β^2) that affect the response, y. Tests for σ_α^2 and σ_β^2 do make sense even when $\sigma_{\alpha\beta}^2$ has been shown to be greater than zero.

For the fixed-effects model following a nonsignificant test on the AB interaction, we can test for main effects due to factors A and B by using

$$F = \frac{MSA}{MSE} \quad \text{and} \quad F = \frac{MSB}{MSE}$$

TABLE 16.8   A Comparison of Appropriate Interaction Tests for Fixed- and Random-Effects Models

Fixed-Effects Model	Random-Effects Model
H_0: $\alpha\beta_{11} = \alpha\beta_{12} = \cdots = \alpha\beta_{ab} = 0$	H_0: $\sigma_{\alpha\beta}^2 = 0$
H_a: At least one $\alpha\beta_{ij}$ differs from the rest	H_a: $\sigma_{\alpha\beta}^2 > 0$
T.S.: $F = \dfrac{MSAB}{MSE}$	T.S.: $F = \dfrac{MSAB}{MSE}$
R.R.: Based on $df_1 = (a - 1)(b - 1)$, $df_2 = ab(r - 1)$	R.R.: Same

TABLE 16.9   Tests for an $a \times b$ Factorial Experiment with Replication: Random-Effects Model

	Factor A		Factor B	
H_0: $\sigma_\alpha^2 = 0$			H_0: $\sigma_\beta^2 = 0$	
H_a: $\sigma_\alpha^2 > 0$			H_a: $\sigma_\beta^2 > 0$	
T.S.: $F = \dfrac{MSA}{MSAB}$			T.S.: $F = \dfrac{MSB}{MSAB}$	
R.R.: Based on $df_1 = (a-1)$, $df_2 = (a-1)(b-1)$			R.R.: Based on $df_1 = (b-1)$, $df_2 = (a-1)(b-1)$	

respectively. As we see from the expected mean squares column of Table 16.7, no matter what the result of the test H_0: $\sigma_{\alpha\beta}^2 = 0$, we can form an F test for the components σ_α^2 and σ_β^2 using the test procedures shown in Table 16.9. Note that the test statistics differ from those used in the fixed-effects case, where the denominator of all F statistics is MSE.

tests, main effects

EXAMPLE 16.2   Consider the experimental situation described for the data of Table 16.4, where we were concerned with the effects of analysts and subjects on the DNA concentration of plaque in 18-to-20-year-old females. Now suppose that the plaque collection from each woman was subdivided into six samples and each analyst (unknown to him or her) made a DNA concentration determination on two samples for each subject. These data are recorded in Table 16.10 (in units of 10 μg).

Perform an analysis of variance for this experiment. Conduct all tests with $\alpha = .05$, and draw your conclusions.

TABLE 16.10   DNA Concentration Data, Example 16.2

Analyst, Factor B	Subject, Factor A			Totals
	1	2	3	
1	13.2	10.6	8.5	63.3
	12.3	9.8	8.9	
	25.5	20.4	17.4	
2	12.5	9.6	7.9	62.0
	12.9	10.7	8.4	
	25.4	20.3	16.3	
3	13.0	9.9	8.3	62.5
	12.4	10.3	8.6	
	25.4	20.2	16.9	
Totals	76.3	60.9	50.6	187.8

Solution

Using the shortcut formulas of Chapter 14, we obtain the sums of squares as follows:

$$TSS = \sum y_{ijk}^2 - \frac{G^2}{n} = 2017.38 - \frac{(187.8)^2}{18}$$

$$= 2017.38 - 1959.38 = 58$$

$$SSA = \frac{\sum A_i^2}{6} - \frac{G^2}{n} = 2015.14 - 1959.38 = 55.76$$

$$SSB = \frac{\sum B_j^2}{6} - \frac{G^2}{n} = 1959.52 - 1959.38 = .14$$

$$SSAB = \frac{\sum (AB)_{ij}^2}{2} - SSA - SSB - \frac{G^2}{n}$$

$$= 2015.46 - 55.76 - .14 - 1959.38 = .18$$

$$SSE = TSS - SSA - SSB - SSAB = 1.92$$

Our results are summarized in an analysis of variance table, Table 16.11.

We can proceed with appropriate statistical tests, using the results presented in the AOV table. For the AB interaction, we have:

H_0: $\sigma_{\alpha\beta}^2 = 0$

H_a: $\sigma_{\alpha\beta}^2 > 0$

T.S.: $F = \dfrac{MSAB}{MSE} = \dfrac{.05}{.21} = .24$

R.R.: For $\alpha = .05$, we will reject H_0 if F exceeds 3.63, the critical value for a = .05, $df_1 = 4$, and $df_2 = 9$.

Conclusion: There is insufficient evidence to reject H_0. There does not appear to be a significant variability in DNA concentrations due to combinations of analysts and subjects.

For factor B we have:

H_0: $\sigma_\beta^2 = 0$

H_a: $\sigma_\beta^2 > 0$

TABLE 16.11   AVO Table for the Data of Example 16.2

Source	SS	df	MS	EMS
A	55.76	2	27.88	$\sigma_\varepsilon^2 + 2\sigma_{\alpha\beta}^2 + 6\sigma_\alpha^2$
B	.14	2	.07	$\sigma_\varepsilon^2 + 2\sigma_{\alpha\beta}^2 + 6\sigma_\beta^2$
AB	.18	4	.05	$\sigma_\varepsilon^2 + 2\sigma_{\alpha\beta}^2$
Error	1.92	9	.21	σ_ε^2
Totals	58.00	17		

T.S.: $F = \dfrac{MSB}{MSAB} = \dfrac{.07}{.05} = 1.4$

R.R.: For $\alpha = .05$, we will reject H_0 if F exceeds 6.94, the critical value for a = .05, $df_1 = 2$, and $df_2 = 4$.

Conclusion: There is insufficient evidence to indicate a significant variability in DNA determinations from analyst to analyst.

Note that if $\sigma_{\alpha\beta}^2 = 0$, both MSAB and MSE are unbiased estimates of σ_ε^2, and MSB has expectation equal to $\sigma_\varepsilon^2 + 6\sigma_\beta^2$. It can be argued that when MSAB is small relative to MSE, it would be wise to pool the sum of squares for AB and for error to form a combined estimate of σ^2. We can illustrate this procedure for our data. The pooled mean square based on 13 degrees of freedom is

MS$_{pooled}$ $MS_{pooled} = \dfrac{SSAB + SSE}{4 + 9} = \dfrac{2.10}{13} = .16$

and can be used to test for main effects. The F test for H_0: $\sigma_\beta^2 = 0$ would be

$F = \dfrac{MSB}{MS_{pooled}} = \dfrac{.07}{.16} = .44$

Comparing the computed value of F to the critical value 3.81 for a = .05, $df_1 = 2$, and $df_2 = 13$, we see that there is insufficient evidence to reject H_0. Since this is the same conclusion we reached by using $F = MSB/MSAB$, we recommend pooling only in cases in which MSAB is considerably less than MSE.

The test for factor A follows.

H_0: $\sigma_\alpha^2 = 0$

H_a: $\sigma_\alpha^2 > 0$

T.S.: $F = \dfrac{MSA}{MSE} = \dfrac{27.88}{.21} = 132.76$

R.R.: For $\alpha = .05$, we will reject H_0 if F exceeds 4.26, the critical value based on a = .05, $df_1 = 2$, and $df_2 = 9$.

Conclusion: Since the observed value of F is much larger than 4.26, we reject H_0 and conclude that there is a significant variability in DNA concentrations in plaque collections from females between the ages of 18 and 20 years.

In this section we have compared a random-effects model to a fixed-effects model for the completely randomized design and for the $a \times b$ factorial experiment with r observations per cell. This study has been in no way exhaustive, but it has shown that there are alternatives to a fixed-effects model. A more detailed study of the random-effects model would certainly include factorial experiments with more than two factors

nested sampling   and the **nested sampling experiment** of Section 16.6. For the latter design, levels of factor

experiment B are nested (rather than cross-classified) within levels of factor A. For example, in considering the potency of a chemical, we could sample different manufacturing plants, batches of chemicals within a plant, and determinations within a batch. Note that the factor "batches" is not cross-classified with the factor "plants" since, for example, batch 1 for plant 1 is different from batch 1 for plant 2.

In Section 16.4 we will consider extending the results of this section to include a mixed model for an $a \times b$ factorial experiment.

EXERCISES

16.3   Refer to Example 16.2. Suppose that only one observation was made by an analyst on a plaque sample from each subject. Taking the first observation for each factor-level combination, we have the following sample data.

Analyst	Subject 1	2	3	Totals
1	13.2	10.6	8.5	32.3
2	12.5	9.6	7.9	30.0
3	13.0	9.9	8.3	31.2
Totals	38.7	30.1	24.7	93.5

a. Write an appropriate linear statistical model identifying all terms in the model.
b. Write down the expected mean squares.

16.4   Refer to Exercise 16.3. Perform an analysis for variance. Use $\alpha = .05$ for all tests.

16.5   Officials of a marketing research corporation were interested in studying the effect of a new promotional campaign for an improved brand of D-cell batteries. The study was conducted in a random sample of four standard metropolitan statistical areas (SMSAs), which had outlet stores for a random sample of three chain stores (selected from a large list of grocery, drug, and department stores). Sales volumes (in dollars) were recorded for a random sample of two weeks following the promotional campaign in the designated areas. These data are shown below.

Chain Store	SMSA 1	2	3	4
1	98	149	79	340
	112	126	61	302
2	87	96	119	125
	75	138	104	133
3	140	159	169	460
	190	185	150	420

a. Write an appropriate linear statistical model. List the assumptions and identify terms.
b. Perform an analysis of variance, showing expected mean squares. Use $\alpha = .05$.

<table>
<tr><td>16.4</td></tr>
</table>

A TWO-FACTOR EXPERIMENT, ONE FACTOR FIXED AND ONE RANDOM: A MIXED-EFFECTS MODELS

In Section 16.3 we compared the analysis of variance tables for fixed- and random-effects models for a two-factor experiment resulting from an $a \times b$ factorial experiment. **mixed model** Suppose, however, that we have a **mixed-effects model**, where one factor (A) is fixed and the other is random. For example, in Section 16.3 we considered an experiment to examine the effects of different subjects and different analysts on the DNA content of plaque. If the three subjects were selected at random and if the three analysts chosen were the only analysts of interest, we would have a mixed model with fixed analysts and random subjects.

The model for the $a \times b$ factorial experiment is the same as that given in Section 16.3 except that there are different assumptions.

$$y_{ijk} = \mu + \alpha_i + \beta_j + \alpha\beta_{ij} + \varepsilon_{ijk}$$

where we use the following assumptions:

assumptions
1. μ is an overall unknown mean.
2. α_i is a fixed effect corresponding to the ith level of factor A.
3. β_j is a random effect due to the jth level of factor B. β_j is normally distributed, with mean 0 and variance σ_β^2.
4. $\alpha\beta_{ij}$ is a random effect due to the ith level of A and jth level of B. $\alpha\beta_{ij}$ is normally distributed, with mean 0 and variance $\sigma_{\alpha\beta}^2$.
5. The β_js and $\alpha\beta_{ij}$s are all independent.

Using these assumptions, the analysis of variance table (incorporating Table 16.7) for a fixed, random, or mixed model in a two-factor experiment with replication is as shown in Table 16.12.

The expected mean squares column of Table 16.12 can be helpful in determining appropriate tests of significance. The test for $\sigma_{\alpha\beta}^2$ is the same in the mixed model as in the random-effects model.

test for $\sigma_{\alpha\beta}^2$

H_0: $\sigma_{\alpha\beta}^2 = 0$

H_a: $\sigma_{\alpha\beta}^2 > 0$

TABLE 16.12   AOV for an $a \times b$ Factorial Experiment with r Observations per Cell

				EMS		
Source	SS	df	MS	Fixed Effects	Random Effects	Mixed Effects A Fixed, B Random
A	SSA	$a - 1$	MSA	$\sigma_\varepsilon^2 + rb\theta_A$	$\sigma_\varepsilon^2 + r\sigma_{\alpha\beta}^2 + br\sigma_\alpha^2$	$\sigma_\varepsilon^2 + r\sigma_{\alpha\beta}^2 + rb\theta_A$
B	SSB	$b - 1$	MSB	$\sigma_\varepsilon^2 + ra\theta_B$	$\sigma_\varepsilon^2 + r\sigma_{\alpha\beta}^2 + ar\sigma_\beta^2$	$\sigma_\varepsilon^2 + ar\sigma_\beta^2$
AB	SSAB	$(a-1)(b-1)$	MSAB	$\sigma_\varepsilon^2 + r\theta_{AB}$	$\sigma_\varepsilon^2 + r\sigma_{\alpha\beta}^2$	$\sigma_\varepsilon^2 + r\sigma_{\alpha\beta}^2$
Error	SSE	$ab(r-1)$	MSE	σ_ε^2	σ_ε^2	σ_ε^2
Totals	TSS	$abr - 1$				

T.S.: $F = \dfrac{MSAB}{MSE}$

R.R.: Based on $df_1 = (a - 1)(b - 1)$ and $df_2 = ab(r - 1)$

No matter what the results of our tests for $\sigma_{\alpha\beta}^2$, we could proceed to use the following tests for factors A and B, which follow from entries in the expected mean squares column of Table 16.12. For factor A we have

test, factor A H_0: $\alpha_1 = \alpha_2 = \cdots = \alpha_a = 0$

H_a: At least one of the αs differs from the rest.

T.S.: $F = \dfrac{MSA}{MSAB}$

R.R.: Based on $df_1 = (a - 1)$ and $df_2 = (a - 1)(b - 1)$

For factor B we have

test, factor B H_0: $\sigma_\beta^2 = 0$

H_a: $\sigma_\beta^2 > 0$

T.S.: $F = \dfrac{MSB}{MSE}$

R.R.: Based on $df_1 = (b - 1)$ and $df_2 = ab(r - 1)$

The analysis of variance procedure outlined for a mixed-effects model in a two-factor experiment can be illustrated for a randomized block design, where blocks are assumed to be random.

EXAMPLE 16.3   A corporation is interested in comparing two different sunscreens (s_1 and s_2) for protecting the skin of persons who want to avoid burning or additional tanning while exposed to the sun. A random sample of females (ages 20–25 years) agreed to participate in the study. For each person two 1″ × 1″ squares were marked off on either side of the back, under the shoulder but above the small of the back. Sunscreen s_1 was then randomly assigned to the two squares on one side of the back, with s_2 assigned to the other two squares. A reading based on the color of skin in a square was made prior to the application of a fixed amount of the assigned sunscreen, and then again after application and exposure to the sun for a two-hour period. The data recorded in Table 16.13 are differences (postexposure minus preexposure) for the persons in the study. A high response indicates more burning.

Use these data to run an analysis of variance and draw conclusions.

Solution   We can compute the sums of squares for the sources of variability in the AOV table using the usual shortcut formulas.

$$TSS = \sum y_{ijk}^2 - \frac{G^2}{n}$$

$$= 2771.6 - \frac{(299.4)^2}{40} = 2771.6 - 2241.01 = 530.59$$

TABLE 16.13   Data (Differences) for the Sunscreen Experiment, Example 16.3

Sunscreen, A	Persons, B										Totals
	1	2	3	4	5	6	7	8	9	10	
	8.2	3.6	10.7	3.9	12.9	5.5	9.1	13.7	8.1	2.5	
1	7.6	3.5	10.3	4.4	12.1	5.9	9.7	13.2	8.7	2.8	156.4
	15.8	7.1	21.0	8.3	25.0	11.4	18.8	26.9	16.8	5.3	
	6.1	4.3	9.6	2.3	12.4	4.8	8.3	12.9	8.0	2.1	
2	6.8	4.7	9.2	2.5	12.8	4.0	8.6	13.6	7.5	2.5	143.0
	12.9	9.0	18.8	4.8	25.2	8.8	16.9	26.5	15.5	4.6	
Totals	28.7	16.1	39.8	13.1	50.2	20.2	35.7	53.4	32.3	9.9	299.4

$$SSA = \frac{\sum A_i^2}{20} - \frac{G^2}{n} = 2245.50 - 2241.01 = 4.49$$

$$SSB = \frac{\sum B_j^2}{4} - \frac{G^2}{n} = 2758.50 - 2241.01 = 517.49$$

$$SSAB = \frac{\sum (AB)_{ij}^2}{2} - SSA - SSB - \frac{G^2}{n}$$

$$= 2768.96 - 4.49 - 517.49 - 2241.01 = 5.97$$

Then by subtraction, SSE = TSS − SSA − SSB − SSAB = 2.64.

Substituting $a = 2$, $b = 10$, and $r = 2$ into an AOV table similar to that shown in Table 16.12, we have the results shown in Table 16.14.

A test for the random component $\alpha\beta_{ij}$ is as follows:

H_0: $\sigma_{\alpha\beta}^2 = 0$

H_a: $\sigma_{\alpha\beta}^2 > 0$

T.S.: $F = \dfrac{MSAB}{MSE} = \dfrac{.66}{.13} = 5.08$

R.R.: For $\alpha = .05$, we will reject H_0 if the computed value of F exceeds 2.39, the value in Table 6 for a = .05, $df_1 = 9$, and $df_2 = 20$.

Conclusion: Since 5.08 exceeds 2.39, we reject H_0 and conclude that $\sigma_{\alpha\beta}^2 > 0$; that is, there is a significant source of random variation due to the combination of

TABLE 16.14   AOV Table for the Data of Example 16.3

Source	SS	df	MS	EMS Mixed Model
A	4.49	1	4.49	$\sigma_\varepsilon^2 + 2\sigma_{\alpha\beta}^2 + 20\theta_A$
B	517.49	9	57.50	$\sigma_\varepsilon^2 + 4\sigma_\beta^2$
AB	5.97	9	.66	$\sigma_\varepsilon^2 + 2\sigma_{\alpha\beta}^2$
Error	2.64	20	.13	σ_ε^2
Totals	530.59	39		

the ith level of A (sunscreens) and the jth level of B (persons). We would infer from this that one sunscreen may not necessarily be better than the other for all persons.

Even with this significant F test for $\sigma_{\alpha\beta}^2$, we can proceed to test for effects due to factors A and B separately. For factor B we have

H_0: $\sigma_\beta^2 = 0$

H_a: $\sigma_\beta^2 > 0$

T.S.: $F = \dfrac{MSB}{MSE} = \dfrac{57.50}{.13} = 442.31$

R.R.: For $\alpha = .05$, we will reject H_0 if F exceeds 2.39, the value in Table 6 for a = .05, $df_1 = 9$, and $df_2 = 20$.

Conclusion: Since 442.31 exceeds 2.39, we reject H_0 and conclude that $\sigma_\beta^2 > 0$. Thus there is a significant source of random variation due to variability from person to person.

For factor A we have

H_0: $\alpha_1 = \alpha_2 = 0$

H_a: $\alpha_1 \neq \alpha_2$

T.S.: $F = \dfrac{MSA}{MSAB} = \dfrac{4.49}{.66} = 6.80$

R.R.: For $\alpha = .05$, we will reject H_0 if F exceeds 5.12, the value in Table 6 for a = .05, $df_1 = 1$, and $df_2 = 9$.

Conclusion: Since $6.80 > 5.12$, we reject H_0 and conclude that the mean response (post minus pre) differs for the two sunscreens. Since $\bar{y}_{s1} = 156.4/20 = 7.82$ and $\bar{y}_{s2} = 143/20 = 7.15$, we would conclude that s_2 offers more protection on the average than s_1. However, as noted previously, there are significant sources of variability due to persons and the combination of persons with sunscreens.

This discussion of mixed models, which has been illustrated for the $a \times b$ factorial with r observations per cell, provides only a brief introduction to the study of mixed models. Indeed, we could spend one or more quarters of study at the graduate level covering topics appropriate for mixed models. For more advanced work, we could examine factorial experiments with three or more factors (some random, others fixed). In addition, when examining the effect of two factors (both fixed effects) on a response while blocking on a **split-plot design**   third factor (which is random), a **split-plot design** becomes an important alternative to a factorial experiment that is laid off in a randomized block design. The difference between a split-plot design and the factorial experiment set off in a randomized block design lies in the method of applying treatments (factor-level combinations) to experimental units. For each block, levels of factor 1 are randomly assigned to experimental units. Then levels of the second factor are randomly assigned to subunits within each level of factor 1. This randomization is quite different from the randomization used in a factorial experiment that is laid off in a randomized block design. A discussion of this topic is presented in Section 16.6.

EXERCISES

 16.6 Prior to conducting a clinical trial that involves a subjective evaluation of a patient's progress, the participating physicians are asked to agree on certain criteria for reaching an evaluation. To examine the consistency in their evaluations before the initiation of a particular clinical trial, a pilot study was conducted on four patients who had been treated with a drug that was to be included in the trial. Each of the five physicians who were to participate in the study was asked to evaluate (on a 0-to-10-point scale) the degree of cure after a two-week treatment period. Since the clinical evaluations of a patient's cure were to be based on the results of a bacterial culture analysis, each physician analyzed two cultures from each patient. This feature was unknown to the physicians, who were merely told they would be analyzing eight separate bacterial cultures. The evaluations based on these cultures are recorded here.

a. Treating physicians as fixed and patients as random, write an appropriate linear statistical model. Identify all terms in the model.

b. Show the expected mean squares column in the AOV table.

	Patient			
Physician	1	2	3	4
1	7.2	4.2	9.5	5.4
	9.6	3.5	9.3	3.9
2	8.5	2.9	8.8	6.3
	9.6	3.3	9.2	6.0
3	9.1	1.8	7.6	6.1
	8.6	2.4	7.1	5.6
4	8.2	3.6	7.3	5.0
	9.0	4.4	7.0	5.4
5	7.8	3.7	9.2	6.5
	8.0	3.9	8.3	6.9

16.7 Refer to Exercise 16.6. Perform an analysis of variance. Draw your conclusions, using $\alpha = .05$.

16.5 RULES FOR OBTAINING EXPECTED MEAN SQUARES

We discussed the AOVs for one- and two-factor experiments for fixed-effects models in Chapter 14 and for random or mixed models earlier in this chapter. We will see in this section that for any k-factors experiment of data, with r observations per factor-level combination, it is possible to write expected mean squares for all main effects and interactions for fixed, random, or mixed models using some rather simple rules. *The importance of these rules is that, having written down the expected mean squares for an unfamiliar experimental design, we often can construct appropriate F tests.* The assumptions for the fixed and random models will be the same as we have used in describing fixed, random, and mixed models in previous sections.

classifying
interactions

Two rules for **classifying interactions** as fixed or random effects are needed before we can proceed with the rules for obtaining expected mean squares.

RULES FOR THE
CLASSIFICATION OF
INTERACTIONS

1. If a fixed effect interacts with another fixed effect, the resulting interaction term is a fixed effect.
2. If a random effect interacts with another effect (fixed or random), the resulting interaction term is a random component.

EXAMPLE 16.4

Consider a 3×6 factorial with two observations per factor-level combination. Classify the AB interaction as fixed or random for the following situations:

a. A and B are both fixed effects.
b. A is fixed and B is random.
c. A and B are both random.

Solution

We apply the rules for classifying interactions.

a. AB is a fixed effect since A (fixed) interacts with B (fixed).
b. AB is a random component since A (fixed) interacts with B (random).
c. AB is random since A (random) interacts with B (random).

EXAMPLE 16.5

Consider a factorial experiment in the factors A, B, and C. Classify the AB, AC, BC, and ABC interactions as fixed or random when A and B are fixed effects and C is random.

Solution

We apply the classification rules.

AB is fixed; A (fixed) interacts with B (fixed).
AC is random; A (fixed) interacts with C (random).
BC is random; B (fixed) interacts with C (random).
ABC is random; A (fixed) interacts with BC (random).

mean square table

Before we state the rules for determining expected mean squares, it is convenient to construct a **mean square table**. These steps are summarized here and illustrated for an $a \times b$ factorial experiment with factor A random, B fixed, and r observations for each factor-level combination of A and B.

STEPS IN
CONSTRUCTING A
MEAN SQUARE
TABLE

1. Write the model for the experiment. For an $a \times b$ factorial experiment, the model is

$$y_{ijk} = \mu + \alpha_i + \beta_j + \alpha\beta_{ij} + \varepsilon_{k[ij]}$$

Note: We use brackets in the ε-term to indicate that there are $k = 1, 2, \ldots, r$ observations for each factor-level combination of factors A and B (i.e., for each choice of i, j).

2. Construct a table with each term in the model (except μ) forming a row heading. This table takes the following form for our example.

α_i
β_j
$\alpha\beta_{ij}$
$\varepsilon_{k(ij)}$

3. Form a column in the table for each subscript in the model.

	i	j	k
α_i			
β_j			
$\alpha\beta_{ij}$			
$\varepsilon_{k(ij)}$			

4. Above each column heading indicate whether the subscript corresponds to a fixed (F) or random (R) effect.

5. Also indicate the number levels for each subscript. Additions to the table from step 4 and 5 are shown here for our example.

| | a | b | r |
| | R | F | R |
	i	j	k
α_i			
β_j			
$\alpha\beta_{ij}$			
$\varepsilon_{k(ij)}$			

6. For each row in a given column, enter the number of levels associated with the column subscript, unless the row term contains the column subscript.

| | a | b | r |
| | R | F | R |
	i	j	k
α_i		b	r
β_j	a		r
$\alpha\beta_{ij}$			r
$\varepsilon_{k(ij)}$			

7. Examine the terms of the model, which are listed in the first column of the table. For each term with brackets in the subscript, place a 1 under the column(s) with a subscript included in the brackets.

	a	b	r
	R	F	R
	i	j	k
α_i		b	r
β_j	a		r
$\alpha\beta_{ij}$			r
$\varepsilon_{k(ij)}$	1	1	

8. Fill in the remaining cells of a column with a 0 if the column is headed by an F; fill in the remaining cells of a column headed by an R with a 1.

	a	b	r
	R	F	R
	i	j	k
α_i	1	b	r
β_j	a	0	r
$\alpha\beta_{ij}$	1	0	r
$\varepsilon_{k(ij)}$	1	1	1

This is the *mean square table* used for computing the expected mean squares for a two-factor experiment with factor A random, factor B fixed, and r observations per factor-level combination of A and B.

It's easy to compute expected mean squares once you have the mean square table. These rules are listed here.

RULES FOR OBTAINING AN EMS USING A MEAN SQUARE TABLE

1. Examine the subscript(s) of the term.
2. Eliminate any row in the mean square table that doesn't have the subscript(s).
3. Cover each column of the table headed by a nonbracketed subscript of the term.
4. Multiply the remaining, uncovered entries in each row to obtain the coefficients of terms in the expected mean square.

EXAMPLE 16.6

Compute $E(MSA)$ for a two-factor experiment with a levels of factor A (random), b levels of factor B (fixed), and r observations per factor-level combination.

Solution

Refer to the mean square table above. We note that α_i has the subscript i; thus we eliminate the second row of the table. Covering column 1 (which is headed by i), we

TABLE 16.15   Mean Square Table for Computing $E(MSA)$

	a R i	b F j	r R k	Product of Remaining Entries	Term EMS
α_i	1	b	r	br	$br\sigma_\alpha^2$
β_j	a	0	r	—	
$\alpha\beta_{ij}$	1	0	r	0	
$\varepsilon_{k[ij]}$	1	1	1	1	σ_ε^2

multiply the remaining entries. Table 16.15 shows these steps, the multiplication, and the terms of the EMS.

Using the last column of Table 16.15, we have

$$E(MSA) = \sigma_\varepsilon^2 + br\sigma_\alpha^2$$

The computation $E(MSB)$ follows in a similar manner. The term β_j has a subscript j, so we eliminate the second column (which is headed by j) and the first row (which contains no j) of the mean square table. The remaining entries are multiplied to obtain the coefficients of the expected mean square (see Table 16.16).

Hence

$$E(MSB) = \sigma_\varepsilon^2 + r\sigma_{\alpha\beta}^2 + ar\theta_B$$

where θ_B is a constant of the form

$$\theta_B = \frac{\sum_j \beta_j^2}{b-1}$$

EXAMPLE 16.7   a. Set up the mean square table for computing expected mean squares for a two-factor experiment with factor A fixed, factor B random, and r observations for each factor-level combination of A and B.
   b. Compute $E(MSA)$.

Solution   a. The model for this experimental situation is

$$y_{ijk} = \mu + \alpha_i + \beta_j + \alpha\beta_{ij} + \varepsilon_{k[ij]}$$

The corresponding mean square table is shown in Table 16.17.
   b. $E(MSA)$ is found as shown in Table 16.18.

TABLE 16.16   Mean Square Table for Computing $E(MSB)$

	a R i	b F j	r R k	Product of Remaining Entries	Term EMS
α_i	1	b	r	—	
β_j	a	0	r	ar	$ar\theta_\beta$
$\alpha\beta_{ij}$	1	0	r	r	$r\sigma_{\alpha\beta}^2$
$\varepsilon_{k[ij]}$	1	1	1	1	σ_ε^2

TABLE 16.17   Mean Square Table for Example 16.7

	a	b	r
	F	R	R
	i	j	k
α_i	0	b	r
β_j	a	1	r
$\alpha\beta_{ij}$	0	1	r
$\varepsilon_{k[ij]}$	1	1	1

TABLE 16.18   Computations for $E(MSA)$

	a	b	r	Product of Remaining Entries	Term EMS
	F	R	R		
	i	j	k		
α_i	0	b	r	br	$br\theta_\alpha$
β_j	a	1	r	—	
$\alpha\beta_{ij}$	0	1	r	r	$r\sigma^2_{\alpha\beta}$
$\varepsilon_{k[ij]}$	1	1	1	1	σ^2_ε

Thus $E(MSA) = \sigma^2_\varepsilon + r\sigma^2_{\alpha\beta} + br\theta_A$ where $\theta_A = \sum_i \alpha_i^2/(a-1)$. This agrees with what we obtained in the previous example for the fixed factor (B) with appropriate changes in notation.

Previously we have been concerned with only fixed-effects models. For these models the test statistics are always formed using the affected mean square in the numerator divided by MSE. But for random and mixed models the test statistics are not always the same. The test statistic for interaction, F equals MSAB/MSE, is the same for the fixed, random, and mixed models; but the F tests for factors A and B do change depending on the assumptions for α_i and β_j. For example, the F test for factor A is MSA/MSE for A fixed, B fixed, and for A random, B fixed. In constrast, the F test for factor A is MSA/MSAB for A fixed, B random, and for A random, B random. So you can see the importance of knowing the expected mean squares for random and mixed models.

There's a special case of the two-factor experiment that should be mentioned: when there is only one observation ($r = 1$) per factor-level combination of A and B, and we have the model

$$y_{ij} = \mu + \alpha_i + \beta_j + \alpha\beta_{ij} + \varepsilon_{ij}$$

You can see from the abbreviated AOV table of Table 16.19 that there are no degrees of freedom for error, so there is no test for interaction and depending on the model there may not be a valid test for the main effects. The only possible remedy is for the situation in which one can assume that there is no interaction between factors A and B, in which case all main effects can be tested using the mean square error from the model

$$y_{ij} = \mu + \alpha_i + \beta_j + \varepsilon_{ij}$$

TABLE 16.19   Abbreviated AOV Table, Two-Factor Experiment $(r = 1)$

Source	df
A	$a - 1$
B	$b - 1$
AB	$(a - 1)(b - 1)$
Error	—
Total	$ab - 1$

regardless of whether we have a fixed, random, or mixed model. When the no-interaction assumption is not reasonable, the experiment must allow for replication $(r > 1)$ at the factor-level combinations of A and B in order to obtain valid tests of the interaction and main effects.

These same rules that we used for the two-factor experiment can also be used for more complicated experiments and although the rules may seem a bit cumbersome, with practice they are quite easy to use. We'll give one more example using a three-factor experiment. For additional details regarding assumptions, derivations, and more complicated applications, see Hicks (1973).

EXAMPLE 16.8   Give the expected mean squares for a $3 \times 5 \times 2$ factorial experiment with $r = 4$ observations per factor-level combination. Treat factors A and B as fixed and factor C as random.

Solution   The complete model for this experiment is given here along with the corresponding mean square table:

$$y_{ijkl} = \mu + \alpha_i + \beta_j + \gamma_k + \alpha\beta_{ij} + \alpha\gamma_{ik} + \beta\gamma_{jk} + \alpha\beta\gamma_{ijk} + \varepsilon_{l(ijk)}$$

We'll set up the mean square table for general values of a, b, c, and r and then substitute later. The mean square table using the rules of pages 773–775 is shown

TABLE 16.20   Mean Square Table for Example 16.8

	a	b	c	r
	F	F	R	R
	i	j	k	l
α_i	0	b	c	r
β_j	a	0	c	r
γ_k	a	b	1	r
$\alpha\beta_{ij}$	0	0	c	r
$\alpha\gamma_{ik}$	0	b	1	r
$\beta\gamma_{jk}$	a	0	1	r
$\alpha\beta\gamma_{ijk}$	0	0	1	r
$\varepsilon_{l(ijk)}$	1	1	1	1

TABLE 16.21   Computations for $E(MSA)$, Example 16.8

	a F i	b F j	c R k	r R l	Product of Remaining Entries	Term EMS
α_i	0	b	c	r	bcr	$bcr\theta_A$
β_j	a	0	c	r	—	
γ_k	a	b	1	r	—	
$\alpha\beta_{ij}$	0	0	c	r	0	
$\alpha\gamma_{ik}$	0	b	1	r	br	$br\sigma^2_{\alpha\gamma}$
$\beta\gamma_{jk}$	a	0	1	r	—	
$\alpha\beta\gamma_{ijk}$	0	0	1	r	0	
$\varepsilon_{l(ijk)}$	1	1	1	1	1	σ^2_ε

in Table 16.20. The expected mean squares for the model terms can be obtained by applying the EMS rules to this mean square table. For example, for $E(MSA)$, we have the uncovered entries in Table 16.21.

From the last column of this table we have

$$E(MSA) = \sigma^2_\varepsilon + br\sigma^2_{\alpha\gamma} + bcr\theta_A$$

Substituting $a = 3$, $b = 5$, $c = 2$, and $r = 4$, this becomes

$$E(MSA) = \sigma^2_\varepsilon + 20\sigma^2_{\alpha\gamma} + 40\theta_A$$

where

$$\theta_A = \sum_i \alpha_i^2/2$$

Similarly, the expected mean squares for factors B and C can be shown to be

$$E(MSB) = \sigma^2_\varepsilon + ar\sigma^2_{\beta\gamma} + acr\theta_B$$
$$= \sigma^2_\varepsilon + 12\sigma^2_{\beta\gamma} + 24\theta_B$$

and

$$E(MSC) = \sigma^2_\varepsilon + abr\sigma^2_\gamma$$
$$= \sigma^2_\varepsilon + 60\sigma^2_\gamma$$

The table for computing the expected mean square for the AB interaction is shown in Table 16.22. The expected mean square for MSAB is

$$E(MSAB) = \sigma^2_\varepsilon + r\sigma^2_{\alpha\beta\gamma} + cr\theta_{AB}$$
$$= \sigma^2_\varepsilon + 4\sigma^2_{\alpha\beta\gamma} + 8\theta_{AB}$$

where

$$\theta_{AB} = \sum_{ij} \frac{\alpha\beta_{ij}^2}{8}$$

TABLE 16.22   Computations for E(MSAB), Example 16.8

	a	b	c	r	Product of Remaining Entries	Term EMS
	F	F	R	R		
	i	j	k	l		
α_i	0	b	c	r	—	
β_j	a	0	c	r	—	
γ_k	a	b	1	r	—	
$\alpha\beta_{ij}$	0	0	c	r	cr	$cr\theta_{AB}$
$\alpha\gamma_{ik}$	0	b	1	r	—	
$\beta\gamma_{jk}$	a	0	1	r	—	
$\alpha\beta\gamma_{ijk}$	0	0	1	r	r	$r\sigma^2_{\alpha\beta\gamma}$
$\varepsilon_{l(ijk)}$	1	1	1	1	1	σ^2_ε

After application of the EMS rules to the AC and BC interactions, we obtain

$$E(MSAC) = \sigma^2_\varepsilon + br\sigma^2_{\alpha\gamma}$$
$$= \sigma^2_\varepsilon + 20\sigma^2_{\alpha\gamma}$$

and

$$E(MSBC) = \sigma^2_\varepsilon + ar\sigma^2_{\beta\gamma}$$
$$= \sigma^2_\varepsilon + 12\sigma^2_{\beta\gamma}$$

Similarly, it can be shown that MSABC and MSE have expectations

$$E(MSABC) = \sigma^2_\varepsilon + r\sigma^2_{\alpha\beta\gamma}$$
$$= \sigma^2_\varepsilon + 4\sigma^2_{\alpha\beta\gamma}$$

and

$$E(MSE) = \sigma^2_\varepsilon$$

A summary of the expected mean squares, which we have computed using the EMS rules of pp. 773–775 for the $3 \times 5 \times 2$ factorial with $r = 4$ observations per cell, is shown in Table 16.23.

TABLE 16.23   Partial AOV for Example 16.8

Source	df	EMS
A	$a - 1$	$\sigma^2_\varepsilon + br\sigma^2_{\alpha\gamma} + bcr\theta_A$
B	$b - 1$	$\sigma^2_\varepsilon + ar\sigma^2_{\beta\gamma} + acr\theta_B$
C	$c - 1$	$\sigma^2_\varepsilon + abr\sigma^2_\gamma$
AB	$(a - 1)(b - 1)$	$\sigma^2_\varepsilon + r\sigma^2_{\alpha\beta\gamma} + cr\theta_{AB}$
AC	$(a - 1)(c - 1)$	$\sigma^2_\varepsilon + br\sigma^2_{\alpha\gamma}$
BC	$(b - 1)(c - 1)$	$\sigma^2_\varepsilon + ar\sigma^2_{\beta\gamma}$
ABC	$(a - 1)(b - 1)(c - 1)$	$\sigma^2_\varepsilon + r\sigma^2_{\alpha\beta\gamma}$
Error	$abc(r - 1)$	σ^2_ε

EXAMPLE 16.9

Refer to Example 16.8. Give an appropriate F statistic for

$$H_0: \ \theta_A = 0$$

and

$$H_0: \ \sigma_{\beta\gamma}^2 = 0$$

Solution

Using the expected mean squares listed in Table 16.23, it is clear that the test statistic for $H_0: \ \theta_A = 0$ is $F = \text{MSA}/\text{MSAC}$; the test statistic for $H_0: \ \sigma_{\beta\gamma}^2 = 0$ is $F = \text{MSBC}/\text{MSE}$.

We can always obtain valid tests for all sources of variability in fixed-effects models, but this is not true for some random- and mixed-effects models. Some researchers have developed approximate F tests that can be constructed for sources of variability in random- or mixed-effects models where no valid F test is available (see, for example, Satterthwaite (1946) and Welch (1956)). We will not cover this topic in this text since these tests can become quite involved and are open to some controversy.

For some random and mixed models, the objective of the researcher might include **estimation and prediction** **estimation** of the variances for random effects and **prediction** of the average value of y. We briefly discussed estimation of the expected value of y and how we use the expected mean squares to obtain the variance of our estimates for a random-effects model. Estimation of the average value of y for a mixed model is more difficult and beyond the scope of this text.

estimates of variance components   Another use of expected mean squares is in the **estimation of the variances** associated with random effects in the model. For example, in a random-effects model for a one-factor experiment of t treatments and r observations per treatment, the expected mean squares for treatments and error are $\sigma_\varepsilon^2 + r\sigma_\alpha^2$ and σ_ε^2, respectively. As before, MSE is our estimate of σ_ε^2. Similarly, since MST estimates $\sigma_\varepsilon^2 + r\sigma_\alpha^2$, by substituting MSE for σ_ε^2, we can equate MST to the expected mean square for treatments and solve for σ_α^2. The solution, $(\text{MST} - \text{MSE})/r$, is an unbiased estimate of the variance of the treatments' source of variability in a one-factor experiment. This procedure of equating mean squares to expected mean squares can be used for obtaining estimates of variance components (variances of random effects) in random- and mixed-effects models for balanced designs. The problem of variance component estimation for unbalanced designs is a difficult one and is beyond the scope of this text. If you are interested in this topic, we refer you to three references for additional reading on the subject of variance component estimation: Hicks (1973), Mendenhall (1968), and Searle (1971).

16.6   NESTED SAMPLING AND THE SPLIT-PLOT DESIGN

Sometimes in an experiment one factor is "nested" within another. This can be illustrated with the following example. A pharmaceutical company conducted tests to determine the stability of its product (under room-temperature conditions) at a specific point in time. Two manufacturing sites were used. At each site, a random sample of three batches of the product was obtained and additional random samples of 10 different tablets were obtained from each batch. The design can be represented as shown in Figure 16.1.

FIGURE 16.1

Two-Factor
Experiment with
Batches Nested
in Sites

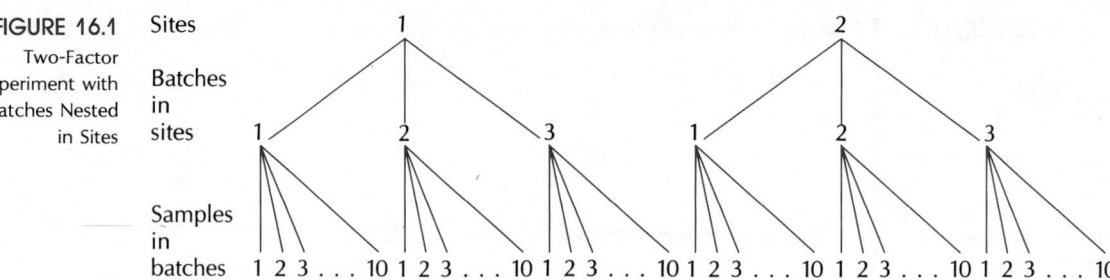

Although this might look like the usual two-factor experiment with sites (factor A) and batches (factor B), it should be noted that the three batches taken from site 1 are different from the three batches taken from site 2. In this sense, factor B (batches) is said to be *nested* in factor A (sites). For this experimental situation it will be impossible to evaluate the effect of the interaction of factor B with factor A, because each level of factor B does not appear with each level of factor A, as would happen with a factorial arrangement of factors A and B. Here the three batches within a site are unique to that site.

The general model for a two-factor experiment (r observations per cell) where factor B is nested in factor A can be written as

$$y_{ijk} = \mu + \alpha_i + \beta_{j(i)} + \varepsilon_{ijk} \qquad \begin{aligned} i &= 1, 2, \ldots, a \\ j &= 1, 2, \ldots, b \\ k &= 1, 2, \ldots, r \end{aligned}$$

Note that this model is similar to the model for the two-factor experiment of Section 16.3, except that there is no interaction term $\alpha\beta_{ij}$ and the term for factor B $\beta_{j(i)}$ is subscripted to denote the jth level of factor B is nested in the ith level of factor A. The analysis of variance table for this design is shown in Table 16.24.

Three of the more common situations are shown in Table 16.24 with the expected mean squares. Note in particular, that

1.   The F test for factor B is always

$$F = \frac{\text{MSB(A)}}{\text{MSE}}$$

TABLE 16.24   AOV Table for a Two-Factor Experiment (r Observations per Cell) with Factor B Nested in Factor A

				EMS		
Source	SS	df	MS	Fixed	Mixed (A Fixed)	Random
A	SSA	$a - 1$	MSA	$\sigma_\varepsilon^2 + rb\theta_A$	$\sigma_\varepsilon^2 + r\sigma_\beta^2 + rb\theta_A$	$\sigma_\varepsilon^2 + r\sigma_\beta^2 + rb\sigma_\alpha^2$
B(A)	SSB(A)	$a(b - 1)$	MSB(A)	$\sigma_\varepsilon^2 + r\theta_B$	$\sigma_\varepsilon^2 + r\sigma_\beta^2$	$\sigma_\varepsilon^2 + r\sigma_\beta^2$
Error	SSE	$ab(r - 1)$	MSE	σ_ε^2	σ_ε^2	σ_ε^2
Total	TSS	$abr - 1$				

2. The F test for factor A in the fixed-effects model is

$$F = \frac{MSA}{MSE}$$

For the random- and mixed-effects model, however, the corresponding test for factor A is

$$F = \frac{MSA}{MSB(A)}$$

3. When $r = 1$, there is no test for factor B, but we can test for factor A in the random- and mixed-effects model using

$$F = \frac{MSA}{MSB(A)}$$

EXAMPLE 16.10

An experiment was conducted to determine the content uniformity of film-coated tablets produced for a cardiovascular drug used to lower blood pressure. A random sample of three batches was obtained from each of two blending sites; within each batch a random sample of five tablets was assayed to determine content uniformity. These data are shown here:

Site	1			2		
Batches within each site	1	2	3	1	2	3
Tablets	5.03	4.64	5.10	5.05	5.46	4.90
within	5.10	4.73	5.15	4.96	5.15	4.95
each	5.25	4.82	5.20	5.12	5.18	4.86
batch	4.98	4.95	5.08	5.12	5.18	4.86
	5.05	5.06	5.14	5.05	5.11	5.07

a. Run an analysis of variance.
b. Is there evidence to indicate batch-to-batch variability in content uniformity? Does the F test run depend on whether we assume batches fixed or random?
c. Draw conclusions about batch.

Solution

a. For these data we have $a = 2$ sites, $b = 3$ batches within each blender, and $r = 5$ tablets per batch. For these data, the analysis of variance table is shown here. (See following output.)

Source	SS	df	MS	F
A	.01825	1	.01825	.16
$B(A)$	.45401	4	.11350	9.39
Error	.29020	24	.01209	
Total	.76246	29		

b., c. The F test for batches is

$$F = \frac{MSB(A)}{MSE} = 9.39$$

based on $df_1 = 4$ and $df_2 = 24$ degrees of freedom. Since the observed value of F, 9.39, exceeds the tabled value of F for $\alpha = .05$, we conclude that there is considerable batch-to-batch variability in content uniformity of tablets. This test does not depend on whether the batches are random.

```
CONTENT UNIFORMITY OF FILM-COATED TABLETS
               INPUT DATA

      OBS    SITE    BATCH    CONTENT

       1      1        1       5.03
       2      1        1       5.10
       3      1        1       5.25
       4      1        1       4.98
       5      1        1       5.05
       6      1        2       4.64
       7      1        2       4.73
       8      1        2       4.82
       9      1        2       4.95
      10      1        2       5.06
      11      1        3       5.10
      12      1        3       5.15
      13      1        3       5.20
      14      1        3       5.08
      15      1        3       5.14
      16      2        1       5.05
      17      2        1       4.96
      18      2        1       5.12
      19      2        1       5.12
      20      2        1       5.05
      21      2        2       5.46
      22      2        2       5.15
      23      2        2       5.18
      24      2        2       5.18
      25      2        2       5.11
      26      2        3       4.90
      27      2        3       4.95
      28      2        3       4.86
      29      2        3       4.86
      30      2        3       5.07

     N=      30
```

```
CONTENT UNIFORMITY OF FILM-COATED TABLETS

       ANALYSIS OF VARIANCE PROCEDURE

          CLASS LEVEL INFORMATION

          CLASS    LEVELS    VALUES

          SITE        2       1 2

          BATCH       3       1 2 3

    NUMBER OF OBSERVATIONS IN DATA SET = 30
```

```
                    CONTENT UNIFORMITY OF FILM-COATED TABLETS

                        ANALYSIS OF VARIANCE PROCEDURE
```

DEPENDENT VARIABLE: CONTENT

SOURCE	DF	SUM OF SQUARES	MEAN SQUARE	F VALUE	PR > F	R-SQUARE	C.V.
MODEL	5	0.47226667	0.09445333	7.81	0.0002	0.619393	2.1803
ERROR	24	0.29020000	0.01209167		ROOT MSE		CONTENT MEAN
CORRECTED TOTAL	29	0.76246667			0.10996211		5.04333333

```
SOURCE                    DF              ANOVA SS       F VALUE      PR > F
SITE                      1              0.01825333        1.51       0.2311
BATCH(SITE)               4              0.45401333        9.39       0.0001



TESTS OF HYPOTHESES USING THE ANOVA MS FOR BATCH(SITE) AS AN ERROR TERM
SOURCE                    DF              ANOVA SS       F VALUE      PR > F
SITE                      1              0.01825333        0.16       0.7089
```

By now you may have realized that a whole new series of experimental designs have opened up with the introduction of nested effects. Thinking beyond the two-factor design, one could imagine a general multifactor design with factor A, factor B nested in levels of factor A, factor C nested in levels of A and B, and so on. The analysis of variance table for a three-factor nested design with all factors random is shown in Table 16.25.

Other extensions of these designs are possible as well. For example, one could have a three-factor experiment where factors A and B are cross-classified but factor C is nested within levels of factors A and B. This would be an example of a *partially nested design*.

Suppose that a marketing research firm is responsible for sampling potential customers to obtain their opinions on two products (A_1 and A_2) in four geographic areas of the country ($B_1, \ldots B_4$). A random sample of six stores selling product A_i is obtained in each geographic area. For each store selected for product A_i in geographic area B_j, 10 people are interviewed concerning product i. For this design, factor C (stores) would be nested in levels of factors A (products) and B (geographic areas) and there would be $r = 10$ observations (opinions) for each level of factor C (stores) nested in levels of factors A and B.

The possibilities of nested and partial nested designs are seemingly endless, but, unfortunately, we will not have an opportunity to examine them here. The interested reader should refer to Winer (1971) and Mendenhall (1968) for a more extensive treatment of this topic. We will, however, consider one very popular design that is similar to a partially nested design. It is called the *split-plot design* because it had its origin in agriculture experimentation. We will illustrate its use with an example.

The yields of three different varieties of soybeans are to be compared under two different levels of fertilizer application. If we were interested in getting (say) $r = 2$ observations at each combination of fertilizer and variety of soybeans, we would need 12 equal-sized plots. Taking fertilizers as factor A and varieties as a treatment factor T, one possible

TABLE 16.25   AOV for a Three-Factor Nested Design—All Factors Random (r Observations per Cell)

Source	SS	df	MS	EMS
A	SSA	$a - 1$	MSA	$\sigma_\varepsilon^2 + r\sigma_\gamma^2 + rc\sigma_\beta^2 + rcb\sigma_\alpha^2$
$B(A)$	SSB(A)	$a(b - 1)$	MSB(A)	$\sigma_\varepsilon^2 + r\sigma_\gamma^2 + rc\sigma_\beta^2$
$C(A, B)$	SSC(A, B)	$ab(c - 1)$	MSC(A, B)	$\sigma_\varepsilon^2 + r\sigma_\gamma^2$
Error	SSE	$abc(r - 1)$	MSE	σ_ε^2
Total	TSS	$abcr - 1$		

FIGURE 16.2

Split-Plot Design

A_1 Wholeplot 1	A_2 Wholeplot 2	A_2 Wholeplot 3	A_1 Wholeplot 4
T_2	T_3	T_1	T_3
T_1	T_2	T_3	T_1
T_3	T_1	T_2	T_2

design would be the standard 2×3 factorial experiment with $r = 2$ observations per factor-level combination. But, since the application of fertilizer to a plot occurs when the soil is being prepared for planting, it would be difficult (logistically) to first apply fertilizer A_1 to six of the plots dictated by the factorial arrangement of factors A and T and then fertilizer A_2 to the other six plots before planting the required varieties of soybeans in each plot.

A design that would be easier to execute would have each fertilizer applied to two larger "wholeplots" and then the varieties of soybeans planted in three "subplots" (equal in size to the plots of the previous design) within each wholeplot. A design of this type appears in Figure 16.2.

This design is called a split-plot design, and with this design there is a two-stage randomization. First, levels of factor A (fertilizers) are randomly assigned to the wholeplots; second, the levels of factor T (soybeans) are randomly assigned to the subplots within a wholeplot (see Figure 16.3). Using this design, it would be much easier to prepare the soil and to apply the appropriate fertilizer to the larger wholeplots and then to plant varieties of soybeans in the subplots, rather than to prepare the soil and to apply fertilizer to the

FIGURE 16.3

Two-Stage
Randomization for
a Split-Plot Design

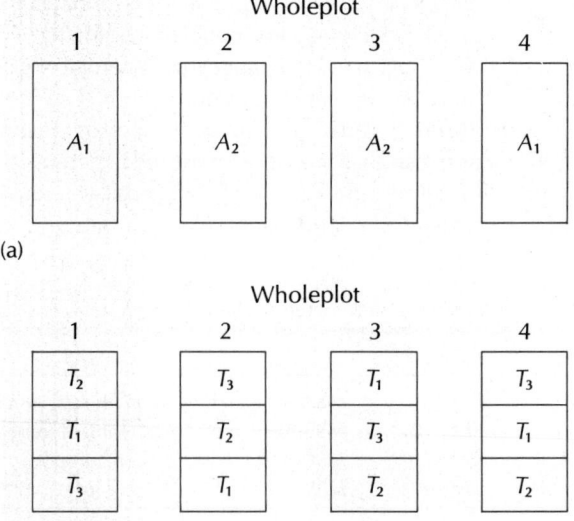

FIGURE 16.4

A Two-Factor
Split-Plot Design
Laid Out
in Blocks

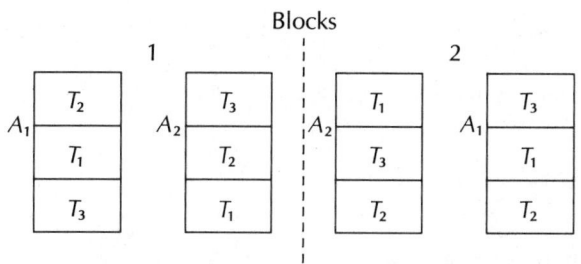

subplots and then to plant soybeans in the subplots, as would be the case for a standard 2 × 3 factorial experiment.

A variation on this design introduces a *blocking factor* (such as farms). So for our example, there may be $b = 2$ farms with $a = 2$ wholeplots per farm and $t = 3$ subplots per wholeplot. This design is shown in Figure 16.4.

The model for this more general two-factor split-plot design laid off in b blocks is as follows:

$$y_{ijk} = \mu + \alpha_i + \beta_j + \alpha\beta_{ij} + \tau_k + \alpha\tau_{ik} + \varepsilon_{ijk}$$

where y_{ijk} denotes the measurement receiving the ith level of factor A and the kth level of Factor T in the jth block. The parameters α_i, τ_k, and $\alpha\tau_{ik}$ are the usual main effects and interaction parameters for a two-factor experiment, whereas β_j is the effect due to block j and $\alpha\beta_{ij}$ is the interaction between the ith level of factor A and the jth block. The analysis corresponding to this model is shown in Table 16.26. Here we assume factors A and T are fixed effects, whereas blocks are random.

The sums of squares for the sources of variability listed in Table 16.26 can be obtained using the general shortcut formulas for main effects and interactions in a factorial experiment. Using these expected mean squares, we can obtain a valid F-test for factor A in the whole plot portion of the analysis and for factor T and the AT interaction in the subplot portion. These are shown here. Note that no test is made for the variability due to blocks.

TABLE 16.26   AOV for a Two-Factor Split-Plot Design Laid Off in Blocks
(A, T Fixed; Blocks Random)

Source	SS	df	EMS
Between wholeplots			
Blocks	SSB	$b - 1$	$\sigma_\varepsilon^2 + at\sigma_\beta^2$
A	SSA	$a - 1$	$\sigma_\varepsilon^2 + t\sigma_{\alpha\beta}^2 + bt\theta_A$
AB (wholeplot error)	SSAB	$(a - 1)(b - 1)$	$\sigma_\varepsilon^2 + t\sigma_{\alpha\beta}^2$
Within wholeplots			
T	SST	$(t - 1)$	$\sigma_\varepsilon^2 + ab\theta_T$
AT	SSAT	$(a - 1)(t - 1)$	$\sigma_\varepsilon^2 + b\theta_{AT}$
Subplot error	SSE	$a(b - 1)(t - 1)$	σ_ε^2
Totals	TSS	$abt - 1$	

Wholeplot analysis

H_0: $\theta_A = 0$ (or, equivalently, H_0: all $\alpha_i = 0$), $F = \dfrac{\text{MSA}}{\text{MSAB}}$

Subplot analysis

H_0: $\theta_{AT} = 0$ (or, equivalently, H_0: all $\alpha\tau_{ik} = 0$), $F = \dfrac{\text{MSAT}}{\text{MSE}}$

H_0: $\theta_T = 0$ (or, equivalently, H_0: all $\tau_k = 0$), $F = \dfrac{\text{MST}}{\text{MSE}}$

EXAMPLE 16.11

Soybean yields (in bushels per subplot unit) are shown here for a two-factor split-plot design laid off in $b = 3$ blocks. Fertilizers (factor A) were applied at random to the wholeplot units within each farm. Soybean varieties (factor T) were then randomly allocated to the subplots within each wholeplot. Conduct an analysis of variance using these sample data. Give an approximate p-value for each test.

	1			2			3	
	Fertilizers			Fertilizers			Fertilizers	
	1	2		2	1		1	2
Varieties			Varieties			Varieties		
1	10.6	10.9	2	11.9	11.5	3	9.5	9.8
2	11.4	11.7	3	12.6	12.1	1	8.1	8.2
3	11.8	12.4	1	11.6	10.8	2	8.7	9.3

Solution

For these data with $a = 2$, $b = 3$, and $t = 3$, the sums of squares are as shown (see TYPE III SS column in the following computer output)

$$\text{SSA} = 0.841$$
$$\text{SSB} = 29.934$$
$$\text{SSAB} = 0.0597$$
$$\text{SST} = 4.955$$
$$\text{SSAT} = 0.021$$
$$\text{SSE} = 0.598$$
$$\text{TSS} = 35.325$$

```
                              GENERAL LINEAR MODELS PROCEDURE

DEPENDENT VARIABLE: RESPONSE

SOURCE            DF      SUM OF SQUARES      MEAN SQUARE      F VALUE      PR > F        R-SQUARE              C.V.

MODEL             9        34.72666667        3.85851852       51.59       0.0001       0.983062            2.5519

ERROR             8         0.59833333        0.07479167                   ROOT MSE            RESPONSE MEAN

CORRECTED TOTAL  17        35.32500000                                     0.27348065            10.71666667


SOURCE            DF         TYPE I SS     F VALUE     PR > F       DF       TYPE III SS     F VALUE     PR > F
FERT              1        0.84500000      11.30       0.0099        1       0.84072727      11.24       0.0100
BLOCK             2       28.86333333     192.96       0.0001        2      29.93366667     200.11       0.0001
BLOCK*FERT        2        0.04333333       0.29       0.7560        2       0.05966667       0.40       0.6837
VARIETY           2        4.95450000      33.12       0.0001        2       4.95450000      33.12       0.0001
VARIETY*FERT      2        0.02050000       0.14       0.8739        2       0.02050000       0.14       0.8739


TESTS OF HYPOTHESES USING THE TYPE III MS FOR BLOCK*FERT AS AN ERROR TERM

SOURCE            DF        TYPE III SS     F VALUE     PR > F

FERT              1        0.84072727      28.18       0.0337
```

The analysis of variance table using Type III sums of squares is shown here.

Source	ss	df	ms	F	p-value
Between wholeplots					
Blocks	29.934	2	14.967	—	—
A	0.841	1	0.841	28.18	0.0337
AB (wholeplots)	0.060	2	0.030	—	—
Within wholeplots					
T	4.955	2	2.477	33.12	0.0001
AT	0.021	2	0.010	0.14	0.8739
Subplot error	0.598	8	0.075	—	—
Totals	35.325	17			

The distinction between this two-factor split-plot design and the standard two-factor experiments discussed in Chapter 14 lies in the randomization. In a split-plot design, there are two stages to the randomization process; first, levels of factor A are randomized to the wholeplots within each block, and then levels of factor B are randomized to the subplot units within each wholeplot of every block. In contrast, for a two-factor experiment laid off in a randomized block design (see Section 14.3), the randomization is a one-step procedure; treatments (factor-level combinations of the two factors) are randomized to the experimental units in each block. Much more has been done with split-plot designs than we will be able to cover in this text. The interested reader is referred to Steel and Torrie (1980), Snedecor and Cochran (1980), Winer (1971), and Mendenhall (1968), where the analyses for the basic split-plot design and more complicated variations on this design are discussed.

| 16.7 | SUMMARY |

Fixed, random, and mixed models are easily distinguished if we think in terms of the general linear model. The fixed-effects model relates a response to $k \geq 1$ independent variables and one random component, while a random-effects model is a general linear model with $k = 0$ and more than one random component. The mixed model, a combination of the fixed- and random-effects models, relates a response to $k \geq 1$ independent variables and more than one random component.

The application of random-effects models to experimental situations was illustrated for the completely randomized design and for the $a \times b$ factorial experiment. Similarities were noted between tests of significance in an analysis of variance for a random-effects model and for the corresponding fixed-effects model. Inferences resulting from an analysis of variance for a mixed model were illustrated using the $a \times b$ factorial experiment.

Unfortunately, in an introductory course, only a limited amount of time can be devoted to a discussion of random- and mixed-effects models. To expand our discussion in the text, the results of Section 16.5 are useful in developing the expected mean squares for sources of variability in the analysis of variance table for balanced designs. Using these expectations we can then attempt to construct appropriate test statistics for evaluating the significance of any of the fixed or random effects in the model.

The hardest part in any of these problems involving random- or mixed-effects models arises from trying to estimate $E(y)$, with an appropriate confidence interval for a random-effects model and the average value of y at some level or combination of levels for fixed effects in a mixed model. In Exercise 16.2 we illustrated how to obtain an estimate of $E(y)$ for a random-effects model and how to construct an approximate confidence interval. The problem becomes even more complicated for mixed models and hence is discussed in more advanced studies.

The final topics covered in this chapter were nested designs and split-plot designs. A brief introduction showed several variations on the basic factorial experiments discussed in Chapter 14 and in earlier sections of this chapter. The designs presented represent only a few of the more common designs possible when considering nested effects in a multi-factor experimental setting. The interested reader should consult the references at the end of this chapter to pursue these topics in more detail.

EXERCISES

16.8 Distinguish between inferences related to θ_A (when factor A is fixed) and σ_α^2 (when factor A is random).

16.9 Refer to Example 16.8. Can valid test statistics be formulated to test for each of the sources of variability? List the valid tests and the corresponding test statistics.

16.10 Consider a factorial experiment with a levels of factor A, b levels of factor B, c levels of factor C, and r observations per factor-level combination.
   a. Write down the expected mean squares for all sources of variability in an analysis of variance table, treating A, B, and C as fixed effects.
   b. Note appropriate tests for all sources.

16.11 Refer to Exercise 16.10.

EXERCISES

a. Repeat part (a), but now consider factors *A*, *B*, and *C* as random effects.

b. Write down the test statistics for appropriate *F* tests for sources of variability.

16.12 Refer to Exercise 16.10. Suppose now that factor *A* is fixed and factors *B* and *C* are random.

a. Write down expected mean squares for all sources in an AOV table.

b. Indicate the test statistics for those sources of variability that can be tested using an exact *F* test.

 16.13 The civil engineering department at a university was awarded a large grant to study the campus traffic problems and to recommend alternative solutions. One small phase of the study involved obtaining daily counts on the number of cars crossing, but not making use of, the campus facilities. To do this, a team of volunteers was stationed at each entrance to monitor simultaneously the license number and the time of entrance or exit for each car passing through the checkpoint. By comparing lists for all checkpoints and allowing a reasonable time for cars to traverse the campus, the teams were able to determine the number of cars crossing but not using the campus facilities during the 8:00 AM to 5:00 PM time period. A random sample of six weeks throughout the academic year was used, with two midweek days selected for study in the weeks sampled. The traffic volume data appear below.

Week 1	Week 2	Week 3	Week 4	Week 5	Week 6
680	438	539	264	693	530
618	520	600	198	646	575

a. Write an appropriate linear statistical model. Identify all terms in the model.

b. Perform an analysis of variance, indicating expected mean squares. Use $\alpha = .05$.

16.14 Refer to Exercise 16.13. Estimate the average number of cars crossing but not using the campus facilities for a midweek day of a randomly selected week and give an approximate confidence interval. (Hint: Refer to Exercise 16.2.)

16.15 Refer to Exercise 16.6. Suppose the five physicians chosen for the pilot study were considered to be a random sample from many possible physicians.

a. Write an appropriate model. Indicate how the assumptions for this model differ from those of part (a) in Exercise 16.6.

b. Compare the AOV table and conclusions drawn here to those of Exercise 16.7.

16.16 Refer to Exercise 16.15.

a. Which model and analysis seem to be more appropriate?

b. Might you also consider a fixed-effects model? Why or why not?

16.17 Refer to Exercise 14.3. Suppose that we consider the five investigators as a random sample from a population of all possible investigators for the rocket propellant experiment.

a. Write an appropriate linear statistical model, identifying all terms and listing your assumptions.

b. Perform an analysis of variance. Include an expected mean squares column in the analysis of variance table.

16.18 Refer to Exercise 16.17. Indicate the differences in the hypothesis under test and differences in the conclusions drawn for the fixed and random effects.

16.19 Obtain the expected mean squares for the experiment described in Example 14.4 if we assume that the drivers were selected at random from the many possible drivers in a large city.

16.20 Obtain expected mean squares for the experiment described in Exercise 14.9, assuming that both rows and columns are random effects.

16.21   Refer to the data of Exercise 14.30.
   a. Give the expected mean squares under the assumption that investigators were selected at random from a large group of similarly qualified persons throughout the country.
   b. Indicate how your analysis and conclusions would change from those in Exercise 14.31 under the assumption of part (a).

16.22   Refer to the data of Example 16.2.
   a. Using the expected mean squares, give formulas for estimates of σ_ε^2, σ_γ^2, σ_β^2, and σ_α^2.
   b. Using the data of Example 16.2 and the formulas of part (a), find estimates for all the variance components.

16.23   Refer to Exercise 16.5. Use the expected mean squares to obtain estimates of all the variance components.

 16.24   Core soil samples are taken in each of six locations within a territory being investigated for surface mining of bituminous coal. Each of the core samples is divided into four subsamples for separate analyses of the sulfur content of the sample.
   a. Identify the design and give a model for this experimental setting.
   b. Give the sources of variability and degrees of freedom for an AOV.

16.25   The sample data for Exercise 16.24 are shown here. Run an AOV and draw conclusions.

	Analyses			
Location	1	2	3	4
1	15.2	16.8	17.5	16.2
2	13.1	13.8	12.6	12.9
3	17.5	17.1	16.7	16.5
4	18.3	18.4	18.6	17.9
5	12.8	13.6	14.2	14.0
6	13.5	13.9	13.6	14.1

16.26   Tablet hardness is one comparative measure for different formulations of the same drug product; some combinations of ingredients (in addition to the active drug) in a formulation give rise to harder tablets than do other combinations. Suppose that three batches of a formulation are to be examined; 10 different 1-pound samples of tablets are obtained from each batch, and six tablets are tested from each of the 1-pound samples.
   a. Identify the design.
   b. Give an appropriate model with assumptions.
   c. Give the sources of variability and degrees of freedom for an AOV.

16.27   Refer to Exercise 16.26. Given that the sums of squares are as listed here, perform an analysis of variance and draw conclusions about the tablet hardness data for the formulation under study.

Source	SS
Between batches	2200.8
Between samples within a batch	1650.4
Between tablets within a sample	90.3

17 EXPERIMENTS WITH REPEATED MEASURES (optional)

INTRODUCTION

In all the experimental situations discussed so far in this text (except for the paired difference experiment), we have assumed that only one observation is taken on each experimental unit. For example, in an experiment to compare the effects of three different cardiovascular compounds on blood pressure, we could use a single-factor design where n_1 patients are assigned to compound 1, n_2 to compound 2, and n_3 to compound 3. Then the model would be

$$y_{ij} = \mu + \alpha_i + \varepsilon_{ij}$$

where α_i is the (fixed or random) effect due to compound i and ε_{ij} is the random effect associated with patient j treated with compound i. For this design, we would get one measurement (y_{ij}) for each patient.

The practicalities of most applied research settings make it mandatory from a cost and efficiency standpoint to obtain more than one observation per experiment unit. For example, in conducting clinical research, it is often difficult to find patients who have the condition to be studied *and* who are willing to participate in a clinical trial. Hence, it is important to obtain as much information as possible once a suitable number of patients have been located. In this chapter, we will consider several different experimental settings involving one or more factors and repeated measures. Rather than obtaining just one observation per patient, as suggested in the design with n_i patients treated with drug product i, we could obtain t different measurements corresponding to t different time points following administration of the assigned treatment. This experimental setting is shown in Table 17.1.

TABLE 17.1   Repeated Time Points for Each Patient

Compound	Time Period			
	1	2	$\cdots$	t
1	y_{111}	y_{112}	$\cdots$	y_{11t}
	$\vdots$	$\vdots$		$\vdots$
	y_{1n_11}	y_{1n_12}	$\cdots$	y_{1n_1t}
2	y_{211}	y_{212}	$\cdots$	y_{21t}
	$\vdots$	$\vdots$		$\vdots$
	y_{2n_21}	y_{2n_22}	$\cdots$	y_{2n_2t}
3	y_{311}	y_{312}	$\cdots$	y_{31t}
	$\vdots$	$\vdots$		$\vdots$
	y_{3n_31}	y_{3n_32}	$\cdots$	y_{3n_3t}

In Table 17.1, y_{ijk} denotes the observation at the time k for the jth patient on compound i. Note that we are getting $t > 1$ observations per patient, rather than only 1. The methods of this chapter can be used to analyze these data, as well as those in several other experimental situations. In this chapter, we will focus on experimental settings in the pharmaceutical industry. Similar, comparable applications can be found for these designs in the R & D and manufacturing operations of most major industries, such as the automotive, chemical, and aerospace industries.

17.2   SINGLE-FACTOR EXPERIMENTS WITH REPEATED MEASURES

In the previous section we discussed some reasons why one might want to get more than one observation per patient. Another reason for obtaining more than one observation per patient is that frequently the variability *among* or *between* patients is much greater than the variability *within* a patient. If this is the case, it might be better to block on patients and to give each patient each treatment. Then the comparison among compounds is a within-patient comparison, rather than a comparison between patients, as would be the case with the single-factor experiment with n_i different patients assigned to compound i. A single-factor design that reflects this within-patient emphasis is shown in Table 17.2.

With this design, n patients would be treated separately with each of the three compounds. Presumably, the order of treatment would be randomized and there would be a

TABLE 17.2   A Within-Patient Comparison of Compounds 1, 2, and 3

Compound	Patient			
	1	2	$\cdots$	n
1	y_{11}	y_{12}	$\cdots$	y_{1n}
2	y_{21}	y_{22}	$\cdots$	y_{2n}
3	y_{31}	y_{32}	$\cdots$	y_{3n}

sufficiently long "washout" period between the treatments, so that the results from one compound would not affect the results for another compound.

Here again, we are obtaining more than one observation per patient and presumably getting more useful information about the three drug products in question. One model for this experimental setting is

$$y_{ij} = \mu + \alpha_i + \pi_j + \varepsilon_{ij}$$

Note that this model looks like any other single-factor experimental setting with a compounds and n patients. However, the assumptions are different because we are obtaining more than one observation per patient. For this model, we make the following assumptions:

1. α_i is a constant and $\sum \alpha_i = 0$.
2. The π_j are independent and normally distributed $(0, \sigma_\pi^2)$.
3. The ε_{ij} are independent of the π_j.
4. The ε_{ij} are independent and normally distributed $(0, \sigma_\varepsilon^2)$.

From these assumptions it can be shown that the variance of y_{ij} is $\sigma_\pi^2 + \sigma_\varepsilon^2$ and the co-variance for any two observations from patient j is constant. These assumptions give rise to a variance-covariance matrix for the y_{ij}, which exhibits *compound symmetry*. For example, with $a = 3$, and $n = 2$, the variance-covariance for the y_{ij} would appear as

$$\mathbf{V}_{6 \times 6} = \begin{bmatrix} E & F \\ F & E \end{bmatrix}$$

where

$$\mathbf{E}_{3 \times 3} = (\sigma_\pi^2 + \sigma_\varepsilon^2) \begin{bmatrix} 1 & \rho & \rho \\ \rho & 1 & \rho \\ \rho & \rho & 1 \end{bmatrix}$$

and

$$\mathbf{F}_{3 \times 3} = \begin{bmatrix} 0 & 0 & 0 \\ 0 & 0 & 0 \\ 0 & 0 & 0 \end{bmatrix}$$

Note that with **compound** symmetry, two observations on the same patient are correlated, whereas observations on two different patients are not. The analysis of variance for the experimental design being discussed and this set of assumptions is shown in Table 17.3. This AOV looks all too familiar. When the assumptions hold, and hence when compound symmetry holds, the statistical test on factor A ($F = MSA/MSE$) is appropriate. But there are

TABLE 17.3   AOV For the Experimental Setting Depicted in Table 17.2

Source	SS	df	EMS (A Fixed, Patients Random)
Between patients	SSP	$n - 1$	$\sigma_\varepsilon^2 + a\sigma_\pi^2$
Within patients			
A	SSA	$a - 1$	$\sigma_\varepsilon^2 + n\theta_A$
Error	SSE	$(a - 1)(n - 1)$	σ_ε^2
Totals			

some other more general conditions* that also lead to a valid F test for factor A using $F =$ MSA/MSE. How restrictive are these assumptions and how can we tell when the test is appropriate?

There are no easy answers to these questions since there are no simple tests to check for compound symmetry. The general conditions (called the Huynh–Feldt conditions) under which the F test for factor A is valid are often not met because observations on the same patient taken closely in time are more highly correlated than are observations taken farther apart in time. So be careful about this. In general, when the variance-covariance matrix does not follow a pattern of compound symmetry, the F test for factor A has a positive bias, which allows rejection of H_0: all $\alpha_i = 0$ more often than is indicated by the critical F values.

From a practical standpoint, the best thing to do in a given experimental setting is to make certain that there is sufficient time between applications of the treatment to allow washout (or elimination) of the previous treatment and to make certain that the design is applied in only those situations where the disease is relatively stable, so that following treatment and washout, each patient (or experimental unit) is essentially the same as prior to receiving treatment. For example, even when studying the effect of blood pressure–lowering drugs, we would expect the hypertension to be stable enough that the patients would return to their predrug level blood pressures after washout of the first assigned compound before receiving the second assigned compound, and so on.

In Section 17.3, more will be said about how to judge whether the underlying assumptions for the test hold, and if they do not, how to proceed. For further information on this topic, refer to higher-level textbooks covering repeated-measures experiments in detail (for example, Winer (1962).

17.3   TWO-FACTOR EXPERIMENTS WITH REPEATED MEASURES ON ONE OF THE FACTORS

We can extend our discussion of repeated measures experiments to two-factor settings. For example, in comparing the blood pressure–lowering effects of cardiovascular compounds, we could randomize the patients so that n different patients receive each of the three compounds. But rather than having each patient receive each compound, we could take multiple measurements across time for each patient. For example, we might be interested in obtaining blood pressure readings immediately prior to receiving a single dose of the assigned compound and then every 15 minutes for the first hour and hourly thereafter for the next six hours. The data for this type of setting are depicted in Table 17.1. Note that this is a two-factor experiment (compounds and time) with repeated measures taken over one of the two factors (time).

The model that we will use for a two-factor experiment comparing a levels of factor A (compounds), having n patients per level of factor A, and b levels of factor B (time) is

$$y_{ijk} = \mu + \alpha_i + \pi_{j(i)} + \beta_k + \alpha\beta_{ik} + \varepsilon_{ijk}$$

* *JASA* 1970 and *SAS Statistics Users Guide.*

TABLE 17.4   Analysis of Variance for a Two-Factor Experiment, Repeated Measures on One Factor

Source	SS	df	EMS (A, B Fixed; Patients Random)
Between patients			
A	SSA	$a - 1$	$\sigma_\varepsilon^2 + b\sigma_\pi^2 + nb\theta_A$
Patients in A	SSP(A)	$a(n - 1)$	$\sigma_\varepsilon^2 + b\sigma_\pi^2$
Within patients			
B	SSB	$b - 1$	$\sigma_\varepsilon^2 + na\theta_B$
AB	SSAB	$(a - 1)(b - 1)$	$\sigma_\varepsilon^2 + n\theta_{AB}$
Error	SSE	$a(b - 1)(n - 1)$	σ_ε^2
Totals	TSS	$abn - 1$	

where α_i, β_k, and $\alpha\beta_{ik}$ are fixed effects corresponding to main effects for factor A (compounds), factor B (times), and their interaction, respectively. The term $\pi_{j(i)}$ denotes a random effect due to the jth patient in the ith level of factor A. We assume the $\pi_{j(i)}$ are independent and normally distributed $(0, \sigma_\pi^2)$.

Based on these assumptions, we have the analysis of variance shown in Table 17.4. Based on Table 17.4, it is clear that the following tests can be performed

1. $H_0: \theta_{AB} = 0$

$$F = \frac{\text{MSAB}}{\text{MSE}}$$

2. $H_0: \theta_B = 0$

$$F = \frac{\text{MSB}}{\text{MSE}}$$

3. $H_0: \theta_A = 0$

$$F = \frac{\text{MSA}}{\text{MSP(A)}}$$

EXAMPLE 17.1   Ten subjects agreed to participate in a study to examine the concentration of drug in the bloodstream for two different dosage forms (capsule and tablet) of the same product following a single dose. Presumably, within limits, the higher the concentration, the more effective the drug product. Five subjects were allocated at random to the capsule form and the other five to the tablet form. Subjects fasted from 8:00 PM of the night prior to starting the study until four hours following ingestion of the assigned dose form (at 8:00 AM of the next day). Blood samples (15 ml) were obtained immediately preceding the assigned dose and at 0.5, 1, 2, 3, and 4 hours after dosing, and were analyzed for the concentration of the drug product in the bloodstream. These data (in ng/ml) are shown here

	Tablet Time							Capsule Time					
Subject	0	0.5	1	2	3	4	*Subject*	0	0.5	1	2	3	4
1	0	50	75	120	60	30	1	0	30	55	80	130	65
2	0	40	80	135	70	40	2	0	25	50	75	125	60
3	0	55	75	125	85	50	3	0	35	65	85	140	85
4	0	70	85	140	90	40	4	0	45	70	90	145	80
5	0	60	90	150	95	50	5	0	50	75	95	160	90

a. Plot the mean sample data (response versus time) for each of the dose forms. Do the dose forms seem to have different availability patterns across time?

b. **Run a repeated measures analysis of variance.**

Solution

a.

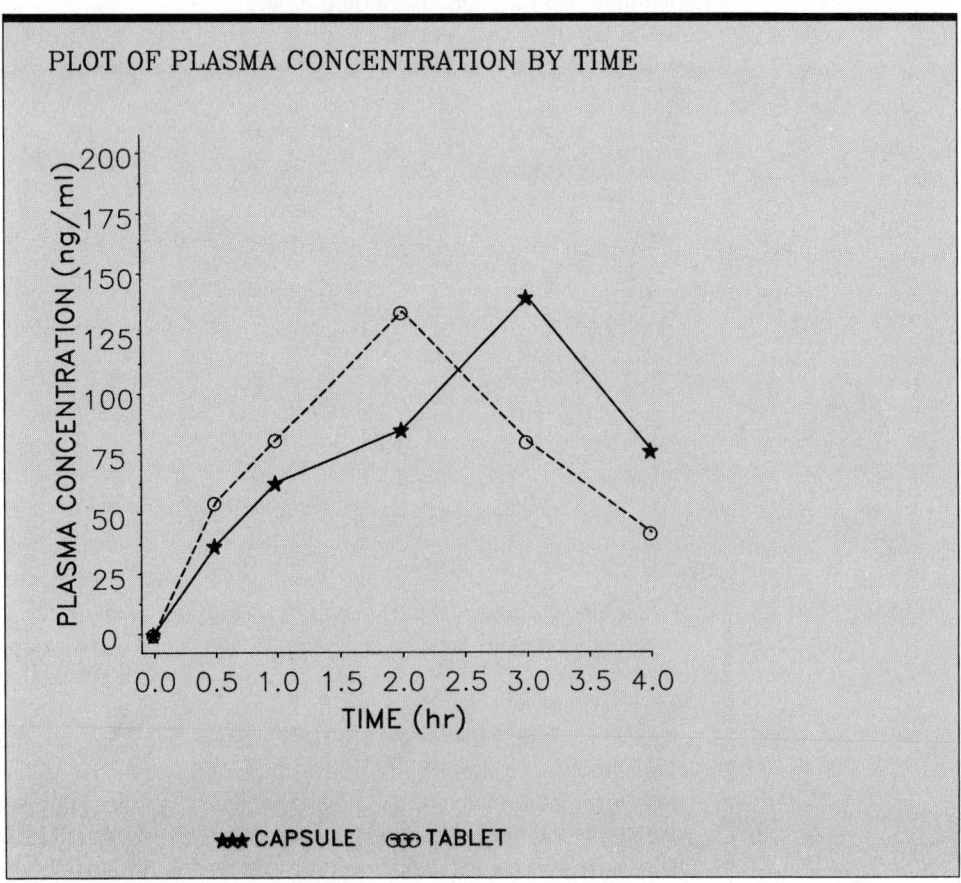

PLOT OF PLASMA CONCENTRATION BY TIME

★★★ CAPSULE   ⊙⊙⊙ TABLET

b.

Source	ss	df	MS	F	p-value
Between patients					
Formulation	33.75	1	33.75	0.08	0.7810
Patients in formulation	3266.67	8	408.33		
Within patients					
Time	86692.08	5	17,338.42	424.61	0.0001
Time × formulation	19478.75	5	3,895.75	95.41	0.0001
Error	1633.33	40	40.83		

Conclusion: There is evidence (based on the significant time × formulation inter-action) that the two formulations have different availability (concentration) patterns across time.

The F test for factor A is based on between-subject effects and hence is *not* affected by the repeated measures on factor B. But, the F-ratios for the within-patients effects are affected and, as with the one-factor experiment with repeated measures, we must worry about the conditions under which these F tests are appropriate. If compound symmetry of the variance-covariance matrix for the y_{ijk}s holds, then we can apply these tests; also if the Huynh–Feldt conditions alluded to previously hold, then we can apply these F tests. Some have suggested (Greenhouse and Geisser (1959); Huynh and Feldt (1970) that "adjusted" F-values be used to determine the statistical significance of a repeated measures F test when there is some departure from the underlying conditions for that test. The adjustments recommended by the various authors follow the same pattern. A quantity epsilon is defined as a multiplicative adjustment factor for the numerator and denominator degrees of freedom for the F test in question. This epsilon (which we'll denote by e) is not to be confused with the random error term ε in our models. For most of these adjustments, the multiplicative factor e ranges between 0 and 1, taking on a value of 1 when the underlying conditions for a valid F test are met and smaller values as the degree of depar-ture from those conditions increases. A value of e having been determined for a given situation, the computed F statistic is compared to the critical value for an F distribution with numerator and denominator degrees of freedom multiplied by e.

The ideas behind the adjustment can be seen if we use the experimental setting for Table 17.4 as the basis for discussion. Here we have a two-factor experiment with repeated measures on the second factor (B). The F tests for the within-patient effects, B and AB shown in Table 17.4 are valid provided the Huynh–Feldt conditions.

For a given experiment, we compute a value of e and adjust the degrees of freedom for the F test by multiplying df_1 and df_2 by e. So, to run a test of H_0: $\theta_{AB} = 0$, a value of e is computed from the sample data and the computed F statistic

$$F = \frac{MSAB}{MSE}$$

is compared to a critical value of F_α based on $df_1 = e(a-1)(b-1)$ and $df_2 = ea(b-1)(n-1)$. Note that when e = 1, the underlying conditions hold and we have the original, recommended degrees of freedom, $df_1 = (a - 1)(b - 1)$ and $df_2 = a(b - 1)(n - 1)$.

In experimental situations where repeated measures data are to be analyzed and where you have access to SAS, you can use PROC GLM to compute revised p-values for two different adjustments to the degrees of freedom. The first adjustment, proposed by Greenhouse and Geisser (1959), uses a sample estimate of e. This adjustment, labeled "G-G" in the SAS output, has been shown, in simulation studies, to be ultraconservative, because the actual p-value may be much smaller than that indicated by the p-value using the G-G adjustment. The second adjustment factor (proposed by Huynh and Feldt (1970)) is based on a different formula for e. Once again, however, an estimate of this adjustment factor is computed from the sample data. The degrees of freedom for critical values of the F statistics are then adjusted using the estimate of e. This adjustment is labeled "H-F" in the PROC GLM output. Although the Greenhouse–Geiser e and Huynh–Feldt e both must be in the interval $0 < e \leq 1$, the H-F estimate of e can sometimes be greater than 1. In these situations, a value of $e = 1$ is used in determining the appropriate degrees of freedom for the F test.

EXAMPLE 17.2   Refer to the output for Example 17.1.

a. Locate the estimated values for the Greenhouse–Geisser adjustment factor and the Huynh–Feldt adjustment factor.
b. Are the conclusions for the tests on time effects and the time formulation interaction affected by these adjustments?

GENERAL LINEAR MODELS PROCEDURE

DEPENDENT VARIABLE: CONC

SOURCE	DF	SUM OF SQUARES	MEAN SQUARE	F VALUE	PR > F	R-SQUARE	C.V.
MODEL	19	109471.25000000	5761.64473684	141.10	0.0001	0.985299	9.6698
ERROR	40	1633.33333333	40.83333333		ROOT MSE		CONC MEAN
CORRECTED TOTAL	59	111104.58333333			6.39009650		66.08333333

SOURCE	DF	TYPE I SS	F VALUE	PR > F	DF	TYPE III SS	F VALUE	PR > F
FORM	1	33.75000000	0.83	0.3687	1	33.75000000	0.83	0.3687
PATIENT(FORM)	8	3266.66666667	10.00	0.0001	8	3266.66666667	10.00	0.0001
TIME	5	86692.08333333	424.61	0.0001	5	86692.08333333	424.61	0.0001
FORM*TIME	5	19478.75000000	95.41	0.0001	5	19478.75000000	95.41	0.0001

TESTS OF HYPOTHESES USING THE TYPE III MS FOR PATIENT(FORM) AS AN ERROR TERM

SOURCE	DF	TYPE III SS	F VALUE	PR > F
FORM	1	33.75000000	0.08	0.7810

GENERAL LINEAR MODELS PROCEDURE

TESTS OF HYPOTHESES FOR BETWEEN SUBJECTS EFFECTS

SOURCE	DF	TYPE III SS	MEAN SQUARE	F VALUE	PR > F
FORM	1	33.75000000	33.75000000	0.08	0.7810
ERROR	8	3266.66666667	408.33333333		

GENERAL LINEAR MODELS PROCEDURE

UNIVARIATE TESTS OF HYPOTHESES FOR WITHIN SUBJECT EFFECTS

SOURCE	DF	TYPE III SS	MEAN SQUARE	F VALUE	PR > F	ADJUSTED PR > F G - G	H - F
TIME	5	86692.08333333	17338.41666667	424.61	0.0001	0.0001	0.0001
TIME*FORM	5	19478.75000000	3895.75000000	95.41	0.0001	0.0001	0.0001
ERROR(TIME)	40	1633.33333333	40.83333333				

GREENHOUSE-GEISSER EPSILON = 0.5571
HUYNH-FELDT EPSILON = 0.9916

Solution

a. The Greenhouse–Geisser estimate of e is 0.5571 and the Huynh–Feldt estimate of e is 0.9916.

b. Time effects: F tests based on the G-G adjustment and on the H-F adjustment yield p-values of 0.0001 and 0.0001, respectively, the same as the original F tests; hence, the adjustments do not change the original conclusion.

Time × Formulation interaction: As with the F test on time, the adjustments did not change the p-values or the conclusions drawn from the original tests.

In conclusion, if you have access to SAS when doing an analysis of variance for a repeated measures experiment, it would be wise to check the effects of adjustments to F tests on the factors affected by repeated measures. If the conclusions based on the original ($e = 1$) test differ from those based on the H-F or G-G adjustment, we recommend adhering to the conclusions based on the less conservative H-F adjustment.

EXERCISES

17.1 A study was run to compare the effects of group therapy over time on patients who had been tested for depression and categorized into one of four levels of depression based on the Hamilton Depression Scale. There were 24 patients, with six patients per group, and each patient participated in the group therapy and was tested daily for a period of two weeks (14 times). Give the sources of variability and degrees of freedom for a repeated measures AOV.

17.2 Draw conclusions for Exercise 17.1 based on the sums of squares listed in the table below.

	SS
Between patients	
Depression group	962
Patients in group	582
Within Patients	
Time	165
Time × group	120
Error	65

17.3 An antihistamine is frequently studied using a model to examine its effectiveness (compared to a placebo) in inhibiting a positive skin reaction to a known allergen. Consider the following situation. Individuals are screened to find 20 subjects who demonstrate sensitivity to the allergen to be used in the study. The 20 subjects are then randomly assigned to one of two treatment groups (the known antihistamine and an identical-appearing placebo), with 10 subjects per group. At the start of the study, a baseline (predrug) sensitivity reading is obtained, and then each patient begins taking the assigned medication for three days. Then skin sensitivity readings are taken at 1, 2, 3, 4, and 8 hours following the first dose. The percent inhibition of the skin sensitivity reaction (reduction in swelling area where the allergen is applied, compared to baseline) is shown here for each of the 20 patients.

		Percent Inhibition Time (hr)				
		1	2	3	4	8
Treatment	patient					
1	1	10.5	28.2	15.3	43.0	29.0
	2	41.2	25.3	27.8	28.0	53.2
	3	43.0	20.8	29.3	5.2	26.5
	4	61.4	61.6	62.8	43.8	19.6
	5	5.0	28.2	31.6	19.5	2.3
	6	−10.2	27.2	38.1	35.5	18.0
	7	−12.9	22.1	34.0	43.4	34.2
	8	27.1	26.5	38.8	28.5	17.4
	9	13.0	19.7	23.5	29.4	39.6
	10	28.9	26.1	11.2	18.1	16.5
2	1	3.0	9.3	1.0	15.0	3.0
	2	− 1.5	−10.1	20.2	18.3	13.5
	3	10.8	20.6	28.3	25.2	15.8
	4	15.3	19.8	25.4	31.3	21.7
	5	8.7	8.0	17.5	26.6	16.4
	6	− 4.6	5.8	12.7	15.6	29.6
	7	−16.6	28.4	32.7	34.4	15.8
	8	9.4	15.7	22.7	29.8	23.2
	9	−19.3	15.7	21.7	30.4	26.1
	10	−12.8	12.3	0.1	21.3	10.6

(A negative value means there was an increase in swelling, compared to baseline.)

a. Compare means and standard deviations by time period for each group.

b. Plot these data showing mean percent inhibition by time for each treatment group. Does the antihistamine group appear to differ from the placebo group?

17.4 Refer to the data from Exercise 17.3. Give a model for this design and run a repeated measures analysis of variance to compare the two treatment groups. Do the analysis of variance results agree with your intuition based on the plot of Exercise 17.3?

17.5 Another question that may be asked relates to the onset of antihistaminic activity. How might you define onset? For each of the treatment groups, use a t test to determine the test time at which there is a significant reduction from baseline. What do these results suggest?

17.4   CROSSOVER DESIGNS

We will now consider an extension to the single-factor experiment discussed in Section 17.2. Recall that in Table 17.2 we presented data for an experimental situation where each of n patients received the same three compounds in a random order. A Latin square

TABLE 17.5   A 3 × 3 Latin Square Design

Sequence		Factor B (Periods) 1	2	3
1	n	A_1	A_2	A_3
2	n	A_2	A_3	A_1
3	n	A_3	A_1	A_2

arrangement of the compounds is an experimental design that provides the same advantages as the single-factor experiment with repeated measures (namely, multiple observations per patient and a within-patient comparison of the treatments) while offering some protection that patients didn't change with time.

A 3 × 3 Latin square design for this experimental situation is shown in Table 17.5. The design itself is called a three-period crossover design.

With this design, $3n$ patients are randomly assigned to the sequences (rows) of the design, n to each sequence. The periods correspond to the order in which the compounds are taken. The model for this design is

$$y_{ijkl} = \mu + \delta_k + \pi_{l(k)} + \alpha_i + \beta_j + \alpha\beta_{ij}^* + \varepsilon_{ijkl}$$

where δ_k is the fixed effect for the kth sequence and $\pi_{l(k)}$ is the random effect of the lth patient in sequence k; α_i and β_j are the fixed effects for compounds (factor A) and periods (factor B). The reason for the asterisk on the interaction term will be discussed later. The analysis of variance for this design (three-period crossover design) is shown in Table 17.6.

The sums of squares, degrees of freedom, and expected mean squares for the between-patient effects and main effects for factors A and B in the within-patient portion of the analysis of variance are straightforward, but you will notice that there is an asterisk on the AB interaction term and that this interaction term has only two, rather than four, degrees of freedom. Actually, the missing two degrees of freedom are in the sum of squares due to sequences. In fact, it can be shown that the AB interaction sum of squares is equal to

$$SSAB = SSSeq + SSAB^*$$

We will use this identity to compute $SSAB^*$.

TABLE 17.6   Analysis of Variance for a Three-Period Crossover Design

Source	SS	df	EMS (A, B Fixed; Patients Random)
Between patients			
Sequence	SSSeq	2	$\sigma_\varepsilon^2 + 3\sigma_\pi^2 + 3n\theta_{Seq}$
Patients in sequences	SSP(Seq)	$3(n-1)$	$\sigma_\varepsilon^2 + 3\sigma_\pi^2$
Within patients			
A (compounds)	SSA	2	$\sigma_\varepsilon^2 + 3n\theta_A$
B (periods)	SSB	2	$\sigma_\varepsilon^2 + 3n\theta_B$
AB*	SSAB*	2	$\sigma_\varepsilon^2 + n\theta_{AB}$
Error	SSE	$3(2)(n-1)$	σ_ε^2
Totals	TSS	$9n-1$	

The test that we run in the analysis of variance will give us partial information about the *AB* interaction; actually, we are testing the within-patient portion of that interaction.

EXAMPLE 17.3   Twelve males volunteered to participate in a study to compare the durations of effect of three different formulations of a drug product. Formulation 1 was a 50-mg tablet, formulation 2 was a 100-mg tablet, and formulation 3 was a sustained-release formulation capsule. A three-period crossover design was used, with four volunteers assigned to each of the three treatment sequences. On each treatment day, volunteers were given their assigned formulation and were observed to determine the duration of effect (blood pressure lowering). There was a one-week washout between each treatment period of the study. The sample data are shown here.

			Period		
			1	2	3
Sequence	1	$n = 4$	A_1	A_2	A_3
2		$n = 4$	A_2	A_3	A_1
3		$n = 4$	A_3	A_1	A_2

	Patient	Period		
Sequence	in Sequence	1	2	3
	1	1.5	2.2	3.4
1	2	2.0	2.6	3.1
	3	1.6	2.7	3.2
	4	1.1	2.3	2.9
	1	2.5	3.5	1.9
2	2	2.8	3.1	1.5
	3	2.7	2.9	2.4
	4	2.4	2.6	2.3
	1	3.3	1.9	2.7
3	2	3.1	1.6	2.5
	3	3.6	2.3	2.2
	4	3.0	2.5	2.0

Solution   Based on the analysis of variance shown in the accompanying computer output, there is a hint of a period by treatment interaction ($p = .0853$), which appears negligible in the presence of a highly significant treatment effect ($p = .0001$). This is borne out in the plot of mean durations versus period for the three sequences shown here. As can be seen, the longest durations on the average were observed with formulation 3 followed by formulation 2 and then 1.

When there are only two compounds to be examined, the Latin square arrangement, called a two-period crossover design, would have $2n$ patients randomly assigned to the two sequences, n to each sequence. The two-period crossover design is shown in Table 17.7.

GENERAL LINEAR MODELS PROCEDURE

DEPENDENT VARIABLE: CONC

SOURCE	DF	SUM OF SQUARES	MEAN SQUARE	F VALUE	PR > F	R-SQUARE	C.V.
MODEL	17	11.08638889	0.65214052	5.69	0.0003	0.843089	13.5579
ERROR	18	2.06333333	0.11462963		ROOT MSE		CONC MEAN
CORRECTED TOTAL	35	13.14972222			0.33856998		2.49722222

SOURCE	DF	TYPE I SS	F VALUE	PR > F	DF	TYPE III SS	F VALUE	PR > F
SEQUENCE	2	0.23388889	1.02	0.3804	0	0.00000000	.	.
PATIENT(SEQUENCE)	9	0.66916667	0.65	0.7425	9	0.66916667	0.65	0.7425
TREAT	2	9.51722222	41.51	0.0001	2	9.51722222	41.51	0.0001
PERIOD	2	0.01722222	0.08	0.9279	2	0.01722222	0.08	0.9279
PERIOD*TREAT	2	0.64888889	2.83	0.0853	2	0.64888889	2.83	0.0853

TESTS OF HYPOTHESES USING THE TYPE I MS FOR PATIENT(SEQUENCE) AS AN ERROR TERM

SOURCE	DF	TYPE I SS	F VALUE	PR > F
SEQUENCE	2	0.23388889	1.57	0.2595

PLOT OF MEAN DURATION VERSUS PERIOD OF STUDY

SEQUENCE 111 1   222 2   333 3

TABLE 17.7   Layout for a Two-Period Crossover Design

Sequence		Factor B (Periods)	
		1	2
1	n	A_1	A_2
2	n	A_2	A_1

TABLE 17.8   AOV Table for a Two-Period Crossover Design

Source	ss	df	EMS (A, B Fixed; Patients Random)
Between patients			
Sequences	SSSeq	1	$\sigma_\varepsilon^2 + 2\sigma_\pi^2 + 2n\theta_{Seq}$
Patients			
in sequences	SSP(Seq)	2(n − 1)	$\sigma_\varepsilon^2 + 2\sigma_\pi^2$
Within patients			
A	SSA	1	$\sigma_\varepsilon^2 + 2n\theta_A$
B	SSB	1	$\sigma_\varepsilon^2 + 2n\theta_B$
Error	SSE	2(n − 1)	σ_ε^2
Totals	TSS	4n − 1	

The corresponding analysis of variance for the model is

$$y_{ijkl} = \mu + \delta_k + \pi_{l(k)} + \alpha_i + \beta_j + \varepsilon_{ijkl}$$

where δ_k is the fixed effect due to sequence k, and α_i and β_j are the fixed effects due to treatment i and period j. As before, $\pi_{l(k)}$ represents the lth person in sequence k.

Note there is no AB interaction term in this model. We must assume this interaction is negligible; otherwise the design is inappropriate because there are no degrees of freedom available for testing the significance of the AB interaction. The AOV table for a two-period crossover design is shown in Table 17.8.

There are many other extensions to the repeated measures designs discussed in this chapter. For example, one could combine the concept of repeated measures on the same factor illustrated in Table 17.4 with the crossover design. Such a plan is illustrated in Table 17.9. Thus, rather than taking one observation per patient within each period, we would take observations at t different time points. For example, we could measure blood pressure every 15 minutes for the first hour following treatment with compound i, and then hourly for the next seven hours. This would be done in each of the periods for a total of 10 blood pressure measurements on each patient in each time period.

TABLE 17.9   Two-Period Crossover Design with Repeated Measures with a Period

	Period			
	1		2	
Sequence	Time		Time	
	1 2 ··· t		1 2 ··· t	
1	A_1		A_2	
2	A_2		A_1	

Although we won't give the analysis of variance for this extension to the repeated measures experiments discussed in this chapter, and will not cover other more complicated repeated measures designs, we want you to be aware of the wealth of possible designs that are available if you are willing to take more than one observation per experimental unit. The interested reader is referred to Winer (1971, Chapters 4, 7, and 9) and Buncher (1981, Chapter 7).

<table>
<tr><td>17.5</td><td></td></tr>
</table>

SUMMARY

In this chapter, we have discussed some of the initial concepts and designs associated with repeated measures experiments. We introduced single- and two-factor experiments, analyses for these experiments, and the special topics of two and three-period crossover designs. These methods are only a beginning, however. Rather than presenting an exhaustive, detailed account of the subject, we have looked at these few situations to see the applicability and utility of some of the repeated measures designs and procedures. Facility in designing and analyzing such experiments can be gained only after more detailed coverage of repeated measures topics through additional reading and course work.

EXERCISES

 17.6   An investigational drug product was studied under sleep laboratory conditions to determine its effect on duration of sleep. A group of 16 patients willing to participate in the study were randomly assigned to one of two drug sequences; eight were to receive the investigational drug in period 1 and an identical-appearing placebo in period 2, and the remaining eight patients were to receive the treatment in the reverse order.
a. Identify the design.
b. Give a model for this design.
c. State the assumptions that might affect the appropriateness of this design.

17.7   Sleep duration data (in hours/night) are shown for the patients of Exercise 17.6

	Patient	Period 1	Period 2
Sequence 1	1	8.6	8.0
	2	7.5	7.1
	3	8.3	7.4
	4	8.4	7.3
	5	6.4	6.4
	6	6.9	6.8
	7	6.5	6.1
	8	6.0	5.7

(continued)

	Patient	Period 1	Period 2
Sequence 2	9	7.3	7.9
	10	7.5	7.6
	11	6.4	6.3
	12	6.8	7.5
	13	7.1	7.7
	14	8.2	8.6
	15	7.2	7.8
	16	6.7	6.9

Sequence 1 received the investigational drug first and placebo second; the reverse order applied to sequence 2.

a. Compute means and standard errors per sequence, per period.

b. Plot these data to show what happened during the study. Does the investigational drug appear to affect sleep duration? In what way?

c. Run a repeated measures analysis of variance for this design. Draw conclusions. Does the analysis of variance confirm your impressions in part (b)?

17.8 Refer to Exercise 17.6. Suppose we ignore the order in which the patients received the treatments. Count the number of patients who had higher sleep duration on the investigational drug than on placebo.

a. Suggest another simple test for assessing the effectiveness of the investigational drug.

b. Give a p-value for the test of part (a).

17.9 Refer to Exercise 17.6. Suppose the sleep durations for period 2 of sequence 1 were as follows:

8.5
7.6
8.5
8.3
7.2
7.0
6.4
6.1

a. Plot the study data for both sequences.

b. Does the design still seem to be appropriate? Is there a possible explanation for what happened?

17.10 Refer to Exercise 17.9. In spite of the results from period 2, we can still get a between-patient comparison of the treatment groups if we use the period 1 results only. Suggest an appropriate test, run the test, and give the p-value for your test. Draw a conclusion.

17.11 Many of us have been exposed to advertising related to the "bioavailability" of generic and brand-name formulations of the same drug product. One way to compare the bioavailability of two formulations of a drug product is to compare areas under the concentration curve (AUC) for subjects treated with both formulations. For example, the shaded area in Figure 17.1 represents the AUC for a patient treated with a single dose of a drug. A two-period crossover design was used to compare the bioavailability of a brand-name (A_1) and generic version (A_2) of a weight-reducing agent. A random sample of six subjects

EXERCISES

FIGURE 17.1

AUC for a Patient
Treated with a
Single Dose of Drug,
Exercise 17.11

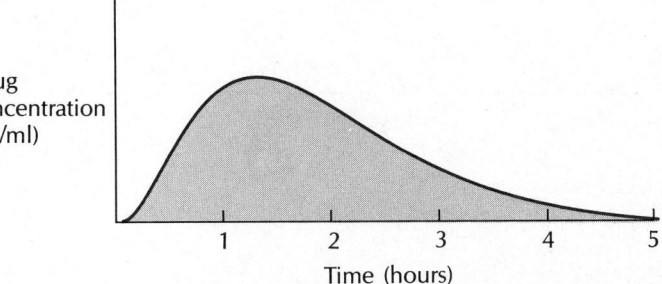

Drug
concentration
(ng/ml)

Time (hours)

was assigned to sequence 1 (A_2, then A_1), and another six subjects to sequence 2 (A_1, then A_2). The AUCs for these patients are shown here.

Sequence	Patients in Sequence	Period 1	Period 2
1	1	80.2	40.4
	2	79.1	38.5
	3	85.1	54.3
	4	108.4	96.6
	5	41.2	66.5
	6	72.7	78.2
2	1	74.6	35.6
	2	125.3	67.3
	3	145.5	90.8
	4	86.7	86.5
	5	107.8	103.1
	6	79.7	83.9

   a. Plot the formulation means (AUC) by period for each sequence.
   b. Is there evidence of a period effect?
   c. Do the formulations appear to differ relative to AUC?

17.12 Refer to Exercise 17.11. Run an analysis of variance for a two-period crossover design. Does your analysis confirm the intuition you expressed in Exercise 17.11?

17.13 Refer to Exercise 17.11. Compare the mean AUCs for the two formulations using *only* the period 1 data. Does this analysis confirm the analysis of Exercise 17.12? Why or why not might the analysis of Exercise 17.12 be more suitable than the "parallel" analysis of this exercise?

17.14 A study was conducted to demonstrate the effectiveness of an investigational drug product in reducing the number of epileptic seizures in patients who have not been helped by standard therapy. Thirty patients participated in the study, with 15 randomized to the drug treatment group and 15 to the placebo group. Patient demographic data are displayed here.

| | | Group | |
		Investigational Drug ($n_1 = 15$)	Placebo ($n_2 = 15$)
Age (yrs)	Mean ($\pm$ SD)	37.2 ($\pm$ 10.5)	39.5 ($\pm$ 9.6)
	Range	19–68	21–65
Sex	M	20	16
	F	10	14
Duration of illness (yrs)	Mean ($\pm$ SD)	10.7 ($\pm$ 6.5)	11.5 ($\pm$ 7.3)
	Range	1–18	1–26

a. Do the groups appear to be comparable related to these demographic variables?

b. Are the mean ages or durations of illness different? How would you make this comparison?

c. How might you compare the sex distributions of the two groups?

17.15 The seizure data for the study of Exercise 17.14 are shown here. Note that we have baseline seizure rates as well as seizure rates for five months while on therapy.

a. Plot the mean seizure rates by month for the two groups. Does the investigational drug appear to work?

b. Run a repeated measures AOV and draw conclusions.

| | | | Time (months) | | | | |
Group	Patient	Baseline	1	2	3	4	5
Drug	1	15	11	10	6	5	3
	2	13	6	5	1	2	1
	3	12	8	3	0	3	0
	4	18	4	2	3	1	2
	5	30	15	14	10	8	20
	6	14	7	9	3	4	1
	7	25	12	18	13	10	6
	8	22	21	18	16	17	25
	9	23	17	14	10	7	1
	10	14	2	1	0	0	0
	11	15	4	5	6	3	2
	12	17	8	7	8	2	6
	13	26	13	10	9	7	4
	14	28	2	1	3	1	3
	15	29	27	29	25	24	22
Placebo	1	16	15	18	14	13	12
	2	18	14	13	12	10	15
	3	14	10	5	4	6	7
	4	19	15	16	9	12	15
	5	12	10	14	16	17	12
	6	11	13	8	7	6	11
	7	31	32	30	21	24	20
	8	32	35	34	31	20	24
	9	21	20	18	15	16	18

(continued)

EXERCISES

Group	Patient	Baseline	Time (months)				
			1	2	3	4	5
	10	26	22	23	21	15	14
	11	13	10	14	12	8	6
	12	17	15	10	3	2	3
	13	18	16	12	14	13	11
	14	23	15	14	18	19	20
	15	10	8	11	10	9	6

17.16 Refer to the data of Exercise 17.15.

a. Consider the change in seizure rate from baseline to the five-month reading. Compare the two groups using these data. Do you reach a similar conclusion?

b. Since seizure rates can be quite variable, some people might compare the maximum change for patients in the two groups. Do these data support your previous conclusions?

17.17 Gasoline efficiency ratings were obtained on a random sample of 12 automobiles, six each of two different models. These ratings were taken at five different times for each of the 12 automobiles.

a. Compute the mean efficiencies for each model at each time point, and plot these data.

b. Draw conclusions from the analysis of variance.

c. What effects, if any, do the correction factors have on the within model comparisons in the analysis of variance shown here.

```
                    LISTING OF DATA

OBS    MODEL    CAR    TIME1    TIME2    TIME3    TIME4    TIME5

 1       1       1     1.43     1.47     1.39     1.40     1.44
 2       1       2     1.50     1.41     1.51     1.53     1.41
 3       1       3     1.79     1.88     1.89     2.00     1.90
 4       1       4     1.87     1.78     2.00     2.00     2.11
 5       1       5     1.85     1.89     1.93     1.86     1.81
 6       1       6     1.89     1.66     1.78     1.77     1.67
 7       2       7     1.63     1.62     1.64     1.63     1.53
 8       2       8     1.81     1.83     1.84     1.83     1.86
 9       2       9     2.25     2.10     2.34     2.27     2.32
10       2      10     1.79     1.80     1.92     2.03     2.02
11       2      11     2.11     2.00     2.33     2.46     2.35
12       2      12     2.10     2.03     2.00     2.09     1.87
```

REPEATED MEASURES ANALYSIS OF VARIANCE

GENERAL LINEAR MODELS PROCEDURE

TESTS OF HYPOTHESES FOR BETWEEN SUBJECTS EFFECTS

SOURCE	DF	TYPE III SS	MEAN SQUARE	F VALUE	PR > F
MODEL	1	0.95760667	0.95760667	3.38	0.0960
ERROR	10	2.83722667	0.28372267		

REPEATED MEASURES ANALYSIS OF VARIANCE

GENERAL LINEAR MODELS PROCEDURE

UNIVARIATE TESTS OF HYPOTHESES FOR WITHIN SUBJECT EFFECTS

SOURCE	DF	TYPE III SS	MEAN SQUARE	F VALUE	PR > F	ADJUSTED PR > F G - G	H - F
TIME	4	0.09579333	0.02394833	3.03	0.0285	0.0719	0.0512
TIME*MODEL	4	0.01182667	0.00295667	0.37	0.8260	0.6906	0.7528
ERROR(TIME)	40	0.31654000	0.00791350				

GREENHOUSE-GEISSER EPSILON = 0.4943
HUYNH-FELDT EPSILON = 0.6770

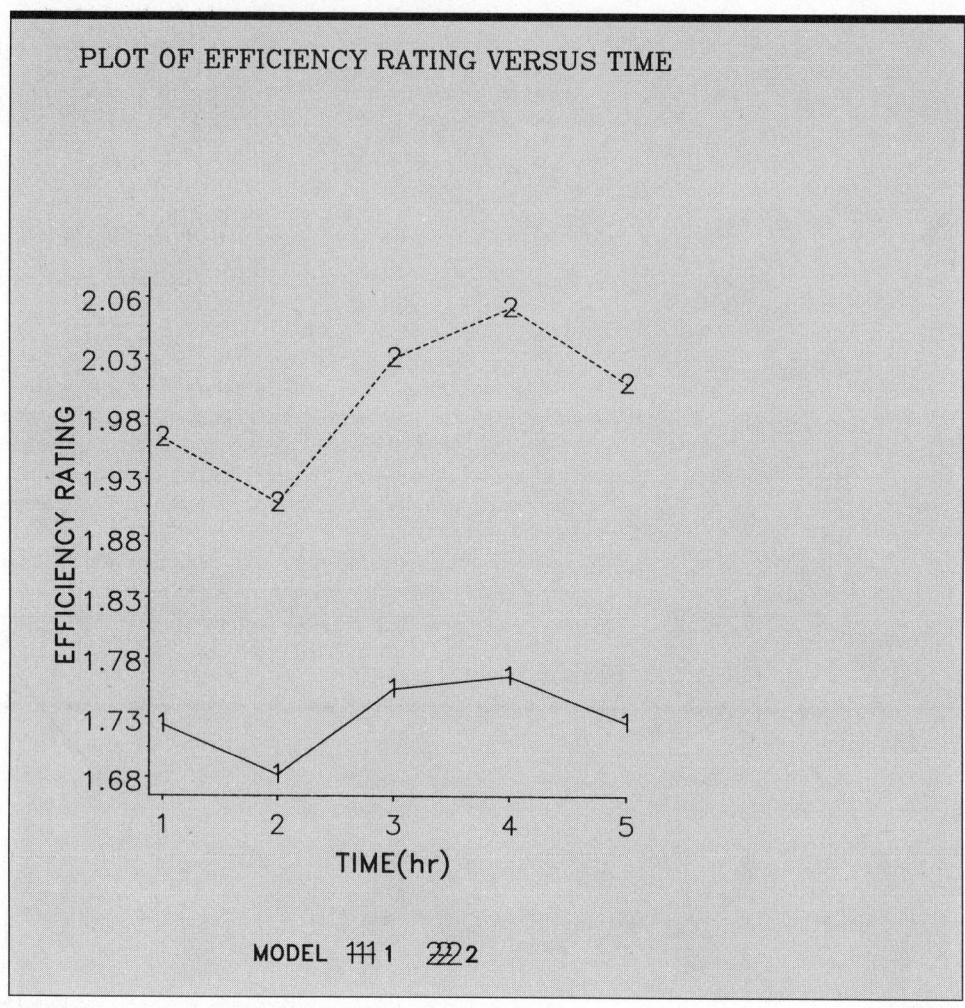

18 THE ANALYSIS OF COVARIANCE

18.1 INTRODUCTION

The analysis of covariance is a procedure for comparing treatment means that incorporates information on a quantitative variable x. This topic appears in most statistical methods textbooks, but it has been our experience that many students become so engrossed in the computational formulas that they rarely understand the problems involved. We will approach the subject in a different way and try to avoid becoming too involved with formulas. Since the analysis of covariance combines features of the analysis of variance and regression, we will make use of a general linear model. By referring to and building on our work with general linear models in preceding chapters, we can more easily understand the topic of covariance analysis. We begin our presentation with a completely randomized design.

18.2 A COMPLETELY RANDOMIZED DESIGN WITH ONE COVARIATE

A completely randomized design is used to compare t population means. To do this, we obtain a random sample of n_i observations on the variable y in the ith population ($i = 1, 2, \ldots, t$). Now in addition to measuring the response variable y on each experimental unit, we measure a second variable x, often called a *covariable*, or a **covariate**. For example, in studying the effects of different methods of reinforcement on the reading achievement levels of eight-year-old children, we could measure not only the final achievement level y for each child but also the prestudy reading performance level x. Ultimately we would

813

want to make comparisons among the different methods while taking into account information on both y and x.

It should be noted that x can be thought of as an independent variable, but unlike most situations discussed in previous chapters, we cannot control the value of x (as we controlled settings of temperature or pressure) prior to observing the variable. In spite of this, we may still write a model for the completely randomized design treating the covariate as an independent variable.

For purposes of illustration we will consider the problem of comparing $t = 2$ treatments from a completely randomized design with one covariable. Later we'll show how these results can be extended to include $t > 2$ treatments and one or more covariables.

EXAMPLE 18.1

An investigator is interested in comparing two drug products (A and B) in overweight female volunteers. The experiment calls for 20 randomly selected subjects who are at least 25% overweight. Ten of these women are to be randomly assigned to product 1 and the remaining 10 of product 2. The response of interest is a score on a rating scale used to measure the mood of a subject. To obtain a score, a subject must complete a checklist indicating how each of 50 adjectives describes her mood at that time. From this checklist we can obtain an anxiety-tension score, a danger-hostility score, an active score (measuring such factors as alertness and energy), and many others.

On the study day, all 20 volunteers are required to complete the checklist at 8 AM. Then each subject is given the prescribed medication (1 or 2). Each subject is required to complete the checklist again at 10 AM.

Write a model relating a subject's 10-AM checklist score y to the two independent variables "drug product" and "8-AM (predrug) checklist score." Interpret the βs.

Solution

For this experimental situation we have one qualitative independent variable (drug) and one quantitative independent variable. Letting x_1 denote the checklist score at 8 AM, the model is

$$y = \beta_0 + \beta_1 x_1 + \beta_2 x_2 + \beta_3 x_1 x_2 + \varepsilon$$

where

$x_1 = $ checklist score at 8 AM

$x_2 = 1$ if product 2 $x_2 = 0$ otherwise

The expected value of y for our model is

$$E(y) = \beta_0 + \beta_1 x_1 + \beta_2 x_2 + \beta_3 x_1 x_2$$

Substituting $x_2 = 0$ and $x_2 = 1$, respectively, for drug products 1 and 2, we have the expected value of y, the adjective checklist score at 10 AM.

Product 1: $E(y) = \beta_0 + \beta_1 x_1$
Product 2: $E(y) = (\beta_0 + \beta_2) + (\beta_1 + \beta_3)x_1$

Thus β_0 and β_1 are the intercept and slope, respectively, defining the linear relationship between y and x_1 for product 1. Since $\beta_0 + \beta_2$ and $\beta_1 + \beta_3$ represent the corresponding intercept and slope, respectively, for product 2, β_2 is the difference between the intercepts for lines 1 and 2, whereas β_3 is the difference between the slopes for the two lines.

The major assumptions in an analysis of covariance are (1) the treatments' regression equations are linear in the covariable, and (2) the linear regressions for the different treatments are parallel.

The objective of an analysis of covariance is to compare the treatment means after adjusting for differences among the treatments due to differences in the covariable levels for the treatment groups. The **adjusted treatment means** are found by predicting y for each treatment group corresponding to $x = \bar{x}$, the mean value of the covariable across all treatment groups.

adjusted treatment means

EXAMPLE 18.2   The data for Example 18.1 are shown here.

	Drug 1		Drug 2	
	8 AM	10 AM	8 AM	10 AM
	x_1	y	x_1	y
	5	20	7	19
	10	23	12	26
	12	30	27	33
	9	25	24	35
	23	34	18	30
	21	40	22	31
	14	27	26	34
	18	38	21	28
	6	24	14	23
	13	31	9	22

a. Write the model for an analysis of covariance.
b. Use the computer output shown here to give the linear regression equations for both treatment groups.
c. Compute the sample mean anxiety scores for the two products; also compute the adjusted treatment means.
d. Does there appear to be a difference between the adjusted treatment means?

ANALYSIS OF COVARIANCE
GENERAL LINEAR MODELS PROCEDURE

DEPENDENT VARIABLE: Y

SOURCE	DF	SUM OF SQUARES	MEAN SQUARE	F VALUE	PR > F	R-SQUARE	C.V.
MODEL	2	546.22979465	273.11489733	37.96	0.0001	0.817037	9.3627
ERROR	17	122.32020535	7.19530620		ROOT MSE		Y MEAN
CORRECTED TOTAL	19	668.55000000			2.68240679		28.65000000

SOURCE	DF	TYPE I SS	F VALUE	PR > F	DF	TYPE III SS	F VALUE	PR > F
X1	1	430.92383794	59.89	0.0001	1	540.17979465	75.07	0.0001
X2	1	115.30595671	16.03	0.0009	1	115.30595671	16.03	0.0009

PARAMETER	ESTIMATE	T FOR HO: PARAMETER=0	PR > ¦T¦	STD ERROR OF ESTIMATE
INTERCEPT	18.35999493	12.15	0.0001	1.51153263
X1	0.82748130	8.66	0.0001	0.09550226
X2	-5.15465839	-4.00	0.0009	1.28765245

OBSERVATION	OBSERVED VALUE	PREDICTED VALUE	RESIDUAL
1	20.00000000	22.49740145	-2.49740145
2	24.00000000	23.32488275	0.67511725
3	25.00000000	25.80732666	-0.80732666
4	23.00000000	26.63480796	-3.63480796
5	30.00000000	28.28977057	1.71022943
6	31.00000000	29.11725187	1.88274813
7	27.00000000	29.94473317	-2.94473317
8	38.00000000	33.25465839	4.74534161
9	40.00000000	35.73710229	4.26289771
10	34.00000000	37.39206490	-3.39206490
11	19.00000000	18.99770567	0.00229433
12	22.00000000	20.65266827	1.34733173
13	26.00000000	23.13511218	2.86488782
14	23.00000000	24.79007479	-1.79007479
15	30.00000000	28.10000000	1.90000000
16	28.00000000	30.58244391	-2.58244391
17	31.00000000	31.40992521	-0.40992521
18	35.00000000	33.06488782	1.93511218
19	34.00000000	34.71985042	-0.71985042
20	33.00000000	35.54733173	-2.54733173

SUM OF RESIDUALS	0.00000000
SUM OF SQUARED RESIDUALS	122.32020535
SUM OF SQUARED RESIDUALS - ERROR SS	-0.00000000
FIRST ORDER AUTOCORRELATION	-0.20370243
DURBIN-WATSON D	2.30336716

a. The analysis of covariance model for this situation is

$$y = \beta_0 + \beta_1 x_1 + \beta_2 x_2 + \varepsilon$$

where

x_1 = prestudy checklist score

and

$x_2 = 1$ if product 2

$= 0$ if product 1

The separate regression equations are:

Product 1: $y = \beta_0 + \beta_1 x_1 + \varepsilon$
Product 2: $y = (\beta_0 + \beta_2) + \beta_1 x_1 + \varepsilon$

Hence β_1 is the common slope and β_2 is the difference in intercepts.

b. The least squares prediction equations for the two products can be obtained from the parameter estimates shown in the output:

Product 1: $\hat{y} = 18.360 + 0.827 x_1$
Product 2: $\hat{y} = (18.360 - 5.155) + 0.827 x_1$
$= 13.205 + 0.827 x_1$

These prediction equations are shown in Figure 18.1.

FIGURE 18.1

Analysis of Covariance
Predicted Line

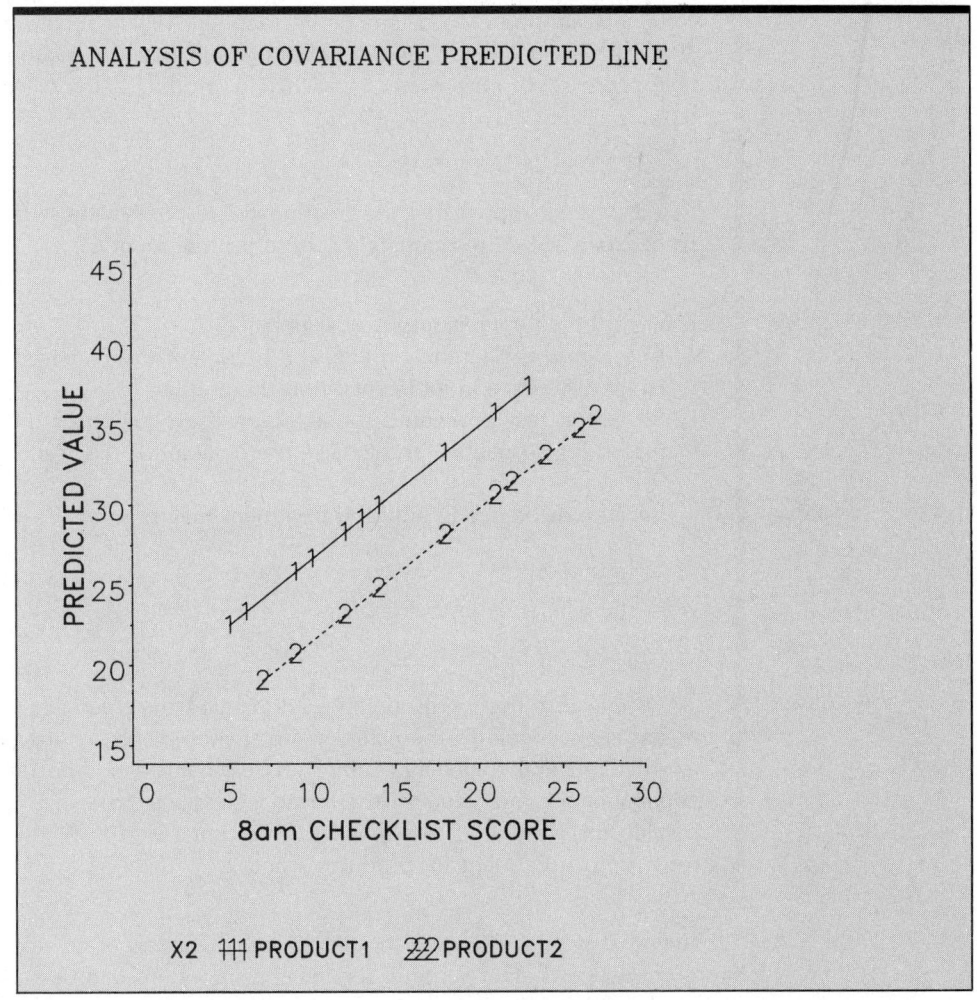

ANALYSIS OF COVARIANCE PREDICTED LINE

X2   ‡‡‡ PRODUCT1   ƨƨƨ PRODUCT2

c. From the output shown here we have the following sample means.

	Product 1	Product 2	Overall
y	29.2	28.1	28.65
x_1	13.1	18.0	15.55

In order to get the adjusted treatment means, we substitute the mean pretreatment value ($\bar{x}_1 = 15.55$) into the separate prediction equations.

Adjusted treatment mean, product 1:

$$\hat{y} = 18.360 + 0.827(15.55) = 31.220$$

Adjusted treatment mean, product 2:

$$\hat{y} = 13.205 + 0.827(15.55) = 26.065$$

Note that although the difference in treatment means for products 1 and 2 is only $29.2 - 28.1 = 1.1$, when we adjust for the fact that the mean pretreatment (covariable) score is 13.1 for product 1 and 18.0 for product 2, the difference between A and B is

$$31.220 - 26.065 = 5.155$$

d. It does appear that the treatment 1 response (adjusted) for the covariable is higher than the corresponding response for treatment 2.

EXAMPLE 18.3

Refer to the computer output of Example 18.2.
a. Compare the difference in adjusted treatment means based on your calculations to the difference in intercepts from the output.
b. Suggest a test procedure for assessing the equality of the adjusted treatment means for products 1 and 2. Give the *p*-value for this test.

Solution

a. The difference in adjusted treatment means is

$$31.220 - 26.065 = 5.155$$

while the difference in intercepts is

$$18.360 - 13.205 = 5.155$$

b. Because the regression lines are parallel, the *difference* in adjusted treatment means is also equal to the difference in the intercepts for the two regressions. Thus a formal test of the equality of adjusted treatments would be identical to a test of the equality of the intercepts of the linear regressions. From the computer output, the *p*-value for $H_0: \beta_2 = 0$ (difference in intercepts) is 0.0009; there is a difference in response to treatment for products 1 and 2.

Now that we have worked through a simple analysis of covariance we should stand back and assess whether some of the initial assumptions can be supported. We'll review this for the data of Example 18.2.

EXAMPLE 18.4

a. Plot the data of Example 18.2 by treatment group. Is there evidence of lack of fit of a linear regression model? Can we test for lack of fit?
b. Is there evidence to indicate nonparallel linear regressions? Can we test for this? How?

Solution

The sample data are displayed in Figure 18.2 by treatment group.
a. Recall that in order to test for lack of fit, we must be able to calculate a sum of squares due to pure experimental error. Since there are no replications of *y* values at any of the settings of the independent variables x_1 and x_2, we cannot test for lack of fit of the linear model.
b. Under the assumption that a linear regression model adequately fits the data for the two drugs, we could test for parallelism of the two regression lines using

FIGURE 18.2

Analysis of
Covariance,
Example 18.4

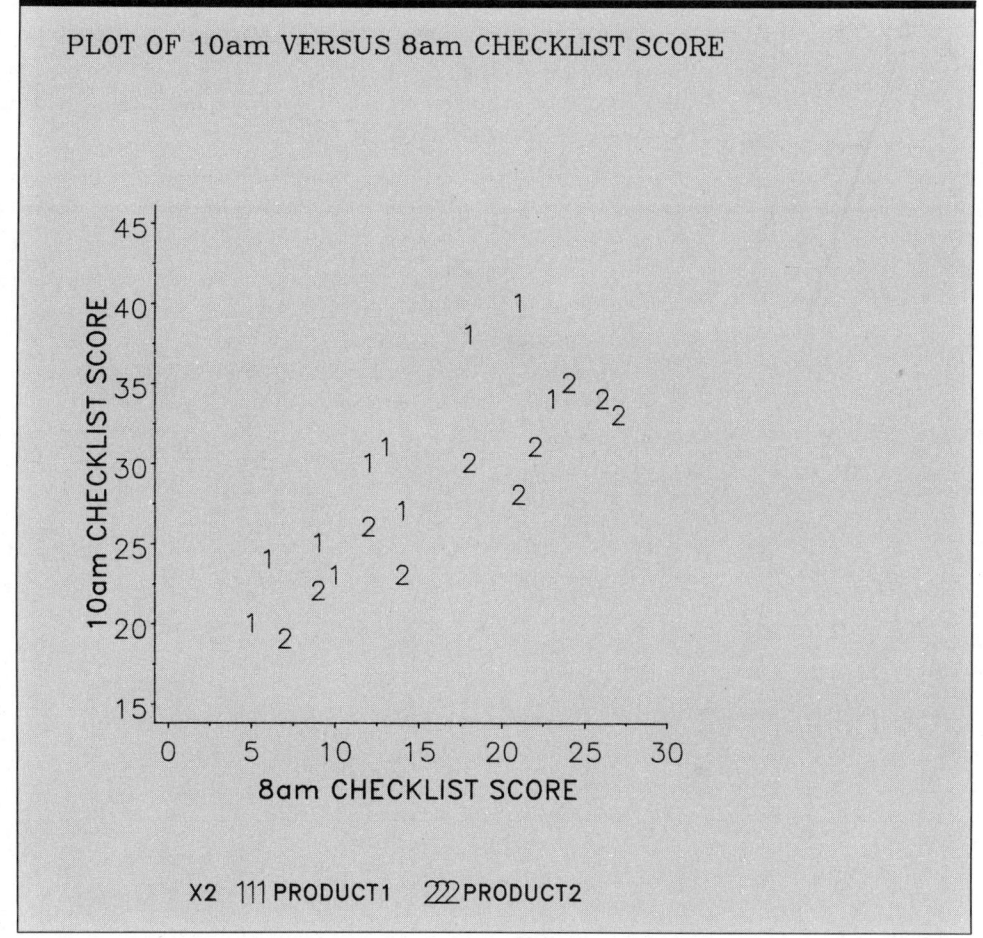

the model

$$y = \beta_0 + \beta_1 x_1 + \beta_2 x_2 + \beta_3 x_1 x_2 + \varepsilon$$

where

$$x_1 = \text{covariable}$$

and

$$x_2 = \begin{cases} 0 & \text{if product 1} \\ 1 & \text{if product 2} \end{cases}$$

For this model (substituting $x_2 = 0$ and 1) it is easy to see that β_3 is the difference in slope for the two regression lines

Product 1: $(x_2 = 0)$: $y = \beta_0 + \beta_1 x_1 + \varepsilon$

Product 2: $(x_2 = 1)$: $y = (\beta_0 + \beta_2) + (\beta_1 + \beta_3)x_1$

Hence a test for parallelism of the linear regression lines can be done using H_0: $\beta_3 = 0$. Based on the output shown here, the t-value for this test is -1.34, with probability greater than .1991. (These values are read from the computer output.) Since this probability is greater than .05, we have insufficient evidence to reject the null hypothesis of parallelism for the two lines.

```
                          ANALYSIS OF COVARIANCE
                       GENERAL LINEAR MODELS PROCEDURE

DEPENDENT VARIABLE: Y

SOURCE              DF    SUM OF SQUARES    MEAN SQUARE    F VALUE    PR > F    R-SQUARE       C.V.

MODEL                3     558.56687443    186.18895814     27.09    0.0001   0.835490      9.1512

ERROR               16     109.98312557      6.87394535               ROOT MSE          Y MEAN

CORRECTED TOTAL     19     668.55000000                              2.62182100     28.65000000


SOURCE              DF      TYPE I SS    F VALUE    PR > F     DF      TYPE III SS    F VALUE    PR > F

X1                   1    430.92383794      62.69    0.0001      1    312.89948313      45.52    0.0001
X2                   1    115.30595671      16.77    0.0008      1      1.21055917       0.18    0.6803
X1*X2                1     12.33707978       1.79    0.1991      1     12.33707978       1.79    0.1991


                                    T FOR HO:     PR > |T|      STD ERROR OF
PARAMETER           ESTIMATE       PARAMETER=0                   ESTIMATE

INTERCEPT         16.42262086          7.94       0.0001         2.06736784
X1                 0.97537245          6.75       0.0001         0.14456764
X2                -1.31392521         -0.42       0.6803         3.13098272
X1*X2             -0.25363332         -1.34       0.1991         0.18932291


OBSERVATION          OBSERVED         PREDICTED        RESIDUAL
                      VALUE            VALUE

     1            20.00000000       21.29948313      -1.29948313
     2            24.00000000       22.27485558       1.72514442
     3            25.00000000       25.20097294      -0.20097294
     4            23.00000000       26.17634539      -3.17634539
     5            30.00000000       28.12709030       1.87290970
     6            31.00000000       29.10246275       1.89753725
     7            27.00000000       30.07783521      -3.07783521
     8            38.00000000       33.97932502       4.02067498
     9            40.00000000       36.90544238       3.09455762
    10            34.00000000       38.85618729      -4.85618729
    11            19.00000000       20.16086957      -1.16086957
    12            22.00000000       21.60434783       0.39565217
    13            26.00000000       23.76956522       2.23043478
    14            23.00000000       25.21304348      -2.21304348
    15            30.00000000       28.10000000       1.90000000
    16            28.00000000       30.26521739      -2.26521739
    17            31.00000000       30.98695652       0.01304348
    18            35.00000000       32.43043478       2.56956522
    19            34.00000000       33.87391304       0.12608696
    20            33.00000000       34.59565217      -1.59565217


                          ANALYSIS OF COVARIANCE
                       GENERAL LINEAR MODELS PROCEDURE

DEPENDENT VARIABLE: Y

          SUM OF RESIDUALS                            0.00000000
          SUM OF SQUARED RESIDUALS                  109.98312557
          SUM OF SQUARED RESIDUALS - ERROR SS        -0.00000000
          FIRST ORDER AUTOCORRELATION                -0.29462211
          DURBIN-WATSON D                             2.55074048
```

Suppose we are interested in comparing $t = 3$ treatments with only one covariable. Using the notation of the general linear model, we let x_1 denote the covariate. Then we can write the model

$$y = \beta_0 + \beta_1 x_1 + \beta_2 x_2 + \beta_3 x_3 + \beta_4 x_1 x_2 + \beta_5 x_1 x_3 + \varepsilon$$

TABLE 18.1   Expected Values for the Model
$y = \beta_0 + \beta_1 x_1 + \beta_2 x_2 + \beta_3 x_3 + \beta_4 x_1 x_2 + \beta_5 x_1 x_3 + \varepsilon$

Treatment	Expected Values of y
1	$\beta_0 + \beta_1 x_1$
2	$(\beta_0 + \beta_2) + (\beta_1 + \beta_4) x_1$
3	$(\beta_0 + \beta_3) + (\beta_1 + \beta_5) x_1$

where

$$x_2 = 1 \text{ if treatment 2} \qquad x_2 = 0 \text{ otherwise}$$

$$x_3 = 1 \text{ if treatment 3} \qquad x_3 = 0 \text{ otherwise}$$

From our discussions in Chapters 12 and 14 we recognize this model as a general linear model relating a response y to a quantitative variable (the covariate x_1) and a qualitative variable (treatments). The βs of the model can be interpreted using Table 18.1. As can be seen from the table of expected values, the model defines a straight line for each of the three treatments. The intercepts and slopes for the three lines are as indicated in Table 18.1.

assumptions   Typically, for a completely randomized design with a single covariate, the response y is assumed to be linearly related to the covariate, and the slope of the straight-line relationship is assumed to be the same for all treatments. While we have already written the model to relate y linearly to x_1, we need not be bound by these assumptions, for it is possible to write the model with higher-order terms in x_1 (provided, of course, there are enough different values of x_1 recorded in the sample data). At present, however, we will illustrate a covariance analysis with a model that is linear in the covariate x_1.

tests   The assumption of constant slope of the three treatment groups can be tested using the null hypothesis

$$H_0: \beta_4 = \beta_5 = 0 \text{ (the slopes are identical)}$$

against the alternative hypothesis that at least one of the slopes differs from the rest. With our original model as the complete model and the model

reduced model $$y = \beta_0 + \beta_1 x_1 + \beta_2 x_2 + \beta_3 x_3 + \varepsilon$$

as the reduced model, the mean square drop based on two degrees of freedom would be the numerator in an F test of the null hypothesis

$$H_0: \beta_4 = \beta_5 = 0$$

If there is insufficient evidence to reject the hypothesis of equality of the slopes ($\beta_4 = \beta_5 = 0$), we use the reduced model to describe the experimental situation. Under this model, the straight-line relationship between y and the covariate x_1 would have expectations

treatment 1: $\beta_0 + \beta_1 x_1$

treatment 2: $(\beta_0 + \beta_2) + \beta_1 x_1$

treatment 3: $(\beta_0 + \beta_3) + \beta_1 x_1$

The model for a covariance analysis in a completely randomized design is not usually written in terms of the parameters in a general linear model. In other sources you might see the model written as

$$y_{ij} = \alpha_i + \beta x_{ij} + \varepsilon_{ij} \qquad (i = 1, 2, \ldots, t; j = 1, 2, \ldots, n_i)$$

Note that we assume that the relationship between the response y and the covariate x is linear with the same slope but different intercepts across treatment groups. The reduced model illustrated for our $t = 3$ example,

$$y = \beta_0 + \beta_1 x_1 + \beta_2 x_2 + \beta_3 x_3 + \varepsilon$$

is the general linear model analogy to this same situation.

Referring to our example, when all three treatments have the same slope (i.e., $\beta_4 = \beta_5 = 0$ in the complete model), a test of the equality of treatment means can be made using the sum of squares due to treatments adjusted for the covariate. We do this by fitting a complete and a reduced model, using the model

$$y = \beta_0 + \beta_1 x_1 + \beta_2 x_2 + \beta_3 x_3 + \varepsilon$$

for the null hypothesis $H_0: \beta_2 = \beta_3 = 0$. The sum of squares drop $(SSE_2 - SSE_1)$ will be the sum of squares due to treatments adjusted for the covariate. The test statistic $F = MS_{drop}/MSE$ is based on $df_1 = $ two degrees of freedom (the number of βs set equal to zero under H_0).

EXERCISES

18.1   Consider a completely randomized design for $t = 5$ treatments, with a single covariate x_1 and six observations per treatment. Write the complete general linear model under the assumption that the response y is linearly related to the covariate x_1 for each treatment. Identify the parameters in your model.

18.2   Refer to Exercise 18.1. Indicate the relationships among the parameters of the model for the following cases; show a graph for each case.
 a. The lines are not parallel.
 b. The lines are parallel, but do not coincide.
 c. The lines are coincident.

18.3   Refer to Exercise 18.1. How would you test for parallelism among the straight lines for the 5 treatment groups? Identify how you would obtain the test statistic. What are the degrees of freedom associated with the test statistic?

18.4   Refer to Exercise 18.1. Assume the lines are parallel. Give the test for adjusted treatment means. How would you estimate the mean response for treatment 1 with $x_1 = 5$?

18.5   Perform an analysis of covariance for the data shown. (Hint: When people refer to performing an analysis of covariance, they usually mean to assume a linear relationship, then test for parallelism, and then test for adjusted treatment means assuming parallelism.) The data are given here. Use $\alpha = .05$.

	Treatment				
	1		2		3
x	y	x	y	x	y
26	100	24	118	37	124
35	150	28	134	31	95
28	106	29	138	14	60
21	95	32	147	27	86
29	113	36	165	18	68
34	144	35	159	25	81

| 18.3 | THE EXTRAPOLATION PROBLEM |

In the previous section we discussed how to compare two (or more) treatments from a completely randomized design with one covariable. If the regression equations for the treatments are linear in the covariable and parallel, we said we could compare the treatments using the adjusted treatment means. However, as with most methods, the analysis of covariance methods should not be used blindly. Even if the linearity and parallelism assumptions hold, we can have problems if the values of the covariable do not have considerable overlap for the treatment groups. We will illustrate this with an example.

Suppose that we are interested in comparing self-esteem scores for alcoholics and drug addicts. A sample of nine alcoholics and a sample of nine drug addicts were obtained and for each individual, we obtained his or her self-esteem score and age. These data are shown in Table 18.2.

If we blindly followed the analysis of covariance procedures without looking at the data, we would likely find the regression equations for alcoholics and addicts to be reasonably linear and parallel and the test for adjusted treatments to be significant. This is all shown in the accompanying output.

TABLE 18.2   Self-Esteem Scores and Ages for a Sample of Alcoholics and Drug Addicts

Alcoholics		Drug Addicts	
Self-Esteem	Age	Self-Esteem	Age
25	15	20	30
22	17	17	31
24	18	18	33
20	19	15	35
21	21	14	36
17	22	15	37
14	23	12	38
16	24	10	40
15	25	11	41

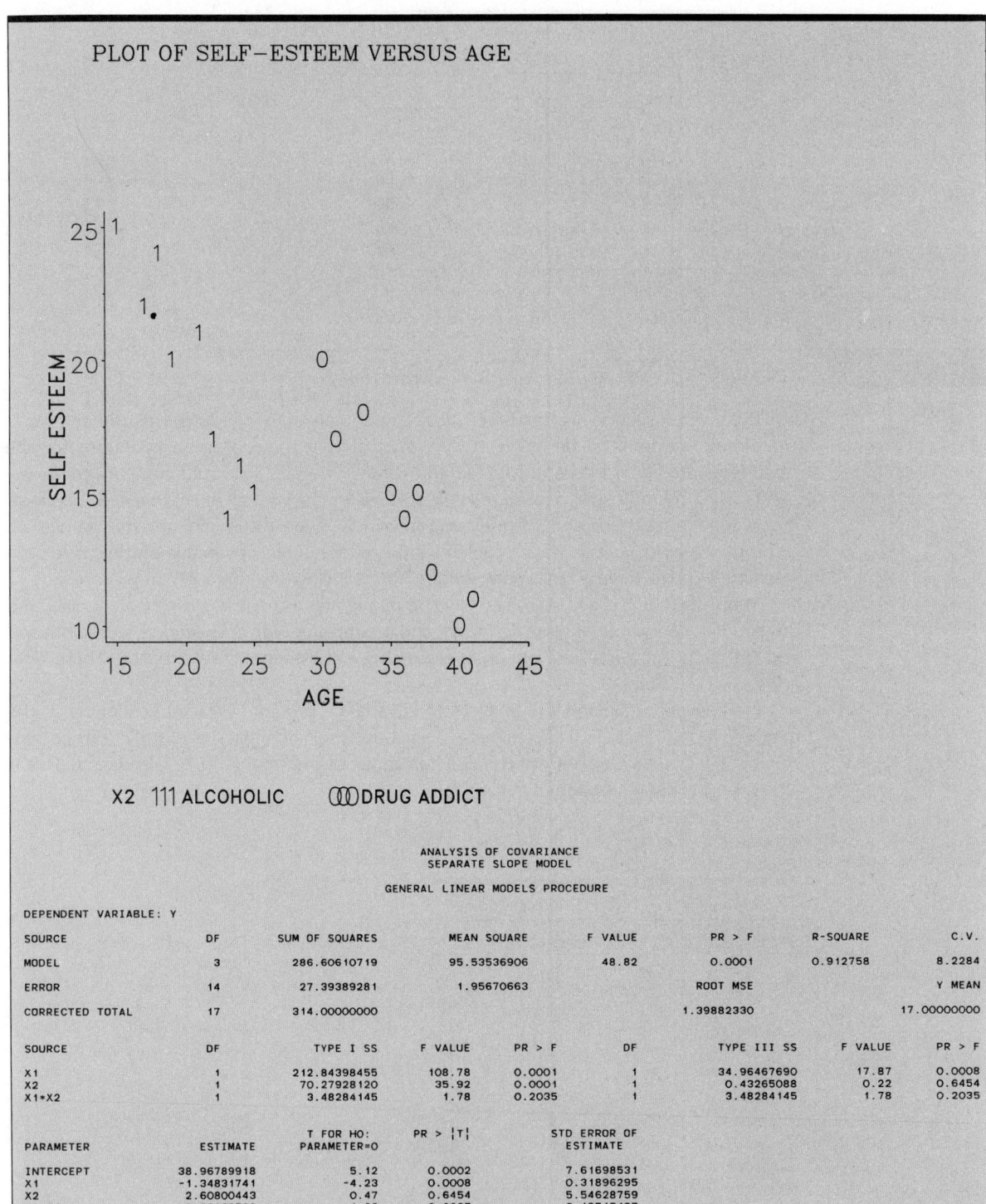

PLOT OF SELF—ESTEEM VERSUS AGE

X2 111 ALCOHOLIC   000 DRUG ADDICT

ANALYSIS OF COVARIANCE
SEPARATE SLOPE MODEL

GENERAL LINEAR MODELS PROCEDURE

DEPENDENT VARIABLE: Y

SOURCE	DF	SUM OF SQUARES	MEAN SQUARE	F VALUE	PR > F	R-SQUARE	C.V.
MODEL	3	286.60610719	95.53536906	48.82	0.0001	0.912758	8.2284
ERROR	14	27.39389281	1.95670663		ROOT MSE		Y MEAN
CORRECTED TOTAL	17	314.00000000			1.39882330		17.00000000

SOURCE	DF	TYPE I SS	F VALUE	PR > F	DF	TYPE III SS	F VALUE	PR > F
X1	1	212.84398455	108.78	0.0001	1	34.96467690	17.87	0.0008
X2	1	70.27928120	35.92	0.0001	1	0.43265088	0.22	0.6454
X1*X2	1	3.48284145	1.78	0.2035	1	3.48284145	1.78	0.2035

PARAMETER	ESTIMATE	T FOR H0: PARAMETER=0	PR > |T|	STD ERROR OF ESTIMATE
INTERCEPT	38.96789918	5.12	0.0002	7.61698531
X1	-1.34831741	-4.23	0.0008	0.31896295
X2	2.60800443	0.47	0.6454	5.54628759
X1*X2	0.26036560	1.33	0.2035	0.19515497

```
                        ANALYSIS OF COVARIANCE
                          COMMON SLOPE MODEL

                     GENERAL LINEAR MODELS PROCEDURE
```

DEPENDENT VARIABLE: Y

SOURCE	DF	SUM OF SQUARES	MEAN SQUARE	F VALUE	PR > F	R-SQUARE	C.V.
MODEL	2	283.12326574	141.56163287	68.77	0.0001	0.901666	8.4396
ERROR	15	30.87673426	2.05844895		ROOT MSE		Y MEAN
CORRECTED TOTAL	17	314.00000000			1.43472957		17.00000000

SOURCE	DF	TYPE I SS	F VALUE	PR > F	DF	TYPE III SS	F VALUE	PR > F
X1	1	212.84398455	103.40	0.0001	1	185.12326574	89.93	0.0001
X2	1	70.27928120	34.14	0.0001	1	70.27928120	34.14	0.0001

PARAMETER	ESTIMATE	T FOR H0: PARAMETER=0	PR > |T|	STD ERROR OF ESTIMATE
INTERCEPT	28.92404838	24.33	0.0001	1.18877463
X1	-0.94290288	-9.48	0.0001	0.09942750
X2	9.68641053	5.84	0.0001	1.65775088

Do alcoholics and drug addicts really have different self-esteem scores? One possible explanation for the difference in scores is that we're dealing with two different age groups; the alcoholics sampled ranged in age from 15 to 25 years, whereas the drug addicts were between the ages of 30 and 41. This difference in ages for the two groups is borne out in the scatter plot shown in Figure 18.3.

The mean ages for the two groups are 20.44 and 35.67 years, respectively, while the combined mean age is 28.06 years. Note that the combined mean is outside the age range for each of the separate samples. We have no information about self-esteem scores for drug addicts less than 30 years of age and no information about self-esteem scores for alcoholics above the age of 25. Hence, it would be inappropriate to compare the predicted

FIGURE 18.3

Analysis of Covariance Plot of y versus x_1

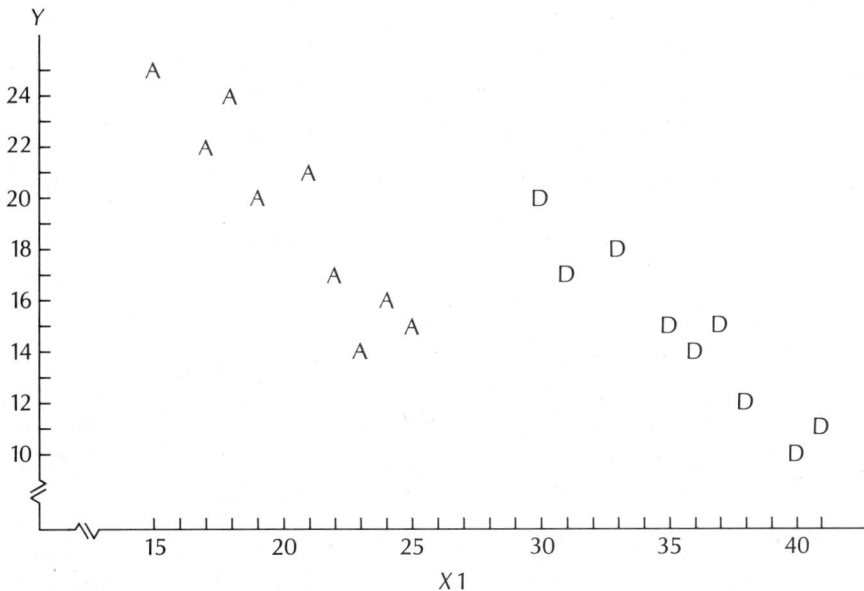

self-esteem scores at the "adjusted" age (28.06) since this involves an extrapolation beyond the ages observed for the separate samples. For this example, it would be difficult to make any comparison between the alcoholics and drug addicts because of the age differences and other possible (unmeasured) differences between the two groups.

So, don't forget to look at your data. The potential for extrapolation, although not as obvious as for our example, should become apparent with plots of the data. Then you can avoid using an analysis of covariance to make comparisons of adjusted treatment means when the adjustment (or, in fact, any comparison) may be inappropriate. These same problems can occur with the extensions of these methods to include more than one covariable and more complicated experimental designs; it's just more difficult to detect the problem.

18.4   MULTIPLE COVARIATES AND MORE COMPLICATED DESIGNS

The sample procedures discussed in Section 18.2 can also be applied to completely randomized designs with one or more covariates. Including more than one covariate in the model merely means that we have more than one quantitative independent variable in our model. For example, we might wish to compare the social status y of several different occupational groups while incorporating information on the number of years x_1 of formal education beyond high school and the income level x_2 of each individual in a group. As mentioned previously, we need not restrict ourselves to linear terms in the covariate(s). Thus we might have a response related to two covariates (x_1 and x_2) and $t = 3$ treatments using the model

$$y = \beta_0 + \beta_1 x_1 + \beta_2 x_1^2 + \beta_3 x_2 + \beta_4 x_3 + \beta_5 x_4 + \beta_6 x_1 x_3$$
$$+ \beta_7 x_1 x_4 + \beta_8 x_1^2 x_3 + \beta_9 x_1^2 x_4 + \beta_{10} x_2 x_3 + \beta_{11} x_2 x_4 + \varepsilon$$

where

$$x_3 = 1 \text{ if treatment 2} \qquad x_3 = 0 \text{ otherwise}$$
$$x_4 = 1 \text{ if treatment 3} \qquad x_4 = 0 \text{ otherwise}$$

We can readily obtain an interpretation of the βs by using a table of expected values similar to Table 18.1.

An analysis of covariance for more complicated designs can also be obtained using general linear model methodology. Consider an analysis of covariance for a randomized block design. For simplicity, we will assume that there are two blocks, three treatments, one covariate x_1, and more than one observation per cell.

EXAMPLE 18.5   Write the model for the experimental situation described in the previous paragraph, assuming the response is linearly related to x_1 for each treatment. Identify the parameters in your model.

Solution   The model is

$$y = \beta_0 + \beta_1 x_1 + \beta_2 x_2 + \beta_3 x_3 + \beta_4 x_4 + \beta_5 x_1 x_2 + \beta_6 x_1 x_3 + \beta_7 x_1 x_4 + \varepsilon$$

TABLE 18.3   Expected Values for the Randomized Block Design with One Covariate, Example 18.5

Block	Treatment	Expected values
	1	$\beta_0 + \beta_1 x_1$
1	2	$(\beta_0 + \beta_3) + (\beta_1 + \beta_6)x_1$
	3	$(\beta_0 + \beta_4) + (\beta_1 + \beta_7)x_1$
	1	$(\beta_0 + \beta_2) + (\beta_1 + \beta_5)x_1$
2	2	$(\beta_0 + \beta_2 + \beta_3) + (\beta_1 + \beta_5 + \beta_6)x_1$
	3	$(\beta_0 + \beta_2 + \beta_4) + (\beta_1 + \beta_5 + \beta_7)x_1$

where

$x_2 = 1$ if block 2 $x_2 = 0$ otherwise

$x_3 = 1$ if treatment 2 $x_3 = 0$ otherwise

$x_4 = 1$ if treatment 3 $x_4 = 0$ otherwise

We immediately recognize this as a model relating a response y to a quantitative variable x_1 and two qualitative variables: blocks and treatments. An interpretation of the βs in the model is readily obtained from the table of expected values shown in Table 18.3.

The model we formulated in Example 18.5 provides for a linear relationship between y and x_1 for each of the treatments in each block, but it allows for differences among intercepts and slopes. If we wanted to test for the equality of the slopes across treatments and blocks, we would use the null hypothesis

$$H_0: \beta_5 = \beta_6 = \beta_7 = 0$$

If there is insufficient evidence to reject H_0, we would proceed with the reduced model (obtained by setting $\beta_5 = \beta_6 = \beta_7 = 0$ in our model).

$$y = \beta_0 + \beta_1 x_1 + \beta_2 x_2 + \beta_3 x_3 + \beta_4 x_4 + \varepsilon$$

A test for differences among treatments adjusted for the covariate could be obtained by fitting a complete and a reduced model for the null hypothesis

$$H_0: \beta_3 = \beta_4 = 0$$

EXERCISES

18.6   Write a model for a 4 × 4 Latin square design with one covariate x_1. Assume that the response is linearly related to the covariate. Identify the parameters in the model.

18.7   Refer to Exercise 18.6.

a. Indicate how you would test for parallelism among the different straight lines. How many degrees of freedom would the F test have?

b. Indicate how you would perform a test for the effects of treatments adjusted for the covariate.

18.8   Refer to Exercise 18.6. Write a complete model assuming that the response is a second-order function of the covariate x_1. Can you identify parameters in the model? How would you test for parallelism of the second-order model?

18.5   SUMMARY

In this chapter we presented a procedure called the analysis of covariance. Here, for each value of y, we also observe a value of a concomitant variable x. This second variable, called a covariate, is recognized as an uncontrolled quantitative independent variable. Because of this fact we can formulate models using general linear model methodology of previous chapters.

In most situations when reference is made to an analysis of covariance, it is assumed that the response is linearly related to the covariate x, with the slope of the line the same for all treatment groups. Then a test for treatments adjusted for the covariate is performed. Actually, many people run analyses of covariance without checking the assumptions of parallelism. Rather than trying to force a particular model onto an experimental situation, it would be much better to postulate a reasonable (not necessarily linear) model relating the response y to the covariate x through the design used. Then by knowing the meanings of the parameters in the model, we can postulate hypotheses concerning the parameters and test these hypotheses by fitting complete and reduced models.

EXERCISES

    18.9   An investigator studied the effects of three different antidepressants (A, B, and C) on patient ratings of depression. To do this, patients were stratified into six age–sex combinations. From a random sample of three patients from each stratum, the experimenter randomly allocated the three antidepressants. On the day the study was to be initiated, a baseline (pretreatment) depression scale rating was obtained from each patient. The assigned thereapy was then administered and maintained for one week. At this time, a second rating (posttreatment) was obtained from each patient. The pre- and posttreatment ratings appear below (higher score indicates more depression).

Block	Sex	Age (Years)	Pretreatment A	B	C	Posttreatment A	B	C
1	F	<20	48	36	31	21	25	17
2	F	20–40	43	31	28	22	21	19
3	F	>40	44	35	29	18	24	18
4	M	<20	42	38	29	26	20	17
5	M	20–40	37	34	28	21	24	15
6	M	>40	41	36	26	18	24	19

a. Identify the experimental design.
b. Write a first-order model relating the posttreatment response y to the pretreatment rating x_1 for each treatment.

18.10 Refer to Exercise 18.9.

    a. Use a computer program to fit the model of part (b) of Exercise 18.8. Use $\alpha = .05$.

    b. Test for parallelism of the lines.

    c. Assuming that the lines are parallel, test for differences in treatment means adjusted for the covariate. Use $\alpha = .05$.

18.11 Refer to Exercises 18.9 and 18.10.

    a. Assuming parallelism of the response lines, perform a test for block differences adjusted for the covariate. Use $\alpha = .05$.

    b. How might you partition the block sum of squares into five meaningful single-degree-of-freedom sums of squares?

    c. Write a model and perform the tests suggested in part (b). Use $\alpha = .05$.

18.12 A random sample of 10 measurements was selected from each of two populations. In addition to measuring a response variable y on each experimental unit, a second variable x was also measured. The sample data appear in the accompanying table.

Population 1		Population 2	
x	y	x	y
30	165	24	180
27	170	31	169
20	130	20	171
21	156	26	161
33	167	20	180
29	151	25	170
27	165	22	169
25	162	30	160
28	169	24	178
32	173	21	182

    a. Plot the sample data.

    b. Write a first-order model relating the response y to the covariate x.

18.13 a. Fit the model of Exercise 18.12(b)

    b. Test for parallelism. Use $\alpha = .05$.

    c. What other tests are appropriate?

DATA MANAGEMENT AND REPORT PREPARATION

Over the past 18 chapters, we've discussed particular statistical methods, how those methods are applied to specific data sets, and how findings from statistical analyses presented in the form of computer output are interpreted. We have not concentrated on the processing steps that one follows between the time the data are received and the time the data are available in computer-readable form for analysis, nor have we discussed the form and content of the report that summarizes the results of a statistical analysis. In this chapter we consider the data-processing steps and statistical report writing. The chapter is not a complete manual with all the tools required; rather, it is an overview—what a manager should know about these steps. As an example, the chapter reflects standard procedures in the pharmaceutical industry, which is highly regulated. Procedures differ somewhat in other industries.

19.1 PREPARING DATA FOR STATISTICAL ANALYSIS

We begin with a discussion of the steps involved in processing data from a study. In practice, these steps may consume 75% of the total effort from the receipt of the raw data to the presentation of results from the analysis. What are these steps, why are they so important, and why are they so time-consuming?

To answer these questions, let's list the major data-processing steps in the cycle, which begins with receipt of the data and ends when the statistical analysis begins. Then we'll discuss each step separately.

STEPS IN PREPARING DATA FOR ANALYSIS

1. Receiving the raw data source
2. Creating the data base from the raw data source
3. Editing the data base
4. Correcting and clarifying the raw data source
5. Finalizing the data base
6. Creating data files from the data base

raw data source

1. *Receiving the raw data source.* For each study that is to be summarized and analyzed, the data arrive in some form, which we'll refer to as the **raw data source**. For a clinical trial, the raw data source is usually case report forms, sheets of $8\frac{1}{2}$- by 11-inch paper that have been used to record study data for each patient entered into the study. For other types of studies, the raw data source may be sheets of paper from a laboratory notebook, a magnetic tape (or any other form of machine-readable data), hand-tabulations, and so on.

data trail

It is important to retain the raw data source, since it is the beginning of the **data trail**, which leads from the raw data to the conclusions drawn from a study. Many consulting operations involved with the analysis and summarization of many different studies keep a log that contains vital information related to the study and raw data source. General information contained in a study log is shown below:

LOG FOR STUDY DATA

1. Date received, and from whom
2. Study investigator
3. Statistician (and others) assigned
4. Brief description of study
5. Treatments (compounds, preparations, etc.) studied
6. Raw data source
7. Response(s) measured
8. Reference number for study
9. Estimated (actual) completion date
10. Other pertinent information

Later, when the study has been analyzed and results have been communicated, additional information can be added to the log on how the study results were communicated, where these results are recorded, what data files have been saved, and where these files are stored.

2. *Creating the data base from the raw data source.* For most studies that are scheduled for a statistical analysis, a machine-readable data base is created. The steps taken to create the data base and the eventual form of the data base vary from one operation to another, depending on the software systems to be used in the statistical analysis. However, we can give a few guidelines based on the form of the entry system.

When the data are to be typed at a terminal, the raw data are first checked for legibility. Any illegible numbers or letters or other problems should be brought to the attention of the study coordinator. Then a coding guide that assigns column numbers and variable names to the data is filled out. Certain codes for missing values (for example, not available) are also defined here. Also, it is helpful to give a brief description of each variable. The **machine-readable data base** data file keyed in at the terminal is referred to as the **machine-readable data base**. A listing of the data base should be obtained and checked carefully against the raw data source. Any errors should be corrected at the terminal and verified against an updated listing.

Sometimes data are received in machine-readable form. For these situations the magnetic tape or disk file is considered to be the data base. You must, however, have a coding guide to "read" the data base. Using the coding guide, obtain a listing of the data base and check it *carefully* to see that all numbers and characters look reasonable and that proper formats were used to create the file. Any problems that arise must be resolved before proceeding further.

Some data sets are so small that it is not necessary to create a machine-readable data file from the raw data source. Instead, calculations are performed by hand or the data are entered into an electronic calculator. For these situations check any calculations to see that they make sense. Don't believe everything you see; redoing the calculations is not a bad idea.

3. *Editing the data base.* The types of edits done and the completeness of the editing process really depend on the type of study and how concerned you are about the accuracy and completeness of the data prior to the analysis. For example, in using SAS files it is wise to examine the minimum, maximum, and frequency distribution for each variable to make certain nothing looks unreasonable.

logic checks Certain other checks should be made. Plot the data and look for problems. Also, certain **logic checks** should be done depending on the structure of the data. If, for example, data are recorded for patients at several different visits, then the data recorded for visit 2 can't be earlier than the data for visit 1; similarly, if a patient is lost to follow-up after visit 2, we can't have any data for that patient at later visits.

For small data sets we can do these data edits by hand, but for large data sets the job may be too time-consuming and tedious. If machine editing is required, look for a software system that allows the user to specify certain data edits. Even so, for more complicated edits and logic checks it may be necessary to have a customized edit program written in order to machine edit the data. This programming chore can be a time-consuming step; plan for this well in advance of the receipt of the data.

4. *Correcting and clarifying the raw data source.* Questions frequently arise concerning the legibility or accuracy of the raw data during any one of the steps from the receipt of the raw data to the communication of the results from the statistical analysis. We have found it helpful to keep a list of these problems or discrepancies in order to define the data trail for a study. If a correction (or clarification) is required to the raw data source, this should be indicated on the form and the appropriate change made to the raw data source. If no correction is required, this should be indicated on the form as well. Keep in mind that the machine-readable data base should be changed to reflect any changes made to the raw data source.

5. *Finalizing the data base.* You may have been led to believe that all data for a study arrive at one time. This, of course, is not always the case. For example, with a marketing survey, different geographic locations may be surveyed at different times and hence those responsible for data processing do not receive all the data at one time. All these subsets of data, however, must be processed through the cycles required to create, edit, and cor-

rect the data base. Eventually the study is declared complete and the data is processed into the data base. At this time the data base should be reviewed again and final corrections made before beginning the analysis. The reason for this is that, for large data sets, the analysis and summarization chores take considerable human labor and computer time. It's better to agree on a final data base analysis than to have to repeat all analyses on a changed data base at a later date.

6. *Creating data files from the data base.* Generally there are one or two sets of data files created from the machine-readable data base. The first set, referred to as **original files** reflects the basic structure of the data base. A listing of the files is checked against the data base listing to verify that the variables have been read with correct formats and missing value codes have been retained. For some studies the original files are actually used for editing the data base.

original files

A second set of data files, called **work files**, may be created from the original files. Work files are designed to facilitate the analysis. They may require restructuring of the original files, a selection of important variables, or the creation or addition of new variables by insertion, computation, or transformation. A listing of the work files is checked against that of the original files to ensure proper restructuring and variable selection. Computed and transformed variables are checked by hand calculations to verify the program code.

work files

If original and work files are SAS data sets, you should utilize the documentation features provided by SAS. At the time an SAS data set is created, a descriptive label for the data set, of up to 40 characters, should be assigned. The label can be stored with the data set imprinted wherever the contents procedure is used to print the data set's contents. All variables can be given descriptive names, up to eight characters in length, which are meaningful to those involved in the project. In addition, variable labels up to 40 characters in length can be used to provide additional information. Title statements can be included in the SAS code to identify the project and describe each job. For each file, a listing (proc print) and a dictionary (proc contents) can be retained.

For files created from the data base using other software packages, you should utilize the labeling and documentation features available in the computer program.

Even if appropriate statistical methods are appliied to data, the conclusions drawn from the study are only as good as the data on which they are based. So you be the judge. The amount of time you should spend on these data-processing chores before analysis really depends on the nature of the study, the quality of the raw data source, and how confident you want to be about the completeness and accuracy of the data.

19.2   GUIDELINES FOR A STATISTICAL ANALYSIS AND REPORT

In this section we briefly discuss a few guidelines for a statistical analysis and some important elements of a statistical report for communicating results. The statistical analysis of a large study can usually be broken down into three types of analyses: (1) preliminary analyses, (2) primary analyses, and (3) backup analyses.

preliminary analyses

The **preliminary analyses**, which are often descriptive or graphic, familiarize the statistician with the data and provide a foundation for all subsequent analyses. These analyses may include frequency distributions, histograms, descriptive statistics, an examination of comparability of the treatment groups, correlations, or univariate and bivariate plots.

primary analyses   **Primary analyses** are those used to address the objectives of the study and the
backup analyses   analyses on which conclusions are drawn. **Backup analyses** include alternate methods for
examining the data that confirm the results of the primary analyses; they may also include
new statistical methods that are not as readily accepted as the more standard methods.
Several guidelines for analyses are presented below:

**PRELIMINARY,
PRIMARY, AND
BACKUP ANALYSES**

1. Analyses should be performed with software that has been extensively tested.
2. Computer output should be labeled to reflect which study is analyzed, what
   subjects (animals, patients, etc.) are used in the analysis, and a brief description of the analysis preferred. For example, TITLE statements in SAS are very
   helpful.
3. Variable labels and value labels (e.g., 0 = none, 1 = mild) should appear on
   the output.
4. A list of the data used in each analysis should be provided.
5. The output for all analyses should be checked *carefully*. Did the job run successfully? Are the sample sizes, means, and degrees of freedom correct? Other
   checks may be necessary as well.
6. All preliminary, primary, and backup analyses that provide the informational
   base from which study conclusions are drawn should be saved.

After the statistical analysis is completed, conclusions must be drawn and the results
communicated to the intended audience. Sometimes it is necessary to communicate these
results as a formal written statistical report. A general outline for a statistical report that
we have found useful and informative is listed below:

**GENERAL OUTLINE
FOR A STATISTICAL
REPORT**

1. Summary
2. Introduction
3. Experimental design and study procedures
4. Descriptive statistics
5. Statistical methodology
6. Results and conclusions
7. Discussion
8. Data listings

19.3   DOCUMENTATION AND STORAGE OF RESULTS

The final part of this cycle of data processing, analysis, and summarization concerns the
documentation and storage of results. For formal statistical analyses that are subject to
careful scrutiny by others, it is important to provide detailed documentation for all data
processing and the statistical analyses so that the data trail is clear and the data base or

work files readily accessible. Then the reviewer can follow what has been done, redo it, or extend the analyses. The elements of a documentation and storage file depend on the particular setting in which you work. The contents for a general documentation storage file are shown below:

STUDY DOCUMENTATION AND STORAGE FILE

1. Statistical report
2. Study description
3. Random code (used to assign subjects to treatment groups)
4. Important correspondence
5. File creation information
6. Preliminary, primary, and backup analyses
7. Raw data source
8. A data management sheet, which includes the log as well as information on the storage of the data files

SUMMARY

It is hoped that the information presented in this chapter has opened your eyes and broadened your perspective on the data processing and statistical analysis. Most textbooks assume that the data are ready for analysis when received or displayed. In practice, however, much work is required to prepare the data for analysis and to write a report. The material presented here gives the flavor of what is being done in the pharmaceutical industry. Because many other businesses or experimental settings are less highly regulated, it may be possible to relax some of the steps that have been outlined here. The important point is that you should actively consider whether these steps are required for a study; don't just ignore them.

Besides discussing the steps required to prepare data for analysis, we have presented a few guidelines for a statistical analysis, a general outline for a statistical report, and the contents of a documentation and storage system. Remember, these are only examples of what can be done. Other variations are certainly reasonable and appropriate.

APPENDIX: STATISTICAL TABLES

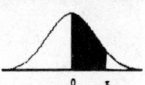

TABLE 1   Normal Curve Areas

z	.00	.01	.02	.03	.04	.05	.06	.07	.08	.09
0.00	.0000	.0040	.0080	.0120	.0160	.0199	.0239	.0279	.0319	.0359
0.10	.0398	.0438	.0478	.0517	.0557	.0596	.0636	.0675	.0714	.0753
0.20	.0793	.0832	.0871	.0910	.0948	.0987	.1026	.1064	.1103	.1141
0.30	.1179	.1217	.1255	.1293	.1331	.1368	.1406	.1443	.1480	.1517
0.40	.1554	.1591	.1628	.1664	.1700	.1736	.1772	.1808	.1844	.1879
0.50	.1915	.1950	.1985	.2019	.2054	.2088	.2123	.2157	.2190	.2224
0.60	.2257	.2291	.2324	.2357	.2389	.2422	.2454	.2486	.2517	.2549
0.70	.2580	.2611	.2642	.2673	.2704	.2734	.2764	.2794	.2823	.2852
0.80	.2881	.2910	.2939	.2967	.2995	.3023	.3051	.3078	.3106	.3133
0.90	.3159	.3186	.3212	.3238	.3264	.3289	.3315	.3340	.3365	.3389
1.00	.3413	.3438	.3461	.3485	.3508	.3531	.3554	.3577	.3599	.3621
1.10	.3643	.3665	.3686	.3708	.3729	.3749	.3770	.3790	.3810	.3830
1.20	.3849	.3869	.3888	.3907	.3925	.3944	.3962	.3980	.3997	.4015
1.30	.4032	.4049	.4066	.4082	.4099	.4115	.4131	.4147	.4162	.4177
1.40	.4192	.4207	.4222	.4236	.4251	.4265	.4279	.4292	.4306	.4319
1.50	.4332	.4345	.4357	.4370	.4382	.4394	.4406	.4418	.4429	.4441
1.60	.4452	.4463	.4474	.4484	.4495	.4505	.4515	.4525	.4535	.4545
1.70	.4554	.4564	.4573	.4582	.4591	.4599	.4608	.4616	.4625	.4633
1.80	.4641	.4649	.4656	.4664	.4671	.4678	.4686	.4693	.4699	.4706
1.90	.4713	.4719	.4726	.4732	.4738	.4744	.4750	.4756	.4761	.4767
2.00	.4772	.4778	.4783	.4788	.4793	.4798	.4803	.4808	.4812	.4817
2.10	.4821	.4826	.4830	.4834	.4838	.4842	.4846	.4850	.4854	.4857
2.20	.4861	.4864	.4868	.4871	.4875	.4878	.4881	.4884	.4887	.4890
2.30	.4893	.4896	.4898	.4901	.4904	.4906	.4909	.4911	.4913	.4916
2.40	.4918	.4920	.4922	.4925	.4927	.4929	.4931	.4932	.4934	.4936
2.50	.4938	.4940	.4941	.4943	.4945	.4946	.4948	.4949	.4951	.4952
2.60	.4953	.4955	.4956	.4957	.4959	.4960	.4961	.4962	.4963	.4964
2.70	.4965	.4966	.4967	.4968	.4969	.4970	.4971	.4972	.4973	.4974
2.80	.4974	.4975	.4976	.4977	.4977	.4978	.4979	.4979	.4980	.4981
2.90	.4981	.4982	.4982	.4983	.4984	.4984	.4985	.4985	.4986	.4986
3.00	.4987	.4987	.4987	.4988	.4988	.4989	.4989	.4989	.4990	.4990

z	Area
3.50	.49976737
4.00	.49996833
4.50	.49999660
5.00	.49999971

Source: Computed by P. J. Hildebrand.

APPENDIX   STATISTICAL TABLES

TABLE 2   Upper-tail Areas for the Normal Curve

z	.00	.01	.02	.03	.04	.05	.06	.07	.08	.09
0.00	.5000	.4960	.4920	.4880	.4840	.4801	.4761	.4721	.4681	.4641
0.10	.4602	.4562	.4522	.4483	.4443	.4404	.4364	.4325	.4286	.4247
0.20	.4207	.4168	.4129	.4090	.4052	.4013	.3974	.3936	.3897	.3859
0.30	.3821	.3783	.3745	.3707	.3669	.3632	.3594	.3557	.3520	.3483
0.40	.3446	.3409	.3372	.3336	.3300	.3264	.3228	.3192	.3156	.3121
0.50	.3085	.3050	.3015	.2981	.2946	.2912	.2877	.2843	.2810	.2776
0.60	.2743	.2709	.2676	.2643	.2611	.2578	.2546	.2514	.2483	.2451
0.70	.2420	.2389	.2358	.2327	.2296	.2266	.2236	.2206	.2177	.2148
0.80	.2119	.2090	.2061	.2033	.2005	.1977	.1949	.1922	.1894	.1867
0.90	.1841	.1814	.1788	.1762	.1736	.1711	.1685	.1660	.1635	.1611
1.00	.1587	.1562	.1539	.1515	.1492	.1469	.1446	.1423	.1401	.1379
1.10	.1357	.1335	.1314	.1292	.1271	.1251	.1230	.1210	.1190	.1170
1.20	.1151	.1131	.1112	.1093	.1075	.1056	.1038	.1020	.1003	.0985
1.30	.0968	.0951	.0934	.0918	.0901	.0885	.0869	.0853	.0838	.0823
1.40	.0808	.0793	.0778	.0764	.0749	.0735	.0721	.0708	.0694	.0681
1.50	.0668	.0655	.0643	.0630	.0618	.0606	.0594	.0582	.0571	.0559
1.60	.0548	.0537	.0526	.0516	.0505	.0495	.0485	.0475	.0465	.0455
1.70	.0446	.0436	.0427	.0418	.0409	.0401	.0392	.0384	.0375	.0367
1.80	.0359	.0351	.0344	.0336	.0329	.0322	.0314	.0307	.0301	.0294
1.90	.0287	.0281	.0274	.0268	.0262	.0256	.0250	.0244	.0239	.0233
2.00	.0228	.0222	.0217	.0212	.0207	.0202	.0197	.0192	.0188	.0183
2.10	.0179	.0174	.0170	.0166	.0162	.0158	.0154	.0150	.0146	.0143
2.20	.0139	.0136	.0132	.0129	.0125	.0122	.0119	.0116	.0113	.0110
2.30	.0107	.0104	.0102	.0099	.0096	.0094	.0091	.0089	.0087	.0084
2.40	.0082	.0080	.0078	.0075	.0073	.0071	.0069	.0068	.0066	.0064
2.50	.0062	.0060	.0059	.0057	.0055	.0054	.0052	.0051	.0049	.0048
2.60	.0047	.0045	.0044	.0043	.0041	.0040	.0039	.0038	.0037	.0036
2.70	.0035	.0034	.0033	.0032	.0031	.0030	.0029	.0028	.0027	.0026
2.80	.0026	.0025	.0024	.0023	.0023	.0022	.0021	.0021	.0020	.0019
2.90	.0019	.0018	.0018	.0017	.0016	.0016	.0015	.0015	.0014	.0014
3.00	.0013	.0013	.0013	.0012	.0012	.0011	.0011	.0011	.0010	.0010

z	Area
3.500	.00023263
4.000	.00003167
4.500	.00000340
5.000	.00000029

Source: Computed by J. W. Stegeman using SAS.

TABLE 3   Critical Values of T_L and T_U for the Wilcoxon Rank Sum Test: Independent Samples *(test statistic is rank sum associated with smaller sample (if equal sample sizes, either rank sum can be used))*

a. $\alpha = .025$ one-tailed; $\alpha = .05$ two-tailed

n_2 \ n_1	3		4		5		6		7		8		9		10	
	T_L	T_U	T_L	T_U	T_L	T_U	T_L	T_U	T_L	T_U	T_L	T_U	T_L	T_U	T_L	T_U
3	5	16	6	18	6	21	7	23	7	26	8	28	8	31	9	33
4	6	18	11	25	12	28	12	32	13	35	14	38	15	41	16	44
5	6	21	12	28	18	37	19	41	20	45	21	49	22	53	24	56
6	7	23	12	32	19	41	26	52	28	56	29	61	31	65	32	70
7	7	26	13	35	20	45	28	56	37	68	39	73	41	78	43	83
8	8	28	14	38	21	49	29	61	39	73	49	87	51	93	54	98
9	8	31	15	41	22	53	31	65	41	78	51	93	63	108	66	114
10	9	33	16	44	24	56	32	70	43	83	54	98	66	114	79	131

b. $\alpha = .05$ one-tailed; $\alpha = .10$ two-tailed

n_2 \ n_1	3		4		5		6		7		8		9		10	
	T_L	T_U	T_L	T_U	T_L	T_U	T_L	T_U	T_L	T_U	T_L	T_U	T_L	T_U	T_L	T_U
3	6	15	7	17	7	20	8	22	9	24	9	27	10	29	11	31
4	7	17	12	24	13	27	14	30	15	33	16	36	17	39	18	42
5	7	20	13	27	19	36	20	40	22	43	24	46	25	50	26	54
6	8	22	14	30	20	40	28	50	30	54	32	58	33	63	35	67
7	9	24	15	33	22	43	30	54	39	66	41	71	43	76	46	80
8	9	27	16	36	24	46	32	58	41	71	52	84	54	90	57	95
9	10	29	17	39	25	50	33	63	43	76	54	90	66	105	69	111
10	11	31	18	42	26	54	35	67	46	80	57	95	69	111	83	127

Source: From F. Wilcoxon and R. A. Wilcox, *Some Rapid Approximate Statistical Procedures* (Pearl River, N.Y. Lederle Laboratories, 1964), pp. 20–23. Reproduced with the permission of American Cyanamid Company.

APPENDIX   STATISTICAL TABLES

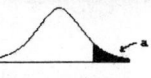

TABLE 4   Percentage Points of the *t* Distribution

df	a = .1	a = .05	a = .025	a = .01	a = .005	a = .001
1	3.078	6.314	12.706	31.821	63.657	318.309
2	1.886	2.920	4.303	6.965	9.925	22.327
3	1.638	2.353	3.182	4.541	5.841	10.215
4	1.533	2.132	2.776	3.747	4.604	7.173
5	1.476	2.015	2.571	3.365	4.032	5.893
6	1.440	1.943	2.447	3.143	3.707	5.208
7	1.415	1.895	2.365	2.998	3.499	4.785
8	1.397	1.860	2.306	2.896	3.355	4.501
9	1.383	1.833	2.262	2.821	3.250	4.297
10	1.372	1.812	2.228	2.764	3.169	4.144
11	1.363	1.796	2.201	2.718	3.106	4.025
12	1.356	1.782	2.179	2.681	3.055	3.930
13	1.350	1.771	2.160	2.650	3.012	3.852
14	1.345	1.761	2.145	2.624	2.977	3.787
15	1.341	1.753	2.131	2.602	2.947	3.733
16	1.337	1.746	2.120	2.583	2.921	3.686
17	1.333	1.740	2.110	2.567	2.898	3.646
18	1.330	1.734	2.101	2.552	2.878	3.610
19	1.328	1.729	2.093	2.539	2.861	3.579
20	1.325	1.725	2.086	2.528	2.845	3.552
21	1.323	1.721	2.080	2.518	2.831	3.527
22	1.321	1.717	2.074	2.508	2.819	3.505
23	1.319	1.714	2.069	2.500	2.807	3.485
24	1.318	1.711	2.064	2.492	2.797	3.467
25	1.316	1.708	2.060	2.485	2.787	3.450
26	1.315	1.706	2.056	2.479	2.779	3.435
27	1.314	1.703	2.052	2.473	2.771	3.421
28	1.313	1.701	2.048	2.467	2.763	3.408
29	1.311	1.699	2.045	2.462	2.756	3.396
30	1.310	1.697	2.042	2.457	2.750	3.385
40	1.303	1.684	2.021	2.423	2.704	3.307
60	1.296	1.671	2.000	2.390	2.660	3.232
120	1.289	1.658	1.980	2.358	2.617	3.160
240	1.285	1.651	1.970	2.342	2.596	3.125
inf.	1.282	1.645	1.960	2.326	2.576	3.090

Source: Computed by P. J. Hildebrand.

STATISTICAL TABLES

TABLE 5 Percentage Points of the Chi-square Distribution

df	a = .999	a = .995	a = .99	a = .975	a = .95	a = .9
1	.000002	.000039	.000157	.000982	.003932	.01579
2	.002001	.01003	.02010	.05064	.1026	.2107
3	.02430	.07172	.1148	.2158	.3518	.5844
4	.09080	.2070	.2971	.4844	.7107	1.064
5	.2102	.4117	.5543	.8312	1.145	1.610
6	.3811	.6757	.8721	1.237	1.635	2.204
7	.5985	.9893	1.239	1.690	2.167	2.833
8	.8571	1.344	1.646	2.180	2.733	3.490
9	1.152	1.735	2.088	2.700	3.325	4.168
10	1.479	2.156	2.558	3.247	3.940	4.865
11	1.834	2.603	3.053	3.816	4.575	5.578
12	2.214	3.074	3.571	4.404	5.226	6.304
13	2.617	3.565	4.107	5.009	5.892	7.042
14	3.041	4.075	4.660	5.629	6.571	7.790
15	3.483	4.601	5.229	6.262	7.261	8.547
16	3.942	5.142	5.812	6.908	7.962	9.312
17	4.416	5.697	6.408	7.564	8.672	10.09
18	4.905	6.265	7.015	8.231	9.390	10.86
19	5.407	6.844	7.633	8.907	10.12	11.65
20	5.921	7.434	8.260	9.591	10.85	12.44
21	6.447	8.034	8.897	10.28	11.59	13.24
22	6.983	8.643	9.542	10.98	12.34	14.04
23	7.529	9.260	10.20	11.69	13.09	14.85
24	8.085	9.886	10.86	12.40	13.85	15.66
25	8.649	10.52	11.52	13.12	14.61	16.47
26	9.222	11.16	12.20	13.84	15.38	17.29
27	9.803	11.81	12.88	14.57	16.15	18.11
28	10.39	12.46	13.56	15.31	16.93	18.94
29	10.99	13.12	14.26	16.06	17.71	19.77
30	11.59	13.79	14.95	16.79	18.49	20.60
40	17.92	20.71	22.16	24.43	26.51	29.05
50	24.67	27.99	29.71	32.36	34.76	37.69
60	31.74	35.53	37.48	40.48	43.19	46.46
70	39.04	43.28	45.44	48.76	51.74	55.33
80	46.52	51.17	53.54	57.15	60.39	64.28
90	54.16	59.20	61.75	65.65	69.13	73.29
100	61.92	67.33	70.06	74.22	77.93	82.36
120	77.76	83.85	86.92	91.57	95.70	100.62
240	177.95	187.32	191.99	198.98	205.14	212.39

TABLE 5 (cont'd)

a = .1	a = .05	a = .025	a = .01	a = .005	a = .001	df
2.706	3.841	5.024	6.635	7.879	10.83	1
4.605	5.991	7.378	9.210	10.60	13.82	2
6.251	7.815	9.348	11.34	12.84	16.27	3
7.779	9.488	11.14	13.28	14.86	18.47	4
9.236	11.07	12.83	15.09	16.75	20.52	5
10.64	12.59	14.45	16.81	18.55	22.46	6
12.02	14.07	16.01	18.48	20.28	24.32	7
13.36	15.51	17.53	20.09	21.95	26.12	8
14.68	16.92	19.02	21.67	23.59	27.88	9
15.99	18.31	20.48	23.21	25.19	29.59	10
17.28	19.68	21.92	24.72	26.76	31.27	11
18.55	21.03	23.34	26.22	28.30	32.91	12
19.81	22.36	24.74	27.69	29.82	34.53	13
21.06	23.68	26.12	29.14	31.32	36.12	14
22.31	25.00	27.49	30.58	32.80	37.70	15
23.54	26.30	28.85	32.00	34.27	39.25	16
24.77	27.59	30.19	33.41	35.72	40.79	17
25.99	28.87	31.53	34.81	37.16	42.31	18
27.20	30.14	32.85	36.19	38.58	43.82	19
28.41	31.41	34.17	37.57	40.00	45.31	20
29.62	32.67	35.48	38.93	41.40	46.80	21
30.81	33.92	36.78	40.29	42.80	48.27	22
32.01	35.17	38.08	41.64	44.18	49.73	23
33.20	36.42	39.36	42.98	45.56	51.18	24
34.38	37.65	40.65	44.31	46.93	52.62	25
35.56	38.89	41.92	45.64	48.29	54.05	26
36.74	40.11	43.19	46.96	49.65	55.48	27
37.92	41.34	44.46	48.28	50.99	56.89	28
39.09	42.56	45.72	49.59	52.34	58.30	29
40.26	43.77	46.98	50.89	53.67	59.70	30
51.81	55.76	59.34	63.69	66.77	73.40	40
63.17	67.50	71.42	76.15	79.49	86.66	50
74.40	79.08	83.30	88.38	91.95	99.61	60
85.53	90.53	95.02	100.43	104.21	112.32	70
96.58	101.88	106.63	112.33	116.32	124.84	80
107.57	113.15	118.14	124.12	128.30	137.21	90
118.50	124.34	129.56	135.81	140.17	149.45	100
140.23	146.57	152.21	158.95	163.65	173.62	120
268.47	277.14	284.80	293.89	300.18	313.44	240

Source: Computed by P. J. Hildebrand.

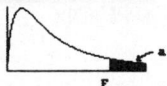

TABLE 6 Percentage Points of the *F* Distribution (df$_2$ Between 1 and 6)

df$_2$	a	\multicolumn{10}{c}{df$_1$}									
		1	2	3	4	5	6	7	8	9	10
1	.25	5.83	7.50	8.20	8.58	8.82	8.98	9.10	9.19	9.26	9.32
	.10	39.86	49.50	53.59	55.83	57.24	58.20	58.91	59.44	59.86	60.19
	.05	161.4	199.5	215.7	224.6	230.2	234.0	236.8	238.9	240.5	241.9
	.025	647.8	799.5	864.2	899.6	921.8	937.1	948.2	956.7	963.3	968.6
	.01	4052	5000	5403	5625	5764	5859	5928	5981	6022	6056
2	.25	2.57	3.00	3.15	3.23	3.28	3.31	3.34	3.35	3.37	3.38
	.10	8.53	9.00	9.16	9.24	9.29	9.33	9.35	9.37	9.38	9.39
	.05	18.51	19.00	19.16	19.25	19.30	19.33	19.35	19.37	19.38	19.40
	.025	38.51	39.00	39.17	39.25	39.30	39.33	39.36	39.37	39.39	39.40
	.01	98.50	99.00	99.17	99.25	99.30	99.33	99.36	99.37	99.39	99.40
	.005	198.5	199.0	199.2	199.2	199.3	199.3	199.4	199.4	199.4	199.4
	.001	998.5	999.0	999.2	999.2	999.3	999.3	999.4	999.4	999.4	999.4
3	.25	2.02	2.28	2.36	2.39	2.41	2.42	2.43	2.44	2.44	2.44
	.10	5.54	5.46	5.39	5.34	5.31	5.28	5.27	5.25	5.24	5.23
	.05	10.13	9.55	9.28	9.12	9.01	8.94	8.89	8.85	8.81	8.79
	.025	17.44	16.04	15.44	15.10	14.88	14.73	14.62	14.54	14.47	14.42
	.01	34.12	30.82	29.46	28.71	28.24	27.91	27.67	27.49	27.35	27.23
	.005	55.55	49.80	47.47	46.19	45.39	44.84	44.43	44.13	43.88	43.69
	.001	167.0	148.5	141.1	137.1	134.6	132.8	131.6	130.6	129.9	129.2
4	.25	1.81	2.00	2.05	2.06	2.07	2.08	2.08	2.08	2.08	2.08
	.10	4.54	4.32	4.19	4.11	4.05	4.01	3.98	3.95	3.94	3.92
	.05	7.71	6.94	6.59	6.39	6.26	6.16	6.09	6.04	6.00	5.96
	.025	12.22	10.65	9.98	9.60	9.36	9.20	9.07	8.98	8.90	8.84
	.01	21.20	18.00	16.69	15.98	15.52	15.21	14.98	14.80	14.66	14.55
	.005	31.33	26.28	24.26	23.15	22.46	21.97	21.62	21.35	21.14	20.97
	.001	74.14	61.25	56.18	53.44	51.71	50.53	49.66	49.00	48.47	48.05
5	.25	1.69	1.85	1.88	1.89	1.89	1.89	1.89	1.89	1.89	1.89
	.10	4.06	3.78	3.62	3.52	3.45	3.40	3.37	3.34	3.32	3.30
	.05	6.61	5.79	5.41	5.19	5.05	4.95	4.88	4.82	4.77	4.74
	.025	10.01	8.43	7.76	7.39	7.15	6.98	6.85	6.76	6.68	6.62
	.01	16.26	13.27	12.06	11.39	10.97	10.67	10.46	10.29	10.16	10.05
	.005	22.78	18.31	16.53	15.56	14.94	14.51	14.20	13.96	13.77	13.62
	.001	47.18	37.12	33.20	31.09	29.75	28.83	28.16	27.65	27.24	26.92
6	.25	1.62	1.76	1.78	1.79	1.79	1.78	1.78	1.78	1.77	1.77
	.10	3.78	3.46	3.29	3.18	3.11	3.05	3.01	2.98	2.96	2.94
	.05	5.99	5.14	4.76	4.53	4.39	4.28	4.21	4.15	4.10	4.06
	.025	8.81	7.26	6.60	6.23	5.99	5.82	5.70	5.60	5.52	5.46
	.01	13.75	10.92	9.78	9.15	8.75	8.47	8.26	8.10	7.98	7.87
	.005	18.63	14.54	12.92	12.03	11.46	11.07	10.79	10.57	10.39	10.25
	.001	35.51	27.00	23.70	21.92	20.80	20.03	19.46	19.03	18.69	18.41

APPENDIX   STATISTICAL TABLES

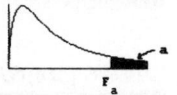

TABLE 6 (cont'd)

					df_1							
12	15	20	24	30	40	60	120	240	inf.	a	df_2
9.41	9.49	9.58	9.63	9.67	9.71	9.76	9.80	9.83	9.85	.25	1
60.71	61.22	61.74	62.00	62.26	62.53	62.79	63.06	63.19	63.33	.10	
243.9	245.9	248.0	249.1	250.1	251.1	252.2	253.3	253.8	254.3	.05	
976.7	984.9	993.1	997.2	1001	1006	1010	1014	1016	1018	.025	
6106	6157	6209	6235	6261	6287	6313	6339	6353	6366	.01	
3.39	3.41	3.43	3.43	3.44	3.45	3.46	3.47	3.47	3.48	.25	2
9.41	9.42	9.44	9.45	9.46	9.47	9.47	9.48	9.49	9.49	.10	
19.41	19.43	19.45	19.45	19.46	19.47	19.48	19.49	19.49	19.50	.05	
39.41	39.43	39.45	39.46	39.46	39.47	39.48	39.49	39.49	39.50	.025	
99.42	99.43	99.45	99.46	99.47	99.47	99.48	99.49	99.50	99.50	.01	
199.4	199.4	199.4	199.5	199.5	199.5	199.5	199.5	199.5	199.5	.005	
999.4	999.4	999.4	999.5	999.5	999.5	999.5	999.5	999.5	999.5	.001	
2.45	2.46	2.46	2.46	2.47	2.47	2.47	2.47	2.47	2.47	.25	3
5.22	5.20	5.18	5.18	5.17	5.16	5.15	5.14	5.14	5.13	.10	
8.74	8.70	8.66	8.64	8.62	8.59	8.57	8.55	8.54	8.53	.05	
14.34	14.25	14.17	14.12	14.08	14.04	13.99	13.95	13.92	13.90	.025	
27.05	26.87	26.69	26.60	26.50	26.41	26.32	26.22	26.17	26.13	.01	
43.39	43.08	42.78	42.62	42.47	42.31	42.15	41.99	41.91	41.83	.005	
128.3	127.4	126.4	125.9	125.4	125.0	124.5	124.0	123.7	123.5	.001	
2.08	2.08	2.08	2.08	2.08	2.08	2.08	2.08	2.08	2.08	.25	4
3.90	3.87	3.84	3.83	3.82	3.80	3.79	3.78	3.77	3.76	.10	
5.91	5.86	5.80	5.77	5.75	5.72	5.69	5.66	5.64	5.63	.05	
8.75	8.66	8.56	8.51	8.46	8.41	8.36	8.31	8.28	8.26	.025	
14.37	14.20	14.02	13.93	13.84	13.75	13.65	13.56	13.51	13.46	.01	
20.70	20.44	20.17	20.03	19.89	19.75	19.61	19.47	19.40	19.32	.005	
47.41	46.76	46.10	45.77	45.43	45.09	44.75	44.40	44.23	44.05	.001	
1.89	1.89	1.88	1.88	1.88	1.88	1.87	1.87	1.87	1.87	.25	5
3.27	3.24	3.21	3.19	3.17	3.16	3.14	3.12	3.11	3.10	.10	
4.68	4.62	4.56	4.53	4.50	4.46	4.43	4.40	4.38	4.36	.05	
6.52	6.43	6.33	6.28	6.23	6.18	6.12	6.07	6.04	6.02	.025	
9.89	9.72	9.55	9.47	9.38	9.29	9.20	9.11	9.07	9.02	.01	
13.38	13.15	12.90	12.78	12.66	12.53	12.40	12.27	12.21	12.14	.005	
26.42	25.91	25.39	25.13	24.87	24.60	24.33	24.06	23.92	23.79	.001	
1.77	1.76	1.76	1.75	1.75	1.75	1.74	1.74	1.74	1.74	.25	6
2.90	2.87	2.84	2.82	2.80	2.78	2.76	2.74	2.73	2.72	.10	
4.00	3.94	3.87	3.84	3.81	3.77	3.74	3.70	3.69	3.67	.05	
5.37	5.27	5.17	5.12	5.07	5.01	4.96	4.90	4.88	4.85	.025	
7.72	7.56	7.40	7.31	7.23	7.14	7.06	6.97	6.92	6.88	.01	
10.03	9.81	9.59	9.47	9.36	9.24	9.12	9.00	8.94	8.88	.005	
17.99	17.56	17.12	16.90	16.67	16.44	16.21	15.98	15.86	15.75	.001	

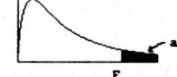

TABLE 6 (cont'd) Percentage Points of the F Distribution (df$_2$ Between 7 and 12)

df$_2$	a	1	2	3	4	5	6	7	8	9	10
							df$_1$				
7	.25	1.57	1.70	1.72	1.72	1.71	1.71	1.70	1.70	1.69	1.69
	.10	3.59	3.26	3.07	2.96	2.88	2.83	2.78	2.75	2.72	2.70
	.05	5.59	4.74	4.35	4.12	3.97	3.87	3.79	3.73	3.68	3.64
	.025	8.07	6.54	5.89	5.52	5.29	5.12	4.99	4.90	4.82	4.76
	.01	12.25	9.55	8.45	7.85	7.46	7.19	6.99	6.84	6.72	6.62
	.005	16.24	12.40	10.88	10.05	9.52	9.16	8.89	8.68	8.51	8.38
	.001	29.25	21.69	18.77	17.20	16.21	15.52	15.02	14.63	14.33	14.08
8	.25	1.54	1.66	1.67	1.66	1.66	1.65	1.64	1.64	1.63	1.63
	.10	3.46	3.11	2.92	2.81	2.73	2.67	2.62	2.59	2.56	2.54
	.05	5.32	4.46	4.07	3.84	3.69	3.58	3.50	3.44	3.39	3.35
	.025	7.57	6.06	5.42	5.05	4.82	4.65	4.53	4.43	4.36	4.30
	.01	11.26	8.65	7.59	7.01	6.63	6.37	6.18	6.03	5.91	5.81
	.005	14.69	11.04	9.60	8.81	8.30	7.95	7.69	7.50	7.34	7.21
	.001	25.41	18.49	15.83	14.39	13.48	12.86	12.40	12.05	11.77	11.54
9	.25	1.51	1.62	1.63	1.63	1.62	1.61	1.60	1.60	1.59	1.59
	.10	3.36	3.01	2.81	2.69	2.61	2.55	2.51	2.47	2.44	2.42
	.05	5.12	4.26	3.86	3.63	3.48	3.37	3.29	3.23	3.18	3.14
	.025	7.21	5.71	5.08	4.72	4.48	4.32	4.20	4.10	4.03	3.96
	.01	10.56	8.02	6.99	6.42	6.06	5.80	5.61	5.47	5.35	5.26
	.005	13.61	10.11	8.72	7.96	7.47	7.13	6.88	6.69	6.54	6.42
	.001	22.86	16.39	13.90	12.56	11.71	11.13	10.70	10.37	10.11	9.89
10	.25	1.49	1.60	1.60	1.59	1.59	1.58	1.57	1.56	1.56	1.55
	.10	3.29	2.92	2.73	2.61	2.52	2.46	2.41	2.38	2.35	2.32
	.05	4.96	4.10	3.71	3.48	3.33	3.22	3.14	3.07	3.02	2.98
	.025	6.94	5.46	4.83	4.47	4.24	4.07	3.95	3.85	3.78	3.72
	.01	10.04	7.56	6.55	5.99	5.64	5.39	5.20	5.06	4.94	4.85
	.005	12.83	9.43	8.08	7.34	6.87	6.54	6.30	6.12	5.97	5.85
	.001	21.04	14.91	12.55	11.28	10.48	9.93	9.52	9.20	8.96	8.75
11	.25	1.47	1.58	1.58	1.57	1.56	1.55	1.54	1.53	1.53	1.52
	.10	3.23	2.86	2.66	2.54	2.45	2.39	2.34	2.30	2.27	2.25
	.05	4.84	3.98	3.59	3.36	3.20	3.09	3.01	2.95	2.90	2.85
	.025	6.72	5.26	4.63	4.28	4.04	3.88	3.76	3.66	3.59	3.53
	.01	9.65	7.21	6.22	5.67	5.32	5.07	4.89	4.74	4.63	4.54
	.005	12.23	8.91	7.60	6.88	6.42	6.10	5.86	5.68	5.54	5.42
	.001	19.69	13.81	11.56	10.35	9.58	9.05	8.66	8.35	8.12	7.92
12	.25	1.46	1.56	1.56	1.55	1.54	1.53	1.52	1.51	1.51	1.50
	.10	3.18	2.81	2.61	2.48	2.39	2.33	2.28	2.24	2.21	2.19
	.05	4.75	3.89	3.49	3.26	3.11	3.00	2.91	2.85	2.80	2.75
	.025	6.55	5.10	4.47	4.12	3.89	3.73	3.61	3.51	3.44	3.37
	.01	9.33	6.93	5.95	5.41	5.06	4.82	4.64	4.50	4.39	4.30
	.005	11.75	8.51	7.23	6.52	6.07	5.76	5.52	5.35	5.20	5.09
	.001	18.64	12.97	10.80	9.63	8.89	8.38	8.00	7.71	7.48	7.29

TABLE 6 (cont'd)

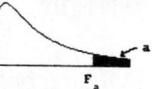

12	15	20	24	30	40	60	120	240	inf.	a	df₂
1.68	1.68	1.67	1.67	1.66	1.66	1.65	1.65	1.65	1.65	.25	7
2.67	2.63	2.59	2.58	2.56	2.54	2.51	2.49	2.48	2.47	.10	
3.57	3.51	3.44	3.41	3.38	3.34	3.30	3.27	3.25	3.23	.05	
4.67	4.57	4.47	4.41	4.36	4.31	4.25	4.20	4.17	4.14	.025	
6.47	6.31	6.16	6.07	5.99	5.91	5.82	5.74	5.69	5.65	.01	
8.18	7.97	7.75	7.64	7.53	7.42	7.31	7.19	7.13	7.08	.005	
13.71	13.32	12.93	12.73	12.53	12.33	12.12	11.91	11.80	11.70	.001	
1.62	1.62	1.61	1.60	1.60	1.59	1.59	1.58	1.58	1.58	.25	8
2.50	2.46	2.42	2.40	2.38	2.36	2.34	2.32	2.30	2.29	.10	
3.28	3.22	3.15	3.12	3.08	3.04	3.01	2.97	2.95	2.93	.05	
4.20	4.10	4.00	3.95	3.89	3.84	3.78	3.73	3.70	3.67	.025	
5.67	5.52	5.36	5.28	5.20	5.12	5.03	4.95	4.90	4.86	.01	
7.01	6.81	6.61	6.50	6.40	6.29	6.18	6.06	6.01	5.95	.005	
11.19	10.84	10.48	10.30	10.11	9.92	9.73	9.53	9.43	9.33	.001	
1.58	1.57	1.56	1.56	1.55	1.54	1.64	1.53	1.53	1.53	.25	9
2.38	2.34	2.30	2.28	2.25	2.23	2.21	2.18	2.17	2.16	.10	
3.07	3.01	2.94	2.90	2.86	2.83	2.79	2.75	2.73	2.71	.05	
3.87	3.77	3.67	3.61	3.56	3.51	3.45	3.39	3.36	3.33	.025	
5.11	4.96	4.81	4.73	4.65	4.57	4.48	4.40	4.35	4.31	.01	
6.23	6.03	5.83	5.73	5.62	5.52	5.41	5.30	5.24	5.19	.005	
9.57	9.24	8.90	8.72	8.55	8.37	8.19	8.00	7.91	7.81	.001	
1.54	1.53	1.52	1.52	1.51	1.51	1.50	1.49	1.49	1.48	.25	10
2.28	2.24	2.20	2.18	2.16	2.13	2.11	2.08	2.07	2.06	.10	
2.91	2.85	2.77	2.74	2.70	2.66	2.62	2.58	2.56	2.54	.05	
3.62	3.52	3.42	3.37	3.31	3.26	3.20	3.14	3.11	3.08	.025	
4.71	4.56	4.41	4.33	4.25	4.17	4.08	4.00	3.95	3.91	.01	
5.66	5.47	5.27	5.17	5.07	4.97	4.86	4.75	4.69	4.64	.005	
8.45	8.13	7.80	7.64	7.47	7.30	7.12	6.94	6.85	6.76	.001	
1.51	1.50	1.49	1.49	1.48	1.47	1.47	1.46	1.45	1.45	.25	11
2.21	2.17	2.12	2.10	2.08	2.05	2.03	2.00	1.99	1.97	.10	
2.79	2.72	2.65	2.61	2.57	2.53	2.49	2.45	2.43	2.40	.05	
3.43	3.33	3.23	3.17	3.12	3.06	3.00	2.94	2.91	2.88	.025	
4.40	4.25	4.10	4.02	3.94	3.86	3.78	3.69	3.65	3.60	.01	
5.24	5.05	4.86	4.76	4.65	4.55	4.45	4.34	4.28	4.23	.005	
7.63	7.32	7.01	6.85	6.68	6.52	6.35	6.18	6.09	6.00	.001	
1.49	1.48	1.47	1.46	1.45	1.45	1.44	1.43	1.43	1.42	.25	12
2.15	2.10	2.06	2.04	2.01	1.99	1.96	1.93	1.92	1.90	.10	
2.69	2.62	2.54	2.51	2.47	2.43	2.38	2.34	2.32	2.30	.05	
3.28	3.18	3.07	3.02	2.96	2.91	2.85	2.79	2.76	2.72	.025	
4.16	4.01	3.86	3.78	3.70	3.62	3.54	3.45	3.41	3.36	.01	
4.91	4.72	4.53	4.43	4.33	4.23	4.12	4.01	3.96	3.90	.005	
7.00	6.71	6.40	6.25	6.09	5.93	5.76	5.59	5.51	5.42	.001	

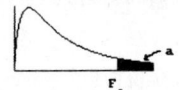

TABLE 6 (cont'd)   Percentage Points of the F Distribution (df$_2$ Between 13 and 18)

| df$_2$ | a | \multicolumn{10}{c}{df$_1$} |
		1	2	3	4	5	6	7	8	9	10
13	.25	1.45	1.55	1.55	1.53	1.52	1.51	1.50	1.49	1.49	1.48
	.10	3.14	2.76	2.56	2.43	2.35	2.28	2.23	2.20	2.16	2.14
	.05	4.67	3.81	3.41	3.18	3.03	2.92	2.83	2.77	2.71	2.67
	.025	6.41	4.97	4.35	4.00	3.77	3.60	3.48	3.39	3.31	3.25
	.01	9.07	6.70	5.74	5.21	4.86	4.62	4.44	4.30	4.19	4.10
	.005	11.37	8.19	6.93	6.23	5.79	5.48	5.25	5.08	4.94	4.82
	.001	17.82	12.31	10.21	9.07	8.35	7.86	7.49	7.21	6.98	6.80
14	.25	1.44	1.53	1.53	1.52	1.51	1.50	1.49	1.48	1.47	1.46
	.10	3.10	2.73	2.52	2.39	2.31	2.24	2.19	2.15	2.12	2.10
	.05	4.60	3.74	3.34	3.11	2.96	2.85	2.76	2.70	2.65	2.60
	.025	6.30	4.86	4.24	3.89	3.66	3.50	3.38	3.29	3.21	3.15
	.01	8.86	6.51	5.56	5.04	4.69	4.46	4.28	4.14	4.03	3.94
	.005	11.06	7.92	6.68	6.00	5.56	5.26	5.03	4.86	4.72	4.60
	.001	17.14	11.78	9.73	8.62	7.92	7.44	7.08	6.80	6.58	6.40
15	.25	1.43	1.52	1.52	1.51	1.49	1.48	1.47	1.46	1.46	1.45
	.10	3.07	2.70	2.49	2.36	2.27	2.21	2.16	2.12	2.09	2.06
	.05	4.54	3.68	3.29	3.06	2.90	2.79	2.71	2.64	2.59	2.54
	.025	6.20	4.77	4.15	3.80	3.58	3.41	3.29	3.20	3.12	3.06
	.01	8.68	6.36	5.42	4.89	4.56	4.32	4.14	4.00	3.89	3.80
	.005	10.80	7.70	6.48	5.80	5.37	5.07	4.85	4.67	4.54	4.42
	.001	16.59	11.34	9.34	8.25	7.57	7.09	6.74	6.47	6.26	6.08
16	.25	1.42	1.51	1.51	1.50	1.48	1.47	1.46	1.45	1.44	1.44
	.10	3.05	2.67	2.46	2.33	2.24	2.18	2.13	2.09	2.06	2.03
	.05	4.49	3.63	3.24	3.01	2.85	2.74	2.66	2.59	2.54	2.49
	.025	6.12	4.69	4.08	3.73	3.50	3.34	3.22	3.12	3.05	2.99
	.01	8.53	6.23	5.29	4.77	4.44	4.20	4.03	3.89	3.78	3.69
	.005	10.58	7.51	6.30	5.64	5.21	4.91	4.69	4.52	4.38	4.27
	.001	16.12	10.97	9.01	7.94	7.27	6.80	6.46	6.19	5.98	5.81
17	.25	1.42	1.51	1.50	1.49	1.47	1.46	1.45	1.44	1.43	1.43
	.10	3.03	2.64	2.44	2.31	2.22	2.15	2.10	2.06	2.03	2.00
	.05	4.45	3.59	3.20	2.96	2.81	2.70	2.61	2.55	2.49	2.45
	.025	6.04	4.62	4.01	3.66	3.44	3.28	3.16	3.06	2.98	2.92
	.01	8.40	6.11	5.18	4.67	4.34	4.10	3.93	3.79	3.68	3.59
	.005	10.38	7.35	6.16	5.50	5.07	4.78	4.56	4.39	4.25	4.14
	.001	15.72	10.66	8.73	7.68	7.02	6.56	6.22	5.96	5.75	5.58
18	.25	1.41	1.50	1.49	1.48	1.46	1.45	1.44	1.43	1.42	1.42
	.10	3.01	2.62	2.42	2.29	2.20	2.13	2.08	2.04	2.00	1.98
	.05	4.41	3.55	3.16	2.93	2.77	2.66	2.58	2.51	2.46	2.41
	.025	5.98	4.56	3.95	3.61	3.38	3.22	3.10	3.01	2.93	2.87
	.01	8.29	6.01	5.09	4.58	4.25	4.01	3.84	3.71	3.60	3.51
	.005	10.22	7.21	6.03	5.37	4.96	4.66	4.44	4.28	4.14	4.03
	.001	15.38	10.39	8.49	7.46	6.81	6.35	6.02	5.76	5.56	5.39

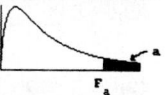

12	15	20	24	30	40	60	120	240	inf.	a	df₂

					df₁						
1.47	1.46	1.45	1.44	1.43	1.42	1.42	1.41	1.40	1.40	.25	13
2.10	2.05	2.01	1.98	1.96	1.93	1.90	1.88	1.86	1.85	.10	
2.60	2.53	2.46	2.42	2.38	2.34	2.30	2.25	2.23	2.21	.05	
3.15	3.05	2.95	2.89	2.84	2.78	2.72	2.66	2.63	2.60	.025	
3.96	3.82	3.66	3.59	3.51	3.43	3.34	3.25	3.21	3.17	.01	
4.64	4.46	4.27	4.17	4.07	3.97	3.87	3.76	3.70	3.65	.005	
6.52	6.23	5.93	5.78	5.63	5.47	5.30	5.14	5.05	4.97	.001	
1.45	1.44	1.43	1.42	1.41	1.41	1.40	1.39	1.38	1.38	.25	14
2.05	2.01	1.96	1.94	1.91	1.89	1.86	1.83	1.81	1.80	.10	
2.53	2.46	2.39	2.35	2.31	2.27	2.22	2.18	2.15	2.13	.05	
3.05	2.95	2.84	2.79	2.73	2.67	2.61	2.55	2.52	2.49	.025	
3.80	3.66	3.51	3.43	3.35	3.27	3.18	3.09	3.05	3.00	.01	
4.43	4.25	4.06	3.96	3.86	3.76	3.66	3.55	3.49	3.44	.005	
6.13	5.85	5.56	5.41	5.25	5.10	4.94	4.77	4.69	4.60	.001	
1.44	1.43	1.41	1.41	1.40	1.39	1.38	1.37	1.36	1.36	.25	15
2.02	1.97	1.92	1.90	1.87	1.85	1.82	1.79	1.77	1.76	.10	
2.48	2.40	2.33	2.29	2.25	2.20	2.16	2.11	2.09	2.07	.05	
2.96	2.86	2.76	2.70	2.64	2.59	2.52	2.46	2.43	2.40	.025	
3.67	3.52	3.37	3.29	3.21	3.13	3.05	2.96	2.91	2.87	.01	
4.25	4.07	3.88	3.79	3.69	3.58	3.48	3.37	3.32	3.26	.005	
5.81	5.54	5.25	5.10	4.95	4.80	4.64	4.47	4.39	4.31	.001	
1.43	1.41	1.40	1.39	1.38	1.37	1.36	1.35	1.35	1.34	.25	16
1.99	1.94	1.89	1.87	1.84	1.81	1.78	1.75	1.73	1.72	.10	
2.42	2.35	2.28	2.24	2.19	2.15	2.11	2.06	2.03	2.01	.05	
2.89	2.79	2.68	2.63	2.57	2.51	2.45	2.38	2.35	2.32	.025	
3.55	3.41	3.26	3.18	3.10	3.02	2.93	2.84	2.80	2.75	.01	
4.10	3.92	3.73	3.64	3.54	3.44	3.33	3.22	3.17	3.11	.005	
5.55	5.27	4.99	4.85	4.70	4.54	4.39	4.23	4.14	4.06	.001	
1.41	1.40	1.39	1.38	1.37	1.36	1.35	1.34	1.33	1.33	.25	17
1.96	1.91	1.86	1.84	1.81	1.78	1.75	1.72	1.70	1.69	.10	
2.38	2.31	2.23	2.19	2.15	2.10	2.06	2.01	1.99	1.96	.05	
2.82	2.72	2.62	2.56	2.50	2.44	2.38	2.32	2.28	2.25	.025	
3.46	3.31	3.16	3.08	3.00	2.92	2.83	2.75	2.70	2.65	.01	
3.97	3.79	3.61	3.51	3.41	3.31	3.21	3.10	3.04	2.98	.005	
5.32	5.05	4.78	4.63	4.48	4.33	4.18	4.02	3.93	3.85	.001	
1.40	1.39	1.38	1.37	1.36	1.35	1.34	1.33	1.32	1.32	.25	18
1.93	1.89	1.84	1.81	1.78	1.75	1.72	1.69	1.67	1.66	.10	
2.34	2.27	2.19	2.15	2.11	2.06	2.02	1.97	1.94	1.92	.05	
2.77	2.67	2.56	2.50	2.44	2.38	2.32	2.26	2.22	2.19	.025	
3.37	3.23	3.08	3.00	2.92	2.84	2.75	2.66	2.61	2.57	.01	
3.86	3.68	3.50	3.40	3.30	3.20	3.10	2.99	2.93	2.87	.005	
5.13	4.87	4.59	4.45	4.30	4.15	4.00	3.84	3.75	3.67	.001	

TABLE 6 (cont'd)   Percentage Points of the F Distribution (df$_2$ Between 19 and 24)

df$_2$	a	\multicolumn{10}{c}{df$_1$}									
		1	2	3	4	5	6	7	8	9	10
19	.25	1.41	1.49	1.49	1.47	1.46	1.44	1.43	1.42	1.41	1.41
	.10	2.99	2.61	2.40	2.27	2.18	2.11	2.06	2.02	1.98	1.96
	.05	4.38	3.52	3.13	2.90	2.74	2.63	2.54	2.48	2.42	2.38
	.025	5.92	4.51	3.90	3.56	3.33	3.17	3.05	2.96	2.88	2.82
	.01	8.18	5.93	5.01	4.50	4.17	3.94	3.77	3.63	3.52	3.43
	.005	10.07	7.09	5.92	5.27	4.85	4.56	4.34	4.18	4.04	3.93
	.001	15.08	10.16	8.28	7.27	6.62	6.18	5.85	5.59	5.39	5.22
20	.25	1.40	1.49	1.48	1.47	1.45	1.44	1.43	1.42	1.41	1.40
	.10	2.97	2.59	2.38	2.25	2.16	2.09	2.04	2.00	1.96	1.94
	.05	4.35	3.49	3.10	2.87	2.71	2.60	2.51	2.45	2.39	2.35
	.025	5.87	4.46	3.86	3.51	3.29	3.13	3.01	2.91	2.84	2.77
	.01	8.10	5.85	4.94	4.43	4.10	3.87	3.70	3.56	3.46	3.37
	.005	9.94	6.99	5.82	5.17	4.76	4.47	4.26	4.09	3.96	3.85
	.001	14.82	9.95	8.10	7.10	6.46	6.02	5.69	5.44	5.24	5.08
21	.25	1.40	1.48	1.48	1.46	1.44	1.43	1.42	1.41	1.40	1.39
	.10	2.96	2.57	2.36	2.23	2.14	2.08	2.02	1.98	1.95	1.92
	.05	4.32	3.47	3.07	2.84	2.68	2.57	2.49	2.42	2.37	2.32
	.025	5.83	4.42	3.82	3.48	3.25	3.09	2.97	2.87	2.80	2.73
	.01	8.02	5.78	4.87	4.37	4.04	3.81	3.64	3.51	3.40	3.31
	.005	9.83	6.89	5.73	5.09	4.68	4.39	4.18	4.01	3.88	3.77
	.001	14.59	9.77	7.94	6.95	6.32	5.88	5.56	5.31	5.11	4.95
22	.25	1.40	1.48	1.47	1.45	1.44	1.42	1.41	1.40	1.39	1.39
	.10	2.95	2.56	2.35	2.22	2.13	2.06	2.01	1.97	1.93	1.90
	.05	4.30	3.44	3.05	2.82	2.66	2.55	2.46	2.40	2.34	2.30
	.025	5.79	4.38	3.78	3.44	3.22	3.05	2.93	2.84	2.76	2.70
	.01	7.95	5.72	4.82	4.31	3.99	3.76	3.59	3.45	3.35	3.26
	.005	9.73	6.81	5.65	5.02	4.61	4.32	4.11	3.94	3.81	3.70
	.001	14.38	9.61	7.80	6.81	6.19	5.76	5.44	5.19	4.99	4.83
23	.25	1.39	1.47	1.47	1.45	1.43	1.42	1.41	1.40	1.39	1.38
	.10	2.94	2.55	2.34	2.21	2.11	2.05	1.99	1.95	1.92	1.89
	.05	4.28	3.42	3.03	2.80	2.64	2.53	2.44	2.37	2.32	2.27
	.025	5.75	4.35	3.75	3.41	3.18	3.02	2.90	2.81	2.73	2.67
	.01	7.88	5.66	4.76	4.26	3.94	3.71	3.54	3.41	3.30	3.21
	.005	9.63	6.73	5.58	4.95	4.54	4.26	4.05	3.88	3.75	3.64
	.001	14.20	9.47	7.67	6.70	6.08	5.65	5.33	5.09	4.89	4.73
24	.25	1.39	1.47	1.46	1.44	1.43	1.41	1.40	1.39	1.38	1.38
	.10	2.93	2.54	2.33	2.19	2.10	2.04	1.98	1.94	1.91	1.88
	.05	4.26	3.40	3.01	2.78	2.62	2.51	2.42	2.36	2.30	2.25
	.025	5.72	4.32	3.72	3.38	3.15	2.99	2.87	2.78	2.70	2.64
	.01	7.82	5.61	4.72	4.22	3.90	3.67	3.50	3.36	3.26	3.17
	.005	9.55	6.66	5.52	4.89	4.49	4.20	3.99	3.83	3.69	3.59
	.001	14.03	9.34	7.55	6.59	5.98	5.55	5.23	4.99	4.80	4.64

APPENDIX · STATISTICAL TABLES

TABLE 6 (cont'd)

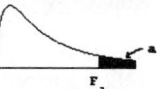

				df_1							
12	15	20	24	30	40	60	120	240	inf.	a	df_2
1.40	1.38	1.37	1.36	1.35	1.34	1.33	1.32	1.31	1.30	.25	19
1.91	1.86	1.81	1.79	1.76	1.73	1.70	1.67	1.65	1.63	.10	
2.31	2.23	2.16	2.11	2.07	2.03	1.98	1.93	1.90	1.88	.05	
2.72	2.62	2.51	2.45	2.39	2.33	2.27	2.20	2.17	2.13	.025	
3.30	3.15	3.00	2.92	2.84	2.76	2.67	2.58	2.54	2.49	.01	
3.76	3.59	3.40	3.31	3.21	3.11	3.00	2.89	2.83	2.78	.005	
4.97	4.70	4.43	4.29	4.14	3.99	3.84	3.68	3.60	3.51	.001	
1.39	1.37	1.36	1.35	1.34	1.33	1.32	1.31	1.30	1.29	.25	20
1.89	1.84	1.79	1.77	1.74	1.71	1.68	1.64	1.63	1.61	.10	
2.28	2.20	2.12	2.08	2.04	1.99	1.95	1.90	1.87	1.84	.05	
2.68	2.57	2.46	2.41	2.35	2.29	2.22	2.16	2.12	2.09	.025	
3.23	3.09	2.94	2.86	2.78	2.69	2.61	2.52	2.47	2.42	.01	
3.68	3.50	3.32	3.22	3.12	3.02	2.92	2.81	2.75	2.69	.005	
4.82	4.56	4.29	4.15	4.00	3.86	3.70	3.54	3.46	3.38	.001	
1.38	1.37	1.35	1.34	1.33	1.32	1.31	1.30	1.29	1.28	.25	21
1.87	1.83	1.78	1.75	1.72	1.69	1.66	1.62	1.60	1.59	.10	
2.25	2.18	2.10	2.05	2.01	1.96	1.92	1.87	1.84	1.81	.05	
2.64	2.53	2.42	2.37	2.31	2.25	2.18	2.11	2.08	2.04	.025	
3.17	3.03	2.88	2.80	2.72	2.64	2.55	2.46	2.41	2.36	.01	
3.60	3.43	3.24	3.15	3.05	2.95	2.84	2.73	2.67	2.61	.005	
4.70	4.44	4.17	4.03	3.88	3.74	3.58	3.42	3.34	3.26	.001	
1.37	1.36	1.34	1.33	1.32	1.31	1.30	1.29	1.28	1.28	.25	22
1.86	1.81	1.76	1.73	1.70	1.67	1.64	1.60	1.59	1.57	.10	
2.23	2.15	2.07	2.03	1.98	1.94	1.89	1.84	1.81	1.78	.05	
2.60	2.50	2.39	2.33	2.27	2.21	2.14	2.08	2.04	2.00	.025	
3.12	2.98	2.83	2.75	2.67	2.58	2.50	2.40	2.35	2.31	.01	
3.54	3.36	3.18	3.08	2.98	2.88	2.77	2.66	2.60	2.55	.005	
4.58	4.33	4.06	3.92	3.78	3.63	3.48	3.32	3.23	3.15	.001	
1.37	1.35	1.34	1.33	1.32	1.31	1.30	1.28	1.28	1.27	.25	23
1.84	1.80	1.74	1.72	1.69	1.66	1.62	1.59	1.57	1.55	.10	
2.20	2.13	2.05	2.01	1.96	1.91	1.86	1.81	1.79	1.76	.05	
2.57	2.47	2.36	2.30	2.24	2.18	2.11	2.04	2.01	1.97	.025	
3.07	2.93	2.78	2.70	2.62	2.54	2.45	2.35	2.31	2.26	.01	
3.47	3.30	3.12	3.02	2.92	2.82	2.71	2.60	2.54	2.48	.005	
4.48	4.23	3.96	3.82	3.68	3.53	3.38	3.22	3.14	3.05	.001	
1.36	1.35	1.33	1.32	1.31	1.30	1.29	1.28	1.27	1.26	.25	24
1.83	1.78	1.73	1.70	1.67	1.64	1.61	1.57	1.55	1.53	.10	
2.18	2.11	2.03	1.98	1.94	1.89	1.84	1.79	1.76	1.73	.05	
2.54	2.44	2.33	2.27	2.21	2.15	2.08	2.01	1.97	1.94	.025	
3.03	2.89	2.74	2.66	2.58	2.49	2.40	2.31	2.26	2.21	.01	
3.42	3.25	3.06	2.97	2.87	2.77	2.66	2.55	2.49	2.43	.005	
4.39	4.14	3.87	3.74	3.59	3.45	3.29	3.14	3.05	2.97	.001	

TABLE 6 (cont'd) Percentage Points of the F Distribution (df$_2$ Between 25 and 30)

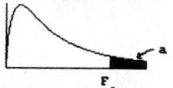

| df$_2$ | a | \multicolumn{10}{c}{df$_1$} |
		1	2	3	4	5	6	7	8	9	10
25	.25	1.39	1.47	1.46	1.44	1.42	1.41	1.40	1.39	1.38	1.37
	.10	2.92	2.53	2.32	2.18	2.09	2.02	1.97	1.93	1.89	1.87
	.05	4.24	3.39	2.99	2.76	2.60	2.49	2.40	2.34	2.28	2.24
	.025	5.69	4.29	3.69	3.35	3.13	2.97	2.85	2.75	2.68	2.61
	.01	7.77	5.57	4.68	4.18	3.85	3.63	3.46	3.32	3.22	3.13
	.005	9.48	6.60	5.46	4.84	4.43	4.15	3.94	3.78	3.64	3.54
	.001	13.88	9.22	7.45	6.49	5.89	5.46	5.15	4.91	4.71	4.56
26	.25	1.38	1.46	1.45	1.44	1.42	1.41	1.39	1.38	1.37	1.37
	.10	2.91	2.52	2.31	2.17	2.08	2.01	1.96	1.92	1.88	1.86
	.05	4.23	3.37	2.98	2.74	2.59	2.47	2.39	2.32	2.27	2.22
	.025	5.66	4.27	3.67	3.33	3.10	2.94	2.82	2.73	2.65	2.59
	.01	7.72	5.53	4.64	4.14	3.82	3.59	3.42	3.29	3.18	3.09
	.005	9.41	6.54	5.41	4.79	4.38	4.10	3.89	3.73	3.60	3.49
	.001	13.74	9.12	7.36	6.41	5.80	5.38	5.07	4.83	4.64	4.48
27	.25	1.38	1.46	1.45	1.43	1.42	1.40	1.39	1.38	1.37	1.36
	.10	2.90	2.51	2.30	2.17	2.07	2.00	1.95	1.91	1.87	1.85
	.05	4.21	3.35	2.96	2.73	2.57	2.46	2.37	2.31	2.25	2.20
	.025	5.63	4.24	3.65	3.31	3.08	2.92	2.80	2.71	2.63	2.57
	.01	7.68	5.49	4.60	4.11	3.78	3.56	3.39	3.26	3.15	3.06
	.005	9.34	6.49	5.36	4.74	4.34	4.06	3.85	3.69	3.56	3.45
	.001	13.61	9.02	7.27	6.33	5.73	5.31	5.00	4.76	4.57	4.41
28	.25	1.38	1.46	1.45	1.43	1.41	1.40	1.39	1.38	1.37	1.36
	.10	2.89	2.50	2.29	2.16	2.06	2.00	1.94	1.90	1.87	1.84
	.05	4.20	3.34	2.95	2.71	2.56	2.45	2.36	2.29	2.24	2.19
	.025	5.61	4.22	3.63	3.29	3.06	2.90	2.78	2.69	2.61	2.55
	.01	7.64	5.45	4.57	4.07	3.75	3.53	3.36	3.23	3.12	3.03
	.005	9.28	6.44	5.32	4.70	4.30	4.02	3.81	3.65	3.52	3.41
	.001	13.50	8.93	7.19	6.25	5.66	5.24	4.93	4.69	4.50	4.35
29	.25	1.38	1.45	1.45	1.43	1.41	1.40	1.38	1.37	1.36	1.35
	.10	2.89	2.50	2.28	2.15	2.06	1.99	1.93	1.89	1.86	1.83
	.05	4.18	3.33	2.93	2.70	2.55	2.43	2.35	2.28	2.22	2.18
	.025	5.59	4.20	3.61	3.27	3.04	2.88	2.76	2.67	2.59	2.53
	.01	7.60	5.42	4.54	4.04	3.73	3.50	3.33	3.20	3.09	3.00
	.005	9.23	6.40	5.28	4.66	4.26	3.98	3.77	3.61	3.48	3.38
	.001	13.39	8.85	7.12	6.19	5.59	5.18	4.87	4.64	4.45	4.29
30	.25	1.38	1.45	1.44	1.42	1.41	1.39	1.38	1.37	1.36	1.35
	.10	2.88	2.49	2.28	2.14	2.05	1.98	1.93	1.88	1.85	1.82
	.05	4.17	3.32	2.92	2.69	2.53	2.42	2.33	2.27	2.21	2.16
	.025	5.57	4.18	3.59	3.25	3.03	2.87	2.75	2.65	2.57	2.51
	.01	7.56	5.39	4.51	4.02	3.70	3.47	3.30	3.17	3.07	2.98
	.005	9.18	6.35	5.24	4.62	4.23	3.95	3.74	3.58	3.45	3.34
	.001	13.29	8.77	7.05	6.12	5.53	5.12	4.82	4.58	4.39	4.24

APPENDIX   STATISTICAL TABLES

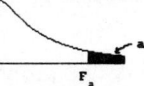

TABLE 6 (cont'd)

					df₁						
12	15	20	24	30	40	60	120	240	inf.	a	df₂
1.36	1.34	1.33	1.32	1.31	1.29	1.28	1.27	1.26	1.25	.25	25
1.82	1.77	1.72	1.69	1.66	1.63	1.59	1.56	1.54	1.52	.10	
2.16	2.09	2.01	1.96	1.92	1.87	1.82	1.77	1.74	1.71	.05	
2.51	2.41	2.30	2.24	2.18	2.12	2.05	1.98	1.94	1.91	.025	
2.99	2.85	2.70	2.62	2.54	2.45	2.36	2.27	2.22	2.17	.01	
3.37	3.20	3.01	2.92	2.82	2.72	2.61	2.50	2.44	2.38	.005	
4.31	4.06	3.79	3.66	3.52	3.37	3.22	3.06	2.98	2.89	.001	
1.35	1.34	1.32	1.31	1.30	1.29	1.28	1.26	1.26	1.25	.25	26
1.81	1.76	1.71	1.68	1.65	1.61	1.58	1.54	1.52	1.50	.10	
2.15	2.07	1.99	1.95	1.90	1.85	1.80	1.75	1.72	1.69	.05	
2.49	2.39	2.28	2.22	2.16	2.09	2.03	1.95	1.92	1.88	.025	
2.96	2.81	2.66	2.58	2.50	2.42	2.33	2.23	2.18	2.13	.01	
3.33	3.15	2.97	2.87	2.77	2.67	2.56	2.45	2.39	2.33	.005	
4.24	3.99	3.72	3.59	3.44	3.30	3.15	2.99	2.90	2.82	.001	
1.35	1.33	1.32	1.31	1.30	1.28	1.27	1.26	1.25	1.24	.25	27
1.80	1.75	1.70	1.67	1.64	1.60	1.57	1.53	1.51	1.49	.10	
2.13	2.06	1.97	1.93	1.88	1.84	1.79	1.73	1.70	1.67	.05	
2.47	2.36	2.25	2.19	2.13	2.07	2.00	1.93	1.89	1.85	.025	
2.93	2.78	2.63	2.55	2.47	2.38	2.29	2.20	2.15	2.10	.01	
3.28	3.11	2.93	2.83	2.73	2.63	2.52	2.41	2.35	2.29	.005	
4.17	3.92	3.66	3.52	3.38	3.23	3.08	2.92	2.84	2.75	.001	
1.34	1.33	1.31	1.30	1.29	1.28	1.27	1.25	1.24	1.24	.25	28
1.79	1.74	1.69	1.66	1.63	1.59	1.56	1.52	1.50	1.48	.10	
2.12	2.04	1.96	1.91	1.87	1.82	1.77	1.71	1.68	1.65	.05	
2.45	2.34	2.23	2.17	2.11	2.05	1.98	1.91	1.87	1.83	.025	
2.90	2.75	2.60	2.52	2.44	2.35	2.26	2.17	2.12	2.06	.01	
3.25	3.07	2.89	2.79	2.69	2.59	2.48	2.37	2.31	2.25	.005	
4.11	3.86	3.60	3.46	3.32	3.18	3.02	2.86	2.78	2.69	.001	
1.34	1.32	1.31	1.30	1.29	1.27	1.26	1.25	1.24	1.23	.25	29
1.78	1.73	1.68	1.65	1.62	1.58	1.55	1.51	1.49	1.47	.10	
2.10	2.03	1.94	1.90	1.85	1.81	1.75	1.70	1.67	1.64	.05	
2.43	2.32	2.21	2.15	2.09	2.03	1.96	1.89	1.85	1.81	.025	
2.87	2.73	2.57	2.49	2.41	2.33	2.23	2.14	2.09	2.03	.01	
3.21	3.04	2.86	2.76	2.66	2.56	2.45	2.33	2.27	2.21	.005	
4.05	3.80	3.54	3.41	3.27	3.12	2.97	2.81	2.73	2.64	.001	
1.34	1.32	1.30	1.29	1.28	1.27	1.26	1.24	1.23	1.23	.25	30
1.77	1.72	1.67	1.64	1.61	1.57	1.54	1.50	1.48	1.46	.10	
2.09	2.01	1.93	1.89	1.84	1.79	1.74	1.68	1.65	1.62	.05	
2.41	2.31	2.20	2.14	2.07	2.01	1.94	1.87	1.83	1.79	.025	
2.84	2.70	2.55	2.47	2.39	2.30	2.21	2.11	2.06	2.01	.01	
3.18	3.01	2.82	2.73	2.63	2.52	2.42	2.30	2.24	2.18	.005	
4.00	3.75	3.49	3.36	3.22	3.07	2.92	2.76	2.68	2.59	.001	

TABLE 6 (cont'd)   Percentage Points of the F Distribution (df$_2$ at Least 40)

df$_2$	a	1	2	3	4	5	6	7	8	9	10
40	.25	1.36	1.44	1.42	1.40	1.39	1.37	1.36	1.35	1.34	1.33
	.10	2.84	2.44	2.23	2.09	2.00	1.93	1.87	1.83	1.79	1.76
	.05	4.08	3.23	2.84	2.61	2.45	2.34	2.25	2.18	2.12	2.08
	.025	5.42	4.05	3.46	3.13	2.90	2.74	2.62	2.53	2.45	2.39
	.01	7.31	5.18	4.31	3.83	3.51	3.29	3.12	2.99	2.89	2.80
	.005	8.83	6.07	4.98	4.37	3.99	3.71	3.51	3.35	3.22	3.12
	.001	12.61	8.25	6.59	5.70	5.13	4.73	4.44	4.21	4.02	3.87
60	.25	1.35	1.42	1.41	1.38	1.37	1.35	1.33	1.32	1.31	1.30
	.10	2.79	2.39	2.18	2.04	1.95	1.87	1.82	1.77	1.74	1.71
	.05	4.00	3.15	2.76	2.53	2.37	2.25	2.17	2.10	2.04	1.99
	.025	5.29	3.93	3.34	3.01	2.79	2.63	2.51	2.41	2.33	2.27
	.01	7.08	4.98	4.13	3.65	3.34	3.12	2.95	2.82	2.72	2.63
	.005	8.49	5.79	4.73	4.14	3.76	3.49	3.29	3.13	3.01	2.90
	.001	11.97	7.77	6.17	5.31	4.76	4.37	4.09	3.86	3.69	3.54
90	.25	1.34	1.41	1.39	1.37	1.35	1.33	1.32	1.31	1.30	1.29
	.10	2.76	2.36	2.15	2.01	1.91	1.84	1.78	1.74	1.70	1.67
	.05	3.95	3.10	2.71	2.47	2.32	2.20	2.11	2.04	1.99	1.94
	.025	5.20	3.84	3.26	2.93	2.71	2.55	2.43	2.34	2.26	2.19
	.01	6.93	4.85	4.01	3.53	3.23	3.01	2.84	2.72	2.61	2.52
	.005	8.28	5.62	4.57	3.99	3.62	3.35	3.15	3.00	2.87	2.77
	.001	11.57	7.47	5.91	5.06	4.53	4.15	3.87	3.65	3.48	3.34
120	.25	1.34	1.40	1.39	1.37	1.35	1.33	1.31	1.30	1.29	1.28
	.10	2.75	2.35	2.13	1.99	1.90	1.82	1.77	1.72	1.68	1.65
	.05	3.92	3.07	2.68	2.45	2.29	2.18	2.09	2.02	1.96	1.91
	.025	5.15	3.80	3.23	2.89	2.67	2.52	2.39	2.30	2.22	2.16
	.01	6.85	4.79	3.95	3.48	3.17	2.96	2.79	2.66	2.56	2.47
	.005	8.18	5.54	4.50	3.92	3.55	3.28	3.09	2.93	2.81	2.71
	.001	11.38	7.32	5.78	4.95	4.42	4.04	3.77	3.55	3.38	3.24
240	.25	1.33	1.39	1.38	1.36	1.34	1.32	1.30	1.29	1.27	1.27
	.10	2.73	2.32	2.10	1.97	1.87	1.80	1.74	1.70	1.65	1.63
	.05	3.88	3.03	2.64	2.41	2.25	2.14	2.04	1.98	1.92	1.87
	.025	5.09	3.75	3.17	2.84	2.62	2.46	2.34	2.25	2.17	2.10
	.01	6.74	4.69	3.86	3.40	3.09	2.88	2.71	2.59	2.48	2.40
	.005	8.03	5.42	4.38	3.82	3.45	3.19	2.99	2.84	2.71	2.61
	.001	11.10	7.11	5.60	4.78	4.25	3.89	3.62	3.41	3.24	3.09
inf.	.25	1.32	1.39	1.37	1.35	1.33	1.31	1.29	1.28	1.27	1.25
	.10	2.71	2.30	2.08	1.94	1.85	1.77	1.72	1.67	1.63	1.60
	.05	3.84	3.00	2.60	2.37	2.21	2.10	2.01	1.94	1.88	1.83
	.025	5.02	3.69	3.12	2.79	2.57	2.41	2.29	2.19	2.11	2.05
	.01	6.63	4.61	3.78	3.32	3.02	2.80	2.64	2.51	2.41	2.32
	.005	7.88	5.30	4.28	3.72	3.35	3.09	2.90	2.74	2.62	2.52
	.001	10.83	6.91	5.42	4.62	4.10	3.74	3.47	3.27	3.10	2.96

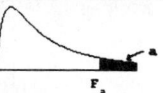

TABLE 6 (cont'd)

12	15	20	24	30	40	60	120	240	inf.	a	df$_2$
					df$_1$						
1.31	1.30	1.28	1.26	1.25	1.24	1.22	1.21	1.20	1.19	.25	40
1.71	1.66	1.61	1.57	1.54	1.51	1.47	1.42	1.40	1.38	.10	
2.00	1.92	1.84	1.79	1.74	1.69	1.64	1.58	1.54	1.51	.05	
2.29	2.18	2.07	2.01	1.94	1.88	1.80	1.72	1.68	1.64	.025	
2.66	2.52	2.37	2.29	2.20	2.11	2.02	1.92	1.86	1.80	.01	
2.95	2.78	2.60	2.50	2.40	2.30	2.18	2.06	2.00	1.93	.005	
3.64	3.40	3.14	3.01	2.87	2.73	2.57	2.41	2.32	2.23	.001	
1.29	1.27	1.25	1.24	1.22	1.21	1.19	1.17	1.16	1.15	.25	60
1.66	1.60	1.54	1.51	1.48	1.44	1.40	1.35	1.32	1.29	.10	
1.92	1.84	1.75	1.70	1.65	1.59	1.53	1.47	1.43	1.39	.05	
2.17	2.06	1.94	1.88	1.82	1.74	1.67	1.58	1.53	1.48	.025	
2.50	2.35	2.20	2.12	2.03	1.94	1.84	1.73	1.67	1.60	.01	
2.74	2.57	2.39	2.29	2.19	2.08	1.96	1.83	1.76	1.69	.005	
3.32	3.08	2.83	2.69	2.55	2.41	2.25	2.08	1.99	1.89	.001	
1.27	1.25	1.23	1.22	1.20	1.19	1.17	1.15	1.13	1.12	.25	90
1.62	1.56	1.50	1.47	1.43	1.39	1.35	1.29	1.26	1.23	.10	
1.86	1.78	1.69	1.64	1.59	1.53	1.46	1.39	1.35	1.30	.05	
2.09	1.98	1.86	1.80	1.73	1.66	1.58	1.48	1.43	1.37	.025	
2.39	2.24	2.09	2.00	1.92	1.82	1.72	1.60	1.53	1.46	.01	
2.61	2.44	2.25	2.15	2.05	1.94	1.82	1.68	1.61	1.52	.005	
3.11	2.88	2.63	2.50	2.36	2.21	2.05	1.87	1.77	1.66	.001	
1.26	1.24	1.22	1.21	1.19	1.18	1.16	1.13	1.12	1.10	.25	120
1.60	1.55	1.48	1.45	1.41	1.37	1.32	1.26	1.23	1.19	.10	
1.83	1.75	1.66	1.61	1.55	1.50	1.43	1.35	1.31	1.25	.05	
2.05	1.94	1.82	1.76	1.69	1.61	1.53	1.43	1.38	1.31	.025	
2.34	2.19	2.03	1.95	1.86	1.76	1.66	1.53	1.46	1.38	.01	
2.54	2.37	2.19	2.09	1.98	1.87	1.75	1.61	1.52	1.43	.005	
3.02	2.78	2.53	2.40	2.26	2.11	1.95	1.77	1.66	1.54	.001	
1.25	1.23	1.21	1.19	1.18	1.16	1.14	1.11	1.09	1.07	.25	240
1.57	1.52	1.45	1.42	1.38	1.33	1.28	1.22	1.18	1.13	.10	
1.79	1.71	1.61	1.56	1.51	1.44	1.37	1.29	1.24	1.17	.05	
2.00	1.89	1.77	1.70	1.63	1.55	1.46	1.35	1.29	1.21	.025	
2.26	2.11	1.96	1.87	1.78	1.68	1.57	1.43	1.35	1.25	.01	
2.45	2.28	2.09	1.99	1.89	1.77	1.64	1.49	1.40	1.28	.005	
2.88	2.65	2.40	2.26	2.12	1.97	1.80	1.61	1.49	1.35	.001	
1.24	1.22	1.19	1.18	1.16	1.14	1.12	1.08	1.06	1.00	.25	inf.
1.55	1.49	1.42	1.38	1.34	1.30	1.24	1.17	1.12	1.00	.10	
1.75	1.67	1.57	1.52	1.46	1.39	1.32	1.22	1.15	1.00	.05	
1.94	1.83	1.71	1.64	1.57	1.48	1.39	1.27	1.19	1.00	.025	
2.18	2.04	1.88	1.79	1.70	1.59	1.47	1.32	1.22	1.00	.01	
2.36	2.19	2.00	1.90	1.79	1.67	1.53	1.36	1.25	1.00	.005	
2.74	2.51	2.27	2.13	1.99	1.84	1.66	1.45	1.31	1.00	.001	

Source: Computed by P. J. Hildebrand.

TABLE 7   Poisson Probabilities (μ Between .1 and 4.0)

y	.1	.2	.3	.4	μ .5	.6	.7	.8	.9	1.0
0	.9048	.8187	.7408	.6703	.6065	.5488	.4966	.4493	.4066	.3679
1	.0905	.1637	.2222	.2681	.3033	.3293	.3476	.3595	.3659	.3679
2	.0045	.0164	.0333	.0536	.0758	.0988	.1217	.1438	.1647	.1839
3	.0002	.0011	.0033	.0072	.0126	.0198	.0284	.0383	.0494	.0613
4	.0000	.0001	.0003	.0007	.0016	.0030	.0050	.0077	.0111	.0153
5	.0000	.0000	.0000	.0001	.0002	.0004	.0007	.0012	.0020	.0031
6	.0000	.0000	.0000	.0000	.0000	.0000	.0001	.0002	.0003	.0005

y	1.1	1.2	1.3	1.4	μ 1.5	1.6	1.7	1.8	1.9	2.0
0	.3329	.3012	.2725	.2466	.2231	.2019	.1827	.1653	.1496	.1353
1	.3662	.3614	.3543	.3452	.3347	.3230	.3106	.2975	.2842	.2707
2	.2014	.2169	.2303	.2417	.2510	.2584	.2640	.2678	.2700	.2707
3	.0738	.0867	.0998	.1128	.1255	.1378	.1496	.1607	.1710	.1804
4	.0203	.0260	.0324	.0395	.0471	.0551	.0636	.0723	.0812	.0902
5	.0045	.0062	.0084	.0111	.0141	.0176	.0216	.0260	.0309	.0361
6	.0008	.0012	.0018	.0026	.0035	.0047	.0061	.0078	.0098	.0120
7	.0001	.0002	.0003	.0005	.0008	.0011	.0015	.0020	.0027	.0034
8	.0000	.0000	.0001	.0001	.0001	.0002	.0003	.0005	.0006	.0009

y	2.1	2.2	2.3	2.4	μ 2.5	2.6	2.7	2.8	2.9	3.0
0	.1225	.1108	.1003	.0907	.0821	.0743	.0672	.0608	.0550	.0498
1	.2572	.2438	.2306	.2177	.2052	.1931	.1815	.1703	.1596	.1494
2	.2700	.2681	.2652	.2613	.2565	.2510	.2450	.2384	.2314	.2240
3	.1890	.1966	.2033	.2090	.2138	.2176	.2205	.2225	.2237	.2240
4	.0992	.1082	.1169	.1254	.1336	.1414	.1488	.1557	.1622	.1680
5	.0417	.0476	.0538	.0602	.0668	.0735	.0804	.0872	.0940	.1008
6	.0146	.0174	.0206	.0241	.0278	.0319	.0362	.0407	.0455	.0504
7	.0044	.0055	.0068	.0083	.0099	.0118	.0139	.0163	.0188	.0216
8	.0011	.0015	.0019	.0025	.0031	.0038	.0047	.0057	.0068	.0081
9	.0003	.0004	.0005	.0007	.0009	.0011	.0014	.0018	.0022	.0027
10	.0001	.0001	.0001	.0002	.0002	.0003	.0004	.0005	.0006	.0008
11	.0000	.0000	.0000	.0000	.0000	.0001	.0001	.0001	.0002	.0002

y	3.1	3.2	3.3	3.4	μ 3.5	3.6	3.7	3.8	3.9	4.0
0	.0450	.0408	.0369	.0334	.0302	.0273	.0247	.0224	.0202	.0183
1	.1397	.1304	.1217	.1135	.1057	.0984	.0915	.0850	.0789	.0733
2	.2165	.2087	.2008	.1929	.1850	.1771	.1692	.1615	.1539	.1465
3	.2237	.2226	.2209	.2186	.2158	.2125	.2087	.2046	.2001	.1954
4	.1733	.1781	.1823	.1858	.1888	.1912	.1931	.1944	.1951	.1954
5	.1075	.1140	.1203	.1264	.1322	.1377	.1429	.1477	.1522	.1563
6	.0555	.0608	.0662	.0716	.0771	.0826	.0881	.0936	.0989	.1042

APPENDIX   STATISTICAL TABLES

TABLE 7 (cont'd)   Poisson Probabilities (μ Between 3.1 and 10.0)

y	3.1	3.2	3.3	3.4	3.5	3.6	3.7	3.8	3.9	4.0
7	.0246	.0278	.0312	.0348	.0385	.0425	.0466	.0508	.0551	.0595
8	.0095	.0111	.0129	.0148	.0169	.0191	.0215	.0241	.0269	.0298
9	.0033	.0040	.0047	.0056	.0066	.0076	.0089	.0102	.0116	.0132
10	.0010	.0013	.0016	.0019	.0023	.0028	.0033	.0039	.0045	.0053
11	.0003	.0004	.0005	.0006	.0007	.0009	.0011	.0013	.0016	.0019
12	.0001	.0001	.0001	.0002	.0002	.0003	.0003	.0004	.0005	.0006
13	.0000	.0000	.0000	.0000	.0001	.0001	.0001	.0001	.0002	.0002

y	4.1	4.2	4.3	4.4	4.5	4.6	4.7	4.8	4.9	5.0
0	.0166	.0150	.0136	.0123	.0111	.0101	.0091	.0082	.0074	.0067
1	.0679	.0630	.0583	.0540	.0500	.0462	.0427	.0395	.0365	.0337
2	.1393	.1323	.1254	.1188	.1125	.1063	.1005	.0948	.0894	.0842
3	.1904	.1852	.1798	.1743	.1687	.1631	.1574	.1517	.1460	.1404
4	.1951	.1944	.1933	.1917	.1898	.1875	.1849	.1820	.1789	.1755
5	.1600	.1633	.1662	.1687	.1708	.1725	.1738	.1747	.1753	.1755
6	.1093	.1143	.1191	.1237	.1281	.1323	.1362	.1398	.1432	.1462
7	.0640	.0686	.0732	.0778	.0824	.0869	.0914	.0959	.1002	.1044
8	.0328	.0360	.0393	.0428	.0463	.0500	.0537	.0575	.0614	.0653
9	.0150	.0168	.0188	.0209	.0232	.0255	.0281	.0307	.0334	.0363
10	.0061	.0071	.0081	.0092	.0104	.0118	.0132	.0147	.0164	.0181
11	.0023	.0027	.0032	.0037	.0043	.0049	.0056	.0064	.0073	.0082
12	.0008	.0009	.0011	.0013	.0016	.0019	.0022	.0026	.0030	.0034
13	.0002	.0003	.0004	.0005	.0006	.0007	.0008	.0009	.0011	.0013
14	.0001	.0001	.0001	.0001	.0002	.0002	.0003	.0003	.0004	.0005
15	.0000	.0000	.0000	.0000	.0001	.0001	.0001	.0001	.0001	.0002

y	5.5	6.0	6.5	7.0	7.5	8.0	8.5	9.0	9.5	10.0
0	.0041	.0025	.0015	.0009	.0006	.0003	.0002	.0001	.0001	.0000
1	.0225	.0149	.0098	.0064	.0041	.0027	.0017	.0011	.0007	.0005
2	.0618	.0446	.0318	.0223	.0156	.0107	.0074	.0050	.0034	.0023
3	.1133	.0892	.0688	.0521	.0389	.0286	.0208	.0150	.0107	.0076
4	.1558	.1339	.1118	.0912	.0729	.0573	.0443	.0337	.0254	.0189
5	.1714	.1606	.1454	.1277	.1094	.0916	.0752	.0607	.0483	.0378
6	.1571	.1606	.1575	.1490	.1367	.1221	.1066	.0911	.0764	.0631
7	.1234	.1377	.1462	.1490	.1465	.1396	.1294	.1171	.1037	.0901
8	.0849	.1033	.1188	.1304	.1373	.1396	.1375	.1318	.1232	.1126
9	.0519	.0688	.0858	.1014	.1144	.1241	.1299	.1318	.1300	.1251
10	.0285	.0413	.0558	.0710	.0858	.0993	.1104	.1186	.1235	.1251
11	.0143	.0225	.0330	.0452	.0585	.0722	.0853	.0970	.1067	.1137
12	.0065	.0113	.0179	.0263	.0366	.0481	.0604	.0728	.0844	.0948
13	.0028	.0052	.0089	.0142	.0211	.0296	.0395	.0504	.0617	.0729
14	.0011	.0022	.0041	.0071	.0113	.0169	.0240	.0324	.0419	.0521
15	.0004	.0009	.0018	.0033	.0057	.0090	.0136	.0194	.0265	.0347

TABLE 7 (cont'd) Poisson Probabilities (μ Between 5.5 and 20.0)

y	5.5	6.0	6.5	7.0	7.5	8.0	8.5	9.0	9.5	10.0
16	.0001	.0003	.0007	.0014	.0026	.0045	.0072	.0109	.0157	.0217
17	.0000	.0001	.0003	.0006	.0012	.0021	.0036	.0058	.0088	.0128
18	.0000	.0000	.0001	.0002	.0005	.0009	.0017	.0029	.0046	.0071
19	.0000	.0000	.0000	.0001	.0002	.0004	.0008	.0014	.0023	.0037
20	.0000	.0000	.0000	.0000	.0001	.0002	.0003	.0006	.0011	.0019
21	.0000	.0000	.0000	.0000	.0000	.0001	.0001	.0003	.0005	.0009
22	.0000	.0000	.0000	.0000	.0000	.0000	.0001	.0001	.0002	.0004
23	.0000	.0000	.0000	.0000	.0000	.0000	.0000	.0000	.0001	.0002

y	11.0	12.0	13.0	14.0	15.0	16.0	17.0	18.0	19.0	20.0
0	.0000	.0000	.0000	.0000	.0000	.0000	.0000	.0000	.0000	.0000
1	.0002	.0001	.0000	.0000	.0000	.0000	.0000	.0000	.0000	.0000
2	.0010	.0004	.0002	.0001	.0000	.0000	.0000	.0000	.0000	.0000
3	.0037	.0018	.0008	.0004	.0002	.0001	.0000	.0000	.0000	.0000
4	.0102	.0053	.0027	.0013	.0006	.0003	.0001	.0001	.0000	.0000
5	.0224	.0127	.0070	.0037	.0019	.0010	.0005	.0002	.0001	.0001
6	.0411	.0255	.0152	.0087	.0048	.0026	.0014	.0007	.0004	.0002
7	.0646	.0437	.0281	.0174	.0104	.0060	.0034	.0019	.0010	.0005
8	.0888	.0655	.0457	.0304	.0194	.0120	.0072	.0042	.0024	.0013
9	.1085	.0874	.0661	.0473	.0324	.0213	.0135	.0083	.0050	.0029
10	.1194	.1048	.0859	.0663	.0486	.0341	.0230	.0150	.0095	.0058
11	.1194	.1144	.1015	.0844	.0663	.0496	.0355	.0245	.0164	.0106
12	.1094	.1144	.1099	.0984	.0829	.0661	.0504	.0368	.0259	.0176
13	.0926	.1056	.1099	.1060	.0956	.0814	.0658	.0509	.0378	.0271
14	.0728	.0905	.1021	.1060	.1024	.0930	.0800	.0655	.0514	.0387
15	.0534	.0724	.0885	.0989	.1024	.0992	.0906	.0786	.0650	.0516
16	.0367	.0543	.0719	.0866	.0960	.0992	.0963	.0884	.0772	.0646
17	.0237	.0383	.0550	.0713	.0847	.0934	.0963	.0936	.0863	.0760
18	.0145	.0255	.0397	.0554	.0706	.0830	.0909	.0936	.0911	.0844
19	.0084	.0161	.0272	.0409	.0557	.0699	.0814	.0887	.0911	.0888
20	.0046	.0097	.0177	.0286	.0418	.0559	.0692	.0798	.0866	.0888
21	.0024	.0055	.0109	.0191	.0299	.0426	.0560	.0684	.0783	.0846
22	.0012	.0030	.0065	.0121	.0204	.0310	.0433	.0560	.0676	.0769
23	.0006	.0016	.0037	.0074	.0133	.0216	.0320	.0438	.0559	.0669
24	.0003	.0008	.0020	.0043	.0083	.0144	.0226	.0328	.0442	.0557
25	.0001	.0004	.0010	.0024	.0050	.0092	.0154	.0237	.0336	.0446
26	.0000	.0002	.0005	.0013	.0029	.0057	.0101	.0164	.0246	.0343
27	.0000	.0001	.0002	.0007	.0016	.0034	.0063	.0109	.0173	.0254
28	.0000	.0000	.0001	.0003	.0009	.0019	.0038	.0070	.0117	.0181
29	.0000	.0000	.0001	.0002	.0004	.0011	.0023	.0044	.0077	.0125
30	.0000	.0000	.0000	.0001	.0002	.0006	.0013	.0026	.0049	.0083
31	.0000	.0000	.0000	.0000	.0001	.0003	.0007	.0015	.0030	.0054
32	.0000	.0000	.0000	.0000	.0001	.0001	.0004	.0009	.0018	.0034
33	.0000	.0000	.0000	.0000	.0000	.0001	.0002	.0005	.0010	.0020

Source: Computed by D. K. Hildebrand.

TABLE 8   Random Numbers

Line/Col.	(1)	(2)	(3)	(4)	(5)	(6)	(7)	(8)	(9)	(10)	(11)	(12)	(13)	(14)
1	10480	15011	01536	02011	81647	91646	69179	14194	62590	36207	20969	99570	91291	90700
2	22368	46573	25595	85393	30995	89198	27982	53402	93965	34095	52666	19174	39615	99505
3	24130	48360	22527	97265	76393	64809	15179	24830	49340	32081	30680	19655	63348	58629
4	42167	93093	06243	61680	07856	16376	39440	53537	71341	57004	00849	74917	97758	16379
5	37570	39975	81837	16656	06121	91782	60468	81305	49684	60672	14110	06927	01263	54613
6	77921	06907	11008	42751	27756	53498	18602	70659	90655	15053	21916	81825	44394	42880
7	99562	72905	56420	69994	98872	31016	71194	18738	44013	48840	63213	21069	10634	12952
8	96301	91977	05463	07972	18876	20922	94595	56869	69014	60045	18425	84903	42508	32307
9	89579	14342	63661	10281	17453	18103	57740	84378	25331	12566	58678	44947	05585	56941
10	85475	36857	53342	53988	53060	59533	38867	62300	08158	17983	16439	11458	18593	64952
11	28918	69578	88231	33276	70997	79936	56865	05859	90106	31595	01547	85590	91610	78188
12	63553	40961	48235	03427	49626	69445	18663	72695	52180	20847	12234	90511	33703	90322
13	09429	93969	52636	92737	88974	33488	36320	17617	30015	08272	84115	27156	30613	74952
14	10365	61129	87529	85689	48237	52267	67689	93394	01511	26358	85104	20285	29975	89868
15	07119	97336	71048	08178	77233	13916	47564	81056	97735	85977	29372	74461	28551	90707
16	51085	12765	51821	51259	77452	16308	60756	92144	49442	53900	70960	63990	75601	40719
17	02368	21382	52404	60268	89368	19885	55322	44819	01188	65255	64835	44919	05944	55157
18	01011	54092	33362	94904	31273	04146	18594	29852	71585	85030	51132	01915	92747	64951
19	52162	53916	46369	58586	23216	14513	83149	98736	23495	64350	94738	17752	35156	35749
20	07056	97628	33787	09998	42698	06691	76988	13602	51851	46104	88916	19509	25625	58104
21	48663	91245	85828	14346	09172	30168	90229	04734	59193	22178	30421	61666	99904	32812
22	54164	58492	22421	74103	47070	25306	76468	26384	58151	06646	21524	15227	96909	44592
23	32639	32363	05597	24200	13363	38005	94342	28728	35806	06912	17012	64161	18296	22851
24	29334	27001	87637	87308	58731	00256	45834	15398	46557	41135	10367	07684	36188	18510
25	02488	33062	28834	07351	19731	92420	60952	61280	50001	67658	32586	86679	50720	94953

Abridged from William H. Beyer, ed., *Handbook of Tables for Probability and Statistics,* 2nd ed. © The Chemical Rubber Co., 1968. Used by permission of CRC Press, Inc.

TABLE 9 Critical Values for the Wilcoxon Signed-rank Test [$n = 5(1)54$]

One-Sided	Two-Sided	$n = 5$	$n = 6$	$n = 7$	$n = 8$	$n = 9$
$p = .1$	$p = .2$	2	3	5	8	10
$p = .05$	$p = .1$	0	2	3	5	8
$p = .025$	$p = .05$		0	2	3	5
$p = .01$	$p = .02$			0	1	3
$p = .005$	$p = .01$				0	1
$p = .0025$	$p = .005$					0
$p = .001$	$p = .002$					

One-Sided	Two-Sided	$n = 15$	$n = 16$	$n = 17$	$n = 18$	$n = 19$
$p = .1$	$p = .2$	36	42	48	55	62
$p = .05$	$p = .1$	30	35	41	47	53
$p = .025$	$p = .05$	25	29	34	40	46
$p = .01$	$p = .02$	19	23	27	32	37
$p = .005$	$p = .01$	15	19	23	27	32
$p = .0025$	$p = .005$	12	15	19	23	27
$p = .001$	$p = .002$	8	11	14	18	21

One-Sided	Two-Sided	$n = 25$	$n = 26$	$n = 27$	$n = 28$	$n = 29$
$p = .1$	$p = .2$	113	124	134	145	157
$p = .05$	$p = .1$	100	110	119	130	140
$p = .025$	$p = .05$	89	98	107	116	126
$p = .01$	$p = .02$	76	84	92	101	110
$p = .005$	$p = .01$	68	75	83	91	100
$p = .0025$	$p = .005$	60	67	74	82	90
$p = .001$	$p = .002$	51	58	64	71	79

One-Sided	Two-Sided	$n = 35$	$n = 36$	$n = 37$	$n = 38$	$n = 39$
$p = .1$	$p = .2$	235	250	265	281	297
$p = .05$	$p = .1$	213	227	241	256	271
$p = .025$	$p = .05$	195	208	221	235	249
$p = .01$	$p = .02$	173	185	198	211	224
$p = .005$	$p = .01$	159	171	182	194	207
$p = .0025$	$p = .005$	146	157	168	180	192
$p = .001$	$p = .002$	131	141	151	162	173

One-Sided	Two-Sided	$n = 45$	$n = 46$	$n = 47$	$n = 48$	$n = 49$
$p = .1$	$p = .2$	402	422	441	462	482
$p = .05$	$p = .1$	371	389	407	426	446
$p = .025$	$p = .05$	343	361	378	396	415
$p = .01$	$p = .02$	312	328	345	362	379
$p = .005$	$p = .01$	291	307	322	339	355
$p = .0025$	$p = .005$	272	287	302	318	334
$p = .001$	$p = .002$	249	263	277	292	307

Source: Computed by P. J. Hildebrand.

TABLE 9 (cont'd)

One-Sided	Two-Sided	$n = 10$	$n = 11$	$n = 12$	$n = 13$	$n = 14$
$p = .1$	$p = .2$	14	17	21	26	31
$p = .05$	$p = .1$	10	13	17	21	25
$p = .025$	$p = .05$	8	10	13	17	21
$p = .01$	$p = .02$	5	7	9	12	15
$p = .005$	$p = .01$	3	5	7	9	12
$p = .0025$	$p = .005$	1	3	5	7	9
$p = .001$	$p = .002$	0	1	2	4	6

One-Sided	Two-Sided	$n = 20$	$n = 21$	$n = 22$	$n = 23$	$n = 24$
$p = .1$	$p = .2$	69	77	86	94	104
$p = .05$	$p = .1$	60	67	75	83	91
$p = .025$	$p = .05$	52	58	65	73	81
$p = .01$	$p = .02$	43	49	55	62	69
$p = .005$	$p = .01$	37	42	48	54	61
$p = .0025$	$p = .005$	32	37	42	48	54
$p = .001$	$p = .002$	26	30	35	40	45

One-Sided	Two-Sided	$n = 30$	$n = 31$	$n = 32$	$n = 33$	$n = 34$
$p = .1$	$p = .2$	169	181	194	207	221
$p = .05$	$p = .1$	151	163	175	187	200
$p = .025$	$p = .05$	137	147	159	170	182
$p = .01$	$p = .02$	120	130	140	151	162
$p = .005$	$p = .01$	109	118	128	138	148
$p = .0025$	$p = .005$	98	107	116	126	136
$p = .001$	$p = .002$	86	94	103	112	121

One-Sided	Two-Sided	$n = 40$	$n = 41$	$n = 42$	$n = 43$	$n = 44$
$p = .1$	$p = .2$	313	330	348	365	384
$p = .05$	$p = .1$	286	302	319	336	353
$p = .025$	$p = .05$	264	279	294	310	327
$p = .01$	$p = .02$	238	252	266	281	296
$p = .005$	$p = .01$	220	233	247	261	276
$p = .0025$	$p = .005$	204	217	230	244	258
$p = .001$	$p = .002$	185	197	209	222	235

One-Sided	Two-Sided	$n = 50$	$n = 51$	$n = 52$	$n = 53$	$n = 54$
$p = .1$	$p = .2$	503	525	547	569	592
$p = .05$	$p = .1$	466	486	507	529	550
$p = .025$	$p = .05$	434	453	473	494	514
$p = .01$	$p = .02$	397	416	434	454	473
$p = .005$	$p = .01$	373	390	408	427	445
$p = .0025$	$p = .005$	350	367	384	402	420
$p = .001$	$p = .002$	323	339	355	372	389

TABLE 10   Percentage Points of the Studentized Range

Error df	α	\multicolumn{10}{c}{t = number of treatment means}
		2	3	4	5	6	7	8	9	10	11
5	.05	3.64	4.60	5.22	5.67	6.03	6.33	6.58	6.80	6.99	7.17
	.01	5.70	6.98	7.80	8.42	8.91	9.32	9.67	9.97	10.24	10.48
6	.05	3.46	4.34	4.90	5.30	5.63	5.90	6.12	6.32	6.49	6.65
	.01	5.24	6.33	7.03	7.56	7.97	8.32	8.61	8.87	9.10	9.30
7	.05	3.34	4.16	4.68	5.06	5.36	5.61	5.82	6.00	6.16	6.30
	.01	4.95	5.92	6.54	7.01	7.37	7.68	7.94	8.17	8.37	8.55
8	.05	3.26	4.04	4.53	4.89	5.17	5.40	5.60	5.77	5.92	6.05
	.01	4.75	5.64	6.20	6.62	6.96	7.24	7.47	7.68	7.86	8.03
9	.05	3.20	3.95	4.41	4.76	5.02	5.24	5.43	5.59	5.74	5.87
	.01	4.60	5.43	5.96	6.35	6.66	6.91	7.13	7.33	7.49	7.65
10	.05	3.15	3.88	4.33	4.65	4.91	5.12	5.30	5.46	5.60	5.72
	.01	4.48	5.27	5.77	6.14	6.43	6.67	6.87	7.05	7.21	7.36
11	.05	3.11	3.82	4.26	4.57	4.82	5.03	5.30	5.35	5.49	5.61
	.01	4.39	5.15	5.62	5.97	6.25	6.48	6.67	6.84	6.99	7.13
12	.05	3.08	3.77	4.20	4.52	4.75	4.95	5.12	5.27	5.39	5.51
	.01	4.32	5.05	5.50	5.84	6.10	6.32	6.51	6.67	6.81	6.94
13	.05	3.06	3.73	4.15	4.45	4.69	4.88	5.05	5.19	5.32	5.43
	.01	4.26	4.96	5.40	5.73	5.98	6.19	6.37	6.53	6.67	6.79
14	.05	3.03	3.70	4.11	4.41	4.64	4.83	4.99	5.13	5.25	5.36
	.01	4.21	4.89	5.32	5.63	5.88	6.08	6.26	6.41	6.54	6.66
15	.05	3.01	3.67	4.08	4.37	4.59	4.78	4.94	5.08	5.20	5.31
	.01	4.17	4.84	5.25	5.56	5.80	5.99	6.16	6.31	6.44	6.55
16	.05	3.00	3.65	4.05	4.33	4.56	4.74	4.90	5.03	5.15	5.26
	.01	4.13	4.79	5.19	5.49	5.72	5.92	6.08	6.22	6.35	6.46
17	.05	2.98	3.63	4.02	4.30	4.52	4.70	4.86	4.99	5.11	5.21
	.01	4.10	4.74	5.14	5.43	5.66	5.85	6.01	6.15	6.27	6.38
18	.05	2.97	3.61	4.00	4.28	4.49	4.67	4.82	4.96	5.07	5.17
	.01	4.07	4.70	5.09	5.38	5.60	5.79	5.94	6.08	6.20	6.31
19	.05	2.96	3.59	3.98	4.25	4.47	4.65	4.79	4.92	5.04	5.14
	.01	4.05	4.67	5.05	5.33	5.55	5.73	5.89	6.02	6.14	6.25
20	.05	2.95	3.58	3.96	4.23	4.45	4.62	4.77	4.90	5.01	5.11
	.01	4.02	4.64	5.02	5.29	5.51	5.69	5.84	5.97	6.09	6.19
24	.05	2.92	3.53	3.90	4.17	4.37	4.54	4.68	4.81	3.92	5.01
	.01	3.96	4.55	4.91	5.17	5.37	5.54	5.69	5.81	5.92	6.02
30	.05	2.89	3.49	3.85	4.10	4.30	4.46	4.60	4.72	4.82	4.92
	.01	3.89	4.45	4.80	5.05	5.24	5.40	5.54	5.65	5.76	5.85
40	.05	2.86	3.44	3.79	4.04	4.23	4.39	4.52	4.63	4.73	4.82
	.01	3.82	4.37	4.70	4.93	5.11	5.26	5.39	5.50	5.60	5.69
60	.05	2.83	3.40	3.74	3.98	4.16	4.31	4.44	4.55	4.65	4.73
	.01	3.76	4.28	4.59	4.82	4.99	5.13	5.25	5.36	5.45	5.53
120	.05	2.80	3.36	3.68	3.92	4.10	4.24	4.36	4.47	4.56	4.64
	.01	3.70	4.20	4.50	4.71	4.87	5.01	5.12	5.21	5.30	5.37
∞	.05	2.77	3.31	3.63	3.86	4.03	4.17	4.29	4.39	4.47	4.55
	.01	3.64	4.12	4.40	4.60	4.76	4.88	4.99	5.08	5.16	5.23

TABLE 10 (cont'd)

12	13	14	15	16	17	18	19	20	α	Error df
				t = number of treatment means						
7.32	7.47	7.60	7.72	7.83	7.93	8.03	8.12	8.21	.05	5
10.70	10.89	11.08	11.24	11.40	11.55	11.68	11.81	11.93	.01	
6.79	6.92	7.03	7.14	7.24	7.34	7.43	7.51	7.59	.05	6
9.48	9.65	9.81	9.95	10.08	10.21	10.32	10.43	10.54	.01	
6.43	6.55	6.66	6.76	6.85	6.94	7.02	7.10	7.17	.05	7
8.71	8.86	9.00	9.12	9.24	9.35	9.46	9.55	9.65	.01	
6.18	6.29	6.39	6.48	6.57	6.65	6.73	6.80	6.87	.05	8
8.18	8.31	8.44	8.55	8.66	8.76	8.85	8.94	9.03	.01	
5.98	6.09	6.19	6.28	6.36	6.44	6.51	6.58	6.64	.05	9
7.78	7.91	8.03	8.13	8.23	8.33	8.41	8.49	8.57	.01	
5.83	5.93	6.03	6.11	6.19	6.27	6.34	6.40	6.47	.05	10
7.49	7.60	7.71	7.81	7.91	7.99	8.08	8.15	8.23	.01	
5.71	5.81	5.90	5.98	6.06	6.13	6.20	6.27	6.33	.05	11
7.25	7.36	7.46	7.56	7.65	7.73	7.81	7.88	7.95	.01	
5.61	5.71	5.80	5.88	5.95	6.02	6.09	6.15	6.21	.05	12
7.06	7.17	7.26	7.36	7.44	7.52	7.59	7.66	7.73	.01	
5.53	5.63	5.71	5.79	5.86	5.93	5.99	6.05	6.11	.05	13
6.90	7.01	7.10	7.19	7.27	7.35	7.42	7.48	7.55	.01	
5.46	5.55	5.64	5.71	5.79	5.85	5.91	5.97	6.03	.05	14
6.77	6.87	6.96	7.05	7.13	7.20	7.27	7.33	7.39	.01	
5.40	5.49	5.57	5.65	5.72	5.78	5.85	5.90	5.96	.05	15
6.66	6.76	6.84	6.93	7.00	7.07	7.14	7.20	7.26	.01	
5.35	5.44	5.52	5.59	5.66	5.73	5.79	5.84	5.90	.05	16
6.56	6.66	6.74	6.82	6.90	6.97	7.03	7.09	7.15	.01	
5.31	5.39	5.47	5.54	5.61	5.67	5.73	5.79	5.84	.05	17
6.48	6.57	6.66	6.73	6.81	6.87	6.94	7.00	7.05	.01	
5.27	5.35	5.43	5.50	5.57	5.63	5.69	5.74	5.79	.05	18
6.41	6.50	6.58	6.65	6.73	6.79	6.85	6.91	6.97	.01	
5.23	5.31	5.39	5.46	5.53	5.59	5.65	5.70	5.75	.05	19
6.34	6.43	6.51	6.58	6.65	6.72	6.78	6.84	6.89	.01	
5.20	5.28	5.36	5.43	5.49	5.55	5.61	5.66	5.71	.05	20
6.28	6.37	6.45	6.52	6.59	6.65	6.71	6.77	6.82	.01	
5.10	5.18	5.25	5.32	5.38	5.44	5.49	5.55	5.59	.05	24
6.11	6.19	6.26	6.33	6.39	6.45	6.51	6.56	6.61	.01	
5.00	5.08	5.15	5.21	5.27	5.33	5.38	5.43	5.47	.05	30
5.93	6.01	6.08	6.14	6.20	6.26	6.31	6.36	6.41	.01	
4.90	4.98	5.04	5.11	5.16	5.22	5.27	5.31	5.36	.05	40
5.76	5.83	5.90	5.96	6.02	6.07	6.12	6.16	6.21	.01	
4.81	4.88	4.94	5.00	5.06	5.11	5.15	5.20	5.24	.05	60
5.60	5.67	5.73	5.78	5.84	5.89	5.93	5.97	6.01	.01	
4.71	4.78	4.84	4.90	4.95	5.00	5.04	5.09	5.13	.05	120
5.44	5.50	5.56	5.61	5.66	5.71	5.75	5.79	5.83	.01	
4.62	4.68	4.74	4.80	4.85	4.89	4.93	4.97	5.01	.05	∞
5.29	5.35	5.40	5.45	5.49	5.54	5.57	5.61	5.65	.01	

TABLE 11   Percentage Points of the Duncan New Multiple Range Test

Error df	α	2	3	4	5	6	7	8	9	10	12	14	16	18	20
1	.05	18.0	18.0	18.0	18.0	18.0	18.0	18.0	18.0	18.0	18.0	18.0	18.0	18.0	18.0
	.01	90.0	90.0	90.0	90.0	90.0	90.0	90.0	90.0	90.0	90.0	90.0	90.0	90.0	90.0
2	.05	6.09	6.09	6.09	6.09	6.09	6.09	6.09	6.09	6.09	6.09	6.09	6.09	6.09	6.09
	.01	14.0	14.0	14.0	14.0	14.0	14.0	14.0	14.0	14.0	14.0	14.0	14.0	14.0	14.0
3	.05	4.50	4.50	4.50	4.50	4.50	4.50	4.50	4.50	4.50	4.50	4.50	4.50	4.50	4.50
	.01	8.26	8.5	8.6	8.7	8.8	8.9	8.9	9.0	9.0	9.0	9.1	9.2	9.3	9.3
4	.05	3.93	4.01	4.02	4.02	4.02	4.02	4.02	4.02	4.02	4.02	4.02	4.02	4.02	4.02
	.01	6.51	6.8	6.9	7.0	7.1	7.1	7.2	7.2	7.3	7.3	7.4	7.4	7.5	7.5
5	.05	3.64	3.74	3.79	3.83	3.83	3.83	3.83	3.83	3.83	3.83	3.83	3.83	3.83	3.83
	.01	5.70	5.96	6.11	6.18	6.26	6.33	6.40	6.44	6.5	6.6	6.6	6.7	6.7	6.8
6	.05	3.46	3.58	3.64	3.68	3.68	3.68	3.68	3.68	3.68	3.68	3.68	3.68	3.68	3.68
	.01	5.24	5.51	5.65	5.73	5.83	5.81	5.95	6.00	6.0	6.1	6.2	6.2	6.3	6.3
7	.05	3.35	3.47	3.54	3.58	3.60	3.61	3.61	3.61	3.61	3.61	3.61	3.61	3.61	3.61
	.01	4.95	5.22	5.37	5.45	5.53	5.61	5.69	5.73	5.8	5.8	5.9	5.9	6.0	6.0
8	.05	3.26	3.39	3.47	3.52	3.55	3.56	3.56	3.56	3.56	3.56	3.56	3.56	3.56	3.56
	.01	4.74	5.00	5.14	5.23	5.32	5.40	5.47	5.51	5.5	5.6	5.7	5.7	5.8	5.8
9	.05	3.20	3.34	3.41	3.47	3.50	3.52	3.52	3.52	3.52	3.52	3.52	3.52	3.52	3.52
	.01	4.60	4.86	4.99	5.08	5.17	5.25	5.32	5.36	5.4	5.5	5.5	5.6	5.7	5.7
10	.05	3.15	3.30	3.37	3.43	3.46	3.47	3.47	3.47	3.47	3.47	3.47	3.47	3.47	3.48
	.01	4.48	4.73	4.88	4.96	5.06	5.13	5.20	5.24	5.28	5.36	5.42	5.48	5.54	5.55
11	.05	3.11	3.27	3.35	3.39	3.43	3.44	3.45	3.46	3.46	3.46	3.46	3.46	3.47	3.48
	.01	4.39	4.63	4.77	4.86	4.94	5.01	5.06	5.12	5.15	5.24	5.28	5.34	5.38	5.39
12	.05	3.08	3.23	3.33	3.36	3.40	3.42	3.44	3.44	3.46	3.46	3.46	3.46	3.47	3.48
	.01	4.32	4.55	4.68	4.76	4.84	4.92	4.96	5.02	5.07	5.13	5.17	5.22	5.23	5.26
13	.05	3.06	3.21	3.30	3.35	3.38	3.41	3.42	3.44	3.45	3.45	3.46	3.46	3.47	3.47
	.01	4.26	4.48	4.62	4.69	4.74	4.84	4.88	4.94	4.98	5.04	5.08	5.13	5.14	5.15
14	.05	3.03	3.18	3.27	3.33	3.37	3.39	3.41	3.42	3.44	3.45	3.46	3.46	3.47	3.47
	.01	4.21	4.42	4.55	4.63	4.70	4.78	4.83	4.87	4.91	4.96	5.00	5.04	5.06	5.07
15	.05	3.01	3.16	3.25	3.31	3.36	3.38	3.40	3.42	3.43	3.44	3.45	3.46	3.47	3.47
	.01	4.17	4.37	4.50	4.58	4.64	4.72	4.77	4.81	4.84	4.90	4.94	4.97	4.99	5.00

The column group header reads: r = number of ordered steps between means

TABLE 11 (cont'd)

| Error | | \ r = number of ordered steps between means | | | | | | | | | | | | | |
df	α	2	3	4	5	6	7	8	9	10	12	14	16	18	20
16	.05	3.00	3.15	3.23	3.30	3.34	3.37	3.39	3.41	3.43	3.44	3.45	3.46	3.47	3.47
	.01	4.13	4.34	4.45	4.54	4.60	4.67	4.72	4.76	4.79	4.84	4.88	4.91	4.93	4.94
17	.05	2.98	3.13	3.22	3.28	3.33	3.36	3.38	3.40	3.42	3.44	3.45	3.46	3.47	3.47
	.01	4.10	4.30	4.41	4.50	4.56	4.63	4.68	4.72	4.75	4.80	4.83	4.86	4.88	4.89
18	.05	2.97	3.12	3.21	3.27	3.32	3.35	3.37	3.39	3.41	3.43	3.45	3.46	3.47	3.47
	.01	4.07	4.27	4.38	4.46	4.53	4.59	4.64	4.68	4.71	4.76	4.79	4.82	4.84	4.85
19	.05	2.96	3.11	3.19	3.26	3.31	3.35	3.37	3.39	3.41	3.43	3.44	3.46	3.47	3.47
	.01	4.05	4.24	4.35	4.43	4.50	4.56	4.61	4.64	4.67	4.72	4.76	4.79	4.81	4.82
20	.05	2.95	3.10	3.18	3.25	3.30	3.34	3.36	3.38	3.40	3.43	3.44	3.46	3.46	3.47
	.01	4.02	4.22	4.33	4.40	4.47	4.53	4.58	4.61	4.65	4.69	4.73	4.76	4.78	4.79
22	.05	2.93	3.08	3.17	3.24	3.29	3.32	3.35	3.37	3.39	3.42	3.44	3.45	3.46	3.47
	.01	3.99	4.17	4.28	4.36	4.42	4.48	4.53	4.57	4.60	4.65	4.68	4.71	4.74	4.75
24	.05	2.92	3.07	3.15	3.22	3.28	3.31	3.34	3.37	3.38	3.41	3.44	3.45	3.46	3.47
	.01	3.96	4.14	4.24	4.33	4.39	4.44	4.49	4.53	4.57	4.62	4.64	4.67	4.70	4.72
26	.05	2.91	3.06	3.14	3.21	3.27	3.30	3.34	3.36	3.38	3.41	3.43	3.45	3.46	3.47
	.01	3.93	4.11	4.21	4.30	4.36	4.41	4.46	4.50	4.53	4.58	4.62	4.65	4.67	4.69
28	.05	2.90	3.04	3.13	3.20	3.26	3.30	3.33	3.35	3.37	3.40	3.43	3.45	3.46	3.47
	.01	3.91	3.08	4.18	4.28	4.34	4.39	4.43	4.47	4.51	4.56	4.60	4.62	4.65	4.67
30	.05	2.89	3.04	3.12	3.20	3.25	3.29	3.32	3.35	3.37	3.40	3.43	3.44	3.46	3.47
	.01	3.89	4.06	4.16	4.22	4.32	4.36	4.41	4.45	4.48	4.54	4.58	4.61	4.63	4.65
40	.05	2.86	3.01	3.10	3.17	3.22	3.27	3.30	3.33	3.35	3.39	3.42	3.44	3.46	3.47
	.01	3.82	3.99	4.10	4.17	4.24	4.30	4.34	4.37	4.41	4.46	4.51	4.54	4.57	4.59
60	.05	2.83	2.98	3.08	3.14	3.20	3.24	3.28	3.31	3.33	3.37	3.40	3.43	3.45	3.47
	.01	3.76	3.92	4.03	4.12	4.17	4.23	4.27	4.31	4.34	4.39	4.44	4.47	4.50	4.53
100	.05	2.80	2.95	3.05	3.12	3.18	3.22	3.26	3.29	3.32	3.36	3.40	3.42	3.45	3.47
	.01	3.71	3.86	3.93	4.06	4.11	4.17	4.21	4.25	4.29	4.35	4.38	4.42	4.45	4.48
∞	.05	2.77	2.92	3.02	3.09	3.15	3.19	3.23	3.26	3.29	3.34	3.38	3.41	3.44	3.47
	.01	3.64	3.80	3.90	3.98	4.04	4.09	4.14	4.17	4.20	4.26	4.31	4.34	4.38	4.41

Reproduced from D. B. Duncan, "Multiple Range and Multiple F Tests," Biometrics, 11: 1–42, 1955. With permission from the Biometric Society and the author.

TABLE 12   Waller–Duncan k Ratio Test Values of t_c for $k = 100$

df_1	df_2												
	6	8	10	12	14	16	18	20	24	30	40	60	120
$F = 1.2$ (a = .913, b = 2.449)													
2–6	*	*	*	*	*	*	*	*	*	*	*	*	*
8	2.91	2.94	2.96	2.97	2.98	2.99	2.99	2.99	3.00	3.00	3.00	3.00	3.00
10	2.93	2.98	3.01	3.04	3.05	3.06	3.07	3.08	3.09	3.10	3.10	3.11	3.12
12	2.95	3.01	3.05	3.08	3.10	3.12	3.13	3.14	3.16	3.17	3.19	3.20	3.21
14	2.96	3.03	3.08	3.12	3.14	3.16	3.18	3.19	3.21	3.23	3.25	3.27	3.29
16	2.97	3.05	3.11	3.15	3.18	3.20	3.22	3.24	3.26	3.28	3.31	3.33	3.36
20	2.99	3.08	3.14	3.19	3.23	3.26	3.28	3.30	3.33	3.37	3.40	3.44	3.47
40	3.02	3.13	3.22	3.29	3.35	3.39	3.43	3.47	3.52	3.58	3.64	3.72	3.79
100	3.04	3.17	3.28	3.36	3.44	3.50	3.55	3.59	3.67	3.76	3.86	3.98	4.11
∞	3.05	3.20	3.32	3.42	3.50	3.58	3.64	3.70	3.80	3.91	4.06	4.24	4.45
$F = 1.4$ (a = .845, b = 1.871)													
2–4	*	*	*	*	*	*	*	*	*	*	*	*	*
6	2.85	2.84	2.83	2.82	2.82	2.81	2.80	2.80	2.79	2.78	2.77	2.75	2.74
8	2.88	2.89	2.90	2.90	2.90	2.89	2.89	2.89	2.88	2.88	2.87	2.86	2.85
10	2.90	2.93	2.94	2.95	2.95	2.96	2.96	2.96	2.95	2.95	2.95	2.94	2.93
12	2.92	2.95	2.98	2.99	3.00	3.00	3.01	3.01	3.01	3.01	3.01	3.00	2.99
14	2.93	2.97	3.00	3.02	3.03	3.04	3.04	3.05	3.05	3.06	3.06	3.05	3.05
16	2.94	2.99	3.02	3.04	3.06	3.07	3.08	3.08	3.09	3.09	3.10	3.10	3.09
20	2.95	3.01	3.05	3.08	3.10	3.11	3.12	3.13	3.14	3.15	3.16	3.16	3.16
40	2.98	3.06	3.12	3.16	3.19	3.22	3.24	3.25	3.28	3.30	3.31	3.32	3.32
100	2.99	3.09	3.16	3.22	3.26	3.29	3.32	3.34	3.38	3.41	3.43	3.45	3.42
∞	3.01	3.12	3.20	3.26	3.31	3.35	3.39	3.42	3.46	3.50	3.53	3.54	3.46
$F = 1.7$ (a = .767, b = 1.558)													
2	*	*	*	*	*	*	*	*	*	*	*	*	*
4	*	*	*	*	*	2.61	2.59	2.58	2.56	2.54	2.52	2.50	2.48
6	2.82	2.79	2.76	2.74	2.72	2.71	2.70	2.69	2.67	2.65	2.63	2.61	2.58
8	2.84	2.83	2.81	2.80	2.78	2.77	2.76	2.75	2.74	2.72	2.70	2.68	2.65
10	2.86	2.86	2.85	2.84	2.83	2.82	2.81	2.80	2.79	2.77	2.75	2.73	2.70
12	2.87	2.88	2.88	2.87	2.86	2.85	2.84	2.84	2.82	2.81	2.79	2.76	2.73
14	2.88	2.90	2.90	2.89	2.89	2.88	2.87	2.86	2.85	2.83	2.81	2.79	2.75
16	2.89	2.91	2.91	2.91	2.90	2.90	2.89	2.89	2.87	2.86	2.84	2.81	2.77
20	2.90	2.93	2.93	2.94	2.93	2.93	2.92	2.92	2.91	2.89	2.87	2.84	2.80
40	2.93	2.97	2.99	3.00	3.00	3.00	3.00	2.99	2.98	2.97	2.94	2.89	2.83
100	2.94	2.99	3.02	3.04	3.05	3.05	3.05	3.05	3.04	3.02	2.98	2.92	2.83
∞	2.95	3.01	3.05	3.07	3.08	3.09	3.09	3.08	3.07	3.05	3.01	2.93	2.81

TABLE 12 (*cont'd*)

	df_2												
df_1	6	8	10	12	14	16	18	20	24	30	40	60	120
F = 2.0 (a = .707, b = 1.414)													
2	*	*	*	*	*	*	*	*	*	*	*	*	*
4	2.74	2.67	2.63	2.59	2.56	2.54	2.52	2.51	2.49	2.46	2.44	2.41	2.39
6	2.79	2.74	2.70	2.67	2.64	2.62	2.60	2.59	2.57	2.54	2.52	2.49	2.46
8	2.81	2.77	2.74	2.71	2.69	2.67	2.65	2.64	2.62	2.59	2.56	2.53	2.49
10	2.83	2.80	2.77	2.74	2.72	2.70	2.69	2.67	2.65	2.62	2.59	2.56	2.52
12	2.84	2.82	2.79	2.77	2.75	2.73	2.71	2.70	2.67	2.64	2.61	2.57	2.53
14	2.85	2.83	2.81	2.79	2.77	2.75	2.73	2.72	2.69	2.66	2.63	2.59	2.54
16	2.85	2.84	2.82	2.80	2.78	2.76	2.74	2.73	2.70	2.67	2.64	2.59	2.54
20	2.86	2.85	2.84	2.82	2.80	2.78	2.77	2.75	2.72	2.69	2.65	2.61	2.55
40	2.88	2.89	2.88	2.86	2.85	2.83	2.81	2.80	2.77	2.73	2.68	2.62	2.55
100	2.89	2.91	2.90	2.89	2.88	2.86	2.84	2.82	2.79	2.75	2.69	2.62	2.53
∞	2.90	2.92	2.92	2.91	2.90	2.88	2.86	2.85	2.81	2.76	2.69	2.61	2.52
F = 2.4 (a = .645, b = 1.309)													
2	*	*	*	*	*	*	*	*	*	*	*	*	2.18
4	2.71	2.63	2.57	2.53	2.49	2.47	2.44	2.43	2.40	2.37	2.34	2.31	2.28
6	2.75	2.68	2.63	2.58	2.55	2.52	2.50	2.48	2.46	2.42	2.39	2.36	2.32
8	2.77	2.71	2.66	2.62	2.59	2.56	2.54	2.52	2.49	2.45	2.42	2.38	2.34
10	2.79	2.73	2.68	2.64	2.61	2.58	2.56	2.54	2.50	2.47	2.43	2.39	2.34
12	2.79	2.74	2.70	2.66	2.62	260	2.57	2.55	2.52	2.48	2.44	2.39	2.35
14	2.80	2.75	2.71	2.67	2.64	2.61	2.58	2.56	2.53	2.49	2.44	2.40	2.35
16	2.81	2.76	2.72	2.68	2.65	2.62	2.59	2.57	2.53	2.49	2.45	2.40	2.34
20	2.82	2.77	2.73	2.69	2.66	2.63	2.60	2.58	2.54	2.50	2.45	2.40	2.34
40	2.83	2.80	2.76	2.72	2.69	2.66	2.63	2.60	2.56	2.51	2.46	2.39	2.33
100	2.84	2.81	2.78	2.74	2.71	2.67	2.64	2.62	2.57	2.51	2.45	2.39	2.32
∞	2.85	2.83	2.79	2.76	2.72	2.68	2.65	2.62	2.57	2.51	2.45	2.38	2.31
F = 3.0 (a = .577, b = 1.225)													
2	*	*	2.41	2.36	2.32	2.29	2.27	2.25	2.22	2.20	2.17	2.14	2.11
4	2.68	2.57	2.50	2.45	2.41	2.38	2.35	2.33	2.30	2.27	2.24	2.20	2.17
6	2.71	2.61	2.54	2.49	2.44	2.41	2.39	2.36	2.33	2.29	2.26	2.22	2.18
8	2.72	2.63	2.56	2.51	2.47	2.43	2.40	2.38	2.34	2.31	2.27	2.22	2.18
10	2.74	2.65	2.58	2.52	2.48	2.44	2.41	2.39	2.35	2.31	2.27	2.22	2.18
12	2.74	2.66	2.59	2.53	2.49	2.45	2.42	2.40	2.36	2.31	2.27	2.22	2.18
14	2.75	2.66	2.60	2.54	2.49	2.46	2.43	2.40	2.36	2.32	2.27	2.22	2.17
16	2.75	2.67	2.60	2.55	2.50	2.46	2.43	2.40	2.36	2.32	2.27	2.22	2.17
20	2.76	2.68	2.61	2.55	2.51	2.47	2.43	2.41	2.36	2.32	2.27	2.22	2.17
40	2.77	2.70	2.63	2.57	2.52	2.48	2.44	2.41	2.37	2.32	2.26	2.21	2.16
100	2.78	2.71	2.64	2.58	2.53	2.49	2.45	2.42	2.37	2.31	2.26	2.21	2.16
∞	2.79	2.71	2.65	2.59	2.53	2.49	2.45	2.42	2.37	2.31	2.26	2.20	2.15

*All differences not significant. $a = 1/F^{1/2}$; $b = [F/(F-1)]^{1/2}$.

From "Corrigenda" by R. Waller and D. Duncan, *Journal of the American Statistical Association,* 67, 1972, 253–55, Table A2. Reproduced by permission of the American Statistical Association.

TABLE 12 (cont'd)

| df$_1$ | \multicolumn{13}{c}{df$_2$} |
	6	8	10	12	14	16	18	20	24	30	40	60	120
\multicolumn{14}{l}{$F = 4.0$ (a = .500, b = 1.155)}
2	2.58	2.44	2.35	2.29	2.25	2.22	2.20	2.18	2.15	2.12	2.09	2.06	2.03
4	2.63	2.50	2.41	2.35	2.30	2.27	2.24	2.22	2.18	2.15	2.12	2.08	2.05
6	2.65	2.52	2.43	2.37	2.32	2.28	2.25	2.23	2.19	2.16	2.12	2.08	2.04
10	2.67	2.55	2.46	2.39	2.34	2.30	2.26	2.24	2.20	2.16	2.12	2.08	2.04
20	2.69	2.57	2.47	2.40	2.35	2.30	2.27	2.24	2.20	2.15	2.11	2.07	2.03
∞	2.71	2.59	2.49	2.42	2.36	2.31	2.27	2.24	2.19	2.15	2.11	2.06	2.02
\multicolumn{14}{l}{$F = 6.0$ (a = .408, b = 1.095)}
2	2.53	2.37	2.27	2.21	2.16	2.13	2.10	2.08	2.05	2.02	1.99	1.96	1.93
4	2.56	2.40	2.30	2.23	2.18	2.14	2.12	2.09	2.06	2.02	1.99	1.96	1.93
6	2.58	2.42	2.31	2.24	2.19	2.15	2.12	2.09	2.06	2.02	1.99	1.95	1.92
10	2.59	2.43	2.32	2.24	2.19	2.15	2.12	2.09	2.06	2.02	1.99	1.95	1.92
20	2.60	2.44	2.32	2.25	2.19	2.15	2.12	2.09	2.05	2.02	1.98	1.95	1.92
∞	2.61	2.44	2.33	2.25	2.19	2.15	2.12	2.09	2.05	2.02	1.98	1.95	1.92
\multicolumn{14}{l}{$F = 10.0$ (a = .316, b = 1.054)}
2	2.48	2.30	2.19	2.12	2.07	2.04	2.01	1.99	1.96	1.93	1.90	1.87	1.85
4	2.49	2.31	2.20	2.13	2.08	2.04	2.01	1.99	1.96	1.93	1.90	1.87	1.84
6	2.50	2.31	2.20	2.13	2.08	2.04	2.01	1.99	1.96	1.93	1.90	1.87	1.84
≥10	2.51	2.32	2.20	2.13	2.08	2.04	2.01	1.99	1.96	1.93	1.90	1.87	1.84
\multicolumn{14}{l}{$F = 25.0$ (a = .200, b = 1.021)}
2–4	2.40	2.20	2.10	2.03	1.99	1.95	1.93	1.91	1.88	1.86	1.83	1.80	1.78
≥6	2.41	2.21	2.10	2.03	1.99	1.95	1.93	1.91	1.88	1.86	1.83	1.80	1.78
\multicolumn{14}{l}{$F = \infty$ (a = 0, b = 1)}
≥2	2.33	2.13	2.03	1.97	1.93	1.90	1.88	1.86	1.84	1.81	1.79	1.76	1.74

TABLE 13   Waller–Duncan k Ratio Test Values of t_c for $k = 500$

df_1	6	8	10	12	14	16	18	20	24	30	40	60	120
						df_2							
$F = 1.2$ (a = .913, b = 2.449)													
2–16	*	*	*	*	*	*	*	*	*	*	*	*	*
20	4.70	4.82	4.89	*	*	*	*	*	*	*	*	*	*
40	4.75	4.91	5.03	5.12	5.20	5.25	5.30	5.34	5.41	5.48	5.55	5.61	5.67
100	4.79	4.98	5.13	5.25	5.34	5.43	5.50	5.56	5.65	5.76	5.89	6.02	6.13
∞	4.81	5.03	5.20	5.34	5.46	5.56	5.65	5.73	5.86	6.02	6.20	6.41	6.56
$F = 1.4$ (a = .845, b = 1.871)													
2–14	*	*	*	*	*	*	*	*	*	*	*	*	*
16	4.61	4.66	4.68	4.69	4.69	4.69	4.69	4.68	4.67	4.65	4.62	4.58	4.53
20	4.64	4.70	4.73	4.75	4.76	4.77	4.77	4.76	4.76	4.74	4.72	4.68	4.62
40	4.68	4.78	4.85	4.89	4.92	4.94	4.96	4.96	4.97	4.97	4.95	4.90	4.81
∞	4.74	4.88	4.99	5.06	5.12	5.17	5.20	5.23	5.26	5.28	5.26	5.16	4.82
$F = 1.7$ (a = .767, b = 1.558)													
2–8	*	*	*	*	*	*	*	*	*	*	*	*	*
10	*	*	*	*	*	*	*	*	*	4.08	4.02	3.95	3.87
12	4.50	4.46	4.42	4.38	4.34	4.30	4.27	4.24	4.19	4.14	4.07	3.99	3.90
20	4.55	4.54	4.52	4.49	4.46	4.43	4.40	4.37	4.32	4.26	4.18	4.08	3.95
40	4.59	4.61	4.61	4.60	4.57	4.55	4.52	4.49	4.44	4.36	4.26	4.12	3.93
∞	4.64	4.69	4.71	4.72	4.71	4.69	4.66	4.63	4.57	4.46	4.31	4.07	3.76
$F = 2.0$ (a = .707, b = 1.414)													
2–6	*	*	*	*	*	*	*	*	*	*	*	*	*
8	*	*	*	*	*	3.98	3.93	3.89	3.83	3.76	3.69	3.60	3.51
10	4.41	4.31	4.22	4.15	4.08	4.03	3.98	3.94	3.88	3.80	3.72	3.63	3.53
20	4.48	4.41	4.34	4.27	4.21	4.16	4.10	4.06	3.98	3.89	3.78	3.65	3.51
40	4.51	4.47	4.41	4.35	4.29	4.23	4.17	4.12	4.03	3.92	3.78	3.62	3.44
∞	4.55	4.53	4.49	4.43	4.37	4.31	4.25	4.19	4.07	3.93	3.75	3.54	3.33
$F = 2.4$ (a = .645, b = 1.309)													
2–8	*	*	*	*	*	*	*	*	*	*	*	*	*
6	*	*	*	*	3.77	3.71	3.65	3.61	3.54	3.47	3.39	3.30	3.22
8	4.31	4.14	4.01	3.91	3.83	3.76	3.70	3.66	3.58	3.50	3.41	3.32	3.22
10	4.33	4.18	4.05	3.95	3.87	3.79	3.73	3.68	3.60	3.51	3.42	3.31	3.21
20	4.39	4.26	4.14	4.04	3.95	3.87	3.80	3.74	3.64	3.53	3.41	3.28	3.15
∞	4.45	4.35	4.25	4.14	4.03	3.94	3.85	3.78	3.64	3.50	3.34	3.18	3.04

*All differences not significant. a $= 1/F^{1/2}$; b $= [F/(F - 1)]^{1/2}$.

From "Corrigenda" by R. Waller and D. Duncan, *Journal of the American Statistical Association*, 67 (1972): 253–255, Table A2. Reproduced by permission of the American Statistical Association.

TABLE 13 (cont'd)

df_1	6	8	10	12	14	16	18	20	24	30	40	60	120

df_2

$F = 3.0$ (a = .577, b = 1.225)

df_1	6	8	10	12	14	16	18	20	24	30	40	60	120
2	*	*	*	*	*	*	*	*	*	*	*	*	*
4	*	*	*	*	*	3.43	3.38	3.33	3.26	3.19	3.12	3.04	2.97
6	4.19	3.95	3.79	3.66	3.56	3.49	3.43	3.37	3.30	3.21	3.13	3.04	2.95
10	4.24	4.02	3.85	3.72	3.62	3.53	3.46	3.40	3.31	3.21	3.12	3.02	2.92
20	4.28	4.08	3.91	3.77	3.65	3.56	3.48	3.41	3.31	3.20	3.09	2.98	2.87
∞	4.33	4.15	3.97	3.82	3.69	3.57	3.48	3.40	3.28	3.15	3.03	2.92	2.82

$F = 4.0$ (a = .500, b = 1.155)

df_1	6	8	10	12	14	16	18	20	24	30	40	60	120
2	*	*	*	*	*	*	*	*	*	*	*	2.81	2.75
4	*	3.74	3.54	3.40	3.30	3.22	3.16	3.11	3.04	2.96	2.89	2.81	2.74
6	4.08	3.78	3.58	3.43	3.32	3.24	3.17	3.12	3.04	2.95	2.87	2.79	2.71
10	4.12	3.83	3.62	3.46	3.34	3.25	3.17	3.11	3.03	2.94	2.85	2.77	2.69
20	4.15	3.86	3.64	3.48	3.35	3.25	3.17	3.10	3.01	2.92	2.83	2.74	2.66
∞	4.19	3.90	3.67	3.49	3.35	3.24	3.15	3.09	2.99	2.89	2.80	2.72	2.65

$F = 6.0$ (a = .408, b = 1.095)

df_1	6	8	10	12	14	16	18	20	24	30	40	60	120
2	*	*	3.28	3.14	3.04	2.97	2.91	2.87	2.81	2.74	2.68	2.62	2.56
4	3.90	3.54	3.32	3.17	3.06	2.98	2.92	2.87	2.80	2.73	2.66	2.60	2.53
6	3.93	3.57	3.33	3.18	3.06	2.98	2.91	2.86	2.79	2.72	2.65	2.58	2.52
10	3.95	3.59	3.34	3.18	3.06	2.97	2.91	2.85	2.78	2.71	2.64	2.57	2.51
20	3.97	3.60	3.35	3.18	3.06	2.97	2.90	2.84	2.77	2.70	2.63	2.56	2.51
∞	3.99	3.62	3.36	3.18	3.05	2.96	2.89	2.83	2.76	2.69	2.62	2.56	2.50

$F = 10.0$ (a = .316, b = 1.054)

df_1	6	8	10	12	14	16	18	20	24	30	40	60	120
2	3.72	3.33	3.10	2.96	2.86	2.79	2.74	2.70	2.64	2.58	2.52	2.47	2.42
4	3.75	3.35	3.11	2.96	2.86	2.79	2.73	2.69	2.63	2.57	2.51	2.46	2.41
10	3.78	3.36	3.11	2.96	2.85	2.78	2.72	2.68	2.62	2.56	2.50	2.45	2.40
20	3.79	3.36	3.11	2.96	2.85	2.78	2.72	2.68	2.62	2.56	2.50	2.45	2.40
∞	3.80	3.37	3.11	2.95	2.85	2.77	2.72	2.67	2.61	2.56	2.50	2.45	2.40

$F = 25.0$ (a = .200, b = 1.021)

df_1	6	8	10	12	14	16	18	20	24	30	40	60	120
2	3.55	3.14	2.92	2.79	2.70	2.64	2.59	2.56	2.51	2.46	2.41	2.36	2.32
10	3.57	3.14	2.92	2.79	2.70	2.64	2.59	2.55	2.50	2.45	2.41	2.36	2.32
∞	3.57	3.14	2.92	2.78	2.70	2.63	2.59	2.55	2.50	2.45	2.41	2.36	2.32

$F = \infty$ (a = 0, b = 1)

df_1	6	8	10	12	14	16	18	20	24	30	40	60	120
≥2	3.39	3.00	2.80	2.69	2.61	2.55	2.51	2.48	2.44	2.39	2.35	2.31	2.27

TABLE 14 Percentage Points of $F_{max} = s^2_{max}/s^2_{min}$

Upper 5% points

df$_2$ \ t	2	3	4	5	6	7	8	9	10	11	12
2	39.0	87.5	142	202	266	333	403	475	550	626	704
3	15.4	27.8	39.2	50.7	62.0	72.9	83.5	93.9	104	114	124
4	9.60	15.5	20.6	25.2	29.5	33.6	37.5	41.1	44.6	48.0	51.4
5	7.15	10.8	13.7	16.3	18.7	20.8	22.9	24.7	26.5	28.2	29.9
6	5.82	8.38	10.4	12.1	13.7	15.0	16.3	17.5	18.6	19.7	20.7
7	4.99	6.94	8.44	9.70	10.8	11.8	12.7	13.5	14.3	15.1	15.8
8	4.43	6.00	7.18	8.12	9.03	9.78	10.5	11.1	11.7	12.2	12.7
9	4.03	5.34	6.31	7.11	7.80	8.41	8.95	9.45	9.91	10.3	10.7
10	3.72	4.85	5.67	6.34	6.92	7.42	7.87	8.28	8.66	9.01	9.34
12	3.28	4.16	4.79	5.30	5.72	6.09	6.42	6.72	7.00	7.25	7.48
15	2.86	3.54	4.01	4.37	4.68	4.95	5.19	5.40	5.59	5.77	5.93
20	2.46	2.95	3.29	3.54	3.76	3.94	4.10	4.24	4.37	4.49	4.59
30	2.07	2.40	2.61	2.78	2.91	3.02	3.12	3.21	3.29	3.36	3.39
60	1.67	1.85	1.96	2.04	2.11	2.17	2.22	2.26	2.30	2.33	2.36
∞	1.00	1.00	1.00	1.00	1.00	1.00	1.00	1.00	1.00	1.00	1.00

Upper 1% points

df$_2$ \ t	2	3	4	5	6	7	8	9	10	11	12
2	199	448	729	1036	1362	1705	2063	2432	2813	3204	3605
3	47.5	85	120	151	184	21(6)	24(9)	28(1)	31(0)	33(7)	36(1)
4	23.2	37	49	59	69	79	89	97	106	113	120
5	14.9	22	28	33	38	42	46	50	54	57	60
6	11.1	15.5	19.1	22	25	27	30	32	34	36	37
7	8.89	12.1	14.5	16.5	18.4	20	22	23	24	26	27
8	7.50	9.9	11.7	13.2	14.5	15.8	16.6	17.9	18.9	19.8	21
9	6.54	8.5	9.9	11.1	12.1	13.1	13.9	14.7	15.3	16.0	16.6
10	5.85	7.4	8.6	9.6	10.4	11.1	11.8	12.4	12.9	13.4	13.9
12	4.91	6.1	6.9	7.6	8.2	8.7	9.1	9.5	9.9	10.2	10.6
15	4.07	4.9	5.5	6.0	6.4	6.7	7.1	7.3	7.5	7.8	8.0
20	3.32	3.8	4.3	4.6	4.9	5.1	5.3	5.5	5.6	5.8	5.9
30	2.63	3.0	3.3	3.4	3.6	3.7	3.8	3.9	4.0	4.1	4.2
60	1.96	2.2	2.3	2.4	2.4	2.5	2.5	2.6	2.6	2.7	2.7
∞	1.00	1.0	1.0	1.0	1.0	1.0	1.0	1.0	1.0	1.0	1.0

s^2_{max} is the largest and s^2_{min} the smallest in a set of t independent mean squares, each based on df$_2$ = n − 1 degrees of freedom. Values in the column t = 2 and in the rows df$_2$ = 2 and ∞ are exact. Elsewhere the third digit may be in error by a few units for the 5% points and several units for the 1% points. The third digit figures in parentheses for df$_2$ = 3 are the most uncertain.
From *Biometrika Tables for Statisticians*, 3rd ed., Vol. 1, edited by E. S. Pearson and H. O. Hartley (New York: Cambridge University Press, 1966), Table, p. 202. Reproduced by permission of the *Biometrika Trustees*.

TABLE 15 Values of 2 arcsin $\sqrt{\hat{\pi}}$

$\hat{\pi}$		$\hat{\pi}$		$\hat{\pi}$		$\hat{\pi}$		$\hat{\pi}$	
.001	.0633	.041	.4078	.36	1.2870	.76	2.1177	.971	2.7993
.002	.0895	.042	.4128	.37	1.3078	.77	2.1412	.972	2.8053
.003	.1096	.043	.4178	.38	1.3284	.78	2.1652	.973	2.8115
.004	.1266	.044	.4227	.39	1.3490	.79	2.1895	.974	2.8177
.005	.1415	.045	.4275	.40	1.3694	.80	2.2143	.975	2.8240
.006	.1551	.046	.4323	.41	1.3898	.81	2.2395	.976	2.8305
.007	.1675	.047	.4371	.42	1.4101	.82	2.2653	.977	2.8371
.008	.1791	.048	.4418	.43	1.4303	.83	2.2916	.978	2.8438
.009	.1900	.049	.4464	.44	1.4505	.84	2.3186	.979	2.8507
.010	.2003	.050	.4510	.45	1.4706	.85	2.3462	.980	2.8578
.011	.2101	.06	.4949	.46	1.4907	.86	2.3746	.981	2.8650
.012	.2195	.07	.5355	.47	1.5108	.87	2.4039	.982	2.8725
.013	.2285	.08	.5735	.48	1.5308	.88	2.4341	.983	2.8801
.014	.2372	.09	.6094	.49	1.5508	.89	2.4655	.984	2.8879
.015	.2456	.10	.6435	.50	1.5708	.90	2.4981	.985	2.8960
.016	.2537	.11	.6761	.51	1.5908	.91	2.5322	.986	2.9044
.017	.2615	.12	.7075	.52	1.6108	.92	2.5681	.987	2.9131
.018	.2691	.13	.7377	.53	1.6308	.93	2.6062	.988	2.9221
.019	.2766	.14	.7670	.54	1.6509	.94	2.6467	.989	2.9315
.020	.2838	.15	.7954	.55	1.6710	.95	2.6906	.990	2.9413
.021	.2909	.16	.8230	.56	1.6911	.951	2.6952	.991	2.9516
.022	.2978	.17	.8500	.57	1.7113	.952	2.6998	.992	2.9625
.023	.3045	.18	.8763	.58	1.7315	.953	2.7045	.993	2.9741
.024	.3111	.19	.9021	.59	1.7518	.954	2.7093	.994	2.9865
.025	.3176	.20	.9273	.60	1.7722	.955	2.7141	.995	3.0001
.026	.3239	.21	.9521	.61	1.7926	.956	2.7189	.996	3.0150
.027	.3301	.22	.9764	.62	1.8132	.957	2.7238	.997	3.0320
.028	.3363	.23	1.0004	.63	1.8338	.958	2.7288	.998	3.0521
.029	.3423	.24	1.0239	.64	1.8338	.959	2.7338	.999	3.0783
.030	.3482	.25	1.0472	.65	1.8546	.960	2.7389		
.031	.3540	.26	1.0701	.66	1.8965	.961	2.7440		
.032	.3597	.27	1.0928	.67	1.9177	.962	2.7492		
.033	.3654	.28	1.1152	.68	1.9391	.963	2.7545		
.034	.3709	.29	1.1374	.69	1.9606	.964	2.7598		
.035	.3764	.30	1.1593	.70	1.9823	.965	2.7652		
.036	.3818	.31	1.1810	.71	2.0042	.966	2.7707		
.037	.3871	.32	1.2025	.72	2.0264	.967	2.7762		
.038	.3924	.33	1.2239	.73	2.0488	.968	2.7819		
.039	.3976	.34	1.2451	.74	2.0715	.969	2.7876		
.040	.4027	.35	1.2661	.75	2.0944	.970	2.7934		

From *Experimental Design: Procedures for the Behavioral Sciences,* by Roger E. Kirk. Copyright © 1968 by Wadsworth Publishing Company, Inc. Reprinted by permission of the publisher, Brooks/Cole Publishing Company, Belmont, Calif.

TABLE 16   Values of d_n Used in Control Limits for μ

n	d_n
2	1.128
3	1.693
4	2.059
5	2.326
6	2.534
7	2.704
8	2.847
9	2.970
10	3.078
11	3.173
12	3.258

TABLE 17   Values of D_n and D_n' Used in Control Limits for Product Variability

Number of Observations in Sample, n	D_n	D_n'
2	0	3.267
3	0	2.575
4	0	2.282
5	0	2.115
6	0	2.004
7	0.076	1.924
8	0.136	1.864
9	0.184	1.816
10	0.223	1.777
11	0.256	1.744
12	0.284	1.716
13	0.308	1.692
14	0.329	1.671
15	0.348	1.652
16	0.364	1.636
17	0.379	1.621
18	0.392	1.608
19	0.404	1.596
20	0.414	1.586
21	0.425	1.575
22	0.434	1.566
23	0.443	1.557
24	0.452	1.548
25	0.459	1.541
Over 25	$3/\sqrt{n}$	$3/\sqrt{n}$

REFERENCES

Books and Articles

Anderson, M. J., R. C. Lamb, C. H. Mekelsen, and R. L. Wiscombe (1974). Feeding liquid whey to dairy cattle. *Journal of Dairy Science* 57.

Andrews, D. F., and A. M. Herzberg (1985). *Data: A collection of problems from many fields for the student and research worker.* New York: Springer-Verlag.

Bancroft, T. A. (1968). *Topics in intermediate statistical methods.* Vol. 1. Ames, Iowa: Iowa State University Press.

Barr, A. J., J. H. Goodnight, J. P. Sall, and J. T. Helwig (1976). *A user's guide to SAS 76.* Raleigh, N.C.: SAS Institute, Inc.

Bishop, Y. M. M., S. E. Fineberg, and P. W. Holland (1975). *Discrete multivariate analysis: theory and practice.* Cambridge, Mass.: MIT Press.

Boardman, T. J., and D. R. Moffitt (1971). Graphical Monte Carlo type 1 error rates for multiple comparison procedures. *Biometrics* 27: 738–744.

Bock, R. D. (1975). *Multivariate statistical methods in behavioral research.* Chap. 8. New York: McGraw-Hill.

Brown, S. S., Healy, M. J. R., and Kearns, M. (1981). Report on the interlaboratory trial of the reference method for the determination of total calcium in serum. *Journal of Clinical Chemistry and Clinical Biochemistry* 19: 395–426.

Brownlee, K. A. (1965). *Statistical theory and methodology in science and engineering.* 2nd ed. New York: Wiley.

Carmer, S. G., and M. R. Swanson (1973). An evaluation of ten pairwise multiple comparison procedures by Monte Carlo methods. *Journal of the American Statistical Association* 68: 66–74.

Carter, R. L. (1981). Restricted maximum likelihood estimation of bias and reliability in the comparison of several measuring methods. *Biometrics* 37: 733–741.

Chew, V. (1976). *Comparing treatment means—A compendium.* Technical report no. 105. Agricultural Research Service of the U.S. Department of Agriculture and the Department of Statistics, University of Florida.

Cochran, W. G. (1954). Some methods for strengthening the common χ^2 test. *Biometrics* 10: 417–451.

REFERENCES

————(1964). Approximate significance levels of the Behrens–Fisher test. *Biometrics* 20: 191–195.

Cochran, W. G., and G. M. Cox (1957). *Experimental design.* 2nd ed. New York: Wiley.

Conover, W. J. (1980). *Practical nonparametric statistics.* 2nd ed. New York: Wiley.

Conover, W. J., and R. L. Iman (1981). Rank transformations as a bridge between parametric and nonparametric statistics. *The American Statistician* 35: 124–133.

Cornell, J. A. (1971). A review of multiple comparison procedures for comparing a set of *k* population means. *Proceedings of the Soil and Crop Science Society of Florida* 31: 92–97.

Cox, D. R. (1970). *The analysis of binary data.* New York: Halsted Press.

CRC standard mathematical tables (1961). 12th ed. Cleveland, Ohio: The Chemical Rubber Co.

Davies, O. L. (1956). *The design and analysis of experiments.* New York: Hafner.

Dixon, W. J., and F. J. Massey, Jr. (1969). *Introduction to statistical analysis.* 3rd ed. New York: McGraw-Hill.

Draper, N. R., and H. Smith (1981). *Applied regression analysis.* 2nd ed. New York: Wiley.

Duncan, A. J. (1959). *Quality control and industrial statistics.* 2nd ed. Chap. 21. Homewood, Ill.: Irwin.

Duncan, D. B. (1955). Multiple range and multiple *F* tests. *Biometrics* 11: 1–42.

————(1957). Multiple range tests for correlated and heteroscedastic means. *Biometrics* 13: 164–176.

————(1975). *T* tests and intervals for comparisons suggested by the data. *Biometrics* 31: 339–359.

Dunnett, C. W. (1955). A multiple comparison procedure for comparing several treatments with a control. *Journal of the American Statistical Association* 50: 1096–1121.

————(1980a). Pairwise multiple comparisons in the homogeneous variance, unequal sample size case. *Journal of the American Statistical Association* 75: 789–795.

————(1980b). Pairwise multiple comparisons in the unequal variance case. *Journal of the American Statistical Association* 75: 796–800.

Fisher, R. A. (1949). *The design of experiments.* Edinburgh: Oliver and Boyd.

Forthofer, R. N., and G. G. Koch (1973). An analysis of compounded functions of categorical data. *Biometrics* 29: 143–157.

Forthofer, R. N., and R. G. Lehnen (1981). *Public program analysis: A new categorical data approach.* Belmont, Calif.: Wadsworth.

Forthofer, R. N., C. F. Starmer, and J. E. Grizzle (1971). A program for the analysis of categorical data by linear models. *Journal of Biomedical Systems* 2: 3–48.

Gaines, P. A., H. J. Kaselman, and J. C. Rogan (1981). Simultaneous pairwise multiple comparison procedures for means where sample sizes are unequal. *Psychological Bulletin* 90: 594–598.

Geisser, S., and S. W. Greenhouse (1958). An extension of Box's results on the use of the *F* distribution in multivariate analysis. *Annals of Mathematical Statistics* 29: 885–891.

Goodman, L. A. (1971). The analysis of multidimensional contingency tables: stepwise procedures and direct estimation methods for building models for multiple classifications. *Technometrics* 13: 33–61.

Graybill, F. A. (1976). *Theory and application of the linear model.* Boston, Mass.: Duxbury Press.

Greenhouse, S. W., and S. Geisser (1959). On methods in the analysis of profile data. *Psychometrika* 24: 95–112.

Grizzle, J. E., C. F. Starmer, and G. G. Koch (1969). Analysis of categorical data by linear models. *Biometrics* 25: 489–504.

Grizzle, J. E., and O. D. Williams (1972). Log linear models and tests of independence for contingency tables. *Biometrics* 28: 137–156.

Haberman, S. J. (1973). *C-tab analysis of multidimensional contingency tables by log-linear models.* Ann Arbor, Mich.: National Educational Resources, Inc.

Harnett, D. L. (1970). *Introduction to statistical methods.* Reading, Mass.: Addison-Wesley.

Harter, H. L. (1957). Error rates and sample sizes for range tests in multiple comparisons. *Biometrics* 13: 511–598.

Hicks, C. R. (1973). *Fundamental concepts in experimental design.* 2nd ed. New York: Holt, Rinehart and Winston.

Hildebrand, D. K., and L. Ott (1987). *Statistical thinking for managers.* 2nd ed. Boston, Mass.: Duxbury Press.

Hogg, R. V., and A. J. Craig (1978). *Introduction to mathematical statistics.* 4th ed. New York: Macmillan.

Hollander, M., and D. A. Wolfe (1973). *Nonparametric statistical methods.* New York: Wiley.

Hurlburt, R. J., and D. K. Spiegel (1976). Dependence of *F* ratios sharing a common denominator mean square. *American Statistician* 30: 74–78.

Huynh, H., and L. S. Feldt (1970). Conditions under which mean sqaure ratios in repeated measurement designs have fixed *F*-distributions. *Journal of the American Statistical Association* 65: 1582–1589.

Kirk, R. E. (1968). *Experimental design: Procedures for the behavioral sciences.* Belmont, Calif.: Brooks/Cole.

Kramer, C. Y. (1956). Extension of multiple range tests to group means with unequal numbers of replications. *Biometrics* 12: 307–310.

——— (1957). Extension of multiple range tests to group correlated adjusted means. *Biometrics* 13: 13–18.

Kritzer, H. (1979). Approaches to the analysis of complex contingency tables: A guide for the perplexed. *Sociological Methods and Research* 7: 305–329.

Ku, H. H., R. N. Varner, and S. Kullback (1971). Analysis of multidimensional contingency tables. *Journal of the American Statistical Association* 66: 55–64.

REFERENCES

Landis, J. R., M. M. Cooper, G. G. Koch, and J. Kennedy (1978). Computer program for testing average partial association in three-way contingency tables (Parcat). *Computer Programs in Biomedicine* 9: 223–246.

Landis, J. R., E. R. Heyman, and G. G. Koch (1978). Average partial association in three-way contingency tables: A review and discussion of alternative tests. *International Statistical Review* 46: 237–544.

Landis, J. R., W. M. Stanish, J. L. Freeman, and G. G. Koch (1976). A computer program for the generalized chi-square analysis of categorical data using weighted least squares (Gencat). *Computer Programs in Biomedicine* 6: 196–231.

Landis, J. R., W. M. Stanish, and G. G. Koch (1976). A computer program for the generalized chi-square analysis of categorical data using weighted least squares to compute Wald statistics (Gencat). *Biostatistics Technical Report No. 8.* Ann Arbor: Department of Biostatistics, University of Michigan.

Mallows, C. L. (1973). Some comments on Cp. *Technometrics* 15: 661–675.

Mantel, N., and W. Haenszel (1959). Statistical aspects of the analysis of data from retrospective studies of disease. *Journal of the National Cancer Institute* 22: 719–748.

Mendenhall, W. (1968). *Introduction to linear models and the design and analysis of experiments.* Belmont, Calif.: Wadsworth.

————(1983). Introduction to probability and statistics. 6th ed. Boston, Mass.: Duxbury Press.

Miller, R. G., Jr. (1964). A trustworthy jackknife. *Annals of Mathematical Statistics* 35: 1594–1605.

Neave, H. R. (1966). A development of Tukey's quick test for location. *Journal of the American Statistical Association* 61: 897–1262.

————(1975). A quick and simple technique for general slippage problems. *Journal of the American Statistical Association* 70: 721–726.

Omstead, P. S., and J. W. Tukey (1947). A corner test for association. *Annals of Mathematical Statistics* 18: 495–513.

Ostle, B. (1963). *Statistics in research.* 2nd ed. Ames, Iowa: Iowa State University Press.

Ott, L., R. Larson, and W. Mendenhall (1987). *Statistics: A tool for the social sciences.* 4th ed. Boston, Mass.: Duxbury Press.

Ott, L., and W. Mendenhall (1985). *Understanding statistics.* 4th ed. Boston, Mass.: Duxbury Press.

Pearson, E. S., and H. O. Hartley (1942). The probability integral of the range in samples of n observations from a normal population. *Biometrics* 32: 301–310.

————(1943). Tables of the probability integral of the Studentized range. *Biometrics* 33: 89–99.

————(1966). *Biometrika tables for statisticians.* 3rd ed. Vol. 1. London: Cambridge University Press.

Rowe, N. H., R. H. Anderson, and L. A. Wanninger (1974). Effects of read-to-eat breakfast cereals on dental caries experience in adolescent children: A three-year study. *Journal of Dental Research* 53: 33.

REFERENCES

Satterthwaite, F. E. (1946). An approximate distribution of estimates of variance components. *Biometric Bulletin* 2: 110–114.

Scheffé, H. (1953). A method for judging all contrasts in an analysis of variance. *Biometrika* 40: 87–104.

———(1959). *The analysis of variance*. New York: Wiley.

Searle, S. R. (1971). *Linear models*. New York: Wiley.

Siegel, S. (1956). *Nonparametric statistics for the behavioral sciences*. New York: McGraw-Hill.

Snedecor, G. W., and W. G. Cochran (1980). *Statistical methods*. 7th ed. Ames, Iowa: Iowa State University Press.

Snowdon, C. T., and B. A. Sanderson (1974). Lead pica produced in rats. *Science* 183: 92–94.

Sokal, R. R., and F. J. Rohlf (1969). *Biometry*. San Francisco: Freeman.

Stedl, J. L. (1981). The effect of sample size on the true α level with Funcat—A simulation study. *Proceedings of the Sixth Annual SAS User's Group*. Cary, N.C.: SAS Institute, Inc.

Steel, R. G. D., and J. H. Torrie (1980). *Principles and procedures of statistics,* 2nd ed. New York: McGraw-Hill.

"Student" (1908). The probable error of a mean. *Biometrika* 6: 1–25.

Sugiura, N., and M. Otake (1974). An extension of the Mantel-Haenszel procedure to $k \, 2 \times c$ contingency tables and the relation to the logit model. *Communications in Statistics* 3(9): 829–842.

Tanur, J. M., F. Mosteller, W. H. Kruskal, R. S. Pieters, and G. R. Rising (1972). *Statistics: A guide to the unknown*. San Francisco: Holden-Day.

Tukey, J. W. (1953). *The problem of multiple comparisons*. Mimeographed. Princeton, N.J.: Princeton University.

———(1959). A quick compact two-sample test to Duckworth's specifications. *Technometrics* 1: 31–48.

Waller, R. A., and D. B. Duncan (1969). A Bayes rule for the symmetric multiple comparison problem. *Journal of the American Statistical Association* 64: 1484–1503.

———(1972). Corrigenda. *Journal of the American Statistical Association* 67: 253–255.

Welch, B. L. (1956). On linear combinations of several variances. *Journal of the American Statistical Association* 51: 132–148.

Wilcoxon, F. and R. A. Wilcox (1964). *Some rapid approximate statistical procedures*. Pearl River, N.Y.: Lederle Laboratories.

Winer, B. J. (1962). *Statistical principles in experimental design*. New York: McGraw-Hill.

Zahn, D. A. (1974). *Documentation for Contab—A computer program to aid in the analysis of multidimensional contingency tables using log-linear models*. FSU statistical report M292. Tallahassee: Department of Statistics, Florida State University.

Statistical Software Packages

BMDP

Dixon, W. J. (1983). BMDP Statistical Software. Berkeley: University of California Press.

Minitab

Ryan, T. A., Joiner, B. L., and Ryan, B. F. (1985). Minitab Handbook, N. Scituate, Mass.: Duxbury Press.

SAS

SAS Institute (1983). SAS Introductory Guide. Cary, N.C.: SAS Institute.

SPSS

Norusis, M. J. (1983). SPSS* Introductory Statistics Guide. New York: McGraw-Hill.

ANSWERS TO SELECTED EXERCISES

Chapter 2

2.1
a.

College	No. of Majors	Percentage
Agriculture	1,500	5.7
Arts and science	11,000	41.5
Business administration	7,000	26.4
Education	2,000	7.5
Engineering	5,000	18.9

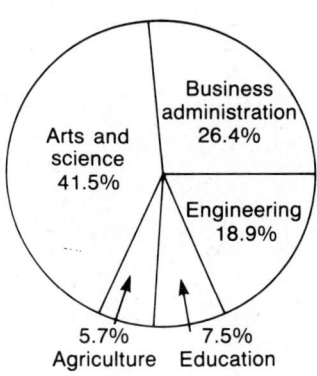

b.

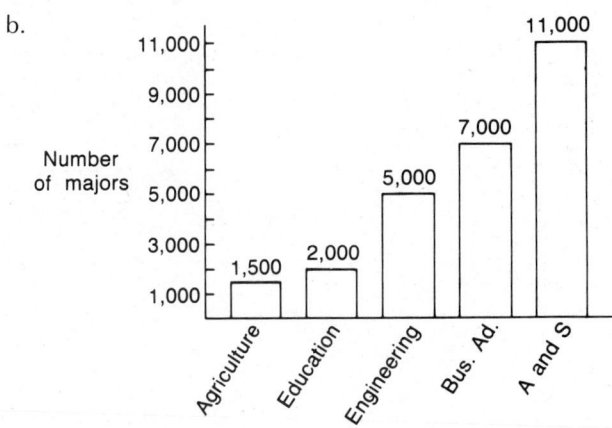

2.2 a. No

b.

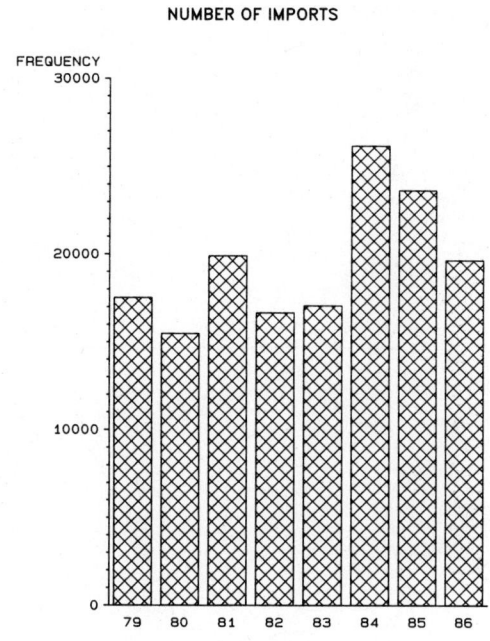

NUMBER OF IMPORTS

2.3 Pie chart is better.

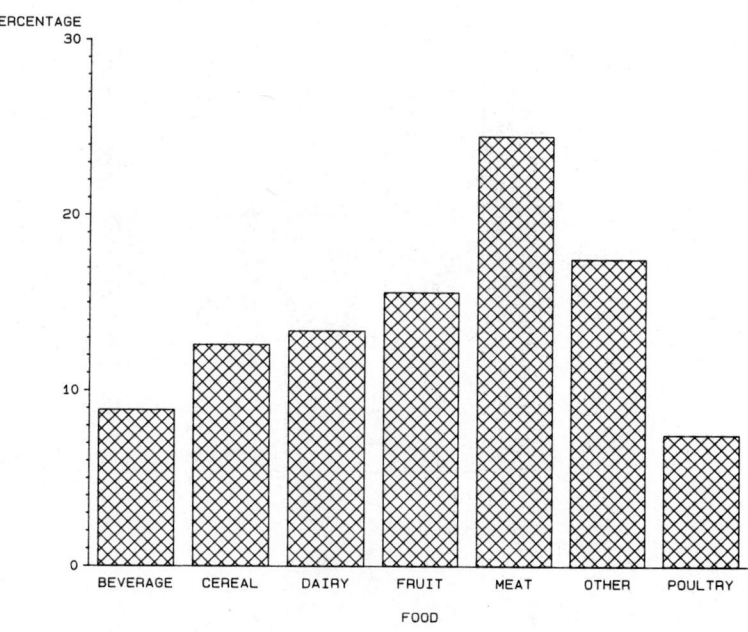

WHERE OUR FOOD DOLLARS GO

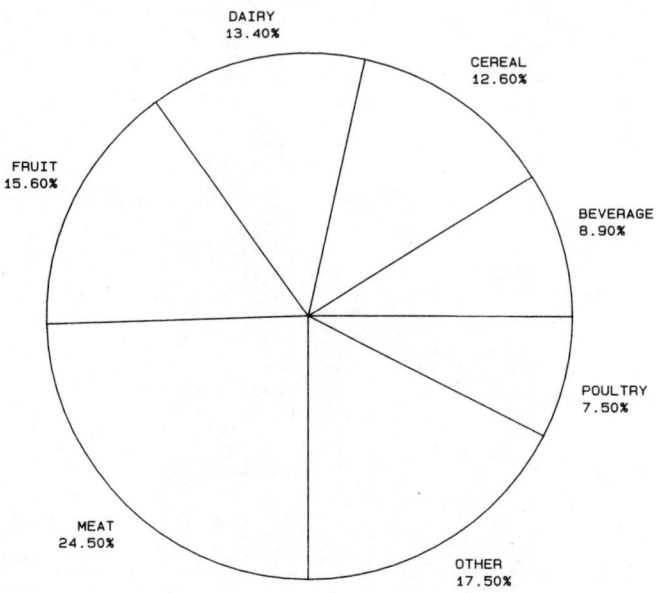

WHERE OUR FOOD DOLLARS GO

2.4

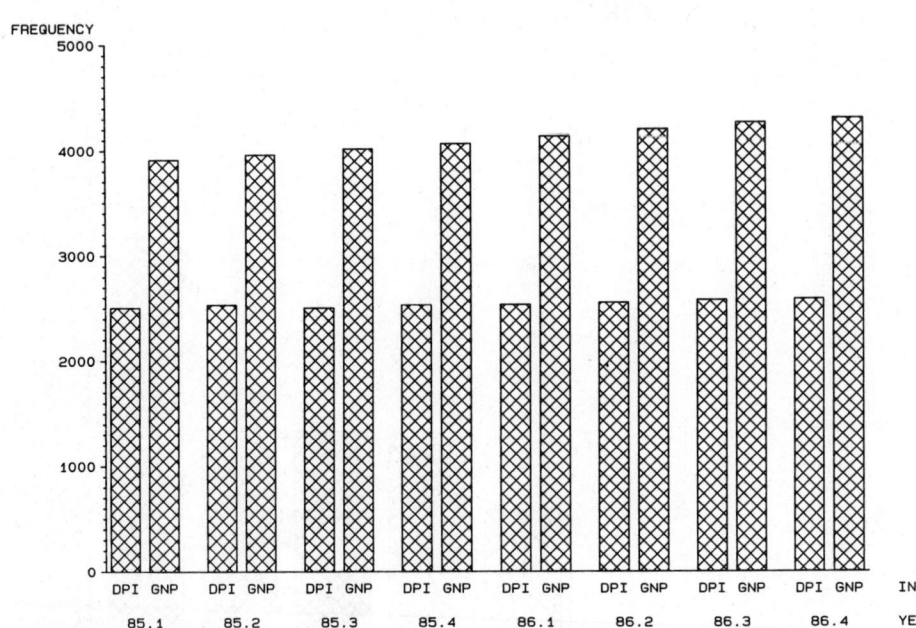

DPI —VS— GNP (in billions of dollars)

2.5 a. Range = 1.05 − .72 = .33

b. Class	Class Intervals	Class Frequency	Relative Frequency
1	.705– .755	2	2/25
2	.755– .805	4	4/25
3	.805– .855	8	8/25
4	.855– .905	4	4/25
5	.905– .955	4	4/25
6	.955–1.005	2	2/25
7	1.005–1.055	1	1/25

d. $\frac{7}{25}$

2.7

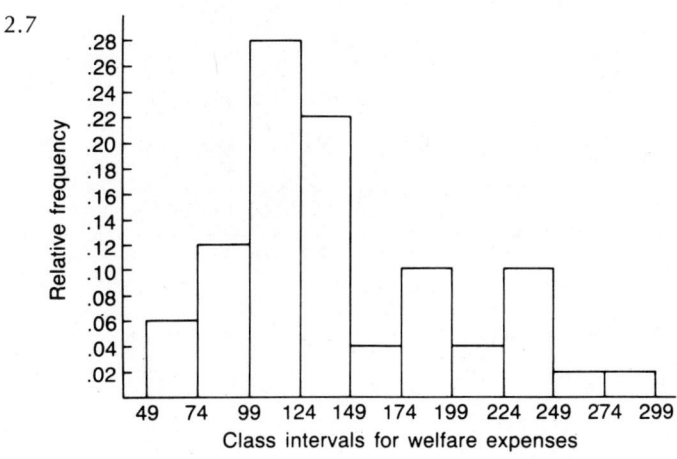

Dollars	Relative Frequency
50– 74	.06
75– 99	.12
100–124	.28
125–149	.22
150–174	.04
175–199	.1
200–224	.04
225–249	.1
250–274	.02
275–299	.02

2.8

CLASS	FREQUENCY	PERCENT	CUMULATIVE FREQUENCY	CUMULATIVE PERCENT
22.5-24.5	1	4.8	1	4.8
26.5-28.5	1	4.8	2	9.5
28.5-30.5	10	47.6	12	57.1
30.5-32.5	9	42.9	21	100.0

2.9

RELATIVE FREQUENCY HISTOGRAM

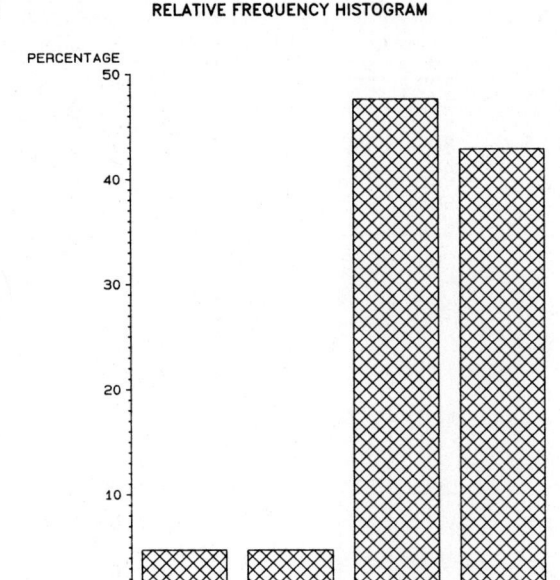

2.10

PLOT OF CLASS*NUMS   LEGEND: A = 1 OBS, B = 2 OBS, ETC.

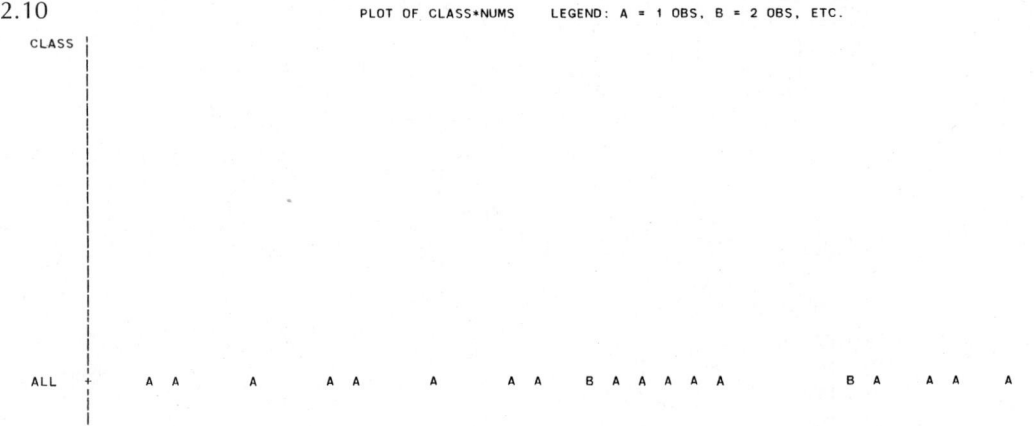

ANSWERS TO SELECTED EXERCISES

2.11
```
STEM LEAF                   #
  4  5                      1
  4  144                    3
  3  55689                  5
  3  0                      1
  2  556789                 6
  2  23                     2
  1  569                    3
  1  2                      1
  0  89                     2
     ----+----+----+----+
```

2.12
```
STEM LEAF                   #
  3  6                      1
  3  12                     2
  2  79                     2
  2  0144                   4
  1  55688                  5
  1  023344                 6
  0  6678                   4
  0  1234                   4
     ----+----+----+---+
MULTIPLY STEM.LEAF BY 10++01
```

2.13 The stem-and-leaf plot because it displays the magnitudes of the values as well as the frequencies.

FREQUENCY HISTOGRAM

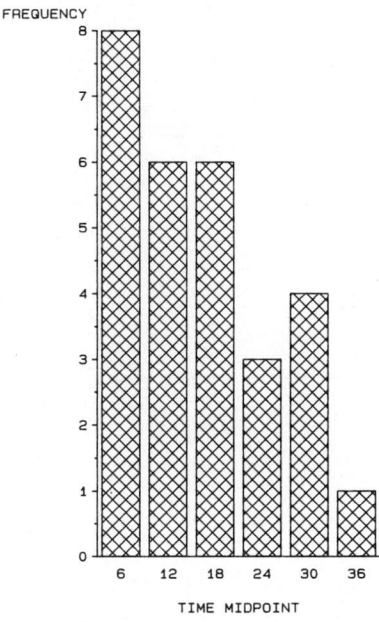

2.14 For the years 1967–1983, the difference between males and females has remained relatively constant, whereas there appears to have been a dramatic drop in verbal scores for both sexes from 1970 on, with more of a drop in females' scores.

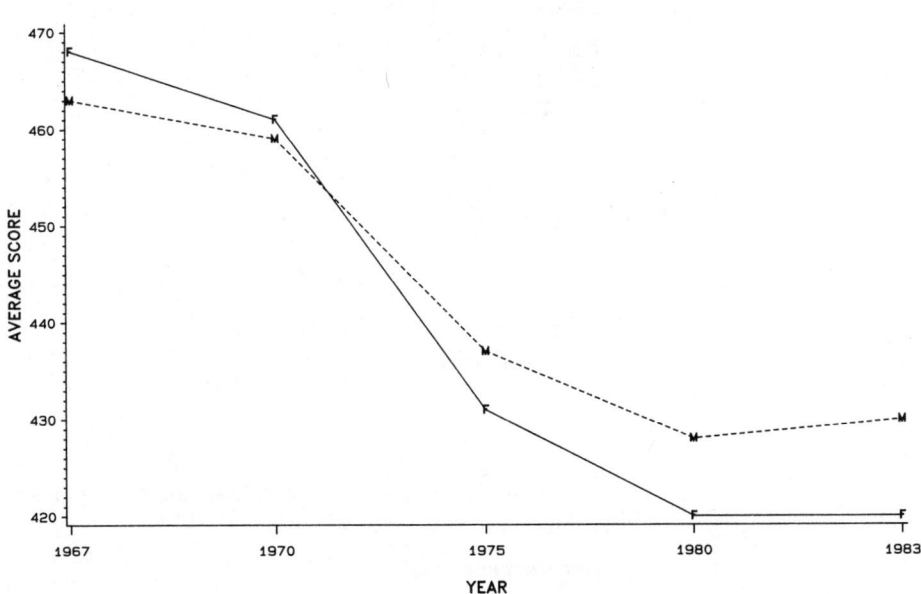

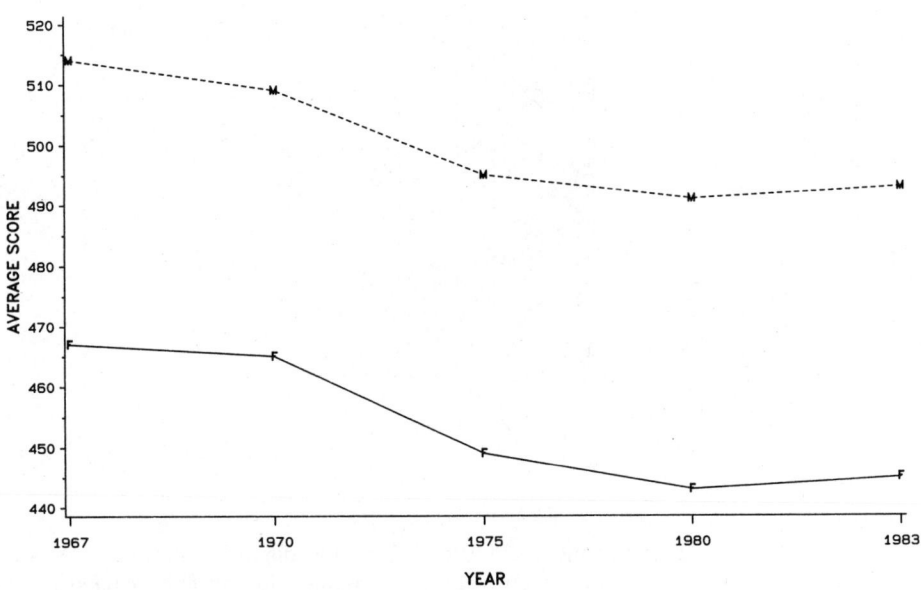

ANSWERS TO SELECTED EXERCISES

2.16 Stem-and-leaf plot for telephone data

```
3 │ 50
4 │ 50  70  70  80  80  80
5 │ 00  10  20  20  30  30  30  40  40  40  40  40  50  50  50  60  60  60  60  60  70  70  70  70  70  70  70  80  80  80  80  90  90
6 │ 00  10  10  10  20  20  30  50  50
7 │ 20
```

2.18 The frequency histogram and the stem-and-leaf plot for these data have the same shape.

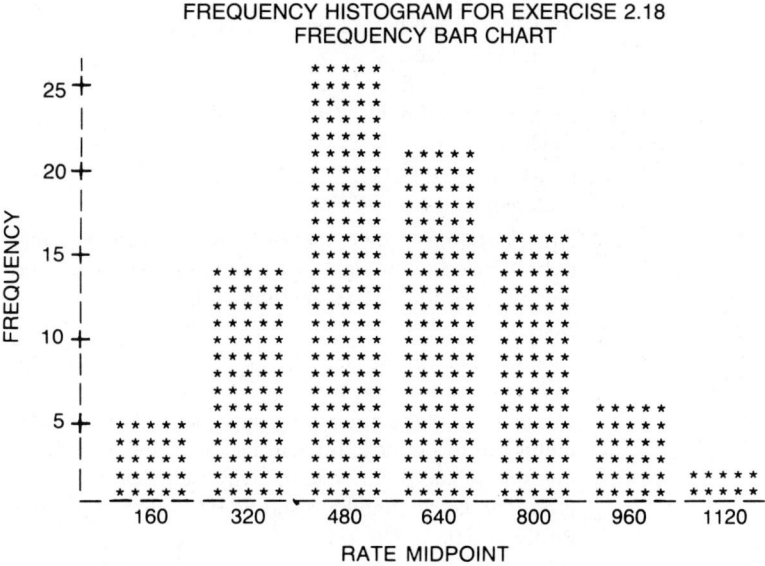

FREQUENCY HISTOGRAM FOR EXERCISE 2.18
FREQUENCY BAR CHART

2.19 Mean = 15.2
Median = 14.5
Mode = 18

2.20 Mean = 17.7
Median = 14.5
Mode = 18

2.21 14.8

2.22 Mode = 37.4
Median = 36.23
Mean = 36.03

2.23 Mode = 5
Median = 5.5
Mean = 6

2.24 Mean = 7.6
Median = 7.4
Mode = 7

2.25 Median = 906
Mean = 970.1

2.26 Mean = 8.716
Median = 8.675

2.27 Mean = 8.466
Median = 8.6

2.29 a. The rounded measurements are

2.10	1.98	1.99	2.77
2.47	2.70	2.10	2.36
1.75	2.43	2.67	1.80
2.94	2.17	2.65	3.09
1.69	2.80	2.06	2.20
2.75	2.82	2.55	2.93
2.82	2.38	2.22	1.85
2.52	2.68	2.92	2.28
2.77	2.39	3.05	1.96

b. Mode = 2.10, 2.77, 2.82   c. Median = (2.43 + 2.47)/2 = 2.45
d. Mean = 2.43

2.31 Since 10 of the 15 measurements are less than the sample mean, the distribution appears to be skewed to the right. The sample median is a more appropriate measure of central tendency for these data; median = 61.61.

2.33 Average mean = 21.5
Average median = 21.5
Average mode = 11.3
The average of the three net group means and the mean of the complete set of measurements are the same. This will be true whenever the groups have the same number of measurements. But, due to the way the mode and median are computed, the average of group modes and medians should differ from the overall median and mode. This is true for these data.

2.35 Data set one: Data set two: Data set three:
$\bar{y} = 2$ $\bar{y} = .02$ $\bar{y} = 1002$
$s^2 = 1; s = 1$ $s^2 = .0001; s = .01$ $s^2 = 1; s = 1$
For data sets one and two, s is the same for both exercises. For data set three, s is not the same for both exercises. The formula, $s = \sqrt{\sum(y_i - \bar{y})^2/(n - 1)}$ appears to be more accurate. Be careful when using the shortcut formula for s, because of rounding errors that may occur.

2.37 Urban 30%
Suburban 50%
Rural 20%

2.38 Resigned 24%
Transferred 26.4%
Retired 49.6%

2.39 <29   20%
30–39   24%
40–49   24%
>50   32%

2.40 a. Hospital A   22
              B   17.8
              C   26
              D   30
c. Hospital B

2.41 a. Yes

b.

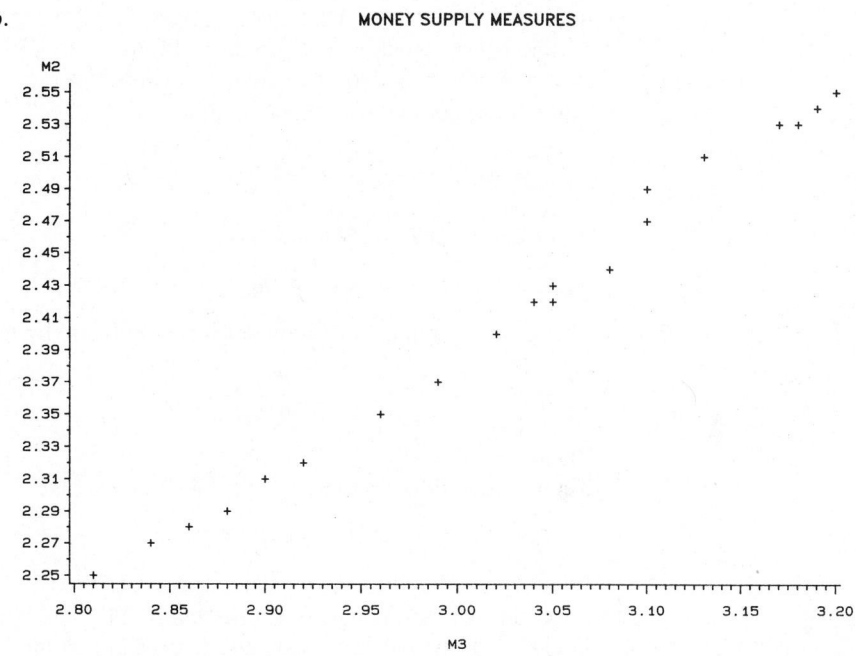

MONEY SUPPLY MEASURES

2.42 M2 and M3 move simultaneously and seem to reflect the same changes in the money supply.

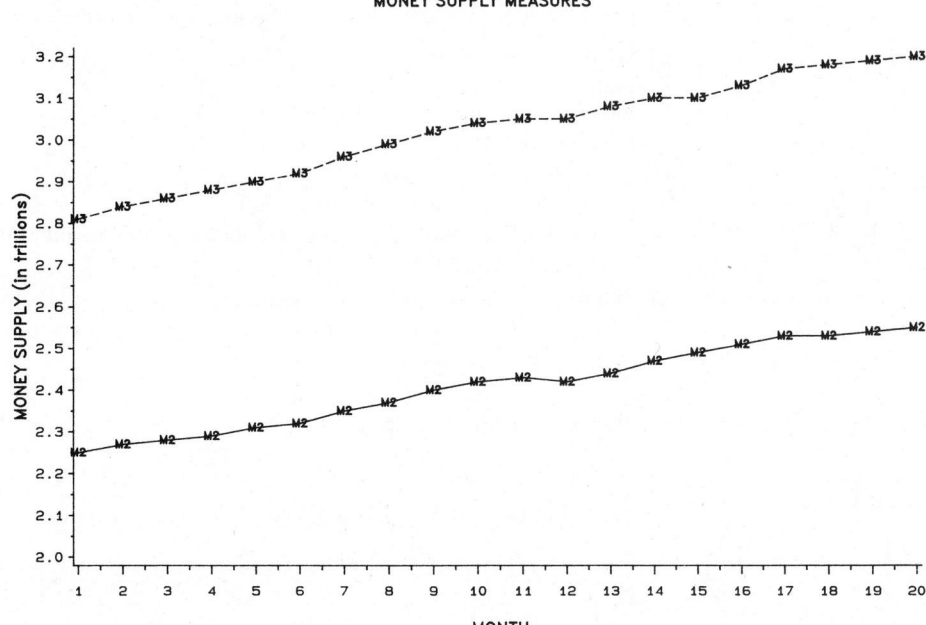

MONEY SUPPLY MEASURES

2.43 b. $s^2 = \frac{1}{4}[37 - (11)^2/5] = 3.20$; $s = \sqrt{3.20} = 1.79$

2.45 The 50 times are arranged in order of magnitude below:
4,4,5,5,5,5,6,6,7,8,9,9,9,10,11,11,12,12,12,13,14,14,15,15,15,
16,16,16,17,17,17,18,19,21,21,21,22,23,25,25,25,26,27,27,29,29,33,33,34
a. Mode = 5; median = 15; mean = 15.96   b. $s \approx$ range/4 = $(34 - 4)/4 = 7.5$
c. $s = 8.39$ d. No, they are not mound-shaped.

2.47 The formula for the mean in coded units is

$$\bar{y} = \frac{\sum fy_c}{n} = \frac{-108}{100} = -1.08$$

The mean for the original units is

$$\bar{y} = w\bar{y}_c + m = 0.1(-1.08) + 4.4 = 4.292$$

The sample variance (in coded units) can be found by the following formula:

$$s_c^2 = \frac{1}{n-1}\left[\sum fy_c^2 - \frac{(\sum fy_c)^2}{n}\right] = \frac{1}{99}\left[1066 - \frac{(-108)^2}{100}\right] = 9.59$$

Then $s_c = \sqrt{9.59} = 3.10$.
The standard deviation for the original chick data is

$$s = ws_c = 0.1(3.10) = 0.31$$

These answers agree with those in Exercise 2.46.

2.49 a. Mean = .65; median = .55; mode = .5   b. Mean = 1.21; median = .55;
mode = .5; median and mode are same as in part (a)

2.51 The coded measurements are 7,5,5,6,5,4,3,9,12,9. The mean and standard deviation
for the coded measurements are, respectively, 6.5 and 2.76. Then $\bar{y} = \bar{y}_c/10 = .65$ and
$s = s_c/10 = .276$ or .28.

2.53 From the box plot we can obtain the following information:

lower quartile, $Q_1 \approx 525$
upper quartile, $Q_3 \approx 574$
interquartile range, $Q_3 - Q_1 \approx 574 - 525 = 49$
median, $M = 560$

2.55 a. Mode = 2.5; median = $L + (w/f_m)(.5n - cf_b) = 5.5 + (2/13)(45 - 35) = 7.04$
b. Mean = $\sum fy/n = 747/90 = 8.3$
c. Since the distribution is skewed to the right, the median provides a better measure of
the center of the distribution.

2.57 The formula for the mean in coded units is

$$\bar{y}_c = \frac{\sum fy_c}{n} = \frac{-7}{90} = -0.078$$

The mean in the original units is

$$\bar{y} = w\bar{y}_c + m = 2(-0.078) + 8.5 = 8.34$$

The formula for the standard deviation in coded units is

$$s_c^2 = \frac{1}{n-1}\left[\sum fy_c^2 - \frac{(\sum fy_c)^2}{n}\right] = \frac{1}{89}\left[665 - \frac{(-7)^2}{90}\right] = 7.4658$$

Then $s_c = \sqrt{7.4658} = 2.73$.
The standard deviation for the original murder rate data is

$$s = ws_c = 2(2.73) = 5.46$$

2.59 a. Median $= .5 + (1/185)(250 - 170) = .93$; mode $= 1.0$
   b. Mean $= \sum fy/500 = 663/500 = 1.33$
   c. The distribution is skewed to the right.

2.61 Stem-and-leaf plot of blood donors

   a. 2 | 50 60 70 74 95
      3 | 01 08 10 15 15 20 25 32 33 34 34 56 68 70 86

   b. Median location $= \dfrac{20 + 1}{2} = 10.5$; median $= \dfrac{315 + 320}{2} = 317.5$

   Quartile location $= \dfrac{10 + 1}{2} = 5.5$

   Lower quartile, $Q_1 = \dfrac{295 + 301}{2} = 298$

   Upper quartile, $Q_3 = \dfrac{334 + 334}{2} = 334$

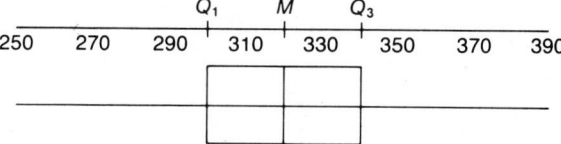

2.63 a. For the south: $\overline{y} = \dfrac{187.7}{10} = 18.8$

   For the north: $\overline{y} = \dfrac{297}{9} = 33$

   For the west: $\overline{y} = \dfrac{103.7}{8} = 13$

   b. $\overline{y} = \dfrac{10(18.8) + 9(33) + 8(13)}{27} = 21.8$ c. $\overline{y} = \dfrac{588.4}{27} = 21.8$

2.65 a. Old policy:

   $$\overline{y} = \frac{69}{15} = 4.6$$

   $$s^2 = \frac{1}{14}\left[413 - \frac{(69)^2}{15}\right] = 6.8; \ s = 2.6$$

   New policy:

   $$\overline{y} = \frac{34}{15} = 2.3$$

   $$s^2 = \frac{1}{14}\left[226 - \frac{(34)^2}{15}\right] = 10.6; \ s = 3.3$$

   b. Exercise for student

2.67 b.

	Plants	Arrests
Mean	677,794.6	95.0
10% trimmed mean	420,497.6	73.9
20% trimmed mean	142,327.6	59.7

2.68 Scatterplot

2.69

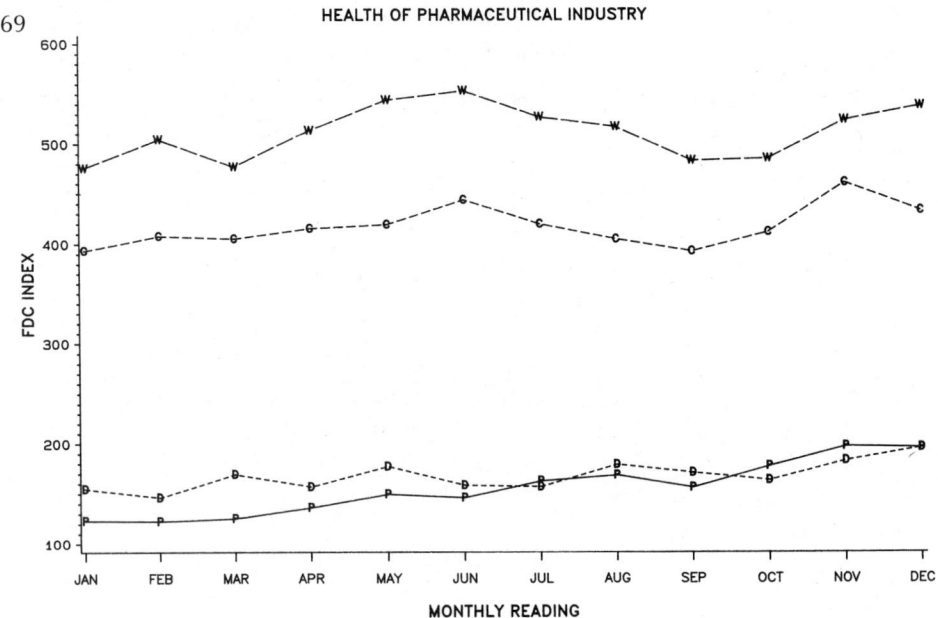

2.70

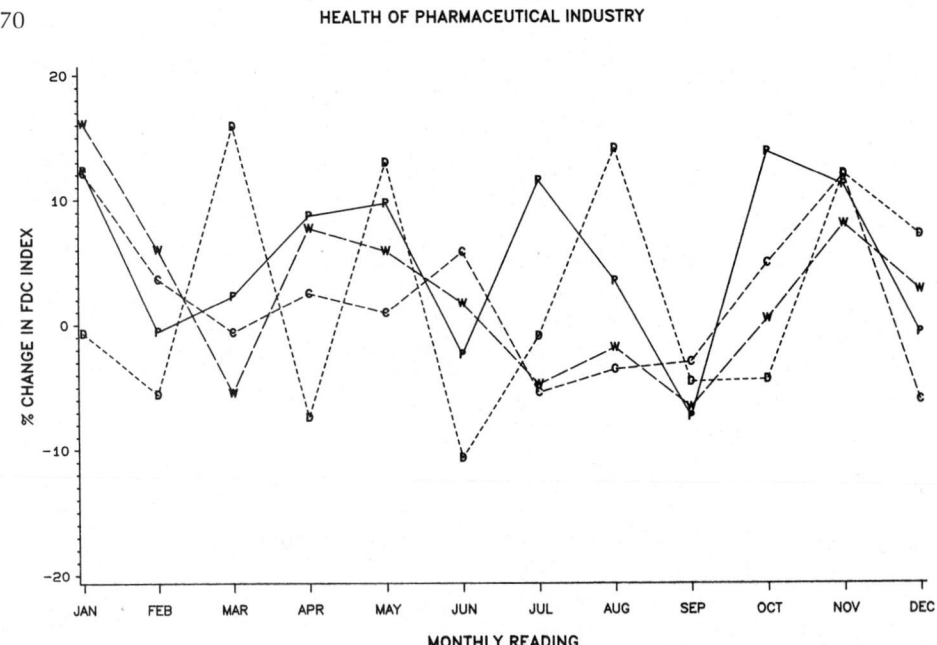

2.71 a. 141.125
b. 73.209
c. To reflect the relative importance of the component companies

Chapter 3

3.1 a. Personal or subjective probability   b. Relative frequency concept of probability
c. Classical interpretation of probability.

3.3 a. $P(A) = 0.65$, $P(B) = 0.40$
b. $P(A \cap B) = 0.30$
c. $P(A \cup B) = 0.75$

3.4 No, because $P(A \cap B) = 0.30 \neq 0$

3.5 There are 12 possible schedules.

3.6 b. $P(\text{both OK}) = 0.30$
c. $P(\text{one OK, one shaky}) = 0.60$
d. $P(\text{both shaky}) = 0.10$

3.7 $P(A) = .436$, $P(B) = .291$, $P(A \cap B) = .109$

3.8 a. No
b. $P(B \mid A) = .25$, $P(B \mid \overline{A}) = .323$

3.9 b. $P(A) = \frac{3}{8}$, $P(B) = \frac{7}{8}$, $P(A \cap B) = \frac{3}{8}$, $P(A \mid B) = \frac{3}{7}$
c. No

3.10 a. $P(A) = 0.63$, $P(B) = 0.36$, $P(C) = 0.14$
b. $P(A \mid B) = 0.61$, $P(A \mid \overline{B}) = 0.64$, $P(\overline{B} \mid C) = 1.0$
c. $P(A \cup B) = 0.77$, $P(A \cap C) = 0.0$, $P(B \cap C) = 0.0$

3.11 a. 0.49
b. 0.91

3.12 a. 0.665
b. 0.27
c. 0.065

3.13 To calculate the actual probabilities, use the binomial formula $P(y)$ for $y = 0, 1, 2, 3$ with the probability of a head being .5.

$$P(0) = .001; \ P(1) = .01; \ P(2) = .045; \ P(3) = .12$$

3.15 A. a. $P(0) = .028$ b. $P(4) = .198$
c. $P(y \le 4) = P(0) + P(1) + P(2) + P(3) + P(4) = .847$
d. $P(10) = .000006$

B. $P(y \le 100) = \sum_{y=0}^{100} \frac{1000!}{y!(1000-y)!} .6^y .4^{1000-y}$

3.16 a. 0.201
b. 0.3456
c. 0.204

3.17 a. 0.8263
b. 0.1737
c. 0.9993
d. 0.0086

3.18 a. 0.194
    b. 0.264
    c. 0.276
    d. 0.599

3.19 No; not identical trials

3.21 Exercise for student.

3.23 a. 0.475
    b. 0.025
    c. 0.95
    d. 84.2

3.25 a. .4332   b. .4641

3.27 1.645, −1.645

3.29 a. .025   b. .0136   c. .0021   d. .2327

3.31 a. .0336   b. Since 55 is 2.67 standard deviations above $\mu = 39$ and the probability of observing a value of 55 or greater is .0038, we would probably conclude that the voucher had been lost or misplaced.

3.35 No

3.37

150	729	611	584	255
465	143	127	323	225
483	368	213	270	062
399	695	540	330	110
069	409	539	015	564

3.39 —IRANDOM 50 INTEGERS BETWEEN 1 and 1000, PUT IN C1.

458.	247.	797.	795.	543.	211.	481.	379.	707.	130.
826.	435.	548.	48.	938.	184.	526.	640.	306.	31.
633.	972.	163.	825.	275.	612.	564.	22.	854.	17.
691.	239.	907.	896.	854.	17.	105.	491.	219.	710.
520.	388.	494.	582.	695.	290.	747.	612.	723.	806.

—STOP

3.41 The sampling distribution of the sample sum will be approximately normal, with mean $= 16(60) = 960$ and standard deviation $= 5\sqrt{16} = 20$. Observing a measurement more than 70 units (3.5 standard deviations) away from the mean (960) of a normal distribution is very improbable.

3.43 a. .7462   b. .1587   c. .0188

3.45 a. 125 ± 32 should contain approximately 68% of the weeks; 125 ± 64 should contain approximately 95% of the weeks; 125 ± 96 should contain approximately all of the weeks.
    b. .1379

3.47 a. .99   b. .38

3.49 a. Approximately normal with mean $= 3.7$ and standard deviation $= .06$
    b. Approximately normal with mean $= 3.7$ and standard deviation $= .03$
    c. Essentially 0

3.51 a. Approximately normal with mean $= 1300$ and standard deviation $= 35.78$
    b. 25th percentile $= 1275.85$; 75th percentile $= 1324.15$

3.53 a. 10   b. .0571, .0057

3.57 a. Essentially zero   b. Essentially zero

3.59 2.63.

ANSWERS TO SELECTED EXERCISES

3.61 No, it has a mean breaking strength higher than the old fabric.

3.63 .03078; the probability that only one or less of the five is selected is 3 changes in 100. Thus this improbable occurrence is a rare event. Having actually observed this event, we could draw one of two conclusions. Either we have observed an unlikely event, or the board is currently admitting with a probability less than .7. We would tend to accept the latter conclusion.

3.65 a. .484   b. .0717   c. .4562

3.66 a. y = # of individuals (out of 1000) planning to buy a drink
   b. No
   c. Normal approximation

3.67 a. 0.005
   b. 0.9044

3.68 a. $P(y \leq 25) \approx 0.00$
   b. No

3.71 Mean = 3.50 = median

3.73 Mean = 3.2, median = 3.0

Chapter 4

4.1 a. 101.95–108.05
   b. 100.99–109.01

4.3 a. The width of the interval will be narrower by $1/\sqrt{2}$ times previous width.
   b. The width of the interval will cut in half (for example, $1/\sqrt{4}$ times previous width).

4.4 107.66–112.34 mg

4.5 3.99–6.41%

4.6 25.56–$27.24

4.7 Point estimate of μ is 3.2; 3.2 ± .176.

4.9 850 ± 25.3

4.11 430 ± 17.11

4.12 a. 188,752,3007
   b. By 4

4.16 Using the range/4 as an estimate for σ, we have $n \approx 1537$.

4.18 $n = 44$

4.20 $\bar{y}_c = 109.3$; UCL = 110.56; LCL = 108.04

4.22 With $\Delta = 4$, $z_\alpha = 1.645$, $z_\beta = 1.28$ we obtain $n = 5.01 \approx 6$.

4.24 a. Reject H_0; $\mu > 38$
   b. No; type II error is possible only when H_0 is not rejected.

4.25 Power = .8107, .9997, and 1.0000

4.26 Reject H_0; $\mu > 2.0$

4.27 Power = .1831, .5643, .8568, and .9980

4.28 a. 9.7 ± 2.21
   b. We are 95% confident that this interval captures the population mean speed for all fourth-grade students.
   c. 9.7 ± 2.76

4.30 H_0: $\mu = 80$. H_a: $\mu \neq 80$. T.S.: $t = (84.2 - .80)/(12.22/\sqrt{10}) = 1.09$. $p > .20$

4.32 $t = -.31$. There is insufficient evidence to indicate that $\mu < 5.0$.

4.34 a. H_0: $\mu = 5.2$. H_a: $\mu < 5.2$. T.S.: $z = (5.0 - 5.2)/(.70/\sqrt{50}) = -2.02$. R.R.: For $\alpha = .05$, reject H_0 if $z < -1.645$

   b. Since $z < -1.645$, we reject H_0 and conclude that the mean dissolved oxygen count is less than 5.2 ppm.

4.36 $z = -2.36$, $p = .0091$

4.38 430 ± 22.53

4.40 a. 30.51 ± 4.09 b. 30.51 ± 5.39

4.42 22 ± 1.04

4.44 $98.4 \pm .04$

4.46 a. H_a: $\mu \neq 29$ b. $p \approx .48$

4.48 $.76 \pm .04$

4.50 58 ± 4.08

4.52 $\beta = 0$ for $\mu = 40$; $\beta = .0301$ for $\mu = 38$; $\beta = .3192$ for $\mu = 36$; $\beta = .8624$ for $\mu = .34$

4.54 a. 0.76200 b. $.98$

   c. 0.72044 to 0.080356. We are 98% confident that the population mean proportion of patients per hospital with group medical insurance lies in the interval $.72044$ to $.80356$.

4.56 $z = 2.39$; $p = .0168$

4.58 a. The t distribution is appropriate when the sample is selected from a population with a mound-shaped distribution. In this case, since $s > \bar{y}$ the distribution will be skewed, and a confidence interval for μ based on t would be inappropriate.

   b. The sample median could be used to estimate the center of the distribution.

4.60 $1,537$

4.63 17.48–22.52%

4.65 515.5–604.5 mg

4.66 658

4.67 a. $p < .001$

   b. $12,562.5$ pints

Chapter 5

5.1 $t = 10.71$; reject H_0: $\mu_1 - \mu_2 = 0$

5.3 $s = 4.87$, $t = 1.91$, $.025 < p < .05$

5.5 No, the distribution will be skewed since $s > \bar{y}$ for the magnesium data.

5.7 a. No

   b. $t = 4.24$, $p < .001$

      $t' = 4.23$, $p < .001$

5.9 -6487.7 to -5976.3

5.10 b. -22.4 to 2.4

   c. -41.4 to -20.6 does not include zero

5.11 b.

Age	99% C.I.
9	−5.2 to 41.2
13	−1.7 to 43.7
17	2.6 to 47.4

5.13 a. Plumber 1: $\bar{y} = 88.8$, $s = 7.9$
Plumber 2: $\bar{y} = 108.9$, $s = 8.7$

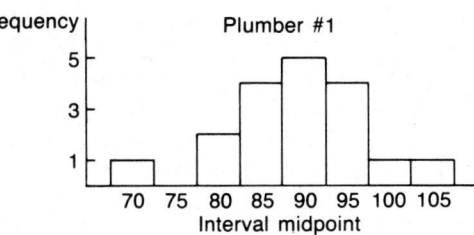

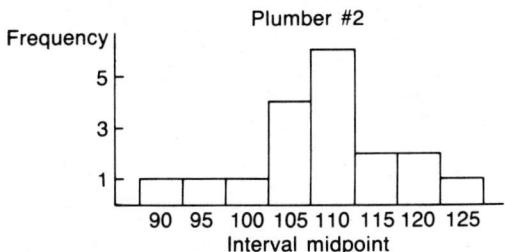

b. Since both graphs show a single, peaked, roughly symmetrical distribution, a t test appears to be appropriate.

5.15 a. $z = 2.92$; reject H_0
b. Two-sample t test if 2 populations have a common variance.

5.19 a.

	Plumber 1	Plumber 2
Largest	102.6	126.8
Smallest	71.4	90.2

$C = 15 + 11 = 26$. At $\alpha = .05$ we reject H_0
b. The t test, rank sum test and Tukey–Duckworth test all had the same conclusion, to reject H_0.

5.21 Exercise for student

5.23 $t = 4.91$

5.25 865

5.27 a. $p = .0001$. We reject H_0—the hypothesis that the population mean scores for those not exposed and those exposed to a minority environment are identical.
b. The conclusion here is the same as the conclusion of Exercise 5.24. In this case, it does not make a difference if you use a t test or a Wilcoxon's signed rank test.

5.29 $T = 16$; we reject H_0

5.31 1.3 ± 3.35; yes, since a mean score of zero is included in the interval.

5.33 $n = 43.8 \approx 44$

5.35 $6.16 \pm .88$

5.37 4.9 ± 2.59

5.39 $t = 2.12$; reject

5.41 8.1 ± 5.43

5.43 A 95% confidence interval on $\mu_1 - \mu_2$ is -23.40 ± 12.02.

5.45 a. There are not enough observations to draw any conclusions concerning normality. There is no suggestion that the population variances are different.

5.47 a. $p < .001$; we are rejecting the hypothesis that there is no difference between the average symptom scores of patients on drug A and those on the placebo.

   b. Differences existing at the end of the study may be due to baseline differences. To guard against baseline differences, both treatment groups should have the same type of patients. For example, one treatment group should not include all the severe patients.

5.55 a. HMOs in general have a shorter length of stay.

   b. $t = 14.93$, $p < .001$

Chapter 6

6.1 $\chi^2 = 5.28$; insufficient evidence to reject H_0

6.2 $\chi^2 = 9.08$; insufficient evidence to reject H_0. The goodness-of-fit test is insensitive to changes in the cell probabilities under H_0.

6.3 $\chi^2 = 49.007$, $p < .001$

6.4 $\chi^2 = 6.95$; insufficient evidence to reject H_0

6.6 $\chi^2 = 6.00$; $p = .05$; we reject H_0

6.8 $\chi^2 = 123.60$; $p < .001$.

6.10 $\chi^2 = 2.40$; insufficient to reject H_0

6.11 Yes

6.12 a. Yes

   b. Yes, $n\pi = 16.7$ and $n(1 - \pi) = 33.3$

   c. $0.246 - 0.515$. Increase the sample size.

6.13 a. 0.274–0.326

   b. 0.255–0.305

   c. 0.226–0.274

   d. 0.197–0.243

   e. 0.187–0.233

6.15 a. A histogram using % response as the vertical axis

6.16 a. A table listed in descending order of proof of illiteracy

6.17 a. Normal approximation is valid

   b. Change for breakfast is statistically significant. One would have to decide whether a 2% increase (on a national basis) represents a relatively large shift.

6.19 a. $\hat{\pi} = .37$; half-width of a 95% confidence interval is .024

   b. 8955

6.21 Exercise for student

6.23

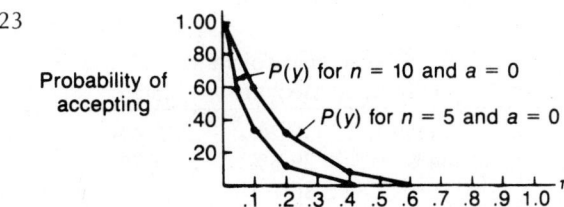

Probability of accepting — $P(y)$ for $n = 10$ and $a = 0$ — $P(y)$ for $n = 5$ and $a = 0$

6.25 Yes, none of the percentages was outside the range 0 to 30.

6.27 $.025 \pm .169$

6.29 $z = 2.3$; $p = .011$

6.31 a. $z = 3.43$, $p < .001$
   b. What was the amount of hair growth? What side effects were observed? What characteristics distinguish responders from nonresponders?

6.32 a. $z = 7.91$, $p < .001$
   b. If cocaine has a similar effect in humans, it is a *very* dangerous drug, even more so than heroine.

6.33 For $\pi = .002$, $P(y \le 2) = .42319$
   For $\pi = .003$, $P(y \le 2) = .173578$

6.35 a. $.0025$
   b. $.9826$
   c. $.9975$

6.37 Use the Poisson approximation. $P(y \ge 1) = .3935$

6.39 $\chi^2 = 26.80$; we reject H_0 and conclude that the number of shutdowns per day does not follow a common Poisson distribution.

6.41 a. $E_{11} = 11$; $E_{12} = 11$; $E_{21} = 9$; $E_{22} = 9$
   b. H_0: The fact that a sheep is a responder or non-responder is independent of treatment.
      $\chi^2 = 6.46$; we reject H_0 and conclude that the two variables are dependent, and that whether or not a sheep is a responder depends on the treatment received.

6.43 The results are the same.

6.44 a. No
   b. $\chi^2 = 1.75$, insufficient evidence to reject H_0

6.48 $\chi^2 = 32.96$, $p < .001$

6.51 $\chi^2_{MH} = 11.55$; $p = .0007$; we reject the null hypothesis that there is no difference, on the average, in males and females for number of alcohol-related arrests.

6.53 $< \$20,000$ 57%
       $\$20,000-\$40,000$ 31%
        $> \$40,000$ 12%
   $\chi^2 = 27.22$; $p < .001$; we reject H_0 that the two variables are independent.

6.55 $\chi^2 = 6.43$; insufficient evidence to reject H_0

6.57 $\chi^2 = 17.07$; since our computed value is greater than the critical value, we will reject H_0 and conclude that there is a relationship between the two variables.

6.58 $0.64-0.68$

6.64 $p = .0475$

6.66 $z = .85$; insufficient evidence to reject H_0: $\pi = .3$

6.68 a.

University	Academic Tolerance			Political Ideology	
	Low	Medium	High	Left	Right
1	22.4%	30.2%	47.4%	61.2%	38.8%
2	44%	29.6%	26.4%	50.9%	49.1%
3	35%	36%	29%	12%	88%

b. $\chi^2_{CMMA} = 39.82$; $p < .001$

6.70 In Exercise 6.69, when we collapsed the global outcome categories we could not reject H_0 that the treatment groups were the same. When the full data are used we can reject H_0 and conclude that differences exist between treatment groups. The lesson that can be learned from these results is that loss of information can distort test results.

6.72 a. $z = -4.12$; we reject H_0 that there is no difference between the sensitivities of the two fuses.

b. $\chi^2 = 16.67$; the relationship between the z test and the chi-square test of independence is that, except for rounding errors, z^2 is equal to χ^2 for df $= 1$.

6.74 $\chi^2 = 13.57$; reject H_0.

6.76 a. $z = -2.86$. At the $\alpha = .05$ level we reject H_0 that there is no difference in the number of acceptable tablets between formulations one and two. Thus the proportion of acceptable tablets differs for the two formulations.

b. $\chi^2 = 8$; we reject H_0 that the number of acceptable/unacceptable tablets is independent of the formulation; that is, the proportion of acceptable tablets depends on the formulation used.

c. The relationship between the z test and the chi-square test of independence is that except for rounding errors, z^2 is equal to χ^2 for df $= 1$.

6.78 We can analyze the data by running a chi-square test of independence. With $\chi^2_9 = 59.64$, we reject H_0 at the $\alpha = .05$ level and conclude that age and regularity of seat belt usage are not independent.

6.80 a. The critical assumption is the normality of the differences. Although there are only 12 observations, the data do not appear to be skewed.

b. The t test for H_0: $\mu_d = 0$ versus H_0: $\mu_d > 0$ has $t = 21.39$ and $p < .001$. We can conclude that the entry-level blood pressure and the follow-up blood pressure were not the same. With so large a t value, this conclusion does not depend heavily on whether or not there is a minor departure from normality.

6.81 a. Yes

b. $\chi^2 = 373.45$, df $= 33$, $p < .001$

6.82 a. $>$ one week: 0.431 to 0.609

$\leq$ one week: 0.036 to 0.304

b. No

c. $\hat{\pi} = .447$; 0.367 to 0.527

6.84 a. No

b. $\chi^2 = 21.63$, df $= 4$, $p < .001$

6.85 a. $\chi^2 = 23.46$, df $= 8$, $p < .005$

6.87 $\chi^2 = 87.94$, df $= 10$; reject H_0

ANSWERS TO SELECTED EXERCISES

Chapter 7

7.1 $.0017 < \sigma^2 < .0038$

7.3 $2.0434 < \sigma^2 < 15.7067$

7.5   No; a t test would have been inappropriate because the two samples did not have a common variance.

7.7   a. $10.4203 < \sigma^2 < 58.7436$

      b. The critical assumption is that the random sample is drawn from a normal population.

7.9   a. 22.5 mg to 27.5 mg

      b. A graph of the data shows a single-peak mound-shaped distribution that supports the normality assumption.

      c. For testing H_0: $\sigma^2 = (27.5 - 22.5/'4)^2 = 1.5625$, $\chi^2_{29} = 40.058$; we cannot reject H_0.

7.11 $3.586 < \sigma^2 < 13.226$.

      The 95% confidence interval supports the test findings of the consumer group since $\sigma^2 = 4$ is in the interval. The test of Example 7.2 does not have much power to detect an increase in σ^2 of 25% over the claimed value because the confidence interval is relatively wide.

7.13 We wish to test H_0: $\mu_1 - \mu_2 = 0$ versus the alternative H_a: $\mu_1 - \mu_2 \neq 0$. The test statistic, t, is -7.29 and the p-value for the test is less than $.01$. Therefore we will conclude that differences in the average returns for the two portfolios do exist. The method we used was the test for $\mu_1 - \mu_2$, which assumes independent samples, normality, and equal variances. From the information given in Exercise 7.12 we know that we are dealing with independent samples. Also in 7 12, we did not reject the hypothesis for equal variances. There is not enough information to make a decision concerning normality.

7.15 $73.35 < \sigma^2 < 315.47$; $8.56 < \sigma < 17.76$

Chapter 8

8.1   a.

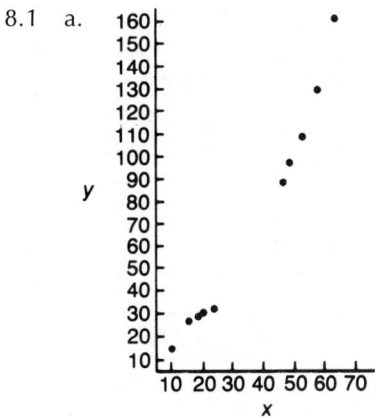

      b. Exercise for student   c. Exercise for student

8.3 a. 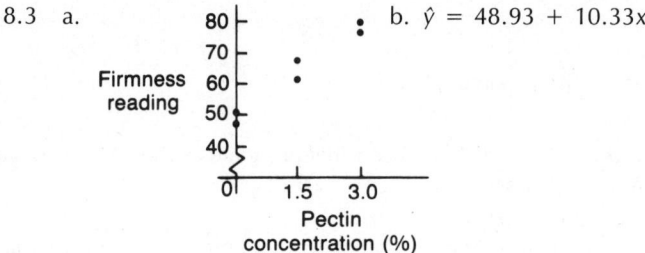 b. $\hat{y} = 48.93 + 10.33x$

8.5 a. 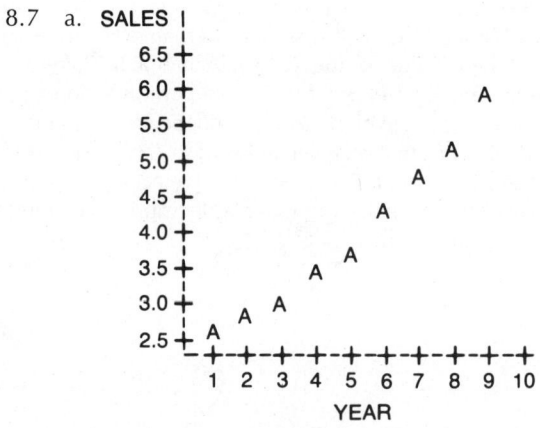 b. $\hat{y} = 8.46 - .14x$

8.7 a.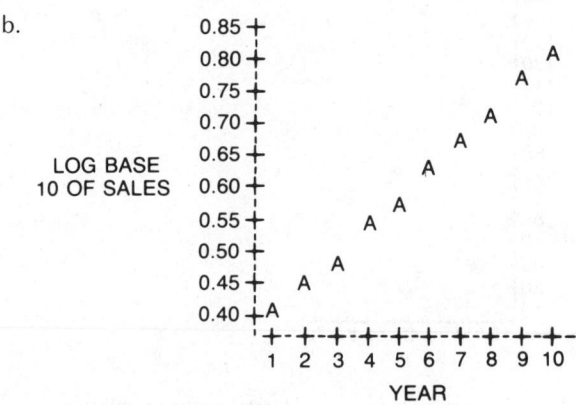

b.

c. Log sales vs. year

8.9   Using the regression equation with sales as the dependent variable, sales in year 11 would be 6.604. Using the regression equation with log sales as the dependent variable, sales in year 11 would be 7.140. Looking at the graph of sales versus year, the forecast of 7.140 for sales in year 11 appears to be more plausible.

8.11 a.

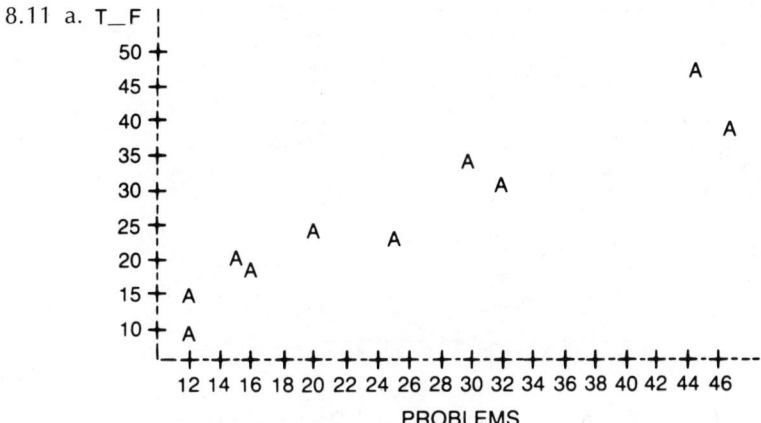

b. $r = .948$

8.13 a.

Amplitude vs Distance

No, y decreases as x decreases
b. $r_s = -.95$

8.15 a.   b. $\gamma = .8903$

Expenditure vs City size

8.17 a. $y = \beta_0 + \beta_1 x_1 + \beta_2 x_2 + \varepsilon$
      b. $y = \beta_0 + \beta_1 x_1 + \beta_2 x_1^2 + \beta_3 x_2 + \beta_4 x_2^2 + \varepsilon$
          $y = \beta_0 + \beta_1 x_1 + \beta_2 x_2 + \beta_3 x_1 x_2 + \varepsilon$

8.19

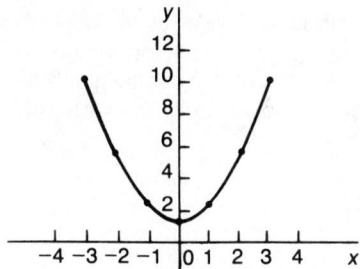

8.21

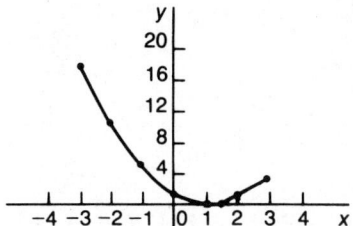

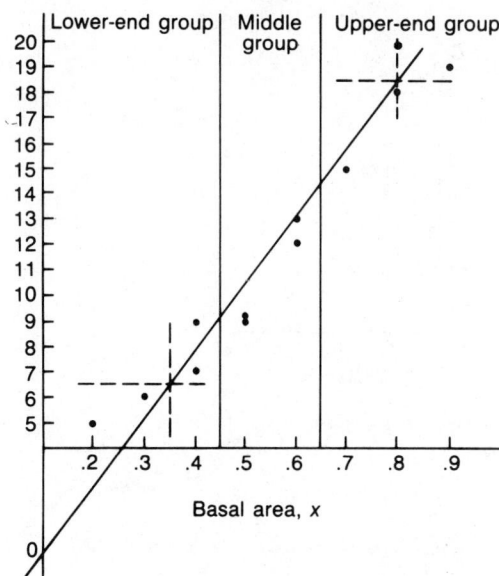

Basal area, x

8.23 The approximate linear regression equation is as follows:

$$\hat{y} = 26.67x \qquad (\hat{\beta}_0 = 0)$$

This approximate linear regression line appears to overestimate the regression line in the output.

8.25 a. $\beta_1 x_1$, first degree   b. Fourth-order
   $\beta_2 x_1^2$, second degree
   $\beta_3 x_1^3$, third degree
   $\beta_4 x_1^4$, fourth degree

8.27 a. $\beta_1 x_1$, first degree
$\beta_2 x_1^2$, second degree
$\beta_3 x_2$, first degree
$\beta_4 x_1 x_2$, second degree

b. This model is neither a first- nor second-order model. The first-order model for two independent variables is

$$y = \beta_0 + \beta_1 x_1 + \beta_2 x_2 + \varepsilon$$

while the second-order model is

$$y = \beta_0 + \beta_1 x_1 + \beta_2 x_2 + \beta_3 x_1^2 + \beta_4 x_2^2 + \beta_5 x_1 x_2 + \varepsilon$$

8.29 a. Model: $y = 1.2 + x$

x	y
-1	0.2
-0.5	0.7
0	1.2
0.5	1.7
1.0	2.2

b. Model: $y = 1.2 + x + .4x^2$

x	y
-1.0	0.6
-0.5	0.8
0	1.2
0.5	1.8
1.0	2.6

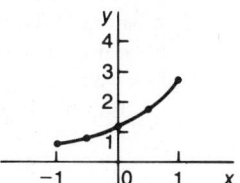

c. Model: $y = 1.2 + x + .4x^2 + .6x^3$

x	y
-1.0	0
-0.5	0.725
0	1.2
0.5	1.875
1.0	3.2

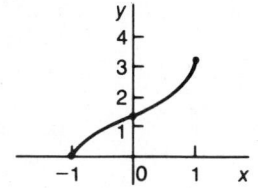

8.31 $y = \beta_0 + \beta_1 x + \varepsilon$

8.33 $y = \beta_0 + \beta_1 x + \beta_2 x^2 + \varepsilon$

8.35

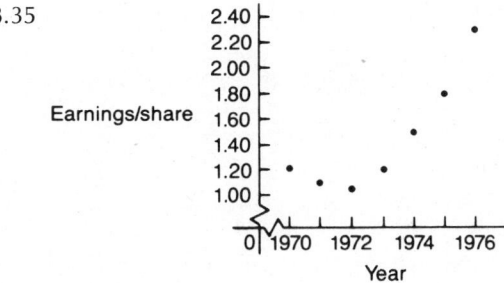

8.37 a.

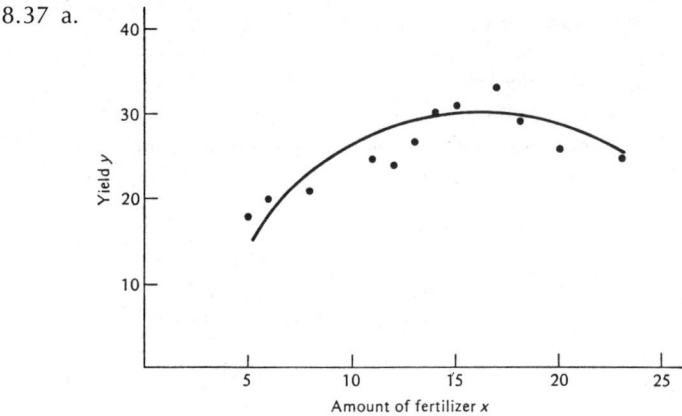

$$y = \beta_0 + \beta_1 x_1 + \beta_2 x_1^2 + \beta_3 x_2 + \beta_4 x_2^2 + \beta_5 x_1 x_2 + \varepsilon$$

b. $y = \beta_0 + \beta_1 x + \beta_2 x^2 + \varepsilon$ c. $r_s = .67$

8.39 $r = .961$

8.41 Since $r > 0$, a positive linear relationship exists between the two test results. One factor that might contribute to the less than complete agreement between the two test results would be the familiarity of the students with the chosen letter. For example, suppose the first letter was a B and the second letter a Z. The students would probably have more trouble thinking of words that start with a Z than a B.

8.43 a.

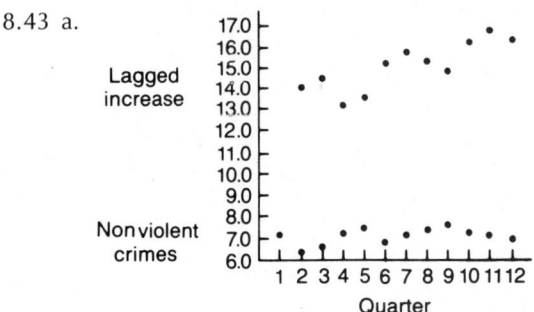

b. $r = .007$

8.44 a.

DEP VARIABLE: CARTONS

ANALYSIS OF VARIANCE

SOURCE	DF	SUM OF SQUARES	MEAN SQUARE	F VALUE	PROB>F
MODEL	1	705.56200	705.56200	84.794	0.0001
ERROR	13	108.17134	8.32087207		
C TOTAL	14	813.73333			

ROOT MSE	2.884592	R-SQUARE	0.8671	
DEP MEAN	22.53333	ADJ R-SQ	0.8568	
C.V.	12.80144			

PARAMETER ESTIMATES

VARIABLE	DF	PARAMETER ESTIMATE	STANDARD ERROR	T FOR H0: PARAMETER=0	PROB > |T|
INTERCEP	1	6.07911554	1.93588368	3.140	0.0078
HOURS	1	0.21406181	0.02324642	9.208	0.0001

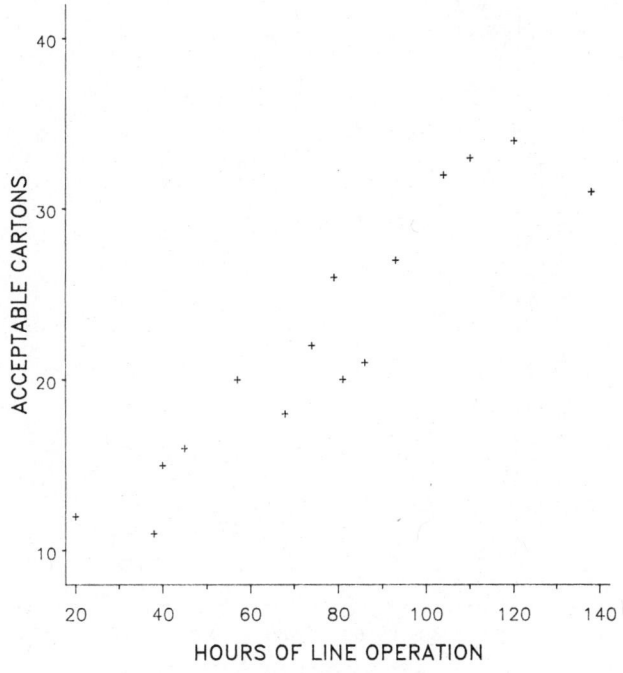

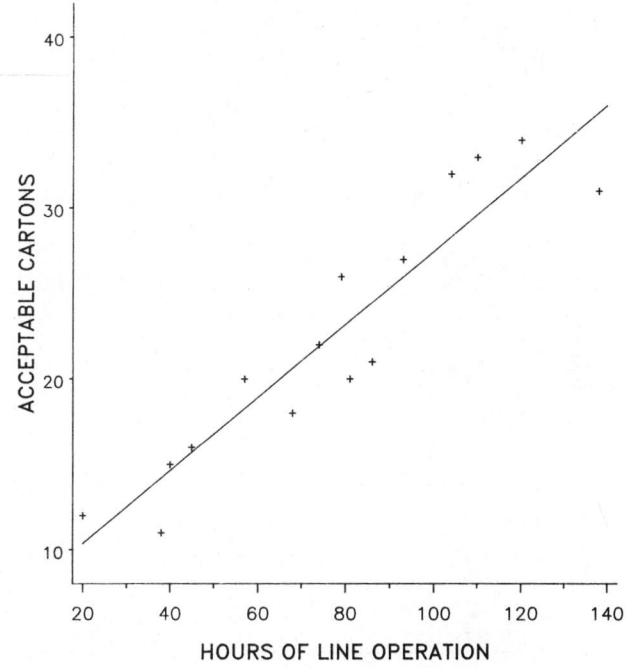

b. $\hat{y} = 6.08 + 0.21x$

8.45 a.

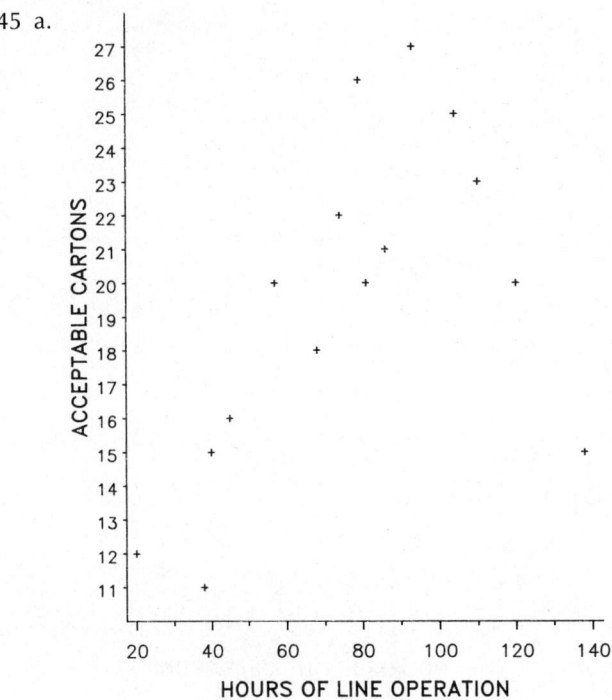

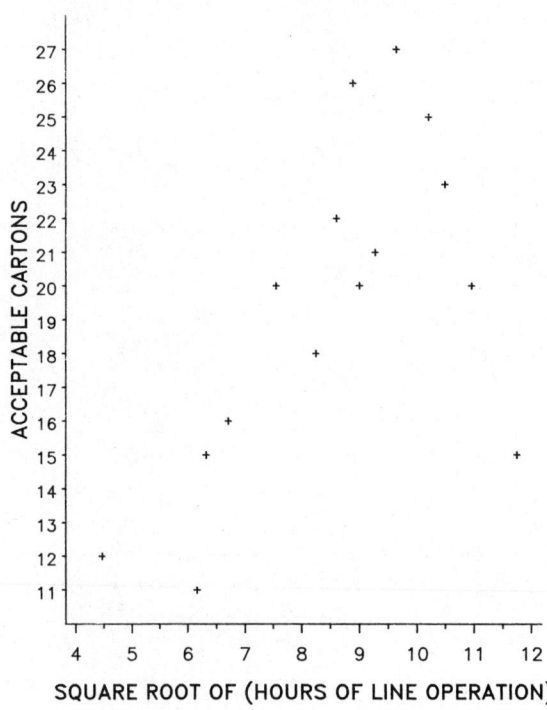

ANSWERS TO SELECTED EXERCISES

```
1   TITLE1 ' 8.8';
1   DATA REG;
4       INPUT LOCATION CARTONS HOURS;
4       X2=HOURS**.5;
4       CARDS;
4       1   12  20
4       2   11  38
4       3   15  40
4       4   16  45
4       5   20  57
4       6   18  68
4       7   22  74
4       8   26  79
4       9   20  81
4       10  21  86
4       11  27  93
4       12  25  104
4       13  23  110
4       14  20  120
4       15  15  138
4       ;
1   PROC REG;
4       MODEL CARTONS=X2;
```

DEP VARIABLE: CARTONS

ANALYSIS OF VARIANCE

SOURCE	DF	SUM OF SQUARES	MEAN SQUARE	F VALUE	PROB>F
MODEL	1	118.85538	118.85538	7.195	0.0188
ERROR	13	214.74462	16.51881672		
C TOTAL	14	333.60000			

ROOT MSE	4.064335	R-SQUARE	0.3563	
DEP MEAN	19.4	ADJ R-SQ	0.3068	
C.V.	20.95018			

PARAMETER ESTIMATES

VARIABLE	DF	PARAMETER ESTIMATE	STANDARD ERROR	T FOR H0: PARAMETER=0	PROB > |T|
INTERCEP	1	6.97313075	4.75014884	1.468	0.1659
X2	1	1.45330964	0.54179899	2.682	0.0188

b. No. $\sqrt{x}$

8.47

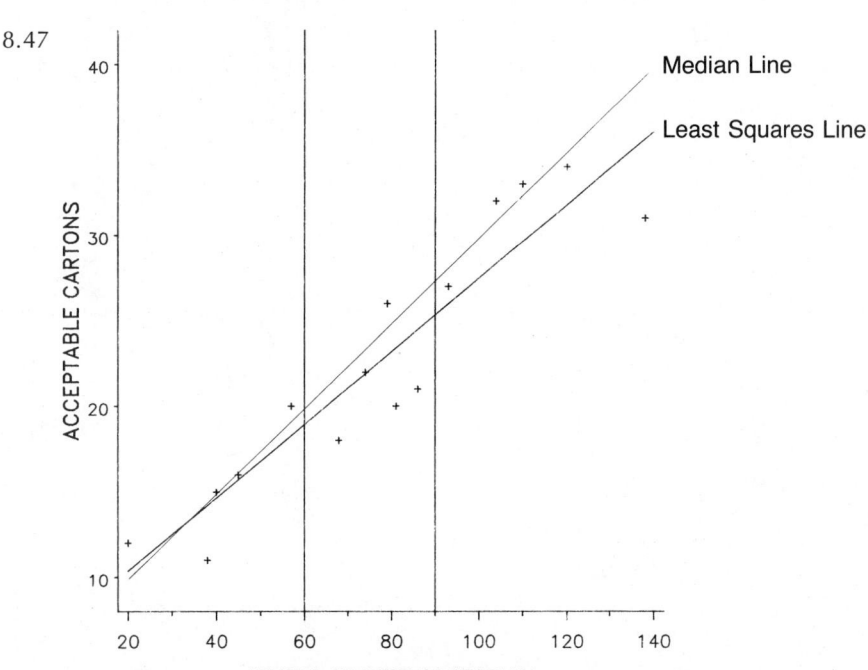

8.48 a. Yes
      b. $\hat{y} = 13.90 + 0.56x$
      c. .962

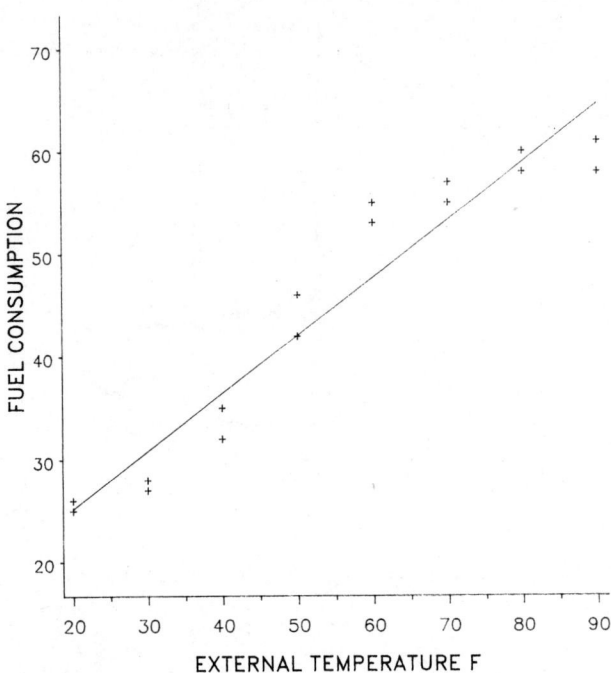

8.49 .984

8.50 a.

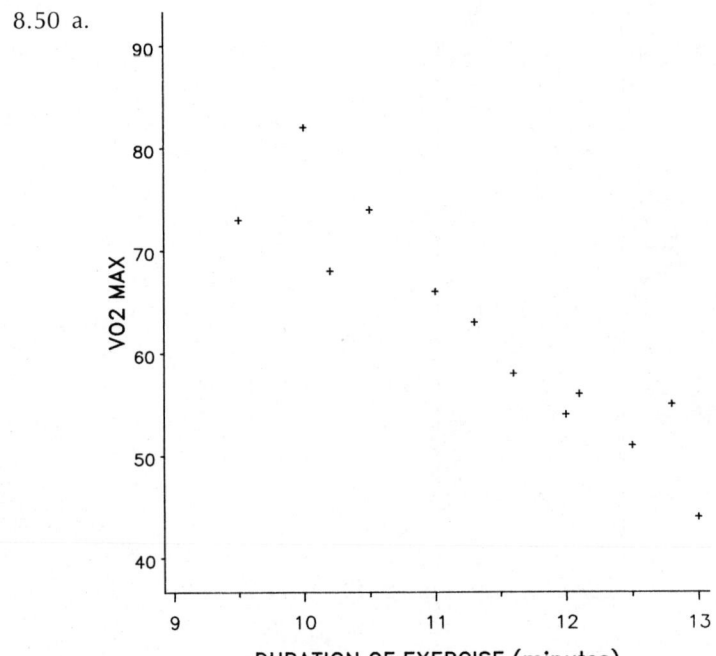

ANSWERS TO SELECTED EXERCISES

b. Yes
c. $\hat{y} = 162.57 - 8.84x$

DEP VARIABLE: VO2MAX

ANALYSIS OF VARIANCE

SOURCE	DF	SUM OF SQUARES	MEAN SQUARE	F VALUE	PROB>F
MODEL	1	1141.36689	1141.36689	61.156	0.0001
ERROR	10	186.63311	18.66331108		
C TOTAL	11	1328.00000			

ROOT MSE	4.320105	R-SQUARE	0.8595
DEP MEAN	62	ADJ R-SQ	0.8454
C.V.	6.967912		

PARAMETER ESTIMATES

VARIABLE	DF	PARAMETER ESTIMATE	STANDARD ERROR	T FOR H0: PARAMETER=0	PROB > |T|
INTERCEP	1	162.56583	12.92006797	12.582	0.0001
EXERCISE	1	-8.84095189	1.13052648	-7.820	0.0001

8.51 a. $\hat{y} = 179.12 - 10.40x$

```
4     ;
1  PROC REG;
4     MODEL VO2MAX=EXERCISE;
```

DEP VARIABLE: VO2MAX

ANALYSIS OF VARIANCE

SOURCE	DF	SUM OF SQUARES	MEAN SQUARE	F VALUE	PROB>F
MODEL	1	1579.07362	1579.07362	36.503	0.0001
ERROR	10	432.59305	43.25930491		
C TOTAL	11	2011.66667			

ROOT MSE	6.577181	R-SQUARE	0.7850
DEP MEAN	60.83333	ADJ R-SQ	0.7635
C.V.	10.8118		

PARAMETER ESTIMATES

VARIABLE	DF	PARAMETER ESTIMATE	STANDARD ERROR	T FOR H0: PARAMETER=0	PROB > |T|
INTERCEP	1	179.12087	19.67026538	9.106	0.0001
EXERCISE	1	-10.39890430	1.72117949	-6.042	0.0001

b. $r^2 = .785$

Chapter 9

9.1 a. $\hat{y} = 48.93 + 10.33x$ b. 5.6983
    c. .7957

9.3 a.

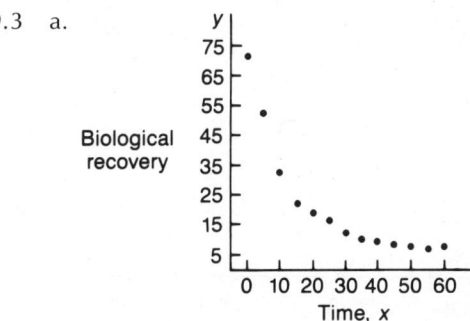

b.

Biological Recovery (%)	Log (%)	Biological Recovery (%)	Log (%)
70.6	1.85	10.0	1.00
52.0	1.72	9.1	.96
33.4	1.52	8.3	.92
22.0	1.34	7.9	.90
18.3	1.26	7.7	.89
15.1	1.18	7.7	.89
13.0	1.11		

9.5 $t = -9.640$; we reject the null hypothesis that $B_1 = 0$ at the $\alpha = .05$ level.

9.7 a.

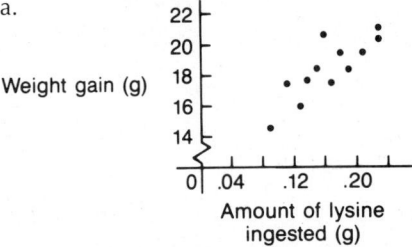

From the graph, a linear model appears to be appropriate.

b. $\hat{y} = 12.509 + 35.828x$

9.9 a. Since there is a nonzero, fixed percentage of lysine mixed into all feed, the mean response for no lysine eaten (β_0) should be zero. Thus, it would not make sense to give any physical interpretation to the estimate of β_0.

b. The model $y = \beta_1 x + \varepsilon$ will force the estimated regression line to pass through the origin. The model $y = \beta_0 + \beta_1 + \varepsilon$ will not necessarily pass through the origin.

9.12

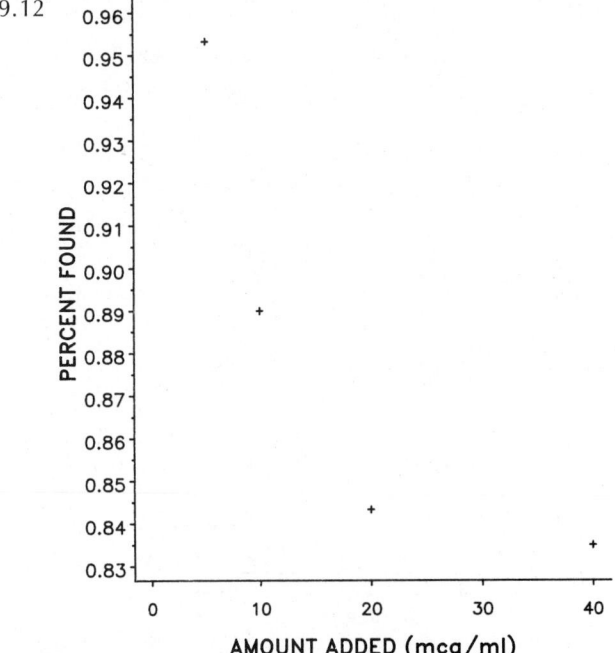

9.13 a.

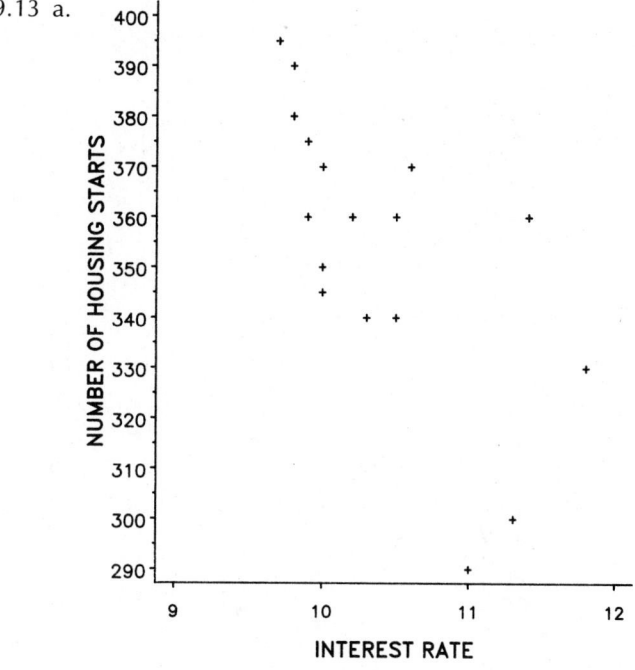

b. $\hat{y} = 672.886 - 30.654x$
c. Yes

9.14 10.2%: 381.7
      9.5%: 360.2

9.15 $c = 14$; at the $\alpha = .01$ level we reject the null hypothesis that the two variables are not correlated and conclude that there is a significant positive correlation between plaque weight and DNA.

9.17 a.

Btu, y

Weight, x

b. $c = 12$; we reject H_0 that the two variables are not correlated at the $\alpha = .05$ level.

9.19 1.059 to 1.103

9.21 16.889 to 21.743

9.23 a. $\hat{y} = -1.7333 + 1.3167x$
      b. $t = 6.342$; reject H_0: $\beta_1 = 0$ and conclude that $\beta_1 > 0$.

9.25 a.

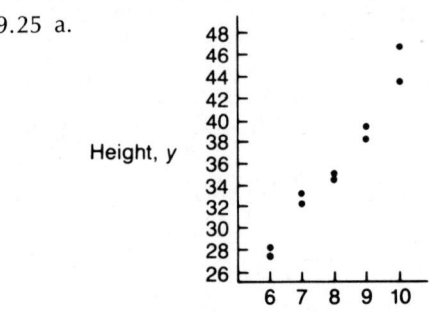

b. $\hat{y} = 3.370 + 4.065x$

c. RESIDUALS

```
           |
    2.5  +
    2.0  +
    1.5  +        A
    1.0  +
    0.5  +   A    A
         |  A
    0.0  + A
         |  A
   -0.5  +                    A       A
   -1.0  +           A
         |           A
   -1.5  +
   -2.0  +                A
         -+---+---+---+---+---+-
        27.760 31.825 35.890 39.955 44.02
              PREDICTED VALUE
```

The plot of the residuals and the predicted values suggests that we may need additional terms in our model.

9.27 $F = 8.27$; we reject H_0 that a linear regression model is appropriate at the $\alpha = .05$ level.

9.29 a. $\hat{x} = 14.47$ b. 11.08 to 17.63

9.31 From the output:
For 50% $\hat{x} = 2.36703$, (0.79139, 4.44615)
For 75% $\hat{x} = 5.86145$, (2.20339, 12.88173)

9.33 $\hat{x}_U = 5.23$, $\hat{x}_L = 4.33$

9.36 a. $\hat{y} = 804.85 + 2.34x$; β_0 and β_1 significantly different from zero.

9.37 $\hat{y} = 2.62x$

DEP VARIABLE: MANHOURS

ANALYSIS OF VARIANCE

SOURCE	DF	SUM OF SQUARES	MEAN SQUARE	F VALUE	PROB>F
MODEL	1	784019888	784019888	4057.357	0.0001
ERROR	15	2898512.31	193234.15		
U TOTAL	16	786918400			

ROOT MSE	439.5841	R-SQUARE	0.9963	
DEP MEAN	6532.5	ADJ R-SQ	0.9961	
C.V.	6.729186			

NOTE: NO INTERCEPT TERM IS USED. R-SQUARE IS REDEFINED.

PARAMETER ESTIMATES

VARIABLE	DF	PARAMETER ESTIMATE	STANDARD ERROR	T FOR HO: PARAMETER=0	PROB > |
ORDERS	1	2.61914428	0.04111855	63.697	0.00

9.38 $\hat{y} = -4.505 + 3.075x$; $t = 15.59$; reject H_0: $\beta_1 = 0$

9.40 a. $\hat{y} = 3.211 + .468x$

b. RESIDUALS

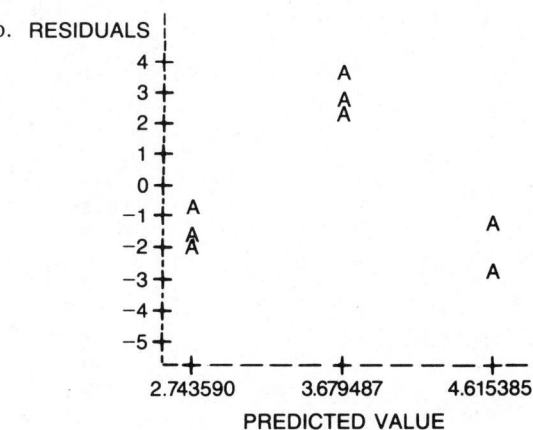

The plot of the residuals and the predicted values suggest that we may need additional terms in our model.

9.42 a.

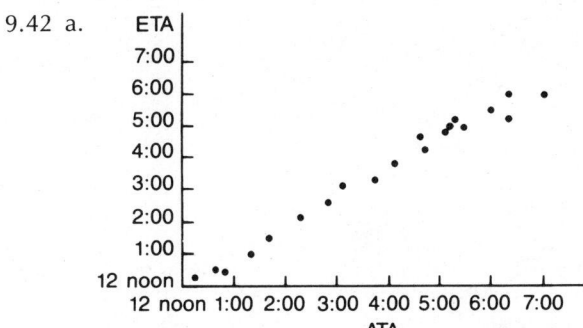

b. Yes, from the graph we can see that a linear relationship appears to be appropriate.

9.44 Exercise for student.

9.46 a. Yes $\hat{y} = 10.333 + .267x$

b. No, the plots show no obvious pattern of lack of fit.

9.48 a. $y_i = \beta_0 + \beta_1 x_i + \varepsilon_i$ $\hat{y} = -663.9 + 144.482x$

b. $t = 4.95$; $p = .0001$; we reject H_0 and conclude that β_0 is significantly different from zero.

c. Exercise for student

9.49 b. .72
c. Yes

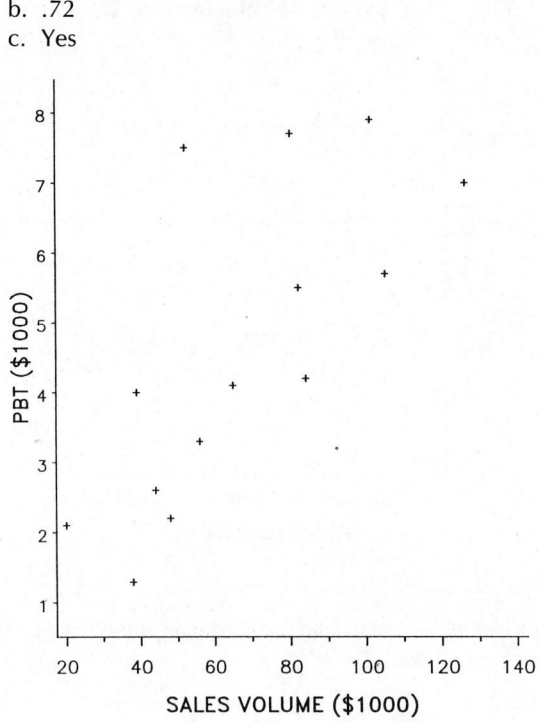

9.51

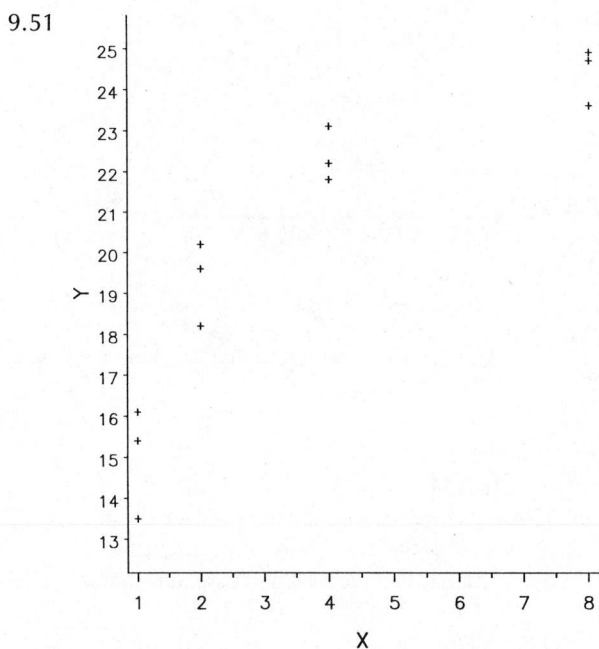

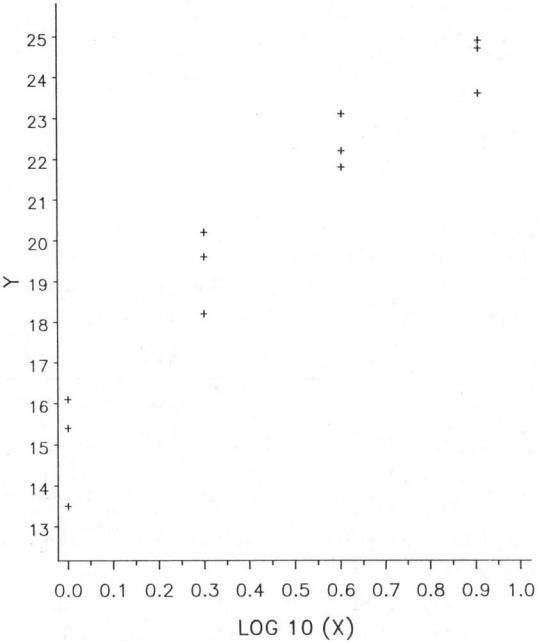

9.52 a. $\hat{y} = 15.81 + 1.19x$
b. $s_\varepsilon^2 = 3.58$

9.53 a. $\hat{y} = 15.59 + 10.36 \log_{10} x$
b. $s_\varepsilon^2 = 1.16$

9.54 $x = 5$ (21.98, 23.70)
$x = 9$ (24.25, 26.73)

9.56 a. $\hat{y} = 41.4 + 28.2 \log_{10} x$
c. $p < .001$

9.57 0.05

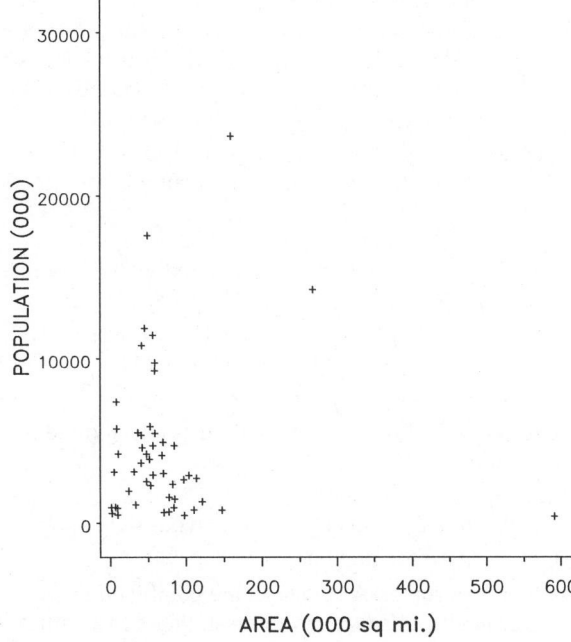

9.58 b. $\hat{y} = 2.3765 + .0005x$
c. $p = .0126$
9.59 $.92763; p = .0077$

Chapter 10

10.1 a. Exercise for student
b. $F_{4,45} = 13.96$; we accept H_0 at the $\alpha = .05$ level and conclude that no differences exist between the mean nicotine tar content of the five brands.

10.3

Source	SS	df	MS	F	p
Location	.1924	3	.0641	1.005	>.05
Error	1.2755	20	.0638		
Totals	1.4679	23			

10.5 $H' = 10.01$; since the observed value of H' does not exceed 13.2767, we cannot reject H_0 at $\alpha = .01$.

10.7 A non-parametric alternative is the Kruskal–Wallis test. $H' = 9.95$; we reject H_0 at the $\alpha = .05$ level. This is the same conclusion as that of Exercise 10.6. We can conclude that differences do exist between diet I and diet II.

10.9 $H' = 2.26$; we accept H_0 at the $\alpha = .05$ level. This test confirms the results of Exercise 10.8. If the results differed, the non-parametric approach would be the analysis to use.

10.11 a. $\chi_3^2 = 26.62; p < .001$.
b. $F_{3,28} = 55.67; p = .0001$. Both tests indicate rejection of the null hypothesis that all means are equal.

10.12 The analysis of variance is less sensitive to deviations from normality and equality of variances among the groups when the sample sizes are equal. Since the sample sizes are not equal in this exercise we choose to do a Kruskal–Wallis test. $\chi_3^2 = 21.15$; $p < .001$. We reject the null hypothesis.

10.14 a. $y_{ij} = \mu + \alpha_i + \varepsilon_{ij}; i = 1, 2, 3; j = 1, 2, \ldots, n_i; n_1 = 6, n_2 = 5, n_3 = 4$
$y_{ij} = j$th sample measurement for ith nutrient
$\mu = $ overall mean
$\alpha_i = $ effect of ith nutrient
$\varepsilon_{ij} = $ random error associated with jth observation from ith nutrient
b. $F_{2,12} = 24.31; p < .001$

10.16 a. $y_{ij} = \mu + \alpha_i + \varepsilon_{ij}; i = 1, 2, 3, 4; j = 1, 2, \ldots, 8$
b. $F_{3,28} = 11.05; p < .001$; reject H_0

10.18 $F_{2,33} = 6.65; p < .01$

10.20 $\chi_2^2 = 9.996; p < .01$. Both tests rejected the null hypothesis that all means are equal at $\alpha = .01$.

10.22 a. $F_{3,20} = 10.99; p < .001$
b. At least one of the pillow types differs from the others.

10.24 Exercise for student

10.26 If you run an ANOVA on the combined ranks of the data, the conclusions will be the same as a Kruskal–Wallis test. (Refer to an article by Conover and Iman, *The American*

Statistician, August 1981.) If we do this for the data of 10.26 we obtain $F_{4,25} = 46.08$, $p < .001$. These results are the same as the results of 10.26. We can conclude that differences do exist among the 5 diets.

10.28 a. $F_{4,20} = 6.03$; $p < .005$

b. We reject H_0, that the days of the week are the same.

10.30 If we replace 9.8 with 15.8 we obtain the following by running an ANOVA on the combined ranks: $F_{4,25} = 46.08$; $p < .001$. This conclusion is the same as a Kruskal–Wallis test. (See Exercise 10.26.) In some cases you may wish to run both a Kruskal–Wallis test and an analysis of variance. If an extreme value exists in the data it may affect the outcome of the analysis of variance. A Kruskal–Wallis test is not sensitive to extreme values because we are looking at the ranks of the data.

Chapter 11

11.1 a. Yes   b. No

11.3 $\hat{l}_1 = -2\bar{y}_1 - \bar{y}_2 + \bar{y}_4 + 2\bar{y}_5$
$\hat{l}_2 = 2\bar{y}_1 - \bar{y}_2 - 2\bar{y}_3 - \bar{y}_4 + 2\bar{y}_5$
$\hat{l}_3 = -\bar{y}_1 + 2\bar{y}_2 - 2\bar{y}_4 + \bar{y}_5$
$\hat{l}_4 = \bar{y}_1 - 4\bar{y}_2 + 6\bar{y}_3 - 4\bar{y}_4 + \bar{y}_5$

11.5 $F_{4,45} = 15.069$; $p < .001$; reject H_0 that all means are equal.

11.7 The analysis of variance of Exercise 11.5 is identical to the analysis of this exercise; $F_{4,45} = 15.069$

11.9 a. $\hat{l} = -2.11$; reject H_0 that $l = 0$ b. $\hat{l} = -.12$; cannot reject H_0: $l = 0$ at $\alpha = .05$ c. $\hat{l} = .46$; cannot reject H_0: $l = 0$ at $\alpha = .05$ d. $\hat{l} = .90$; cannot reject H_0: $l = 0$ at $\alpha = .05$

11.11 a. LSD = 4.708; A   B   C

b. $W = 5.734$; A   B   C. Conclusions are the same.

c. A   B   C. Conclusions are the same.

11.13 $W_2 = 3.538$; $W_3 = 3.714$; 3   2   1

11.15 LSD = .83

$$4 \quad \underline{3 \quad 2} \quad 1 \quad 5$$

11.17 Fisher's LSD procedure. Summary: 4   3   2   1. Per-comparison error rate is controlled.

11.19 LSD = 12.72. Summary: $\underline{1 \quad 2}$ 3

11.21 $F_{2,147} = 85.88$; we reject H_0 at the $\alpha = .05$ level. We conclude that differences exist among the methods of grass seed preparation.

11.23 $F_{3,16} = 4.35$; we reject H_0 at the $\alpha = .05$ level. We conclude that differences exist among the gasoline blends.

11.25 $\hat{l}_3 = 6.7$, $s = 15.84$; we cannot reject H_0, that $l_3 = 0$ at the $\alpha = .05$ level.

11.27 $W = 10.3$

Blend

$$1 \quad \underline{2 \quad 3 \quad 4}$$

11.29 a.

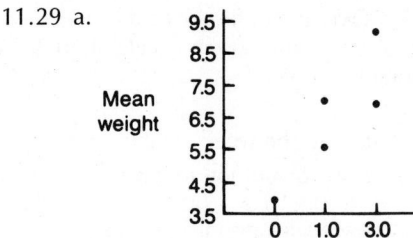

b. Additive A   Additive B

    $\hat{y} = 4.76 + .74x$ $\hat{y} = 5.81 + 1.19x$

Chapter 12

12.1   a. $y = \beta_0 + \beta_1 x_1 + \beta_2 x_2 + \beta_3 x_3 + \varepsilon$

     b. The general linear model for relating a response y to three independent quantitative variables would be the same as the model in part (a).

12.3 $y = \beta_0 + \beta_1 x_1 + \beta_2 x_2 + \beta_3 x_3 + \beta_4 x_4 + \beta_5 x_5 + \beta_6 x_6 + \beta_7 x_7 + \beta_8 x_8 + \beta_9 x_9 + \varepsilon$

     where $x_1 = x_1$ $x_4 = x_2^2$ $x_7 = x_1 x_2$

                 $x_2 = x_1^2$ $x_5 = x_3$ $x_8 = x_1 x_3$

                  $x_3 = x_2$ $x_6 = x_3^2$ $x_9 = x_2 x_3$

12.5   a.

Location		1	
Treatment	1	2	3
	$E(y) = \beta_0$	$E(y) = \beta_0 + \beta_1$	$E(y) = \beta_0 + \beta_2$

Location		2	
Treatment	1	2	3
	$E(y) = \beta_0 + \beta_3$	$E(y) = \beta_0 + \beta_1 + \beta_3$	$E(y) = \beta_0 + \beta_2 + \beta_2$

     β_0—mean of treatment 1, location 1

     β_1—difference of the mean of treatment 2 and the mean of treatment 1 for a given location

     β_2—difference of the mean of treatment 3 and the mean of treatment 1 for a given location

     β_3—difference of the mean of treatment 1, location 1, and the mean of treatment 1, location 2, for a given treatment

     b. $\beta_0 + \beta_1 + \beta_3 - (\beta_0 + \beta_2 + \beta_3) = \beta_1 - \beta_2$. It is the same for location 1.

     c. Exercise for student

12.7   a.

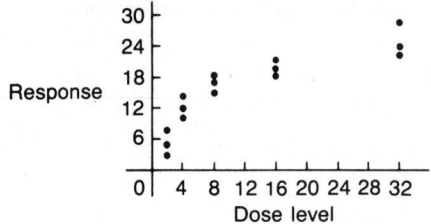

ANSWERS TO SELECTED EXERCISES

b. $\hat{y} = 8.67 + 0.58x$ c. A quadratic model

d. The quadratic model. The quadratic model has a much lower SSE than the linear model. The quadratic term is significant, $p = .0062$. The plot of the predicted values versus the residuals for the linear model has more of an upward trend than the same plot for the quadratic model.

12.9 a.

b. Quadratic model

c. The quadratic model. For the quadratic model the plot of the predicted values versus the residuals does not show any problems. The same plot for the linear model does show a trend. The R-squares for the quadratic and the cubic models are almost equivalent. However, the cubic term of the cubic model is not significant, $p = .1794$. All terms of the quadratic model are significant.

12.11 $\hat{y} = -19.41 + 29.10x_1 + 3.29x_2$
$\hat{y} = -23.81 + 31.30x_1 + 3.84x_2 - 0.28x_1x_2$
The model without the interaction term is more appropriate since the interaction term is not significant, $p = .3697$

12.13 a. $\hat{y} = 1.85 - 0.035x + .0003x^2$
b. $s_{\varepsilon} = .023$; $s_{\hat{\beta}_1} = .0013$; $s_{\hat{\beta}_2} = .00002$

12.15 a. β_0 would be the biological recovery at time zero.
b. No, the biological recovery is not zero at time zero. If we used this model we would be assuming that regression through the origin would be appropriate.

12.17 a. $r_{yx_1} = .856$; $r_{yx_2} = .870$; $r_{x_1x_2} = .894$
b. $r_{yx_1}^2 + r_{yx_2}^2 = 1.49 \neq .788 = R_{y \cdot x_1 x_2}^2$. The independent variables x_1 and x_2 are correlated since $R_{y \cdot x_1 x_2}^2 \neq r_{yx_1}^2 + r_{yx_2}^2$.

12.19 a. $\hat{y} = -2.781 + 1.048x$
b. $p = .0001$; we reject H_0 that $\beta_1 = 0$.
c. $R^2 = .8989$; $R = .948$; $r = .948$

12.21 The prediction band would be wider for a higher degree of certainty. The bands would be smaller for a degree of less certainty.

12.23 (16.889, 21.743); The 95% prediction interval computed in this exercise is wider than the confidence interval computed in Exercise 12.20 because we are predicting the value of a random variable instead of estimating a parameter.

12.25 $F_{2,8} = 30.795$; reject H_0 that $\beta_2 = \beta_3 = 0$ at $\alpha = .05$.

12.27 a. $\hat{y} = 48.93 + 10.33x$ b. .79599   c. 2.3881

12.29 a. $\hat{y} = 1.854 - .036x + .0003x^2$ b. .0006
c. .0170, .0013, .00002

12.31 $1.854 \pm .038$

12.36 $R^2 = 81$

12.38 $p = .0176$; we reject H_0: $\beta_2 = 0$

12.40 a. $\hat{y} = -44.013 + 5.087x_1 + 3.019x_2 + .205x_3 - 5.409x_4$
    (1.970)   (2.204)   (0.373)   (3.850)
  b. $R^2 = .9629; 22.123$

12.42 a. $y = \beta_0 + \beta_1x_1 + \beta_2x_2 + \varepsilon$
  b. $y = \beta_0 + \beta_1x_1 + \beta_2x_1^2 + \beta_3x_2 + \beta_4x_2^2 + \beta_5x_1x_2 + \varepsilon$

12.44 a. $y = \beta_0 + \beta_1x_1 + \beta_2x_2 + \varepsilon$
  b. $\hat{y} = 1.04 + 1.59x_1 + .22x_2$ c. 2.720

12.46 a. $\hat{y} = -144.312 + 25.764x - 0.686x^2$
  b. $p = .0197$; we can reject H_0, that $\beta_2 = 0$. The quadratic term for this model is significant, thus a quadratic model is appropriate.

12.55 a. $\mathbf{A}' = \begin{bmatrix} 2 & 3 \\ 1 & 2 \end{bmatrix}$ $\mathbf{B}' = \begin{bmatrix} 2 & 1 \\ 1 & 1 \end{bmatrix}$

  b. $\mathbf{A}^{-1} = \begin{bmatrix} 2 & -1 \\ -3 & 2 \end{bmatrix}$

12.57 a. $\begin{bmatrix} 3 & 0 & 2 \\ 0 & 2 & 0 \\ 2 & 0 & 2 \end{bmatrix}$ b. 4

12.59 a. $\begin{bmatrix} 1 & 1 & 1 \\ 1 & 2 & 3 \end{bmatrix}$

  b. $\begin{bmatrix} 3 & 6 \\ 6 & 14 \end{bmatrix}$

12.61 $\begin{bmatrix} \frac{4}{3} \\ 1 \end{bmatrix}$

12.63 a. 4   b. $\begin{bmatrix} -\frac{1}{4} & 1 & -\frac{1}{4} \\ -\frac{3}{4} & 0 & \frac{1}{4} \\ \frac{2}{4} & -1 & \frac{2}{4} \end{bmatrix}$

Chapter 13

13.1 Exercise for student

13.3 If we run a best subset regression procedure with the C_p criterion, the following variables would be included in the model: Age, Income, and Sex ($C_p = 3.00$).

13.5 Exercise for student

13.7 $\hat{y} = 288.06 + 402.53x_1 - 0.34x_2 - 1.99x_1x_2$,   where $x_1 = $ log (dose)   and $x_2 = $ weight. From examination of the residual plots, it can be seen that there is some lack of fit for this model; the residuals appear to be larger for smaller values of $\hat{y}$ or equivalently for smaller values of weighs (x_2) or log dose (x_1).

13.9 Based on the observed versus predicted plots and the residual plots, the second model seems to provide a better fit to the data. The first model does a poor job predicting the larger concentrations and misses the peak concentration. The second model does a better job at fitting the peak and fitting the larger concentrations while still doing an adequate job at the lower concentrations observed at the initial and late post-dose times.

13.11 There appears to be no evidence of serial correlation for the difference model. A plot of the residuals versus the month bears this out.

13.13 a. The prediction equation is as follows:

$$\hat{y} = -29.88 + 2.33x_1 + 3.82x_2 - 0.20x_1x_2$$

By examination of the residual plots, the model appears to fit the data.

b. 1. Zero expectation: The model appears to fit the data thus this assumption should hold.

2. Constant variance: As can be seen from the plot of the residuals versus the predicted values, there appears to be no problem with the assumption of constant variance.

3. Normality: From examination of the plots, no outliers appear to be present. Also a stem-and-leaf plot of the residuals does not indicate any departure from normality.

4. Independence: The Durbin–Watson statistic is 1.27. Since this value is less than 1.5, we should suspect positive serial correlation. One solution to this problem could be to run an analysis on the first differences of the variables.

13.15 As can be seen from the residual plot, the model appears to be underestimating y for small values of y and overestimating y for larger values of y. There is no evidence from this plot that any of the model assumptions have been violated. Other plots that should be considered are the residuals versus the independent variables, a histogram or stem-and-leaf plot of the residuals, and a plot of the residuals versus time if appropriate. The Durbin–Watson test for serial correlation should also be run.

13.17 a. $y = \beta_0 + \beta_1x_1 + \beta_2x_2 + \beta_3x_3 + \beta_4x_1x_2 + \beta_5x_1x_3 + \varepsilon$

$x_1 = $ log of drug dose $\beta_0 = y$-intercept for product A regression line
$x_2 = 1$ if product B $\beta_1 = $ slope of product A regression line
$x_2 = 0$ otherwise $\beta_2 = $ difference in y-intercepts for products B and A
$x_3 = 1$ if product C $\beta_3 = $ difference in y-intercepts for products C and A
$x_3 = 0$ otherwise $\beta_4 = $ difference in slopes for products B and A
  $\beta_5 = $ difference in slopes for products C and A

b. $y = \beta_0 + \beta_1x_1 + \beta_4x_1x_2 + \beta_5x_1x_3 + \varepsilon$

13.19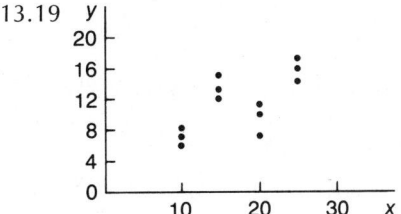

An appropriate polynomial model would be

$$y = \beta_0 = \beta_1x + \beta_2x^2 + \beta_3x^3 + \varepsilon$$

13.21 a. $\hat{y} = -91.00 + 19.217x - 1.163x^2 + .023x^3$

b. The model of Exercise 13.20 appears to fit the data well. The model of Exercise 13.21 fits the data well except for one point that appears to be underestimated.

13.23 a.

Dose Level	Log_{10} Dose
2	.30103
4	.60206
8	.90309
16	1.20412
32	1.50515

b. $\hat{y} = 1.20 + 16.17x_1$

c. The results in part (b) agree with the answers in the computer printout.

d. The linear and quadratic models of Exercise 13.22 have values of the Durbin–Watson statistic less than 1.5, suggesting possible serial correlation. The residual plots for these two models also suggest a serial correlation. In contrast, the model with the logarithmic transformation provides a much better fit to the sample data. This model has the highest value for R^2, and the Durbin–Watson statistic is larger.

13.25 a. No. There are only six data points and the total degree of freedom is only five; thus, we could not fit a model with all of these terms.

b. $F_{1,1} = 1.127$; at the $\alpha = .05$ level, we have insufficient evidence to indicate a lack of fit for the model.

13.27

Comparison	χ_1^2	p
A vs. B	1.7884	.1811
A vs. C	.1118	.7381
A vs. D	1.0444	.3068
B vs. D	5.5914	.0180
C vs. D	1.8671	.1718

13.29 $\hat{y} = 60.48 - 0.71x_1 + 0.003x_1^2 + 8.88x_2$

$\hat{y} = 42.28 - 0.42x_1 + 0.002x_1^2 + 69.54x_2 - 0.95x_1x_2 + .003x_1^2x_2$

13.31 The model given would not be more appropriate. The linear model of Exercise 13.30 gives the best fit of the data. The residual plots bear this out as well as the high R^2 value.

13.33 a. y = number of sales per month

x_1 = price per gallon

x_2 = interest rate

$x_3 = \begin{cases} 1 \text{ if standard} \\ 0 \text{ if luxury} \end{cases}$

b. The prediction equation for the complete model is as follows:
For $x_3 = 0$ (luxury)

$$\hat{y} = 40.567 + 12.201x_1 - 4.595x_1^2 - 5.328x_2 + 0.154x_2^2$$

For $x_3 = 1$ (standard)

$$\hat{y} = -59.903 + 297.394x_1 - 120.608x_1^2 - 12.976x_2 + 0.372x_2^2$$

c. The regression coefficients for the two models appear to be quite different. However, care has to be taken in interpreting these differences since the interaction terms have relatively large standard errors. Perhaps we should consider a model having fewer interaction terms involving the dummy variable, x_3, but more interaction terms involving the two qualitative independent variables, x_1 and x_2.

13.35 One problem may be the presence of nonconstant variance. The plot of the residuals versus the predicted values shows that as $\hat{y}$ increases the residual also increases. A transformation of the data could help to stabilize the variance.

13.37 a. $\hat{y} = 44.1824 - 0.4940x + 0.0014x^2$

b. $F_{3,9} = 1.2$; since our value does not exceed the critical F value, we have insufficient evidence to indicate a lack of fit for our model.

c. Examination of the residual plots does not reveal any problems with the model assumptions.

13.39 $F_{2,10} = 5.45$; we reject H_0 at the $\alpha = .05$ level and conclude that there are additional polynomial terms that should be included in the model.

13.41 Test for $\beta_3 = 0$; $t = 4.69$; $p = .0001$

13.43 a.

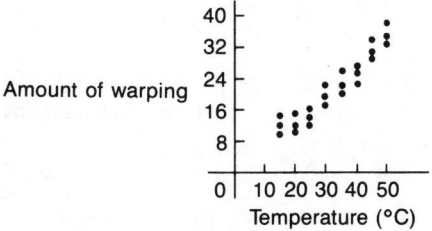

b. $\hat{y} = -1.5397 + 0.7063x$

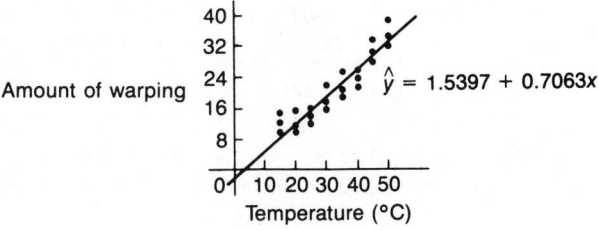

c. $\hat{y} = 9.1786 - 0.0468x + 0.0116x^2$

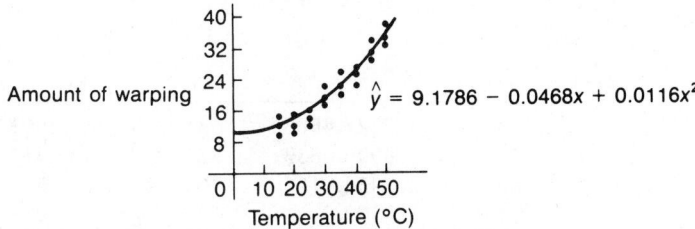

Both equations give a good fit to the data, but the quadratic one fits the lower values of the independent variable better than the linear model.

d. Using the linear equation, $\hat{y} = 17.5304$. Using the equation for the quadratic model, $\hat{y} = 16.3714$.

Chapter 14

14.1 a. Treatments $F = 5.62$

      Blocks $F = 351.56$

b. 160.3

14.3   a. Blocks are investigators and treatments are mixtures.
       b. Mixtures are applied at random to experimental units within an investigator, with
          the stipulation that each mixture appears once in an investigator.

14.5

Source	SS	df	MS	F	p
Year	.220	1	.220	6.47	.03
Curricula	21.260	10	2.126	62.53	<.01
Error	.340	10	.034		
Totals	21.820	21			

14.7   a. $F = 5.505, p \approx .01$
       b. $F = 6.914, p < .001$; it seems appropriate to block on people.

14.8   a. Formulation 1 appears different from formulations 2 and 3.

14.9   a. $y_{ijk} = \mu + a_i + \beta_j + \gamma_k + \varepsilon_{ijk}$;
          $i = 1, 2, 3, 4; j = 1, 2, 3, 4; k = 1, 2, 3, 4$

       b.

Source	SS	df	MS	F	p
Rows	.00085	3	.000283	2.26	.18
Columns	.01235	3	.004117	32.94	<.01
Treatments	.48015	3	.16005	1280.4	<.01
Error	.00075	6	.000125		
Totals	.49410	15			

We cannot reject the null hypothesis that rows are equal but we can reject the
hypotheses that columns and treatments are equal.

14.11 Exercise for student

14.13 a. Randomized block design

14.18 A 3×4 factorial experiment

14.20 a. Yes   b. $t = 5.31, p < .001$

14.22 a.

Source	SS	df	MS	F
A	2227.4583	3	742.4861	17.56
B	3996.0833	2	1998.0416	47.24
AB	456.9166	6	76.1528	1.80
Error	507.5	12	42.29167	
Total	7187.9583	23		

       b. The p-value for the F test on the variety-by-treatment interaction is 0.1815. Thus,
          we have insufficient evidence to indicate an interaction between the two main
          effects. We may now proceed with the tests for main effects.
       c. The p-values listed for both varieties and pesticides are very small. Hence, we
          would conclude that there are significant differences among the varieties and among
          pesticides.
       d. The results from the output differ slightly from those in the SAS output and Example
          14.9 because of rounding-off errors. However, all three solutions result in the same
          conclusions.

ANSWERS TO SELECTED EXERCISES

14.24 a.

Source	SS	df	MS
Copper	171856.19	3	57285.396
Manganese	3745496.19	3	1248498.7
Interaction	9994.56	9	1110.5069
Totals	3927346.94	15	

We cannot test for interaction.

b. $y = \beta_0 + \beta_1 x_1 + \beta_2 x_1^2 + \beta_3 x_1^3 + \beta_4 x_2 + \beta_5 x_2^2 + \beta_6 x_2^3 + \varepsilon$

14.26 a.

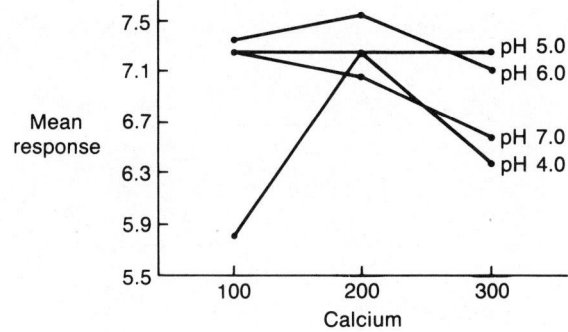

	Sample Means Calcium		
pH	100	200	300
4.0	5.8	7.3	6.4
5.0	7.3	7.3	7.3
6.0	7.4	7.6	7.2
7.0	7.3	7.1	6.6

b. $y_{ijk} = \mu + \alpha_i + \beta_j + \alpha\beta_{ij} + \varepsilon_{ijk}$; $i = 1, 2, 3, 4$; $j = 1, 2, 3$; $k = 1, 2, 3$

c.

Source	SS	df	MS	F	p
pH	4.4608	3	1.4869	21.93	<.01
Calcium	1.4672	2	.7336	10.82	<.01
Interaction	3.2550	6	.5425	8.00	<.01
Error	1.6267	24	.0678		
Totals	10.8097	35			

14.28 Summary: M_n

20	50	80	110
1506.5	1948.25	2470	2767.5

14.30 a. $y_{ijk} = \mu + \alpha_i + \beta_j + \alpha\beta_{ij} + \varepsilon_{ijk}$; μ = overall mean; α_i = effect of investigator i, $i = 1, 2, 3$; β_j = effect of treatment j, $j = 1, 2, \ldots, 9$; $\alpha\beta_{ij}$ = interaction between investigator i and treatment j; ε_{ijk} = error, $k = 1, 2, \ldots, 5$

b.

Source	SS	df	MS	F	p
Investigators	.3628	2	.1814	7.37	<.01
Treatments	3.5117	8	.4390	17.85	<.01
Interaction	.5012	16	.0313	1.27	.23
Error	2.6520	108	.0246		
Totals	7.0277	134			

14.32 a. Completely randomized design; $y_{ij} = \mu + \alpha_i + \varepsilon_{ij}$; $i = 1, 2, 3, 4$; $j = 1, 2, \ldots, 7$

b.

Source	SS	df	MS	F	p
Class	892.7143	3	297.5714	11.055	<.01
Error	646.0000	24	26.9167		
Totals	1538.7143	27			

14.34 a. Randomized block design; $y_{ij} = \mu + \alpha_i + \beta_j + \varepsilon_{ij}$; $i = 1, 2, \ldots, 5$; $j = 1, 2, \ldots, 5$

b.

Source	SS	df	MS	F	p
Temperature	16.4216	4	4.1054	42.92	<.01
Pane	7.1176	4	1.7794	18.60	<.01
Error	1.5304	16	.09565		
Totals	25.0696	24			

14.36 a. $4 \times 3 \times 3$ factorial

b. $y_{ijk} = \mu + \alpha_i + \beta_j + \gamma_k + \alpha\beta_{ij} + \alpha\gamma_{ik} + \beta\gamma_{jk} + \varepsilon_{ijk}$; $i = 1, 2, 3, 4$; $j = 1, 2, 3$; $k = 1, 2, 3$

c.

Source	SS	df	MS	F	p
Main effects					
ph	4.4608	3	1.4869	28.48	<.01
calcium	1.4672	2	.7336	14.05	<.01
grove	.0089	2	.0044	.08	.92
Interaction					
pH-calcium	3.2550	6	.5425	10.39	<.01
pH-grove	.4000	6	.0667	1.28	.34
calcium-grove	.5911	4	.1478	2.83	.07
Error	.6267	12	.0522		
Totals	10.8097	35			

14.38 a. $3 \times 2 \times 2$ factorial laid off in 6 blocks

Chapter 15

15.1   a. The complete and reduced models for testing treatments are as follows:
      complete model: $y_{ij} = \mu + \alpha_i + \beta_j + \varepsilon_{ij}$
      reduced model: $y_{ij} = \mu + \beta_j + \varepsilon_{ij}$
  b. The analysis of variance table for testing the effect of treatments is shown below.

Source	SS	df	MS	F	p
Blocks	2.61464912	4	——	——	——
Treatments (adj.)	140.80083333	3	46.93361111	904.41411	0.0001
Error	0.5708333	11	0.05189394		
Total	143.98631579	18			

Since our computed value of F greatly exceeds the critical value and our value of p is so small, we would reject H_0 and conclude that one of the adjusted treatment means differs significantly from the rest.

  c. The complete model for testing blocks would be the same complete model as in section (a). The reduced model is

$$y_{ij} = \mu + \alpha_i + \varepsilon_{ij}$$

  d. The AOV table for testing blocks would be

Source	SS	df	MS	F	p
Blocks (adj.)	2.11266667	4	0.52816666	10.17781	.0014
Treatments	141.30281579	3	——	——	——
Error	0.57083333	11	.05189394		
Total	143.98631579	18			

Our computed value of F also exceeds its critical value; hence, we will conclude that at least one of the dairies (blocks) differs significantly from the rest.

15.3

Source	SS	df	MS	F	p
Plots	386.89	3	128.96	27.44	<.01
Insecticides	1932.14	2	966.07	205.55	<.01
Error	23.49	5	4.70		
Totals	2342.52	10			

15.5   From a one-way analysis of variance on treatments:

    TSS = 143.9863
    SST = 141.3028
    SSE = TSS − SST = 2.6835 = SSE_2

From Exercise 15.1, we know that $SSE_1 = 0.5708$. Thus,

    $SS_{drop} = SSE_2 - SSE_1 = 2.6835 - 0.5708 = 2.1127$

This figure agrees with that shown in the Type III sum of squares column of the computer output in Exercise 15.1.

15.7 a. AOV for testing mixtures:

Source	SS	df	MS	F	p
Investigators	8910.03	4			
Mixtures (adj.)	210,236.35	3	70,078.78	1014.16	<.01
Error	760.15	11	69.10		
Totals	219,906.53	18			

AOV for testing investigators:

Source	SS	df	MS	F	p
Investigators (adj.)	344.60	4	86.15	1.25	>.05
Mixtures	218,801.78	3			
Error	760.15	11	69.10		
Totals	219,906.53	18			

b. Use complete and reduced models excluding this observation also.

15.9 The AOV table for testing the effect of treatment adjusted for rows and columns would be:

Source	SS	df	MS	F
Rows	22.3285	4	———	———
Columns (adjusted for rows	2.1340	4	———	———
Treatments (adjusted for rows and columns)	165.4943	4	41.3736	315.35
Error	1.4432	11	0.1312	
Total	191.4000	23		

The F test for treatments adjusted by rows and columns is significant; i.e., the adjusted treatment means differ significantly.

Source	SS	F
Rows (adjusted for columns and treatments)	14.3668	27.38
Columns (adjusted for rows and treatments)	0.9428	1.80

The F test for rows adjusted by columns and treatments is significant; however, the other test for columns is not significant. Hence, investigators are significantly different, but days are not. Our conclusions in this exercise agree exactly with those in Exercise 15.8.

15.11

Source	SS	df	MS	F	p
Operator (adjusted for model and blend)	7.492	3	2.497	0.70	>.05
Model (adjusted for operator and blend)	695.087	3	231.696	64.68	<.01
Blend (adjusted for operator and model)	82.297	3	27.432	7.66	>.05
Error	17.908	5	3.582		

15.13

Source	SS	df	MS	F	p
Persons	1034.8	9			
Treatments (adj.)	1747.056	5	349.411	11.55	<.01
Error	453.611	15	30.241		
Totals	3235.467	29			

15.15 Same answer as Exercise 15.13

15.17 Use the Type III (adjusted) sum of squares for blocks.

15.19 $W = 17.72$; a summary of the procedure follows.

3N   3E   2N   3S   1W   3W   2S   1N   1E   2E   1S   2W

15.21 a. An appropriate dependent variable would be the number of defects.

b. $y_{ijk} = \mu + \alpha_i + \beta_j + \gamma_{k(i)} + \alpha\beta_{ij} + \beta\gamma_{jk(i)} + \varepsilon_{ijk}$

μ: An overall mean that is an unknown constant

α_i: An effect due to training i: $i = 1, 2$

β_j: An effect due to inspector j: $j = 1, 2$

$\gamma_{k(i)}$: An effect due to assembly line k in training i: $k = 1, 2, 3$

ε_{ijk}: A random error associated with training i, inspector j and assembly line k

c.

Source	df
Training	1
Line (Training)	4
Inspector	1
Inspector × Training	1
Inspector × Line (Training)	4
Error	12
Total	23

15.23 a. No, the model does not change. We can analyze the data by fitting complete and reduced models.

b. Using the output for the complete and reduced models, we can test for the interaction Inspector × Line (Train) and also for the interaction Inspector × Train. For Inspector × Line (Train): $F_{4,10} = 0.06$; we cannot reject H_0 at the $\alpha = .05$ level, and we conclude that the interaction is not significant. For Inspector × Train: $F_{1,14} = 0.14$; we cannot reject H_0 at the $\alpha = .05$ level, and we conclude that the interaction is not significant.

15.25 A dependent variable that would account for both major and minor defects would be the ratio of major defects to minor defects. You may want to transform the data before calculating the ratios since there are several values for major defects.

Chapter 16

16.1 a. $y_{ij} = \mu + \alpha_i + \varepsilon_{ij}; i = 1, 2, \ldots, 5; j = 1, 2, 3, 4$
$y_{ij} = j$th observation from vat i
μ = overall unknown mean
α_i = random effect due to the ith vat; α_i is normally distributed with mean 0 and variance σ_α^2
ε_{ij} = random error associated with response j on ith vat

b.

Source	SS	df	MS	EMS	F
Vats	11.948	4	2.987	$\sigma^2 + 4\sigma_\alpha^2$	32.53
Error	1.3775	15	.09183	σ^2	
Totals	13.3255	19			

16.3 a. $y_{ij} = \mu + \alpha_i + \beta_j + \varepsilon_{ij}; i = 1, 2, 3; j = 1, 2, 3$
y_{ij} = observation from jth subject in analyst i
μ = overall unknown concentration mean
α_i = random effect due to the ith analyst
β_j = random effect due to the jth subject
ε_{ij} = random error associated with response in analyst i, subject j

b.

Source	EMS
Analysts	$\sigma^2 + 3\sigma_\alpha^2$
Subjects	$\sigma^2 + 3\sigma_\beta^2$
Error	σ^2

16.5 a. $y_{ijk} = \mu + \alpha_i + \beta_j + \alpha\beta_{ij} + \varepsilon_{ijk}; i = 1, 2, 3, 4; j = 1, 2, 3$
y_{ijk} = response (sales volume) for the kth observation at the ith level of SMSA and jth level of chain store
μ = overall unknown sales volume mean
α_i = random effect due to the ith SMSA
β_j = random effect due to the jth store
$\alpha\beta_{ij}$ = random effect due to the ith level of SMSA and the jth level of store
ε_{ijk} = random error associated with kth response at SMSA i, store j

b.

Source	SS	df	MS	EMS	F	p
SMSA	136,644.125	3	45,548.042	$\sigma^2 + 2\sigma_{\alpha\beta}^2 + 6\sigma_\alpha^2$	5.049	<.05
Store	62,973.000	2	31,486.5	$\sigma^2 + 2\sigma_{\alpha\beta}^2 + 8\sigma_\beta^2$	3.490	>.05
Interaction	54,127.000	6	9021.1667	$\sigma^2 + 2\sigma_{\alpha\beta}^2$	22.03	<.05
Error	4913.500	12	409.4583	σ^2		
Totals	258,657.625	23				

16.7

Source	SS	df	MS	EMS	F	p
Physicians	3.8115	4	.9529	$\sigma^2 + 2\sigma_{\alpha\beta}^2 + 8\theta_A$	.71	>.05
Patients	180.13275	3	60.0442	$\sigma^2 + 10\sigma_B^2$	173.41	<.05
Interaction	16.1585	12	1.3465	$\sigma^2 + 2\sigma_{\alpha\beta}^2$	3.89	<.05
Error	6.925	20	.34625	σ^2		
Totals	207.02775	39				

16.9 Yes

$H_0: \theta_A = 0; F = MSA/MSAC$

$H_0: \theta_B = 0; F = MSB/MSBC$

$H_0: \sigma_\gamma^2 = 0; F = MSC/MSE$

$H_0: \theta_{AB} = 0; F = MSAB/MSABC$

$H_0: \sigma_{\alpha\gamma}^2 = 0; F = MSAC/MSE$

$H_0: \sigma_{\beta\gamma}^2 = 0; F = MSBC/MSE$

$H_0: \sigma_{\alpha\beta\gamma}^2 = 0; F = MSABC/MSE$

16.11 a.

Source	EMS
A	$\sigma^2 + r\sigma_{ABC}^2 + cr\sigma_{AB}^2 + br\sigma_{AC}^2 + bcr\sigma_A^2$
B	$\sigma^2 + r\sigma_{ABC}^2 + cr\sigma_{AB}^2 + ar\sigma_{BC}^2 + acr\sigma_B^2$
C	$\sigma^2 + r\sigma_{ABC}^2 + br\sigma_{AC}^2 + ar\sigma_{BC}^2 + abr\sigma_C^2$
AB	$\sigma^2 + r\sigma_{ABC}^2 + cr\sigma_{AB}^2$
AC	$\sigma^2 + r\sigma_{ABC}^2 + br\sigma_{AC}^2$
BC	$\sigma^2 + r\sigma_{ABC}^2 + ar\sigma_{BC}^2$
ABC	$\sigma^2 + r\sigma_{ABC}^2$
Error	σ^2

b.

Source	F Test	df_1	df_2
ABC	MSABC/MSE	$(a-1)(b-1)(c-1)$	$abc(r-1)$
BC	MSBC/MSABC	$(b-1)(c-1)$	$(a-1)(b-1)(c-1)$
AC	MSAC/MSABC	$(a-1)(c-1)$	$(a-1)(b-1)(c-1)$
AB	MSAB/MSABC	$(a-1)(b-1)$	$(a-1)(b-1)(c-1)$

There are no exact F tests for σ_A^2, σ_B^2, and σ_C^2.

16.13 a. $y_{ij} = \mu + \alpha_i + \varepsilon_{ij}$; $i = 1, 2, \ldots, 6$; $j = 1, 2$

$y_{ij} = j$th observation from week i

μ = overall unknown mean

α_i = random effect due to the ith week

ε_{ij} = random error associated with response j at week i

b.

Source	SS	df	MS	EMS	F	p
Weeks	255,089.42	5	51,017.88	$\sigma^2 + 2\sigma_\alpha^2$	26.76	<.05
Error	11,439.5	6	1906.58	σ^2		
Totals	266,528.92	11				

16.15 a. $y_{ijk} = \mu + \alpha_i + \beta_j + \alpha\beta_{ij} + \varepsilon_{ijk}$; $i = 1, 2, \ldots, 5$; $j = 1, 2, 3, 4$; $k = 1, 2$; in this model α_i is a random effect due to the ith physician, while in Exercise 16.6 it was a fixed effect. In this model α_i, β_j, and $\alpha\beta_{ij}$ are all independent.

b.
Source	SS	df	MS	EMS	F	p
Physicians	3.8115	4	.9529	$\sigma^2 + 2\sigma_{\alpha\beta}^2 + 8\sigma_\alpha^2$	.71	>.05
Patients	180.1328	3	60.0442	$\sigma^2 + 2\sigma_{\alpha\beta}^2 + 10\sigma_\beta^2$	44.59	<.05
Interaction	16.1585	12	1.3465	$\sigma^2 + 2\sigma_{\alpha\beta}^2$	3.89	
Error	6.9250	20	.34625	σ^2		
Totals	207.0278	39				

16.17 a. $y_{ij} = \mu + \alpha_i + \beta_j + \varepsilon_{ij}$; $i = 1, 2, 3, 4$; $j = 1, 2, \ldots, 5$

  y_{ij} = observation from jth investigator in mixture i

  μ = overall unknown mean

  α_i = fixed effect due to ith mixture

  β_j = random effect due to jth investigator; β_j is normally distributed with mean 0 and variance σ_β^2; the β s are independent

  ε_{ij} = random error associated with response in mixture i, investigator j; the ε_{ij} are independent

b.
Source	SS	df	MS	EMS	F	p
Mixtures	261,260.95	3	87,086.983	$\sigma^2 + 5\theta_A$	1264.73	<.05
Investigators	452.50	4	113.125	$\sigma^2 + 4\sigma_\beta^2$	1.64	>.05
Error	826.30	12	68.858	σ^2		
Totals	262,539.75					

16.19
Source	EMS
Rows	$\sigma^2 + 4\sigma_\alpha^2$
Columns	$\sigma^2 + 4\sigma_\beta^2$
Treatments	$\sigma^2 + 4\theta_c$
Error	σ^2

16.21 a.
Source	EMS
Investigators	$\sigma^2 + 45\sigma^2$
Treatments	$\sigma^2 + 5\sigma_{\alpha\beta}^2 + 15\theta_B$
Interaction	$\sigma^2 + 5\sigma_{\alpha\beta}^2$
Error	σ^2

b. In the F test for differences among treatments in this exercise, the T.S. is $MST/MS_{\text{interaction}}$ with $df_1 = 8$ and $df_2 = 16$. In Exercise 15.18 the T.S. was MST/MSE with $df_1 = 8$ and $df_2 = 108$. If the test for investigators was significant in this exercise, we would conclude that σ_α^2 is greater than 0, whereas in Exercise 15.18 the conclusion would be that the means are unequal.

16.23 $\hat{\sigma}_\epsilon^2 = 409.46$; $\hat{\sigma}_\beta^2 = 2808.17$; $\hat{\sigma}_{\alpha\beta}^2 = 4305.85$; $\hat{\sigma}_\alpha^2 = 6087.81$

Chapter 17

17.2

Source	SS	df
Between patients		
Depression group	962	3
Patients in group	582	20
Within patients		
Time	165	13
Timex group	120	39
Error	65	260

17.3

Treatment	Time	Mean	Std. Dev.
1	1	20.70	23.98
	2	28.57	12.00
	3	31.24	14.30
	4	29.44	12.65
	8	25.63	14.26
2	1	−0.76	12.26
	2	12.55	10.43
	3	18.23	10.83
	4	24.79	6.91
	8	17.57	7.83

17.6   a. 2 period crossover design
       c. No interaction between period and drug

17.8   a. Binomial or paired t-test ignoring order.

Chapter 18

18.1 $y = \beta_0 + \beta_1 x_1 + \beta_2 x_2 + \beta_3 x_3 + \beta_4 x_4 + \beta_5 x_5 + \beta_6 x_1 x_2 + \beta_7 x_1 x_3 + \beta_8 x_1 x_4 + \beta_9 x_1 x_5 + \varepsilon$

x_1 = covariate
x_2 = 1 if treatment 2 is applied
x_2 = 0 otherwise
x_3 = 1 if treatment 3 is applied
x_3 = 0 otherwise

x_4 = 1 if treatment 4 is applied
x_4 = 0 otherwise
x_5 = 1 if treatment 5 is applied
x_5 = 0 otherwise

18.3   The assumption of constant slope for the 5 treatment groups can be tested using the following null hypothesis:

$H_0: \beta_6 = \beta_7 = \beta_8 = \beta_9 = 0$

The complete model is given in Exercise 17.1 and the reduced model is $y = \beta_0 + \beta_1 x_1 + \beta_2 x_2 + \beta_3 x_3 + \beta_4 x_4 + \beta_5 x_5 + \varepsilon$. The test statistic is $F = MS_{drop}/MSE_1$, with $df_1 = 4$ and $df_2 = 20$.

18.5   b. Test for parallelism: $F_{2,12} = 4.133$. The test for adjusted treatment means is not valid since the slopes are not constant for the three groups.
       c. Yes, the plot suggests the lines are not parallel.

18.7 a. Obtain SSE_1 from the complete model described in Exercise 18.6. Obtain SSE_2 from the reduced model:

$$y = \beta_0 + \beta_1 x_1 + \beta_2 x_2 + \beta_3 x_3 + \beta_4 x_4 + \beta_5 x_5 + \beta_6 x_6 + \beta_7 x_7 + \beta_8 x_8 + \beta_9 x_9 + \beta_{10} x_{10} + \varepsilon$$

$SSE_{drop} = SSE_2 - SSE_1$; $MS_{drop} = SS_{drop}/9$; $F = MS_{drop}/MSE_1$, with $df_1 = 9$ and $df_2 = $ degrees of freedom for error in the complete model

b. Use the reduced model described in part (a) as the complete model and set $\beta_8 = \beta_9 = \beta_{10}$ equal to 0 for the reduced model.

18.9 a. Randomized block design with antidepressants as treatments and age-sex combinations as blocks.

b. $y = \beta_0 + \beta_1 x_1 + \beta_2 x_2 + \beta_3 x_3 + \beta_4 x_4 + \beta_5 x_5 + \beta_6 x_6 + \beta_7 x_7 + \beta_8 x_8$
$+ \beta_9 x_1 x_2 + \beta_{10} x_1 x_3 + \beta_{11} x_1 x_4 + \beta_{12} x_1 x_5 + \beta_{13} x_1 x_6 + \beta_{14} x_1 x_7 + \beta_{15} x_1 x_8 + \varepsilon$

18.11 a. $F_{5,9} = .174$; cannot reject H_0 that the blocks are equal.

b. Use 5 orthogonal contrasts:
(1) males versus females
(2) <20 versus 20–40
(3) <20 and 20–40 versus 2(>40)
(4) interaction between sex and first 2 age groups
(5) interaction between sex and all age groups

c. $y = \beta_0 + \beta_1 x_1 + \beta_2 x_2 + \beta_3 x_3 + \beta_4 x_4 + \beta_5 x_5 + \beta_6 x_6 + \beta_7 x_4 x_5 + \beta_8 x_4 x_6 + \varepsilon$

$x_1 =$ covariate
$x_2 = 1$ if treatment B
$x_2 = 0$ otherwise
$x_3 = 1$ if treatment C
$x_3 = 0$ otherwise
$x_4 = 1$ if female
$x_4 = -1$ if male
$x_5 = 1$ if less than 20
$x_5 = -1$ if 20–40
$x_5 = 0$ if greater than 40
$x_6 = 1$ if less than 20 or 20–40
$x_6 = -2$ if greater than 40

Source	SS	df
x_1 (covariate)	25.799	1
x_2 } (treatments)	59.260	1
x_3	8.224	1
x_4 (sex)	.183	1
x_5 } (age)	4.66	1
x_6	1.292	1
x_7 } (age × sex)	.114	1
x_8	.386	1
Error	68.581	9
Totals	168.500	17

Although the single-degree-of-freedom sums of squares are given in the AOV table, no additional tests need be performed since the tests for blocks (sex, age, age × sex) of part (a) failed to show any significant effects

18.13 a.

Source	SS	df	MS	F	p
x_1	.294	1	.294	.004	>.05
x_2	726.744	1	726.744	10.865	<.01
$x_1 x_2$	825.577	1	825.577	12.343	<.01
Error	1070.185	16	66.887		
Totals	2622.8				

b. $F_{1,16} = 12.343$; reject H_0 that the lines are parallel.

c. The test for a difference between the 2 populations is not valid because the lines are not parallel. The assumptions for a convariance analysis are not met.

INDEX

INDEX

INDEX

INDEX